Advances in Materials Technology for Fossil Power Plants

Proceedings from the Fifth International Conference
October 3–5, 2007
Marco Island, Florida, USA

Editors:
R. Viswanathan
D. Gandy
K. Coleman

Sponsored By:

EPRI Report Number 1016250

First printing, March 2008

Comments, criticisms, and suggestions are invited, and should be forwarded to EPRI.

ISBN-13: 978-0-87170-869-4
SAN: 204-7586

Distributed by ASM International®
Materials Park, OH 44073-0002
www.asminternational.org

Printed in the United States of America

Multiple copy reprints of individual articles are available from Technical Department, ASM International.

Conference Organizing Committee

Kent Coleman
EPRI, United States, Chairman

David Gandy
EPRI, United States

R. Viswanathan
EPRI, United States

Doug Arrell
Siemens, United States

Fred Glaser
U.S. Department of Energy, United States

Steve Goodstine
Alstom Power, United States

Jeff Hawk
General Electric, United States

Mario Marrocco
Ohio Coal Development Office, United States

Bob Purgert
Energy Industries of Ohio, United States

Robert Romanosky
U.S. Department of Energy, United States

John Shingledecker
Oak Ridge National Laboratory, United States

Greg Stanko
Foster Wheeler, United States

Jim Tanzosh
Babcock & Wilcox, United States

Brian Vitalis
Babcock Power, United States

Patricia Rawls
NETL, United States

International Advisory Board

Baldev Raj
Indira Gandhi Centre for Atomic Research, India

B.S. Rao
National Thermal Power Corporation, Ltd., India

Patricia Rawls
NETL, United States

Raman Singh
Monash University, Australia

Fred Starr
European Technology Development, Ltd., United Kingdom

Marc Staubli
Alstom Power, Switzerland

A. Strang
Alstom Power (retired), United Kingdom

Jose Antonio Tagle
Iberdrola, S.A., Spain

Yasuhiko Tanaka
Japan Steel Works, Japan

Zishan Xie
Institute for Science and Technology, China

Contents

Section 4: Oxidation

Section 5: Creep/Life Management

Section 6: Welding

Section 7: Oxy Fuel

Section 8: Reference Information

Preface

Several factors have renewed worldwide interest in advanced, high-efficiency coal power plants during the last decade. These factors include the abundance of coal and the need to maintain a viable coal option, as well as reduced fuel costs, emissions, and waste from power plants. The net thermal efficiency of fossil plants has improved from 33% high-heating value (HHV) in the case of the aging fleet of "subcritical plants" to nearly 42% HHV for supercritical plants operating under steam conditions of 1100 °F/3600 psi (593 °C/25 MPa). To boost efficiencies above 45% HHV, research and development projects have been carried out in Europe, the United States, and Japan on ultrasupercritical (USC) power plants that operate at steam conditions of 1300 °F/4000 psi (700 °C/28 MPa) and above. In Europe, in-plant demonstrations of prototype components have begun. In the United States, a five-year project on materials for USC plants has recently been completed, and new initiatives are underway.

The key enabling technology that drives high-efficiency power plants is the development of advanced materials and coatings with considerable increases in creep strength and corrosion resistance over traditional alloys. Major strides have been made in 9–12% chromium (Cr) ferritic steels containing cobalt (Co), tungsten (W), and other elements for both boilers and steam turbines that are capable of operating at temperatures of up to 1150 °F (625 °C). To operate beyond this limit, vastly improved austenitic steels for tubing applications such as HR3C, NF 709, Super 304 H, 347 HFG, and many others have been developed. For temperatures exceeding 1200 °F (650 °C), nickel-based alloys such as Inconel 740, Haynes 230 (a modified version of IN 617), HR6W, and others have been developed and evaluated.

The conference was the fifth in a series of conferences held every three years by the Electric Power Research Institute (EPRI), on the subject of materials for advanced power plants. Previous conferences were held in London (UK), San Sebastian (Spain), Swansea (Wales), and Hilton Head (United States). The present conference was intended to continue to promote information exchange between scientists and engineers on an international level.

The assistance of Stacey Burnett in all phases of organizing the conference and in compiling this publication is gratefully acknowledged. Financial support for the conference was provided by EPRI Program P87, "Fossil Materials and Repair."

R. Viswanathan, D. Gandy, and K. Coleman
November 2007

Advances in Materials Technology for Fossil Power Plants
Proceedings from the Fifth International Conference
R. Viswanathan, D. Gandy, K. Coleman, editors, p 1-15

DOI: 10.1361/cp2007epri0001

U.S. Program on Materials Technology for Ultrasupercritical Coal-Fired Boilers

R.Viswanathan, EPRI
R.Purgert, Energy Industries of Ohio
S.Goodstine, Alstom Power
J.Tanzosh, Babcock & Wilcox
G.Stanko, Foster Wheeler Development Corporation
J.P. Shingledecker, Oak Ridge National Laboratory
B.Vitalis, Riley Power

Abstract

One of the pathways for achieving the goal of utilizing the available large quantities of indigenous coal, at the same time reducing emissions, is by increasing the efficiency of power plants by utilizing much higher steam conditions. The US Ultra-Supercritical Steam (USC) Project funded by US Department of Energy (DOE) and the Ohio Coal Development Office (OCDO) promises to increase the efficiency of pulverized coal-fired power plants by as much as nine percentage points, with an associated reduction of CO2 emissions by about 22% compared to current subcritical steam power plants, by increasing the operating temperature and pressure to 760°C (1400°F) and 35 MPa (5000 psi), respectively. Preliminary analysis has shown such a plant to be economically viable. The current project primarily focuses on developing the materials technology needed to achieve these conditions in the boiler. The scope of the materials evaluation includes mechanical properties, steam-side oxidation and fireside corrosion studies, weldability and fabricability evaluations, and review of applicable design codes and standards. These evaluations are nearly completed, and have provided the confidence that currently-available materials can meet the challenge. While this paper deals with boiler materials, parallel work on turbine materials is also in progress. These results are not presented here in the interest of brevity.

Introduction and Background

In the 21st Century, the world faces the critical challenge of providing abundant, cheap electricity to meet the needs of a growing global population, while at the same time, preserving environmental values. Most studies of this issue conclude that a robust portfolio of generation technologies and fuels should be developed to assure that the United States will have adequate electricity supplies in a variety of possible future scenarios. Traditional methods of coal combustion emit pollutants (including CO2) at high levels relative to other generation options. Maintaining coal as a generation option will require methods for addressing these environmental issues.

This project, through a government/industry consortium, is undertaking a five-year effort to evaluate and develop materials technologies that allow the use of advanced steam cycles in coal-based power plants. These advanced cycles, with target steam temperatures up to 760°C (1400°F) will increase the efficiency of coal-fired boilers and reduce emissions substantially.

Worldwide, more than a dozen plants are operating at steam conditions close to 593°C (1100°F)/27 MPa (4000 psi and plant operation at 620°C (1150°F) appears to be a near-term possibility.(1) Research, development and demonstration programs have been underway in Europe and Japan aimed at materials capable of withstanding steam conditions up to 650°C (1200°F), and over the next decade to 700°C (1300°F). It is imperative that the US boiler manufacturers match those capabilities in this advanced technology area. This is one of the objectives of the US DOE/OCDO project. Furthermore, materials technology is generic in nature and cuts across many energy systems operating at high temperatures. Consequently, materials technology developments for high-temperature applications are expected to be of enduring value. In the near term, materials developed for USC plants can be confidently used for retrofit applications in currently-operating plants to increase their reliability.

The project objectives are addressed through the eight tasks listed below:

Task 1. Conceptual Design and Economic Analysis
Task 2. Mechanical Properties of Advanced Alloys
Task 3. Steamside Oxidation and Resistance
Task 4. Fireside Corrosion Resistance
Task 5. Weldability
Task 6. Fabricability
Task 7. Coatings
Task 8. Design Data and Rules

The current scope of the project is nearly complete. Highlights from the results achieved so far are summarized in this paper.

Results

Conceptual Design and Economic Analysis

Based on temperature/pressure calculations performed for various sections of a conceptual boiler(2-4), possible materials selection have been performed from a creep strength point of view. Haynes® 230, Inconel® 740 and CCA 617* were selected for the highest-temperature, heavy-section applications as well as tubular components; the austenitics HR 6W and SUPER 304H were selected for tubular applications; and ferritic alloy SAVE 12 was selected for heavy sections at lower temperatures. Alloys T92, T23 and HCM 12 were considered for application in waterwall tubing. The compositions and intended applications of the alloys are shown in Table1.

---

* CCA617: **C**ontrolled **C**ompositional **A**nalysis, a modified composition of Alloy 617

Table 1
Candidate Alloys

Alloy	Nominal Composition	Developer	Application
Haynes 230	57Ni-22Cr-14W-2Mo-La	Haynes	P, SH/RH Tubes
INCO 740	50Ni-25Cr-20Co-2Ti-2Nb-V-Al	Special Metals	P, SH/RH Tubes
CCA 617	55Ni-22Cr-.3W-8Mo-11Co-Al	VDM	P, SH/RH Tubes
HR6W	43Ni-23Cr-6W-Nb-Ti-B	Sumitomo	SH/RH Tubes
Super 304H	18Cr-8Ni-W-Nb-N	Sumitomo	SH/RH Tubes
Save 12	12Cr-W-Co-V-Nb-N		P
T92	9Cr-2W-Mo-V-Nb-N	Nippon Steel	WW Tubes
T23	2-1/4Cr-1.5W-V	Sumitomo	WW Tubes
HCM12	12Cr-1Mo-1W-V-Nb	----	WW Tubes
P - pipe			

A tentative bill of materials for the conceptual boiler is shown in Fig 1. Overall, the feasibility of designing an ultra-supercritical 750 MW boiler operating at 760°C (1400°F)/35 MPa (5000 psi) throttle steam conditions with existing material technology is encouraging. It was estimated that such a plant will increase the plant from

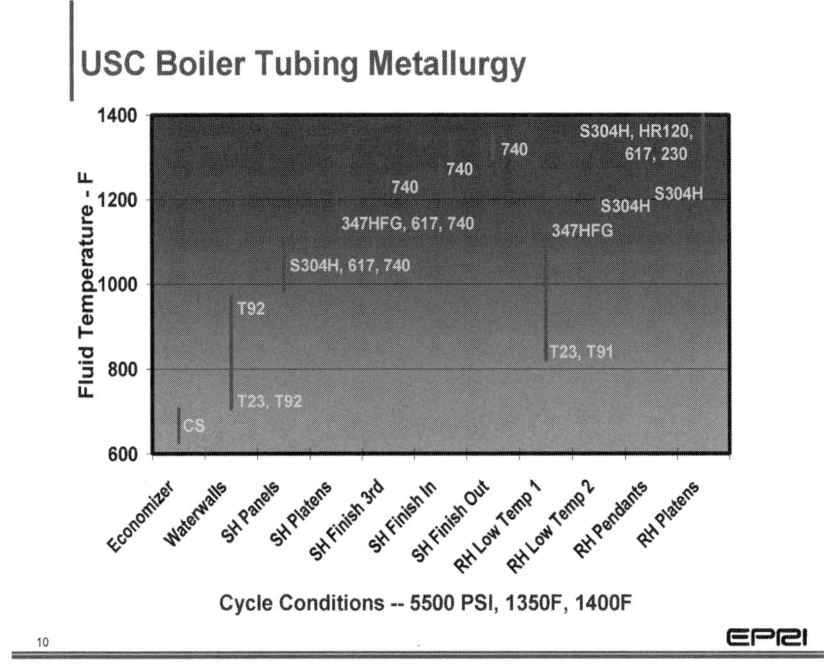

Figure 1
Alloy choices for components of a conceptual USC boiler

the efficiency from an average of 35% (HHV)* (current domestic aging sub-critical steam fleet) to approximately 45% (HHV).(2,3). For a double reheat configuration the efficiency will reach 47%. Expressed in terms of European parameters, this could be as high as 52% LHV*.

Based on a 20-year break-even consideration, an assumed capacity factor of 80%, and coal cost of $1.42/G J. ($1.50/MBtu), the fuel cost savings over the 20-year period, are sufficient to allow an ultra-supercritical plant to be cost competitive even if the total plant capital cost is 12 to 15% more than a comparable scale facility built using conventional subcritical boiler and cycle design (5,6),. Balance-of-plant (BOP) costs are expected to be lower than those for existing boiler and cycle designs due to smaller coal handling and pollution control systems, resulting from the improved plant efficiencies. As a result of these reductions in fuel and BOP costs and the smaller plant size due o increasd efficiency, boiler and steam turbine capital costs can be permitted to be higher compared to subcritical plant. If a potential "carbon tax" were also taken into account, the advantage of the USC plant becomes even more substantial. These economic calculations are being up dated

An alternative design consideration that mitigates the high cost of the superalloys is a design with increased throttle temperature without a major increase in throttle pressure. As mentioned earlier, the costs involved in increasing steam temperature are lower than for raising steam pressure. Reducing throttle pressure may allow the more extensive use of standard austenitic materials, while maintaining the plant efficiency gains due to higher throttle temperature. This approach of using lower pressure (in combination with the better than expected material properties observed for some alloys and the lower stresses calculated from the reference stress approach((7) as described later), favors the possibility of achieving the initial goal of 760°C (1400°F) throttle steam.

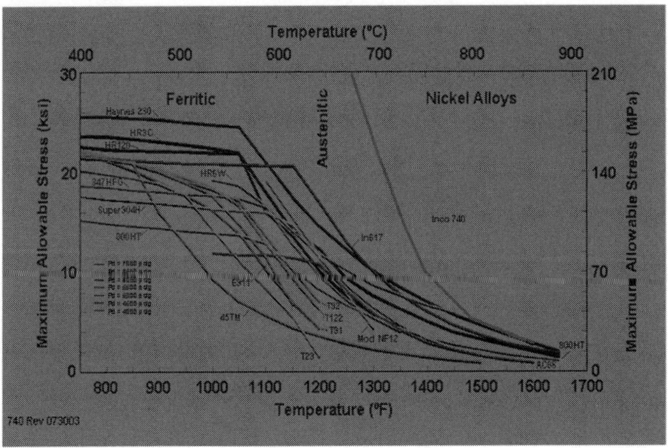

Figure 2
Allowable stresses for various classes of alloys(8)

---

* HHV and LHV: Higher Heating Value and Lower Heating Value, respectively (related to the assumption regarding recovery of heat of vaporization of moisture). Efficiency calculations were based on typical US plant siting and operating practices.

Mechanical Properties

Figure 2 is a plot of the allowable stresses as a function of temperature used to compare and determine the temperature capabilities of various classes of alloys.(8) The figure also shows the actual stresses at several steam pressures.(8) The nickel-based alloys Inconel 740, Haynes 230, Inconel 625, Inconel 617, and HR 120 have much higher temperature capability, in decreasing order as listed compared to austenitic steels, followed by the ferritic steels. Purely from the creep strength point of view, at a pressure of 35 MPa (5000 psi) for a 5 cm (2 in) by 1.25cm (0.5 in.) tube (stress 60 MPa or 8.7 ksi), ferritic steels are useful up to about 620°C (1150°F) (metal temperature) and austenitic steels up to about 675°C(1250°F). At metal temperatures higher than this, Ni-based alloys are required.

Creep rupture testing has exceeded 30000hr for SAVE 12, Super 304H, Haynes 230, CCA617, and Inconel 740. Creep-rupture tests on HR6W tubing are complete, and the results confirm earlier predictions (based on short-term data) that indicated the alloy had poorer creep strength than expected from literature. Figure 3 compares the expected and measured 100,000 hour rupture strength as a function of temperature for HR6W. The measured values are 20 to 30% lower than expected. When compared to Super 304H, the creep strength of HR6W is inferior below ~675°C while it shows an improvement over Super 304H above 675°C.

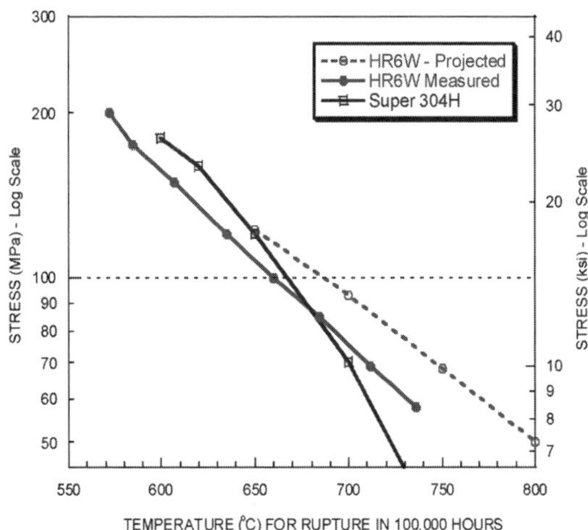

Figure 3
100,000 hour rupture life for austenitic tubing (extrapolations based on data up to ~20,000 hours). Measured strength of HR6W is significantly lower than projected values from material supplier.

Creep tests on CCA617 have exceeded expectations in the lower temperature regime (650°C to 700°C), but above 750°C the creep strength was indistinguishable from that of the standard Inconel 617. Microstructural studies suggest that this strength difference is due to the precipitation of gamma prime, but at 750°C these precipitates were found to coarsen significantly with time, and at 800°C contributed little to the hardening of the material. (9) This suggests that at 750°C and above, no long-term strength advantaged is expected for CCA617 due to gamma prime precipitation. Figure 4 shows the extrapolated 100,000 hour rupture strengths of 617 and CCA617. The 100 MPa-100,000 hour rupture strength for CCA617 is estimated at 725°C compared to ~700°C for 617.

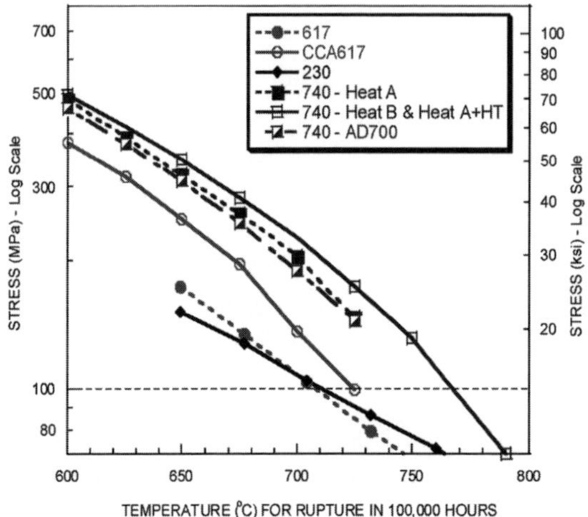

Figure 4
100,000 hour rupture life for nickel-based tubing and plates (extrapolations based on data up to ~23,000 hours). CCA617 shows improved strength compared to 617 (725°C at 100MPa). Inconel 740 rupture strength shows heat-to-heat and heat-treatment variations. All the data suggests better strength than AD700 program reported values.

Creep testing of Inconel 740 is being performed on both tubing and thick plate. Tests on two heats of material and two heat-treatments have shown some data scatter for rupture life as shown in figure 4. All material was given a standard 800°C-16hr heat-treatment before testing. Heat 'A' shows behavior similar (although slightly improved) to that reported on the AD700 program(10). However when Heat 'A' was subjected to an additional solution heat-treatment, its rupture behavior followed that of Heat 'B' which was significantly better than that reported by the AD700 program. Heat 'B' has a 100MPa-100,000hr estimated rupture life greater than 760°C, which meets the aggressive strength requirements of this project. For ASME code qualification, data is required on a minimum of three heats of material. A third heat of 740 has been obtained for the required testing.

To evaluate the fabrication and welding processes developed on the program, testing is being performed on welds, weldments, and cold-bent tubes. Cross-weld creep tests on HR6W and on Super 304H have shown that 'overmatching' filler metals have superior creep strength compared to the base metal with most ruptures in the base metal at expected base metal lives. Cross-weld tests on Haynes 230 tubes and thin plates show the weld metal to be approximately 80% the strength of the base metal with all cross-weld failures in the weldments. Current cross-weld creep studies are focused on thick plate weldments in CCA617 (specimens up to 37mm in thickness), tube weldments in 617, dissimilar metal welds (DMW) in CCA617/Super 304H, and plate weldments in Inconel 740. The cross-weld tests on Inconel 740 are in support of the welding development efforts on this alloy and include matching 740 filler metal and Nimonic 263 filler metal. To examine cold-work effects, pressurized creep tests on tube bends have been performed on Haynes 230 and HR6W. At relatively high levels of cold work (20 to 35% outer fiber strain) after 8,000 hours in test, neither alloy showed large reductions in rupture life at 775°C indicating the ASME code rules for cold-work will be conservative for the solid solution nickel-based alloys. Tests on Inconel 740 tubes bends are currently underway to examine cold-work effects in age-hardenable nickel-based alloys.

In support of design rules and model development, LCF, thermal fatigue (thermal shock experiments), and creep-fatigue tests are being performed on some alloys. These data will be used primarily to determine the effect thermal transients will have on thick-section components. This is important because the current experience for most boilers is with ferritic steels which have different physical properties compared to the nickel-based alloys.

Steamside Oxidation Tests

Steamside oxidation testing has been completed at 650°C (1 and 17 atm); 700°C (17 atm); and 800°C (1 and 17 atm). The oxidation rates primarily followed a parabolic rate law, and the rate constants derived were found to be in agreement with literature data (where this information is available)(11). Among the ferritic steels, two new steels: MARB2 (developed by NIMS) and VM12 (developed by V&M Tubes) displayed the lowest weight gains/metal loss, whereas all other 2-9% steels were subject to rapid oxidation. Ferritic steels that showed the lowest weight loss also exhibited the lowest tendency for their oxide to spall. Clear evidence of spallation was observed for alloys T23, T91 and T92. Ferritic material MARB2 and austenitic material 304H displayed the formation of oxide islands, indicative of the development of a non-protective oxide at some location. The highest tendency among the austenitics for scale exfoliation was found in 347HFG. Shot blasting was found to be beneficial in both Super 304H and in 347HFG steel. The oxidation behavior of some nominally 12%Cr steels (for example, T122) was very variable, with anomalous temperature dependence apparently resulting from the fact that their ability to form a consistently-protective scale was unusually sensitive to minor changes in factors such as levels of minor alloying elements, surface condition, and test variables. However, one nominally 12%Cr steel (VM12) was able to reliably form a protective scale, and performed as well as some austenitic and Ni-based alloys.

No spallation was observed from the austenitic steels or Ni-based alloys in the 650°C (1200°F) tests. Of the austenitic and Ni-based alloys, the Co-containing alloys Nimonic 263, CCA 617 and Inconel 740 exhibited the best oxidation behavior. All of the austenitic and Ni-based materials tested formed a dense layer of chromium oxide that will result in low oxidation rates at this temperature. The Ni-based alloys that contained between 0.5 and 1.3% Al formed near-surface intergranular aluminum oxide penetrations beneath the external (protective) chromia scales, which effectively increased the thickness of load-bearing section lost.

Results from tests at 650 and 600°C in 1 atm steam, shown in Fig. 5 indicate that the oxidation susceptibility appears to be independent of Cr level, once a threshold level of about 10% is reached.

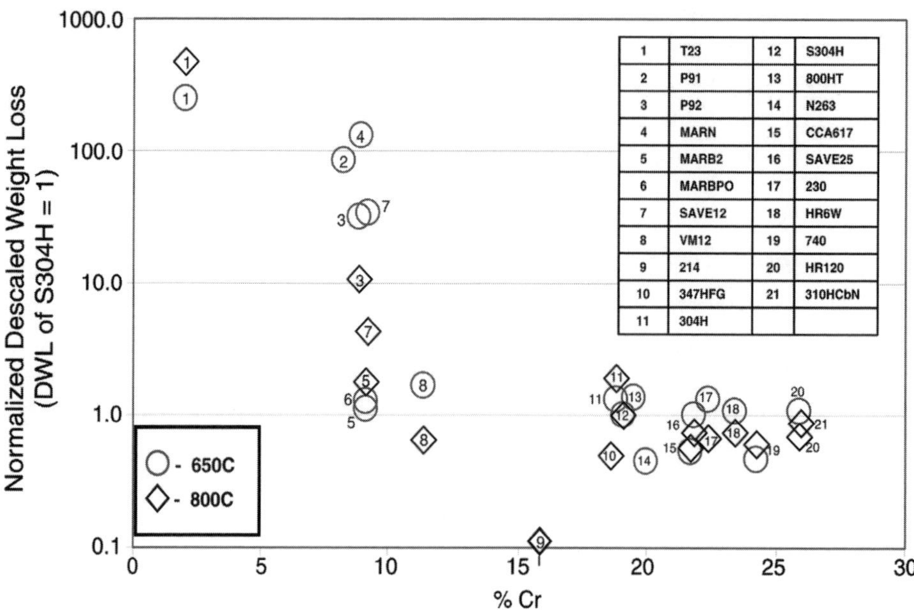

Figure 5
Normalized weight loss of sample after steamside oxidation 1000 hr at 800°C(11)

Among the diffusion coatings tested, SiCr and FeCr coatings performed much better than AlCr coatings. Based on calculated oxidation rates, it appears that the life of austenitic and Ni-based tubing was unlikely to be controlled by steamside oxidation rates rather than fireside corrosion. The primary concern with respect to steamside oxidation is oxide exfoliation.

Fireside Corrosion

The objective of this task is to evaluate the relative resistance of various candidate alloys to fireside corrosion over the full temperature range expected in an USC steam plant.(12,13) The testing involved the corrosive environments representative of three different domestic coal types: Eastern (E), Midwestern (MW), and Western (W). Also, three types of testing were undertaken: (1) laboratory testing simulating conditions expected at the waterwalls and superheater/reheaters, using appropriate deposit compositions and gas mixtures; (2) steam loops formed by welding together spool pieces of the various materials and inserted into the superheater circuit to be exposed to much higher temperature boiler conditions; and (3) retractable air-cooled probes inserted inside operating boilers. Steam loops are hydrotested prior to installation in a boiler, which burns high sulfur coal.(13,14) USC boiler design operating conditions are simulated by throttling the steam flow to achieve metal temperatures up to 760°C (1400°F). Operating parameters during the test are monitored remotely. Physical monitoring of the tubes is accomplished during outages at the plant by means of diameter measurements for hot corrosion wastage, and photographs of the surface condition.

Results of laboratory tests, In the case of the waterwall conditions, specimens tested under MW and E conditions exhibited more wastage than those tested under W-coal conditions (see Fig. 6(12)). The difference in the amount of wastage between the MW/E and W conditions was greater at

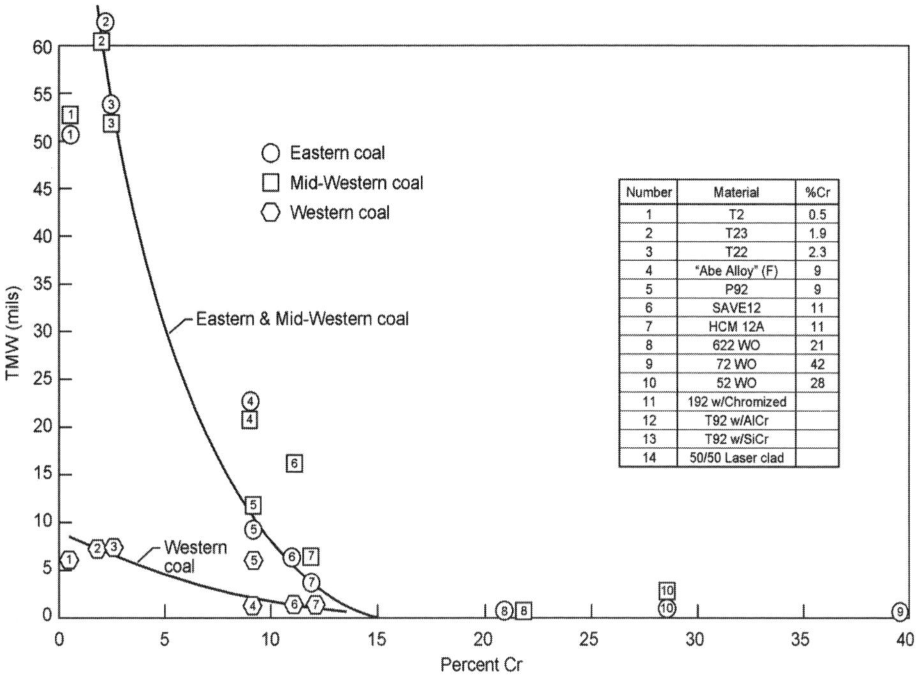

Figure 6
Total metal wastage for alloys under waterwall conditions at 525°C (975°F), 1000 hr(12)

454°C (850°F) and 525°C (975°F) than at 593°C (1100°F). The MARB2 and SAVE12 alloys (which contain Co) exhibited higher wastage rates than did the P92 and HCM12A alloys at all three temperatures.

The Inconel-622, -52, and -72 weld overlays and laser clad 50/50 displayed significant improvement in corrosion resistance compared to the wrought alloys at all temperatures, although the Inconel-622 overlay started to show more attack at 593°C (1100°F) compared to -52 and -72. With regard to the diffusion coatings (on T92), they also exhibited significantly improved corrosion resistance compared to the wrought alloys. The chromized coating displayed the best performance, followed by the SiCr and the AlCr. As in the SH/RH tests, the AlCr coating exhibited the most subsurface attack.

The results for superheater/reheater corrosion so far confirm that:

As noted by others, of US Midwestern and Eastern coals are much more corrosive than the environment typical of a Western coal.

Corrosion behavior was a function of Cr level, with corrosion decreasing rapidly as the chromium level increased to 22-27%, and then leveling off.

Fe-Ni-Cr and Ni-Fe-Cr alloys HR3C (53Fe-25Cr-20Ni), 353 (36Fe-35Ni- 25Cr-1Si), 120 (37Ni-35Fe-25Cr), and HR6W (40Ni-25Fe-23Cr-6W) performed better than the Ni-based alloys tested.

Of the Ni-based alloys, Inconel 740 exhibited better corrosion resistance than Haynes 230 and CCA617.

Weld overlays Inconel-72 (42% Cr) and Inconel-52 (28% Cr) performed better than Inconel-622 (21% Cr).

Diffusion coatings of Cr and Cr-Si displayed better corrosion resistance than Al-Cr and were comparable to weld overlays, but because they are thinner will be breached sooner.

Of the weld overlays, Inconel-72 (typically 42 Cr) and Inconel-52 (typically 28 Cr) performed better than Inconel-622 (typically 21 Cr) at all temperatures, with 72 performing better than the 52. At 650°C (1200°C) and 704°C (1300°C), the 72 and 52 were better than the wrought alloys in the 22-27 Cr range. With regard to the diffusion coating (on Super 304H), the FeCr and SiCr compositions performed better than the AlCr, which exhibited the most extensive subsurface penetration. The FeCr and SiCr coatings were comparable to the weld overlays but, because they are thinner, will be breached sooner.

Weldability Studies

The scope of this task includes study of six alloys: SAVE12, Super 304H, HR6W, Haynes 230, Inconel 740, CCA 617. Two product forms, tubing and pipe or plate, are being studied, and welding procedures for 15 materials/product forms/welding processes combinations are being developed. Welding procedures are being developed for three dissimilar metal weld configurations, and weldments for each of the combinations are being evaluated15).

Preliminary results indicate that submerged arc welding (SAW), a high deposition rate process favored by boilermakers for thick sections, might not be feasible for Ni-based materials; tests on Haynes 230 and Inconel 740 have been unsuccessful. A 7.5cm (3 in.) thick plate of Haynes 230 was successfully welded using pulsed gas metal arc welding (GMAW) technique,.(16) Hot wire gas tungsten arc welding tests conducted on 2-inch thick Inconel 740 plates using a 25% Helium/75% Argon shielding gas and matching filler wire gave encouraging results. There were very few microfissures compared to earlier samples using a pulsed gas metal arc process. Bend and tensile tests will be conducted for further evaluation of this plate in an attempt to fully qualify a welding procedure.

An orbital gas-tungsten arc welding (GTAW) process has been qualified for Super 304H and test specimens are being fabricated. While attempts to weld tubing using an automatic GMAW process were unsuccessful, type 347 filler produced acceptable welds. An orbital GTA process was qualified for tubing of alloy CCA 617, and a SAW process was qualified for plate and test specimens are being fabricated. Attempts to perform shielded metal arc welding (SMAW) using matching filler were unsuccessful due to slag control problems with CCA 617 electrodes. On the other hand, successful SMAW welds were achieved using conventional 617 electrodes. Collaboration is being pursued with the alloy vendor (Special Metals Corp.) in view of their experience in welding 2.5 cm (1 in) thick plates of Inconel 740 using GMAW.

Fabricability Studies

The primary objective of this task is to conduct fabrication studies on the alloys of interest and to assess the effects of fabrication on material properties, so that potential fabrication problems may be identified.(17) Experience in welding, machining, cutting, boring and grinding of Haynes 230, Inconel 740, HR6W, CCA 617 and Super304H stainless steel has been gained in the course of fabrication of two steam test loops. Protective weld overlay claddings using alloys Inconel 52, Inconel 72 and Inconel 622 were successfully applied to the tube sections that form the test loops. Field welding was additionally demonstrated during installation of the test loops in the boiler. Samples of strained material required for characterizing the recrystallization/precipitation behavior of USC steam alloys were made by controlled straining (ranging from 0 to 50%) of special tapered tube specimens. Fabrication of multiple U-bends from Haynes 230 and HR6 W. tubing (2 in. OD X 0.4 in. MW) was successfully demonstrated using production equipment. Tube U-bends with strains of 15%, 20% and 35% were produced as shown in Fig 7.

USC Alloy -Participant -	Evaluation Planned	Results & Status	Comments	Follow -on Activities
Alloy 230 - B&W -	Cold U -Bend Trials	13.3% 20% & 33.3%	Good; Some ovality	P-Creep at ORNL
	Tubewelding Trials	625 Filler; 622,52,72 WOL	Good; Field Test Loops Fabricated	Fabricate Demo Article
	Machining Trials	Completed	Good; Field Test Loops Fabricated	Fabricate Demo Article
	Strain/Rx/Pptn Trials	1500°F & 1600°F	100,250,400 & 550 hrs	Assess Microstructures
	Swaging Trials	Planned Spring 2005	Machined tubes to 0.2" Wall	Assess Microstructures
	Demo Article Fab	Planned Spring 2005	Design Completed	Fabricate; Display TBD

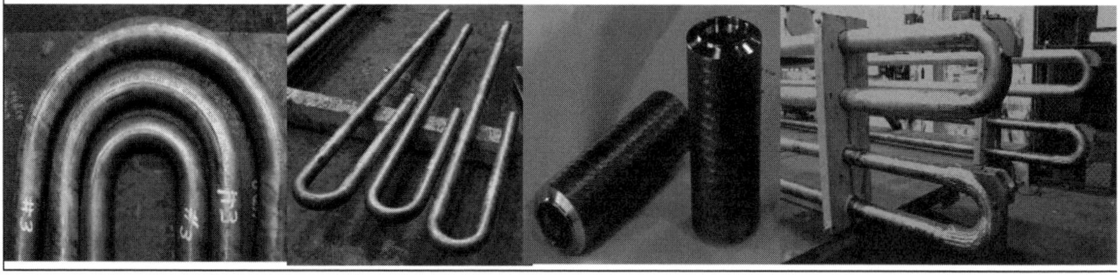

Figure 7
Current Status of Workplan Activities – Alloy 230

Swaging trials of two of the USC steam alloys have been completed. Data on recrystallization behavior, phase precipitation and dissolution in the Ni-based alloys are being compiled to assist in the understanding and development of fabrication procedures. Summary of results for one of Alloy 230 are shown in the table in Figure 7.(17),for illustrating the type of data being gathered.

As a demonstration of the fabrication capabilities achieved in the course of this project, a mock up section of a header was fabricated, see Figure 8. The mock up illustrates capabilities with respect to fabrication of CCA 617 alloy into the header shape by bending of plate; girth welding;

seam welding; socket welding machining; swaging; hole drilling; and dissimilar welding between CCA 617 and Super304H and T91 tubing.

Fabrication Process

Forming

- Press forming of headers and piping
- Bending of tubing
- Swaging of tube ends

Machining

- Weld grooves for header and pipe longitudinal and circumferential seams
- Socket weld grooves for tube-to-header joints
- Weld grooves for tube circumferential seams

Welding

- Submerged arc welding (SAW) for header and pipe longitudinal and circumferential seam
- Gas tungsten arc welding (GTAW) for tube-to-tube joints
- Shielded metal arc (SMAW) and gas tungsten arc welding (GTAW) for tube-to-header socket joints
- Dissimilar welds

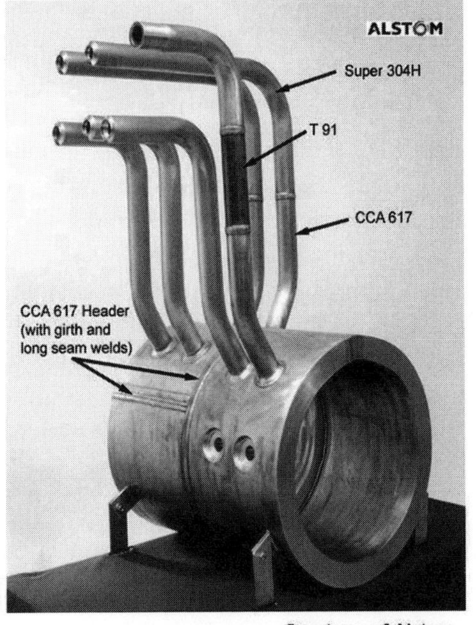

Courtesy of Alstom

Figure 8
Demonstration of fabrication capabilities with alloys of interest for an USC steam boiler

Coatings

Several specimens with claddings, sprayed coatings [cold spray, high-velocity oxygen/fuel (HVOF)] and diffusion chromium, chromium-silicon, chromium-aluminum coatings have been prepared using commercially-viable processing.(4) Results show that ferritic steels benefit most from coatings; austenitic steels may benefit; while Ni-based alloys are not likely to need coatings at all. Process scale-up activities are being pursued. The scale-up tests for the chromizing process have been completed, and the evaluation of results revealed excellent reproducibility for both types of materials, Super304H and T 92. Development of parameters for depositing HVOF and cold spraying techniques for 50Ni/50Cr coatings also has been completed. Optimal parameters have been selected and used to coat Haynes 230 tubes. For the other coating processes of interest, parameters for deposition of 50Ni/50Cr by plasma-transferred arc and laser cladding are being measured.

Design Data and Rules

A new set of design equations has been developed for cylindrical components (tubes, pipes, headers, shells and drums). Since these are based on analysis, they are less conservative than those used in conventional design practice.(7) The impact of the new design equations will be that a consistent failure criterion will be applied to all sizes and types of cylindrical sections. This criterion incorporates a limit load approach in the time-independent regime, and a reference stress approach in the time-dependent regime. It is expected that these proposed rules will permit the use of thinner-walled components than would be permitted under the current ASME and European rules, without compromising component reliability and safety. It is estimated that in ultra-supercritical boilers, where expensive materials are required, a 12% reduction of the cost of boiler pressure parts can be achieved. Also, thinner-walled components would be less subject to thermal fatigue, and therefore the plant would be less susceptible to damage from cyclic operation. This less conservative approach also would permit use of materials with lower creep strengths under a given set of conditions, thus offering a wider selection of alloys for use at high temperatures. Proposed revisions to Section I of the ASME code to permit the use of simplified and more technically-defensible design equations were submitted to Subcommittee I, and accepted by them in September 2004. Subsequent to that, the revisions were included in the Main Committee ballot.

Summary of Accomplishments

- **Developed a USC plant heat balance design which will reduce fuel consumption and emissions by 30% compared to current state-of-the-art subcritical cycle. Developed two conceptual designs of USC boilers outlining component design and material test conditions. Performed plant cost and feasibility study showing that the cost of the advanced alloys will be more than offset by savings in balance of plant equipment cost and fuel savings within a 20-year period.**
- **Alloys with the creep strength required to withstand the target temperatures and stresses at different locations in the boiler have been identified and tested for times exceeding 30,000 hours. Tensile tests, fatigue tests, and other mechanical property tests have been completed.**
- **Laboratory steamside oxidation testing of various alloys and coatings has been completed, allowing relative ranking of materials at appropriate exposure temperatures.**
- **Laboratory and field-exposure corrosion testing of various alloys and coatings has been completed, allowing relative ranking of materials at appropriate exposure temperatures and coal ash types.**
- **Developed practical welding procedures for some alloys and identified difficulties with others.**
- **Subjected alloys to common fabrication processes and produced prototype assemblies.**
- **Identified external and internal coatings and methods, prepared samples for oxidation and corrosion testing, and evaluated the results.**
- **Proposed a more accurate and less conservative design formula which was adopted into the ASME B&PV Code Section I; Advanced the knowledge base of the boiler manufacturers and improved their competitive position vis a vis capabilities in Europe and Japan.**

Acknowledgements

The sponsorship and guidance of this work by Dr. Romanosky, Patrecia Rawls of NETLand Mario Maracco and Bob Brown of OCDO is gratefully acknowledged. The project manager is Robert Purgert of Energy Industries of Ohio(EIO). The technical results reported are based on contributions by numerous investigators including Paul Weitzel, Mark Palkes, George Booras, Robert Swindeman, Jeff Sarver, Ian Perrin, John Sanders, Mike Borden, Walt Mohn,, John

Fishburn, and many others. Their contributions are gratefully acknowledged. The project consortium consists of ALSTOM, Babcock &Wilcox, Fosterwheeler, Riley Power and EPRI, Energy Industries of Ohio is the project manager. Detailed results from the entire project have been summarized in the phase 1 Final report.

References

[1] R. Viswanathan, A.F. Armor and G. Booras, "A Critical Look at Supercritical Plants", Power Magazine, April 2004, p 42-49.

[2] M. Palkes, "Task 1A, Conceptual Design – ALSTOM Approach, "Boiler Materials for Ultra-supercritical Coal Power Plants", DOE Grant DE-FG-26-01NT41175, OCDO Grant D-00-20, Topical Report USC T-3, February 2003.

[3] "Task 1 B, Conceptual Design – Babcock and Wilcox Approach, "Boiler Materials for Ultra-supercritical Coal Power Plants", DOE Grant DE-FG26-01NT41175, OCDO Grant D-00-20, Topical Report USC T-3, February 2003.

[4] S.L. Goodstine and D.C. Nava, "Use of Surface Modification of Alloys for Ultra-supercritical Coal-fired Boilers", in Proc. of the 4th International Conference on Advances in Materials Technology for Fossil Power Plants, Hilton Head Island, SC, October 2004, ASM International, in press.

[5] G. Booras, "Task 1 C, Economic Analysis", Boiler Materials for Ultra-supercritical Coal Power Plants, DOE Grant DE-FG26-01NT41175, OCDO Grant D-00-20, Topical Report USC T-1, February 2003.

[6] G. Booras, N. Holt and R. Viswanathan, "Economic Analysis of New Coal-fired Generation Options", in Proc. of the 4th International Conference on Advances in Materials Technology for Fossil Power Plants, Hilton Head Island, SC, October 2004, ASM International, in press.

[7] John D. Fishburn, and I. Perrin, "A Single Technically Consistent Design Formula for the Thickness of Cylindrical Sections Under Internal Pressure", PVP2005-71026, Proceedings of PVP2005, 2005 ASME Pressure Vessels and Piping Division Conference, July 17-21, 2005, Denver, Colorado.

[8] B. Vitalis, Private Communication to R. Viswanathan, Riley Power Inc., 2003.

[9] J.P. Shingledecker, R.W. Swindeman, Q. Wu, and V.K. Vasudevan, "Creep Strength of High-Temperature Alloys for Ultra-supercritical Steam Boilers". Proceedings to the 4th International Conference on Advances in Materials Technology for Fossil Power Plants, October 25-28, 2004. ASM International, in press.

[10] R. Blum, R.W. Vanstone. "Materials Development for Boilers and Steam Turbines Operating at 700°C." *Parsons 2003 - Proceedings of the Sixth International Charles Parsons Turbine Conference*, Institute of Materials, Minerals, and Mining, London, 2003. 489-510.

[11] J.M. Sarver and J.M. Tanzosh, "An Evaluation of the Steamside Oxidation of Candidate USC Materials at 650oC and 800oC, in Proc. of the 4th International Conference on Advances in Materials Technology for Fossil Power Plants, Hilton Head Island, SC, October 2004, ASM International, in press.

[12] G. Stanko, unpublished work, Foster Wheeler Inc., 2004.

[13] D.K. McDonald and E.S. Robitz, "Coal Ash Corrosion Resistant Material Testing Program", in Proc. of the 4th International Conference on Advances in Materials Technology for Fossil Power Plants, Hilton Head Island, SC, October 2004, ASM International, in press.

[14] J.M. Tanzosh, D.J. DeVault and W.R. Mohn, "Engineering Design and Fabrication of USC Test Loops, in Proc. of the 4th International Conference on Advances in Materials Technology for Fossil Power Plants, Hilton Head Island, SC, October 2004, ASM International, in press.

[15] M.P. Borden, "Weldability of Materials for USC Boiler Application", in Proc. of the 4th International Conference on Advances in Materials Technology for Fossil Power Plants, Hilton Head Island, SC, October 2004, ASM International, in press.

[16] J. Sanders, Babcock and Wilcox and F.L. Flower, Haynes Corp., unpublished work, 2004.

[17] W.R. Mohn and J.M. Tanzosh, "Considerations in Fabricating USC Boiler Components from Advanced High Temperature Materials, in Proc. of the 4th International Conference on Advances in Materials Technology for Fossil Power Plants, Hilton Head Island, SC, October 2004.

Advances in Materials Technology for Fossil Power Plants
Proceedings from the Fifth International Conference
R. Viswanathan, D. Gandy, K. Coleman, editors, p 16-28

DOI: 10.1361/cp2007epri0016

Creep-Rupture Behavior and Recrystallization in Cold-Bent Boiler Tubing for USC Applications

John P. Shingledecker
Oak Ridge National Laboratory
Oak Ridge, TN, 37831-6155
shingledecjp@ornl.gov

Abstract

Creep-rupture experiments were conducted on candidate Ultrasupercritical (USC) alloy tubes to evaluate the effects of cold-work and recrystallization during high-temperature service. These creep tests were performed by internally pressurizing cold-bent boiler tubes at 775°C for times up to 8000 hours. The bends were fabricated with cold-work levels beyond the current ASME Boiler and Pressure Vessel (ASME B&PV) Code Section I limits for austenitic stainless steels. Destructive metallographic evaluation of the crept tube bends was used to determine the effects of cold-work and the degree of recrystallization. The metallographic analysis combined with an evaluation of the creep and rupture data suggest that solid-solution strengthened nickel-based alloys can be fabricated for high-temperature service at USC conditions utilizing levels of cold-work higher than the current allowed levels for austenitic stainless steels.

Introduction

Increased efficiency and decreased emissions of pulverized coal-fired boilers can be achieved by increasing steam temperatures and pressures [1]. An advanced Ultrasupercritical (USC) steam boiler with steam parameters of 760°C and 35MPa could reach efficiencies of 50%, which would cause a decrease in CO_2 emissions of about 22% compared to standard subcritical boiler technology in the United States [2]. The critical barrier to reaching steam parameters of 760°C is the limitations of current boiler materials which can be fabricated and put into service for long times (>100,000 hours) in such an aggressive environment [3]. To meet this challenge, the U.S. Department of Energy (DOE) and the Ohio Coal Development Office (OCDO) is sponsoring the U.S. USC Steam Boiler Consortium, which is made up of the U.S. boiler manufactures, the Electric Power Research Institute (EPRI), Energy Industries of Ohio (EIO), and Oak Ridge National Laboratory (ORNL) [2]. The objective of this consortium is to identify, evaluate, and qualify the materials needed for operating a USC steam boiler at a target steam condition of 760°C and 35MPa.

During the fabrication of boiler components, it is necessary to deform (bend, flatten, forge, etc.) materials. The level of deformation or strain imparted during fabrication is in some cases limited by the ASME Boiler and Pressure Vessel Code (B&PV), or it is controlled by the fabrication and heat-treatement schedules of the manufacturer. The rules and methodology for low-alloy steels and stainless steels are well developed, but for advanced USC boilers, nickel-based alloys will be required. Rules do not exist for this class of materials; thus, ORNL, as part of the U.S. USC consortium, is studying the effect of cold-work on the creep behavior for some of the candidate nickel-based alloys through evaluation of cold-bent tubing.

Background

For a tube bend with a nominal outside radius of r and a nominal bending radius to the centerline of the tube of R, Section I of the ASME B&PV Code (PG-19) defines the degree of cold-work or cold strain for austenitic materials as [4]:

$$\%\text{Strain} = 100r/R \qquad (1)$$

Table PG-19 [4] defines the maximum forming strain as a function of design temperature for each material and specifies the minimum heat-treatment (annealing) temperature required if the allowed forming strain is exceeded. The basis for the rules in this table are based on two metallurgical phenomena associated with cold-strained austenitic material, recrystallization and ductility impingement.

Recrystallization occurs when a cold-worked material is exposed at high-temperature for sufficient time that the strain energy of cold-work is relieved by the formation of new fine grains through the processes of recovery, recrystallization, and grain growth. If this happens during service, the creep resistance of the alloy will be greatly reduced due to the dynamic recovery process and an increase in unpinned grain boundaries. For high-temperature service, this limit is generally 10% in Table PG-19.

Ductility impingement [5] is used to describe a phenomenon typically observed in higher-strength stainless steels which derive their creep strength from carbide precipitation. The distribution of carbides in stainless steels such as 347H is initially governed by the nucleation sites available for precipitation. The MX (NbC) precipitates in these alloys are highly misfit with the lattice so precipitation typically starts in regions of the lattice which can accommodate large strains such as dislocations or other precipitates. These sites are relatively few in annealed material, so carbide distribution is uniform and carbide size is large (growth is favored over nucleation). However, cold-straining introduces a large number of dislocations within the grains. In this case, a fine distribution of intragranular carbides will form on these dislocations during high-temperature exposure, which has been shown to greatly increase the creep resistance of the alloy [6, 7]. However, fine intragranular precipitation also causes creep deformation to concentrate in the grain boundary regions resulting in low stress-rupture ductility, where grain boundary cracking is the failure mode [8]. Ductility impingement is not well understood, but it is thought to have caused early failures in cold-worked 347H stainless steel and has lead to the rules in PG-19.

Experimental Procedure

The chemistry and product details for the materials in this study are shown in Table 1. HR6W is a Ni-Fe-Cr austenitic alloy with additions of Nb and W which provide strengthening through the precipitation of carbides and fine laves phase [9]. Haynes 230 is an ASME B&PV code approved solid solution strengthened nickel-based alloy [10].

Table 1. Materials – Chemistry in wt%

Material	C	Si	Mn	Cr	Ni	W	Fe	Mo	Co	Other
HR6W Tube Product chemistry	0.07	0.25	1.00	23.44	44.79	6	24.03			0.12Ti 0.3Nb
Haynes 230 Tube Nominal Chemistry **Maximum*	0.10	0.4	0.5	22	57	14	3*	2	5*	0.3 Al 0.02 La 0.015B*

Tubes, nominal dimensions 50.8mm O.D. and 10.2mm wall thickness, were bent in the solution annealed condition using standard manufacturing processes to strain levels (equation 1) of 35% by Riley Power, USA for HR6W, and 20% and 33.3% for Haynes 230 by The Babcock & Wilcox Company, USA.

Test specimens were fabricated by sectioning the tube bends into a J-type arrangement where end caps (fabricated from Haynes 230 bar stock) were welded (manual GTAW, 617 filler) onto the ends of the bend as shown in Figure 1 with a straight section at one end. The internal volume of the tube was partially filled with 304 stainless steel balls to minimize the explosive force of the tube if rupture occurred. A pressure-stem (Haynes alloy ® 25) was welded to the end cap which protruded from the furnace. This was welded to standard stainless steel high-pressure (autoclave) tubing.

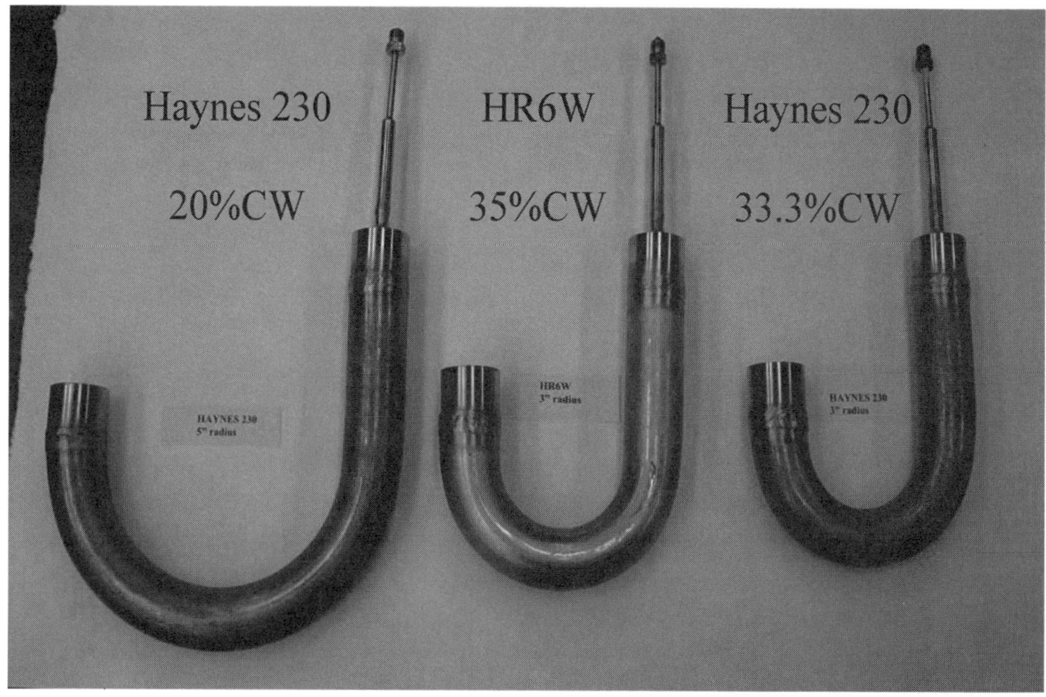

Figure 1. Pressurized tube-bend test specimens

Internal pressure (Ar-5%He gas) was applied and controlled via a high-pressure oil-less pump and pressure-gauges in line with each specimen. All three specimens were tested at the same time in the furnace, Figure 2. Temperature was controlled by thermocouples tied directly to the specimens and six thermocouples per specimen were used to monitor the temperature over the entire tube. Temperature and pressure were recorded on a digital data acquisition system and showed that for the duration of the test, the pressure was +/-25psi and the temperature was +/-2°C of the intended target. To measure creep deformation, the tests were periodically interrupted and the tube bends were removed for measurement by a coordinate measuring machine. Test conditions were 775°C-4,200psi for the HR6W and 775°C-5,600 psi for the Haynes 230 tubes.

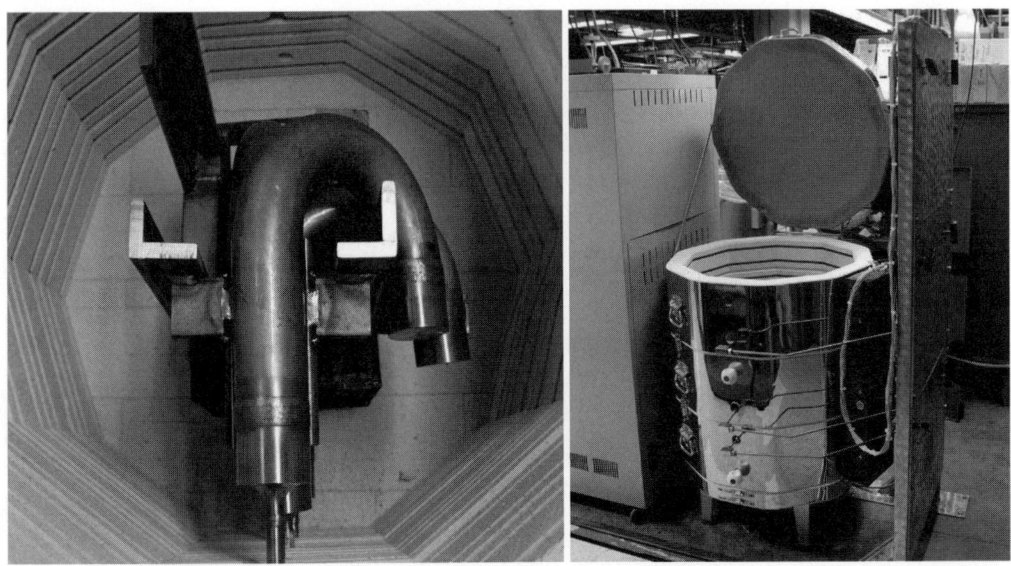

Figure 2. ORNL's high-pressure creep testing facility. Images show USC cold-bent tubes (left) in large furnace (right)

Results

HR6W

The HR6W test ruptured after 7,108.9 hours. The rupture occurred at the apex (90° location) of the bend 45 to 60 degrees from the extrados (EX) as shown in Figure 3. Destructive microstructural analysis showed extensive grain boundary cavitation throughout the tube and failure initiation on the inner diameter (I.D.) of the tube, Figure 4, although extensive cavitation was also observed on the outer diameter (O.D.) of the EX. Extensive early stage recrystallization was observed throughout the tube but was not uniquely associated with the heavily cavitated regions of the tube.

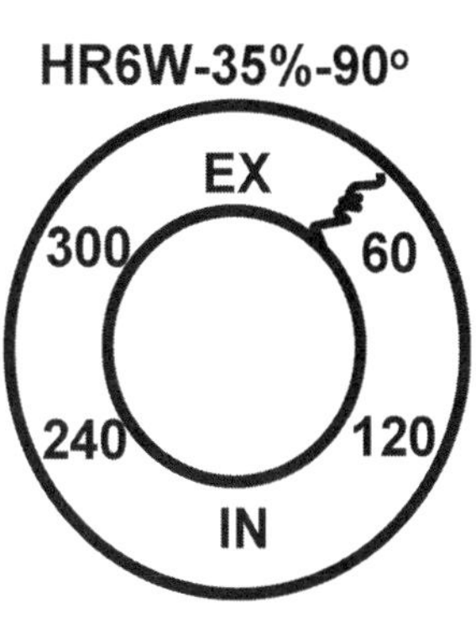

Figure 3. Location and macro image of HR6W rupture

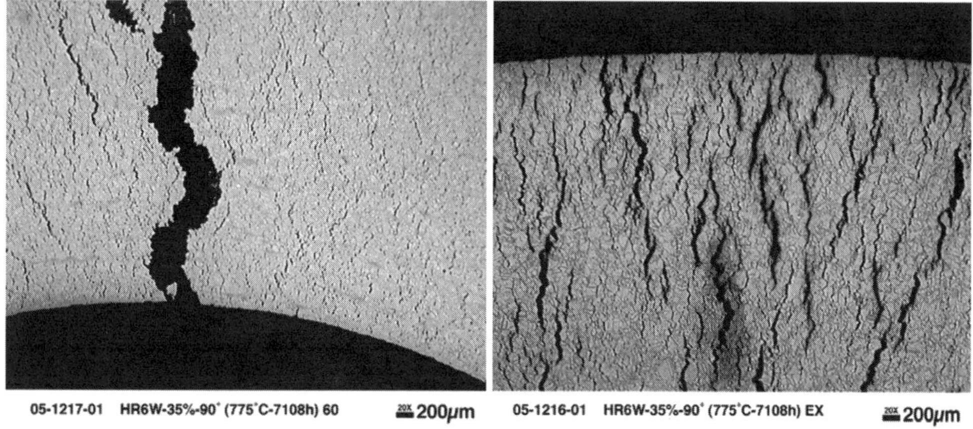

Figure 4. Optical micrographs from apex (90° location) of bend. Intergranular cavitation initiating in the I.D. region 45-60 degrees from the EX (left) caused failure, and extensive cavitation was also observed on the O.D. at the EX (right).

Haynes 230

The Haynes 230 tests were stopped after 8,020.0 hours and destructively evaluated as rupture had not occurred. Material was removed and metallographic mounts were prepared from the apex of the bend (90° location). These specimens were examined for evidence of recrystallization or cavitation around the tube from the extrados (EX) to the intrados (IN). For the tube with 33.3% cold-work, extensive cavitation and some recrystallization was observed at the 60° rotation near the I.D., Figure 5, where the HR6W tube bend ruptured. For the tube with 20% cold-work, a few possible cavities were found at the same location but this could not be confirmed because of uncertainties due to metallographic polishing and etching, Figure 6, and no recrystallization was observed. Very minor recrystallization was observed near the O.D. surface of the EX, figure 7, but no cavities were found and no other evidence of recrystallization was found around the entire tube.

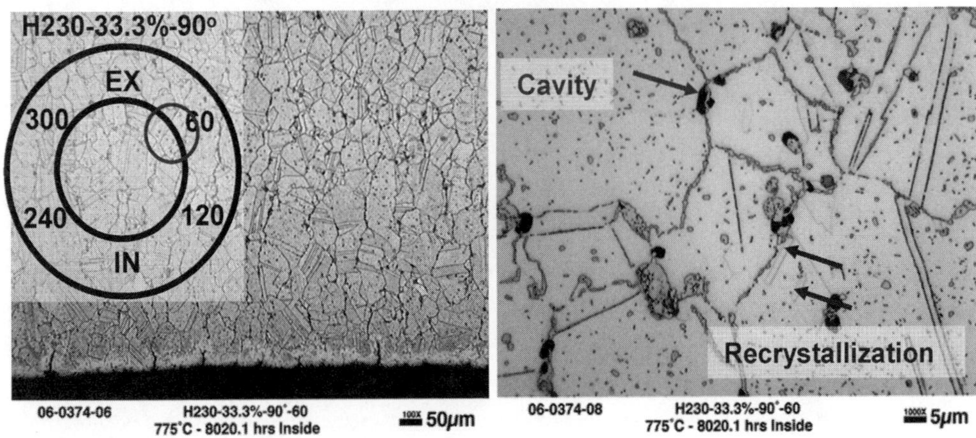

Figure 5. Location and optical micrographs showing extensive cavitation and recrystallization in creep-tested Haynes 230 with 33.3% cold work

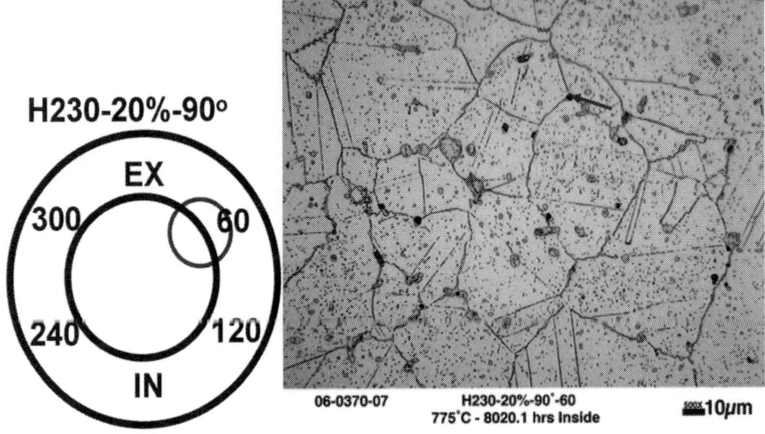

Figure 6. Location and optical micrograph of 20% cold-worked specimen showing possible cavitation

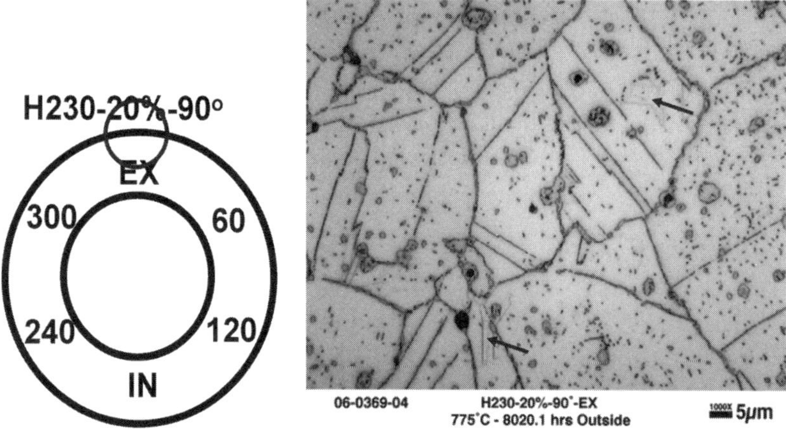

Figure 7. Location and optical micrograph of 20% cold-worked specimen showing very minor recrystallization (arrows) near the O.D. EX surface (recrystallization was not observed at any other locations)

Creep strain and ductility results

To evaluate the creep strain in the tubes, the hoop strain was defined by the initial diameter (d_o) and the final diameter (d) as:

$$\text{Hoop Strain (\%)} = 100(d - d_o)/d_o \tag{2}$$

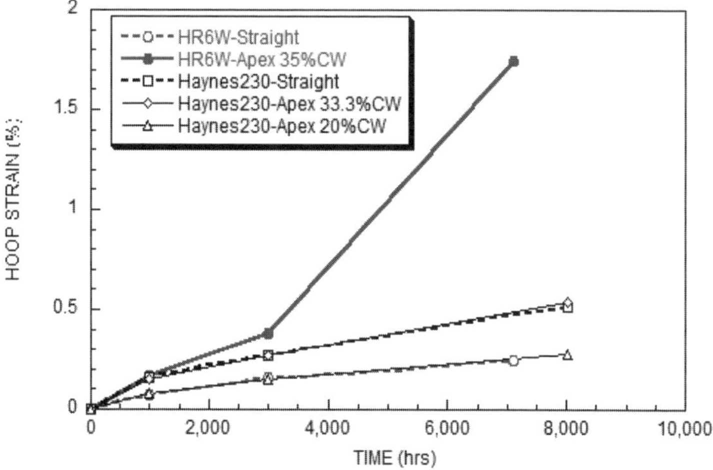

Figure 8. Interrupted hoop strain measurements for HR6W and Haynes 230 tube bends

The results for the interrupted strain measurements for the straight portion of each tube (no cold-work) and for the hoop strain at the apex of the tube bend (maximum cold-strain) are plotted in Figure 8. For the HR6W with 35% cold-work, the strain at rupture was 1.75% compared with an average straight tube strain of only 0.25%. For the Haynes 230, stopped before rupture, the hoop strain in the apex was equal to that of the straight tube in the case of the 33.3% cold-worked material, and it was slightly less for the material with 20% cold-work.

Discussion

To rationalize the recrystallization observed in the tube bends, the data of Mohn and Stanko [11] were plotted for onset of recrystallization in 10,000 hours as a function of cold-strain level in Figure 9, since this time is close to that of HR6W rupture time and stoppage of the Haynes 230 tests. The higher strain levels (35 and 33.3%) lay in the region beyond the onset of recrystallization for both materials, so recrystallization during the test is expected. However for the 20% strained Haynes 230, a temperature of 775°C is at the estimated point of onset for Haynes 230.

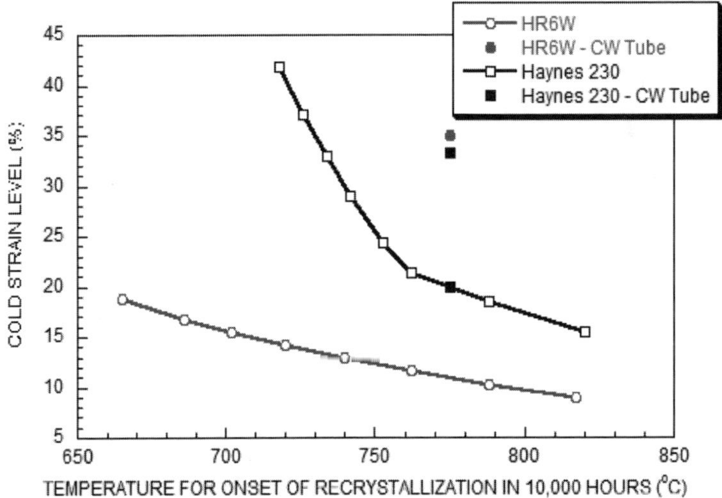

Figure 9. Temperature for the onset of recrystallization as a function of cold-work for HR6W and Haynes 230 based on ref. [11]

Evaluation of the creep damage within the tubes was performed by estimating the stress in the tube bends utilizing a simplified reference stress approach for the case of a straight tube and for a thin wall torus (tube bend under internal pressure) [12]. These values were then compared to the calculated time to rupture curves for the alloys based on experimental data produced as part of the USC steam boiler consortium for the specific heats used in the study [13]. A plot of these data is shown in Figure 10.

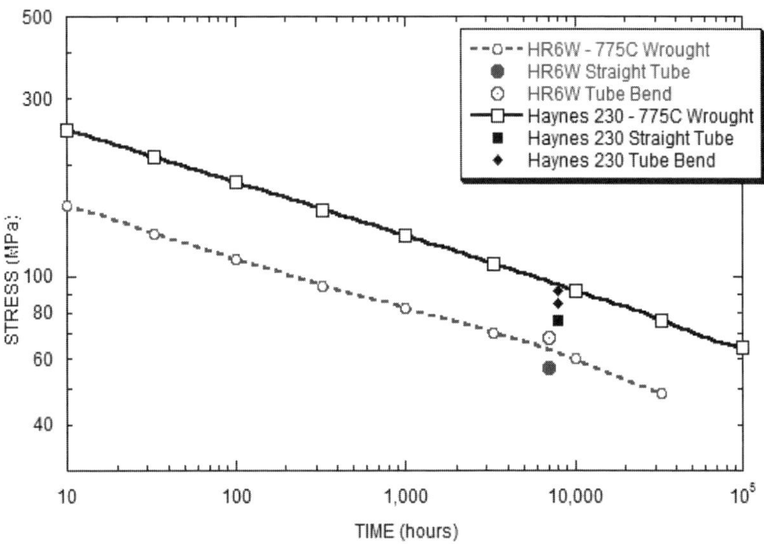

Figure 10. Estimated time to rupture curves at 775°C for HR6W and Haynes 230, and final tube bend examination point.

In the case of the HR6W straight tube, the rupture occurred at a stress reduction of approximately 10% that of the estimated rupture stress for the given time, but for the calculated tube bend stress the life exceeded expectations. This indicated that the observed recrystallization, although extensive, was a secondary factor in determining rupture life as the stress in the bent region of the tube was sufficiently high to cause rupture. The measured creep strains after 1,000 and 3,000 hours also show that the HR6W was creeping significantly faster in the bent region due to a higher stress. However, the extensive O.D. cracking at the EX in Figure 4, which did not cause failure, suggests that cold work did cause significant creep damage which in the long-term (for a lower stress longer-time test) may cause a premature failure.

For the Haynes 230 straight tube stress, the estimated life at the test condition was approximately 35,000 hours so the tests were stopped at 0.25 consumed life which corresponds to 80% of the stress required for rupture at the stoppage time. If the bend is taken into account, the 20% and 33.3% cold-worked tubes were stopped at 0.5 and 0.8 consumed life which corresponds to 90% and 96% of the stress required for rupture, respectively.

For the 20% strained tube, the absence of cavitation or widespread recrystallized regions is consistent with expectations that Haynes 230 is quite resistant to recrystallization for this condition. Examination of the creep strain accumulation (figure 8) in the tube bend shows some reduction in observed strain compared to the straight tube strain, but this difference is only about a factor of two. Furthermore, both regions show equivalent creep strain rates after 1,000 hours. This shows that although there are some differences in initial strain rate (geometric or material dependent) the steady-state creep behavior of the regions is equivalent, thus a ductility impingement mechanism, categorized by a sharp decrease in creep rate, is not observed.

For the 33.3% cold-strained region, the result is similar to the HR6W. Although recrystallization is observed, the life of the tube has not been reduced since 80% of creep life or 96% of the creep stress has been consumed without failure. The extensive cavitation is typical of a material containing significant creep damage.

Conclusions

Pressurized creep-rupture tests on cold-bent tubes were conducted on two candidate USC nickel-based alloys, HR6W and Haynes 230, to evaluate the role of cold-work and recrystallization on the creep-rupture life of the alloys. Tests were conducted to ~8,000 hours and the significant findings of this work were:

1. The austenitic alloy HR6W behaves similarly to other stainless steels, readily recrystallizing at high levels of cold strain (35%) during high-temperature testing. Interestingly, analysis of the rupture results showed the observed recrystallization in the tubes played a minor role in causing rupture as the small bend radius raised the stress sufficiently in the tube bend to cause failure. However, microstructural evidence for extensive creep damage in the areas of highest cold strain indicated that longer-term testing may cause a premature failure.

2. Haynes 230 was more resistant to recrystallization than HR6W.

3. For Haynes 230 strained to 33.3%, recrystallization occurred during testing but analysis of the rupture predictions and microstructural creep damage indicated that the recrystallization did not cause a large reduction in creep strength.

4. For Haynes 230 strained to 20%, little evidence of creep damage or recrystallization was observed as predicted. Creep strain measurement of the bent region and the straight tube region showed minor differences in initial creep rate, but minimum creep rates were comparable for the two regions. This suggested a ductility impingement mechanism was not operating in this alloy.

5. Based on the Haynes 230 test results, solid solution strengthened nickel-based alloys appear less susceptible to premature failure in service due to cold-work than traditional stainless steels. An upper limit of 20% strain for service temperatures at USC conditions is proposed.

Acknowledgments

Research was supported by the U.S. Department of Energy, Office of Fossil Energy, Advanced Research Materials Program, under Contract DE-AC05-00OR22725 with UT-Battelle, LLC. This work was performed in cooperation and with the support of the U.S. DOE/OCDO USC Steam Boiler Consortium. Special thanks Brian Sparks and Tom Geer of ORNL for their work on the experimental study and to Mike Brady who reviewed this manuscript.

References

1. R. Viswanathan, R. Purgert, U. Rao. "Materials for Ultra-Supercritical Coal-Fired Power Plant Boilers." *Materials for Advanced Power Engineering 2002, Proceedings Part II.* Forschungszentrum Julich GmbH, 2002. 1109-1129

2. R. Viswananthan, J.F. Henry, J. Tanzosh, G. Stanko, J. Shingledecker, B. Vitalis, R. Purgert. "U.S. Program on Materials Technology for Ultra-Supercritical Coal Power Plants." *Journal of Materials Engineering and Performance.* Vol. 14 (3) June 2005. 281-292

3. J.P. Shingledecker, I.G. Wright. "Evaluation of the Materials Technology Required for a 760°C Power Steam Boiler." *Proceedings to the 8th Liege Conference on Materials for Advanced Power Engineering 2006.* Forschungszentrum Jülich GmbH (2006) pp. 107-120

4. 2004 ASME Boiler and Pressure Vessel Code. Section I – Rules for Construction of Power Boilers, PG-19. © 2004 The American Society of Mechanical Engineers.

5. 2004 ASME Boiler and Pressure Vessel Code. Section II – Materials, Part D, Appendix A. © 2004 The American Society of Mechanical Engineers.

6. E.Evangelista, D.D'Angelo, L.Kloc, A.Rosen, S.Spigarelli. "Cold Work Effect on Particle Strengthening in Creep of AISI 347 Stainless Steel." *D. Coutsouradis et al. (eds.), Materials for Advanced Power Engineering, Part I.* © 1994 Kluwer Academic Publishers. 485-494

7. I.Ben-Haroe, A.Rosen, I.W.Hall. "Evolution of microstructure of AISI 347 stainless steel during heat treatement." *Materials Science and Technology*. July 1993, Vol. 9, 620-626

8. K.J.Irvine, J.D.Murray, F.B. Pickering. "The effect of heat-treatment and microstructure on the high-temperature ductility of 18%Cr-12%Ni-1%Nb steels." *Journal of The Iron and Steel Institute.* October 1960, 166-179

9. Y. Sawaragi, Y. Hayase, K. Yoshikawa. "Development of an Austenitic Alloy with High Elevated Temperature Strength and Superior Corrosion Resistance for Superheater Tubing of Ultra Super Critical Boilers." *Proceedings of the International Conference on Stainless Steels,* 1991, Chiba, ISIJ.

10. Haynes ® 230 ® Alloy product brochure. *H-3000G* © 2000, Haynes International, Inc.

11. W.Mohn, G.Stanko. *This proceedings*

12. *Boiler Materials for Ultrasupercritical Coal Power Plants – Task 8, An overview of Reference Stress Approach,* NETL/DOE/OCDO, 2003. USC T-6.

13. J.P.Shingledecker. *Unpublished research, 2007.*

Advances in Materials Technology for Fossil Power Plants
Proceedings from the Fifth International Conference
R. Viswanathan, D. Gandy, K. Coleman, editors, p 29-45

DOI: 10.1361/cp2007epri0029

Refurbishment of Aged PC Power Plants with Advanced USC Technology

Masafumi Fukuda*
Hideyuki Sone **
Eiji Saito ***
Yoshinori Tanaka ****
Takeo Takahashi *****
Akira Shiibashi ******
Jun Iwasaki *******
Shinichi Takano ********
Sakae Izumi *********

*Sophia University, Tokyo, Japan
**The Institute of Applied Energy, Tokyo, Japan
***Hitachi Ltd., Hitachi, Japan
****Mitsubishi Heavy Industries, Ltd., Takasago, Japan
*****Toshiba Corporation, Tokyo, Japan
******Mitsubishi Heavy Industries, Ltd., Yokohama, Japan
*******Babcock Hitachi K. K., Tokyo, Japan
********Ishikawajima Harima Heavy Industries, Ltd., Tokyo, Japan
*********Fuji Electric Systems Co. Ltd, Kawasaki, Japan

Abstract

The capacity of PC power plants in Japan rose to 35GW in 2004. The most current plants have a 600 deg-C class steam temperature and a net thermal efficiency of approximately 42% (HHV). Older plants, which were built in the '70s and early '80s, will reach the point where they will need to be rebuilt or refurbished in the near future. The steam temperatures of the older plants are 538 deg-C or 566 deg-C. We have done a case study on the refurbishment of one of these plants with the advanced USC technology that uses a 700 deg-C class steam temperature in order to increase the thermal efficiency and to reduce CO_2 emissions.

The model plant studied for refurbishing has a 24.1MPa/538 deg-C /538 deg-C steam condition. We studied three possible systems for the refurbishing. The first was a double reheat system with 35MPa/700 deg-C /720 deg-C /720 deg-C steam conditions, the second one was a single reheat 25MPa/700 deg-C/720 deg-C system, the last one was a single reheat 24.1MPa/610 deg-C/720 deg-C system. In addition to these, the most current technology system with 600 deg-C main and reheat temperatures was studied for comparison.

The study showed that the advanced USC Technology is suitable for refurbishing old plants. It is economical and environmentally-friendly because it can reuse many of the parts from the old plants and the thermal efficiency is much higher than the current 600 deg-C plants. Therefore, CO_2 reduction is achieved economically through refurbishment.

1. Introduction

Since the "Oil Shock" during the '70s, many PC power plants have been built in Japan to establish a more secure energy supply system. Figure 1-1 shows the history of the capacity of PC power plants. The capacity rose to 35GW in 2004 and 1/4 of the electricity in Japan today, is produced by coal. However, the climate change brought about by CO_2 emissions is a well known problem. The pressure on PC power plants to reduce CO_2 emissions is much larger than before. The most current plants have a 600 deg-C class steam temperature and the net thermal efficiency is approximately 42% (HHV). This means that the CO_2 emission of the most current plants is 10% less than those of the plants from the '70's. It is believed that future coal power plants will reduce CO_2 emissions even further.

There are many plans for CO_2 reduction. The US government is promoting the FutureGen project that aims at a 95% CO_2 reduction with CO_2 recovery using IGCC and CO_2 sequestration technology. In Japan, we have the IGCC project, which is working to achieve higher thermal efficiency than the latest PC plants in order to reduce CO_2 emissions. Raising the steam temperature of PC plants is another option.

As seen in figure 1-1 there are many older PC plants, which were built in the '70s and early '80s, which will reach the point where they will need to be rebuilt or refurbished in the near future. Almost all Japanese PC plants already have environmental protection equipment like DeNOx,

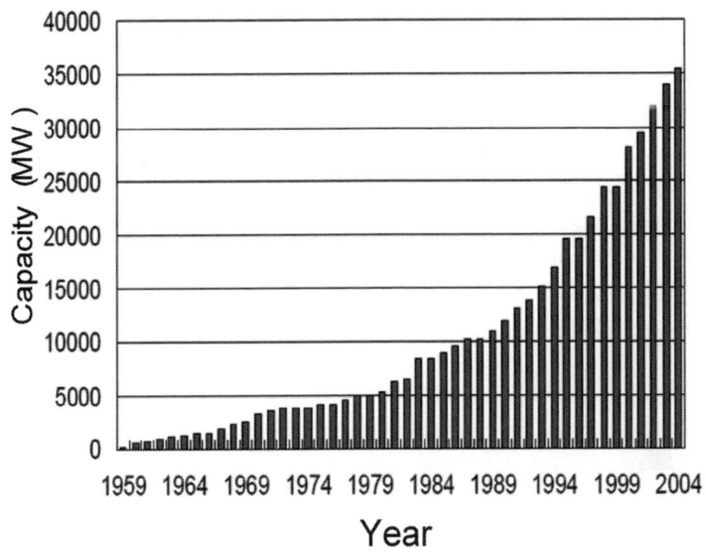

Fig. 1-1 Capacity of PC Plants in Japan

DeSOx, and EP which make up a large part of the plants. If we can refurbish the old plants with the higher steam temperature technology it is possible to reduce CO_2 at a much lower cost than if we used the other technologies because we can reuse the expensive environmental protection equipment and some other cold parts from the existing plants.

This paper presents the study of the refurbishment of a PC power plant with a 700 deg-C class steam condition, using the advanced USC technology.

2. Selection of a model plant

At first we selected a model plant to be refurbished. Our earlier study showed that, for the 700 deg-C class temperature steam condition refurbishment, a plant with super critical steam pressure is more suitable than one with sub-critical steam pressure, and that the plant capacity should be more than 400MW. Fig. 2-1 shows the number of existing PC plants and their steam condition and capacity. Some plants from the early '80s have super critical steam pressure and a capacity of over 400MW. One advantage to refurbishing these plants using the 700 deg-C class technology is that we can reuse many parts from the old plants. Additionally, in around 2020 when the 700 deg-C class technology is expected to be ready, these plants will be over 40 years old and considered ready to be refurbished.

Therefore, we selected a model plant with the following parameters.

Model Plant Super Critical Single Reheat

Net Plant Output	500 MW
Main Steam Pressure	24.1 MPa

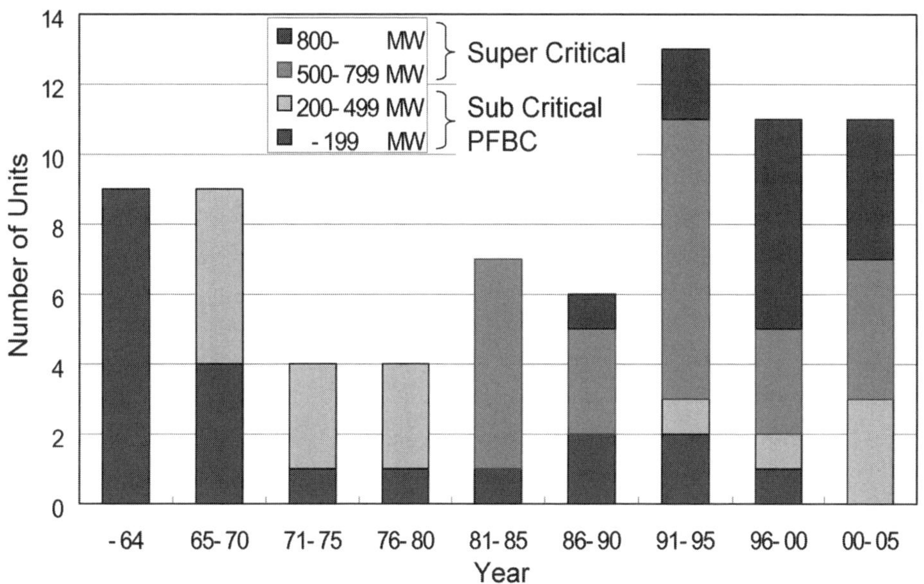

Fig. 2-1 Number of Existing Plants and Steam Condition

Main Steam Temperature	538 deg-C
Reheat Steam Temperature	538 deg-C

3. Case Studies

We set up 4 case studies for up-grading the plant. The net plant output is 500MW for all cases.

Case A	Double Reheat	
	Main Steam Pressure	35 MPa
	Main Steam Temperature	700 deg-C
	1st Reheat Steam Temperature	720 deg-C
	2nd Reheat Steam Temperature	720 deg-C
Case B	Single Reheat	
	Main Steam Pressure	25 MPa
	Main Steam Temperature	700 deg-C
	Reheat Steam Temperature	720 deg-C
Case C	Single Reheat	
	Main Steam Pressure	24.1 MPa
	Main Steam Temperature	610 deg-C
	Reheat Steam Temperature	720 deg-C
Case D	Single Reheat	
	Main Steam Pressure	25 MPa
	Main Steam Temperature	600 deg-C
	Reheat Steam Temperature	600 deg-C

Case A had a 700 deg-C class double reheat condition. It was expected to have the highest level of thermal efficiency improvement. But, it would be necessary to change a single reheat plant to a double reheat plant.

Case B had a 700 deg-C class single reheat condition. It was expected to have a fairly high level of thermal efficiency improvement and not require extreme remodeling.

Case C had a 700 deg-C class temperature only in the reheat system. The temperature of the main steam system is kept at 610 deg-C. This configuration would enable us to choose ferritic materials for the main steam system. It was expected to have a good level of thermal efficiency improvement and require only light remodeling without the heavy use of Ni-based alloys.

Case D has the most current steam condition. It was chosen to compare the 700 deg-C technology with the latest 600 deg-C technology.

Table 4-1-1 Boiler Materials

Parts	Candidate Materials	
Heat Exchanger Tube	KaSUS310J2TB(NF709R)	(22Cr-25Ni-Mo-Nb-N)
	KaSUS310J1TB(HR3C)	(25Cr-20Ni-Nb-N)
	HR6W	(23Cr-40Ni-6W-Ti-Nb)
	SB167 UNS N06617 (Alloy617)	(52Ni-22Cr-13Co-9Mo-Ti-Al)
	CCA617	(55Ni-22Cr-3W-8Mo-11Co-Al)
	SB622 UNS N06230 (Haynes230)	(57Ni-22Cr-14W-2Mo-La)
	Alloy740	(50Ni-25Cr-20Co-2Ti-2Nb-V-Al)
	Nimonic263	(50Ni-20Cr-20Co-6Mo-2Ti-Al)
Header Pipe	KaSUS304J1HTB(Super304H)	(18Cr-9Ni-3Cu-Nb-N)
	KaSUS321J2HTB(TEMPALOY AA-1)	(18Cr-10Ni-3Cu-Ti-Nb)
	KaSUS347J1TB(XA704)	(18Cr-9Ni-W-Nb)
	KaSUS310J2TB(NF709R)	(22Cr-25Ni-Mo-Nb-N)
	KaSUS310J1TB(HR3C)	(25Cr-20Ni-Nb-N)
	HR6W	(23Cr-40Ni-6W-Ti-Nb)
	SB167 UNS N06617 (Alloy617)	(52Ni-22Cr-13Co-9Mo-Ti-Al)
	CCA617	(55Ni-22Cr-3W-8Mo-11Co-Al)
	SB622 UNS N06230 (Haynes230)	(57Ni-22Cr-14W-2Mo-La)
	Alloy740	(50Ni-25Cr-20Co-2Ti-2Nb-V-Al)
	Nimonic263	(50Ni-20Cr-20Co-6Mo-2Ti-Al)
Header, Pipe (<650℃)	Advanced 9Cr-W-Co Steel	
	· 9Cr-3W-3Co Steel	(9Cr-3W-3Co-Nb-V-B)
	· LowC-9Cr-2.4W-1.8Co Steel	(0.03C-9Cr-2.4W-1.8Co-Nb-V-B)

Table 4-2-1 Turbine Materials

Components	Parts	Candidate Materials
VHPT HPT IPT (700~650℃)	Rotor	LTES
		USC141
		FENIX-700
		IN625, IN617
		12Cr Steel
	Blade	U500, U520, IN-X750, IN713C, M252
		USC141
	Nozzle	MAR-M509, X-45, IN713C
		USC141
	Inner Casing	LTES (Cast)
		IN625, 617 (Cast)
		Austenitic Cast Steel
		12Cr Cast Steel
	Outer Casing	12Cr Cast Steel
	Bolt	LTES
		USC141
		U500, Waspaloy, etc

4. Potential Materials *

4-1 Boiler Materials

Ni or Ni-Fe-based alloys were selected for the 700 deg-C class parts. Austenitic steels were considered for use for the parts under 680 deg-C. We also considered an advanced 9Cr steel (Abe alloy) for the parts under 650 deg-C to reduce the overall cost.

4-2 Turbine Materials

We considered four Ni-based alloys for the turbine rotors. LTES(1) is a Ni-based alloy that has been developed by Mitsubishi Heavy Industries, Ltd. This alloy was developed to have a thermal expansion coefficient similar to 12Cr steel, so it conforms well to conventional steels (Fig. 4-2-1). USC141 was developed by Hitachi and has characteristics similar to LTES. FENIX-700(2), which has superior long-term stability at 700 deg-C was developed from IN706 by Hitachi (Fig. 4-2-2). The other materials are conventional steam or gas turbine materials or materials for the chemical industry.

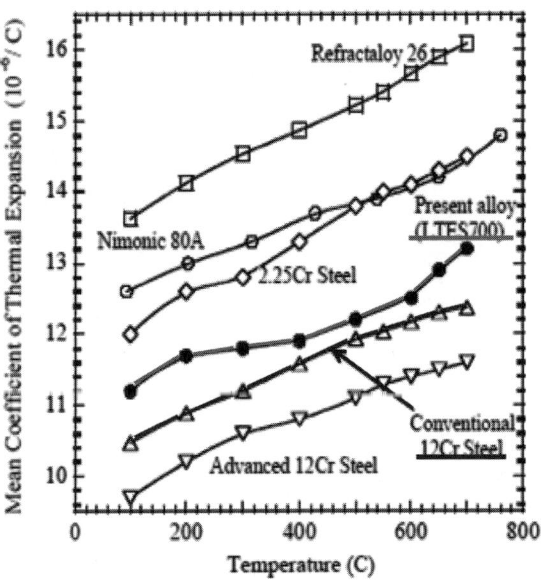

Fig. 4-2-1 Coefficient of Thermal Expansion

* Product names mentioned herein may be trademarks of their respective companies

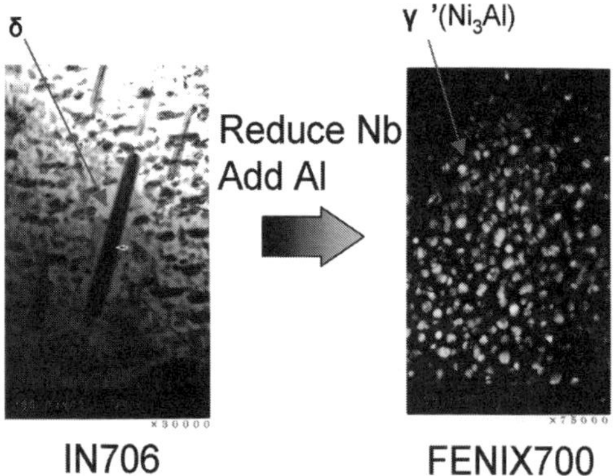

Fig. 4-2-2 FENIX700

5. Turbine Design Concepts

We used three turbine design concepts for the case study. They are 'Welded Rotor', 'Divided Turbine' and 'Cooling'.

5-1 Welded Rotor

It is necessary to use Ni or Ni-Fe-based wrought material for rotors of 700 deg-C class steam turbines unless advanced cooling technology is used. However, it is difficult to make large Ni or Ni-Fe-based wrought parts and it is assumed that the largest possible size of a part is around 10 tons. Turbine rotors for large capacity power stations usually weigh 20 to 50 tons. The 'Welded Rotor' concept was proposed to make large rotors by welding Ni or Ni-Fe-based wrought material and steel. Fig. 5-1-1 shows a typical welded rotor.

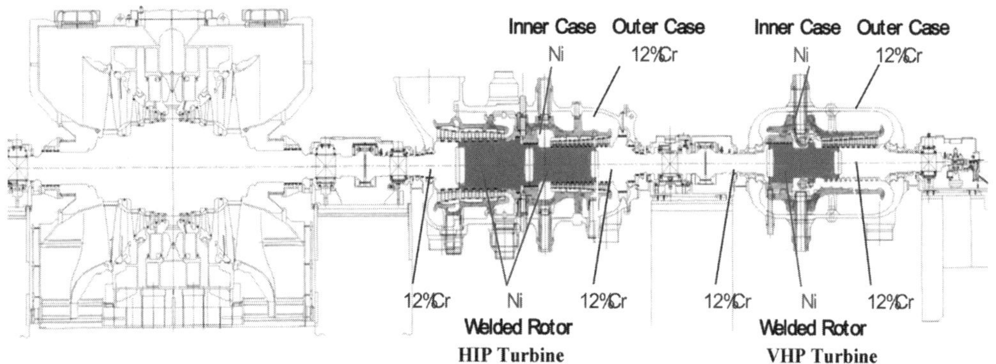

Fig. 5-1-1 Welded Rotor

5-2 Divided Turbine

In this concept, the turbine is divided into two small turbines so that one of them is small enough to be made of Ni or Ni-Fe-based materials. Fig. 5-2-1 shows an example of the concept, in which a high-pressure turbine (HPT) is divided into HPT1 and HPT2. HPT1 is made of Ni or Ni-Fe-based materials and the weight of the rotor is about 10 tons.

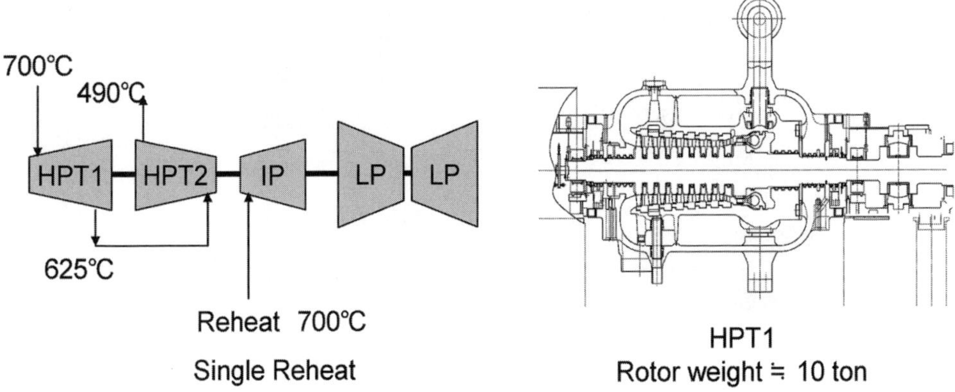

Fig. 5-2-1 Divided Turbine

5-3 Cooling

By introducing the gas turbine cooling technology, we can reduce the use of Ni or Ni-Fe-based materials [3]. Fig. 5-3-1 shows an intermediate pressure turbine (IPT) that is cooled. Ni-based alloy is used only for the blades and some stationary parts around the steam inlet.

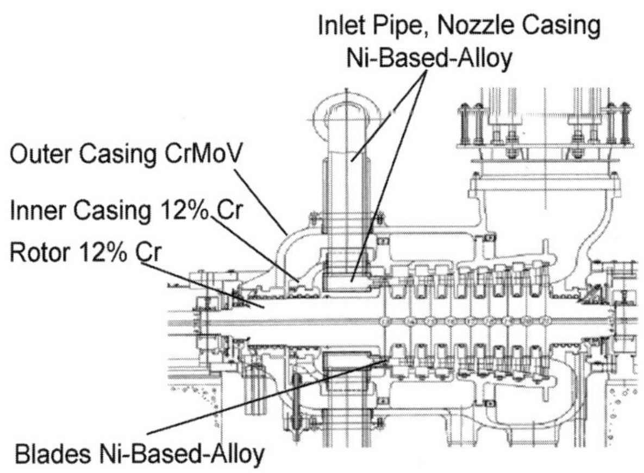

Fig. 5-3-1 Cooled Turbine

6. Case Study

6-1 Case A

Fig. 6-1-1 shows the system configuration of Case A. A typical double reheat steam cycle was used and the pressure and temperature were raised to a 35MPa 700 deg-C main steam condition. Reheat systems have a 720 deg-C steam temperature.

The selected materials for Case A are shown in Fig. 6-1-2. Ni or Ni-Fe-based alloys were chosen for a part of the super heaters and reheaters, the large steam pipes and the valves going from the boiler to the turbines, and a part of the turbine rotors and casings. The turbine rotors consist of

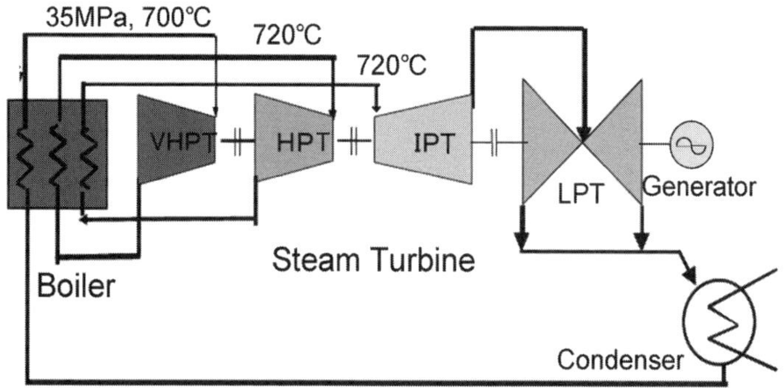

Fig. 6-1-1 System Configuration of Case A

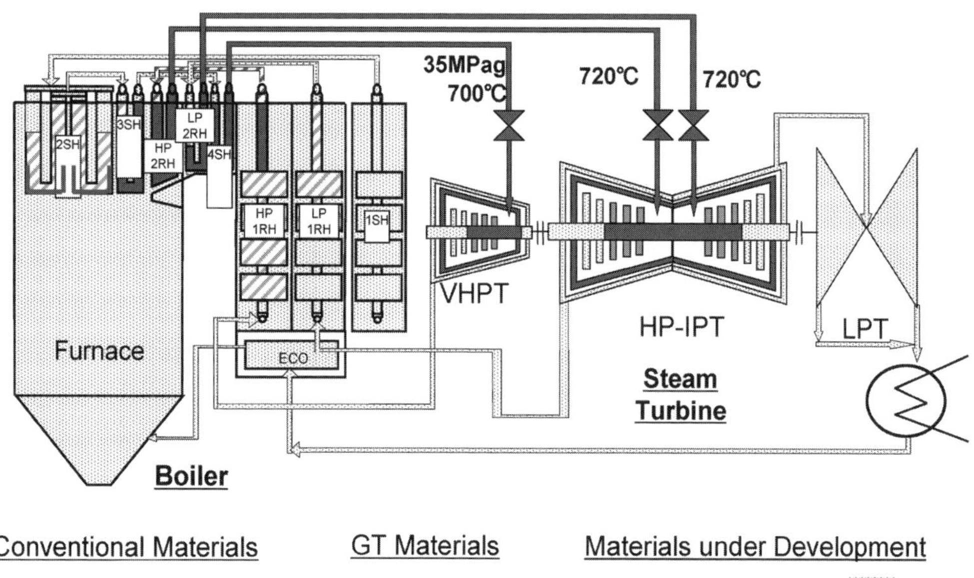

Fig. 6-1-2 Selected Materials for Case A

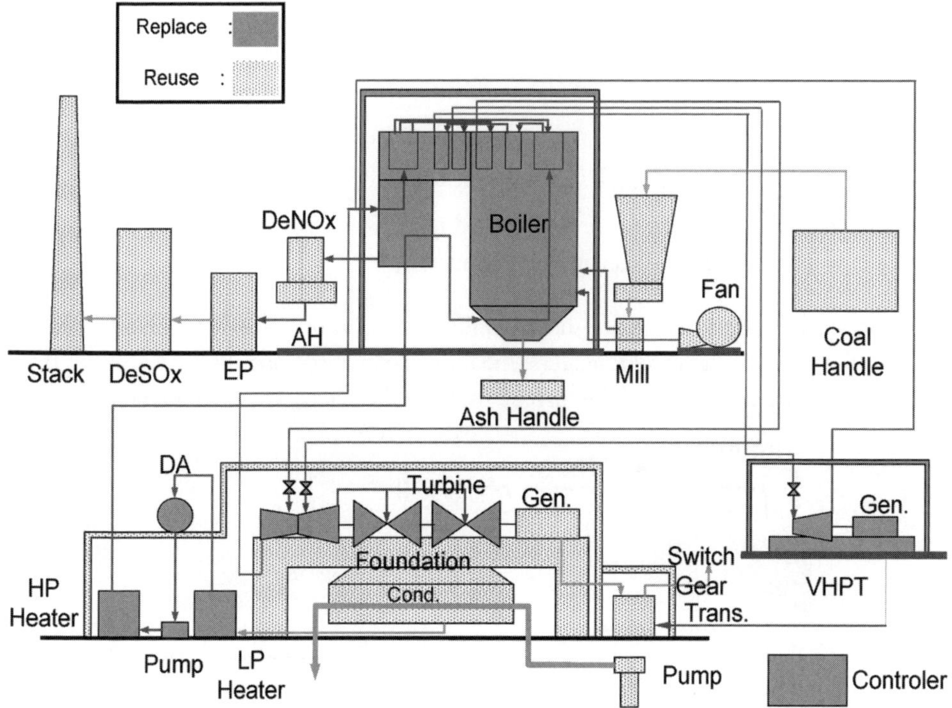

Fig. 6-1-3 Scope of Replacement of Case A

Ni-based alloy and 12Cr Steel, which are welded together. One of the potential materials for the welded rotors is LTES. The turbine nozzles and blades for the high temperature stages use Ni-based materials that are being used for gas turbines. The steam pipes that connect the reheaters are made of ferritic or austenitic steels that are being developed.

Fig. 6-1-3 shows the parts that were replaced in the remodeling. In this case, the boiler was completely replaced and a very high-pressure turbine (VHPT) was added because the original single-reheat system was changed to a double reheat system. The turbines, heaters, and the pipes between the boiler and the turbines were replaced, as well. In spite of the heavy remodeling, we can see that a large portion of the plant can be reused.

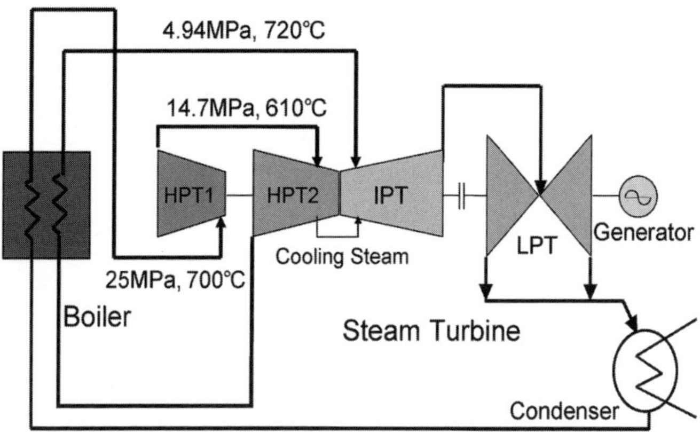

Fig. 6-2-1 System Configuration of Case B

6-2 Case B

Fig. 6-2-1 shows the system configuration of Case B. A high pressure turbine (HPT) was divided into two turbines, HPT1 and HPT2. HPT1, which has 25MPa and a 700 deg-C inlet steam and a 610 deg-C exit steam condition, is the high temperature part of the HPT. HPT1 is a relatively small turbine because the pressure ratio is only 1.7. A trial design of HPT1 with 10 ton class rotor material is shown on Fig. 5-2-1. The HPT2 uses conventional steam turbine materials as the inlet temperature is 610 deg-C. Because the 720-deg-C IPT is cooled, the rotor and the casings of HPT2 and IPT are made of conventional steam turbine materials.

The selected materials for Case B are shown in Fig. 6-2-2. Ni or Ni-Fe-based alloys were chosen for a part of the super heaters and a reheater, the large steam pipes and the valves going from the boiler to the HPT1 and IPT, and a rotor and a casing for HPT1. A potential rotor material for HPT1 is FENIX700. The turbine nozzles and blades for the high temperature stages use Ni-based materials that are being used for gas turbines. The steam pipes that connect the superheaters and the reheaters are made of ferritic or austenitic steels that are currently being developed.

Fig. 6-2-3 shows the parts that are replaced in the remodeling. In this case, almost all of the heat exchangers of the boiler are replaced, but the original furnace and the steel structures are reused. Because of the divided configuration of the HPT, almost all of the turbines and their related parts are replaced. The number of replaced parts is reduced if the welded rotor concept similar to Case A is applied to the HPT. High-pressure heaters, and the pipes from the boiler to the turbines are also replaced. We can reuse more boiler parts in Case B than in Case A because Case B uses a

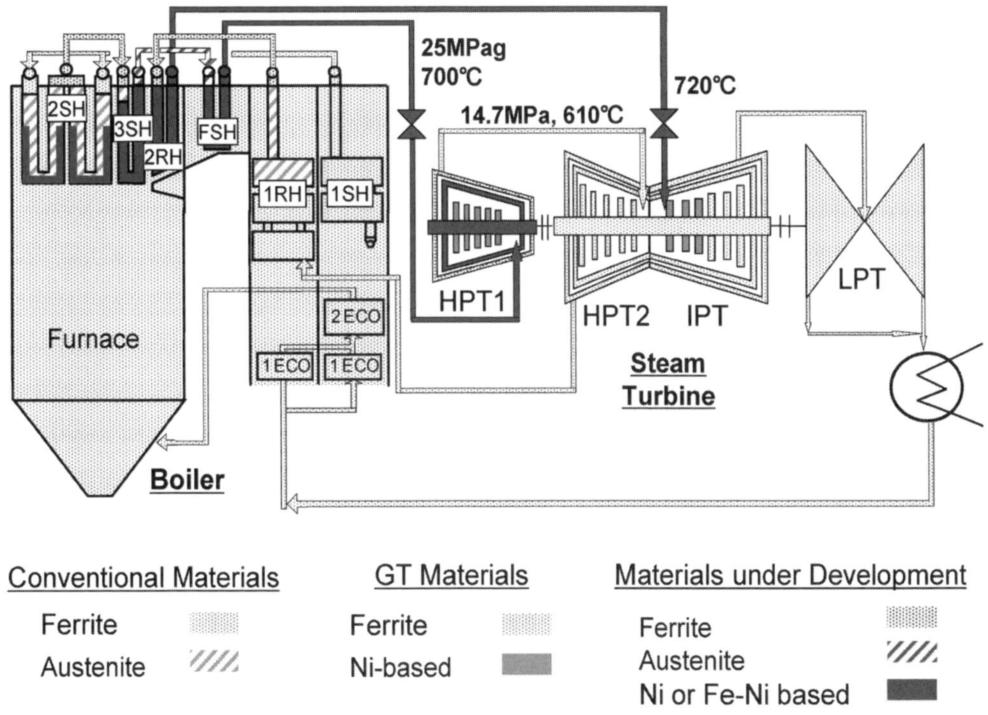

Fig. 6-2-2 Selected Materials for Case B

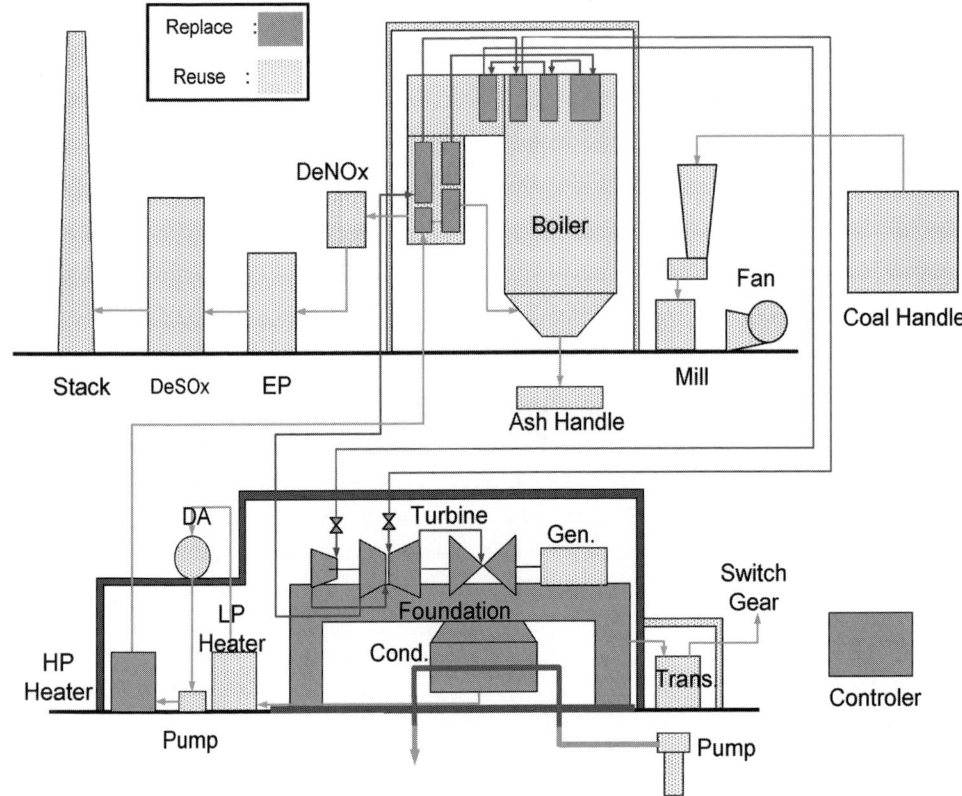

Fig. 6-2-3 Scope of Replacement of Case B

single reheat configuration in the remodeling.

6-3 Case C

Fig. 6-3-1 shows the system configuration of Case C. Case C has a typical single reheat system. The main steam temperature is kept to 610 deg-C to avoid using Ni-based alloys in the main steam system. The 720 deg-C IPT is cooled.

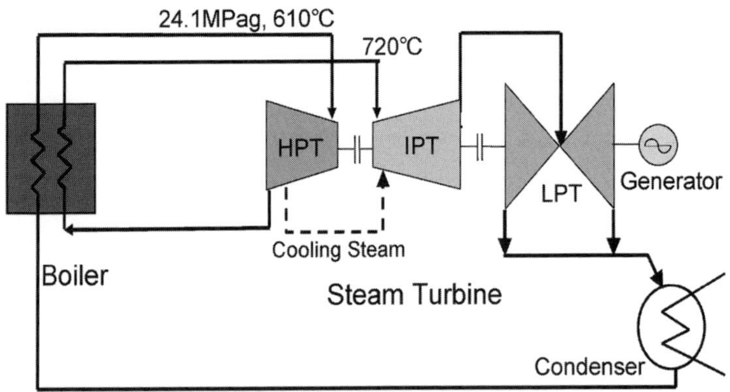

Fig. 6-3-1 System Configuration of Case C

The selected materials for Case C are shown in Fig. 6-3-2. Ni or Ni-Fe-based alloys were chosen for one of the reheaters and the large steam pipe and valve going from the reheater to the IPT.

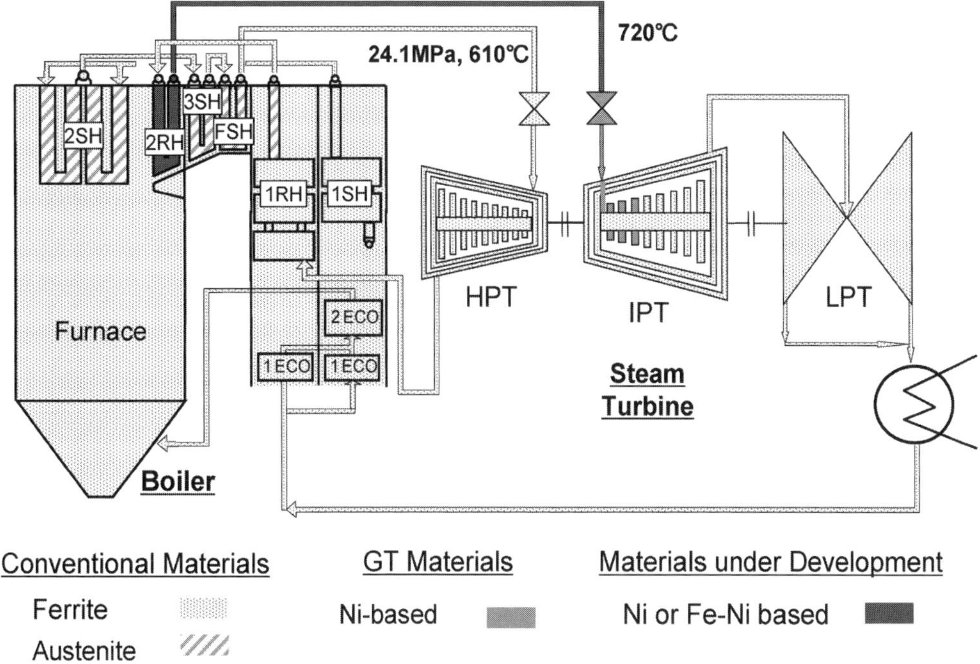

Fig. 6-3-2 Selected Materials for Case C

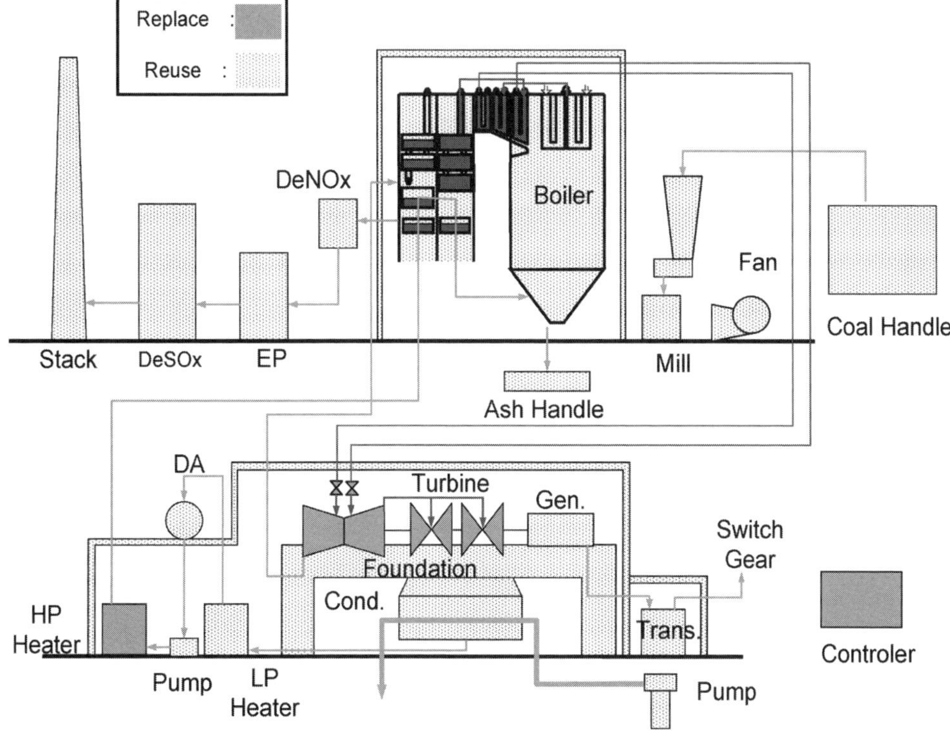

Fig. 6-3-3 Scope of Replacement of Case C

The IPT rotor and casings, except for the steam inlet portion, don't need Ni or Ni-Fe-based alloys because of the cooling. Only the turbine nozzles and blades for the high temperature stages use Ni-based materials that have been used for gas turbines for many years.

Fig. 6-3-3 shows the parts that are replaced in the remodeling. We can see that a lot of parts can be reused. In this case, some of the boiler's heat exchangers are replaced, but the existing furnace and the steel structures are reused. Even though they are replaced, a large amount of the heat exchangers use conventional ferritic or austenitic steel. Although the turbines are replaced, the turbine foundation, condenser, generator and the turbine building can be reused. High-pressure heaters and the pipes from the boiler to the turbines are also replaced. The main steam pipe from the boiler to the HPT uses conventional ferritic steel.

7. Thermal efficiency and CO_2 emissions

7-1 Thermal efficiency

Fig. 7-1-1 shows the estimated plant thermal efficiency (Net, HHV) of each case after refurbishment. The thermal efficiency of the original plant is about 40%. Case D is a reference case, which uses the current technology. Main and reheat steam temperatures are 600 deg-C and the system configuration is a single reheat cycle. Compared to Case D, Case A has a 3.8 point advantage. Case A has the best thermal efficiency, 46%, and requires relatively heavy remodeling. The thermal efficiency of Case B is 44.3% and 2.1 points better than that of Case D and 1.7 points less than that of Case A. As Case B has a single reheat cycle, the remodeling is not as extensive as in Case A. The thermal efficiency of Case C is 43.4%. Because Case C employs a 610 deg-C main steam temperature and a 720 deg-C reheat steam temperature, the thermal efficiency advantage of Case C is less than Case A and B. However, it doesn't need any Ni or Ni-Fe-based alloy parts in the main steam system.

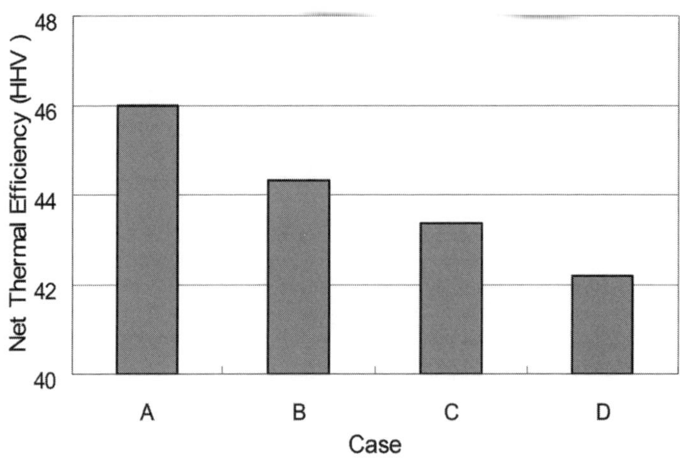

Fig. 7-1-1 Estimated Thermal Efficiency

7-2 CO_2 Emissions

The CO_2 emission in each case study was calculated based on its thermal efficiency. Fig 7-2-1 shows the CO_2 emissions based on the entire life of the plant, which includes CO_2 emissions during construction and the CO_2 emissions equivalent to the methane leaks from coal mines. Case A has 20% less CO_2 emissions than sub-critical PC plants. Case A and IGCC have almost the same CO_2 emissions. IGCC is a project that aims at a 46% thermal efficiency in Japan.

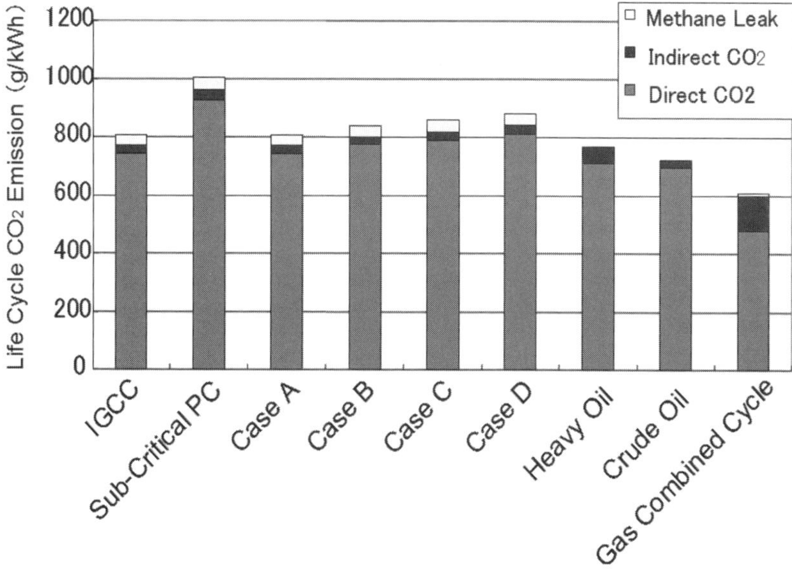

Fig. 7-2-1 CO_2 Emissions

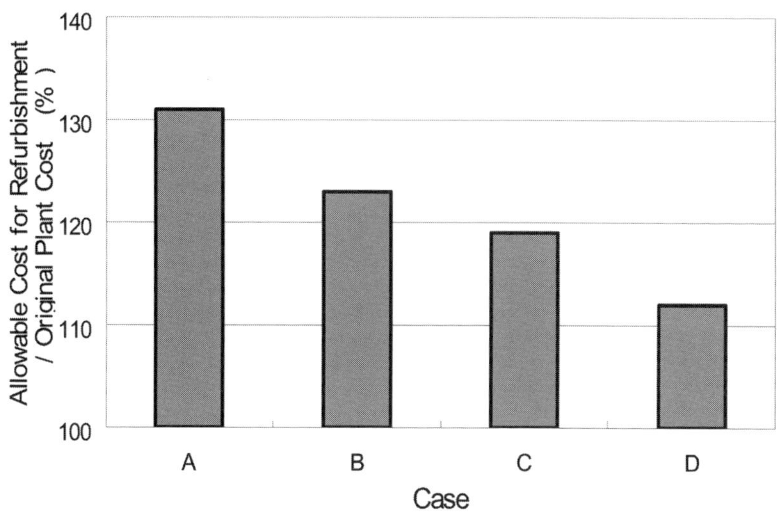

Fig. 8-1 Justifiable Cost for Refurbishment

8. Economics

Assuming the same cost of electricity and an additional life span equal to the one the original plant had, we calculated the justifiable cost for the refurbishment in each case. Because the rate of thermal efficiency is so much higher in each case, than that of the original plant, the justifiable cost for the refurbishment is much higher than the cost to build an original plant (Fig. 8-1). As mentioned earlier, we can reuse relatively cold parts in the refurbishment. The amount of the parts that can be reused is 50 - 60% of the original plant. This means that we can spend the justifiable cost for the refurbishment to remodel the relatively hot parts that make up 40 - 50% of the original plant.

9. Conclusion

We will have many old PC power plants in Japan in the near future. Our case study demonstrated the refurbishment of a 538 deg-C class super critical pressure old plant with the 700 deg-C class advanced USC technology and showed the following results.

It is possible to increase the net thermal efficiency (HHV) of the plant from 40% to 46% by using a double reheat cycle (35MPa / 700 C / 720 C / 720 C) for the refurbishment. In addition, a considerable number of parts from the original plant can be reused. Therefore, a 15% CO_2 reduction can be achieved economically.

If a single reheat cycle (25MPa / 700 C / 720 C) is used for the refurbishment, the achieved thermal efficiency becomes 44.3% and the scope of reused parts becomes larger than that of the double reheat cycle.

Increasing only the reheat steam temperature to 700 deg-C class is a reasonable option for the refurbishment. The updated steam condition is a single reheat 24.1MPa / 610 deg-C / 720 deg-C. In this case we can reuse a lot of parts from the original plant although the thermal efficiency is only 43.4%. In addition, we can use conventional ferritic steel for the main steam system.

Therefore, the study showed that the Advanced USC Technology is suitable for refurbishing old plants. It is economical and environmentally-friendly because it can reuse many parts from the old plant and the thermal efficiency is much higher than the most current 600 deg-C plants. Therefore, CO_2 reduction is achieved economically through the refurbishment.

References

1. M. Fukuda, et al., Materials and Design for Advanced High Temperature Steam Turbines, *Advances in Materials Technology for Fossil Power Plants,* (2005), pp. 491-505.

2. R. Yamamoto, et al., Development of Wrought NI-Based Superalloy with Low Thermal Expansion for 700C Steam Turbines, *Advances in Materials Technology for Fossil Power Plants*, (2005), pp. 623-637.

3. S. Imano, et al., Modification of Ni-Fe base Superalloy for Steam Turbine Applications, *Advances in Materials Technology for Fossil Power Plants*, (2005), pp. 575-586.

Advances in Materials Technology for Fossil Power Plants
Proceedings from the Fifth International Conference
R. Viswanathan, D. Gandy, K. Coleman, editors, p 46-58

DOI: 10.1361/cp2007epri0046

The Development of Electric Power and High Temperature Materials Application in China–An Overview

Fusheng Lin
Shanghai Power Equipment Research Institute, Shanghai 200240, China
Shichang Cheng
Central Iron and Steel Research Institute, Beijing, Beijing 100081, China
Xishan Xie
University of Science and Technology Beijing, Beijing 100083, China

Abstract

The rapid development of Chinese economy (recently in the order of 10%/year) is requiring sustainable growth of power generation to meet its demand. In more than half century after the foundation of People's Republic of China, the Chinese power industry has reached a high level. Up to now, the total installed capacity of electricity and annual overall electricity generation have both jumped to the 2nd position in the world, just next to United States. A historical review and forecast of China electricity demand to the year of 2010 and 2020 will be introduced.

Chinese power plants as well as those worldwide are facing to increase thermal efficiency and to decrease the emission of CO_2, SO_X and NO_X. According to the national resources of coal and electricity market requirements in the future 15 years power generation especially the ultra-super-critical (USC) power plants with the steam temperature up to 600°C or higher will get a rapid development. The first two series of 2×1000MW USC power units with the steam parameters 600°C, 26.25MPa have been put into service in November and December 2006 respectively. In recent years more than 30 USC power units will be installed in China.

USC power plant development will adopt a variety of qualified high temperature materials for boiler and turbine manufacturing. Among those materials the modified 9-12%Cr ferritic steels, Ni-Cr austenitic steels and a part of nickel-base superalloys have been paid special attention in Chinese materials market.

Introduction

With the rapid development of Chinese national economy and the rising living standards of the people, more and more electricity is demanded in China, which actually speeds up the development of electrical power and power equipment manufacturing industry. Chinese power units as well as those worldwide are facing to increase thermal efficiency and to decrease the emission of CO_2, SO_X and NO_X. All these have promoted the research and development of gas-steam combined cycle, supercritical and ultra-supercritical power units. The efficiency of thermal power units depends on the steam parameters, i.e. steam temperature and pressure. However, these parameters are limited by high temperature materials, which need to be researched and developed furthermore.

The Electric Power Development in China

Installed Capacity and Total Electricity Generated

In the year 1882, the first power generating unit with installed capacity of 16 horsepower (11.76kW) was set up. Since then, by 1949, it had got up to 1.85GW; by the end of 1965, 15.08GW; by the end of 1980, 65.87GW; by the end of 2000, 319.32GW; by the end of 2005, 517.18GW (in which fossil fired power plant occupied 75.67%, water power plants 22.7%, nuclear power plants 1.32%, wind power plants 0.20%) [1]. The average annual increase rate of installed capacity during 1950-1965 is 14.01%, while during 1966-1980, it is 10.33%, and during 1980-2000, 8.21% and during 2001-2006, 11.75%. In recent years, due to nationwide shortage of electricity, power plant construction has been speed up. The newly added installed capacity in 2004 was 50.98GW, and in 2005, 74.8GW, in 2006, 104.82GW [2]. At present, the currently being built power plants have a total installed capacity of 250GW.

The total electricity generated in 1952 is 7300 GWh, and in 1980, 300.6 TWh; in 2000, 1368.5 TWh; in 2005, 2497.5 TWh (in which fossil fired power 81.83%, water power 15.87%, nuclear power 2.13%). During 1953-1965, the average annual increase of generated electricity is 18.67%, while during 1966-1980, 10.46%; during 1981-2000, 7.87%; during 2001-2006, 12.90%. The total electricity generated in 2006 was 2834.4 TWh. The increase of installed capacities and electricity generated in the past years are listed in Fig.1 [1][3][4]. The structure of installed capacity and power generated in 2006 are listed in Fig. 2 and 3.

Unit Capacity and Parameter

In 1950's, fossil fired power equipment was manufactured in China. At the beginning, the generating units are only intermediate pressure ones with capacities of about 6-12MW, 3.5MPa/435°C, then high pressure units with capacities of 25-100MW, 9.0MPa/535°C and super-high pressure units were gradually developed. Sub-critical units of 300MW with the parameters of 16.5MPa/550°C/550°C (changed into 16.5MPa/535°C/535°C after modification) were put into operation in 1970's. Licensed sub-critical units of 300-600MW with 17.0MPa/538°C/566°C were put into service in 1980's. Supercritical units of 600MW with 24.1MPa/538°C/566°C and 800MW with

25MPa/545°C/545°C were operated in 1990's. Supercritical units of 900MW with 25MPa/538°C/566°C and 600MW with 24.2MPa/566°C/566°C were put into service in 2000's and 1000MW with 26.25MPa/600°C/600°C in November, 2006. Unit capacity and steam parameter developments in China are listed in Fig. 4 and 5.

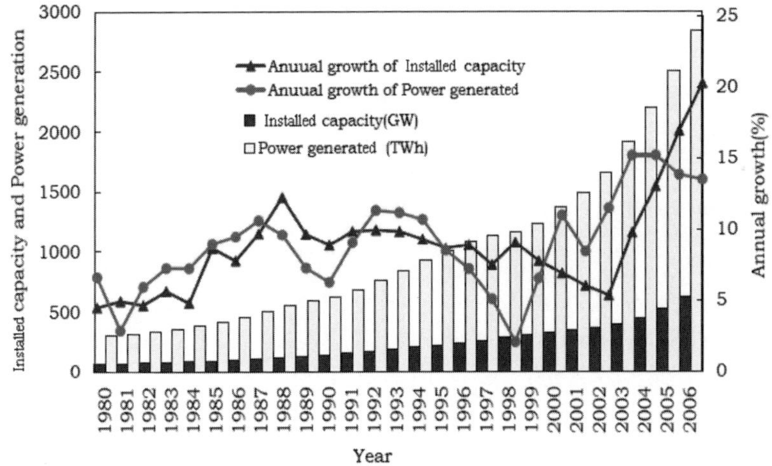

Fig.1 Installed capacity, electricity generated and its annual growth rate

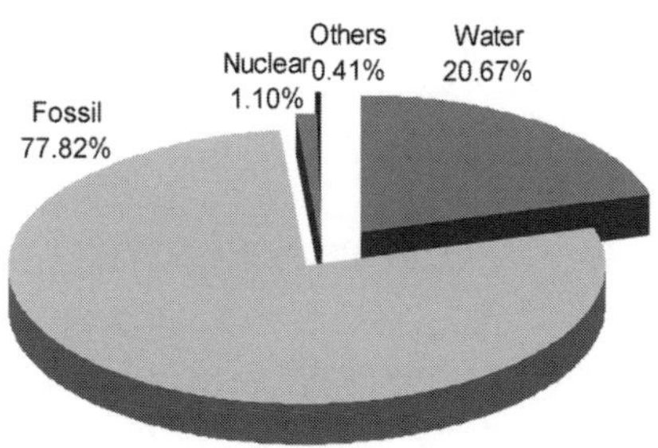

Fig.2 The structure of installed capacity in 2006

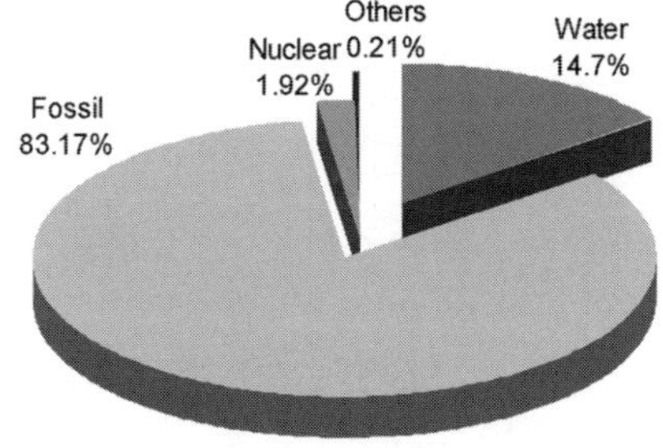

Fig.3 The structure of power generated in 2006

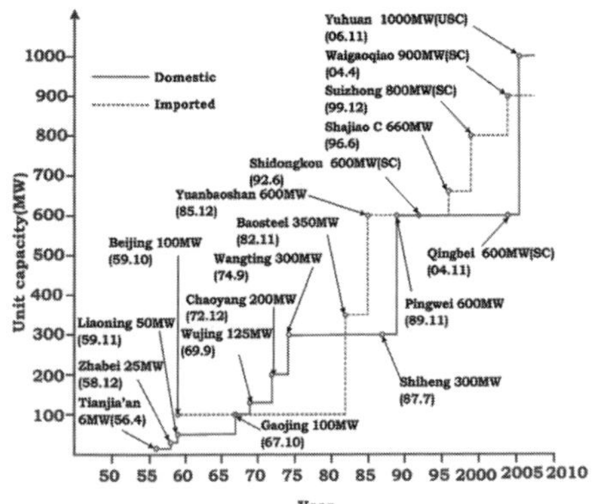

Fig.4 The development of unit capacity

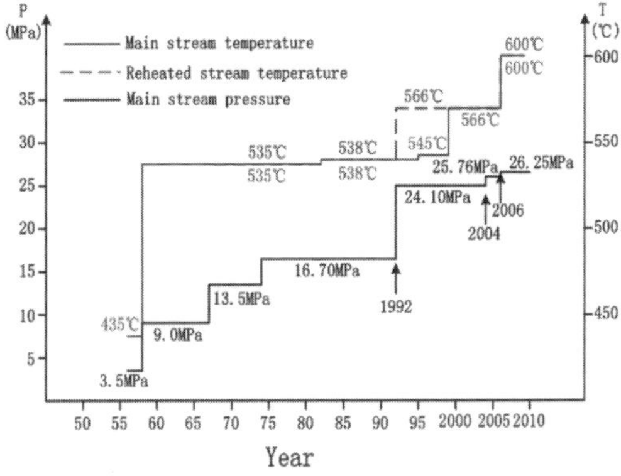

Fig.5 The development of unit parameter

Forecast on Electric Power Development in China

The GDP of China in 2000 is the quadruple of that of 1980. Also, it realized quick development during 2001-2006. The GDP in 2001-2006 and its growth rate are listed in Fig. 6. The demand of electric power from people living level increases very quickly because of the improvement of people life. In 2005, power consumption of urban and rural people living level is 283.8 TWh , with an increase of 16.19% and 8% higher than last year. However, this part of power consumption only occupied 11.4% of the total. Power consumption per person was 217kWh. The demand of the electric power will be still growing in the future.

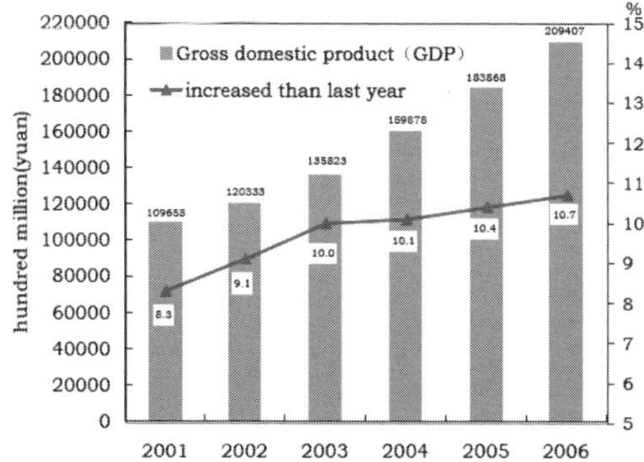

Fig.6 The GDP in 2001-2006 and its growth rate

Forecast on power generation, installed capacity and its structure for 2010 and 2020 are shown in table 1 and 2.

Tab.1 Forecast on power generation and installed capacity for 2010 and 2020

Year	Actual value of 2000	Actual value of 2005	2010	2020
Power generation (TWh)	1368.5	2497.5	3810	6050
Annual growth (%)		12.79	8.81	4.73
Installed capacity (GW)	319.32	517.18	852	1330
Annual growth (%)		10.12	10.50	4.55

Tab.2 Structure of installed capacity (GW) for 2010 and 2020 (in%)

Year	Actual value of 2000		Actual value of 2005		2010		2020	
Item	Capacity	%	Capacity	%	Capacity	%	Capacity	%
Installed capacity	319.32	100	517.18	100.0	852	100	1330	100
Water	79.35	24.8	117.39	22.7	190	22.3	300	22.6
Coal	221.15	69.3	367.70	71.1	583.5	68.5	838.2	63.0
Oil	15.4	4.8	16.59	3.2	18	2.1	18	1.4
Gas	0.96	0.3	7.09	1.4	36	4.2	72	5.4
Nuclear	2.1	0.7	6.85	1.3	13.70	1.6	40	3.0
New energy	0.36	0.1	1.57	0.3	10.8	1.3	61.8	4.6

Power Equipment Manufacturing

Chinese power equipment manufacturing industry started from 1953. The first 6MW fossil fired power unit was produced in Shanghai in 1955 and the first 25MW fossil fired power unit produced in Harbin. In 1974, the Dongfang manufacturing base was established nearby Chengdu in Sichuan (Fig.7). Except above mentioned 3 big manufacturing bases, there are also several medium and small manufacturers in Beijing, Wuhan and etc. The first 100MW high pressure, 125MW and 200MW super-high pressure and 300MW sub-critical power unit were put into operation at Gaojing Power Plant in 1967, at Wujin Power Plant in 1969 and at Chaoyang Power Plant in 1972 respectively. The first 300MW and 600MW sub-critical units with licensed technology from CE and Westinghouse were put in service at Shiheng Power Plant in 1987 and in Pingwei Power Plant in 1989 respectively. The first 600MW unit jointly manufactured with foreign companies was put into operation at Qingbei Power Plant in 2004. The first 1000MW ultra super-critical power units were put into operation at Yuhuan Power Plant in November 2006.

The demand of electricity increases sharply and the power equipment production develops quickly as shown in Fig. 8 [5-10]. The output of power equipment keeps rising year after year. The output of power equipment is almost double in 2004 with the comparison of 2003 and it reached 110GW in 2006.

Fig. 7 Three big manufacturing bases of power equipments in China

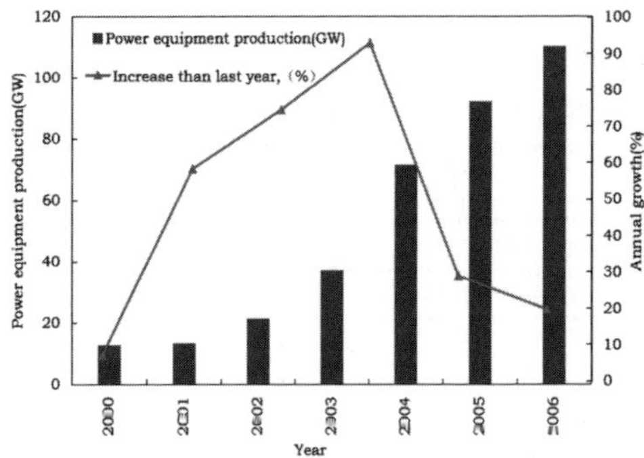

Fig. 8 Power equipment production and its annual growth rate

Equipments for electric generation with water, fossil-fired, gas, nuclear and wind force will be developed simultaneously in China. From the first 800KW water power generating unit in China to the present 700MW water power generating unit manufactured independently, which was put into service on July 8, 2007. The fossil power units have developed; from the first 6MW intermediate pressure fossil-fired power generating set to the present ultimate 300MW large circulating fluidized bed units and 1000MW large ultra supercritical fossil fired power units. For gas fired power units, 53 gas turbine combined cycle generating sets of class F and E have been contracted by Harbin, Dongfang and Shanghai three big manufacturers jointly with foreign gas turbine manufacturers, some of them have been already put in service. In 1980's, self-designed and manufactured nuclear power units of 300MW came out, and the manufacture of 650MW nuclear power units also have been done and 1000MW nuclear power units are just in manufacturing process. China will make 4 nuclear power sets AP1000 of third-generation in collaborating with Westinghouse, USA and the first AP1000 unit will be put in service soon. The wind power in China develops very rapidly too. Now the manufacturers in China are able to produce 1.5MW wind power units.

Future development of power equipments will concentrate on large-scale ultra supercritical fossil fired power units, large air-cooling units, large CFB boilers, large nuclear power units, large gas turbine combined cycle generating sets, large water power and water pumping energy storage units, large wind power units as well as new energy source power generation technology and clean coal combustion technology etc.

The Development of Supercritical and Ultra Supercritical Power Plants in China

Energy Conservation and Environmental Protection Policy of Chinese Government Promote the Quick Development of Supercritical and USC Power Plants

Chinese government pays more attention to save energy, reduce consumption and protect environment and has made a target that unit GDP energy consumption should be reduced 20% and the emission of main pollutants should be cut down 10% in the years of 2006 - 2010. In respect of the electrical power structure, our strategy is: developing water power to a large scale, optimizing fossil fired power generation, promoting nuclear power development, using natural gas at a certain degree and encouraging new energy resource exploitation. In respect of fossil fired power, China will shut down the small units with high pollution and high energy consumption and encourages to make supercritical and ultra supercritical units of above 600MW and large circulation fluidized bed units and to push forward the application of IGCC technology. Chinese government has required to shut down the small fossil fired power plants with the total capacity of 50GW in the near future. For environmental improvement by the year of 2010, power units with desulphurization devices will reach 300 GW, the emission of SO_2 will be lowered to a level of 2.7g/kWh. In recent years, the coal consumption of current supply in China came down step by step and they were 392, 370, 366 and 356g/kWh in the year of 2000, 2005, 2006 and 1-6 months of 2007[1][2][3][11].

Coal resources are comparatively rich in China. The coal-fired power plants will still occupy the most important position among the total power supply. Energy conservation and environmental protection policy promote the quick development of supercritical units and ultra supercritical units in China.

The Development of Supercritical and Ultra Supercritical Power Plants in China

In June, 1992, No.1 units of Shidongkou Power Plant was put into service. Thus the adoption of supercritical electric power units started in China. More than 10 supercritical units were successfully imported in the following 10 years. Chinese power equipment manufacturing enterprises worked together with foreign companies and began to make supercritical units in 2002 and ultra supercritical units in 2004. The first 600MW Chinese made super-critical unit was put into operation at Qingbei Power Plant in 2004. The first 1000MW ultra super-critical power units were put into operation at Yuhuan Power Plant in November, 2006. By the end of June, 2007, power equipment manufacturing enterprises have contracted 204×600MW super-critical power units, in which 103 units were already delivered, and contracted 36×600MW ultra supercritical power units, in which 3 units have been already delivered and contracted 46×1000MW ultra supercritical power units, in which 6 units were already delivered, 4×1000MW ultra supercritical power units have been put in operation.

After half year operation of the first 2×1000MW ultra super-critical power units at Yuhuan Power Plant in China. Various performance indexes of the units have reached designed target [12]. Among them, boiler efficiency is 93.88%, the heat consumption of the steam turbine is 7295.8kJ/kWh,. At fixed load, the coal consumption of the power generation is 270.6g/kWh, the coal consumption of the current supply is 283.2g/kWh, the thermal efficiency of the unit is 45.4%, the emission of NOx is 270mg/m^3 and the emission of SO_2 is 17.6mg/m^3 (See Fig.9, No.1 ultra super-critical power units at Yuhuan Power Plant).

Fig.9 No.1 ultra super-critical power units at Yuhuan Power Plant

The net increase of installed capacity in China will be 700GW from 2007 to 2020. In consideration of the additional 100GW caused by shutting down those aged or high power consumption units and newly added installed capacity would mainly be of coal fired power units (about 500GW). Although IGCC is an advanced coal fired power generation technology, it will not be popularized in the near future due to technical reasons. In order to raise power generation efficiency and effectively reduces pollution, it is estimated that approximately 70% coal fired power units newly installed between

2011 and 2020 will be ultra supercritical ones. The ultra super-critical power units will continuously increase after 2020. It can be seen that the ultra super-critical power units have very big market in China and the demand will exceed 350 units till 2020.

High Temperature Materials Application and Development for Supercritical and Ultra Supercritical Units

Materials Application for Supercritical Units and Ultra Supercritical Units

Since 1980s, fossil fired USC development program has been widely executed in Japan, Europe and USA, which based on ferritic steel series of 9~12%Cr and austenitic steel series 18Cr-9Ni and 25Cr-20Ni, aims to develop new ferritic steels with high strength and low price to substitute original austenitic steels, and new austenitic steels with higher heat-resisting strength and corrosion resistance compared with original austenitic steels. Especially the 9~12%Cr ferritic steel series have been extensively developed by adding alloying elements such as Mo, W, V, Nb(Ta), Cu, Co, Ni, N, B and Re etc. New steels of (9~12)CrMoVNbN, (9~12)CrMoWVNb(B)N and (9~12)CrMoWCoVNb(B)N series developed on the basis of (9~12)CrMoV steels are possible to be used at 600°C, 620°C and 650°C. Superalloys such as Inconel 740, Inconel 625, Alloy 617, Nimonic 263, Haynes 230 etc. are now extensively being studied for 37.5MPa, 700°C ultra supercritical power units in European communities and 38.5MPa, 760°C ultra supercritical power units in USA.

Tab. 3 Materials for SC and USC Units

Component	566/566°C	600/600°C
Headers/Steam Pipes	P91, P23	P92, P122, P91
Superheaters/Reheaters	T91, T92,TP304H,347H	TP347HFG,Super304H TP310HNbN
HP/IP Rotors	1.25Cr1MoV 10Cr1MoVNbN	10Cr1MoVNbN 10Cr1Mo1WVNbN 10Cr0.5Mo1.8WVNbN
HP/IP Inner Cylinder	1.25Cr1MoV 9.5Cr1MoVNbN	9.5Cr1MoVNbN 10Cr1Mo0.8WVNbN
Valve Casing	1.25Cr1MoV 9.5Cr1MoVNbN	9.5Cr1MoVNbN 10Cr1Mo0.8WVNbN
High Temp. Buckets	10.5Cr1Mo1WNiVNbN 10Cr0.7Mo1.8W3.2CoVNbNB	10.5Cr1Mo1WNiVNbN 10Cr0.7Mo1.8W3.2CoVNbNB Nimonic80A , R26
High Temp. Bolts	10.5CrMoVNbNB 10Cr0.7Mo1.8W3.2CoVNbNB	10.5CrMoVNbNB 10Cr0.7Mo1.8W3.2CoVNbNB Nimonic80A,R26 GH4145
LP Rotors	3.5NiCrMoV	3.5NiCrMoV, High -purity 3.5NiCrMoV

At present time, the manufacturers in China have cooperated with foreign companies to produce USC power units. As boiler manufacturing enterprises cooperate with

Mitsubishi, Alstom and Babcock-Hitachi. Steam turbine manufacturing enterprises cooperate with Siemens, Hitachi, Toshiba and Mitsubishi. So the applications of high temperature materials are not unified. Materials applied for supercritical units and ultra supercritical units in China are listed in Tab.3.

High Temperature Materials Demand of Supercritical and Ultra Supercritical Power plants in China

Up to now the key high temperature materials for supercritical and ultra supercritical power units are still mainly imported from foreign countries. China requires to make the key high temperature materials by our national enterprises. For making seamless tubes and large pipes, China enterprises will be equipped with more 35MN, 60MN and one 350MN extrusion machines in the near future.

Large-scale castings for 1000MW USC steam turbine already can be made in China. Large-scale forging have been also trial-produced. High-purity low alloy steel forging for low pressure rotor and 12%Cr high-quality forged rotor for 1000MW USC steam turbine high pressure rotor are made in China (see Fig.10 and Fig.11).

Fig.10 High-purity low alloy steel forging for 1000MW USC steam turbine low pressure rotor (forged from 292-ton ingot)

Fig.11 12%Cr high-quality forging for 1000MW USC steam turbine high pressure rotor (forged from 79-ton ingot)

High Temperature Materials Research and Development for Ultra Supercritical Units

Most of high temperature materials research and development for USC power units are concentrated in research institutes and universities such as Central Iron & Steel Research Institute (CISRI), Institute of Metal Research (IMR), University of Science & Technology Beijing (USTBeijing) in close cooperation with Shanghai Power Equipment Research Institute, Xian Thermal Power Research Institute and also with steel plants, heavy machineries, boiler and turbine makers [13].

Central Iron & Steel Research Institute invented multi-alloying low alloy heat resisting steel G102 (12Cr2MoWVTiB) in 1960's which had been successfully used in large quality for superheaters and reheaters at the temperatures below 580°C. Medium Cr-content heat resisting steel G106 (10Cr5MoWVTiB) was developed in 1970's and used at the temperature below 600°C.

CISRI is still doing the research work on modification of several heat-resisting steels and alloys [14]. Several typical R&D projects are as follows:

1) G107 (25Cr-20Ni type): Chemical composition adjustment on the base of HR3C for improvement of stress rupture strength;

2) G108 (18Cr-8Ni type): Alloying element modification on the base of Super 304H for elimination intergranular corrosion;

3) G109 (25Cr-20Ni type): To develop an austenitic steel with high temperature strengths equivalent to NF 709R;

4) G110 (Fe-Ni base type alloy with 25%Cr): γ' strengthened Fe-Ni base superalloy for superheaters at 700°C.

Institute of Metal Research, Chinese Academy of Sciences developed a γ' strengthened Fe-Ni base superalloy with Nb, Ti and Al (Ni-33Fe-19Cr-2Mo-1Nb-1Ti-0.4Al) without Co for superheater tube material since 1969. This superalloy designated as GH2984 was successfully produced in 1978 and used as superheater tubes for marine application. After 10-year service GH2984 tubes were well in shape without any damage and which can be continuously used [15].Fig.12 shows the stress-rupture curves at 650°C, 700°C and 750°C in comparison with other superalloys. It can be seen that stress rupture strength for 10^5h at 700°C can be higher than 100MPa to meet the superheater material requirement for long-term service at 700°C.

GH2984 characterizes with high oxidation and corrosion resistance because of its high content of Chromium (18-20%Cr), which can form a tight and adherent Cr_2O_3 film, that guarantees this alloy having better oxidation resistance. The average oxidation rate of GH2984 at 700°C is 0.0058g/mm^2·h only. Hot corrosion resistance of GH2984 was studied by crucible method with the medium of 25%NaCl+75%Na_2SO_4 at 650-820°C. The weight loss at 700 and 750°C is 1.29 and 1.63 mg/cm^2·h respectively.

Structure stability study of GH2984 has been carried out at 700°C from 1000 to 18000h. After standard heat treatment the spherical γ' average size of about 20μm and an amount of 5.74%wt uniformly distributed in γ-matrix. After 18000h long time aging at 700°C, the radius of γ' particles increases linearly with the third root of aging time. The amount of γ' gradually increases to 7.25%wt. The strengthening effect can be still kept due to the increase of γ' fraction and compensates the loss of strength caused by γ' growth.GH2984 not only characterizes with good strengths, oxidation and corrosion resistance but also easy formability for tube making. GH2984 is a promising 700°C tube candidate for superheaters and reheaters of USC boilers.

Inconel 740 (Ni-25Cr-20Co-0.5Mo-2Nb-1.6Ti-1.1Al) has been developed on the base of Nimonic 263 by Special Metals Corporation (SMC Huntington), USA for USC boiler superheater and reheater materials at the temperatures above 750°C. Inconel 740 characterizes with the highest stress-rupture strength for 10^5h at 750°C among today's commercial superalloys and also with excellent oxidation and corrosion resistance at high temperatures [16,17]. It is considered to be used in Europe for most advanced fossil power plant at the steam temperature above 700°C. Fig.13 shows the stress-rupture curves in the temperature range of 700-800°C. It can be seen that the stress-rupture strength of Inconel 740 for 10^5h is higher than the demand of 100MPa. The corrosion resistance can also meet the requirement of ≤2mm cross-section loss in 2×10^5h.

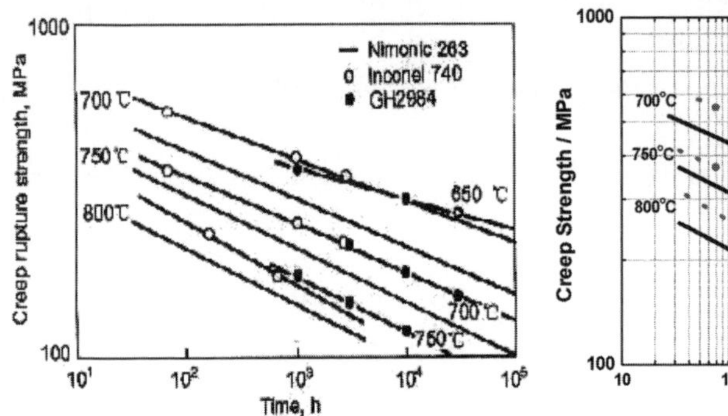

Fig.12 Stress ruptures properties of GH2984, Inconel 740 and Nimonic 263 alloys [15]

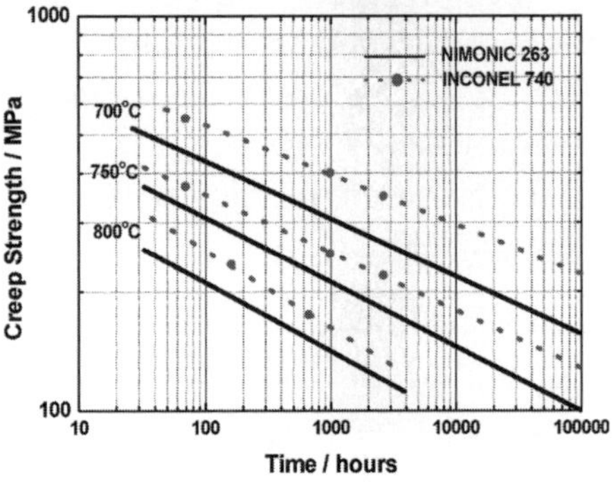

Fig.13 Comparison of stress rupture strength of Inconel 740 with Nimonic 263 [17]

One of the most important factors for Inconel 740 is the structure stability at the temperatures above 750°C. University of Science & Technology Beijing has worked together with SMC(Huntington) USA to study the structure stability at the temperatures 700, 725, 750, 800°C till 5000hrs and at 850°C for 1000hrs. Long term experimental results show that the structure instability of Inconel 740 includes γ′ coagulation, γ′→η transformation and G phase formation [17]. Structure stability study shows that Inconel 740 keeps good structure stability during prolonged aging at temperatures under 725°C. Structure stability of Inconel 740 at the temperature above 750°C should be improved.

Thermodynamic calculation and structure stability improvement study show that Al, Ti and Nb contents have an obvious effect on precipitation behavior of γ′ and η phase. The improvement of structure stability of Inconel 740 can be achieved by the adjustment of Al, Ti and/or Si level in the alloy. The modified alloys (see Fig.14) exhibit more stable structure at 750°C aging till 5018h, at 800°C aging for 5000h and at 850°C aging for 1000h as increasing more stable γ′ fraction, retarding the transformation of γ′ to η and/or elimination G phase formation. The detail research results can be found in recent other paper [18]. It can be concluded that Inconel 740 is a most perspective candidate for USC boiler tubing materials at the temperature above 750°C.

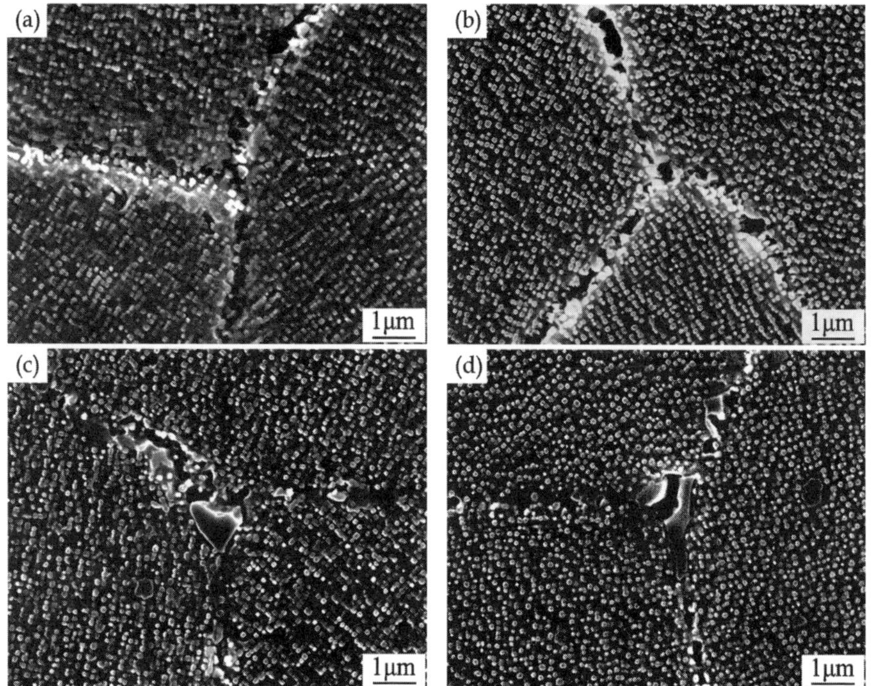

Fig. 14 SEM images of 4 modified INCONEL 740 alloys aged at 750°C for 5018h (Heat 1 (a), Heat 2 (b), Heat 3 (c), Heat 4 (d) [18]

Conclusions

(1) The rapid development of Chinese economy promotes the growth of power generation. Coal fire fossil electric power has occupied the most important position not only to-day but also in the future 20 years.

(2) Chinese fossil power plants are facing to increase thermal efficiency and to decrease the emission of CO_2, SO_X and NO_X. Ultra-super-critical (USC) power plants with the steam temperature at 600°C (or above) will get a rapid development.

(3) Chinese USC power plants development will adopt a variety of qualified high temperature materials for boiler and turbine manufacturing.

(4) The research and development of high temperature materials for USC power plants in China with the steam temperature up to 600-700°C have been also taken in consideration.

References

[1] Statistical Data of National Power Industry in 2005. *CEC Department of Statistics and Information*, Aug, (2006).

[2] Statistical Data of National Power Industry in 2006. *China Electricity Council,* Feb, (2007).

[3] Historical Development Indexs of Chinese Power Industry. State power information network. (2005).

[4] Handbook of Power Generating Equipments. *Management Centre of Electric Power.* Apr, (2006).

[5]- [10] Statistical Communique of National Economy and Social Development in 2001-2006. *National Bureau of Statistics of China*. Feb, (2002-2007).

[11] The Brief Production Situation of National Power Industry during 1-6 Months of 2007. *China Electricity Council.* Jul, (2007).

[12] the Technical Indexes of First Chinese USC Power Plant with 1000MW Units. *International Energy Network.* Jun, (2007).

[13] Z. Q. Hu and J. T. Guo, "Development of High Temperature Materials for Energy Market in China," *Materials for Advanced Power Engineering 2006,* Energy Technology (2006), p.189.

[14] CISRI Internal Report (unpublished).

[15] J. T. Guo, X. K. Du, "A Superheater Tube Superalloy GH 2984 with excellent properties," *Acta Mtallurgica Sinica,* 41(2005), p. 1221.

[16] S. J. Patel, "Introduction to INCONEL 740: An Alloy Designed for Superheater Tubing in Coal-Fired Ultra Supercritical Boilers", *Acta Mtallurgica Sinica* (English Letters), 18 (2005), p. 479.

[17] S. Zhao, J. Dong, X. Xie, G. D. Smith and S. J. Patel, "Thermal Stability study on a New Ni-Cr-Co-Mo-Nb-Ti-Al Superalloy," *Superalloys 2004,* TMS (2004), p. 63.

[18] X. Xie, S. Zhao, J. Dong, G. D. Smith, B. A. Baker and S. J. Patel, "A New Improvement of Inconel 740 for USC Power Plants," *Proceedings of the 5th International Conference on Advances in Materials Technology for Fossil Power Plants,* Marco Island Florida USA (in press).

Advances in Materials Technology for Fossil Power Plants
Proceedings from the Fifth International Conference
R. Viswanathan, D. Gandy, K. Coleman, editors, p 59-81

DOI: 10.1361/cp2007epri0059

Consideration of Weld Behavior in Design of High Temperature Components

K. Maile, A. Klenk, M. Bauer and E. Roos
Materialprüfungsanstalt Universität Stuttgart
Pfaffenwaldring 32
70569 Stuttgart
Germany

Abstract

This paper describes the steps necessary for consideration of weld behavior in order to be used in modern design procedures. Specific behavior of similar and dissimilar welds in the creep regime are described as well as procedures and criteria to be used for the assessment of welded joints.

1 Introduction

Due to the increase of the capacity of modern power plants, coming along with the increase of the steam parameters, the design of components (e. g. piping, header) is approaching more and more the design limits. In addition the need for larger pipe dimensions raise questions on the application of longitudinally welded pipes. Due to this, the need for consideration of the behavior of welded joints in high temperature components is increasing. Furthermore, new design approaches -design by analysis- i. e. inelastic finite element analyses, are more and more used instead of conventional design, based on stress limits given in standards or code of practices.

In order to consider the weld behavior properly and to build a base for consideration of welds in inelastic analyses, the following steps are necessary:

- Understanding the failure mechanism of welded joints under long term creep loading depending on the materials used.
- Specific design of welded pipes.
- Extrapolation of creep data of welded joints.
- Derivation of weld creep strength factors.
- Development of procedures to introduce the behavior of welded joints and heat affected zones in inelastic finite element analyses.

- Consideration of transferability of weld creep strength and weld creep strength factors on components.

2 Failure mechanisms of welded joints in the creep range

2.1 Ferritic steels – creep rupture

Due to the heat input during welding, microstructural changes take place in a small area – depending on the welding process, welding geometry and material type – besides the fusion line. If the A_1 temperature is exceeded in that area, phase transformations lead to changes in the microstructure of ferritic steels, which can clearly be seen using optical microscope, that are called heat affected zone (HAZ). Thus, the HAZ does not represent the optimal microstructure and precipitation characteristics that can be found in the unaffected base material (BM) needed for optimized creep resistance. Furthermore, according to standards and manufacturers specifications, the unaffected base material has to be subjected to a specific heat treatment for maximum creep strength.

Dividing up the HAZ into three different zones is a state of the art classification of that area. The coarse grain zone near the fusion line (HAZ1), the area of the outer HAZ (intercritical zone – HAZ3) and an area lying in between with a medium grain size (fine grain zone - HAZ2). These areas show different creep strength and different creep rupture deformation values. In real life of course, each zone has a fluent transition into the next zone.

In general, the coarse grain zone shows relatively high creep strength but rather low creep rupture elongation. The intercritical zone and the transition region into the unaffected base material show a significantly lower creep strength than the unaffected base material. A hardness profile crossweld usually has a characteristic shape – a maximum hardness in the weld metal and a steep decrease in the HAZ with a hardness trough in the area of the outer HAZ (transition to the uninfluenced base material respectively). The different strength and deformation behaviour in the HAZ (stress and strain redistribution) and the related constraint effects as well as the influence of the surrounding material layers, like weld metal (WM) and BM, result in a complex, interacting and multiaxial stress situation in the weld. This particular stress situation in combination with the rather small creep resistance of the HAZ3 results in general in premature creep damage in the intercritical heat affected zone, depending however on the loading and temperature conditions. In case of long term creep loading perpendicular to the weld, a change of fracture location from WM to HAZ3 can be observed and therefore often premature failure of the component.

In (1) creep tests on crossweld specimens of different base materials were evaluated. These investigations clearly show the great impact of the base material on the failure behavior of the welded specimens. According to Figure 1, unalloyed steels show the smallest difference between the creep rupture strength of the welded joint compared to the parent material's creep strength, whereas bainitic and in particular martensitic steels show a stronger decrease in creep strength after welding. Especially welded joints of martensitic steels, due to their complex microstructure and the related specific heat treatment, are subjected to the greatest decrease in creep strength as a result of the effects described above. Furthermore, a strong influence of the service temperature can be found. It strikes out, that at the maximum service temperatures of the new 9% Cr-steels the factor $R_{m/t/\delta}$ welded joint / $R_{m/t/\delta}$ base material reaches a value of 0.5, see Figure 1.

This factor is of great importance with regard to design and loading of welded components and will be discussed in detail in chapter 4.

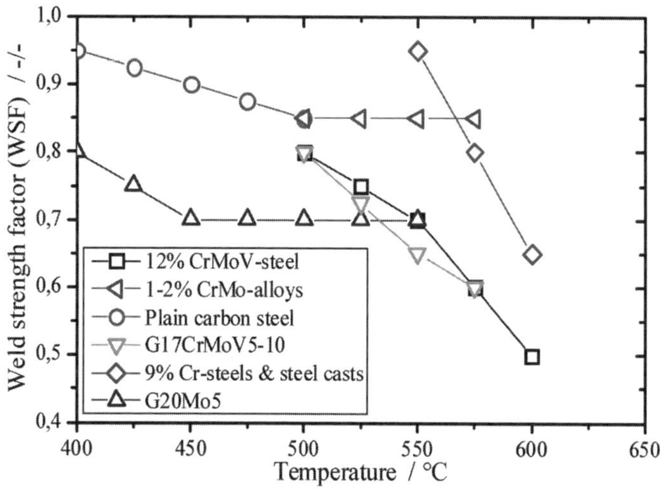

Figure 1: Weld strength factors for 100.000 h creep rupture strength (1).

The negative effects on the microstructure and the precipitation characteristics described above can be removed/diminished by performing a complete additional heat treatment (quenching and tempering) of the welded component according to standards or recommended procedures by the steel manufacturer. By doing so, the microstructure of the HAZ is retransformed into the base material's optimized microstructure for maximum creep resistance. If a matching weld metal, with equivalent creep strength comparing to the parent material is used, such components can show a service performance similar to seamless components.

For longitudinally welded pipes there is an alternative technology available, if a complete additional annealing of the component post welding is not possible e. g. due to expense limitations or technical restrictions. The coupling of the submerged arc (SAW) welding process with subsequent continuous austenizing – by inductive heating - of the whole pipe. Tempering can be done in a separate furnace afterwards. Current research on this new technique within the scope of an AVIF research project showed very promising results (2). Creep tests on P91 (X10CrMoVNb9-1) crossweld specimens, extracted from pipes which were subjected to the heat treatment described above, with running times up to 16.000 h showed, that the change in fracture location and thus premature failure could be avoided. The determined creep rupture data lies well within the scatter band of the base material. There are indications for similar performance of

P92 (X10CrWMoVNb9-2) crossweld specimens (fracture location in the base material at running times of up to 5.000 h and testing temperature of 650 °C) (2).

Challenges to meet of this technique are the relatively short austenizing time and the possibility of a too low temperature (below A_{C3}) on the inside of the pipes due to the inductive heating.

Therefore productive measures have to be taken by the pipe manufacturer to guarantee a complete austenizing even on the inside of the pipes. Continuous control and documentation – e. g. by temperature measurements on the pipe's inner surface – seem mandatory for quality assurance.

The evaluation of the microstructural results showed that despite the low austenizing time of only approximately 2 minutes, a positive effect on the microstructure can be found. The results of creep tests with simulated HAZ showed further potential for improvement by further increasing the normalizing temperature, yet limits with regard to height and normalizing time have to be respected. Furthermore it has to be noticed, that a modification of the pipe geometry (wall thickness, pipe diameter) and of the alloy design could require further adjustments.

2.2 Austenitic steels and alloys

In austenitic steels or alloys respectively (e. g. Nickel alloys) the γ-area is stabilized up to room temperature by adding Ni and Cr, i. e. no phase transformation occurs during the cooling process from the melt. For that reason a HAZ like in the case of ferritic steels can not observed, a change of fracture location going along with premature creep failure does not occur.

Nevertheless, changes in the microstructure in the heat affected zone can occur, like e. g. coarsening of particles, precipitation of new types of particles, dissolution of particles and recrystallization/relaxation effects and grain coarsening. These changes can have a minor impact to the creep strength in the long term area.

2.3 Dissimilar welds

Components, made of a large number of different materials, are used to built a modern power or process plant. In general the materials are chosen due to their specific material behavior e. g. creep or corrosion resistance, machinability etc. Of course, cost effectiveness and the need to reduce the exhaust of greenhouse gases are also major design parameters. Unfortunately, this inherits dissimilar welds which are often a challenge in design and fabrication. Dissimilar welds presented in Table 1 are common in power and process plants.

Three major failure mechanisms have to be considered when designing dissimilar welds, as depicted in Figure 2. The next paragraph gives a short introduction on the three major failure mechanisms found in dissimilar welds under high temperature loading. Service experience and a large number of research projects (e. g. (3)) have shown that dissimilar welds often fail in the region of the material/zone with the worst creep resistance.

This can exemplarily be seen in Figure 3, showing the plastic strain accumulation in a crossweld specimen under creep loading. Here, a high alloyed ferritic steel (X20CrMoV12-1) is welded to a high alloyed austenitic steel Alloy 800 (X10NiCrAlTi32-20) using an austenitic filler material (NiCr16FeMn). The buffer layer is necessary to minimize the effect of different thermal elongation coefficients and thus to prevent from hot cracking in the weld metal.

The coefficient of thermal expansion of the weld metal is chosen between the ones of the base materials to reduce the danger of thermal fatigue cracking in service and during start-ups. Nevertheless a strain concentration in the HAZ of the ferritic base material could be observed after a relative short testing time, which lead, later on, to a failure in that region.

Table 1: Common dissimilar welds

Base material A	**Base material B**	**Weld Metal**
low alloy ferritic steel (e. g. 0.6 Mo or 1CrMoV-steel)	low alloy ferritic steel (e. g. P22, 1Cr-cast steel)	matching A or B
low alloy ferritic steel (e. g. P22, 1Cr-cast steel)	high alloy ferritic steel (e. g. 12 Cr-steel X20, P91, P92, VM12)	matching A or B or Ni based WM
high alloy ferritic steel (e. g. 12 Cr steel X20, P91, P92, VM12)	high alloy austenitic steel (e. g. SS304, SS316, HR3C, Alloy 800)	austenite
low alloy ferritic steel (e. g. P22, 1CrMoV)	high alloy austenitic steel (e. g. SS304,)	Ni based WM
high alloy ferritic steel (P92, VM12)	Ni based alloy (e. g. Alloy 617)	Ni based WM

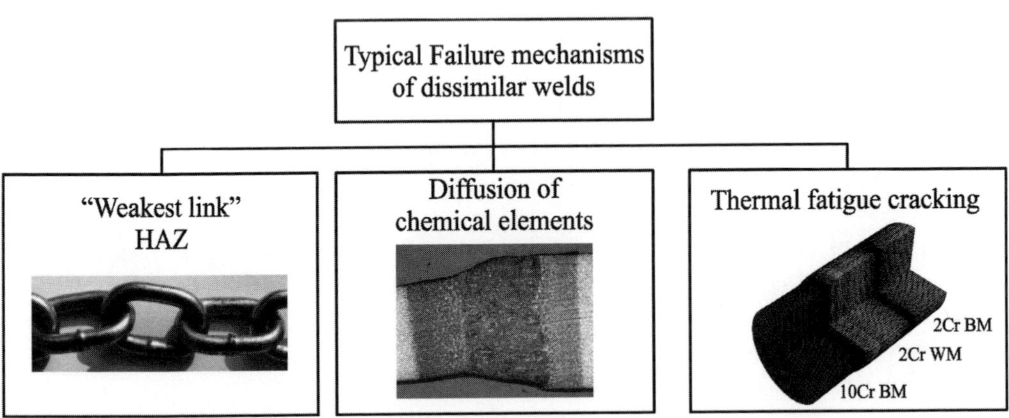

Figure 2: Overview on typical failure mechanisms of dissimilar welds

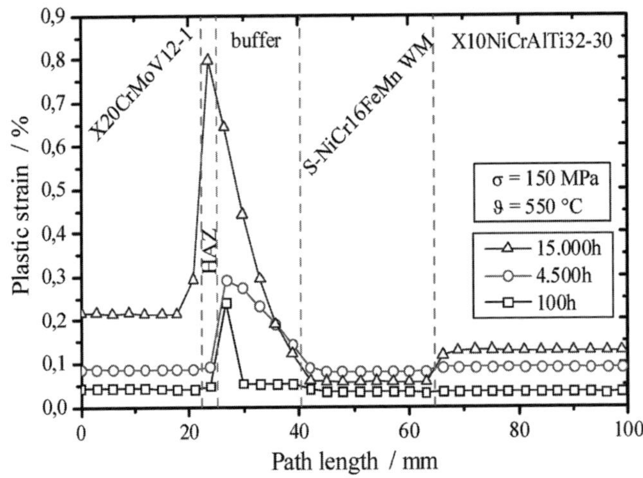

Figure 3: Plastic strain accumulation for different loading times of a crossweld specimen (3).

The second failure mechanism mentioned above is the depletion of chemical elements, especially C and Cr, in the transition zone. Investigations carried out in the framework of the COST 505 project (4), (5) show, as the results of microprobe analysis, the carbon and chromium profiles across the transition zone of a dissimilar weld (here 1CrMoV-cast steel (GS-17CrMoV5-11) vs. 12CrMoV-steel (X20CrMoV12-1)), see Figure 4 and Figure 5. It can be seen from Figure 4, that in the as welded state, the carbon content is more or less constant over the whole transition zone, i. e. there is no carbon migration during welding. If the joint is exposed to high temperatures and long times, as shown in Figure 5, a decarburized zone develops due to carbon depletion in the coarse grained HAZ of the low alloyed cast steel. In general this comes along with a severe decrease in strength and thus premature failure.

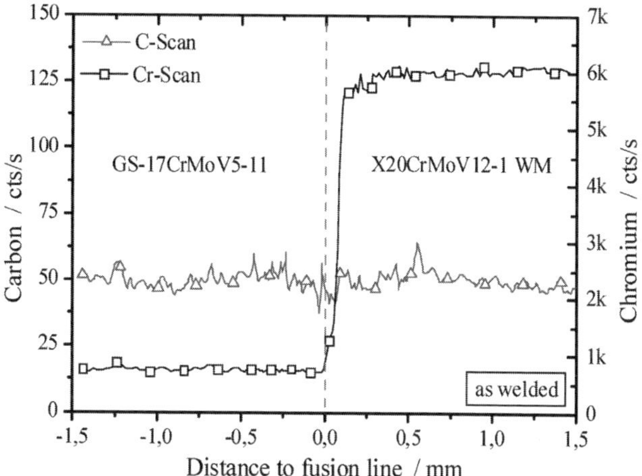

Figure 4: C- and Cr distribution at the interface 1% Cr-BM / 12% Cr-WM in the as welded condition

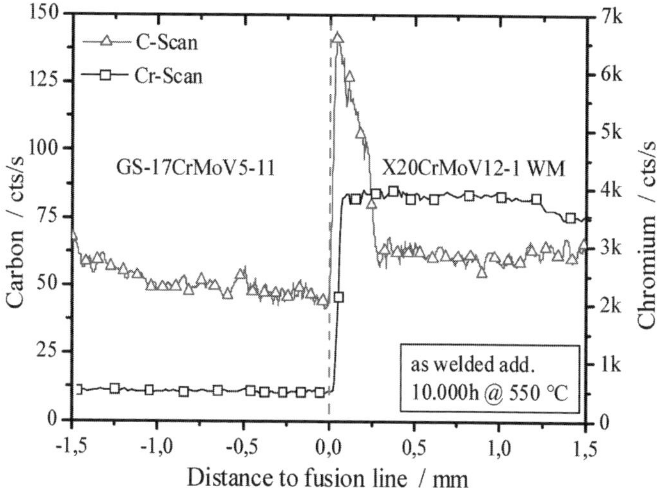

Figure 5: C- and Cr distribution at the interface 1% Cr-BM / 12% Cr-WM in the as welded condition plus an additional long time exposure at 550 °C for 10.000 h

Finally, the influence of the thermal expansion of the different materials has a major influence on failure of dissimilar welds. As it can be seen in Figure 6, high stresses result in the region of the fusion line from the different material expansion during service and - a fact that is often neglected – during post weld heat treatment. Here, a state of the art PWHT is simulated, starting at RT, heating up to 780 °C, two hours of holding time and cooling down to RT at a moderate cooling rate.

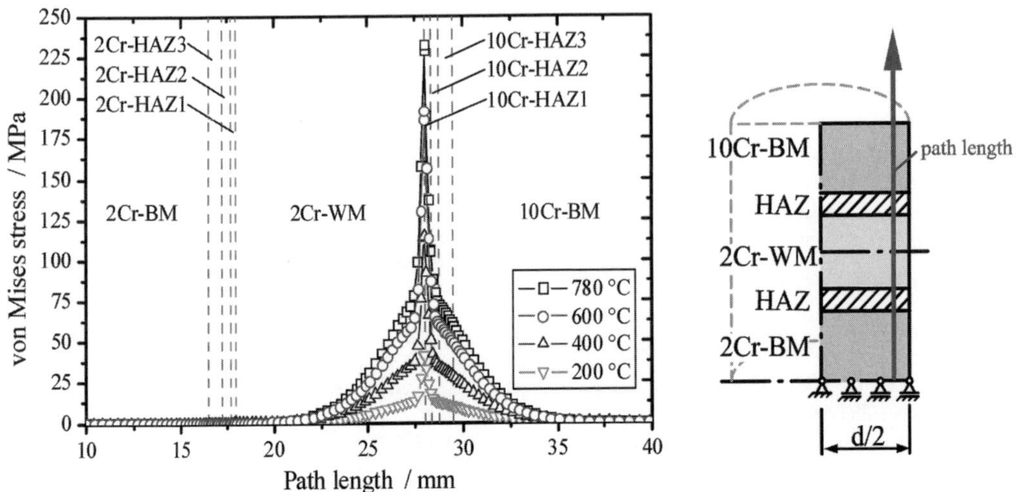

Figure 6: Von Mises stress in a crossweld specimen resulting from thermal expansion in a dissimilar weld, $\Delta\vartheta = 760$ °C

Of course the stress peaks relieve in a relative short period of time due to stress redistribution and relaxation, but nevertheless, the damage and respectively plastic strain accumulated within that period of time is still present, although relatively small, see Figure 7.

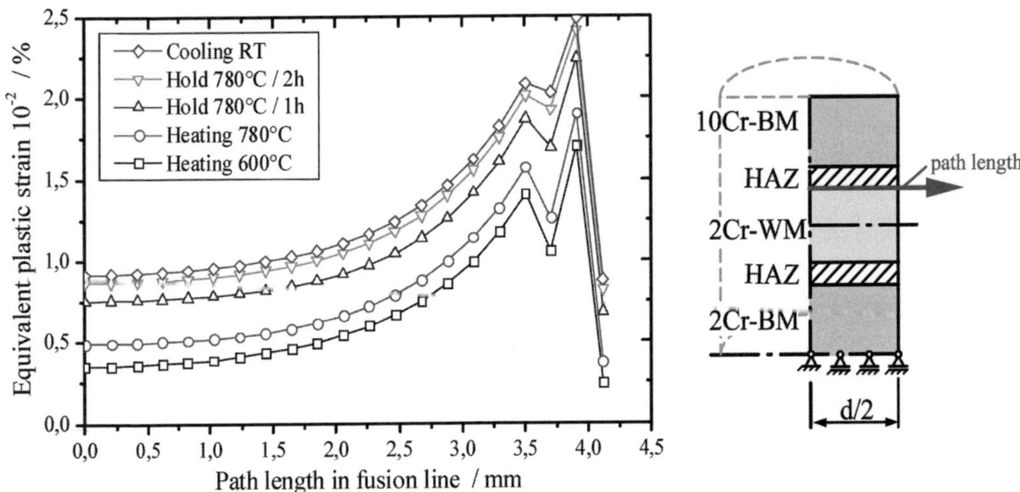

Figure 7: Equivalent plastic strain of a crossweld specimen in the fusion line 2Cr-WM / 10Cr--BM resulting from thermal loading

In case of a (thermo) cyclic loading of the dissimilar weld, these effects described above are even more severe and can lead to thermal fatigue cracking.

2.4 Further failure mechanisms

In addition to the previously mentioned creep failure in the HAZ, the following damage mechanisms can also occur:

- Intergranular corrosion (IGC), observed mainly if non stabilized steels are subjected to appropriate media (e. g. flue gas)

- Intergranular cracking during welding and after post weld heat treatment (stress relief cracking), predominantly in the coarse grain zone of low alloyed heat resistant steels when vanadium is present.

- Intergranular cracking after service loading (stress relief cracking) in case of nickel alloys.

All these mechanisms can be the origin of long term creep damage.

3 Transferability of results obtained with crossweld creep specimens to components

Certain requirements have to be considered when transferring results/strength parameters of base material creep tests, and in particular, of crossweld specimens to components. On the first hand, it is required that the microstructure of the lab specimen (i. e. microstructure and heat treatment) is representative for the component. On the second hand, loading and stress condition are of particular importance as they affect the local, time-dependent relaxation behavior. Furthermore, stationary service conditions have to be assumed, i.e. relaxation processes are not influenced by cyclic loading.

3.1 Thin walled smooth pipes under internal pressure

At temperatures with technically significant creep and internal pressure loading, a stress and strain redistribution can be observed in thin walled pipes due to an inhomogeneous stress distribution after loading. The stress peak causes the material on the inside of the pipe to creep faster than the material on the outer wall, resulting in a stress redistribution. Therefore, after a relatively short loading time, an almost homogeneous stress distribution over the wall thickness can be observed in thin walled pipes under internal pressure. Due to this constant stress distribution, an uniaxially loaded creep specimen can be regarded as representative with regard to the transferability of the material behavior. However, it has to be mentioned, that the creep sample does not exactly represent the creep deformation behavior in the component because the multiaxiality of the stress state has a major impact and does not exist in the lab specimen to such a degree (unless in the case of necking). Hence, the component will show lower creep strain values than the small scale specimen. Furthermore, a constant stress level can be assumed in the component, whereas in the specimen, the cross-section area decreases permanently and thus the resulting (true) stress increases – leading to a larger strain accumulation and a shorter time to fracture.

3.2 Thin walled smooth pipes with circumferential seams under internal pressure

Girth welds are not fully loaded in case of internal pressure loading, as the maximum principle stress is parallel to the seam in circumferential direction. Consequently, the HAZ is only loaded locally besides a stabilizing effect by surrounding weld and base material. As mentioned above, the stress/strain characteristics of the fully loaded welded joint has only to be taken into account here, if the second principle stress in axial direction is increased, e. g. by an additionally superimposed axial loading and reaches stress values comparable to fully loaded joints. Hence the stress/strain characteristics of the lab specimens can be adopted directly, if an almost complete stress and expansion redistribution - as described above - took place and an almost homogeneous stress situation is present. Nevertheless, the HAZ with its bad creep resistance represents an area of premature damage. Creep tests on large samples that are loaded vertically to the seam, showed that a premature damage does occur in the area of the HAZ, but this does not affect the overall rupture time of the component (6).

3.3 Thin walled smooth pipes with longitudinal seams under internal pressure

In this case, the seam is fully loaded, because it is orientated vertically to the maximum circumferential stress. After stress redistribution, an almost homogeneous stress situation, comparable to the one in a crossweld creep specimen can be found. Thus the parameters of this sample can be transferred directly. It is assumed that the redistribution of stress and strain in the crossweld specimen corresponds to the one in the component and the size effects don't play a role in the material behavior due to the thin walled pipe. Design should be based on material specific, time and temperature depending weld strength factors instead of a uniform weld reduction factor.

3.4 Thick walled components with welded seams under internal pressure

With an increasing outer to inner diameter ratio u – pipes becoming more and more thick walled – a time dependent, inhomogeneous stress distribution can be observed over the wall, even after long operation times, see Figure 8.

Transfer of material properties gathered from creep test is possible if analytical methods are used which correspond to a representative stress found in the cross section. This can easily be done for homogeneous cross-section areas, however with increasing wall thickness the assessment gets more and more conservative, as the local stress redistributions are only partly considered. For that reason, inelastic finite element (FE) calculations show in general lower stresses (and hence smaller wall thicknesses), due to advanced creep laws taking into account local stress and strain redistributions.

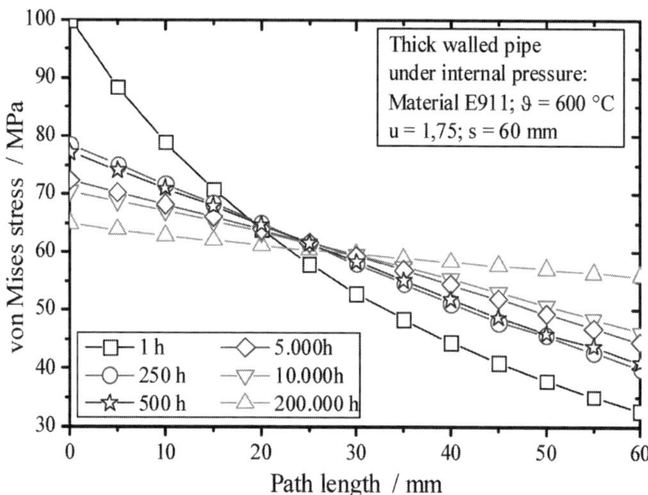

Figure 8: Stress distribution over the wall (thick walled pipe under internal pressure)

However, particular problems arise, if the material is not homogeneous over the cross-section area. This is especially true for welded joints (in the worst case WM, HAZ and BM in the same cross-section) due to the reasons mentioned before.

This leads, depending on the orientation of the zones with different creep behavior to the direction of maximum principle stress, to a different stress situation over the cross-section. As already mentioned above, a specific stress situation can be found in creep tested crossweld specimens that results from the size effect and the orientation of the HAZ perpendicular to the loading direction.

Therefore, the transferability of small scale material properties to thick walled parts based on analytical methods is difficult, if the orientation of the weld to the loading direction is different than in the lab test (in general perpendicular). Main reason for this is the disregard of constraint effects and support of surrounding material layers by analytical methods.

Detailed information on the failure behavior can be obtained by inelastic FE analyses. The stress and strain behavior in the welded joint of a longitudinal welded pipe depends on time, pipe geometry (diameter, wall thickness, weld seam geometry, weld metal), loading condition (internal pressure, additional axial loading) and the material properties (BM, HAZ, WM).

By using the finite element method (FEM) and corresponding advanced inelastic material laws the loading condition in components can be determined (7).

Figure 9 shows the multiaxiality of the stress state q in a longitudinally welded E911 thick walled pipe in radial direction in the intercritical HAZ after 15.000 h of simulation time. According to (8) the multiaxiality q can be defined as follows:

$$q = \frac{1}{\sqrt{3} \cdot h} = \frac{1}{\sqrt{3}} \cdot \frac{\sigma_{mises}}{\sigma_{hydro}} \tag{1}$$

with σ_{mises} representing the von Mises equivalent stress, given in principle stresses below:

$$\sigma_{mises} = \frac{1}{\sqrt{2}} \cdot \sqrt{(\sigma_1 - \sigma_2)^2 + (\sigma_2 - \sigma_3)^2 + (\sigma_3 - \sigma_1)^2} \tag{2}$$

and σ_{hydro} as the hydrostatic stress (pressure):

$$\sigma_{hydro} = \frac{1}{3} \cdot (\sigma_1 + \sigma_2 + \sigma_3) \tag{3}$$

The most severe state of multiaxiality for creep damage can be found below the surface close to the outer diameter of the pipe, as shown in Figure 9.

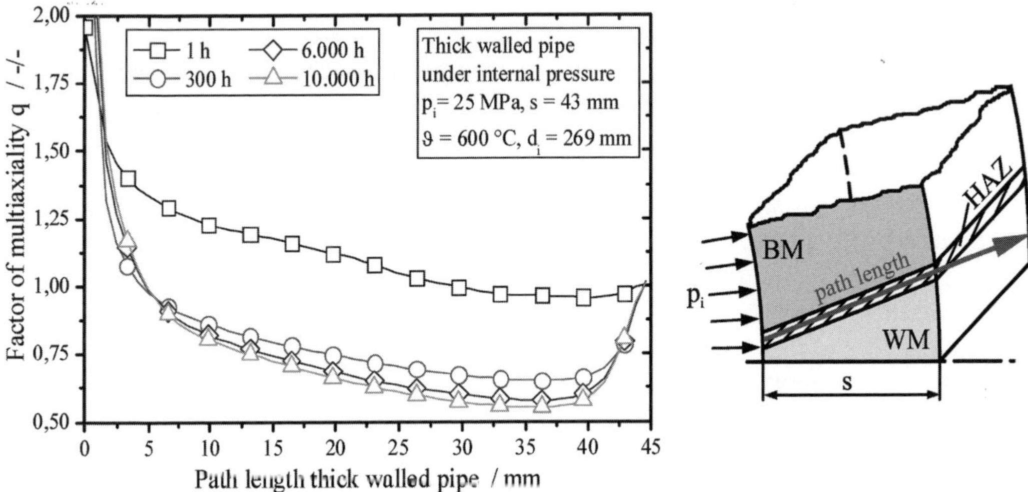

Figure 9: Multiaxiality q in the HAZ3 of a longitudinal welded E911 pipe under internal pressure

The state of multiaxiality in the HAZ of the crossweld specimen, see Figure 10, correspond to those in the HAZ of a longitudinal welded pipe under internal pressure to a large extend, compare Figure 9. Therefore – as stated before - the transferability of material parameters from lab specimens to welded components is possible, provided that the loading direction is perpendicular to the weld (fully loaded seams).

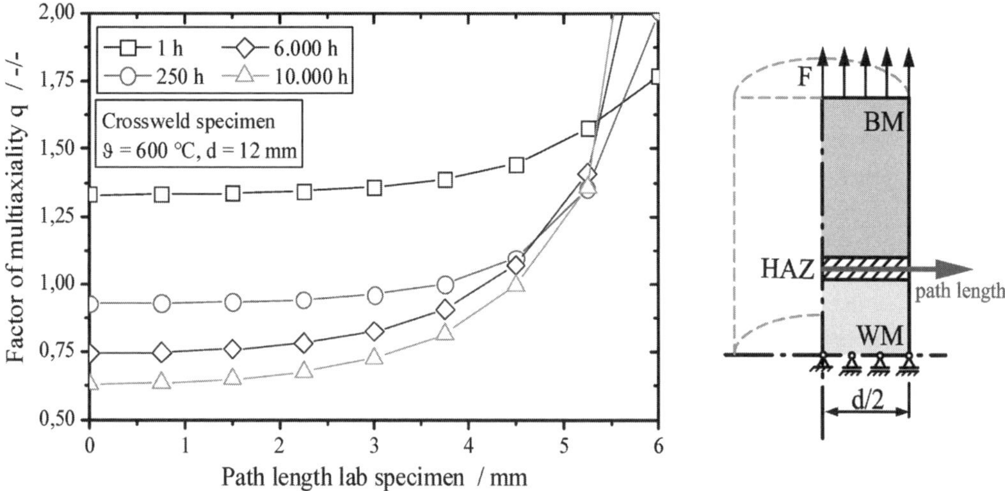

Figure 10: Multiaxiality q in the HAZ3 of a crossweld specimen

The damage evolution at 15.000 h simulation time in the HAZ3 of longitudinally welded pipes is depicted in Figure 11 for various wall thicknesses but constant internal pressures. It can be seen, that for this geometry, the damage has a minimum close to the inside of the pipes.

The constitutive equations used for these FE simulations inherit a damage parameter D or better a damage rate $\dot{D}$, governed by equation (4). The exponents A, n and m are approximated in a phenomenological way using genetic algorithms and simplex functions - to creep data.

$$\dot{D} = 10^{A_D} \cdot \sigma^{n_D} \cdot \varepsilon^{m_D} \tag{4}$$

The outer surface shows a relatively low damage potential too, whereas below the surface the maximum damage can be observed. Actually, this means that the damage evolution is initiated by creep right below the outer surface - progresses to the surface in the course of time after already having damaged a great area of the wall thickness.

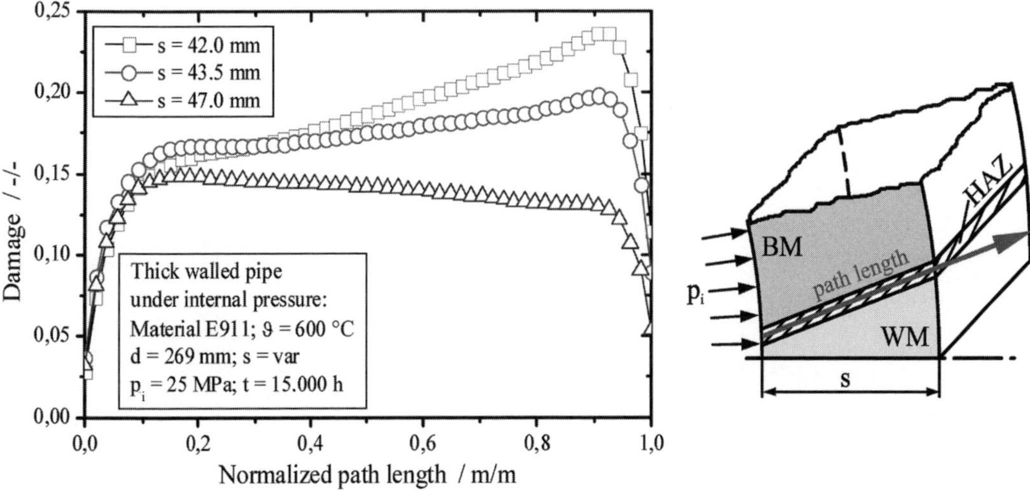

Figure 11: Hypothetic creep damage in the HAZ3 of a longitudinally welded E911 pipe

These effects also have an influence on other constructive elements used in power and process plants, like e. g.:

- Connecting (longitudinal welded) pipes with circumferential seams
- Attaching nozzles or junctions
- Manufacturing of pipe bends or T-pieces

The reasons mentioned above show the advantages of inelastic FEM with adequate creep laws compared to analytical "back of an envelope" calculations by performing stress and strain analysis. Therefore FE simulations are the preferred tool to clarify to what extend results of crossweld specimens can be transferred to large components.

4 Design of welded components according to conventional design rules / design standards

In general, design of welded components in the creep range implies the determination and application of additional safety factors to handle the reduced creep strength of welds.

The heat affected zone represents - as mentioned above - the "weakest" link in welded components of ferritic steels. Figure 1 makes clear that all ferritic steels and especially the new martensitic steels, do show a sensitivity of premature failure in the HAZ in case of fully loaded welds. If a weld strength factor of 0.5 is reached (e. g. 12% Cr steels at 600 °C), even not fully loaded welds like hoop welds in pressurized pipes fail prematurely. This situation can even get worse, if additional loads like axial forces or bending moments are present, that superimpose the stresses of the internal pressure. In this case, local creep failure can occur even for larger weld strength factors. For that reason the design of welded components is of particular importance.

Primarily it has to be stated, that design on the basis of base material parameters only - with additional safety factors is not adequate. As it can be seen from Figure 1, the weld strength reduction factor concerning creep strength has to be determined material dependant. Furthermore it has to be emphasized, that damage and failure in the HAZ is a local damage process, concentrated in a very small area with resulting in local deformations. Thus state of the art deformation criteria, like the 1% or 2% strain criteria in the base metal aren't sufficient to prevent from HAZ failure. The corresponding DIN EN 13480-3, paragraph 5.3.1 specifies an additional reduction of the base material parameter by 20%, if no creep strength values for the welded joints are available. The draft of EN 13445-2:2002/prA1:2006.7 or EN 13445-3:2002/prA1:2006.9 is of similar content: The weld strength factor z is multiplied with the factor z_c (weld creep strength reduction factor) which can take the values listed below:

- z_c is determined by tests of limited durations on cross weld specimens according to Annex C of EN 13445-2
- $z_c = 1$, if the creep values of the welded joints are not lower than the lower accepted scatter band (- 20 %) of specified mean values of the base material
- $z_c < 1$, if these conditions are not met
- $z_c = 0.8$, otherwise

It has to be concluded from Figure 1, that for high alloyed steels at high service temperatures, weld strength values down to 50 % of the base material's mean value are possible. Hence both DIN EN 13480-3 and the draft of EN 13445-3:2002/prA1:2006.9 are not conservative – in particular cases failure can occur. Thus the material-independent determination of safety parameters is not adequate to the situation. The application of material specific, time and temperature dependent weld strength factor instead of a standard reduction factor would improve the material utilization and the design accuracy (1). Furthermore solutions/recommendations should be provided for the 200.000 h extrapolation problem already known from homogeneous base materials and even more critical in the case of welded components.

5 Recommendations for FE-analysis of welded components in the creep range

In typical technical applications, it is often sufficient to describe the primary and secondary creep state because the tertiary creep regime is not generally of interest for design of these components. In this case a modified Garofalo (9) creep model can be used to describe the behavior of the component under high temperature loading:

$$\varepsilon_{cr} = \varepsilon_{cr,max} \cdot h(t) + \dot{\varepsilon}_{cr,min} \cdot t = X_1 \cdot (1 - e^{-X_2 \cdot t}) + X_3 \cdot t \tag{5}$$

In general, the parameters X_1, X_2 and X_3 are stress dependant functions whose constants are fitted to creep test results phenomenologically:

$$X_1 = A \cdot e^{B \cdot \sigma};\ X_2 = C \cdot \sigma^D;\ X_3 = E \cdot e^{F \cdot \sigma} \tag{6}$$

If consideration of advanced life time assessment and modeling of damage and failure behavior is planned, the inelastic finite element calculations must be based on constitutive models covering all three creep stages. In this case, a modified creep law of Graham and Walles (10) is recommended. In its general form, the Graham-Walles material law reads as follows:

$$\dot{\varepsilon}_{cr} = A_1 \cdot \sigma^{n_1} \cdot \varepsilon_{cr}{}^{m_1} + A_2 \cdot \sigma^{n_2} \cdot \varepsilon_{cr}{}^{m_2} + A_3 \cdot \sigma^{n_3} \cdot \varepsilon_{cr}{}^{m_3} \tag{7}$$

An advantage of this modified creep law is the implementation of a strain hardening rule, which allows a better description of load level changes than the time hardening version. Using this material model, it is possible to calculate the local stress states and the time dependent deformation of specimens and components. The tertiary creep behavior can also be represented by an effective stress concept with a damage parameter D. As mentioned above, the material laws are fitted phenomenological to creep data from uniaxially loaded creep tests. Therefore an additional parameter quantifying the multiaxiality of the stress state in real components is required. The constitutive formulation of the modified Graham-Walles creep law is given below

$$\dot{\varepsilon} = 10^{A_1} \cdot \left(\frac{\sigma_{mises}}{(1-D)}\right)^{n_1} \cdot \varepsilon_{eq}{}^{m_1} + 10^{A_2} \cdot \left(\frac{\sigma_{mises}}{(1-D)}\right)^{n_2} \cdot \varepsilon_{eq}{}^{m_2} \tag{8}$$

Here, the damage rate is a function of the multiaxiality q of the stress state as defined in eq. (1):

$$\dot{D} = 10^{A_{D1}} \cdot \left(\left(\frac{\sqrt{3}}{q}\right)^{\alpha} \cdot \sigma_{mises}\right)^{n_{D1}} \cdot \tilde{\varepsilon}^{m_{D1}} + 10^{A_{D2}} \cdot \left(\left(\frac{\sqrt{3}}{q}\right)^{\alpha} \cdot \sigma_{mises}\right)^{n_{D2}} \cdot \tilde{\varepsilon}^{m_{D2}} \tag{9}$$

with a strain contribution given in equation (10)

$$\tilde{\varepsilon} = \sqrt{\langle\varepsilon_1\rangle^2 + \langle\varepsilon_2\rangle^2 + \langle\varepsilon_3\rangle^2} \,. \tag{10}$$

Current research (11), (12) shows that it is of great importance to implement the primary creep range within the constitutive equations in order to obtain conservative results. Therefore a simple Norton creep law disqualifies for advanced life time assessments and modeling of damage and failure behavior of welded components in the creep range. Furthermore it might be necessary to adopt the material law to the specific melt of the component to be analyzed, if sample creep tests of the material show a big deviation from the fitting data (e. g. ECCC mean value curves, etc.).
In order to obtain most accurate results, a state of the art five material model –as shown in Figure 12 - should be used to model the weld.

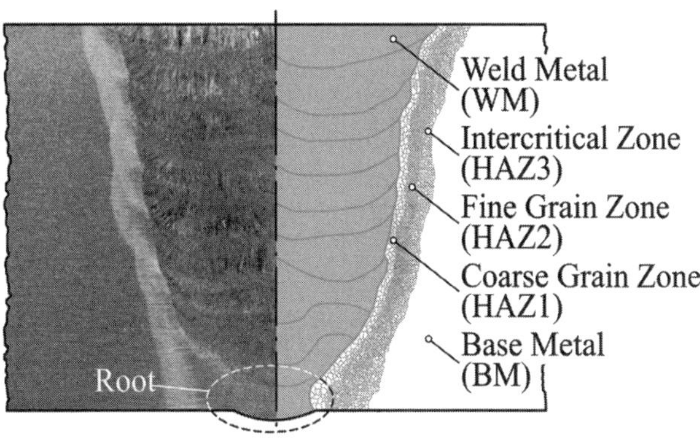

Figure 12: Five zone weld discretization

Hence it is necessary to have reliable creep data for the different heat affected zones to fit the material laws. Therefore base metal specimens are thermally treated according to the temperature cycle monitored during welding, see Figure 13. This simulation can be either done on a Gleeble thermal-mechanical testing machine or by controlled inductive heating and oil cooling (13). As the temperature distribution in the specimen in a Gleeble thermal simulation is only constant in a relatively small area (in general 20 mm) HAZ1 and HAZ2 simulations always result in a adjacent HAZ3. This requires a special shape of the specimens to avoid failure in that region. Nevertheless the temperature cycle can be realized much easier on a Gleeble system and requires an experienced operator in case of simulating with inductive heating and cooling. The peak temperatures of the simulation is dependant on the material, welding technique, component size etc.. The peak temperatures given in Figure 13 are used to simulate the HAZ of an E911 base material and submerged arc welding.

In Figure 14 exemplarily results for creep tests on HAZ simulated material are presented. Again, the poor creep resistance of the HAZ3 simulated material can be clearly seen. At least five different stress levels should be examined (for each material, resp. zone) for best results in fitting the creep laws. Due to the long testing times and high costs respectively, low stress levels (below 75 MPa) are often omitted in research programs but are extremely important for an accurate FE-analysis.

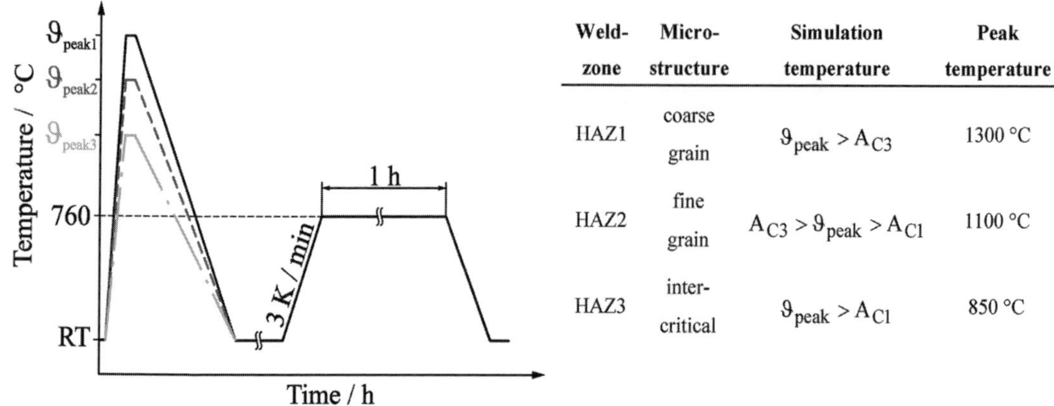

Weld-zone	Micro-structure	Simulation temperature	Peak temperature
HAZ1	coarse grain	$\vartheta_{peak} > A_{C3}$	1300 °C
HAZ2	fine grain	$A_{C3} > \vartheta_{peak} > A_{C1}$	1100 °C
HAZ3	inter-critical	$\vartheta_{peak} > A_{C1}$	850 °C

Figure 13: Thermal simulation of heat affected zone material and peak temperatures for HAZ simulation (E911 BM, SAW)

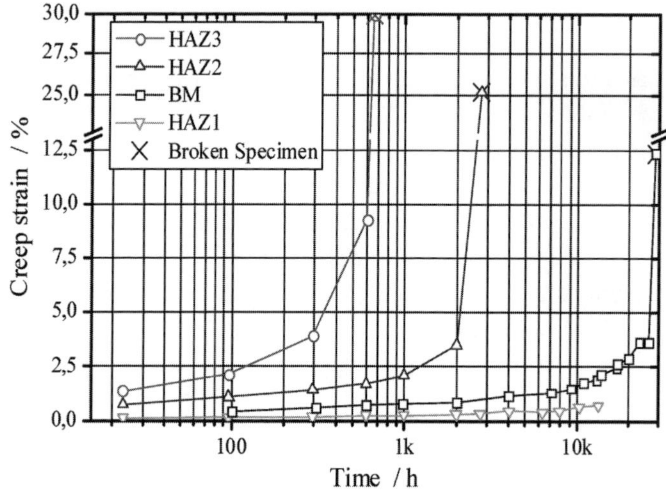

Figure 14: Creep strain characteristics for BM and HAZ

The need for the implementation of the factor of multiaxiality of the stress state and the coupling with damage evolution in the constitutive equations is demonstrated in the following. Results of numerical simulations of P91 hollow cylinders under internal pressure (14), (15) are depicted in Figure 14. Based on the stress state only, the inner surface of a pressurized pipe or, as shown in this case, a hollow cylinder under internal pressure would be the position of failure.

Experience with real components (here pipes/hollow cylinders under internal pressure) shows, that failure mostly initiate at the outer surface. This is the region with critical states of multiaxiality leading to creep cavities formation. By introducing the multiaxiality q as shown in equation (9), failure location of components can be predicted more realistic as shown in Figure 16.

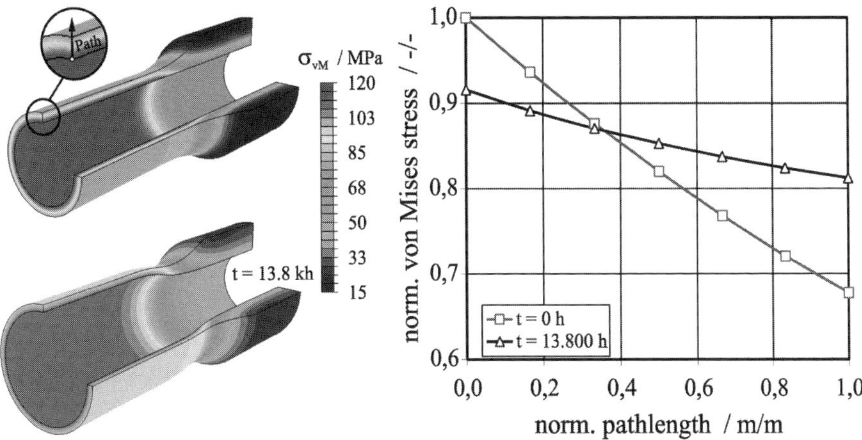

Figure 15: Change in von Mises stress distribution with pure pressure (t = 0 h) and after 13800 h of creep exposure; the deformation is shown with a scale of 20:1

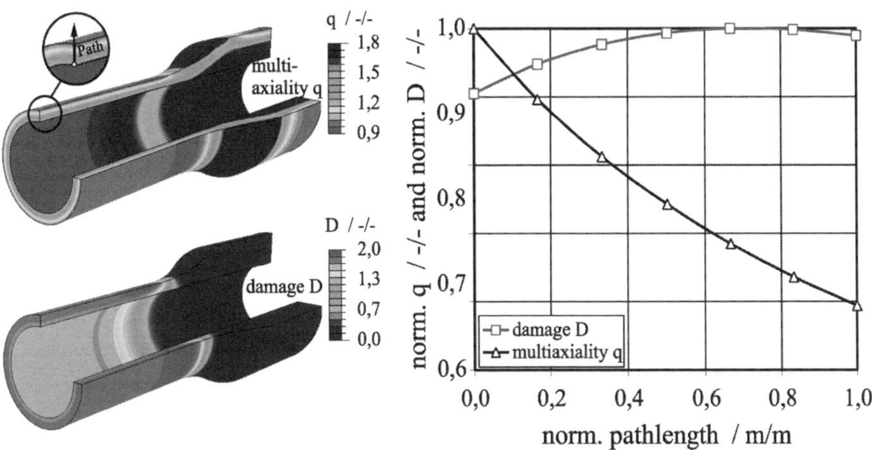

Figure 16: Distribution of the quotient of multiaxiality q and damage D within a smooth hollow cylinder specimen made of P91 under internal pressure of p = 255 bar after 13800 h at 600 °C

6 Availability of data or extrapolation of creep strength parameters of welded joints

The availability of data regarding creep strength behavior of welded joints is directly linked to the extrapolation to obtain long term values. State of the art and knowledge allows an extrapolation of the factor of 3 for base material parameters based on the rupture time.

However, the new martensitic materials have shown that this factor in connection with a relatively small number of experiments may imply dramatic corrections of the extrapolated data to lower values if – after a certain amount of time - an enhanced data basis with long term experiments exists.

It can be assumed, that in case of steels, receiving an optimized creep strength due to a complex microstructure and precipitation characteristics, thermodynamic processes negatively influence the creep strength - dependent on the temperature - in the long term. Thus, the formation of new precipitations in the temperature range above 600 °C, like e. g. the Z-phase and the corresponding solution of other particles, causes a significant drop in creep strength for long test times ("S-shaped creep curve").

In these cases extrapolation of creep data is not conservative and always requires the verification through long term tests.

In case of creep strength of welded joints, this effect goes along with the change in fracture location. Therefore, the extrapolation of data obtained from crossweld specimens inherits further risks. The facts just mentioned also have to be considered if the HAZ (metallographic notch) is removed by an additional annealing post welding, especially if no adequate heat treatment can be performed due to the reasons mentioned in (1). The application of the previously mentioned evaluation factors in the standards, like e. g. in EN 13445-3:2002/prA1:2006.9 have to be restricted in that perspective. If at a relatively low temperature no change in fracture location is observed and thus z_c is defined with 1.0, it is still questionable if this is also valid in the long term range. Even if z_c is set to 0.8, this value has to be verified with long-term creep data. Exemplarily a data set (base metal and welded joints) for E911 at 600 °C is shown in Figure 17. The welded joints A129 SAW and A129 MAW1 show a change in fracture location at approximately 4.500 h. The difference between the rupture curve of the welded joints and the rupture curve of the base material increases with increasing rupture times.

An extrapolation to 100.000 h seems problematic in this case, as there is no information available which precipitations in the HAZ are thermodynamically instable and which phase modifications do occur in this time range.

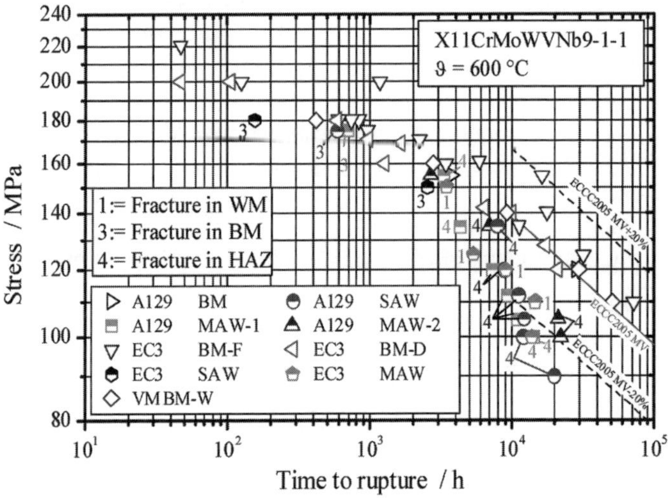

Figure 17: Creep rupture strength for different melts and weld procedures of base material and crossweld specimen

From the results presented so far, it can be stated that a verified data base is mandatory for the economical and safe design of welded components. An experimental data base comprising the following key issues is recommended for steels planned to be used in welded joints:

- Three test temperatures (operation temperature, +/- 30 °C).
- Testing time: at least 1/3 of the desired life time, in general 1.000 h, 3.000 h, 10.000 h, 30.000 h, 60.000 h, 100.000 h, etc.).
- Extrapolation for fracture location in the HAZ is only meaningful, if additional long term tests show no further decrease in creep strength. If this requirement is fulfilled, an extrapolation factor of 3 is feasible. An evaluation method is proposed in (13) which uses crossweld data after change in fracture location and - if available - additional simulated HAZ data for an appropriate assessment. In case of no long term data available, a conservative assessment taking into account the possibility of further decrease of the weld strength factor for long times is required.
- If the fracture location is not in the HAZ, a Larson-Miller (LM) evaluation is recommended. Additional experiments at higher temperatures have to be conducted for values of the LM – parameter greater or equal the ones of the extrapolation point. The purpose of these experiments is to verify that no change of fracture location into the HAZ does occur. Testing time has to exceed 10 kh and no microstructural changes may happen due to the chosen test temperature.

As mentioned above, data of creep tests on welded joints is not included in the specific material standards. Therefore, common research projects (e. g. (16)) are at present the source of information and knowledge for design purposes. Even if an additional annealing is performed on welded components (compare paragraph 2.1, (2)) the sustainability of the heat treatment in the long term range has to be verified by a corresponding research program.

7 Conclusions

Fully loaded seams in pressurized components operating in the creep range require different strategies in design and monitoring. This is due to the particular failure behavior in the HAZ under creep loading which leads - for a number of materials discussed in this paper - to premature fracture in the HAZ. Until the verification, that an additional heat treatment (quench and temper) post welding compensates this particular failure behavior, special measures have to be taken, in order to guarantee a safe operation.

These measures primarily consist in the determination of material specific, time and temperature dependent weld strength factors instead of standard reduction factors. Those factors are not included in the established standards and have to be determined experimentally – similarly to the design parameters of the base material.

Due to the high risk potential in case of failure of fully loaded components, the monitoring has to be optimized based on additional instrumentation and inelastic numeric analyses to determine the state of damage of a component.

Acknowledgments

Parts of the investigations presented in this paper are supported by the Research Association of the Iron and Steel Manufacturing Industry (AVIF) in Germany under contract No. A221 & A129. The support is highly acknowledged. Furthermore the investigations are supported by the COST group – mainly in the framework of COST 536 and by the members of the Work Group W1 of the German Creep Group. These contributions are truly appreciated.

References

(1) Schubert, J., A. Klenk and K. Maile: Determination of weld strength factors for the creep rupture strength of welded joints. International Conference on Creep and Fracture in High Temperature Components – Design and Life Assessment Issues, European Creep Collaborative Committee (ECCC), IOM London, 12-14 September 2005

(2) AVIF Forschungsvorhaben A 196 „Längsnahtgeschweißte Rohre", Abschlussbericht, MPA Universität Stuttgart, 2006

(3) Maile, K, D. Blind and K. Schneider: Analyse des Zeitstandverhaltens von nichtgleichartigen Schweißverbindungen für den Anlagenbau, DGM und DVM Vortrags- und Diskussionstagung, Bad Nauheim, 1989

(4) COST 505, Project A6: Investigations of welded joints of dissimilar steels for high temperature application in steam turbines, IWS Technical University of Graz, 1989

(5) COST 505, Project D18: The influence of welding and stress relieving on the diffusion and precipitation process in welded transition joints of 1% Cr-steels with 12% Cr-steels, MPA Stuttgart, 1989, IWS Technical University of Graz, 1989

(6) Theofel, H. and K. Maile: Einfluss der Wärmeeinbringung beim Schweißen auf die Zeitstandfestigkeit und das Kriechverhalten von Schweißverbindungen, untersucht an warmfesten CrMoV Stählen mit 1% und 12% Cr. Abschlußbericht zum Forschungsvorhaben AiF-Nr. 5710, MPA Stuttgart 1987

(7) Bauer, M., A. Klenk, K. Maile and E. Roos: Numerical Investigations on Optimisation of Weld Creep Performance in Martensitic Steels, 8th International Seminar: Numerical Analysis of Weldability, 25. - 27. September 2006, Graz - Seggau Austria

(8) Clausmeyer, H., K. Kussmaul and E. Roos: Influence of Stress State on the failure behavior of cracked components made of steel, Appl. Mech. Rev. Vol. 44, 2, ASME, 1991.

(9) Granacher, J. and A. Pfenning: Kriechgleichungen II – rechnergestützte Beschreibung des Kriechverhaltens ausgewählter hochwarmfester Legierungen. FVV Forschungsvorhaben Nr. 432, 1991

(10) Graham, A. and K. F. A. Walles: Relationships between Long- and Short-Time Creep and Tensile Properties of a Commercial Alloy, Journal of The Iron and Steel Institute, 179, pp. 104-121, 1955.

(11) Optimierung der Schweißverbindung zur Verhinderung des vorzeitigen Kriechversagens in der WEZ über die gezielte Festigkeitsauswahl des Schweißgutes (Mismatch), AVIF Forschungsvorhaben A221, MPA Stuttgart, 2004

(12) Maile, K. and E. Roos: Forschungsvorhaben A229 - Schädigungsentwicklung 3 - Kriterien zur Schädigungsbeurteilung von Hochtemperaturbauteilen aus martensitischen 9-11 % Cr-Stählen

(13) ECCC Recommendations, Volume III Part 2: Data acceptability criteria and data generation: Creep Data for Welds, Issue 3, Eds. B. Buchmayr, A. Klenk, European Creep Collaborative Committee, 2005

(14) Klenk, A., M. Rauch and K. Maile: Influence of stress state on creep damage development in components, Proceedings Int. Conf. on Plant Life Extension, Cambridge, England, 2004.

(15) Rauch, M. et al.: Creep damage development in martensitic 9Cr steels, Proceedings Int. Conf. on Plant Life Extension, Cambridge, England, 2004.

(16) Nachweis der Langzeiteigenschaften von Schweißverbindungen moderner Stähle für den Einsatz in Dampferzeugern im Bereich bis 620°C, Abschlussbericht zum Forschungsvorhaben A129 (AVIF), MPA Universität Stuttgart, 2002

Advances in Materials Technology for Fossil Power Plants
Proceedings from the Fifth International Conference
R. Viswanathan, D. Gandy, K. Coleman, editors, p 82-91

DOI: 10.1361/cp2007epri0082

UltraGen: a Proposed Initiative by EPRI to Advance Deployment of Ultra-Supercritical Pulverized Coal Power Plant Technology with Near-Zero Emissions and CO_2 Capture and Storage

John Wheeldon, Jack Parkes, and Des Dillon
Electric Power Research Institute,
Palo Alto, California

UltraGen is an initiative proposed by EPRI to accelerate the deployment and commercialization of clean, efficient, ultra-supercritical pulverized coal (USC PC) power plants that are capable of meeting any future CO_2 emissions regulations while still generating competitively-priced electricity. In addition to reducing CO_2, these advanced systems will have to achieve near-zero emissions of criteria pollutants (SO_2, NO_X, and filterable and condensable particulate) and hazardous air pollutants such as mercury.

An essential step in reducing emissions is raising PC generating efficiency to lower the coal fed for a given power output. The most significant method to increase efficiency is to progress to advanced USC steam conditions. Exceptionally strong high-nickel boiler and steam turbine materials currently under evaluation by the US-DOE program are expected ultimately to allow main steam temperatures of up to 1400°F. At this temperature, compared to a sub-critical PC unit, generating efficiency will be up to 10 percentage points higher with the potential to lower CO_2 emissions per MWh by 25 percent for any given coal as illustrated in Figure 1[1]. Producing less CO_2 per MWh in turn lowers the cost of capture and storage by reducing the size of the equipment required and, for example, the auxiliary power required for compression.

Figure 2 shows the trend of cost of electricity (COE) with generating efficiency [2]. For plants without CO_2 capture there is little effect upon COE. For plants with CO_2 capture there is a clear reduction - the COE for the advanced USC PC plant with double reheat being around 10 percent lower than that of a sub-critical plant. Based on these observations and conclusions from supporting studies, the UltraGen strategy to capture and store CO_2 most economically is to advance improvements concurrently in both power generation, and capture and storage technologies. However, if these technologies are to be commercially available once CO_2 capture is required, then they must be deployed now so that they can evolve and improve through design and operational experience.

It is expected that the UltraGen plants will be commercial projects incorporating advanced technology demonstration features. They will be on economic dispatch satisfying the demand for competitively-priced electricity, with main steam temperatures between 1100°F to 1400°F, improved emission controls, and the capture and storage of CO_2. The first UltraGen project will be at the lower end of the temperature range and subsequent projects will have progressively higher temperatures. Part of the UltraGen work program will be to identify how to fund the differential cost incurred by the host utility in demonstrating these improved features

Background to UltraGen

Current advanced supercritical PC plants have main steam temperatures of around 1100°F and use ferritic steels for the high-temperature steam tubing and headers. Although these materials can operate at higher temperatures, the maximum projected temperature is from 1160 to 1180°F. To accelerate progress to higher steam temperatures, the transition from ferritic to high-nickel alloy materials should be made as soon as possible. Discussions with equipment suppliers suggest that within three years sufficient information will be available to design a plant with main steam temperatures up to 1290°F. The design of a 1400°F plant is several years away and will require additional testing of materials and components in a power plant test facility.

To limit the potential for increased fireside corrosion in the early UltraGen plants, the fuels proposed have low-sulfur and low-chlorine contents. Candidate fuels would include:

- Internationally-traded bituminous coal and Utah bituminous coal.
- Sub-bituminous coal: the coal most widely used for new US PC plant designs.
- Some lignites may also qualify.

In addition to higher steam temperatures, environmental controls also have to be improved and progressed to near-zero-emissions, and the CO_2 capture plant has to be integrated with the power island. Each project will advance improvements in some if not all of these areas.

Based upon these requirements the UltraGen program will consist of the following demonstration projects.

- **UltraGen I** with main steam temperatures up to 1120°F.
- **UltraGen II** with main steam temperatures up to 1290°F.
- **ComTes-1400** to test out materials and components for use in UltraGen III.
- **UltraGen III** with main steam temperatures up to 1400°F.

The plant with the lowest cost of electricity and the lowest emissions incorporating CO_2 capture and storage will be UltraGen III. Although this cannot be built today, working towards this objective can start now by progressively demonstrating the technology features that will be incorporated in UltraGen III. Active participation in the program will contribute to the stage-wise development of the combustion-based solution to reduce emissions of greenhouse gases.

UltraGen I

UltraGen I is likely to operate with a main steam temperature of around 1120°F and use currently available, commercially proven ferritic steels. It will control emissions at a level lower than that achieved by currently operating plants and it will demonstrate a CO_2 capture plant sending 1-million tons per year to storage or for enhanced oil recovery. The capture plant will process approximately 25 percent of the flue gas flow and be around 20 times the size of the largest plant operating on a coal-fired unit.

If its design features were incorporated with a project currently in the planning stages, UltraGen I could come into service around 2012. Major design features are presented in Table 1 and in Figure 3.

- The net output after CO_2 capture is 800 MW, matching the median size of USC PC plants currently proposed for deployment in the US.

Table 1. Performance Parameters for UltraGen I, II, and III

	UltraGen I	UltraGen II	UltraGen III
Net output after capture, MW	800	600	600
Net output before capture, MW	850 to 900	650 to 700	630 to 670
Efficiency before capture, % (HHV)	39	42 to 44	45 to 48
Main steam temperature, °F	1120	1290	1400
High-temperature material	Ferritic	High nickel	High nickel
Flue gas slip stream flow, % (1)	25 (2)	50 or 100 (3)	50 or 100 (3)
SO_2, lb/MBtu (lb/MWh)	0.03 (0.25)	0.01 (0.080)	0.01 (0.080)
NO_X, lb/MBtu (lb/MWh)	0.03 (0.25)	0.01 (0.080)	0.01 (0.080)
Total particulate, lb/MBtu (lb/MWh)	0.010 (0.088)	0.008 (0.064)	< 0.008 (0.064)
Mercury, percent capture	90	Greater than 90	Greater than 90
Differential cost, $million	250 to 350	TBD	TBD
Projected earliest in-service date	2012	2015	2021

(1) To post-combustion CO_2 capture plant (2) To achieve 1-million tons/year

(3) 50 percent results in CO_2 emissions of around 900 lb/MWh. 100 percent results in up to 3.8 million tons/year capture.

- The HHV efficiency will be around 39 percent (8,750 Btu/kWh) depending upon the coal used and whether single or double reheat is employed. The steam conditions are expected to be 3,750 psia/1120°F/1130°F using ferritic alloys for the high-temperature tubing and headers.
- 25 percent of the flue gas is routed to a post-combustion CO_2 capture plant designed to capture at least 90 percent of the inlet CO_2. Using the best available solvents and effective heat integration of the two plants, the energy required to capture and compress the CO_2 is estimated to be between 50 and 100 MW.
- The emissions of SO_2 and NO_X are reduced to 0.03 lb/MBtu (0.25 lb/MWh), levels approximately half those of the cleanest plant currently permitted in the US.
- Total particulate (condensable and filterable) are reduced to 0.010 lb/MBtu (0.088 lb/MWh), which is 30 percent lower than the cleanest plant currently permitted in the US.
- 90 percent of the mercury is captured, which is comparable to the cleanest plant currently permitted in the US. Improvements in capture efficiency are considered premature at this time given the current status of the technology.

UltraGen I incorporates improvements required to progress to UltraGen II and ultimately to UltraGen III. In addition, individual projects could be included at existing sites or on new projects. For example, a utility may have an interest in testing out improved controls for NO_X but not for SO_2 or CO_2.

UltraGen II

The thermal and economic performance of UltraGen II will be an advancement on that of UltraGen I. The main steam temperature will be raised to 1290°F through the use of high-nickel alloys and emission will be reduced further to near-zero levels, considered to be below 10 ppmv.

Two options for CO_2 capture are possible; capturing 50 percent to lower emissions to be similar to those of a natural gas combined cycle; and capturing 90 percent to store up to 3.5 million tons per year.

If as suggested the design data becomes available by 2010 and a host is identified, UltraGen II could come into service around 2015. Major design features are presented in Table 1 and in Figure 4.

- The net output after CO_2 capture is 600 MW. This is at the lower end of the range for USC PC plants currently proposed for commercial deployment in the US and is selected to limit the differential cost arising from the use of high-nickel alloys.
- The HHV efficiency before capture will be from 42 to 44 percent (7,770 to 8,140 Btu/kWh) depending upon the coal used, the main steam temperature selected, and whether single or double reheat is used.
- The main steam temperature will be up to 1290°F at around 4,100 psia using high-nickel alloys for the high temperature tubing and headers.
- 50 or 100 percent of the flue gas is routed to a post-combustion CO_2 capture plant designed to capture at least 90 percent of the inlet CO_2. It is anticipated that improvements in post combustion CO_2 capture technology and plant integration will be made lowering the energy consumed in capturing and compressing the CO_2 relative to UltraGen I. The parasitic energy required is estimated to be between 50 and 100 MW, respectively, for the two capture efficiencies.
- Alternatively, depending upon its state of development, the UltraGen II plant could be built as an oxy-combustor.
- The emissions of SO_2 and NO_X are reduced to 0.01 lb/MBtu (0.080 lb/MWh) a factor of 3 lower than for UltraGen I. Achieving this will level will depend upon the success of UltraGen I
- Total particulate (condensable and filterable) is reduced to 0.008 lb/MBtu (0.064 lb/MWh), 20 percent lower than for UltraGen I.
- Mercury capture is increased to a value greater than for UltraGen I, but the exact value will depend upon progress made in mercury control and measurement technology.

As with UltraGen I, aspects of UltraGen II might also be carried out as separate projects thus spreading costs and any perceived risks. Retrofit applications to existing plants may also be adopted for the near-zero emission and CO_2 capture technologies applications. After UltraGen I additional USC plants may be built with similar steam conditions but with higher net outputs.

ComTes-1400

This concept has been pioneered by the European Union AD-700 program where critical parts of a high temperature steam circuit are incorporated into an operating coal-fired power plant. Components that would be tested include:

- Evaporator and superheater panels.
- Headers, manifolds, and high-energy piping.
- By-pass, safety, and throttle valving.

In addition, the program would include boiler and steam turbine design studies investigating measures to reduce the amount of high-nickel alloy required and how best to integrate the CO_2 capture and storage systems with the power plant to achieve the lowest cost of electricity.

If planning for the test facility was to commence today with design work starting in three years time, it could come into service in 2011. Following a 5-year test program with project planning commencing in 2016, UltraGen III could come into service in 2021.

UltraGen III

The major design features are presented in Table 1 and in Figure 5.

- The net output after CO_2 capture is 600 MW.
- The HHV efficiency before capture will be from 45 to 48 percent (7,120 to 7,600 Btu/kWh) depending upon the coal used, the main steam temperature selected, and whether single or double reheat is used.
- The main steam temperature will be up to 1400°F at around 4250 psia using high-nickel alloys for the high temperature tubing and headers.
- Either 50 or 100 percent of the flue gas is routed to a post-combustion CO_2 capture plant designed to capture at least 90 percent of the inlet CO_2. The energy required is estimated to be 30 and 70 MW, respectively.
- Alternatively, depending upon its state of development, the UltraGen III plant could be built as an oxy-combustor.
- The emissions of SO_2 and NO_X are reduced to 0.01 lb/MBtu (0.080 lb/MWh).
- Total particulate (condensable and filterable) is reduced to below the 0.008 lb/MBtu (0.064 lb/MWh) level established for UltraGen II.
- Mercury capture is increased to a value greater than for UltraGen II, but the exact value will depend upon progress made in mercury control and measurement technology.

UltraGen III will include all technological improvements identified by the UltraGen Program. After UltraGen III additional USC plants may be built with similar steam conditions but with higher net outputs.

Estimated Costs for UltraGen Projects

Compared to currently proposed commercial USC PC plants, the cost of the UltraGen projects will be higher. Estimated differential costs between a standard USC plant and UltraGen I incorporating all additional features are in the range \$250 to \$350 million (excluding pipeline costs). These are made up as follows:

- Capital cost for additional capacity to provide power for CO_2 capture.
- Capital and operating cost for improved emissions control.
- Capital and operating costs for CO_2 capture plant.

The incremental costs for UltraGen II are being prepared by EPRI in conjunction with an Architect Engineer and boiler manufacturer; information will become available in 2008. Tasks included in the study are as follows:

- Establish the thermal and economic performance for a 750-MW USC PC boiler fired with sub-bituminous coal and single and double reheat.

- Establish the design for the USC PC plant fired with Pittsburgh #8 and identify the most-cost effective means of reducing SO_2 to limit degradation of the CO_2 capture solvent.
- Establish the design for the USC PC plant fired with lignite and assess the thermal and economic impact of using a commercially-available dryer to dry the fuel.

Separate engineering and economic evaluation studies are being carried out by EPRI to assess how best to integrate the PC plant with a CO_2 capture plant.

The costs for UltraGen III will be obtained as part of the planning work included in the ComTes-1400 program.

Closing Comments

The proposed UltraGen Program is an ambitious step that allows pulverized coal technology to play a role in reducing CO_2 emissions in a cost-effective manner. This objective cannot be achieved economically today and must be worked towards in a structured demonstration program supported by appropriate testing and evaluation activities to lower costs and mitigate the associated risks. Whilst the costs of UltraGen are substantial they are considered to be outweighed by the benefits. The UltraGen concept is based on deploying commercial projects incorporating upgrades to thermal and environmental performance. To encourage such deployment a host must be found and measures taken so that they do not bear the differential cost of these upgrades. Part of the EPRI effort will be to identify how best to fund this cost differential.

A single project will not provide the solution, rather a series of projects is required each building off the one before. In this way UltraGen I will help resolve critical issues that currently form a barrier to deploying UltraGen II, which in its turn will remove barriers to the deployment of UltraGen III. In this way a clear deployment path is established allowing performance considered unachievable today to become the norm in the future.

References

1. "The Challenge of Carbon Capture" CO_2 Capture and Storage, EPRI Journal, Spring 2007, Page 15.
2. *Program on Technology Innovation: Evaluation of Amine-Based, Post-Combustion CO_2 Capture Plants*, Palo Alto California. EPRI Report 1011402, November 2005.

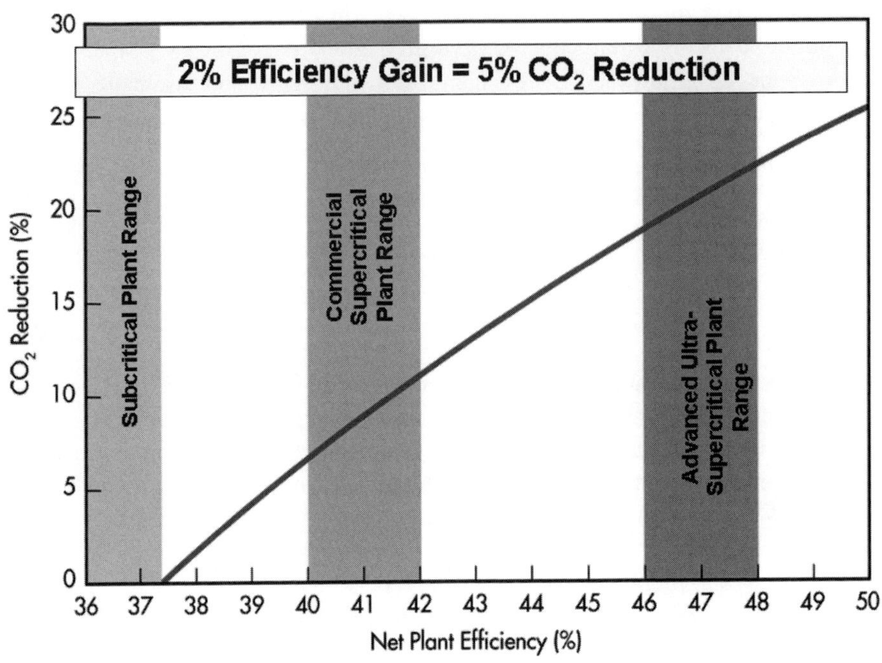

Figure 1. **PC Plant efficiency versus CO_2 Reduction**

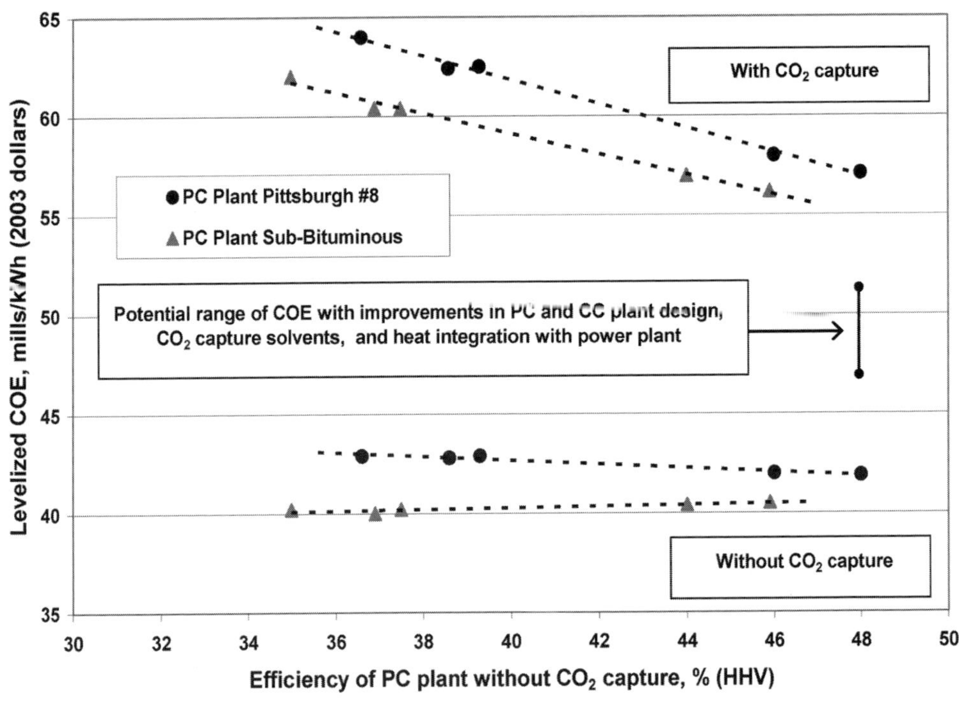

Figure 2. Variation of Cost of Electricity with Generating Efficiency for PC Plants With and Without CO_2 capture

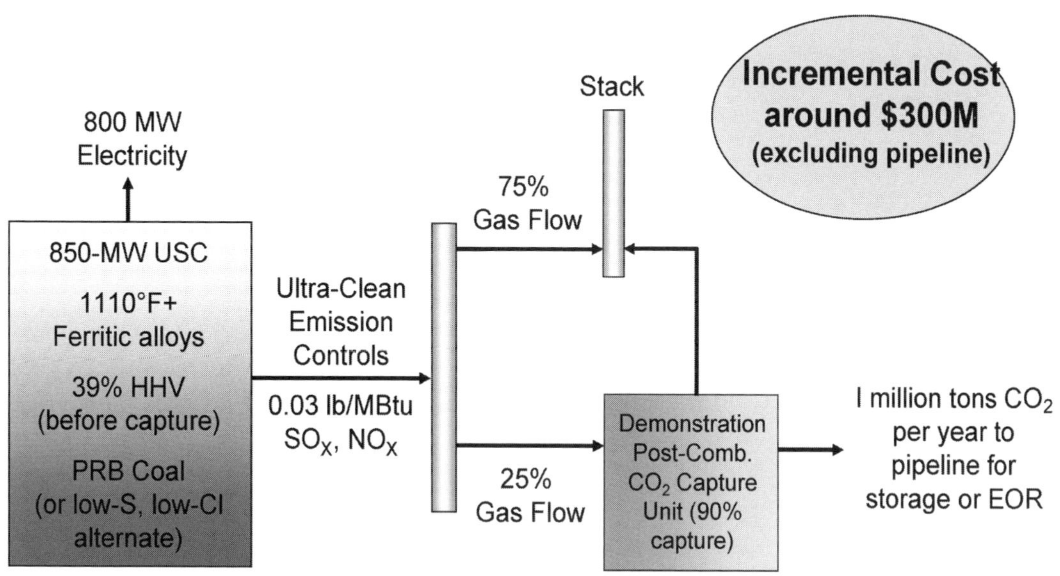

Figure 3. Design Parameters for UltraGen I

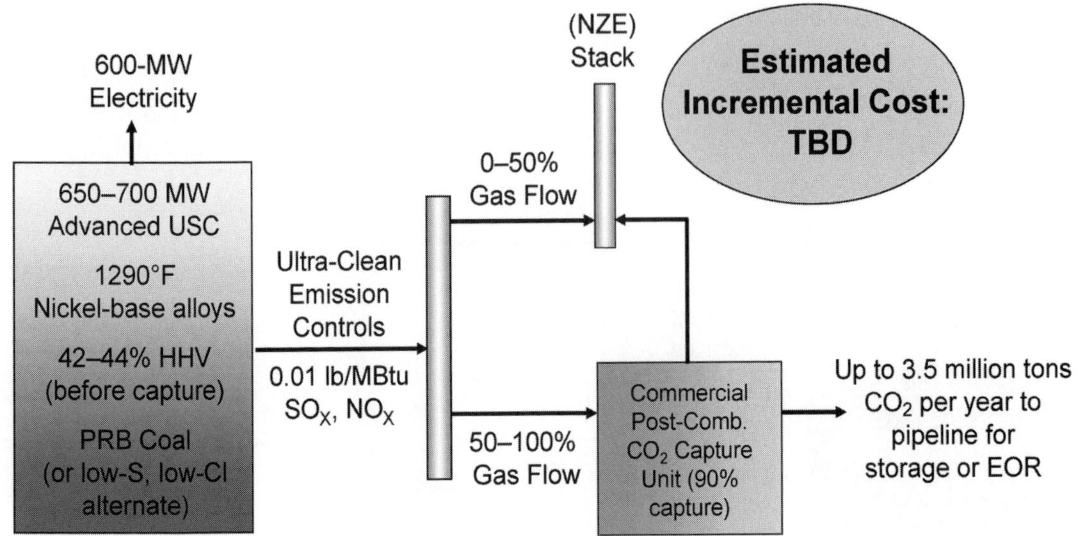

Figure 4. Design Parameters for UltraGen II

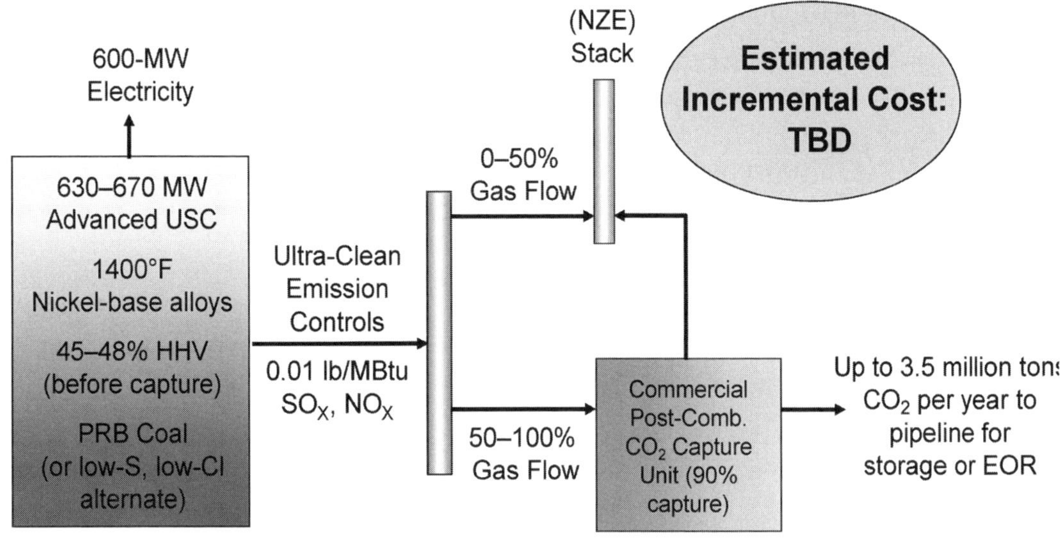

Figure 5. Design Parameters for UltraGen III

Advances in Materials Technology for Fossil Power Plants
Proceedings from the Fifth International Conference
R. Viswanathan, D. Gandy, K. Coleman, editors, p 92-106

DOI: 10.1361/cp2007epri0092

Feasibility of MARBN Steel for Application to Thick Section Boiler Components in USC Power Plant at 650 °C

F. Abe*
M. Tabuchi*
H. Semba**
M. Igarashi[+]
M. Yoshizawa[+]
N. Komai[++]
A. Fujita[+++]

* National Institute for Materials Science (NIMS),
1-2-1 Sengen, Tsukuba 305-0047, Japan
** National Institute for Materials Science, now Sumitomo Metal Industries, Ltd.
now 1-8 Fuso-cho, Amagasaki 660-0891, Japan
[+] Sumitomo Metal Industries
1-8 Fuso-cho, Amagasaki, Hyogo 660-0891, Japan
[++] Mitsubishi Heavy Industries
5-717-1, Fukahori-machi, Nagasaki 851-0392, Japan
+++ Mitsubishi Heavy Industries
2-1-1, Arai-cho, Shinhama, Takasago 676-8686, Japan

Abstract

A 9Cr-3W-3Co-VNbBN steel, designated MARBN (**MAR**tensitic 9Cr steel strengthened by **B**oron and **N**itrides), has been alloy-designed and subjected to long-term creep and oxidation tests for application to thick section boiler components in USC power plant at 650°C. The stabilization of lath martensitic microstructure in the vicinity of prior austenite grain boundaries (PAGBs) is essential for the improvement of long-term creep strength. This can be achieved by the combined addition of 140ppm boron and 80ppm nitrogen without any formation of boron nitrides during normalizing at high temperature. The addition of small amount of boron reduces the rate of Ostwald ripening of $M_{23}C_6$ carbides in the vicinity of PAGBs during creep, resulting in stabilization of martensitic microstructure. The stabilization of martensitic microstructure retards the onset of acceleration creep, resulting in a decrease in minimum creep rate and an increase in creep life. The addition of small amount of nitrogen causes the precipitation of fine MX, which further decreases the creep rates in the transient region. The addition of boron also suppresses the Type IV creep-fracture in welded joints by suppressing grain refinement in heat affected zone. The formation of protective Cr_2O_3 scale is achieved on the surface of 9Cr steel by several methods, such as pre-oxidation treatment in Ar gas, Cr shot-peening and coating of thin layer of Ni-Cr alloy, which significantly improves the oxidation resistance of 9Cr steel in steam at 650°C. Production of a large diameter and thick section pipe and also fabrication of welds of the pipe have successfully been performed from a 3 ton ingot of MARBN.

Introduction

Energy security combined with lower carbon dioxide emissions is increasingly quoted to protect global environment in the 21st century. Coal is an abundant, low cost resource used for electric power generation. However, traditional coal-fired power plants have been emitting CO_2 at high levels relative to other electric power generation options. Adoption of ultra supercritical (USC) power plants with increased steam parameters significantly improves efficiency, which reduces fuel consumption and the emissions of CO_2.

Materials development projects for advanced ultra supercritical (A-USC) power plants with steam conditions of 700°C and above have been already initiated to gain net efficiency higher than 50% at Thermie AD700 project aiming at 700°C in Europe (1) and at DOE Vision 21 project aiming initially at 760°C in the US (2). These projects involve the replacement of 9-12%Cr martensitic steels by nickel base superalloys for the highest temperature components. However, it should be noted that nickel base superalloys are much more expensive than austenitic and ferritic/martensitic steels. In Japan, we have been discussing the feasibility of nickel base superalloys, austenitic steels and 9-12Cr martensitic steels for components with steam temperatures above 700°C, at around 700°C and up to 650°C, respectively. The combination of nickel base superalloys, austenitic steels and 9-12Cr martensitic steels makes it possible to construct highly efficient and economically viable A-USC power plant. The increase in steam parameters exceeding 600°C requires extensive R&D of advanced ferritic/martensitic steels with sufficient long-term creep rupture strength higher than conventional ones.

This paper describes feasibilities of MARBN steel (3), which is a **MAR**tensitic 9Cr steel strengthened by **B**oron and **N**itrides and was developed by NIMS in corporation with private companies in Japan, for application to thick section boiler components in USC power plant with steam temperature of 650°C.

Critical issues for development of ferritic steels

Critical issues for the development of ferritic steels for thick section boiler components, such as main steam pipe and header, with steam condition of 650°C are the improvement of following properties.

(1) long-term creep strength (resistance to degradation in long-term creep strength)
(2) oxidation resistance in high-temperature steam
(3) resistance to Type IV cracking strength loss in welded Joints
(4) thermal cycling capabilities (resistance to creep-fatigue damage)

Usually, the creep rupture strength of 100MPa at operating temperature and 10^5 h is a target value for the development of new steels used under creep conditions. At present, we have no deciding criterion for oxidation resistance. We think that candidate steels for boiler components operating at 650°C should exhibit oxidation resistance in steam at 650°C better than that of Gr.91 in steam at 600°C, because Gr.91 are being now used for long duration in power plants operating at 600°C. In terms of creep strength of welded joints, the weld strength reduction factor (WSRF) of 10^5 h creep rupture strength should be larger than 0.75 for new steels at 650°C, because the WSRF for Gr.91 is 0.75 at 600°C (4).

Experimental

9Cr-3W-3Co-0.2V-0.05Nb steels with 140 ppm boron but different nitrogen contents were used to investigate the effect of nitrogen on creep strength of 9Cr boron steel base metal. The chemical compositions of the steels are given in Table 1 (5). The steels were basically prepared by vacuum induction melting to 50 kg ingots. Hot forging and hot rolling were carried out to produce plates of 20 mm in thickness. Creep tests were carried out at 650°C for up to about 3 x10^4 h under constant load condition, using specimens of 10 mm in gauge diameter and 50 mm in gauge length.

Table 1 Chemical compositions and heat treatment conditions of 9Cr-3W-3Co-VNb steels with 0.014% (140 ppm) boron but different nitrogen contents.

(mass %)

	C	Si	Mn	Cr	W	Co	V	Nb	N	B	normalizing	tempering
0.0015N	0.076	0.30	0.51	9.00	3.02	3.02	0.19	0.053	0.0015	0.0132	1150°C x 1 h	770°C x 4 h
0.0034N	0.078	0.30	0.51	8.99	2.91	3.01	0.19	0.049	0.0034	0.0139	1080°C x 1 h	800°C x 1 h
0.0079N	0.078	0.31	0.49	8.88	2.85	3.00	0.20	0.051	0.0079	0.0135	1150°C x 1 h	770°C x 4 h
0.030N	0.078	0.30	0.51	9.08	3.05	3.03	0.20	0.055	0.0300	0.0150	1150°C x 1 h	770°C x 4 h
0.065N	0.081	0.31	0.51	8.90	3.07	3.00	0.20	0.054	0.0650	0.0144	1150°C x 1 h	770°C x 4 h

Another series of 9Cr-3W-3Co-0.2V-0.05Nb steels with different contents of boron and nitrogen and conventional steel P92, whose chemical compositions are given in Table 2, were used to investigate the effect of boron and nitrogen on creep strength of welded joints. Plates of about 30 mm thickness were subjected to multi-layers Gas Tungsten Arc (GTA) welding. The post weld heat treatment (PWHT) was carried out for each specimen including the base metal at 740°C for 4.7 h.
The concentration of boron in $M_{23}C_6$ precipitates was analyzed by field emission scanning Auger electron spectroscopy. The details of FE-AES analysis were described elsewhere (6).

Table 2 Chemical compositions and heat treatment conditions of 9Cr-3W-3Co-VNb steels with different boron and nitrogen contents and conventional steel P92.

(mass %)

	C	Si	Mn	Cr	W	Mo	Co	V	Nb	N	B	Normalizing	Tempering
47B-17N	0.079	0.30	0.48	8.77	2.93	-	2.91	0.18	0.046	0.0017	0.0047	1080°C x 1 h	800°C x 1 h
90B-14N	0.074	0.30	0.48	8.93	3.13	-	2.92	0.18	0.046	0.0014	0.009	1080°C x 1 h	800°C x 1 h
130B-15N	0.077	0.30	0.49	8.97	2.87	-	2.91	0.18	0.046	0.0015	0.013	1080°C x 1 h	800°C x 1 h
180B-11N	0.078	0.30	0.49	8.91	2.85	-	2.91	0.18	0.047	0.0011	0.018	1080°C x 1 h	800°C x 1 h
160B-85N	0.079	0.31	0.51	8.81	3.05	-	3.10	0.20	0.055	0.0085	0.016	1150°C x 1 h	770°C x 4 h
P92	0.09	0.16	0.47	8.72	1.87	0.45	-	0.21	0.06	0.050	0.002	1070°C x 1 h	780°C x 1 h

Results and discussion

Long-term stability of base metal

Combination of boron-strengthening and MX nitride-strengthening. The effect of nitrogen addition on the creep rupture strength of 9Cr-3W-3Co-0.2V-0.05Nb steel containing 140 ppm boron (Table 1) are shown in Fig. 1 (3, 5). The creep rupture strength significantly increases with increasing nitrogen content from 0.0015 to 0.0079 %, but the excess addition of nitrogen as high as 0.030 and 0.065% (300 and 650 ppm) decreases the creep rupture strength. The excess addition of nitrogen of 0.030 and 0.065% causes the formation of boron nitrides during normalizing heat treatment. The solubility product for boron nitrides in 9 to 12Cr steels at normalizing temperatures of 1050 to 1150°C is given by

$$\log [\%B] = -2.45 \log [\%N] - 6.81, \quad (1)$$

where [%B] and [%N] are the concentration of boron and nitrogen in mass% (7). At a boron concentration of 140 ppm, 95 ppm nitrogen can dissolve in the matrix without any formation of boron nitrides during normalizing. The present results indicate that boron and nitrogen in solid solution after normalizing is very effective for the improvement of creep rupture strength. But the formation of boron nitrides reduces the strengthening effects.

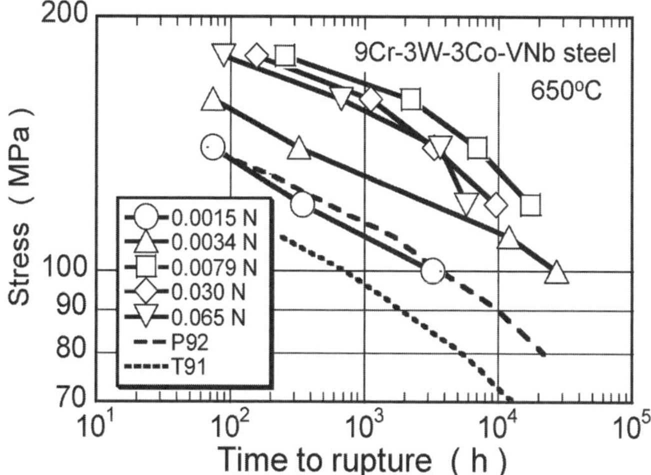

Figure 1 Effect of nitrogen on creep rupture strength of 9Cr-3W-3Co-VNb steel with 140 ppm boron at 650°C.

Enough creep ductility is also required for thick section components in USC plant, because the thermal cycling capabilities of thick section components would be severely restricted by fatigue damage. Yamaguchi and co-workers (8) have carried out creep-fatigue tests at 650°C for various kinds of ferritic steels and found the boron addition is also advantageous for the improvement of creep-fatigue life. The creep-fatigue life is proportional to the reduction of area in creep rupture test, namely, proportional to the creep ductility but not proportional to the creep strength.

Figures 2(a) and 2(b) show the effect of nitrogen addition on the creep deformation behavior of the steel at 650°C and 120 MPa. The addition of nitrogen causes a rapid decrease in creep rate with time and also with strain in the transient region. It should be noted that the creep rate in the transient region is the same among the three steels containing different nitrogen of 79, 300 and 650 ppm (Fig.2(a)). The onset of acceleration creep is retarded up to longer time by the addition of 79 ppm nitrogen but it shifts to earlier times with increasing nitrogen content above 79 ppm. In Fig. 2(b), the onset of acceleration creep takes place at a high strain of 0.045 in the very low nitrogen 0.0015N steel as shown by the arrow, while it takes place at a low strain of 0.007 in the 0.0079N, 0.030N and 0.065N steels. This suggests that the addition of nitrogen promotes the heterogeneity in creep deformation. The acceleration of creep rate by strain $d\ln\dot{\varepsilon}/d\varepsilon$ after reaching a minimum creep rate is evaluated from the slope of creep rate versus strain curves to be 32.2, 121.1, 113.0 and 90.7 for the 0.0015N, 0.0079N, 0.030N and 0.065N steels, respectively.

The minimum creep rate steeply decreases with increasing nitrogen content up to 79 ppm but it turns to increase slowly above 79 ppm nitrogen, as shown in Fig.3. This reflects the nitrogen content dependence of creep rupture strength shown in Fig.1.

In order to make clear the reason why the creep rates in the transient region are the same among the three steels containing nitrogen of 79, 300 and 650 ppm (Fig. 2(a)), Fig.4 shows

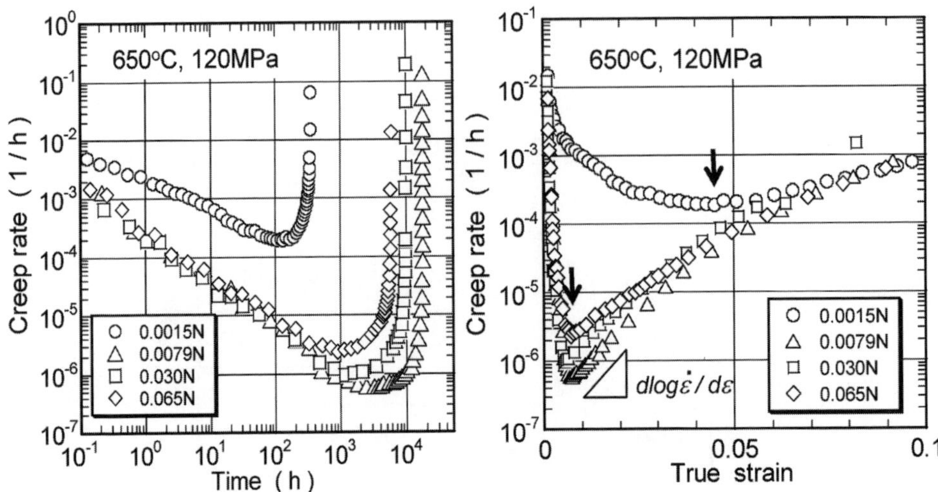

Figure 2. Effect of nitrogen on creep rate versus time curves and creep rate versus strain curves of the steels at 650°C and 120 MPa.

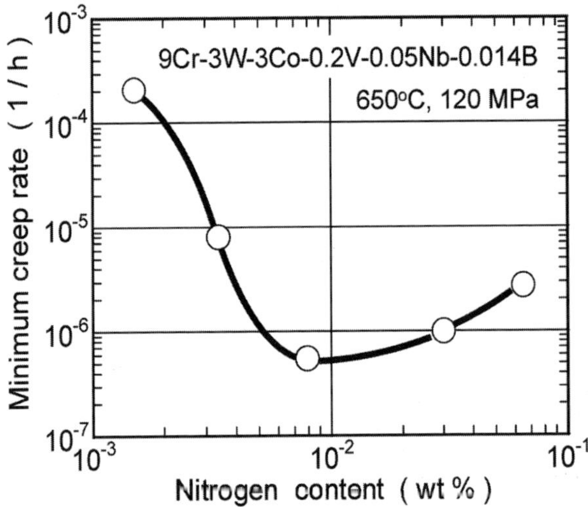

Figure 3. Effect of nitrogen content on minimum creep rate of the steel with 140 ppm boron at 650°C and 120 MPa.

the amount of dissolved nitrogen and precipitated nitrogen in the 0.0015N, 0.0079N and 0.065N steels after tempering. In the low nitrogen 0.0015N and 0.0079N steels, most of nitrogen is in solution after tempering. But in the high nitrogen 0.065N steel, most of nitrogen has already precipitated during tempering. Dissolved nitrogen can precipitate as fine MX carbonitrides during creep at 650°C. Indeed, very fine vanadium-rich MX carbonitrides were observed to have precipitated in the 0.0079N steel during aging for 1000 h at 650°C (5). Fine MX carbonitrides precipitated during creep are responsible for the significant decrease in creep rate in the transient region shown in Fig. 2. It should be noted that the dissolved nitrogen concentration is roughly the same between the 0.079N and 0.065N steels after tempering. This results in the same the creep rates in the transient region among the steels containing nitrogen above 79 ppm.

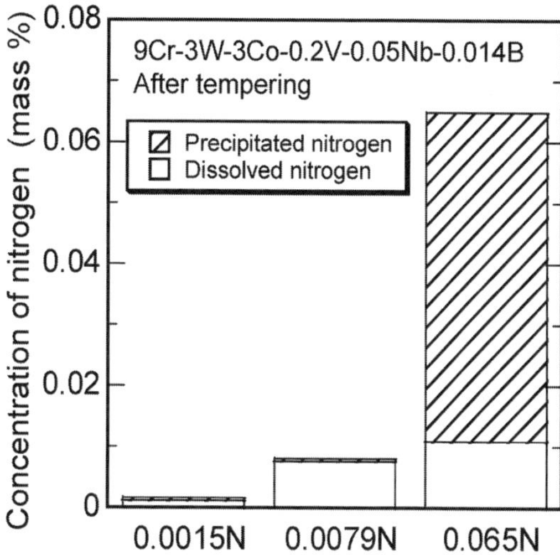

Figure 4. Dissolved and precipitated nitrogen concentration in the 0.0015N, 0.0079N and 0.065N steels after tempering.

Next, let us consider the reason why the onset of acceleration creep takes place at earlier times in the high nitrogen steels (Fig. 2(a)). We have revealed that in the 0.0034N steel with 140 ppm boron, the fine distribution of $M_{23}C_6$ carbides along prior grain boundaries (PAGBs) is still maintained during exposure at 650°C and that boron is enriched in $M_{23}C_6$ carbides near PAGBs (5). The fine distribution of $M_{23}C_6$ carbides is also observed in the steel containing no boron after tempering, but extensive coarsening takes place during exposure at 650°C. This indicates that boron reduces the rate of Ostwald ripening of $M_{23}C_6$ carbides near PAGBs during creep at 650°C. The fine distribution of $M_{23}C_6$ carbides near PAGBs retards the onset of acceleration creep, resulting in lower minimum creep rate and longer time to rupture. Figure 5 compares the enrichment of boron in $M_{23}C_6$ carbides between the 0.0015N and 0.065N steels. In the 0.0015N steel, no boron nitride formed during normalizing and hence most of boron can contribute to the enrichment in $M_{23}C_6$ carbides near PAGBs. On the other hand, in the 0.065N steel, large amount of boron nitrides formed during normalizing. Therefore, most of soluble boron is consumed to form boron nitrides and only a slight amount of boron contributes to the enrichment in $M_{23}C_6$ carbides. This is a reason why the onset of acceleration creep takes place at earlier time in the high nitrogen steel.

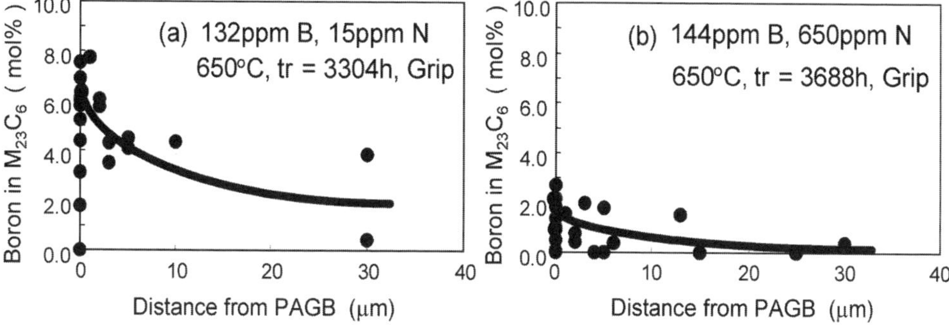

Figure 5. Boron content in $M_{23}C_6$ carbides in the steels after creep rupture testing at 650°C for 3000-4000 h, as a function of distance from prior austenite grain boundary.

Formation process of $M_{23}(CB)_6$. In iron-carbon-boron ternary phase diagram at 900°C, $Fe_{23}(CB)_6$ appears at alloy compositions containing several % boron (9). Because the boron content in the present steels is only 140ppm in maximum, $Fe_{23}(CB)_6$ is difficult to form in the matrix of present steels. Therefore, we estimate segregation of boron at grain boundaries. Using a binding energy of 62.7kJ/mol (10) reported for Type 316 stainless steel, Fig.6 shows the concentration of boron at grain boundaries as a function of temperature. At a normalizing temperature of 1100°C, segregation of several % boron can be achieved at grain boundaries. During tempering at 800°C and subsequent creep at 650°C, the segregation is more significant because of lower temperatures.

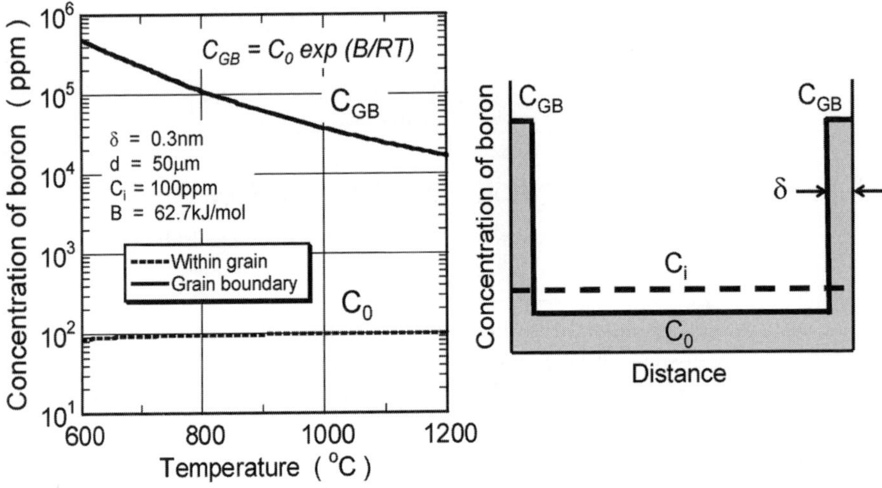

Figure 6 Estimation of grain boundary segregation of boron.

Based on the estimation of grain boundary segregation of boron described above, we think that at first during normalizing heat treatment at 1100°C, only grain boundary segregation of boron takes place but no precipitation of $M_{23}C_6$ carbide because of high temperature, as shown in Fig. 7. During tempering at 800°C, the precipitation of $M_{23}C_6$ carbide can take

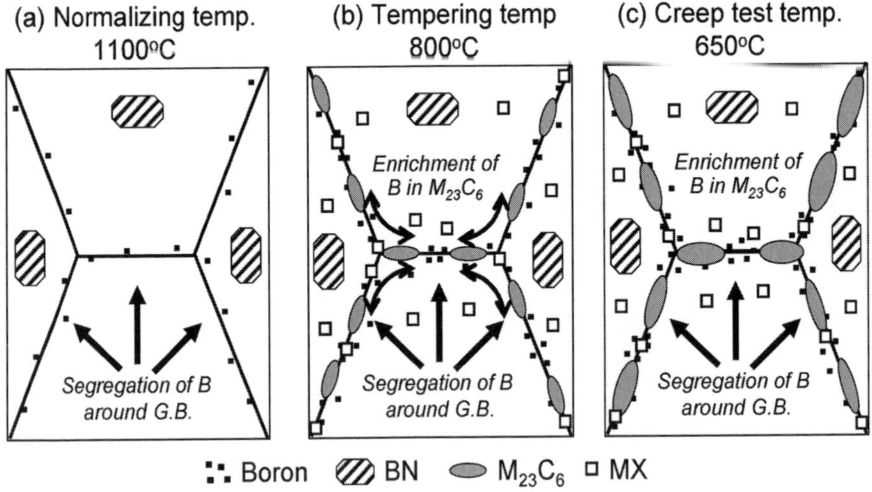

Figure 7. Enrichment process of boron into $M_{23}C_6$ during heat treatment.

place preferentially at grain boundaries and lath-block boundaries. Because the grain boundary segregation of boron is achieved, $M_{23}(CB)_6$ can form by enrichment of boron into the $M_{23}C_6$ carbides. During creep at 650°C, the enrichment of boron in $M_{23}(CB)_6$ is more significant because of lower temperature.

When the steel contains high nitrogen contents, boron nitrides can form during normalizing, consuming most of soluble boron. This significantly reduces the segregation of boron at grain boundaries and also reduces the enrichment of boron in $M_{23}C_6$, as shown in Fig.5.

Reduction of coarsening rate of $M_{23}(CB)_6$. According to the mechanism of Ostwald ripening, the main factors affecting the coarsening rate of precipitate particles are diffusion coefficient, solid solubility and interfacial energy (11). However, we have no experimental result showing a significant decrease in diffusion coefficient and solid solubility by the addition of small amount of boron. Furthermore, we have no evidence showing the enrichment of boron at the interface between $M_{23}C_6$ carbide particle and alloy matrix. Therefore, these parameters could be excluded from the main explanation of the reduction of coarsening rate of $M_{23}C_6$ carbides by boron. It should be noted that Ostwald ripening in solid matrix requires accommodation of local volume change around a growing particle, because specific volume of carbide is larger than that of the alloy matrix. As a small carbide goes into solution, carbon atoms take up interstitial sites in the matrix and vacancies are created at the carbide interface, as shown in Fig.8. Then vacancies migrate through the matrix and arrive at a growing carbide interface, which accommodates local volume change. If boron atoms occupy vacancies around growing carbide particle, local volume change cannot be accommodated. This causes the reduction of coarsening rate. We think the main effect due to boron is to occupy vacancies around growing $M_{23}C_6$ carbide particle.

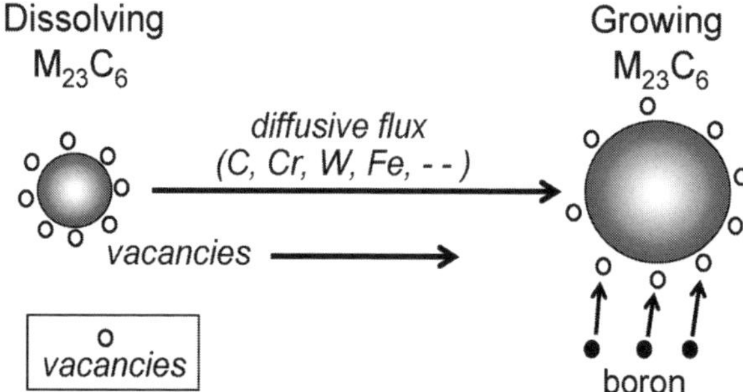

Figure 8 Accommodation of local volume change around a growing $M_{23}C_6$ particle.

Long-term stability of welded joints and HAZ microstructure

Figure 9 compares the creep rupture data between the base metals and welded joints for 9Cr-3W-3Co-VNb steels containing different boron and nitrogen contents and P92 (Table 2) at 650°C (12). Although P92 contains 20 ppm boron, the degradation of creep rupture strength in welded joints is quite large at long times due to Type IV fracture. This suggests that even if the steel contains boron, it does not necessarily suppress the Type IV fracture. The degradation in creep rupture strength of welded joints is negligibly small in the 9Cr-3W-3Co-VNb steel even if it contains only 47ppm boron but no addition of nitrogen (Fig.9(a)) and also even if it contains 160 ppm boron and 85 ppm nitrogen (Fig. 9(e)).

We have revealed that multi-axial stress condition in the fine-grained HAZ with lower creep strength, resulting from mechanical constrain effect by the surrounding weld metal and base metal with higher creep strength, is essential for the brittle Type IV fracture. Therefore, we have systematically examined the effect of boron and nitrogen contents on grain refinement behavior for the 9Cr steel during heating at temperatures between 900 and 1100°C. The results are summarized in Fig. 10, overlapping on the composition diagram for the formation of BN during normalizing heat treatment. No addition of boron causes the grain refinement at around Ac_3 temperature, irrespective of nitrogen content. The addition of boron and nitrogen without any formation of BN during normalizing causes no grain refinement, while the steels with the formation of BN during normalizing produce the fine-grained microstructure. The present results suggest that soluble boron is essential for the suppression of grain refinement during heating.

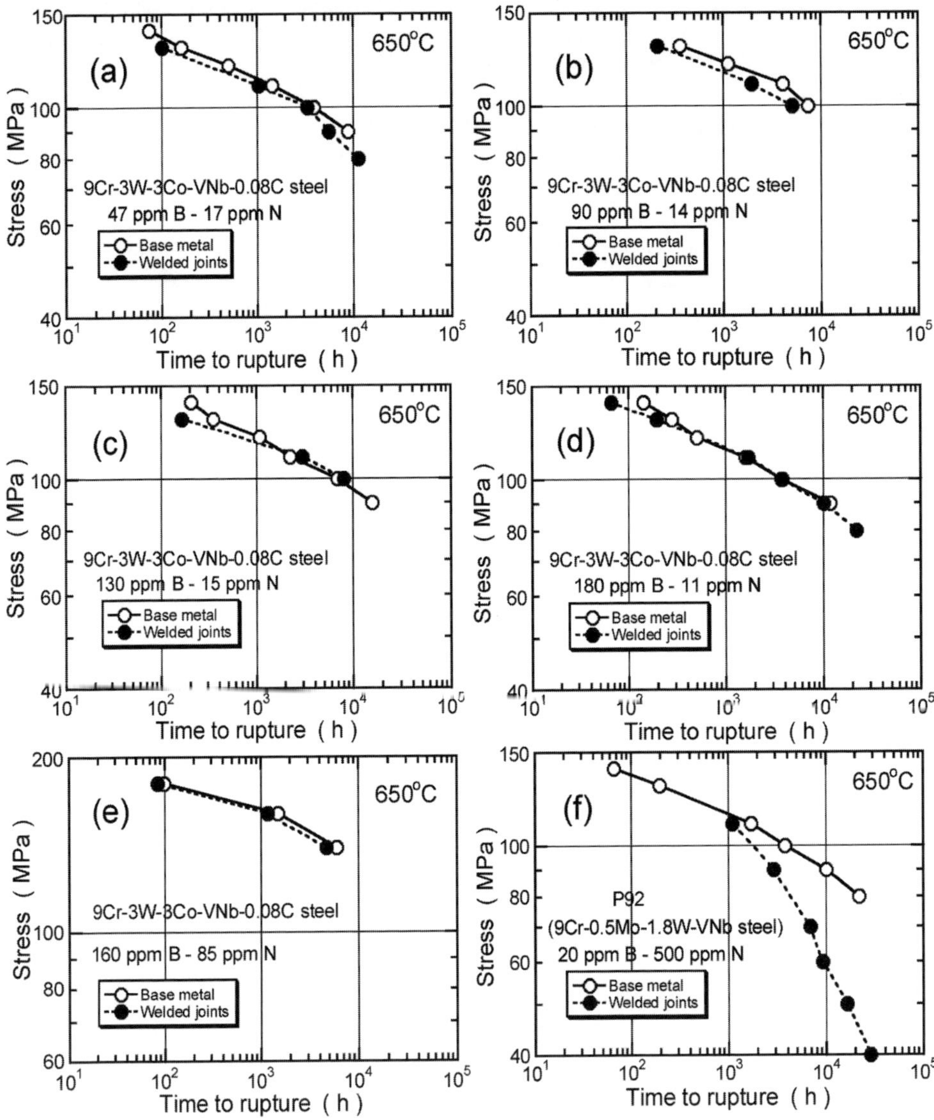

Figure 9 Creep rupture data for base metals and welded joints of the steels at 650°C.

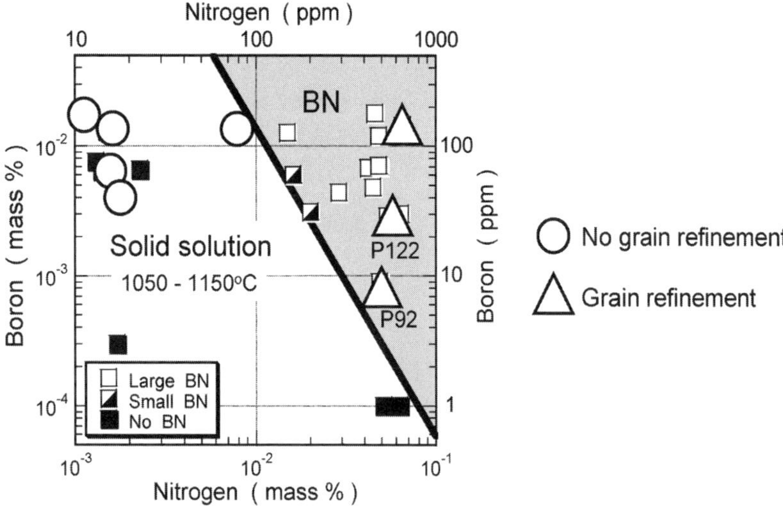

Figure 10 Grain refinement and no grain refinement in 9 to 12Cr steels during heating at 900 to 1100°C.

Oxidation resistance in steam

Figure 11 compares the weight gain due to oxidation in steam at 650°C among various 9 to 12Cr steels (13). The sheet specimens having a size of 10x20x2 mm were cut from bulk materials, which were already heat treated, ground on a SiC paper of 320 grit, rinsed in acetone and then supplied to the oxidation test in steam at 650°C. The weight gain of P92 and P122 in steam at 650°C is much larger than that of T91 at 600°C. This suggests that existing steels, even in 12Cr steel P122, cannot satisfy the present criterion for oxidation resistance that candidate steels for boiler components operating at 650°C should exhibit oxidation resistance in steam at 650°C better than that of Gr.91 in steam at 600°C. It should

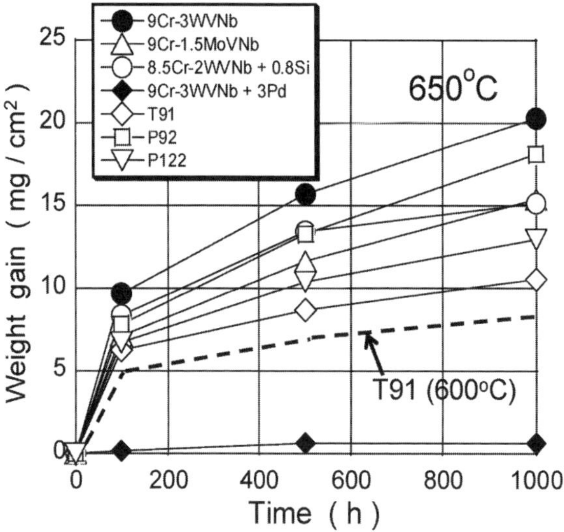

Figure 11 Weight gain of the steels in steam at 650°C, comparing with that of T91 at 600°C.

be noted that the weight gain of 9Cr-3WVNb steel with 3% Pd is significantly lower at 650°C than that of T91 at 600°C (14). Thin scale of Cr-rich oxide, presumably Cr_2O_3, forms on the specimen surface of 9Cr-3WVNb steel with 3%Pd in steam at 650°C, while thick scale consisting of magnetite in the outer layer and Fe-Cr spinel oxides in the inner layer forms on the surface of the other steels including T91. No evidence is found for the formation of any oxide containing Pd on the surface of 9Cr-3WVNb steel with 3%Pd. Based on the results in Fig.11, the formation of protective Cr_2O_3 scale is essential for the development of oxidation-resistant ferritic steels for USC boilers at 650°C.

The formation of protective Cr-rich oxide scale is achieved by the combination of Si addition and pre-oxidation treatment in argon gas (15), by the combination of shot-peening of Cr and pre-oxidation treatment in air at 700°C (16), and by coating of Ni-20Cr and Ni-50Cr thin layers (17). The oxidation test was carried out for 9Cr-3WVNb steel containing different Si contents of 0 to 0.8% in steam at 650°C after pre-oxidation treatment in Ar gas. In the condition of no pre-oxidation treatment, the addition of Si decreases the weight gain of the steel in steam but the effect of Si is not large. The pre-oxidation treatment in Ar gas further decreases the weight gain during subsequent oxidation in steam. This is more significant with increasing Si concentration and with increasing pre-oxidation time, as shown in Fig.12. For example, the combination of 0.3%Si addition and pre-oxidation treatment in Ar gas at 650°C for 100h satisfies the present criterion for oxidation resistance in steam at 650°C.

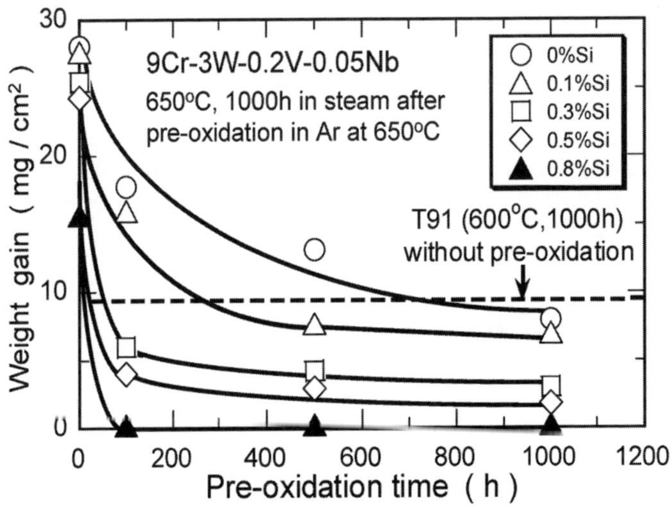

Figure 12 Weight gain of 9Cr-3W-0.2V-0.05Nb steel containing different Si contents by oxidation in steam at 650°C for 1000h, as a function of pre-oxidation time in Ar gas at 650°C.

The thin scale of Cr-rich oxides formed during pre-oxidation treatment is stable during subsequent oxidation in steam at 650°C, as shown in Fig.13. The breakaway in the weight gain curve is not observed for the specimen with pre-oxidation treatment.

The resistance to exfoliation of Cr_2O_3 scale is another requirement. At first we have tried to examine the resistance to exfoliation of Cr_2O_3 scale by thermal cycling test in steam after pre-oxidation treatment (16). The increase in weight gain in steam was substantially the same between thermal cycling test and continuous oxidation test. This indicates that the Cr_2O_3 scale formed during pre-oxidation is stable and highly resistant to exfoliation. Next, we have tried to cross-cut the Cr_2O_3 scale by a sharp edge after pre-oxidation and then supplied to oxidation test in steam at 650°C for 1000h. After cutting the Cr_2O_3 scale, the steel surface was directly exposed to steam. But no evidence was found for any exfoliation or breakaway of surrounding Cr_2O_3 scale during subsequent steam oxidation. Again this indicates that the Cr_2O_3 scale formed during pre-oxidation is very stable and highly resistant to exfoliation. It should be noted that at present there is no standard test method for the assessment of resistant to exfoliation.

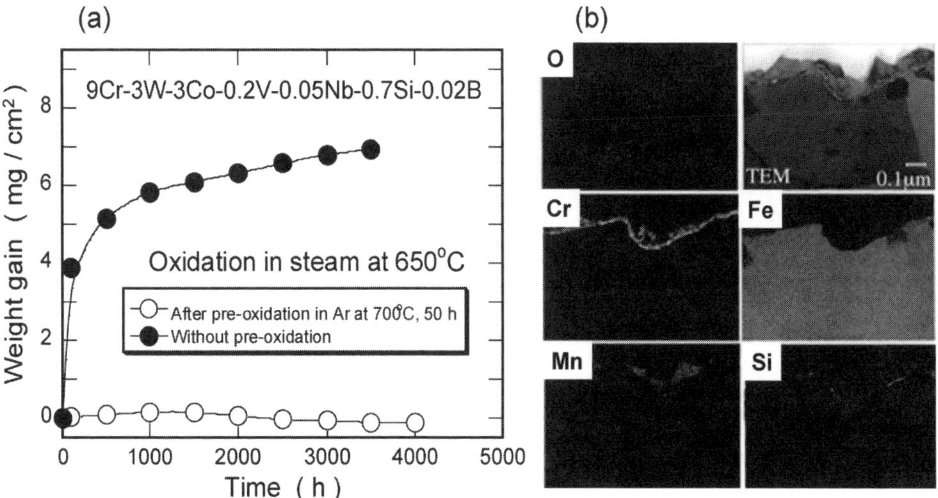

Figure 13 (a) Weight gain of 9Cr-3W-3Co-VNbSi steel (MARB2) in steam at 650°C and (b) formation of protective Cr_2O_3 scale during pre-oxidation in argon at 700°C for 50 h.

Production of pipe and fabrication of welds

The production of a large diameter and thick section pipe and subsequent fabrication of circumference welds of the pipe have successfully been performed from a 3 ton ingot of MARBN. The chemical compositions of MARBN are given in Table 3. From the viewpoints of enhancing creep rupture strength for base metal (Fig.1) and welded joints (Fig.8) and of no formation of BN during normalizing heat treatment (Fig.9), the boron and nitrogen contents were set to be 0.013% (130ppm) boron and 0.007% (70ppm) nitrogen. The residual impurity Al was minimized, because it consumes soluble nitrogen to form AlN during creep, which degrades the creep strength (18). A 3 ton ingot, having a size of 785mm in diameter and 580mm in height, was prepared by vacuum melting. The ingot was successfully hot-forged to a pipe, which was heat treated and machined to a size of 470 mm in outer diameter, 65 mm in thickness and 1300 mm in length. The pipe was normalized at 1100°C for 3h followed by air cooling and then tempered at 780°C for 4 h followed by air cooling. Figure 14 shows the appearance of pipe after machining. The creep specimens were taken from the outside, centre and inside parts in the wall thickness and subjected to the creep rupture testing at 650°C. The results are shown in Fig.15, together with those for the small melts of 0.0034N and 0.0079N steels in Table 1. Although the creep rupture data for the MARBN pipe are located just below those for the 0.0079N steel because of a little lower nitrogen content, the MARBN pipe material exhibits substantially the same creep rupture strength as that of the small melt of 0.0079N steel in Table 1. We are now continuing long-term creep rupture tests.

Table 3 Chemical compositions of MARBN pipe.

(mass %)

	C	Si	Mn	Cr	W	Co	V	Nb	N	B	sol-Al
MARBN Pipe	0.08	0.33	0.51	9.03	2.79	3.01	0.19	0.056	0.0071	0.013	0.001

Figure 14 Appearance of MARBN pipe,
470 mm in outer diameter, 65 mm in thickness and 1300 mm in length.

Circumference welding of the MARBN pipe has successfully been carried out using gas tungsten-arc welding (GTAW) process and using two kinds of filler wires; Ni-base AWS ER NICr-3 (Alloy 82) and matching filler wires. The preheating temperature was kept to be 150°C or more to avoid low-temperature cracking, while the interpass temperature was below 200°C to avoid high-temperature cracking. After welding, the welded joints were subjected to a postweld heat treatment (PWHT) at 740°C for 4 h. Figure 16 shows circumference TIG welding of the MARBN pipe and appearance of welded joints after PWHT, with Ni-base AWS ER NICr-3 and matching filler wires. The creep rupture testing of welded joints is being now carried out.

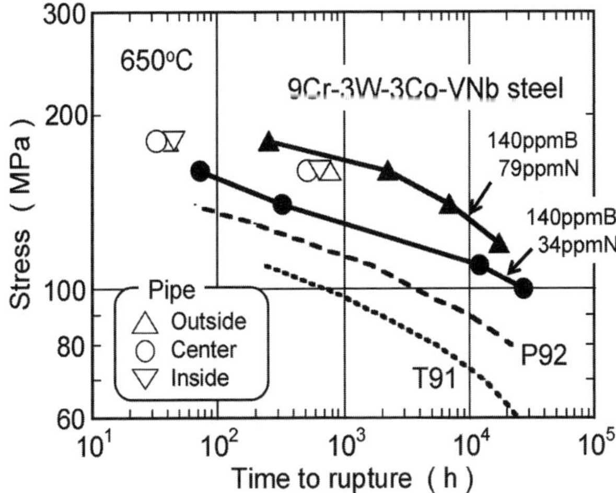

Figure 15 Creep rupture data for MARBN pipe,
together with 0.0034N and 0.0079N steels, T91 and P92 at 650°C.

Figure 16 Circumference TIG welding of MARBN pipe and appearance of welded pipes after PWHT, with Ni-base AWS ER NICr-3 and matching filler wires.

Conclusions

(1) A highly creep-resistant martensitic 9Cr steel (MARBN) has been alloy-designed on the base of microstructure stabilization, especially in the vicinity of PAGB. Creep-strengthening by nitrides as well as boron and no formation of boron nitrides during normalizing heat treatment cause a significant improvement of creep rupture strength. The creep rupture strength at 650°C and 10^5 h is estimated to be 80 to 90 MPa, which is much higher than that of existing high strength steels P92 and P122.

(2) The addition of boron without any formation of boron nitrides during normalizing heat treatment suppresses grain refinement in HAZ of welded joints and hence suppresses the Type IV fracture in welded joints. This satisfies the criterion that the weld strength reduction factor (WSRF) of 10^5 h creep rupture strength should be larger than 0.75 at 650°C. It should be noted that the optimization of weld metal will be required in future.

(3) The formation of protective Cr-rich scale is achieved on the surface of 9Cr steel by the pre-treatments, such as pre-oxidation treatment and coating. This significantly improves the oxidation resistance of 9Cr steel in steam.

(4) The production of a large diameter and thick section pipe having a size of 470 mm in outer diameter, 65 mm in thickness and 1300 mm in length and also fabrication of circumference welds of the pipe have successfully been performed from a 3 ton ingot of MARBN.

Acknowledgement

The authors would like to thank all members participating with Ultra-Steel Project at NIMS. They are also grateful to members of the Creep Group at NIMS for their sincere efforts.

References

1. R. Blum and R. W. Vanstone, "Materials Development for Boilers and Steam Turbines Operating at 700°C", Proc. of 8th Liege Conference on Materials for Advanced Power Engineering 2006, Liege, Belgium, (2006) pp.41-60.

2. R. Viswanathan, J. F. Henry, J. Tanzosh, G. Stanko, J. Shingledecker and B. Vitalis, "U.S. Program on Materials Technology for Ultrasupercritical Steam Coal-fired Power Plants", as in (1) pp.893-915.
3. F. Abe, "Metallurgy for Long-term Stabilization of Ferritic Steels for Thick Section Boiler Components in USC Power Plant at 650°C", as in (1) pp.965-980.
4. M. Tabuchi and Y. Takahashi, "Evaluation of Creep Strength Reduction Factors for Welded Joints of Modified 9Cr-1Mo steel (P91)", Proc. of 2006 ASME PVP Conference, Vancouver, Canada, PVP2006-ICPVT11-93350 (2006).
5. H. Semba and F. Abe, "Alloy design and creep strength of advanced 9%Cr USC boiler steels containing high concentration of boron", Energy Materials, **1** (2006) 238-244.
6. T. Horiuchi, M. Igarashi, F. Abe, "Improved Utilization of Added B in 9Cr Heat-Resistant Steels Containing W", ISIJ International, **42** (2002) S67-S71.
7. K. Sakuraya, H. Okada and F. Abe, "BN type inclusions formed in high Cr ferritic heat resistant steel" Energy Materials, **1** (2006) 158-166.
8. M. Kimura, K. Kobayashi and K. Yamaguchi, "Creep and fatigue properties of newly developed ferritic heat-resisting steels for ultra super critical (USC) power plants", Materials Science Research Intern., **9** (2003) 50-54.
9. M. Lucco Borlera and G. Pradelli, "Iron-carbon-boron ternary phase diagram", Met. Ital., **59** (1967) 907.
10. L. Karsson and H. Norden, "Non-equilibrium grain boundary segregation of boron in austenitic stainless steels-II. Fine segregation behaviour" Acta Metall., **36** (1988) 13-24.
11. M. Y. Wey, T. Sakuma and T. Nishizawa, 'Growth of alloy carbide particles in austenite', Trans. JIM, **22** (1981) 733-742.
12. F. Abe, M. Tabuchi, M. Kondo and H. Okada, "Suppression of Type IV fracture in welded joints of advanced ferritic power plant steels - effect of boron and nitrogen", Materials at High Temperatures **23** (2006) 145-154.
13. H. Kutsumi, T. Itagaki and F. Abe, "Improvement of Steam Oxidation Resistance for Ferritic Heat Resistant Steels", Proceedings of 7th Liege Conference on Materials for Advanced Power Engineering 2002, edited by J. Lecomte-Beckers, M. Carton, F. Schubert and P. J. Ennis, Liege, Belgium (2002) pp.1629-1638.
14. T. Itagaki, H. Kutsumi, H. Haruyama, M. Igarashi and F. Abe, "Steam Oxidation of High-Chromium Ferritic Steels Containing Palladium", Corrosion, **61** (2005) 307-316.
15. H. Kutsumi, H. Haruyama and F. Abe, "Application of the Pre-Oxidation Treatment in Ar Gas to the NIMS High-Strength Steels", Proceedings of 4th International Conference on Advances in Materials Technology for Fossil Power Plants, Hilton Head Island, SC, USA (2004) pp.463-471.
16. H. Haruyama, H. Kutsumi, S. Kuroda and F. Abe, "Effect of Shot Peening and Pre-Oxidation Treatment in Air on Steam Oxidation Resistance of Mod.9Cr-1Mo Steel", as in (15) pp.412-419.
17. T. Sundararajan, S. Kuroda and F. Abe, "Steam Oxidation of 80Ni-20Cr High-Velocity Oxyfuel Coating on 9Cr-1Mo steel: Diffusion-Induced Phase Transformation in the Substrate Adjacent to the Coating", Metallurgical and Materials Transactions A, **36A** (2005) 2165-2174.
18. F. Abe, H. Tanaka and M. Murata, "Impurity Effects on Heat-to-Heat Variation in Creep Life for Some Heat Resistant Steels", Proceedings of BALTICA VII International Conference on Life Management and Maintenance for Power Plants, Helsinki, 12-14 June 2007 (2007) pp.171-184.

Advances in Materials Technology for Fossil Power Plants
Proceedings from the Fifth International Conference
R. Viswanathan, D. Gandy, K. Coleman, editors, p 107-118

DOI: 10.1361/cp2007epri0107

Materials Solutions for Advanced Steam Power Plants

D. L. Klarstrom
L. M. Pike
Haynes International, Inc.
1020 W. Park Avenue
Kokomo, IN 46904-9013

Abstract

Significant research efforts are being performed in Europe, Japan and the U.S. to develop the technology to increase the steam temperature in fossil power plants in order to achieve greater efficiency and reduce the amount of greenhouse gases emitted. The realization of these advanced steam power plants will require the use of nickel-based superalloys having the required combination of high temperature creep strength, oxidation resistance, thermal fatigue resistance, thermal stability and fabricability. HAYNES® 230® and 282® alloys are two alloys that meet all of these criteria. The metallurgical characteristics of each alloy are described in detail, and the relevant high temperature properties are presented and discussed in terms of potential use in advanced steam power plants.

Introduction

A number of efforts have been established in Europe, Japan and the United States to develop more efficient fossil fuel power plants by increasing steam temperatures and pressures (1). These so-called ultrasupercritical steam power plants will operate with steam temperatures in the 700-760°C (1300-1400°F) temperature range with pressures above 24 MPa (3.5 ksi). Such conditions are well beyond the capabilities of current carbon and stainless steels, and, hence, the use of nickel-base alloys will be required in order to satisfy the long term service goals.

In the initial survey of materials, HAYNES® 230® alloy was identified as a candidate for a variety of components due to its high ASME Code allowable stresses in the temperature range of interest. Because the alloy is solid-solution strengthened, its use is facilitated since no special aging heat treatment is required after fabrication to engender strength. The alloy is noted for its excellent high temperature creep strength and oxidation resistance, and it is probably the most thermally stable solid-solution strengthened alloy currently available (2).

INCONEL® alloy 740, an age-hardenable nickel-base alloy, was also identified as a candidate material because of its potentially higher allowable stresses. However, the alloy is not thermally stable in the temperature range of interest, which may limit its use (2). Recently, a new age-hardenable superalloy, HAYNES 282® alloy, was commercially introduced which offers the potential of even higher allowable stress levels than alloy 740 in addition to superior thermal stability (3). The new alloy has creep strength competitive with R-41 alloy, but it offers much better formability and weldability.

Physical Metallurgy

The compositions of 230 and 282 alloys are shown in Table 1 along with those of other alloys used for comparisons. Among the nickel-based solid-solution strengthened alloys, 230 alloy is unique because it relies primarily on tungsten as the major strengthening element. Tungsten was selected because it diffuses more slowly in nickel than molybdenum, and, therefore, it enhances high temperature creep strength (2). This solid-solution strengthened alloy also relies on the formation of chromium-rich $M_{23}C_6$ carbides which precipitate on and immobilize dislocations formed during high temperature creep. A small addition of boron is used to improve the lattice mis-match between the carbide and the face-centered cubic matrix, which improves the carbide stability. The alloy also forms large, tungsten-rich M_6C carbides during ingot solidification that get distributed throughout the microstructure during hotworking operations. Due to their high tungsten content, these carbides do not dissolve easily, and they serve as important second-phase obstacles for controlling the grain size of the alloy. The alloy is typically annealed in the temperature range of 1177 to 1232°C(2150 to 2250°F) to develop a grain size of ASTM 4-5. This grain size provides the best combination of creep and fatigue strength.

Table 1. Nominal Compositions of 230, 282 and Comparative Alloys, Weight %

Alloy	Ni	Co	Fe	Cr	Mo	W	Mn	Si	Al	Ti	C	B	Others
Solid-Solutioned Strengthened													
230	57	5*	3*	22	2	14	0.5	0.4	0.5*	0.1*	0.10	0.015*	La-0.02
X	47	1.5	18	22	9	0.6	1*	1*	0.5*	0.15*	0.10	0.008*	-
625	62	1*	5*	21	9	-	0.5*	0.5*	0.4*	0.4*	0.10*	-	-
617	54	12.5	1	22	9	-	-	-	1.2	0.3	0.07	-	-
Age-Hardened													
282	57	10	1.5*	20	8.5	-	0.3*	0.15*	1.5	2.1	0.06	0.005	-
263	52	20	0.7*	20	6	-	0.4	0.2	0.6*	2.4*	0.06	0.005*	Al+Ti=2.6
Waspaloy	58	13.5	2*	19	4.3	-	0.1*	0.15*	1.5	3	0.08	0.006	Zr-0.05
R-41	52	11	5*	19	10	-	0.1*	0.5*	1.5	3.1	0.09	0.006	Zr-0.07*
740	48	20	0.7	25	0.5	-	0.3	0.5	0.9	1.8	0.03	-	Nb-2

*Maximum

The age-hardenable alloys all depend on the formation of gamma-prime, $Ni_3(Al, Ti)$, for their strength. The (Al + Ti) content for 282 alloy is higher than for 263 alloy, which ensures that its high temperature creep strength will be better. Furthermore, its addition of 8.5% Mo in combination with the (Al + Ti) of 3.6% provides creep properties which exceed those of Waspaloy and which are competitive with R-41 alloy. The alloy is typically annealed in the temperature range of 1107 to 1149°C (2025 to 2100°F) to obtain a grain size of ASTM 4 - 4 ½ for optimum resistance to creep and low cycle fatigue. It is then given a two-step heat treatment consisting of 1010°C (1850°F)/2 hrs./AC + 788°C (1450°F)/8 hrs./AC. The first step is carried out at a temperature above the gamma-prime solvus temperature of 997°C(1827°F) primarily to form $M_{23}C_6$ carbides at the grain boundaries in the preferred morphology to resist grain boundary sliding during creep. The second step results in the formation of the matrix strengthening gamma-prime precipitates. Because the alloy contains titanium, a small quantity of titanium-rich MC carbides and carbonitrides can be found scattered throughout the microstructure.

Mechanical Properties

Creep Strength

The low strain creep strength of 230 alloy is quite good among the solid-solutioned strengthened alloys. This is illustrated in Figure 1 for a creep strain of 1% for plate.

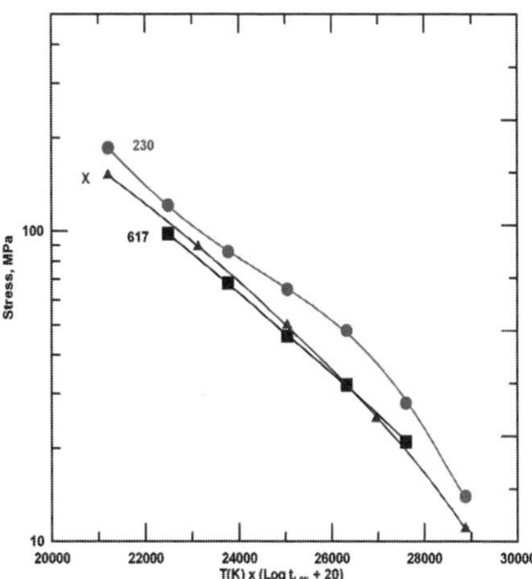

Figure 1. Comparison of 1% creep strengths of 230, X and 617 alloys (plate).

The alloy shows an advantage over the entire range shown, which makes it an attractive candidate for use in ultrasupercritical steam plants.

Among the age-hardened alloys, 740 alloy has been carefully examined because its creep strength is much higher than the solid-solution strengthened alloys. However, the introduction of HAYNES 282 alloy has provided an alloy with even better high temperature strength. This is shown in Figure 2, which compares the 1000 hour rupture strengths in a Larson-Miller format over the temperature range of 650-816°C (1200-1500°F). From the figure, it can be seen that 282 alloy provides a significant advantage over most of the range, and especially in the 700-760°C (1300°-1400°F) range of interest.

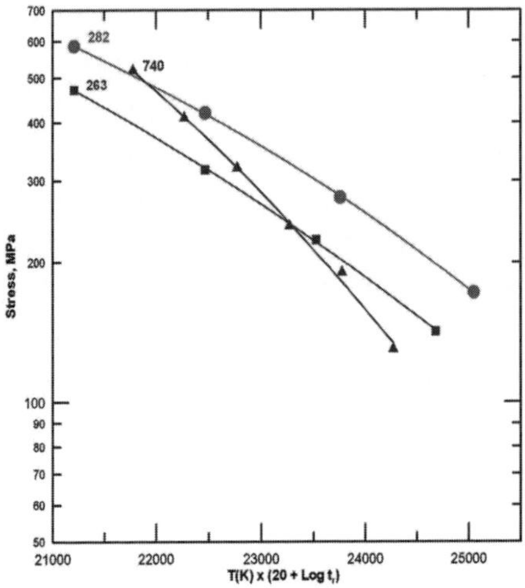

Figure 2. Comparison of the 1000-hour stress rupture strengths of plate and bar.

Of particular note, is that the advantage of 740 alloy over 263 alloy declines rather rapidly within this temperature range, and at 760°C, 740 alloy is actually weaker than 263 alloy. This is probably due to the fact that 740 alloy contains no addition of molybdenum to strengthen the matrix as does 263 alloy.

Fatigue Properties

Among the solid-solution strengthened alloys, 230 alloy possesses outstanding fatigue strength. This is illustrated in Figure 3 for fully reversed, strain-controlled, low cycle

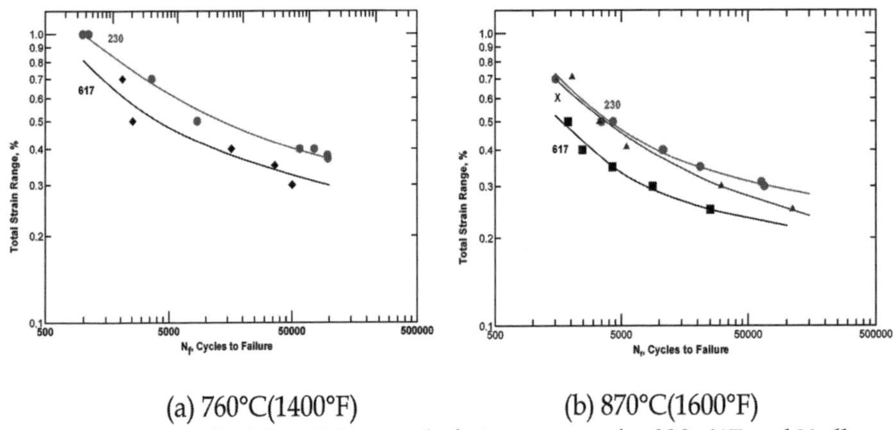

(a) 760°C(1400°F) (b) 870°C(1600°F)

Figure 3. Strain controlled (R=-1) low-cycle fatigue curves for 230, 617 and X alloys.

fatigue at 760°C(1400°F) and 870°C(1600°F) at a frequency of 0.33 Hz (5) for plate. The alloy has also been shown to possess excellent resistance to thermal fatigue (6). It was found to be superior to 617 alloy as shown in Figure 4 for plate samples. The advantage of 230 was also found to hold for welded samples. Resistance to thermal fatigue would be an important consideration for plant shutdown operations.

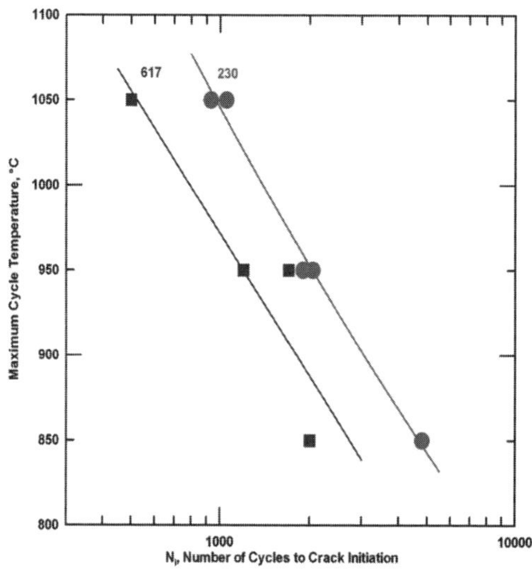

Figure 4. Thermal fatigue crack initiation as a function of T_{max} for 230 and 617 alloys (5).

HAYNES 282 alloy has also been found to possess excellent resistance to low-cycle fatigue (7). A comparison of 282 and 263 alloys is shown in Figure 5 for strain-controlled LCF at 815°C and a frequency of 0.33Hz. The thermal fatigue behavior

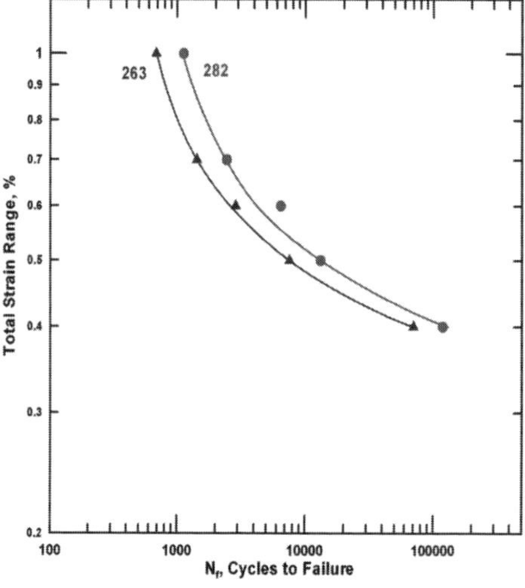

Figure 5. Comparison of the LCF lives of 282 and 263 alloys at 815°C(1500°F) (sheet).

of 282 has not yet been investigated. However, based on its low coefficient of thermal expansion, it would be expected to be very good. A comparison of the mean coefficients of thermal expansion of 282, 263 and 740 alloys is presented in Figure 6.

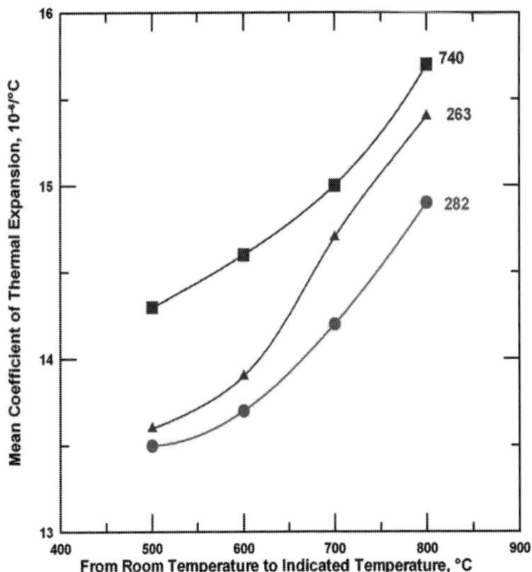

Figure 6. Mean coefficients of thermal expansion for 263, 282 and 740 alloys.

From the figure, it can be seen that 282 alloy has the lowest expansion coefficient among the three alloys. Therefore, one would expect that it would have thermal fatigue resistance superior to 263 and 740 alloys .

Oxidation Resistance

HAYNES 230 alloy was formulated to have excellent resistance to oxidation with additions of 22% chromium in a nickel base along with minor additions of manganese, silicon and lanthanum (2). During oxidation, the alloy develops a manganese-chromium spinel oxide which is very protective. Recent studies relavent to its use in ultrasupercritical steam power plants have shown that the alloy exhibited excellent resistance to oxidation in moist air (3% H_2O) over the 650-800°C(1200-1470°F) temperature range (8). It was judged to have better oxidation resistance than 263, 617 and 740 alloy which were included in the test program.

The oxidation resistance of HAYNES 282 alloy has not yet been evaluated for use in ultrasuperacritical steam power plants. Based on its composition, one would expect it to have more than acceptable oxidation resistance. For example, one would expect that it would be similar to 263 alloy which has been shown to possess excellent resistance to steam oxidation at 650°C(1200°F) and 800°C(1470°F) (9).

Thermal Stability

Thermal stability is an important issue for consideration in ultrasupercritical steam power plants since they will be in operation for many decades. It is of concern from a number of aspects. First, it is of primary importance for the performance of repair and overhaul operations which will most likely involve welding accompanied by some mechanical adjustments of the affected components. Second, loss of low temperature ductility during service exposure enacts a penalty on resistance to thermal fatigue. Third, service exposures reduce the low temperature fracture toughness of most alloys. Thus, issues of safety arise.

In all of these aspects, HAYNES 230 alloy has demonstrated excellent behavior. It has been shown, for example, that the alloy retains a high level of toughness after long term exposures in the 650-870°C(1200-1600°F) temperature range as illustrated in Figure (7) (10).

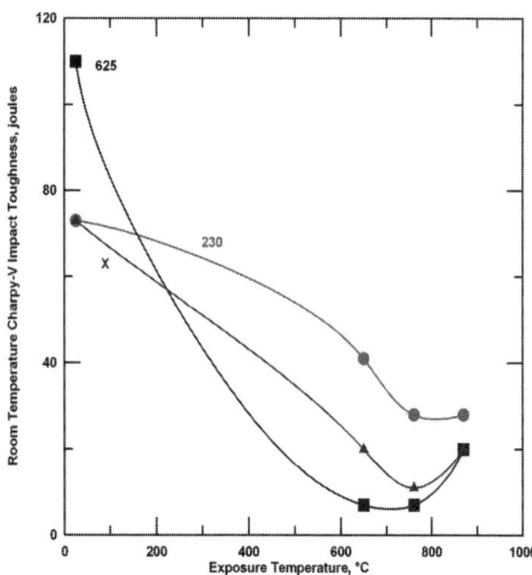

Figure 7. Room temperature Charpy-V impact toughness after 8,000 hours of exposure at various temperatures for plate samples.

The ability of the alloy to be weld repaired in the service exposed condition was demonstrated in a study performed on plates that had been given a thermal exposure of 760°C(1400°F)/1500 hrs. (11). The plates were TIG welded using 230-W® filler metal and tested in accordance with the requirements of Section IX of the ASME Boiler and Pressure Vessel Code. All of the results met these requirements.

With regard to the effects of ductility loss on low temperature, low cycle fatigue life, Figure 8 demonstrates the reduction of 425°C(800°F) LCF life at a total strain range of 0.65% resulting from a simulated service exposure at 760°C(1400°F)/1000 hours. These results are significant because thermal stresses become tensile at low temperatures during cooling. The debit in ductility becomes important at temperatures below approximately 650°C(1200°F). All of the alloys exhibit some loss of LCF life. However, it can be appreciated from the figure that 230 alloy has better LCF life in the aged condition than either of the other alloys in the annealed condition.

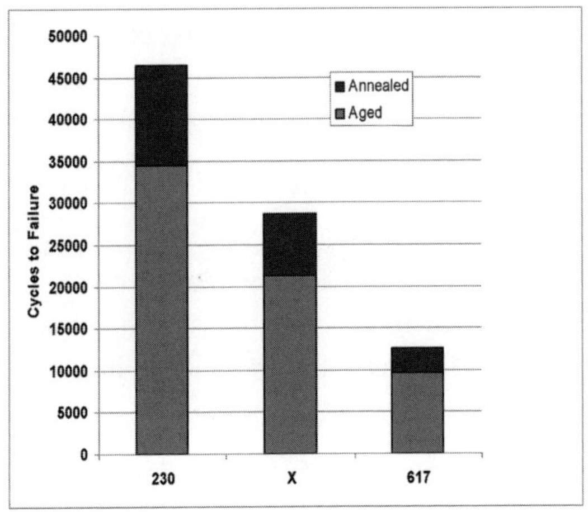

Figure 8. Effect of aging at 760°C(1400°F)/1000 Hrs. on the 425°C(800°F) LCF life of several alloys at a total strain range of 0.65% for plate samples.

The thermal stability of HAYNES 282 alloy has not yet been extensively studied. However, preliminary data indicates that it has good retention of ductility on thermal exposure. Figure 9 summarizes the effects of exposures of 8,000 hours at temperatures of 760°C and 870°C on room temperature tensile elongation of 282 plate. The levels of retained ductility would appear to be quite satisfactory. Metallographic examination

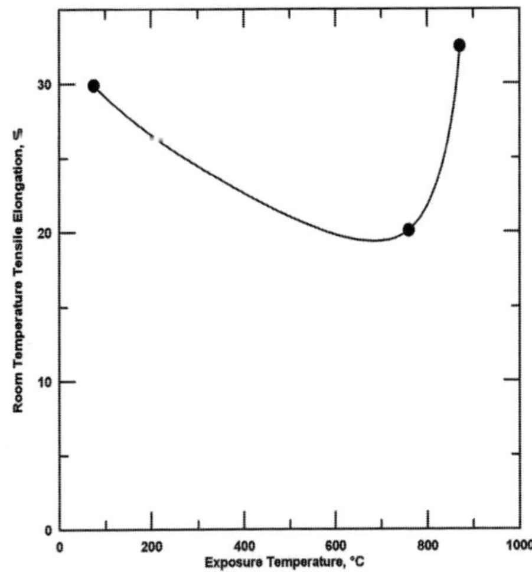

Figure 9. Room temperature tensile elongation for 282 plate samples exposed for 8,000 hours at 760°C(1400°C) and 870°C(1600°F)

of the exposed samples indicated that the relatively minor ductility drop was associated with the precipitation of carbides. No evidence of η-phase (Ni_3Ti) or TCP phases such as σ- or μ-phase, which are associated with large ductility losses, were observed in the microstructures.

Fabricability

In the annealed condition, 230 and 282 alloys both possess high levels of formability. A summary of the room temperature tensile properties of annealed plate is given in Table 2. Both alloys have reasonably low levels of yield strength along with high levels of ductility, indicating that typical forming operations should not be a problem. The coldworking characterictics of the two alloys along with other alloys for comparison

Table 2. Room temperature tensile properties of annealed plate.

Alloy	0.2% Yield Strength, MPa	Ultimate Tensile Strength, MPa	Elongation, %
230	375	840	48
282	394	831	57

are shown in Figure 10. From the figure, it can be seen that the two alloys tend to work harden more than X alloy, but less than 625 alloy.

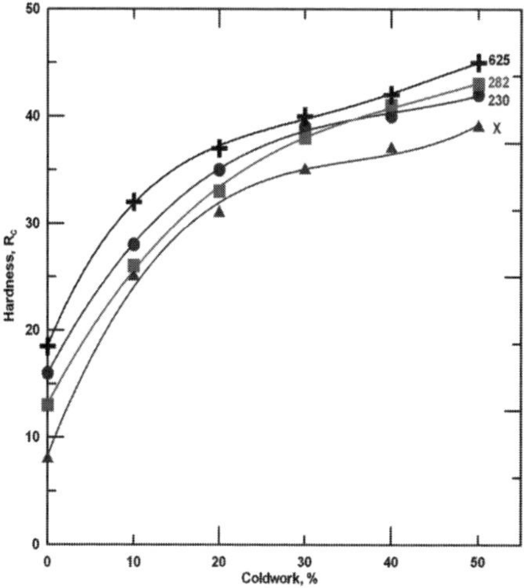

Figure 10. Hardness as a function of coldwork for 230, 282, 625 and X alloys.

Weldability is another important property to consider for fabrications. The weldability of 230 alloy for the TIG and MIG welding processes has been investigated in detail (11).

Due to the presence of boron in the basemetal, a boron-free filler metal, 230-W alloy, is used to avoid solidification hot cracking. Using this filler metal, it was demonstrated that plates up to 38 mm (1.5 in.) could be made without encountering weld metal microfissuring. Later studies have shown that plates up to 76 mm (3 in.) could be successfully welded (12).

HAYNES 282 alloy was developed to have improved resistance to strain-age cracking, an important problem affecting certain age-hardenable alloys (4). Using a test developed in the late 1960's referred to as the controlled heating rate test (CHRT) (13), the alloy was shown to be much more resistant to strain-age cracking than Waspaloy or R-41 alloys, which have comparable high temperature creep strength. This is illustrated in Figure 11. Studies using standard TIG and MIG welding processes have been conducted at Haynes International to further examine the welding characteristics of the alloy. Plates of the alloy 12 mm (0.5 in.) thick were welded by both processes without difficulty using a matching composition filler metal. Bend bars were were tested in the as-welded condition, and all samples successfully passed R=2t, 180° bends.

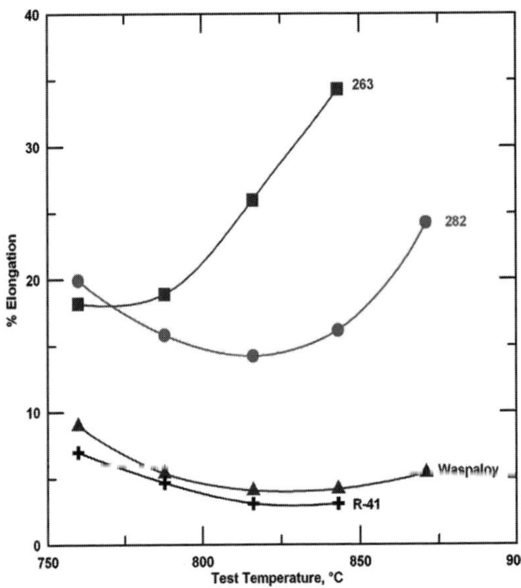

Figure 11. Controlled heating rate test results for 263, 282, Waspaloy and R-41 alloys.

Concluding Remarks

The realization of advanced ultrasupercritical steam fossil power plants will require the use of special nickel-base alloys. Both HAYNES 230 and 282 alloys are prime candidates for such service because they satisfy the main material requirements. HAYNES 230 alloy has been included in the materials evaluation efforts, and investigation of the properties of the alloy is ongoing. HAYNES 282 alloy represents a new material that has high temperature strength capabilities beyond those of 740 alloy, which has previously been touted as having the highest strength capabilities. In addition to its creep strength, its fatigue resistance, oxidation resistance, and excellent fabrication characteristics make it a very attractive candidate. Therefore, it is believed that a serious effort should be undertaken to thoroughly evaluate 282 alloy for use in ultrasupercritical steam power plants.

References

1. "A critical look at supercritical power plants," R. Viswanathan, A. F. Armor and G. Borres, *Power*, April, 2004, p. 42.

2. D. L. Klarstrom, *Materials Design Approaches and Experiences*, Warrendale, PA: TMS, 2001, pp. 297-307.

3. S. Zhao, et al., *Superalloys 2004*, Warrendale, PA: TMS, 2004, pp. 63-72.

4. L. M. Pike, "HAYNES® 282™ alloy – A New Wrought Superalloy Designed for Improved Creep Strength and Fabricability," Paper No. GT2006-91204, presented at ASME Turbo Expo 2006, Barcelona, Spain (May 2006).

5. S. K. Srivastava and D. L. Klarstrom, "The LCF Behavior of Several Solid Solution Strengthened Alloys Used in Gas Turbine Engines," Paper No. 90-GT-80, presented at ASME Turbo Expo 1990, Brussels, Belgium (June, 1990).

6. "The Thermal Fatigue Behavior of Combustor Alloys IN 617 and HAYNES 230 Before and After Welding," F. Meyer-Olbersleben, N. Kasik, B. Ilschner, and F. Rézaï-Aria, Met. and Matls. Trans. A, Vol. 30A, April, 1999, 981-989.

7. L. M. Pike, "Low-Cycle Fatigue Behavior of HAYNES® 282® Alloy and Other Wrought Gamma-Prime Strengthened Alloys," Paper No. GT2007-2867, presented at ASME Turbo Expo 2007, Montreal, Canada (May, 2007).

8. G. R. Holcomb, et al., "Oxidation of Advanced Steam Turbine Alloys," Paper No. 06453, presented at NACE Corrosion Expo 2006, San Diego, CA (March, 2006).

9. J. M. Sarver and J. M. Tanzosh, "Steamside Oxidation Behavior of Candidate USC Materials at 650°C and 800°C, presented at the 8th Ultra-Steel Workshop, Tsukuba, Japan (July, 2004).

10. D. L. Klarstrom, "Thermal Stability of a Ni-Cr-W-Mo Alloy," Paper No. 407, presented at NACE Corrosion 1994, Baltimore, MD (February, 1994).

11. "Weldability Studies of HAYNES® 230® Alloy," S. C. Ernst, Welding J., Vol. 73, April, 1994, pp 80s-89s.

12. R. Viswanathan, et al., "Materials for Ultrasupercritical Coal-Fired Power Plant Boilers," Proc. 4th Intl. Conf. on Advances in Materials for Fossil Fuel Power Plants, ASM International, 2004, pp 3-19.

13. "Ranking the Resistance of Wrought Superalloys to Strain-Age Cracking," M. D. Rowe, Welding J., Vol. 85, February, 2006, pp. 27s-34s.

Advances in Materials Technology for Fossil Power Plants
Proceedings from the Fifth International Conference
R. Viswanathan, D. Gandy, K. Coleman, editors, p 119-128

DOI: 10.1361/cp2007epri0119

Creep Properties of Carbon and Nitrogen Free Austenitic Alloys for USC Power Plants

S.Muneki
H.Okubo and F.Abe
National Institute for Materials Science
1-2-1 Sengen Tsukuba
Ibaraki Japan

Abstract:

Various carbon and nitrogen free martensitic alloys were produced for the application which required long time creep properties at high temperatures. But they were easy transformed to austenite phase before the creep tests because of low Ac1 temperature. In this paper, a new attempt has been demonstrated using carbon and nitrogen free austenitic alloys strengthened by intermetallic compounds. We choose Fe-12Ni-9Co-10W-9Cr-0.005B based alloy. Furthermore, we discussed about creep characteristics among the wide range of the testing conditions more over 700°C and steam oxidation resistance to confirm the possibility of the alloys for the future USC power plants under the severe environments.

1.Introduction

High Cr ferritic steels, such as 9Cr-1MoVNb steel have successfully been used for large diameter and thick section boiler components, such as main steam pipe and header in USC boilers in fossil-fired power plants. Recent trend to utilization of clean energy leading to protection of global environment has been accelerating application of ultra super critical boilers, which efficiency in power generation than in conventional ones, and thus release less amount of carbon dioxide, etc. The USC power plants requires heat resistant materials with improved creep rupture strength at elevated temperatures over 650°C because of increase in operating temperature and pressure of the steam used[1,2]. Since the metastable microstructure of the martensitic alloys[3] in which the transformation temperatures are very low below 700°C is easily generated by the DSS operation of recent USC power plants, it is difficult that these materials are applied as large components. So we selected carbon and nitrogen free austenitic alloys.

In this study the effect of aluminum and silicon addition to the Fe-12Ni-9Co-10W-9Cr-0.005B alloy on the creep properties and oxidation resistance at elevated temperatures more over 700°C.

2. Experimental procedure

Table 1 shows chemical composition of alloys used in this study. They were melted 10kg ingots in a vacuum induction furnace. Creep rupture tests were conducted in the temperature range between 700°C and 800°C and in the applied stress range between 300MPa and 60MPa using a 6mm diameter and 30mm gage length tensile specimen as solution treated condition at 1000°C for 30minutes.

Table 1 Chemical composition of alloys used (mass%).

Alloy	C	N	Ni	Co	W	Cr	Ti	Al	B	Si	Fe
Base	0.0003	0.0006	12.06	9.04	10.10	8.66	0.16	0.09	0.0051	0.04	Bal.
3Al	0.004	0.0010	11.71	8.77	9.98	8.50	0.20	3.08	0.0050	0.01	Bal.
3Si	0.002	0.0012	11.75	8.85	9.85	8.73	0.18	0.13	0.0051	2.95	Bal.

The creep test was started immediately in which the test piece reached at the creep test temperature. In the steam oxidation test, the plate specimen of 2mm thickness, 10mm width and 20mm length was used. Test temperatures were 700°C and 750°C, and it was done to longest time until 8,466h. Water used in this study was controlled dissolved oxygen decreased under 10ppb, and electric conductivity under 8.0μS/m.

3. Results and discussion

3.1 Creep properties at 700°C

Fig.1 shows relationship between creep rate and time of three alloys crept at 700°C with 200MPa and 150MPa. An initial creep rate of 3Si alloy was about 5 x 10^{-2} (1/h). Change of creep rate was very small. An initial creep rate of base alloy decreased about 1/10 at 8 x 10^{-3} (1/h). Creep rate gradually decreased with increasing time in the transition creep region, and then reached at the minimum creep rate at about 1 x 10^{-3} (1/h). After that creep rate did not change between 1h and about 10h, and then slightly increased with increasing time. Time to rupture of base alloy increased more over 10 times from that of 3Si alloy. Furthermore creep resistance of 3Al alloy remarkably increased. When the applied stress decreased to 150MPa an initial creep rate of 3Si alloy decreased to 1 x 10^{-2} (1/h). Time to rupture increased about 10 times. Creep rates of base and 3Al alloys were similar values at about 2 x 10^{-3} (1/h). Creep rates of two alloys decreased with increasing time as similar slope. Minimum creep rate of base alloy indicated at about 100h. Creep rate increased to the acceleration creep region after incubation time between 100h and about 1,000h. Time to rupture of base alloy was 2,027h. Creep rate of 3Al alloy further deceased than that of base alloy. Time to rupture of 3Al alloy remarkably increased to about 7,000h.

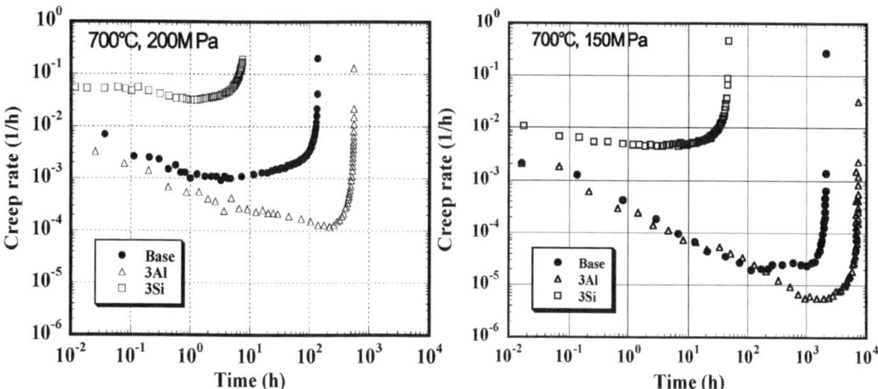

Fig.1 Relationship between creep rate and time of three alloys crept at 700°C with 200MPa and 150MPa.

Fig.2 shows change of ruptured elongation and reduction of area of three alloys crept at 700°C. Ruptured elongation of 3Si alloy changed 130% to 70%. Elongation of other two showed relatively high more over 40% until the time at 1,000h. And elongation of the longest time at about 37,000h of 3Al alloy showed about 20%.

While change of reduction of area of 3Si alloy kept the highest value more over 80% until about 4,300h. Reduction of area of 3Al and base alloy showed higher value more over 60% until about 7,000h.

In conventional high Cr ferritic heat resistant steels initial hardness before the creep test was controlled by tempering at HV 200-220. As a result fine and a large amount of carbides and nitrides were distributed into the microstructure. They have been maintained long term stability during creep tests.

While in case of carbon and nitrogen free alloys intermetallic compounds distributed into the microstructure before and during the creep test.

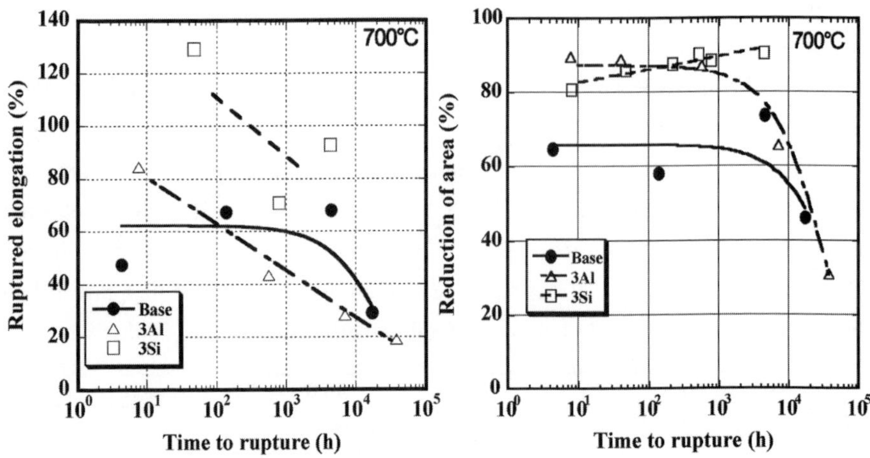

Fig.2 Ruptured elongation and reduction of area of three alloys crept at 700°C.

Fig.3 shows ruptured hardness and stress dependence of three alloys crept at 700°C. Hardness of 3Al alloy did not change and indicated the highest value at about HV400 until about 37,000h Hardness of 3Si alloy showed higher value between HV300 and 250. Hardness of base alloy changed HV250 to HV200 as similar value of initial hardness of the conventional high Cr ferritic heat resistant steel.

Stress dependence of 3Al alloy was the best creep resistance among three alloys. Time to rupture changed 7.7h in 300MPa to about 37,400h in 100MPa. The time of base alloy changed 4.4h in 300MPa to about 17,100h in 100MPa. But 3Si alloy indicated poor creep resistance in the same stress level.

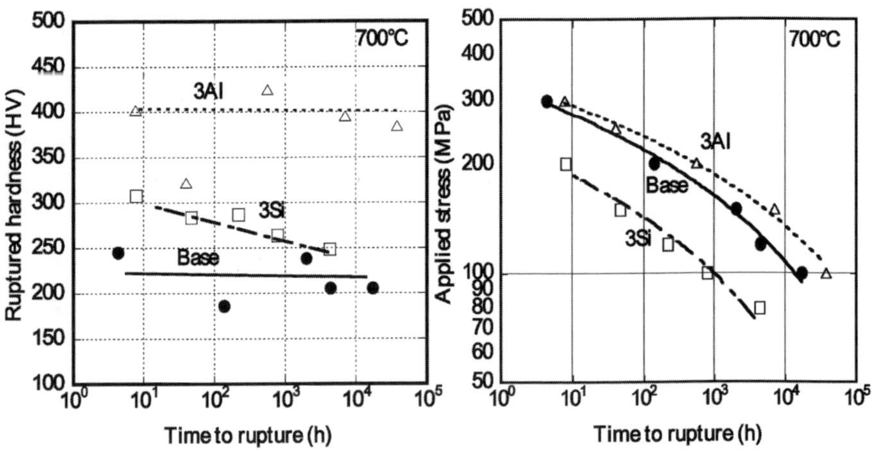

Fig.3 Ruptured hardness and stress dependence of three alloys crept at 700°C.

3.2 Creep properties at 750°C and 800°C

Fig.4 shows relationship between creep rate and time of three alloys crept at 750°C and 800°C with 100MPa. Initial creep rates of 3Si and 3Al alloys were similar values at about 1 x 10-2 and 6 x 10-3 (1/h), in respectively. But creep rate decreased very small in 3Si alloy. And the alloy ruptured at about 85h. Creep rate of 3Al alloy considerable decreased with increasing the time until about 2,000h. And the alloy ruptured at about 6,100h. All three alloys indicated typical creep deformation behavior even at 800°C in 100MPa. 3Si alloy ruptured very short time at about 10h. But time to rupture of base and 3Al alloys remarkable increased more over 10 times. Time to rupture of base and 3Al alloys was about 250h and 500h, in respectively.

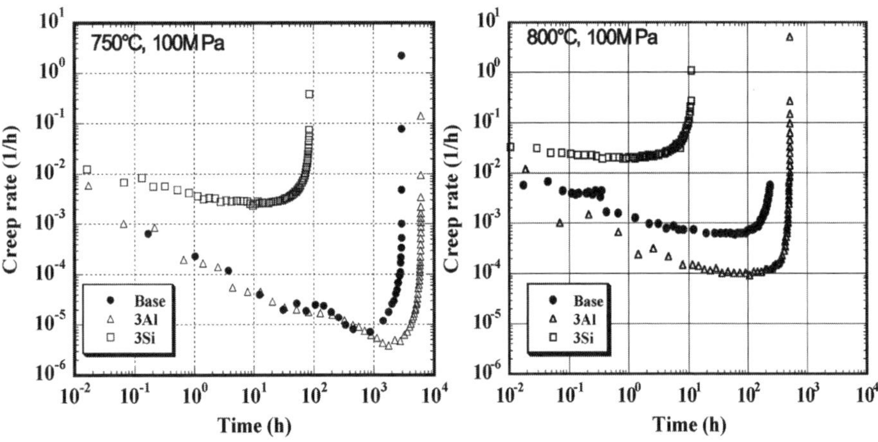

Fig.4 Relationship between creep rate and time of three alloys crept at 750°C and 800°C with 100MPa.

Fig.5 shows change of ruptured elongation at 750°C. Ruptured elongation of 3Si alloy indicated the highest value more over 88% until 2,360h. Elongation of 3Al and base alloys showed higher value more over 37% until 6,000h. Elongation of base alloy ruptured at about 46,500h was 18.3%.

Reduction of area of 3Si and 3Al alloys kept higher values more over 80%. Reduction of area of base alloy gradually decreased with increasing the time to rupture. And the reduction of area of base alloy ruptured at about 46,500h was 2.7%.

Fig.6 shows change of ruptured hardness and stress dependence at 750°C. Hardness of 3Si and base alloys did not change with increasing time to rupture. But hardness of 3Al alloy indicated HV260 at about 1.0h in the ruptured time. Then the hardness gradually increased with increasing the time and reached at HV380. It will be precipitation hardening during the creep test.

Stress dependence of 3Al alloy exhibited the best creep resistance even at 750°C. Time to rupture changed 0.8h in 300MPa to 6,100h in 100MPa. The time of base alloy changed 28h in 200MPa to about 46,500h in 60MPa.

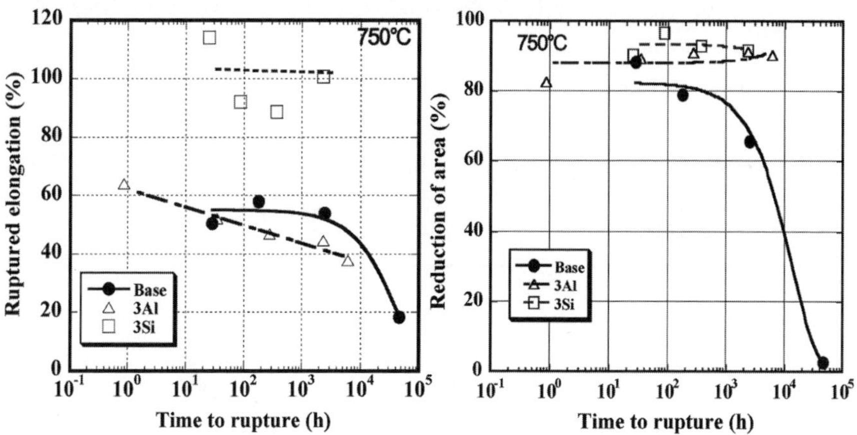

Fig.5 Ruptured elongation and reduction of area of three alloys crept at 750°C.

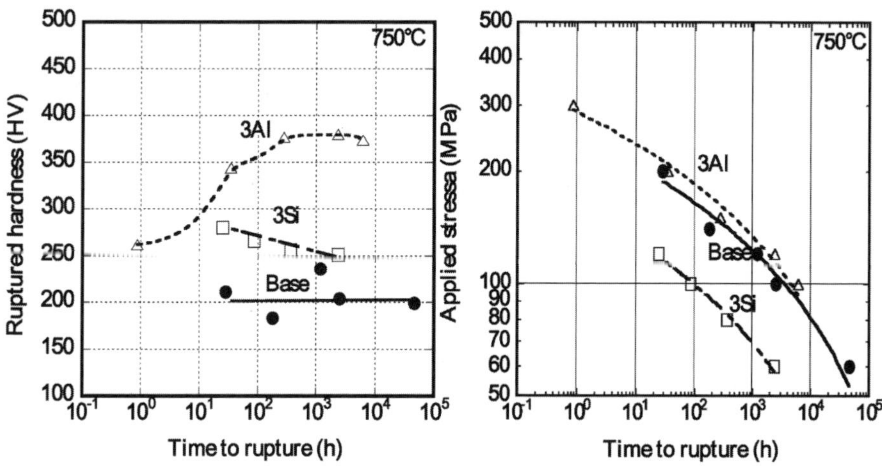

Fig.6 Ruptured hardness and stress dependence of three alloys crept at 750°C.

3.3 Precipitates distribution

Distribution and morphology changes of precipitates during creep tests were investigated by field emission SEM. This observation consists of four fields in backscattered electron images at the gage portion from the crept specimen.

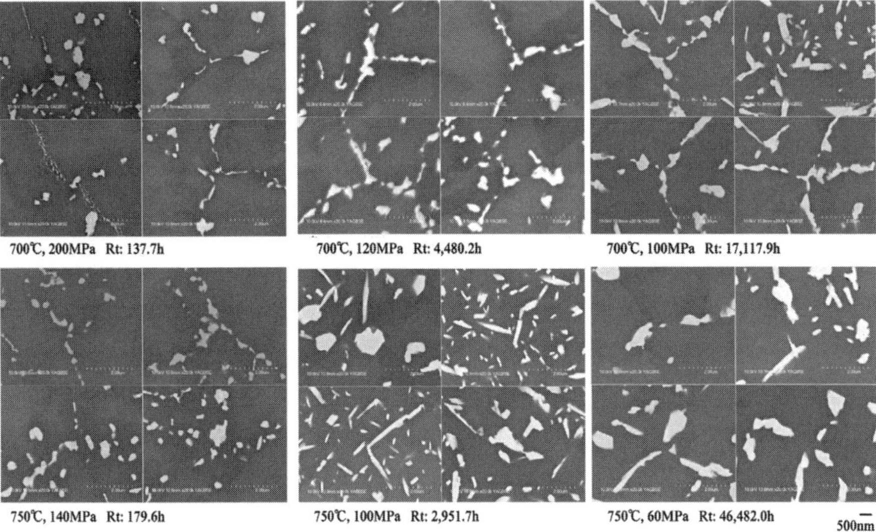

Fig. 7 Backscattered electron micrographs showing the effect of time to rupture of base alloy crept at 700°C and 750°C.

Fig.7 showed back scattered electron images of base alloy crept at 700°C and 750°C. Precipitates more over 1μm observed at the grain boundaries and inside the grain even at 137h in ruptured time of 700°C crept specimen. White shining particles into the image identified intermetallic compounds such as Fe_2W and Fe_7W_6. With increasing rupture time size of precipitates enormously increased, especially at grain boundaries.

Fig. 8 shows an example of 3Al alloy. As compared that of base alloy precipitates into the grain distributed uniform and very fine. These fine precipitates do not identify stll now. It might be some intermetallic compounds consists of Ni and Al such as Ni_3Al. Cohesion and coalescence of precipitates recognized at 558h in short time ruptured sample.

Fig.9 shows an example of 3Si alloy. Globular and large precipitates were observed from very short time ruptured sample. And needlelike precipitates were also observed. These differences of shape and distribution of precipitates seemed to bring about lowering of the creep strength.

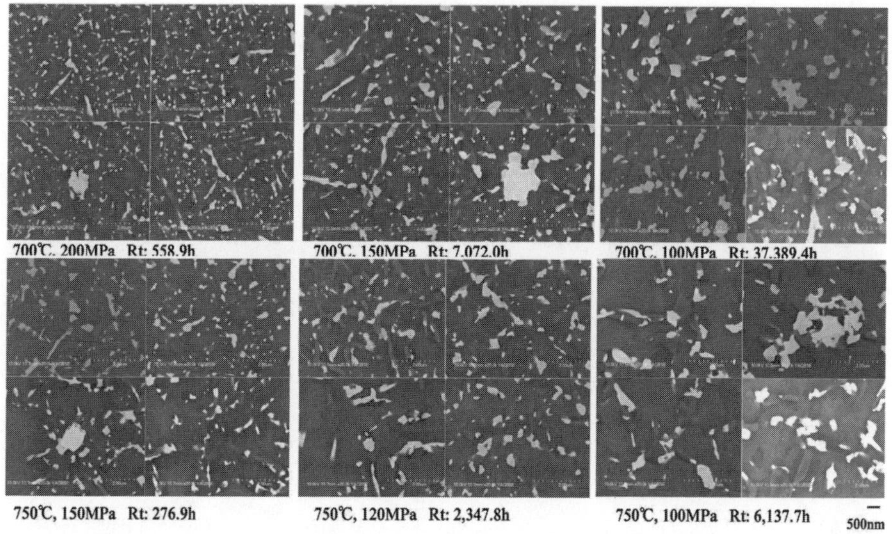

Fig. 8 Backscattered electron micrographs showing the effect of time to rupture of 3Al alloy crept at 700°C and 750°C.

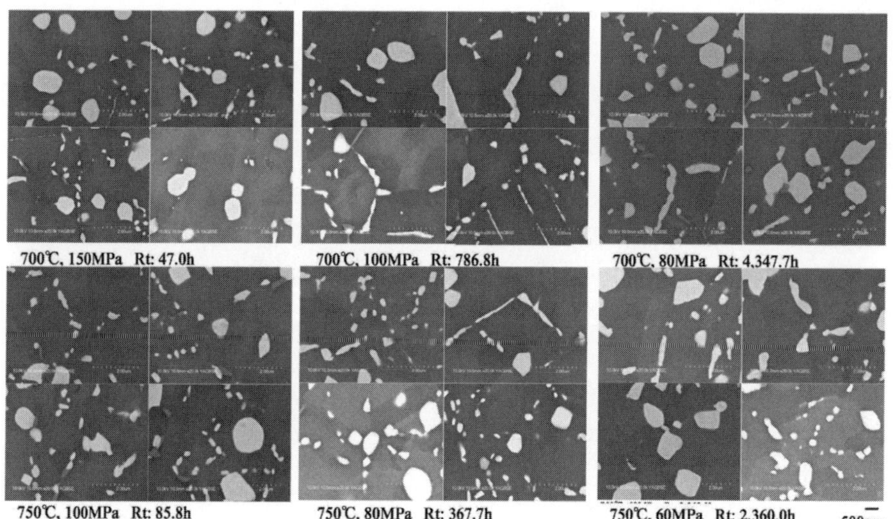

Fig. 9 Backscattered electron micrographs showing the effect of time to rupture of 3Si alloy crept at 700°C and 750°C.

3. 4 steam oxidation resistance

Fig.10 shows change of mass gain in three alloys steam oxidized at 700°C and 750°C.

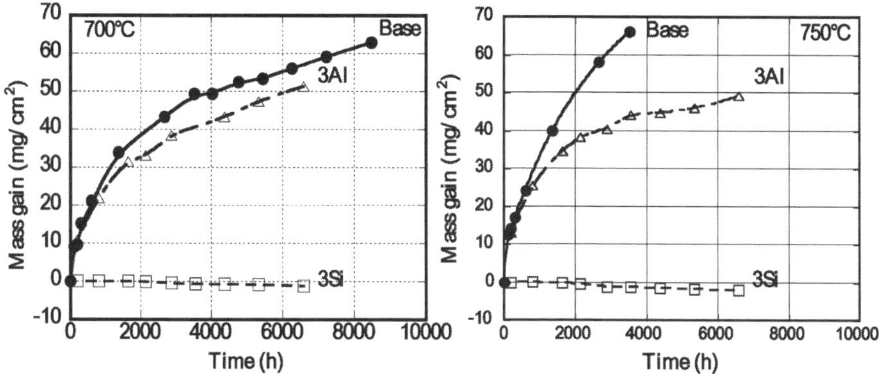

Fig.10 Mass gain of three alloys steam oxidized at 700°C and 750°C.

Mass gain of base alloy at 700°C showed parabolic increase until 8466h. 3Al alloy was also similar slope until 6,579h. On the other hand mass gain of 3Si alloy did not increase until 6,579h This alloy showed good steam oxidation resistance at 700°C, while steam oxidation resistance at 750°C of base and 3Al alloys showed slightly different behavior. Steam oxidation increase of base alloy very rapidly increase occurred in very short time. Steam oxidation increase of 3Al alloy became gentle after 2,000h in exposure time. Change of mass gain in 3Si alloy did not appeared until 6,579h. This alloy exhibited good steam oxidation resistance even at 750°C.

3.5 Possibility of three alloys in creep strength at high temperatures

Fig.11 shows attainability of creep strength used by Larson-Miller parameter in three alloys crept at 700 to 800°C. Creep strength of all three alloys changed in wide stress range. Creep strength of 3Al alloy at 700°C for 100,000h was estimated with 100MPa. And the strength of base alloy was estimated with 85MPa. Best combination of creep strength and steam oxidation resistance at 700 and 750°C was the Fe-12Ni-9Co-10W-9Cr-3Al-0.2Ti-0.005B alloy.

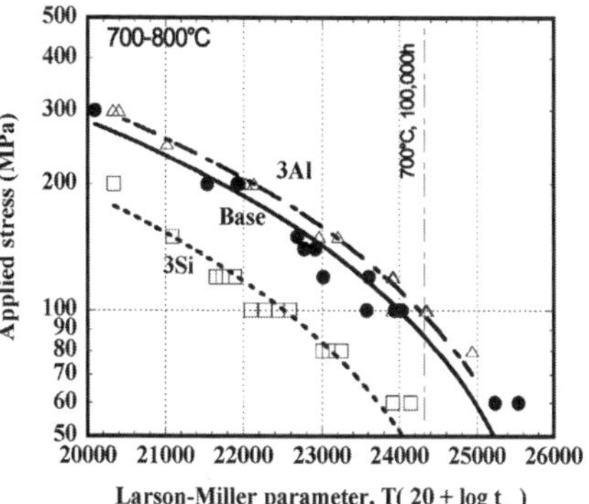

Fig.11 Attainability of three alloys on the creep strength at 700°C for 100,000h.

However, it will be necessary to improve steam oxidation resistance.

4 Conclusion

Effect of 3% Al and Si addition to the Fe-12Ni-9Co-10W-9Cr-0.2Ti-0.1Al-0.005B austenitic alloy on the creep properties and steam oxidation resistance more over 700°C have been investigated. Among three alloys creep strength of 3Al alloy was estimated with 100MPa in the creep condition of 700°C for 100,000h. Steam oxidation behavior of 3Si alloy was not recognized until 6,579h in exposure time. In three alloys, the best combination of creep strength and steam oxidation resistance has not yet been got. It will be necessary to improve steam oxidation resistance of 3Al alloy.

Acknowledgement

The creep tests were performed by the Creep Group in Materials Information Technology Station, NIMS.
The author is grateful to member of the Creep Group and Mrs. Moriiwaand Mrs. Ishitsuka for their sincere efforts.

References

1) F.Masuyama: Proc.5th Int.Conf. on Advances in Materials technology for Fossil Power Plants, (2004),35
2) R.Blum, R.W.Vanstone and C.M-Gouze: ibid (2004),116
3) S.Muneki, H.Okubo, H.Okada, M.Igarashi and F.Abe: Proc. 6th Int. Charles Parsons Turbine Conf. (2003),569

Advances in Materials Technology for Fossil Power Plants
Proceedings from the Fifth International Conference
R. Viswanathan, D. Gandy, K. Coleman, editors, p 129-139

DOI: 10.1361/cp2007epri0129

Development of High Strength HCMA (1.25Cr-0.4Mo-Nb-V) Steel Tube

T. Nakashima[1]
K. Miyata[2], H. Hirata[2], M. Igarashi[2]
A. Iseda[3]
Sumitomo Metal Industreis, Ltd.
[1]Wakayama steel works, 1850 Minato, Wakayama640-8555, Japan
[2]Corporate Research and Development Laboratories
[3]Tokyo Head Office

ABSTRACT

Improvement of thermal efficiency of new power plants by increasing temperature and pressure of boilers has led us to the development of high creep strength steels in the last 10 years.
HCMA is the new steel with base composition of 1.25Cr-0.4Mo-Nb-V-Nd, which has been developed by examining the effects of alloying elements on microstructures, creep strength, weldability, and ductility. The microstructure of the HCMA is controlled to tempered bainite with low carbon content and the Vickers hardness value in HAZ is less than 350Hv to allow the application without preheating and post weld heat treatment. The HCMA tube materials were prepared in commercial tube mills. It has been demonstrated that the allowable stress of the HCMA steel tube is 1.3 times higher than those of conventional 1%Cr boiler tubing steels in the temperatures range of 430 to 530°C. It is noted that creep ductility has been drastically improved by the suitable amount of Nd (Neodymium)-bearing. The steam oxidation resistance and hot corrosion resistance of the HCMA have been proved to be the same level of the conventional 1%Cr and 2%Cr steels. It is concluded that the HCMA has a practical capability to be used for steam generator tubing from the aspect of good fabricability and very high strength.
This paper deals with the concept of material design and results on industrial products.

1. Introduction

The power generation industry worldwide has identified a need for the development of coal-fired boilers operating at much higher efficiencies than the current generation of supercritical plants. This increased efficiency is expected to be achieved using ultra supercritical steam condition, which will require the development of new materials.
So far, new steels such as T92 [1] were developed and are used for USC power plants. However, though 1%Cr steels with higher strength such as NF1H and HCMV3 [2] were developed, these steels have not been commercially installed for practical boiler of USC power plant. In order to meet the requirement of practical use, the new steel named "HCMA" has been developed.

Newly developed HCMA steel put emphasis on properties of higher tensile and creep strength for practical use, steam oxidation resistance and hot corrosion resistance, and thermal conductivity were considered. In the following, the concept of material design, the manufacturing process, properties of the products and welding joints are described.

2. The concept of Material design

The composition of HCMA steel has been optimized to provide:

- The ultimate tensile strength ≧ 570MPa, and yield strength ≧ 390MPa,
- The allowable tensile strength 1.3 times higher than those of conventional 1%Cr boiler tubing steels in the temperature range of 430 to 530°C,
- The toughness in a satisfactory range for the practical application,
- The resistance to stress relief (SR) cracking,
- The weldability that allows the application without preheating and post weld heat treatment,
- The best compromise in creep ductility and creep strength.

Fig.1 shows a concept of HCMA. It is known that the full bainite structure leads to higher creep strength as well as tensile strength in the low alloy steels. However, the microstructures and mechanical properties of the bainite are very sensitive to heat treatment in 1%Cr steels [3-5]. Therefore the chemistry of HCMA is chosen to expand the stable bainite region.

In addition, (V, Nb, Mo) C precipitation and solute-Mo in the tempered bainite matrix [6] are optimized to obtain the strength target. The toughness and the resistance to SR cracking have been improved by suppression of grain coarsening with a pinning effect of Ti(C, N). With reducing C-content, the maximum hardness value in HAZ is reduced to less than 350Hv, resulting in the hardness criteria for application without preheating and post weld heat treatment. The previous studies pointed that the full bainite structure tended to reduce the creep ductility. However, we have demonstrated that a suitable addition of Nd drastically improved the creep ductility. The resulting chemical composition of HCMA is shown in Table1.

Table 1: Specified chemical composition of HCMA steel tube

	Composition											
	C	Si	Mn	P	S	Cr	Mo	V	Nb	S.Al	B	Others
Spec.	0.04/ 0.12	≦ 0.50	0.10/ 0.90	≦ 0.025	≦ 0.010	1.05/ 1.45	0.30/ 0.50	0.05/ 0.20	0.02/ 0.08	≦ 0.030	0.001/ 0.006	Nd

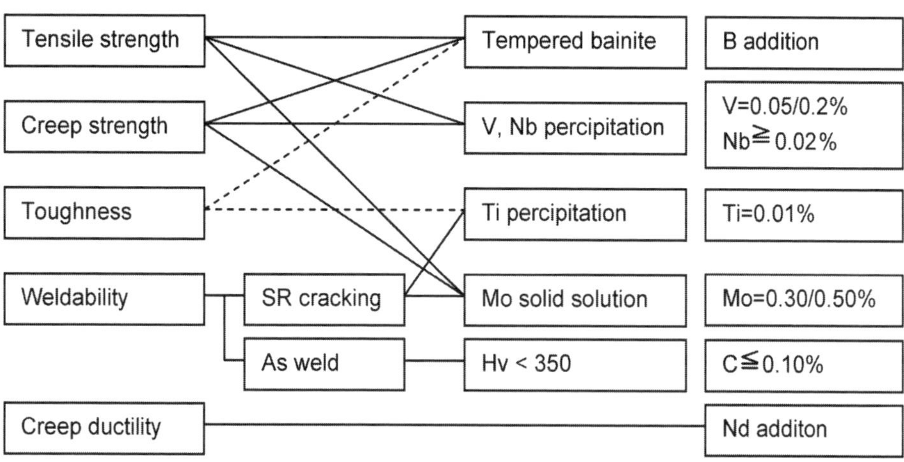

Figure 1. The concept of HCMA design

3. Test materials

Test materials were hot finished tube products through the process shown in table 2. Table 3 shows the chemical composition of test tubes.

Table 2: Trial tube manufacturing process

Heat	Product (O.D. x WT)	Steel making process	Tube manufacturing process	Heat treatment
A	φ63.5xt9.0mm	210ton Converter	Mannesmann-mandrel mill process	975°C normalized and 720°C tempered
B, C	φ50.8xt8.0mm	150kg VIM	Hot extrusion process	

Table 3: chemical composition of trial tubes (mass%)

	Composition										
Heat	C	Si	Mn	P	S	Cr	Mo	V	Nb	S.Al	B
A	0.06	0.26	0.70	0.007	0.002	1.11	0.43	0.10	0.05	0.004	0.002
B	0.07	0.22	0.70	0.008	0.001	1.26	0.44	0.10	0.05	0.002	0.004
C	0.06	0.24	0.72	0.009	0.001	1.26	0.44	0.10	0.05	0.004	0.002

4. Results

4.1 Mechanical properties

Fig.2 shows tensile properties of HCMA steel tubes. It is specified that tensile strength is more than 570MPa and yield strength is more than 390MPa to realize higher allowable stress at 500°C below.

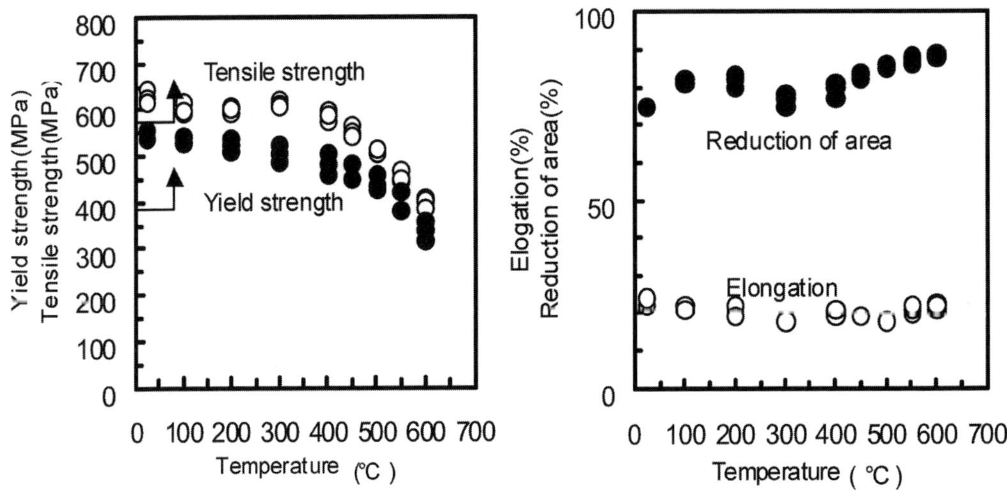

Figure 2. Tensile properties at elevated temperatures (Specimens: ϕ6mm GL=30mm)

Fig.3 shows creep rupture properties of HCMA steel tubes. No marked degradation in rupture strength occurred while the test was running for about 40,000hr.

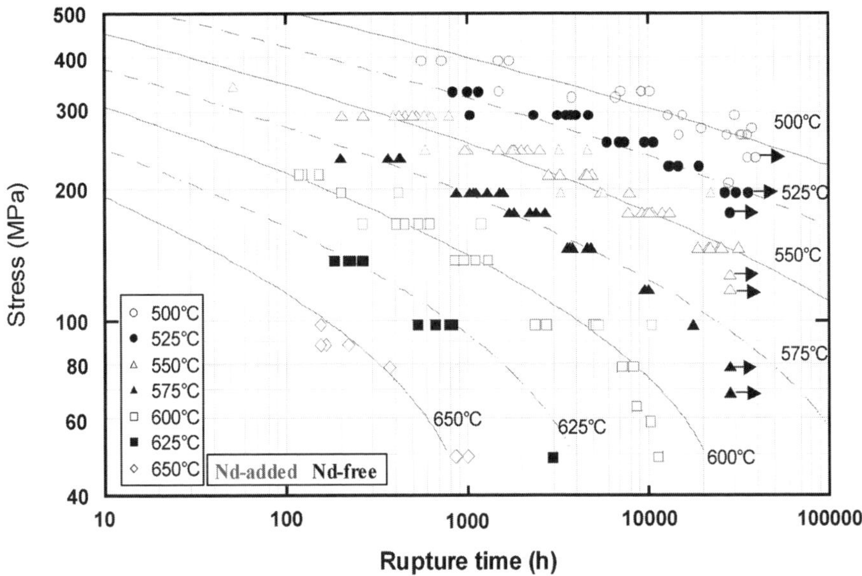

Figure 3. Creep rupture properties of HCMA steel including lab products

Fig.4 shows the allowable stress of HCMA steel tube compared with the curves of conventional 1%Cr and 2%Cr steel tubes. The allowable stress in the temperature range of 500°C and below was calculated by using the data of tensile properties. On the other hand, the allowable stress in the temperature range of 500°C above was calculated by using the data of creep properties. The allowable stress of HCMA steel tube is 1.3 times higher than those of conventional 1%Cr steel tubes in the temperature range of 430 to 530°C.

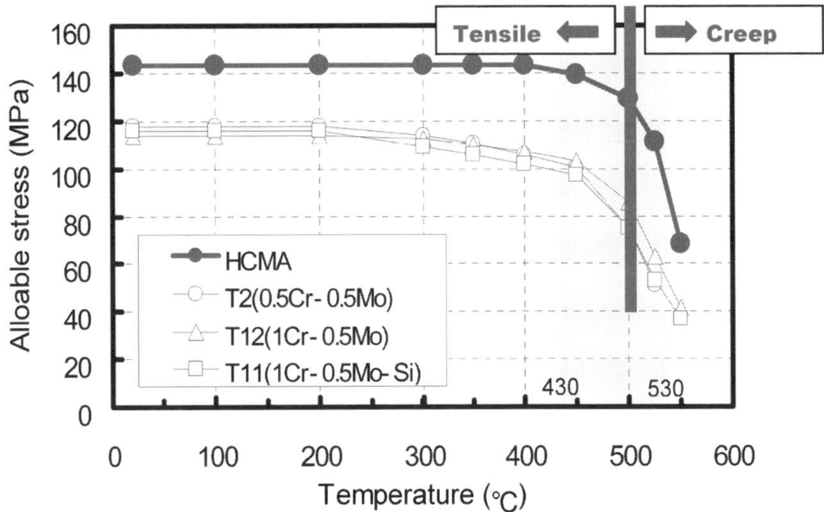

Figure 4. Comparison of allowable stress

4.2 Microstructure

The HCMA steel tubes are composed of fully tempered bainite as shown in Fig.5.

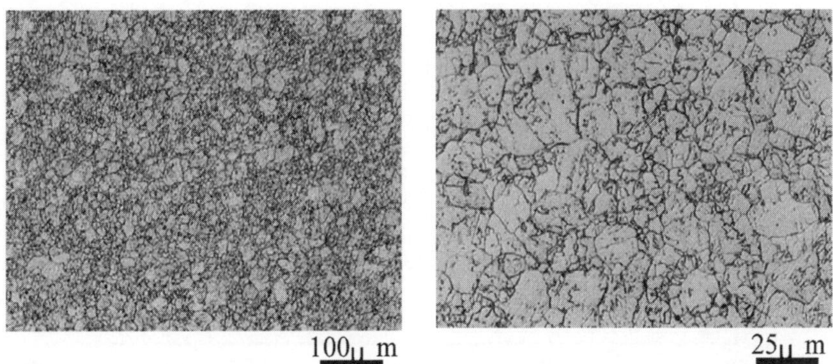

Figure 5. Optical microstructure of fully tempered bainite (Heat A)

Fig.6 shows TEM images of the specimen crept for 5863h at 550°C with 196MPa. Fine MX-type (M: V, Nb, Mo X: C, N) carbides precipitate in grains and interfere with the motion of dislocations. Partially spherical $M_{23}C_6$ carbides precipitate on grain boundaries. These carbides are supposed to have influence on higher tensile strength as well as creep strength.

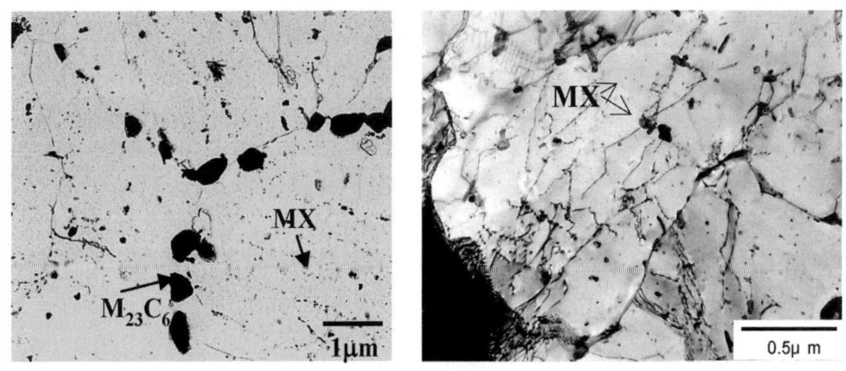

(a) Extracted replica sample (b) Thin film sample

Figure 6. TEM images of the specimen crept for 5863h at 550°C with 196MPa

Fig. 7 shows the effect of Nd on the microstructural evaluation around grain boundaries during long time creep. In Nd-free steel, creep strain was locarized around grain boundaries, resulting in loss of creep ductility. On the other hand, the steel with Nd has been deformed uniformly without localized recovery around grain boundaries and improved creep ductility. Nd element adhered to impurity such as Sulfer and is supposed to strengthen grain boundaries. The adhesion was identified as Nd-oxysulfide by EDX.

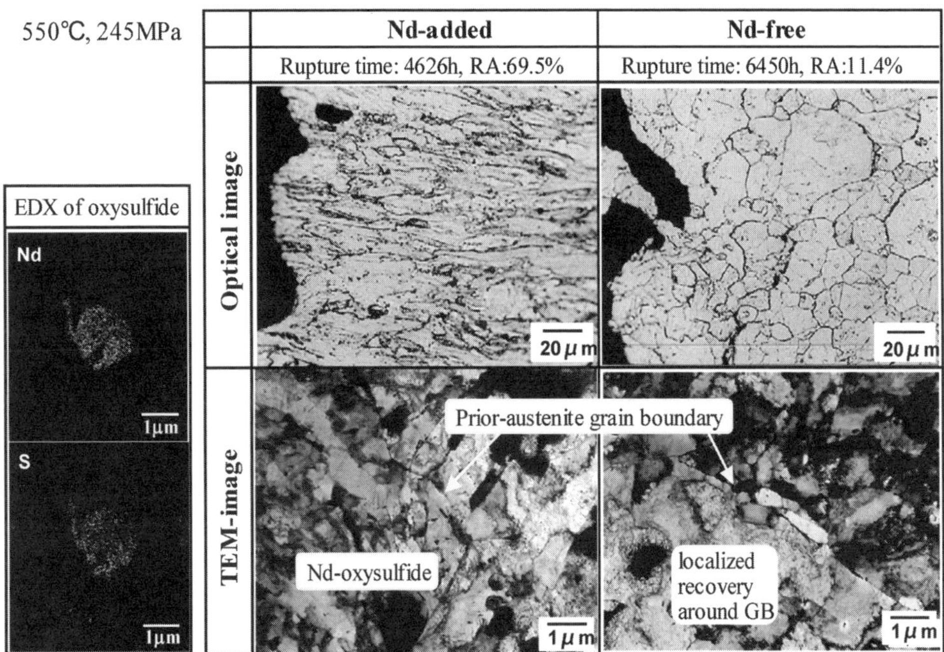

Figure 7. Optical and TEM images of the specimens crept at 550°C with 245MPa

4.3 Toughness after heating

Fig.8 shows the Charpy impact values after heating at 500°C and 550°C for up to 3000hr. The values are still high with little degradation enough to allow the steel to be used as a boiler tube.

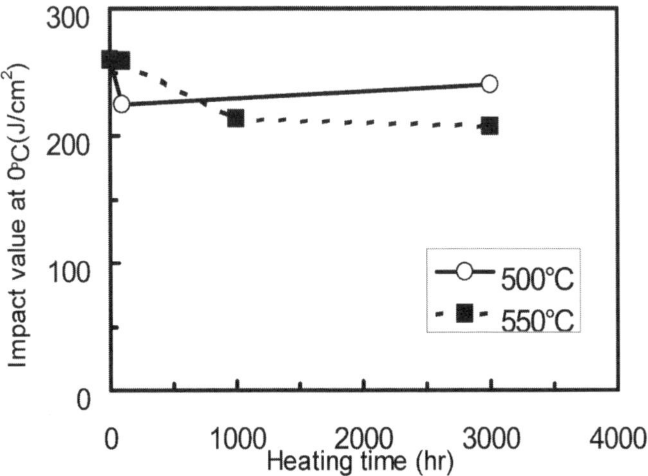

Figure 8. Charpy impact values after heating (Specimens: 5mm sub-size 2mmV-notch)

4.4 Steam and hot corrosion resistance

Fig.9 shows the comparison of steam resistance after 1000 hour heating between HCMA and conventional steels. The steam oxidation properties of HCMA are similar to those of conventional 1%Cr and 2%Cr steels. The oxide scale of HCMA is composed of relatively homogeneous two oxide layers. The internal layer was identified as Fe-Cr oxide with a spinal structure. The external layer is Fe oxide with a hematite structure.

Fig.10 shows the comparison of hot corrosion resistance after 20hour heating in synthetic ash between HCMA and conventional steels. The hot corrosion properties of HCMA are similar to those of conventional 1%Cr and 2%Cr steels.

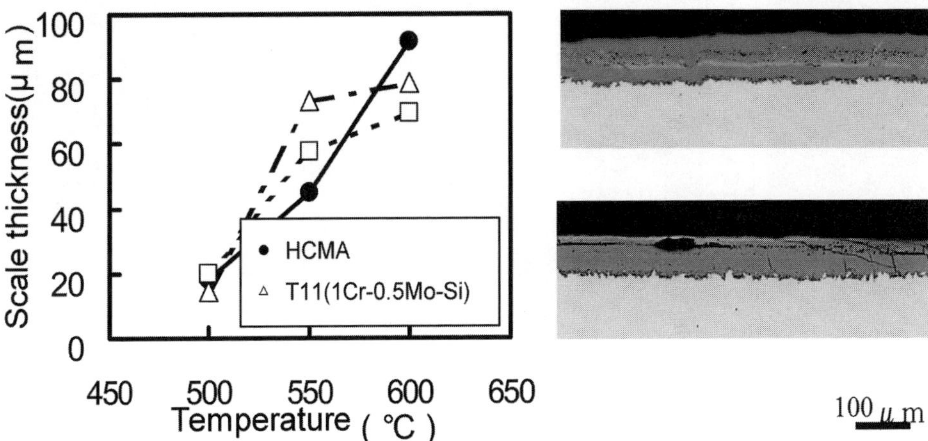

Figure 9. In steam oxidation resistance after 1000 hour heating

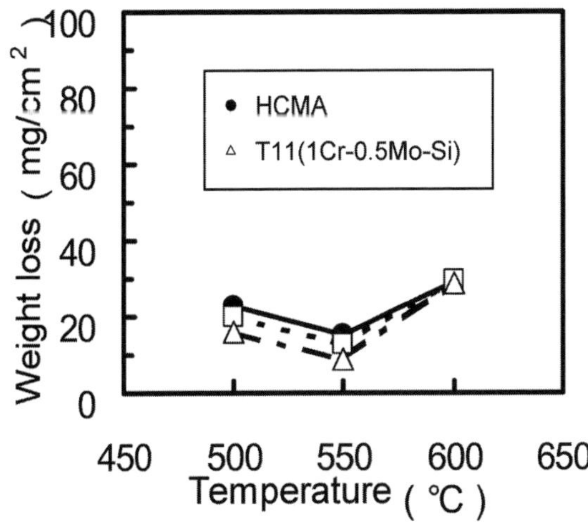

Figure 10. Hot corrosion resistance after in 20 hour heating in synthetic ash
Gas: 0.2%SO_2-5%O_2-15%CO_2-Bal.N_2
Ash: 37.5mol%Na_2SO_4-37.5mol%K_2SO_4-25mol%Fe_2O_3

4.5 Weldability

Welding test was conducted on the HCMA steel tube by using gas tungsten arc welding (GTAW). Chemical compositions of filler wire and welding conditions are shown in table 4 and in table 5 respectively.

Table 4: Chemical composition of the welding wire (mass%)

	Composition							
Welding consumable	**C**	**Si**	**Mn**	**Ni**	**Cr**	**Mo**	**W**	**Nb**
T-HCM2S	0.04	0.50	0.49	0.49	2.19	0.10	1.59	0.040

Table 5: welding conditions

Welding method	Layer	Welding current (A)	Welding voltage (V)	Welding speed (cm/min)	Heat imput (kJ/cm)	Pre-heating	Post weld heat treatment
GTAW	1 to 9	120 - 150	13 - 15	8 - 10	11.7 -13.5	None	None

Fig.11 shows the results of side bend tests. No defects such as weld cold cracking were observed after bending without preheating.

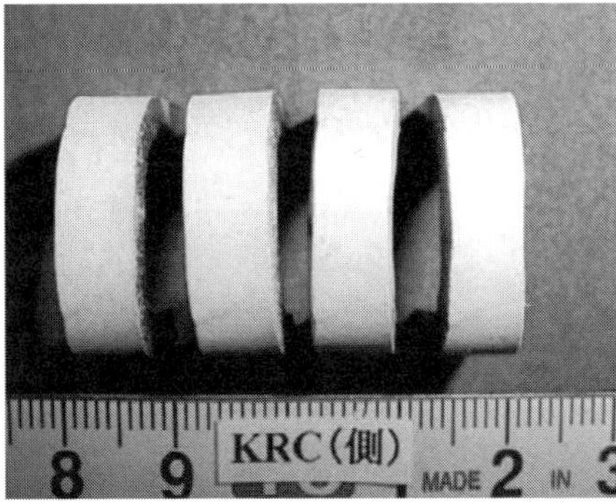

Figure 11. Appearance of the weld joints after the side-bend test

Fig.12 shows the tensile strength of the weld joints compared with those of base metals shown in Fig.2. The all test specimens were fractured at base metal and tensile properties of the weld joints were equal to those of base metals at room temperature, 450°C and 550°C.

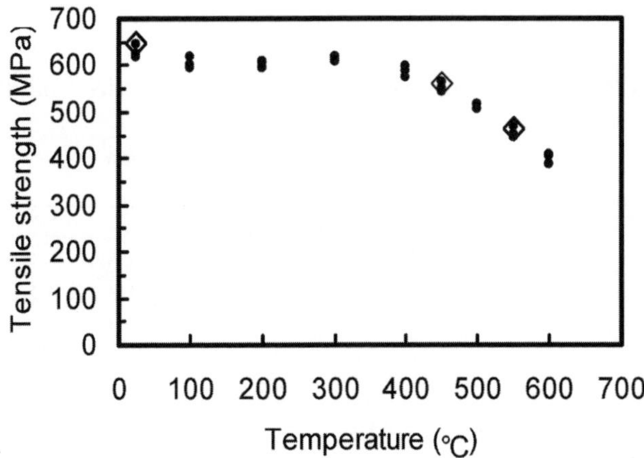

Figure 12. Tensile strength of the weld joints

Fig.13 shows the hardness distribution of the weld joint. The maximum hardness value in HAZ was below 350Hv without post weld heat treatment.

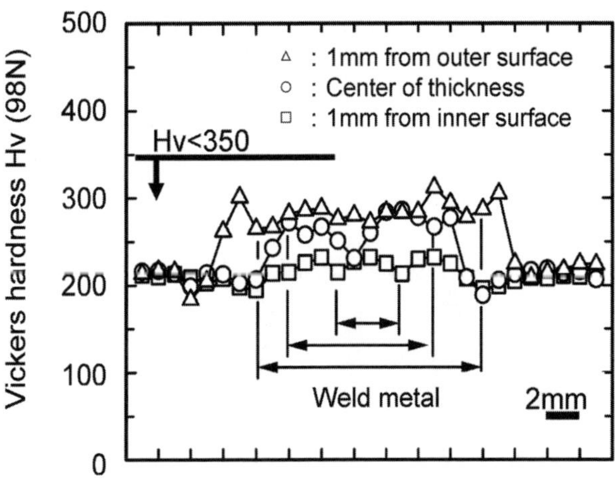

Figure 13. Hardness distribution of the welded joints

5. Conclusion

The new 1.25Cr-0.4Mo-Nb-V steel tube named "HCMA" has been developed. HCMA steel tube has high tensile strength at elevated temperatures. The allowable stress is dominated at 500°C and below by tensile property and at 500°C above by creep property. The allowable stress of HCMA steel tube is 1.3 times higher than those of conventional 1%Cr steel tubes in the temperature range of 430 to 530°C. The steam oxidation resistance and hot corrosion resistance have been proved to be the same level of conventional 1%Cr and 2%Cr steels. HCMA steel tubes have been installed and evaluated in a practical boiler for a long-time filed exposure test.

References

1. M. Ohgami et al.: *Tetsu to Hagane*, vol.76 (1990), p. 1124. [magazine article]
2. H. Yamada et al.: *THE THERMAL AND NUCLEAR POWER*, vol.52 (2001), p. 1217. [magazine article]
3. E. Miyoshi and T. Okada: *Tetsu to Hagane*, vol.11 (1965), p. 2110. [magazine article]
4. K. Tanosaki: *Tetsu to Hagane*, vol.64 (1978), p. 3603. [magazine article]
5. T. Yukitoshi and K. Nishida: *Tetsu to Hagane*, vol.59 (1973), p. 47. [magazine article]
6. K. Miyata and Y. Sawaragi: *ISIJ International*, vol.41 (2001), p. 281. [magazine article]

Advances in Materials Technology for Fossil Power Plants
Proceedings from the Fifth International Conference
R. Viswanathan, D. Gandy, K. Coleman, editors, p 140-152

DOI: 10.1361/cp2007epri0140

Experience with the Use of P91 Steel and Development of Tools for Component Integrity/ Life Assessment

Ahmed Shibli, David Robertson, European Technology Development, Surrey, UK
ashibli@etd1.co.uk

Abstract

Martensitic steel P91 with higher creep strength was first introduced as thick section components in power plants some 18 years ago. However, more recently a number of failures have been experienced in both thick and thin section components and this has given rise to re-appraisal of this steel. Thick section components are generally known to have failed due to Type IV cracking. Furthermore, due to the restructuring of the electricity industry worldwide many of the existing steam plant are now required to operate in cycling mode and this requires the use of materials with high resistance to *thermal fatigue*. Here high strength P91 is assumed to offer an additional benefit in that the *reduced section thickness* increases pipework flexibility and reduces the level of through wall temperature gradients in thick section components. Because of this envisaged benefit a number of operators/owners of the existing plant, especially in the UK, have been substituting these new higher strength steels for the older materials, especially when a plant is moved from base load to cyclic operation. There has also been a perceived advantage of *higher steam side oxidation resistance* of superheater tubes made from high Cr steels. For the Heat Recovery Steam Generators (HRSGs) used in Combined Cycle Gas Turbines (CCGTs) there is a requirement to produce compact size units and thus high strength steels are used to make smaller size components. This paper discusses these issues and compares the envisaged benefits with the actual plant experience and more recent R&D findings.

In view of these incidents of cracking and failures it is important to develop life assessment tools for components made from P91 steel. ETD has been working on this through a 'multi-client project' and this aspect will be discussed in this paper.

Keywords: **P91, T91, creep, cyclic operation, power plant, steels, integrity assessment**

1 Introduction

The 9Cr martensitic steels were developed as a result of the demand for ferritic steels with higher creep strength. To date, there has been a somewhat limited in-plant experience of their long-term performance, especially with thick section components such as headers and steam pipework. However, judging from earlier experience with lower grade ferritic alloys, and limited experience with these newer materials, the long-term performance of components fabricated from 9Cr martensitic steels needs careful observation. Indeed, a recent study conducted by European Technology Development [1] showed that there have been some surprising reports of premature failures of P91 weldments in plant, but as usual the causes of such failures have been complex. Nevertheless, one has to guard against the possibility that the situation for weldments could become more critical as the plants reach longer life, particularly for the plants that are operating in cyclic mode. The ETD study referred to above has shown that Type IV failure in welded structures of 9-12 Cr martensitic steels is now becoming a real concern for power plant operators. In the absence of any reliable plant monitoring or inspection techniques and lack of integrity/ life assessment tools and methodologies for this type of material, the concern is how to predict and prevent early and potentially violent, and in some cases life-threatening, failures. This paper discusses the effects of creep and fatigue on the medium to long-term performance of these steels and shows typical cases of plant failures. It further discusses the issues involved in inspecting and monitoring 9-12Cr martensitic steel components and in developing the integrity and life assessment tools.

2 Type IV cracking and thermal fatigue issues

The introduction in the 1950s of the 0.5CrMoV, 1.25Cr0.5Mo and 2.25Cr1Mo ferritic steels allowed an increase in steam temperature from about 450°C to 540-568°C with a corresponding increase in thermal efficiency. However, these materials were found to be susceptible to several forms of weld cracking. While some types of cracking were a function of welding and material properties, Type IV cracking (in the fine grain or inter-critical HAZ adjacent to the base metal) was a more persistent and life-limiting type. Type IV has in fact become a major maintenance issue for today's power plants using CrMoV type steels.

It was recognised in the 1980s that P91 martensitic steel, like all ferritic steels, is potentially susceptible to Type IV cracking [e.g. 2]. The maintenance consequences of Type IV cracking of the 1.25Cr0.5Mo and 2.25Cr1Mo ferritic steels did not become apparent for about a decade after these alloys were deployed. The deployment of P91 in utility power boilers as thick section components began first in the UK in about 1989/1990. Like the low alloy ferritic steels before it, P91 was introduced on the basis of the performance of its base metal strength and it is only in the last eight years or so that attention has moved to the performance of its weldments, the weakest link in the chain.

Market forces and competition now dictate that power plants are ‘cycled’ on a daily basis to follow load demands and maximise profits. This means that the materials used have not only to show satisfactory creep performance but they must also be tolerant to thermal fatigue. As in the case of creep, the solution is sought both in design and materials innovations and developments. Thus cycling is driving a redesign of existing and new plants and the use of new materials of construction. To avoid thermal fatigue, caused by frequent cycling, one solution is to utilise higher strength alloy steels for pressure vessel construction. Lower thickness components require less time to reach thermal equilibrium and are thus expected to be less vulnerable to thermo-mechanical damage. However, as will be shown later from a limited amount of research work, preliminary indications are that the Type IV region of the weldment may not perform as well as expected under the creep-fatigue interaction regime.

A survey of worldwide plant failures was carried out in the study by ETD [1]. This was a challenging task as plant operators are generally reluctant to discuss cracking incidents/failures. Nevertheless, some information was available from recently published literature while ETD managed to obtain some more from plant operators. This is detailed in [1], but some typical failures are described here.

2.1 Dissimilar metal weld failures

In the USA, a leak occurred in a main steam-line piping weld between P91 pipe and a 1¼Cr1Mo¼V control valve casting in a HRSG after less than 5000h operation [3], as shown in Figure 1. The components had been welded using 2¼Cr1Mo filler. Failure analysis showed that creep-dominated fracture had occurred through the narrow partially decarburised zone in the 2CrMo weld metal immediately adjacent to the fusion boundary on the P91 side of joint. During service, strain was concentrated in this weak zone. The investigators reported that welding and heat treatment had been properly performed, and that the thermal cycling of the plant played no significant influence in the failure. The investigation did not identify an axial stress of sufficient magnitude to cause such a premature failure, but there was an obvious design issue in that the 1½ inch wall thickness piping had been welded to the 3 inch thick valve body, which meant that the thinnest part of the 2CrMo weld metal was subjected to the same nominal axial stress as the P91 pipe. The recommended modification was to insert a conical P91 transition piece between the pipe and casting in order to reduce the axial stress on the 2CrMo weld metal to the same level as that seen by the 1CrMoV valve. The investigators also suggested that creep relaxation of system stresses had been hindered due to the high creep strength of P91 at the application temperature.

2.2 Type IV weldment failures

Six known incidents of cracking and failures occurred at the West Burton power plant in the UK. As usual, after a few years of confidentiality, information on these failures has been published and the details have been well documented [4,5,6,7]. Four of these failures occurred in 'bottle' type joints, and one in a header end cap, while advanced cracking was found in another header end cap at the same plant. The first bottle failure occurred after only 20,000 hours service, while the end cap failure occurred after a service of only 36,000 hours. In all cases, the nominal operating temperature was 565°C. It should be noted that these were the longest serving thick wall components in the UK and perhaps elsewhere, as the West Burton Power Station was the first one to get replacement components of P91 when this retrofit programme started in the UK in 1980s.

As [4,5] show the failure was attributed to overtempering of the P91 base material, which was inferred from the unusually low hardness values of the failed casts (although this could not be traced back to the heat treatment records due to their non-availability). Another contributing factor was low nitrogen to aluminium ratio (<2) and high aluminium content (>0.02 wt%, but within the ASME specification), which was thought to result in the formation of aluminium nitrides, leaving less nitrogen available for formation of the MX carbo-nitride creep strengthening precipitates. In the case of the header end cap, the design involving a sharp change of section near the weld also contributed to the cracking and failure. However, such a design had been successfully used in P22 vessels for a long time and was in accordance with the British Standards design specifications.

More recently, RWEnpower (UK) found extensive cracking on branch, stub and attachment welds on a P91 boiler superheater outlet header in one of their coal-fired plants

[8]. Over 100 stub welds were reportedly affected by the cracking. This was a retrofit header installed on a 500MW unit in 1992. The cracking had occurred after only 58,000 hours of service. The outlet steam conditions were 568°C and 165barg and it was reported that the boiler had operated within normal temperature and pressure conditions. Full inspection was undertaken on this header after a check of the manufacturing records had shown that low N:Al ratio casts were present. All investigated cracking was found to be Type IV and almost all cracks were reported to be on the P91 header barrel sides of the welds. The N:Al ratio in the affected casts ranged from 1.4 to 2.8, the lower ratio casts showing widespread cracking.

RWEnpower reported that no external system stresses were expected and the cracking was considered to be due to the hoop stress on the header [8]. The temperature distribution along the header varied but could not explain the cracking distribution, although it was expected to have some effect. It was concluded that the problem had occurred both due to overtempering and the low N:Al ratio. It is interesting, however, to note that a recent publication by E.ON-UK [9] states that their research shows that the creep strength of the parent metal has no bearing on the cross-weld rupture strength when failure occurs at the Type IV position.

2.3 Thermal fatigue cracking of tube-to-header welds

Thermal fatigue cracking of tube-to-header welds in HRSGs has been reported in the USA and elsewhere [e.g. 10, 11]. An example is shown in Figure 2. Cyclic operation of the HRSGs resulted in large cyclic bending stresses at the attachment welds, as a result of susceptible designs. Although the thermal fatigue cracking was due to the thermal cycling and design issues, there have been cases where the damage was exacerbated by the presence of soft zones in the tube material adjacent to the weld. These soft zones were produced as a result of overtempering during PWHT. The soft zones would also be expected to affect the creep performance of the welds in the longer-term.

3 Discussion of failure mechanisms and research findings

3.1 Creep rupture and Type IV failures

As is well known, the introduction of P91 to the power plant industry was based on the rupture strength of the base metal. Only short-term data were available at the start of the use of this steel. However, this situation is now changing. Some of the work carried out as early as 1980s showed that the cross-weld rupture strength of this steel fell much below that of the base metal. Recently, a number of other publications on 9-12Cr martensitic steels have confirmed this view [e.g. 12, 13]. Allen *et al* [12], through their work on E911 steel (9Cr-1Mo-1W) have shown that the weld strength reduction factor for this steel (tested at 625°C) could be as high as 40% when the failure was in the Type IV position. Extrapolation to 100,000 hours showed that the strength reduction factor could be 50%, which compares unfavourably with the 20% strength reduction factor for the traditional low strength P22 steel tested at 550°C [14].

With respect to the work from [12] quoted above, the criticism can be that tests carried out on cross-weld specimens were conducted at higher temperature (625°C) than the service temperatures, which in the case of replacement P91 material is between 540 to 568°C and in new plants in Denmark, Japan and Germany is in the region of 580 to 600°C. However, some work at 600°C carried out in the project SOTA [15] also showed that P91 specimens tested in creep crack growth mode were more vulnerable to Type IV cracking than P22 specimens tested at 550°C (in the temperature region at which 2.25Cr1Mo steel is used in power plant). Work carried out at 625°C on cross-weld specimens in the project HIDA [16] for test durations of up to 4,000 hours showed a strength drop, from base metal, of 35%. These tests had shown Type IV cracking for test durations of over 2,500 hours.

The UK experience with low alloy ferritic steels used in power plants shows that most of the Type IV failures in these steels appeared only after 50,000 hours of service, i.e. the Type IV failures were a medium to long-term service phenomenon. In the UK, where the experience with the use of thick section P91 has been the longest (over 60,000 hours to date), at least six failures and over 100 incidences of cracking have been experienced, some only after a service of 20,000 hours, in spite of the fact that the service temperature 568°C was only modest for this steel and the amount of material in service is also very low compared with the conventional ferritic steels.

3.2 Creep-fatigue cracking

The most common problem experienced as a result of plant cycling is thermal fatigue damage. This can manifest either in the form of cracking of an individual component or by the mechanical failure of structures. Cracking of a component is attributed to severe thermal gradients arising from excessive steam to metal and through-wall temperature differences associated with rapid rates of change of steam temperatures, as generally observed during start-up, shut down and load changes.

Materials behave in a complex way when both creep and fatigue mechanisms are present. The mechanisms usually act synergistically to cause premature failure. Creep strains can reduce fatigue life and fatigue strains can reduce creep life. The American Society of Mechanical Engineers (ASME) recognised the effects of interaction and provides guidance on the interaction between creep and fatigue and its effect on the life expectancies of materials (ASME Boiler and Pressure Vessel Code, Case N-47). Figure 3 (based on ASME N-47) demonstrates the interaction and consequences of creep and fatigue for P22 (2.25Cr1Mo).

The creep-fatigue interactions are not currently well-defined, and the limit line as shown by the 'elbow' line in Figure 3 represents the design life limit, expressed in fraction of material creep life and fatigue life. This limit line is used to establish the effect of combining the two mechanisms and demonstrates how they act together to exacerbate the effect of the individual processes. The line is a highly conservative representation of the phenomenon. It does, however, serve to demonstrate the effect of the interaction.

By way of an example, consider a component originally designed for say 10,000 cycles, which might have been designed to operate in a unit which two-shifts on a daily basis over 30 years. Assume also that the component operates in the creep range and was designed for 150,000 hours operation. If the unit were to operate on a base load regime, it will of necessity, accrue some thermal cycles, probably of the order of 1,000 over the projected life. The thick 'continuous base load operation' line indicates the effective operation of the component. The actual component life is given by the point where it intersects the 'elbow' line. This shows a reduction of the component life to about 75% of its predicted creep life. Similarly, if the component operates on a two-shifting unit with 300 cycles per year whilst operating in the creep range, the actual life may be as low as 40% of the anticipated fatigue life, as shown by the intersection of the 'continuous two shift operation' line. Where operational cycling is introduced on a former base load unit, it can be seen that the residual life can be greatly reduced to between 40% and 60% of the original design life due to the combined effects of creep and fatigue.

The key implication is that older units designed for base-load operation and used in this capacity over many years are very susceptible to component failure when they are eventually forced to cycle regularly. This logic obviously applies to components made from both the traditional ferritic steels and the new 9-12Cr martensitic steels. Thus, while increases in failure rates due to cycling may not be noted immediately, critical components will eventually start to fail. Shorter component life expectancies will result in higher plant EFOR (equivalent forced outage rate), longer scheduled outages, and/ or higher capital and maintenance costs to replace components at or near the end of their service lives. In addition, it may result in reduced total plant life or more capital to extend the life of the plant.

A relatively recent creep and high temperature fatigue crack growth study in the European Commission funded project HIDA [16] on welded P91 and P22 components has suggested that the creep-fatigue interaction (studied at 625°C) in P91 components containing welds could be even more severe than P22 (studied at 565°C). These tests were conducted at constant temperature but the load was cycled. These tests were carried out on large feature specimens (both circumferential and seam welded pipes) and on laboratory specimens of the fracture mechanics type and both have shown the same effect, i.e. that cycling can be more harmful to P91 than to P22. Full details of this work are described in a number of recent publications [17, 18]. The work on the P91 cross-weld rupture specimens tested in the HIDA programme has shown that when fracture occurs in the Type IV position the creep ductility of these specimens can be as low as 1 to 2%. This is much lower than the over 4% creep ductility of P22 cross-weld specimens tested at 565°C and failed at the Type IV position. It is thought that this may explain the more severe effect of creep-fatigue interaction on P91 weldments.

Mohrmann et al [19] carried out low cycle fatigue (LCF) tests with and without hold times at room temperature and 600°C on welded E911 (9Cr-1Mo-1W) specimens. These tests showed a decrease in lifetime by a factor of about two compared with the base material. The hold times further reduced the lifetime. Similarly, Bicego et al [20] carried out LCF tests at 600°C on cross-weld specimens of P91, P92 and E911 steels and found a similar life reduction of about 2 compared with the base metal.

The above studies show that: a) creep-fatigue interaction can have a more severe effect on 9Cr martensitic steel weldments than on the parent metal, and b) the creep-fatigue interaction effect on P91 weldments can be more severe than on P22 weldments. There is also independent evidence that the Type IV position in P91 can be vulnerable to cracking/failure. This work by Tabuchi et al [21] has shown that $M_{23}C_6$ precipitates and Laves phases form faster at the fine grain HAZ region in 9Cr martensitic type steels (compared with the other regions of the weldment) and this makes the Type IV position in these steels very vulnerable. This is because matrix-strengthening elements such as Cr, Mo and W are depleted from the matrix to form these precipitates, which are not as effective in imparting creep strength.

All that this means is that in cycling duty plant P91 thick section components are more likely to fail than in base load plant. From the known plant experience there is already a suspicion that this may be the case. However, a systematic database of failures now needs to be built up to confirm this or otherwise.

4. Life assessment issues

One problem with Type IV failure in welded high temperature steel structures has been that the cracking starts sub-surface and cannot always be detected by more economical and time-saving procedures such as MPI, metallographic replication, etc. The cracking only emerges on the surface near the last leg of its journey, thus making the structures potentially unsafe. P91 steel has further problems in that unlike the low alloy ferritic steels operating in the creep regime, the more easily visible changes such as spheroidisation and break down of the microstructure at the scale that can be seen under an optical microscope does not occur in P91. This then means that the much more costly transmission electron microscopy (TEM) has to be used to identify changes in the microstructure. Even then the changes may not be so obvious to a non-specialist, as it is mainly the changes in the precipitates and dislocation density that adversely affect the life of these steels.

However, recent systematic research and study, particularly in Europe and Japan, of the inspection and monitoring of component microstructural damage, cavitation, cracking and failure are giving indications of potential new inspection, monitoring and integrity assessment techniques that may be used successfully on 9-12Cr steels. ETD has surveyed these studies in depth for its recently completed project on 'Plant and R&D Experience with the Use of New Steels' [1], and has now launched a new Group Sponsored Project aimed at testing, validating and verifying the more promising NDE techniques and the relationship of the damage detected to the component life. The new project will then make recommendations in 2008 for the most suitable techniques and parameters needed for their successful application for early stage damage detection, component monitoring, inspection and integrity/ life assessment. Some of the recent NDE developments are discussed below.

4.1 Use of Atomic Force Microscopy for on-site cavitation measurement

Recent Japanese studies [22] of systematic changes in the creep cavitation behaviour of P91 have revealed that it may be possible to use cavitation observed under a scanning electron microscope (SEM), or even under a light optical microscope (LOM) to estimate the life consumption rate (LCR). Figure 4 by these authors shows the number density of cavities as a function of creep life consumed. ETD are studying at present the use of a portable Scanning Force Microscope (Figure 5) for on-site measurement of cavitation to the magnification and resolution comparable with SEM.

4.2 Use of extraction replicas and TEM for component life assessment

The observation of microstructural damage is widely used in inspection, monitoring and life management of low alloy ferritic steels employed in high temperature plant. The damage evolution is material-dependent, and requires confirmation from inspection data (replicas, dye penetrant, ultrasonic, etc). For most low-alloy steels, compilation of inspection data has been used to establish guidelines for life management. However, in the case of P91 the study of microstructural deterioration is not so convenient as, unlike the low alloy steels, no spheroidisation of the microstructure takes place. The high creep resistance of these steels is based on their martensitic microstructure, carbide and nitride precipitates, dislocation (subgrain) structure and Laves phase. Precipitates like $M_{23}C_6$ and MX, i.e. V(C, N), form during the standard heat treatment, while precipitates like M_2X and Laves and Z phases form during service at high temperature. $M_{23}C_6$ coarsens at service temperature while the fine MX precipitates are much more stable at typical service temperatures. The presence of these features could be used as a guideline to determine the safe life. This could be established by means of TEM and extraction replicas applied to the components. Material for examination can also be obtained from very small specimens (~1mm thickness) that can be cut out of thick section components almost 'non-destructively' using practically non-invasive techniques.

4.3 Detection of creep damage using ultrasonic inspection

Recently, creep cavitation in uniaxial creep tests and creep crack growth tests on 9-12Cr steels was investigated using an intelligent phased array ultrasonic inspection system [23]. The system was found to be capable of detecting the formation of cavities at the crack tip and a crack length of about 0.1mm. The correlation between the detected and actual crack length (as measured destructively) showed excellent agreement. These researchers found the phased array ultrasonic system to be effective in detecting cracks and cavities in creep related tests with acceptable accuracy.

4.4 Hardness monitoring as a life assessment tool

In systematic research carried out recently, creep test specimens were subjected to hardness measurements and a new hardness-based method was proposed for the life assessment of the HAZ of P91 welded joints [22]. The hardness in the crept specimens of both base metal and weldments changed in a similar manner with creep time, and dropped dramatically in the later stage of creep life. This work also showed a linear relationship between hardness and creep life fraction of 0.2 to 0.9 for the base metal and weldments. The hardness technique employing a statistical approach was proposed for non-destructive creep life assessment method for P91 welded joints. This can be particularly helpful as the portable hardness testers are liable to a bigger margin of error than the conventional laboratory based testers. However, together with the Japanese collaborators ETD are now looking at the refinement of the portable hardness testers (Figure 6) for more precise and reliable measurement of in-situ hardness.

5. Conclusions

Both the research findings and the plant experience so far show that the Type IV weldment region of the 9Cr martensitic steels (P91, P92) may be prone to cracking when used under creep and especially under creep-fatigue plant cyclic conditions. This vulnerability to cracking may be similar to, or even higher than, that experienced in the past in the case of lower alloy steels. The significantly higher strength of this steel allows components to be made to reduced wall thickness and this has been considered to be beneficial for its use in cycling plant due to lower thermal gradients across the wall. However, preliminary indications are that the creep-fatigue interaction effect on the welded structures made from this steel type may have a more adverse effect than on the traditional low alloy steels used in power plant boilers. This means that greater attention needs to be paid to inspection and monitoring of components made from these steels. In this regard, there are a number of NDE techniques that show promise, although further testing, validation and verification are required which is now being conducted in a multi-client or group sponsored project.

References

[1] Shibli I A, Robertson D G, "Review of the use of new high strength steels in conventional and HRSG boilers: R&D and plant experience", ETD Report No: 1045-gsp-40, October 2006.

[2] Middleton C, Metcalfe E, "A review of laboratory Type IV cracking data in high chromium ferritic steels", Paper C386/027, Published in IMechE Proceedings, London, 1990.

[3] Henry J F, Fishburn J D, "Investigation of leak in a main steamline piping weld joining P91 piping to a 1.25Cr1Mo0.25V control valve using 2¼Cr-1Mo filler metal: Causes and implications for the use of new high strength steels", Proc. Intl. Seminar "Cyclic Operation of Heat Recovery Steam Generators", 24 June 2005, London, UK, Published by European Technology Development, etd@etd1.co.uk

[4] Brett S J, Allen D J, Pacey J, "Failure of a modified 9Cr header endplate", Proc. Conf. "Case Histories in Failure Investigation", Milan, Sept. 1999, pp.873-884.

[5] Allen D J, Brett S J, "Premature failure of a P91 header endcap weld: minimising the risks of additional failures", Proc. Conf. "Case Histories in Failure Investigation", Milan, Sept. 1999, pp.133-143.

[6] Brett S J, "Identification of weak thick section modified 9Cr forgings in service", Proc. Swansea Creep Conference, Swansea, UK, April 2001.

[7] Brett S J, "The creep strength of weak thick section modified 9Cr forgings", Proc. Conf. "Baltica V", Vol.1, June 2001.

[8] Brett S J, "In-Service Type IV Cracking in a Modified 9Cr (Grade 91) Header", Proc. ECCC Creep Conference, London, Sep. 2005, Proceedings published by ETD Ltd. etd@etd1.co.uk

[9] Allen D J, Harvey B, Brett S J, "FOURCRACK - An investigation of the creep performance of advanced high alloy steel welds", Proc. ECCC Creep Conference, Sep. 2005, Proceedings published by ETD Ltd. etd@etd1.co.uk

[10] Pearson M, Anderson R W, "Measurement of damaging thermal transients in F-Class horizontal HRSGs", Proc. Intl. Seminar "Cyclic Operation of Heat Recovery Steam Generators", 24 June 2005, London, UK, Published by European Technology Development, etd@etd1.co.uk

[11] Lutfi bin Samsudin M, "Experience with HRSG T91 tube failures", Proc. Intl. Seminar "Industry & Research Experience in the Use of P91/T91 in HRSG and Conventional Boilers", 7-8 Dec 2005, London, UK, Published by European Technology Development, etd@etd1.co.uk

[12] Allen D J, Fleming A, "Creep performance of similar and dissimilar E911 steel weldments for advanced high temperature plant", Proc. 5th Charles Parsons Conference on "Advance materials for 21st century turbines and power plant", 3-7 July 2000, Churchill College, Cambridge, UK, pp.276-290.

[13] Orr J, Buchanan L W, Everson H, "The commercial development and evaluation of E911, a strong 9% CrMoNbVWN steel for boiler tubes and headers", Published in the Proceedings of the Conference "Advanced heat resistant steel for power generation", Editors: R Viswanathan, J Nutting, San Sabastian, Spain, Publisher: IOM Publications, London, UK.

[14] Etienne C F, Heerings J H, "Evaluation of the influence of welding on creep resistance (strength reduction factor and lifetime reduction factor)", IIW doc. IX01725-93, TNO, Appledoorn, The Netherlands.

[15] Shibli I A, "Creep and fatigue crack growth in P91 weldments", Proceedings of the Swansea Creep Conference, Swansea, UK, April 2001.

[16] Shibli I A, "Overview of the HIDA project", Proc. 2nd International HIDA Conference on "Advances in Defect Assessment in High Temperature Plant", 4-6 Oct. 2000, Stuttgart, Germany.

[17] Shibli I A, Le Mat Hamata N, Gampe U, Nikbin K, "The effect of low frequency cycling on creep crack growth in welded P22 and P91 pipe tests", Proc. 2nd International HIDA Conference on "Advances in Defect Assessment in High Temperature Plant", 4-6 Oct. 2000, Stuttgart, Germany.

[18] Le Mat Hamata N, Shibli I A, "Creep crack growth of seam-welded P22 and P91 pipes", Proc. 2nd International HIDA Conference on "Advances in Defect Assessment in High Temperature Plant", 4-6 Oct. 2000, Stuttgart, Germany.

[19] Mohrmann R, Hollstein T, Westerheide R, "Modelling of low-cycle fatigue behaviour of the steel E911", Published in "Materials for advanced power engineering 1998", Proceedings of the 6th Liege Conference, Volume 5, Part 1.

[20] Bicego V, Bontempi P, Mariani R, Taylor N, "Fatigue behaviour of modified 9Cr steels, base and welds", Published in "Materials for advanced power engineering 1998", Proceedings of the 6th Liege Conference, Volume 5, Part 1.

[21] Tabuchi M, Watanabe T, Kubo K, Matsui M, Kinugawa J, Abe F, "Creep crack growth behaviour in HAZ of weldments for W containing high Cr steel", Proc. 2nd International HIDA Conference on "Advances in Defect Assessment in High Temperature Plant", 4-6 Oct. 2000, Stuttgart, Germany.

[22] Masuyama F, "Integrity and life assessment of P91 components", Proc. Intl. Seminar "Industry & research experience in the use of P/T91 in HRSGs/boilers", Published by European Technology Development, etd@etd1.co.uk

[23] Chan-Seo Jeong, Si-Yeon Bae, Byeong-Soo Lim, "The effect of cavity on the creep fracture in P92 and P122 steels and its detection by intelligent phased array ultrasonic", School of Mechanical Engineering, Sungkyunkwan University, Kyonggi-do, Korea.

Fig. 1: Cracking of 2Cr weld metal adjacent to the weld toe on P91 side of joint [3].

Fig. 2: Thermal fatigue failure at tube-to-header weld [11]; detail of oxide-filled thermal fatigue crack close to failure location.

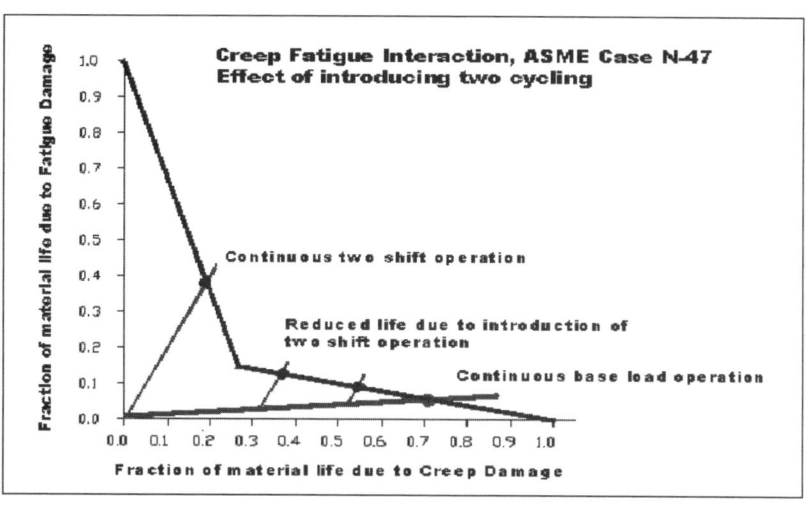

Fig. 3: Interaction and consequences of creep and fatigue (based on ASME N-47) for a typical power plant steel (2.25Cr1Mo).

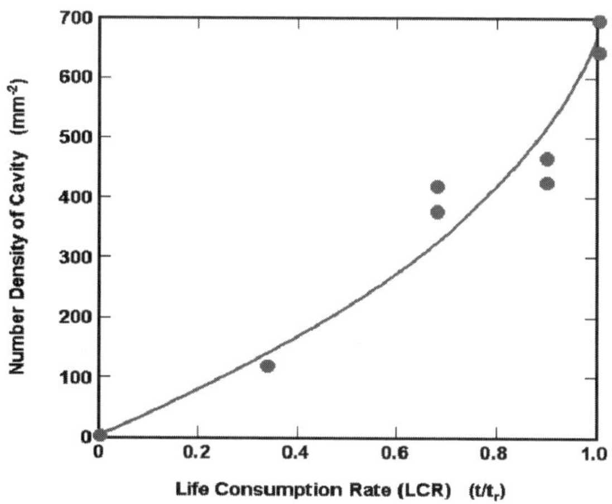

Fig. 4: Number density of cavity versus life (LCR) of T91 tube base metal [22].

Fig. 5: Scanning Force Microscope. From top left hand (clockwise) – Lab version, portable version, portable version hand held, portable version in use on a pipe.

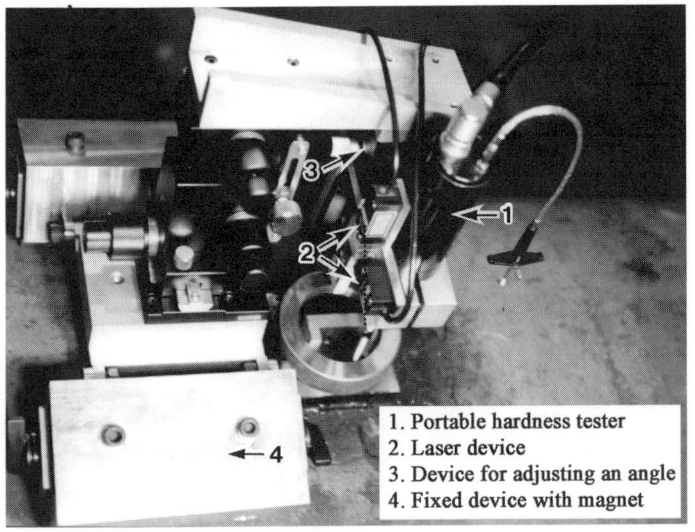

Fig. 6: Laser guided portable hardness tester for precise measurement

Advances in Materials Technology for Fossil Power Plants
Proceedings from the Fifth International Conference
R. Viswanathan, D. Gandy, K. Coleman, editors, p 153-167

DOI: 10.1361/cp2007epri0153

INVESTIGATING THE STRAIN LIMITS OF COLD FORMED, HIGH TEMPERATURE AUSTENITIC MATERIALS FOR FABRICATING USC BOILER COMPONENTS

Walt Mohn
The Babcock & Wilcox Company
20 South Van Buren Avenue
Barberton, OH 4203-0351

Greg Stanko
Foster Wheeler Development Corporation
12 Peach Tree Hill Road
Livingston, NJ 07039

Abstract

The construction of highly efficient Ultra Supercritical (USC) boiler systems to operate with steam temperatures up to 760°C (1400°F) and with steam pressures up to 34.5 MPa (5000 psi) will require the use of advanced high temperature, high strength materials. As part of a 5-year project to qualify advanced boiler materials for USC power plants, a number of austenitic materials have been selected for further development and use in USC boiler systems, including alloys 230, 740, CCA 617, HR6W, and Super 304H. In one task of this project, boiler fabrication guidelines appropriate for the use of these alloys were investigated. Because it is recognized that cold formed and mechanically strained austenitic materials can degrade in material creep strength, a study to investigate the limits of strain and temperature exposure for the USC alloys was undertaken. An objective of this work was to determine for each USC alloy a relationship between the level of cold strain and the conditions of time and temperature that will cause recrystallization and significant microstructural change. The ultimate goal of this work was to determine limits of strain, due to cold forming, that can be tolerated before heat treatment is required, similar to those limits provided for the austenitic materials (e.g., 300-series stainless steels, alloy 800H) in Table PG-19 in Section I of the ASME Boiler and Pressure Vessel Code. This paper will describe the technical approach for 1) preparing specimens having discrete cold strains ranging from about 1 to 40 percent, 2) exposing these strained specimens for selected times at various elevated temperatures, 3) identifying the on-set of recrystallization in the microstructures of the exposed specimens, and 4) establishing a useful engineering method to predict conditions for the on-set of recrystallization in the USC alloys using the experimental results.

Introduction

The fabricability of an alloy is generally understood to mean the relative ease that it can be formed, machined, joined, heat treated, or otherwise processed to produce a component with the desired shape, properties and characteristics. In fabricating alloys for the construction of fossil-fueled commercial utility boilers, the effects of various metalworking processes on alloy microstructure and properties can have a significant impact on boiler design and configuration, operating service life, total cost, and reliability.

Cold forming operations performed during the manufacture of austenitic alloy pressure parts, such as superheater tube U-bends produced by the Babcock & Wilcox Company (B&W), as shown in Figure 1, may cause impaired service performance when the component operates in the creep range, typically above 538°C (1000°F).

Figure 1 Cold formed alloy 230 U-bends (with respective outer fiber strains of 33.3%, 20%, and 13.3%) produced from 50.8 mm OD X 10.2 mm MW tubes in the Babcock & Wilcox Company's (B&W) manufacturing facility in Cambridge, Ontario Canada.

At such temperatures, recrystallization to a finer grain size can lead to an increase in creep rate and a decrease in rupture strength. The result can lead to premature failure of the cold formed component. However, heat treatment at specified temperatures after cold forming, which can significantly increase manufacturing costs, will restore the desired properties of the material and will minimize the threat of premature failure of the component due to recrystallization during service. The major variables affecting the recrystallization kinetics are 1) amount of cold work, 2) temperature, 3) time, and 4) alloy composition.

The commercial alloys that had initially been considered for use in USC power plant fabrication were chosen principally on the basis of their elevated temperature properties and their resistance to hot corrosion and steam oxidation. While it was recognized that the austenitic alloys selected for USC boiler construction could be produced by commercial suppliers with the desired properties in the form of bar, plate, and tubing, it was still necessary to investigate their fabricability to ensure that subsequent processing (e.g., welding and heat treatment, hot and cold bending) used in the manufacture or repair of USC boiler systems would not degrade those properties. Accordingly, an investigation in the USC Boiler Materials Development project was conducted to evaluate the fabricability of the high-temperature, corrosion-resistant alloys selected for use in building USC power plants. As part of this effort, the response of the USC alloys to mechanical deformation and subsequent thermal treatment was systematically explored. Determining the conditions which cause the on-set of recrystallization in each USC alloy is of great interest in the fabrication of boiler components, as recrystallization has long been known to have a significant affect on creep properties of austenitic materials. Very fine austenitic grain sizes that typically develop as recrystallization progresses can cause a significant creep-weakened condition for these alloys. This phenomenon is more fully described in the Nonmandatory Appendices (Appendix A-370) of ASME Section II Part D.[1] Hence, it is important to understand and determine limits of strain, due to cold forming, that can be tolerated before heat treatment is required, similar to those limits provided for the austenitic materials (e.g., 300-series stainless steels, alloy 800H) in Table PG-19 in Section I of the ASME Boiler and Pressure Vessel Code.[2]

In addition to studying the effects of recrystallization, the precipitation of carbides, gamma prime and other phases were also characterized for the USC alloys which exhibit this behavior. It was recognized that understanding the formation of such phases was important since, for precipitation strengthened alloys (such as alloy 740) and for some solution strengthened alloys (such as alloy 230), strain due to cold working can accelerate the aging reaction, causing not only the early appearance of precipitates, but also affecting their size and distribution.[3] This response could degrade the tensile and creep properties of a boiler pressure component, leading to a reduction in design service life.

Technical Approach – Recrystallization Studies

Although the effects of recrystallization on the microstructure and properties of the USC alloys have not yet been fully characterized, it was assumed for this analysis that recrystallization in these materials would have similar deleterious effects on their performance as has been observed in other austenitic alloys. Accordingly, the on-set of recrystallization in each USC alloy microstructure was established as a conservative criterion in this study for determining the limits of cold strain that can be tolerated without heat treatment.

It is well known that recrystallization, like creep, is a diffusion-controlled process with an Arrhenius relationship that is dependent on both time and temperature exposure and which is a function of the amount of strain induced in the material during cold working operations.[4] Accordingly, the Larson-Miller Parameter (LMP) can be used to correlate the level of cold strain with the conditions of time and temperature that will cause the on-set of recrystallization.[5] In this work, specimens of USC materials with a known amount of strain were measured for hardness and then thermally exposed at various selected temperatures and times that represent both post-fabrication heat treatment options, as well as boiler operating conditions. Following exposure, the hardness of these specimens was again measured, and the microstructure was examined and characterized to determine the thermal conditions for the on-set of recrystallization. For different levels of cold strain, temperature/time combinations experimentally determined to cause the on-set of recrystallization were used to calculate the LMP. Cold strains associated with conditions that caused recrystallization were then plotted against the LMP. For strain levels of 15%, 20%, 30%, and 40%, tables showing different predicted time and temperature combinations to cause the on-set of recrystallization were also prepared.

To produce alloy test specimens with a known amount of cold strain, tapered tubes were machined from solution annealed USC tube materials (nominally 50.8 mm OD X 10.2 mm MW) that were procured for the project. Controlled straining of the tapered tube specimens, conducted in a load frame by Foster Wheeler Development Corporation (FWDC), shown in Figure 2, produced materials with engineering strains typically ranging from 0.1% to about 40% without necking. Graduated fiduciary marks initially scribed on the surface of the tapered tube specimens, as shown in Figure 3, provided indices for the accurate measurement of strains at locations throughout the gage length, as shown in Figure 4.

Following controlled straining, each tapered tube was carefully cut longitudinally into eight identical strips. These strips were, in turn, cut in half at the center (the maximum strain location) to produce sixteen thermal exposure specimens having strains along their length ranging from 0.1% to about 40%. Strained specimens of each USC alloy were then heated in a furnace at selected temperatures and times. After thermal exposure, the longitudinal cross-sectional microstructures of specimens were metallographically prepared and microscopically examined. The strain corresponding to the onset of recrystallization could then be determined by position along the length of the exposed specimen. The longitudinal cross-sectional microstructure of solution annealed alloy 230 having very little strain (0.1%) is shown in Figure 5. The longitudinal cross-sectional microstructure of alloy 230 having about 41.8% is shown in Figure 6.

Figure 2 **Controlled straining of the tapered tensile specimens, conducted in a load frame by Foster Wheeler Development Corporation (FWDC), produced materials with engineering strains typically ranging from 0.1% to about 40% without necking.**

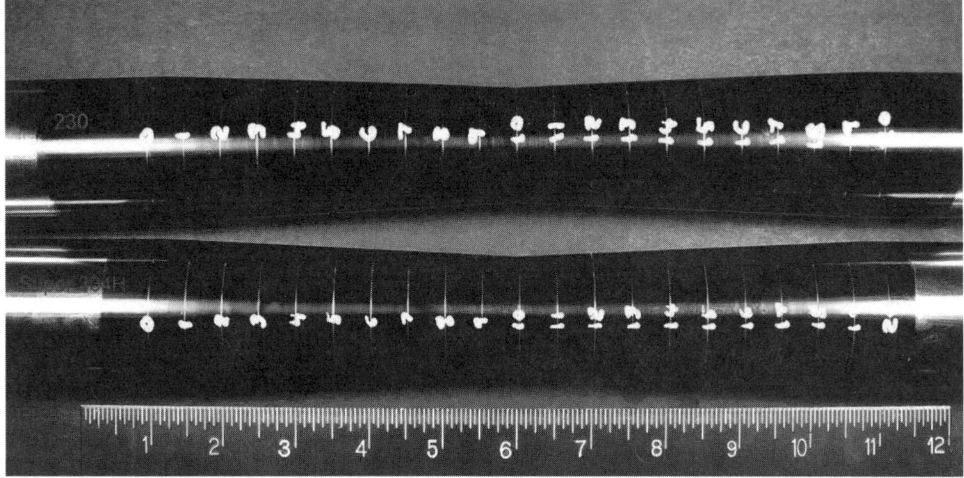

Figure 3 **Tapered tube specimens with scribed fiduciary marks as they appeared before controlled straining at FWDC.**

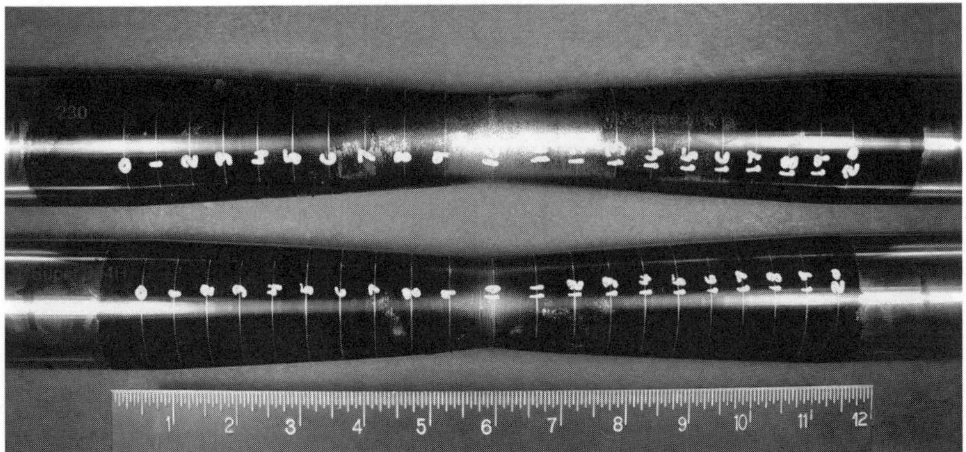

Figure 4 **Tapered tube specimens after straining in a load frame to produce materials with strains ranging from 0.1 % (at the extremities) to about 40% (at the center).**

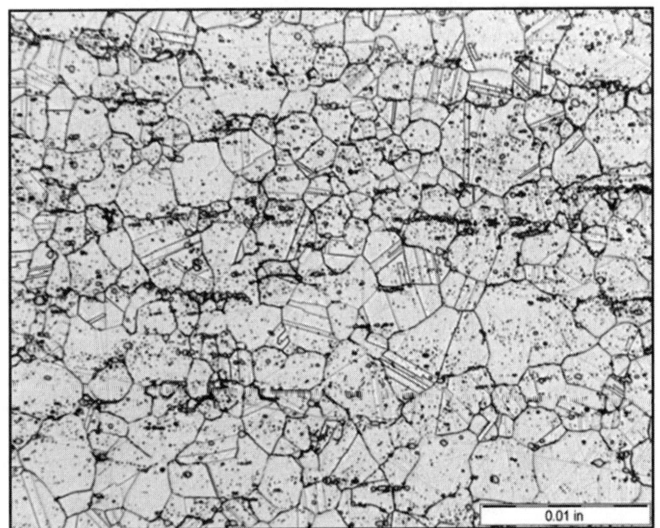

Figure 5 **Longitudinal cross-sectional microstructure of an alloy 230 specimen taken from the outer extremity of the solution annealed tapered tube after controlled straining. Strain at that location was calculated to be 0.1%. Average microhardness at that location was measured to be 94 HRB (205 HB).**

Etchant:
10 ml nitric acid
10 ml acetic acid
15 ml hydrochloric acid
2-5 drops glycerol

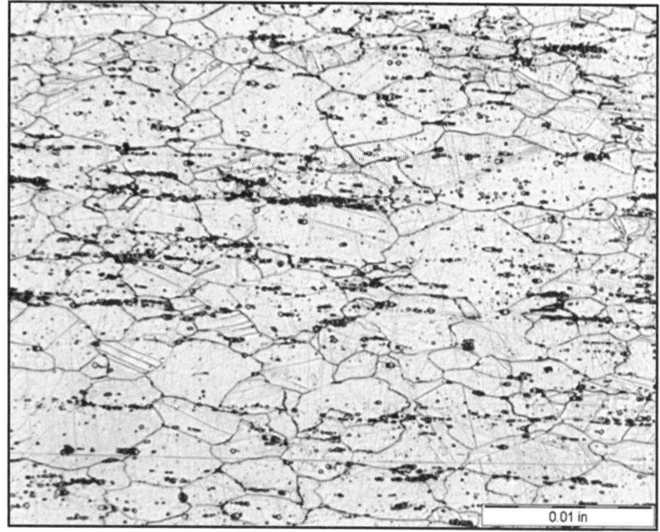

Figure 6 **Longitudinal cross-sectional microstructure of an alloy 230 specimen taken from the center of the solution annealed tapered tube after controlled straining. Strain at that location was calculated to be 41.8%. Average microhardness at that location was measured to be 40 HRC.**

Etchant:
10 ml nitric acid
10 ml acetic acid
15 ml hydrochloric acid
2-5 drops glycerol

Results and Discussion

Although recrystallization studies were conducted for all materials selected for the USC Project, results only for alloy 230 are presented in this paper. The methodology for analyzing and assessing the results for the other alloys, including 740, CCA 617, HR6W, and Super 304H was similar to that employed for the analysis and assessment of alloy 230. Complete results of the recrystallization studies for all USC alloys will be published elsewhere.

The microstructures of two strained alloy 230 specimens, each exhibiting characteristics of the on-set of recrystallization resulting from separate thermal exposures, are shown in Figures 7 and 8. Table 1 presents the six different thermal exposures (temperatures and times) that were conducted in this study, along with the corresponding strain levels at which the on-set of recrystallization was determined through microstructural examination. For each recorded strain level in Table 1, the LMP was calculated (see equation 1) using the corresponding exposure temperatures and times. For these calculations, a Larson-Miller constant of C=19.7 was used, an optimized value previously determined for alloy 230 in long-term creep studies.[6]

$$LMP = T_k (C + \log t) = (273 + T) (C + \log t) \quad (1)$$

Where T_k = Absolute Temperature in kelvins = 273 + Exposure Temperature T in degrees Celsius

C = optimized Larson-Miller constant determined for alloy 230 = 19.7

t = time of thermal exposure in hours

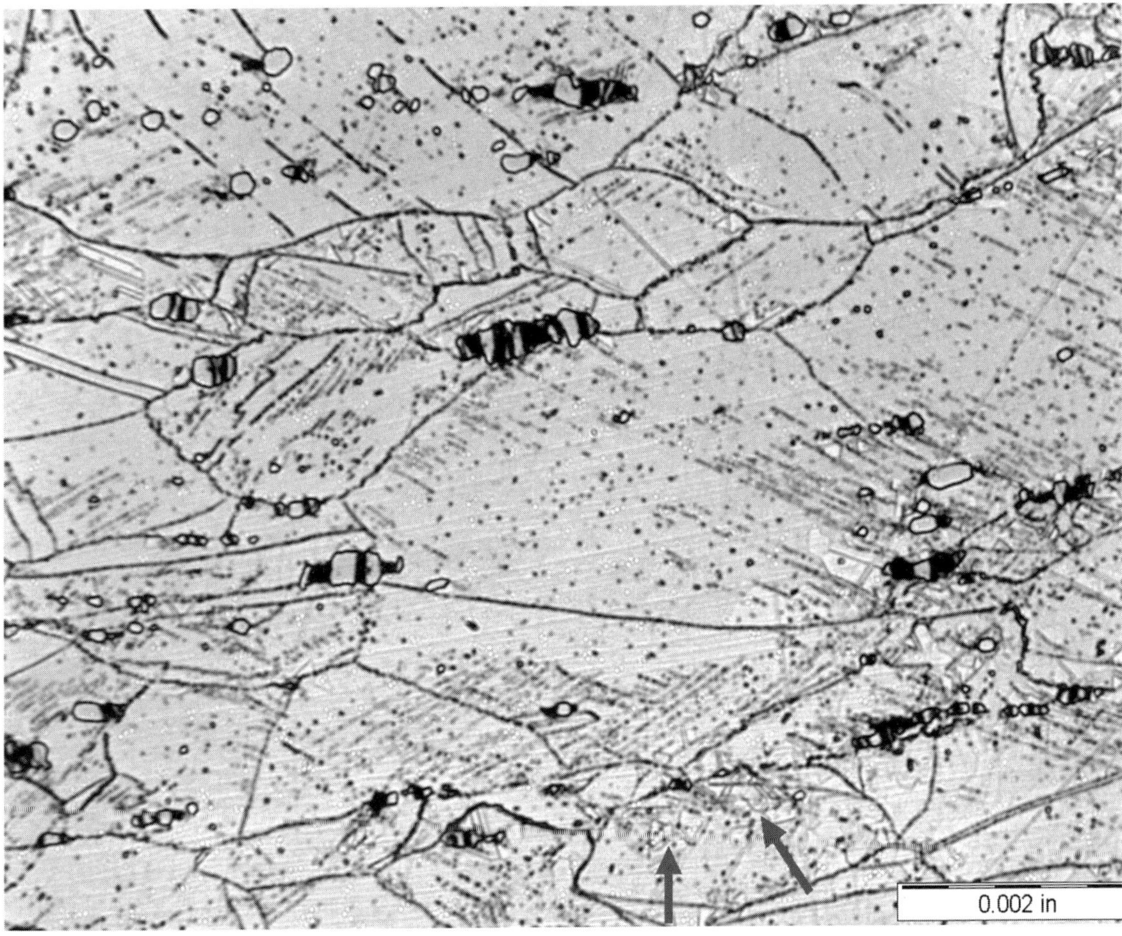

Figure 7 **Longitudinal cross-sectional microstructure of an alloy 230 specimen with cold strain of 41.8% after thermal exposure at 1500°F for 100 hours. Arrows point to features that indicate the initiation of recrystallization. Measured hardness was 35 HRC (327 HB).**

Etchant:
10 ml nitric acid
10 ml acetic acid
15 ml hydrochloric acid
2-5 drops glycerol

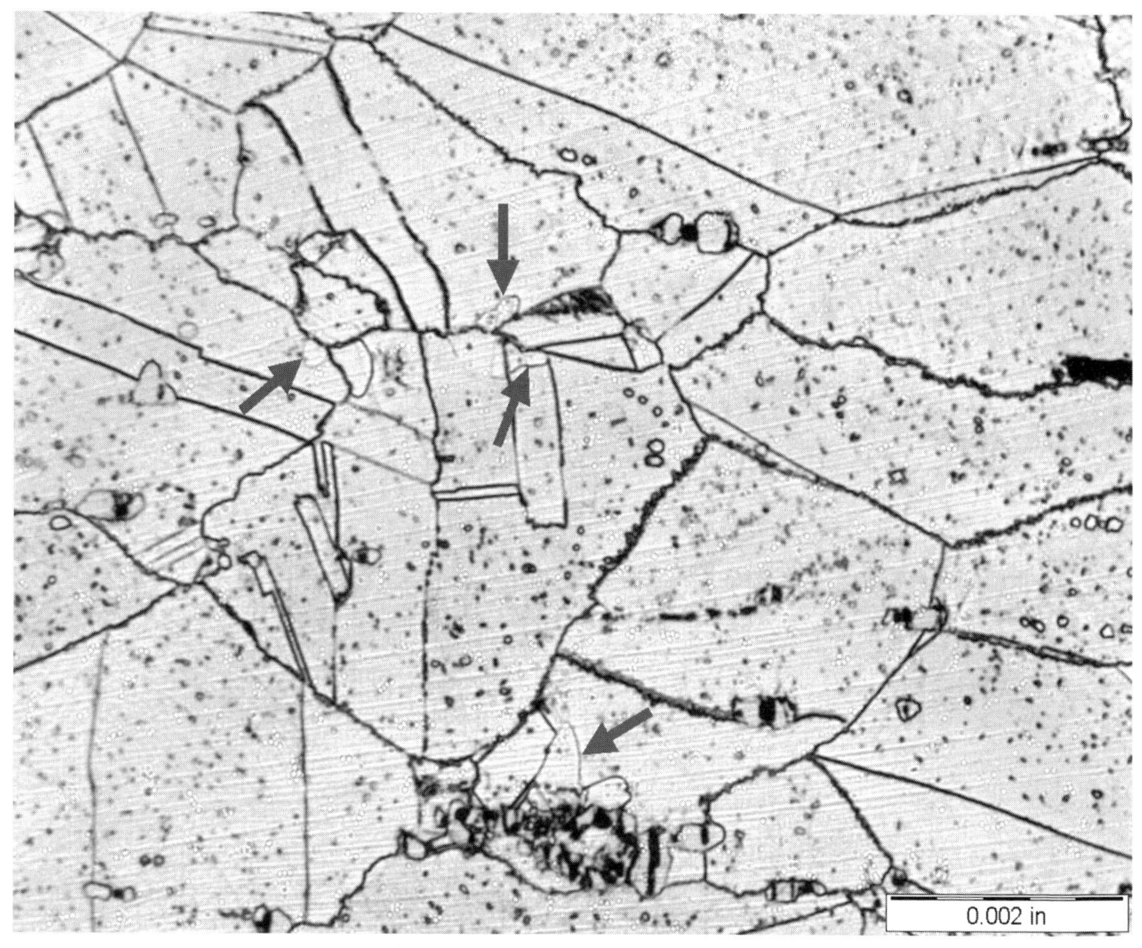

Figure 8 **Longitudinal cross-sectional microstructure of an alloy 230 specimen with cold strain of 20.5% after thermal exposure at 1500°F for 700 hours. Arrows point to features that indicate the initiation of recrystallization. Measured hardness was 31 HRC (294 HB).**

Etchant:
10 ml nitric acid
10 ml acetic acid
15 ml hydrochloric acid
2-5 drops glycerol

TABLE 1

Summary of Thermal Exposures and Cold Strain Levels at which Initial Recrystallization was Apparent in the Microstructure with Corresponding Calculations of Larson-Miller Parameter

Alloy 230 Thermal Exposure	Cold Strain Level (%) Corresponding with the On-set of Recrystallization after Thermal Exposure	Calculated Larson-Miller Parameter X 10^{-3} (kelvins, C=19.7)
816°C (1500°F) – 100 hours	41.8	23.631
816°C (1500°F) – 700 hours	20.5	24.552
843°C (1550°F) – 124 hours	25.0	24.321
871°C (1600°F) – 100 hours	20.5	24.824
871°C (1600°F) – 1000 hours	15.5	25.969
899°C (1650°F) – 12.5 hours	23.6	24.374

Using Microsoft Excel, the values of strain corresponding to the on-set of recrystallization in Table 1 were plotted on a semi-logarithmic scale against the calculated values of the LMP. A second-order polynomial curve, defined by Equation 2, was then fitted to the data points, as shown in Figure 9.

$$y = 7.3308x^2 - 374.52x + 4797.9 \tag{2}$$

Where y = strain level to cause the on-set of recrystallization = ε_{Rx}
x = Larson-Miller Parameter (LMP)

In a practical sense, the equation and curve relate a range of strains (ε_{Rx}) - corresponding to the on-set of recrystallization - to the LMP. In this analysis, the cold strains that are defined by the curve extend from about 14% to 42%. Thus for any given ε_{Rx} within this range, the corresponding LMP can be determined and used to calculate the time necessary to cause the on-set of recrystallization for any selected exposure temperature.

It should be noted that the R^2 value for the polynomial curve fit, shown in the upper right-hand corner of Figure 9, is actually the square of the correlation coefficient. The correlation coefficient, R, gives a measure of the reliability of the fitted curve's relationship between the *x* and *y* values. A value of R = 1 indicates an exact relationship between *x* and *y*. Values of R close to 1 indicate excellent reliability. If the correlation coefficient is relatively far away from 1, the predictions based on the fitted curve will be less reliable.

Using Equation 2 for the fitted curve in Figure 9, values of the LMP were calculated for recrystallization strain levels of 15%, 20%, 30%, and 40%. For the on-set of recrystallization to occur at 15% cold strain, the LMP was calculated to be 25,280. Using this value of the LMP and a constant C=19.7, the time for on-set of recrystallization to occur at 15% cold strain was calculated for five selected exposure temperatures, including 648°C (1200°F), 704°C (1300°F), 732°C (1350°F), 760°C (1400°F), and 816°C (1500°F), and results are summarized in Table 2. Similar calculations were done for the on-set recrystallization with cold strains of 20%, 30%, and 40%, and results are respectively summarized in Tables 3, 4, and 5.

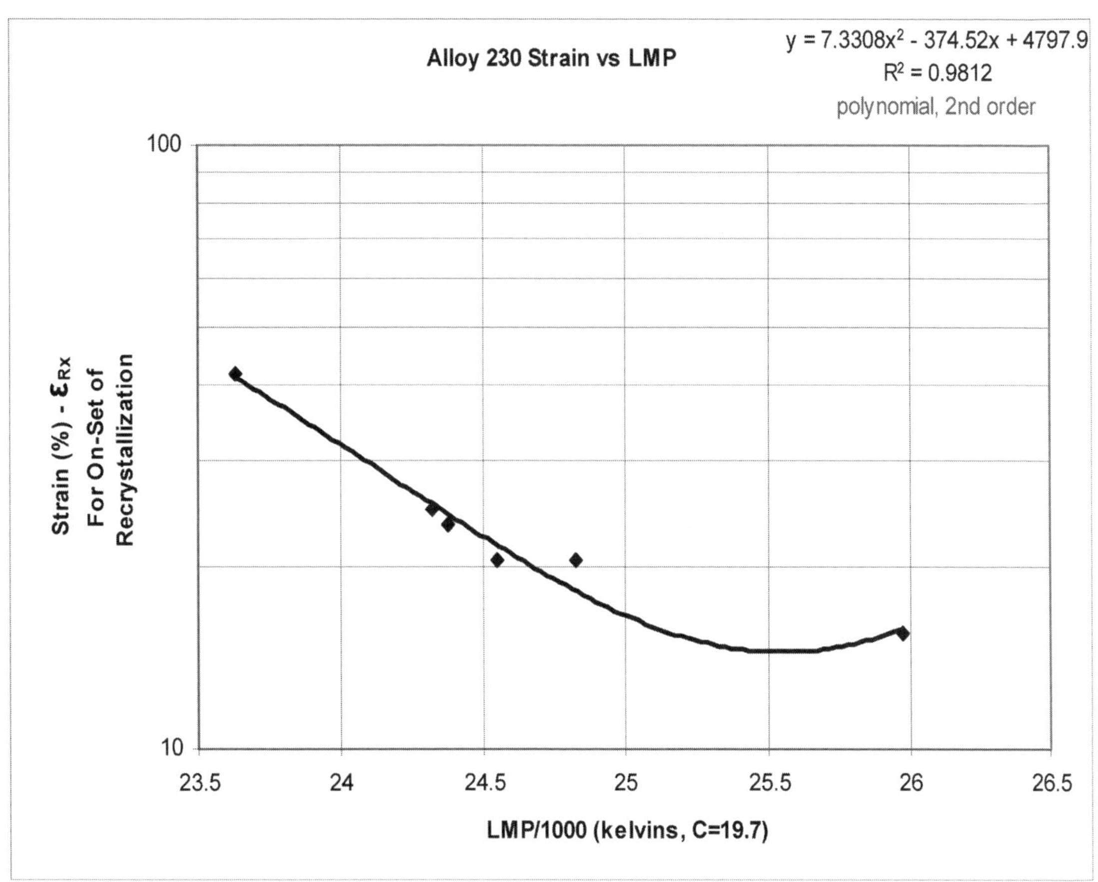

Figure 9 Cold strain levels corresponding to the on-set of recrystallization (ε_{Rx}) for thermally exposed alloy 230 specimens are plotted on a semi-logarithmic scale against the Larson-Miller Parameter (LMP) X 10^{-3}. A second order polynomial curve fit to the data has been incorporated.

TABLE 2

15% Cold Strain*

Calculated Time at Selected Exposure Temperatures for the On-Set of Recrystallization in Alloy 230 Materials Cold Strained 15%

Temperature	Calculated Exposure Time for Rx (hrs) (LMP/1000=25.28, C=19.7)
648°C (1200°F)	> 1,000,000
704°C (1300°F)	> 1,000,000
732°C (1350°F)	284,596
760°C (1400°F)	59,212
816°C (1500°F)	3,266

* LMP/1000 = 25.28 at 15% cold strain, calculated using Equation 2 for fitted curve in Figure 9

TABLE 3

20% Cold Strain*

Calculated Time at Selected Exposure Temperatures for the On-Set of Recrystallization in Alloy 230 Materials Cold Strained 20%

Temperature	Calculated Exposure Time for Rx (hrs) (LMP/1000=24.68, C=19.7)
648°C (1200°F)	> 1,000,000
704°C (1300°F)	363,918
732°C (1350°F)	71,980
760°C (1400°F)	15,545
816°C (1500°F)	918

* LMP/1000 = 24.68 at 20% cold strain, calculated using Equation 2 for fitted curve in Figure 9

TABLE 4

30% Cold Strain*

Calculated Time at Selected Exposure Temperatures for the On-Set of Recrystallization in Alloy 230 Materials Cold Strained 30%

Temperature	Calculated Exposure Time for Rx (hrs) (LMP/1000=24.09, C=19.7)
648°C (1200°F)	> 1,000,000
704°C (1300°F)	90,597
732°C (1350°F)	18,627
760°C (1400°F)	4,173
816°C (1500°F)	264

* LMP/1000 = 24.09 at 30% cold strain, calculated using Equation 2 for fitted curve in Figure 9

TABLE 5

40% Cold Strain*

Calculated Time at Selected Exposure Temperatures for the On-Set of Recrystallization in Alloy 230 Materials Cold Strained 40%

Temperature	Calculated Exposure Time for Rx (hrs) (LMP/1000=23.68, C=19.7)
648°C (1200°F)	> 1,000,000
704°C (1300°F)	34,472
732°C (1350°F)	7,281
760°C (1400°F)	1,673
816°C (1500°F)	111

* LMP/1000 = 23.68 at 40% cold strain, calculated using Equation 2 for fitted curve in Figure 9

To assess the exposure times for on-set of recrystallization predicted to occur at the four strain levels, a typical power plant design lifetime of 250,000 hours was used as a benchmark. Also, in current USC power plant designs that have been developed in the USC Project, alloy 230 is intended to operate up to 704°C (1300°F). The results shown in Table 2 indicate that, for operating metal temperatures up to about 732°C (1350°F), 15% cold strain can be tolerated in alloy 230 forming operations without having to heat treat the formed components. That is, the on-set of recrystallization (and possible loss of material properties) in formed components having 15% residual strain would not be expected at metal temperatures up to 732°C (1350°F) during the design lifetime of a commercial utility powerplant. Similarly, the results shown in Table 3 indicate that, for operating metal temperatures up to about 704°C (1300°F), 20% cold strain can be tolerated in alloy 230 forming operations without having to heat treat the formed components.

Because maximum metal temperatures of alloy 230 components in a USC powerplant are anticipated to be in the vicinity of 704°C (1300°F), the data presented in Tables 4 and 5 indicate that components produced with cold strains greater than 20% will likely have to be heat treated, since the calculated times for recrystallization with such strains at 704°C (1300°F) are less than the 250,000-hour design lifetime for commercial utility powerplants.

Conclusions and Plans for Additional Work

Based on the results of this analysis, 20% cold strain is a conservative limit for cold formed alloy 230 components without heat treatment which are intended for operation at USC service temperatures up to 704°C (1300°F). The procedures employed in this work to predict conditions for the on-set of recrystallization using the Larson-Miller Parameter appear to be a viable method for determining the practical strain limits for cold formed high temperature materials considered for use in USC power plants. To confirm this finding, additional furnace exposures at selected temperatures of cold strained alloy 230 specimens for durations up to 10,000 hours are planned to verify the Larson-Miller extrapolations. Similar long-term thermal exposure testing is also planned for the other USC austenitic alloys, including alloys 740, CCA 617, HR6W, and Super 304H.

Acknowledgements

The authors would like to thank Patricia Rawls and Bob Romanosky of the National Energy Technical Laboratory, Fred Glaser of the Department of Energy, Bob Brown and Mario Marrocco of the Ohio Coal Development Office, Bob Purgert of Energy Industries of Ohio, and R. Viswanathan of the Electric Power Research Institute for their support and guidance in the USC Boiler Materials Development Project, partially funded under U.S. DOE Contract No. DE-FG26-01NT41175 and OCDO Contract No. D-00-20. The constructive recommendations and technical contributions of other participating consortium team members in this project from B&W, FWDC, Alstom Power, and Riley Power are gratefully acknowledged. Special thanks to John Shingledecker of ORNL and to Jim Tanzosh of B&W for their assistance in the analysis of technical data and in the review of the manuscript.

References

1. ASME Section II, Part D, Nonmanditory Appendix A – Metallurgical Phenomenon, Section A-370, pg.813, 2004.
2. ASME Section I, Table PG-19, pg.10, 2004.
3. R. Viswanathan and W.T. Bakker, "Materials for Boilers in Ultra Supercritical Power Plants," Proceedings of 2000 International Joint Power Generation Conference, Miami Beach, FL, July 23-26, 2000, ASME.
4. J.G. Byrne, Recovery, Recrystallization, and Grain Growth, The MacMillan Company, New York, NY, 1965.
5. Cedric W. Richards, Engineering Materials Science, Chap. 8, pgs.302-310, Wadsworth Publishing Co., Belmont, CA 1961.
6. J. Shingledecker, ORNL, Private Communications, March 2007.

Advances in Materials Technology for Fossil Power Plants
Proceedings from the Fifth International Conference
R. Viswanathan, D. Gandy, K. Coleman, editors, p 168-184

DOI: 10.1361/cp2007epri0168

CREEP PROPERTIES AND STRENGTHENING MECHANISMS IN 23Cr-45Ni-7W (HR6W) ALLOY AND Ni-BASE SUPERALLOYS FOR 700°C A-USC BOILERS

H. Semba
H. Okada
M. Igarashi

Sumitomo Metal Industries, Ltd
Corporate Research and Development Laboratories,
1-8, Fuso-cho, Amagasaki, 660-0891, Japan

Abstract

Establishment of materials technologies on piping and tubing for advanced ultra super critical (A-USC) plants operated at steam temperatures above 700°C is a critical issue to achieve its hard target. 23Cr-45Ni-7W alloy (HR6W) has been developed in Japan, originally as a high strength tubing material for 650°C USC boilers. In order to clarify the capability of HR6W as a material applied to A-USC plants, creep properties and strengthening mechanisms of HR6W were investigated in comparison with γ'-strengthened Alloy 617. It has been revealed that the amount of added W is intimately correlated with precipitation amount of Laves phase and thus it is a crucial factor controlling creep strength. Stability of long term creep strength and superior creep rupture ductility have been proved by creep rupture tests at 650-800°C up to 60000h. The 10^5h extrapolated creep rupture strengths are estimated to be 88MPa at 700°C and 64MPa at 750°C. Microstructural stability closely related with long term creep strength and toughness has also been confirmed by microstructural observations after creep tests and aging. Creep rupture strength of Alloy 617 has been found to be much higher than that of HR6W at 700 and 750°C, while comparable at 800°C. A thermodynamic calculation along with microstructural observation indicates that the amount of Laves phase in HR6W gradually decreases with increasing temperature, while that of γ' in Alloy 617 rapidly decreases with increasing temperature, and almost dissolves at 800°C. This may lead to an abrupt drop in creep strength of Alloy 617 above 750°C. Capability of HR6W as a material for A-USC plants was discussed in terms of creep properties, microstructural stability and other reported mechanical properties including creep-fatigue resistance. It can be concluded that HR6W is a promising candidate for piping and tubing in A-USC plants.

Introduction

Advanced ultra super critical (A-USC) power plants operated at steam temperatures of 700-760°C are the most promising technology to improve the efficiency of fossil power plants for reducing their CO_2 emission. R&D projects on A-USC plants and relevant materials technologies have been launched in Europe and the US [1][2]. In Japan, development of advanced heat resistant steels and alloys for USC plants with steam temperatures of 600-650°C has successfully progressed [3] and a new project on A-USC plants and further innovations in materials technology will be initiated soon based on the established alloy design and application technologies so far [4][5]. High strength 9-12%Cr ferritic steels are favorable and widely used for thick sections such as main steam pipes in 'conventional' USC boilers with steam temperatures at around 600°C. However, it is difficult to apply these ferritic steels to the components exposed above 650°C in A-USC plants in terms of creep strength and steam oxidation resistance. 18-25%Cr austenitic steels also have limitations on use as thick wall pipes from the standpoints of thermal fatigue resistance due to higher thermal expansion coefficient. Therefore, Alloy 617 and other Ni-base superalloys having superior creep strength at around 700°C due to precipitation of γ' and lower thermal expansion coefficient than austenitic steels, are candidate materials for thick section components as well as tubing in A-USC plants. However, it seems that these γ'-hardened Ni-base superalloys have poor ductility, toughness, fatigue resistance and workability in addition to higher material costs.

23Cr-45Ni-7W alloy (HR6W) has been developed for a tubing alloy of near future USC plants with steam temperatures above 650°C in order to overcome the above disadvantages of conventional heat resistant steels and alloys [6][7][8]. It is strengthened by Laves phase without γ'. It has recently been shown that HR6W thick wall pipe with 457mm in diameter and 60mm in thickness was successfully manufactured [9][10]. This paper describes alloy design, creep properties and strengthening mechanisms of HR6W compared with γ'-strengthened Alloy 617, and than capability of HR6W as a material applied to A-USC plants is discussed.

Alloy Design of HR6W

HR6W with a nominal composition of 0.08C-23Cr-45Ni-7W-0.1Ti-0.2Nb-B has been developed, considering with creep strength, corrosion resistance and microstructural stability at around 700-800°C [6][7][8]. A combination of Nb and N in austenitic steels, employing precipitation strengthening of NbCrN and solid-solution strengthening of N is capable for increasing creep strength as shown in HR3C (25Cr-20Ni-0.45Nb-0.25N) steel. It was, however, found that addition of Mo or W to high nitrogen bearing austenitic steels aiming for further improvement of creep strength causes precipitation of blocky nitrides, leading to degradation of long-term creep strength and toughness [11]. Therefore, a chemical composition of HR6W has been optimized without N in austenitic alloys containing more than 20%Cr from the viewpoint of corrosion resistance.

Fig. 1 shows calculated phase diagrams in a Cr-Ni-Fe-Mo quaternary system. Although addition of Mo is effective for improving high temperature strength of austenitic steels and alloys, these phase diagrams show that precipitation of σ phase is inevitable in a Cr-Ni-Fe-Mo system with more than 20%Cr. On the other hand, addition of W to a Cr-Ni-Fe system containing more than 20%Cr stabilizes Laves phase and does not cause precipitation of σ phase with appropriate amount of Ni as shown in Fig. 2. It is also found that the two-phase region of γ+Laves is

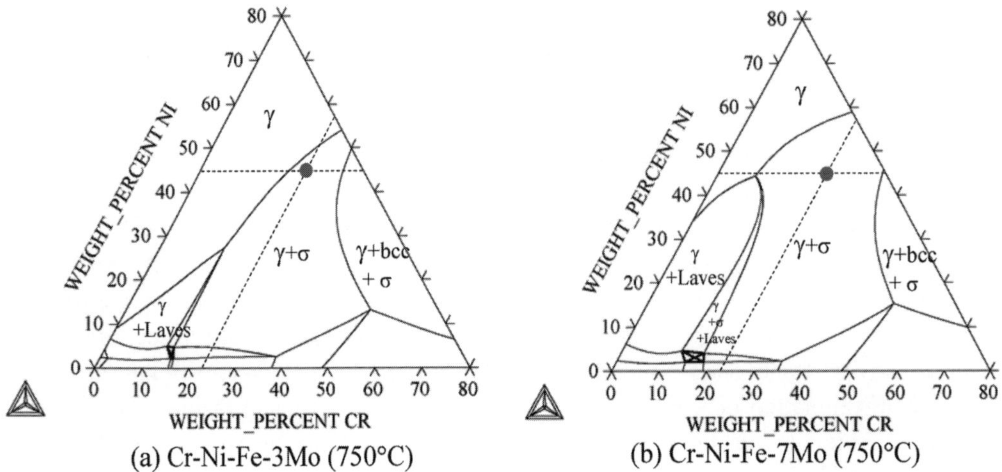

(a) Cr-Ni-Fe-3Mo (750°C) (b) Cr-Ni-Fe-7Mo (750°C)

Fig. 1 Phase diagrams of Cr-Ni-Fe-Mo quaternary system computed by Thermo-Calc.

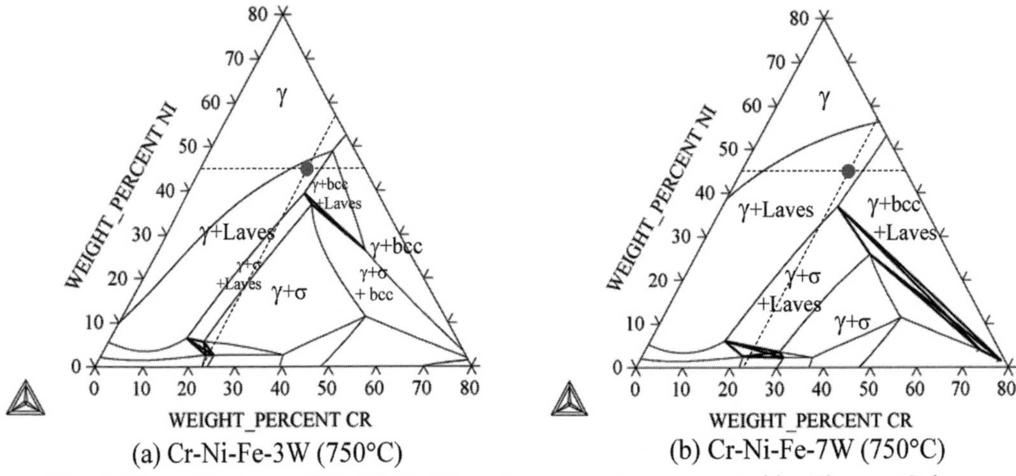

(a) Cr-Ni-Fe-3W (750°C) (b) Cr-Ni-Fe-7W (750°C)

Fig. 2 Phase diagrams of Cr-Ni-Fe-W quaternary system computed by Thermo-Calc.

extended in the Cr-Ni-Fe-7W system compared with the Cr-Ni-Fe-3W system. These calculations indicate that the addition of adequate amount of W to austenitic alloys can enhance precipitation strengthening of Laves phase without instability of austenite matrix. Indeed, precipitation of σ phase has been identified in 0.08C-23Cr-42/45Ni-3/5Mo alloys after aging at 750°C, while only Laves phase has been observed in addition to $M_{23}C_6$ carbides in 0.08C-23Cr-42/45Ni-5/7W alloys, resulting in higher creep strength and better toughness than those Mo bearing alloys as shown Figs. 3 to 5 [6]. A small amount of Ti, Nb and B is also added to HR6W. These elements enhance uniform dispersion of fine $M_{23}C_6$ carbides [12], leading to increasing precipitation strengthening of $M_{23}C_6$. It has been reported that dislocations formed around TiC and NbC, precipitating prior to $M_{23}C_6$ during an initial stage of creep deformation, enhanced uniform dispersion of fine $M_{23}C_6$ [13]. Fig. 6 shows an effect of Cr content on the amount of precipitates in 0.08C-10/40Cr-45Ni-

7W-0.1Ti-0.2Nb alloys computed by Thermo-Calc. It is found that the maximum amount of Laves phase and $M_{23}C_6$ is obtained with about 23%Cr. Hence, the Cr content of HR6W has been decided as 23% considering with the balance between phase stability and corrosion resistance. Calculated phase fraction of HR6W is shown in Fig. 7. It is seen that Laves phase precipitates above 800°C. This suggests that strengthening of Laves phase sufficiently contributes to creep resistance at 700-800°C.

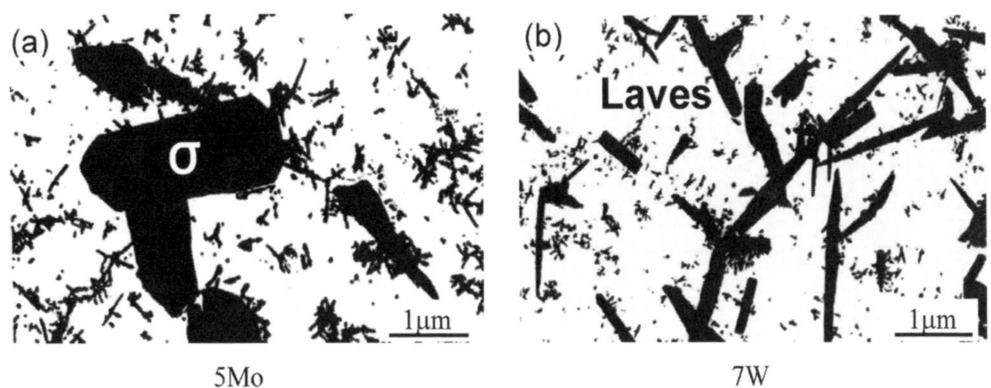

Fig. 3 TEM images of extracted replicas of 0.08C-23Cr-44Ni alloys aged at 750°C for 3000h.

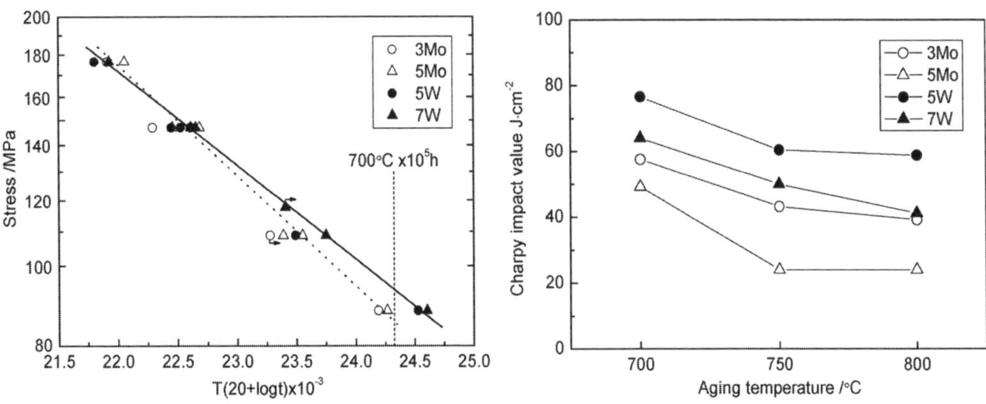

Fig. 4 Effects of Mo and W on creep strength in 0.08C-23Cr-42/45Ni alloys.

Fig. 5 Effects of Mo and W on toughness in 0.08C-23Cr-42/45Ni alloys aged for 3000h.

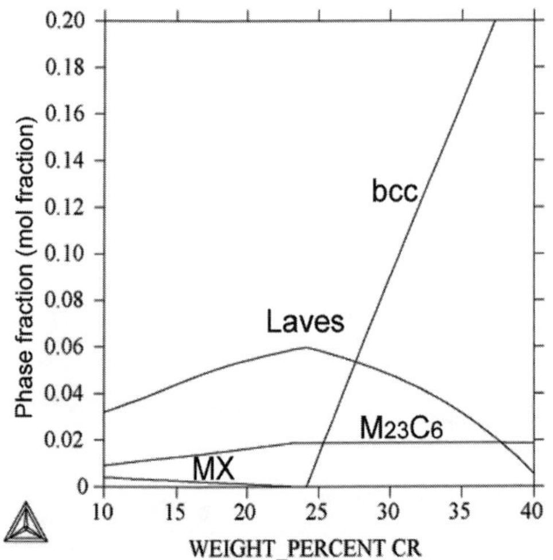

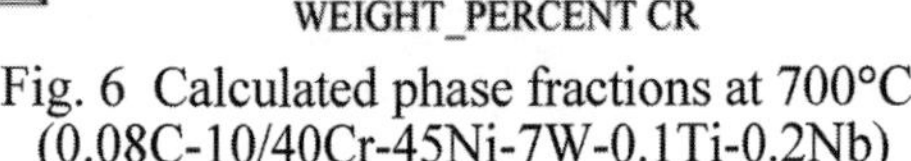
Fig. 6 Calculated phase fractions at 700°C. (0.08C-10/40Cr-45Ni-7W-0.1Ti-0.2Nb)

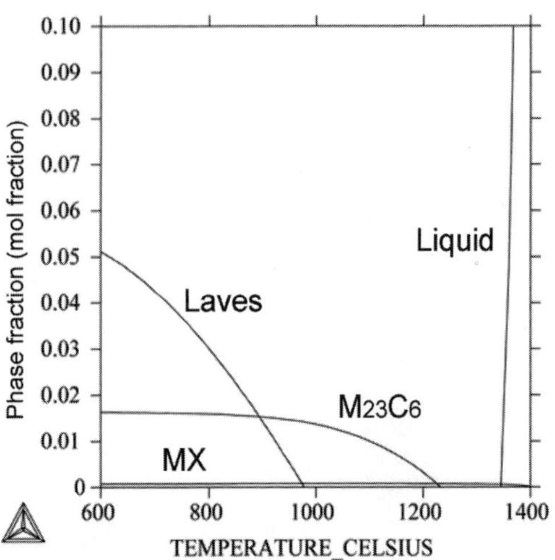

Fig. 7 Calculated phase fractions of HR6W. (0.08C-23Cr-45Ni-7W-0.1Ti-0.2Nb)

Creep Properties and Strength Mechanisms of HR6W

Creep Properties

Creep rupture strength of the developed HR6W tubes and plates are shown in Fig. 8. Typical size of the tubes tested is 50.8mm in diameter and 8mm in thickness, manufactured by cold drawing after hot extrusion. Long-term creep rupture tests have been performed and progressed up to 60000h at 650-800°C. It has been proved that the average creep strength curves at 650-800°C are stable by long-term creep tests over 10000h. The 10^5h extrapolated creep rupture strengths of HR6W by Larson-Miller method are estimated to be 88MPa at 700°C, 64MPa at 750°C and 46MPa at 800°C, respectively. The creep rupture strength of HR6W at 800°C is comparable with that of Alloy 617 [14]. Fig. 9 shows creep rupture elongations of HR6W. It is noted that HR6W demonstrates superior rupture ductility, compared with Alloy 617 and other Ni-base superalloys strengthened by γ' phase even in the long-term tests. Fig. 10 shows creep curves of HR6W at 750°C, 108MPa (ruptured at 1047.4h). Creep rate decreases about one order of magnitude in the primary region, then gently increases, leading to tertiary creep. While the onset strain of tertiary creep is around 0.01, the tertiary stage is relatively long, resulting in sufficient creep strain of 0.4. Good creep rupture ductility of HR6W should result from this deformation behavior.

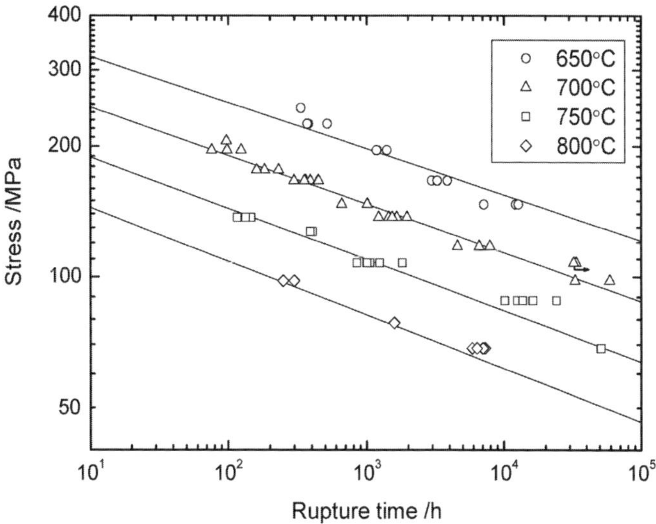

Fig. 8 Creep rupture strength of HR6W tubes and plates.

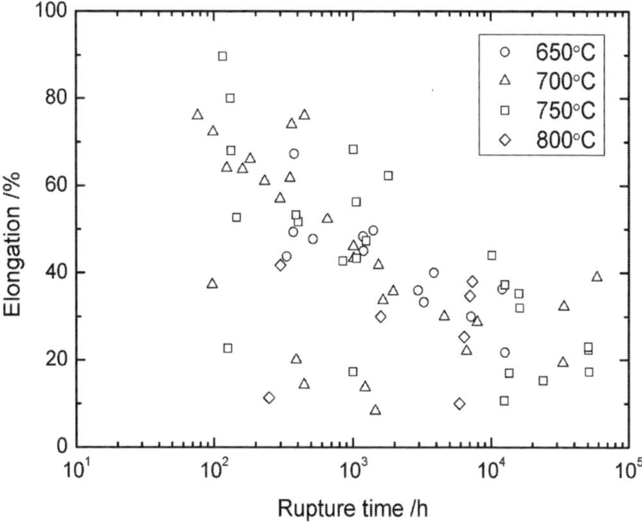

Fig. 9 Creep rupture elongation of HR6W tubes and plates.

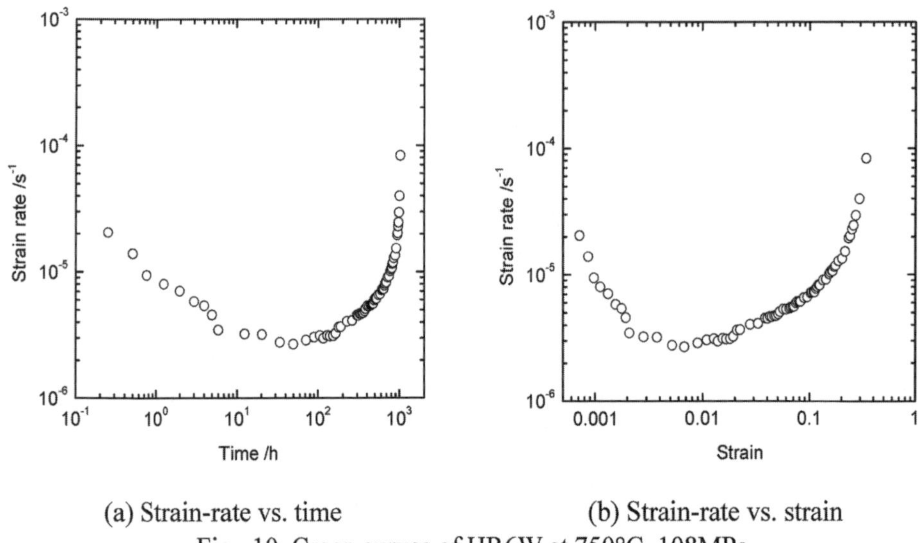

(a) Strain-rate vs. time (b) Strain-rate vs. strain

Fig. 10 Creep curves of HR6W at 750°C, 108MPa.

Strengthening Mechanisms

Results of chemical analysis of extracted residues have revealed that about 3mass% of added W exists as solution and the rest exists as precipitates despite the amount of W content after aging at 750°C for 3000h. This suggests that about 3mass% of W contributes to solid solution strengthening. Microstructural observations of the aged HR6W plats clarified that precipitates are mainly Laves and $M_{23}C_6$. σ and other brittle intermetallic phases have not been identified. It is considered that the entire precipitated W exists in Laves and $M_{23}C_6$, as also shown in the computed equilibrium compositions of constituting phases (Table 1). Fig. 11 shows creep rupture strength of the 0.08C-23.1Cr-42.8Ni-6.6W-Ti-Nb and 0.09C-23.0Cr-36.6Ni-3.1W-Ti-Nb alloys. Much precipitation of Laves phase was observed in addition to $M_{23}C_6$ in the former, while precipitates were mainly $M_{23}C_6$ and few Laves was observed in the latter. It is found the creep rupture strength of the 6.6W steel is much higher than the 3.1W steel especially in the long-term region. This may mainly due to precipitation of Laves phase, suggesting that it increases creep rupture strength about one order of magnitude. TEM observations of the crept specimen have revealed that fine dispersion of Laves phase in HR6W has been kept even after long-term creep deformation for 58798h at 700°C (Fig. 12 (a)). Laves phase slightly coarsens at higher temperatures as shown in Figs. 12 (b) and (c). It is, however, noted that the amount of Laves phase does not decrease rapidly, as indicated by computed results (Fig. 7). Furthermore, coarse and brittle intermetallic phases, such as σ, have not been identified. Those may lead to relatively stable long-term creep rupture strength as shown in Fig. 8. Fine dispersion of $M_{23}C_6$ inside grains which precipitates coherently and acts as effective obstacles against dislocation motion has been confirmed in the same crept samples as well as that covering grain boundaries. A small amount of Ti, Nb and B is considered to enhancing this fine precipitation of $M_{23}C_6$ as mentioned above. Strengthening by precipitates on grain boundaries, for instance, γ' in Alloy 80A and carbides in Alloy 617 has been reported [15][16]. Creep resistance increases with increasing a grain boundary covering ratio. $M_{23}C_6$ densely covers grain boundaries in HR6W during creep deformation. This may also contribute to creep resistance. It is thus concluded that in HR6W the long-term creep strength has been achieved by fine dispersion of Laves phase together with $M_{23}C_6$ and solid solution strengthening of W. Superior microstructural stability results in sufficient creep rupture ductility and toughness after high temperature exposure, and also contributes to suppressing degradation of long-term creep strength.

Table 1. Equilibrium compositions of matrix, Laves and $M_{23}C_6$ in HR6W at 750°C computed by Thermo-Calc.

(mass%)

Phase						
Matrix	C	Cr	Ni	W	Ti	Nb
	0.0007	22.52	45.95	3.14	0.07	0.17
Laves	W	Fe	Cr			
	63.05	18.62	18.32			
$M_{23}C_6$	Cr	W	Fe	C		
	71.30	19.28	0.044	0.049		

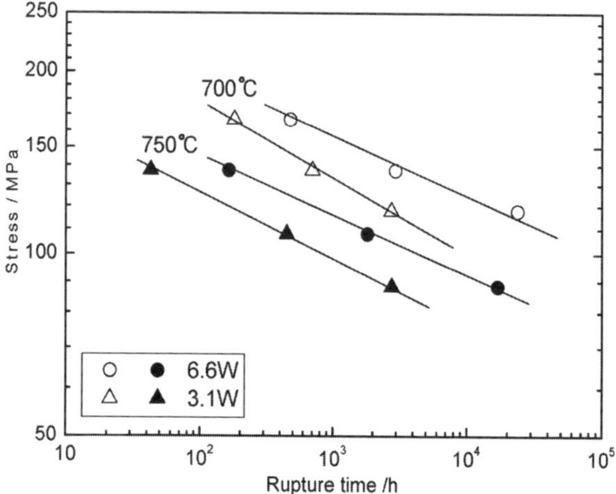

Fig.11 Effect of Laves phase on creep rupture strength.

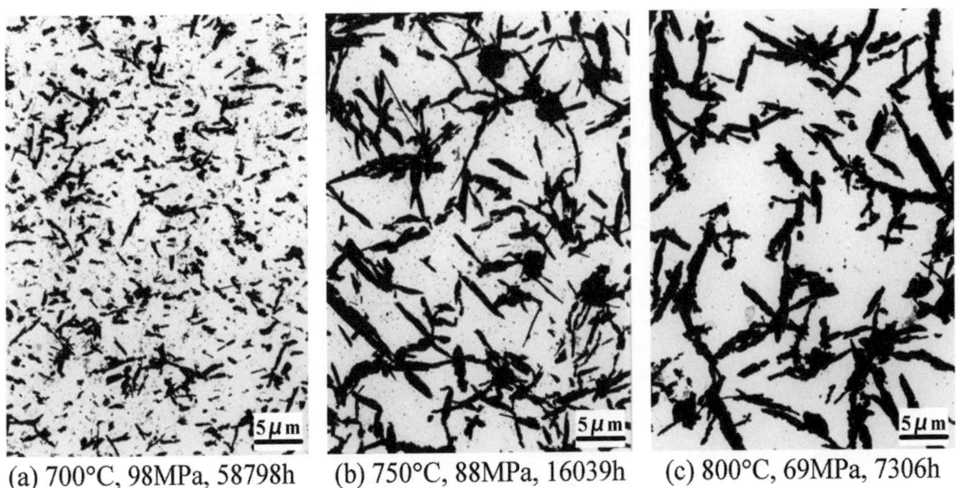

(a) 700°C, 98MPa, 58798h (b) 750°C, 88MPa, 16039h (c) 800°C, 69MPa, 7306h

Fig. 12 TEM images of extracted replica for HR6W crept at 700, 750 and 800°C.

Effect of W Content on Creep Strength

Laves phase is important strengthening determinant as described in the previous section. Therefore, the effect of W content, which is closely related to precipitation amount of Laves phase, has been investigated. Table 2 shows creep rupture time of the 0.08C-23.6C-44.9Ni-7.2W and 0.08C-23.2Cr-45.7Ni-5.5W alloys. Note that rupture times of the 5.5W alloy is extremely shorter than the 7.2W alloy. In order to clarify the precipitation behavior of both the alloys, chemical analysis of extracted residues was carried out. Figs. 13 and 14 show the amount of precipitated W and Cr after aging at 700°C. It is found that precipitation of W in the 7.2W alloy is much faster than the 5.5W alloy and the total amount of precipitated W after aging for 10000h is much larger than the 5.5W alloy. Although precipitated W is contained in Laves and $M_{23}C_6$, it mainly exists as Laves, as indicated by calculated phase fractions and equilibrium compositions of HR6W (Fig. 7 and Table 1). Precipitated Cr, which mainly corresponds to precipitation of $M_{23}C_6$, increases in the early stage of aging as shown in Fig. 14. It is, therefore, considered that the difference in the amount of precipitated W after aging for 10000h between the 7.2W and 5.5W alloys corresponds to the difference in the amount of Laves phase. A thermodynamic calculation of equilibrium phase fractions also supports these results. The equilibrium phase fraction of Laves phase in the 5.5W alloy is 38% less than that in the 7.2W alloy (Fig. 15). These differences in Laves precipitation have been confirmed by the extracted replicas from the crept samples of the 7.2W and 5.5W alloys shown in Fig. 16. Although there is little difference in the dispersion of $M_{23}C_6$ along grain boundaries and inside grains between these two alloys, it is found that Laves phase in the 5.5W alloy is coarser and much less than that in the 7.2W alloy. These results indicate that shorter creep lives in the 5.5W alloy is mainly due to the difference in the precipitation behavior of Laves phase, suggesting the W content in HR6W should be carefully controlled. Some discrepancies in creep strength of HR6W between the experiments and the database have been reported [2]. This might be involved in these effects of W.

Table 2 Creep rupture time of the 7.2W and 5.5W alloys (hours).

	700°C, 98MPa	750°C, 88MPa	800°C, 69MPa
0.08C-23.6Cr-44.9Ni-7.2W	58798	16039	7306
0.08C-23.2Cr-45.7Ni-5.5W	19137	3088	2658

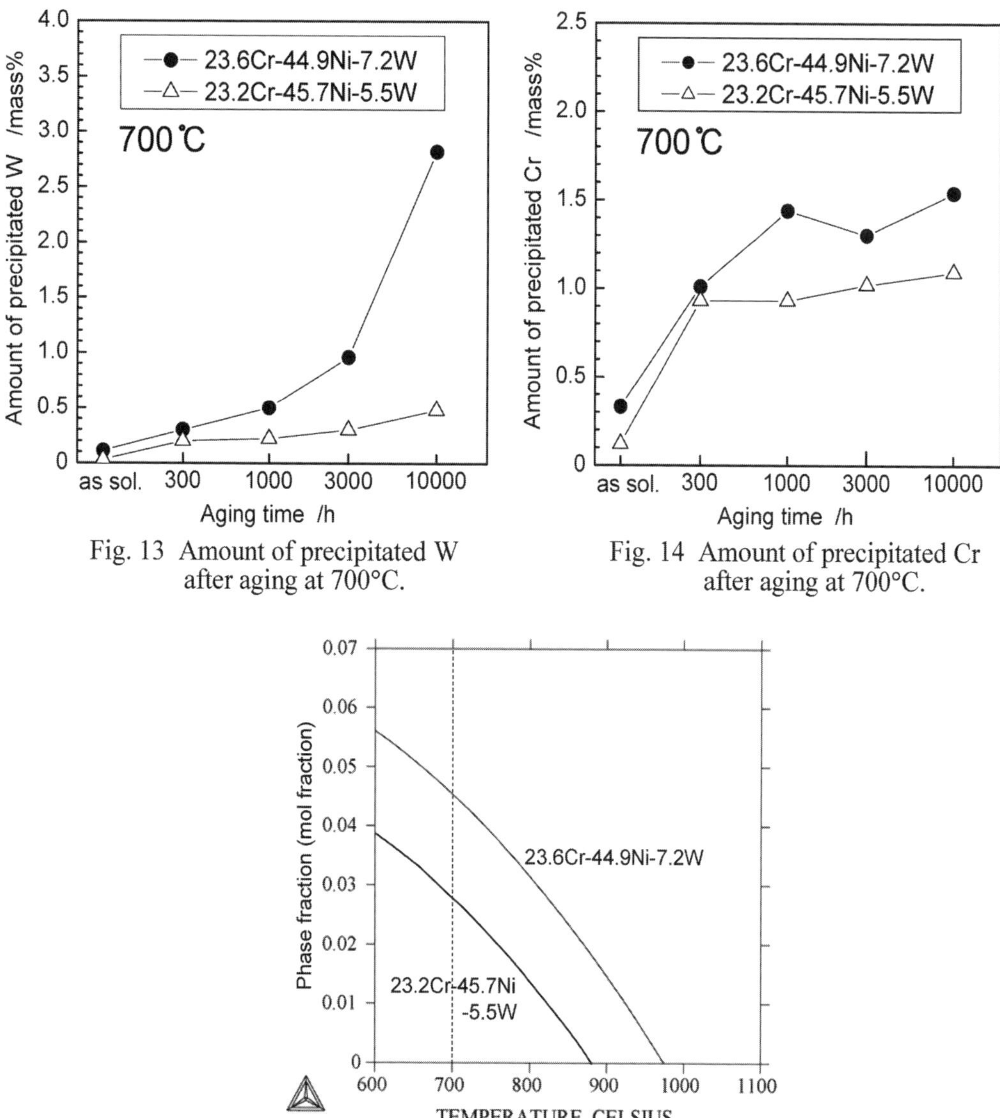

Fig. 13 Amount of precipitated W after aging at 700°C.

Fig. 14 Amount of precipitated Cr after aging at 700°C.

Fig. 15 Calculated phase fraction of Laves.

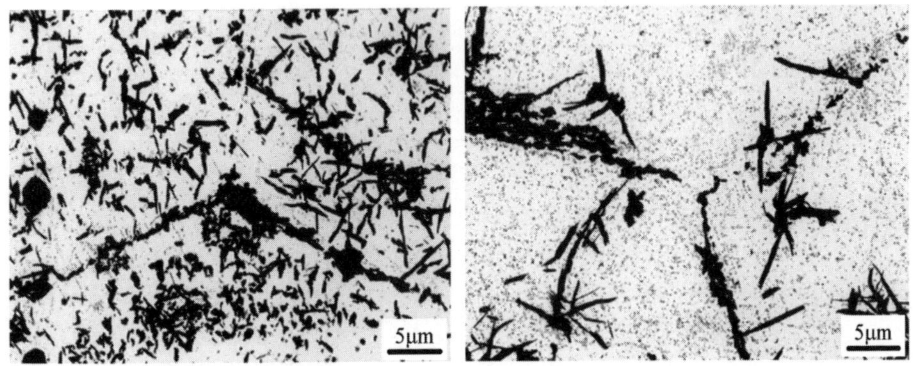

(a) 7.7W alloy, 58798h (b) 5.5W alloy, 19137h

Fig. 16 TEM images of extracted replica for the 7.2W and 5.5W alloys crept at 700°C, 98MPa.

Creep Strength and Microstructure of Alloy 617

Creep strength and deformation behavior of Alloy 617, which is strengthened by γ' phase, were investigated using laboratory vacuum induction melted and forged plates followed by cold rolling and solution annealing. Creep rupture strength is shown in Fig. 17. It can be seen that creep rupture strength of Alloy 617 is much higher than HR6W at 700 and 750°C. However, it significantly falls down above 750°C and is comparable with HR6W at 800°C. Fig. 18 shows a strain-rate versus strain curve of Alloy 617 at 750°C, 180MPa (ruptured at 1256h). This testing stress was decided to be ruptured at a similar time to HR6W discussed in the previous section (Fig. 10). Note that onset strain of tertiary creep is about 0.004, and then creep rate rapidly increases, resulting in the small total creep strain of 0.07 in contrast to that of HR6W. Equilibrium phase fractions and compositions were calculated and are shown in Fig. 19 and Table 3. After this computation, γ', $M_{23}C_6$ and M_6C precipitate and most of the added Co and Mo dissolve in the matrix at 750°C. Indeed, they have been identified in the crept specimens at 750°C. This calculation shows that the amount of γ' phase rapidly decreases with increasing temperature and it almost dissolves at 800°C. It is also found that an increase in the amount of M_6C entails a decrease in that of $M_{23}C_6$ with increasing temperature.

Figs. 20 and 21 show the TEM images of extracted replica taken from grip portions of the crept samples at 750 and 800°C. Therefore, the following results do not involve the effect of applied stress. Only γ' phase was extracted by the non-aqueous electrolyte extraction method. It can be seen that γ' at 800°C is coarser and much less than that at 750°C. Average diameter and number density were measured by image analysis of these TEM micrographs as shown in Table 4. Volume fraction was obtained with these measured values, assuming uniform spherical particles. Particle spacing, that is, mean free path (MFP) was calculated with equation (1).

$$MFP = \frac{4r}{3} \cdot \frac{1-f}{f} \quad (1)$$

where r is a radius of particles and f is a volume fraction of dispersed particles [17]. It is found that the observed volume fraction of γ' at 800°C is very small, corresponding with the calculation shown in Fig. 19. The resultant mean free path at 800°C is an order of magnitude larger than that at 750°C. These results indicate that creep strength of Alloy 617 in the temperature range of 700-800°C drastically falls down as the amount of γ' decreases and its morphology changes with increasing temperature.

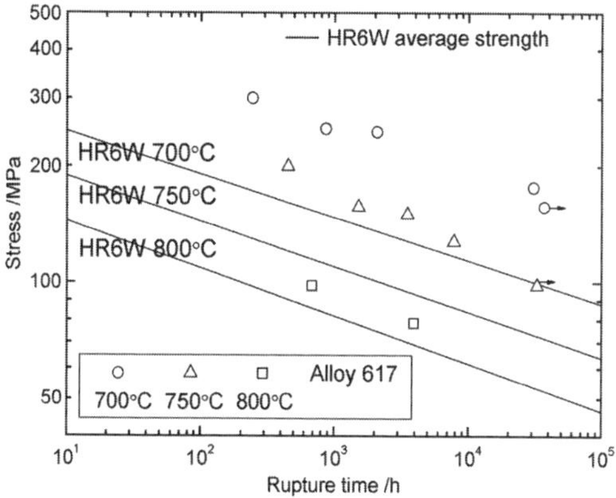

Fig. 17 Creep rupture strength of Alloy 617.

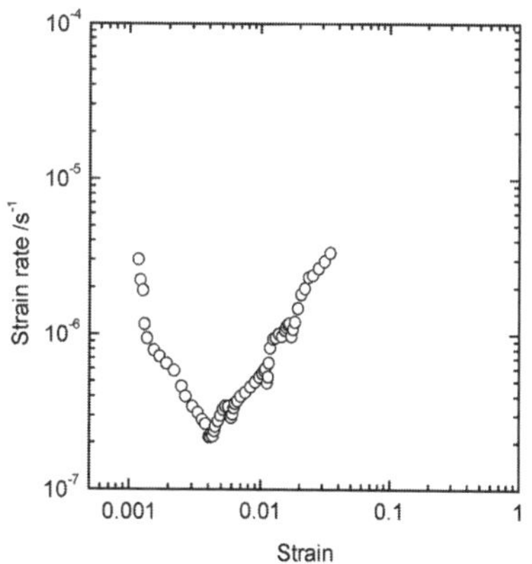

Fig. 18 Creep-rate vs. strain curve at 750°C, 180MPa.

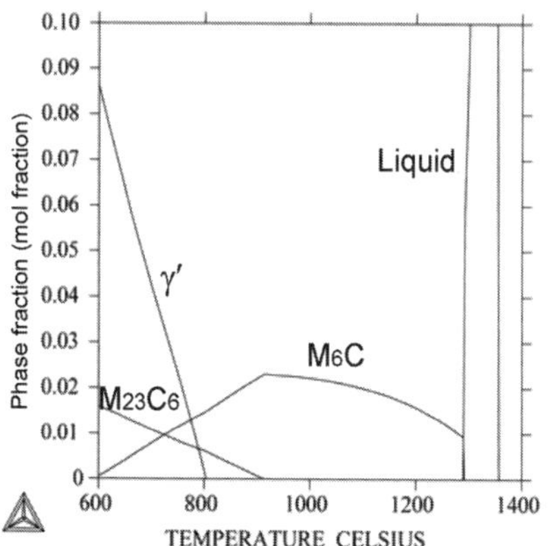

Fig. 19 Calculated phase fractions of Alloy 617. (Ni-0.07C-22Cr-12Co-9Mo-1.2Al-0.3Ti-1.5Fe)

Table 3 Equilibrium compositions of matrix, γ', $M_{23}C_6$ and M_6C at 750°C computed by Thermo-Calc.

(mass%)

Phase						
Matrix	Ni	Cr	Co	Mo	Fe	Al
	53.6	22.2	12.4	8.5	1.5	1.1
γ'	Ni	Al	Ti	Co	Cr	Mo
	77.1	7.1	6.9	4.5	2.6	1.4
$M_{23}C_6$	Cr	Mo	Ni	Co	Fe	C
	66.1	20	6.1	2.5	0.2	5.1
M_6C	Mo	Ni	Cr	Co	C	
	56.8	22.7	14.1	2.9	2.6	

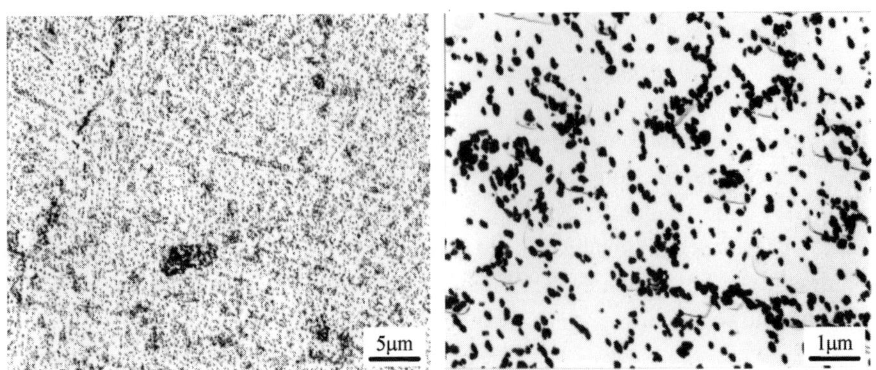

Fig. 20 TEM images of extracted replica for grip portion of the crept specimen exposed at 750°C for 7770h.

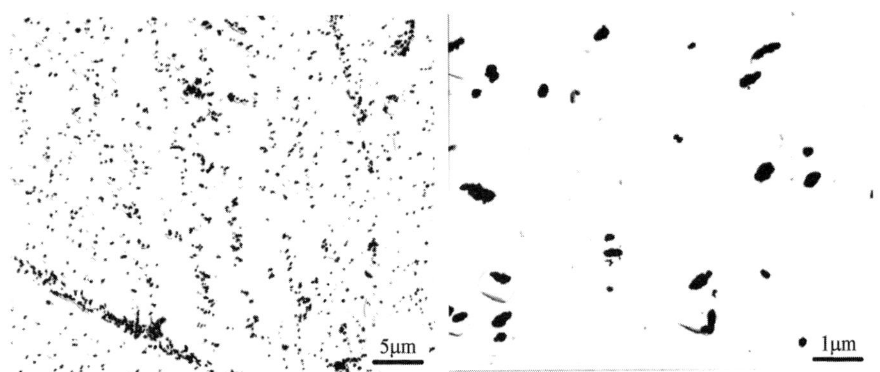

Fig. 21 TEM images of extracted replica for grip portion of the crept specimen exposed at 800°C for 3928h.

Table 4 Morphology of γ' after exposure at 750 and 800°C

	Average diameter (nm)	Number density (m^{-2})	Volume fraction	Mean free path (m)
750°C, 7770h	168	5.6×10^{12}	0.033	3.3×10^{-6}
800°C, 3928h	225	8.2×10^{11}	0.004	3.4×10^{-5}

Capability of HR6W as an A-USC Boiler Alloy

Thermal expansion coefficients of various steels and alloys have been measured as shown in Fig. 22. These of 18-25%Cr austenitic stainless steels are about 50% higher than that of the Gr.91 ferritic steel. Large thermal expansion is disadvantageous for thick wall piping in terms of thermal fatigue resistance. On the other hand, thermal expansion coefficients of HR6W is about 15% lower than the austenitic steels though they are slightly higher than Alloy 617. Fig. 23

shows Charpy impact values of HR6W after aging at 650-800°C for 300-10000h. It is found that HR6W shows sufficient impact values even after long-term aging. This is in contrast with the fact that the toughness of Alloy 617 after aging drops significantly [9]. The stable microstructure in HR6W mentioned above should result in sufficient toughness. Creep-fatigue tests for HR6W and Alloy 617 have been carried out at 700°C [18]. It has been revealed that creep-fatigue life of HR6W under the CP (slow-fast) type strain waveform is much longer than that of Alloy 617, resulting from its higher creep ductility compared with Alloy 617 (see the reference [18]). Gleeble testing has revealed that hot workability of HR6W is much better than Alloy 617 and it is comparable with SUS316H below 1200°C, although its zero ductility temperature is lower than SUS316H [9]. Indeed, a thick wall pipe of HR6W, which is 457mm in diameter, 60mm in thickness and 3.5m in length, has been successfully manufactured by the Erhart Push Bench press method from a 4.9ton ingot melted in a electric furnace [9][10]. This trial production has shown that hot workability of HR6W is sufficient for manufacturing thick wall piping. Creep and microstructural properties of this produced pipe has being under investigation. Considering these results in addition to excellent creep properties and microstructural stability mentioned above, it is concluded that HR6W is a promising candidate for piping and tubing in USC plants operated at steam temperatures above 650°C. Current creep rupture strength of HR6W is slightly lower than the requirements for structural components exposed at the highest temperature and pressure in A-USC plants with operating steam temperatures of 700-760°C. However, it is considered that HR6W can be applied to piping and tubing except these components used under the severest steam conditions and can contribute to reducing materials related costs instead of Ni-base superalloys. For instance, hot reheat pipes, in which steam pressure is relatively low, might be possible components that HR6W can be applied. Further research and development on improving creep strength of HR6W should lead to successful development of A-USC plants.

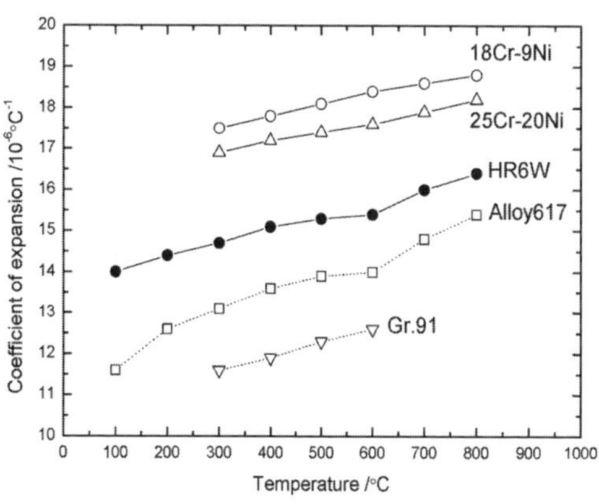

Fig. 22 Comparison of thermal expansion coefficient.

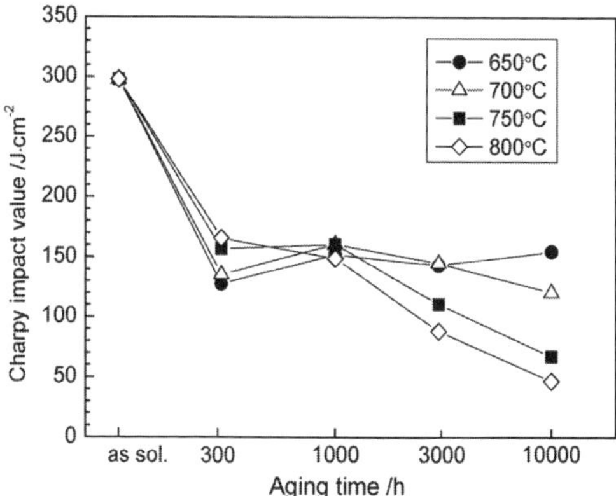

Fig. 23 Charpy impact values at 0°C after aging at 650-800°C for 300-10000h.

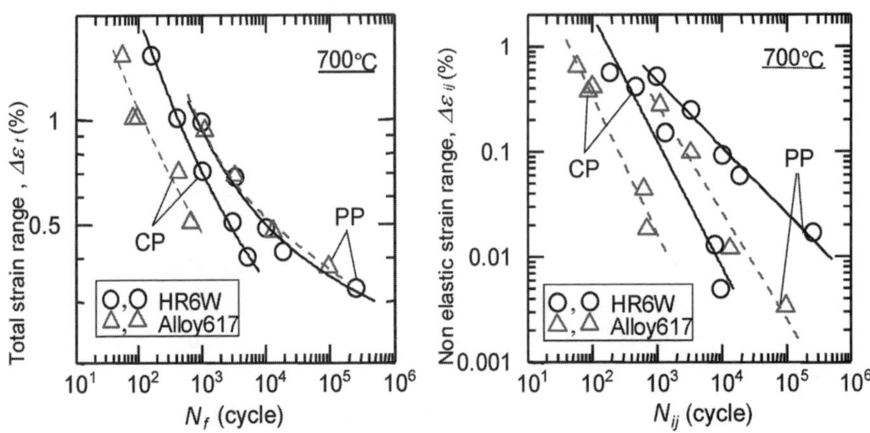

Fig. 24 High temperature creep-fatigue properties of HR6W and Alloy 617 [18].
Specimen: 10mm in diameter and 25mm in gauge length,
Strain PP (fast 0.8%/s-fast 0.8%/s), CP (slow 0.01%/s-fast 0.8%/s)

Conclusions

HR6W with a nominal composition of 0.08C-23Cr-45Ni-7W-0.1Ti-0.2Nb-B has been developed, considering with creep strength, corrosion resistance and microstructural stability at around 700-800°C. It is strengthened by precipitates of Laves and dissolved W in the matrix in addition to fine dispersion of $M_{23}C_6$. It has been clarified that the amount of added W is intimately correlated with precipitation amount of Laves phase, and thus it is a crucial factor controlling creep strength. Stability of long term creep strength and superior creep rupture ductility have been proved by creep rupture tests at 650-800°C up to 60000h. The 10^5h extrapolated creep rupture strengths of HR6W are estimated to be 88MPa at 700°C and 64MPa at 750°C. Microstructural stability closely related with long term creep strength and toughness has also been confirmed by microstructural observations after creep tests and aging. Creep rupture strength of Alloy 617 has been found to be much higher than that of HR6W at 700 and 750°C, while comparable at 800°C. A thermodynamic calculation along with microstructural observation indicates that the amount of Laves phase in HR6W gradually decreases with increasing temperature, while that of γ' in Alloy 617 rapidly decreases with increasing temperature, and almost dissolves at 800°C. This may lead to an abrupt drop in creep strength of Alloy 617 above 750°C. Capability of HR6W as an A-USC plant material has been shown. It can be concluded that HR6W is a promising candidate for piping and tubing in A-USC plants.

References

1. R. Blum and R. W. Vanstone, Materials for Advanced Power Engineering 2006, Proceedings of the 8th Liege Conference (2006), pp.41-60.

2. R. Viswanathan and R. Purgert, Proceedings of 8th International Conference on Creep and Fatigue at Elevated Temperatures, July 22-26, 2007, CREEP2007-26826.

3. F. Masuyama, Materials for Advanced Power Engineering 2006, Proceedings of the 8th Liege Conference (2006), pp.175-187.

4. J. Iwasaki, A. Shiibashi, S. Takano. T. Sato, H. Okada and F. Abe, The Thermal and Nuclear Power, 58(2007), pp.649-655.

5. T. Fujikawa, M. Fukuda, E. Saito, T. Takahashi and S. Izumi, The Thermal and Nuclear Power, 58(2007), pp.656-662.

6. Y. Sawaragi, Y.Hayase and Y. Yoshikawa, Proceedings of International Conference on Stainless Steels, Chiba, ISIJ (1991), p.633

7. H. Semba, M. Igarashi, Y. Yamadera, A. Iseda and Y. Sawaragi, Report of the 123rd Committee on Heat-Resisting Materials and Alloys, JSPS, 44 (2003), pp. 119-127.

8. M. Igarashi, H. Semba and H. Okada, Proceedings of 8th Workshop on the Innovative Structural Materials for Infrastructure in 21st Century, NIMS, Tsukuba, Japan (2004), pp.194-199.

9. H. Okada, M. Igarashi, K. Ogawa, Y. Noguchi, H. Matsuo and S. Yamamoto, CAMP-ISIJ, 19(2006), p.1230.

10. A. Iseda, H. Okada, H. Semba, M. Igarashi and Y. Sawaragi, Proceedings of Symposium on Heat Resistant Steels and Alloys for USC power Plants 2007, July 2-6, 2007, Seoul, Korea, pp.229-237.

11. Y. Sawaragi, H. Teranishi, K. Yoshikawa and N. Otsuka, *Tetsu-to-Hagané*, 70 (1984). p.s1409.

12. T. Shinoda, T. Mimino, K. Kinoshita and I. Minegishi, *Tetsu-to-Hagané*, 54 (1968). p.1472.

13. R. Tanaka and T. Shinoda, *Zairyo*, 21 (1972), p.198.

14. ECCC Creep Datasheets 2005

15. A. M. ElBatahgy, T. Matsuo and M. Kikuchi, Report of the 123rd Committee on Heat-Resisting Materials and Alloys, JSPS, 30 (1989), pp. 41-49.

16. R. Ishii, Y. Terada, T. Matsuo and M. Kikuchi, Report of the 123rd Committee on Heat-Resisting Materials and Alloys, JSPS, 30 (1989), pp. 159-168.

17. J.W. Martin, *Micromechanisms in particle-hardened alloys*, Cambridge University Press (1980), p.41.

18. Y. Noguchi, M. Miyahara, H. Okada, M. Igarashi and K. Ogawa, Proceedings of 8th International Conference on Creep and Fatigue at Elevated Temperatures, July 22-26, 2007, CREEP2007-26471.

Advances in Materials Technology for Fossil Power Plants
Proceedings from the Fifth International Conference
R. Viswanathan, D. Gandy, K. Coleman, editors, p 185-196

DOI: 10.1361/cp2007epri0185

LONG-TERM CREEP PROPERTIES AND MICROSTRUCTURE OF SUPER304H, TP347HFG AND HR3C FOR ADVANCED USC BOILERS

A. Iseda [1)]
H. Okada [2)]
H. Semba [2)]
M. Igarashi [2)]
Sumitomo Metal Industries, Ltd
1) Tube and Pipe Company, 8-11, Harumi 1-chome, Chuo-ku, Tokyo, 104-6111, Japan
2) Corporate Research and Development Laboratories,
1-8, Fuso-cho, Amagasaki, 660-0891, Japan

Abstract

SUPER304H (18Cr-9Ni-3Cu-Nb-N, ASME CC2328) and TP347HFG (18Cr-12Ni-Nb, ASME SA213) ,which are applied fine-grained microstructure, have been developed for high strength and high steam oxidation resistant steel tubes. The longest creep rupture tests at 600°C for 85,426h of SUPER304H and at 700°C for 55,858h of TP347HFG prove to keep stable strength and microstructure with very few amount of σ phase and no other brittle phases when compared with conventional austenitic stainless steels. HR3C (25Cr-20Ni-Nb-N, ASME CC2115) have been developed for high strength and high corrosion resistant steel tube used in severe corrosion parts of recent USC boilers with steam temperature around 600°C. The longest creep test at 700°C and 69MPa for 88,362h of HR3C confirms high enough and stable creep strengths and microstructure at 600-800 °C. These steel tubes have been installed in Eddystone No.3 USC power plant as superheater and reheater tubes since 1991, and have been removed and investigated microstructure after long-term service exposure. This paper describes up-dated long-term creep rupture properties of the steels and microstructural changes after long-term creep rupture and aging. Three steel tubes have been successfully applied as standard materials for superheater and reheater tubes in newly built USC boilers in the world.

1. Introduction

Recently, many USC boilers with steam temperature about 600 °C and pressure about 25MPa have been built and successfully operated in Japan and Europe since 1995, and have been newly built and planed all over the world. In these high temperature and pressure USC boilers, some advanced materials have been developed and successfully applied all over the world (1)(2).

Among them, SUPER304H, TP347HFG and HR3C are the most well known standard materials as superheater and reheater tubes for recent USC boilers.

Table 1 shows nominal compositions, standards and properties of three steels comparing with conventional ASME SA213 TP347H. SUPER304H (ASME CC2328) and TP347HFG (ASME SA213) have been developed adopting the thermo-mechanical process as shown in Fig.1 in order to achieve both fine-grained microstructure and high creep rupture strength (3)(4).

Table 1 Nominal Compositions, Standards and Properties of Developed Steels

	Conventioal TP347H	TP347HFG	SUPER304H	HR3C
Composition	18Cr-12Ni-0.9Nb	18Cr-12Ni-0.9Nb	18Cr-9Ni-3Cu-0.5Nb	25Cr-20Ni-0.4Nb-N
ASME	SA213	SA213	CC2328	CC2115
Japanese METI	KA-SUSTP347HTB	-	KA-SUS304J1HTB	KA-SUS310J1TB
ASTM Grain Size (example)	6	9.5	8.5	5
Allowable Stress (MPa) 650°C / 700°C	54 / 32	61 / 33	78 / 47	71 / 40
Steam Oxidation (1000h, μm)[1] 650°C / 700°C	27 / 40	15 / 20	19 / 25	<2 / <2

Note 1) Steam oxidation scale (inner layer)

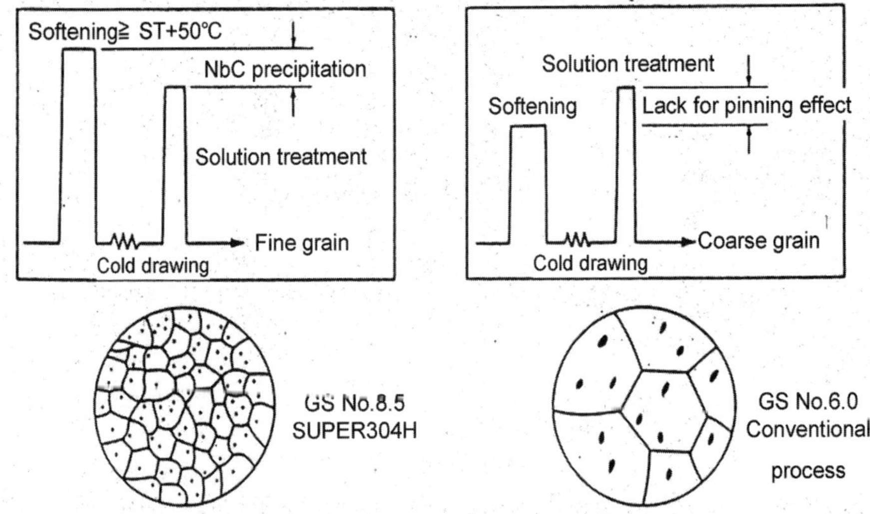

Fig. 1 Thermo-Mechanical Process for Fine-Grained Microstructure

This thermo-mechanical process adopts the 50°C higher softening treatment than final solution treatment before cold drawing. At final solution treatment, precipitated NbC is necessary to keep the pinning effect of grain-growth. Therefore, carbon and niobium (columbium) contents of SUPER304H and TP347HFG are optimized and strictly controlled to keep fine-grained microstructure with ASTM grain size No. 7-10. Excellent steam oxidation resistance of the fine-grained microstructure has been proved by long-term service exposure tests (2)(3). Both steels are applied precipitation strengthening of finely dispersed NbCrN during creep and prevention of σ phase degradation at the long-term creep stage. In addition, SUPER304H is strengthened by high nitrogen content with 0.15 mass% for tensile region and strengthened by finely dispersed Cu-rich phase for creep region (3).

HR3C (ASME CC2115, TP310HCbN) has higher Cr content with 25% than 18Cr austenitic stainless steels. Therefore, fine-grained microstructure is not required for HR3C in order to achieve enough steam oxidation resistance (2). Creep strengthening of HR3C with 25Cr-20Ni-0.4Nb-0.2N is achieved by finely dispersed $M_{23}C_6$ and NbCrN during creep and also preventing σ phase precipitation at the long-term creep stage (5).

Recently, SUPER304H, TP347HFG and HR3C steel tubes have been widely used in newly built and near future planned USC boilers as superheater and reheater all over the world. In order to assure the reliability of strength and microstructure for the developed steels, this paper describes up-dated long-term creep rupture and aging properties.

2. Long-Term Creep Rupture Properties

2-1. SUPER304H

Fig.2 shows creep rupture properties of SUPER304H steel tubes. New data are added as solid marks.

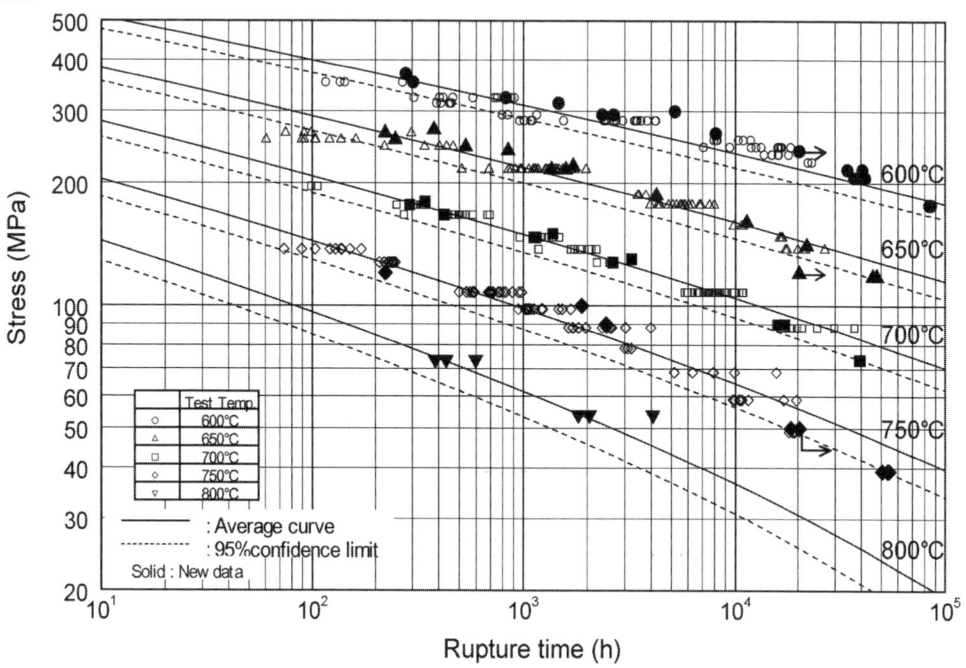

Fig. 2 Creep Rupture Properties of SUPER304H

Average and 95% minimum curves at 600-800°C are calculated by Larson-Miller parametric method using all data. Creep rupture strengths keep stable for long-term creep stage over 50,000h at each temperature. The longest creep rupture at 600°C and 177MPa is 85,426.7h which is enough for proving stable strength of SUPER304H.

Fig.3 shows creep rupture ductilities of SUPER304H. Creep rupture ductilities are enough high and do not show significant drop even after long-term creep rupture over 50,000h.

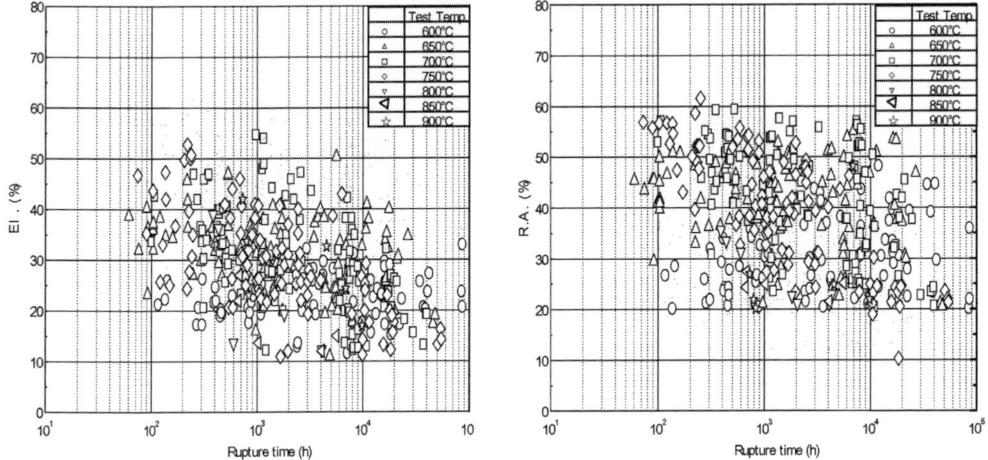

Fig.3 Creep Rupture Ductilities of SUPER304H

2-2. TP347HFG

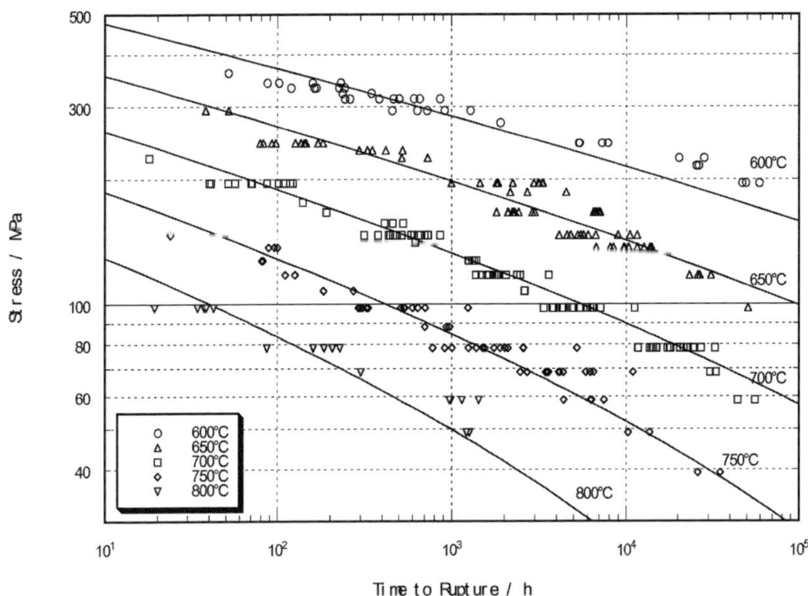

Fig.4 Creep Rupture Properties of TP347HFG

Fig.4 shows creep rupture properties of TP347HFG. Average curves at 600-800°C are calculated by Larson-Miller parametric method using all data. Creep rupture strengths keep stable for long-term creep stage over 50,000h at each temperature. The longest data at 700°C and 59MPa are 55,858.1h which is enough to prove stable strength of TP347HFG.

2-3. HR3C

Fig.5 shows creep rupture properties of HR3C. New data are added as solid marks. Average and 95% minimum curves at 600-800°C are calculated by Larson-Miller parametric method using all data. Creep rupture strengths keep stable for long-term creep stage over 50,000h. The longest creep rupture at 700°C and 69MPa is 88,362.7h which is enough for proving stable strength of HR3C.

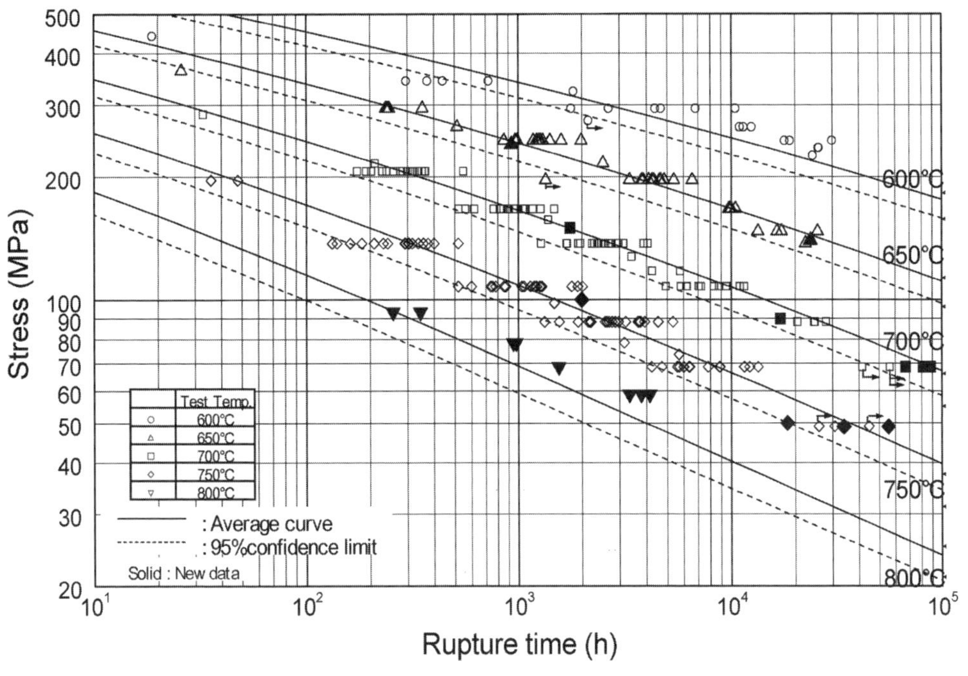

Fig.5 Creep Rupture Properties of HR3C

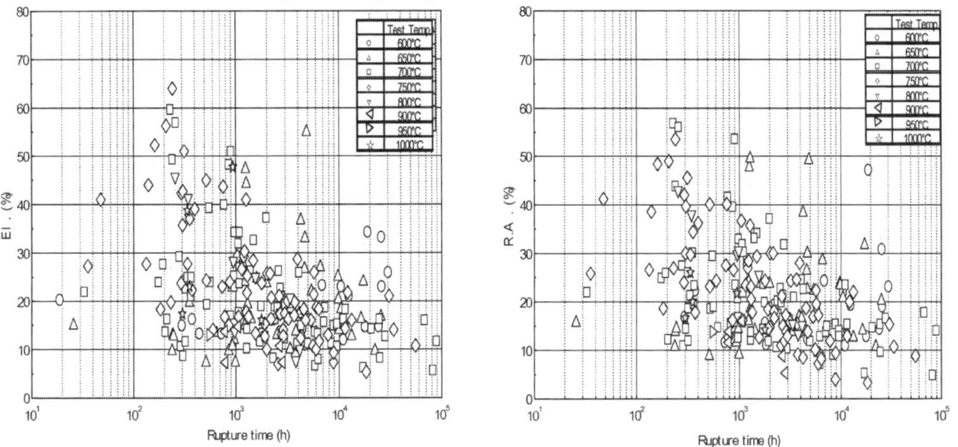

Fig.6 Creep Rupture Ductilities of HR3C

Fig.6 shows creep rupture ductilities of HR3C. Creep rupture ductilities are enough high and do not show significant drop even after long-term creep rupture at 700°C for 88,362.7h.

2-3. Extrapolated Long-Term Creep Strengths

Based on the up-dated long-term creep rupture data, extrapolated 100,000h creep rupture strengths are calculated by Larson-Miller parametric method and listed in Table 2.

Table 2 Extrapolated 100,000h Creep Rupture Strengths (MPa)

Steel	SUPER304H			TP347HFG			HR3C		
Temperature (°C)	600	650	700	600	650	700	600	650	700
100,000h Creep Rupture Strength, Ave.	178	115	70.3	159	99.3	57.3	176	111	67.1
Allowable Stress/0.67	181	116	70.0	154	91.0	49.3	176	103	61.7
100,000h Creep Rupture Strength, Min.	163	104	62.1	142	86.9	48.3	158	98.1	58.4
Allowable Stress/0.8	151	98.0	58.6	129	76.3	41.2	148	86.0	51.7

The up-dated 100,000h strengths for three steels are proved to keep stable and satisfy the average and the minimum strengths from ASME allowable stress values.

3. Toughness and Structural Changes of HR3C after Long-Term Aging

Fig.7 shows Charpy impact values at 0°C after aging at 500-750°C up to 30,000h. Although the impact values after aging drop to one tenth of virgin material, such drop occurs at the short-term aging and then the impact values after long-term aging have enough for a practical use.

Fig.8 shows optical micrographs of HR3C after long-term aging. A lot of precipitates are observed within grains and along with grain boundary. No σ phase like precipitates are not observed even after long-term aging.

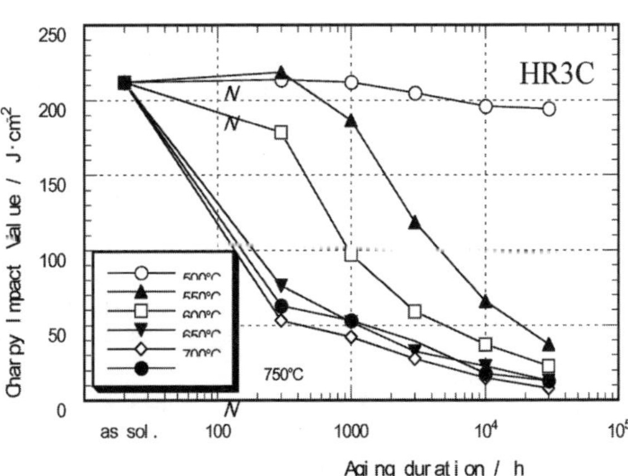

Fig.7 Charpy Impact Values at 0°C after Aging of HR3C

Amount of extracted residues after aging of HR3C are analyzed and shown in Fig.9. Precipitation has been still continuing at 650°C over 10,000h and almost saturated at 750°C over 300h. Therefore, the drop of Charpy impact values attributes to the precipitation behavior at each temperature and seems to stop anymore, because precipitation is over and remarkable brittle phase is not observed.

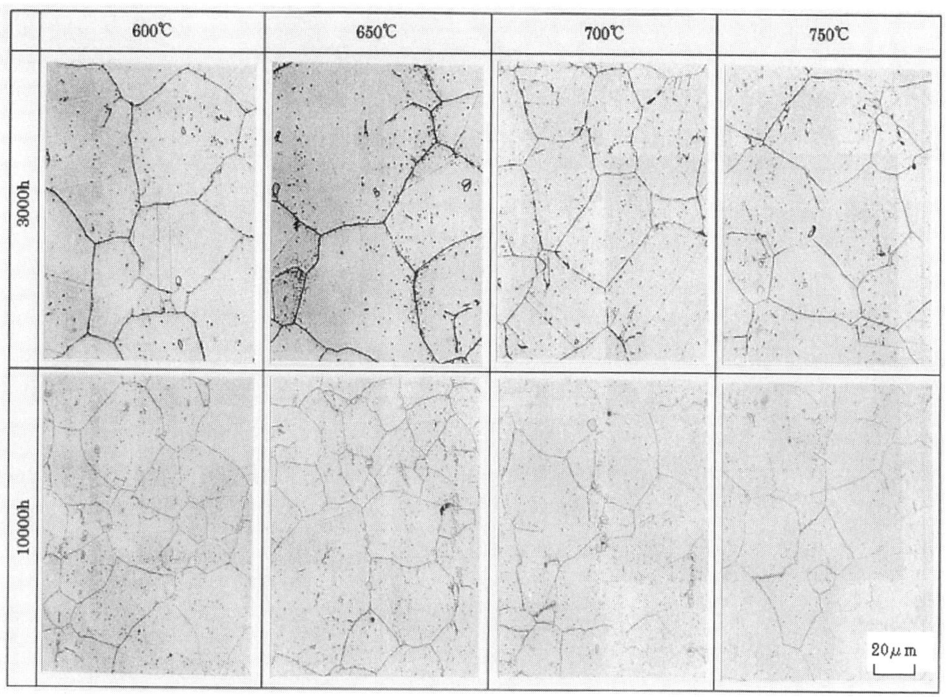

Fig.8 Optical Micrographs after Long-Term Aging of HR3C

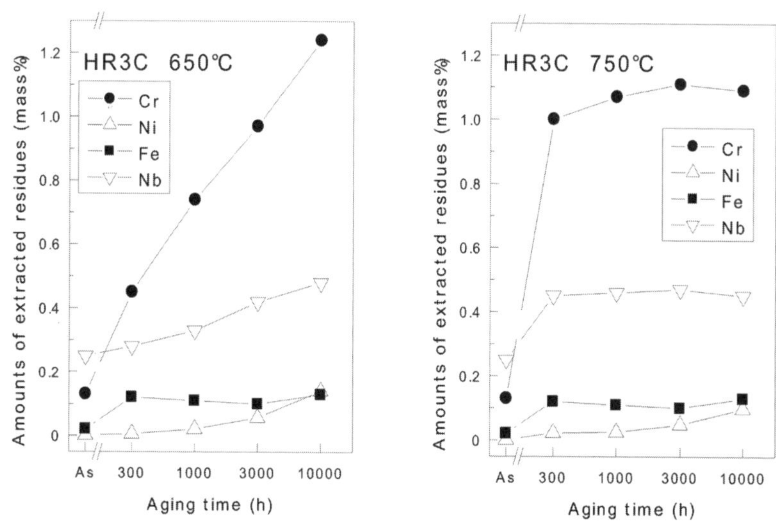

Fig.9 Analysis of Extracted Residues after Long-Term Aging of HR3C

In Fig.9, extracted Fe which seems to be included mainly in $M_{23}C_6$ is low and does not increase at long-term aging. This is another evidence that σ phase does not precipitate significantly, because Fe is a main element of σ phase with consist of ($Fe_{0.5}Cr_{0.5}$) normally. Conclusively,

HR3C has enough toughness and stable macrostructure without significant σ phase precipitation even after long-term aging.

3. Microstructural Stabilities after Long-Term Creep Rupture

3-1. SUPER304H

Fig.10 and Fig. 11 show optical micrographs and TEM images of extracted replicas of SUPER304H respectively after the longest creep rupture at each temperature. Many precipitates are observed in grains and along grain boundaries. These precipitates are mainly identified as $M_{23}C_6$ and finely dispersed NbCrN, which are effective in long-term creep strength. Although a few and very small σ phase is observed and identified along grain boundaries, amount of σ phase precipitation is significantly smaller than those of conventional 304H and 321H type steels (6)(7).

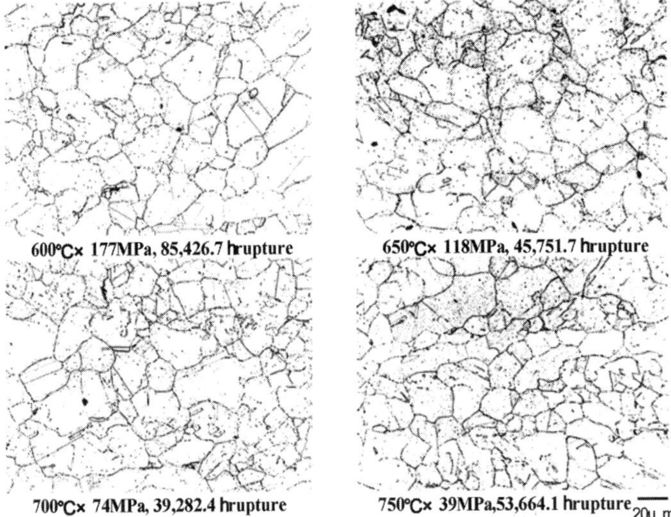

Fig.10 Optical Micrographs after Longest Creep Rupture of SUPER304H

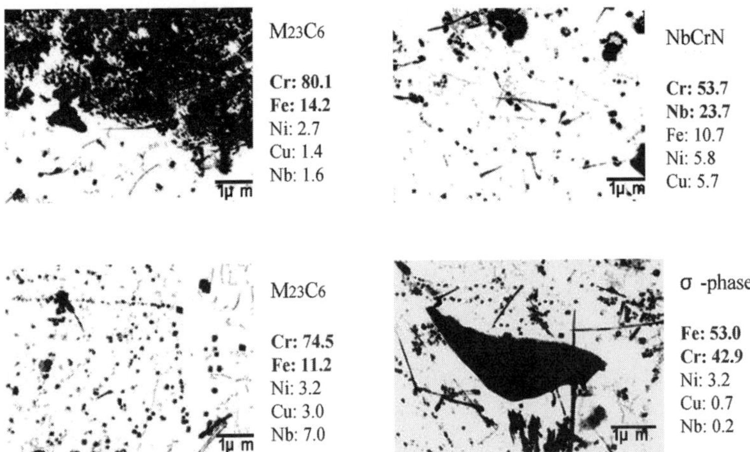

Fig.11 TEM Images of Extracted Replicas after Longest Creep Rupture at 600°C and 177MPa for 85,426.7h of SUPER304H

3-2. HR3C

Fig.12 and Fig. 13 show optical micrographs and TEM images of extracted replicas for HR3C respectively after the longest creep rupture at each temperature. Many precipitates are observed in grains and along grain boundaries. These are mainly identified as $M_{23}C_6$ and finely dispersed NbCrN, which are effective in long-term creep strength. Although very small σ phase and G phase are observed and identified along grain boundaries, total amounts of σ phase and G phase precipitation seem to be smaller than those of conventional 304H and 316H type steels (6)(8).

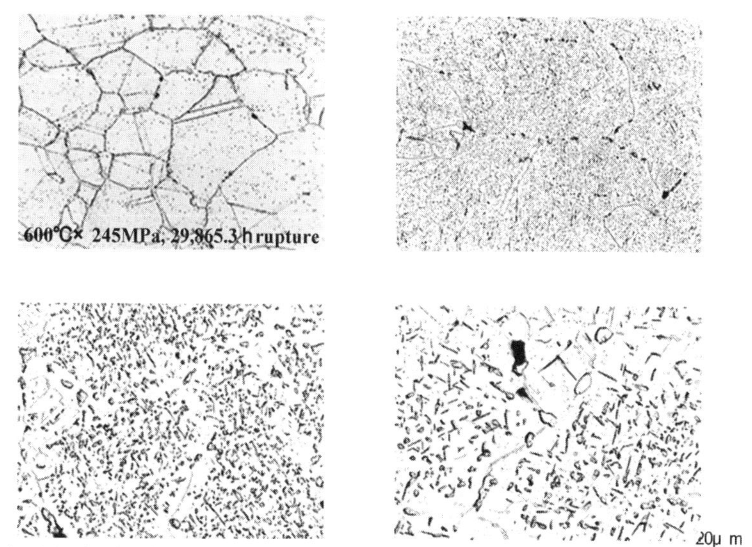

Fig.12 Optical Micrographs after Longest Creep Rupture of HR3C

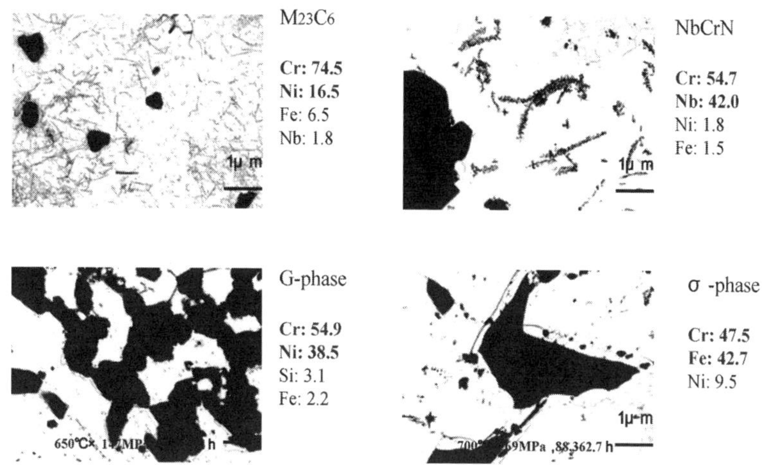

Fig.13 TEM Images of Extracted Replicas after Longest Creep Rupture of HR3C

3-3. Precipitation Behavior of SUPER304H and HR3C

Time-Temperature Precipitation diagrams of SUPER304H and HR3C are estimated and shown in Fig.14 and Fig.15 respectively by means of microstructural observation of the long-term crept and aged samples. For comparisons, TTP diagrams of conventional 304H and 321H steels reported by NIMS are shown in Fig.16 and 17 respectively (6)(7). In the laboratory crept samples, significant σ phase is not observed for SUPER304H as mentioned the above. In the case of HR3C, precipitation curve of σ phase is not settled because remarkable σ phase is not observed. Comparing with conventional 304H and 321H steels, σ phase precipitation behavior of SUPER304H and HR3C is proved to be few and prevented. Therefore, it is concluded that phase stabilities of SUPER304H and HR3C are good enough and reliability of long-term creep rupture strength and aging toughness are supported based on the long-term creep rupture data over 85,000h for SUPER304H and over 88,000h for HR3C.

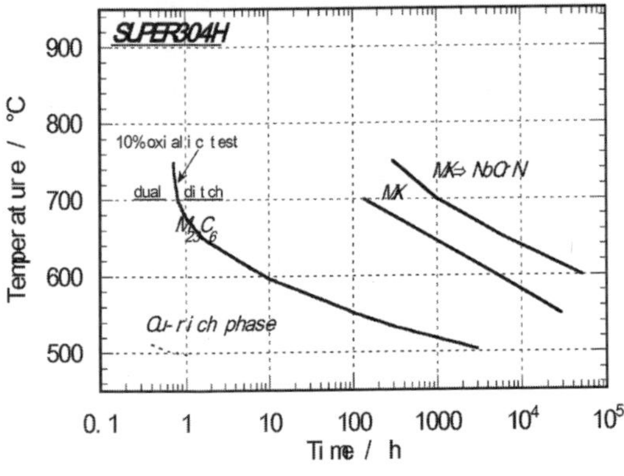

Fig.14 Estimated TTP Diagram of SUPER304H

Fig.15 Estimated TTP Diagram of HR3C

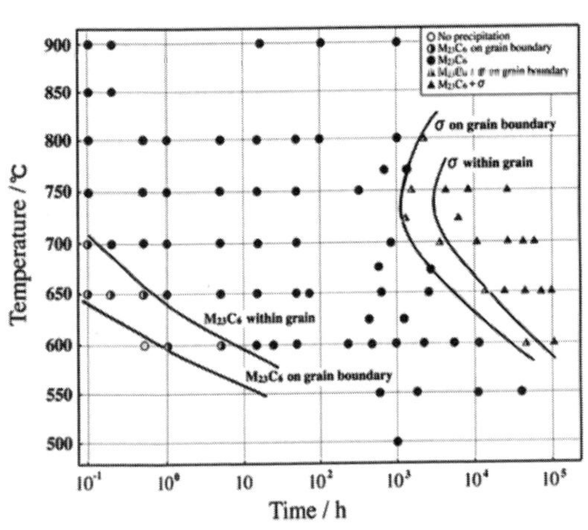

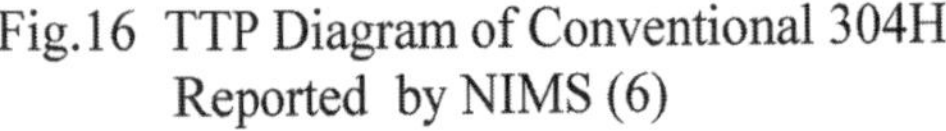

Fig.16 TTP Diagram of Conventional 304H Reported by NIMS (6)

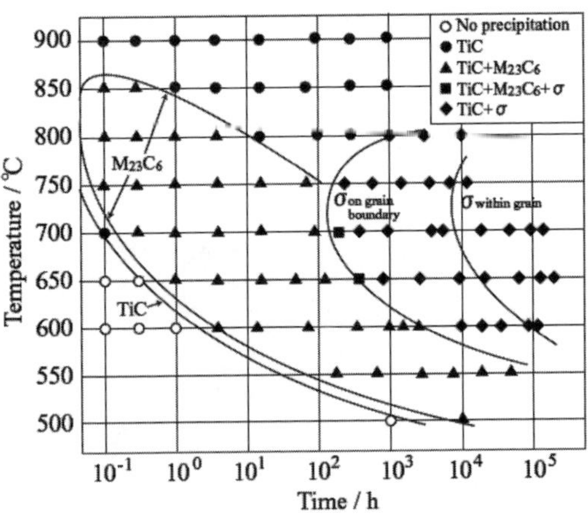

Fig.17 TTP Diagram of Conventional 321H Reported by NIMS (7)

SUPER304H, HR3C and TP347HFG steel tubes have been installed in Eddystone No.3 USC power plant as superheater and reheater tubes since 1991, and have been removed and investigated microstructure. Total service time is 75,075h operating at 650°C and 35MPa. The stable microstructure and the strength reliability of SUPER304H, TP347HFG and HR3C after long-term service in the practical USC boiler are discussed in detail and proved definitely (9).

4. Conclusion

SUPER304H (18Cr-9Ni-3Cu-Nb-N, ASME CC2328) and TP347HFG (18Cr-12Ni-Nb, ASME SA213) ,which are applied fine-grained microstructure, have been developed and widely applied for superheater and reheater tubes in newly built USC boilers since 1995. In order to verify the long-term creep rupture strengths, up-dated data, witch include the longest creep rupture at 600°C for 85,426h of SUPER304H and at 700°C for 55,858h of TP347HFG, are collected and analyzed by Larson-Miller parametric method using all data. The extrapolated 100,000h creep rupture strengths of both steels are still high enough and satisfy ASME allowable stresses. The stable microstructure and the phase stability withoutσ phase are confirmed after long-term creep rupture over 85,000h, which definitely supports the long-term creep rupture strength.

HR3C (25Cr-20Ni-Nb-N, ASME CC2115) have been developed for high strength and high corrosion resistant steel tube and used in a severe corrosion parts of recent USC boilers. The up-dated long-term creep rupture data including the longest creep rupture at 700°C for 88,362h of HR3C confirm that the extrapolated 100,000h creep rupture strengths are high enough and satisfy ASME allowable stresses. The microstructural investigation including the longest crept specimen supports the stability of finely dispersed precipitates with preventing significantσ phase precipitation and proves the reliability of long-term creep rupture strength. The long-term aging tests up to 30,000h at 500-750°C confirm enough toughness and stable microstructure.

The above three steel tubes have been installed in Eddystone No.3 USC power plant as superheater and reheater tubes since 1991, and have been investigated and proved the stable microstructure and strengths. Sumitomo Metals' SUPER304H, TP347HFG and HR3C steel tubes have been successfully applied as standard materials for superheater and reheater tubes in newly built USC power plants in the world.

References

1. R. Viswanathan, Proceedings of the 7th Liege Conference, Part II(2002), 1009.

2. A. Iseda, Proceedings of the 14th Conference on the Electric Power Supply Industry CEPSI 2002 Fukuoka, Japan (November 5-8, 2002), Paper No.IT-7, pp350-355.

3. Y. Sawaragi, K. Ogawa, S. Kato, A. Natori and S. Hirano, The Sumitomo Search, vol.48 (1992), pp50-58.

4. H. Teranishi, Y. Sawaragi, M. Kubota and Y. Hayase, The Sumitomo Search, vol.38 (1989), pp63-74.

5. Y. Sawaragi, H. Teranishi, A. Iseda, K. Yoshikawa, The Sumitomo Search, vol.44 (1990), pp146-158.

6. NIMS Creep Data Sheet, "Metallographic Atlas of Long-Term Crept Materials No.M-1" (2002).

7. NIMS Creep Data Sheet, "Metallographic Atlas of Long-Term Crept Materials No.M-3" (2005).

8. NIMS Creep Data Sheet, "Metallographic Atlas of Long-Term Crept Materials No.M-2" (2003).

9. H. Okada, M. Igarashi, S.Yamamoto, O. Miyahara, A. Iseda, N. Komai and F. Masuyama, Proceedings of 8th Conference on Creep Fatigue at Elevated Temperatures, San Antonio, TX, USA (July 22-26, 2007), Paper No. PVP2007-26561.

Advances in Materials Technology for Fossil Power Plants
Proceedings from the Fifth International Conference
R. Viswanathan, D. Gandy, K. Coleman, editors, p 197-207

DOI: 10.1361/cp2007epri0197

Prediction of the Loss of Precipitation Strengthening in Modern 9-12% Cr Steels – A Numerical Approach

I. Holzer[1]
E. Kozeschnik[1,2]
H. Cerjak[1]
[1] Institute for Materials Science, Welding and Forming,
Graz University of Technology, Kopernikusgasse 24, A-8010 Graz, Austria.
Ph. 43-316-873-7181 Fax 43-316-873-7187
[2] Materials Center Leoben Forschungsgesellschaft mbH,
Franz-Josef Straße 13, A-8700 Leoben, Austria.
Ph. 43-3842-459221 Fax 43-3842-459225

Abstract

The creep resistance of 9-12% Cr steels is significantly influenced by the presence and stability of different precipitate populations. Numerous secondary phases grow, coarsen and, sometimes, dissolve again during heat treatment and service. Based on the software package MatCalc, the evolution of these precipitates during the thermal treatment of the COST 522 steel CB8 is simulated from the cooling process after cast solidification to heat treatment and service up to the aspired service life time of 100.000h. On basis of the results obtained from these simulations in combination with a newly implemented model for evaluation of the maximum threshold stress by particle strengthening, the strengthening effect of each individual precipitate phase, as well as the combined effect of all phases is evaluated – a quantification of the influence of Z-Phase formation on the long-term creep behaviour is thus made possible. This opens a wide field of application for alloy development and leads to a better understanding of the evolution of microstructural components as well as the mechanical properties of these complex alloys.

Keywords: Microstructure evolution, heat treatment, precipitation kinetics, simulation, particle strengthening, threshold stress.

1 Introduction

Great efforts were undertaken in the field of material development for thermal power plant applications during the last decades to produce materials with optimized mechanical properties, especially superior creep strength. These materials are characterized by a microstructure with a dense distribution of precipitates, tight subgrain structure and high dislocation density within the subgrains as shown in Figure 1b, which exhibits superior resistance against plastic deformation even at high temperatures and long service times. Such a behaviour is achieved by strong pinning forces on dislocations as well as on grain and subgrain boundaries caused by precipitates, which leads to a stabilisation of the microstructure.

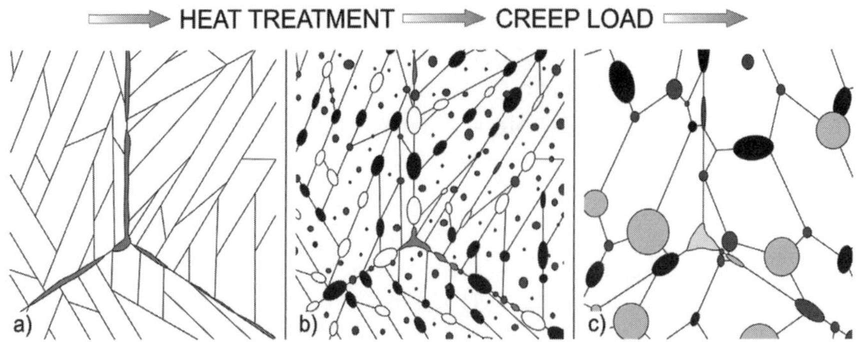

Figure 1: Schematic evolution of the microstructure during heat treatment and service. Note that all images relate to the same length scale.

These interactions, strongly influenced by particle evolution, lead to the described microstructure with superior creep properties, which have been investigated in numerous experimental and theoretical studies (1-8). In the following, precipitation strengthening in modern 9-12% Cr steels will be shortly outlined and discussed with regard to the contribution to creep strength.

2 Precipitation strengthening in ferritic/martensitic 9-12% Cr steels

As pointed out in the introduction, precipitates act as efficient obstacles for dislocation movement as well as microstructure stabilizing components, interacting with grain and subgrain boundaries (see also ref. (9)). The precipitate – dislocation interaction will be outlined in the following with regard to strength contribution and effect of creep behavior.

2.1 Precipitate – dislocation interaction

As pointed out by McLean (10), precipitates and dislocations can interact in one of the following ways:

1. A dislocation can pass coherent precipitates by cutting (breaking) the precipitate. A stacking fault is left in the precipitate.

2. A dislocation can pass precipitates by bending between them and closing the bent lines to loops. A dislocation is left around the by-passed precipitate (Orowan mechanism).

3. A dislocation can pass the precipitate by climbing.

4. A dislocation can drag the precipitates with it. This mechanism is possible only for very small precipitates. The velocity-determining factor in this case is the mobility of the dragged precipitates.

The operating precipitate - dislocation interaction mechanism is depending on a number of factors, among which the availability of glide planes, the height of the local forces and the hardness of the precipitates are most important. Due to the physical nature of the four processes, mechanisms 1 and 2 are considerably faster than mechanisms 3 and 4. If the latter are the operating creep mechanisms, the creep rate is significantly lower than the creep rate based on mechanisms 1 and 2. For details, the interested reader is referred to references (10-14).

2.2 Back stress of precipitates and influence on creep strength

If an external force is acting on a microstructure, part of the external driving pressure σ_{ex} is counteracted by heterogeneous internal microstructural constituents, such as precipitates and interfaces. Consequently, not the entire external load can be assumed to represent the driving force for plastic deformation processes such as creep. Only the part of the external stress σ_{ex}, which exceeds the amount of inner stress σ_i from the counteracting microstructure, effectively contributes to the deformation process. Since the inner stress reduces the effect of the external stress, this inner stress is commonly denoted as back-stress and the approach is known as the back-stress concept. The effective creep stress σ_{eff} can be expressed as

$$\sigma_{eff} = \sigma_{ex} - \sigma_i \,. \tag{1}$$

In a recent treatment by Dimmler (15), the inner stress σ_i has been expressed as a superposition of individual contributions from dislocations and precipitates. An advanced approach is shown in the following, where also the contribution from subgrain boundaries is taken into account. Thus, the inner stress can be expressed as

$$\sigma_i = M\tau_i = M(\tau_{disl} + \tau_{prec} + \tau_{sgb}), \tag{2}$$

where M is the Taylor factor (usually between 2 and 3, see ref. (15)) and τ is the shear stress. The subscripts in the bracket term denote contributions from dislocations, precipitates and subgrain boundaries, respectively. When taking into account the inner stress, the general Norton creep law (see, e.g., reference (16)) can be rewritten as

$$\dot{\varepsilon} = A \cdot (\sigma_{ex} - \sigma_i)^n = A \cdot \sigma_{eff}^n, \tag{3}$$

where A and n are constants. Assuming that subgrain size and dislocation density remain more or less constant during most of the service, the inner stress or back stress of the microstructure and, therefore, the mechanical behaviour of the alloy will be mostly influenced by the precipitate evolution.

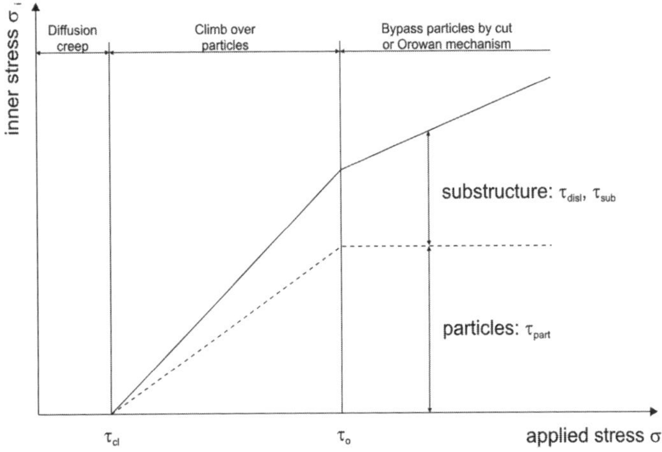

Figure 2: Inner stress σ_i depending on the back-stress of particles (τ_{part}), substructure (τ_{disl}, τ_{sub}) and applied stress σ (see also ref. (15))

From Figure 2, it can be seen that, depending on the applied stress, threshold stresses can be defined where the particle bypass mechanism for mobile dislocations change. At stresses higher than τ_{cl}, dislocations climb over particles and above τ_o, dislocations overcome the particles by cutting or by the Orowan mechanism. The latter represents the maximum back-stress caused by hard precipitates and can be expressed according to Ashby as (17)

$$\tau_{part} = C\frac{Gb}{\lambda}\ln\left(\frac{\zeta}{r_0}\right), \tag{4}$$

where C is a constant (C=0.159 for screw dislocations and C=0.227 for edge dislocations), G is the shear modulus, b is the Burgers vector, λ is the mean particle distance, r_0 is the 'inner cut-off radius' and ζ the 'outer cut-off radius' of the dislocation. Dimmler (15) showed that the transition from viscous glide to the power law breakdown creep in 9-12% Cr steels is linked to this mechanism change of particle bypassing. In the following, the dependence of this maximum back-stress as function of several individual precipitate phases will be examined on the example of the COST alloy CB8, a typical representative of a 9-12% Cr ferritic/martensitic steel.

3 The COST alloy CB8

The experimental test alloy CB8 has been designed in the COST[1] program 522. It is a typical 9-12% Cr steel for cast application, which showed excellent creep properties in short-term tests. For this reason, the variant CB8 has been extensively investigated also at longer times. However, a significant drop in creep strength has been observed after approximately 10,000 hours of service exposure. The chemical analysis of the steel CB8 is given in Table 1. The time temperature regime of the production process with the additional service at 650°C is shown in Figure 3.

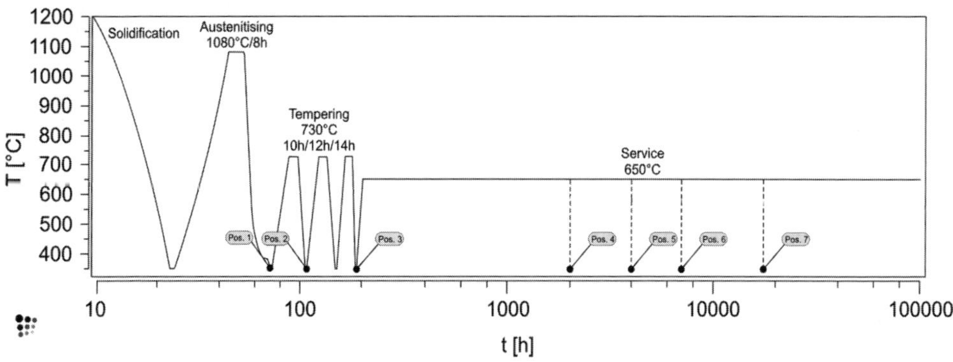

Figure 3: Simulated time-temperature regime of the whole lifetime of COST alloy CB8 with defined specimen positions for experimental characterization

At defined positions during the heat treatment (position 1-3) and service at 650°C (position 4-7), specimens were taken to characterize the present microstructure. Table 2 summarizes the

[1] COST (COoperation in the field of Science and Technology): European collaborative project to develop improved high temperature materials for thermal power plants.

experimental observed evolution of phase fraction (f), mean particle radius (R) and number density (N) of MX, VN and $M_{23}C_6$ precipitates (from Sonderegger (18) and Plimon (19)).

Table 1: Chemical composition of steel CB8 – heat 173 (in wt%)

C	Si	Mn	Cr	Ni	Mo	W	V	Nb	Co	Al	B (ppm)	N
0.17	0.27	0.2	10.72	0.16	1.40	-	0.21	0.060	2.92	0.028	112	0.0319

Table 2: Overview of evolution of phase fraction (f), mean particle radius (R) and number density (N) of MX, VN and $M_{23}C_6$ precipitates during heat treatment and service (18)

		MX			VN			$M_{23}C_6$		
Pos.	time [h]	f [%]	R [nm]	N [m^{-3}]	f [%]	R [nm]	N [m^{-3}]	f [%]	R [nm]	N [m^{-3}]
1	67.5	0.07	31.5	$3.07 \cdot 10^{18}$	-	-	-	-	-	-
2	100	0.12	23.0	$8.61 \cdot 10^{18}$	-	-	-	2.55	50.0	$2.33 \cdot 10^{19}$
3	180	0.18	27.0	$9.2 \cdot 10^{19}$	-	-	-	2.28	54.0	$1.77 \cdot 10^{19}$
4	2,000	-	-	-	0.230	21.0	$3.07 \cdot 10^{19}$	2.19	36.5	$5.55 \cdot 10^{19}$
5	4,000	-	-	-	0.200	39.0	$4.33 \cdot 10^{18}$	2.13	57.5	$1.60 \cdot 10^{19}$
6	7,000	-	-	-	0.300	28.0	$1.64 \cdot 10^{19}$	4.13	69.5	$1.42 \cdot 10^{19}$
7	16,000	-	-	-	0.060	34.5	$1.62 \cdot 10^{18}$	2.03	61.5	$1.11 \cdot 10^{19}$

4 Results and Discussion of Numerical Simulation of Precipitation Evolution and Strengthening

For the following numerical simulation of the precipitate evolution, the software MatCalc[2] (20) is used. The underlying theory and model implementation is described in refs. (21-23).

4.1 *Kinetic simulation*

In the present simulation, the precipitate phases $M_{23}C_6$, M_7C_3, MX, Laves and the modified Z-phase are taken into account in accordance with experimental findings. The experiments show that two types of MX precipitates are present in this steel, i.e. a vanadium- and nitrogen-rich phase (VN) and a niobium- and carbon-rich phase (NbC). During the simulation, all precipitates interact with each other by exchanging atoms with the matrix phase. The kinetics of this process is controlled by the multi-component diffusivities of all elements, which are available through kinetic databases, such as the mobility database of the software package DICTRA (24). The thermodynamic parameters for calculation of the chemical potentials are taken from the TCFE3 database (25) with some modifications specific to this type of steel (26). Apart from accurate thermodynamic and kinetic data, a very important input parameter for the simulation is the type of heterogeneous nucleation site for each of the precipitate phases, which can be grain boundaries, subgrain boundaries, dislocations, grain boundary edges and/or grain boundary corners. For the present simulation, the nucleation site for each precipitate is defined according to the experimental observations summarized in ref. (18).

---

[2] http://matcalc.tugraz.at

The simulated temperature profile (top image in Figure 4) is defined according to the time-temperature regime shown in Figure 3. The simulation starts at 1400°C, which is slightly below the solidus temperature of this alloy during the casting procedure. It is assumed that all elements are homogeneously distributed in the matrix at this time and no precipitates exist. The material then cools linearly to a temperature of 350°C. This temperature corresponds to the observed austenite to martensite start temperature. In the simulation, it is assumed that this transformation occurs instantaneously and the parent and target phases have identical chemical composition. It is further assumed that no diffusive processes and, consequently, no precipitation occurs below this temperature. At this point, the matrix phase is changed from face centered cubic (fcc) austenite to body centered cubic (bcc) ferrite structure.[3] In the next step, the material is reheated for austenitization. At the experimentally observed transformation temperature of 847 °C, the ferrite matrix is changed to austenite again. After austenitization and the following cooling phase, the transformation to martensite/ferrite is assumed to take place again at 350°C. After three annealing quality heat treatments, service at 650°C for 100,000 hours is simulated.

The three lower plots in Figure 4 display the evolution of the phase fraction, mean precipitate radius and number density of each precipitate type, calculated by the software MatCalc. The simulation results are in good agreement with the obtained experimental data from Table 2. After about 10,000h, a depletion of VN in the matrix is observed until they completely dissolve at about 20,000h. This dissolution of VN is linked to the appearance of Z-phase, which is observed in the investigated specimens. As mentioned above, growth, coarsening and dissolution of secondary phases strongly influence the precipitation strengthening in engineering materials during their lifetime, which will be discussed in the following.

---

[3] Since no separate thermodynamic description is available for the bct martensite phase, the bct phase is substituted by the bcc phase in the simulations.

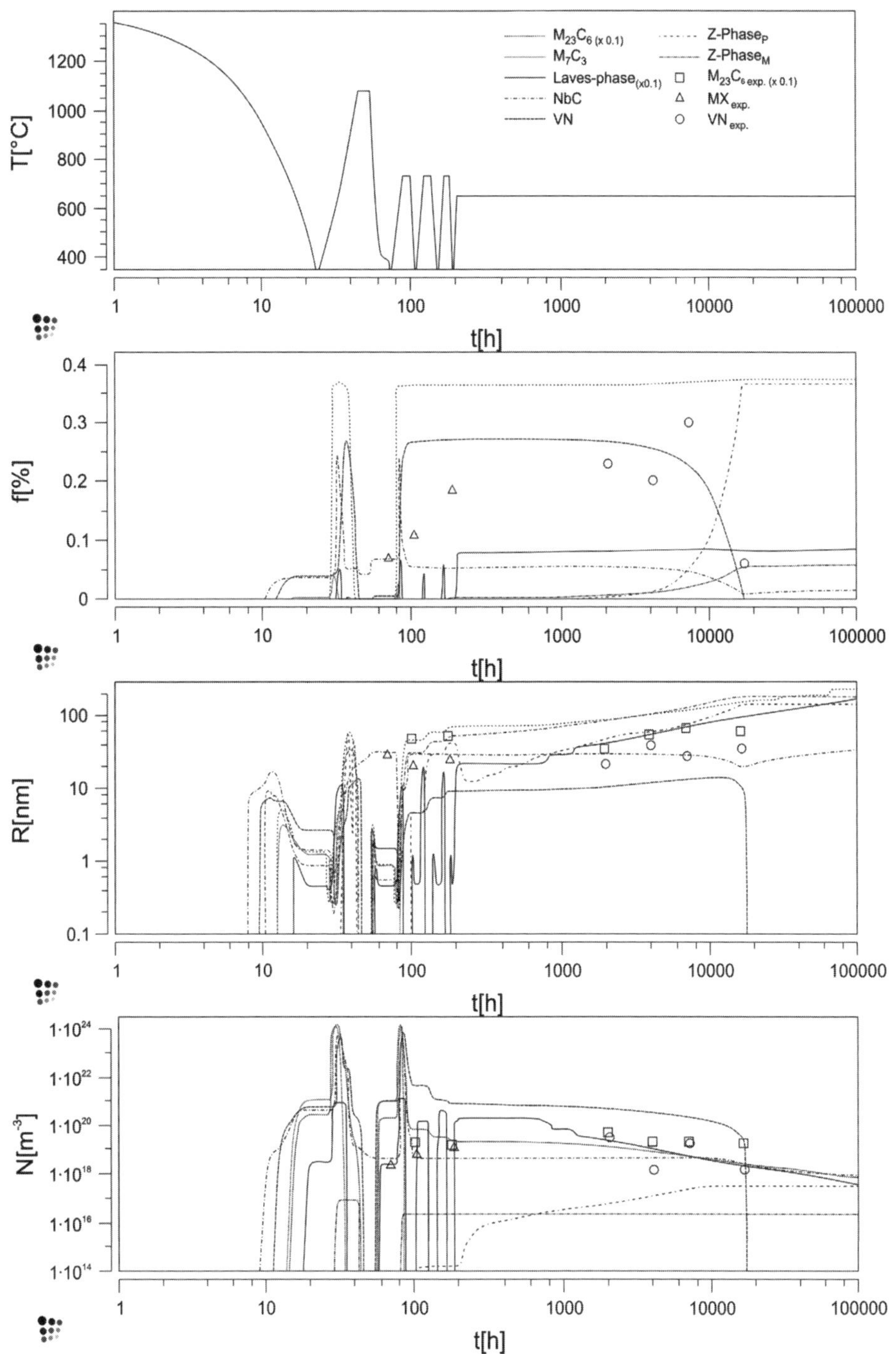

Figure 4: Kinetic simulation of the precipitate evolution in CB8 during heat treatment and service using MatCalc (f…phase fraction, R…mean precipitation radius and N…number density). The phase fractions of $M_{23}C_6$ and Laves-phase are multiplied by a factor of 1/10 to give a better visual representation of the results. Experimental values are taken from Table 2.

4.2 Determination of precipitation strengthening during service of CB8

Following the argumentation in section 2.2 the maximum obstacle effect of precipitates can be evaluated with equation (4). For a complex alloy with several precipitate populations, an assumed outer cut-off radius of two times the precipitate radius and an inner cut-off radius of two times the Burger's vector with b ~ 0,25 nm, equation (4) takes the form

$$\tau_O = CGb\sqrt{\frac{\pi \cdot \Sigma n_i}{6}} \ln\left(\frac{\Sigma r_i^4}{b \cdot \Sigma r_i^3}\right), \tag{5}$$

where n_i is the number density and r_i the radius of each precipitate class. The constants in equation (5) are assumed to be C=0.19 and G=62.3 GPa (converted from the data in ref. (27)). Figure 5 shows the predicted evolution of the back-stress during service of the COST alloy CB8 at 650°C calculated with the model outlined above. The graphs show the predicted contributions of each phase separately as well as the total back-stress including the combined effect of all precipitates with and without the effect of Z-phase.

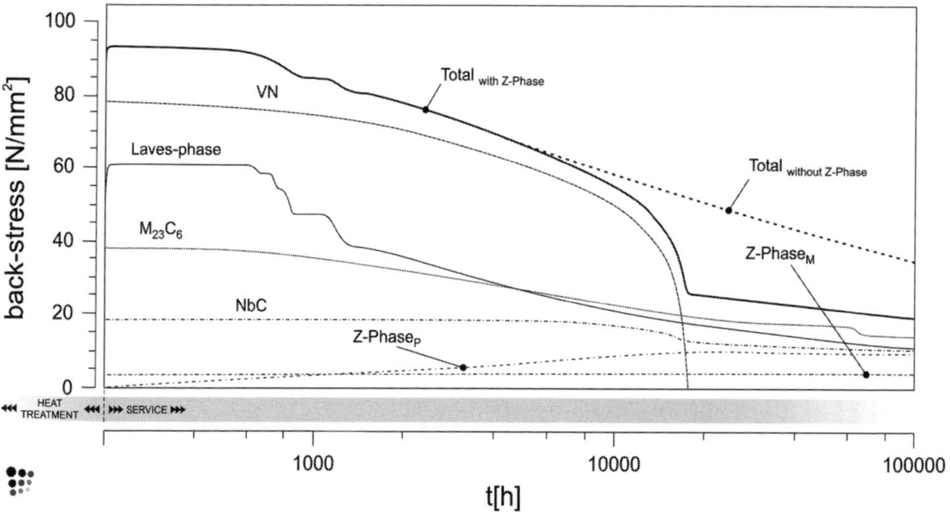

Figure 5: Predicted degradation of the back-stress during service of CB8 including the modified Z-phase contribution (solid line) and artificially suppressing this phase (dashed line). Subscript P denotes precipitates of Z-Phase nucleated on existing VN particles, the subscript M denotes Z-Phase precipitates nucleated in the matrix.

In the as-received condition, after all heat treatments, the total strength contribution from precipitation hardening is estimated to be in the order of 90 MPa[4]. Due to the inevitable effect of Ostwald ripening (coarsening), the density of precipitates is reduced, which is reflected in the gradual decrease of the total back-stress up to times of approximately 10,000 hours. When the modified Z-phase is included, the total back-stress shows a drastic reduction between 10,000 and 20,000 hours. This effect is due to the enhanced nucleation and subsequent growth of the Z-phase, which causes dissolution of the VN precipitates. In the later stages, the

[4] This quantity varies strongly with selection of input values and should therefore not be considered in terms of absolute values.

decrease of back-stress continues in a steady manner again, however, on a much lower level than before. Comparison of the curves for the integrated back-stress in Figure 5 indicates that Z-phase formation causes an additional back-stress reduction at 100,000 hours of approximately 20 MPa – compare the solid and dashed line of the total back-stress. This effect is assumed to be responsible for the drop in creep strength of various different ferritic/martensitic creep resistant 9-12% Cr-steels during long-term creep exposure (9,18).

5 Outlook

Since precipitation strengthening is a key mechanism for improving mechanical properties of creep resistant materials, the introduced model opens a wide field in material development. It is shown that the strengthening contribution from precipitates can be predicted for the entire lifetime of an engineering material. The strengthening effect of MX particles and the deteriorating influence of Z-Phase due to dissolution of MX particles is shown. Therefore, an explanation for the sudden collapse of creep strength of this steel can be given. However, for a profound understanding of all microstructural processes of influence on mechanical properties of such materials, further work has to be done. An advanced model to consider also the influence of the substructure on the generated back-stress of a microstructure will be object of future works.

6 Acknowledgement

This work was part of the Austrian research cooperation "ARGE ACCEPT – COST 536" and was supported by the Austrian Research Promotion Agency Ltd. (FFG) which is gratefully acknowledged. Financial support by the Österreichische Forschungsförderungsgesellschaft mbH, the Province of Styria, the Steirische Wirtschaftsförderungsgesellschaft mbH and the Municipality of Leoben under the frame of the Austrian Kplus Programme in the projects SP16 and SP19 is gratefully acknowledged.

References

1 R. Lagneborg, *Effect of grain size and precipitation of carbides on creep properties in Fe-20%Cr-35%Ni alloys*, J. of Iron and Steel Institute, 1503-1506, (1969).

2 K.E. Amin, J.E. Dorn, *Creep of a dispersion strengthened steel*, Acta Metall., Vol. 7, 1429-1434, (1969).

3 F.R.N. Nabarro, *Grain Size, Stress, and Creep in Polycrystalline Solids*, Physics of the Solid State, Vol. 42, 1456-1459, (2000).

4 J. Eliasson, A. Gustafson, R. Sandström, *Kinetic Modelling of the influence of Particles on Creep Strength*, Key Eng. Mater., Vols. 171-174, 277-284, (2000).

5 K. Maruyama, K. Sawada, J. Koike, *Strengthening Mechanisms of Creep Resistant Tempered Martensitic Steel*, ISIJ Int., Vol. 41, 641-653, (2001).

6 K. Sawada, K. Kubo, F. Abe, *Creep behaviour and stability of MX precipitates at high temperature in 9Cr-0,5Mo-1,8W-VNb steel*, Mater. Sci. Eng., Vol. A 319-321, 784-787, (2001).

7 F. Kauffmann, G. Zies, D. Willer, C. Scheu, K. Maile, K.H. Mayer, S. Straub, *Microstructural Investigation of the Boron containing TAF Steel and the Correlation to the Creep strength*, 31. MPA-Seminar *Werkstoff- und Bauteilverhalten in der Energie- & Anlagentechnik*, Stuttgart, 13.-14. Oktober 2005.

8 A. Kostka, K.-G. Tak, R.J. Helling, Y. Estrin, G. Eggeler, *On the contribution of carbides and micrograin boundaries to the creep strength of tempered martensite ferritic steels*, Acta Mater., Vol. 55, 539-550, (2007).

9 E. Kozeschnik, I. Holzer, Precipitation during heat treatment and service – Characterisation, simulation and strength contribution (Chapter 12), in: *Creep resistant steels*, edited by F. Abe, T.-U. Kern and R. Viswanathan, Woodhead Publishing, Cambridge (2007), in press.

10 M. McLean, *On the threshold stress for dislocation creep in particle strengthened alloys*, Acta metal. Vol. 33, 545-556, (1985).

11 L.M. Brown, R.K. Ham, in *Strengthening Methods in Crystals* (edited by A. Kelly and R.B. Nicholson), Elsevier, Amsterdam (1971).

12 R. Lagneborg, *Bypassing of dislocations past particles by a climb mechanism*, Scripta Metallurgica Vol. 7, 605-614, (1973).

13 J.D. Verhoeven, *Fundamentals of Physical Metallurgy*, John Wiley & Sons, New York, (1975).

14 E. Arzt, M.F. Ashby, *Threshold stresses in materials containing dispersed particles*, Scripta Metallurgica Vol. 16, 1285-1290, (1982).

15 G. Dimmler, PhD thesis, *Quantification of creep resistance and creep fracture strength of 9-12%Cr steel on microstructural basis*, Graz University of Technology, 2003 (in German).

16 J. Čadek, *Creep in metallic materials*, Elsevier, (1988).

17 M. Ashby, *The theory of the critical shear stress and work hardening of dispersion-hardened crystals*, G.S. Ansell, T.D. Cooper and F.V. Lenel (Eds), *Metallurgical Society Conference*, Vol. 47, Gordon and Breach, New York, 143-205, (1968).

18 B. Sonderegger, PhD thesis, *Characterisation of the Substructure of Modern Power Plant Steels using the EBSD-Method*, Graz University of Technology, 2005 (in German).

19 S.W. Plimon, Master thesis, *Simulation of an industrial heat treatment and accompanying microstructural investigation of a modern 9-12% chromium steel*, Graz University of Technology, 2004 (in German).

20 E. Kozeschnik and B. Buchmayr, Mathematical Modelling of Weld Phenomena 5 (London, Institute of Materials, Book, 734, 2001), 349-361.

21 J. Svoboda, F. D. Fischer, P. Fratzl, and E. Kozeschnik, Modelling of kinetics in multi-component multi-phase systems with spherical precipitates I. – Theory, *Mater. Sci. Eng. A*, Vol 385 (No. 1-2), 2004, p 166-174.

22 E. Kozeschnik, J. Svoboda, P. Fratzl, and F. D. Fischer, Modelling of kinetics in multi-component multi-phase systems with spherical precipitates II. – Numerical solution and application, *Mater. Sci. Eng. A*, Vol 385 (No. 1-2), 2004, p 157-165.

23 E. Kozeschnik, J. Svoboda, and F. D. Fischer, Modified evolution equations for the precipitation kinetics of complex phases in multi-component systems, *CALPHAD*, Vol 28 (No. 4), 2005, p 379-382.

24 J.O. Andersson, L. Höglund, B. Jönsson, and J. Ågren, Computer simulations of multicomponent diffusional transformations in steel, in: *Fundamentals and Applications of Ternary Diffusion*, G.R. Purdy (ed.), Pergamon Press, New York, NY, 1990, p 153-163.

25 TCFE3 thermodynamic database, Thermo-Calc Software AB, Stockholm, Sweden, 1992-2004.

26 J. Rajek, PhD thesis, *Computer simulation of precipitation kinetics in solid metals and application to the complex power plant steel CB8*, Graz University of Technology, (2005).

27 G. Guntz, M. Julien, G. Kottmann, F. Pellicani, A. Pouilly and J.C. Vaillant, *The T 91 Book – Ferritic tubes and pipe for high temperature use in boilers*, Vallourec Industries – France, 1991.

Advances in Materials Technology for Fossil Power Plants
Proceedings from the Fifth International Conference
R. Viswanathan, D. Gandy, K. Coleman, editors, p 208-219

DOI: 10.1361/cp2007epri0208

VM12 – A NEW 12%CR STEEL FOR APPLICATION AT HIGH TEMPERATURE IN ADVANCED POWER PLANTS - STATUS OF DEVELOPMENT -

J. Gabrel
Vallourec & Mannesmann Research Center, France
C. Zakine
Bo. Lefebvre
B. Vandenberghe
Vallourec & Mannesmann Tubes (V&M Tubes), France

Abstract

The T/P91 and T/P92 steel grades were developed as a result of a demand of high creep strength for advanced power plants. Nevertheless, their operating temperature range is limited by their oxidation performance which is lower compared with usual 12%Cr steels or austenitic steels.

Moreover, the new designed power plants require higher pressure and temperature in order to improve efficiency and reduce harmful emissions.

For these reasons, Vallourec and Mannesmann have recently developed a new 12%Cr steel which combines good creep resistance and high steam-side oxidation resistance. This new steel, with a chromium content of 12% and with other additional elements such as cobalt, tungsten and boron, is named VM12.

Manufacturing of this grade has been successfully demonstrated by production of several laboratory and industrial heats and rolling of tubes and pipes in several sizes using different rolling processes. This paper summarizes the results of the investigations on base material, including creep tests and high temperature oxidation behavior, but also presents mechanical properties after welding, cold bending and hot induction bending.

Introduction

The development of steels like T/P911 and T/P92 has achieved an increase in creep rupture strength of 10 to 25% in comparison with modified 9% chromium steel T/P91. Their use allows operating temperatures up to about 610°C (1130°F) for live steam and 620°C (1150°F) for hot reheat steam [1 - 3] limited by their oxidation and corrosion resistance.

Worldwide activities aim at further improvement of plant efficiency by continuing to increase the steam temperature. New ferritic steels are also needed for use at higher operating temperature between 610 and 650°C (1130° and 1200°F).

Alloy Design – Input Data

The aim of the R&D project initiated by Vallourec & Mannesmann Tubes (V&M) was to develop a new 12% Cr steel with improved oxidation resistance to allow operating temperatures of up to 650°C (1200°F) [4, 5]. The chromium content needs to be higher than 11% in order to meet the requirement of good oxidation resistance. In addition the strength properties of the new steel should be close to the best ferritic steel T/P92 and the processing similar to that of the well-known T/P91.

In order to obtain good creep resistance, Vanadium, Columbium (Niobium) and Nitrogen have been introduced in quantities similar to the 9% Cr-steels in order to achieve precipitation strengthening by MX particles. Other strengthening elements such as molybdenum, tungsten and boron have also been added. This steel is a tungsten bearing steel with 1.4% W and only 0.3% Mo, using the strengthening effect of Laves phase precipitation similar to T/P92.

Finally to obtain a fully martensitic structure from 9 to 11% Cr the nickel equivalent has to be increased simultaneously. This has been done by adding Cobalt.

Tube and Pipe Production

Based on the positive results from laboratory heats [6], the industrialization of the new steel grade was launched under the name VM12. Table 1 gives the different processes and products. Whatever the tube or pipe sizes and processes, no problems occurred during production. The casting and rolling revealed comparable behavior to T/P91.

Process	*Steel plants*	*Quantity tons*	*Round bar*	*Mills*	*Process*	*Products*	
						mm	*inches*
Electric Arc Furnace - Vacuum degassing	1	20	*Ingot Forging*	A	*continuous mandrel rolling*	*Tubes 60.3 x 8.8*	*2.4 x 0.35 in.*
				E	*push bench*	*Tubes 51 x 4*	*2 x 0.16 in.*
			Ingot	B	*pierce and pilger rolling*	*Pipes 406.4 x 35*	*16 x 1.4 in.*
	2	80	*Continuous Casting + Forging*	A	*continuous mandrel rolling*	*Tubes 38 x 6.3 - 38 x 7.1 44.5 x 10 - 51 x 4*	*1.5 x 0.25 – 1.5 x 0.28 in. 1.75 x 0.4 - 2 x 0.16 in.*
				E	*push bench*	*Tubes 51 x 4*	*2 x 0.16 in.*
				C	*continuous mandrel rolling*	*Tubes 140 x 12*	*5.5 x 0.47 in.*
	3	65	*Ingot*	D	*pierce and draw rolling*	*Pipes 460 x 80*	*18 x 3.1 in.*
				B	*pierce and pilger rolling*	*Pipes 460 x 60*	*18 x 2.3 in.*

Table 1: Status of the industrial production of VM12 grade

Heat Treatment

VM12 steel is used in the normalized and tempered condition. The typical heat treatment after hot rolling is:

- Normalization: 1040-1080°C / air cooling (1905-1975°F)
- Tempering: 750-800°C / air cooling (1380-1470°F)

Depending on chemical composition, A_{C1} temperature was found to be between 820°C (1510°F) and 830°C(1525°F), and A_{C3} around 890°C (1635°F).
The figure 1 shows the complete CCT diagram.

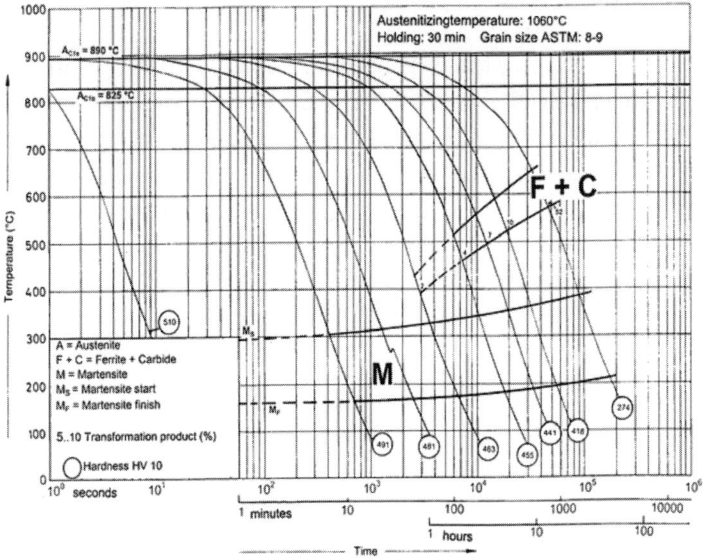

Figure 1: CCT diagram of VM12 steel

Chemical Composition

The chemical composition of the first three VM12 industrial heats is given in table 2 and compared with grade T92 ASTM A213.

Grades		C %	Mn %	P %	S %	Si %	Cr %	W %	Mo %	V %	Cb %	Co %	N %	B %	Al %	Ni %	Cu %
A213	Min	0.07	0.30	-	-	-	8.50	1.5	0.30	0.15	0.04		0.030	0.001	-	-	
T92	Max	0.13	0.60	0.020	0.010	0.50	9.50	2.00	0.60	0.25	0.09		0.070	0.006	0.04	0.40	
Industrial Heat 1		0.115	0.35	0.018	0.001	0.49	11.50	1.50	0.29	0.26	0.050	1.62	0.065	0.0049	0.008	0.29	0.080
Industrial Heat 2		0.106	0.32	0.015	0.002	0.45	11.25	1.44	0.24	0.26	0.047	1.55	0.055	0.0044	0.012	0.26	0.045
Industrial Heat 3		0.110	0.18	0.018	0.002	0.48	11.25	1.44	0.27	0.23	0.051	1.50	0.053	0.0035	0.010	0.23	0.080

Table 2: Chemical composition of the different heats (mass - %)

Structure:

The structure is composed of tempered martensite.

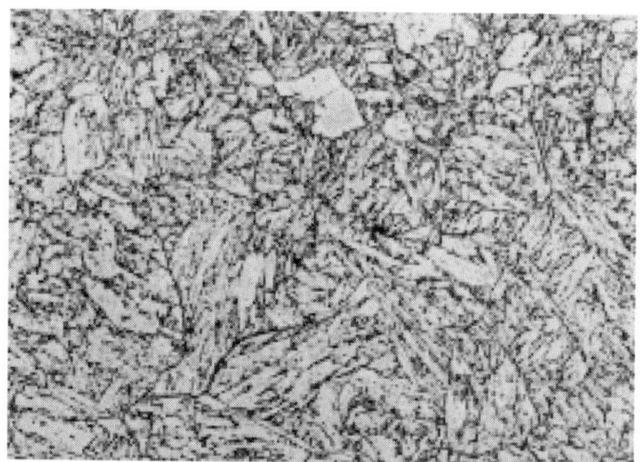

Figure 2: Microstructure (pipe 406.4 x 35 mm) after normalization and tempering

As an example, the chromium and nickel equivalents have been calculated and plotted in Figure 3, thus allowing a rough estimation of the phase stability.

All VM12 heats are located near the martensite area where the Cr_{Eq} is maximum for this structure. For comparison, Figure 3 also shows the location of the typical V&M T92. According to this figure, the VM12 heats may show some δ-ferrite. However no δ-ferrite was observed on tubing products after final heat treatment.

A small amount of δ-ferrite (less than 2%) may be present on thick products. This low amount of δ-ferrite has no influence on the rolling behavior and mechanical properties.

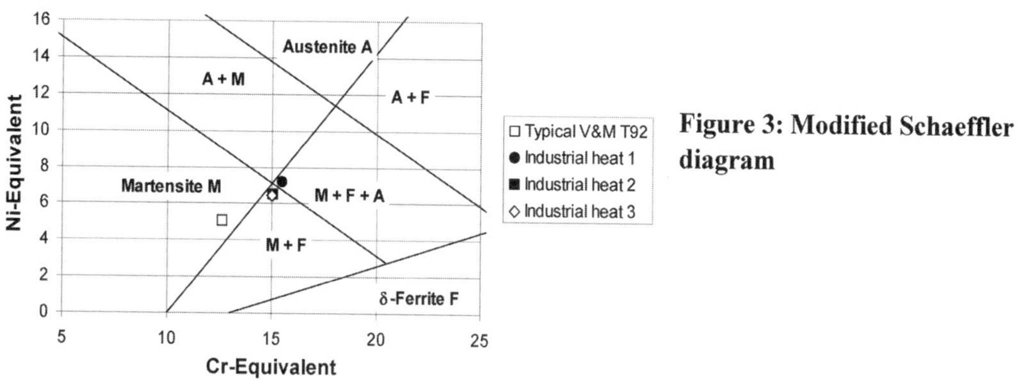

Figure 3: Modified Schaeffler diagram

$$Cr_{Eq} = Cr + 2Si + 1.5Mo + 5V + 5.5Al + 1.75\ Cb + 1.5Ti + 0.75W$$
$$Ni_{Eq} = Ni + Co + 0.5Mn + 30C + 0.3Cu + 25\ N$$

Steam side corrosion

Inner steam oxidation is the cause of various problems in power plants, such as:

- The increase in the metal temperature during operating due to the formation of an insulating layer of oxide between the internal tube surface and the steam.
- The formation of oxide layers reducing the tube thickness and increasing stress.
- Exfoliation of the oxide layers, blocking tubes or causing erosion inside the turbine.

Steam-side oxidation resistance is a limitation of the design temperature of 9%Cr steels. In spite of the fact that T/P92 steel has good mechanical properties at elevated temperatures, its steam oxidation resistance is similar to T91. Consequently the operating temperature of such steels reached its limit at around 610°C (1130°F).

Improved oxidation behavior was one of the key requirements for the new VM12 steel.

Therefore oxidation tests were performed at 600°C (1110°F) and 650°C (1200°F) in pure water vapor, using test facilities at Ecole des Mines-Douai, France [7]. Corrosion damage was measured using the mass losses obtained after a reducing descaling process.

A testing time of 8,000 h was reached. As an example the results at 650°C are shown in Figure 4.

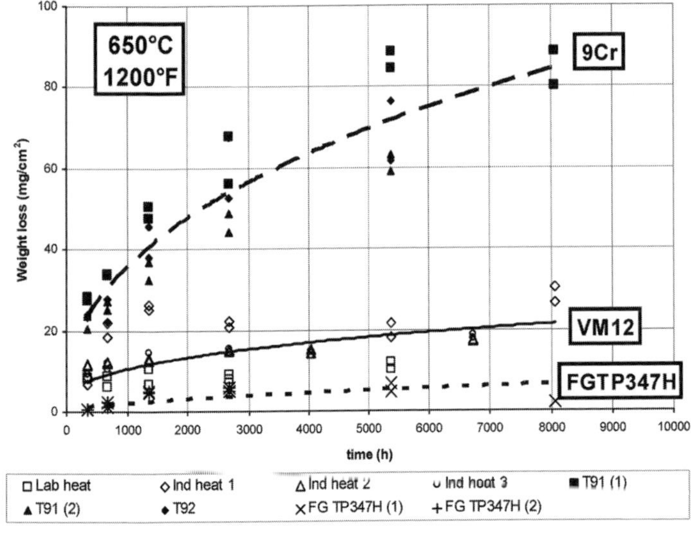

Figure 4: Steam-side oxidation at 650°C (1200°F)

In addition to the VM12 steel, T91, T92 and FG TP347H steels were also plotted for comparison.

The weight change measurements clearly indicate better oxidation resistance of the new VM12 steel compared to T91 and also to T92.

Moreover, the results are close to those of austenitic steel FG TP347H.

For all test times and temperatures, two different types of corroded surfaces were observed:

- The first one is characterized by a relatively homogeneous two-oxide layer: The scale thickness grows with temperature and time of exposure. A typical example is T/P91.
- The second one shows localized nodules of corrosion products beneath which the steel is attacked: this type is best represented by FG TP347H. In the locations of the nodules, the oxide thickness increases with time but simultaneously the corrosion attack expands onto the whole surface.

Additional tests on VM12 were performed in an argon atmosphere with 50 vol.-% steam in a second laboratory (Forschungszentrum-Jülich, Germany). More recently Alstom Germany as part of the COST project and Babcock & Wilcox in the framework of the DOE program [8] and [9] have confirmed the excellent steam oxidation resistance of VM12 independently of the specific test conditions.

Mechanical Properties

Tensile Properties

Yield and tensile strength values at room and elevated temperatures are plotted in Figure 5. The tensile properties of VM12 are similar to those of T/P92.

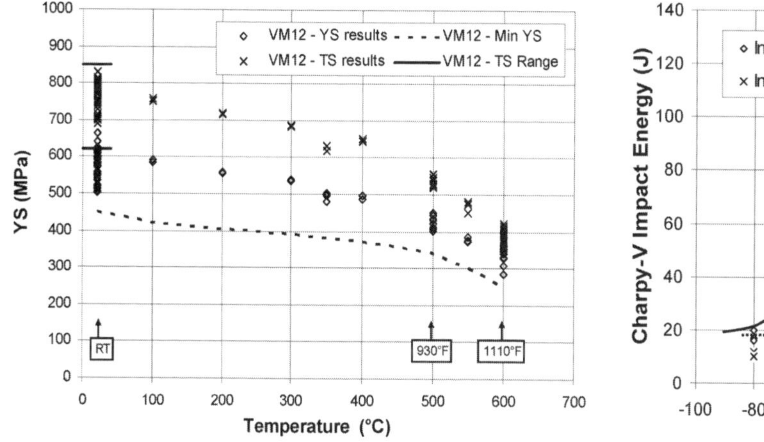

Figure 5: Yield and tensile strength
Figure 6: Charpy-V notch results

Impact tests

As illustrated in figure 6 both VM12 tube and pipe from the first industrial heat show good impact properties. For example at room temperature the absorbed energies are about 120 J (tube) and 100 J (pipe).

Creep tests

A creep testing program is in progress at temperatures of 525, 550, 575, 600, 625, 650 and 675°C (from 980 to 1250°F). Presently the longest tests are now greater than 25,000 hours and the accumulated testing time is close to 1,400,000 hours.

The current statuses at 600°C (1110°F) and 625°C (1160°F) are given in figure 7. A first estimation gives a mean rupture stress at 600°C for 100,000 hours at 100 MPa (14.5 ksi). In comparison with the assessment performed in 2005 by the European Creep Collaborative Committee (ECCC) for T/P92 (113 MPa-16.4 ksi) and T/P911 (98 MPa-14.2 ksi), VM12 seems to behave more like T/P911 than T/P92. However, the current status of testing does not allow a final judgment to be made on creep strength.

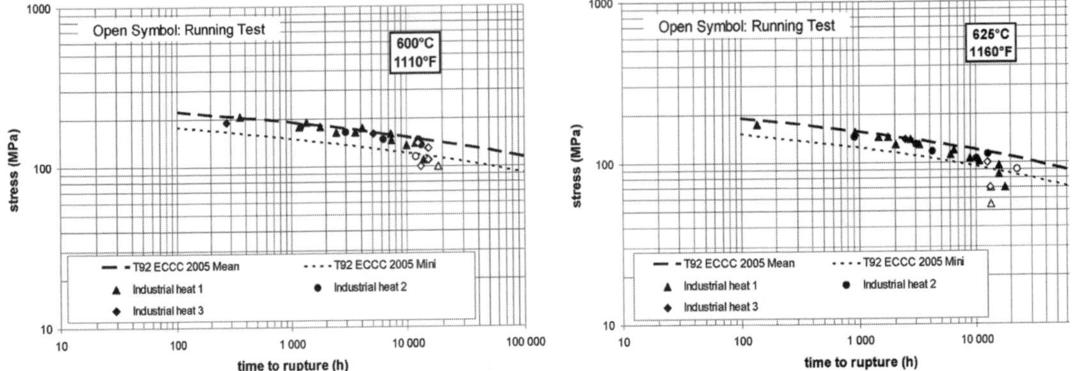

Figure 7: Creep test results at 600 and 625°C (1110 and 1160°F)

Processing

Welding

Welding consumables (coated electrode, solid wire and flux) and welding procedures have been developed by partner companies [10].
During weldability tests, neither hot cracks nor end crater cracks have been observed. Also the material did not show any susceptibility to stress relief cracking.

Main conclusions from weld simulations are:

- The lowest toughness and highest hardness can be expected in the coarse grained HAZ
- Fulfilling the hardness criteria (<350HV) seems easily achievable with a post-welding heat treatment at 770°C (1420°F).

Firstly, industrial welds (SMAW and SAW) were performed on 406.4 x 35 mm pipe from the first industrial heat. After post welding heat treatment performed at 770°C (1420°F) / 2 hours, the microstructure and mechanical properties (tensile, impact, hardness and creep) showed satisfactory results. Figure 6 shows the status of the creep tests on the SMA-weld at 625°C (1160°F).

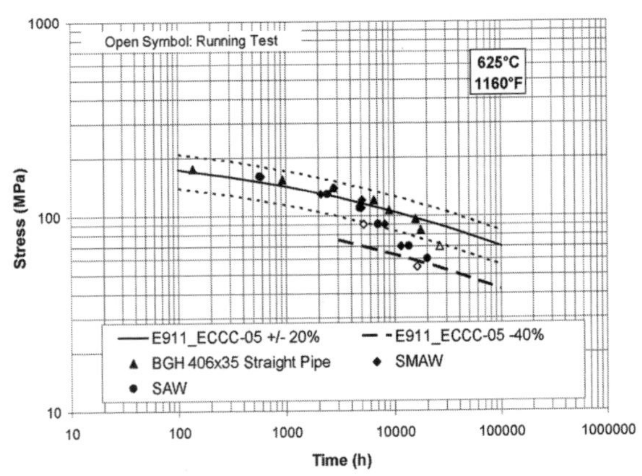

Figure 8: Cross-weld creep tests

on SMAW and SAW joint

Induction bending

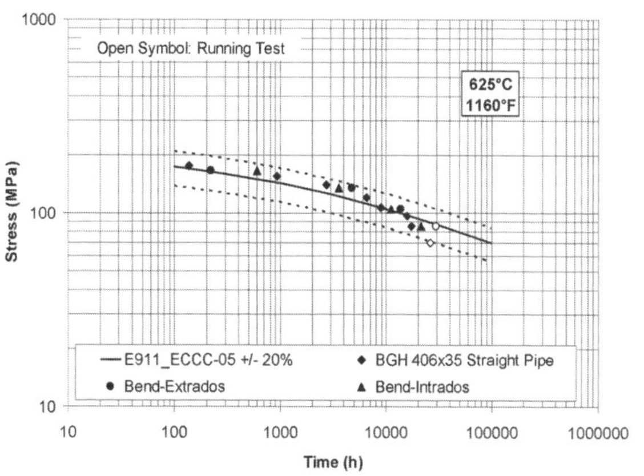

Figure 9: Creep results after induction bending, normalizing and tempering

An induction bend has been made from the same pipe used for welding trials. After complete heat treatment (normalization and tempering), microstructure and mechanical properties are very similar to the parent pipe as illustrated by Figure 9.

Cold bending

For this purpose we worked with two partner companies to perform cold bending qualifications. Thanks to their facilities (Schwarzewirtz machines 100DB with back thrust jack - figure 8), cold bends were performed on 38 x 6.3 mm and 60.3 x 8.8 mm tubes with a radius of between 3 and 1.3 times the outside diameter. After bending without any problem, each bent tube was inspected using the dye penetrant method. No external defects were detected. The behavior during bending was very comparable to that of T91 or T92. The hardness and microstructure were satisfactory in the as bent condition and after stress relief treatment (SRT) at 750°C (1380°F).

Figure 10: Cold bending

Principally, no recrystallization has been observed demonstrating the very stable structure of the VM12 steel. Moreover this observation was confirmed after cold deformation of up to 35% followed by SRT. Creep tests after bending and cold deformation show similar results to those obtained on the parent tube.

Use of VM12 in Neurath Power Plant

Thanks to its excellent steam-side oxidation resistance and good processing, VM12 has been recently chosen in the superheater and reheater of the German project Neurath Units F and G (Table 3). This world most powerfull lignite-fired power plant (1100MW per unit) should be on

the grid in 2010. The supercritical steam parameters are 290b/600°C/620°C – 4,200 psi/1110°F/1150°F which lead to an efficiency of 43%.
Although the development of the steel grade is not yet complete, a TÜV approval has been received for this application under the designation VM12-SHC.

Process	Steel plants	Quantity tons	Round bar	Mills	Process	Tube sizes mm
Electric Arc Furnace - Vacuum degassing	2	900 (2006-2007)	Continuous Casting + Forging	A	continuous mandrel rolling	44.5 x 8.8 44.5 x 10.0
				E	push bench	51 x 4

Table 3: Status of the industrial production of VM12-SHC grade in progress for Neurath Units F and G

Unit F has at present been delivered and the second unit G is in progress. The results confirm the excellent fabricability and good mechanical properties (Table 4) of the VM12-SHC.

		YS MPa	TS MPa	E %	Impact energy J		
VM12-SHC	Min	450	620	19	≥ 40		
	Max		850				
Neurath Project	Min	502	739	19	57	56	59
	Max	639	814	24.1	135	133	141
	Average	585	775	20.5	94	96	96

Table 4: Mechanical properties at room temperature

Some other projects are also booked or planned. The VM12-SHC grade has not only been appropriated for new projects but it could be used for retrofitting or maintenance.

Outlook

In order to improve creep properties and to keep the excellent oxidation resistance, three new laboratory heats with slight modification of the chemical composition have been produced.
In comparison with the previous VM12-SHC the main modifications are (Table 5):

- For the laboratory heat A: lower Co (0.88% instead of 1.4) compensated by higher C (0.16% instead of 0.12), lower Nb (0.038% instead of 0.055).
- For the laboratory heat B, in addition to the previous modifications, lower Cr at 10.65%.
- For the laboratory heat C, in addition to the previous modifications, lower Co at 0.60%

Whatever the heat, the tensile and impact properties are similar to VM12SHC. Moreover the 3 new laboratory heats, and in particular the heats A and B, have better toughness.-
At present, the steam oxidation results are located in the same scatter-band as the previous VM12-SHC in spite of the decrease in Cr content.

Grades		*C* %	*Mn* %	*P* %	*S* %	*Si* %	*Cr* %	*W* %	*Mo* %	*V* %	*Cb* %	*Co* %	*N* %	*B* %	*Al* %	*Ni* %	*Cu* %
VM12-	*Min*	*0.10*	*0.15*	-	-	*0.40*	*11.0*	*1.30*	*0.20*	*0.20*	*0.03*	*1.40*	*0.030*	*0.0030*	-	*0.10*	-
SHC	*Max*	*0.14*	*0.45*	*0.020*	*0.010*	*0.60*	*12.0*	*1.70*	*0.40*	*0.30*	*0.08*	*1.80*	*0.070*	*0.006*	*0.02*	*0.40*	*0.25*
Heat A		*0.158*	*0.42*	*0.005*	*0.001*	*0.49*	*11.36*	*1.46*	*0.31*	*0.25*	*0.038*	*0.88*	*0.042*	*0.004*	*0.007*	*0.23*	*0.022*
Heat B		*0.156*	*0.42*	*0.004*	*0.001*	*0.49*	*10.62*	*1.46*	*0.31*	*0.25*	*0.038*	*0.89*	*0.039*	*0.005*	*0.015*	*0.23*	*0.022*
Heat C		*0.160*	*0.42*	*0.005*	*0.001*	*0.49*	*10.71*	*1.46*	*0.31*	*0.25*	*0.039*	*0.61*	*0.040*	*0.0047*	*0.008*	*0.23*	*0.021*
Industrial heat		*0.14*	*0.48*	*0.019*	*0.005*	*0.41*	*11.16*	*1.46*	*0.28*	*0.23*	*0.042*	*0.89*	*0.037*	*0.0046*	*0.005*	*0.18*	*0.014*

Table 5: Chemical composition of the modified VM12

The creep tests on both laboratory heats B and C, which have reached now 17 000 h, show better behavior at 650°C (1200°F) than the VM12-SHC. Nevertheless, the tests are too short to allow a reliable judgment (Figure 11).

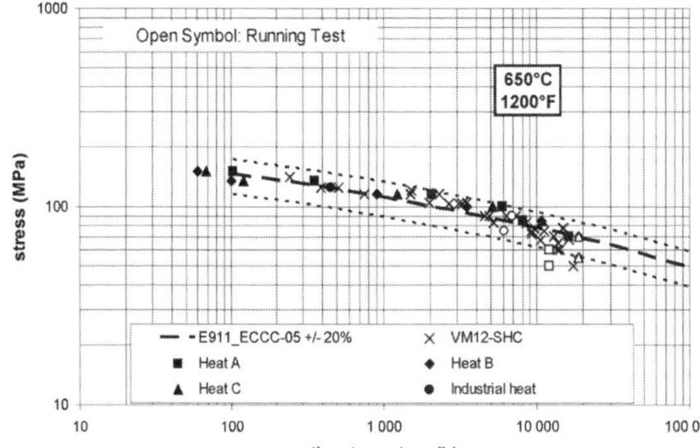

Figure 11: Comparison between modified VM12 and VM12-SHC

Casting and rolling (pipe 406.4 x 50 mm) were checked on an industrial heat with chemistry close to the laboratory heat A (Table 5). The behavior is similar to the one of VM12-SHC.
The first mechanical properties are satisfactory.

Conclusion

Regarding VM12, the results on industrial tubular products confirm the first results obtained with laboratory heats.
Thanks to a good adjustment of the chemical composition, tubes and pipes have been rolled without any trouble using the same mill procedures used for T/P91.
The structure and mechanical properties at room and elevated temperatures are similar to those of T/P92. With industrial heats, we also get excellent steam-side oxidation behavior at 650°C (1200°F); it is close to that of austenitic steels.
Welding and bending trials with boiler maker partners show that the recommendations defined for 91 and 92 grades will be similar for the new grade, especially for welding, cold and induction bending operations.

The creep properties seem similar to those of T/P911 and are not as high as initially expected in comparison with TP92. However the creep results are clearly higher than for other 12%Cr steels with comparable oxidation resistance.

All these results allowed us to put this grade under the name VM12-SHC on the market for some specific German power plant projects. Moreover VM12 is still under development and should be improve in the future.

Acknowledgement

Part of this work was performed within the framework of the European COST Action 536 with funding by the German BMWA under Contract No. 0327705F

Bibliographie

[1] W. Bendick, F. Deshayes, K. Haarmann and J-C Vaillant, EPRI Conf. Advanced Heat Resistant Steels for Power Generation, San Sebastian, Spain, 27-29 April, 1998

[2] D. Richardot, J.-C. Vaillant, A. Arbab and W. Bendick, The T92/P92 Book. Vallourec & Mannesmann Tubes, 2000

[3] A. Arbab, J.-C. Vaillant, and B. Vandenberghe, 3rd EPRI Conference "Advances in Material Technology for Fossil Power Plants", 5-6 April, 2001, Swansea, UK, ed. by R. Viswanathan et al., The Institute of Materials, London, 2001, pp. 99-112

[4] J. Gabrel, W. Bendick, JC. Vaillant, B. Vandenberghe and Bo. Lefebvre, 4th Int. Conference on advances in Materials Technology for fossil Power Plants, "VM12-A new 12%Cr steel for boiler tubes, headers and steam pipes in ultra-super- critical power plants.", 15-28 Oct. 2004, 10 pages

[5] B. Vandenberghe, J. Gabrel, JC. Vaillant and Bo. Lefebvre, AFIAP conference, "Development of a new 12%Cr steel for tubes and pipes in power plants with steam temperatures up to 650°C", Paris 28-30 Sept 2004, 12 pages

[6] W. Bendick, J. Gabrel, JC. Vaillant and B. Vandenberghe, Liège conference, "Development of a new 12%Cr steel for tubes and pipes in power plants with steam temperatures up to 650°C", Liege 30 Sept- 2 Oct. 2002, 10 pages

[7] V. Lepingle, G. Louis, D. Petelot, B. Lefebvre, B. Vandenberghe, « Long term exposure of new 12% Cr boiler steels in steam high temperature ». Eurocorr 2005 - Lisbon - 4-8 Sept 2005

[8] J.M. Sarver, J.M. Tanzosh, "Preliminary results from steam oxidation tests performed on candidate materials for ultra supercritical boiler" EPRI International Conference on Materials and Corrosion Experience for Fossil Power Plants, Charleston, 2003

[9] J.M. Sarver, J.M. Tanzosh, 4th Int. Conference on advances in Materials Technology for fossil Power Plants, "An Evaluation of the Steamside Oxidation of Candidate USC Materials at 650°C and 800°C", 15-28 Oct. 2004

[10] J. Vekeman, A. Dhooge, S. Huysemans, B. Vandenberghe and C. Jochum, Weldability and high temperature behaviour of 12%Cr steel for tubes and pipes in

power plants with steam temperatures up to 650°C, Liège conference - pp. 1369-1380

Advances in Materials Technology for Fossil Power Plants
Proceedings from the Fifth International Conference
R. Viswanathan, D. Gandy, K. Coleman, editors, p 220-230

DOI: 10.1361/cp2007epri0220

A NEW IMPROVEMENT OF INCONEL ALLOY 740 FOR USC POWER PLANTS

Xishan Xie, Shuangqun Zhao, Jianxin Dong
University of Science and Technology Beijing, Beijing 100083, China
G. D. Smith, B. A. Baker, S. L. Patel
Special Metals Corporation, Huntington , WV 25705, USA

Abstract

A new nickel base superalloy, INCONEL® alloy740, is under development for the application of ultra-supercritical (USC) boilers above 750°C. This alloy can fulfill the requirements of long time high temperature stress rupture strength (100MPa for 10^5hrs) and corrosion resistance (2mm/2×10^5hrs) allowance. Experimental results show that the most important structure changes at elevated temperatures are γ' coarsening, γ' to η transformation and G phase formation. A further improvement for more strengthening effect and structure stability of INCONEL alloy 740 has been conducted. On the basis of thermodynamic calculation, small adjustment of several alloying elements has been adopted and some modifications of INCONEL alloy 740 were designed and melted. Long time structure stability has been studied to 5000hrs at 750 and 800°C, and 1000hrs at 850°C. The mechanical property and oxidation resistant tests have been also conducted for the modified alloys in comparison with original INCONEL alloy 740. The preliminary experimental results show that minimal modification can improve the stress rupture strength and also maintain corrosion resistance. Microstructure examination indicates the thermal stability of the alloy is improved. INCONEL alloy 740 is promising to be used as USC boiler at 750°C or higher temperatures.

Introduction

Worldwide electricity development is requiring power plants to raise thermal efficiency and to reduce SO_X, NO_X and CO_2 emissions. A key solution is to raise the steam parameters. Recent ultra-supercritical steam conditions up to 35 MPa and 700°C, are being planned by both the European THERMIE AD700 and the German MARCKO DE2 projects, will enhance the efficiencies of coal-fired boilers to about 50% and lead to superheater/reheater midwall temperatures as high as 740°C to 760°C as well. At these temperatures and pressures, the superheater and reheater materials will therefore be required which have a high creep rupture strength (100MPa/10^5h) at temperatures of about 750°C, together with high corrosion resistance (≤2mm cross-section loss in 2×10^5h). For long-term service at these high pressures and temperatures, neither solid solution

® INCONEL is trademark of the Special Metals Corporation group of companies.

solution strengthening austenitic stainless steels nor a currently available nickel-base superalloy, such as NIMONIC® alloy 263, INCONEL alloy 617, CCA617, INCONEL alloy 690 and INCONEL alloy 671, can meet the requirements for stress rupture strength and corrosion resistance [1-3]. A new precipitation-hardenable nickel base superalloy, designated as INCONEL alloy 740, is under development at Special Metals Corporation (Huntington) for USC power plant application at temperatures above 750°C [4, 5]. The nominal chemical composition is (in wt.%): C 0.03, Cr 25.0, Co 20.0, Mo 0.5, Al 0.9, Ti 1.8, Nb 2.0, Fe 0.7, Mn 0.3, Si 0.5, Ni balance. In collaboration with Special Metals Corporation, we have investigated mechanical properties, strengthening mechanism, structural stability and corrosion resistance of this new alloy at high temperatures [6-8]. So far, all the studies prove that this alloy successfully surpasses the established strength design target as mentioned above and exhibits favorable resistance to coal ash corrosion in laboratory tests designed to simulate boiler conditions as well. This paper briefly introduces the important results of structure stability of the alloy during long-term aging at elevated temperatures. On the basis of thermodynamic calculation, the methods for structure stability improvement have been suggested and four experimental heats of modified alloys have been produced at Beijing and Huntington. The mechanical properties and corrosion resistance of modified alloys were evaluated at elevated temperature, and their structure stabilities were examined at 750, 800 and 850°C as well.

Microstructure Stability of INCONEL 740

Typical SEM and TEM structures of INCONEL alloy 740 at standard heat treatment condition (1150°C annealing+800°C/16h aging) are shown in Fig.1. SEM structure (Fig.1a) indicates that carbides mainly precipitated at grain boundaries and partially in the grains. The exact and detailed results of the precipitates were analyzed by physical and chemical phase analyses, which indicate that these precipitates are γ' (12.980wt%), MC (0.183wt%), $M_{23}C_6$ (0.115wt%) and G phase (0.054wt%). The MC and $M_{23}C_6$ are (Nb,Ti)C and $Cr_{23}C_6$ type carbides, respectively. As shown in Fig. 1b, the uniformly

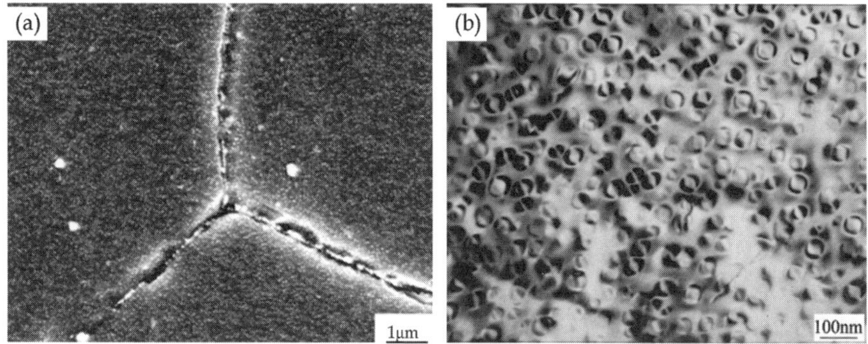

Fig.1 SEM and TEM images of INCONEL 740 at standard heat-treatment condition

dispersive fine γ' precipitates are observed in the matrix, which contributes the main strengthening effect for this new alloy. The formation of cubic γ' morphology can be attributed to the medium degree of γ-γ' lattice mismatch.

Fig.2 shows SEM images of the alloy aged at 704 and 725°C for 4000h. The morphology and distribution of the precipitates in the samples are quite similar as in Fig. 1. Very fine γ' precipitates distribute in the grains. The large particles (MC) distribute in matrix at random. The grain boundary precipitates can be obviously observed after long time exposure at 704°C. Detailed phase analyses indicate that the major carbides MC, $M_{23}C_6$ mixed with a few of G phase. The MC carbide forms from the liquid and it is stable at high temperatures. $M_{23}C_6$ mainly forms at grain boundaries and have no significant variation with prolonged aging time. However, after aging for 4000 h at 725°C, a small quantity of needle-like and blocky η precipitates have formed mainly nearby grain boundaries (Fig. 2b).

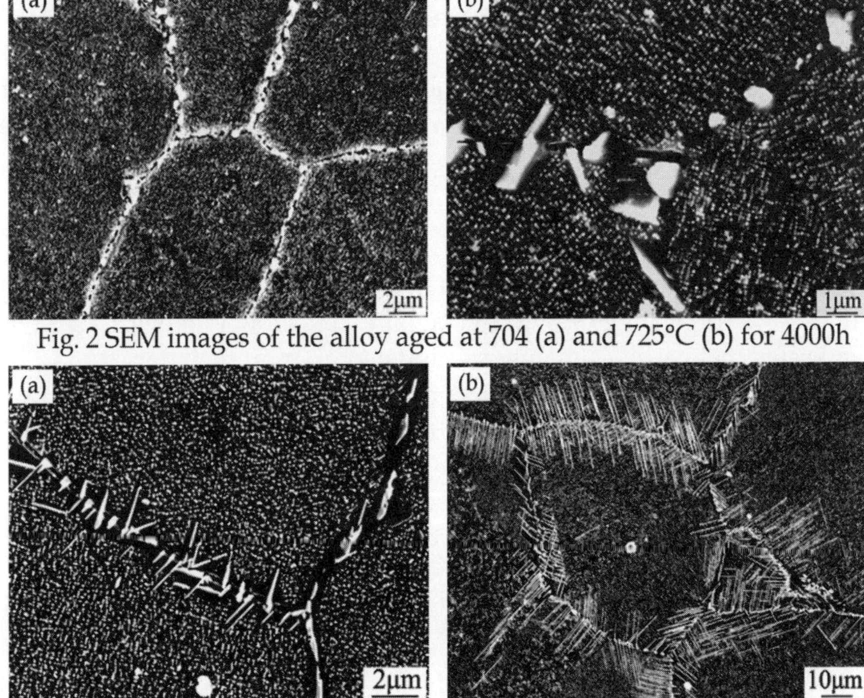

Fig. 2 SEM images of the alloy aged at 704 (a) and 725°C (b) for 4000h

Fig. 3 SEM images of the alloy aged at 750°C for 1000 (a) and 5000h (b)

Fig. 3 shows SEM images of the alloy aged at 750°C for 1000 and 5000h. The same as in Fig. 2b, the significant structure change is η-phase formation besides the growth of γ' particles. After 1000h aging (Fig. 3a), a small amount of η phase, especially in the area close to the grain boundaries, has been observed in the sample. The quantity and size of η precipitates rapidly increase with aging time. By 5000h (Fig. 3b), not only η phase initiates at grain boundaries but also a significant amount of η phase forms in the grains as Widmanstätten pattern structure. In addition, there are γ'-free zones surrounding η plates and nearby the grain boundaries. It indicates that the η phase forms at the expense of γ' phase.

TEM γ' morphologies of the alloy aged at 750°C for 1000 and 5000h are shown in Fig. 4. With an increase in aging time, γ' precipitates grow rapidly. The cubic morphology of γ' precipitates indicates no remarkable change in shape during long time aging. The formation of cubic γ' morphology can be attributed to the medium degree of γ-γ' lattice mismatch. The average radius of γ' particles is about 40 and 60nm.

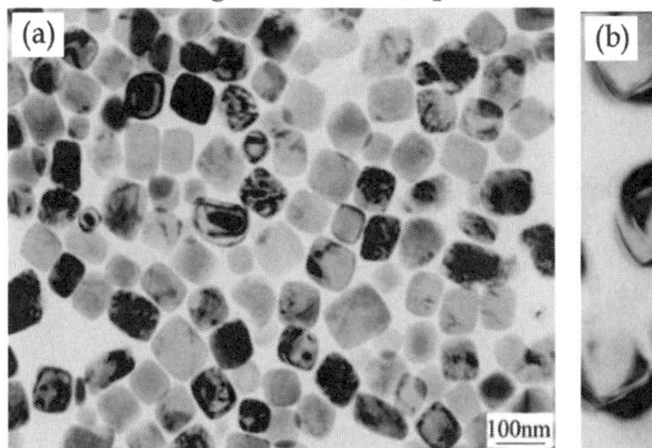

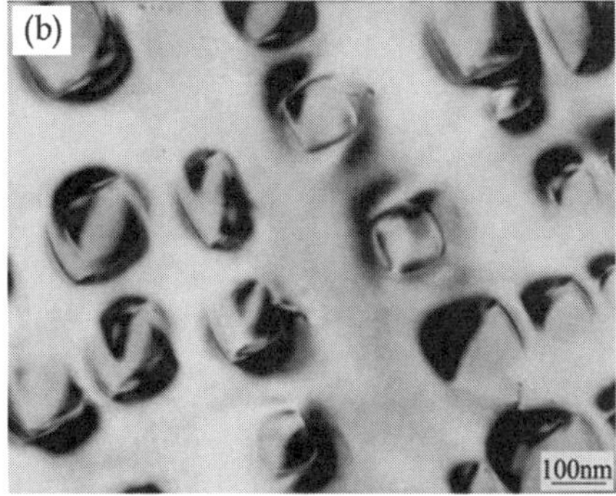

Fig. 4 γ' morphology of the alloy aged at 750°C for 1000 (a) and 5000h (b)

The coarsening behavior of γ' particles was examined as a function of temperatures at 704, 725 and 760°C, and as a function of aging times up to 4000h for the alloy. The relationship between average radius of γ' particles with various aging time at three temperatures are given in Fig. 5. It is clear that the relationships between effective radius $\bar{r}^3$ and aging time t at different aging temperatures are linear, suggesting that the coarsening of γ' particles in this new superalloy containing not too high volume fraction of γ' follows typical $\bar{r}^3 \propto t$ kinetics of diffusion-controlled particle growth. It is also an evidence in Fig. 5 that the γ' particle coarsens more rapidly as the aging temperature increases to 760°C.

In order to study the η precipitation behavior, the samples as solution-annealed were aged at different high temperatures to investigate when the η phase forms. After aging at 725°C for 4000 h, 800°C for 100h, 900°C for 16h and 950°C for 26h, a small amount of η phase formed at grain boundaries and/or in matrix. In addition, the η phase has not formed in the sample aged at 1000°C for 140h. According to these results, the TTT-diagram for η phase formation can be given in Fig. 6. Though no exact results at temperatures above 950°C, the formation trend of η phase is also given in accordance to the thermodynamics calculation with nominal composition, which indicates that the highest stable temperature of η phase might be 1118°C. According to these results, it is likely that the η phase forms in the temperature range of 700~1118°C. The peak temperature of η phase formation is about 900°C.

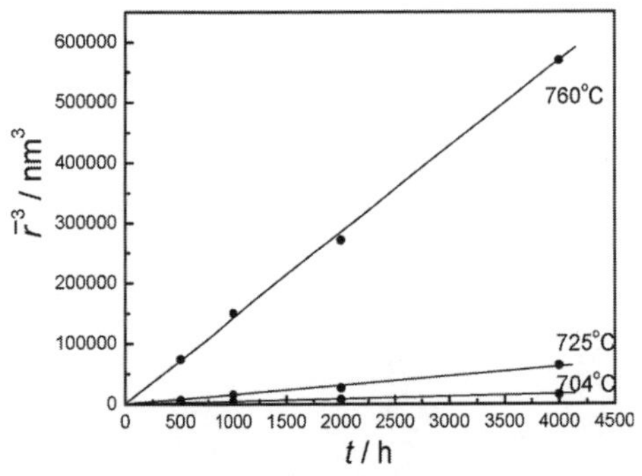

Fig. 5 Coarsening of γ′ phase in the alloy

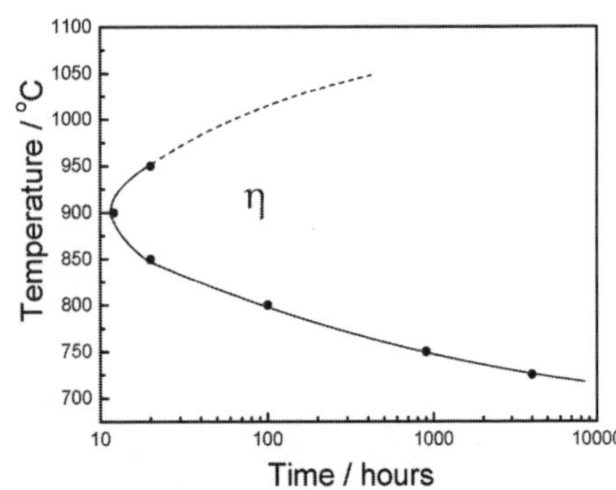

Fig. 6 TTT curve of η phase in the alloy

The weight fractions of precipitates in the alloy aged at 704 and 760°C for 1000 and 2000h obtained by physical and chemical phase analyses are given in Tab. 1. The results indicate that the precipitates in this alloy are γ′(+η), MC, $M_{23}C_6$ and G phase. Because of the similar electrochemical property, the γ′ phase and the η phase could not be differentiated by this method. The weight fraction of γ′ precipitates in samples aged at 760°C is total fractions of γ′ and η. The actual fraction of γ′ phase in the samples aged at 760°C is less than that of at 704°C, because a certain amount of γ′ has transformed to η. The fractions of γ′(+η), MC and $M_{23}C_6$ have no large changes with increasing time at temperatures 704 and 760°C. Phase analyses also indicate that the chemical composition of γ′ precipitate is $Ni_3(Ti,Al,Nb)$ and G phase is $A_6B_{16}Si_7$-type $(Nb,Ti)_6(Ni,Co)_{16}Si_7$. It is worth to mention that the fractions of G phase in the samples aged at 760°C are much higher than those of at 704°C.

Tab. 1 Fraction of precipitates in the alloy (wt%)

Aging	γ′(+η)	MC	$M_{23}C_6$	G
704°C / 1000h	16.629	0.154	0.139	0.046
704°C / 2000h	16.835	0.151	0.151	0.063
760°C / 1000h	14.364	0.161	0.170	0.336
760°C / 2000h	14.633	0.154	0.217	0.471

Thermodynamic calculation for structure stability improvement

Experimental results show that main structure instabilities of the alloy above 750°C aging are rapid coarsening of γ′, η phase transformation, and a certain amount of G phase formation at grain boundaries. Consequently, the structure stability improvement of this alloy should be concentrated on the precipitation behavior of γ′, η and G phase.

Thermodynamic calculation was carried out to predict the phase stability and phase fractions in the alloy using Thermo-Calc software, version M, accompanied with 14-element Ni-database (not including Si and Mn) [9]. The calculated phase diagram of the alloy under equilibrium condition is shown in Fig. 7, in which all predicted equilibrium phases and their weight fractions at each temperature are given. Thermodynamic calculation (Fig. 7) reveals that the equilibrium phases above 600°C include γ, γ′, η, MC and $M_{23}C_6$, which were observed in the alloy, and σ phase as well, which was not observed and may form after longer exposure time. The liquidus, solidus, σ, η and γ′ solvus temperatures of the alloy are about 1369, 1293, 699, 1118 and 850°C, respectively. The predicted solidification temperature range (76°C) of this alloy is in good agreement with the experimental value (74°C).

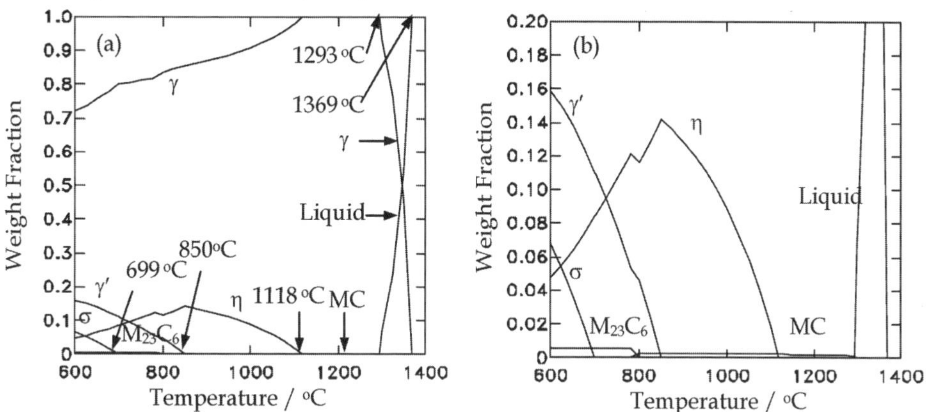

Fig. 7 Phase calculation results (a) of the INCONEL 740 and its partial magnification (b)

Both the physical and chemical phase analyses results and thermodynamic calculation results show the chemical composition of γ′ phase is $Ni_3(Al,Ti,Nb)$, therefore, phase computation should be concentrated on Al, Ti and Nb. Fig. 8a gives the calculated results of the effect of Al content on the weight fraction and solution temperature of γ′ phase. With the increase of Al content, the weight fraction of γ′ phase increases obviously. For example, when the content of Al is 0.7%, 0.9%, 1.1% and 1.5%, the weight percentage of γ′ phase is 1.7%, 7.7%, 13.5% and 23.7% at 750°C, respectively. Simultaneously, the precipitation temperature of γ′ phase is also increase with the content of Al and they are 775°C, 850°C, 904°C and 987°C, respectively. However, the changes of Ti in a certain range (0.8~2.4wt%) and Nb in the range of 1.6~2.4wt% almost have no big influence on γ′ precipitation. It should be considered that these calculations are not accuracy enough, but it shows the tendency.

Phase computations also reveal that the effect of Ti on η phase precipitation is much more than that of Nb. The effect of Ti on η phase precipitation behavior is shown in Fig. 8b. The maximum percentage of η phase in the alloy (Ti 1.8% - nominal composition) will reach to 14.16% at 850°C. When the content of Ti decreases to 0.8%, the maximum percentage of η phase will be 7.33% at 861°C and the precipitation temperatures are limited in the range of 744~1011°C.

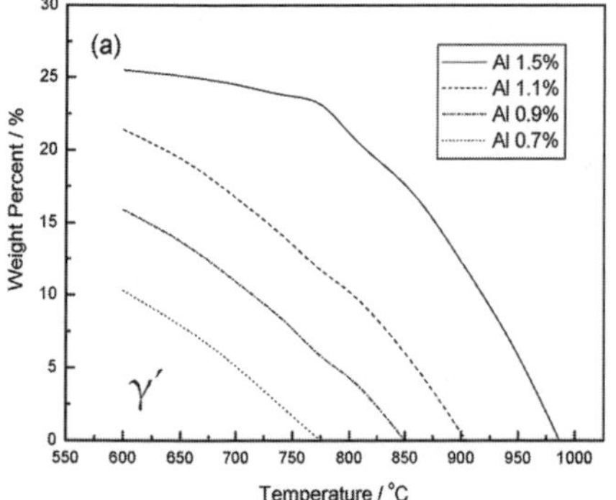

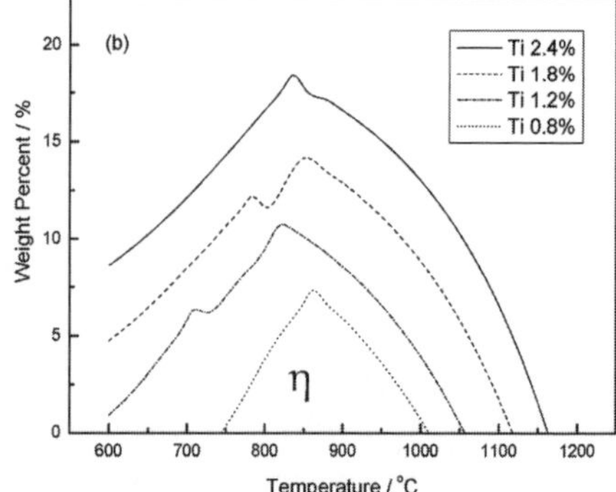

Fig. 8 Calculated curves of the effect of Al (a) on γ′ phase precipitation behavior and the effect of Ti (b) on η phase precipitation behavior

The existence of large blocky G phase at grain boundaries at 760°C may develop embrittlement and will degrade mechanical properties of the alloy, thus the formation of G phase should be restrained. However, due to the lack of Si in the thermodynamic database, the formation of G phase cannot be speculated by Thermo-Calc software.

Experimental results on modified alloys

According to thermodynamic calculation results, 4 modified alloys were made to investigate the improvement of structure stability. Their chemical compositions are listed in Tab. 2. Heat 1 and Heat 2 were made in Beijing to adjust the contents of Al, Ti and to control Si content to a lower level in the alloys. Heat 3 and Heat 4 were made in Huntington to adjust the contents of Al, Ti only. The manufacturing process and heat-treatment conditions of 4 modified alloys are same as former INCONEL alloy 740. The mechanical properties and coal ash/flue gas corrosion resistance of modified alloys were evaluated. All 4 modified alloys were solution-annealed and aged at 750°C for times up to 5018h, respectively. The samples of Heat 1 and Heat 2 were also aged at 800°C for 5000h and at 850°C for 1000h. The microstructures of all samples were examined by using SEM.

The stress rupture curves of INCONEL alloy 740 are shown in Fig. 9a. It indicates that the stress rupture strength of INCONEL alloy 740 at 750°C for 10^5hrs is already above the target stress 100MPa. The γ′ fractions of modified alloys are higher than that of original INCONEL alloy 740. The stress rupture strengths of modified alloys are expected to be higher than that of original INCONEL alloy 740. Long time stress rupture tests are conducted in progress. The cross-sectional metal loss results of the initial laboratory corrosion resistance tests are presented in Fig. 9b for 750°C to 5018hrs. The test condition is similar as the former. Linear extrapolation of the 750°C data for modified alloy 740 would predict that these alloys would meet the requirement of 2 mm / 200,000 hours corrosion allowance criteria in this test environment.

Tab. 2 Chemical compositions of 4 modified alloys (wt.%)

Heat	C	Cr	Co	Nb	Ti	Al	Fe	Mn	Si	Mo	Ni
1	0.016	25.45	20.37	1.99	0.67	1.21	0.066	0.001	0.009	0.001	Bal
2	0.014	25.8	20.4	1.99	1.38	1.25	0.07	0.001	0.011	0.01	Bal
3	0.042	24.5	19.8	2.2	1.12	1.73	0.6	0.29	0.5	0.5	Bal
4	0.031	24.5	19.9	2.37	1.15	1.41	0.1	0.28	0.5	0.5	Bal

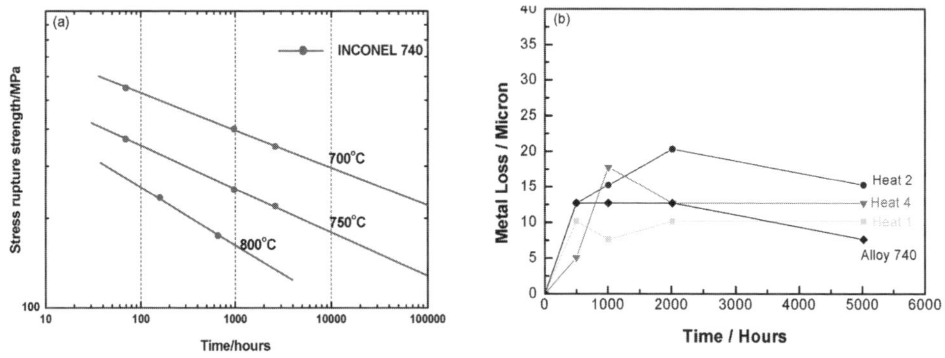

Fig. 9 Stress rupture curves of INCONEL740 (a) and coal ash corrosion resistance comparison between INCONEL 740 and modified alloys (b)

Fig. 10 a and b shows the SEM images of modified alloys Heat 1 and Heat 2 aged at 750°C for 5018h respectively. The strengthening phase γ' has been discerned clearly and a small amount of carbides was also found in the grains and at grain boundaries. In comparison with the previous results of INCONEL 740 (see Fig. 3b), neither η phase in the grains nor G phase at grain boundaries can be found in the modified alloys Heat 1 and Heat 2 (see Fig. 10a and 10b) after long time aging for 5018h at 750°C. Fig. 10 c and d shows the SEM images of modified alloys Heat 3 and Heat 4 aged at 750°C for 5018h respectively. The same as in Fig. 10a and 10b, no η phase was found but blocky carbides and G phase was found at the grain boundary.

Fig. 11 and Fig. 12 show the SEM images of modified alloys Heat 1 and Heat 2 aged at 800°C for 5000h and aged at 850°C for 1000h, respectively. The similar as 750°C aging, there are also no η phase and G phase form in the grains or at the grain boundaries. However, the coarsening of γ' particles is obviously. The more detailed studies on the precipitates, corrosion resistance of these 4 modified alloys are still in progress.

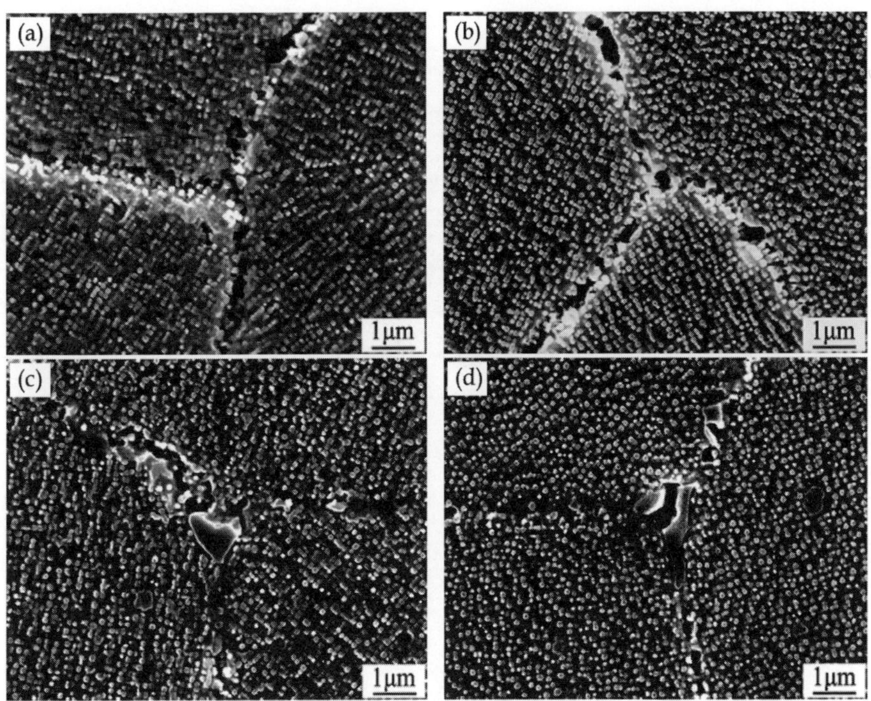

Fig. 10 SEM images of 4 modified alloys aged at 750°C for 5018h (Heat 1 (a), Heat 2 (b), Heat 3 (c), Heat 4 (d))

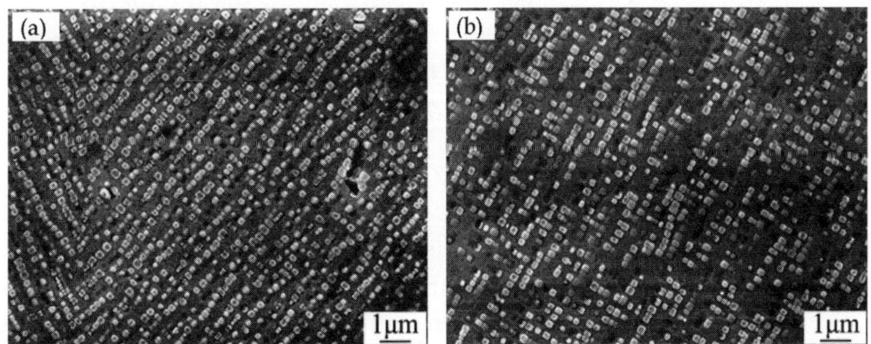

Fig. 11 SEM images of the alloy aged at 800°C for 5000h (Heat 1 (a), Heat 2 (b))

Thermodynamic calculation was also made to evaluate the precipitates of modified alloys. Table 3 gives the results of phase computation. The weight percentages of γ' phase are higher than 15% in 4 modified alloys. A minimum percentage of 15% was deemed acceptable to gain the stress rupture strength target. The solution temperatures of γ' phase in 4 modified alloys all have a large increase. Furthermore, both the precipitation temperature scope and percentage of η phase decrease clearly. These results are in accordance with experimental results as shown in Fig. 10.

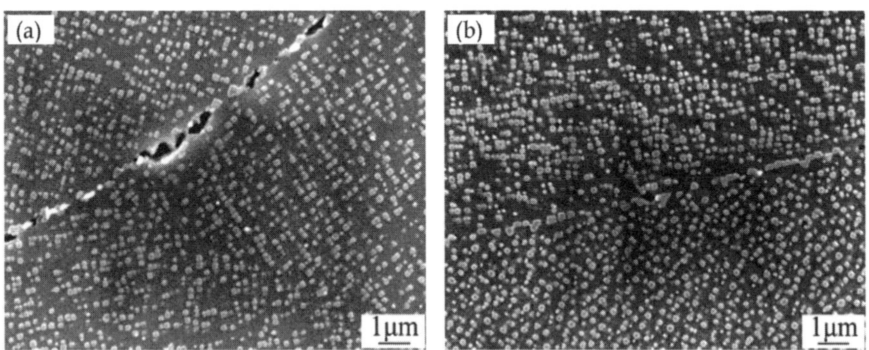

Fig. 12 SEM images of the alloy aged at 850°C for 1000h (Heat 1 (a), Heat 2 (b))

Table 3 Phase computation results of 4 modified alloys

Heat	Content of γ′(η)/%		γ′ Solution Temperature/°C	η Solution Temperature/°C	σ Solution Temperature/°C
	700°C	750°C			
1	15.2	14.1	930	852-1014	652
2	20.3	19.1	934	749-1100	725
3	22.9	21.9	1018	950-1072	728
4	21.3	19.6	962	824-1079	699
NC*	10.9(8.5)	7.7(10.7)	850	<1118	699

* Nominal composition of INCONEL 740

Conclusions

INCONEL alloy 740 keeps a good microstructural stability during prolonged aging at temperatures under 725°C. The main structure instabilities of INCONEL alloy 740 during thermal exposure include the γ′ coarsening, large amount of η formation and certain fraction of G phase existence at grain boundaries after long-term aging above 750°C.

Thermodynamic calculation results show that Al, Ti and Nb contents have an obvious effect on precipitation behavior of γ′ and η phases. The improvement of structure stability of INCONEL alloy 740 can be achieved by the adjustment of Al, Ti and/or Si level in the alloy.

The modified alloys exhibit more stable structure at 750°C aging till 5018h, at 800°C aging for 5000h, and at 850°C aging for 1000h as increasing more stable γ' fraction, retarding the transformation of γ' to η and/or eliminating G phase formation.

Acknowledgments

The University authors give their appreciation to Special Metals Corporation, Huntington WV, USA for funding this project and providing the experimental materials.

References

1. G. D. Smith and H. W. Sizek, "Introduction of an Advanced Superheater Alloy for Coal-Fired Boilers," Paper No. 256, presented at the NACE Annual Conference and Exposition, Orland, FL (March 2000).

2. P. Castello, V. Guttmann, N. Farr and G. D. Smith, "Laboratory Simulated Fuel-Ash Corrosion of Superheater Tubes in Coal-Fired Ultra-Supercritical-Boilers," *Corrosion and Materials*, 51 (2000): p. 786.

3. S. J. Patel, "Introduction to INCONEL Alloy 740: An Alloy Designed for Superheat Tubing Coal-fired Ultra Supercritical Boilers," *Acta Metallurgica Sinica*, 18 (2005): p. 479.

4. INCONEL alloy 740 Bulletin, Special Metals Corporation, Huntington, WV (2004).

5. B. A. Baker, "A New Alloy Designed for Superheater Tubing in Coal-Fired Ultra Supercritical Boilers," *Superalloys 718, 625, 706 and Derivatives 2005*, TMS (2005), p. 601.

6. S. Zhao, X. Xie, G. D. Smith, and S. J. Patel, "Microstructural Stability and Mechanical Properties of a New Nickel-Based Superalloy," *Mater Sci Eng A*, 355 (2003), p. 96.

7. S. Zhao, J. Dong, X. Xie, G. D. Smith and S. J. Patel, "Thermal Stability Study on a New Ni-Cr-Co-Mo-Nb-Ti-Al Superalloy," *Superalloys 2004*, TMS (2004), p. 63.

8. X. Xie, S. Zhao, J. Dong, G. D. Smith and S. J. Patel, "An Investigation of Structure Stability and Its Improvement on New Developed Ni-Cr-Co-Mo-Nb-Ti-Al Superalloy," *Mater Sci Forum*, 475 (2005), p. 613.

9. B, Sundman, B, Jansson and J. O. Anderson, "The Thermo-Calc databank system," *CALPHAD*, 9 (1985), p. 153.

Advances in Materials Technology for Fossil Power Plants
Proceedings from the Fifth International Conference
R. Viswanathan, D. Gandy, K. Coleman, editors, p 231-259

DOI: 10.1361/cp2007epri0231

MATERIALS QUALIFICATION FOR 700 °C POWER PLANTS

Qiurong Chen, Georg-Nikolaus Stamatelopolous, Andreas Helmrich
Alstom Power Boiler GmbH
Augsburger Str. 712
D 70327 Stuttgart, Germany

Josef Heinemann
UTP Schweißmaterial GmbH
Elsässer Strasse 10
D 79189 Bad Krozingen, Germany

K. Maile, A. Klenk
Materialpruefungsanstalt Universitaet Stuttgart
Pfaffenwaldring 32
D 70569 Stuttgart, Germany

Abstract

Components in 700 °C power plants which are subjected to highest temperature and loading will be made of Ni-based alloys. Since a couple of years material development for boiler and turbine components operating in this temperature regime is ongoing. This paper describes investigations of components (e.g. tubing, membrane walls, thick walled components) made from Ni-based alloys. Results of qualification programs for boiler components including welded joints showed the applicability of candidate material Alloy 617. Similar programs and investigations on Alloy 263 and Alloy 740 are under way. Investigations and experiments aimed to optimize and qualify welding consumables have been done in order to transfer the knowledge gained to the manufacturing of components. Long term qualification of the materials is necessary not only with respect to creep behavior but also investigations on deformation capability after long-term ageing needs to be investigated.

1 Introduction

Despite strong efforts especially in Europe to strengthen the development and activation of renewable energy sources, coal fired plants will contribute a significant part to the supply of electric power in the near future. Technical progress in this field must be aimed to make this contribution as efficient as possible with respect to resources and emissions. One of the most important aims is therefore to increase the efficiency and reduce mainly CO_2-emissions. Higher efficiency means higher steam temperature and therefore materials development for higher temperatures are a key-issue. Currently plants are under design and construction realizing the next temperature step with superheater temperatures up to 610 °C and reheater temperatures up to 625 °C. However, the most significant efficiency increase will be achieved with temperatures up to 750 °C.

To realize these option a couple of European (e.g. the THERMIE-programs) and national research projects have been launched, e.g. Cooretec-Projects, Table 1, in Germany. These projects comprise development and qualification of materials for boiler and turbine parts as well as the development of non-destructive testing methods for austenitic materials and Nickel based alloys as well as for their welded joints.

Table 1: Overview on Cooretec research projects

Cooretec (CO_2-reduction technologies) for future power plants

Boiler and Piping

DE-1	Corrosion behaviour of materials for 720°C plan	3,70 M€	2007-2010
DE-2	Reheater materials after cold forming	0,40 M€	2006-2009
	Development of Technology and Qualification of longitudinally welded pipes made of Nickel based alloys		Planned 2007
DE-4	Strength and Deformation Behavior of Tubes, Piping and Forgings made from Ni based alloys (Alloy 617, Alloy 263)	2,07 M€	2007-2011

Steam turbine

DT-1	'COST 536 Material development for 600-620°C	3,90 M€	2005-2009
DT-2	Hard coatings	-	-
DT-3	Qualification of 10%Cr-materials (620°C), Dissimilar welds 2Cr-Ni-based alloy/ 10Cr - Ni based alloy (720°), Numerical work	1,50M€	2007-2010
DT-4	Lifetime concepts and assessment using fracture mechanics methods for high efficiency steam power plants AG TURBO	1,20M€	2007-2009

TD-1	NDE of thickwalled components made of Ni-based alloys and welded joints	1,50 M€	09.06-08.09

These projects are based on long-term experiences mainly for Alloy 617 and other austenitic materials gained during the development of High Temperature Reactors (HTR) /1/. A couple of R&D project resulted in sufficient material database and the basics for the extension of design codes for materials and components in the HTR. These studies were focused on temperatures in the range of 750 °C to 900 °C. The project MARCKO-DE2 was aimed to optimize Alloy 617 for the temperature range 700 °C to 750 °C. A controlled chemistry specification was developed and preliminary qualification of this material and welded joints were demonstrated /2, 3/. According to the current state of the art candidate materials for temperatures above 700 °C for boiler tubes and piping are Alloy 617, Alloy 263 and Alloy 740 /2,3,4/.

2 Technology for 700°C plants

The worldwide abundance of coal guarantees a stable fuel supply for the next centuries and contributes to the selection of pulverized coal-fired power plant technology as the preferred option for many utilities. Additional reasons for a preference of this technology include:

- the reliability of a technology which has evolved in incremental steps,
- high levels of availability and operational flexibility,
- high levels of fuel flexibility ensuring a secure and economic fuel source,
- high efficiencies and low generation costs.

Despite its long history, coal-fired power plant technology is still evolving, and efficiency has been improved significantly since the late 1980s and will be improved still further (Figure 1).

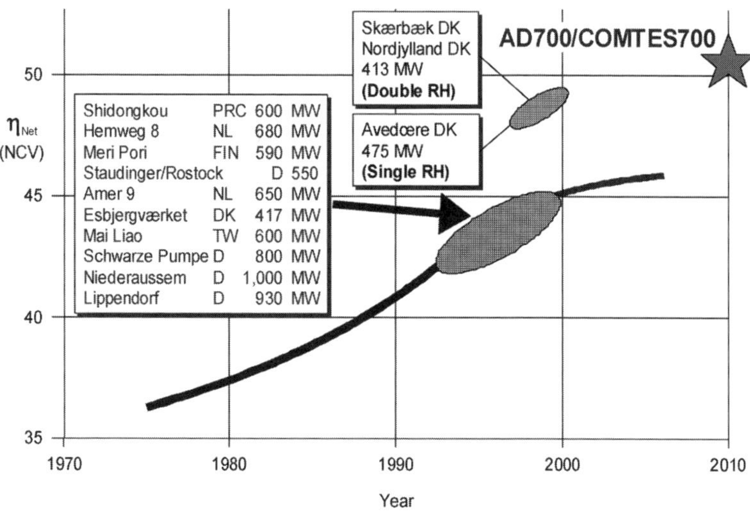

Figure 1: Evolving coal-fired power plant technology

An improvement in the thermal efficiency of the plant does not only reduce the fuel costs but also reduces the release of SO_2, NO_x and CO_2 emissions. In comparison with the world average efficiency, the state of the art power plant can reduce the CO_2 emissions by approximately 35 %

(Figure 2). In order to increase the power plant efficiency, supercritical and ultra-supercritical technology with continuously higher steam parameters are used which will represent also the major development trend in the near future (Figure 3). The supercritical cycle is realized by means of the well known and since the 1960s commercially used once-through technology. This avoids the introduction of a totally new kind of clean coal technology, which may carry some substantial risks. Only material selection and dimensioning of the pressure parts have to be adapted to the new requirements, and the design and construction have to consider the limitations on the material side.

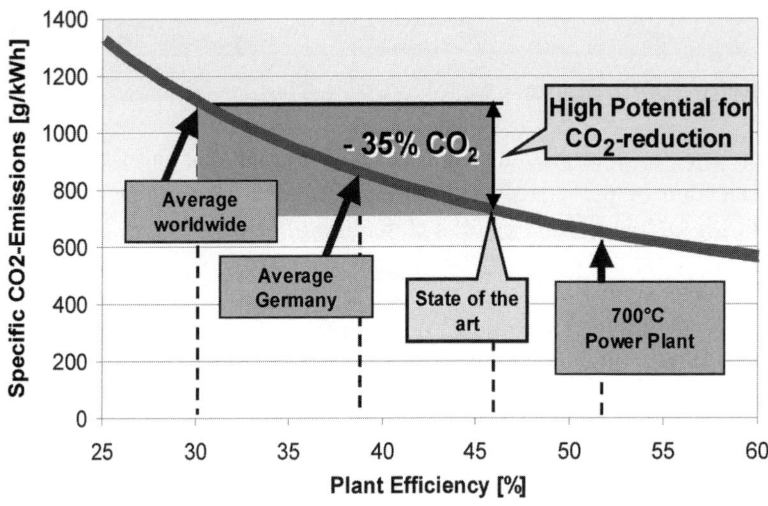

Figure 2: Evolving coal-fired power plant technology

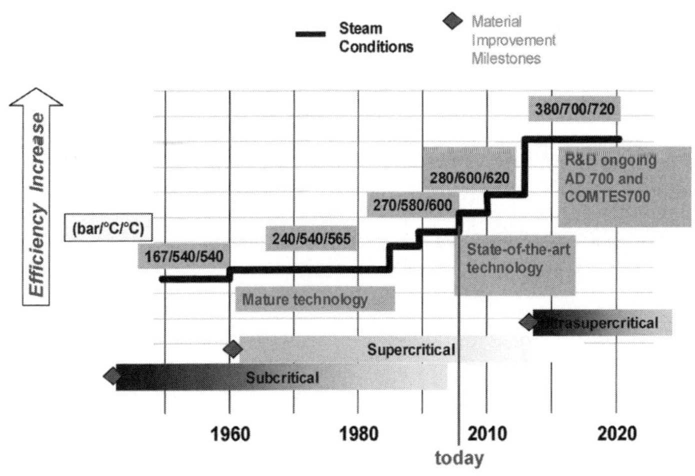

Figure 3: Major developments in the near future

Currently in different R & D programs investigations are performed for an advanced power plant with the maximum steam parameters of 380 bar/700 °C/720 °C which are significantly higher in comparison with the state-of-the-art power plants. The efficiency of the advanced power plants will be increased from the recent range of 43 - 47 % to above 50 %. Due to the higher thermal efficiency of the power plants, the fuel consumption and emissions will be reduced by almost 15 % compared with the best plants currently available.

2.1 Design of the 700 °C boiler

A typical bituminous coal fired boiler with the 700 °C technology is shown in Figure 4. The boiler is designed as a typical tower type boiler. Based on the experiences for the supercritical power plants, a tower type boiler gives the most uniform distribution of the flue gas especially with respect to temperature profile in the superheater and reheater sections. This is a big advantage for a boiler operated at very high steam parameters because the material design margins at high steam conditions are very limited.

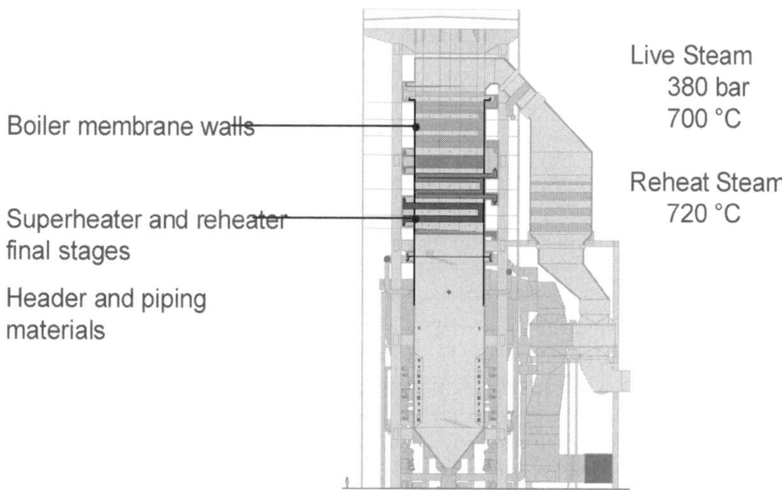

Figure 4: Critical components in 700°C-boilers

Membrane walls enclose the furnace with spiral tubing in the high heat flux area and vertical tubing in the convective heating surface area. The membrane wall of the lower furnace area is designed as evaporator. For this part, spiral tubing is applied. With this layout, the number of parallel tubes and the tube dimensions can be chosen in such a way that sufficient cooling of the tubes can be assured over the whole load range. The big advantage of the spiral tubing, which is especially important for a 700 °C power plant, is that the energy imbalance of the parallel tubes at the evaporator can be minimized due to the even heat pickup of the individual tubes. The separator is arranged directly after the spiral tubing of the membrane walls. The vertical tubing part of the membrane wall is designed as first superheater stage.

The convective heating surfaces are arranged in a conventional and proven way. From bottom to top (with decreasing flue gas temperature) the heat exchangers are arranged as follows: second superheater (grid), final superheater, final reheater, third superheater, first reheater and economizer.

2.2 *Materials of the 700 °C boiler*

The critical components of the boiler that applies the 700 °C technology are the membrane walls, the final superheater and reheater stages as well as the thick-walled components, mainly the high-pressure outlet headers and the piping to the turbine.

Several boiler materials with improved properties have been developed in recent years. The use of the materials for the boiler with advanced steam parameters will be first limited by the relevant creep strength specifications. The application of the materials for membrane walls and thick-walled components has to consider additionally the manufacturing aspects. For superheater and reheater heating surfaces, further restriction results from the risk of flue gas side high-temperature corrosion and steam-side oxidation. For the 700°C power plants, new materials with higher strength, higher corrosion resistance and good weldability and formability are required.

2.2.1 Membrane walls: The membrane wall is a critical component for the 700 °C boiler. Table 2 shows the chemical composition of the candidate wall materials in comparison with conventional ones.

Table 2: Chemical composition of membrane wall materials

Material	C	Cr	Mo	W	Ti	Co	Others
2 – 2.5 % Cr-steels:							
T23	0.04 - 0.10	1.9 - 2.6	0.05 - 0.30	1.45 - 1.75	-	-	V, Nb, N, B
T24	0.05 - 0.10	2.2 - 2.6	0.9 - 1.1	-	0.05 - 0.10	-	V, N, B
9 – 12 % Cr-steels:							
T91	0.08 – 0.12	8.0 – 9.5	0.85 – 1.05	-	-	-	V, Nb, N
T92	0.07 – 0.13	8.5 – 9.5	0.3 – 0.6	1.5 – 2.0	-	-	V, Nb, N, B
VM12	0.10 – 0.14	11.0 – 12.0	0.2 – 0.4	1.3 – 1.7	-	1.4 – 1.8	V, Nb, N, B
HCM12	max. 0.14	11.0 - 13.0	0.8 – 1.2	0.8 - 1.2	-	-	V, Nb
Ni-based alloys:							
Alloy 617 mod.	0.05 - 0.08	21.0 - 23.0	8.0 - 10.0	-	0.3 - 0.5	11.0 - 13.0	Ni, Al, Cu, N, B

The conventional membrane wall materials used so far are low-alloyed steels like 13CrMo45 (T12), T23 and T24, which are characterized by an excellent workability for the manufacturing of the membrane walls, because a post-weld heat treatment is for these materials not necessary. For 700 °C boilers, the maximum metal temperature will reach approximately 560 °C to 600 °C in the membrane wall, which is very high in comparison with the state-of-the art power plant. The creep strength of the candidate materials should be high enough in order to have less creep strains at high temperatures and to have a reasonable wall thickness. Another critical point is the oxidation resistance of the steels. Due to higher metal temperatures and high heat fluxes, the magnetite deposits on the inside of the tube grow rapidly which could raise the temperature further and cause creep damage, if the tube material does not have sufficient oxidation resistance. Due to the requirements of the high creep strength and oxidation resistance, the low-alloyed steels with 2 - 2.25 % chromium content (such as T23 and T24) can be used only up to a limited temperature level. Austenitic steels have high creep rupture strength and high steam oxidation resistance. But the austenitic steels cannot be used for membrane walls due to the poor physical properties (too big thermal expansion coefficient).

The candidate materials can be first selected from the group of martensitic 9 – 12 % chromium steels (T91, T92, VM12, HCM12). In the COST and MARCKO700 programs, it has been demonstrated that the material T91, T92 and VM12 can be fabricated into membrane walls. However, a post-weld heat treatment is required for these materials. Among the group of martensitic 9 - 12 % chromium steels, the material T92 has the highest creep rupture strength. The oxidation resistance of this material is also sufficiently high in the relevant temperature range. This material will be used as preferred steel for the membrane walls of a 700 °C boiler.

Another candidate material among the 9-12% chromium steels could be the material HCM12 under certain circumstances. This material has been manufactured without a post-weld heat treatment and has been or is being tested in the AD700 (phase 2) and COMTES700 projects. The operational experiences and the oxidation resistance results have been very promising. Unfortunately, it was realized in 2005 that a major reduction in long-term creep rupture strength at temperatures above 550 °C must be foreseen for the steel HCM12. In order to make the application of this material possible, the long-term creep properties have to be clarified.

Another group of candidate materials for the membrane walls are the nickel-based alloys for which a post-weld heat treatment is not necessary. A well-examined nickel alloy is the material Alloy 617. The alloy is used in the solution-treated condition and has the advantage of being relatively easily worked. Other advantages of this material are very high creep rupture strengths, high corrosion and oxidation resistance and similar heat expansion coefficients as martensitic steels. In the MARCKO DE2 project, this material has been further optimized in the chemical composition to achieve higher creep strength values (the new version called Alloy 617 mod.) for temperatures up to 750 °C. However, Ni-based alloys are much more expensive. Before such a choice is taken, a careful analysis of the whole boiler economy is needed. In the AD700 (phase 2) and COMTES700 projects, evaporator panels made of HCM12 and Alloy 617 or Alloy 617 mod. have been manufactured and installed for in-plant exposure tests. Figure 5 shows the application limit (maximum possible wall temperatures) for different membrane materials.

2.2.2 Superheater and reheater tubes: For superheater and reheater tubes, the creep strength of the used materials should be high enough at the relevant temperature range. In addition to this requirement, the corrosion resistance on the flue gas side and the oxidation resistance on the steam side have to be taken into consideration. The oxide layer on the inside of the tube leads to higher material temperatures and could cause creep damage. External high-temperature corrosion on the flue gas side reduces the effective wall thickness of the tubes.

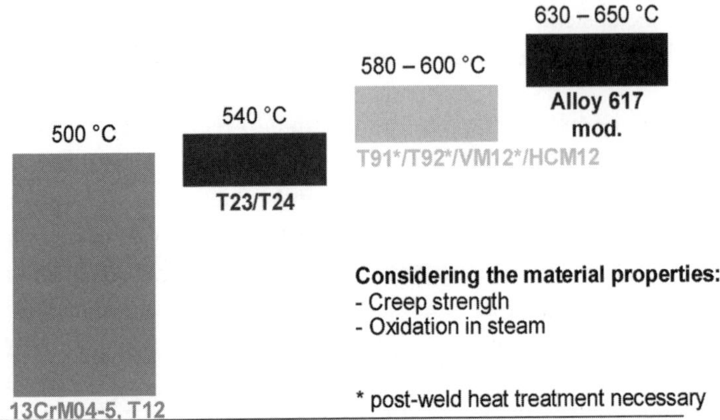

Figure 5: Materials and material application limit (maximum possible wall temperature) for 700°C- membrane walls

Table 3 gives the chemical composition of some austenitic steels and Ni-based alloys for superheater and reheater tubes at high temperatures.

Table 3: Chemical composition of superheater and reheater materials

Steel	Cr	Ni	Mo	W	Ti	Co	Others
Austenitic steels:							
Super304H	17.0 - 19.0	7.5 - 10.5	-	-	-	-	Cu, Nb, N, B
Tempaloy AA-1	17.5 - 19.5	9.0 - 12.0	-	-	0.10 - 0.25	-	Cu, Nb, B
XA704	17.0 – 20.0	8.0 – 11.0	-	1.5 – 2.6	-	-	V, Nb, N
TP347HFG	17.0 - 20.0	9.0 - 13.0	-	-	-	-	Nb+Ta
NF709R	21.5 - 23.0	22.0 - 28.0	1.0 – 2.0	-	max. 0.20	-	Nb, N, B
HR3C	23.0 - 27.0	17.0 - 23.0	-	-	-	-	Nb, N
DMV 310N	24.0 – 26.0	17.0 – 23.0	-	-	-	-	Nb, N
Tempaloy A-3	21.0 - 23.0	14.5 - 16.5	-	-	-	-	Nb, N, B
Alloy 174	22.5	25.0	-	3.6	-	1.5	Cu, N, Nb
HR6W	21.0 – 25.0	40.0 – 55.0	-	4.0 – 8.0	max. 0.20	-	Nb
Ni-based alloy:							
Alloy 617 mod.	21.0 - 23.0	Bal.	8.0 - 10.0	-	0.30 - 0.50	11.0 - 13.0	Al, Cu, N, B
Alloy 740	25.0	Bal.	0.50	-	1.8	20.0	Al, Nb

Higher chromium content has generally higher corrosion and oxidation resistance. Therefore, austenitic steels with higher chromium content will be preferred at higher temperatures. In order to further increase the oxidation resistance on the steam side, additional methods (fine grained structure, shot-peening surface treatment) are used.

New austenitic steel called Alloy 174 (Sanicro 25) has been developed in the AD700 project. This material shows a big improvement in creep rupture strength compared with the best commercially available austenitic steels. Ongoing steam oxidation and flue gas corrosion test demonstrate properties comparable to or slightly better than those obtained for a large variety of 22 – 25 % Cr austenitic steels.

Above certain temperature level, the austenitic steels cannot be used any more due to insufficient creep rupture strength. At such temperature levels, Ni-based alloys have to be selected. The well-examined Alloy 617 and the optimized version Alloy 617 mod. should be preferred used. For this alloy, long-term material properties are known. A good workability has been demonstrated in different R & D programs.

In the AD700 project, a new nickel-based alloy called Alloy 740 has been developed. The development was made on the basis of the material Alloy 263 with the consideration of the creep strength and corrosion resistance. Improvements of the coal ash corrosion resistance were developed in a series of coal ash corrosion tests at different temperatures. The new alloy with the resultant optimized chemical composition is a nickel chromium cobalt alloy, which is age-hardenable by the precipitation of gamma prime but also benefits from solid solution hardening. In-plant tests with Ni-based alloys Alloy 740, Alloy 617 and Alloy 617 mod. as well as austenitic steels such as Alloy 174 are ongoing in the Esbjerg 720 °C Test Rig and COMTES700 project. Figure 6 shows the application limit (maximum possible wall temperatures) for different superheater and reheater tubing materials.

2.2.3 Thick-walled components: For the thick-walled components and piping, there are two goals for the material development. An improved martensitic 9 – 12 % Cr steel is desirable to expand the present temperature range up to approx. 630 °C – 640 °C. A Ni-based alloy with a 100000 h creep rupture strength of at least 110 MPa at 715 °C is needed to allow construction of the final superheater outlet headers and live steam lines with acceptable wall thicknesses. Additional to the creep rupture strength requirement, the steam oxidation resistance of the material has to be taken into consideration. Table 4Table 4 gives the chemical composition of the header and pipe materials.

The conventional materials for the pipe and header application in the 600 °C power plants have been the steel P91, E911 and P92. So far no martensitic materials have demonstrated long-term creep rupture strength better than the steel P92. Some potentially interesting new martensitic steels are under development with the aim of higher creep rupture strength values than P92 at the NIMS in Japan and under the European COST536 program. But for all new developments both long term material properties and the workability have to be investigated in more detail.

For the high steam conditions, nickel-based alloys are needed. Besides high creep rupture strength, the Ni-based alloys have excellent steam oxidation resistance. For acceptable wall thicknesses, the material Alloy 617 or the optimized Alloy 617 mod. should be preferably used. In the COMTES700 project, headers and pipe lines from Alloy 617 mod. which were machined from forged bars have been installed for the in-plant testing.

Table 4: Chemical composition of header and pipe materials

Steel	C	Cr	Mo	W	Co	Others
Martensitic steels:						
P91	0.08 - 0.12	8.0 - 9.5	0.85 - 1.05	-	-	V, Nb, N
E911	0.09 - 0.13	8.5 - 9.5	0.90 - 1.10	0.9 - 1.1	-	V, Nb, N, B
P92	0.07 - 0.13	8.5 - 9.5	0.30 - 0.60	1.5 - 2.0	-	V, Nb, N, B
Ni-based alloys:						
Alloy 617 mod.	0.05 - 0.08	21.0 - 23.0	8.0 - 10.0	-	11.0 - 13.0	Ni, Al, Ti, Cu, N, B
Alloy 263	0.04 – 0.08	19.0 – 21.0	5.6 – 6.1		19.0 – 21.0	Ni, Al, Ti, Cu, B

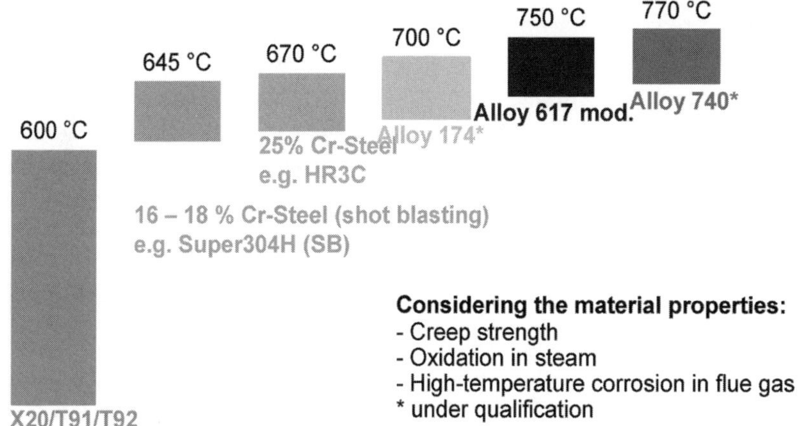

Figure 6: Materials and material application limits (maximum possible wall temperature) for 700°C reheater and superheater tubes

The age-hardenable Ni-based material Alloy 263, which had limited and short time stress rupture data are being upgraded for the power plant applications. This material meets the requirements of high creep strength for the final superheater outlet headers and live steam lines at 700 °C steam temperature. However long term creep data and demonstration of fabricability are still needed. Alloy 263 is under investigation in the AD700 project and will be qualified in the German COORETEC program. Figure 7 shows the application limit (maximum possible wall temperatures) for the header and pipe materials.

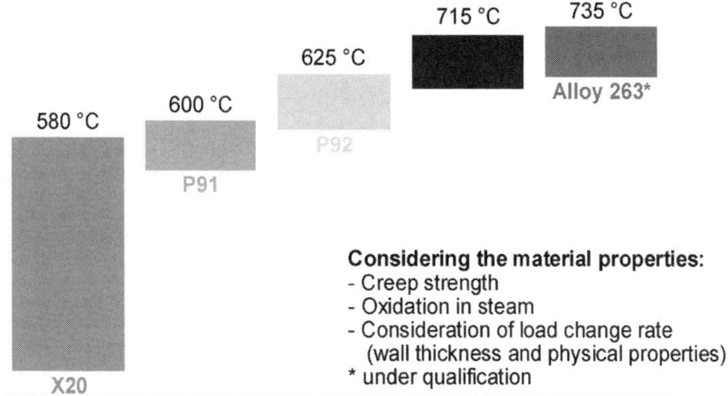

Figure 7: Materials and material application limits (maximum possible wall temperature) for header and pipe materials

3 Qualification of components made of Ni based alloys

3.1 Characterization of base materials

The application of Ni-based alloys is one main issue in the strategy for the realization of 700° C Power Plants. According to actual design studies, the Ni-based alloys Alloy 617 mod., Alloy 263 and Alloy 740 will be used for the final stages related to the components SH- and RH tubes, header and piping and as potential candidate material for the final stages of the membrane walls. For safe application based on code requirements, there is a need of qualifications in order to investigate and to prove material behavior and fabricability. Basically, for the characterization of the material properties and the application as boiler component, the specific chemical composition as shown in Table 5Table 5 has to be considered.

Derived from the chemical composition, the Ni-based alloys can in principle be classified in the following two types:

Alloy 617 mod.: The mechanism of strengthening is characterized by the mechanism of solid solution hardening mainly caused by Mo /5/. Based on that, the specified material properties as well as the application for operation correspond to the solution heat-treated condition. This benefit of higher creep rupture strength is based on controlled chemical composition mainly based on the elements Al, Ti, B and C /3/.

Alloy 263, Alloy 740: The effort of strengthening is based on the mechanism of precipitation hardening with fine distributed precipitations of the intermetallic phase gamma prime but also benefits from solid solution hardening /6/. This effect results basically in higher creep rupture behavior in comparison with Alloy 617 mod.

In order to improve oxidation and corrosion behavior for application of SH-/RH-tubes, Alloy 740 is specified with higher Cr-content. The specified material properties and the application for operation correspond to the precipitation hardened condition following the solution heat-treated condition during fabrication of the semi-finished product or after welding. The development for power plant application was started in the framework of the European R & D-Project Thermie and in the United States in the framework of the R & D Project DOE.

Actual data of creep structure strength are shown in Table 6Table 6 for comparison. Typical microstructures of Alloy 617 mod. and Alloy 740 (precipitation hardened) is shown in Fig. 8.

Table 5: Specific Chemical Composition of Ni-based Alloys

Material	C	Cr	Ni	Mo	Nb	Ti	Others
Alloy 617 mod.	0,05 - 0,08	21.0 - 23.0	Rem.	8.0 – 10.0	----	0.30 – 0.50	N. max. 0,05 Co = 11. – 13.0 Al = 0.80 – 1.30 B = 0,002-0,005
Alloy 740	0,03	25.0	Rem.	0.5	2.0	1.8	Co = 20.0 Al = 0.9
Alloy 263	0,04 - 0,08	19.0 21.0	Rem.	5,6 – 6,1	----	1.9 – 2.4	Co = 20 Al = 0,3 – 0,6 Al + Ti: max. 2,8

Table 6: 100 000 h Creep Rupture Strength at 700°C

Material	Alloy 617	Alloy 617 mod.	Alloy 740	Alloy 263
Creep strength 100.000 h 700°C (N/mm²)	95	119	190	182
Standard	VdTÜV 485 (09.2001)	Evaluation TÜV-Rheinland (01.2004)	Manufacturer's Data	Manufacturer's Data

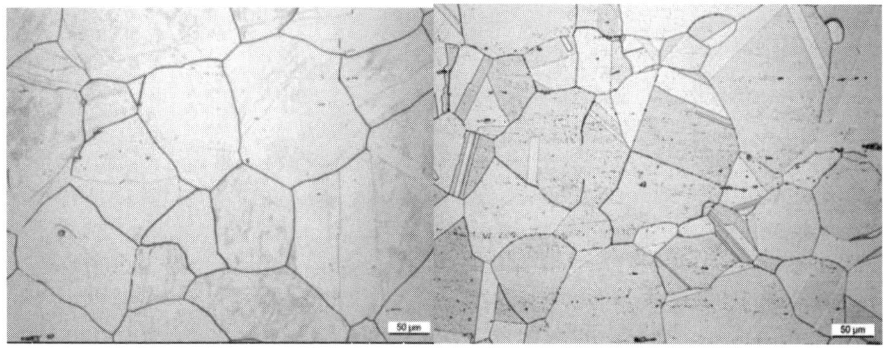

a) Alloy 617 mod. Tube OD 38mm x 6,3 mm   b) Alloy 740, Tube OD 38 mm x 6,3 mm

Figure 8: Typical microstructures of Alloy 617 mod. (a) and Alloy 740 precipitation hard. (b)

3.2 Material behavior

For use of the Ni-based alloys for the particular boiler components, the necessary material behavior in dependence of the design temperature has to be considered. These are short-term properties and creep properties of base metal and weld, corrosion and oxidation resistance, industrial fabricability and application of Non-Destructive-Testing (NDT). Table 7Table 7 gives an overview of the necessary behaviors that have to be proved and verified.

Table 7: Requirements for 700°C-materials

Component	**Header / Piping**	**SH - / RH-tubes**	**Membrane Wall**
Material	**Alloy 617 mod., Alloy 263**	**Alloy 617 mod., Alloy 740**	**Alloy 617 mod.**
Short-term Properties (base metal, weld)	X	X	X
Creep properties (base metal, weld)	X	X	X
Oxidation Resistance	X	X	X
Corrosion Resistance	---	X	X
Fabricability Welding Bending	 X X (inductive)	 X X (cold)	 X X (cold)
NDT	X Ultrasonic testing (UT) on butt welds	--- NDT available	--- NDT available

3.3 Qualification activities - Overview

For qualification and approval of the Ni-based alloys several different R & D projects have been initiated. According EN-Standard, from a formal point of view, short- and long-term properties have to be proved for base metal, weld deposit and welded joint. That means for approval based on a European Material Data Sheet (EMDS), long term properties have to be proved by creep tests on several heats with running time more than 30.000 h.

Corrosion and oxidation behavior will be proved previously in laboratories but also on test loops or test facilities in boilers with increased steam temperatures corresponding to the intended use in 700° C Power Plants. Fabricability will be demonstrated by fabrication of components based on usual boiler fabrication practice at workshop and site. Pre-conditions for fabrication related to the particular bending and welding process are Bending Procedure Qualifications and Welding Procedure Qualifications according to EN-code, approved by a Notified Body. For installation in a boiler, entire fabrication and final inspection will be done following code-requirements under supervision of a Notified Body in order to get approval for operation by authorities.

Table 8Table 8 gives an actual status of initiated European R&D projects based on PED. Intensive activities for investigation and qualification of all Ni-based alloys are also ongoing in the framework for the R&D project DOE in the USA.

Table 8: European R&D projects - overview

Boiler component / Material	**Header, Piping**	**SH-/RH-tubes**	**Membrane Wall**
Alloy 617 mod.	Marcko DE 2 Marcko 700 Esbjerg 700/720°C Comtes 700	Marcko DE 2 Esbjerg 700/720° C Comtes 700	Marcko 700 Thermie Comtes 700
Alloy 740		Thermie Esbjerg 700/720° C Comtes 700 Weisweiler GKM	
Alloy 263	Thermie, Cooretec DE4		
Ni-based alloys UT-Test	Cooretec DE1		

3.4 Fabrication and design criteria

The investigations and qualifications in the framework of the above mentioned R&D-projects are based on criteria as described below.

For **fabrication**, the following processes are part of the qualification programs:

- **SH-/RH-tubes:** Butt Welds, TIG, Cold bending
- **Header, Piping:** Butt Welds, root pass TIG, filler and cover pass SMAW and SAW, Nozzle welds, with full penetration, root pass TIG – automatic from inside without filler metal, filler and cover pass SAW, Inductive bending with subsequent complete new solution heat treatment and mechanical testing.
- **Membrane walls:** Tube-fin -tube welds, SAW

For **design,** some criteria are specified in addition:

- Cold and inductive bending are each based on VGB-rules with increased dimensional requirements /7, 8/
- Full penetration weld of nozzle welds as shown in Figure 9a
- Tube-fin-tube welds with limitation of the remaining gap for SAW-welding based on VdTÜV-requirements /9/ as shown in Figure 9Figure 8b.

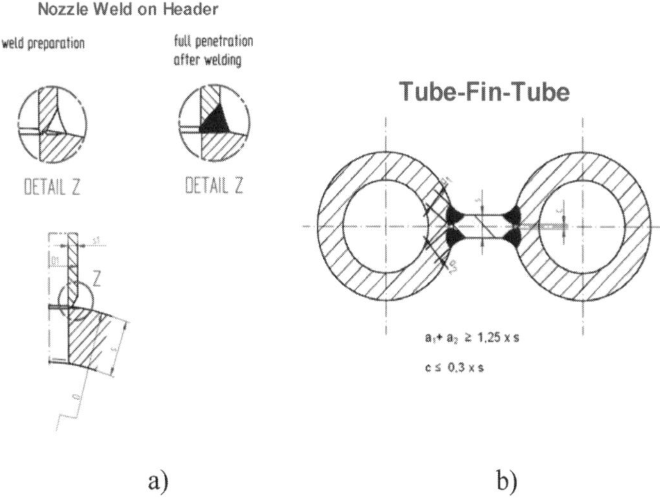

a) b)

Figure 8: Nozzle weld geometry

3.5 Results of qualifications

In context with different R&D-projects, qualifications have been done following EN-Code. Additional criteria and requirements are as mentioned before.

3.5.1 Welding

Butt welds: For Alloy 617 mod., butt welds on pipes and tubes have been carried out without heat treatment after welding, Figure 9 10 and Figure 10 11. For Alloy 740, butt welds on tubes have been done with subsequent precipitation hardening. For NDT, x-ray test and penetration test (PT) was applied. Mechanical tests and examination of macro/micro structure have been done as follows:

- Tensile tests at room temperature
- Hot tensile tests at higher temperatures
- Bend test
- Impact tests (weld and HAZ)
- Macro section with hardness test
- Micro section

From the butt welds, creep tests using crossweld specimens are ongoing or will start. Qualification on Alloy 263 with similar filler will start in the framework of Cooretec DE4 in 2008.

Header, nozzle welds: In the framework of Marcko 700 for Alloy 617 mod., headers with nozzles as full penetration weld have been qualified. After welding, PT-test was done. On macro and micro sections, conditions of full penetration, layer build up, hardness and microstructure have been checked, Figure 12.

Membrane walls: Membrane walls as test panels for qualification of Alloy 617 mod. have been fabricated in the framework of Thermie and Marcko 700. After welding NDT (PT) was done. On macro and micro sections layer build-, geometry of weld (remaining gap) and microstructure have been checked, Figure 13. Details of Welding Procedure Specifications (WPS) and results of qualification welds are shown in Table 9Table 9 and Figures 10 - 13.

Pipe OD 400 x 50 mm, Welding Process TIG/SMAW

5 mm

100 µm

Macro Section **Micro Section**

Figure 9: Butt weld Qualification (Pipe)

Table 9: Details of Welding Procedure Specification

Kind of weld	Pipe / Header, Butt Weld	Tube, Butt Weld	Tube, Butt Weld	Header, Nozzle Weld	Membrane Wall, Tube-Fin-Tube
Material	Alloy 617 mod.	Alloy 617 mod.	Alloy 740	Alloy 617 mod.	Alloy 617 mod.
Welding Process	GTAW /SMAW	GTAW	GTAW	GTAW / SAW	SAW
Filler Metal	Alloy 617 mod.	Alloy 617 mod.	Alloy 617 mod.	- / Alloy 617 mod.	Alloy 617 mod.
Weld position	Tube axis horizontal, fixed	Tube axis horizontal and	Tube axis horizontal and	Nozzle axis vertical	Tube axis horizontal
Weld preparation	U-groove	V-groove	V-groove	V-groove	no machining
Interpass temperature	max. 150° C	max. 150° C	max. 150° C	max. 150° C	not
Heat treatment	no	no	800° C/4 h	no	no

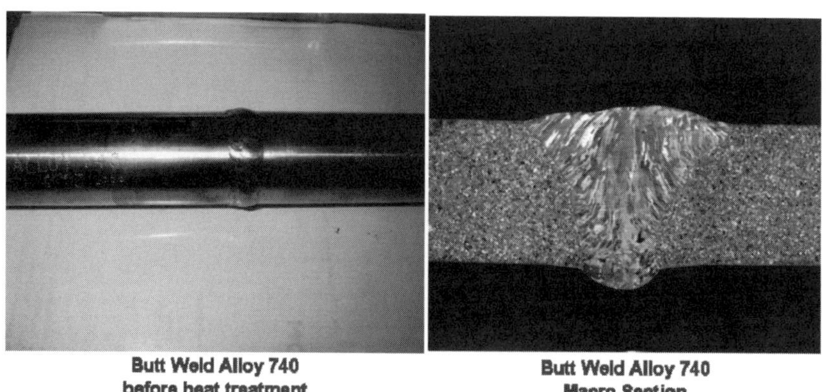

Butt Weld Alloy 740
before heat treatment

Butt Weld Alloy 740
Macro Section

Figure 10: Butt weld Qualification (Tube)

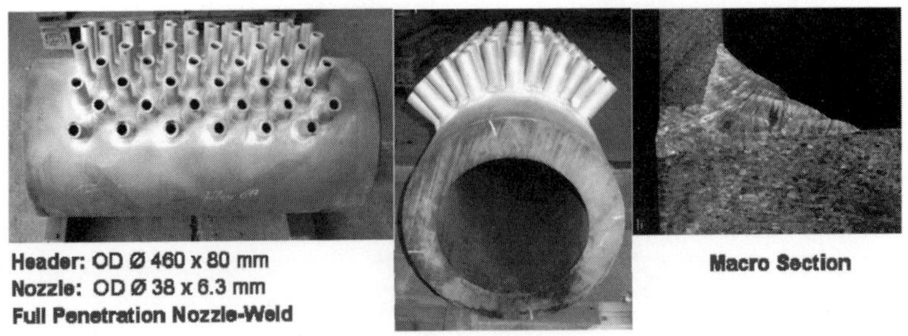

Figure 11: Nozzle Weld Qualification

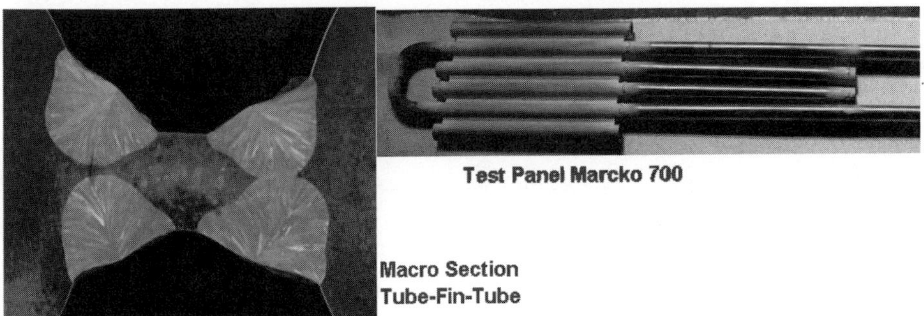

Figure 12: Membrane Wall Weld Qualification

3.5.2 Bending

For Alloy 617 mod., inductive bending was done in the framework of Marcko DE2, Marcko 700 and Comtes 700. After bending and subsequent new solution heat treatment, PT-testing and dimensional control was carried out. After solution heat treatment the mechanical properties and microstructure have been tested as follows:

- Tensile test at room temperature
- Hot tensile test at higher temperature
- Impact test (neutral zone, extrados, intrados)
- Micro section (neutral zone, extrados, intrados)
- Creep tests from inductive bends are ongoing.

Cold bending on Alloy 617 mod. was qualified done in the framework of Esbjerg 720°C and Comtes 700. Due to code requirements solution, heat treatment is specified for Alloy 617 if deformation rate exceeds 10 %. For that, qualification was done on bends with different deformation rates in order to qualify the variations with and without solution heat treatment after bending.

For NDT after bending, PT testing and dimensional check was carried out. In the case of new solution heat treatment, short-term mechanical testing and check of microstructure was carried out.

3.6 Components fabricated for the test facilities and tests at laboratory

In the framework of R& D-projects several components of Alloy 617 mod. and Alloy 740 have been fabricated for installation in test facilities. An overview is shown in Table 10 10 and examples are shown in Figure 14 and Figure 15. The test panel for the membrane wall test shown in Figure 13Figure 12 is described in Section 5.

Table 10: Components for test facilities and tests at Laboratory

Project / Component	Comtes 700	Esbjerg 720° C	Marcko 700	Thermie	Weisweiler
SH-/RH-tubes (butt welds, cold bending Alloy 617 mod.)	Alloy 740	Alloy 740 Alloy 617 mod.	------	Alloy 617 mod.	Alloy 740 Alloy 617 mod.
Header (nozzle welds, butt welds)	Alloy 617 mod.	------	------	------	-------
Piping (butt weld, inductive bends)	Alloy 617 mod.	------	------	------	------
Membrane Wall	Alloy 617 mod.	------	Alloy 617 mod. (tests at laboratory)	Alloy 617 mod.	

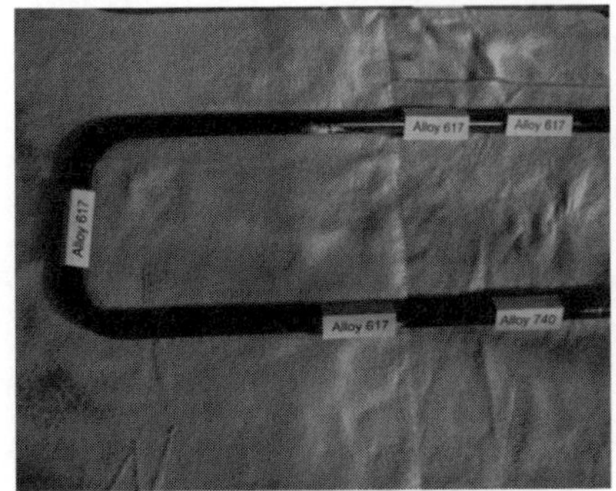

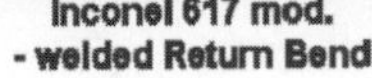

Inconel 617 mod.
- welded Return Bend

Overview before Installation

Figure 13: Welded return bends for Esbjerg 720°C test loop

Header Alloy 617 mod.

Evaporator panel Alloy 617 mod.

Figure 14: Header and Evaporator Panel Alloy 617 mod.

4 Optimization and qualification of welding consumables

4.1 *Basic requirements for consumables and welding of Ni-based alloys*

4.1.1 Qualification of Nickel alloys

The welding consumables are specified in EN 14172 / 18274 and AWS standards A 5.11 / A 5.14. The analyses margins are defined so broadly that it can be reckoned – in case of the occurrence of the minimal or maximal value of particular alloy elements - on a significant modification of the alloy characteristics, as shown in /2/.

In order to fulfill the requirements for the 700°C / 1292 F power plant, different alloy elements have to be limited accordingly or the analysis margins have to narrowed. Conducting the following investigations (welding supplements with equal EN / AWS-qualification), it turned out, that the weld material analysis does fulfill the standards, but elements like Al and Ti do not correspond to the requirements of the power plant operators. The elements Al and Ti are of crucial importance for the oxidation resistance and the long-term creep characteristics.

When employing alloys which are alloyed with elements like Al, Ti and therefore show a high affinity towards oxygen, special preparations regarding processing have to be made. Processes using shielding gas (GTAW – ASME IX, WIG – ISO 857, EN 288 and EN 141) can be regarded as no critical, as the processing is conducted in an inert atmosphere in which alloy elements like e. g. Al and Ti are hardly modified.

As plant constructors do also employ slag-producing processes (ISO 857 / EN 288 arc welding 111 (shielded metal arc welding SMAW), submerged arc welding 1 (SAW), ASME IX – SMAW and SAW), adequate preparations have to be made, in order to prevent a modification of the alloy elements.

For the manufacturer of nickel alloy consumables, utmost care has taken on the composition of the bases materials, which are employed for SMAW and SAW processes. For these processes, base materials are employed which allow a higher level of basicity. By the high level of basicity it can be assured that the percentage of oxygen is being reduced. Thus an oxidation of alloy elements is also considerably minimized.

4.1.2 Processing of nickel alloys in the SMAW and SAW process

Long-time experience in the processing of nickel alloys showed that due to non-adapted welding process data modifications of the weld properties and chemical analysis of the weld arise. Welding has to be conducted with a short electric arc (i.e. electric arc voltage e. g. 22-25 V) and a steep electrode control (80-85°C). The voltage has to be adjusted according to the product informations. The welding has to be performed only in the stringer bead technique. In case of oscillating during welding, the local heat induction increases and hence the burn-off of alloy elements and simultaneously the sensitivity of hot cracking rises. The heat induction in nickel alloys may not exceed 12 kJ/cm, the intermediate layer temperature has to be limited to less than < 150 °C / 302 F.

4.1.3 Development of a welding technology for the SAW process

In /2/ emphasis was placed on the further development of the electrode. As already mentioned the plant constructors widely employ the SAW process. The process is employed for girth welding of piping and headers and for tube-fin-tube-joints. In the SAW process for ferritic-martensitic steels wire diameters of more than 2.4 mm (3/32 inch) are employed. These wire electrode diameters cannot be employed for nickel alloys. If wire electrode diameters of more than 2.4 mm (3/32 inch) are employed for nickel alloys, a significant modification of the alloy composition takes place, i. e. alloy elements burn off and the hot cracking sensitivity increases.

For nickel alloys the wire diameter has to be limited to max. 2.4 mm (3/32 inch).

The objective of the investigation described in the following was to develop a welding technology that ensures that the alloy elements like Al, Ti etc. are only marginally modified. With the wire powder combination UTP UP FX 6170 Co mod. and UTP UP Fx 6170 Co mod., wire diameter 1.6 and 2.0 mm, **multi-layer weldings** are carried out (layer height 25 mm / 1 inch), see Figure 16. The welding parameters were varied correspondingly in order to ensure that the plant constructors get an adequate range of welding process data during the manufacturing of the components.

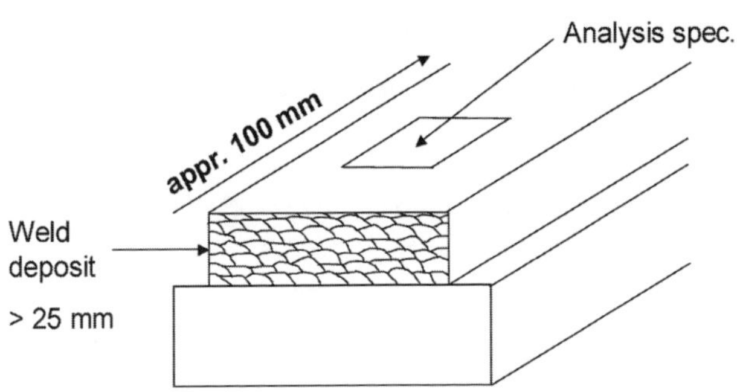

Figure 15: Multi-layer welding

The analyses of the weldings are summarized in Table 11 and 12. As expected for both wire diameters contents of Al and Ti is reduced in the weld. Compensating these elements via the alloying the welding powder is not possible due to metallurgical reasons. Thus, only by means of the welding parameters and the wire diameters these alloy elements can be influenced.

The results observed demonstrate that the wire dimension 2.0 mm should not be employed, as the burn-off is approx. 60 % depending on the welding parameter. As a cause of this elevated burn-off the fusion bath size with a large surface can be mentioned.

Table 11: Chemical composition of weld material for SAW multi-layer welds using UTP UP6170Co mod., wire diameter 1.6 mm and UTP UP FX6170Co mod.

PA	C	Si	Mn	P	S	Cr	Mo	Ni	V	W	Al	Co	Cu	Nb	Ti	Fe
Chg. 35841	0,06	0,04	0,07	0,002	0,002	21,25	8,6	57,16			1,21	11,05	0,01		0,3	0,18
6076.1	0,043	0,36	0,069	0,003	0,001	20,96	8,74	57,47		0,014	0,70	10,99	0,018		0,18	0,44
6076.2	0,045	0,30	0,069	0,004	0,001	20,99	8,71	57,43		0,001	0,79	10,98	0,017		0,20	0,46
6076.3	0,043	0,31	0,067	0,002	0,001	20,85	8,71	57,39		0,011	0,78	10,94	0,015		0,20	0,47
6076.4	0,044	0,32	0,067	0,003	0,001	20,76	8,70	57,26		0,007	0,77	10,87	0,016		0,20	0,41
6076.5	0,047	0,31	0,066	0,003	0,001	20,84	8,67	57,21		0,006	0,85	10,90	0,012		0,22	0,53

Table 12: Chemical composition of weld material for SAW multi-layer welds using UTP UP6170Co mod., wire diameter 2.0 mm and UTP UP FX6170Co mod.

PA	C	Si	Mn	P	S	Cr	Mo	Ni	V	W	Al	Co	Cu	Nb	Ti	Fe
6037.1	0,055	0,74	0,03	0,002	0,001	21,1	8,29	56,69	0,008	0,046	1,2	10,51	0,04	0,011	0,28	0,57
6037.2	0,057	0,75	0,029	0,003	0,001	21,09	8,28	56,75	0,008	0,04	1,21	10,51	0,037	0,012	0,27	0,57
6037.3	0,056	0,75	0,029	0,004	0,001	21,13	8,27	56,74	0,008	0,036	1,22	10,51	0,037	0,011	0,27	0,57
Chg. 131686	0,053	0,03	0,02	0,002	0,002	21,2	8,5	57,42			1,25	10,8	0,01		0,3	0,37
6038.1	0,039	0,45	0,017	0,005	0,001	20,86	8,49	57,9	0,007	0,02	0,51	10,77	0,026	0,002	0,15	0,53
6038.2	0,04	0,39	0,017	0,004	0,002	20,92	8,49	57,83	0,008	0,031	0,61	10,77	0,025	0,002	0,18	0,5
6038.3	0,041	0,35	0,015	0,005	0,002	20,94	8,47	57,8	0,008	0,02	0,66	10,76	0,024	0,003	0,19	0,5

For the wire size 1.6 mm a reduction of the Al-Ti-concentration by approx. 35 % occurs. Using this wire size the requirement of the plant operators is fulfilled, i. e. to guarantee an Al-concentration of 0.8 % within the welding joint. The modified alloy composition welded with the sizes 1.6 and 2.0 mm does not show significant impact on the short-term properties. In the long-term tests the impact of Al-content becomes noticeable. In Figure 17 the result of creep tests on weld material, sampling of the specimen is shown in Figure 18Figure 15, are shown. It is obvious that the TIG-weldings are closest to the base material results. SAW weldment show the most significant reduction. From a test sample girth weld, which was produced using different welds, pure weld metal samples were taken. The samples were chemically analyzed and creep tested. The results of Iso stress tests are compared to base material and to another weld which as a relatively low Al-content of 0.54. It is visible that for SAW-weldments an Al-content of more than 0.7 can be achieved which yield better creep test results.

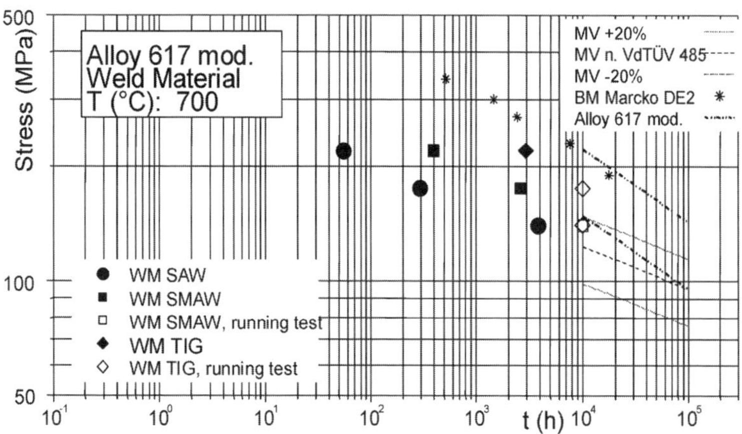

Figure 16: Results of creep rupture tests on weld material

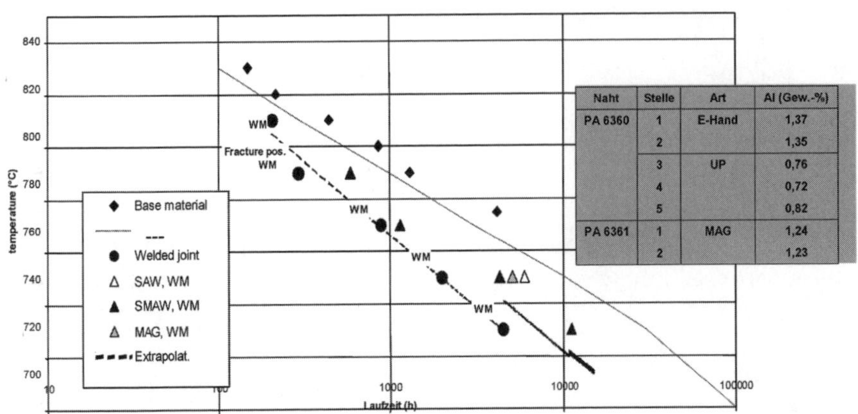

Figure 17: Crossweld tests on Alloy 617 joints, weld material shows different Al-contents

4.1.4 Qualification of consumables

In order to get the permission to employ welding consumables, they have to be submitted to a comprehensive qualification procedure. This qualification is conducted and monitored by an independent organization. In the EN regulations the test scope for welding consumables is defined. In the qualification the welding consumable manufacturer examines at least 3 batches per product type (rod electrode, welding rod, wire electrode, wire powder combination).

In the first step the pure welding good is tested regarding chemical and mechanical-technological characteristics in the short-term area at room temperature as well as in the long-term characteristics. Consequently, the joint weldings on the steels employed in the power plant are manufactured as similar or as dissimilar material joint in the different welding positions (1G=PA, 2G=PC, 3G=PF, 4G=PE). As soon as the complete results of the short-term and long-term tests are available, the independent organization prepares a product specification sheet with material specific data.

A comparable test has also to be conducted by the plant constructor / component manufacturer. In this test the plant constructor proves that he can process this product (welding consumable and base material) in the respective condition. If this test also produces the required result, also the plant constructor receives an accreditation for the manufacturing of welded components.

The corresponding qualification documents of welding consumable and plant constructor are submitted to the plant operators for approval. These qualification documents are required for the approval for the construction of a power plant.

5 Long-term Qualification of materials, welded joints and components subjected to service conditions

In the various investigation programs, see Table 1, long term investigations on creep properties of base materials and welded joints are one of the focal points. For Alloy 617mod. an reasonably good long term data base for base material and welded joints of tubes and pipes obtained by crossweld tests already exists /2, 10/. The results obtained with crossweld specimens from SMAW- and TIG-girth welds of tubes and pipes for temperatures of 700°C and 750°C are within the base material scatter band. Long-term tests longer than 10,000 h testing time show a tendency towards the base material mean values, Figure 19.

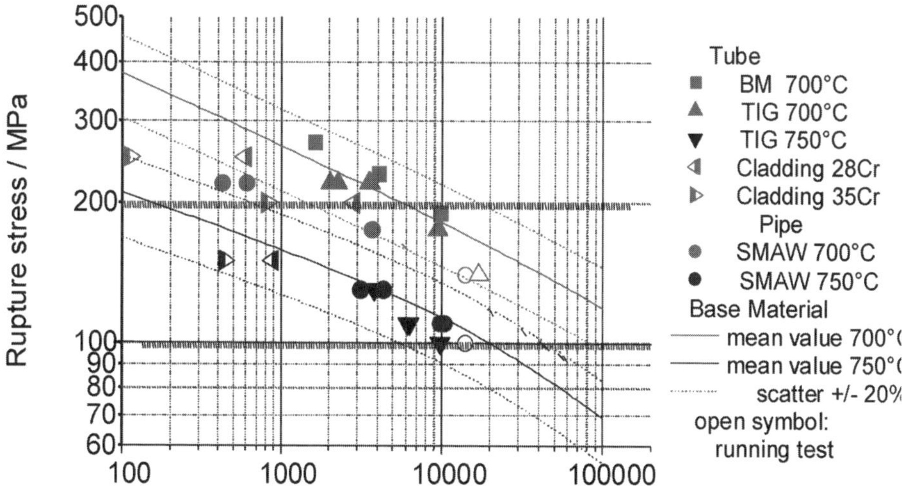

Figure 18: Results on crossweld tests for different similar weldments of Alloy 617mod.

In addition to creep rupture tests, creep tests to develop material models for numerical investigations as well as investigations to prepare a solid design data base and to ensure safety and availability of components, e.g. fracture mechanics investigations are planned. Furthermore investigations on oxidation and corrosion behavior under steam and flue gas conditions are foreseen using special test equipment. As a first step towards the investigation of component behavior under service conditions within the running research project MARCKO 700 /10/ a component like test on a part of the membrane wall is running using panels made from Alloy 617mod., VM12, P92 and T24.

Figure 20 shows the test panel in the test rig. It is instrumented with thermocouples and capacitive strain gauges. Two of the tubes are subjected to steam under near service pressure (250 bar). A special water conditioning and heating system is applied. The two tubes in the center of the panel are subjected to the same pressure using Argon. In addition to internal pressure tensile force perpendicular to the tube axis is applied with hold times of 90 hours. The tests is accompanied by simulation calculation using modern constitutive creep equations including a damage parameter /11/. Results of pre-calculations are shown in Figure 20.

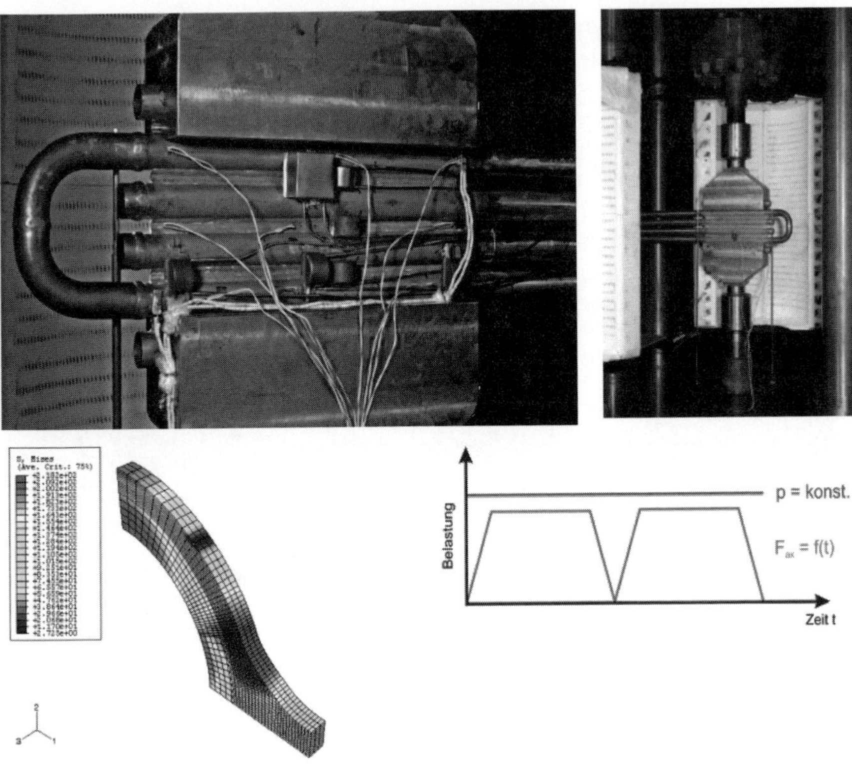

Figure 19: Membrane wall test and numerical simulation of the test panel

6 Summary and conclusions

Based on experiences with field tests and couple of research projects the development of power plants with 700°C technology is well under way. Design and material needs are described and the requirements for the materials used are outlined. Extensive qualification programs are under way and with the first results producibility and weldability of components made of Ni based alloys could be demonstrated.

Although chemical compositions of welding consumables are given in the standards accordingly (EN 14172 / 18274 and AWS A 5.11 /A 5.14) the alloys have to be modified accordingly for the respective applications. In order to meet the requirements of the plant operators, comprehensive tests in advance have to be carried out which can also lead to product modifications. For the processing of nickel alloys close cooperation between manufacturers of welding consumables and component and plant manufacturers is necessary. This considers primarily, the achievement of results that meet the requirements of the operators and secondly that the welding consumables fulfill the qualification according to the EN / AWS standards. For the manufacturing of plant components that are produced in the SAW process, only the wire size 1.6 mm may be employed. With this size only it can be guaranteed that the requirements of the plant operators are met.

In order to ensure safety and availability of components a number of research projects have been initiated. Tests to determine creep properties of weldments running more than 10,000 h show for SMAW and TIG welds that the results are within the base material scatter band. SAW weldings must be carried out using optimized consumables and adapted process parameter to ensure sufficient Al content in the weld deposit.

In the research programs tests under service like conditions are foreseen. As an example the membrane wall test /11/ is described. These investigations will deliver results for long-term behavior under steam conditions in addition to field test.

COMTES700-Funding Notice ...

- ... for publications:

  "The project is funded by the Research Fund for Coal and Steel of the European Commission (project number RFC-CP-04003).

  The driving power for the development of the COMTES700-project was a group of major European power generators named "E_{max} Power Plant Initiative". From this group a consortium was formed ready to support the project with a substantial financial contribution. Members of this consortium are DONG Energy (Denmark), EDF (France), Electrabel (Belgium), EnBW (Germany), E.ON (Germany), PPC (Greece), RWE (Germany), Vattenfall Nordic (Denmark) and Vattenfall Europe (Germany).

  The project is carried out from July 1, 2004, to December 31, 2009, by Alstom Power Boiler GmbH, Hitachi Power Europe GmbH, Burmeister & Wain Energy A/S, Siemens AG, Elsam Engineering A/S and VGB PowerTech e. V.

  The total budget is amounting to 23.2 Million Euro. 70 % thereof is funded by power generators, 26 % by the Research Fund for Coal and Steel and 4 % by the suppliers involved."

- ... short version for presentations:

  "The project is funded by the Research Fund for Coal and Steel of the European Commission (project number RFC-CP-04003) and by a substantial financial contribution of DONG Energy (Denmark), EDF (France), Electrabel (Belgium), EnBW (Germany), E.ON (Germany), PPC (Greece), RWE (Germany), Vattenfall Nordic (Denmark) and Vattenfall Europe (Germany).

  The project is carried out by Alstom Power Boiler GmbH, Hitachi Power Europe GmbH, Burmeister & Wain Energy A/S, Siemens AG, Elsam Engineering A/S and VGB PowerTech e. V."

References

1. Auslegungskriterien für hochtemperaturbelastete metallische und keramische Komponenten sowie des Spannbeton-Reaktordruckbehälters zukünftiger HTR-Anlagen – Teil B: Metallische Komponenten, Forschungsvorhaben des BMFT, KFA Jülich, Abschlussbericht, 08-1988

2. R. U. Husemann, A- Helmrich, J Heinemann, A. Klenk, K. Maile: Applicabiliy of Ni-based welding consumables for boiler tubes and piping in the temperature range up to 720 °C, 4th Int. Conf. On Advances in Materials Technology for Fossil Power Plants, October 25-28, 2004

3. Agarwal, D. C. and U. Brill: Influence of the Tungsten Addition and Content on the Properties of the High Temperature-High strength Ni-Base Alloy 617, 4th Int. Conf. On Advances in Materials Technology for Fossil Power Plants, October 25-28, 2004

4. Shingledecker et al., 4th Int. Conf. On Advances in Materials Technology for Fossil Power Plants, October 25-28, 2004

5. Ulrich Heubner et al. , Nickelwerkstoffe und hochlegierte Sonderstähle, Expert-Verlag, 1993

6. Special Metals, Manufacturer Material Data Sheets

7. VGB – R 501 H; Herstellung sowie Bau- und Montageüberwachung von Dampfkesselanlagen, 2002

8. VGB – R 508 h; Herstellung und Bauüberwachung von Rohrleitungsanlagen in Wärmekraftwerken

9. VdTÜV – Merkblatt 451-68/1, Herstellung gasdicht geschweißter Rohrwände

10. Resarch Project Marcko 700 , Project Description, MPA Stuttgart, 1992

11. K. Maile, A. Klenk, M. Bauer and E. Roos: Consideration of Weld Behavior in Design of High Temperature Components, 5th Int. Conf. On Advances in Materials Technology for Fossil Power Plants, October 3-5, 2007

Advances in Materials Technology for Fossil Power Plants
Proceedings from the Fifth International Conference
R. Viswanathan, D. Gandy, K. Coleman, editors, p 260-270

DOI: 10.1361/cp2007epri0260

Study of Performance Requirements and Construction Rule for 700 Degree-C Class Advanced USC Plant

K. Yoshida
T. Sato
Japan Power Engineering and Inspection Corporation
14-1 Benten-cho, Tsurumi-ku
Yokohama, Japan 230-0044

Abstract

In order to reduce carbon dioxide gas emission, the 700 degree-C class thermal power plant, of which plant efficiency improvement is achieved, has been studied in Japan. The plant technology is based upon advanced materials development, namely developments of high strength alloy, fireside corrosion resistant alloy, steamside oxidation resistant alloy and so on. A part of the materials which were developed are not set forth in the present domestic interpretation of the technical standard. And failure modes, which are estimated in 700 degree-C class A-USC components, are possible to be different from those in the 600 degree-C class USC components. Therefore the failure modes were considered in the 700 degree-C A-USC components and then the performance requirements were led based upon the failure modes. The study was initiated on June 2006 and the draft interpretation will be set forth by March 2011. In the paper the situation of the investigation for failure modes, performance requirements and the draft interpretation are reported.

Introduction

It is important to reduce the amount of carbon dioxide gas emission from thermal power plant, and it is necessary to increase the efficiency of the thermal power generation at present. It is, furthermore, desirable that the increase of the efficiency can be completed by the natural gas or coal which is substituted for oil, because of the decrease of the dependence of the oil on fossil fuel combustion. Natural gas thermal power generation is promoted by the application of the advanced combined cycle power generation system at present in Japan, the turbine inlet gas temperature has reached 1,500 degree-C, and heat efficiency has run into about 51% HHV. It is expected that the heat efficiency reaches about 55% by the future increase of the turbine inlet gas temperature to 1,700 degree-C. On the other hand, there are two way in coal power generation, one is 700 degree-C class advanced ultra supercritical, A-USC, system and the other is the integrated coal gasification combined cycle, IGCC, system. In the latter demonstration plant, as the result that the turbine inlet gas temperature of 1,200 degree-C was applied to the system, the heat efficiency was about 42%. And the heat efficiency of about 50% is predicted because of the turbine inlet gas temperature of 1,500 degree-C. In the former A-USC, it is expected that the heat efficiency of 46% is achieved by the application of double reheat cycle. When the steam temperature is gone up to 800 degree-C in A-USC system, the heat efficiency of about 49% will be realized. As a result, the main steam pressure is 35MPa in 700 degree-C class A-USC, and 38.5MPa in 800 degree-C class, it is possible to design and fabricate the 700 degree-C class components by using the materials developed at present. It is however difficult to design the 800 degree-C class components according to the present materials application, and it is therefore necessary to make a progress on the high temperature material technology.

There are natural gas and coal in the fossil fuel of oil substitution as mentioned above, and the former has the advantage in an environmental viewpoint as compared with the latter, since the carbon dioxide gas emission of the former is about 50% of the latter. However, as natural gas reserves are small as compared with coal deposits, it is necessary to depend on coal at present. Moreover, in coal power, as the thermal efficiency, HHV, higher heating value, of IGCC system exceeds it of A-USC system, the IGCC system has the advantage in an environmental protection. As the kinds of coal are, however, different between both systems, and the A-USC needs are considered in a viewpoint of the power supply stability by power system variety.

As shown above, the A-USC needs are strong and the realization will be expected within ten years in Japan. The components for thermal power generation in Japan are regulated by the Electricity Utilities Industry Law, and it is required that the components shall conform to the technical standard, METI Ordinance No. 51[1]. This technical standard set forth the performance requirements to the thermal power components, and the detailed specification which conforms to the performance requirements is set forth in the interpretation of the technical standard. The materials which can be applied to a main steam pipe of 700 degree-C class A-USC are not however set forth in the interpretation. Although the detailed specification concerning the structural integrity of the components for the failure modes which are predicted up to the 600 degree-C class is set forth in the interpretation, it is not assured that the specification about the structural integrity concerning the failure modes which are predicted in 700 degree-C class is set forth in the interpretation at present. Because, in introducing 500 degree-C class thermal power system into domestic in 1950s, the original technical standard was set forth in the mid of 1950s, and subsequently the technical standard has been technically reviewed and extended so that this can be applied even to 600 degree-C class thermal power components at present. It was started to investigate the interpretation of the technical standard which will ensure the structure integrity of 700 degree-C class A-USC components in June 2006 in these viewpoints.

Materials to 700 degree-C class A-USC

The fundamental design conditions of the 700 degree-C class A-USC currently examined in Japan are shown in Table 2-1. As the main steam temperature is 700 degrees C, and the steam pressure is about 35 MPa, the minimum thickness is set to 138.1mm when the steam pipe is designed by using type 304 stainless steel of which allowable tensile stresses are specified in the interpretation of the technical standard and the outside diameter is 406.4 mm. As the minimum thickness of the P91 steam pipe of 600 degree-C class USC is 62.8 mm in calculation under 26 MPa steam pressure, the above thickness design of A-USC is not practical by the application of the materials which are set forth in the present interpretation. For example, when the main steam pipe is designed by N06617 as candidate material of 700 degree-C class A-USC, the minimum thickness is 67.5mm under 35 MPa pressure, as the result, it is possible to design the main steam piping in 700 degree-C class A-USC. The highest allowable stress material in the present interpretation is just KA-SUS310J3TB stainless steel at 700 degree-C, and when it is applied to the main steam pipe, the minimum thickness is 84.2mm, that is, the steel application is impractical in A-USC design (refer to Table 2-2). It is, therefore, necessary to develop new materials to apply to 700 degree-C class A-USC. Since it is necessary to take the effect of the axial load on piping design into consideration, the needs for materials with a relatively small thermal expansion coefficient is desirable, and nickel base alloys and/or iron-nickel alloys are promising as candidate materials.

Investigation of the technical standard of 700 degree-C class A-USC

ISO/FDIS16528[2] is currently examined as international performance requirement standard of boiler and pressure vessels. After identifying the failure modes predicted to the components under the service condition, and the technical requirements for preventing the failures should be set forth in a technical standard, it is given that the detailed specification shall be constituted according to the technical requirements. On the other hand, the performance requirements for boiler are set forth in the technical standard, METI Ordinance No. 51. The materials, which are applied to 700 degree-C class A-USC, were investigated based on the service condition at first as shown Table 3-1. According to the above ISO/FDIS16528 procedure, the performance requirements for 700 degree-C class A-USC are examined, and the detailed specification investigation was initiated on the assumption that the technical standard will be reset if necessary. A part of the investigation in 2006 fiscal year is reported shown below.

Investigation of Failure Modes and Performance Requirements

Investigation on failure modes was performed, as compared with the failure modes in ISO/FDIS16528, after comparing and contrasting general degradation behaviors estimated by the domestic failure incidents of boiler components and/or assemblies, and the damage mechanisms in ASME-PCC[3]. The failure modes set up from the investigation is shown in Table 3-2. Although the technical standard fundamentally covers all such failure modes, since the detailed specification do not clearly define procedures of the prevention from the failures, it was initiated to materialize the performance requirements corresponding to failure modes. As a result of referring to the technical standard of thermal power plants, the technical standard of nuclear power plants, etc. the draft performance requirements on 11 kinds of failure modes were investigated. In connection with prevention of the brittle fracture, about the procedure for confirming no harmful defect, there is two way, that is, non destructive examination, for example ultrasonic technique, and visual inspection. And moreover, related to the check of suitable fracture toughness, it is possible to exempt fracture toughness testing for materials having superior ductility. At first, the justification of the performance requirements will be proceeded on schedule, and based on the performance requirements the detailed specification will be investigated.

Investigation of the Constitution of Detailed Specification

The fundamental constitution of the detailed specification was defined, that is, material, design, manufacturing, examination and final test, as shown in Table 3-3. These items harmonize to technical requirements in ISO/FDIS16528, and are same to the articles in the interpretation of the technical standard. As the material items are shown in Table 3-4, these items were selected so that they are not left off the constitution referring to both the technical requirements of ISO/FDIS16528 and the regulation matter of the present interpretation of the technical standard. And there are some items which are equivalent to a recommendation matter like consideration of material degradation in service of USC-2160, elsewhere the claims which the specification of the pressure part material of USC-2120 was postulated. Although the detailed specification should consist of only claims postulated, some of recommendation matter is also important for prevention of failure of the components in service. It therefore needs further consideration about the investigation on the measures of a recommendation matter.

Design Margin

The design margin in the interpretation of the technical standard is four to ultimate tensile strength at present. As the allowable stress is generally given through the mean creep rupture strength in creep temperature region in case of thermal power components of 600 degree-C class USC, it can be considered that the design margin is defined to be 1.5 to creep rupture strength for 100,000 hours. The allowable tensile stress basis for the interpretation in a creep temperature region is shown in Table 3-5, and when (a) and (b) in the basis are assumed to be equal, the failure probability in 100,000 hours becomes about 0.04%. It is therefore possible that the allowable stress basis in creep temperature region is rational. Since Favg was introduced in ASME Sec. II, Part D[4], the above design margin might not be sometimes conservative as compared with the Favg. It is possibly necessary to investigate the factor for the candidate materials of 700 degree-C class A-USC.

Candidate Material and Material Properties

Materials for 700 degree-C class components were shown in Table 3-1, candidate materials are not, however, restricted to the materials in Table 3-1. Materials for 650 degree-C class are also candidates for 700 degree-C class A-USC, and some candidate materials are shown in Table 3-6, their allowable tensile stresses up to 750 degree-C should be investigated and defined. Tempered martensitic low alloy steels are under development in candidate materials without the possibility of the long-term creep strength reduction. An iron-nickel alloy was newly developed as one of the candidates for 700 degree-C class main steam piping and hot reheat piping in Japan. Although nickel base alloy has the problem about weldability, their high temperature creep strength is excellent and it possible to adopt the rational design of a main steam pipe by using the alloy. Austenitic stainless steels are leading as a high temperature tubing material, and are already set forth in the interpretation of the technical standard.

Material properties which should be specified for candidate material besides allowable tensile stress are design fatigue curve, modulus of elasticity, thermal expansion coefficient, etc. After instituting the application subject of candidate materials in order to specify above, it should be planned to collect the data tested by samples from possible product form materials, and to analyze them.

Situation of Detailed Specification Investigation

One of the investigation matters, which were carried out as the detailed design specification, is the specification for the prevention of fatigue failure. There is no specification of fatigue evaluation in the interpretation of the technical standard, and the prevention was probably deliberated by the manufacturers. According to EN12952[5], when there is no design life in purchase specifications, it is supposed that 2,000 cold start-ups are assumed and fatigue evaluation is carried out. It is necessary to fix the subjected part of fatigue evaluation, it is, furthermore, important to select the subjected fatigue loading which should be taken into consideration as alternating loads. The kinds of fatigue loading which should be taken into consideration besides cold start-ups are investigated after this.

Investigation about welding of a nickel base alloy is conducted about manufacturing specification. It is relatively difficult to weld a nickel base alloy as compared with welding carbon steel and/or low alloy steel. When the welding procedure, which defects, such as a crack, do not arise in the weld and the properties of the weld heat affected zone are equivalent to base material, can be selected, the weld will meet the performance requirements of the prevention from failures. After investigating the weldability, the mechanical properties and so on, the welding essential variables are, therefore, considered. According to the present investigation for nickel base alloys welding, it is suggested that to establish some supplementary essential variables in the application in high temperature.

The final test includes the hydrostatic test, and the test pressure has been established as a test pressure based on the design margin to ultimate tensile strength. That is, when the design margin is four, 1.5 times pressure of the maximum allowable working pressure has been selected as test pressure which plastic deformation does not occur during the test. When the allowable tensile stress at maximum service temperature is equivalent to one at hydrostatic test temperature and when a yield ratio to ultimate tensile strength is 0.5, about 75% of yield stress is given as membrane stress at the hydrostatic test. It is possible that the test is carried out by proper test pressure as the completion examination of the component. When the design margin in high temperature is, however, 1.5 to creep rupture strength for 100,000 hours, it is possibly necessary to reinvestigate the hydrostatic test pressure. The correction for temperature about the test pressure introduced into the hydrostatic test pressure in ASME Sec.VIII Div.1[6] is rational when maximum allowable temperature does not arrive at creep temperature region, but when the temperature do arrive at creep temperature region, it is necessary to consider the proper test pressure related to the design margin. A concrete investigation of detailed specification is due to be a future subject, based on the present domestic interpretation, to contrast ASME and EN standards, and to consider specification preventing from failures.

Summary

Based on the investigation of the failure modes predicted in 700 degree-C class A-USC, the investigation of the detailed specification has been performed by the schedule to FY2006. The new interpretation of the technical standard of 700 degree-C class A-USC will be set forth according to this investigation and it will lead that 700 degree-C class A-USC is successfully introduced into domestic. This investigation also contributes to the further increase of efficiency of ACC and IGCC system. That is, when turbine exhaust gas temperature increases to higher side, and high temperature steam is supplied from HRSG as compared with current steam, the advanced efficiency may be realized to about 60%.

Acknowledgement

This investigation is sponsored by Nuclear and Industrial Safety Agency, the authors greatly appreciate for the guidance about the master planning by Mr. Norihisa Yuki of Nuclear and Industrial Safety Agency. The RSA committee was established in JAPEIC in order to review and promote the investigation. The authors have great acknowledgement to Prof. Shinsuke Sakai of University of Tokyo who is the chairperson of the committee.

Annex

$$(m - 1.645SEE) \times 0.8 = 0.67m$$
$$\therefore SEE = 0.0987841m$$
$$\therefore 0.67m = m - 0.33m = m - 3.34SEE$$
$$\therefore \Phi(3.34) = 0.999596 = 1 - 0.000404$$

m: average value SEE: standard error of estimate $\Phi(x)$: distribution function

References

1. METI Ordinance No. 51, the technical standards concerning the thermal power plant, March 27, 1997.

2. ISO/FDIS16528, Boiler and pressure vessels.

3. ASME post construction subcommittee on inspection planning agenda, July 26, 2006.

4. ASME Boiler and pressure vessels code, Sec. II, Materials, Part D, Properties.

5. EN12952 Water-tube boilers and auxiliary installations.

6. ASME Boiler and pressure vessels code, Sec. VIII, div. 1, Rule for construction of pressure vessels.

Table 2-1: Design Consideration of 700 deg-C Class A-USC System in Japan

	Single Reheat	Double Reheat
Main Steam Pressure (MPa)	25.0	34.2
Steam Temperature (deg-C)	700/700	700/720/720
Plant Efficiency (%)-HHV	46.0	47.2 - 48.1
Feedwater Temperature (deg-C)	298	322

Table 2-2: Rough Thickness Calculations Results, Design by Rule

Class	Material		Allowable Stress (MPa)	Thickness (mm)
700 deg-C	SUS304TP	Type304	27	138.1
	NCF800HTP	N08810	34	121.6
	KA-SUS310J3TB	SAVE25	60	84.2
	N06617		81	67.5
600 deg-C	KA-STPA28	P91	66	62.8
	KA-STPA29	P92	86	50.8
	KA-SUS310J3TB	SAVE25	116	39.4
	N06617		106	42.6

Table 3-1: Some Candidate Tubes and Piping Materials for 700 deg-C Class A-USC

		Material Specification	
Tubes	Superheater	KA-STBA24J1	2.25Cr-1.7W
		KA-STBA28	T91
		KA-SUS304J1HTB	Super304H
		KA-SUS310J1HTB	HR3C
			HR6W
	Reheater	KA-STBA28	T91
		KA-SUS304J1HTB	Super304H
		KA-SUS310J1HTB	HR3C
			HR6W
Pipings	Main/Reheat Steam (include Connecting Pipe)	STPA22	P12
		KA-STPA24J1	2.25Cr-1.7W
		KA-SUS304J1	Super304H
			HR6W

Table 3-2: Failure Modes for 700 deg-C Class A-USC

	Material[1]	Design[2]	Examination[3]
Corrosion	○		
Erosion	○		○
EAC [4]	○		
Brittle Fracture	○	○	
Ductile Fracture	○	○	
Creep Failure	○	○	
Leak at Flange Joint		△	△
Buckling		△	
Fatigue		○	
Creep-fatigue		△	
Ratcheting		△	
Shake down		△	

[1] Failure Protection due to Material Selection
[2] Failure Protection due to Design Rule
[3] Failure Protection due to Monitoring and/or Examination
[4] EAC: Environmental Assisted Corrosion
[5] ○ : for example, corrosion protection due to superior corrosion resistant material selection under service condition.
[6] △ : for example, flange design based on Sec. VIII Div. 1, Appendix 3

Table 3-3: Draft Contents of Acceptable Approach

700 deg-C Class A-USC		Interpretation of Technical Standard	Technical Requirements of ISO/FDIS16528-1
USC-2000	Material	Article 2.	7.2 Material
USC-3000	Design	Article 3, 4 and 6 to 14.	7.3 Design
USC-4000	Manufacturing	Article 117 to 130.	7.4 Manufacturing
USC-5000	Examination		7.5 Inspection, non-destructive testing and examination
USC-6000	Final Test	Article 5.	7.6 Final Inspection and test

Table 3-4: Draft Contents of USC-2000, Material

	Draft Interpretation	ISO/FDIS16528-1
USC-2100	General Requirements	7.2.1 General
USC-2120	Pressure Retaining Material	7.2.2 Specification of Materials 7.2.2 a), b), d) and e)
USC-2130	Certification of Material	7.2.3 Material Certification
USC-2140	Welding Material	-
USC-2150	Material Identification	7.2.3 Material Certification
USC-2160	Deterioration of Material in Service	7.2.2 Specification of Materials 7.2.2 c)
USC-2200	Material Test Coupons and Specimens	7.2.2 Specification of Materials
USC-2300	Fracture Toughness Requirements	7.2.2 Specification of Materials 7.2.2 a)
USC-2400	Welding Material	-
USC-2500	Examination and Repair	-

Table 3-5: Allowable Stress Basis of Interpretation of Technical Standard

Not to exceed the lowest of followings;
$0.67S_{Ravg}$-10,
$0.8S_{Rmin}$-10, and
$1.0S_C$
[Note] S_{Ravg}-10: average stress to cause rupture at the end of 100,000 hrs.
S_{Rmin}-10: minimum stress to cause rupture at the end of 100,000 hrs.
S_C: average stress to produce a creep rate of 0.01 percents/1,000 hrs.

Table 3-6: Some Candidate Nickel Base Materials for 700 deg-C Class A-USC

		Cr	Ni	Mo
N06617	Alloy 617	20.0-24.0	44.5 min.	8.0-11.0
-	Alloy 263	19.0-21.0	Bal.	5.6- 6.1
-	Alloy 740	25	Bal.	0.5
N06625	Alloy 625	20.0-23.0	58.0min.	8.0-10.0
N06230	Alloy 230	20.0-24.0	Bal.	1.0- 3.0
-	HR6W	24	Bal.	-

-	TEMPALLOY CR30A	30	50	2
N08120	HR120	23.0-27.0	35.0-39.0	2.5max.
N08810	Alloy 800H	19.0-23.0	30.0-35.0	-
N08811	Alloy 800HT	19.0-23.0	30.0-35.0	-
N10276		14.5-16.5	Bal.	15.0-17.0

Advances in Materials Technology for Fossil Power Plants
Proceedings from the Fifth International Conference
R. Viswanathan, D. Gandy, K. Coleman, editors, p 271-280

DOI: 10.1361/cp2007epri0271

Nickel Alloys for High Efficiency Fossil Power Plants

D.C. Agarwal
ThyssenKrupp VDM USA Inc, Houston, Texas

B. Gehrmann
ThyssenKrupp VDM GmbH
Kleffstrasse 23
58762 Altena, Germany

Abstract

In order to satisfy the increasing energy demands of the 21st century via new fossil fuelled power plants and to fulfil the "clean air-blue skies" environmental protection goals, new concepts for these plants have to be developed with higher efficiency and CO_2, sulphur oxides and nitrogen reduction technologies.

With increasing temperatures and pressures, which are the major requirements for increasing the efficiency of the power plants, the needed property requirements with respect to the creep rupture strength and high temperature corrosion resistance as well as other properties , increases for the alloys used in certain components in these plants. Newer and existing nickel alloys appear to be promising candidates for these applications. Developments of new alloys require the detailed study of the following categories of properties:

- Chemical composition, physical and chemical properties
- Mechanical – technological properties (short time, long time dependent properties, aging effects, thermal stability)
- Creep characteristics, fatigue characteristics, fracture mechanics
- Optimization of fabricability procedures for (cold and hot working), welding (properties at heat affected zone and weldments, crack propagation characteristics, ductility, toughness etc.)
- Evaluation of materials- and components properties
- Processing concept for pre-material
- High temperature resistance for oxidation
- Others

This will encompass study and development of materials for membrane walls, materials for headers and piping, materials for reheater and superheater and various other components.

This paper provides a general overview of the existing and newly developed nickel alloys used in the various components of fossil fuelled for high efficient 700°C fired steam power plants.

Nickel alloys for high efficiency fossil power plants

Over the years the operating temperature of fossil fuelled power plants increased with the ultimate goal to improve the overall efficiency. Therefore the property requirement on the construction materials for certain components increased with respect to the creep rupture strength as well as to the oxidation and corrosion resistance. This led to the development of improved materials in nickel based alloys for the proposed high efficiency fossil fired steam power plants. The ultimate aim being achievement of higher efficiency 700°C fossil fired steam power plants.
It is estimated that in order to satisfy the world´s growing electricity demand additional power plant capacities are required of more than 2000 GW in the time period to the year 2020. At the same time the environmental policy and the Kyoto climate protection goals also have to be considered and taken into account. The development of new power plant concepts with higher efficiencies for the energy conversion of fossil fuels and the development of power plants without CO_2-emissions are major tasks and at the same time presents opportunities for developments in newer nickel or cobalt based alloys.
The Federal Ministry of Economics and Labour (BMWA) in Germany have initiated a new developing research concept for the realization of such power plants together with operators, manufacturing companies, research institutions and universities. The name COORETEC of this research activity means CO_2-Reduction TECnologies for fossil fuelled power plants. This initiative will hopefully lead to the increased efficiency while preserving and protecting the environment in two stages:

1) In the short and medium term, a further development of existing and/or newer materials, components and process control for fossil fuelled plants.
2) Beside this, the R&D activities will focus on new power plant concepts in the medium and long term with the ultimate aim of increasing the efficiency and the CO_2-reduction technologies.

The efficiency of coal fired power plants has increased from approximately 18% starting in 1950 up to 40% in the 1990's which is also the average efficiency in Germany of the year 2000. The efficiency of a hard coal fired power plant is approximately 48% and the one of a gas fired power plant approximately 58%.

The targets of the COORETEC research programs are the increase of the efficiencies of a coal fired power plant to a value around 53% and for a gas fired power plant around 65% (see Figure1).

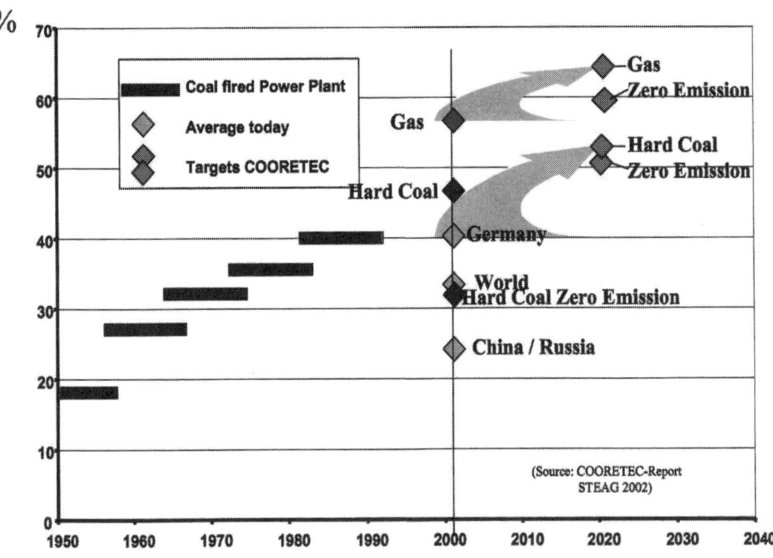

Figure 1: Increase of the efficiency of fossil fired steam power plants over the years [1)].

The figure 2 shows the roadmap of the COORETEC activities. Materials play an important role over the entire time period of development. Investment costs are greatly affected by the cost of specified materials and their thermo-mechanical processing characteristics (weldability and fabricability). The knowledge of the interaction between the operating conditions and the selected material for a component play an important role in the estimation of the technical and economical risks when introducing improved or new power plant concepts. Therefore activities have been started in COORETEC with respect to the need of developing for new and improved materials for fossil fuelled plants. See the propose VISION in Figure 2 below:

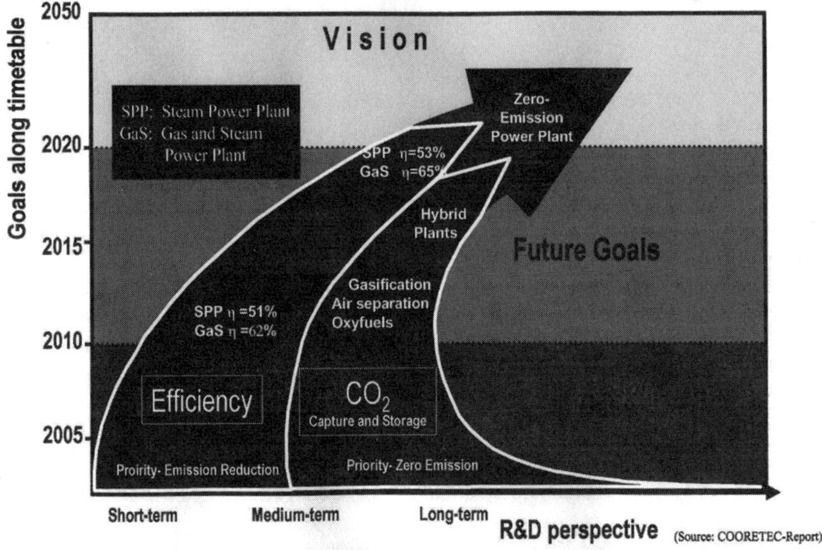

Figure 2: Roadmap of COORETEC [1)].

The producers of the components of fossil fuelled power plants have to qualify the usage of certain materials:

- Approval for the usage of materials according to the corresponding specifications.
- Approval for the usage of materials which have a European material qualification for pressure vessels and singular approvals of materials.

There are many requirements which have to be fulfilled by the specified materials for certain application components and conditions such as: chemical composition, physical and chemical properties, mechanical-technological properties over short time, over long time and with respect to aging.
Very important are creep characteristics, fatigue characteristics as well as fracture mechanics. Approvals are required for working – both cold and hot-, for welding – properties at the heat affected zone and welding deposits and crack propagation susceptibility. The materials- and components properties will need to be fully evaluated. There must exist proven processing concepts/procedures for the pre-material. Beside the mechanical properties, the oxidation and high temperature corrosion resistance are very important as well. Figure 3 shows the conceptual design of a 700 deg C fossil fired power plant.

700 °C- Plant – Future Potential

500 MW

High pressure part	
Steam rating	**343 kg/s** **1235 t/h**
Allowable working pressure (gauge)	**393 bar**
SH-Outlet temperature	**700 °C**
Reheater	
Steam rating	**269 kg/s** **968 t/h**
Allowable working pressure (gauge)	**83 bar**
RH-Outlet temperature	**720 °C**

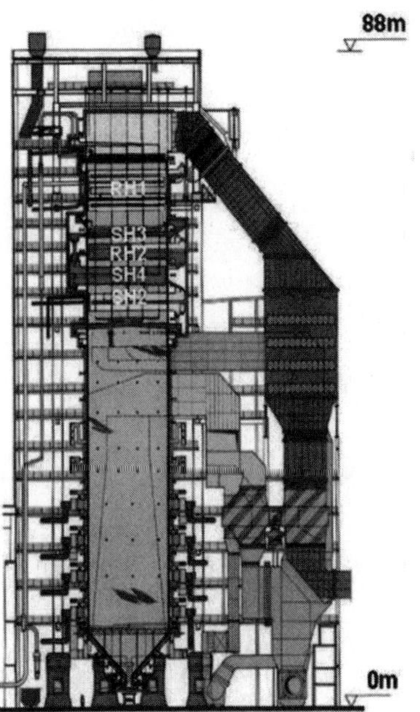

(Source: R.U. Husemann, October 2005, Milano)

Figure 3: Principle design of a 700°C plant [2)].

Materials of Construction

	C	Si	Mn	P	S	Al	Cr	Ni	Mo	V	Nb	N
Min.	0.08	0.20	0.30				8.0		0.85	0.18	0.06	0.03
Max.	0.12	0.50	0.60	0.02	0.01	0.04	9.5	0.4	1.05	0.25	0.10	0.07

Table 1: Chemical composition of the martensitic Cr-steel P 91 (X10CrMoVNb9-1).

The chemical composition of a martensitic Cr-steel P 91 is listed in Table 1. P 91 is a typical alloy that has been used extensively for headers and piping in 600°C plants. However for higher temperatures above 600 deg C its stress-rupture and creep properties are inadequate as shown below in Figure 4.

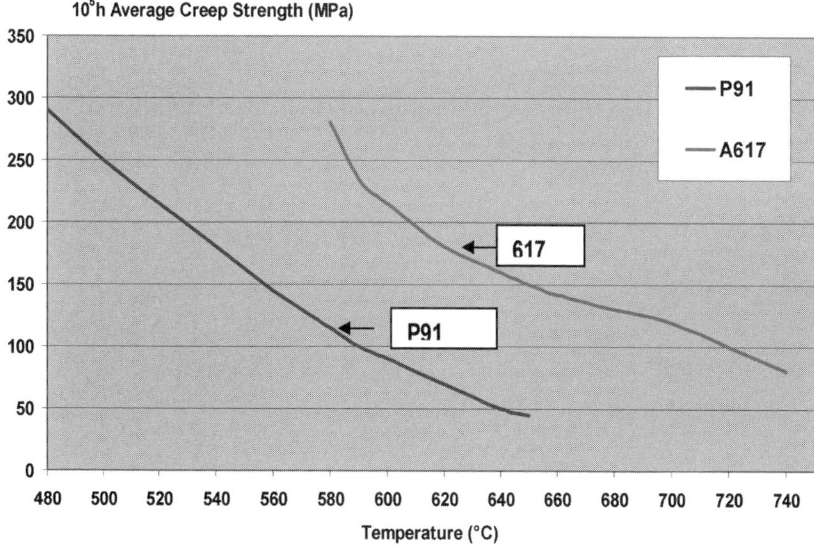

Figure 4: 10^5 hrs. average creep strength of Alloy 617 in comparison to P 91 as function of the temperature.

Figure 4 shows the 10^5 hrs. average creep strength of Alloy 617 in comparison to P 91 as function of the temperature. It can be seen the curve for Alloy 617 is much higher than the curve for P 91. This alloy 617 and its modifications with addition of certain elements like Boron in precisely controlled quantities make it an attractive candidate. Figure 5 shows the improved creep strength of this modified alloy 617 compared to standard alloy 617.

Development of materials for fossil fired steam power plants

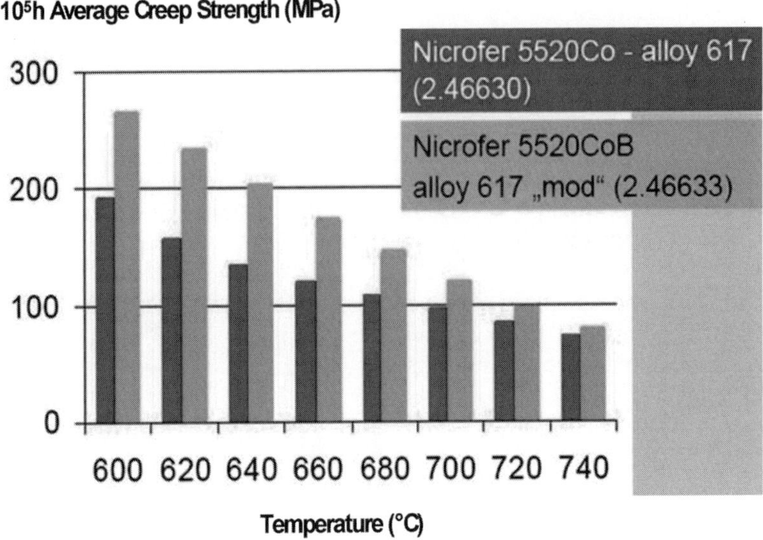

Figure 5: **10^5 hrs. average creep strength of Nicrofer 5520CoB – alloy 617 "mod" in comparison to Nicrofer 5520Co – alloy 617 as function of the temperature.**

Figure 5 shows the 10^5 hrs. average creep strength of the new developed alloy Nicrofer 5520CoB – alloy 617 "mod" in comparison to Nicrofer 5520Co – alloy 617 as function of the temperature.

Ni-alloys for high efficient 700°C fired steam power plants

Alloy 617 / Nicrofer 5520 Co (2.4663)

	Ni	Cr	Fe	C	Mn	Si	Co	Cu	Mo	Al	Ti
Min.	Bal.,	20.0		0.05			10.0		8.0	0.8	
Max.	44.5	24.0	3.0	0.15	1.0	1.0	15.0	0.5	10.0	1.5	0.6

Alloy 617 "mod" / Nicrofer 5520 CoB

	Ni	Cr	Fe	C	Mn	Si	Co	Cu	Mo	Al	Ti	B
Min.	Bal.	21.0		0.05			11.0		8.0	0.8	0.3	0.002
Max.		23.0	1.5	0.08	1.0	1.0	13.0	0.5	10.0	1.3	0.5	0.005

Chemical composition (in wt.-%)

Table 2: Chemical composition of alloy 617 "mod" – Nicrofer 5520CoB and alloy 617 – Nicrofer 5520Co.

The chemical compositions of alloy 617 "mod" – Nicrofer 5520CoB and alloy 617 – Nicrofer 5520Co are listed in Table 2. As is evident from the above chemical composition, the " modified alloy 617" is much more tightly controlled i.e. lower Fe max, lower range for Carbon, higher min. range for cobalt, tighter Al+Ti with small additions of boron. **(ThyssenKrupp VDM has filed a patent application for this modified alloy 617).**

Nicrofer 5520 Co is a nickel-chromium-cobalt-molybdenum alloy with excellent mechanical and creep properties up to 1100°C (2000°F) due to solid solution hardening. As a result of its balanced chemical composition the alloy shows outstanding resistance to high temperature corrosion such as oxidation and carburization under static and cyclic conditions up to temperatures of about 1100°C (2000°F). The alloy good good weldability & fabricability. Nicrofer 5520 Co has a face-centered cubic structure with good metallurgical stability. Its excellent high-temperature strength is achieved by solid-solution hardening. The alloy is not age-hardenable.
However Nicrofer 5520 Co as well as its " modified version" is susceptible to relaxation cracking if new solution-annealed and welded semi-fabricated products are exposed to service temperatures within the temperature range of 550-780°C (1020-1436°F) without a prior postweld stabilizing heat treatment (PWHT) at 980°C (1800°F) for 3 hrs. The heating and cooling rates for such stabilizing heat treatments are not critical.
The subsequent service temperature range within which relaxation cracking may occur extends further to 500-780°C (932-1436°F), if products are reused which have already been in service and which have been repair-welded with matching alloy 617 consumables without a following stabilizing heat treatment at 980°C (1800°F) for 3 hrs.
The " modified alloy 617" is being considered for various components such as: 1) Membrane Walls, 2) Reheater and Superheater tubes (up to 770°C) and 3) Headers and Piping up to design temperatures of 735°C

Alloy C-263 – Nicrofer 5120 CoTi.

Alloy C-263 is another alloy with good potential for use in the high efficiency fossil fired power plants as 1) Reheater and Superheater tubes (up to 770°C) and Headers and Piping up to design temperatures of 735°C

	Ni	Cr	Fe	C	Mn	Si	Cu	Mo	Co	Al	Ti	Al+Ti
Min.	Bal. 51	19.0		0.04				5.6	19.0	0.30	1.90	2.40
Max.		21.0	0.7	0.08	0.6	0.4	0.2	6.1	21.0	0.60	2.40	2.80

Table 3: Chemical composition of alloy alloy C-263 – Nicrofer 5120 CoTi.

The chemical composition of alloy C-263 – Nicrofer 5120 CoTi is listed in Table 3 above. Nicrofer 5120 CoTi is characterized by:

- excellent resistance to oxidation and scaling up to 1000°C (1800°F)
- good mechanical properties and excellent creep values at elevated temperatures
- good weldability without susceptibility to post-weld heat treatment cracking

The high temperature strength of Nicrofer 5120 CoTi is obtained by two strengthening mechanism. The cobalt and molybdenum additions give solid-solution strengthening. The aluminum and titanium additions form precipitates of the γ′-phases Ni3 (Al, Ti) on age-hardening. The cobalt addition also increases the solubility of γ′ above 1100°C (2010°F), thus facilitating hot working despite the high aluminum and titanium contents. In the fully heat-treated condition, the microstructure of Nicrofer 5120 CoTi shows fine discontinuous precipitates of carbides ($M_{23}C_6$) at the grain boundaries. Continuous $M_{23}C_6$ films must be avoided, as this can lead to poor ductility and hot tearing during welding. Correct solution treatment will avoid this effect.

Ni-alloys for high efficient 700°C fired steam power plants

Alloy 617 / Nicrofer 5520 Co and Alloy C263 / Nicrofer 5120 CoTi

- Yield Strength Rp0.2 and Tensile Strength Rm

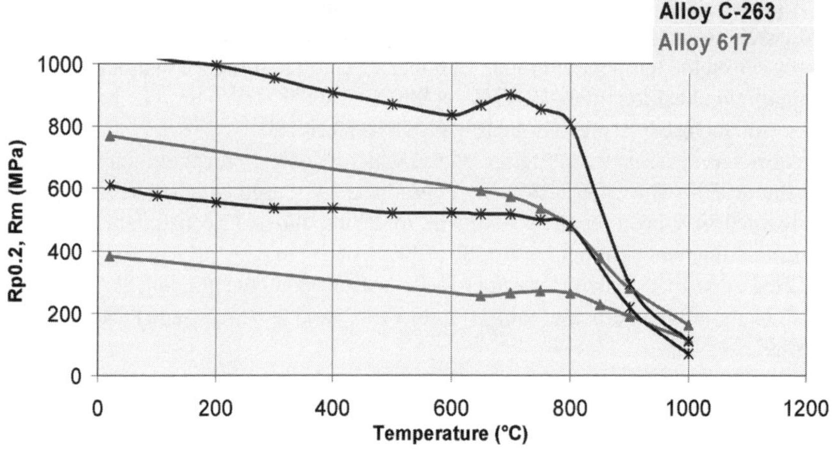

*C-263 in precipitation hardened condition (800°C/8h)

Figure 6: Yield strength Rp0.2 and tensile strength Rm of alloys 617 and C263 as function of the temperature

In the precipitation hardened condition which results after a aging heat treatment of 8 hours at 800°C the yield strength and the tensile strength of alloy C263 are higher than the values for the alloy 617 that can be seen in Figure 6. The drastic drop in these properties in alloy C263 at temperatures higher than 800°C is due to the dis-solution of the age-hardening precipitates.

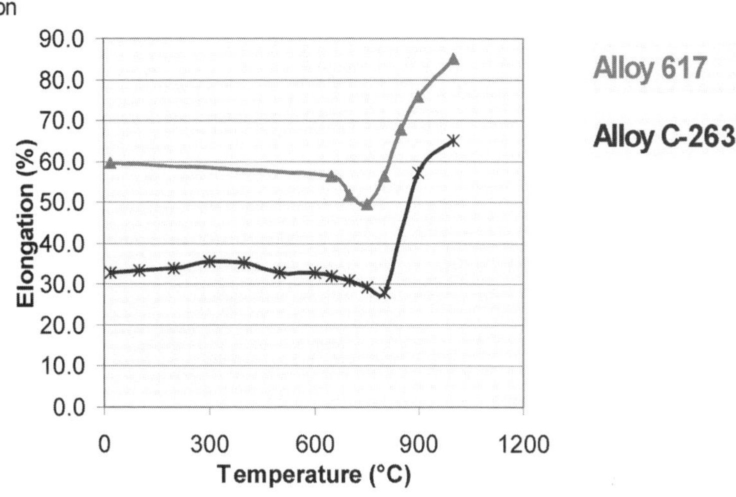

Figure 7: Elongation of alloys 617 and C263 as function of the temperature

The elongation values for the alloys 617 and C-263 as function of the temperature are shown in Figure 7.

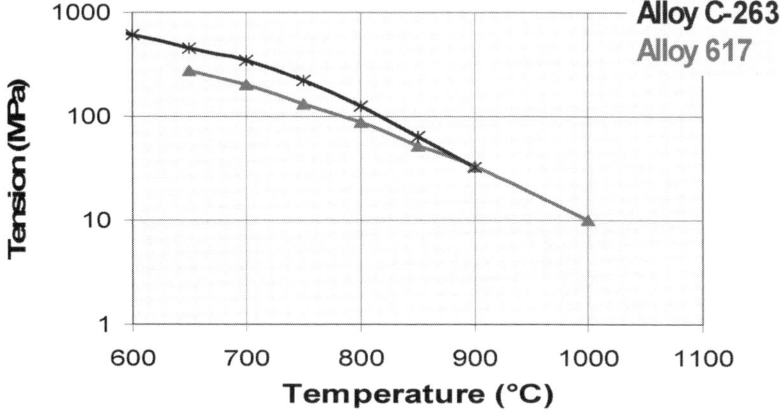

Figure 8: 10.000 hours creep strength of alloys 617 and C-263 as function of the temperature

As discussed before the creep rupture strength of alloy 617 is higher than those of the steels shown before used for components in power plants with lower working temperatures. The Figure 8 shows the curves of creep rupture strength after 10.000 hours for alloys 617 and C263. The one of C263 is even higher than the curve of creep rupture strength of alloy 617, which is the reason why this Ni-alloy is in discussion for certain components of 700°C fossil fired power plants. After 10.000 hours the creep rupture strength of C263 is clear above 100 MPa in the temperature range between 700 and 750°C, which is also the goal for 100000 hours.

The alloys 617 – Nicrofer 5520 Co and C-263 – Nicrofer 5120 CoTi are characterized by their behaviour with respect to very good short-term and long-term mechanical properties with excellent creep values at elevated temperatures, their excellent resistance to oxidation and good weldability. The "modified alloy 617" increases the window of applicability for certain components of 700°C fossil fired power plants.

Summary

In order to satisfy the increasing energy demand and to fulfil the climate protection goals new concepts for fossil fuelled power plants have to be developed with higher efficiency and CO_2-reduction technology. Many research activities in US, Japan and Europe work on this subject. A new one in Germany is COORETEC. With increasing temperature the requirement with respect to the creep rupture strength increases for the used metals for certain components. The existing nickel alloys 617 and C263 are two candidates for this application along with the "modified alloy 617" which provides superior properties in the tests conducted so far.

Acknowledgments

The authors thank Mr. Ralf-Udo Husemann, Laboratory of Hitachi Power Europe GmbH for the permission to use the figure of the principle design of a 700°C power plant (figure 3).

References

1) Source: Cooretec-Report STEAG 2002 (Figures 1 and 2)
2) Source: R.U. Husemann, October 2005, Milano (Figure 3).

Advances in Materials Technology for Fossil Power Plants
Proceedings from the Fifth International Conference
R. Viswanathan, D. Gandy, K. Coleman, editors, p 281-292

DOI: 10.1361/cp2007epri0281

CREEP PROPERTIES OF ADVANCED STEELS FOR HIGH EFFICIENCY POWER PLANTS

I. Charit*, K.L. Murty and C.C. Koch
North Carolina State University
North Carolina State University, Raleigh, NC 27695, USA
*Currently at the University of Idaho, Moscow, ID 83844, USA

Abstract

Driven mainly by the environmental and economic concerns, there is an urgent need for increasing the thermal efficiency of fossil fuel power generation plants, which still languishes at around 32% under current practices. Several programs have been undertaken worldwide to address this issue. One of the immediate options is to increase the steam temperature and pressure (to the supercritical range). However, the current power plant materials appear to have inadequate creep resistance under these demanding conditions along with corrosion/oxidation problems. Hence, to meet these challenges a variety of new steels and stainless steels have been developed in the United States, Japan, and Europe. Alloy design and microstructural design approaches in developing these alloys (ferritic/martensitic, austenitic and oxide-dispersion-strengthened steels) will be briefly reviewed. Further, this paper presents creep data of these steels found in the literature in terms of Larson-Miller parameters (LMP). A detailed account of plausible creep micromechanisms in these advanced steels is also be summarized.

Introduction

Recently gobal warming has been unanimously recognized as a major environmental concern. Greenhouse gas emissions (CO_2, SO_2) from various human activities have been found to be responsible for this. Apart from other major sources (such as, automobiles, aviation), these gases are also produced by thermal power plants all over the world. As the thermal power will continue to provide the greatest contribution to the power mix, it is important that the efficiency of the power plants be increased. The average thermal efficiency (as evaluated from the Rankine steam cycle) of the thermal power plants remains at ~32%. However, enhancing thermal efficiency would yield rich dividends, such as substantial reduction in greenhouse gas emissions, increasing cost-effectiveness of operations and so forth. This increment in efficiency may be achieved through several possible means, such as improving design features of power plant components, using other innovative technologies and using improved structural materials.

The last option has already been embodied in the ultra supecritical steam turbine concept [1]. Increasing the steam temperature and pressure to the ultrasupercritical range (up to 760°C and 35 MPa, respectively) would significantly increase the thermal efficiency (42 to 50%). However to achieve this, the structural materials need to withstand unusually high temperatures and corrosive conditions. Therefore, this option can only be fruitful when structural materials with adequate creep resistance and corrosion/oxidation resistance are developed. For the load-bearing components at higher temperatures, creep is a critical performance-limiting factor. Hence, creep properties of structural materials to be employed in high temperature components of fossil-fuelled power plants are of paramount importance. The purpose of the paper is to highlight various aspects of importance from the point of view of creep in various advanced steels. No attempt will be made for a comprehensive review of the topic.

Several materials have been considered – spanning across various material classes such as steels, superalloys, composites, refractory alloys and so forth. It is further true that different materials can be used in different power plant components. For example, steels can be used for turbine casing and steam supply piping/accessories whereas nickel-based superalloys may find use in superheater tubing, turbine bolting and blading components [2]. In this paper we discuss the creep behavior of advanced steels based on the alloy and microstructural design aspects with an emphasis on enhancing their high temperature capabilities.

Strengthening Mechanisms

There are many strengthening mechanisms which are crucial for imparting strength in the material. The principle of almost all significant strengthening mechanisms is that the motion of dislocations needs to be impeded by some form of obstacles present in the material. These obstacles may be of different characters, such as solute atoms, immobile dislocations, precipitates, dispersoids and/or grain boundaries [3]. Here a brief account of various strengthening mechanisms is presented with particular emphasis on their relative importance to high temperature strength properties. In fact, most of the classical strengthening mechanisms (such as grain size strengthening, strain hardening etc.) that are useful at lower temperatures will vanish or get severely limited at this elevated temperature range. Substitutional solid solution in steels is limited at room temperature (although W and Mo are potent solution strengtheners). Hence, only precipitation and dispersion strengthening (hereafter called together as particle strengthening) mechanisms are the as the major mode of strengthening at higher temperatures.

Creep Fundamentals

Diffusional processes become increasingly important at temperatures above about 0.4-0.5T_m and high temperature deformation behavior can be described by the following phenomenological equation:

$$\dot{\varepsilon} = A\frac{DEb}{kT}\left(\frac{\sigma}{E}\right)^n\left(\frac{b}{d}\right)^p, \quad (1)$$

where $\dot{\varepsilon}$ is the steady-state strain rate, σ is the applied stress, d the grain size, A is a constant dependent on the material and operating mechanism, D is the diffusivity {which is of the functional form, $D_o exp(-Q/RT)$, where D_o is the frequency factor, Q the appropriate activation energy, R the universal gas constant and T the temperature in K}, E the elastic modulus, b the Burgers vector, k the Boltzmann's constant, n the stress exponent and p the inverse grain size exponent. It is important to note that not all the microstructural factors (except grain size) appear in Eq. (1); however, their effect is always present guiding the way the materials respond to the applied stresses at higher temperatures.

Detailed accounts of various creep mechanisms can be found in the literature [4-9]. The creep behavior of pure metals and some alloys is termed as the Class-M (or Class-II) type. A stress exponent of 5 is expected. However, stress exponents of 4-7 generally fit the creep behavior of Class-M alloys. Although there are a number of models describing the Class-M behavior, Weertman's pill-box model continues to be widely accepted [8]. In this model, the dislocations generated by the Frank-Read sources keep on gliding until the edge components of the dislocations are hindered from further movement due to the mutual elastic repulsion, whereas the screw components cross-slip rapidly for mutual annihilation. At high homologous temperatures, diffusion of vacancies become rapid enough for edge dislocations to climb toward each other and in the process, mutually annihilate. Then, Frank-Read sources start operating again. At higher stresses (on the order of 10^{-3}E) creep rates become higher than what is expected from the power law creep, and follow an exponential relation. This is called power law breakdown (PLB) regime. Dislocation mechanisms occurring in the PLB regime are not clearly resolved, but it is thought to be essentially the same as in the power-law regime at lower stresses.

At lower stresses, Newtonian viscous creep mechanisms (n=1) are expected to operate. Depending on the grain size and/or temperature, the grain boundary or lattice diffusivity becomes dominant. At higher temperatures and small grain sizes, Nabarro-Herring (N-H) creep is dominated by the bulk diffusion. N-H creep is grain size dependent (p=2). At even smaller grain sizes or lower temperatures, grain boundary diffusion becomes more significant. Then, a different diffusion creep mechanism, known as Coble creep, operates. This creep mechanism is more grain size sensitive (p=3) than the N-H creep. At very large grain sizes creep rate becomes grain size independent and follows Harper-Dorn creep (H-D) with an activation energy equal to that for volume diffusion.

The creep behavior of many solid solution alloys is termed as Class-A (or Class-I) type as depicted in Figure 1 [5-7,10]. Here solute atoms lock the mobile dislocations in such a way that the glide process becomes sluggish compared to the climb process under a range of creep conditions. As the climb and glide operate in sequence, the slower one i.e. the dislocation glide controls the rate. The activation energy associated with this mechanism is that of solute diffusivity in solid solution (or Darken diffusivity). However, for all practical purposes, the lattice self-diffusivity and Darken diffusivity are comparable, especially in dilute alloys. If the stress is decreased further, the dislocation climb controlled regime would appear. However, diffusional creep may still be observed at further lower stresses. On the other hand, at higher stresses dislocations may break away from the solute atmospheres, thus entering a climb-controlled regime again. This regime is associated with similar characteristics as the climb-controlled regime already discussed (n = 4-7). Following Murty's work [5,7,9], this breakaway stress can be calculated from the equation

$$\sigma_b = \frac{W_m^{\ 2} c_o}{2^{\beta} kTb^3} \tag{2}$$

where W_m is the binding energy between solute atom and the dislocation, c_o is the solute concentration, and β typically ranges between 2-4 depending on the shape of the solute atmosphere.

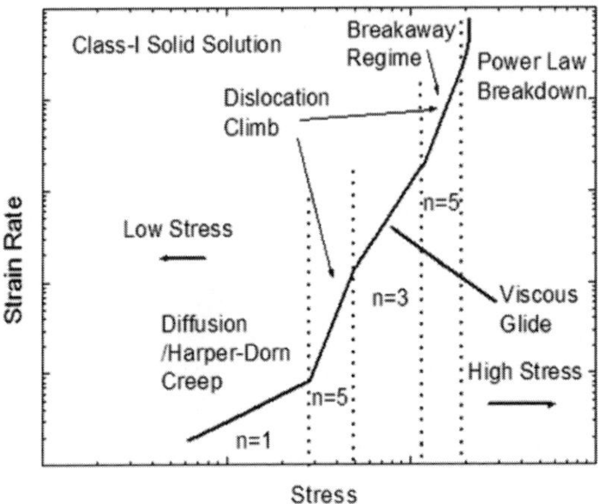

Figure 1: A typical Class-A creep behavior [10].

At even higher stresses, there may appear another regime involving low temperature climb controlled creep with a stress exponent value of n+2 (i.e. 7). This mechanism is associated with the climb processes involving dominance of dislocation core diffusion. However, this is often masked because the power law breakdown regime starts in the near vicinity. Table 2 presents parametric dependencies and constants for relevant deformation mechanisms.

TABLE 1: Parametric dependencies of various diffusion-controlled deformation mechanisms [10]

Mechanism	D	n	p	A
Climb of edge dislocations	D_L	5	0	$\sim 6\times10^7$
Viscous Glide	D_S	3	0	6
Low Temperature Climb	D_C	7	0	2×10^8
Harper-Dorn	D_L	1	0	3×10^{-10}
Nabarro-Herring	D_L	1	2	12
Coble	D_B	1	3	100

(Note. D_L: lattice diffusivity, D_S: solute diffusivity, D_C: dislocation core diffusivity, and D_B: grain boundary diffusivity)

Threshold Stress and Creep Behavior

Of all the strengthening mechanisms, precipitation and/or dispersion strengthening is the most potent strengthening mode known. Fine precipitates with small interparticle spacing are good for resisting dislocation motion. If the precipitates are of incoherent nature, at elevated temperature, attractive interaction between dislocations and precipitate will exist and the dislocation will climb over the particle. Ultimately, a stress will be needed to detach the dislocation away from the particle as shown in Figure 2 [11].

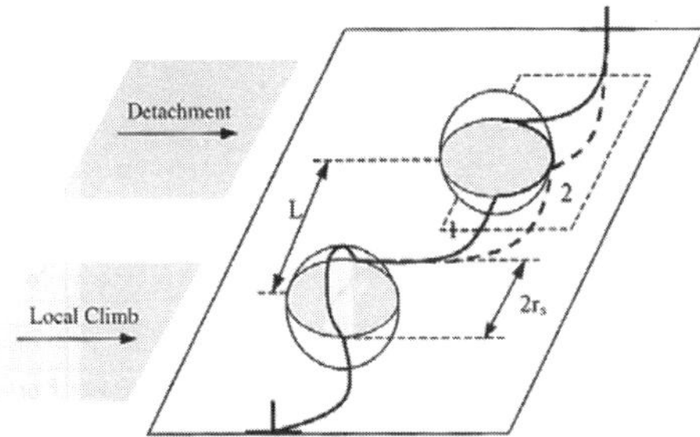

Figure 2. The mechanism of interfacial pinning based on the detachment model [11].

In the presence of threshold stresses, the constitutive equation for creep rate can be expressed by

$$\dot{\varepsilon} = A'(\sigma - \sigma_0)^n, \quad (3)$$

where A' is a constant dependent on material and temperature, σ the applied stress, σ_0 the threshold stress and n the stress exponent [12]. The creep resistance of a material is thought to be superior when threshold stress of a material is high, and below this stress, deformation will be negligible or proceed at sluggish rate.

For other steels, nanoscale MX carbide precipitates are generated, which are responsible for better creep resistance. A study by Taneike et al. [13], attributed the better creep resistance of 9Cr martensitic steel to the presence of fine nanometric carbo-nitrides (5-10 nm) on boundaries, resisting the boundary motion at elevated temperatures and delaying the start of tertiary stage of creep, thus prolonging the creep life. However, in their study discussion on the aspect of particle-dislocation interaction was notably absent. Klueh et al. [14] have shown that fine scale MX nano-precipitates are formed in a 9Cr-martensitic steel. The fine precipitates are shown in Figure 3.

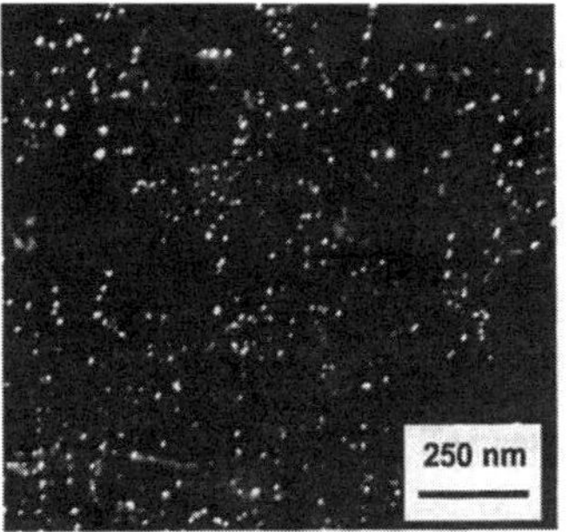

Figure 3. A dark field TEM image of fine precipitates in a 9CrMoNiVN steel [14].

Horton [15] has suggested that grain boundary particles may hinder grain boundary sliding and diffusion can be hindered by the presence of grain boundary particles, and the threshold stress associated with the process gradually becomes larger with increase in particle density on boundaries. Zhang et al. [16] have verified this conjecture in a Fe-15Cr-25Ni alloy.

Generally, the particle coarsening (Ostwald ripening) is given by the following equation,

$$r^m - r_o^m = Kt, \tag{4}$$

where r and r_o are the final and initial mean particle radii, respectively, K is the rate constant, and m is the exponent whose value depends on the controlling coarsening process [17]. Typically, m is 3 for lattice diffusion-controlled coarsening processes. Again, the term 'K' for carbide (M_AC_B) coarsening can be expressed in terms of various material parameters as given by the following equation:

$$K = \frac{8(A+B)E_{p-m}VDu^m}{9ART(u^p - u^m)^2} \tag{5}$$

where E_{p-m} of particle-matrix interfacial energy, V is the molar volume of the carbides, D is the lattice diffusivity of the M atoms in the matrix, T is the temperature in K, and u^p and u^m are the concentration of M atoms in the matrix and carbides, respectively. Mainly, the creep properties of structural materials degrade via the coarsening and/or dissolution of precipitates at elevated temperatures as a function of time until cavities/cracks initiate.

Bhadeshia [18] has described precipitation and coarsening as a simultaneous process. Figure 4 shows that precipitate size grows following an approximate parabolic law, and at longer times it follows a cubic law (as expected for coarsening) when the number density of the particles becomes small due to coarsening. Naturally, if the particles are important for creep resistance in advanced steels, their evolution as a function of time as well as temperature would be of much significance on the related creep properties.

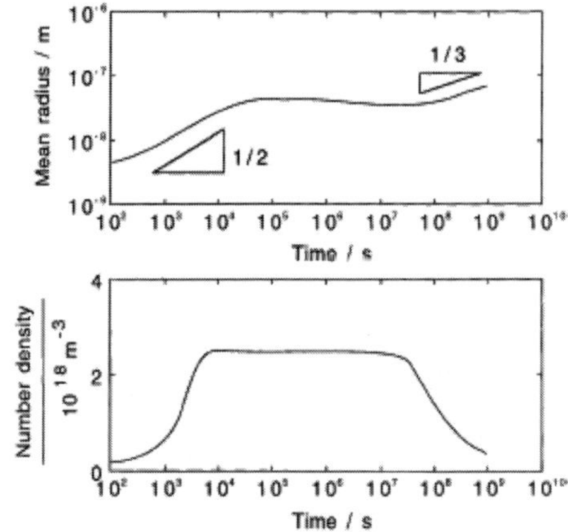

Figure 4. The precipitation and coarsening as an inclusive process in a model Fe-0.05Nb-0.1C (wt.%) steel, starting with a solid solution at 900°C [18].

From the foregoing discussions, it is clear that the potential of dispersion strengthened materials is dependent on how well they can retain strength as a function of temperature. Three factors are important to consider [12]: (a) thermal instability of the particles, (b) a decreasing effectiveness of the dispersoid particles, and (c) the onset of grain boundary processes. However, it is important to note that thermal instability may not be used as a reason whenever there is a decrease in strength at elevated temperatures. Even thermally stable particle dispersions may be unsuitable at high temperatures if there is a lack of sufficient attractive interaction between dislocations and dispersoids. Finally, fine grained structures are often prone to grain boundary sliding and damage formation. When these things start to happen, the strengthening due to particles generally disappears.

Austenitic Stainless Steels

Austenitic stainless steels have traditionally been used for creep-resistant as well as corrosion/oxidation-resistant applications. However, the ultra supercritical power plant requires materials that are much superior compared to the conventional stainless steels. A number of alloys have been developed in recent years. Among them, CF8C Plus (12.5Cr-19.3Ni-4.0Mn-0.3Mo-0.85Nb-0.09C-0.48Si-0.23N-Bal.Fe, in wt.%) presents a noteworthy development. Due to the unique composition, this alloy is fully austenitic following solidification, essentially free of delta-ferrite phases.

The alloying rules used in the development of CF8C Plus have been described by Maziasz and Pollard [19]. The effects were classified as reactant effects, catalytic, inhibitor and interference. The long term creep resistance of CF8C Plus was attributed to the following reasons: a) production of nanoscale precipitates (niobium carbide, not niobium carbonitride) that hinder dislocation movement during creep deformation, b) adjustment of alloy composition making sure long-term precipitate stability at service temperatures, and c) delaying or eliminating the causes of failure (such as, formation of brittle sigma phase). Nitrogen has been added to improve the strength of the austenitic matrix by solid solution strengthening.

For the proposed USC system with a steam temperature of 760°C and a steam pressure of 32.5 MPa, components would have to be designed for at least a stress of 80 MPa. Considering that kind of stress, the corresponding LMP value would be ~24,000 [20]. Assuming a design rupture life of 100,000 h, the maximum allowable temperature (MAT) can be estimated from the following equation:

$$\text{MAT (in } ^{o}\text{C)} = (\text{LMP}_{(\text{at 80 MPa})} / 25) - 273 \qquad (5)$$

The relevant temperature turns out to be about 687°C which falls short of the required temperature. That is why further studies were conducted by adding Cu and W to CF8C Plus base composition. These additions appear to have a beneficial effect on the high temperature mechanical properties. It has been previously noted that aging at 750-800°C of a Cu-containing austenitic steel (Tempaloy) precipitates out fine Cu-rich phases, which in turn add to high temperature creep strength via additional precipitation strengthening [21]. Interestingly, it was further noted that the precipitation of these phases is not influenced by the existing carbides or other precipitating carbides. Hence, there is a possibility of enhancing the high temperature deformation resistance independent of other strengthening modes. Moreover, it is known that W significantly enhances solid solution strengthening of the austenite matrix phase. Therefore, simultaneous addition of Cu and W would lead to much higher creep resistance. Recently, an Al-containing Fe-20Ni-14Cr-2.5Al alloy (austenitic steel, no Ti/V added) has been found to have superior oxidation resistance, and creep-rupture properties exceeding 2000 hours at 750°C and 100 MPa [22].

No study has yet focused exclusively on explaining high temperature creep mechanisms in these advanced austenitic stainless steels. Hence, those are some of the areas that need further study and assessment. Also, the role of microstructural stability on the long-term creep resistance needs to be carefully evaluated.

Ferritic/Martensitic Steels

Ferritic/Martensitic steels constitute an important class of materials for application in the fossil power plants. Various alloying schemes have been proposed and implemented over decades which lead to considerable improvement in high temperature capability of these Cr-containing steels. Klueh [23] has summarized the development of various F-M steels (as shown in Table 2) and has given a detailed rationalization of alloying schemes used.

Table 2: Evolution of F-M Steels used for power generation industry (Klueh, 2005 [23])

Generation/Years	Steel Modification	10^5 hr rupture strength at 600°C (MPa)	Name of Steels	Maximum Temp. (°C)
0 /1940-60	-	40	T22, T9	520-538
1 / 1960-70	Addition of Mo, Nb, V to simple Cr-Mo steels	60	EM-12, HCM9M, HT9, HT91	565
2 / 1970-85	Optimization of C, Nb, V, N	100	HCM12, HCM2S (T23)	593
3 / 1985-95	Partial substitution of W for Mo and addition of Cu, N and B	140	NF616, E911, HCM12A	620
4 / Future	Increase W and addition of Co	180	NF12, SAVE12	650+

Here dilemma of using excess nitrogen as alloying element in the ferritic matrices is discussed. Swada et al. [17] investigated 9Cr ferritic steels (0.002C-9Cr-3W-VNbCoNB) with three different levels of nitrogen (0.049, 0.073 and 0.103wt.%). They noted that an increase in the nitrogen content leads to lowering of creep strength and shorter creep life even if they have finer lath structure and larger prior austenite grains after tempering. There have been two plausible reasons for this effect: (a) Cr_2N phases precipitate at locations where MX nitrides (M – Nb, V; and X – C, N) are to be formed. This effectively reduces the inter-MX particle spacing and thus, the creep strength. (b) Higher coarsening rates of MX nitrides effectively reducing the creep life in the long run.

Figure 5 demonstrates that the creep micromechanisms can undergo transitions as a function of stress – example given for the creep behavior of a P-91 steel (8.1Cr-0.1C-0.9Mo-0.2V-0.07Nb-0.05N-0.3Ni). Here the mechanism with a stress exponent of 1 (presumably diffusional creep) at lower stresses transitions to a mechanism with a stress exponent of 10 (apparent stress exponent, not consistent with parametric dependencies included in Table 1) at higher stresses. However, high stress exponents in particle-containing alloys have often been obtained, and may be explained in terms of a threshold stress generated due to the dislocation-particle interactions (Eq. 3). From this plot, the danger in the blind extrapolation of high stress data to low stresses for predicting creep life becomes apparent, and a much lower creep rate is predicted at lower stresses than the actual rates.

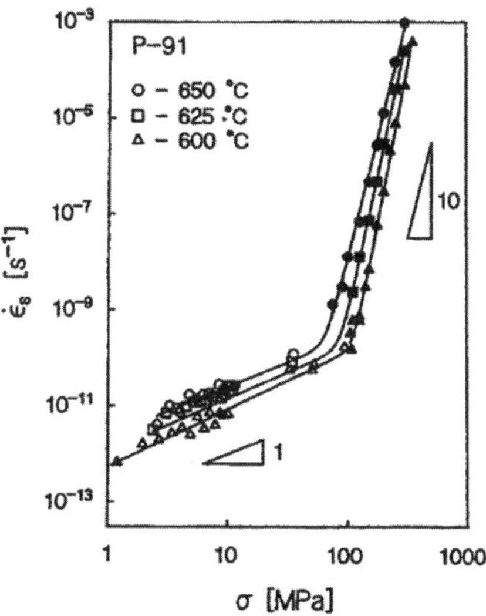

Figure 5. Steady state creep rate vs. stress for a P-91 steel at different temperatures [23].

Oxide Dispersion Strengthened Steels

It was recognized early by the materials community that there is much promise in developing metallic materials with dispersed hard ceramic phases. The initial interest was due to the immense potential of increasing the strength of metals/alloys via dispersion strengthening mechanism (i.e. Orowan bypassing). The development of dispersion strengthened alloys by internal oxidation and the invention of dispersion-strengthened aluminum (SAP) lead to a widespread interest in using this mode of strengthening in other alloy systems as well. Although TD-Ni and TD-Ni-Cr alloys were developed through conventional powder metallurgy techniques, the technical breakthrough in this direction came from Benjamin via the mechanical alloying technique involving high energy ball milling [24]. Several commercial mechanically alloyed Ni-base superalloys (such as MA754, MA758, MA6000), various MA-ODS steels (such as MA956, MA957) and other ODS alloys have since been developed.

Recently, an yttria-containing alloy, 12YWT (known as nanostructured ferritic alloys or NFA), has been developed through mechanical alloying route [25]. This steel possesses very good high temperature creep strength because of the presence of thermally stable, nanoscale oxide dispersoids (diameter < 10 nm, Figure 6a) [25]. Very limited creep (time-dependent plastic deformation) studies have been performed on the MA-ODS steels. The creep rate – stress data for three ODS alloys are shown in Figure 6b [26]. It is interesting to note that the stress exponent (n) is very high compared to what is generally predicted for metallic alloys (see Table 1). This behavior is sometimes typical of various ODS alloys, and is generally attributed to the presence of a threshold stress. It also leads to an overestimation of the creep activation energy. This behavior is again dependent on the precise nature of interaction happening between the nanoscale oxide particles and mobile dislocations. The nature of interface becomes very important in ascertaining the type of threshold stress behavior exhibited by these materials.

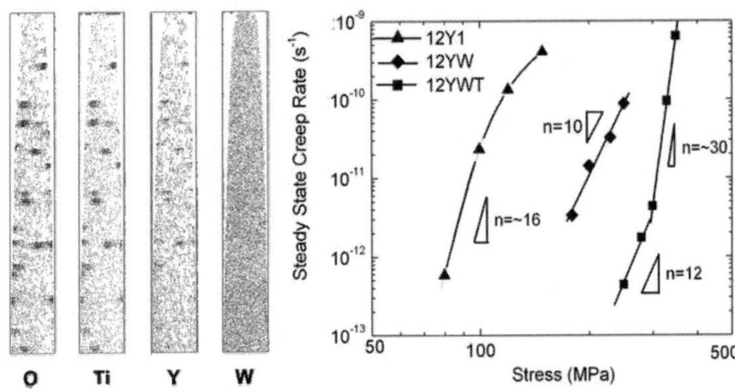

Figure 6. a) The presence of Y-Ti-O nanoclusters in 12YWT alloy (left). b) Steady state creep rate as a function of stress in three different ODS steels at 700°C (right). 12Y1 and 12YW are the ODS alloys without Ti/W and Ti additions, respectively.

With a similar calculation as shown in Eqn. 5, the maximum allowable temperature for 12YWT alloy has been found to be ~1000°C at a stress of 100 MPa (corresponding LMP is 32,000) [26]. Hence, these alloys meet more than what is required for materials in USC plants. Despite their excellent creep properties, their potential for applications in USC plants is limited because of their fabrication difficulties (such as, welding) and expense.

Summary and Conclusion

There are various steels (austenitic, ferritic/martensitic and ODS steels) available but a few meet all the requirements of ultra supercritical power plants. Most important strengthening mechanism that imparts creep resistance is the particle strengthening followed by solid solution strengthening and work hardening. However, microstructural stability is key to achieving long term creep resistance in these advanced steels. Due to the microstructural instability, creep micromechanisms need to be carefully identified. Further, due to the mechanistic transition there lies a danger in the blind extrapolation of high-stress data to low-stress regimes of practical interest.

References

1. R. Viswanathan, R. Purgert and U. Rao, "Materials for Ultrasupercritical Coal-Fired Power Plant Boilers," Materials for Advanced Power Engineering 2002, Proceedings Part II, Forschungszentrum Julich GmbH, 2002, 1109-1129.

2. I.G. Wright, P.J. Maziasz, F.V. Ellis, T.B. Gibbons, and D.A. Woodford, "Materials Issues for Turbines for Operations in Ultra-Supercritical Steam," Procs. Of the 29th Inter. Conf. on Coal Utilization and Fuel Systems, April 18-23, 2004, Clearwater, FL.

3. G.W. Meetham, "Mechanisms for Increasing High Temperature Capability: Part B," Materials and Design, 9 (6) (1988) 308-317.

4. O.D. Sherby and P.M. Burke, "Mechanical Behavior of Crystalline Solids at Elevated Temperatures," Progress in Materials Science 13 (1968) 323.

5. T.G. Langdon, "Dislocations and Creep, in: Proc. of Dislocations and Real Properties of Materials, Institute of Materials, London, UK, 1985, 221.

6. K.L. Murty, "Transitional Creep Mechanisms in Al-5Mg at High Stresses," Scripta Met., 7 (1973) 899.

7. K.L. Murty, "Viscous Creep in Pb-9Sn Alloy," Mater. Sci. Eng., 14 (1974) 169.

8. J. Weertman, "Steady-State Creep of Crystals," J. Applied Physics, 28 (1957) 1185.

9. K.L. Murty, "Transitions in Deformation Mechanisms in Class-A Alloys – Historical Perspective and Recent Applications to Microelectronic Solder and Nuclear Core Materials," Procs. of 7th Inter. Conf. on Creep and Fracture of Engineering Materials and Structures, J.C. Earthman and F.A. Mohamed (eds.),1997, 69-78.

10. I. Charit and K.L. Murty, "High Temperature Mechanical Properties of Molybdenum Solid Solution Alloys," Procs. ICAPP'06, Reno, NV, USA, June 4-8, 2006, Paper 6030.

11. E. Arzt, and M.F. Ashby, "Threshold Stresses in Materials Containing Dispersed Particles," Scripta Metall. 16 (1982) 1285.

12. E. Arzt and J. Rosler, "High Temperature Deformation of Dispersion Strengthened Aluminum Alloys", Dispersion strengthened aluminum alloys, edited by Y.-W. Kim and W.M. Griffith, 1988, p. 31.

13. M. Taneike, F. Abe and K. Sawada, "Creep-strengthening of steel at high tempeartures using nano-sized carbonitride dispersions", Nature, 424 (2003) 294.

14. R.L. Klueh, N. Hashimoto, and P.J. Maziasz, "Development of new nano-particle-strengthened martensitic steels", Scripta Mater., 53 (2005) 275.

15. C.A.P. Horton, "Some observations of grain boundary sliding in the presence of second phase particles", Acta Metall., 20 (1972) 477.

16. J.S. Zhang et al., "Grain boundary precipitation strengthening in high temperature creep of Fe-15Cr-25Ni alloys", Scripta Met., 23 (1989) 547.

17. K. Sawada, M. Taneike, K. Kimura and F. Abe, "Effect of nitrogen content on microstructural aspects and creep behavior in extremely low carbon 9Cr heat resistant steel," ISIJ Inter., 44 (2004) 1243.

18. H.K.D.H. Bhadeshia, "Design of Ferritic Creep Resistant Steels," ISIJ Inter., 41 (2001) 626.

19. P. Maziasz and M. Pollard, "High Temperature Cast Stainless Steel," Advanced Materials and Processes, 161 (2003) 57.

20. J.P. Shingledecker, P.J. Masiasz, N.D. Evans, and M.J. Pollard "Creep Behavior of a New Cast Austenitic Alloy," International J. Pressure Vessel and Piping, 84 (2007) 21-28.

21. A. Tohyama and Y. Minami, "Development of the High Temperature Materials for Ultra Super Critical Boilers," in: Advanced Heat Resistant Steel for Power Generation, R. Viswanathan and J. Nutting (eds.), 1999, 294.

22. Y. Yamamoto, M.P. Brady, Z.P. Lu, P.J. Maziasz, C.T. Liu, B.A. Pint, K.L. More, H.M. Meyer, E.A. Payazant, "Creep-Resistant, Al_2O_3-forming Austenitic Stainless Steels," Science, 316 (2007) 433.

23. Klueh, R.L., "Elevated Temperature Ferritic and Martensitic Steels and Their Application to Future Nuclear Reactors", International Materials Review, 50 (2005) 287.

24. J.S. Benjamin, "Dispersion Strengthened Superalloys by Mechanical Alloying," Metallurgical Transactions, 1 (1970) 2943.

25. M.K. Miller, E.A. Kenik, L. Heatherly, D.T. Hoelzer, P.J. Maziasz, "Atom Probe Tomography of Nanoscale Particles in ODS Ferritic Alloys," Materials Science and Engineering A, 353 (2003) 140.

26. I. Kim, B.Y. Choi, C.Y. Kang, T. Okuda, P. Maziasz and K. Miyahara, "Effect of Ti and W on the Mechanical Properties of 12%Cr Base Mechanical-Alloyed Nano-Sized ODS Ferritic Alloys," ISIJ Inter., 43 (2003) 1640.

Advances in Materials Technology for Fossil Power Plants
Proceedings from the Fifth International Conference
R. Viswanathan, D. Gandy, K. Coleman, editors, p 293-302

DOI: 10.1361/cp2007epri0293

Life Management of Creep Strength Enhanced Ferritic Steels —Solutions for the Performance of Grade 91 Steel

Kent Coleman

Project Summary

Recent in service experience has demonstrated that cracking can occur in CSEF steels very early in life. Indeed it is apparent that 'irregularities' during fabrication can result in components entering service with properties that are substantially deficient when compared to an "average" material. These issues have caused serious concern among users because of the obvious implications to safety of personnel and reliability of equipment. The present program – a collaborative effort between the Electric Power Research Institute and Structural Integrity Associates, Inc.- specifies a workscope to address critical issues associated with the use of these materials. The issues to be addressed range from material procurement, shop fabrication, field erection and the appropriate quality assurance procedures to be applied during each of these phases of implementation, to the in service behavior of both base metal and weld metal, with a particular emphasis on the provision of a comprehensive strategy for life prediction and optimization of maintenance. In addition to the core technical deliverables, the program will be designed to facilitate the transfer of the latest information on these materials through regular participant workshops, in which both the information generated through the program as well as information reflecting world-wide utility experience will be reviewed and discussed to insure optimal direction of program efforts.

Finally, it should be noted that a considerable amount of research has been conducted over the past decade by EPRI and SI. A list of EPRI reports developed over the last decade for Grade 91 is provided in Appendix A. The present program has been developed to accelerate the level of knowledge for Grade 91 as there are simply too many issues to address under EPRI base funding in the Fossil Materials & Repair program (P87).

1. Introduction & Background

It is common knowledge among utility engineers that a small but significant amount of the Grade 91 material installed as critical pressure part components in US power plants has been processed in such a way that the elevated temperature performance of the material likely will not meet the expectations of the designers of those components. Recent inspections at a number of new and operating plants, both standard fossil-fuel fired and heat-recovery type steam generators, have uncovered evidence in multiple components of "soft" material with an undesirable condition of microstructure, indicating that proper controls were not maintained during some phase of the manufacturing or erection of the components. The deficient material has involved both weldments and base metal independent of any weld. In most of the cases the available documentation has not been adequate to allow a determination of how the deficient condition developed, so that it has not been possible to determine accurately the extent of the deficiencies without resorting to costly inspection. Of equal concern is the fact that there is not sufficient information available at present to accurately characterize the deficient material when it is found, so that it is necessary to make certain very conservative assumptions regarding the properties of the deficient material – e.g., that the deficient material is "like" Grade 22 – in making decisions regarding the future operation of the plant.

For the plant operators this clearly is an unacceptable situation, and in response to concerns the Electric Power Research Institute and Structural Integrity Associates, Inc. have joined forces to initiate a broad engineering study of the Creep-Strength Enhanced Ferritic steels. The purpose of this study will be to provide critical information on all aspects of the use of this unique class of steels. (Because of material availability and the number of existing problems, the study will concentrate initially on Grade 91; however, the results of the study ultimately will apply to all of the CSEF steels currently in the ASME B&PV Code and other international codes, including Grades 92, 23, and 911.)

2. Objectives

The study will have two principle objectives, with a number of tasks to be completed to achieve each of these objectives. *The first of the two objectives will focus on what might be called "front-end" issues, that is, those issues that pertain to how the material is ordered, how it is processed, how the quality control is maintained during processing, and how the material should be inspected in the shop and the field to determine its condition prior to or soon after initial operation.* The tasks falling under this objective will include a determination of the most effective methods for finding and characterizing deficient material that will be or already has been installed in plants. This will entail identifying best practices with regard to inspection strategy (i.e., which components and at what level of detail), inspection methods (e.g., hardness testing and metallographic replication, the appropriate surface preparation for each, etc.), and post-inspection activity (i.e., sample removal, engineering analysis, weld repair, heat treatment, etc.). Included in this task will be verification of the utility of new volumetric inspection methods, such as eddy-current based test procedures for the identification of deficient microstructures. A second task falling under this objective, which is a reflection of EPRI/SIA's "life cycle management" approach to the control of these materials, will be to develop the necessary specifications and procedural documents that will enable users to control the quality of the material at every stage of its implementation, from original purchase of the material by a manufacturer, through the manufacturing and construction phases, to life management once the material has been installed. The intent here is to ensure that deficient material never is installed. Another major task for this first objective will be the development of recommendations for the successful weld repair of components that are found to contain flaws or deficiencies.

The second major objective of the study will be to address the broad issue of the life management of these steels once they are placed in service, recognizing that these steels are inherently unstable and will degrade at predictable rates during "normal" operation at elevated temperatures. One of the most important tasks of this phase of the study will be to characterize as fully as possible the elevated temperature mechanical behavior (i.e., creep deformation, creep-fatigue damage, creep crack growth, etc.) of not only material that has been properly processed, but, of equal importance, deficient material in all of the variations of deficiency that may exist.

One critical step in this characterization will be the development of a master record, or atlas of microstructures, which will correlate a specific condition of microstructure with a range of hardness and creep behavior. One of the values of such an atlas will be the ability to compare information obtained during field inspections to the data contained in the atlas as a basis for a qualitative assessment of material serviceability. In characterizing the material behavior, it will be the intent to treat separately the base metal and weld metal, since it is well understood that weldment properties can vary significantly from those of the base material and that in many cases it is the weldment properties that will govern the life of the component.

Also included as part of this phase of the study will be: (1) a validation of new testing and assessment techniques that have been applied to these steels, including small specimen creep indentation testing and the use of elevated temperature strain gauges for direct measurement of on-load strain accumulation, and (2) a review of the relative merits of various life prediction methodologies, including the more common parametric methods, the Omega approach, the Monkman-Grant relationship, etc. Upon completion of the study it is intended that a "User's Handbook" for these steels will be published, which will provide the necessary information to successfully control all major aspects of the use of these steels.

3. Scope of Work/Task Descriptions

The scope of this contract/amendment includes the following tasks;

- Task 1: Initial Material Integrity—Purchase Instructions, Materials and Processing Specifications, Inspection Protocols, and Post-Sampling Protocols
- Task 2: Performance In Service: Elevated Temperature Behavior and Life Assessment Issues

3.1 Task 1: Initial Material Integrity: Purchase Instructions, Materials and Processing Specifications, Inspection Protocols, and Post-Sampling Protocols

3.1.1 <u>Development of Field Inspection Protocol</u>

Review the strengths and limitations of existing inspection techniques as they are applied worldwide and develop recommendations for best practice. This would include an assessment of both equipment (e.g., different types of portable hardness testers as they compare to standard laboratory hardness testers) and methods (e.g., mechanical vs electrolytic polishing for replicas), as well as recommended practice for surface preparation.

For the evaluation of the field hardness testers, it is anticipated that "confidence bands" would be developed reflecting the relative degree of accuracy of each type of tester compared to laboratory standards. In addition research of issues related to hardness conversions, with particular attention given to the disagreements that exist between some published tables regarding conversion from Brinell to Vickers (or Diamond Pyramid Hardness). This work will build upon the earlier EPRI work on Optimal Hardness for P91 Weldments—1004702.

Validate New Inspection Technique(s): At present the non-destructive test methods available for assessing Grade 91 material that has been installed are limited to traditional ultrasonic or radiographic techniques, together with hardness testing and metallographic replication. These methods are limited in their ability to interrogate the microstructural condition of the material below the surface, which is a critical factor in determining how the material will behave in service. To address this problem, an eddy-current based test method has been developed that in trial inspections on test coupons has proven capable of distinguishing "good" from "bad" microstructures at depths of at least 1". Validation of this test method will be conducted to verify that it can consistently identify deficient microstructures and to determine what limitations there may be to the effective use of the technique (e.g., the effect of decarburization).

Establish inspection priorities: develop recommended practice for determining scope of inspection based on available background information, inspection time and budget, level of user risk acceptance, etc. This would include identification of which components in a system should be inspected, where on a given component the inspection should focus, and the appropriate inspection techniques to be used. It is anticipated that the basis for these inspection priorities will reflect, at least in part, the broad base of inspection experience developed by material suppliers, OEMs, users, and inspection companies, as well as the improved understanding of loading conditions that develop in critical components that has emerged from the application of the more advanced stress analysis techniques.

Develop list of actions to be taken, with their purpose, when evidence of deficient material is uncovered during inspection, to include the following:

- confirmation of inspection results through more detailed non-destructive examination (e.g., replication to confirm hardness results; eddy-current based testing to determine extent of deficient material, etc.)
- sample removal
  - non-destructive: "scoop" type sample
  - destructive : boat or plug type samples
- engineering analysis to estimate life of deficient material under actual operating conditions
- monitoring of strain accumulation in service: etched grid pattern, elevated temperature strain gauges

3.1.2 Development of Specifications and Processing Standards for New Material

Purchase instructions, to include restrictions on chemistry (base and weld metal), hardness and heat treatment of material from the mill exceeding those contained in the material specifications, to provide additional life margin during operation.

Manufacturing and processing specifications, to include restrictions on heat treatment and welding practice not addressed in the Codes (eg., control of as-welded material prior to PWHT), and identifying quality control requirements that must be satisfied to insure material integrity (e.g., recommendations on thermocouple type and placement for temperature monitoring during heat treatment) .

Guidelines for the qualification and monitoring of suppliers of components fabricated from the CSEF steels (e.g., verification of furnace calibrations, verification of personnel qualifications, etc.).

3.1.3 Development of Post-Sampling Protocol for the CSEF steels

A recommended practice for removal of samples for destructive analysis and repair of site, including procedures for proper control of all heating operations will be generated.

Additionally, a recommended practice for defect/crack verification and removal through excavation and subsequent weld repair of site, including procedures for proper control of all heating operations will be established.

3.2 Task 2: Performance in Service: Elevated Temperature Behavior and Life Assessment Issues

3.2.1 Development of an "Atlas of Microstructures" correlating microstructure with hardness and creep strength. The intent of this important effort will be to characterize both base metal and, on a more selective basis, HAZ and weld metal structures. The task will involve selection of appropriate batch chemistries representing different levels of inherent material strength (e.g., heats with low and high levels of various elements that can affect long-term behavior, such as nitrogen, columbium, aluminum, nickel, etc.), fabrication methods (e.g., levels of cold work), etc., followed by applying selected heat treatment schedules to allow creation of a large number of samples representing the full range of deficient structures that can be produced during improper processing and testing those samples under conditions that approximate those to which the material will be exposed during service. The test conditions will be carefully selected to insure that the damage formed during testing is similar to the damage that will form in service to minimize inaccuracies in extrapolation of the test results. Where microstructural information has been obtained from other EPRI projects (see references), it will be included as appropriate.

The Atlas of Microstructurs will be directly applicable in accurate characterization of structures in plant components as well as providing a methodology for making judgments of creep strength. The relevance of these judgments will be based on relationships developed between factors such as hardness, composition, phase analysis etc.

3.2.2 Data compilation: creep-fatigue and creep crack growth. The purpose of this task will be to review all available literature pertaining both to crack growth at elevated temperature and to material performance under conditions involving the combined effect of creep and fatigue. All information deemed useful for the purpose of developing an assessment methodology will collected and stored in a database that will be made available to participants. In addition, to the extent that sufficient funding is available, where the data is inadequate to permit accurate characterization of crack growth behavior under a given set of conditions, testing will be conducted under the appropriate conditions to supplement the existing data and thereby improve the accuracy of the assessment methodology. Creep-fatigue ata is already being assembled in an EPRI technology innovation project for Grade 91. Where applicable, this information will be further developed and shared with this program.

3.2.3 Validation of Creep-Indentation Testing as a Means of Assessing the Creep Behavior of the CSEF Steels. This technique has been used successfully in Europe to provide quantitative information on the creep behavior of the CSEF steels, which then has been used in analyses to more accurately estimate component life. Trials conducted by SIA have demonstrated excellent correspondence between the strain rates measured using the creep-indentation test and the material condition (i.e., hardness and microstructure). It will be the intent of this effort to fully validate the data generated by the test, establishing the correspondence between the information generated by this test and the information generated in standard tensile type creep tests using uniaxial test specimens.

3.2.4 Validation of the Use of Elevated Temperature Strain Gauges for On-Load Strain Monitoring. Such gauges permit the direct monitoring of strain accumulation at critical sites on a component (as defined by experience and/or analysis) during operation. By monitoring the rate of strain accumulation of acceptable or deficient material in an area of high stress on a component, a more accurate assessment can be made of the time to rupture of that material. This will permit direct verification of analysis results and will both avoid unnecessary replacement of deficient material where the operating conditions are not severe and will insure timely replacement of deficient material where the operating conditions are particularly rigorous. This technology has been used extensively abroad, and it will be the intent of this effort to import the technology and make the necessary refinements for use in the US as a cost-effective method of component life monitoring.

3.2.5 Establish Life Prediction Methodologies for Grade 91. Various methods have been used for estimating the life of a component fabricated from one of the CSEF steels under a given set of operating conditions, ranging from a simple parametric treatment of a data set (e.g., Larson-Miller), to the Omega method, to the Monkman-Grant relationship. The purpose of this task will be to conduct a critical review of the most frequently-used of the predictive methods, comparing the relative accuracy of the results obtained from each method to the longest-term data available for Grade 91, which now exceeds 100,000 hours. The results of the review will be incorporated into recommendations for a life assessment methodology for these materials that will reflect the improved understanding of material behavior generated by this study. A guideline will be developed to provide life assessment protocol for Grade 91 materials.

4 Project Deliverables

- Recommended Practice for Determining the Scope of Inspection.
- Practice for Assessment of In-Service Materials To Identify Deficient Materials
- Materials Specification and Processing Standard for Procurement of New Grade 91 Materials.
- Atlas of Grade 91 Microstructures Correlated to Hardness and Creep Strength.
- Assessment of creep-indentation testing techniques and Elevated Temperature Strain Monitoring.
- Remaining Life Guideline for CSEF steels

Information derived from the project will be incorporated into both P87 and P63 to assist member utilities in improved management CSEF steels. Additionally, utilization of this information will enable utilities to enhance power plant safety by identification of substandard materials allowing for immediate replacement or reduced operating temperatures or pressures.

5 Project Schedule

Note--Anticipated start date is September 2007

Project Activities and Schedule		**Month**
1.	User's Steering Committee to Review the Proposed Scope of Work and Make the Necessary Additions, Deletions, or Modifications (Kickoff Meeting)	1
2.	Develop Purchase instructions for Grade 91 Material	1-3

3.	Develop Manufacturing and Processing Specifications for Grade 91 Material	1-6
4.	Assemble a Field Inspection Protocol to Be Applied When Inspecting Installed Grade 91 Components	1-12
	a. Review available field inspection techniques and identify benefits and liabilities of each	
	b. Evaluate strengths/weaknesses of hardness test equipment	
	c. Evaluate the effect of surface preparation on hardness test results	
	d. Assess the volumetric inspection technique	
	e. Establish standard inspection scopes for specific components or classes of component based on plant objectives and budget – assign a relative risk factor for each.	
5.	Create a Microstructural Atlas	1-24
	a. Identify variations in condition within what would be considered an "acceptable" processing range, e.g., material subjected to a high tempering parameter compared to material subjected to a moderate tempering parameter compared to material subjected to a low tempering parameter	
	b. Identify the range of possible deficient heat treatment conditions and obtain specimens representing each condition	
	c. For each condition of interest, characterize the microstructure, including the tempered martensitic sub-structure (or the by-product thereof), to determine precipitate size and distribution as well as the degree of recovery	
	d. Obtain a hardness range for each condition	
	e. Characterize the creep behavior for each condition	
	f. Compile an atlas of optical microstructures that correlates sub-structure, hardness and creep behavior for each distinct condition of interest.	
6.	Assess the Creep Indentation Methodology by Comparison with Uniaxial Test Results	12-24
7.	Assess the Performance of Elevated Temperature Strain Gauges	12-24
8.	Develop a Recommended Method for Life Assessment of Components Fabricated from Grade 91	18-30
	a. Review existing life assessment methods and evaluate strengths/weaknesses in light of the information obtained from previous tasks	
	b. Develop a preferred assessment methodology that reflects	

	new information generated as part of this study	
	c. Review existing information on creep-fatigue damage of Grade 91 material and identify additional testing that would be required to satisfactorily characterize its creep-fatigue behavior	
9.	Prepare a User's Handbook – Months 24-30	30-36
	a. Provide Detailed Information on All Critical Aspects of the Management of Components Fabricated from Grade 91 Steel	
	b. Identify important similarities and differences when comparing other CSEF steels, and make recommendations on further study to extend coverage to the other CSEF steels	

Appendix A
EPRI Grade 91 Reports
August 2007

1004702	Optimal Hardness of P91 Weldments
1011352	Effect of Cold-Work and Heat Treatment on the Elevated-Temperature Rupture Properties of Grade 91 Material
1004516	Performance Review of P/T91Steels
1006590	Guideline for Welding P(T)91 Materials
1004915	Normalization of Grade 91 Welds
1009758	Evaluation of Filler Materials for Transition Weld Joints between Grade 91 to Grade 22 Components
1004916	Development of Advanced Methods for Joining Low-Alloy Steel
1009757	Temperbead Repair Welding of Grade 91 Materials
1009758	Evaluation of Transition Joints between grade 91 and Grade 22 Components
TR-101394	Thick-Section Welding of Modified 9Cr-1Mo (P91) Steel
TR-108971	Review of Type IV Cracking in Piping Welds
1006299	Conference on 9Cr Materials Fabrication and Joining Technologies -- Myrtle Beach
2007 Project	Specifying the Optimum Fabrication Practices for Grade 91 Piping and Tubing.
1004516	Performance Review of P/T91Steels
1004703	Post Forming Heat Treatment of P91 Materials
TR-106856	Properties of Modified 9Cr-1Mo Cast Steel
TR-103617	P91 Steel for Retrofit Headers -- Materials Properties
TR-104845	Creep Behavior of Modified 9% CrMo Cast Steel for Application in Coal-Fired Steam Power Plants
TR-105013	Material Considerations for HRSGs in Gas Turbine Combine Cycle Power Plants
TR-114750	Materials for Ultra Supercritical Fossil Power Plants
1001462 (2001)	Advances in Material Technology for Fossil Power Plants—SWANSEA
1011381 (2004)	Advances in Materials Technology for Fossil Power Plants – Hilton Head
TR-111571	Advanced Heat Resistant Steels for Power Generation

Advances in Materials Technology for Fossil Power Plants
Proceedings from the Fifth International Conference
R. Viswanathan, D. Gandy, K. Coleman, editors, p 303-319

DOI: 10.1361/cp2007epri0303

Microstructure and Mechanical Properties Characteristics of Welded Joints Made of Creep-Resistant Steel with 12% Cr, V, W and Co Additions

Dobrzański J*., Hernas A.**,
Pasternak J.***, Zieliński A.*
* Institute for Ferrous Metallurgy, 39-100 Gliwice, Poland
** Silesian University of Technology, 40-019 Katowice, Poland
*** Boiler Engineering Company RAFAKO S.A., 47-400 Raciborz, Poland

Abstract

This paper shows the test results of thick-walled VM12 steel pipes having 12%Cr-V-W content with Co addition. The main objective of this paper was to verify the welding technology of boiler superheater thick-walled components and present the characteristics of strength and technological properties and microstructure of welded joints made at RAFAKO S.A. The research programme included tests made on the parent material and welded joints, such as main mechanical properties at room temperature, creep resistance, low-cycle fatigue at room temperature and at the temperature of 600°C/1112°F, as well as the evaluation of macro- and microstructure. Additionally, the microstructure stability was evaluated (with the use of LM, SEM and TEM), after fatigue resistance at room and at 600°C/1112°F and following additional annealing at 700°C/1292°F and 1,000h of parent material and welded joints. The investigations of the high temperature creep resistance steels grade VM12 have been performed in the frame of COST 536 Action.

1. Introduction

The existing power plants have to cut down their sulphur compounds emissions (SOx); nitrogen compounds emissions (NOx) and carbon dioxides. In order to meet these requirements the power plants have to be modernised. Such modernisation most often includes improvement of the boiler working parameters, the increase in steam temperature and pressure values (t_0, p_0), which additionally add to the efficiency of a power unit.

To this date in Poland, X20CrMoV121 steel acc. to DIN 17117 Standard as well as coil and header parts of superheaters has been used in the largest power units (360 and 500 MW) for their critical components, i.e. those being operated in the most severe temperature and stress conditions. The calculated pressure of these components is up to 21 MPa, the working temperature of steam superheater coil material ranges from 540 to 580°C (1004-1076°F), with that of headers and collectors from 540 to 560°C (1004-1040°F) [1,2, 4,6].

The modernisation related to the increase of calculated parameters, as well as the further development of coal fired power sector, related to designing and erection of new supercritical boiler units, i.e. temperature up to 620/1148 or even 650°C/1202°F and pressure range from 28 MPa to as high as 35 MPa, contributed to a dynamic growth of 9÷12%Cr ferritic steels with molybdenum, tungsten or cobalt additions and micro-additions of niobium and nitrogen. The development of such steels, in a graphical form, related to temperature 600°C/1112°F and the acquired level of timely resistance to creep for 100,000 hours of operation, is shown in Fig. 1 [3,4].

The large-scale modernisation of the existing large Polish power units, which has been going on lately, as well as the commencement of erection of new power units with supercritical parameters and 440 and 800 MW capacity, enforced the development of their welding as well as the mode of their joining to other steel grades, not only originating from 9÷12%Cr group but also low-alloy steels type Cr-Mo, Cr-Mo-V and multi-component ones at large boiler manufacturers' such as RAFAKO S.A. Simultaneously, research works are being carried out concerning component bending and forming technological processes. These research works have been going on for years jointly with IFM (Institute for Ferrous Metallurgy Gliwice), STU (Silesian Technology University Katowice) and Institute of Welding Gliwice. These institutions are jointly performing research programmes, participating in the European programmes such as COST522, 536 and COST538 [5,6]. Some selected test results, obtained for X12CrCoWVNb12-2-2 (VM12) steel of which similar circumferential welded joints were made, are presented below.

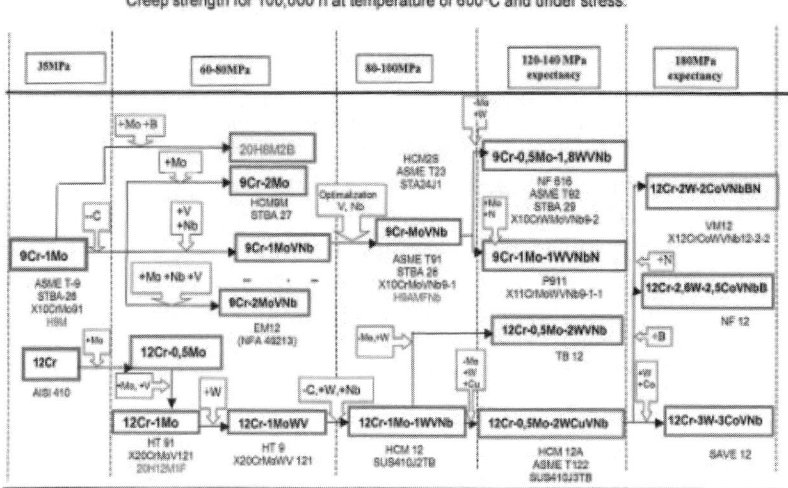

Fig. 1. Development of 9÷12%Cr high-chromium alloyed steels operated in creep conditions.

2. Tested material.

The tested material comprised tubes made of VM12 steel, having the dimensions of ø140x10mm and ø355.6x35mm, to be used for steam superheater header components, manufactured and supply by Vallourec&Mannesman group. The chemical composition of the tube material tested is presented in Table No. 1.

Table. 1. Specification of chemical composition of tested high-chromium tubes made of X12CrCoWVNb12-2-2 (VM12) steel.

Pipes Dimensions $D_z x g_n$ [mm]	Contents, %												
	C	Si	Mn	Cr	Ni	Mo	V	W	Nb	Co	B	N	Ti
Ø140x10mm	0,16	0,52	0,28	10,5	0,36	0,22	0,36	1,15	0,06	0,91	0,004	550 ppm	0,018
Ø355,6x35	0,13	0,48	0,22	11,4	0,19	0,27	0,22	1,30	0,05	1,2	0,003	500 ppm	<0,005

The welding consumable materials for welding VM12 steel were purchased from Thyssen Welding. Within the scope of the joint works over the elaboration of the welding technology at a large boiler manufacturer's in manufacturing conditions – carried out according to COST 536 programme – at RAFAKO S.A. premises the welding technology has been elaborated and similar welded joints for "PC" and "PF" welding positions made. The mechanical tests and material structure evaluations have been performed in order to determine the tube base material properties.

3. X12CrCoWVNb12-2-2 (VM12) steel, base material properties

3.1. Structure

VM12 steel is widely applied in normalised and tempered condition. A typical heat treatment after hot rolling is normalisation at the temperature of 1040-1080°C (1904-1976^{0}F) with air cooling and then tempering at the temperature of 750-800°C (1382-1472^{0}F) and cooling in normal air. Following such heat treatment, the typical obtained structure is tempered martensite, most frequently with small quantity of delta ferrite, as can be seen in Fig. 2.

LM (magnify 500x) SEM (magnify 1000x)

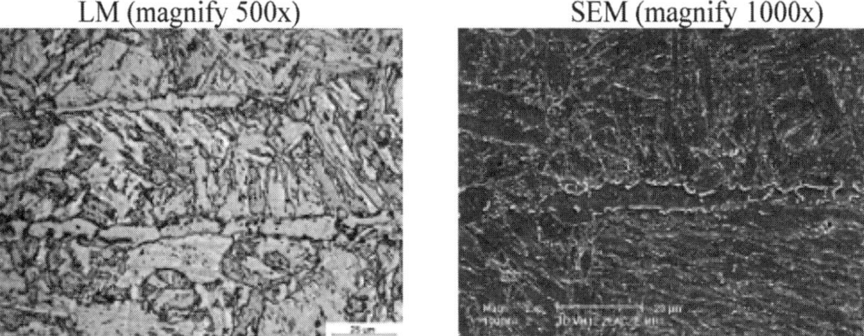

Fig. 2. Microstructure of tested pipe ϕ355.6x35 mm made of X12CrCoWVNb12-2-2 (VM12) steel after normalising and tempering process.

With slower cooling rates, one can expect a structure being a mixture of tempered martensite and bainite, as well as a small quantity of delta ferrite (Fig. 3). In this case the quantity of delta ferrite may not exceed 2%. If this quantity is exceeded, the strength properties and impact strength values may be lowered. If welded joints are made of steel with a too high delta ferrite percentage, one can expect a considerable decrease of impact strength in the individual zone of welded joints. For the material of the tested pipe ϕ355.6x35mm, the transformation temperatures have been determined: A_{c1} at 833°C/1531^{0}F and A_{c3} at 995°C/1823^{0}F. Depending on the chemical composition, the transformation temperature A_{c1} is between 820 and 830°C (1508-1526^{0}F) and A_{c3} above 890°C/1797^{0}F [7].

Similar to P91, P92 and E911 steel, VM12 steel can be characterised by a relatively high initial martensitic transformation temperature M_s, which may reach 400°C/752^{0}F. For the tested pipes ϕ355.6x35mm - M_s at 380°C/716^{0}F have been performed. The steel structure, which is formed as the result of the executed transformation, consists of strip martensite with self-tempering effects. In the tempered martensite with high dislocation density there are mainly $M_{23}C_6$ carbide releases, which include chromium, molybdenum and ferrum. Beside $M_{23}C_6$, there are smaller quantities of iron carbide Fe3C and very fine carbide-nitrides M(C, N). The releases are located mostly on the former austenite grain boundaries, sub-grain boundaries and martensite strips.

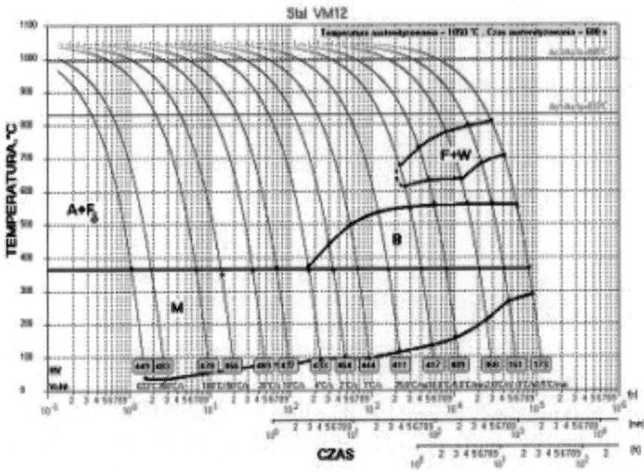

Fig. 3. CTPc curve for the material of tested tube ϕ355.6x35 mm made of X12CrCoWVNb12-2-2 (VM12) steel.

The tests and examinations were carried out on TEM, as well as the results obtained on carbide isolates by means of radiographic phase analysis (Fig. 4). The results achieved are compliant with the test results shows in [8,9].

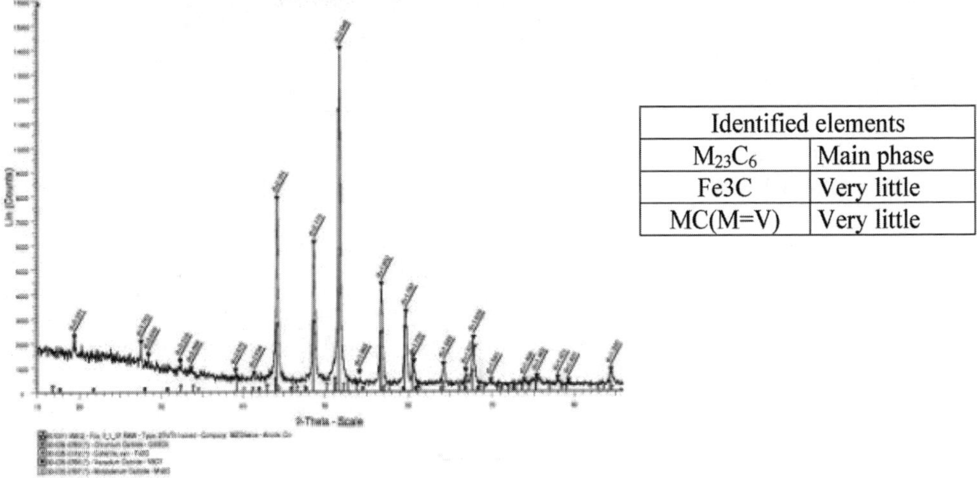

Identified elements	
$M_{23}C_6$	Main phase
Fe3C	Very little
MC(M=V)	Very little

Fig. 4. Test results of qualitative phase analysis of carbide isolates on tube material ϕ355.6x35mm of X12CrCoWVNb12-2-2 (VM12) steel.

3.2. Mechanical properties of the base material

The test results of strength properties at room and elevated temperatures for pipe ϕ355.6x35mm material are shown in Fig. 5. Fig. 6 presents impact test results for the tested thick-walled tube ϕ355.6x35mm in relation to the test temperature.

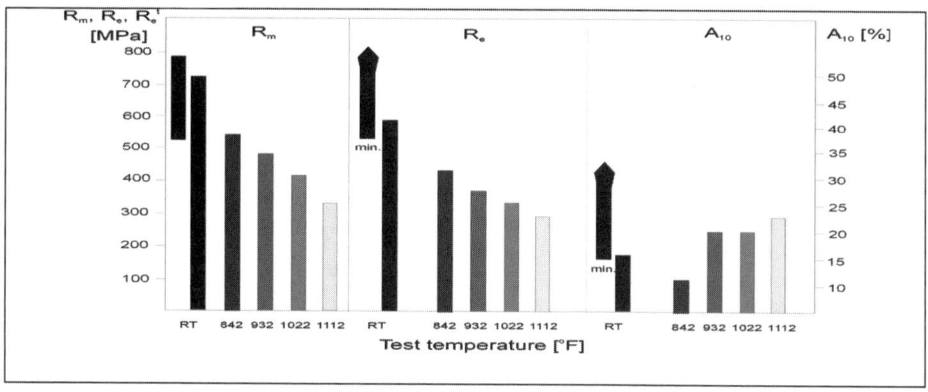

Fig. 5. Test results of strength properties for tube ϕ355.6x35 mm made of X12CrCoWVNb12-2-2 steel in relation to the test temperature.

In Fig. 7 there is a comparison of the obtained impact test results between VM12 steel and X10CrMoVNb9-1 (P91) steel. This comparison confirms that the impact energy of the thick-walled tube made of P91 steel is 250[J], whereas it is only approx. 70[J] for VM12 steel. The impact transition temperatures (ITT), transition into brittle state is respectively approx. "-" 100°C/212^{0}F for P91 steel and only approx. "-" 35°C/95^{0}F for VM12 steel.

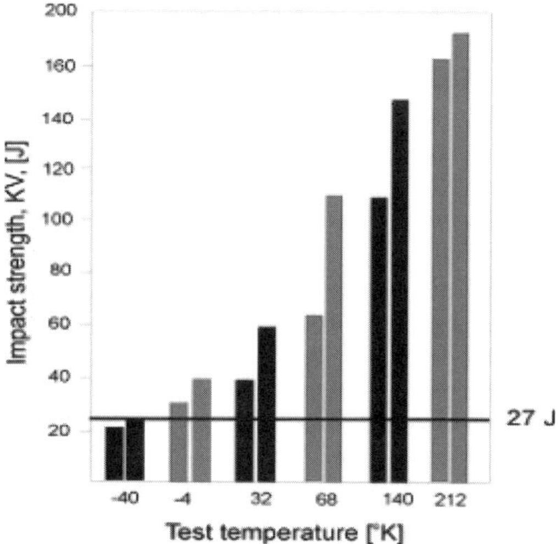

Fig. 6. Test results of impact properties for pipe ϕ355.6x35 mm made of X12CrCoWVNb12-2-2 (VM12) steel in relation to the test temperature.

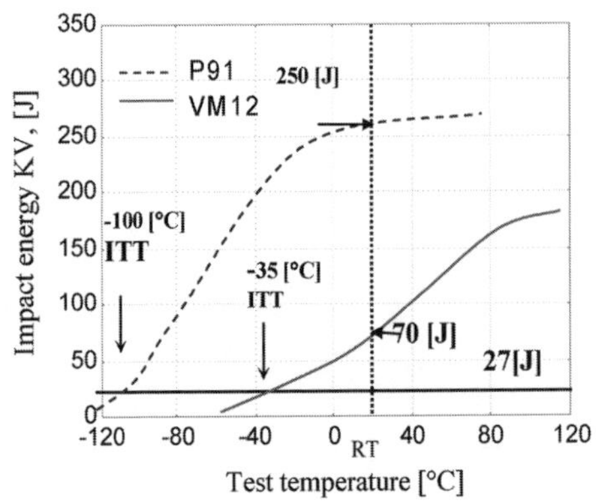

Steel grade	$KCV_{20°C}$[J]	ITT [°C]
X10CrMoVNb9-1 (P91)	250	-100
X12CrCoWVNb12-2-2 (VM12)	70	-35
RT - room temperature ITT - Impact Transition Temperature		

Fig. 7. Comparison of impact properties between X12CrCoWVNb12-2-2 (VM12) steel and X10CrMoVNb9-1 steel (P91).

The test results of creep resistance for the tested 12%Cr steel with Mo, V, W additions and micro-additions of Nb, B and N with various Co content are shown in Fig. 8 in the form of a creep resistance curve at constant test temperature of Tb = 650°C/1202^{0}F. The obtained results of the tests, carried out within COST 522 programme, have enabled to optimise the cobalt addition.

Additionally, creep tests at temperatures 600 and 650°C (1112-1202^{0}F) on pipe ϕ355.6x35 mm material made of VM12 steel at variable test stress levels σ_b from 140 to 180MPa at temperature 600°C/1112^{0}F and from 70 to 150MPa at the temperature of 650°C/1202^{0}F are also being performed. The maximum test duration is approx. 5,000 hours.

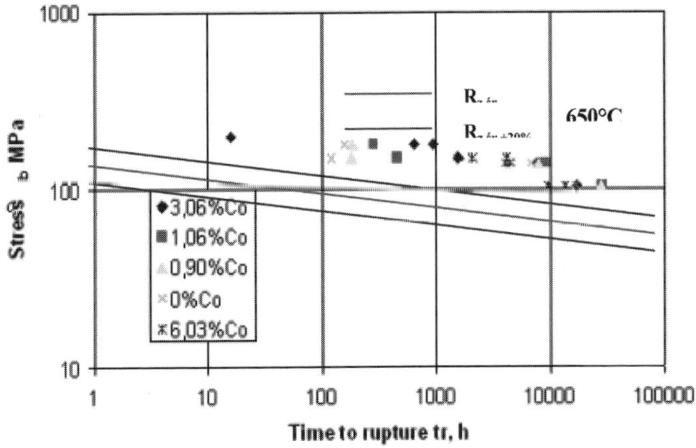

Fig. 8. Creep test results at temperature 650°C/1202^{0}F for 12% Cr steel with Mo, W, V addition and micro-addition of Nb, B and N with various Co content - from 0 to approx. 6%.

The executed low-cycle fatigue tests have shown that the low-cycle strength of VM12 steel are similar to that of other steels belonging to this group, as well as P91, P92 and X20CrMoV12-1, for which the obtained results were determined and compared at room temperature and at 600°C/1112^{0}F. This is shown as an example in Fig. 9 for the total deformation range of $\Delta\varepsilon_c$=0.8%.

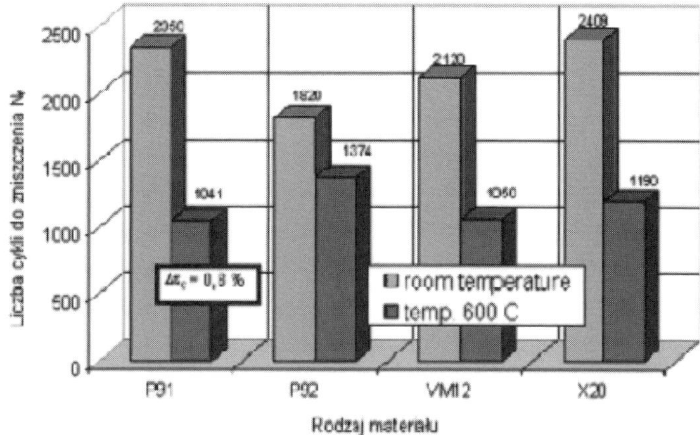

Fig. 9. Low-cycle strength characteristics for P91, P92, VM12 and X20CrMoV121 steels in initial state determine at room and at 600°C temperature, for total deformation range of $\Delta\varepsilon_c$ = 0.8%.

4. Technological requirements and test results of welded joints - after welding and heat treatment

The high-temperature creep resisting steels with tempered martensite structure and containing 9-12%Cr are very difficult in handling when selecting proper welding consumable materials, adequate welding parameters or keeping the welding parameters in the process of welded joint fabrication, particularly in manufacturing conditions. Therefore, welded joints of these steels have to be made with utmost carefulness and strict adherence to the required technological regime of their fabrication. This pertains in particular to the required welding temperature levels and cooling rates, as well as heat treatment parameters after welding.

When welding VM12 steel and in comparison with other steel grades (e.g. P91, P92 or E911), the applied temperature level initiates changes in the structure of fusion penetration zone and HAZ. VM12 steel, like the new martensitic steels already developed with the content of 9-12%Cr (P91, E911, P92), should be welded within the temperature range of approx. 200 to 280°C (392-536^{0}F).

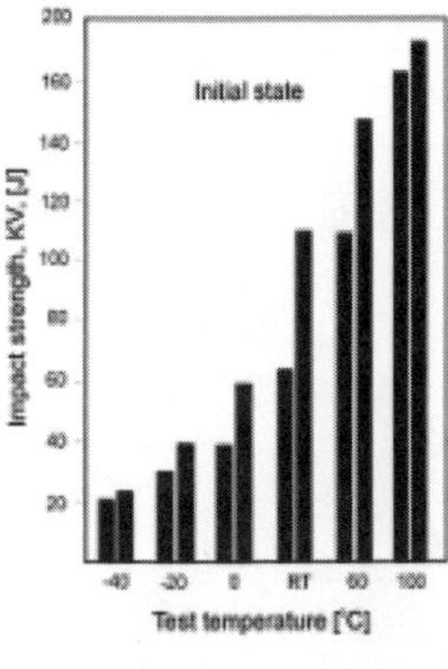

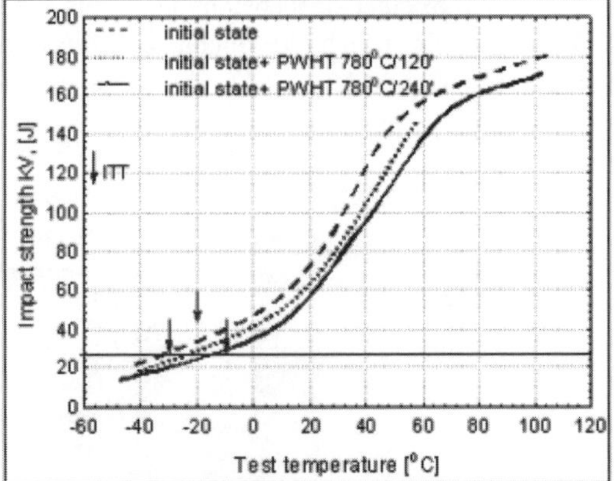

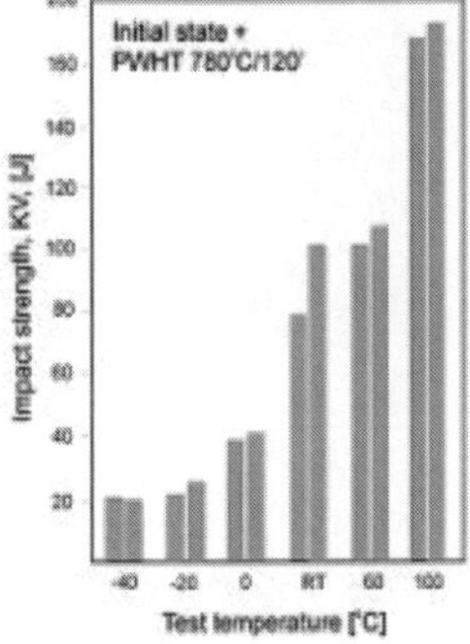

Material Stage	$KCV_{20°C}$	ITT [°C]
IS	70	-35
IS+780°C/120'	50	-10
IS+780°C/240'	60	-20
IS – initial state ITT – Impact Transition Temperature		

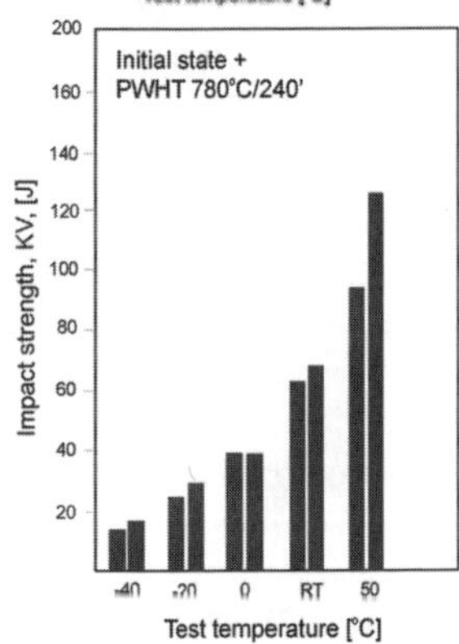

Fig. 10. Influence of PWHT stress relief annealing onto impact strength properties and transition temperature into brittle state alteration.

Owing to the steel martensitic structure, the temperature and technological parameters during welding and heat treatment have to be maintained very carefully. The cooling rate has to be selected in such a way that the martensitic transformation is fully accomplished. VM12 steel can be also characterised by high temperature of the martensitic transformation initial state Ms, i.e. above 350°C/662^{0}F, which depends on the heat / melt chemical composition. As the result of the diffusion processes, during heat treatment following welding there may occur brittleness in the HAZ of the welded joint. Such brittleness may contribute to the proneness of micro-crack occurrence. That is why selection of the appropriate temperature and stress relief annealing time is of crucial importance. Therefore, before the works on elaboration of the welding technology for welded joints made of VM12 steel, the company of Thyssen Welding carried out a thorough verification of welding and heat treatment parameters for the welded joints to be made. The influence of these parameters on impact strength and decrease of

impact transition temperature (ITT) into brittle state are shown in Fig. 10 [10], have been compared to ITT, shows in Fig 7.
After the welding technology verification it was in turn adapted to the manufacturing conditions of a large manufacturer of boiler components. Based on the elaborated welding technologies, similar welded joints of pipe made of VM12 steel with dimensions ϕ140x10 mm and ϕ355.6x35mm, were made for typical "PC" and "PF" welding positions. A graphical presentation of the circumferential welded joints fabrication is shown in Fig. 11.

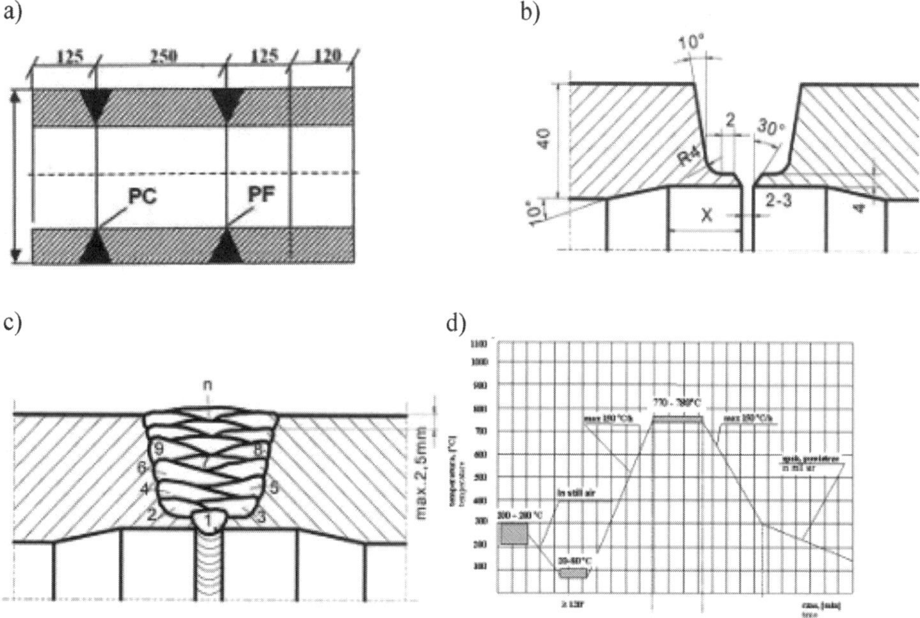

Fig. 11. Fabrication of similar circumferential welded joints on thick-walled tube ϕ355.6x35 mm as an example.

a) welding position and type of welding location,
b) preparation of components for welding,
c) the way and sequence of bead laying,
d) the applied heat treatment parameters.

The welded joints were made with SMAW method (111), manually, with coated electrodes type Thermanit MTS 5 Co T, and root pass made by GTAW method (141) with wire ϕ 2.4mm type Thermanit MTS 5 Co T – on argon shield. All welded joints have been PWHT after welding at parameters 770 – 780^{0}C(1418-1436)/ 120min, as show on Fig. 11d.
The non-destructive examinations were performed on similar welded joints made on pipes ϕ140x10mm and ϕ355.6x35mm in positions PC and PF, on 100% of their length and space.
The applied method and criteria of defects acceptance were in accordance with EN ISO 15614-1 and WBL 439 – Vallourec & Mannesman standard. No defects discontinuities, cracks, hot tears or inadmissible undercuts were found. The results of radiographic examinations, ultrasonic and magnetic particle examinations follows a/m standard.
At Fig. 12 there is present the programme of destructive examinations of similar welded joints made on VM12 steel tubes, as qualification of welding procedures required by EN ISO 15614-1 standard.

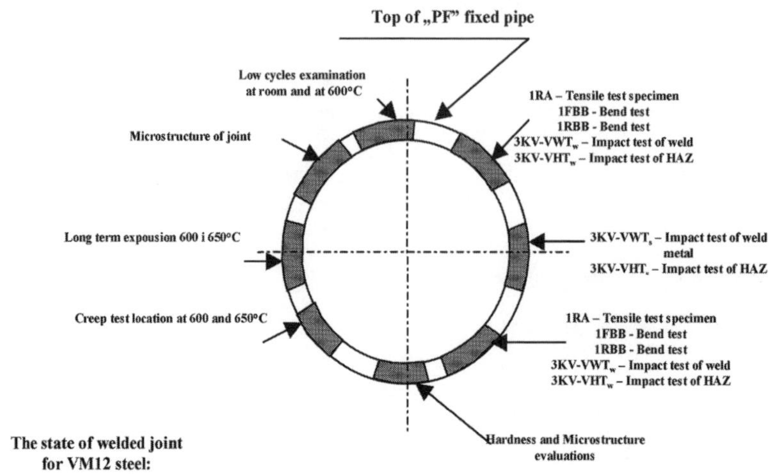

Fig. 12. Programme of destructive examinations of similar welded joints ϕ140x10mm and ϕ355.6x35mm made of steel X12CrCoWVNb12-2-2 (VM12).

4.1. Technological tests performance

Evaluating of welded joints, verification of their technological properties plays a very vital part. An example of the technological examinations is the bending test carried out according to EN ISI 15614-1 standard. The bending angle achieved was measured 130° (face and root tension) without any occurrence of cracks, i.e. The satisfactory result of technological examination was obtained. The results of the test are presented in Fig. 13 as macro-photographs.

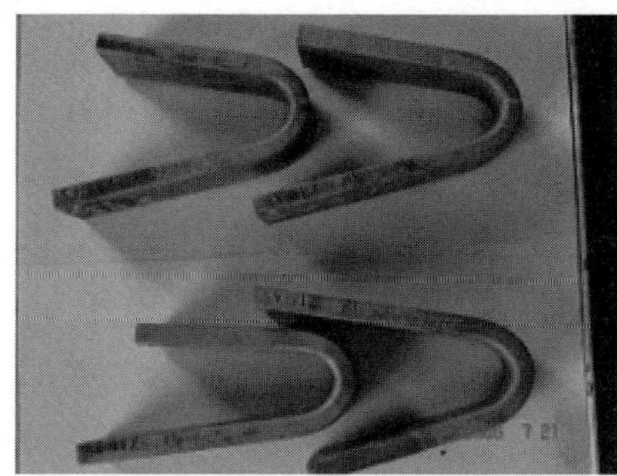

a) b)

Fig. 13. Test results of similar welded joint made of pipe VM12 steel.
a) ϕ355.6x35mm - a front view and b) ϕ140x10mm - a side view.

4.2. Structure and hardness distributions of similar welded joints

The test results of macro and microstructure on the light microscope and scanning electron microscope of the thick-walled tube ϕ355.6x35mm welded joint cross-section are shown in Fig. 14. In the macroscopic photograph of the welded joint cross-section (Fig. 14a), the mode and sequence of bead laying in the welding process can be seen. Fig. 14b shows the microstructure detected in the individual welded joint zones, in the parent material, the heat-affected zone and the weld metal, following welding and heat treatment parameters and chemical composition of specific welds area.

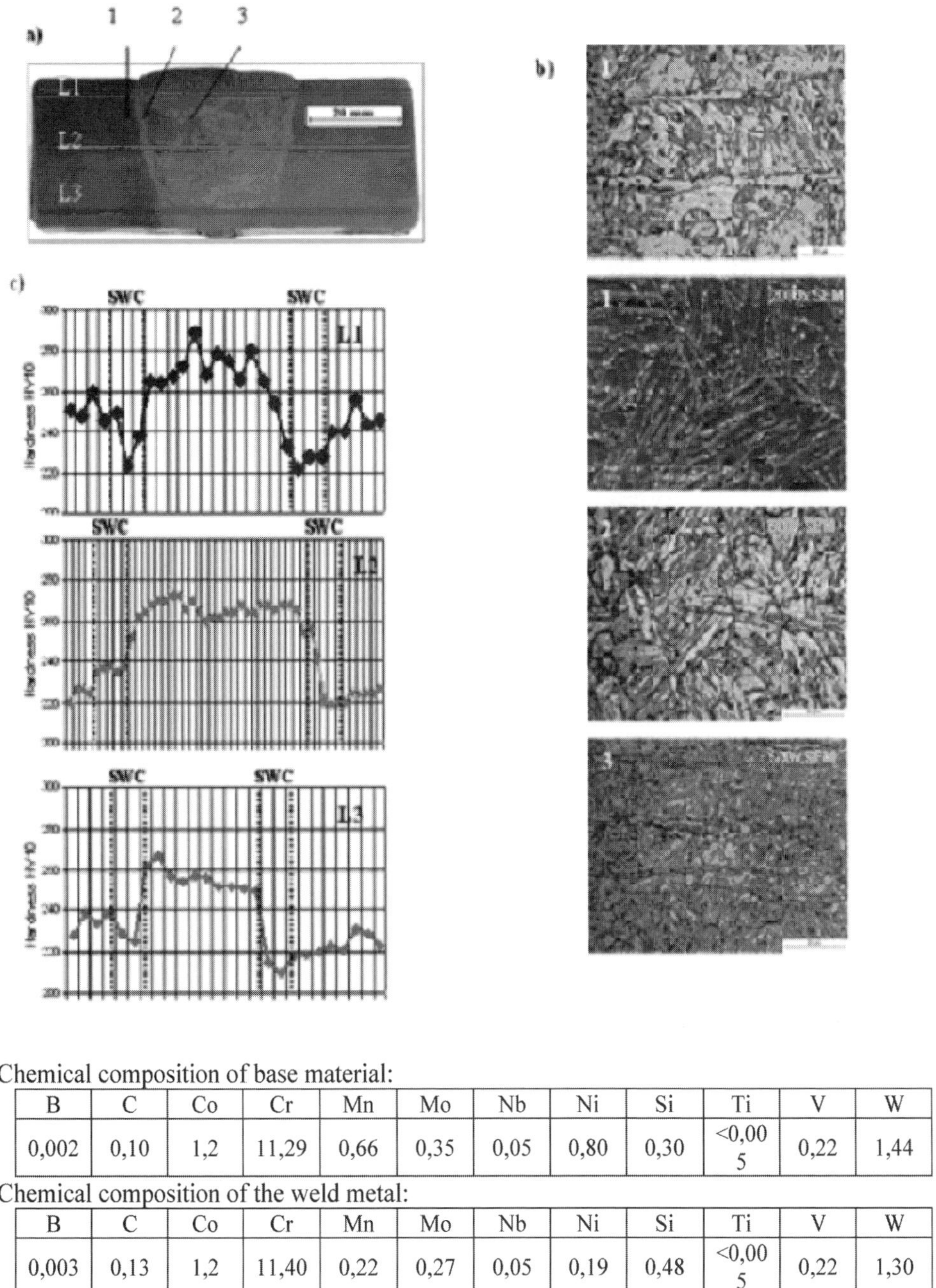

Chemical composition of base material:

B	C	Co	Cr	Mn	Mo	Nb	Ni	Si	Ti	V	W
0,002	0,10	1,2	11,29	0,66	0,35	0,05	0,80	0,30	<0,005	0,22	1,44

Chemical composition of the weld metal:

B	C	Co	Cr	Mn	Mo	Nb	Ni	Si	Ti	V	W
0,003	0,13	1,2	11,40	0,22	0,27	0,05	0,19	0,48	<0,005	0,22	1,30

Fig. 14. Test results of structure, hardness and chemical composition analysis of a similar circumferential welded joint made of thick-walled tube ϕ355.6x35mm material X12CrCoWVNb12-2-2 (VM12) steel.

The structure of the welded joints constitutes of tempered martensite and in the parent material together with elongated delta ferrite grains, whose quantity, depending on the testing location, may reach a few per cent share in the structure. In the HAZ and welded joint material, there are smaller amounts of delta ferrite. In the individual areas of the welded joint, the tempered martensite is slightly different in form and strip sizes, which is manifested in hardness differences between individual zones (Fig. 14c). However, the observed structure and hardness, as well as the chemical composition of the welded joint (Fig. 14) are compliant with the expectation. The hardness meets the requested criteria; lower than 350HV10 (Fig. 14c).

4.3. Mechanical properties of welded joints

The test results of tensile strength, which were carried out on flat test specimens taken out transversely to the welded joint, are shown in Fig. 15. Figure 16 shows the results of impact strength tests, with V-notch cut in parent material, HAZ and welded join respectively.
The tests results of welded joints on several welding method receive in Thyssen Welding are show in table 2 [10], but results of welded joints made by RAFAKO SA are demonstrated in table 3.

Table. 2. Test results of deposited metal of welded joints X12CrCoWVNb12-2-2 (VM12) steel obtained at Thyssen Welding company for various welding methods.

a) Mechanical properties of deposited metal

Welding method	Test temperature [°C]	HT°C/h	$R_{p0,2}$ [MPa]	R_m [MPa]	A_5 [%]	A_v (ISO-V) (+20°C) [J]	A_v (ISO-V) (+50°C) [J]
GTAW	20	770/2	767	906	18	50 23 32	-
SMAW	20	770/2	694	835	6	46 34 42	-
	50	-	-	-	-	-	57 66 58
	600	770/2	335	423	18	-	-
Sub-arc-W	20	770/2	688	819	18	36 48 46	-

a) Properties circumferential welded joint, pipe ϕ406.6x35mm

Test temperature [°C]	PWHT°C/h	$R_{p0,2}$ [MPa]	R_m [MPa]	Rupture point	A_v (ISO-V) [J]	Hardness [HV10]
20	770/4	576	765	Base Material	45 41 31	260
600	770/4	293	383		-	-

Table. 3. Test results of welded joints, X12CrCoWVNb12-2-2 (VM12) steel, pipe ϕ355,6.6x35mm obtained at RAFAKO SA. Base material and circumferential welded joint properties.

Test temperature 20 [°C]	PWHT°C/h	$R_{p0,2}$ [MPa]	R_m [MPa]	Rupture point	A_v (ISO-V) [J]	Hardness [HV10]
Base material	770-780^{0}C/ 2	571, 530	750, 712	BM	98, 78, 108 98, 42, 78	280
Weld "PC" position	770-780^{0}C/ 2	-	766, 791	Base Material	36, 60, 68 26,30, 40	220-298
Weld "PF" position	770-780^{0}C/ 2	-	739, 753		34, 38, 40 34, 40, 112	225-295

The results of tensile strength obtained in Thyssen Welding and RAFAKO SA are comparable and fully meet the requirements for the parent material. However, the obtained impact strength results are relatively low and, with regard to the welded joint material, just slightly higher than the required rupture energy value of 27 [J].

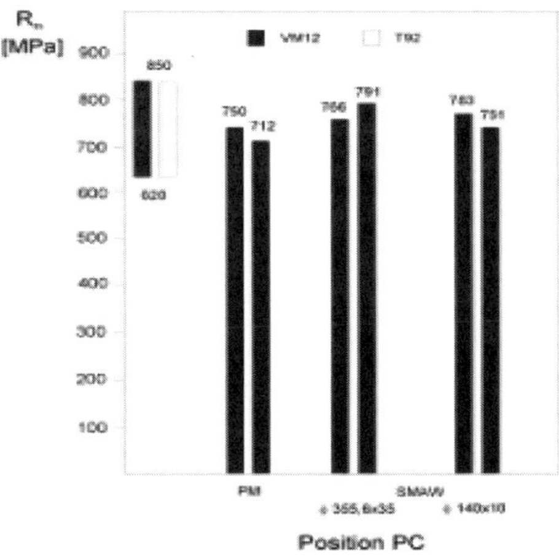

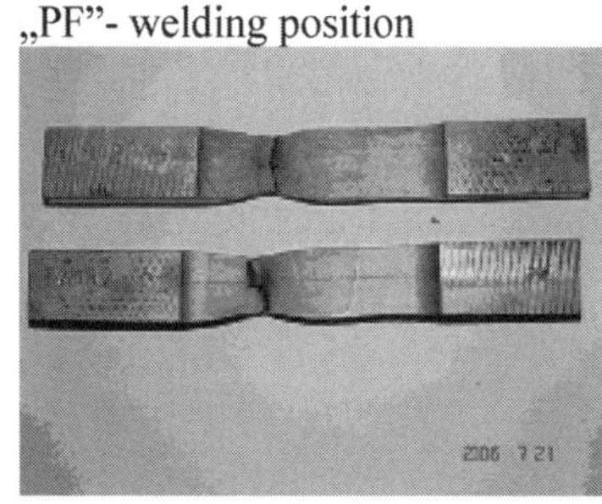

Fig. 15. Results of tensile strength tests for similar circumferential welded joints on pipe ϕ140x10mm and ϕ355.6x35mm made of X12CrCoWVNb12-2-2 (VM12) steel.

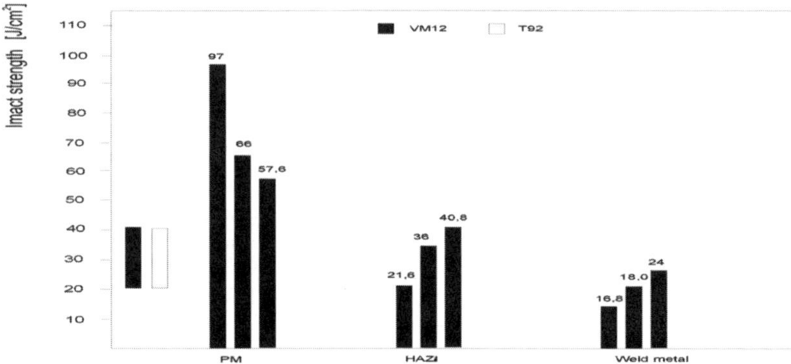

Fig. 16. Results of impact strength tests for similar circumferential welded joints (PM – base material, HAZ and weld metal) of pipe ϕ140x10mm made of X12CrCoWVNb12-2-2 (VM12) steel.

4.4. Assessments of creep resistance results

The creep testing programme included tests at the temperatures of 600 and 650°C (1112-1202^{0}F). The creep tests duration time reached approx. 5,000 h. The creep tests were performed with single-axial tension on test specimens with measurement diameter of d_0 = 5mm. The test results for the similar welded joint ϕ355.6x35mm, in comparison with the test results the welded joint parent material and the requirements for X10CrWMoVNb9-2 (P92) steel, presented in the form of function logσ = f (log t_r), at T_b = const are graphically shown in Fig. 17. Based on the creep tests which have already been executed, it can be concluded that the tested welded joint material does not always comply with the requirements set for VM12 steel, not even for the lower scatter band "–" 20%.

a)

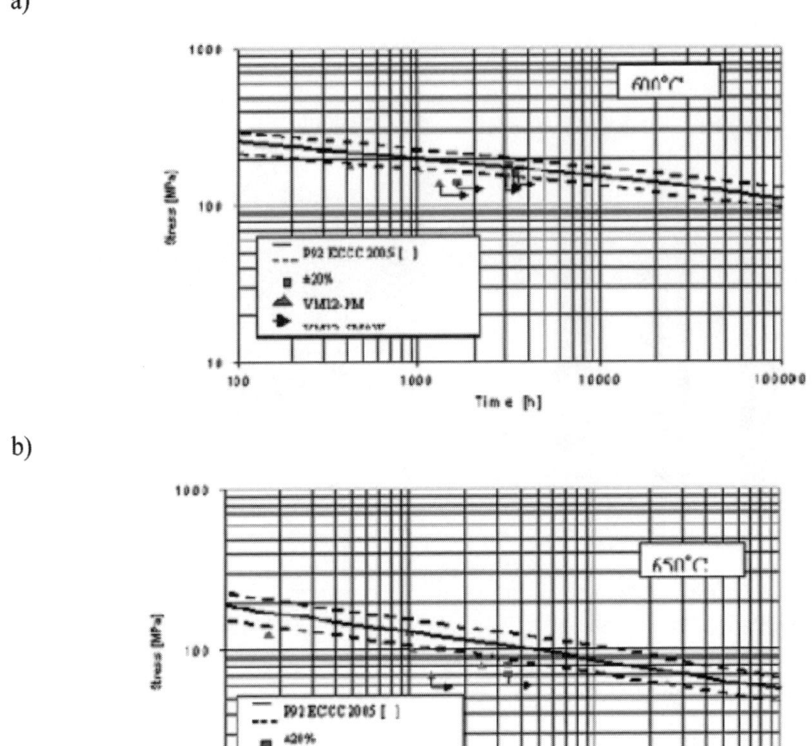

b)

Fig. 17. Comparison of the creep test results of parent material and similar welded joint on pipe ϕ355.6x35mm made of X12CrCoWVNb12-2-2 (VM12) steel with the requirements according to ECCC 2005 for X10CrWMoVNb9-2 (P92) steel grade.

4.5 Objective and results of X12CrCoWVNb12-2-2 steel after long-term annealing

In order to verify the structural stability as a result of long-term influence of the working temperature on welded joints made of VM12 steel, a simulation of the steel long-term annealing at the temperature of 700°C/1292^{0}F and 1,000h was carried out. This corresponds to the superheater tubes working conditions of approximately 100,000 hours at the temperature of 600°C. A comparison of the obtained test results of HV10 hardness for a welded joint made on pipe ϕ355.6x35mm after PWHT and annealing at the temperature of 700°C/1292^{0}F and 1,000h, line L1 are shown in Fig. 18(a) and hardness measured on line L2 see Fig 18(b) [6].

The long-term annealing at such temperature decreased the welded joint hardness by approx. 50HV10. This is due to the changes in microstructure concerning the activation of release processes in the matrix of tempered martensite (Fig. 19) and on martensite – ferrite δ interphase boundary. Additionally, the research works on the release processes development were also carried out. The radiographic analysis, apart from carbides $Cr_{23}C_6$, precipitations also revealed VC, WC and the presence of Laves phase following the long-term annealing. No Z phase, so often described in the literature, was detected.

a)

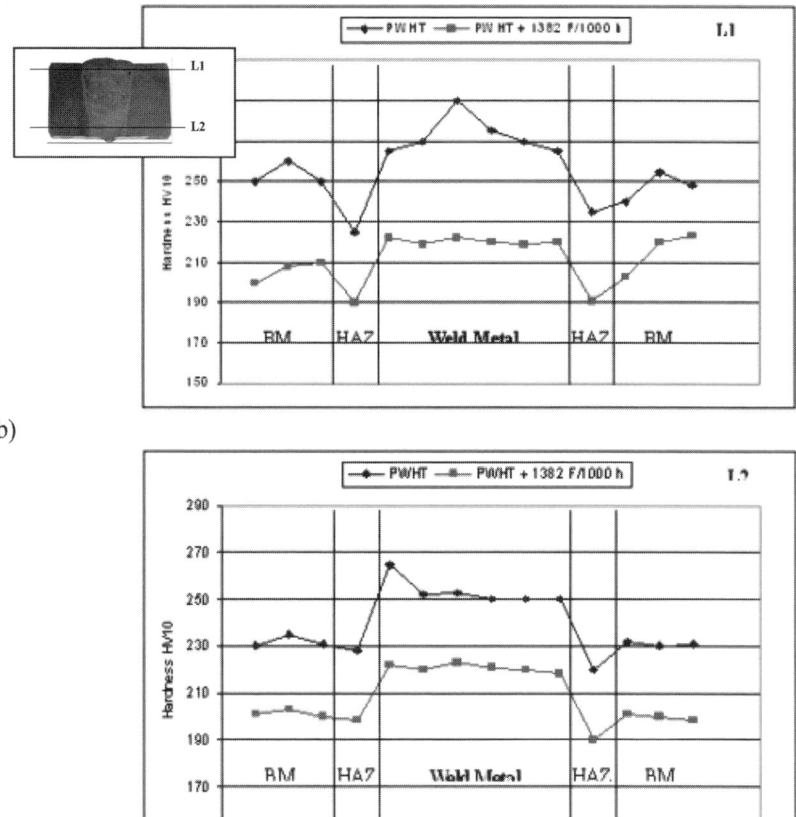

b)

Fig. 18. Welded joints hardness distributions of similar welded joint, pipe ϕ355.6x35mm made of X12CrCoWVNb12-2-2 (VM12) steel after welding and annealing at the temperature of 700°C/ 1,000h - line L1 – Fig. (a) and line L2 – Fig. (b).

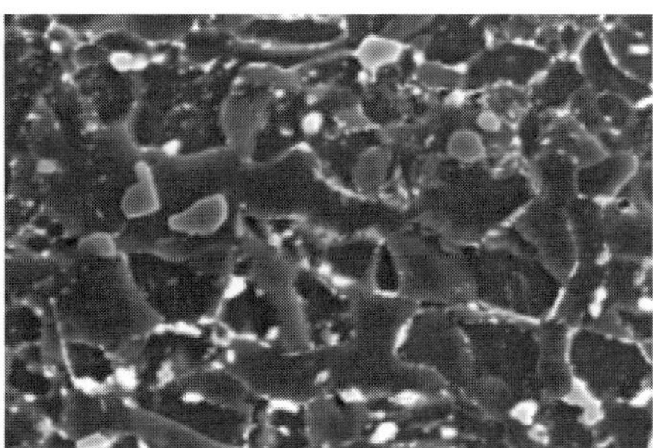

Fig. 19. Microstructure of similar welded joint of pipes Ø355.6x35mm following PWHT and additional annealing at the temperature of 700°C/1292^{0}F and 1,000h.

5. Future objectives

In order to evaluate steel mechanical properties and structure in view of temperature and time parallel influence, long-term annealing at the temperature of Tb = 600 and 650°C (1112-1201^{0}F) is carried out for the same tube material following operation lasting 3,000; 5,000; 10,000hrs and considerably over 10,000h. On the specially prepared materials the examinations regarding the structural changes, development of release processes and changes in the mechanical properties will be executed. The test duration time exceeded 4,500 hours. The obtained results will be presented in next publications.
The test programme within this scope is presented in Table 4. The examinations will focus on the changes in mechanical properties in relation to those in the microstructure.

Table 4. Basic assumption of the test programme regarding the evaluation of stability, mechanical properties and structure of steel owing to simultaneous time and temperature influence.

Base material: X12CrCoWVNb12-2-2 (VM12)	Expose time t_b [h]	1,000	5,000	10,000	30,000
	Test temperature T_b [°C]	Scope of examination: - tensile strength, yield strength, elongation, - impact energy, - structure and hardness investigations			
	600				
	650				

Detailed results of the tests will be presented in the next publications.

6. Conclusions

1. The typical structure for the chemical composition of the tested X12CrCoWVNb12-2-2 steel, after normalising and annealing, is tempered martensite with little amount of delta ferrite. Relatively high temperature of the initial stage of martensitic transformation M_s requires that the course of the welding technology has to be very careful. Cooling of the welded joints has to be carried out below the line of the end of martensitic transformation M_f, which assures martensitic structure in the entire joint area.
2. The obtained strength properties test results of steels tubes X12CrCoWVNb12-2-2 are similar to those of P91 and P92 steel tubes. However, the rupture energy of VM12 steel tube material at room temperature is approx. 70 [J] on average and is considerably lower than that of P91 steel, which reaches approx. 250 [J].
3. The executed examinations of the thick-walled tube circumferential welded joints, in a large boiler manufacturer's conditions, have produced a positive result and are comparable with those obtained at Thyssen Welding Hamm and for the parent material.
The obtained strength test results are compliant with the requirements for the parent material, although the impact strength results are slightly higher than the required rupture energy of 27[J].
4. The executed evaluation of the welded joints in the bending technological test also produced a positive result, in which the angle of 130° was reached without occurrence of cracks.
5. The structure of the homogeneous circumferential welded joints constitutes of tempered martensite with elongated delta ferrite grains, whose quantity is variable depending on the test location and amounts to several % share in the structure. The hardness of the welded joint and HAZ meets the requirement of 350 HV10 after welding and PWHT.

6. The creep resistance at the temperatures of 600 and 650°C (1112-1201^{0}F), based on tests up to approx. 5,000 hours, does not always meet the requirements set for X10CrWMoVNb9-2 steel, being placed in the lower admissible spread limit –20% from the average value. This particularly concerns the test results of homogeneous circumferential welded joint material.
7. Long-term annealing at the temperature of 700°C/1292^{0}F throughout 1,000h resulted in slight lowering of the welded joint hardness by 50HV10. This is the effect of structure changes related to the development of release processes in the tempered martensite matrix and martensite-ferrite delta inter-stage boundary. Apart from carbides $Cr_{23}C_6$, VC and WC releases, the occurrence of Laves phase has been detected.

7. References

1. Dobrzański J., Complex evaluation and further service forecasting of the power boilers pressure part critical elements in creep service. Report IMŻ N-00322/BM/00, unpublished (in Polish).
2. Dobrzański J., The classification method and the technical condition evaluation of the critical elements material of power boilers in creep service made from the 12Cr-1Mo-V, Journal of Materials Processing technology 164-165 (2005) 785-794.
3. Masuyama F.: „Steam Plant Material Developments in Japan". 6th Int. Conf. on Materials for Advanced Power Engineering 1998, Liege, Belgium. ISBN 389336, Forschungszentrum Jülich 1998, p.1087.
4. Dobrzański J., Zieliński A.: Properties and structure of the new martensitic12% Cr steel with tungsten and cobalt for use in ultra supercritical coal fired power plants, Inżynieria Materiałowa nr 3-4 (157-158), str. 134-137, maj-sierpień 2007.
5. Maciosowski A., Dobrzański J., Zieliński A. „Mechanical and creep testing of new steel and welds for critical elements of environmentally friendly Power Plant" Report No3 SPB/COST/96/2006. Cost 536, Gliwice 2006, Poland.
6. Dobrzański J., Pasternak J., Zieliński A. „Properties of Welded Joints of Martensitic Creep Resistance Steels Applied in Polish Power Plants". 3rd Int. Conf. on Integrity of High Temperature Welds. April 2007, London, United Kingdom.
7. Gabral J., Bendick W., Vanderberghe B., Lefabrre B: „Status of development of the VM12 steel for tubular applications in advanced power plants". 8th Int. Conf. on Materials for Advanced Power Engineering 2006, Liege, Belgium. ISBN 389336-436-6, Forschungszentrum Jülich 2006, p.1065.
8. Zielińska-Lipiec A.: „The Analysis of Microstructural Stability of Modified Martensitic Deformation". AGH 2005 Kraków ISSN 0867-8631, Poland.
9. Zielińska-Lipiec A. and other: "Microstructural development of VM12 steel caused by creep deformation at 625°C". 8th Int. Conf. on Materials for Advanced Power Engineering 2006, Liege, Belgium. ISBN 389336-436-6, Forschungszentrum Jülich 2006, p.1077.
10. Gross V., Heuser H., Jochum C. "Neuartige Scheisszusatze fur Bainitishe und Martenzitishe Stahl. Thyssen Welding Proceedings

Advances in Materials Technology for Fossil Power Plants
Proceedings from the Fifth International Conference
R. Viswanathan, D. Gandy, K. Coleman, editors, p 320-337

DOI: 10.1361/cp2007epri0320

ALLOY 33: UPDATE ON FIELD EXPERIENCE IN WATER WALLS AND SUPERHEATERS

Larry Paul
ThyssenKrupp VDM USA, Inc
122 E. Jefferson
Tipton, IN 46072

Abstract

Alloy 33 is a weld overlay material that has generated a lot of interest in the fossil boiler industry. The high chromium content of Alloy 33 has been shown to provide excellent corrosion protection in both waterwall and superheater/reheater tube applications. For waterwall applications, the corrosion resistance has been demonstrated in both laboratory and field tests conducted over the last 5 years. In addition to corrosion resistance, the Alloy 33 has also shown that it is also resistant to cracking (although no material is 100% immune). In the superheater/reheater, the use of spiral clad weld overlay tubes is able to provide resistance to excellent coal ash corrosion. Laboratory and field tests have shown Alloy 33 to have among the best corrosion resistance of all materials studied. The application of Alloy 33 is also easier than other more highly alloyed materials (such as FM-72) and is less expensive. As a result of these favorable experiences, Alloy 33 is now being used commercially to weld overlay both waterwall and superheater/reheater tubes on fossil boilers.

Introduction

The coal-fired boiler environment produces corrosive conditions in both the furnace region and in the boiler convection pass. The use of corrosion resistant weld overlay materials has become widely used to manage the corrosion issues in coal-fired boilers.

In the furnace region the waterwall tubes are attacked by a mixed sulfidation/oxidation mechanism that causes corrosion rates of up to 80 mpy (2 mm/y) (1,2). The highest corrosion rates are seen in boilers that combine low NOx combustion with burning high-sulfur fuels. This results in the formation of corrosive H_2S gases and FeS deposits on and around the boiler tubes. An additional concern for furnace wall tubes is cracking. What has been termed circumferential cracking has been observed in furnace tubes for many years and has been shown to be a corrosion fatigue mechanism (3,4).

Table 1 . Nominal chemical composition of commonly used weld overlay materials.

Alloy	UNS No.	Ni %	Cr %	Mo%	Fe %	Other %
625	N06625	60	22	9	3	3.4Nb
622	N06022	56	22	13	3	3W
309 SS	S30900	13	23	-	62	
FM52	N06052	57	30	-	9	0.5Ti 0.5Al
33	R20033	31	33	1	32	0.45N 0.7Cu
FM72	N06072	56	44	-	0.2	0.6Ti

This cracking mechanism has been mostly restricted to supercritical boilers, which operate at the highest pressures (and therefore the highest tube temperatures).

In the convection pass, coal ash corrosion can attack the re-heater and superheater tubes, the traditional mechanism involves molten salt corrosion caused my molten alkaline iron tri-sulfate. However, some recent studies have also seen other corrosion mechanisms that involve a carburization from of attack.

In both the case of furnace wall tubes and convection pass tubes, the use of a high chromium weld overlay material has provided the best protection. The use of Alloy 622 with controlled chromium content above 22% has been widely used for protection of waterwall tubes. Likewise use the high chromium FM-72 (45% Cr) has been the preferred material to combat coal ash corrosion in convection pass tubes. Because chromium is a key element in resisting corrosion in both of these boiler applications, a new alloy with 33% chromium, know as Alloy 33 is now being used as an overlay material on both waterwall and convection pass tubing (Alloy 33 is also known by the trade name Nicrofer® 3033[1]). Table1 gives the nominal chemical compositions of the most common weld overlay materials now being used in coal fired boilers.

Waterwall Tubing Laboratory Data

The corrosion of waterwall tubes was simulated in the laboratory by exposing coupon samples of weld metal to an atmosphere of is 82.88% N_2 + 10% CO + 5%CO_2 + 2% H_2O + 0.12% H_2S. Various tests were conducted using isothermal, thermal cycling, and alternating oxidizing potentials, details of these tests are reported elsewhere (5,6). An example of the data from this study is shown in Figure 1 and is typical of the results from the laboratory tests. It was clear from the laboratory studies that higher chromium content weld overlay alloys generally showed improved resistance to corrosion in this environment.

[1] Nicrofer® is a trademark of ThyssenKrupp VDM GmbH

Another major finding of the laboratory data was the influence of molybdenum on the corrosion morphology. When alloys with higher molybdenum contents are used for weld overlay, the molybdenum segregates in the interdendritic region of the weld solidification structure. This segregation affects the corrosion morphology. (7) The corrosion morphology Alloy 622 (13%Mo) is compared to Alloy 33 (1%Mo) in Figure 2. It can be seen that that the corrosion, a mixture of oxidation and sulfidation, preferentially follows the dendrite cores in the Alloy 622 weld overlay but this preferential attack is not seen in Alloy 33. Electron microscope images along with corresponding dot maps are shown for Alloy 622 and Alloy 33 in Figures 3 and 4.

It is believed that this corrosion morphology also has an influence on cracking of the overlay materials, as described in the following section.

Waterwall Tubing Field Testing

Extensive laboratory testing showed that higher chromium containing materials generally had better corrosion resistance; in addition molybdenum appeared to offer no benefit to overall corrosion resistance and could even be harmful. In order to demonstrate the corrosion behavior seen in the laboratory, and in particular the excellent corrosion resistance shown by Alloy 33, field testing was initiated. The field tests were also selected to evaluate the cracking resistance of Alloy 33, since the cracking mechanism is quite complex and difficult to reproduce under laboratory conditions.

Multiple boilers were used to evaluate the corrosion and cracking resistance of Alloy 33. The selected boilers represent a wide spectrum of operating conditions and boiler designs. Several boilers where circumferential cracking has occurred on other materials were intentionally selected to ensure a proper evaluation of the Alloy 33. The boilers used in the field testing are shown in Table 2 along with a description of the location and type of sample used.

The field test samples were inspected when a scheduled boiler outage allowed access to the samples. Samples were removed periodically for evaluation when possible. To date the field samples have been in service for 3 years and tests are planned to continue for several more years in order to track the performance over long periods of time.

After 3 years the corrosion resistance of Alloy 33 has been clearly demonstrated in all of the boilers used in the test program. The corrosion rate of Alloy 33 has been less than 1 mpy in all cases. This would translate into a corrosion life of over 70 years for a typical Alloy 33 weld overlay. Figure 5 shows a cross section of a weld overlay removed after 18 months of exposure form Boiler A, this sample was etched to reveal the dendritic structure of the weld overlay. Note that there is a thin uniform and continuous chromium oxide layer on the Alloy 33. Furthermore, the oxidation does not

Table 2. Boilers used for field testing of Alloy 33 weld overlay and location of samples.

Boiler	Boiler Type	Test Panel	Location	Sample Removed
A	Supercritical - Tangential	21″ x 19″	OFA level	18 Months
A	Supercritical - Tangential	21″ x 19″	OFA level	34 Months
B	Supercritical - Tangential	18″ x 240″	OFA level	22 Months
C	Drum (2800 psi) - Tangential	20″ x 120″	10′ above burner	23 Months
D	Supercritical - Tangential	Hand Weld Safe Ends	Between Burner & SOFA	34 Months
E	Supercritical - Opposed Burner	7″ x 21″	Between Burner & SOFA	23 Months

preferentially follow the dendritic structure of the weld, as seen in other high molybdenum, alloys.

Another goal of the field test program was to evaluate the resistance of Alloy 33 to circumferential cracking. There was no observed cracking of the Alloy 33 weld overlay in any of the boilers except one. The one boiler where cracking was observed is not typical regarding neither the weld overlay nor the boiler operational history, as will be shown.

Boiler A - Samples were removed after 18 and 34 months for laboratory inspection. The samples were typical of a field applied weld overlay. The samples were covered with fly ash and deposits, the corrosion scale was predominantly chromium oxide with some evidence of iron sulfide also present, and the weld overlay showed chemistry typical for a weld overlay on alloy steel tubes.

Figure 6 shows an Alloy 33 weld overlay sample removed after 34 months. The bare ends of the tube panel are also evident in this picture. The base tubing was SA213-T11 (1/¼ Cr-½ Mo). Inspection revealed cracking in the bare base metal tubes; these cracks were about 0.020″ (0.5mm) deep and have the typical oxide wedge in the crack with a sulfur "spine" running down the middle of this oxide wedge. This type of circumferential cracking is a form of corrosion fatigue. No cracking was seen in the Alloy 33 overlay directly adjacent to this area. Figure 7 shows micrographs of the cracks in the bare SA213-T11 tubing along side the weld overlay from this same region. One shallow defect was observed that is 0.001″ (0.025mm) deep on the Alloy 33, this is thought to be a weld defect as furthered examination of this overlay at two different independent laboratories found not evidence of cracking of the Alloy 33. Therefore, the Alloy 33 was shown to resist cracking even when in an area known to produce circumferential cracking.

Boiler B - Alloy 33 was tested directly alongside of Alloy 622, both weld overlays were applied at the same time on a 20 foot long full tube panel; 10 tubes were covered with each alloy. This test panel is shown in Figure 8. After 22 months of operation, samples of both alloys were removed and examined from this test panel. Both alloys had very low corrosion rates, however the Alloy 622 developed oxide lobes that were seen to penetrate into the dendrite cores; this is shown in Figure 9.

Development of oxides lobes has been associated with the initiation of cracks in weld overlay materials (4). The association of oxide lobes and the onset of cracking are shown schematically in Figure 10 (after reference 4). During exposure in the boiler environment over time (t) the overlay reacts to form oxide/sulfide scales, which can form lobes in some materials, such as those with high molybdenum. First the oxide lobes are developed (t_1 - t_3) and later these lobes can become linked together (t_4.). Tensile stresses arising from operational variables, such as slag sheds, sliding pressure, etc., will cause theses oxides to crack(t_5.); oxides are ceramic materials which are know to have poor ductility and can crack easily under tensile stresses. These cracks allow sulfur species to penetrate the oxide later and cause the sulfur "spine" observed in these cracks. It is felt that the absence of these oxides lobes on Alloy 33 will lead to improved resistance to circumferential cracking.

Boiler C - Alloy 33 was tested along side Alloy 622 in this lower pressure boiler. This boiler had not experienced any cracking issues of any materials used to date; likewise neither Alloy 33 nor Alloy 622 showed any signs of distress. The corrosion rates were very low for all of the materials. far less than 1 mpy (0.02 mm/y) Chromized tubing was used for many years in this boiler, but had to be replaced after several years when the relatively thin coating was breached on a wide scale basis.

Boiler D - Alloy 33 was used to cover the bare tubes where joint welds were made to join tube panels into the boiler waterwall. These welds were performed by hand in the field and were not typical of automatic overlay welds used to cover most of the boiler tubes. Boiler D had previously tested Alloy 625, 622, FM-52, Type 309 and 312 stainless steels, as well as bare SA2313-T11 tubes. All of these materials have cracked in this boiler. Therefore, this boiler is known to be a particularly aggressive boiler with respect to cracking. Likewise, Alloy 33 hand welds showed evidence of thin tight cracks after 36 months of service in this boiler. Unfortunately, these test samples were placed back into service without obtaining samples for laboratory evaluation.

The experience in Boiler D shows that while Alloy 33 is very resistant to cracking, it is not immune. There are many factors that influence the cracking behavior of a weld overlay in a particular boiler. Figure 11 shows a schematic that describes the material and operational factors that affect the onset of circumferential cracking in a particular boiler. Cracking of weld overlay boiler tubes is a function of both mechanical factors and environmental factors. Mechanical factors include global stress due to boiler design and operating practices as well as local stresses resulting from operating temperature, thermal expansion of the overlay, thermal conductivity of the overlay, and geometry effects due to application and oxidation of the weld overlay (such as described in Figure 10). Environmental factors include coal chemistry, combustion practices, deposition, slagging behavior, temperature, flame stability and anything else that influence the local environment adjacent to the weld overlay.
The complexity of circumferential cracking was one reason why field tests were conducted rather than laboratory tests. After the testing done to date, it is clear that most boilers operate in a regime where cracking concerns are moderate to low but that a small handful of boilers clearly will have problems regardless of tube metallurgy. The most problematic boilers are those with the fireball near to the tube walls, those that burn high sulfur fuels, and those that operate near or above rated furnace capacity.

Boiler E - Alloy 33 was installed as a small panel and was visually inspected after 24 months of operation. There was no evidence of cracking and no visible signs of corrosion on the weld overlay. No samples were taken during this 24 month outage.

Waterwall Tubing Commercial Deployment

With the favorable laboratory and field data for Alloy 33 as a weld overlay for waterwall tubes, a full scale deployment was scheduled at Western Kentucky Energy (WKE) in the spring of 2007 (8-10). This is a drum boiler that recently began burning opportunity fuels, including petroleum coke. This change in fuel along with retrofitting the boiler with low NOx burners resulted in accelerated tube wastage of the SA210-A1 carbon steel boiler tubes. Therefore, a protective weld overlay was required to prevent failure of the pressure parts in the furnace region. Alloy 33 was considered along with several other materials. Because the corrosion resistance of the Alloy 33 was better in the high sulfur environments that the other candidate materials (Alloy 622 and 625) and because of the lower cost, it was selected as the material for this weld overlay project. The lower cost of Alloy 33 allowed WKE to cover more of the furnace region within their allotted budget than would be possible with the other materials.

Approximately 8,000 square feet of the furnace was covered with Alloy 33 overlay, requiring about 40,000 pounds of weld wire. Figure 12 shows the application of the Alloy 33 inside of the boiler at WKE. The weld overlay was applied using multiple automatic welding machines and was completed within 6 weeks.

Superheater and Reheater Tubing Laboratory Data

Because of the high chromium content of Alloy 33 and the ability to apply it to tubes as a weld overlay, interest was developed in using this material for protection from coal ash corrosion in the convection pass. Of coal fired boilers. Laboratory testing demonstrated the excellent corrosion resistance to coal ash corrosion of the Alloy 33 weld overlay (11). The excellent resistance of Alloy 33 to coal ash corrosion is not surprising based on the findings of other researchers.

The benefit of high chromium content in an alloy to combat coal ash corrosion has been recognized for years (12-15). An example of a recent field test which shows the general effect of chromium on reducing coal ash corrosion is shown in Figure 13. The first and obvious conclusion drawn from Figure 13 (and the work of others) is that increasing the chromium content of an alloy increases the resistance to coal ash corrosion. The second observation is that above about 30% Cr there are diminishing effects of further chromium additions to the alloy. The third observation is that there is no data points between 27 and 42% chromium, this is because there have been no real commercially available alloys in this range of chromium composition. Alloy 33, which contains 33% chromium, clearly has sufficient chromium to impart excellent resistance to coal ash corrosion. In addition, the Alloy 33 has excellent weldability and fabrication qualities, far better than the highest chromium alloys such as FM-72.

Superheater and Reheater Tubing Field Testing

Field testing was conducted by Alstom Power by using air-cooled corrosion probes. These air-cooled probes were exposed in the convection pass region of a coal-fired boiler where coal ash corrosion has been observed in the past to be a problem. After 1294 hours of exposure the probes were removed and examined. The Alloy 33 had corrosion resistance roughly equivalent to FM-72 in this test. A cross section of the Alloy 33 and FM-72 samples from this study are shown in Figure 14. This field test is on-going and continues to show that the Alloy 33 has excellent resistance to coal ash corrosion.

Superheater and Reheater Tubing Commercial Deployment

In the summer of 2007, an eastern utility decided upon Alloy 33 as the overlay material for a superheater replacement project. Approximately 7,000 linear feet of tubing was covered with Alloy 33 in by a weld overlay supplier in their shop. This required about 28,000 pounds of Alloy 33 weld wire. A 360° spiral overlay was used that covered the entire outside surface of the tubing; the base tubing was a mixture of SA213—TP347 and SA213-T22. An example of the spiral overlay of Alloy 33 is shown in Figure 15.

This project is still underway as of this writing. However, bending trials were successful and manufacturing procedures were developed for the full scale production and installation of the Alloy 33 superheater tubes.

Acknowledgments

The contributions of Lehigh University in performing the laboratory testing in the low NOx environment was an essential part of the information in this paper, in particular the contributions of Dr. John DuPont, Dr. Arnold Marder, and Ryan Deacon are much appreciated.

References

1. Wate Bakker & S.C. Kung, "Waterwall Corrosion in Coal Fired Boilers a New Culprit: FeS", paper 00246, CORROSION 2000, , NACE, Houston, TX, 2000

2. S.C. Kung & L.D. Paul, "Corrosion of Waterwall Tube Materials in Low NOx Combustion Systems", Paper 65, CORROSION 91, NACE, Houston, TX, 1991

3. H.J. Cialone, I.G. Wright, R.A. Wood, "Circumferential Cracking of Supercritical Boiler Water wall Tubes", Electric Power Research Institute, Palo Alto, CA, 1986

4. K. Luer, J. DuPont, A. Marder, C. Skelonis, "Corrosion Fatigue of Alloy 625 weld claddings in Combustion Environments", Materials at High Temperature, 18 (1), pp 11-19, Science Reviews, 2001

5. Larry Paul, Gregg Clark and Andreas Ossenberg-Engels. Protection of Waterwall Tubes from Corrosion in a Low NOx Coal Fired Boiler, Presented at Coal-Gen, Cincinnati, OH, August 16-18,, 2006

6. Larry Paul, Gregg Clark and Dr. Michael Eckhardt, Laboratory and Field Corrosion Performance of a High Chromium Alloy for Protection of Waterwall Tubes from Corrosion in Low NOx Coal Fired Boilers, Paper 06473, Corrosion 2006, NACE, Houston, TX, 2006

7. R. Deacon, J. DuPont, A. Marder, "High Temperature Corrosion Resistance of Candidate Nickel Base Weld Overlay Alloys in a Low NOx Environment", Materials Science and Engineering A 460-461, pp 392-402, Elsevier, 2007

8. "Alloy 33 Weld Overlay Extends Steel Tube Life and Saves Money", pp13-14, Advanced Materials and Processing, August, 2007

9. "Overlay Your Welds", Power Magazine, P.65, July 2007

10. Larry Paul, Gregg Clark, & Andreas Ossenberg-Engles, "Alloy 33 Weld Overlay Extends Boiler Tube Life and Saves Money" Teresa Hansen, ed., Power Engineering, pp 64-69, September, 2007

11. Larry Paul & Gregg Clark, "Coal Ash Corrosion Resistance of New High Chromium and Chromium-Silicon Alloys", Paper 05453, Corrosion 2005, NACE, Houston, TX, 2005

12. S. Phillips, N. Shinotsuka, K Yamamoto, Y Fukada, "Applications of High Steam Temperature Countermeasures in High Sulfur Coal Fired Boilers", Babcock Hitachi, 2003

13. Babcock & Wilcox Company, Steam- Its Generation and Use, 40th edition, Babcock & Wilcox Company. Barberton, OH 1992, Pages 20-19 to 20-22.

14. J.B. Kitto, "Technology Development for Advanced Pulverized Coal Fired Boilers", BR-1626, presented at Power-Gen International '96, December 4-6, Orlando, FL, 1996.

15. J. L. Blough and W. W. Seitz, "Fireside Corrosion Testing of Candidate Super-Heater Tube Alloys, Coatings, and Cladding Phase II", Foster Wheeler Review and Heat Engineering, Foster Wheeler Development Corporation, 12 Peach Tree Hill Road, Livingston, NJ, 1999.

16. D.K. McDonald, "Coal Ash Corrosion Resistant Testing Program - Evaluation of the First Section Removed in November 2001", Babcock & Wilcox Company, DOE Project DE-FC26-99FT40525, Funded by the Department of Energy (DOE) and the Ohio Coal Development Office/Ohio Department of Development (OCDO/ODOD)

17. Juan Nava and Michael Quinlan, "Surface Technology Solutions for Coal Fired Utility Boilers", Coal-Gen, San Antonio, TX, 2005

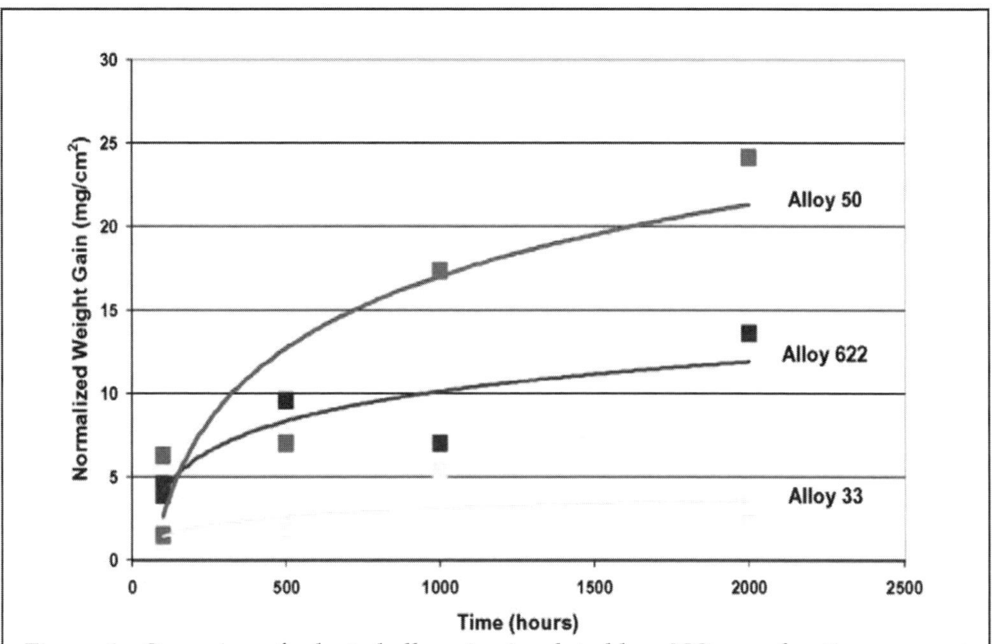

Figure 1. Corrosion of selected alloys in simulated low NOx combustion gases at 500°C (932°F); gas composition is 82.88% N_2 + 10% CO + 5%CO_2 + 2% H_2O + 0.12% H_2S

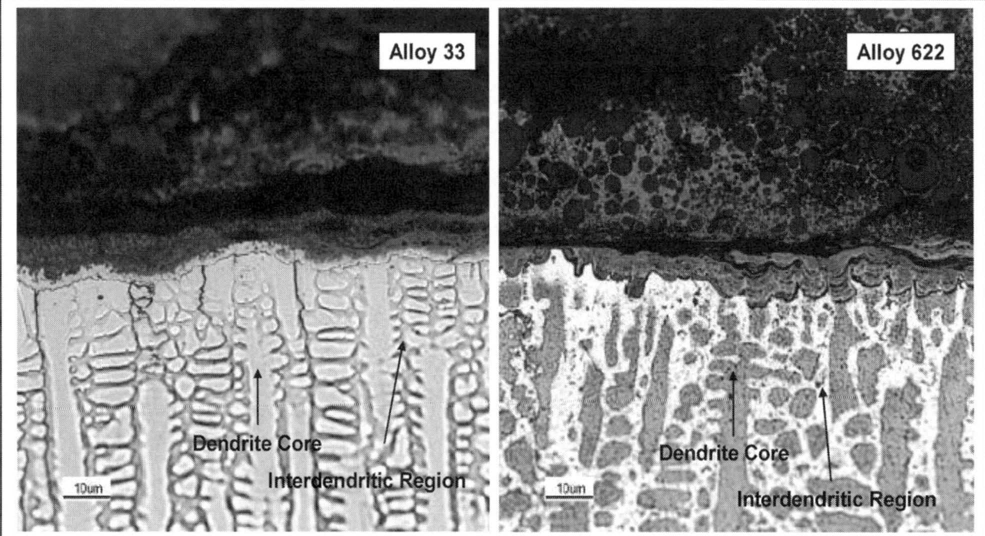

Figure 2. Corrosion morphology of preferential dendritic core attack caused by segregation of molybdenum in Alloy 622 (13% Mo), compared to the more uniform corrosion seen in the low molybdenum Alloy 33 (1%Mo).

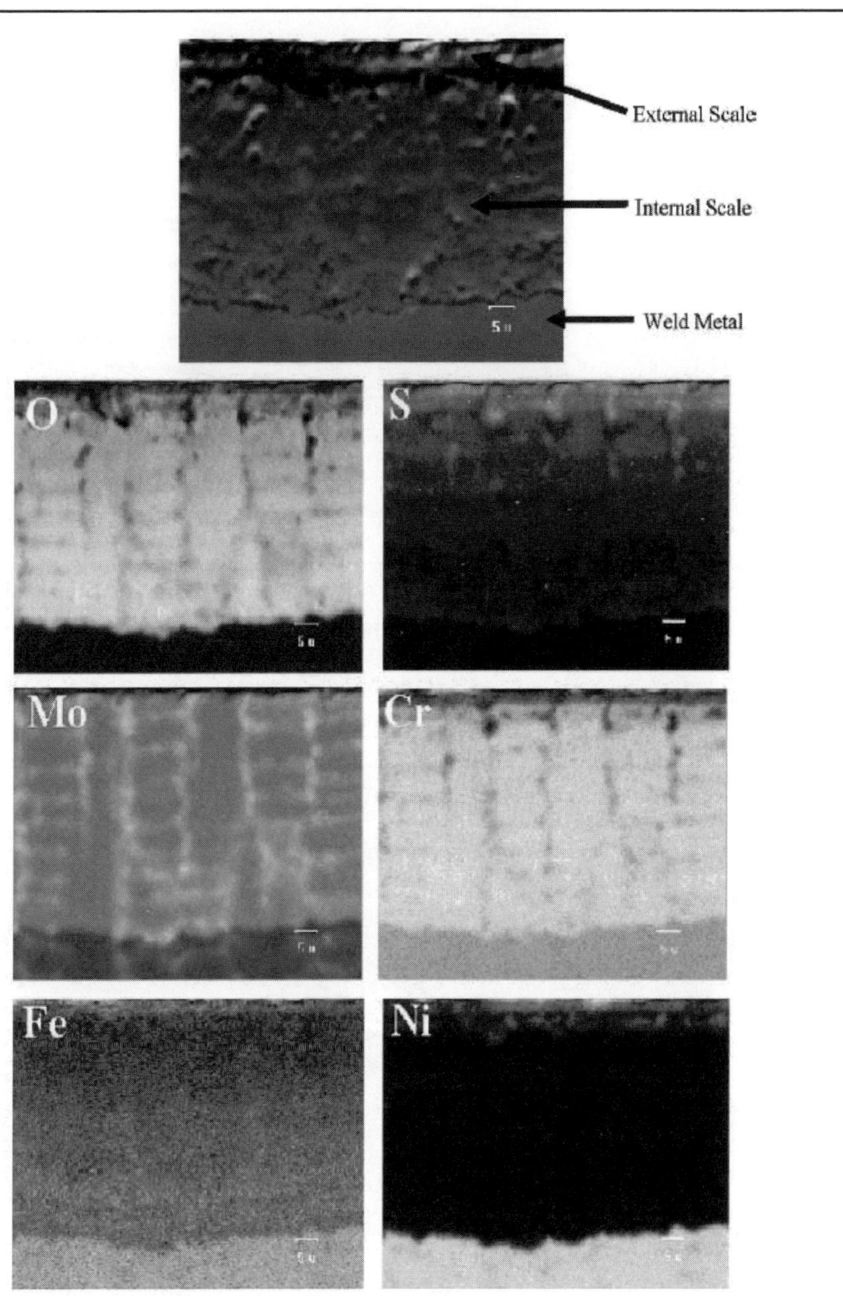

Figure 3. Corrosion morphology of the Alloy 622 weld overlay after 2000 hours of exposure to a simulated low NOx combustion gases. Note that the dendritic structure of the weld deposit is also seen in the corrosion scale with chromium oxides being predominant at the dendrite cores and molybdenum sulfides occurring in the interdendritic region.

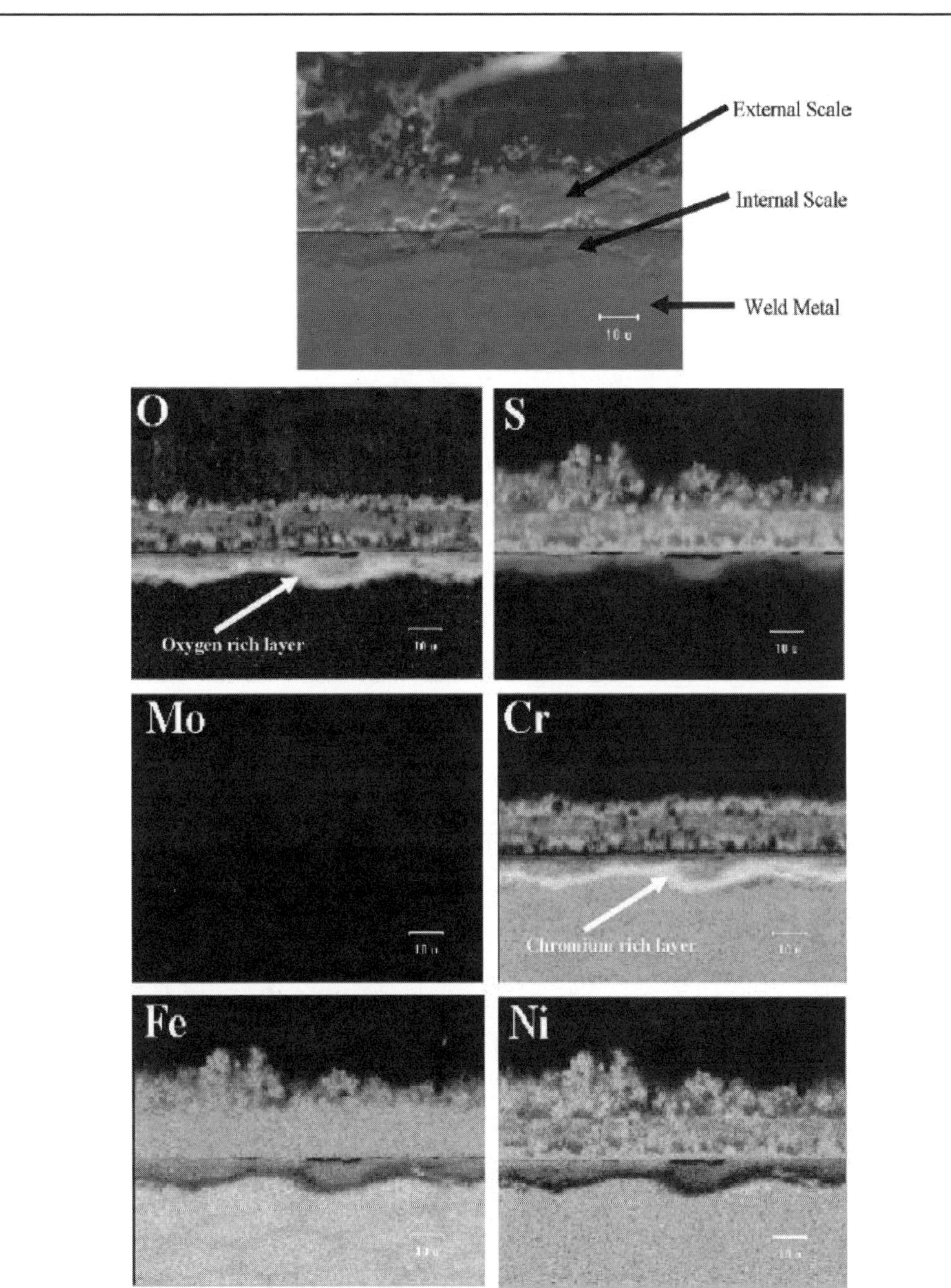

Figure 4 Corrosion morphology of the Alloy 33 weld overlay after 2000 hours of exposure to a simulated low NOx combustion gases. There is a thin uniform chromium oxide scale with no preferential attack as the result of molybdenum segregation.

Figure 5. Corrosion of Alloy 33 after 18 months of exposure in Boiler A. Note the thin uniform and continuous chromium oxide layer and also that oxidation is not preferential with respect to the dendrite structure of the weld overlay.

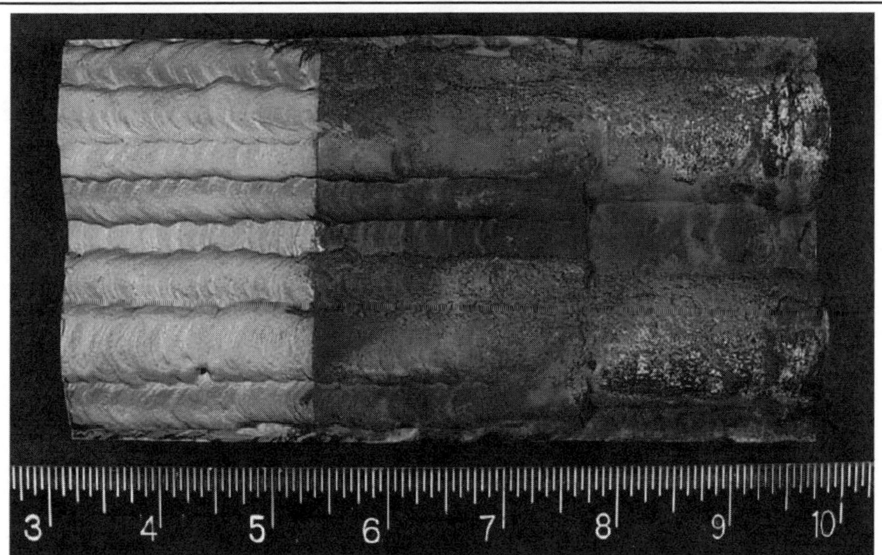

Figure 6. Waterwall tubing sample removed from service after 34 months in Boiler A. The left side of the tubing samples was lightly grit blasted to remove scales and deposits for better visual inspection. Note that the right hand side of the tube samples are not covered in weld overlay; these are bare SA213-T11 tubes.

Figure 7. Cross sections of the tubing samples seen in Figure 6 . The bare SA213-T11 tubes showed fairly deep circumferential cracks that are approximately 0.020″ (0.5mm) deep. The Alloy 33 weld overlay directly adjacent to this region showed no evidence of cracking.

Figure 8. Tube panel installed into Boiler B. 10 tubes were covered with Alloy 33 weld overlay and 10 tubes were covered with Alloy 622 weld overlay; the panel is 20 feet tall.

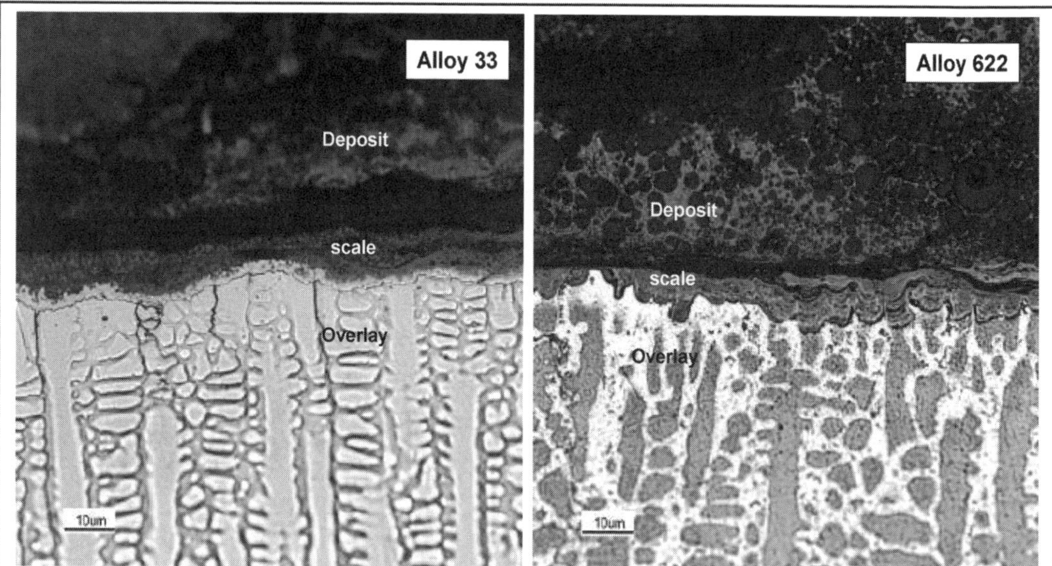

Figure 9. Cross sections of Alloy 33 and Alloy 622 weld overlays after 22 months of service in Boiler B. Note the more uniform scale on Alloy 33 and that the scale on the Alloy 622 overlay is developing lobes that penetrate into the dendrite cores.

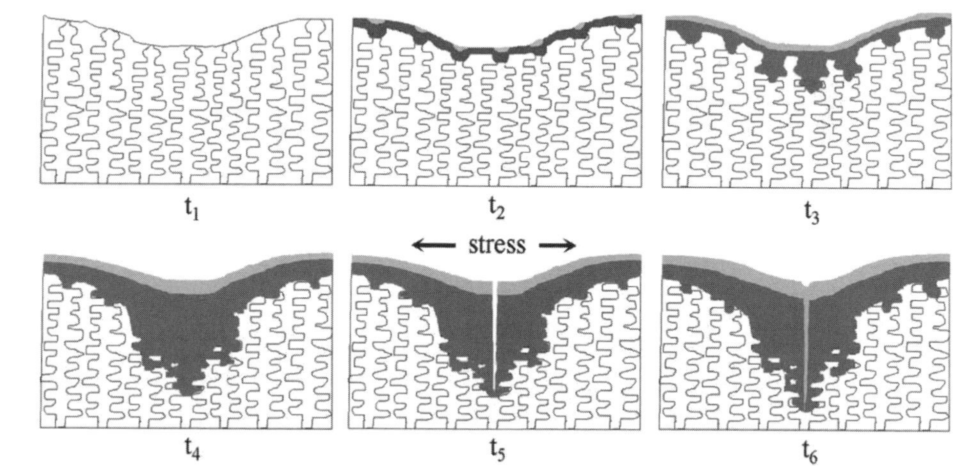

Figure 10. Schematic showing the sequential development over time (t) of oxide/sulfide lobes on a weld overlay exposed in a coal fired boiler. First he oxide lobes are developed (t_1 - t_3), later these lobes can eventually link together (t_4.). Tensile stresses arising from operational variables (such as slag sheds, sliding pressure, etc.) cause the oxide to crack, as oxides have very poor ductility (t_5.). Cracks allow sulfur species to penetrate the oxide later and cause the sulfur "spine" observed in these cracks (after reference 4).

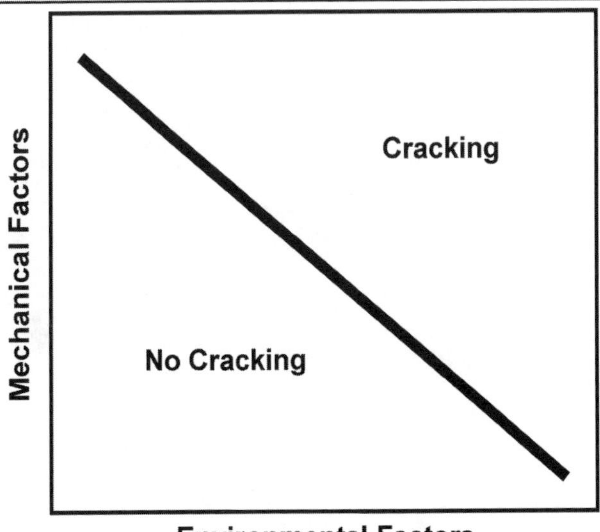

Figure 11. Cracking of weld overlay boiler tubes is a function of both mechanical factors and environmental factors.

Figure 12. Application of the Alloy 33 weld overlay inside of the WKE boiler, approximately 8,000 square feet of the furnace was covered during this project

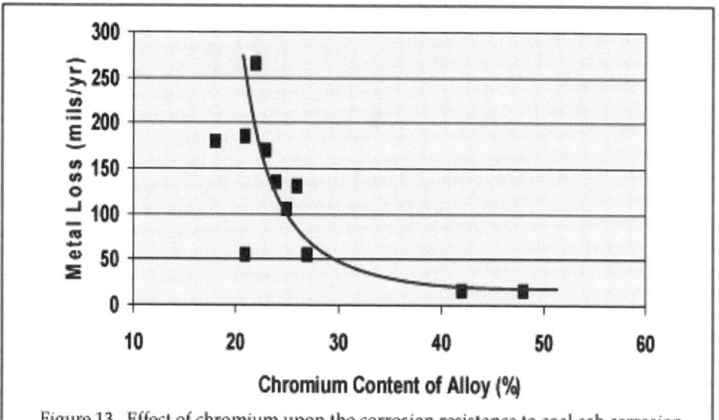

Figure 13. Effect of chromium upon the corrosion resistance to coal ash corrosion, based on field tests in a coal fired boiler burning medium sulfur coal (after reference 13).

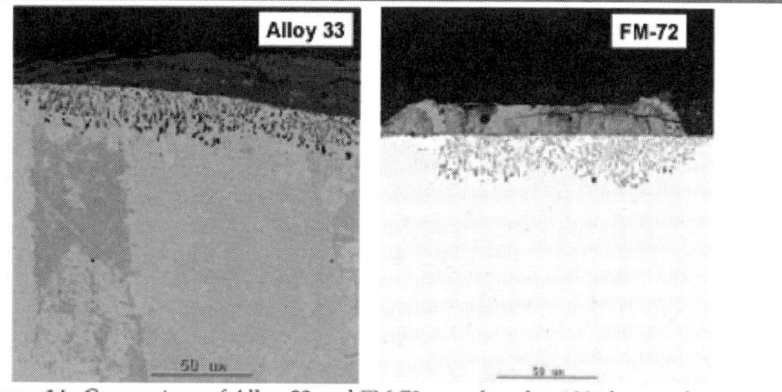

Figure 14. Comparison of Alloy 33 and FM-72 samples after 1294 hours of exposure at 1100-1150°F (538-621°C) in a coal fired boiler. The cross sections both show a chromium oxide scale with a similar thickness and similar amount of sub-scale attack.

Figure 15. Example of a 360° spiral overlay of Alloy 33 on a boiler tube. This is the weld overlay method used for superheater and reheater tubes.

Advances in Materials Technology for Fossil Power Plants
Proceedings from the Fifth International Conference
R. Viswanathan, D. Gandy, K. Coleman, editors, p 338-352

DOI: 10.1361/cp2007epri0338

Material Development and Mechanical Integrity Analysis for Advanced Steam Turbines

Andreas Pirscher* , Brendon Scarlin*, Rod Vanstone**.

* Alstom (Switzerland) Ltd, Baden Switzerland
** Alstom Power, Rugby, England

Abstract

Development activities, which started over 10 years ago within the framework of the COST 522 program, and which are continuing in the **COST 536** Action, are now showing success in terms of the construction of a new generation of steam power plants. These will operate under advanced steam conditions with steam temperatures up to about 1150°F (620°C) and much improved thermal efficiency. The recent successes in terms of placement of orders for such environmentally-friendly power plants in Europe and also in the USA are the result of:

- Development, testing and qualification of 10% Cr steels with improved long term creep properties, for the manufacture of major high temperature components, such as forgings for rotors and castings for turbine and valve casings
- Extensive in-house development, particularly concerning fabrication and weldability of full-sized forged and cast components along with detailed mechanical integrity and fracture mechanics evaluations

These new materials will be employed in several new coal-fired power plants in Europe and particularly in the plant Hempstead in the USA, with live and reheat steam temperatures of 1111°F (599°C) / 1125°F (607°C) ordered by AEP. Details of design concepts and materials selection are described in this paper. This permits steam conditions to be increased to 1112°F (600°C) live steam and 1148°F (620°C) reheat steam. The improved creep properties permit the construction of casings with reduced wall thicknesses, permitting greater thermal flexibility at lower component cost, as well as welded turbine rotors for high temperature application, without the need for cooling in the steam inlet region.

Further major improvements in operating efficiency will be achieved through the introduction of nickel alloys for the major components of steam turbines and boilers. Whereas the European AD700 project aims at a reheat steam temperature of 1328°F (720°C), the US DOE project targets 1400°F (760°C). The **AD700** project has recently been completed and has shown that there are no technical obstacles to the construction of such a steam power plant and engineering tasks are underway prior to the construction of a 550MW demonstration plant. Currently **DOE** activities have clarified the boiler concerns and the contribution to the turbine program is focused on rotor welding, mechanical integrity and steam oxidation resistance.

Introduction

Efficiency improvement in coal-fired steam power plant is made both for economic and environmental reasons [1]. Although some of the recent improvement has been due to better aerodynamic blade profiles and reduction in losses, the major part of the improvement has been due to increased cycle parameters which have been made possible by the introduction of new materials. Whereas for subcritical units low alloyed creep resistant steels could be used for major castings and forgings, 9-12%Cr steels with improved creep strength and oxidation resistance were required for the most highly loaded components such as turbine and valve casings and rotors in supercritical units. This was required to withstand the increased steam conditions. Whereas steels such as the German 12%Cr steel X20 CrMoV121, were adequate for temperatures up to 565°C, improved 9 to 12%Cr steels were required for the high temperature plants

- up to 1112°F (600°C) reheat temperature (typical representative is the plant Lippendorf in operation since 1999) and
- up to 1148°F (620°C) reheat temperature
  - for 50 cycle operation ALSTOM currently has over 13GW of orders for steam turbines with temperatures above 600°C, making use of new materials
  - for 60 cycle operation the most recent order is from AEP for the plant Hempstead with live steam temperature 1111°F (599°C) and reheat steam temperature 1125 °F (607°C).

Figure 1 shows the activities undertaken to qualify and introduce materials for these steam conditions. In COST 501 forged and cast 9 to 10%Cr steels were developed with additions of 1.5%Mo or a combination of 1%Mo and 1%W. They showed much improved creep strength, resistance to embrittlement in operation and weldability. In addition samples from production components were subjected to low cycle fatigue and long term creep testing permitting a statistical evaluation of the results. They are in use at temperatures up to 1112°F (600°C).
The trend to even higher steam conditions was the subject of the COST 522 program which explored the possibilities of stabilizing the tempered martensitic microstructure through addition of small quantities of boron. These are the steels (forged steel FB2 and cast steel CB2) now being employed in orders currently being executed in Europe and the USA.

Further longer term activities target:

- improved 10%Cr steels designed and investigated in COST 536 using programs able to predict the relative amounts of strengthening precipitates generated and microstructural stability and
- known nickel alloys with much improved high temperature properties but also significantly increased costs.
  - The European project Thermie / AD700 aims for a reheat temperature of 1328°F (720°C) through the use of solution treated alloys such as IN617 and IN625.
  - The DOE project aims for a reheat temperature of 1400°F (760°C) using precipitation hardenable nickel alloys, such as C263

Most recent materials development activities

Cast Steels

From a wide range of steels investigated, representing different approaches to producing a high creep strength, the most successful cast steel developed so far in the COST program is designated CB2 [2]. It contains 9%Cr, 1.5%Mo and 1%Co along with 0.01%B. It is essentially free of δ ferrite and its creep strength is provided by a dispersion of fine MX particles along with $M_{23}C_6$ particles which are effectively stabilized by boron. A full mechanical testing program has been carried out including long term creep tests at 1022°F (550°C) to 1202°F (650°C) with test times which have exceeded 94'000h. 52 creep rupture points have been determined and 29 tests are still continuing. On this basis reliable creep strength values can be determined and Figure 2 shows the improvement compared with the previous generation of boron-free cast steels from COST 501. The improvement is more pronounced at the highest temperatures and longest testing times. Two full-sized components have been manufactured and both nondestructively and destructively tested leading to qualification of the castings suppliers in the COST 522/536 programs. The positive effect of boron on the creep properties of 9%Cr steels has been shown in the past, it being particularly effective at lower stresses and rupture times beyond 10'000h [3].

Welding is required for the manufacture of cast components such as turbine casings and valve bodies, both as similar welds (manufacturing and repair) and as dissimilar welds to other components, such as pipes and connecting stubs of forged 9%Cr steels, such as P91 and P92. In general welding leads to a reduction in creep strength due to the non-ideal microstructure generated in the heat affected zone, which persists even after post weld heat treatment.

Different microstructures develop in the heat affected zone (HAZ) as a function of the peak temperature during welding [4]. The coarse grained HAZ is closest to the fusion line and is the zone in which the temperature lies well above Ac3, all carbides are dissolved and grain growth occurs, the fine grained HAZ is the region further from the fusion line, where incomplete solution of carbides occurs and austenite grain growth is limited, the partially transformed intercritical zone (ICHAZ) is the region between Ac3 and Ac1. The zone further away from the weld is only overtempered without the formation of any austenite. Failure at long creep testing times (low stresses) occurs in the fine grained HAZ (FGHAZ) and is referred to as type IV cracking.

The weld reduction factor is defined as the cross-weld creep strength divided by the base material creep strength and is important in determining the application temperature for a particular steel. For previously used cast steels it lies at about 0.65 or below, i.e. 35% loss of creep strength compared with the base material. The experiments of Abe [2] reveal a weld reduction factor at 1202°F (650°C) of 0.64 for P92. On the other hand the 9%Cr steel alloyed with 130ppm B shows no drop in cross-weld creep strength. No FGHAZ is observed, particularly at high boron and low nitrogen content. This fact may be attributed to boron stabilizing the $M_{23}C_6$ particles.

Results of creep tests performed on similar welds of CB2 at 600 and 625°C are shown in Figure 3. It can be seen that there are similar rupture times for weld metal samples and samples taken across the weld. The maximum reduction in creep strength due to the presence of the weld is about 25% for 600°C and 100'000h.

Forged Steels

Similar success has been demonstrated for the forged steel FB2 [5]. A full mechanical testing program has been carried out including long-term creep tests at 550 to 650°C with test durations which have exceeded 65'000h. 74 creep rupture points have been determined and 63 tests are still continuing. On this basis reliable creep strength values can be determined and Figure 4 shows the improvement compared with the previous generation of boron-free forged steels from COST 501. The improvement is much more pronounced at the highest temperatures. Three full-sized components have been manufactured and both nondestructively and destructively tested leading to qualification of the forgings suppliers in the COST 522/536 programs.

In the manufacture of steam turbine rotors ALSTOM welds the high temperature section of 9 to 10%Cr steel to lower temperature sections (e.g. shaft ends) of 1%CrMoV steel. This is preferred to the use of monoblock forgings since it produces a thermally flexible rotor capable of short start-up times and since smaller forgings allow easier nondestructive examination and a wider choice of manufacturers. This has been the practice for more than 30 years. The gradual improvement has been through the use of 9 to 12%Cr steels with steadily improved creep strength values, i.e. from 12%Cr steel (X20CrMoV 12 1) to 10%Cr steels alloyed with Mo and / or W and now to 9-10%Cr steels containing additionally cobalt and boron. The welding techniques and welding consumables have not been changed (submerged arc welding with nominally 5%Cr weld metal). Cross-weld creep tests have been made at 550°C on such a welded joint between FB2 and a 1.5%CrMoV steel. The results are shown in Figure 5 along with results for a weld between a 9-10% Cr steel without boron and a 1%CrMoV steel. As expected, in both cases failures occur on the 1% CrMoV steel side in the HAZ at stress levels which lie even for testing durations of 60'000h only marginally below the minimum value for the 1.5%CrMoV steel base material. The COST536 program is directed towards the further search for 9 to 10%Cr steels, on the basis of new alloying concepts, which may permit operating temperatures to be increased to 1166°F (630°C) to 1202°F (650°C).

Choice of materials and designs for current orders

As an example of the design approach for the latest generation of ultra supercritical (USC) steam turbines, the 1100MW steam turbine with inlet conditions of 595°C/604°C/267bar ordered for the RWE Neurath F/G plants in Germany is shown in Figure 6. This order is one of several received by ALSTOM for new USC steam turbines in Germany to date [6]. In this case machine layout consists of single high pressure (HP), single intermediate pressure (IP) and double low pressure (LP) modules.

HP valves

The HP inlet valves are exposed to the full live steam temperature and pressure over the life of the machine. For the new generation of USC machines a detailed investigation was performed to determine the best selection of material in terms of high temperature properties, cost and thermal flexibility (start-up time). It was clearly demonstrated that CB2 offers significant benefits over conventional 9%Cr steels since its enhanced creep properties allow thinner walled castings with significant weight saving and through-wall thermal stress reduction. For all current and future orders, both for 50 and 60 cycle turbines, high pressure valve casings in CB2 will be used.

HP turbine module design with shrink rings

The HP turbine module is based on a double casing design with bolted outer casing (1.5%Cr casting) and inner casing (9%Cr casting, G-X-12CrMoVNbN9-1) with shrink-rings. This configuration has been used for over 30 years with an excellent service history. The shrink rings allow the inner casing design to be near-symmetric which simplifies casting and machining as well as reducing the deformation under thermal and pressure loading compared to a flanged casing design. The advantages of the shrink ring design compared with a flanged solution increase as the operating conditions become higher. With increasing live steam pressure it is quite straightforward to increase the dimensions and / or number of shrink rings. Since they are located in the area of cooler exhaust steam there is no possibility of thermally induced creep deformation or embrittlement. Figure 7 shows the arrangement of the HP module. In the right-hand view, the outer casing upper half has been removed to show the inner casing and shrink-ring construction.

The rotor is a two-piece construction with the weld located after the sixth blade row. The high temperature forging is procured in 10%CrMoVNbN steel for 50 cycle applications and in steel FB2 for 60 cycles. Depending on the specific turbine output the high temperature forging may cover the HP inlet stages, balance piston and rotor end with journal bearing and coupling. To prevent wire-wooling of the journal surface, a low alloy weld cladding may be introduced over the base material. The low temperature forging is procured in 1% CrMoV steel; a material with which ALSTOM has extensive operational experience.

To improve cycle efficiency an additional HP heater extraction is located towards the end of the HP expansion. This consists of a cast ring in 9% Cr steel located over the extraction slots in the inner casing and held in position by the last shrink ring. Piston rings provide the seal between the extraction ring and outer casing connection. By careful optimization of the inlet blade rows, the use of rotor cooling or the application of a FB2 high temperature forging is not required for live steam temperatures up to 600°C. Thanks to its high creep strength and oxidation resistance the highly successful austenitic stainless steel blading continues to be used for the inlet stages of the HP turbine.

IP valves

Two inlet valve casings are used to regulate steam flow into the inlet spiral. Although the IP valves generally see higher reheat temperatures they must withstand significantly lower pressure loading and are consequently of much thinner walled castings. Material thickness has been standardized to allow the use of a common casing pattern for both 9% Cr steel and CB2 for temperatures up to 620°C.

IP turbine module design

The general arrangement of a double-flow IP turbine for ultra supercritical applications is shown in Figure 8 with the IP-LP crossover piping removed from the upper casing connections. The inner casing consists of an advanced design, which reduces the level of ovalization during operation compared with a standard flanged design. The use of optimized rotor blade groove geometries, front stage blading in nickel-based alloy and FB2 forgings permit the design of IP modules up to 620°C without cooling systems, both for 50 and 60 cycle applications. These would have a negative effect on the machine heat rate and over the machine life result in a significant loss in generating revenue.

High performance blading and advanced sealing

The development of reaction blading profiles by ALSTOM began with the 1000 series in the 1960/70s, progressed through the 8000 series in the 1980s and continued into the 1990s with the HPB (High Performance Blading) series. Current development is focusing on further optimization of the HPB profile as well as refinements in the blade root and tip shroud regions. The control of secondary flows and their interaction with leakage jets offer further areas of potential performance improvement. In Figure 9 a fixed and moving blade of the HPB series are shown. This profile is characterized by its robust shape, excellent vibrational behavior and high performance level.

The reduction of leakage is a major area of potential performance improvement and Alstom have invested heavily in the R&D of new sealing technologies. Areas of leakage loss are the sealing areas at the tip of the moving and fixed blades, as well as the balance piston and rotor-end shaft sealing. In order to reduce leakage on USC turbines, abradable material or brush seals may be used, as appropriate.

Neither for the 50 nor 60 cycle turbines is there any need for cooling of any high temperature components.

Technology and materials for >700°C

Further development of 9 to 12 %Cr steels is underway in Europe in the COST 536 project and promising compositions have been manufactured and are under testing. Nevertheless before such steels can be used for highly loaded components it will be necessary to establish their properties through many years of testing.
Hence in order to achieve further major improvements in efficiency in the short term it is clear that nickel alloys will be required for the fabrication of high temperature components. This includes heavy components such as rotors, inner casings, valve bodies and pipes and thinner section parts such as blades, bolts, and sealing strips. For the heavy components even though materials with the necessary properties are known (typically a creep strength at 1330°F/720°C and 200'000h of 100MPa), they had not previously been manufactured with the sizes required for steam turbines of typically 1000MW capacity.

In Europe the Thermie / AD700 project aimed to demonstrate the ability to produce such large components as forgings, castings and manufactured parts (pipe bends and similar and dissimilar welds). The requirements can be met by solution treatable nickel alloys, such as IN617 and IN625, for which welding can be performed without undue difficulty. Full size similar welds have been produced to simulate pipe welding and manufacturing welding of castings and dissimilar welds have been produced to demonstrate the success of rotor welds, joints between steel castings and nickel alloy castings and joints between valve casings and pipework. The positive experience in manufacturing large components has also demonstrated the ability to make parts of the size required for large plant and has provided clearer figures concerning the costs of the parts, including the part required for processing (such as forming, bending, milling, turning and grinding). Particular development of appropriate NDT techniques was required.
These fabricated parts have been subjected to nondestructive testing with radiographic and ultrasonic techniques and to a full program of mechanical testing. This included tensile, low cycle, creep and fracture mechanics evaluation both in the post weld heat treated condition and after long term thermal exposure to simulate service operation. NDT inspection can detect defects well below the size at which critical crack propagation could occur.

To a large extent conventional designs can be used and there is no requirement for cooling. However Alstom is carrying out a study to assess whether the customer would benefit from a design with a limited amount of cooling. This would permit a reduced usage of high-cost nickel alloys in the turbine. A section through a high pressure turbine suitable for application at 700°C is shown in Figure 10. Nickel alloys are used for

- The rotor, with a dissimilar weld to a steel forging once the temperature has dropped, as a result of steam expansion, to the normal maximum temperature of application of the steel
- The inner casing, again with a weld to a steel casting at lower steam temperature
- Bolts, selected on the basis of matching thermal expansion coefficient and high stress relaxation strength
- Blades, which are milled from bar stock, machining trials have shown how machining costs can be greatly reduced.

A 10%Cr steel or a 1%CrMoV steel is used for the section of the rotors and inner casings at the lower temperature. There are advantages and disadvantages in the choice of steel:

- The choice of a 10%Cr steel permits the dissimilar weld to the nickel alloy to be made at a higher steam temperature so that the nickel alloy forging is limited in size (cost reduction)
- The choice of a 1%CrMoV steel leads to lower stresses during thermal cycling and hence to shorter startup times (matching coefficients of thermal expansion). The reduced thermal stresses also lead to a greater defect tolerance so that the minimum size of defect which must be detected nondestructively is considerably increased.

  To facilitate inspection of the dissimilar rotor weld a steel ring is first welded to the nickel alloy disc, followed by inspection of this weld from both the outside and the root. The weld between the steel ring and the steel disc, at lower temperature is manufactured and tested conventionally.

At the end of the Thermie (AD700) project full-scale demonstration rotor welds were manufactured and nondestructively inspected. Complete mechanical integrity was assured.

Other programs in Europe address additional essential topics such as
- Lifetime calculation methods for nickel alloy components (Cooretec)
- Nondestructive inspection of thick-walled cast, forged and welded nickel alloy components (Cooretec)
- Fabrication and testing of welded joints between nickel alloy and low alloy steel forgings (Marcko)
- Application of coatings and overlays for protection of valve internals and blade surfaces against oxidation and solid particle erosion (Swiss KTI project)
- Surface engineering solutions to extend the lifetime of steam-exposed components and development of sealing technologies (UK DTI project)

On the basis of this work a design study was made to build and operate the COMTES test facility at the steam power plant Scholven in Germany. The steam is taken from the inlet header of the first superheater stage and is led to the evaporator, where it is heated to 600°C. The steam is then heated to 705°C in the test superheater, where it enters the HP bypass test valve or is cooled and mixed with the main superheater steam. The test facility permits demonstration of
- manufacturing of casing by casting and welding
- manufacture of internals including application of wear-resistant coatings
- satisfactory opening and closing behavior
- subsequent nondestructive and destructive inspection.

It was built and commissioned in Summer 2005 and has now completed half of its planned operation.

With these measures it is expected that a thermal efficiency of >50% should be achieved, providing a considerable reduction in cost of electricity. Economic viability of the concept is assured, particularly through measures to minimize the amount of nickel alloy required (use of innovative thinner-walled designs or use of cooling) and also in view of the steadily increasing costs of fossil fuels and the possibility of the introduction of a "CO_2 tax" on lower efficiency plant.

Following on from the AD700 project, further detailed engineering studies are in progress, which it is expected will lead to the ordering of a demonstration plant in 2010, followed by construction and commissioning in 2014.

In the USA a program has been initiated by the DOE, with the aim of demonstrating the basic feasibility of a steam power plant operating at 1400°F (760°C). For this stronger nickel alloys of the precipitation-hardening type are needed. These are readily formable in the solution treated condition and are not necessarily more expensive than the solution hardened alloys of the AD700 project.

A five-year US effort sponsored by DOE and OCDO to develop materials for USC boilers has been in progress for several years with participation of EPRI, ORNL and the US boiler manufacturers. Development of corresponding materials technology for steam turbines has been started in a 3 year programme with participation of Alstom, GE, Siemens, EPRI and ORNL.

The following lists activities, which will be performed within this 3 year project. It will constitute:

- a feasibility study in which technical and economic factors will be addressed
- identification and investigation of potential "show-stoppers"
- limited medium-term experimental testing.

A follow-on programme will be required for full qualification of materials and designs, including long-term testing, component fabrication and pilot plant design, construction and operation.
As a prerequisite the partners agree on the target steam data and plant MW rating. These coincide largely with the targets of the DOE boiler project. It is likely that materials for 760°C application will be of the precipitation-hardened type. Precipitation-hardened materials are likely to be more difficult to cast and weld. Nevertheless a steam temperature of 760°C will provide a higher operating efficiency. Economics studies should compare the situation for steam data of 760°C:

- using 760°C materials without cooling
- or 730°C materials with cooling.

In the first place conventional turbine designs should be assumed, whereby the turbine rotors and inner casings may be either single-piece or welded. At a later stage designs could be optimized to reduce the required amount of expensive nickel alloy.

The following tasks are being pursued:

1) Identify critical turbine components and their requirements

Preliminary conceptual design considerations.
For turbine HP and IP rotors, casings for turbines and valves, stationary and rotating blades, turbine and valve bolts and valve internals firstly define sizes, weights and approximate dimensions.
Evaluate alternate manufacturing routes such as forging, casting, welding, bending, powder metallurgy. Estimate requirements in terms of mechanical and environmental loading.

2) Search for appropriate materials

Identify candidate alloys in terms of mechanical and environmental properties, capability of manufacturing in appropriate shapes and sizes (including weldability, if required) and cost.
Collate available data for these alloys.
Determine whether coatings are needed to protect against steam oxidation and / or solid particle erosion
Identify missing data and formulate programme to fabricate material and provide missing data.
Identify potential "show-stoppers" and formulate program to address these points, e.g. manufacturing capabilities and uniformity of properties of complex alloys, weldability and nondestructive inspection capabilities.

3) Validation of materials suitability

Procure materials and perform medium-term mechanical, exposure and oxidation testing (max.15'000 hours).
Identify and agree activities to evaluate potential "show-stoppers", such as:

- integrity of similar and dissimilar welded joints as appropriate (weldability, nondestructive inspection and properties),
- manufacturing capabilities for large forgings and castings

4) Design

Develop alternative design concepts, including cooling technologies.
Determine permissible start-up / shutdown and cycling numbers and rates
Preliminary investigation of any "show-stoppers".

Areas of Activity of Alstom

1. **Thermodynamic Studies**

This activity started with a review of the preliminary boiler design from the DOE boiler materials programme. The cycle parameters for the study are main steam 348-354 bar, 732°C, reheat steam 70-60 bar, 760°C, condenser pressure 0.085 bar.

2. **Oxidation testing**

Steam exposure tests have been made using Thermo Gravimetric Analysis (TGA) on a selection of nickel alloys and steels at 760 and 800°C. Values of the parabolic rate constant have been determined.

3. **Rotor Welding**

For an intermediate pressure double flow rotor with inlet at 1400°F (760°C) made of C263, a joining weld to IN617 and then a joining weld to a steel may be required. It is likely that the precipitation-hardening alloy such as C263 should be welded in the solution-treated condition to a solution-hardened alloy such as IN617 using an appropriate nickel alloy weld metal. Buttering of the precipitation-hardening nickel alloy may be carried out as required. This weld would receive a post weld heat treatment (PWHT) in order to develop the creep strength of

C263. The IN617 would then be welded in this condition to a steel (1%CrMoV steel is preferred to 10%Cr steel since there is a much smaller mismatch in thermal expansion coefficient). PWHT of this joint would then be appropriate for the steel. Special attention will be paid to avoiding cracking in the nickel – nickel weld (C263 to IN617). An investigation program is in progress.
From the point of view of mechanical integrity an investigation is being performed of the nickel – steel weld. It was decided to investigate primarily this dissimilar weld between rotor sections of IN617 and of 1%CrMoV steel since this will have higher stresses due to the mismatch in thermal expansion coefficient between IN617 and the steel, rather than the weld between C263 and IN617, which matches more closely.

Acknowledgements

The authors would like to thank their partners for support and information provided during the course of the COST project. Thanks are also extended to the national funding bodies, which provided financing for the COST activities.
In the same way acknowledge is given to the EU Commission for funding provided to AD700 and to the project partners who contributed to the technical program.

References

[1] A. Tremmel and D. Hartmann, Efficient steam turbine technology for fossil fuel power plants in economically and ecologically driven markets, VGB Power Tech, 11/2004, p.38-43.
[2] M.Staubli, R.Hanus, T.Weber, K-H.Mayer and T-U.Kern, The European efforts in development of new high temperature casting materials – COST 536, Liege Materials for advanced power engineering, COST Conference, Liege 2006, p.855-870.
[3] F.Abe, Metallurgy for long-term stabilisation of ferritic steels for thick section boiler components in USC power plant at 650°C, Materials for advanced power engineering, COST Conference, Liege 2006, p.965-980.
[4] J.A.Francis, W.Mazur and H.K.D.H.Bhadesia, Type IV cracking in ferritic power plant steels, Materials Science and Technology, 2006, Vol.22, No.12, p.1387 – 1395.
[5] T-U.Kern, M.Staubli, K-H.Mayer, B.Donth, G.Zeiler and A. DiGianfrancesco, The European efforts in development of new high temperature rotor materials – COST 536, Liege Materials for advanced power engineering, COST Conference, Liege 2006, p.843-854.
Madrid 2007
[6] C.Brandt, A.Tremmel and H. Klotz, ALSTOM steam turbine design of world's largest single shaft units in most advanced ultra-supercritical steam power plants, Powergen Europe, Madrid, 2007

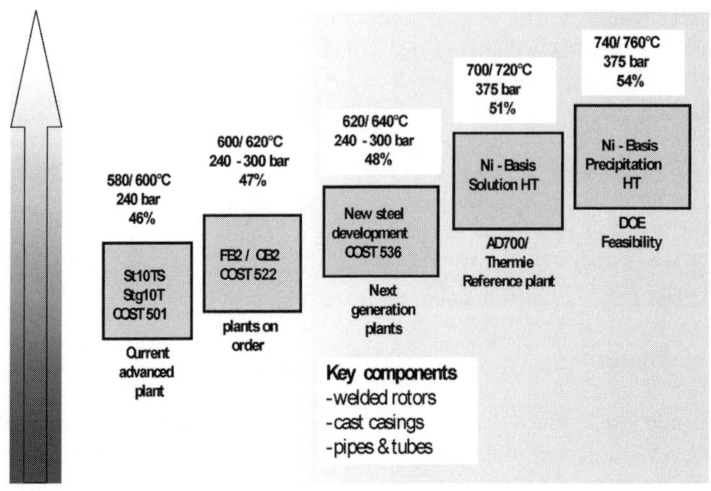

Figure 1: The role of materials development in increasing steam conditions and efficiency

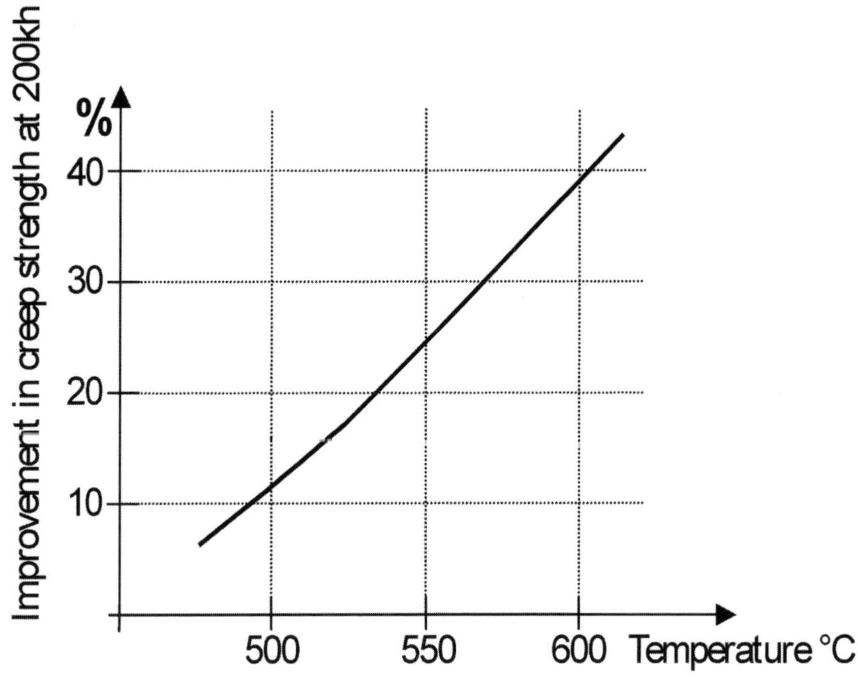

Figure 2: Improvement in creep strength at 200'000h for CB2 compared with boron-free steel

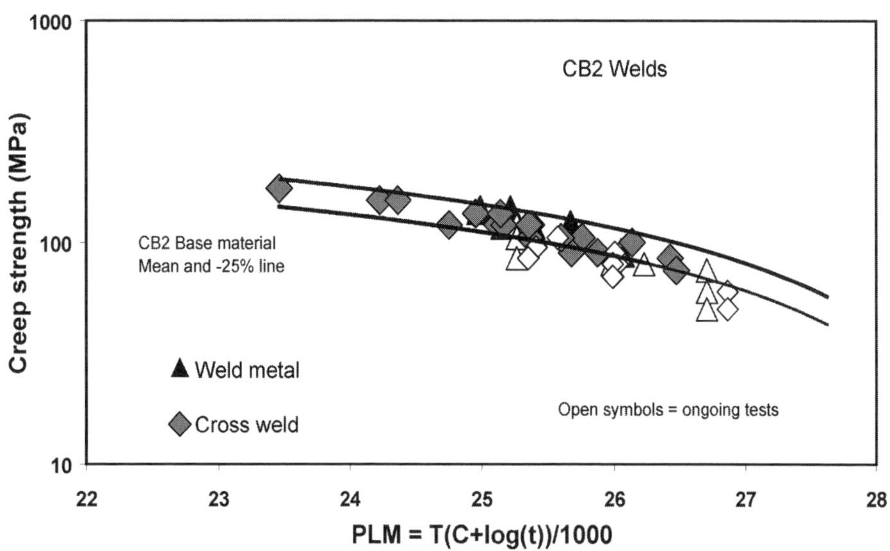

Figure 3: Creep strength as a function of Larson-Miller parameter for CB2 welds.

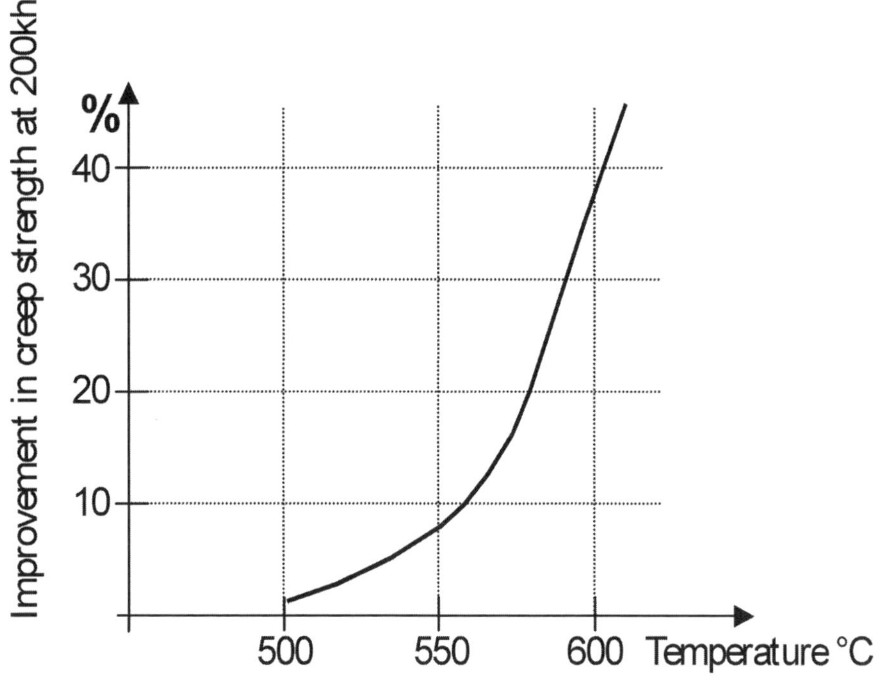

Figure 4: Improvement in creep strength at 200’000h for FB2 compared with boron-free steel.

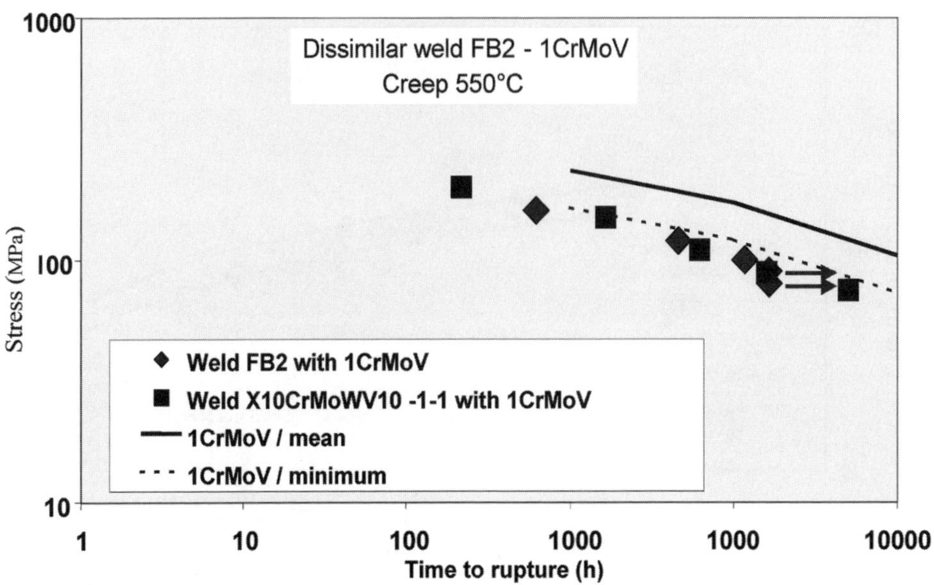

Figure 5: Creep strength of FB2-1%CrMoV steel dissimilar welds at 550°C.

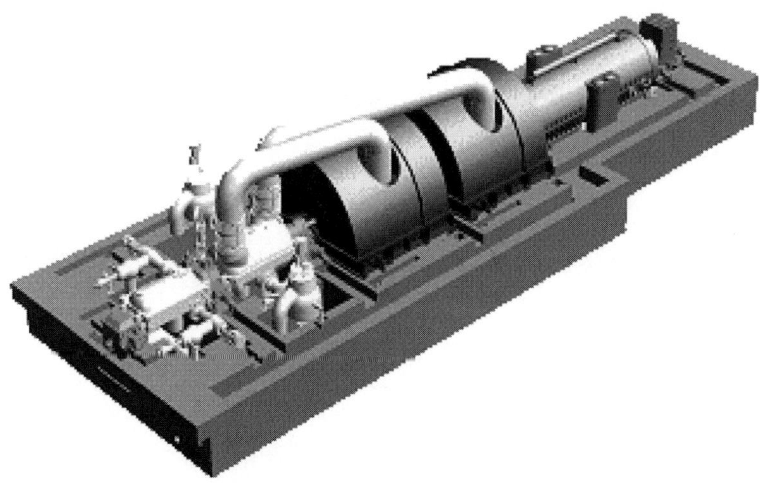

Figure 6: Overall layout of typical 600/620°C steam turbine (single HP turbine, single IP turbine and double LP turbines)

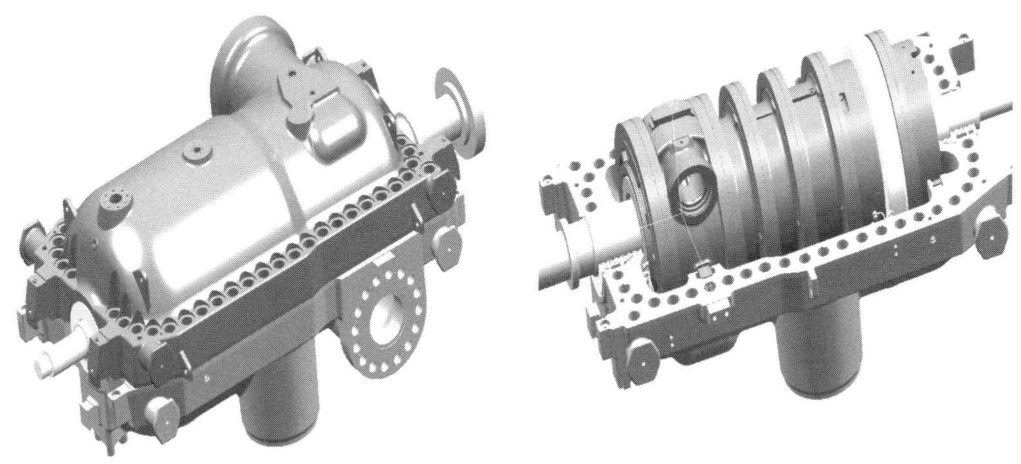

Figure 7: Ultrasupercritical HP module design.

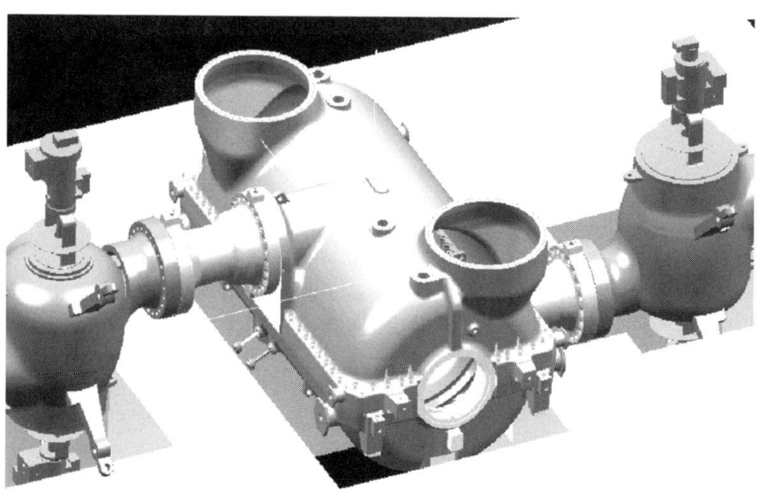

Figure 8: Ultrasupercritical IP module design with inlet valves.

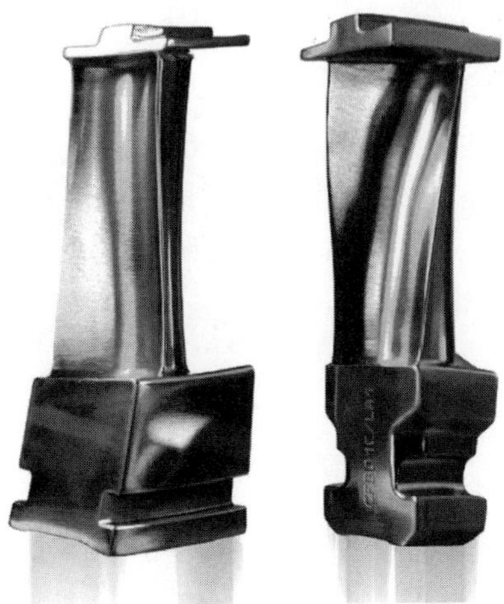

Figure 9: Typical fixed and moving blades of the HPB series.

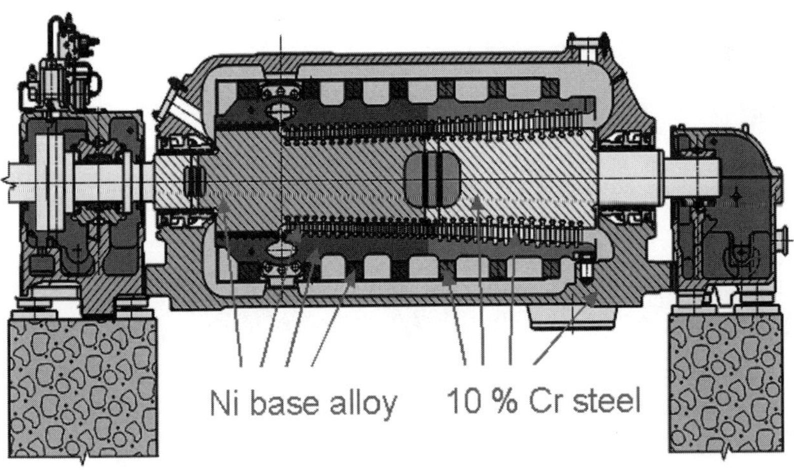

Figure 10: Section through a high pressure rotor suitable for application at 700°C.

Advances in Materials Technology for Fossil Power Plants
Proceedings from the Fifth International Conference
R. Viswanathan, D. Gandy, K. Coleman, editors, p 353-365

DOI: 10.1361/cp2007epri0353

High Chromium Steel Forgings for Steam Turbines at Elevated Temperatures

N. Blaes
B. Donth
D. Bokelmann
Saarschmiede GmbH Freiformschmiede
Bismarckstrasse 57-59
66333 Völklingen, Germany

Abstract

The global trend to high efficiency steam power plants requires ever more steel rotor forgings for operational temperatures at 600°C and above. The European Cost programme was, and still is, concerned with the development of creep resistant 10%-Cr-steels for such applications. Beside gas turbine applications, the steel Cost E is nowadays widely used for steam turbine shafts seeing a strongly rising market share. Saarschmiede has established a fail-safe manufacturing procedure for Cost E rotors with an increasing collectivity of mechanical properties and long term data, which provide reliability for turbine design. Saarschmiede has also started the production of Cost F rotor forgings for HP/IP turbines with delivered weights of up to 44 t . To increase the application temperatures, the benefit of boron additions has been found by investigations in the frame of the Cost programme. For 10%-Cr-steel with boron Saarschmiede has produced full size trial rotors to develop a production procedure. These prototype components are now under testing. New results are presented.

Introduction

The generation of electrical energy using fossil fuels makes a significant contribution to CO_2 emissions. Worldwide, programmes for reducing CO_2 emissions are being initiated. In power stations that burn fossil fuels, one effective way of achieving this is by increasing the degree of efficiency and thereby reducing the specific fuel consumption. Both in Germany and at international level, development projects for increasing efficiency are underway and Saarschmiede is involved in these projects with the focus on rotor manufacturing. The programme Cost 536 aims to increase the steam temperature to 650°C and the programme Thermie/AD 700 targets at steam temperature of 700°C.

Actually, steam temperatures of up to 610°C have been achieved. In the programme Cost 501, 10% Cr steels were developed for rotors in high pressure and medium pressure turbines, and, of these steels, the development type E known as "Cost E" has become most widely accepted. However, type "F", which corresponds to the Japanese steel TMK1, is also increasingly establishing itself on the market.

These steels provide a basis for building power stations with a high level of efficiency and reliable technology.

New developments in the Cost programmes Cost 522 and Cost 536 aim at achieving better long term properties in the turbine rotors through the addition of boron.

Saarschmiede, which is a long-term partner in these development projects, produces the steel Cost E in large quantities and sees a global market for this product.

In the development of boron-alloyed steel, Saarschmiede has manufactured trial rotors and has commenced with production.

Product Portfolio of 10% Cr Steels

In steam power stations using fossil fuels, besides high-pressure and medium pressure shafts of 1% Cr and 2% Cr steel, more and more shafts made of 10% Cr steels are being used. Saarschmiede supplies all leading manufacturers of power stations worldwide with shafts of this type. In the period 2000 to June 2007, a total of 55 shafts made of Cost E (X12CrMoWVNbN1011) and 11 shafts made of Cost F (X14CrMoVNbN10-1) were delivered. Cost E rotors with a delivery weight of up to 45 t and diameters of up to 1280 mm were manufactured. Even larger shafts with a body diameter of 1380 mm and a weight of 44 t have been manufactured from the steel Cost F / TMK1 (1).

Besides the rapid development in the number of shafts sold on the market, there is also a trend towards larger and heavier shafts. This has resulted in special demands on the manufacturing process with regard to ingot production, forging and heat treatment.

Manufacturing

In the course of the development of rotor steels, a large number of small trial melts are created which are examined with regard to their mechanical and creep properties. For good candidate materials, creep tests are carried out over decades since it has been observed that a decrease in the creep strength can still occur after several 10,000 h . Trial shafts are manufactured from the most promising materials in order to verify the creep tests using material manufactured under realistic conditions.

In subsequent series production, care should be taken that the shafts manufactured have the properties that were achieved during the development phase. In particular, the properties should be adjusted so that they are as homogeneous as possible throughout the shaft. To achieve this, one basic requirement is a largely uniform chemical composition and a homogeneous low δ-ferrite microstructure throughout the volume.

With regard to the segregation qualities of large ingots, Saarschmiede generally uses the ESR process for 10% Cr steels for energy engineering. In ESR ingots of up to 175 t in weight and 2300 mm in diameter, a high level of homogeneity of the chemical composition and microstructure is ensured along the whole length of the ingot and through the cross section

(2). In the largest productive ESR facility in the world, 4 electrodes can be simultaneously remelted to form an ingot. (Fig. 1).

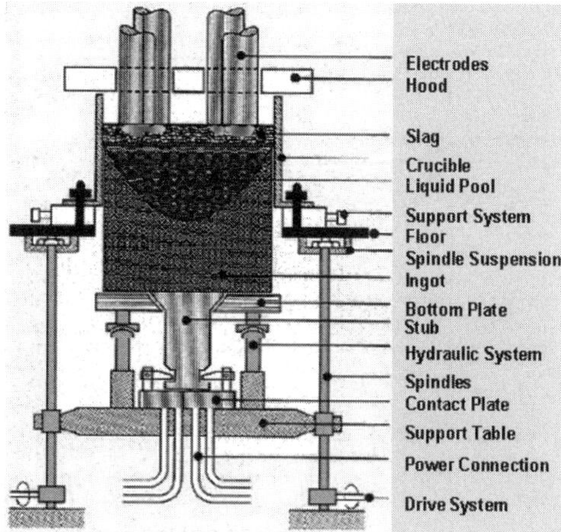

Figure 1. Layout of the ESR Facility

During the whole remelting process, the ESR slag is deoxidized so that oxygen-affine elements such as Al or Si can be adjusted within narrow limits. In order to control the amount of deoxidization, samples for chemical analysis are taken from the liquid steel pool and immediately passed onto the chemical laboratory for analysis. This means that, even during the remelting process, information is directly available regarding the chemical composition along the ingot axis and therefore also through the cross section.

Immediately after remelting, the ESR ingot is transported in hot condition to the neighbouring forging shop and here, it is heated up to forging temperature under controlled conditions. Forging of large high pressure/medium pressure shafts is carried out in one or two upsetting operations (Fig. 2). This ensures sufficient deformation which is important in order to achieve an even, fine-grained microstructure. Furthermore, optimised temperature control is important in order to achieve a small grain size and therefore to allow good ultrasonic inspectability.

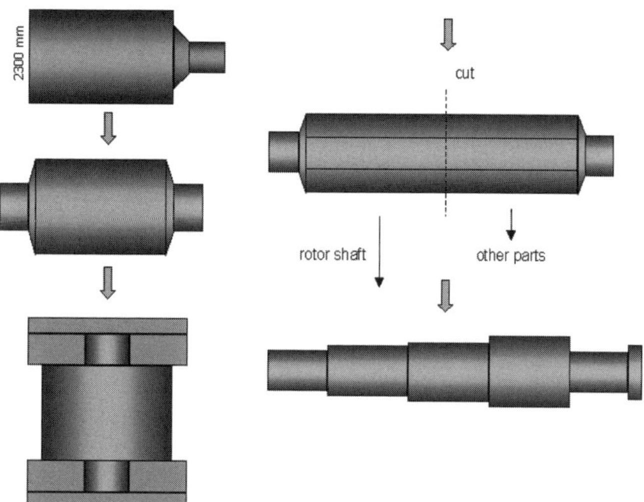

Figure 2. Typical Forging Procedure for 10% Cr-Shafts

The temperature profile in preliminary heat treatment is optimised for good ultrasonic detectability (Fig.3). When the grain size has been evened at austenitizing temperature, time-consuming pearlitic transformation takes place.

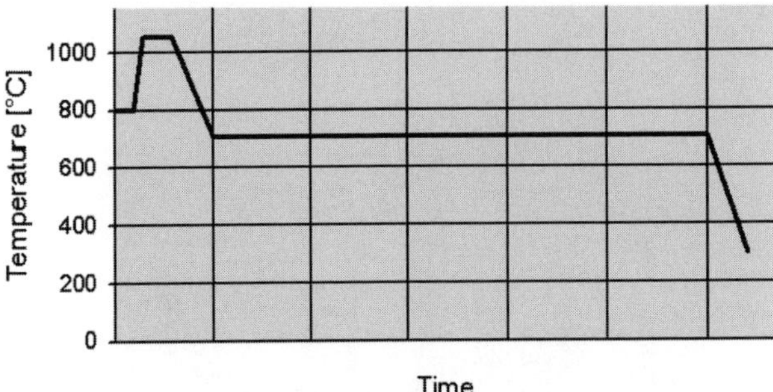

Figure 3. Typical PHT for 10% Cr-Shafts

After that, machining is carried out to achieve the tempering contour which is sufficiently in excess of the final shape. As a rule, at this point, the defect detectability is estimated by determining the Flat Bottom Hole size in the core of the shaft. Hardening is carried out by heating up to austenitizing temperature and cooling in an oil bath (Fig.4). Double annealing is applied in order to avoid residual austenite.

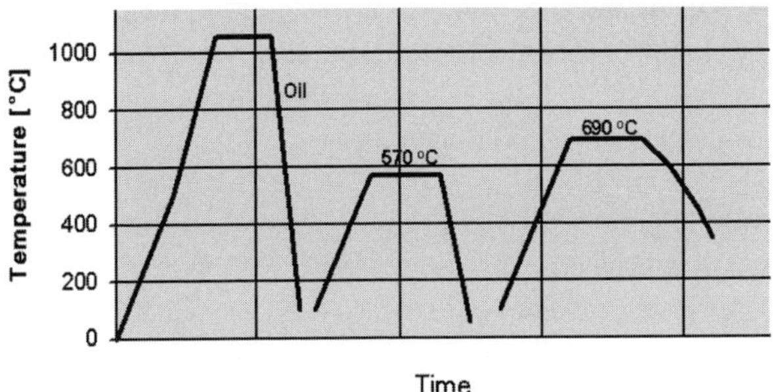

Figure 4. QHT of 10% Cr-Rotors

All turbine shafts are subjected to mechanical testing in order to ensure their suitability for use. The outer segments, radial cores or also an axial core are tested on the shafts. Typical testing locations are shown in Fig. 5.

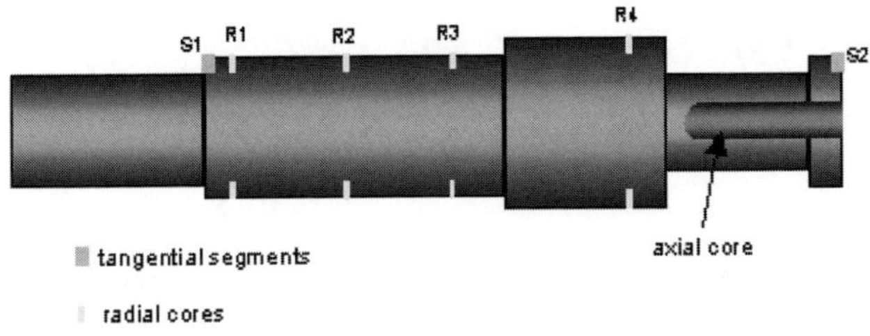

Figure 5. Typical Test Location at a 10% Cr Shaft

Most shafts have been manufactured within a 0.2% yield strength range of between 710 MPa and 790 MPa (Fig. 6). In this range the FATT lies within the range of below 40 °C as Fig. 7

shows. The FATT figures in the radial cores reveal a tendency to higher values. When carrying out the relevant impact strength testing at room temperature, there is a great amount of variance with values of between 40 J and 100 J. Within this variance there is a vague relationship between segment and axial core values (Fig. 8). The great variance can be explained by the FATT close to room temperature. In the transition area, greater variance is normally to be expected.

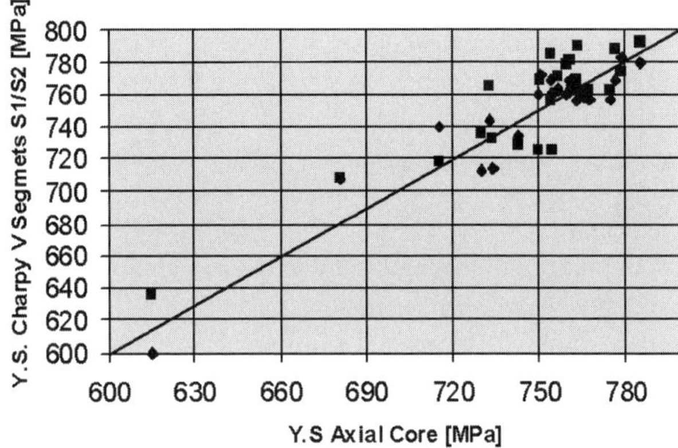

Figure 6. Yield Strength of Segments Versus Axial Core for COST E Shafts

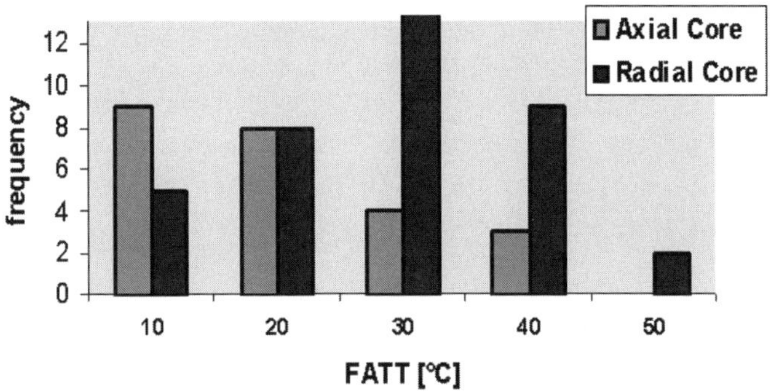

Figure 7. FATT of COST E at Different Sampling Locations

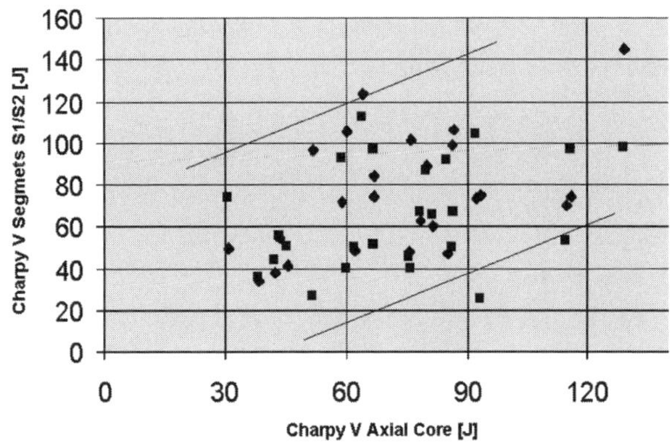

Figure 8. Charpy V Figures of Segments Versus Axial Core for COST E Shafts

The fact that the tangential and radial values are largely uniform is a result of the even temperature control during vertical QHT. It is carried out in gas furnaces with rotational symmetry which are operated with a "digital" control system. With about 100 burners, each individual burner is only operated in two states; either switched off or burning at full power. The total power supplied to the furnace results from the number of burners switched on at the same time i.e. the number of burners running at full power. Burners are switched on and off in cycles in order to provide even temperature distribution. In doing so, all the burners are integrated into a fixed sequence. The total power supplied to the furnace therefore also corresponds to the firing period of a burner. Burning at full power leads to stronger emissions of fumes. The processes of switching on and off lead to strong surges of gas. This causes excellent swirling of the atmosphere throughout the furnace and therefore to homogeneous temperature distribution.

The vertical furnaces of up to 13 m in length are divided into up to 6 zones which can be controlled individually. It is therefore also possible to carry out differential heat treatment whereby different temperatures can be applied to different parts of the shaft.

For annealing, a furnace with revolving atmosphere is also available, in which the whole furnace atmosphere is constantly circulated. Excellent temperature distribution is achieved with this method.

Hardening in oil is carried out in a round basin with constant oil circulation. The circulated oil is pumped through a cooler in order to limit the temperature changes in the oil to a minimum.

The combination of all of these measures means that the shafts are subjected to extremely rotationally symmetrical temperature conditions. If cooling from annealing temperature is sufficiently slow, a symmetrical, very low level of residual stresses is achieved. Special treatment to release stresses is therefore not necessary.

Besides symmetrical temperature control, the rotational symmetry of the properties is reinforced by the manufacturing based on ESR.

A good measure of the uniformity of the forging is provided by the yield strength of radial cores located opposite each other (Fig. 5) The distribution of frequency in Fig. 9 describes the differences in the yield strength of radial cores located opposite each other. Practically all differences are lower than 10 MPa which roughly corresponds to the measuring accuracy. Thus, the shafts have very homogenous properties.

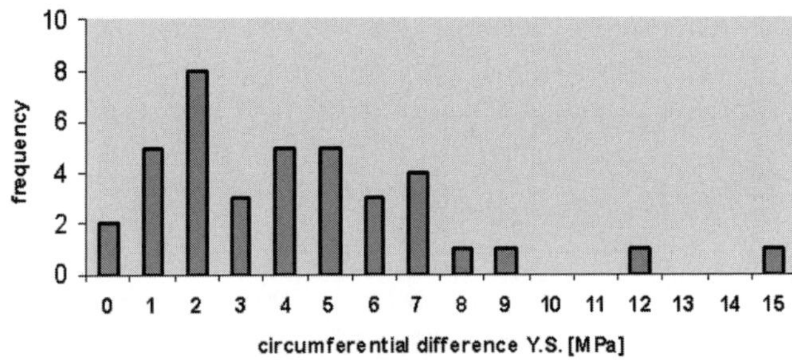

Figure 9. Difference of Yield Strengths between Opposite Radial Cores

The homogeneous microstructure also leads to overall stable behaviour in the thermal stability test. Only slight runout is to be seen and, in particular, slight runout of type C according to SEP 1950 (3). Type C describes the bimetallic behaviour of a rotor and represents the difference between the runout at the end of the hot phase and the runout at the cold end. Fig. 10 shows an example of a thermal stability test in a shaft made of Cost E. The runout according to SEP 1950 corresponds to double the deflection of the centre line. During the temperature increase, clear runout of type A becomes prevalent, resulting from the radiation-based interaction between the furnace and shaft but which is of no importance for later operation. During the temperature evening phase, this A runout recedes. The C runout measured between the hot end and the cold end is very slight.

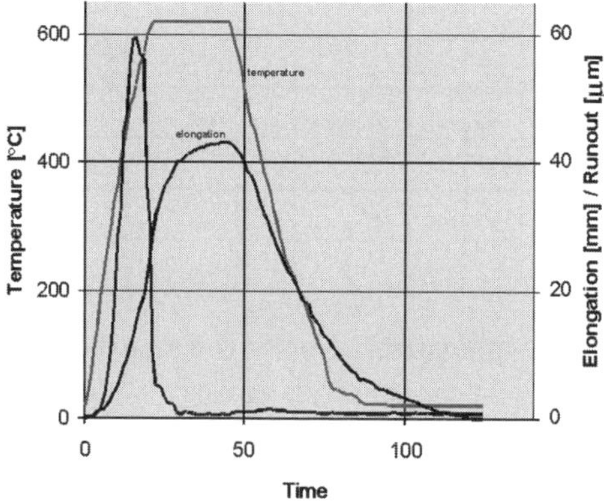

Figure 10. Heat Stability Test Chart

The test is carried out for the Cost E shafts with sensors with a resolution of 0.01 mm. Fig.11 shows a histogram of the C runout achieved in 21 shafts. The high level of symmetry in the properties of the shafts explains the high proportion of shafts which have no runout at all.

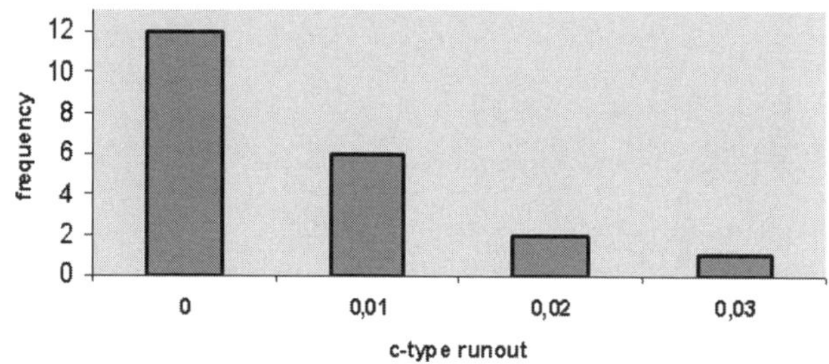

Figure 11. Frequency Distribution of Measured C-Type Runouts

Steel X13CrMoCoVNbNB 9-2-1 (COST-FB2)

Within the European COST programme, several series of trial melts with various alloy concepts were examined. These trials finally led to the steels COST E and COST F which are meanwhile used worldwide. By alloying martensitic 10% Cr steels with boron, it was found that further increases in creep strength could be achieved. This allows a higher application temperature and therefore also an increase of thermal efficiency. In this regard, creep values for up to approx. 80,000 h are available. The most promising type of alloy for application temperatures of up to 620°C is the steel FB2 with approx. 100 ppm boron and 1.2% cobalt. (Table 1) (4)

In order to verify the manufacturability of large production components and to determine the properties, the participants in the COST program decided to manufacture several trial shafts.

The trial shaft manufactured by Saarschmiede (diameter 1200 mm, total length 4000 mm, weight 28000 kg), was produced using the ESR process as is usual for 10% Cr steels. The mean chemical composition is shown in table 1.

Table 1 Chemical Composition of Steel X9 CrMoCoVNbNB 9-2-1 – COST FB2

		C	Si	Mn	P	S	Cr	Mo	Ni	V	Al	As	Cu	Sn	Sb	N	B	Co	Nb
Specification	min.	0.12	0.05	0.30			9.00	1.40	0.10	0.18	0.005					0.015	0.008	1.20	0.045
	max.	0.14	0.10	0.40	0.010	0.005	9.50	1.60	0.20	0.22	0.010	0.012	0.10	0.004	0.001	0.030	0.010	1.30	0.065
ESR Heat		0.14	0.08	0.31	0.006	0.001	9.28	1.51	0.15	0.19	0.007	0.002	0.031	0.004		0.026	0.009	1.32	0.053

As it is possible to take samples for chemical analysis during the remelting process, the boron content can be controlled and kept constant within a range of 20ppm (Fig. 12). At the beginning of the remelting process, first the chemical balance between ESR slag, electrode and molten metal is adjusted. When remelting this boron-alloyed ingot a larger amount of bottom discard resulted (Fig. 12).

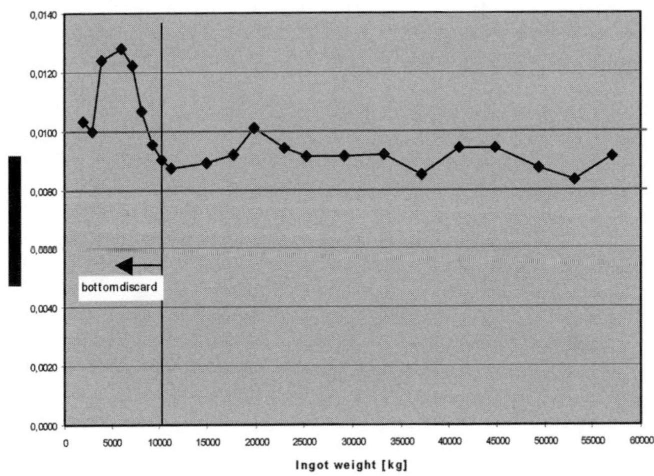

Fig. 12. Development of boron content according to ingot weight

From the ESR ingot manufactured with a 1300 mm diameter and weight of 57 t, a trial shaft was forged in several stretch forging operations and two upsetting operations in a 85 MN hydraulic press (Fig. 13). The forging temperatures were limited to a maximum of 1170°C in order to prevent an excessive in grain growth and to achieve optimum recrystallization behaviour.

When forging boron-alloyed 10% Cr steels, it was observed that in steels with higher boron content, there was a dramatic decrease in ductility (reduction of area) and therefore also a reduction in the forgeability in the temperature range between 1250°C and 1150°C. The utilizable temperature range when forging these steels is, therefore, limited and the upper temperature is restricted to approx. 1150°C. For the steel FB2 with approx. 100 ppm boron, this effect does not occur and so the forgeability at high temperatures is not restricted, Fig. 14.

Fig. 13. FB2 trial rotor made by Saarschmiede

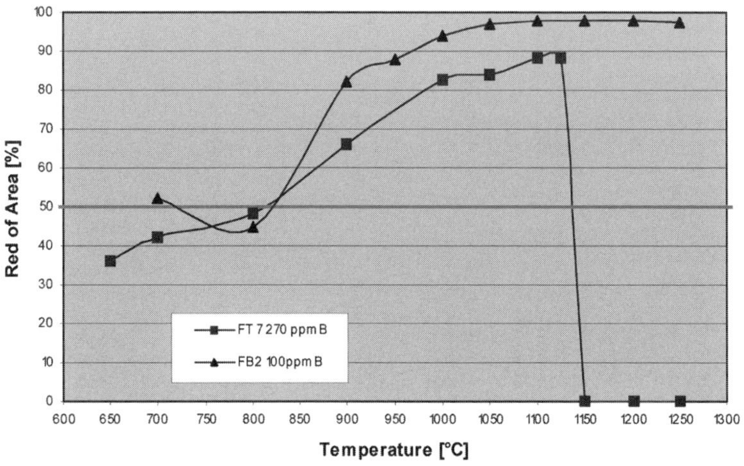

Fig. 14. Test of forgeability for boron alloyed 10% Cr steels

After the forging operation, the forging was subjected to preliminary heat treatment with two martensitic transformations and subsequent annealing in order to achieve a uniform grain microstructure and to improve the characteristics for later ultrasonic inspection. Preliminary heat treatment for boron-alloyed 10% Cr steels with pearlitic transformation is generally not considered since the transformation is a long, slow process and leads to a very incomplete transformation.

The quality heat treatment for adjusting the mechanical properties is carried out by hardening at 1,100°C and double annealing. The aim was to achieve a 0.2 % yield strength of ≥ 700 MPa. Double annealing is performed in order to ensure a totally annealed martensitic microstructure:

1. Austenitizing 1100°C//16h/ water spraying

2. 1st tempering 570°C//22h/ furnace and air cooling

3. 2nd tempering 700°C//22h/ furnace and air cooling

In the state after the preliminary heat treatment and after the quality heat treatment (quenching and tempering), an ultrasonic inspection was carried out in each case (Fig. 15). The trial shaft was free of faults and, after QHT, a good value regarding minimum detectable defect size (MDDS) of 2.0 mm was achieved at a diameter of 1215 mm. This value lies within the expected range and is sufficient for all applications with high pressure turbine shafts in this dimensional range. As figure 15 shows, through QHT, it was possible to improve the MDDS in the body by 0.5 mm compared to the state after preliminary heat treatment. In general, the MDDS in boron-alloyed 10% Cr steels remains a critical value and there has been a deterioration compared to e.g. the steel COST E.

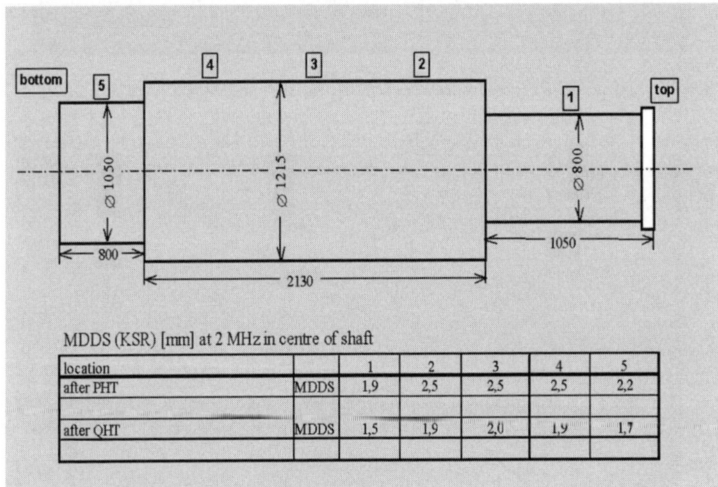

MDDS (KSR) [mm] at 2 MHz in centre of shaft

location		1	2	3	4	5
after PHT	MDDS	1,9	2,5	2,5	2,5	2,2
after QHT	MDDS	1,5	1,9	2,0	1,9	1,7

Fig. 15. NDT results and dimensions

Mechanical properties and microstructure

The trial shaft was subjected to intensive examination in order to determine the mechanical properties (Fig. 16). In order to further examine the creep and creep fatigue behaviour as part of a German research program, a test slice was removed from the large diameter. The homogeneity of the properties between the rim and axial core is remarkable. There is no decrease in the 0.2% yield strength in the axial core and also no reduction in toughness. The creep tests currently being carried out also show very homogeneous behaviour between the rim and the core of the shaft. In the trial running times up till now, no reduction in creep strength or increased creep was determined in the core of the shaft.

In comparison with boron-free 10% Cr steels (COST E), the steel FB2 demonstrates somewhat greater FATT values and lower toughness values. However, these are still sufficiently high to allow use in high pressure shafts and do not present a problem for the fracture mechanical design of the shafts.

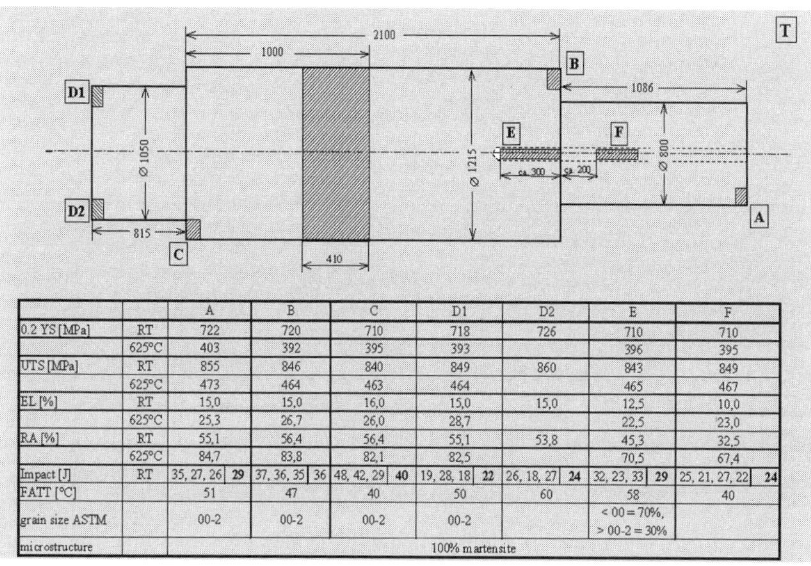

		A	B	C	D1	D2	E	F
0.2 YS [MPa]	RT	722	720	710	718	726	710	710
	625°C	403	392	395	393		396	395
UTS [MPa]	RT	855	846	840	849	860	843	849
	625°C	473	464	463	464		465	467
EL [%]	RT	15,0	15,0	16,0	15,0	15,0	12,5	10,0
	625°C	25,3	26,7	26,0	28,7		22,5	23,0
RA [%]	RT	55,1	56,4	56,4	55,1	53,8	45,3	32,5
	625°C	84,7	83,8	82,1	82,5		70,5	67,4
Impact [J]	RT	35, 27, 26 **29**	37, 36, 35 36	48, 42, 29 **40**	19, 28, 18 **22**	26, 18, 27 **24**	32, 23, 33 **29**	25, 21, 27, 22 **24**
FATT [°C]		51	47	40	50	60	58	40
grain size ASTM		00-2	00-2	00-2	00-2		< 00 = 70%, > 00-2 = 30%	
microstructure		100% martensite						

Fig. 16. Mechanical properties at rim and centre locations of FB2 trial shaft

Metallographic examinations showed that, across the whole cross-section, full hardening with a 100% annealed martensitic microstructure was achieved (Fig. 17). In all sample locations examined, the microstructure was free of δ-ferrite.

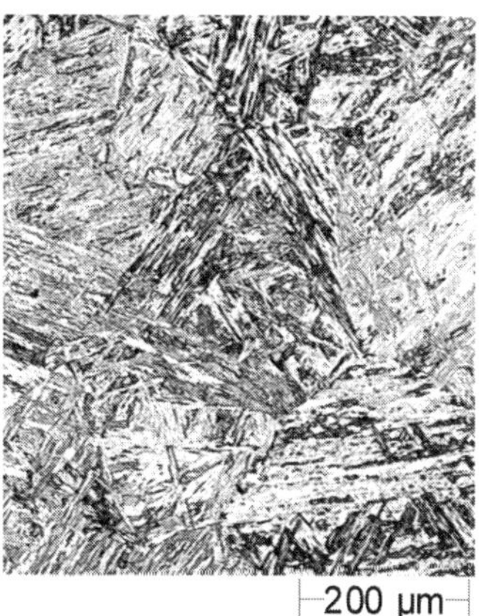

Fig. 17. Microstructure with 100% tempered martensite

In order to examine the homogeneity of the chemical composition and, in particular the distribution of boron across the cross-section of the rotor, samples for chemical analysis were taken from various locations. The results can be seen in fig. 18. The figures of the check analyses show that the chemical composition is very homogeneous through the cross section and there were only slight differences compared to the analysis of the melt. One exception to this is the boron content which systematically proved to have lower values in the check analysis than in the melt analysis. This effect is currently the subject of tests (round robin tests) which are being carried out within the scope of the COST programme.

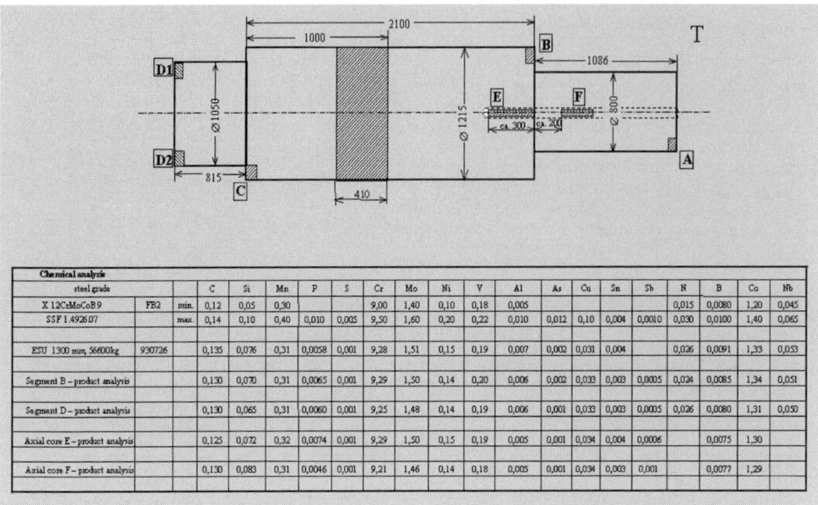

Chemical analysis																				
steel grade			C	Si	Mn	P	S	Cr	Mo	Ni	V	Al	As	Cu	Sn	Sb	N	B	Co	Nb
X 12CrMoCoB 9	FB2	min.	0,12	0,05	0,30			9,00	1,40	0,10	0,18	0,005					0,015	0,0080	1,20	0,045
SSF 1.4926.07		max.	0,14	0,10	0,40	0,010	0,005	9,50	1,60	0,20	0,22	0,010	0,012	0,10	0,004	0,0010	0,030	0,0100	1,40	0,065
ESU 1300 mm, 56600kg	930726		0,135	0,076	0,31	0,0058	0,001	9,28	1,51	0,15	0,19	0,007	0,002	0,031	0,004		0,026	0,0091	1,33	0,053
Segment B – product analysis			0,130	0,070	0,31	0,0065	0,001	9,29	1,50	0,14	0,20	0,006	0,002	0,033	0,003	0,0005	0,024	0,0085	1,34	0,051
Segment D – product analysis			0,130	0,065	0,31	0,0060	0,001	9,25	1,48	0,14	0,19	0,006	0,001	0,033	0,003	0,0005	0,026	0,0080	1,31	0,050
Axial core E – product analysis			0,125	0,072	0,32	0,0074	0,001	9,29	1,50	0,15	0,19	0,005	0,001	0,034	0,004	0,0006		0,0075	1,30	
Axial core F – product analysis			0,130	0,083	0,31	0,0046	0,001	9,21	1,46	0,14	0,18	0,005	0,001	0,034	0,003	0,001		0,0077	1,29	

Fig. 18. Product analysis at different rim and centre locations of FB2 trial shaft

In the meantime, Saarschmiede has commenced with the manufacture of the first production component. This part manufactured for a medium pressure shaft has a diameter of 1090 mm, a length of 2150 mm and a weight (rough machined) of 15150 kg. The quality heat treatment has already been performed and the specified values for ultrasonic inspection with MDDS of around 1.7 mm and the specified mechanical properties could be achieved without difficulty.

Summary

Saarschmiede is an active partner for manufacturers of high efficient power plants by supplying high chromium steels for elevated temperatures. The steels Cost E and Cost F enjoy an increasing market worldwide. These types of steel are produced reliably in rising quantities. Saarschmiede is contributing to research and development projects by the manufacture of trial melts as well as full size trial rotors. Steel FB2 containing boron is a most promising candidate for the next turbine generation. A trial rotor has been produced successfully.

References

1. N. Blaes, D. Bokelmann, P. Braun, B. Donth, G. Weides, Y. Hirakawa, Y. Kadoya, R. Magoshi and M. Tanaka. *Largest turbine rotors ever manufactured from 10% Cr-steels.* 16th International Forgemasters Meeting, Sheffield, 2006.

2. N. Blaes, B. Donth, K.H. Schönfeld and D. Bokelmann. *High Temperature Steel Forgings for Power Generation.* EPRI Fourth International Conference of Advances in Material Technology for Fossil Power Plants, USA, 2004.

3. Stahl-Eisen-Prüfblatt des Vereins Deutscher Eisenhüttenleute. *Warmrundlaufprüfung an Turbinenwellen.* SEP 1950, 3. Ausgabe November 1993.

4. T.-U. Kern, M. Staubli, K.H. Mayer, B. Donth, G. Zeiler and A. DiGianfrancesco. *The European Effort in Development of new high temperature Rotor Materials – COST 536.* Proceedings of the 8th Liége´ Conference, Materials for advanced power engineering, Part II, p. 843-845

Advances in Materials Technology for Fossil Power Plants
Proceedings from the Fifth International Conference
R. Viswanathan, D. Gandy, K. Coleman, editors, p 366-376

DOI: 10.1361/cp2007epri0366

CREEP BEHAVIOUR AND MICROSTRUCTURAL ANALYSIS OF FB2 TRIAL ROTOR STEEL

A. Di Gianfrancesco [1)], L. Cipolla [1)], D. Venditti [1)], S.Neri [2)], M. Calderini [2)]

[1)] Centro Sviluppo Materiali, Via di Castel Romano 100, Roma 00128, Italy
[2)] Società delle Fucine, Viale B. Brin 218, Terni 05100, Italy

Abstract

The development of new ferritic-martensitic steels for rotor applications was one of the main actions in the joint research projects COST 501 and COST 522. During COST 501 several trial compositions of 9-10%Cr steels were tested. In the COST 522 the best candidate, coded FB2 - a 10%Cr steel with additions of Co and B, without W - was selected for scale-up from laboratory trial to full industrial component. Società delle Fucine (SdF) produced a FB2 prototype rotor using a conventional process route, based on ladle furnace and vacuum degassing.
A large creep test programme has been launched to define the creep properties of the full size component and the results are in line with these from laboratory material, in terms of creep resistance as well as ductility. The tests reached more than 30.000 hours and the assessment of creep-rupture tests targeted an improvement of 15-20 MPa to obtain 100.000 creep hours at 600°C with respect to Grade 92.
A programme of microstructural investigation has been recently launched to evaluate the evolution of the structure and to have better knowledge of the role of B as a creep strengthening element.

Keyword: COST 536, 10%Cr Steels, High Temperature Application, Creep strength, material development.

Introduction

The energy production is faced with the introduction of increasingly stringent emission regulations to safeguard health and to preserve the environment for the future generations.
The thermal efficiency is influenced by several factors, but the adoption of supercritical conditions by increasing steam temperatures and pressures plays a key role: very high temperatures and pressures makes mandatory the use of steels suitable for these severe conditions [1]. The increase of steam parameters from 600°C up to 650°C/300bar will generate an efficiency improvement of 8-10% with a corresponding CO_2 reduction [2]. These advanced steam parameters require materials with adequate creep strength and resistance to oxidation. Experience with austenitic materials was unsatisfactory showing considerably restrictions in the operational flexibility of the plants due to the difference in the thermal expansion between austenitic and ferritic components and the consequent stresses. The class of the 9-12% Cr steels offers the highest potential to meet the required creep resistance level for the critical components in steam power plants.
In Europe the main efforts to improve the 9-12%CrMoV steels were concentrated in the COST (**CO**-operation in **S**cience and **T**echnology) Programmes: COST501 (1986-1997), COST 522 (1997-2003). In these programmes new ferritic steels for forging, casting and pipework were developed and characterised to increase the operating steam temperatures from 538-565°C up to

580-600°C. [1-5]. In the running COST 536 Programme (2004-2009) the qualification of these materials are still ongoing.
The HP and IP rotor steels require basically an high creep strength due to the high steam temperature and desire to minimize the use of dovetail cooling steam in Hp and/or IP sections [6].

Materials development for 600°C turbine rotor applications

Within COST 501 a series of advanced steels for HP and IP rotor forgings and castings application as given in Table 1 was qualified.

COST	**Forged Steels**	**C**	**Cr**	**Mo**	**W**	**V**	**Ni**	**Nb**	**N**	**B**	100MPa 100.000h	**Status**
	1CrMoV	0.25	1.0	1.0		0.25					550°C	Long term operating
	12CrMoV	0.23	11.5	1.0		0.25					570°C	Long term operating
501	Type F	0.1	10	1.0	1	0.2	0.7	0.05	0.05		597°C	Operating in plant
501	Type E	0.1	10	1.5		0.2	0.6	0.05	0.05		597°C	Operating in plant
501	Type B	0.2	9.0	1.5		0.2	0.1	0.05	0.02	0.01	620°C	
522	Type FB2 (SdF)	0.13	9.32	1.47		0.2	0.16	0.05	0.019	0.085		Trial rotor manufactured

Table 1: Compositions of improved ferritic steels developed in COST 501 and operating temperature for 100MPa/100.000h of conventional and improved ferritic steels [2]

The steel B and the following FB2 have both addition of B, that it is known to generate a beneficial effect on long-term creep rupture strength.
After the good results obtained in the COST 501 on composition E, F and B a new analysis "FB2" was produced as trial melt. Furthermore the promising properties of the trial melt, at the beginning of the new COST 522, it was decided to scale up FB2 steel to industrial heat in order to manufacture a trial forged rotor. Boehler/Austria has manufactured a full-size rotor forging with a final weight of 17,000kg. The steel making process was Boehler-BEST, consisting in a pouring process with special measures to improve the homogeneity of the ingot [7].

A second proposal for a FB2 trial rotor manufacturing was brought into COST522 by the Italian SdF in Terni; it produced a 52,000kg ingot by conventional steel making (Ladle Furnace and vacuum degassing) with a final rotor weight of 28,000kg [2].
One more trial rotor has been produced by ESR remelting of 57,000kg ingot in Saarschmiede in Europe [8].
Also Doosan Heavy Industry, in the frame of the Korean development program, produced ESR ingot and a forged trial rotor [6].

The manufacture of trial rotor at Società delle Fucine (SdF)

The aim of the investigations is to qualify the different steelmaking processes for this class of Boron containing 10CrMoCoVB alloys. One of the main tasks is to answer the question of how much homogeneity in composition is achievable, and what the properties of the final rotor forging are.

Figure 1 shows the lay-out of the SdF manufacture route: from the melting shop to the final machining. Due to the characterisation work on experimental rotor the component machining was stopped at the step of NDT control after quality heat treatment.

The chemical composition of the SdF trial rotor is based on the FB2 trial melt from COST 501. Table 2 shows the chemical analysis of the cast product. Very good agreement with the aim composition has been obtained with very low content of residual elements.

FB2	C	Si	Mn	P	S	Cr	Mo	Ni	Al	B	Co	N	Nb	V
min	0.12	-	0.3	-	-	9.0	1.45	0.1	-	0.006	1.2	0.015	0.04	0.18
max	0.14	0.006	0.4	0.01	0.005	9.5	1.55	0.2	0.008	0.009	1.4	0.030	0.06	0.22
cast	0.14	0.032	0.32	0.007	0.003	9.1	15	0.14	0.001	0.009	1.23	0.015	0.046	0.2

Table 2: Required and obtained chemical composition of FB2 trial forged rotor
(residual elements: H_2= 1,2ppm; Sb= 0,001%; Sn=0,001%; As=0,006%; Cu=0,035%; W <0.01%)

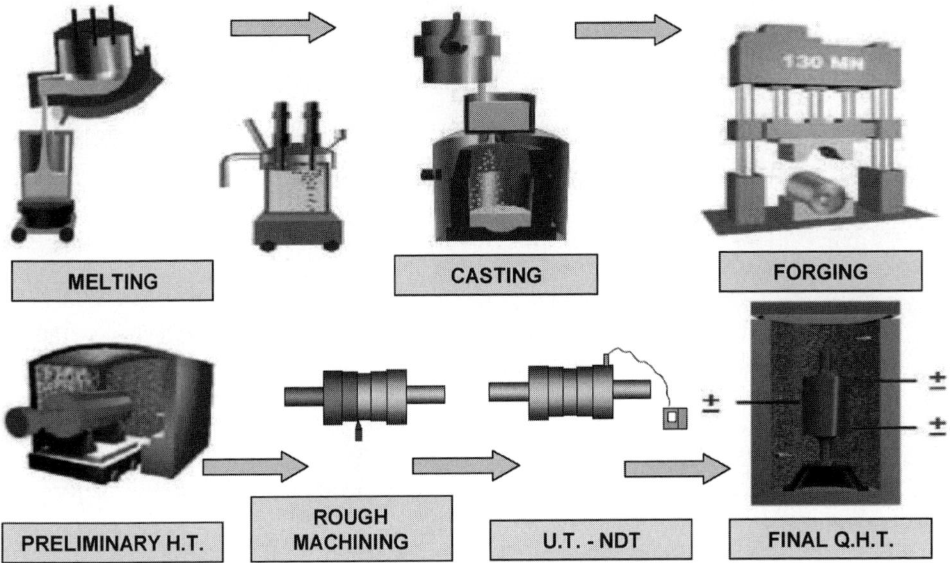

Figure 1: flow chart of rotor manufacturing route at SdF

The final dimensions of the trial component are shown in Figure 2. After forging the trial rotor has been treated as follows:
- Austenitizing: 1100°C/ 17h/oil quenched;
- 1st Tempering: 570°C/ 24h/ air cooled;
- 2nd Tempering: 700°C/ 24h/ air cooled.

The trial component in as treated condition and after preliminary machining has been subject to NDT ultrasonic inspection (US) with a 2MHz source. The maximum defects discovered (flat bottom hole equivalent) in the different positions are summarised in Table 3. Figure 3a show the trial component after final forging and heat treatment and Figure 3b during the NDT control tests.

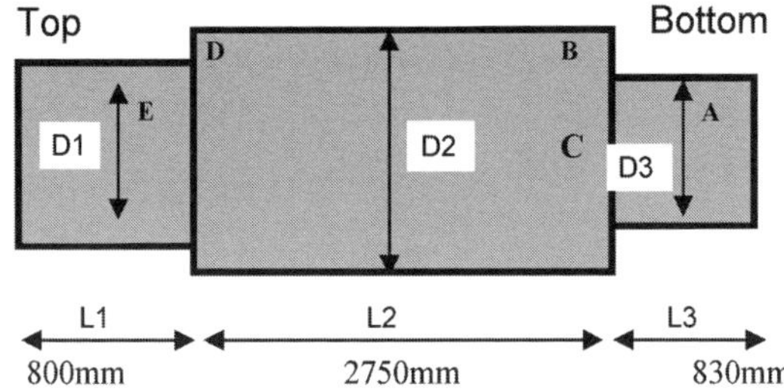

Figure 2: draw of trial rotor

FB2 SdF Trial rotor	Total (L1+L2+L3)	D1	D2	D3
Dimensions (mm)	4380	925	1110	790
Max Defect discovered (mm)		1.0	1.5	1.5

Table 3: Diameter and NDT results

Microstructural and mechanical characterisations

In order to verify the homogeneity of the properties of the trial rotor, the specimens for chemical analysis and mechanical tests have been obtained from different parts of the component (A, B, D, E and C positions from the core (see Figure 2)).

Figure 3: SdF trial forged rotor after heat treatment and during US NDT

The chemical analysis (Table 4) shows a very good homogeneity in the composition of the main alloy elements as well as for the boron and nitrogen content that could be critical for their distribution in a large component processed without remelting.

Position /Element	A	B	D	E	FB2 trial melt
C	0,12	0,13	0,13	0,15	0,13
Cr	9,08	9,13	9,11	9,02	9,32
Mo	1,57	1,59	1,59	1,62	1,47
Ni	0,14	0,14	0,14	0,14	0,16
V	0,21	0,21	0,21	0,22	0,20
Nb	0,054	0,054	0,054	0,060	0,05
B	0,0095	0,0096	0,010	0,012	0,0085
N	0,015	0,016	0,015	0,014	0,019
Co	1,28	1,27	1,27	1,29	0,96

Table 4: chemical composition (wt%) of forged trial rotor in different positions compared with the nominal composition. (Residual elements: H_2= 1.2ppm; Sb= 0.001%; Sn=0.001%; As=0.006%; Cu=0.035%)

The chemical composition of FB2 steel guarantees a fully martensitic structure as shown in the CCT diagram (Figure 4). In fact the microstructural analysis performed at CSM after the final heat treatment (normalizing + tempering) (Figure 5) shows a typical tempered martensitic structure with a 0-2 ASTM grain size. In Figures 6 is showed the FB2 microstructure as obtained by Transmission Electron Microscopy on thin foils. The prior austenite and the martensitic lath boundaries are decorated by a typical precipitation of $M_{23}C_6$ carbides. The microstructure contains a high dislocation density.

Conventional mechanical tests have been performed in different positions and orientation. In Figure 7 the values of the Fracture Appearance Transition Temperature in different positions are presented. Figure 8 shows the results of the tensile test in term of YS, UTS and ductility: elongation and reduction of area. It is possible to note a quite good homogeneity of the obtained values.

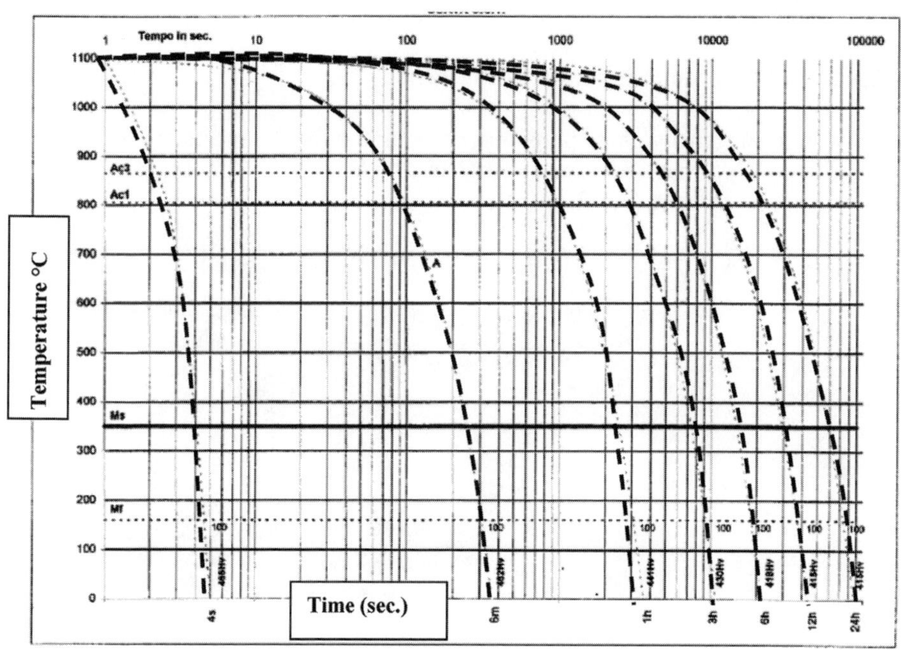

Figure 4: FB2 CCT diagram

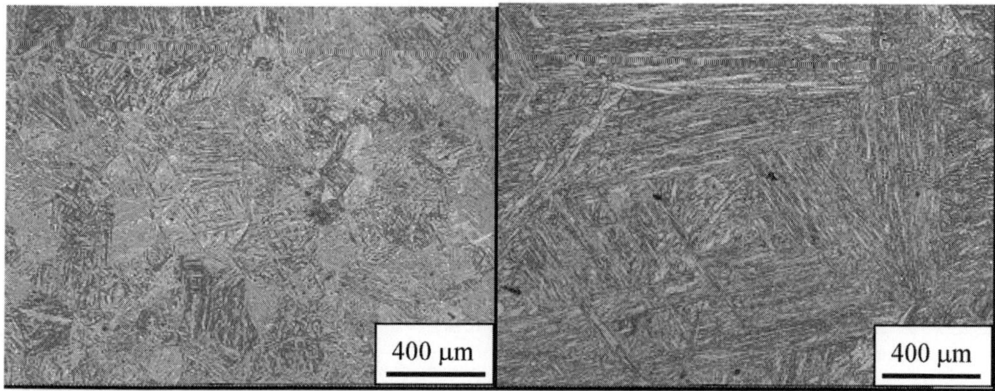

Figure 5: fully tempered martensitic microstructure of trial rotor after quality heat treatment

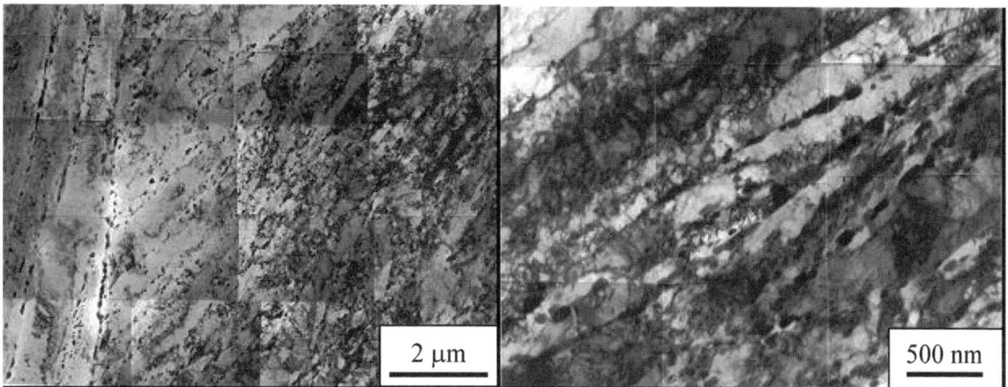

Figure 6: fully tempered martensitic microstructure of trial rotor after quality heat treatment by Transmission Electron Microscopy

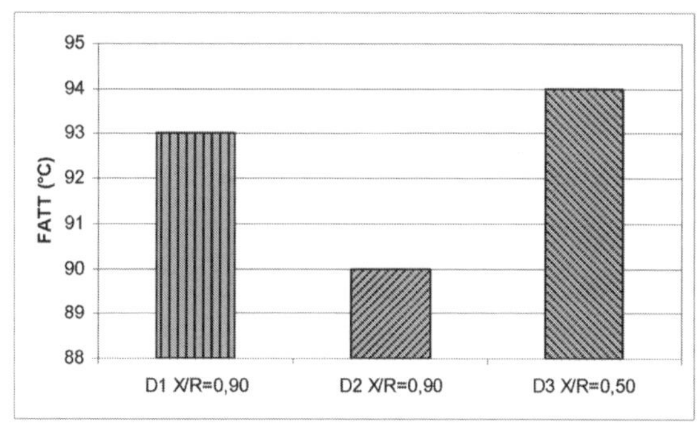

Figure 7: FATT in different positions of the trial rotor

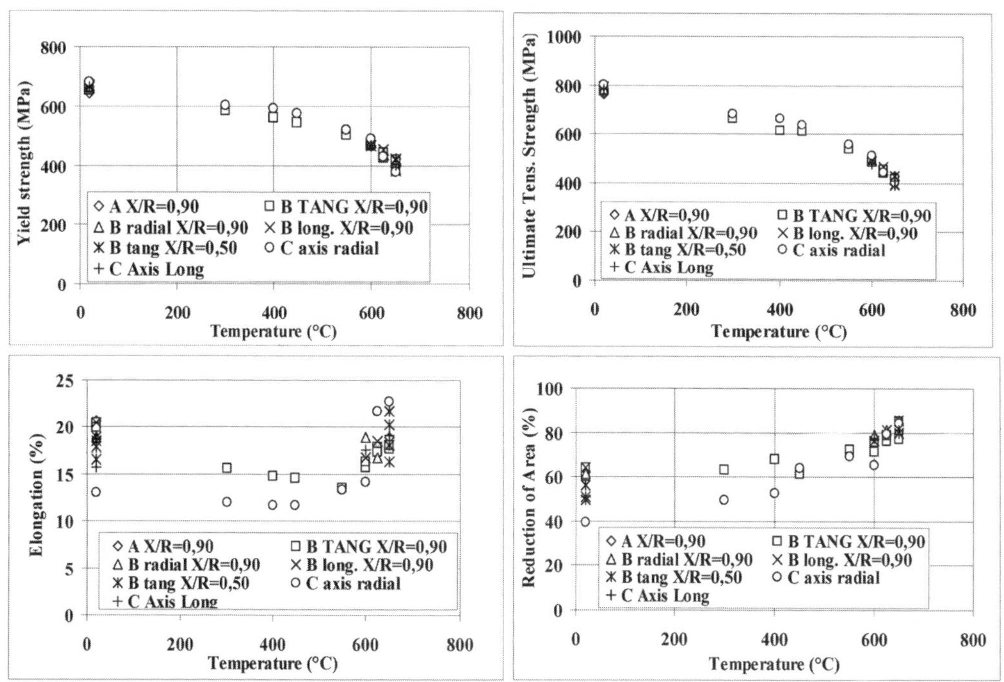

Figure 8: Hot mechanical properties of trial rotor in different positions and specimen orientations

Creep properties

A large creep test programme has been launched at CSM to qualify the trail rotor. Smooth and notched specimens have been machined from positions A, B and C and tested in the temperature range 600-650°C with continuous strain measurements. Some tests have been planned to reach the rupture in 100.000 hours. The creep results obtained on the SdF trial rotor confirm with the behaviour of the 500kg trial melt [9]. Figure 9 summarises the current status of the tests (open points are tests still running) compared with the master curves of COST 501 rotors E, F and B2.

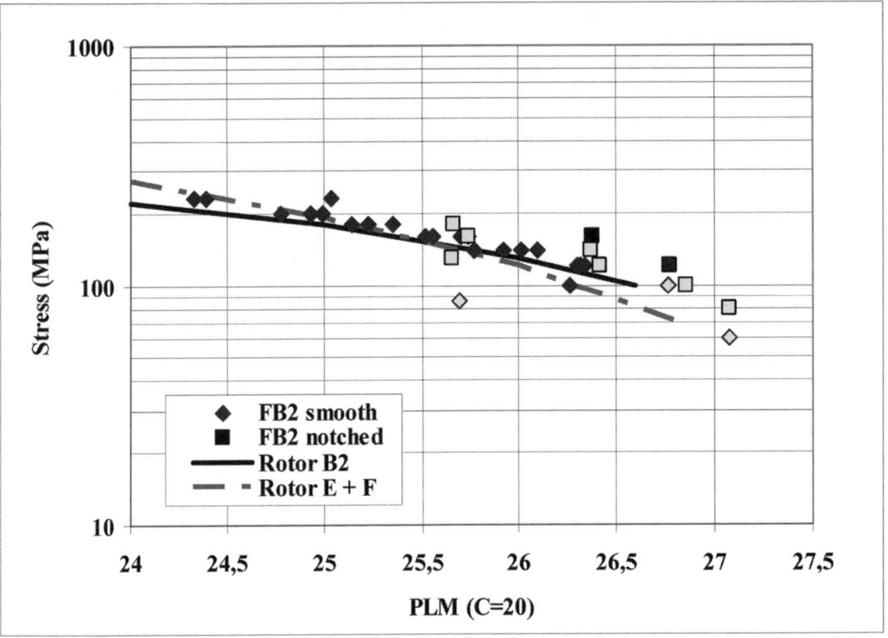

Figure 9: master curve of creep tests of SdF FB2 trial rotor (open point = specimens running) compared with B2 and E + F rotors curves

It is possible to observe that the results of FB2 trial rotor are in the upper band of the previous trial rotors.
A preliminary assessment of the creep results, compared with the recent ECCC assessment for Grades 91 and 92 shows an increase of 15MPa with respect to the Grade 92 at 600°C (Figure 10) [10,11].

Microstructural evolution

In parallel with the mechanical and creep tests a programme to investigate the microstructural evolution of the FB2 steel is started. The thermodynamic tool JMatPro predicts, at equilibrium (Figure 11), the main presence of $M_{23}C_6$ carbides and M(C, N), Laves phase and a little amount of Z-phase. As well known the Z-phase appearance has been identified as the main reason of the drop of the creep properties of all the recently developed 10-12%Cr steels [12].

The first results of microstructural analysis on crept specimens made by SEM+EDS (Energy Dispersion Spettrometry) show the presence of some large precipitates (white circle in Figures 12,13), probably $Fe(Cr,Mo)_2$ Laves phase, as just investigated in all the 9%Cr steels, but not yet the presence of Z-phase particles. STEM investigation are ongoing on these specimens and SADP (Selected Area Diffraction Pattern) will be performed to identify all the phases.

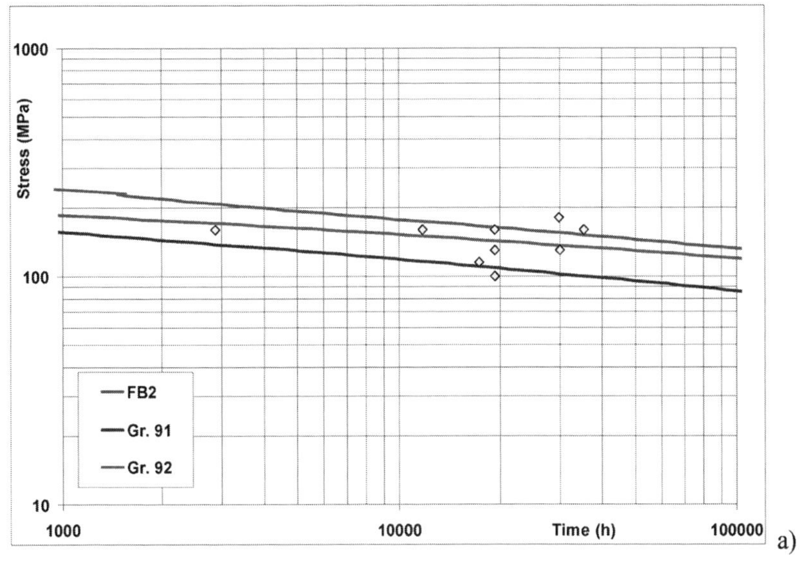

a)

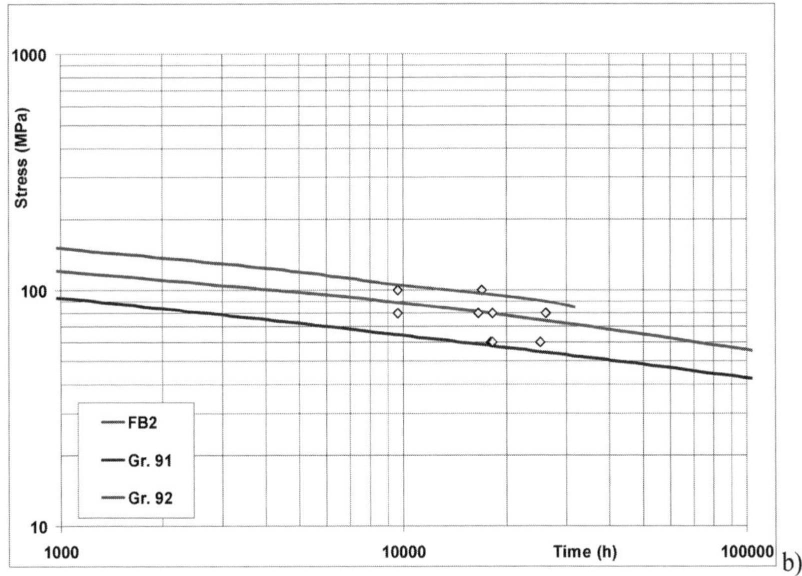

b)

Figure 10: preliminary creep data assessment of FB2 tests compared with Grades 91 and 92: a) 600°C; b) 650°C. (Yellow points are running tests)

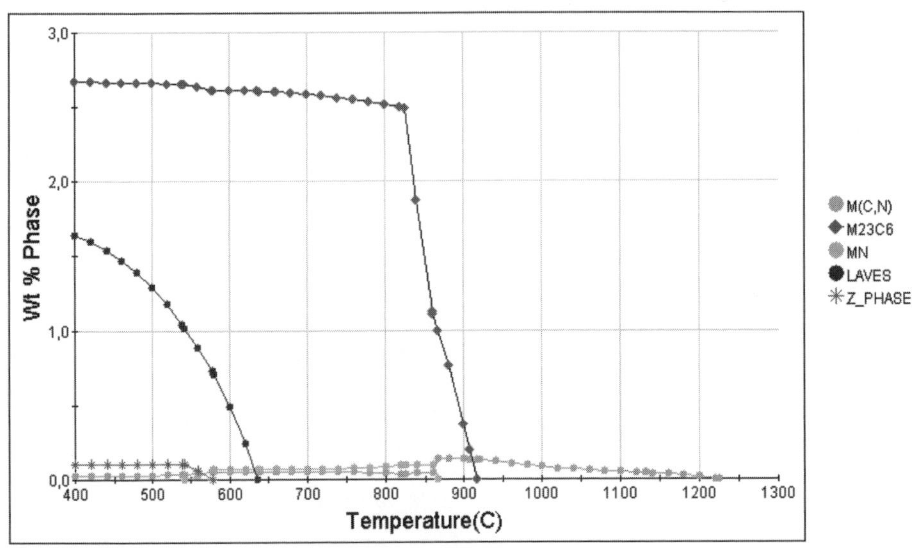

Figure 11: JMatPro equilibrium diagram for FB2 Steel

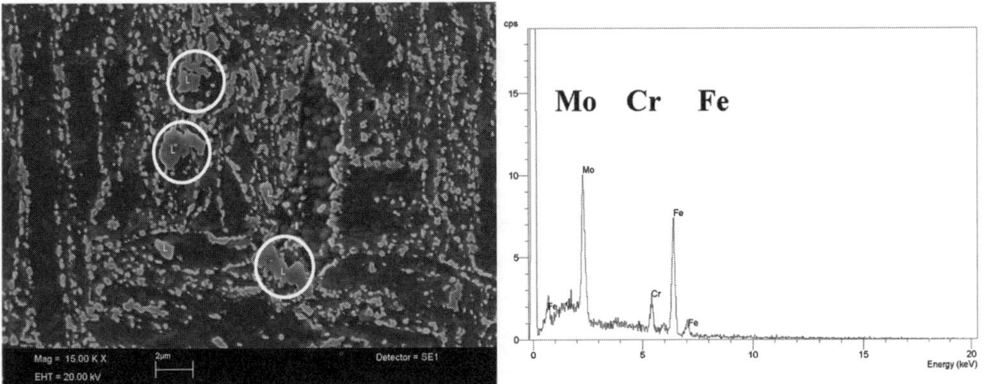

Figure 12: grain boundary precipitation in FB2 specimen crept at 625°C for 23576h by SEM + EDS

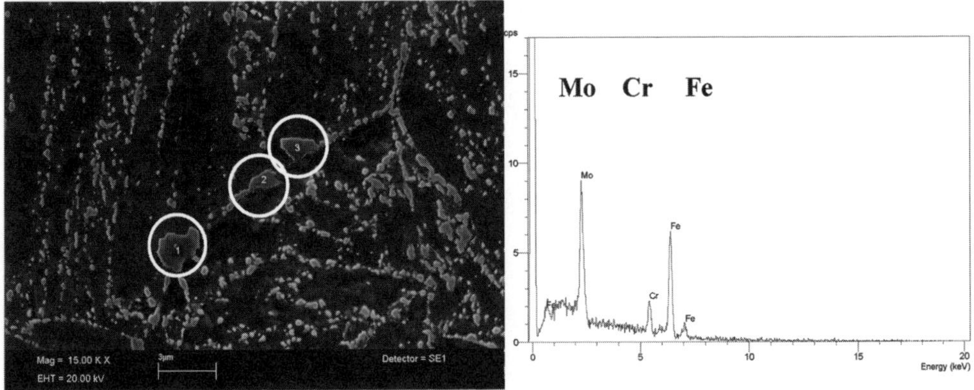

Figure 13: grain boundary precipitation of FB2 specimen crept at 650°C for 10026 hours by SEM + EDS

Summary and conclusions

The result obtained up to now on FB2 trial rotor produced at Società delle Fucine, without ESR ingot remelting, demonstrate that it is possible to realise full scale forged components by conventional process route with very good homogeneity of chemical analysis, mechanical behaviour and creep properties.

In terms of creep behaviour, FB2 seems to offer a significant improvement over the parent E and F type steels and it could be suitable for 625°C applications as forged HP and IP rotors. In fact the experience gained during the last few years seems guarantee that the 9%Cr are more stable materials from the microstructural evolution point of view [12,13], and suggests that the formation of some Z-phase precipitates may not significantly affect the long term creep behaviour of this steel. In fact a very few amount of Z- phase has founded in the main 9%Cr steels, but only after very long simulated service condition [14,15].

Of course these results have to be confirmed with:

- longer creep tests in order to generate a more consistent database, assessment and extrapolation. Tests still running are planned to reach 100.000 hours life;
- the TEM investigation on long term aged specimens in order to know the microstructural evolution of precipitates and in particular to verify the presence of Z-phase. For this item crept specimens will be used.

Acknowledgement

The authors are grateful to their colleagues and partners in the programme COST522 and 536 for their contributions and many discussions during the course of this endeavour. Many thanks are also extended to the COST Management Committee for their guidance in the programme and to the national funding bodies for their financial support of the individual national projects.

REFERENCES

1) R.W. Vanstone, Alloy design and microstructural control for improved 9-12%Cr power plant steels, Annex A, COST 522 Steam Power Plant, Final Report, 1998-2003
2) T.U.Kern, K. Wieghardt, H. Kirchner: Material and design solutions for advanced steam power plants, EPRI Fourth International Conference on Advanced in Materials Technology for Fossil Power Plants; October 25-28, 2004, Hilton Head Island, South Carolina USA
3) B.Scarlin, R.Vanstone, R.Gerdes: Materials developments for Ultrasupercritical Steam Turbines, Ibidem
4) B.Scarlin, T-U.Kern, M.Staubli: The European efforts in material development for 650°C USC Power Plants – COST522, Ibidem
5) Y.Tanaka, T.Azuma, K.Miki: Development of steam turbine rotor forging for high temperature application high temperature steel forgings for power generation, Ibidem
6) S. Ryu, J. Kim: Symposium on heat resistant steels and alloys for USC Power Plants 2007, 3-6 July 2007 Seoul, South Korea.
7) G.Zeiler, W.Meyer, K.Spiradek, J.Wosik: Experiences in manufacturing and long-term mechanical & microstructural testing on 9-12 % chromium steel forgings for power generation plants, EPRI Fourth International Conference on Advanced in Materials

Technology for Fossil Power Plants; October 25-28, 2004, Hilton Head Island, South Carolina USA

8) N.Blaes, B.Donth, K.H.Schönfeld, D.Bokelmann: High temperature steel forgings for power generation, Ibidem
9) T.U.Kern, M.Stabli, K.H.Mayer, B.Donth, G.Zeiler,A.Di Gianfrancesco: The European effort in development of new high temperature rotor materials – COST 536: Int. Conf. Materials for Advanced Power Engineering, 19-21 September 2006 Liege, Belgium
10) L. Cipolla, J. Gabrel: New creep rupture assessment of grade 91: 1st Conference Super High Strength Steel, 2-4 November 2005 Rome, Italy.
11) W.Bendick, J. Gabrel: Assessement of creep rupture strength fr new martensitic 9% Cr steels E911 and T/P92, ECCC Creep Conference September 2005, London, UK
12) J. Hald, H. Danielsen: Z-phase in 9-12%cr steels: EPRI Fourth International Conference on Advanced in Materials Technology for Fossil Power Plants; October 25-28, 2004, Hilton Head Island, South Carolina USA
13) S.Caminada, G.Cumino L.Cipolla, A.Di Gianfrancesco: Long term creep behaviour and microstructural evolution of ASTM grade 91 steel; ibidem
14) K. Sawada, H. Kushima, K. Kimura, M. Tabuchi; ISIJ International, vol.47 (2007) No. 5 pp 733-739,
15) S. Caminada, L. Cipolla, G. Cumino, A. Di Gianfrancesco, D. Venditti; Symposium on heat resistant steels and alloys for USC Power Plants 2007, 3-6 July 2007 Seoul, South Korea.

Advances in Materials Technology for Fossil Power Plants
Proceedings from the Fifth International Conference
R. Viswanathan, D. Gandy, K. Coleman, editors, p 377-390

DOI: 10.1361/cp2007epri0377

Development of Low Thermal Expansion Ni Base Superalloy for Steam Turbine Applications

Takehiro Ohno*, Akihiro Toji*, Toshihiro Uehara*, Gang Bao**,
Takashi Sato**, Jun Sato*** and Shinya Imano***
*Hitachi Metals, Ltd.
2107-2, Yasugi-cho, Yasugi-shi, Shimane-ken 692-8601, Japan
**Babcock-Hitachi K.K. Japan
Takara-machi-Kure-shi, Hiroshiima-ken, 737-0029, Japan
***Hitachi, Ltd.
1-1 Omika-cho, 7-chome, Hitachi-shi, Ibaragi-ken, 319-1292, Japan

Abstract

Recently, operating temperature of USC plants is getting higher and planned up to 700°C. Austenitic superalloys are promising instead of ferritic heat resistant steels because of necessity of high strength at around 650-700°C. In general, austenitic Ni base superalloys indicate higher creep rupture strength than ferritic heat resistant steels, however their thermal expansions are higher, creep rupture ductilities lower, and costs higher. Firstly, to obtain low thermal expansion we have noticed Mo containing superalloy, then investigated the effect of Mo and Co content, the amount of gamma prime phase, Al/Ti ratio in gamma prime phase on thermal expansion, tensile properties, and creep rupture properties, compared with conventional Mo containing alloy252 as a reference. As a result, a developed superalloy, although it contains no Co, combined with modified heat treatment, indicated much higher creep rupture elongation than Alloy252, keeping low thermal expansion and high creep rupture strength close to Alloy252.

Furthermore, Creep-rupture properties at 700°C up to about 20,000h were evaluated for long-term applications, and weldability and mechanical properties of weld joint at 750°C were also investigated for boiler tube applications.

Introduction

Main steam temperature in Ultra-Super Critical (USC) steam power generation plants have reached up to 600°C and is planed to reach higher temperature. In this case, since creep strengths of ferritic heat-resisting steels are thought to be insufficient, applications of γ ' (gamma-prime) precipitation strengthened austenitic superalloys are considered. Austenitic superalloys have sufficient creep strengths, but there are some disadvantages; their thermal expansions are higher, creep-rupture elongations lower, and cost higher than those of ferritic steels.

In order to reduce the thermal expansions of superalloys, the additions of solid-solute alloying elements with low thermal expansions, such as Mo is thought to be effective approach. In conventional superalloys, for example, Alloy 252 contains 10% Mo [(1)] and its thermal expansion is comparatively low and close to that of ferritic steels. Alloy 252, however, contains expensive element: 10% Co, and its creep-rupture elongation is not high enough.

In this study, the effects of Mo, Co, Al/Ti ratio in γ ' phase, the amount of γ ' phase on thermal expansion coefficient and creep-rupture properties in Mo containing γ ' strengthening superalloys including Alloy252 as a reference were investigated to develop an austenitic Ni base superalloy applicable to steam turbines and boiler tubes operated at above 600°C.

Alloy Design

We assumed that the alloys in this study consist of three phases: matrix (γ phase), γ 'phase and carbides. Firstly, we designed each phase separately, then combined three phases with designed amounts of each phase, and calculated chemical compositions of the alloys. Designed compositions and amounts of each phase are shown in Table 1. We have chosen Ni, Cr, Mo and Co as matrix composition, Ni_3 (Al, Ti) asγ ' phase and TiC as carbide. Calculated alloy compositions based on the alloy design in Table1 are shown in Table2.

Alloy No.2 was a base alloy in this experiment, contains 10% Mo, the same amount in Alloy252, but no Co and C was decreased compared with in Alloy252. With No.2 alloy as a base alloy, the amount of C was varied in No.1 alloy, the Al/Ti ratio, in No.31, 32 and 33 alloys, the amount of γ ′ phase, in No.41 alloy, the amount of Mo, in No.51, 52 and 53 alloys. The effect of the amount of Co could be evaluated by comparing No.1 alloy with alloy252.

Table 1 Designed Compositions and Amounts of each phase of Experimental Alloys

Alloy No.	Matrix (γ) Chemical Composition (mass%)				γ ′ ($Ni_3(Al,Ti)$) Amount (mol%)	γ ′ Chemical Composition Atomic Ratio	γ ′ Chemical Composition Al/Ti	TiC Amount (mol%)
	Cr	Mo	Co	Ni				
Alloy252	24	12	12	52	20	$Ni_3(Al_{0.48}Ti_{0.52})$	0.92	1.16
1	24	12	0	64	20	$Ni_3(Al_{0.48}Ti_{0.52})$	0.92	1.16
2	24	12	0	64	20	$Ni_3(Al_{0.48}Ti_{0.52})$	0.92	0.28
31	24	12	0	64	20	$Ni_3(Al_{0.58}Ti_{0.42})$	1.38	0.28
32	24	12	0	64	20	$Ni_3(Al_{0.68}Ti_{0.32})$	2.13	0.28
33	24	12	0	64	20	$Ni_3(Al_{0.78}Ti_{0.22})$	3.55	0.28
41	24	12	0	64	17.5	$Ni_3(Al_{0.58}Ti_{0.42})$	1.38	0.28
51	24	8	0	68	20	$Ni_3(Al_{0.68}Ti_{0.32})$	2.13	0.28
52	24	4	0	72	20	$Ni_3(Al_{0.68}Ti_{0.32})$	2.13	0.28
53	24	0	0	76	20	$Ni_3(Al_{0.68}Ti_{0.32})$	2.13	0.28

Table 2 Calculated Chemical Compositions of Experimental Alloys (mass%)

Alloy No.	C	Cr	Mo	Co	Al	Ti	Ni
Alloy252	0.12	18.9	9.7	10	1.13	2.76	Bal.
1	0.12	18.9	10.1	0	1.06	2.61	Bal.
2	0.04	19.1	9.8	0	1.15	2.71	Bal.
31	0.03	19.1	10.0	0	1.33	1.87	Bal.
32	0.03	19.2	9.9	0	1.56	1.46	Bal.
33	0.03	19.0	10.0	0	1.68	1.03	Bal.
41	0.03	20.0	10.0	0	1.16	1.63	Bal.
51	0.03	19.1	7.0	0	1.51	1.47	Bal.
52	0.03	19.2	4.1	0	1.55	1.51	Bal.
53	0.03	19.3	0	0	1.61	1.53	Bal.

Experimental Procedure

10kg ingots were vacuum induction-melted and hot-forged to 30mm square bars. Specimens were machined from these bars and heat treated as follows.

Solution treatment: 1066°C×4h, air cool

Double aging treatment: 850°C×4h, air cool + 760°C×16h, air cool

Heat treatment conditions described above were determined by several investigations, especially with the intention of improving creep-rupture ductility. Major investigation results are shown in Fig.3 in later section.

Creep-rupture tests were carried out by using smooth and notched combined specimens.

Results and Discussion

Effects of several factors on coefficients of thermal expansion

Fig.1 shows effects of the amount of Co in matrix (γ phase), Al/Ti ratio in γ ' phase, the amount of γ ' phase and the amount of Mo in γ on the coefficients of thermal expansion from RT to 600°C. Numbers shown in Fig.1 indicate the alloy No. in Table 1 and 2.

From Fig.1, followings were found; (1) Co hardly affected the coefficient of thermal expansion, (2) Increasing Al/Ti ratio in γ ' phase above 1.38 increased the coefficient of thermal expansion, (3) Increasing the amount of γ ' phase increased the coefficient of thermal expansion, (4) Increasing the amount of Mo strongly decreased the coefficient of thermal expansion.

Mo is the most effective element in this alloy system to lower the coefficient of thermal expansion. By increasing Mo content to above 18% in γ , the coefficient of thermal expansion decreased to the level close to that of ferritic steels. It was considered, however, a large amount addition of Mo might cause problems in ingot-making and forging processes in actual production. Thus, we decided that appropriate amount of Mo was around 12% in γ .

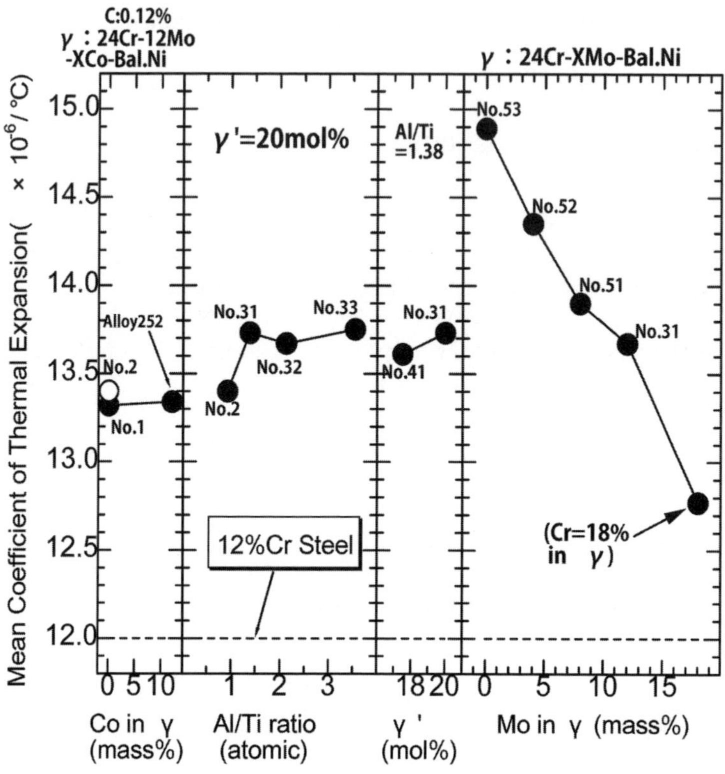

Fig.1 Effects of Several Factors on the Coefficient of Thermal Expansion from Room Temperature to 600°C.

Effects of several factors on high-temperature tensile strength

Fig.2 shows effects of the amount of Co in matrix (γ phase), Al/Ti ratio in γ ' phase, the amount of γ ' phase and the amount of Mo in γ on the tensile strength at 600 °C. The tensile strength at 600°C increased as Co content, the amount of γ ' phase and Mo content in γ increased, and decreased as Al/Ti ratio in γ ' phase increased. All alloys except No.53 showed higher strength than the dotted line in Fig.2, indicating the tensile strength of a representative superalloy Alloy 80A.

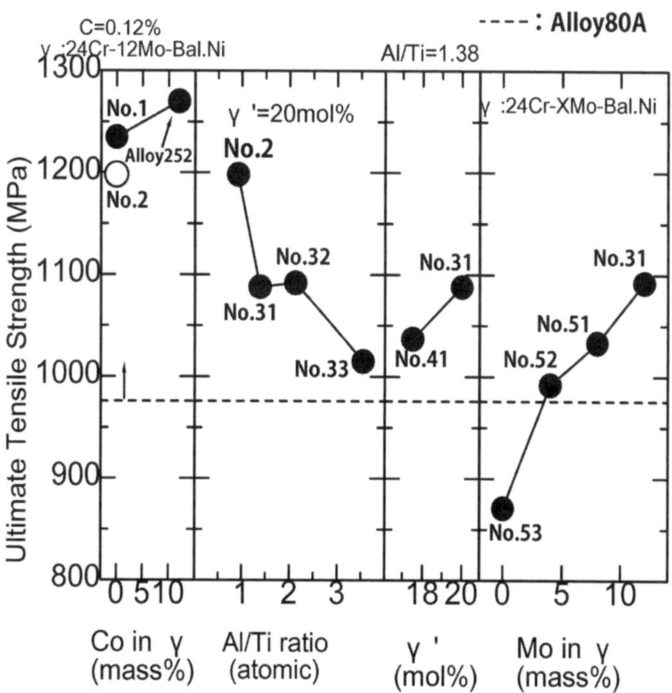

Fig.2 Effects of Several Factors on the Tensile Strength at 600°C

Effects of several factors on creep-rupture properties

Fig.3 shows the effect of the first aging treatment temperatures on creep rupture properties of Alloy No.2. The creep-rupture elongation and reduction of area were increased by the first aging at 850°C.

Fig.4 shows the effects of Co in matrix (γ phase) and C content on creep-rupture properties. The creep-rupture life slightly increased as Co content increased and C content decreased. The creep-rupture ductilities did not affected by these factors strongly.

Fig.5 shows the effects of Al/Ti ratio in γ' phase, the amount of γ' phase and the amount of Mo in γ on creep-rupture properties. The creep-rupture life decreased as Al/Ti ratio in γ' increased, but increased as the amount of γ' phase and Mo content increased. Mo strongly affected the creep-rupture strength. On the other hand, the reduction of area showed reverse tendency. It was indicated, to obtain good creep-rupture ductility above 30% in reduction of area with the addition of 12% Mo in γ, high Al/Ti ratio in γ' above 1.38 and small amount of γ' were effective.

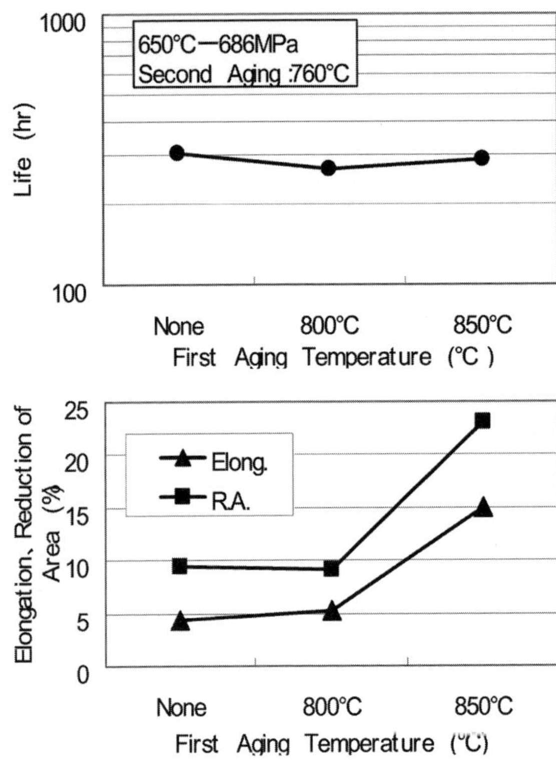

Fig.3 Effects of First Aging Temperature on Creep-rupture Properties of Alloy No.2

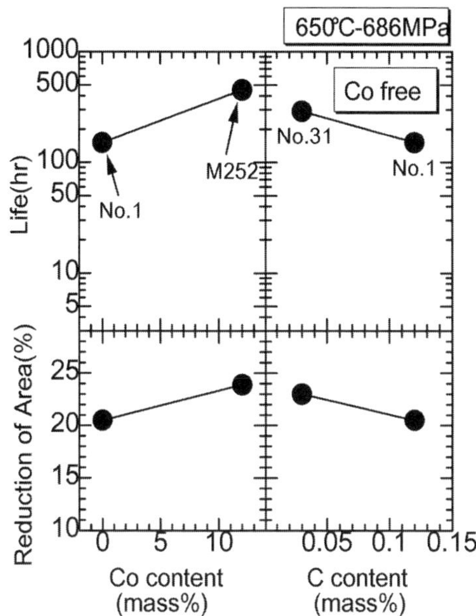

Fig.4 Effects of Co and C Contents on Creep-rupture Properties at 650℃- 637MPa.

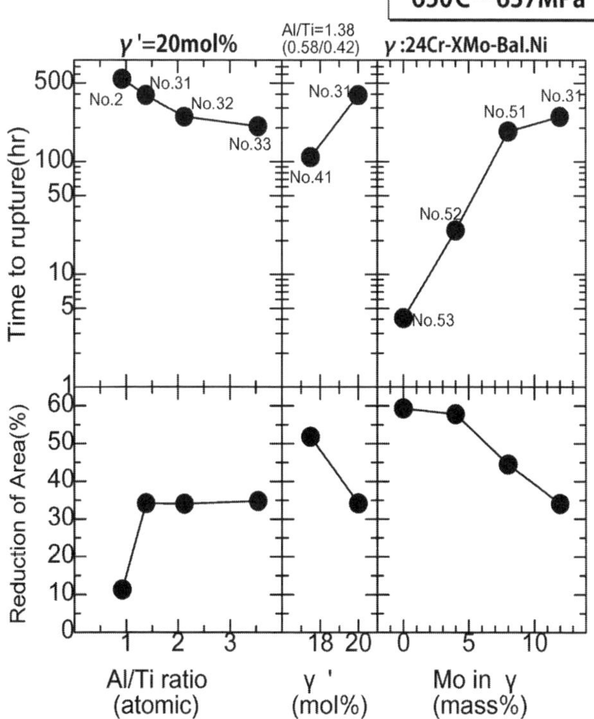

Fig.5 Effects of Several Factors on Creep-rupture Properties at 650℃-637MPa.

Developed alloy USC141

From above results, we have determined several factors as follows; (1) Mo content of 12% in γ (10% in alloy) to obtain low thermal expansion and high creep-rupture strength, (2) Al/Ti ratio in γ ' phase of 1.38 to obtain good balance of tensile strength and creep-rupture ductility, (3) the amount of γ ' phase of 17.5% to obtain good creep-rupture ductility, (4) Co content of no addition. Thus, we selected No.41 alloy for candidate for USC steam turbine applications and named USC141.

Properties of the developed alloy USC141

Photo.1 shows microstructures of USC141 after solution treated and aged indicating uniform precipitation of γ ' particles and grain boundary carbides. Fig. 6 shows creep-rupture strength at 600°C ~ 700°C. Creep-rupture lives up to about 20,000h at 650°C and 700°C were obtained and no significant degradation of creep-rupture strength was indicated. This suggests that USC141 has sufficient creep-rupture strength for long-term applications. Fig. 7 shows Larson-Miller plots of creep-rupture lives of USC141 compared with those of other superalloys and 12%Cr steel. USC141 also revealed good creep- rupture ductilities: over 40% of reductions of area after long-time creep-rupture tests.

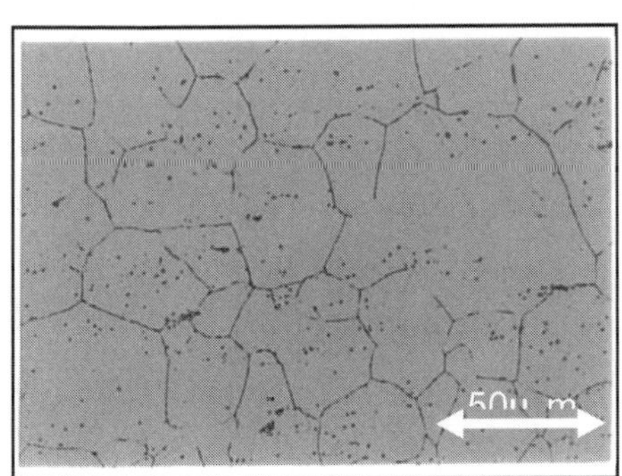
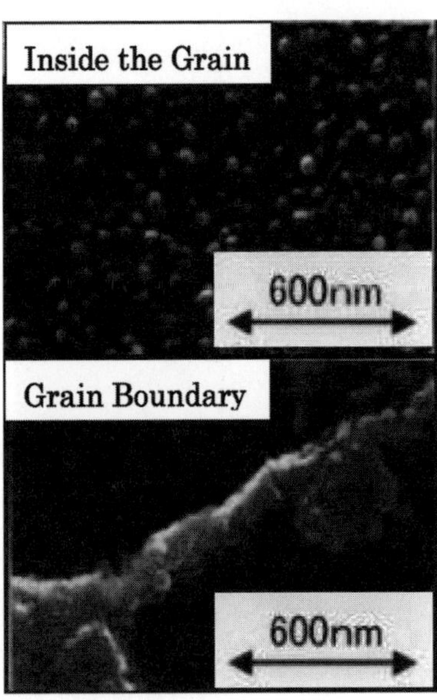

Fhoto.1 Microstructures of USC141 after Heat Treatment

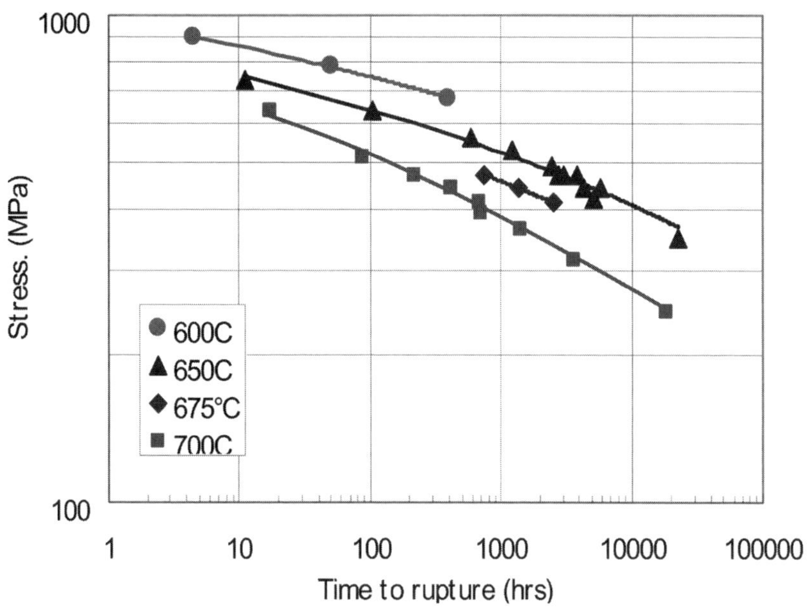

Fig. 6 Creep-rupture Strength of USC141

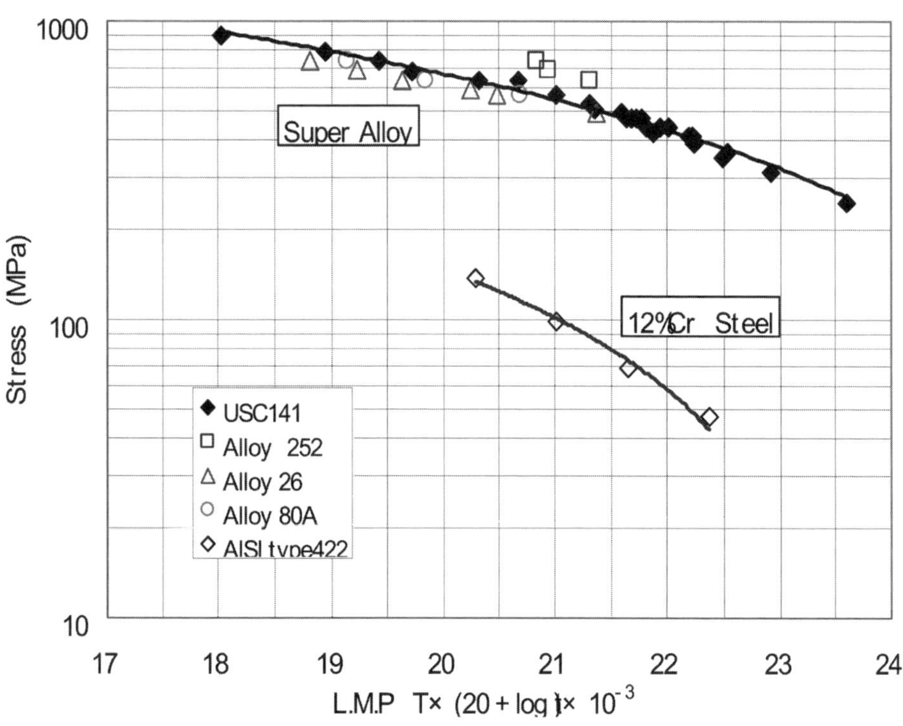

Fig. 7 Creep-rupture strength of USC141 and other alloys by Larson-Miller Plot

Evaluation 0f Weldability of USC141 for Boiler Tube Applications

In order to evaluate USC141 for boiler tube applications, weld joint were manufactured by TIG welding from solution-treated USC141 and Alloy 263(Ni-20Cr-20Co-6Mo-2Ti-Al) as a weld metal, then weldability and mechanical properties of weld joint at 750°C were investigated. The sizes of specimens were 30mm in diameter and 8mm in thickness. After welding, no heat treatment was carried out.

Photo.2 shows an appearance of welded tube specimens and photo.3 shows macrostructure of weldment. No defects were observed. Photo.4 shows a side bending test result, indicating no cracks.

Tensile test results of the weldments at 750°C were shown in Fig. 8. Specimens were ruptured at weld metal. The tensile strength was slightly lower than that of parent metal USC141; it was not a problem as butt-welded boiler tubes. It was enough higher than that of weld metal Alloy 617.

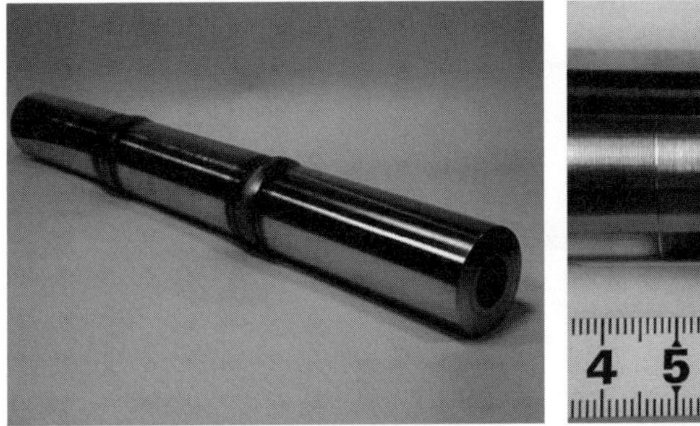

Photo.2 Appearances of a Welded tube Specimen

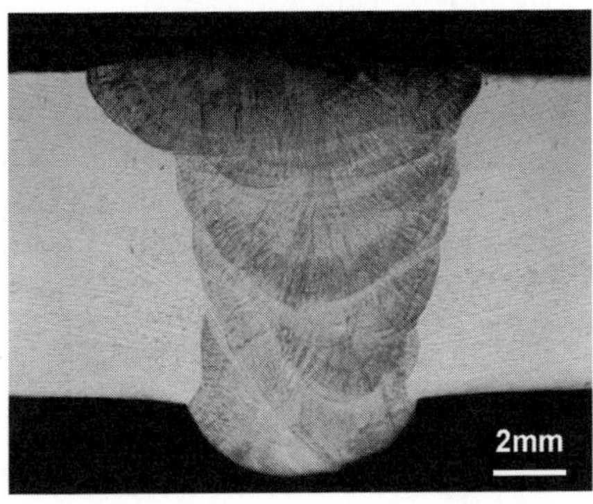

Photo.3 Macrostructure of a Weldment

Photo.4 Side Bending test Result

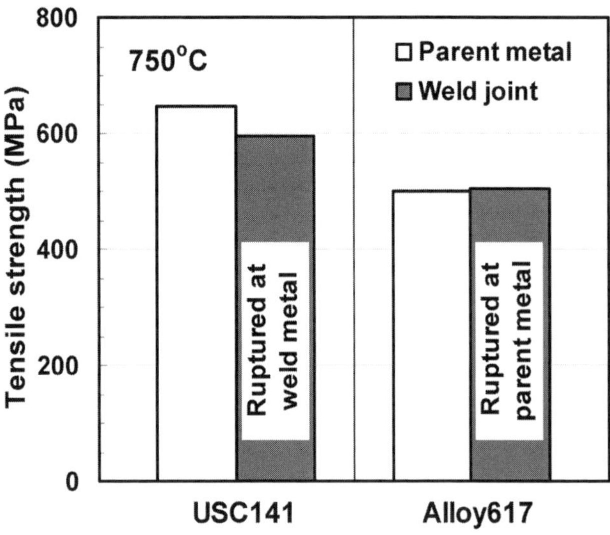

Fig. 8 Tensile Strength of Weldjoints and Parent Metals at 750°C

Conclusions

In order to apply austenitic Ni base superallys for steam turbines operated at around 650°C ~ 700°C, a low thermal expansion superalloy with good balanced creep-rupture properties was developed.

(1) To obtain low thermal expansion, it was found that the addition of Mo was effective and its optimum amount was determined around 10 mass%.

(2) To obtain good balanced creep-rupture properties, Al/Ti ratio in γ ' phase and the amount of γ ' phase were investigated and high Al/Ti ratio and relatively low amount of γ ' phase were chosen especially for good creep-rupture ductility.

(3) Developed alloy USC141 exhibited high creep-rupture strength up to about 20,000h at 700°C.

(4) The weldability of USC141 was evaluated for boiler tube applications and revealed good results.

References

(1).Superalloys II . Edited by C.T.Cims, N.S.Stolof and W.C.Hagel, John Willy & Sons, p588 [book]

Advances in Materials Technology for Fossil Power Plants
Proceedings from the Fifth International Conference
R. Viswanathan, D. Gandy, K. Coleman, editors, p 391-401

DOI: 10.1361/cp2007epri0391

MANUFACTURING EXPERIENCES AND INVESTIGATION OF PROPERTIES OF 12% CR STEEL FORGINGS FOR STEAM TURBINES

M. Mikami
Y. Wakeshima
T. Miyata
K. Kawano
Japan Casting & Forging Corporation
46-59, Sakinohama Nakabaru, Tobata-ku,
Kitakyushu-city 804-8555, Japan

Abstract

Demand of 9-12% chromium steel rotor forgings becomes higher from point of view of environmental protection in coal fired fossil power generations. Japan Casting & Forging Corporation (JCFC) has manufactured 9-12% Cr steel rotor forgings with JCFC's original techniques since 1991. Recently, type E steel developed by European COST program has been trial melted to meet the demand of such high Cr steel forgings in the world. Full size two forgings have been manufactured from approximately 70 ton ingot applying Electro Slag Hot Topping by JCFC (ESHT-J) process. One of the trial forgings has been austenitized at higher temperature in the quality heat treatment to improve long term creep strength. Their productivities and sufficient qualities have been ascertained.

Introduction

12%Cr steel rotor forgings are used in the high and intermediate pressure turbines of advanced coal-fired power generations that achieve higher efficiency and lower emissions than conventional ones. The turbine manufacturers in Europe and Japan have already developed their original advanced type 12%Cr steels for 593°C (1100°F) or higher class turbine rotor shafts. In Europe, type E and F steels has been developed with the COST program and applied to high pressure and intermediate pressure components for a lot of ultra super critical power generations (1)-(9). On the other hand, JCFC has manufactured 9-12%Cr steel rotor forgings since 1991 (10), however, has no experience of manufacturing tungsten alloyed material as a product. Therefore, two trial forgings of type E steel have manufactured by ESHT-J method, and their productivities and qualities have been investigated. Based on our research and experiences to manufacture molybdenum alloyed steels, there is the possibility to improve long term creep rupture strength by raising austenitizing temperature (11). In this paper, the results of manufacturing and investigation of type E steel trial forgings are described.

Manufacturing Process of Trial Forgings

Manufacturing Sequence

Figure 1 shows the manufacturing sequence of the trial forgings. Two forgings were made from an ingot and simultaneously forged as one piece. The two forgings were manufactured with their respective process after the forging was divided into two pieces in the preliminary heat treatment.

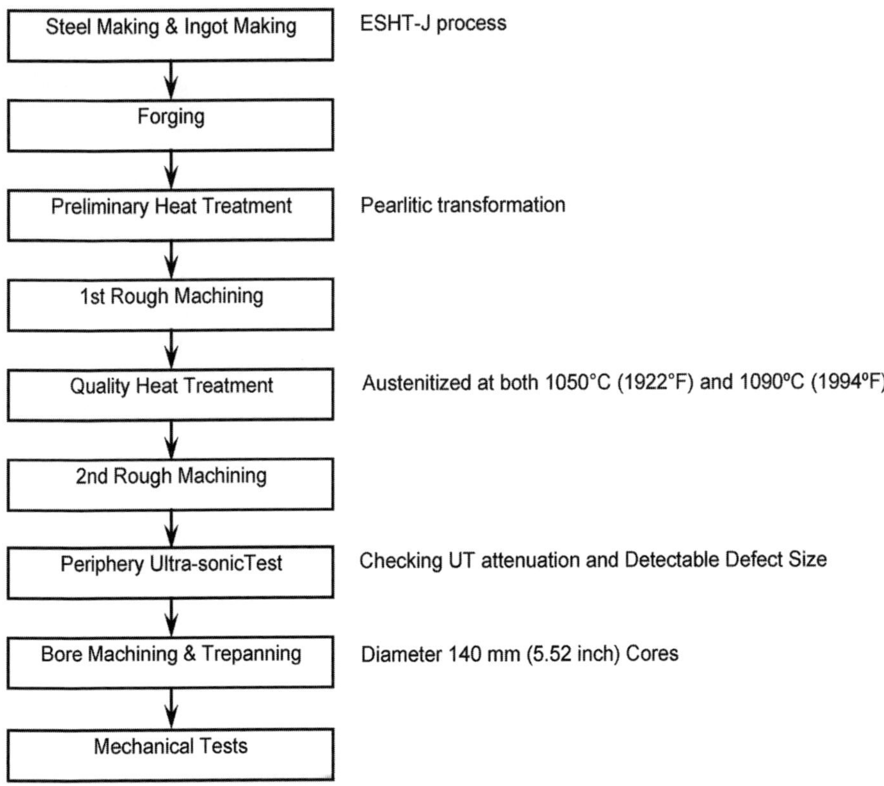

Figure 1. Manufacturing Sequence of Trial Forgings

Steel and Ingot Making

The steel was melted and refined in 100 metric ton Electric Arc Furnace, the molten steel was subsequently transferred to 150 ton Ladle Furnace to perform deoxidizing refining, degassing and adjustment of alloy elements. After the steel making, the molten steel was poured into the mold by the bottom poring method. After that, the molten slag in Ladle Furnace was teemed on the molten steel in the mold. ESHT-J process was performed after the slag pouring to reduce the segregation. Consumable electrodes were subsequently put into the molten slag layer, and the temperature of the hot topping slag and melting rate of the electrodes were controlled by the current. The conditions of ESHT-J operation were considered with the computer simulation of solidification. In case of this trial melting, ESHT-J process was carried out for 19 hours, and 69.5 ton ingot was made by this process.

Forging process

The ingot was worked with 8,000 ton open-die forging press to homogenize solidification structure and to obtain good soundness. The forging sequence is shown in Figure 2.

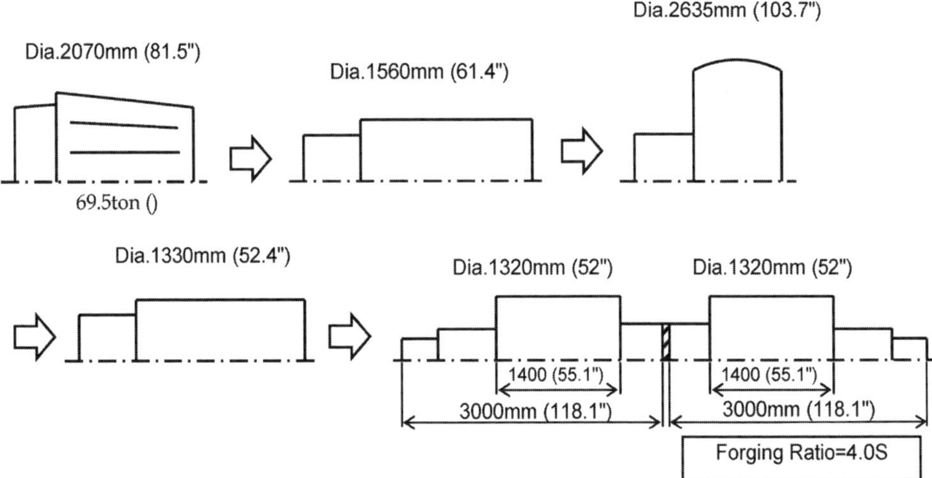

Figure 2. Forging Process

Heat Treatment

The preliminary heat treatment was performed with the optimum heating cycle which enables sufficient pearlitic transformation after forging process.
After the first rough machining, two forgings were austenitized, oil quenched and double tempered with the heat cycle as shown in Figure 3. One of the forgings was austenitized at 1050°C (1922°F) as a standard process in Europe, the other was austenitized at 1090°C (1994°F) as an attempt to improve long-term creep rupture strength. The second tempering processes were carried out at individual temperatures depending on the austenitizing temperatures.

	T_γ	T_{T1}	T_{T2}
Forging A	1050°C (1922°F)	570°C (1058°F)	680°C (1256°F)
Forging B	1090°C (1994°F)	570°C (1058°F)	685°C (1265°F)

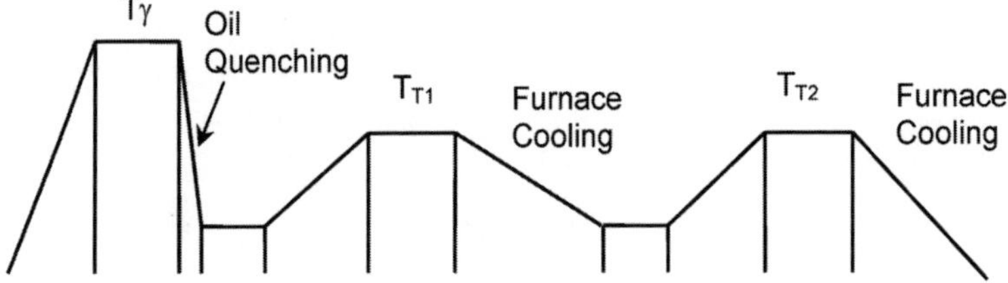

Figure 3. Schematic Heat Cycle for Quality Heat Treatments

Results of Investigation

Ultrasonic Inspection

As the results of periphery UT inspections, no indications were found. Table 1 shows the minimum detectable defect size and ultrasonic attenuation of the forgings. In spite of austenitizing at high temperature, a good UT detectability is still secured on both forging A and forging B.

Table 1. Minimum Defect Size and Ultrasonic Attenuation of Forgings

Forging	Austenitizing Temperature	Diameter [mm] (inch)	Minimum Defect Size [mm] (inch)	Ultrasonic Attenuation (dB/m) 2MHz	5MHz
Forging A	1050°C (1922°F)	1277 (50.3)	0.9 (0.035)	0.4	3.9
Forging B	1090°C (1994°F)	1277 (50.3)	0.8 (0.031)	0.8	5.1

Metallurgical Results of the Forgings

Table 2 shows results of ladle and product analyses at specimen locations. Figure 4, which results of analyses on the samples cut from center cores are rearranged to correspond with height of ingot, shows the change in chemical composition along the axial center of the ingot. Alloy element levels of the forgings do not vary significantly from the bottom to the top of the ingot. ESHT-J process is very effective to obtain homogeneous ingot.

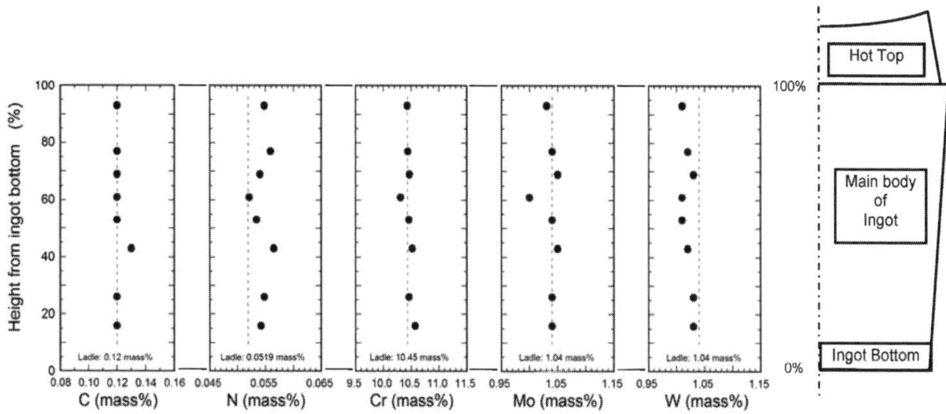

Figure 4. Results of Chemical Analysis Associated with Axial Center of the Ingot

Table 2. Results of Ladle and Product Analysis

Locations		C	Si	Mn	P	S	Ni	Cr	Mo	W	V	Nb	N
Ladle		0.12	0.06	0.43	0.008	0.001	0.75	10.45	1.04	1.04	0.20	0.050	0.052
Forging-A	T1	0.11	0.06	0.42	0.007	0.001	0.75	10.42	1.03	1.03	0.20	0.046	0.052
	T2	0.12	0.06	0.43	0.008	0.001	0.76	10.44	1.04	1.03	0.20	0.047	0.053
	R3	0.12	0.06	0.44	0.008	0.001	0.77	10.51	1.06	1.02	0.20	0.050	0.053
	C3	0.12	0.06	0.43	0.008	0.001	0.76	10.45	1.04	1.02	0.20	0.047	0.053
Forging-B	T1	0.12	0.06	0.43	0.008	0.001	0.76	10.45	1.03	1.03	0.20	0.046	0.053
	T2	0.12	0.06	0.43	0.008	0.001	0.76	10.48	1.05	1.03	0.20	0.048	0.053
	R3	0.12	0.06	0.43	0.008	0.001	0.76	10.48	1.05	1.03	0.20	0.048	0.053
	C3	0.13	0.07	0.44	0.009	0.001	0.78	10.52	1.07	1.03	0.20	0.052	0.054

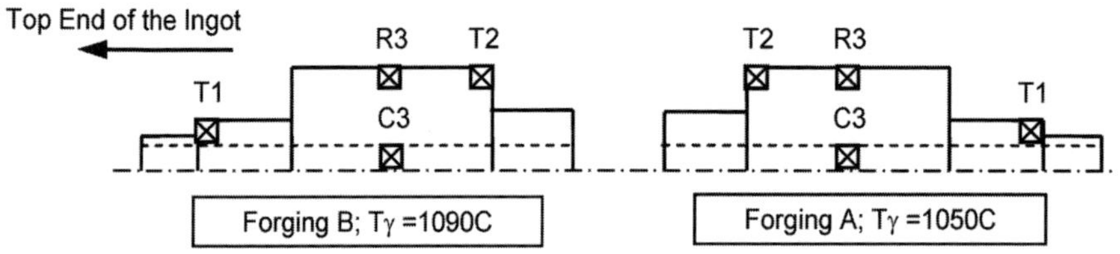

Top End of the Ingot

T1 R3 T2 C3

Forging B; Tγ =1090C

T2 R3 T1 C3

Forging A; Tγ =1050C

Figure 5 shows the microstructures and the austenite grain size number by ASTM method at each sample locations of the forgings. The microstructures are tempered martensite, no delta ferrite and eutectic carbide such as niobium carbo-nitride is observed. The grain sizes of the forgings are enough small to obtain good UT attenuation.

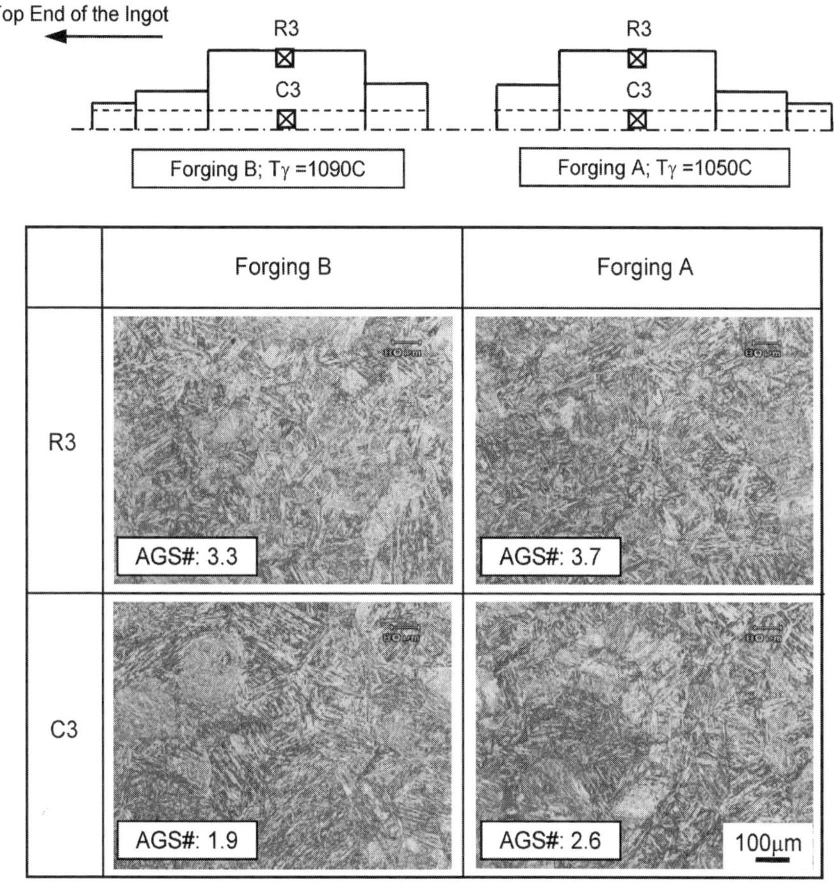

Figure 5. Optical Micrographs of Trial Forgings at Various Sample Locations

Results of Tensile and Charpy Impact Tests

Table 3 shows the results of tensile and Charpy impact tests on the forgings. The tensile strength and the toughness of both forging A and forging B are sufficient levels. According to the test results of the specimens tempered additionally, it is preferable to adjust the 0.2% yield strength level approximately form 700 to 750 N/mm^2 (102 – 109 ksi) to secure higher impact energy.

Table 3. Results of Tensile and Charpy Impact Tests

Forging	Location		Direction	0.02%YS [N/mm²] (ksi)	0.2%YS [N/mm²] (ksi)	TS [N/mm²] (ksi)	El [%]	RA [%]	Impact Energy [J] (ft-lb)			FATT [C] (F)
Forging A	Periphery	T1	Tangential	672 (97)	750 (109)	873 (127)	20.2	60.1	43 (32)	36 (27)	52 (38)	23 (73)
		T2	Tangential	681 (99)	752 (109)	872 (126)	21.4	63.0	35 (26)	46 (34)	50 (37)	25 (77)
		R3	Radial	684 (99)	749 (109)	870 (126)	20.6	58.0	58 (43)	70 (52)	-	20 (68)
	Center	C3	Radial	686 (99)	762 (111)	882 (128)	19.0	54.9	47 (35)	41 (30)	49 (36)	25 (77)
		C3*1	Radial	646 (94)	736 (107)	864 (125)	20.2	58.0	51 (38)	50 (37)	48 (35)	20 (68)
Forging B	Periphery	T1	Tangential	702 (102)	773 (112)	897 (130)	20.2	60.1	35 (26)	31 (23)	31 (23)	30 (86)
		T2	Tangential	696 (101)	780 (113)	896 (130)	19.8	60.1	45 (33)	31 (23)	36 (27)	30 (86)
		R3	Radial	682 (99)	770 (112)	892 (129)	20.0	59.1	63 (46)	53 (39)	-	20 (68)
	Center	C3	Radial	681 (99)	782 (113)	903 (131)	16.0	47.0	43 (32)	31 (23)	29 (21)	27 (81)
		C3*2	Radial	650 (94)	744 (108)	870 (126)	19.4	57.0	48 (35)	75 (55)	56 (41)	20 (68)

*1: Additionally tempered at 680C (1256F) for 12 hours *2: Additionally tempered at 690C (1274F) for 12 hours

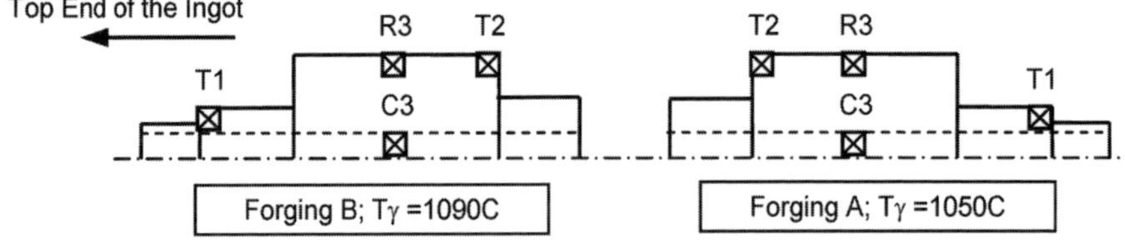

Creep Rupture Tests

Figure 6 shows the creep rupture strength at the center portion of the forgings. Creep rupture tests at 600°C are performed using samples taken from both forgings' center core (sample location: C3). The materials for rupture test are additionally tempered after the trepanning. Testing time of the trial forgings is over 9,000 hours. The rupture strength level of forging B is higher than that of forging A at this time. The creep rupture tests are scheduled to perform beyond 30,000 hours at least to assess the effect of higher austenitizing temperature on the creep rupture strength due to the degradation of long term creep strength.

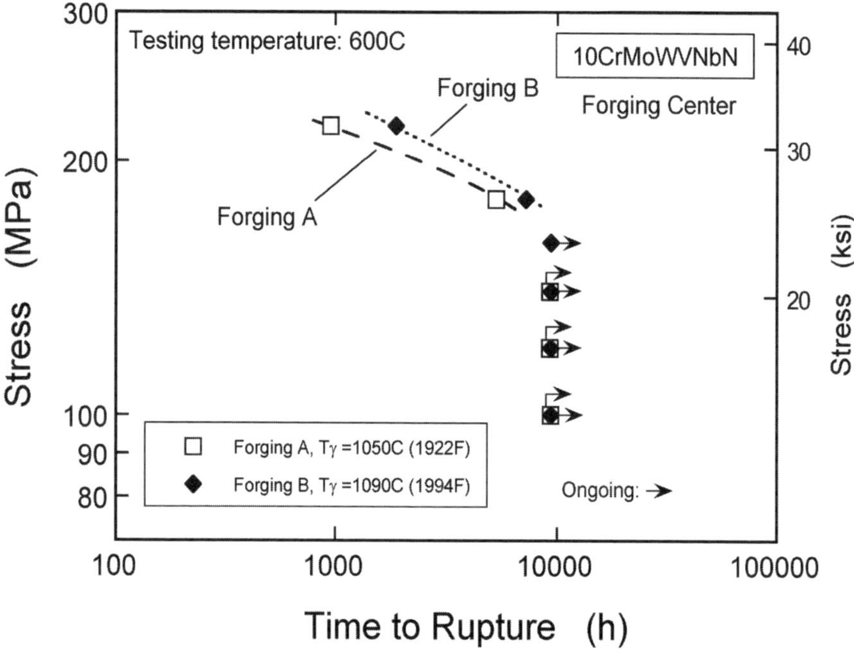

Figure 6. Creep Rupture Strength of Trial Forgings

Microstructural Investigations

Microstructural investigations using a transmission electron microscope were carried out in an attempt to confirm the influence of austenitizing temperature on the microstructure. Figure 7 shows the transmission micrographs of the materials cut from the center cores of both forging A and B. A lot of particles are observed on the subgrain boundaries or in the grains. Energy dispersion X-ray analyses with carbon extraction replicas reveal the presence of fine MX or M_2X type particles in the grains and coarse $M_{23}C_6$ particles on the grain boundaries. The amount of

MX and M_2X particles observed in the forging B is larger than that in the forging A as shown in figure 7. Austenitizing temperature is thought to affect fine MX and M_2X particles since those particles are dissolved by the austenitizing and generated by the subsequent tempering process. Not only creep rupture tests but also further microstructural investigations during creep deformation will be performed to confirm the effectiveness of higher austenitizing temperature.

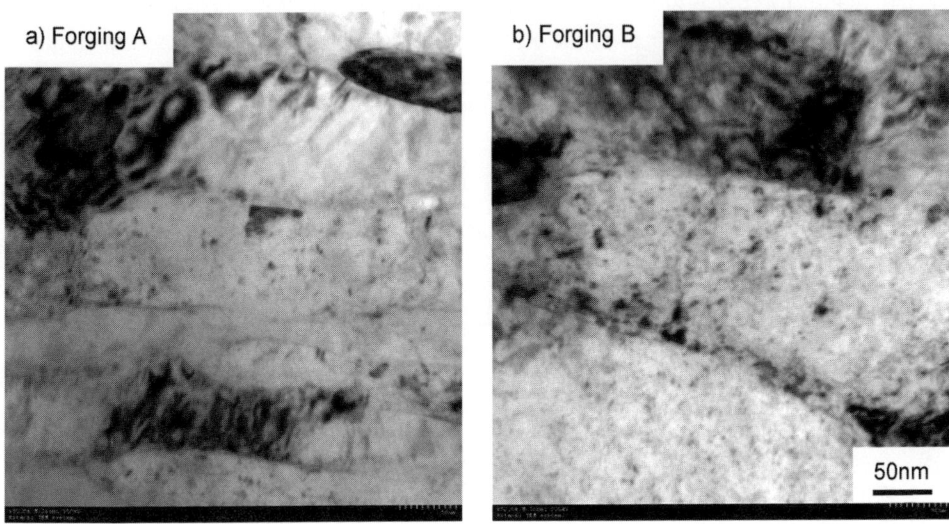

Figure 7. Transmission Micrographs of Trial Forgings at Center Portion

Conclusion

Full scale two trial forgings made from type E steel developed by COST program have been successfully manufactured using JCFC's original techniques. Results are obtained as follows:

1. It has been ascertained that JCFC's original techniques are still effective to type E steel.

2. It was revealed that adjusting tensile strength makes the forgings to secure sufficient toughness even though the forgings are austenitized at 1090°C (1994°F).

3. Higher austenitizing temperature is effective to improve the creep rupture strength as the results of creep rupture tests up to 10,000 hours.

At least 30,000 hours long term creep rupture tests will be performed to confirm the effectiveness of higher austenitizing temperature.

References

1. C. Berger, S. M. Beech, K. H. Mayer, R. B. Scarlin and D. V. Thornton, "High temperature rotor forgings of high strength 10% CrMoV steels," Paper No. 8.1, presented at the 12th International Forgemasters Meeting, Chicago, Illinois (September 1994).

2. M. Taylor, D. V. Thornton and R. W. Vanstone. *Materials for Advanced Power Engineering.* Jülich: Forschungszentrum Jülich GmbH, 1998, pp.297-310.

3. T. –U. Kern, B. Scarlin, R. W. Vanstone and K. H. Mayer. *Materials for Advanced Power Engineering.* Jülich: Forschungszentrum Jülich GmbH, 1998, pp.53-70.

4. K. H. Mayer, T. –U. Kern, K. H. Schönfeld, M. Staubli and E. Tolksdorf. *14th International Forgemasters Meeting Proceedings*, Düsseldorf: Verein Deutscher Eisenhüttenleute, 2000, pp. 277-282.

5. T. -U. Kern, M. Braendle and A. Wiegand. *IFM 2003 JAPAN Proceedings.*: Steel Castings and Forgings Association of Japan, 2003, pp. 238-243.

6. T. -U. Kern, M. Staubli, G. Zeiler, A. Finali and B. Donth. *IFM 2003 JAPAN Proceedings.*: Steel Castings and Forgings Association of Japan, 2003, pp. 244-247.

7. N. Blaes, K. -H. Schöfeld and D. Bokelmann. *IFM 2003 JAPAN Proceedings.*: Steel Castings and Forgings Association of Japan, 2003, pp. 219-226.

8. G. Zeiler, W. Meyer, K. Spiradek and J. Wosik. *Advances in Materials Technology for fossil Power Plants.*: ASM International, 2004, pp. 506-519.

9. N. Blaes, B. Donth, K. -H. Schönfeld and D.Bokelmann. *Advances in Materials Technology for fossil Power Plants.*: ASM International, 2004, pp. 559-574.

10. M. Mikami, K. Morinaka, Y. Okamura, S. Tanimoto, Y. Wakeshima, R. Magoshi, Y. Kadoya and A. Matsuo. *14th International Forgemasters Meeting Proceedings*, Düsseldorf: Verein Deutscher Eisenhüttenleute, 2000, pp. 295-300.

11. Y. Nakamura, M. Mikami, Y. Wakeshima, K. Toriumi, Y. Kadoya and Y. Hirakawa, "Influence of Heat Treatment Conditions on Microstructures of 12%Cr Turbine Rotor", presented at the 16th International Forgemasters Meeting, Sheffield, UK (October 2006), pp. 405-414.

Advances in Materials Technology for Fossil Power Plants
Proceedings from the Fifth International Conference
R. Viswanathan, D. Gandy, K. Coleman, editors, p 402-412

DOI: 10.1361/cp2007epri0402

The Application of Ni-Base Alloy Nimonic 80A for Buckets of USC Steam Turbine in China

Hongwei Shen, Weili Wang
Shanghai Turbine Company Ltd., Shanghai 200240, China
Zhizheng Wang, Lihong Zhang
Special Steel Branch of BAOSTEEL, Shanghai 200940, China
Xishan Xie, Shuhong Fu
University of Science and Technology Beijing, Beijing 100083, China

Abstract

Nimonic 80A, a Ni-base superalloy mainly strengthened by Al and Ti to form γ′-Ni_3 (Al, Ti) precipitation in Ni-Cr solid solution strengthened austenite matrix, has been used in different industries for more than half century (especially for aero-engine application). In consideration of high strengths and corrosion resistance both Shanghai Turbine Company (STC) has adopted Nimonic 80A as bucket material for ultra-super-critical (USC) turbines with the steam parameters of 600°C, 25MPa. First series of two 1000MW USC steam turbines made by Shanghai Turbine Co. were already put in service on the end of 2006.

Large amount of Nimonic 80A with different sizes are produced in Special Steel Branch of BAOSTEEL, Shanghai. Vacuum induction melting and Ar protected atmosphere electro-slag remelting (VIM+PESR) process has been selected for premium quality high strength Nimonic 80A. For higher mechanical properties the alloying element adjustment, optimization of hot deformation and heat treatment followed by detail structure characterization have been done in this paper.

The Chinese premium quality high strength Nimonic 80A can fully fulfill the USC turbine bucket requirements.

Introduction

The rapid development of Chinese economy is requiring sustainable growth of power generation to meet its demand. Chinese power plants as well as those worldwide are facing to increase thermal efficiency and to decrease the emission of SO_x, NO_x and CO_2. According to national resource of coal and electricity market requirements in the future 15 years power generation especially the ultra-super-critical (USC) power plants with the steam temperation up to 600°C or higher will get a rapid development [1]. The first 2

×1000MW USC turbines (see Fig.1) with the steam parameters of 600℃, 25MPa were made by Shanghai Turbine Co. in joint venture with the Siemens Co. Germany and put in service at Yuhuan Plant in Zhejiang province, China on the end of December 2006.

12%Cr heat resisting steel and its modifications [2] are popularly used for steam turbine rotors and buckets of different stages. In consideration of long term (2×10^5hrs) service of USC power plants, neither 12%Cr steels nor advanced modified 9-12%Cr steels can meet the strict demand. In consideration of excellent corrosion resistance and high stress-rupture strength at high temperatures, the premium quality high strength Nimonic 80A has been adopted for our this type of USC turbine with the steam temperature of 600℃, 25MPa.

Fig.1 shows the layout of 1000MW USC steam turbine made by STC, China. Nimonic 80A has been used for several stage buckets (see Fig.2).

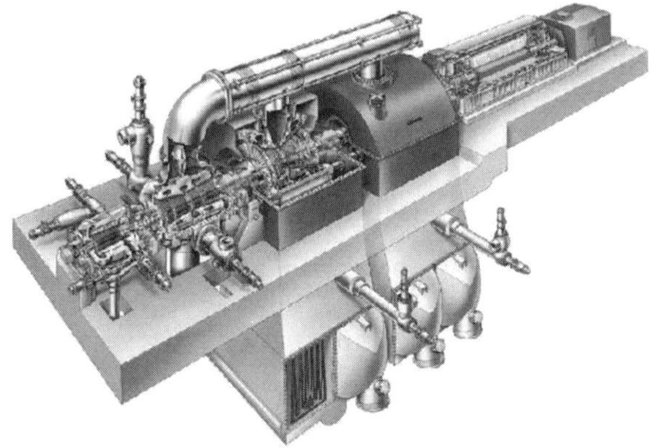

Fig.1 1000MW USC steam turbine produced by STC
and put in service at Yuhuan power plant on the end of 2006

Fig.2 Typical Nimonic 80A buckets for USC turbine

Nimonic 80A has been used for half a century but mainly was used as hot components such as blades in aero-engines for shorter service time (10^4hrs) in comparison with the buckets in steam turbines (10^5hrs). On the other hand the sizes and weight of land base steam turbines are much larger and heavier than that of in the aero-engines. In result of that the technical requirements of USC steam turbine buckets are very strict. Tab. 1 shows the specifications of Nimonic 80A for different applications.

Tab. 1 Mechanical property Requirements of Nimonic 80A for Different Applications

	20°C Tensile			Stress Rupture				Hardness	Impact	Grain
	YS (MPa)	UTS (MPa)	δ (%)	T (°C)	Stress (MPa)	t (hrs)	δ (%)	Brinell (HB)	(J)	(ASTM NO:)
Aerospace MRS 7011	≥620	≥1000	≥ 20	750	310	≥30	-	≥285	-	-
Marine BTWH001-2006	≥800	1200-1400	≥ 15	-	-	-	-	≥330	≥4-5	≥4
Auto EMS210-P	-	-	-	732	380	≥30	≥5	≥270	-	2-7
General ASTM B637	≥620	≥930	≥ 20	760	325	≥23	≥3.5	-	-	-
USC Turbine BTXH042 -2006	≥600	1100-1300	≥ 17	750	310	≥100	≥4	≥260	≥20	≥4

It can be seen that the ultimate tensile strength (1100-1300MPa) for USC turbine application is higher than that of Aerospace MRS 7011 (≥1000MPa). The stress rupture life for USC turbine application at 750°C, 310MPa should be longer than 100hrs, but for Aerospace MRS 7011 requires longer than 30hrs only. Except these requirements the ductilities at room temperature tensile and 750°C stress rupture, impact toughness and the grain size are strictly required.

Tab. 2 shows the chemical composition requirement of Nimonic 80A specification BTXHO-2006 for USC steam turbine application.

It can be seen that the chemical composition in this specification characterizes with a wide ranges especially for the main γ′ strengthening elements of Ti and Al but the sum of Ti+Al should be controlled over 3.50%.

For the purpose to meet all these strict requirements the chemical composition adjustment, metallurgical processing improvement including heat-treatment and structure characterization in combination with mechanical properties determination have been investigated in this paper.

Tab. 2 Chemical composition (wt %) of Nimonic 80A for USC steam turbine application

	C	Ni	Cr	Ti	Al	Ti+ Al	B	S	P	Si	Mn
Spec BTXH042 -2006	0.04 -0.10	Bal	18.00 -21.00	1.80 -2.70	1.00 -1.80	≥ 3.50	≤ 0.008	≤ 0.010	≤ 0.010	≤ 0.30	≤ 1.00

Materials and Experiments

Investigated premium quality high strength Nimonic 80A commercial heats were made by Special Steel Branch of BAOSTEEL, Shanghai, China. Selected good quality raw materials were charged in vacuum induction furnace (VIM) for melting and also vacuum pouring in the forms of electrodes. For keeping homogeneous distribution of Al and Ti in Nimonic 80A ESR ingots the electrodes were remelted by electro-slag refining under argon protected atmosphere (PESR). The PESR ingots were forged via 4500T press and finally deformed via 1500T rotary forging or hot rolling for different sizes of shaped materials.

For optimum chemical composition determination, thermodynamic calculation was carried out to study the effect of Al, Ti and C on γ'-Ni_3 (Al, Ti) and carbide precipitation behavior respectively by using Thermo-Calc software, version M, accompanied with 14 element Ni-database [3]. Selected heats with different combination of Al and Ti were melted by VIM+PESR for structure and mechanical properties comparison.

Grain size and microstructure observation was taken by optical microscopy (OM) and high resolution scanning electron microscopy (SEM). Precipitated phases were electrolytic isolated and identified by X-ray diffraction (XRD) and their weight fractions were determined by micro-chemical analyses.

Mechanical property determination includes hardness, tensile properties, impact toughness and stress-rupture tests at high temperatures.

Thermodynamic Calculation on Phase Formation

Nimonic 80A [4] is based on Ni-Cr solid solution with the main second phase, γ'-Ni_3 (Al, Ti), precipitation hardening and grain-boundary strengthening with a small addition of boron and also a small amount of carbide formation at grain boundaries and in grains. It is important to study the effect of C, Al and Ti on phase formation.

Phase calculation results (see Fig.3) shows that three kind carbides MC, M_7C_3 and $M_{23}C_6$ can exist in Nimonic 80A. In our calculated range of carbon (0.01-0.15%) M_7C_3 and $M_{23}C_6$ can always form, but M_7C_3 precipitates at higher temperature range and $M_{23}C_6$ precipitates at lower temperature range. However, MC carbide can form only at the carbon content higher than 0.3% (see Fig. 3(a)) and its forming temperature is the highest in comparison with M_7C_3 and $M_{23}C_6$, because MC carbide mainly forms directly from the liquid during solidification of the melt. The amount of $M_{23}C_6$, M_7C_3 and MC increases with the increasing of carbon content.

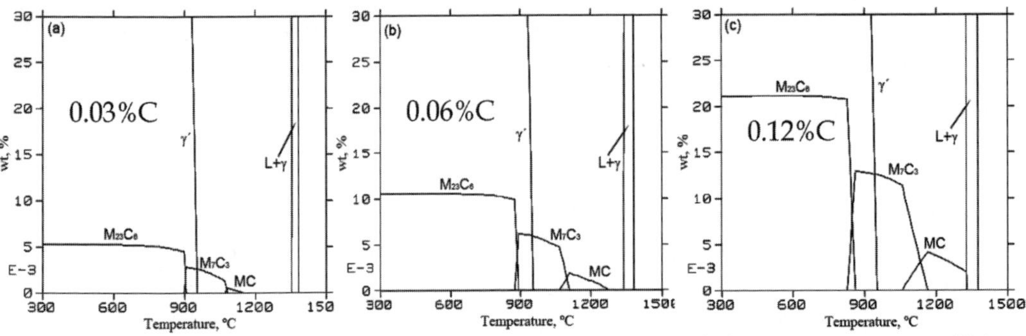

Fig.3 Effect of carbon (0.01-0.15%C) on phase formation behavior in Nimonic 80A

Al and Ti are considered to be the most important strengthening elements for Nimonic 80A. Fig.4 shows the fraction of γ' and its solves temperature are dependent on Al and Ti separately and also on the sum of Al+Ti. The specification requires that the sum of Al+Ti must be higher than 3.5%. Our calculated results show that for more precipitation of γ' the sum of Al and Ti contents should be controlled at the higher levels, if the hot deformation process can be performed at metallurgical processing. For premium quality high strength Nimonic 80A for USC turbine bucket application we require to controlled the carbon content in the range of 0.05-0.06% and keep the Al and Ti to the possible high levels (such as 2.4-2.5%Ti and 1.6-1.7%Al).

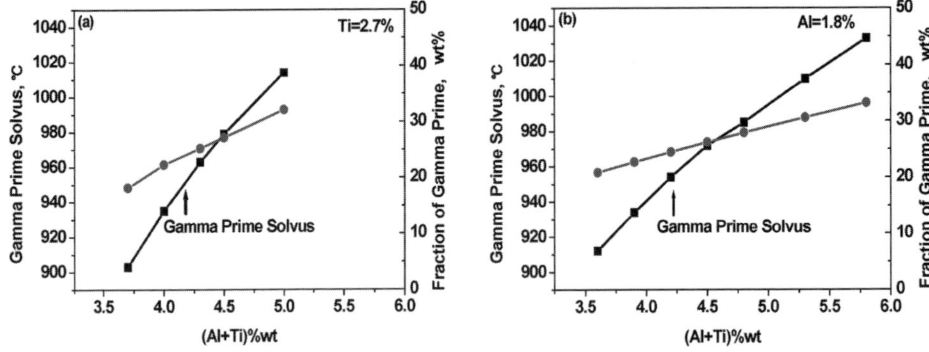

Fig.4 Effect of Al and Ti on fraction and solution temperature of γ' in Nimonic 80A

Production Heats Evaluation

On the base of above mentioned thermodynamic calculation and theoretical analyses six production heats were melted by VIM+PESR and converted to Φ125 mm and Φ90 mm bars. Chemical compositions of six production heats of Nimonic 80A are shown in Tab. 3. It can be seen that the carbon content was controlled at the medium level (0.06%C). Ti was controlled to the highest level (2.5%Ti) and Al was adjusted at the lower level (1.45%Al) for heats 11, 12 and 13 and as high as possible levels at 1.70-1.80%Al for heats 21, 22 and 23. The gas contents of O, N, and H were controlled to the levels as low as possible because of the premium quality requirement. B and Mg were added in the heats for grain boundary strengthening and hot deformation improvement respectively. The average contents of O, N and H of routine production heats are 6, 50, 2ppm respectively and the Mg content is in the range of 40-80ppm (Tab.4).

Tab. 3 Chemical composition (in wt %) of tested Nimonic 80A bars

Alloy No:	C	Mn	Si	S	P	Ni	Cr	Al	Ti	Ti+ Al	B
11	0.056	0.03	0.12	0.003	0.005	bal	19.25	1.45	2.44	3.88	0.0047
12	0.058	0.02	0.09	0.002	0.003	bal	19.22	1.43	2.45	3.89	0.0035
13	0.055	0.03	0.10	0.002	0.005	bal	19.37	1.43	2.48	3.91	0.0033
21	0.061	0.02	0.08	0.002	0.004	bal	19.48	1.74	2.51	4.25	0.0037
22	0.060	0.02	0.09	0.003	0.002	bal	19.47	1.79	2.50	4.29	0.0040
23	0.055	0.04	0.06	0.002	0.005	bal	19.40	1.78	2.50	4.29	0.0033

Tab. 4 The statistic data of O, N, H, and Mg (in ppm) from routine production heats of Nimonic 80A

Melting	O		N		H		Mg	
	Max	Min	Max	Min	Max	Min	Max	Min
VIM+PESR	7.5	5	70	38	4.0	1.0	80	40

Mechanical property data (see Fig. 5) shown that the hardness and tensile strengths are strictly dependent on the sum of Al+Ti. Production data have confirmed our thermodynamic calculation and theoretical analysis that for premium quality high strength Nimonic 80A the optimum contents of C, Ti and Al are 0.06%, 2.5% and 1.7% respectively.

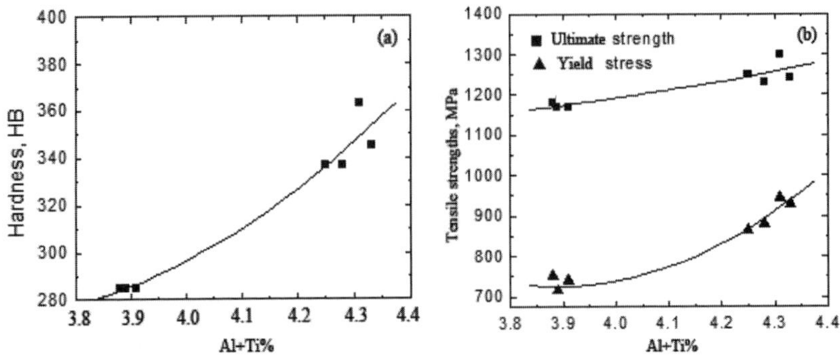

Fig. 5 The effect of Al+Ti (at ~2.5%Ti) on the hardness (a) and tensile strengths (b)

The premium quality Nimonic 80A requires fine grain structure (finer than ASTM 4). Careful hot deformation process has been conducted to convert the ingots to different sizes of shaped products. Fig.6 shows the fine grain structure (finer than ASTM 4) of Φ125mm (Fig.6 a) andΦ90 mm (Fig.6 b) bar products as the typical examples of fine grain Nimonic 80A after rotary forging and hot rolling respectively.

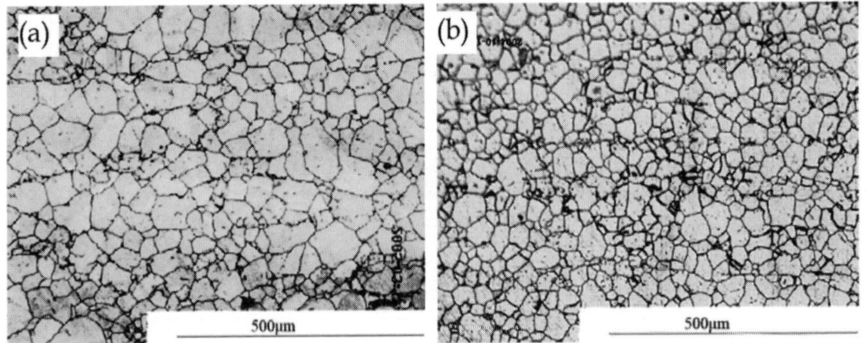

Fig. 6 Grain structure of Nimonic 80A after rotary forging (a) and hot rolling (b)

Heat Treatment

The specification recommends the full treatment of Nimonic 80A as follows: 1050-1080°C/8h/AC+845°C/24h/AC+700°C/16h/AC. For determination of solid solution temperature the effect of temperature on grain size has been investigated as shown in Fig.7. Phase calculation results of Nimonic 80A at 1.8%Al and 2.5%Ti have been shown in Fig.8. In comparison of Fig.7 and Fig.8 it reveals that $M_{23}C_6$ carbide should solute in γ-matrix at the temperature higher than 900°C and M_7C_3 will solute in γ-matrix at higher temperature of 1030°C. The most stable carbide MC (TiC) begins to solute in γ-matrix from 1020°C and should totally solute in γ-matrix above 1145°C. These calculated data can just explain the effect of solid solution temperature on grain size as shown in Fig.7. The grain size of fine grain (ASTM 6-7) Nimonic 80A begins to grow when the most of grain boundary carbides M_7C_3 and $M_{23}C_6$ solute into γ-matrix. The grain size grows rapidly when the most stable carbide MC begins to solute into γ-matrix above 1120°C.

For higher stress rupture strength requirement of Nimonic 80A the solution temperature for Nimonic 80A has been determined.

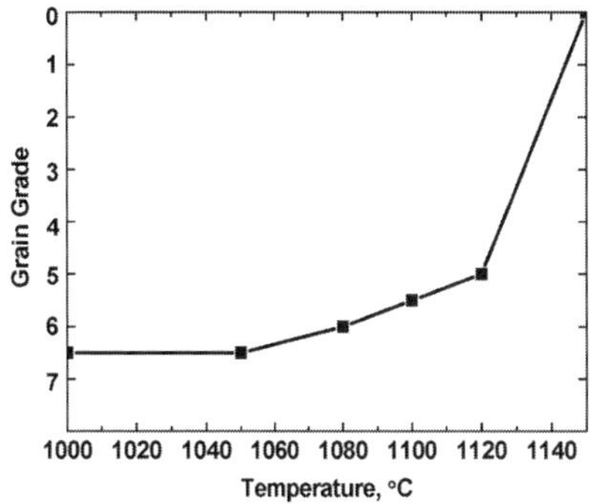

Fig. 7 Effect of solution temperature on grain size of Nimonic 80A

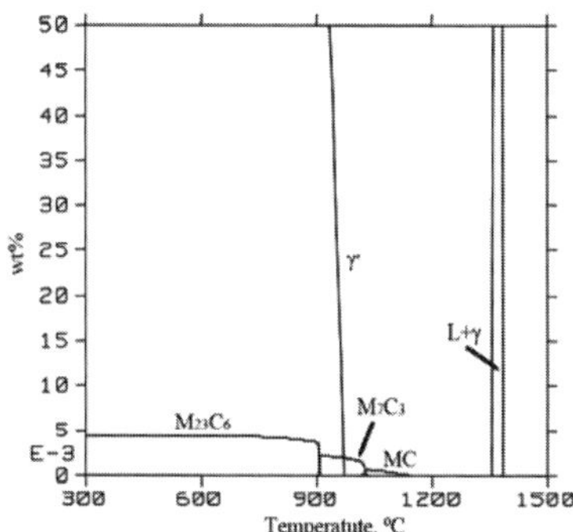

Fig. 8 Calculated phase diagram of Nimonic 80A (1.8%Al, 2.5%Ti)

Structure Characterization

After theoretical calculation and production heats evaluation we have concluded to make the premium quality high strength Nimonic 80A for USC turbine buckets application via qualified double melting ---VIM+PESR and to control the main alloying elements Ti, Al and C chemical composition of specification to meet the optimum target i.e. 2.50%Ti, 1.70%Al and 0.06%C. for fine grain structure requirement we adopt the full heat treatment: 1080°C/8h/AC+845°C/24h/AC+700°C/16h/AC.

Detail microstructure characterization and followed with mechanical properties evaluation have been conducted on a production heat 730-0324 and its chemical composition is shown in Tab. 5.

Tab. 5 Chemical composition (wt %) of investigated heat of Nimonic 80A

Heat	C	Ni	Cr	Ti	Al	Ti+ Al	S	P	B	Mg
730-0324	0.06	Bal	19.93	2.45	1.69	4.14	0.002	0.003	0.0033	0.0045

Typical microstructure of investigated of Nimonic 80A at standard heat treatment condition is shown in Fig.9. It characterizes with a fine grain structure with ASTM grain size No: (see Fig9. a). High resolution SEM images show the fine dispersive γ' particles homogeneously distributed in γ'-matrix (see Fig.9 c) and its average size of γ' is about 50nm. The globular particles of M_7C_3 and $M_{23}C_6$ carbides discontinuously precipitated at grain boundaries as shown in Fig.9 b. The fractions of existing phases are shown in Tab. 6. It can be seen that the premium quality Nimonic 80A characterizes with a higher γ' fraction (17.713%) than that of non-premium quality Nimonic 80A (less than 15%). The premium quality Nimonic 80A contains a certain amount (0.636%) of Cr-rich carbides M_7C_3 and $M_{23}C_6$ and also a small amount (0.036%) of MC carbide.

Tab. 6 Fractions (in wt %) of existing phases in Nimonic 80A at standard heat treatment condition

Phase	γ	γ'	**MC**	$\mathbf{M_7C_3 + M_{23}C_6}$
Fraction (wt %)	81.615	17.713	0.036	0.636

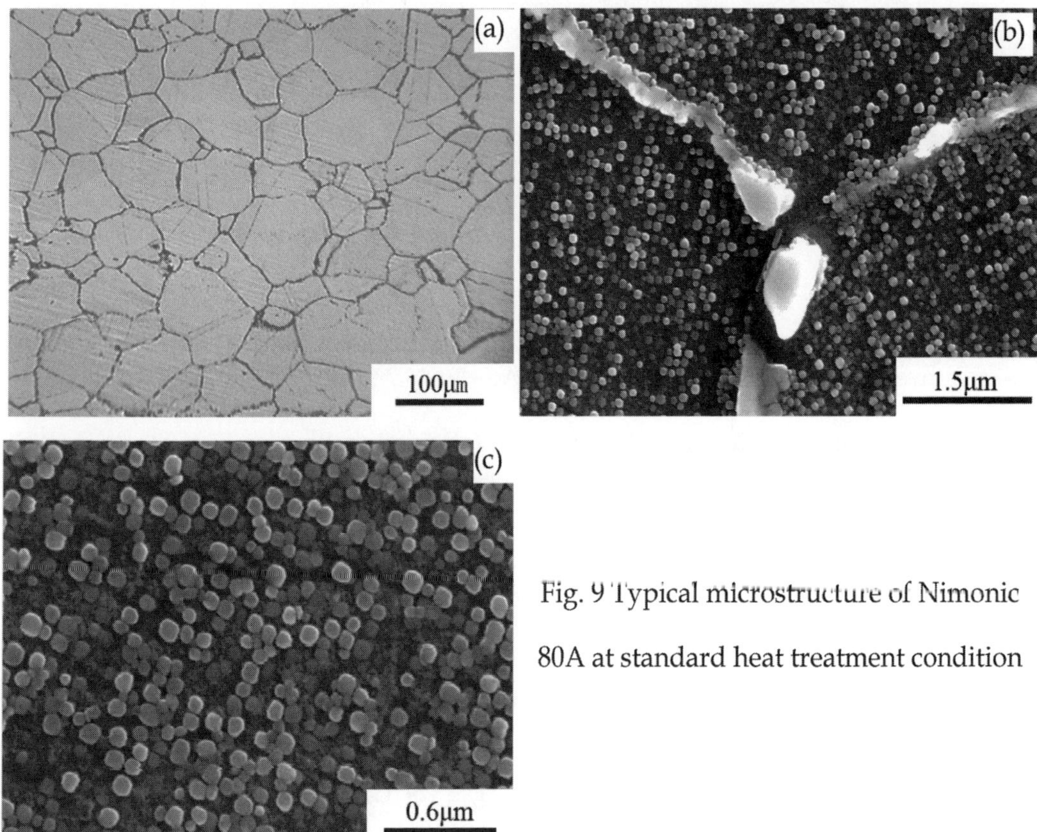

Fig. 9 Typical microstructure of Nimonic 80A at standard heat treatment condition

Mechanical Properties Determination

The mechanical properties including hardness, impact toughness and tensile properties for different sizes of Nimonic 80A from production statistics of routine heats are shown in Tab. 7. It can be seen that fine grain structure (finer than ASTM No: 5) can fulfill the specification requirement (finer than ASTM No: 4). All mechanical properties not only strengths but also ductilities and impact toughness can meet the USC turbine buckets requirement.

Stress rupture properties at 750°C, 310MPa of different sizes of Nimonic 80A bars all can meet the Specification requirement. The long time stress rupture tests and structure stability study are still going on. The test results will be published in the future time.

Tab. 7 Room temperature mechanical properties (average data) of Nimonic 80A bars products

Bar size (mm)	Grain size (ASTM No:)	Hardness (HB)	Impact (J)	Tensile properties			
				UTS (MPa)	YS (MPa)	δ (%)	Ψ (%)
Φ125	7.0-7.5	354	76	1320	945	22.5	39.5
Φ115	7.0-7.5	363	60	1260	905	22.0	35.5
Φ105	7.0-7.5	354	74	1300	965	25.5	40.0
Φ95	7.0-7.5	337	62	1230	830	27.5	35.5

Conclusions

Premium quality fine grain high strength Nimonic 80A of different sizes for USC steam turbine buckets application has been successfully produced in the Special Steel Branch of BAOSTEEL, Shanghai, China. Theoretical analyses, experimental results and production heats evaluation reveal that this premium quality Nimonic 80A can fulfill the demand of power engineering industry for making USC turbine buckets.

Shanghai Turbine Company Ltd. has adopted Nimonic 80A for 1000 MW USC turbine buckets application. First series of 2×1000MW USC turbine was already put in service on the end of December, 2006.

References

[1] F. Lin, S. Cheng and X. Xie, "The Development of Chinese Power Plants and High Temperature Materials Application," *Proceedings of Symposium on Heat Resistant Steels and Alloys for USC Power Plants 2007*, July 3-6, 2007, KIST, Seoul, Korea, p. 32.

[2] K-H Mayer, C. Berger, A. Scholg and Y. Wang, "Improvement of Creep Strength of 650°C Ferritic/Martensitic Super Heat Resistant 11-12%Cr Steels," *Materials for Advanced Power Engineering 2006*, Energietechnik(2006), p. 1053.

[3] B, Sundman, B, Jansson and J. O. Anderson, "The Thermo-Calc databank system," *CALPHAD*, 9 (1985), p. 153.

[4] W. Betteridge and J. Heslop, *The Nimonic Alloys*, 2nd edition, Edward Amkld (1974).

Advances in Materials Technology for Fossil Power Plants
Proceedings from the Fifth International Conference
R. Viswanathan, D. Gandy, K. Coleman, editors, p 413-423

DOI: 10.1361/cp2007epri0413

MATERIALS AND COMPONENT DEVELOPMENT FOR ADVANCED TURBINE SYSTEMS

M. A. Alvin
National Energy Technology Laboratory
626 Cochrans Mill Road
Pittsburgh, PA 15236-0940

F. Pettit, G. Meier, N. Yanar, M. Chyu, D. Mazzotta, W. Slaughter, V. Karaivanov
Mechanical Engineering and Material Science
University of Pittsburgh
Pittsburgh, PA 15260

B. Kang, C. Feng, R. Chen, T-C. Fu
Mechanical and Aerospace Engineering & Civil Engineering
West Virginia University
Morgantown, WV 26506

Abstract

In order to meet the 2010-2020 DOE Fossil Energy goals for Advanced Power Systems, future oxy-fuel and hydrogen-fired turbines will need to be operated at higher temperatures for extended periods of time, in environments that contain substantially higher moisture concentrations in comparison to current commercial natural gas-fired turbines. Development of modified or advanced material systems, combined with aerothermal concepts are currently being addressed in order to achieve successful operation of these land-based engines.

To support the advanced turbine technology development, the National Energy Technology Laboratory (NETL) has initiated a research program effort in collaboration with the University of Pittsburgh (UPitt), and West Virginia University (WVU), working in conjunction with commercial material and coating suppliers as Howmet International and Coatings for Industry (CFI), and test facilities as Westinghouse Plasma Corporation (WPC) and Praxair, to develop advanced material and aerothermal technologies for use in future oxy-fuel and hydrogen-fired turbine applications. Our program efforts and recent results are presented.

Introduction

Increasing the power efficiency of stationary land-based industrial gas turbines will require future turbine inlet temperatures to increase to >1400°C (Table 1). Cooling air will simultaneously need to be controlled to avoid an increase in NO_x emissions. The concept of ultra low or "zero" emission power generation has focused on an oxy-fuel combustion process in which nitrogen is removed in the combustion air, and replaced with steam, thereby preventing the formation of nitrogen oxides. The exhaust stream of the oxy-fuel process is separated into concentrated CO_2 and water.

The development of oxy-fuel gas turbine IGCC plants has two major benefits: (1) the turbine exhaust gas is a highly enriched CO_2 stream with a small amount of excess oxygen. This exhaust gas can be processed to reduce excess oxygen to satisfy sequestration oxygen requirements if needed, then dried and compressed, reducing the cost and plant performance impact of CO_2 removal for sequestration compared to the standard IGCC plant configuration; (2) the syngas, recycle CO_2, and oxygen streams fed to the oxy-fuel turbine combustors will contain very little nitrogen. Thus, NO_x emissions from the gas turbine will be very low, eliminating the need for expensive, low-NO_x combustors and SCR exhaust gas NO_x reduction.

Alternately, hydrogen-fueled combustion turbine systems are conceptually based on a complex cycle composed of a closed Brayton and Rankine cycle. Hydrogen turbines are expected to be a key part of near-term integrated gasification combined cycle (IGCC) plants that can sequester carbon dioxide, such as FutureGen. Hydrogen and oxygen are supplied as the fuel and oxidant respectively to a compressor, and burned in steam. The pressurized steam feed enters a high temperature turbine at 1700°C for stator blades, and 1570°C for rotor blades. Since these elevated temperatures are typically higher than the melting temperatures of alloys used in the construction of the hot gas structural components, the components are cooled by air extracted from the turbine compressor and directed through the cooling passages designed into the component.

Oxidation and hot corrosion in commercial land-based power generation gas turbines are principal concerns, particularly in view of long-term operation (30,000 hrs), as well operation with advanced combustion cycles that potentially target turbine inlet temperatures of >1500°C. Current structural materials as nickel and cobalt-based superalloys cannot withstand temperatures of >950°C. Although single crystal substrates as René N5, CMSX-4, PM2000, and the like, increase the durable operating temperature for the structural airfoil support, operation in the advanced combustion cycles will require development and application of stable thermal barrier coatings (TBCs) with better insulation properties, as well as implementation of acceptable internal component cooling techniques, in order to achieve extended operational life of oxy-fuel and hydrogen fired land-based turbines.

Bond Coat Development and Assessment

As a candidate alloy for advanced turbine blades, the single crystal René N5 matrix was selected in this program for its high temperature creep resistant characteristics, and oxidative stability which results from the formation of a protective external alumina scale. As a combustor liner or disc matrix, the nickel-based superalloy, Haynes 230, was also selected in this program for its high temperature oxidative stability, but primarily for its ability to form an external protective chromia scale. By addressing the response of both metal substrates and their TBC counterparts, information is being generated that potentially could support the use of either alloy at alternate turbine component design locations.

In conjunction with CFI, NETL initiated development of a bond coat system that consisted of metallic-ceramic particles that were held together via an inorganic binder. When applied to René N5 and Haynes 230 coupons manufactured at Howmet International, preliminary bench-scale testing at CFI indicated adherence of the porous ~40-50 μm thick NETL-1 bond coat after being subjected to (1) 24 hrs of static air exposure at 800°C; (2) 10 severe thermal cycles (100°C/minute ramp rate from room temperature to 1000°C, followed by 1 hr dwell time at temperature, then rapid cooling); and 3) 500 hrs of static air testing at 1000°C.

Extended thermal cycle testing was then conducted at 900°C and 1100°C under static conditions for 2,000 hrs. Although the initial NETL-bond coat remained adherent to the base metal substrates at 900°C, the coupons experienced a weight increase indicative of the oxidation throughout the bond coat, as well as along the René N5 and Haynes 230 substrate interfaces. At 1100°C, weight loss of the coupons was observed after 200 hrs, which reflected localized spalling of the coating during both thermal cycling and isothermal testing. Oxidation along the single crystal and superalloy substrate interface was also observed. Similar results were demonstrated when thermal cycling was conducted in the presence of water vapor.

To improve adherence of the bond coat at 1100°C and mitigate oxygen permeation, modifications to the NETL-CFI bond coat were made, subsequently leading to production of a densified metallic coating (Figure 1). Currently extended thermal cycle testing is being conducted to demonstrate the performance of the modified NETL-A1 architecture at 900°C and 1100°C. The rationale for testing at 1100°C is to screen potential alternate bond coat systems, comparing their performance to that of state-of-the-art (SOTA) materials, prior to addressing performance at the advanced turbine operating temperatures shown in Table 1.

Testing has also been initiated at the Westinghouse Plasma Corporation (WPC) facility where rotating (500 rpm), NETL-A1-coated, René N5 tubes are being subjected to external surface temperatures of 1100°C, while being internally cooled with air (Figure 2). After extended exposure (i.e., ~500 hrs), post-test analysis of the tubes will include visual inspection, and measurements to determine thinning of the outer coating. Due to the material and fabrication cost for each tube, post-test destructive characterization will not be performed. Non-destructive evaluation (NDE) techniques, as described in the following section, are planned. Demonstration of the stability and performance of material systems when subjected to an applied heat flux at WPC serves as the basis for extended screening of potential material systems for use in advanced turbine applications, and is being conducted in conjunction with our aerothermal efforts which are also described in this paper.

Efforts have also been focused on the application and adherence of an air plasma sprayed (APS) and electron beam physical vapor deposited (EBPVD), yttria stabilized zirconia (YSZ), top coat along the outer surface of NETL bond coat systems. As shown in Table 2, early failure resulted for the original NETL-1 bond coat system when either the APS or EBPVD YSZ architecture had been subjected to 1100°C thermal cycle testing. Debonding along the metal substrate interface was principally attributed to the adhesion characteristics of the porous as-manufactured NETL-1 bond coat. The surface roughness of the bond coat also lead to irregular columnar growth of the EBPVD YSZ layer (Figure 3). For state-of-the-art CoNiCrAlY-YSZ systems (Figure 4), cycle time-to-failure during 1100°C cyclic oxidation testing was extended. The absence of an applied bond coat system was shown to further promote retention of the YSZ top coat along the surface of the base metal substrates. Extended cycle time-to-failure was observed for APS YSZ on the René N5, an alumina former, while the longest cycle time-to-failure identified to date was EBPVD YSZ on Haynes 230, a chromia former.

Material Diagnostics

Micro-Indentation Testing — Surface Stiffness Response

Recent efforts have been focused on development of a simplified micro-indentation technique (MIT) for monitoring changes in the mechanical properties of bond coat and TBC systems, and ultimately addressing their integrity and life during extended process operation. The MIT consists of a 1.6 mm diameter spherical indenter assembly that is attached to a load cell, which is in turn attached to a PZT actuator. The PZT actuator provides the overall indentation depth measurement which includes the indentation penetration depth and system rigid-body displacement.

Initially, the contact position between the indenter and the surface of the sample is determined for a given load threshold (i.e., 0.1-5N). Subsequently multi-loading and partial-unloading indentation tests are conducted using pre-defined parameters as velocity of the indenter, the penetration depth (i.e., ~10 μm), and the unloading compliance. From the multiple unloading compliance measurements, the effect of system rigid-body displacement is eliminated, and true surface stiffness response of the bond-coated coupons is then determined. Using this technique on flat, polished, 25mm x 25mm x 3mm coupons, a Young's modulus of ~200-210 GPa was determined for Haynes 230, which is in good agreement with literature reported values. The Young's modulus for the single crystal René N5 matrix was identified as ~130-150 GPa. In contrast to defining Young's modulus, an overall surface stiffness response of the coating/substrate architecture was determined due to the irregular morphology of the bond coat surface (1).

When bond-coated René N5 and Haynes 230 coupons were subjected to thermal cycling conditions (25° ↔ 1100°C), the overall surface stiffness response of the coating was shown to decrease with continued cycling (Figure 5). These results strongly correlate with weight change measurements of the coupons after completion of the 100, 200, 300, and 400 thermal cycles. In addition, the greater calculated loss of bond coat stiffness on the Haynes 230 matrix reflected the greater extent of surface crack formations and areas of spallation in comparison to that which resulted along the surface of the thermally cycled, bond-coated, René N5 coupons.

Acousto-Ultrasonics

A nonlinear acousto-ultrasonic technique is also being developed to evaluate interface delamination of bond coatings that are applied to the surface of single crystal and superalloy metal substrates. A greater nonlinear response in the waveform amplitude-dependent characteristics is projected to result when debonding or delamination of the bond coat and/or top coat layers occurs.

The bench-scale test equipment consists of a laser vibrometer and a spectrum analyzer with various ultrasonic transducers, and digitization/acquisition boards that can detect kHz-MHz frequencies. Due to the size of coupons (25mm x 25mm x 3mm), miniature piezoelectric sensors were used. In order to assure that the distance between the pulser and receiver was fixed, both sensors were attached to an acrylic block which served as a sensor bridge. A stretched latex membrane was used as the dry contact interface between the sensor and the coupon surface. A constant sensor contact pressure of 0.6 MPa was applied to the coupon surface, and a series of 30 readings were generated typically along the center of the coupon to assure statistical accuracy of the resulting data.

Testing was performed on (1) uncoated base metal coupons; (2) as-manufactured coated coupons; and (3) coated coupons that had been subjected to 100, 200, 300, and 400 thermal cycles (25° to 1100°C). A wave propagation analysis was conducted using Lamb's solution for an axisymmetric analysis of an elastic half-space due to a point load at the origin. Although the ultrasonic waveform was virtually the same for the bond-coated René N5 coupon after being subjected to 100, 200, 300, and 400, 1100°C thermal cycles, slight shifting in travel time was observed (Figure 6). This implied that material properties as Young's modulus, density, or other geometry changes had occurred. In contrast, marked differences in the waveform and travel time were observed for the bond-coated Haynes 230 coupon. Discontinuity or localized debonding and/or delamination of the coating from the surface of the Haynes 230 matrix was expected to have occurred. Notably, repeatability of the wave characteristics was observed for the thermal-cycled René N5 matrix, implying more overall uniformity and consistency of the residual bond coat.

In conjunction with experimental testing, a finite element method (FEM) for wave propagation simulation was conducted using ABAQUS. Material properties were assumed to be $E = 211$ GPa and $\rho = 8{,}970$ kg/m^3 for Haynes 230, and $E = 213$ GPa and $\rho = 8{,}630$ kg/m^3 for René N5 base metal substrates. FEM time histories of the surface vertical waveform displacement were shown to be comparable to the theoretical solutions for both the arrival time and peak amplitude. In contrast for bond-coated systems, the peak amplitudes and their corresponding arrival times were seen to be clearly influenced by assumed material properties (Figure 7). Similarly when embedded voids were included in modulated sine impulse FEM simulations to reflect delamination of the coating, a strong nonlinear effect from the void with higher harmonics resulted (Figure 8). Correlation of the FEM delamination simulation data with the experimentally generated results for the 100-400 thermally cycled coupons is being undertaken.

Aerothermal-Materials Integration

A computational methodology based on three-dimensional numerical simulation and damage mechanics for predicting thermal-mechanical durability and life of turbine blades is being developed for hydrogen-fired and oxy-fuel applications. Efforts have focused on analysis of stress distributions due to temperature, pressure, and centrifugal loads for a solid NASA E^3 blade model (Figure 9) (2). To complete the thermal-mechanical analysis of the solid substrate, the level of thermal load or boundary condition on the airfoil external surface was determined via 3D CFD simulation modeling using FLUENT. While the heat transfer coefficient was shown to vary strongly along the surface of the airfoil, the projected trends were relatively comparable for airfoils in syngas and hydrogen-fired applications (Table 1). Modeling projected the highest surface temperatures along the leading edge of single crystal airfoils (i.e., CSMX-4; René N5), as well as airfoils with an applied MCrAlY-YSZ TBC (Figure 10). When combined thermal, surface pressure, and centrifugal loads were applied to the model airfoil, the highest stress resulted along the mid-rib and near the root section of the blade (Figure 11).

Currently efforts are being directed to incorporate the finite element analysis (FEA) results into a life prediction model that is capable of evaluating creep evolution damage over the entire airfoil, as well as visualizing the results of the life prediction analysis, and identifying the most critical zones (3). Recent creep damage calculations with a mechanical load comparable to that of future hydrogen-fired systems identified significant short-term impact again along the blade's leading edge, mid-rib, and near the base of the trailing edge (Figure 12). Reducing temperature by 30% over the entire blade projected a significant increase in service operating life (Figure 13).

Acknowledgments

We wish to acknowledge Mr. Richard Dennis, DOE NETL Turbine Technology Manager, for his continued support of this project. The efforts of Mr. James Klotz at CFI, Mr. Bob Grunstra, Mr. Ty Hansen, Mr. Ron Honick, and Mr. Ken Murphy at Howmet International, and Dr. Shyam Dighe and Dr. Ivan Martorell at WPC are gratefully acknowledged.

References

1. C. Feng, M.A. Alvin, B.S.-J. Kang, "A Micro-Indentation Method for Assessment of TBC Bond Coat Systems," Paper No. 351171, presented at the MS&T 2007 Conference, Detroit, MI (September 2007).

2. D. Mazzotta, M.K. Chyu, M.A. Alvin, "Aero-Thermal Characterization of Hydrogen Turbines," Paper No. GT2007-28296, presented at the ASME Turbo Expo 2007, Land, Sea and Air Conference, Montreal, Canada (May 2007).

3. V. Karaivanov, D. Mazzotta, W. Slaughter, M. Chyu, M.A. Alvin, "Three-dimensional Modeling of Creep Damage in Airfoils for Advanced Turbine Systems," Paper No. GT2008-

51278 to be presented at the ASME Turbo Expo 2008, Gas Turbine Technical Congress and Exposition, Berlin, Germany (June 2008).

Table 1: Advanced Turbine Operating Conditions

	Syngas Turbine 2010	Hydrogen Turbine 2015-2020	Oxy-Fuel Turbine 2010	Oxy-Fuel Turbine 2015-2020
Combustor Exhaust Temp, °F (°C)	~+2700 (~+1480)	~+2700 (~+1480)		
Turbine Inlet Temp, °F (°C)	~2500 (~1370)	~2600 (~1425)	~1150 (~620)	~1400 (~760) (HP ~3200 (~1760) (IP
Turbine Exhaust Temp, °F (°C)	~1100 (~595)	~1100 (~595)		
Turbine Inlet Pressure, psig	~265	~300	~450	~1500 (HP) ~625 (IP)
Combustor Exhaust Composition	9.27% CO_2 8.5% H_2O 72.8% N_2 0.8% Ar 8.6% O_2	1.4% CO_2 17.3% H_2O 72.2% N_2 0.9% Ar 8.2% O_2	82% H_2O 17% CO_2 0.1% O_2 1.1% N_2 1% Ar	75-90% H_2O 25-10% CO_2 1.7% O_2, N_2, Ar

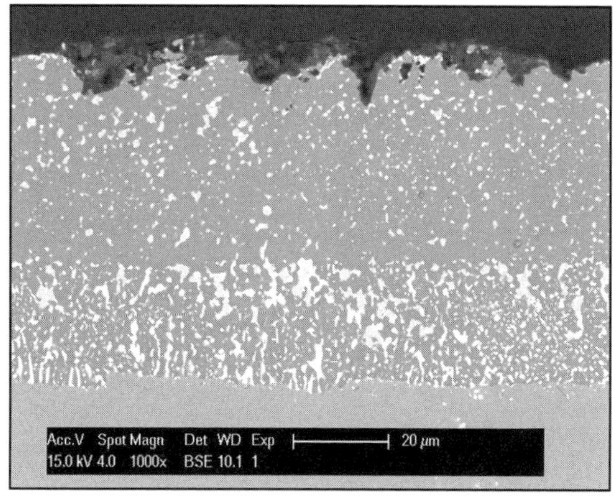

Figure 1. NETL-A1 Bond Coat System

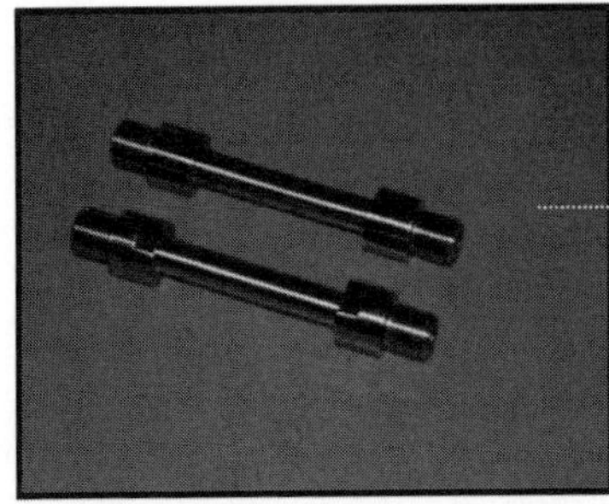

Figure 2. Bond-Coated René N5 Tubes Used for Thermal Flux Testing at the Westinghouse Plasma Corporation Test Facility

Table 2: Cyclic Oxidation Testing of NETL and SOTA TBC Systems at 1100°C
— Cycles to Failure —

Bond Coat + YSZ Top Coat	Substrate	
	René N5	Haynes 230
NETL-1 + APS	20	20
NETL-1 + EBPVD	40	120-140
NETL-A1 + APS		
NETL-A1 + EBPVD		
CoNiCrAlY + APS	200	220-240
CoNiCrAlY + EBPVD		
Pt-Al + EBPVD		
– + APS	240	60
– + EBPVD	320-400	460-480

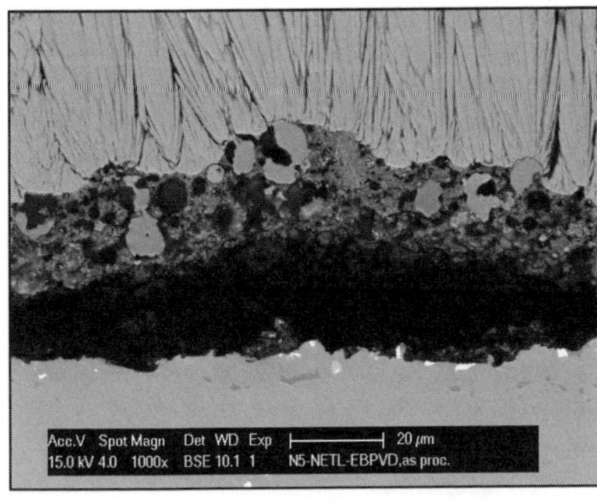

Figure 3. EBPVD YSZ on the NETL-1 Bond Coat

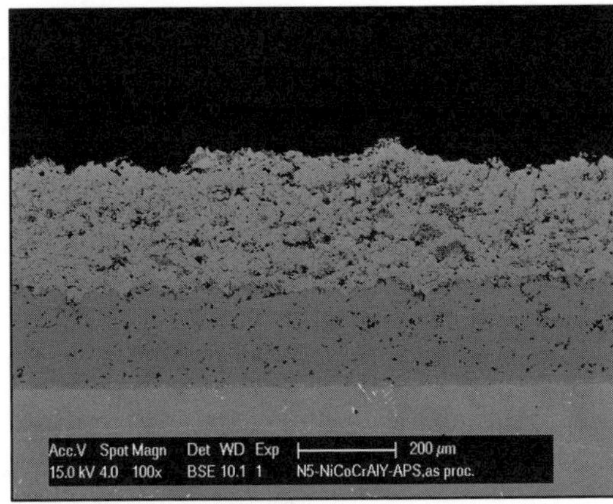

Figure 4. APS YSZ on a CoNiCrAlY Bond Coat

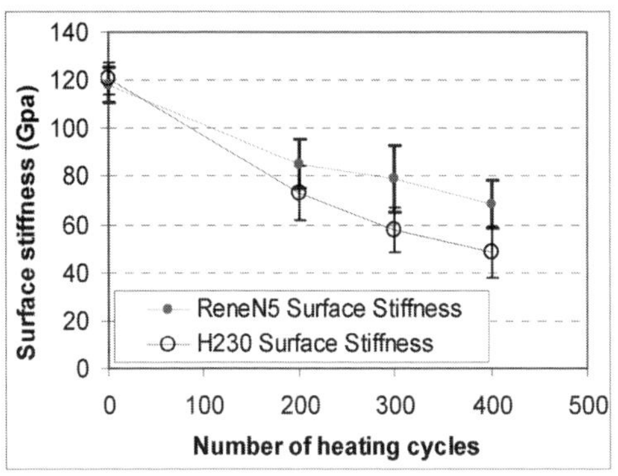

Figure 5. Bond Coat Stiffness Comparison on René N5 and Haynes 230 Alloy as a Function of Extended, High Temperature, Cyclic Oxidation

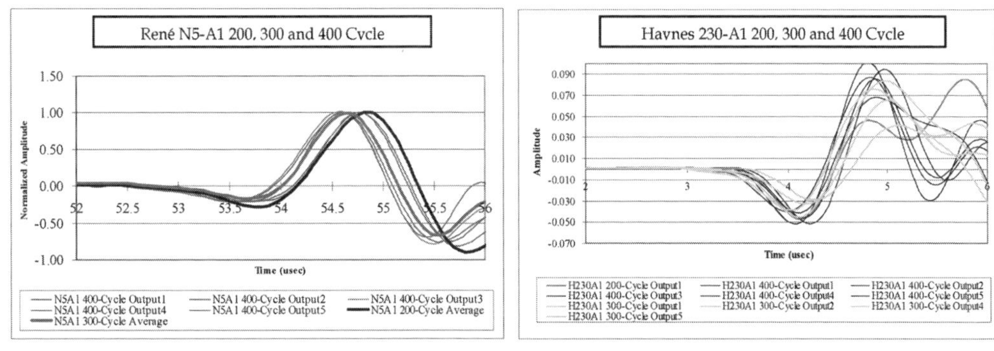

Figure 6. Wavefronts for Bond-Coated René N5 and Haynes 230 Coupons after Exposure to 100-400 Thermal Cycles

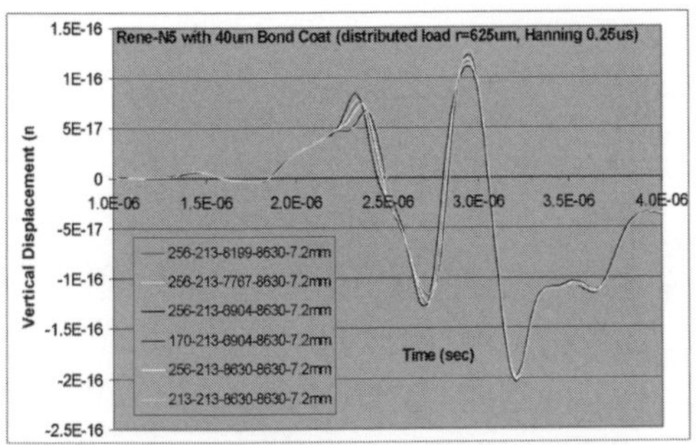

Figure 7. Influence of Bond Coat Properties on Surface Displacement Time History

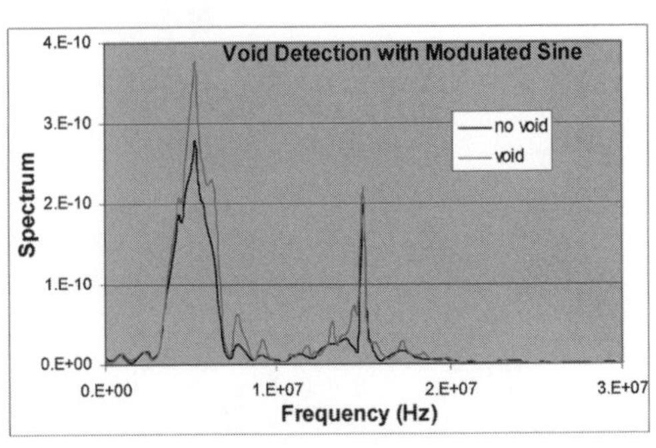

Figure 8. Spectrum of Delaminated TBCs w Modulated Sine Excitations

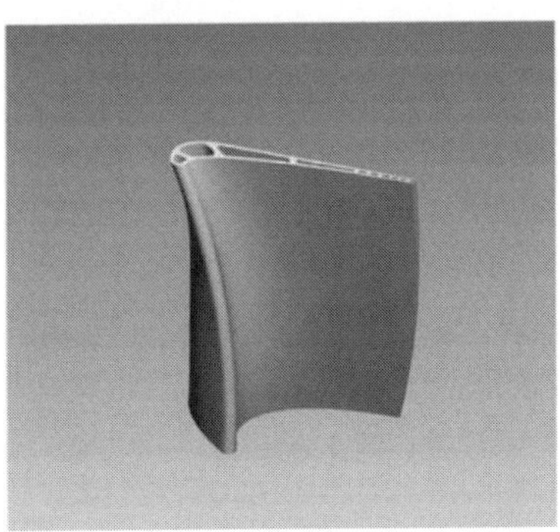

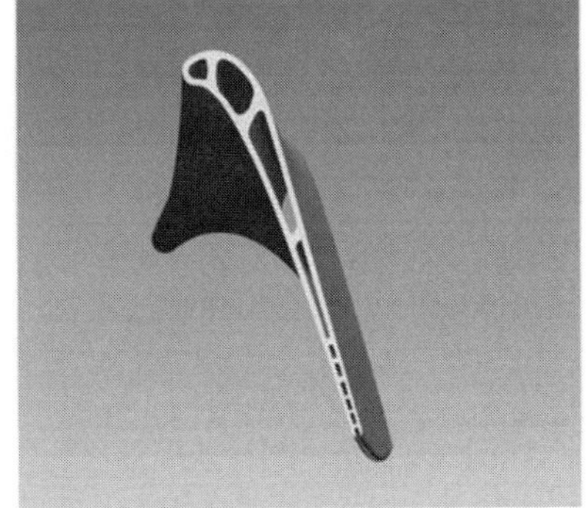

Figure 9. Solid Model of the NASA E³ Blade, Created Using Pro/Engineer and the Finite Element Analysis Application ANSYS

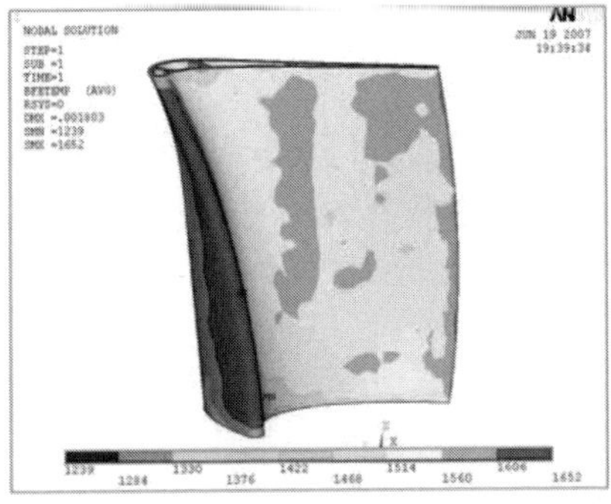

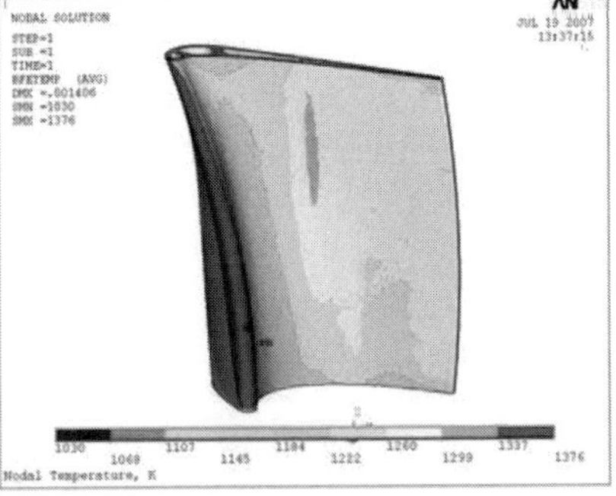

Figure 10. Base Metal and External TBC Surface Contour Temperature Distribution Plots, in Kelvin, for Hydrogen-Fired Airfoil Applications

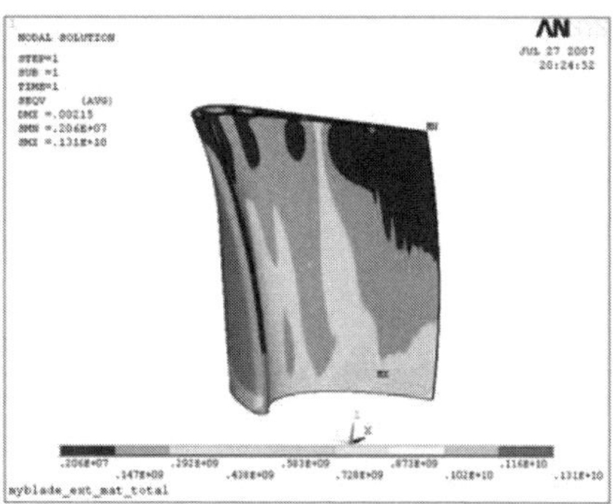

Figure 11. Stress Distribution Plot, in Pascals, Resulting from the Combined Thermal, Surface Pressure, and Centrifugal Loads (Applied Angular Velocity of 15,000 rpm; Rotor Radius of 0.35 m)

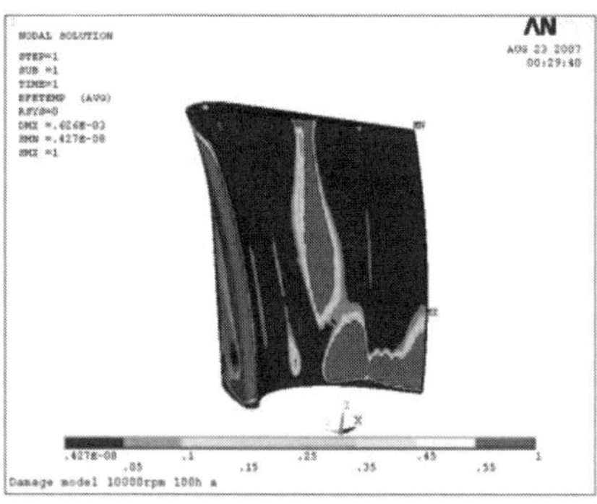

Figure 12. Creep Damage: 100 hrs, 10k rpm

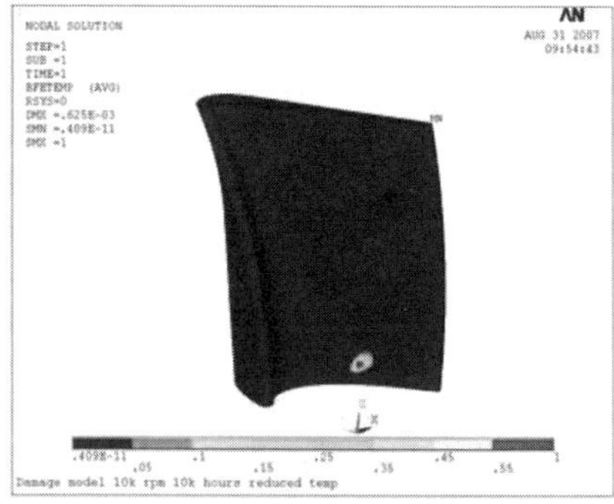

Figure 13. Creep Damage: 10,000 hrs, 10k rpm, Reduced Thermal Load

Advances in Materials Technology for Fossil Power Plants
Proceedings from the Fifth International Conference
R. Viswanathan, D. Gandy, K. Coleman, editors, p 424-433

DOI: 10.1361/cp2007epri0424

Mechanical Properties and Manufacturability of Ni-Fe Base Superalloy (FENIX-700) for A-USC Steam Turbine Rotor Large Forgings

Shinya Imano[1], Jun Sato[1],
Koji Kajikawa[2] and Tatsuya Takahashi[2]

[1]Hitachi Research Laboratory, Hitachi, LTD.;
1-1-7 Omika-cho, Hitachi, Ibaraki, 319-1292, Japan
[2]Muroran Research Laboratory,
The Japan Steel Works, LTD.;
4 Chatsu-Mati, Muroran, Hokkaido, 051-8505, Japan

Abstract

To develop 10-ton class forgings with adequate long-term strength and without segregation defects for A-USC steam turbine rotors, we modified the chemical composition of Alloy706 to improve its microstructure stability and segregation properties. The modified Alloy (FENIX-700) is a γ' phase strengthened alloy without a γ'' phase. Its microstructure stability is superior to Alloy 706 at 700 °C, according to short-term aging tests and phase stability calculations using the CALPHAD method. In this presentation, we present the recent results of various mechanical tests, microstructure observations and manufacturability evaluation tests using FENIX-700. Following is a summary of the results. A trial disk 1-ton class forging of FENIX-700 was manufactured from a double melted ingot. Tensile and creep strength of the forging were equivalent to that of 10-kg class forgings. The trial forging was successful. 10^5h class creep tests were performed using 10kg class forgings and according to the results, the approximate 10^5h creep strength at 700 °C is higher than 100 MPa. The manufacturability of FENIX-700 is better than that of Alloy706, according to segregation tests using a horizontal directional solidification furnace and hot workability tests. Microstructure observation and tensile tests were performed using 10,000h aged specimens (650,700,750 °C). Degradation of tensile strength and yield stress due to coarsening of the γ' phase were observed, but ductility was enhanced by aging. The microstructure stability of FENIX-700 at 700°C was excellent according to microstructure observation of the 10,000h class aged sample and thermodynamic considerations.

Introduction

Ultra super critical steam turbines with main steam temperatures of 700°C or more (A-USC) are now being developed in order to further improve thermal efficiency and reduce CO_2 emissions from fossil power plants. Conventional rotor materials made of ferritic steel are not suitable for A-USC rotors since their maximum temperature is about 650°C; thus, it is necessary to produce the rotors with a high strength Ni base or Ni-Fe base superalloy. Ni-36Fe-16Cr-3Nb-1.7Ti-0.3Al (Alloy706[1)2)]) is advantageous because it is excellent in ingot making, strength, and cost (it contains about 40 weight percent of Fe). Alloy706 is a gas-turbine disk material. However, since Alloy706 suffers from a solidification defect (freckle defect) due to segregation of Nb, it is difficult to make a forged product that weighs more than 10 tons from Alloy706. Reducing Nb, which is a segregation element, would improve the manufacturing properties of the large alloy ingots. However, since Alloy706 is precipitation-strengthened by Ni_3Nb (γ" Phase) and Ni_3Al(γ' Phase)[3)], its strength deteriorates if the Nb concentration is reduced. Also, it is known that harmful phases are precipitated when Alloy706 is subjected to 700°C for a long time[2)], and the Ni-Fe base superalloy is thereby weakened. Thus, as described above, there are problems with respect to the manufacturing properties and high-temperature microstructure stability of the rotor material when manufacturing a steam turbine for main steam temperatures over 700°C. To solve these problems we modified the chemical composition of Alloy706 using the CALPHAD method[4)] and experimental data. In this presentation, we present the recent results of mechanical tests, microstructure observations, and manufacturability evaluation tests using FENIX-700.

Experimental Procedure

Trial forging of 2ton ESR ingot

The trial disk forging of FENIX-700 was manufactured from a double melted ingot (Vacuum induction melting (VIM) followed by electroslag remelting (ESR)). The diameter of the ingot was 450mm, and its weight was about 2 tons. The diameter of the trial disk shape forging was 900mm. Conditions of ingot making and forging were the same as those for Alloy706. Creep and tensile specimens were taken from rim parts and the tensile direction was radial. Table 1 shows the chemical composition of the forging (FENIX-700TF1).

Evaluation of Manufacturability

There is a critical value for macro-segregation ($\varepsilon R^{1.1}{}_C$)[5)] which is a function of cooling rate (ε) and solidification rate (R). To prevent macro-segregation, it is necessary to maintain a $\varepsilon R^{1.1}$ value higher than $\varepsilon R^{1.1}{}_C$ during the solidification process. The critical value for segregation depends on the chemical composition of the alloys. Reducing $\varepsilon R^{1.1}{}_C$ during the ESR or VAR processes effectively avoids segregation. To evaluate the $\varepsilon R^{1.1}{}_C$ of FENIX-700 and compare it to that of Alloy706, segregation tests were performed using the horizontal directional solidification furnace shown in Figure 1. Temperatures were measured at six points, from casting to the end of solidification. Cooling rate (ε) and solidification rate (R) were evaluated from these data at each point. The value of $\varepsilon R^{1.1}$ is evaluated as a function of distance from cooling chill. A macrostructure observation was performed using the specimens after solidification to determine the distance from the cooling chill to the first segregation point. Critical values for segregation ($\varepsilon R^{1.1}{}_C$) are determined using the distance from the cooling chill to the first segregation point and the previous function. Table 2 shows the chemical composition of the ingots after the segregation test. Hot workability was evaluated using THERMECMASTER-Z. Flow stress at 850°C to 1050°C was evaluated as function of strain rate. High temperature ductility was evaluated, according to the results of tensile tests at elevated temperatures.

Mechanical tests after long-term exposure at elevated temperatures

Table 1(FENIX-700AG) shows the chemical composition of the test specimens after long term exposure. These specimens were machined from forged bars (500mm×30m×L), which were made from VIM (50kg) ingots. The forged bars were solution treated at 995°C. After the solution treatment, the forged bars were aged and the mean diameters of the γ' phase were controlled in the range of 50nm to 100nm. After the solution and aging treatments, these specimens were exposed at 650, 700, and 750°C. Exposure times were 4000h and 10000h. Tensile tests and microstructure observations (SEM) were performed using these specimens.

Result and Discussion

Trial forging of 2ton ESR ingot

Figure 2 shows the appearance of the ESR ingot and the trial forging. The trial forging was successful. The grain size was about 3 to 4 after heat treatment. The tensile properties and fatigue properties were as same as those of the 10-kg forging (FENIX-700OR). Figure 3 shows the results of the creep tests. Approximate 10^5h creep strength is more than 100MPa, according to the results of a 10^4h class creep test using the 10-kg forging. The 10^3h class creep rupture strength of the 1-ton forging is higher than that of the 10-kg forging.

Evaluation of Manufacturability

Figure 4 shows the cross-section of specimen Ch.1 after the segregation test. Segregation defects in the Nb-rich region were observed 50mm from the cooling side. The critical values for segregation ($\varepsilon R^{1.1}{}_c$) are evaluated from the cooling rate (ε) and solidification speed (R). Figure 5 shows the relationship between the critical value for segregation and the distance from the cooling side (Ch.1). The critical value for segregation was evaluated from these results. Table 3 shows the critical values for segregation in Ch.1, Ch. 2, and Alloy706[6)]. The ingot size limit depends on the minimum value of $\varepsilon R^{1.1}$ during the remelting process and on the $\varepsilon R^{1.1}{}_c$ value of the material. We also evaluated various $\varepsilon R^{1.1}$ values during ESR processes according to solidification simulations. Using these results, we evaluated the ingot size limitations of FENIX-700 due to macro-segregation. The results of the evaluation can be seen in Figure 6. The ingot size limit for Alloy706 is about 750 mmφ in this condition. The ingot size limit for FENIX-700 is larger than 850 mmφ, because its $\varepsilon R^{1.1}{}_C$ value is smaller than its $\varepsilon R^{1.1}$ value during solidification of the 850 mmφ ingot. We expect that a larger ingot can be made from FENIX-700 by optimizing the ingot making process and making minor modifications to the chemical composition, or by using a VAR process.

Figure 7 shows the flow stress of FENIX-700 at elevated temperatures. The flow stress of FENIX-700 is lower than that of Alloy706, 718, and IN617. Figure 8 shows the results of the high-temperature tensile tests. The tensile elongation and area reduction of FENIX-700 are at low levels up to 850°C because of γ' precipitation, but these values rise to about 100% above 900°C. The area reduction above 850°C is more than 50%. From these and previous results, we conclude that the manufacturability of FENIX-700 is better than that of Alloy706.

Mechanical tests after long-term exposure at elevated temperature

Figure 9 shows tensile elongation, area reduction, and a SEM micrograph for specimens exposed for a long time at elevated temperatures. A coarsening of γ' was obvious at 750°C and a small amount of platelet η phase was observed at the grain boundary. But the spherical γ' phase is the main precipitate in all specimens and the η phase was not observed in optical micrographs of specimens with long-term exposure at 650° and 700°C.

The ductility of these specimens increases with time and temperature. Degradation of ductility by η phase precipitation was not observed. Figure 10 shows the relationship between 0.2% proof stress and 1/λ in the long-term exposed specimens . λ is the mean space of the γ' phase, which was evaluated by SEM image analysis. The variation of 0.2% proof stress is due to coarsening of the γ' phase, because there is a linear relationship between 0.2% proof stress and 1/λ. The effect of η phase precipitation is negligible, because the amount of η phase is very small. The Larson Miller Parameter (LMP) of exposure for 10,000h at 750°C was 24.6(C=20). That value is larger than 24.3, which is equivalent to exposure for 100,000h at 700°C. The η phase is more stable at 750°C than at 700°C, according to thermodynamic calculations using CALPHAD method. It is expected that amount of η phase after exposure for 100,000h at 700°C will be fewer than that of exposure for 10,000h at 750°C. As a result of this investigation, it is concluded that FENIX-700 has adequate microstructure stability at 700°C.

Conclusions

We modified the chemical composition of Alloy706 to improve its microstructure stability and segregation properties, and to develop a 10ton class forging with adequate long-term strength and without segregation defects for A-USC steam turbine rotors. Following are the conclusions according to the results of our investigation of the modified alloy FENIX-700.

1. Trial forging (1ton disk) was successful; creep strength, tensile properties and fatigue property were the same as those of a 10-ton forging.

2. Approximate 10^5h creep strength at 700°C is higher than 100MPa according to 10^4h class creep tests using a 10-ton forging.

3. The limitation on ingot size due to segregation problems is larger for FENIX-700 than for Alloy706, according to segregation tests and ESR process simulations.

4. FENIX-700 has adequate microstructure stability at 700°C for A-USC steam turbine rotors.

Acknowledgements

The authors would like to thank Dr. Takashi Shibata for his advice on ingot making and on the forging process of superalloys.

References

1. C. Berger, J. Granacher, A. Thoma, Proc._Conf. Superalloys 718, 625, 706 and Various Derivatives, 2001, 489-499.

2. H. J. Penkalla, J. Wosik, W. Fischer, F. Schubert, Proc. Conf. Superalloys 718, 625, 706 and Various Derivatives, 2001, 279-290.

3. T. Takahashi,T. Shibata: CAMP-ISIJ VOL.10(1997)- 1414

4. N. Saunders, X. Li, A. P. Miodownik , J-Ph. Proc. Symp. Materials Design Approaches and Experiences, eds. J.-C. Shao et al., 185-197; 2001, Warrendale, PA, TMS.

5. K. Suzuki and T. Miyamoto: Trans. ISIJ, 1978, 18(2), pp.80-89

A. Itoh, S. Suzuki, K. Kajikawa and H. Yamada: JSW technical review(Japanese) 54 (1998) pp.104-112)

Table 1 Chemical Composition of the Specimens(wt%)

	C	Ni	Cr	Fe	Al	Ti	Nb
FENIX-700TF1	0.02	41.3	15.45	Bal.	1.32	1.63	2.1
FENIX700OR	0.03	41.4	15.60	Bal.	1.26	1.70	2.0
FENIX700AG	0.02	42.1	16.10	Bal.	1.11	1.71	2.0

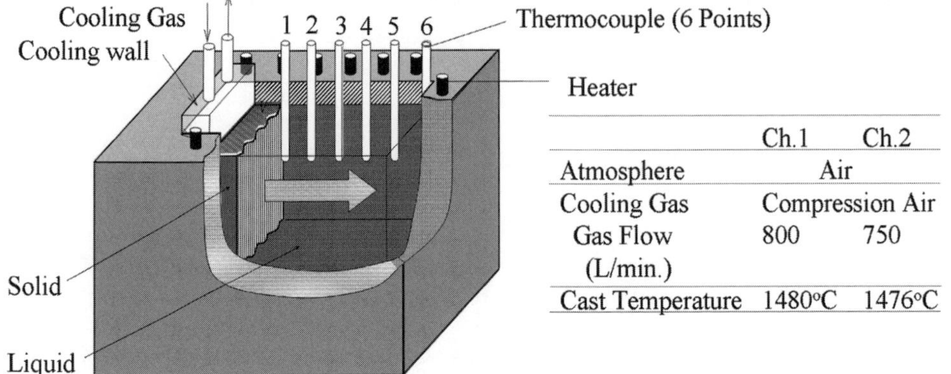

	Ch.1	Ch.2
Atmosphere	Air	
Cooling Gas	Compression Air	
Gas Flow (L/min.)	800	750
Cast Temperature	1480°C	1476°C

Figure 1 Horizontal Directional Solidification Furnace

Table 2 Chemical Composition of Specimens used in Segregation Tests (wt%)

	C	Si	Ni	Cr	Fe	Al	Ti	Nb	O(ppm)	N(ppm)
Ch.1	0.003	0.01	41.16	15.90	35.80	2.10	2.54	2.06	13	102
Ch.2	0.005	0.04	42.70	15.44	36.59	1.23	1.82	2.11	7	72

Appearance of ESR ingot

Appearance of trial forging

Figure 2 Result of the Trial Forging

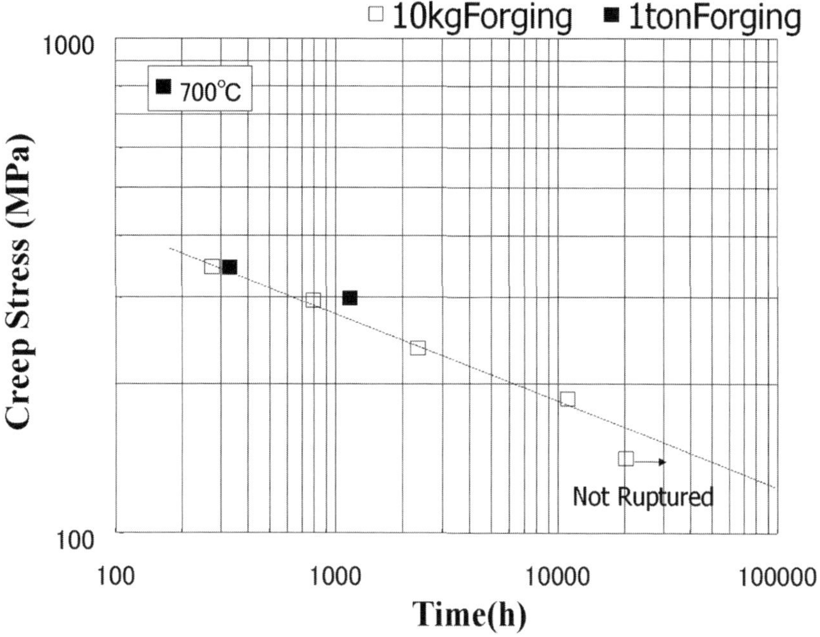

Figure 3 Result of the Creep Rupture Test

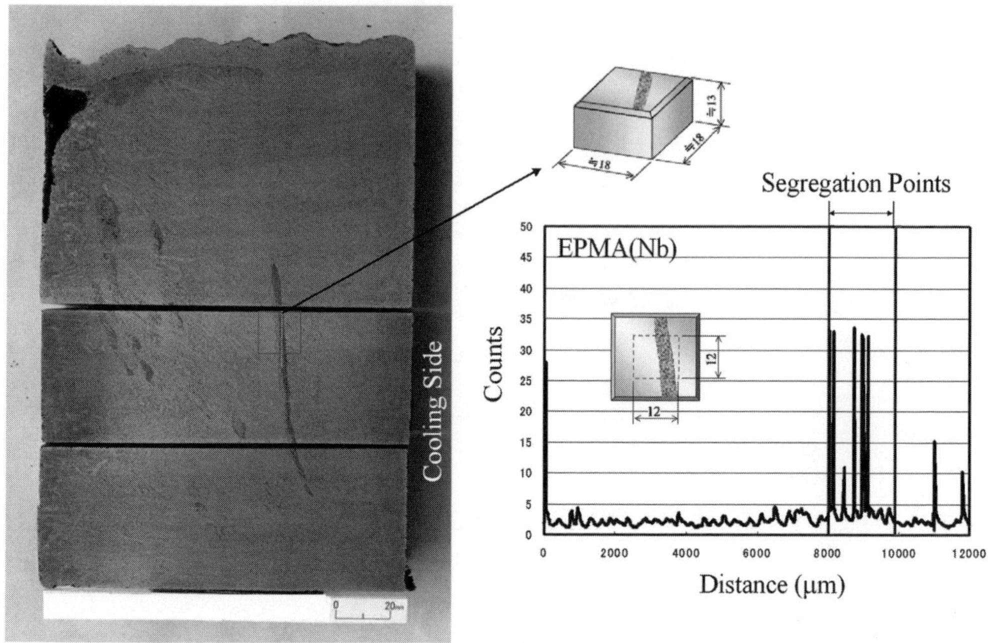

Figure 4 Cross section of specimen Ch.1 after segregation test

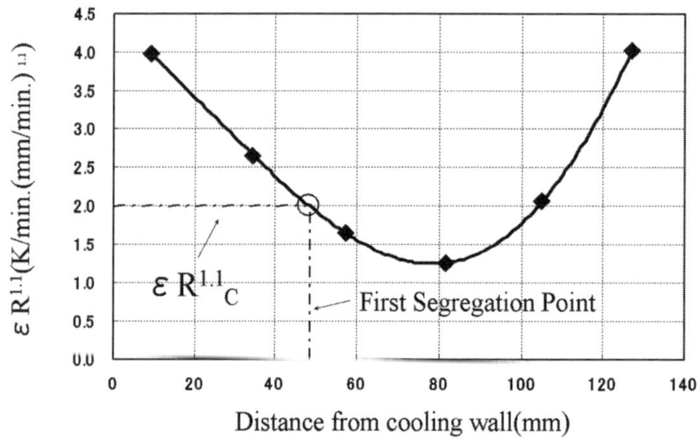

Figure 5 Variation of $\varepsilon R^{1.1}$during solidification(Ch.1)

Table 3 Comparison of $\varepsilon R^{1.1}{}_C$ Value

	Segregation point(mm)**	$\varepsilon R^{1.1}{}_C$
Alloy706	-----	1.60[6)]
Ch.1	50	2.00
Ch.2***	80	1.17

* Distance from cooling wall to first segregation point

** Similar to chemical composition of FENIX-700

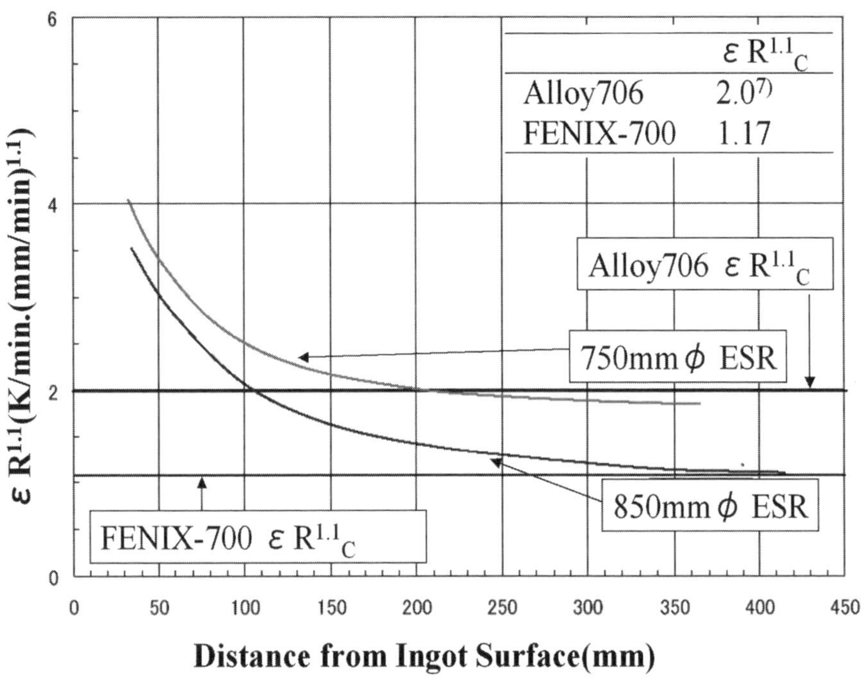

Figure 6 Variation of $\varepsilon R^{1.1}$during remelting process

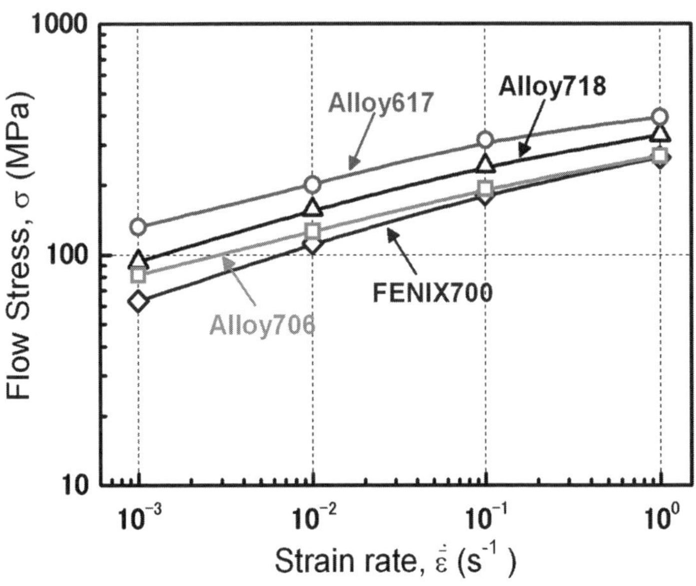

Figure 7 Flow Stress at Elevated Temperature(1050°C)

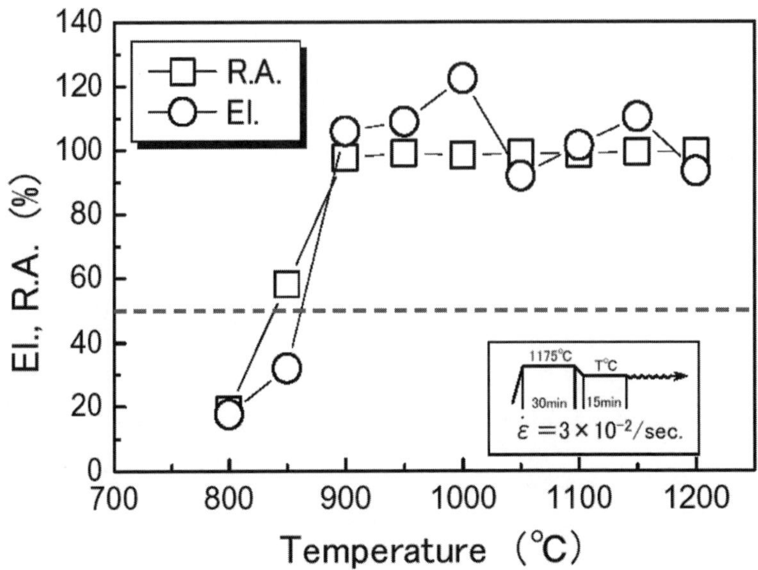

Figure 8 Elongation and Reduction Area at Elevated Temperature

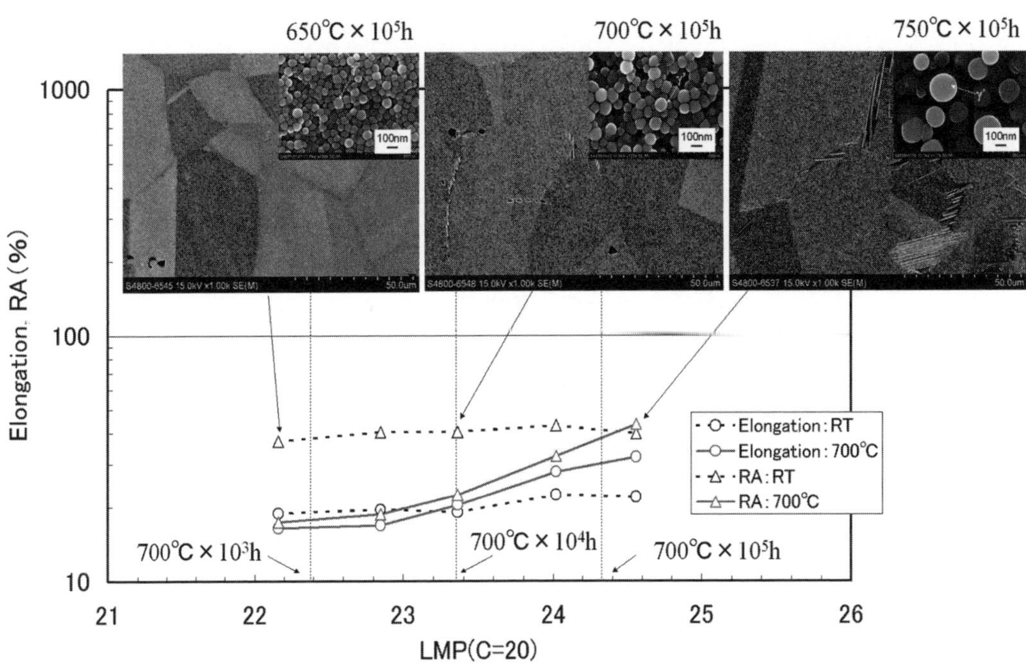

Figure 9 Microstructure and variation of ductility after long time aging

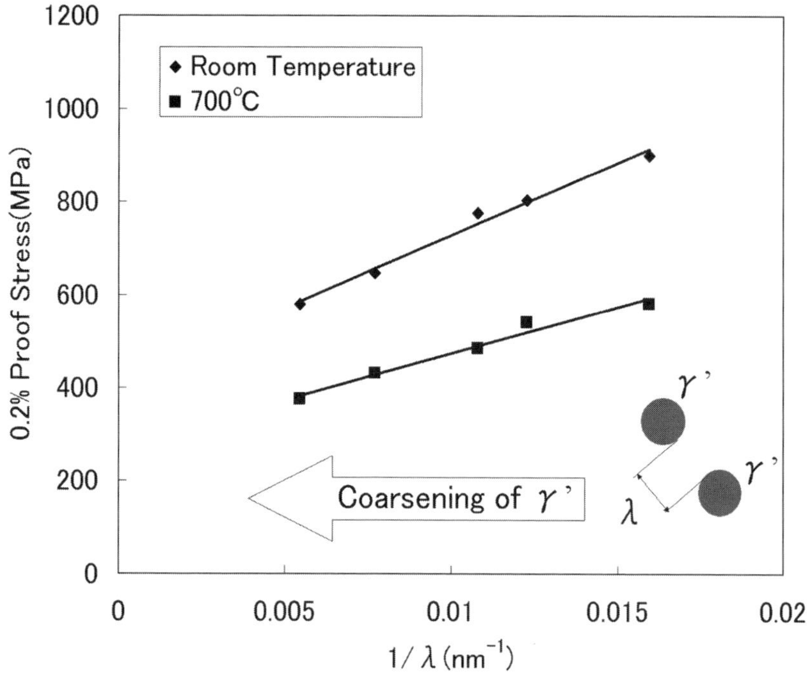

Figure 10 0.2%Proof Stress and Microstructure Parameter(1/ λ)

Advances in Materials Technology for Fossil Power Plants
Proceedings from the Fifth International Conference
R. Viswanathan, D. Gandy, K. Coleman, editors, p 434-446

DOI: 10.1361/cp2007epri0434

Development of Ni-Based Superalloy for Advanced 700°C-Class Steam Turbines

Ryuichi Yamamoto
Takasago Research and Development Center, Mitsubishi Heavy Industries, Ltd.,
2-1-1 Shinhama Arai-cho, Takasago, Hyogo, 676-8686, Japan

Yoshikuni Kadoya
Nagasaki Research and Development Center, Mitsubishi Heavy Industries, Ltd.,
5-717-1, Fukahori-machi, Nagasaki, 851-0392, Japan

Takashi Nakano
Yoshinori Tanaka
Ryotaro Magoshi
Power Systems Headquarters, Mitsubishi Heavy Industries, Ltd.,
2-1-1 Shinhama Arai-cho, Takasago, Hyogo, 676-8686, Japan

Seiji Kurata
Shigeki Ueta
Toshiharu Noda
Research and Development Laboratory, Daido Steel Co., Ltd.,
2-30 Daido-cho, Minami-ku, Nagoya, 457-8545, Japan

Abstract

Advanced 700°C-class steam turbines require using austenitic alloys instead of conventional ferritic 12Cr steels, which are insufficient in creep strength and oxidation resistance above 650°C. Austenitic alloys, however, possess a rather higher coefficient of thermal expansion (CTE) than the 12Cr steels. So far, the authors have examined the effects of alloying elements on CTE and high-temperature strength. Consequently, "LTES700" was developed for steam turbine bolts and blades, which is an Ni-based superalloy with CTE as low as 12Cr steels and high-temperature strength as high as conventional superalloy (Refractaloy 26).

For large parts such as turbine rotors, chemical compositions were refined to increase phase stability and high-temperature strength above 700°C based on "LTES700" while maintaining low CTE. The proof strength of the refined alloy at room temperature was close to that of advanced 12Cr steel rotor material, and the creep rupture strength around 700°C was notably higher than that of 12Cr owing to strengthening by gamma-prime phase precipitates. The phase stability of the refined alloy was confirmed until 5000 hrs heating from 550°C to 700°C. Therefore, the refined alloy is a prospective future material for advanced USC power plants.

1. Introduction

In recent years, growing concern about environmental problems has led to demands regarding CO_2 gas, which has a large greenhouse effect. In order to meet such demands, it is necessary to reduce CO_2 gas emissions by converting fuel energy more efficiently into electric energy, thus reducing the amount of fuel consumed in power plants. For conventional thermal power generation, high temperature and high pressure was promoted to increase the unit capacity and thermal efficiency of power generation, and now, around twenty 600°C-class USC plants are in operation in Japan. In future, there needs to be an increase in advanced USC plants, which are called 'A-USC' and whose steam temperature has been raised to 700°C and above to boost net efficiency and reduce CO_2 emission, in line with increasing fuel consumption and investment costs in Japan, in Europe, and the United States (1, 2, 3).

Ferritic heat-resistant steels are commonly used for 600°C-class steam turbines. In the development of A-USC steam turbines, however, the application of Ni-based superalloys is required for high-temperature components. In general, these alloys have a high coefficient of thermal expansion (CTE). High thermal expansion also imposes careful design considerations associated with axial expansion of the rotor and the differential expansion between the rotating and stationary parts, which may affect turbine performance or construction.

Considering the manufacturing capability requirements of large forgings, Ni-based superalloy-forged components are used only for the high-temperature section of the rotor and are welded to 12Cr or ferritic steel forgings for the lower-temperature section shown in Figure 1 (4). To create a weld turbine rotor, an Ni-based superalloy with low CTE that is similar to ferritic rotor material is required to reduce design considerations.

So far, the authors have examined the effects of alloying elements on CTE and high-temperature strength, and "LTES700" was developed for steam turbine bolts and blades, which is an Ni-based superalloy with CTE as low as 12Cr steels and high-temperature strength as high as conventional superalloy, Refractaloy 26 (4).

Alloy LTES700 has superior properties up to 700°C. In addition, in order to raise generation efficiency, the possibility that the steam temperature may rise to over 700°C should be considered. Alloy LTES700 is strengthened by double precipitation of the gamma-prime [$Ni_3(Al, Ti)$] phase and Laves [$Ni_2(Mo, Cr)$] phase, but it has been ascertained that the Laves phase resolves at around 750°C. So, strengthening by gamma-prime is necessary only in order to maintain high-temperature strength over 700°C. Furthermore, for large parts such as turbine rotors, good manufacturability such as hot workability and weldability is also necessary.

This paper describes an alloy design to develop an austenite alloy, LTES700, which has good manufacturability and CTE as low as 12Cr steels and high-temperature strength as high as conventional superalloy owing to strengthening by the gamma-prime phase only. Furthermore, the material properties of the developed alloy are described. For large parts such as turbine rotors, it is necessary to refine chemical compositions to increase manufacturability and high-temperature strength above 700°C based on alloy LTES700 while keeping low CTE.

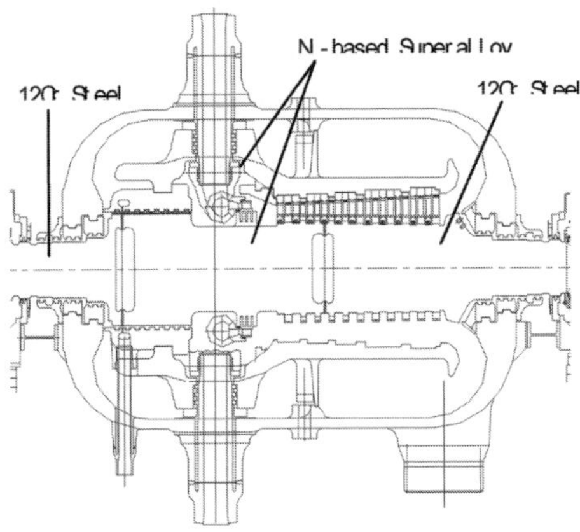

Figure 1. Outline of steam 700°C-class turbine

2. Experimental Procedure

The experimental alloys were melted in a vacuum induction furnace and cast into 50-kg ingots with a circumference of 138 mm and a height of 380 mm. They were forged down into 15-mm round bars. The bars were solution treated at 1100°C for 2 hrs and water cooled, and then aged under various conditions and air cooled. The mean CTE of the experimental alloy from room temperature was measured, and tensile properties at room temperature to 700°C and stress-rupture properties at 600°C to 750°C, tested using a cylindrical combination rupture specimen that is similar to that of Figure 1 in ASTM E 292, were measured.

3. Results and Discussion

3.1. Alloy design

3.1.1 The effect of tungsten amount

In order to precipitate the gamma-prime [$Ni_3(Al, Ti)$] phase only, and to prevent the precipitation of Laves phase, it is necessary to reduce Mo that forms the Laves phase. However, Mo is an effective element to reduce CTE, another element instead of Mo should be substituted to keep CTE as low as alloy LTES700. The authors have clarified that CTE can be predicted by Eq. 1 (4), and W was the same effective element as Mo to reduce CTE.

$$CTE=13.8732 + 7.2764\times10^{-2}\times Cr + 3.751\times10^{-2}\times(Ta+1.95Nb) + 1.9774\times10^{-2}\times Co + 7.3\times10^{-5}\times Co\times Co - 1.835\times10^{-2}\times Al - 7.9532\times10^{-2}\times W - 8.2385\times10^{-2}\times Mo - 1.63381\times10^{-1}\times Ti \quad (1)$$

On the other hand, the higher W content may deteriorate ductility due to the precipitation of alpha-W. Therefore, in order to develop a low-CTE alloy that is strengthened by the gamma-prime phase only, the amount of W was investigated at first, and subsequently, the amount of Mo was investigated in the range without alpha-W precipitation. In order to prevent the precipitation of the Laves phase and keep CTE as low as LTES700, the range of W that influences the precipitation of the alpha-W phase was investigated by replacing Mo with W. Table 1 lists the chemical composition of experimental alloys, and Figure 2 shows the position of LTW3, LTW6, LTW7, LTW8, and LTES700 with Mo and W amounts. The Mo+1/2W amounts of LTW3, LTW6, LTW7, and LTW8 were fixed at 17.7%. In addition, the amount of Al+Ti was increased to cover the strength that would be decreased due to there being no Laves phase.

Table 1 . Chemical composition (mass%) of the designed alloys and LTES700

	Ni	C	Cr	Mo	W	Al	Ti	Mo+1/2W	Al+Ti(at.%)
LTW3	bal.	0.029	12.0	14.2	7.0	1.48	0.91	17.70	4.64
LTW6	bal.	0.032	12.0	12.2	11.0	1.54	0.91	17.70	4.88
LTW7	bal.	0.034	12.1	10.2	15.0	1.55	0.90	17.70	4.99
LTW8	bal.	0.032	12.1	8.2	19.0	1.49	0.92	17.70	4.99
LTW10	bal.	0.031	12.0	6.2	7.0	1.50	0.90	9.70	4.52
LTW13	bal.	0.031	12.0	10.2	7.0	1.47	0.90	13.70	4.57
LTES700	bal.	0.030	12.0	18.0	-	0.9	1.15	18.00	3.49

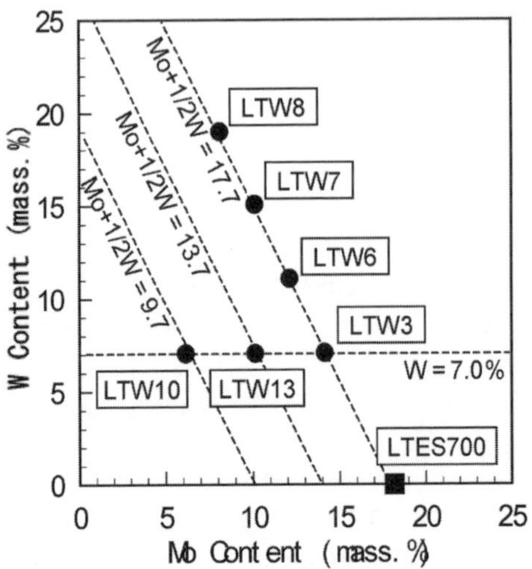

Figure 2. Mo and W content in the designed alloys and LTES700

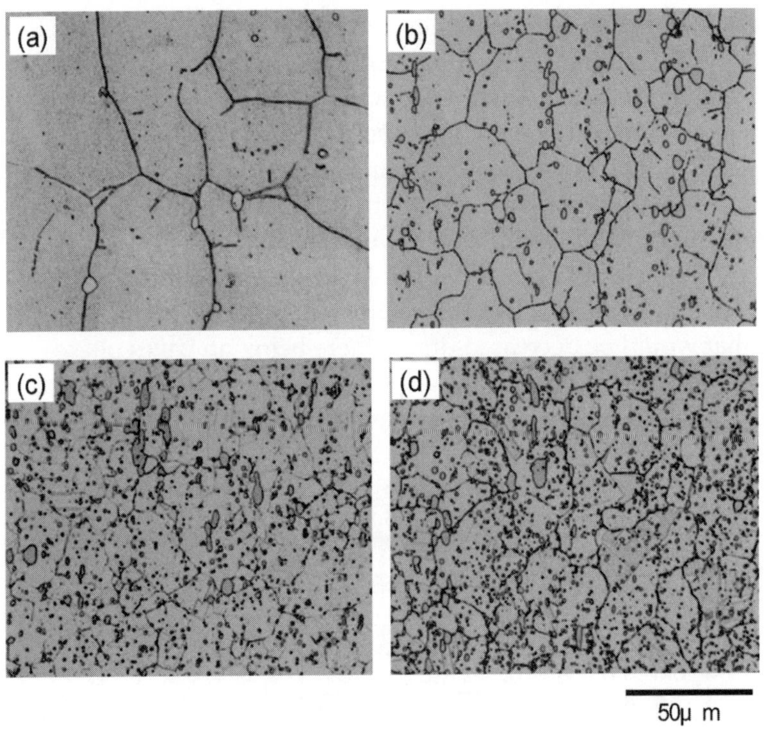

Figure 3. Microstructure of the designed alloys ((a) LTW3 (7%W), (b) LTW6 (11%W), (c) LTW7 (15%W), (d) LTW8 (19%W))

Figure 3 shows optical microstructures of LTW3, LTW6, LTW7, and LTW8. Distinction of phases was carried out by EPMA (electron probe micro analyzer). Figure 4 shows EPMA analysis of LTW8, and Table 2 shows the phases identified by EPMA of LTW3, LTW6, LTW7, and LTW8. LTW6 was confirmed by the existence of an alpha-W phase, and it was determined that the alpha-W phase precipitates in alloys with W over 11%. Alpha-W phase precipitation may deteriorate ductility and toughness, and hot-workability may deteriorate. Therefore, the amount of W must be reduced to less than 11%.

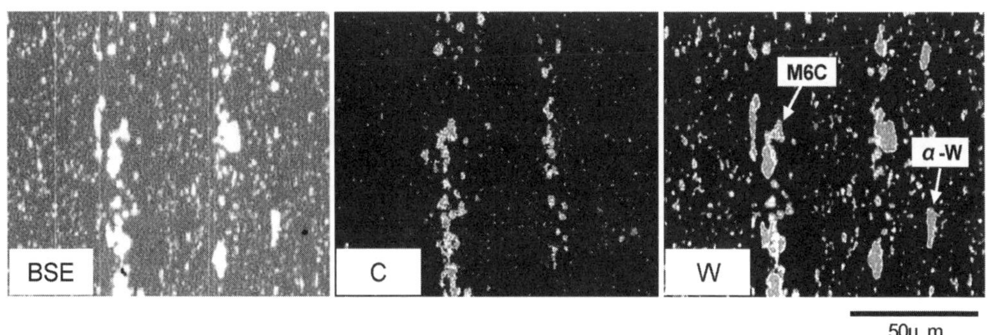

Figure 4. EPMA analysis of LTW8 (19%W))

Table 2. Phase identified by EPMA

	LTW3 (7%W)	LTW6 (11%W)	LTW7 (15%W)	LTW8 (19W)
Carbide	Present	Present	Present	Present
Alpha-W	Not present	Present	Present	Present

3.1.2 The effect of molybdenum amount

To examine the limit of Mo that the Laves phase does not precipitate, LTW10 and LTW13 showed decreased Mo from LTW3, whose W is 7% without the alpha-W phase. The microstructure of LTW3, LTW10, and LTW13 was observed to confirm the precipitation of the Laves phase. Figure 5 shows SEM microstructures, which were etched using used 10% AA (10% acetyl acetone+5% tetramethylammonium chloride+methanol), after aging at 600°C for 1000 hrs to stabilize the Laves phase. The precipitation of the Laves phase for LTW3 and LTW13, 14.2% and 10.2% Mo, was confirmed, but it was not confirmed at all for LTW10, 6.2% Mo. Therefore, it is determined that it is necessary to reduce the amount of Mo to less than 10% to prevent the precipitation of the Laves phase. Since LTW10 has been confirmed regarding phase stability without either the Laves or the alpha-W phase at various temperatures for a long period, LTW10 was designated as LTES700R, wherein R means "refined."

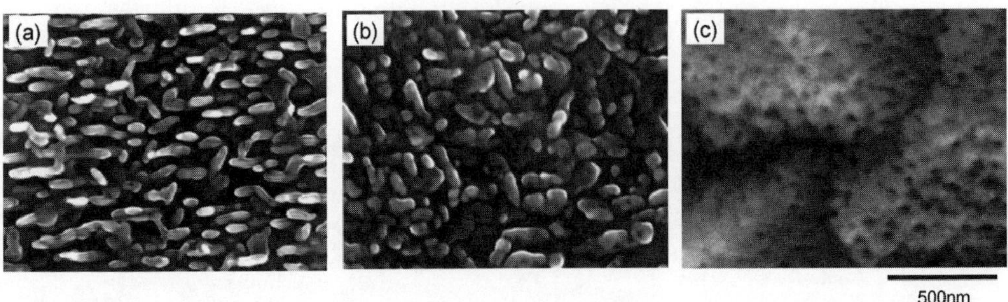

Figure 5. SEM images of aging at 600°C for 1000 hr. (a) LTW3 (14.2%Mo), (b) LTW13 (10.2%Mo), (c) LTW10 (6.2%Mo)

Table 3 shows the composition of alloy LTES700R for material characteristics tests in Chapter 3.2. The amount of Al and Ti is easy to vary at mass production, so the influence of the Ti/Al ratio on the properties is investigated at the same time.

Table 3 . Chemical composition (mass%) of alloy LTES700R

	Ni	C	Cr	Mo	W	Al	Ti	Ti/Al
LTES700R-1	bal.	0.031	12.0	6.2	7.0	1.50	0.90	0.60
LTES700R-2	bal.	0.030	12.0	6.2	7.0	1.66	0.67	0.40

3.2. Characteristics of alloy LTES700R

3.2.1 Precipitation hardening

Hardness after aging at 700°C to 800°C is shown in Figure 6. The hardness of alloy LTES700R is 30 HRC by single aging at 725°C to 775°C. Since the hardness is stable up to 775°C, it is thought that the strength of LTES700R is stable over 700°C. Figure 7 shows hardness after double aging. The hardness rise of 3 HRC after double aging, is higher than after single aging. It is considered that the hardness of alloy LTES700R is stable because it is strengthened by the gamma-prime phase only without the Laves phase.

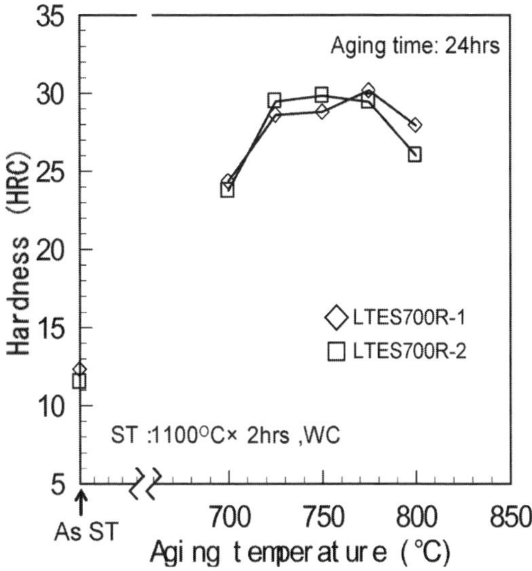

Figure 6. The effect of aging temperature on hardness

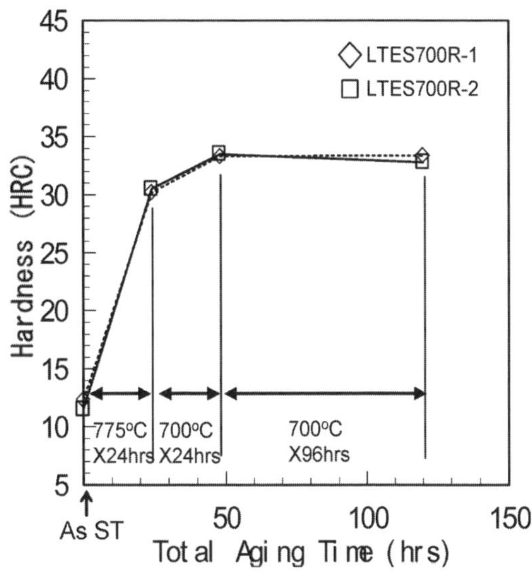

Figure 7. The effect of aging temperature on hardness

3.2.2 Microstructure

Figure 8 shows the microstructure of alloy LTES700R after double aging. Blocky precipitates in the optical photo are the retained eta carbides, M_6C, undissolved in the solution treatment, and the precipitation of the alpha-W phase is not confirmed. On the other hand, the TEM photo shows densely populated gamma-prime phases that are spherical in shape. The precipitation of the Laves phase, which is confirmed in alloy LTES700, is not confirmed in alloy LTES700R.

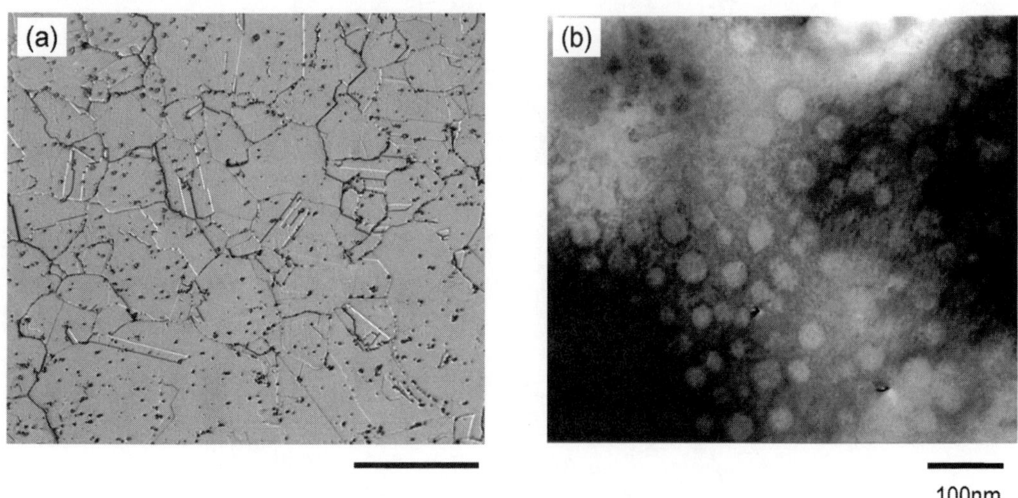

Figure 8. Microstructure of double aged alloy LTES700R
(a) Optical microscope and (b) TEM

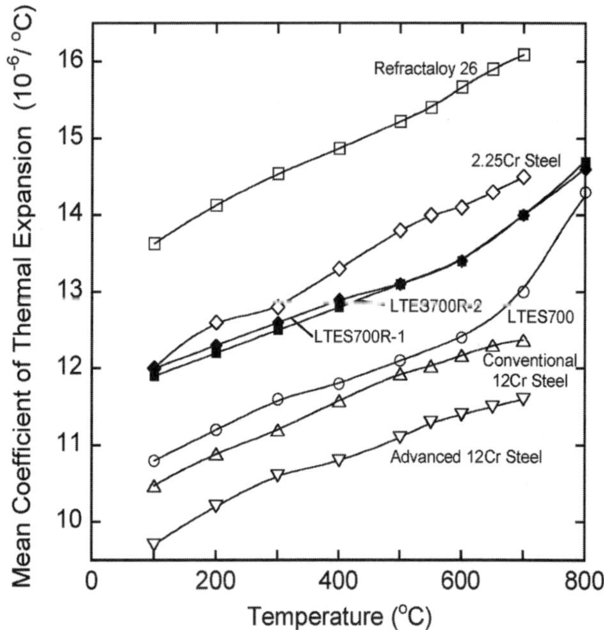

Figure 9. Mean CTE of alloy LTE700R and conventional steels and alloys for steam turbines (4)

3.2.3 Mean coefficient of thermal expansion

The mean coefficients from RT to 800°C are shown in Figure 9 including conventional steels and alloys for steam turbines (4). The CTE of alloy LTES700R is higher than that of alloy LTES700 due to the reduction in the amount of Mo+W. Alloy LTES700R has a coefficient lower than those of 2.25Cr steel and Refractaloy 26 and higher than that those of 12Cr steel.

3.2.4 Mechanical properties

The strength and ductility at RT and elevated temperatures are shown in Figure 10 compared with alloy LTES700 (4) and the advanced 12Cr steel, TMK1 (5). The strength of alloy LTES700R at room temperature is lower than that of alloy LTES700 due to there being no Laves phase; on the other hand, the strength at 700°C becomes near to that of alloy LTES700. Compared with advanced 12Cr steel, TMK1 using as rotor material alloy LTES700R has a similar 0.2% proof strength at room temperature, higher strength at elevated temperature, especially tensile strength, and larger elongation. The strengths of alloy LTES700R at 700°C are higher than those of TMK1 at 600°C.

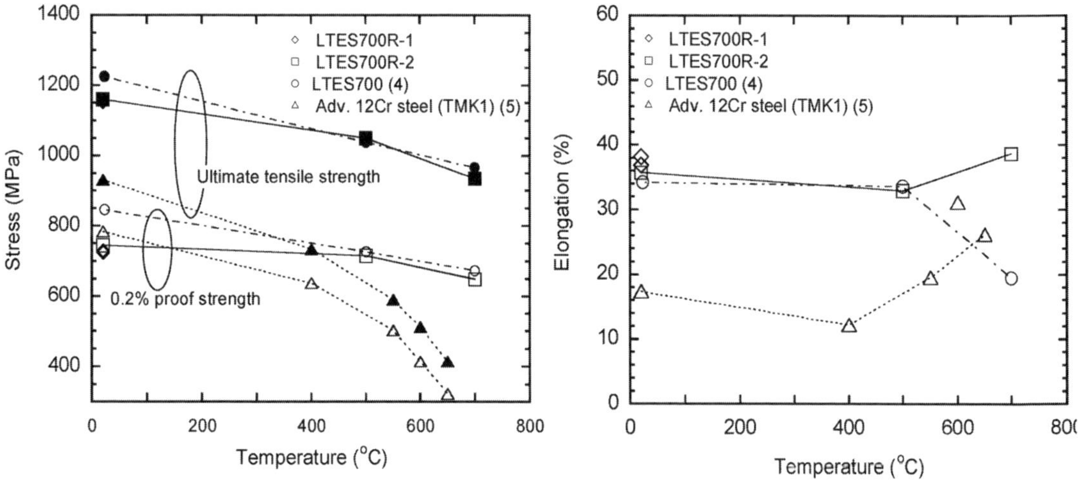

Figure 10 . Tensile properties of alloy LTE700R, LTES700 (4) and advanced 12Cr steel (TMK1) (5)

The stress to the rupture of alloy LTES700R from 600°C to 750°C is plotted against Larson-Miller parameters in Figure 10 compared with the mean data of advanced 12Cr steel MTR10A (6) used as the rotor material for 630°C-class turbines. Notch weakening does not appear. The creep rupture strength of alloy LTES700R is 100°C and above, higher than that of MTR10A, and the 100,000-hr creep rupture strength at 700°C should be above 98 MPa, which is a target of the A-USC turbine rotor material.

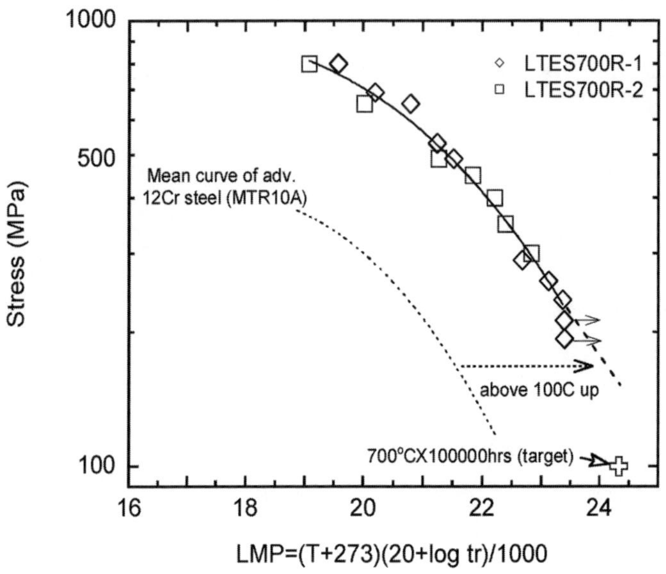

Figure 11 . Creep rupture strength of alloy LTES700R and advanced 12Cr steel (MTR10A)

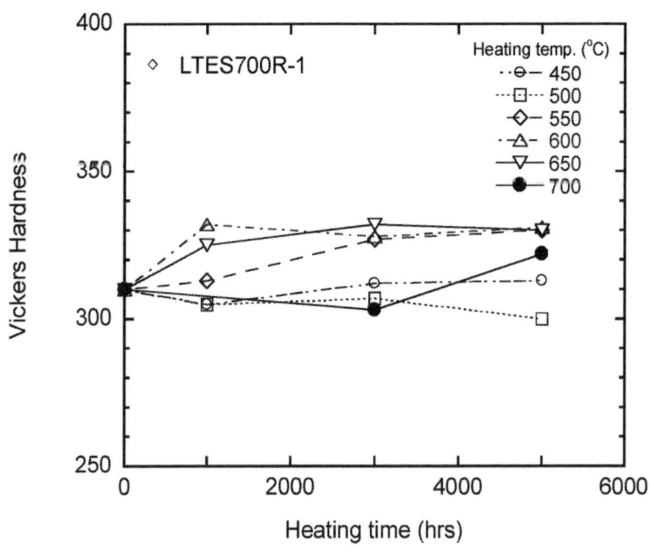

Figure 12 . Relationship between Vickers hardness and aging time from 450°C to 700°C

3.2.5 Phase stability

The relationship between Vickers hardnesses after heating from 450°C to 700°C is shown in Figure 12. The hardness increased by 30 HV after heating above 550°C. Figure 13 shows the TEM microstructure after 1000 hrs of heating at 650°C whose hardness increased by about 30 HV. The gamma-prime phase is only observed on the TEM image; on the other hand, other phases such as the Laves phase is not observed, which is similar to the microstructure after the double aging shown in Figure 7. This indicates that the phase is also stable in alloy LTES700R.

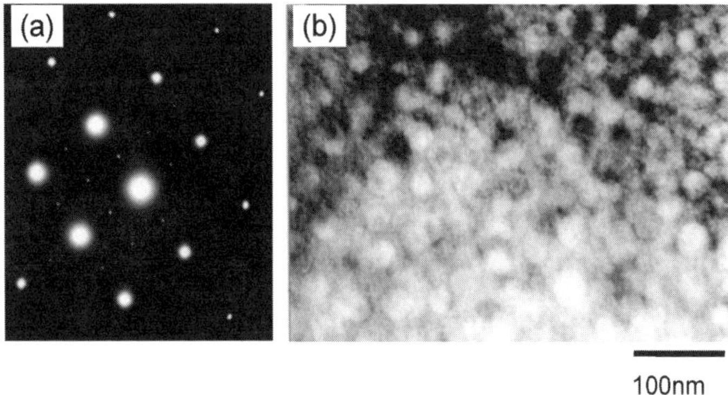

Figure 13 . TEM microstructures of alloy LTES700R after 1000hrs of heating at 650°C. (a) Electron diffraction pattern (B=001), (b) Bright field image

4. Conclusion

(1) Alloy LTES700R strengthened by the gamma-phase was refined from alloy LTES700 by optimizing Mo and W to avoid precipitation of the Laves and alpha-W phases.

(2) CTE of alloy LTES700R is lower than that of 2.25Cr steel and conventional superalloy.

(3) Alloy LTES700R has a similar 0.2% proof strength to the advanced 12Cr steel material at room temperature and higher strength at elevated temperature, especially tensile strength, and larger elongation.

(4) The creep rupture strength of alloy LTES700R is 100°C and above, higher than that of advanced 12Cr steel, and 100,000-hr creep rupture strength at 700°C should be above the target.

(5) The phase stability of alloy LTES700R is confirmed by a heating test from 450°C to 700°C.

The new superalloy, LTES700R, with low thermal expansion will be suitable for large components like turbine rotors, especially weld rotors for A-USC turbines, and it is planned to conduct a trial forging to study castablity and forgability.

References

1. R. Blum and R. W. Vanstone: "MATERIALS DEVELOPMENT FOR BOILERS AND STEAM TURBINES OPERATING AT 700°C," Proceedings of the 8th Liege conference, Liege, Belgium, (2006), 41.

2. R. Viswanathan, J. F. Henry, J. Tanzosh, G. Stanko, J. Shingledecker and B. Vitalis, "U.S. PROGRAM ON MATERIALS TECHNOLOGY FOR USC POWER PLAMTS," Proceeding from the Fourth International Conference on Advanced in Materials Technology for Fossil Power Plants, South Carolina, USA, (2005), 3.

3. M. Sato, M. Yaguchi, Y. Tanaka, J. Iwasaki, M. Fukuda, E. Saito, H. Nakagawa, A. Shiibashi and S. Izumi: "Current R&D Status of Pulverized Coal Fired Power Generation System Using 700°C-class Steam Temperature," The THERMAL AND NUCLEAR POWER, 57 (2006), 89.

4. R. Yamamoto, Y. Kadoya, S. Ueta, T. Noda, R. Magoshi, S. Nishimoto and T. Nakano, ibid 2., 623.

5. Y. Nakabayashi, A. Hizume, Y. Takeda, Y. Takano, T. Fujikawa, H. Yokota, A. Suzuki, S. Kinoshita, M. Koono and T. Tsuchiyama: "ADVANCED 12Cr STEEL ROTOR (TMK1) FOR EPDC'S WAKAMATSU 50MW HIGH-TEMPERATURE TURBINE STEP 1 (593C/593C)," The First International Conference on IMPROVED COAL-FIRED POWER PLANTS, California, USA, (1986), 8.

6. Y. Kagawa, F. Tamura, O. Ishiyama, O. Matsumoto, T. Honjo, T. Tsuchiyama, Y. Manabe, Y. Kadoya, R. Magoshi and H. Kawai: "Development and manufacturing of the next generation of advanced 12%Cr steel rotor for 630°C steam temperature," Proceedings of the 14th International Forgemasters Meeting 2000, Wiesbaden, Germany, (2000), 301.

Advances in Materials Technology for Fossil Power Plants
Proceedings from the Fifth International Conference
R. Viswanathan, D. Gandy, K. Coleman, editors, p 447-470

DOI: 10.1361/cp2007epri0447

NANO-STRUCTURED EROSION RESISTANT COATINGS FOR GAS AND STEAM TURBINES

V. P. "Swami" Swaminathan
TurboMet International
8026 Winter Park
San Antonio, TX 78250

Ronghua Wei
Southwest Research Institute
6220 Culebra Road
San Antonio, TX 78238

David W. Gandy
Electric Power Research Institute
1300 West Harris Blvd
Charlotte, NC 28262

Abstract

Solid particle and liquid particle erosion in the compressor section of gas turbines and steam turbine vanes and blades lead to significant reduction in turbine efficiency over time. This results in increased downtime and operating cost of the power plants. Some of the conventional coatings and erosion protection shields used by the currently available commercial processes have limitations in their temperature and erosion protection capabilities. Under a project funded by the Electric Power Research Institute (EPRI), very hard nano coatings with thickness within 40 microns (about 1.5 mils) have been produced on test samples using a state-of-the-art Plasma Enhanced Magnetron Sputtering (PEMS) technique. Under Phase I of this EPRI project, several coatings were deposited on various substrate alloys for the initial screening tests. Titanium silicon carbonitride nano-composite (TiSiCN), stellite and modified stellite, TiN monolayer coatings and Ti-TiN, Ti-TiSiCN multi-layered coatings have been developed and screening tests completed in this project. The substrate selection is based on some of the alloys currently used in aeroderivative engine compressor blades, land based gas turbine compressor blades and vanes; steam turbine blades and vanes. They include Ti-6Al-4V alloy, 17-4 PH, Custom-450 and Type 403 stainless steels. The PEMS coating technique differs significantly from the conventional techniques such as air plasma spray (APS), low-pressure plasma spray (LPPS), diffusion coatings, chemical or physical vapor deposition (CVD or PVD) used on blades and vanes. PEMS method involves a magnetron sputtering process using a vacuum chamber with an independently generated plasma source from which high current density can be obtained. This method used heavy ion bombardment during coating deposition to increase the coating adhesion and create a highly dense microstructure. A novel method using trimethylsilane gas instead of solid targets was successful in producing this nanocomposite. The PEMS deposited stellite coatings did not show any erosion improvement and dropped from further development. However, other hard coatings developed in this project exhibited excellent solid particle erosion resistance - nearly 25 times higher than the uncoated substrates and 20 times higher than all other nitride coatings produced by traditional commercial processes. This paper covers a brief description of the deposition technology and various properties of the coatings. Hardness indentation and scratch tests were performed to assess the coating adhesion to the substrates. Scanning Electron Microscopy (SEM) with Energy Dispersive Spectroscopy (EDS), and X-Ray diffraction (XRD) analysis were used to study the microstructure and morphology of these coatings. Nanoindentation was conducted to determine the hardness and Young's modulus, while sand erosion tests were conducted to rank the erosion resistance of the coatings produced using several processing variables.

Introduction

Solid particle erosion (SPE) and liquid droplet erosion (LDE) causes severe damage to turbine components such as gas turbine compressor blades and vanes; steam turbine control stage and later stage LP blades. Some examples of such damage are shown in Figure 1. SPE reduces engine efficiency and reliability by eroding the airfoils and potentially leading to catastrophic failures during service. There were several turbine failures attributed such damage in service. In the case of flight engines, it may endanger the lives of the crew especially for flight engines operated in dusty environments [1, 2]. In addition to SPE, liquid droplet erosion (LDE) in the steam path of steam turbines also lead to damage to the nozzles and blades [3]. For a LM600 SPRINT gas turbines using water injection, significant LDE damage was observed on the leading edge of the blades[4]. Similar damage to Frame FA engine R-0 compressor blades was reported and some of the field failures were attributed to such damage. Various coatings have been applied to combat erosion in turbines. Most commonly used are nitride or carbide coatings including single layered TiN, ZrN, CrN and TiAlN; multilayered Cr/CrN and Ti/TiN; and supper-lattice CrN/NbN. Conventional Physical Vapor Deposition (PVD) including magnetron sputtering and cathode arc physical vapor deposition (CAPVD) are the methods used to deposit these coatings. It is known that relatively thick coatings are needed for durable erosion resistance. Rickerby and Burnett [5] observed that thick TiN coatings on both stainless steel and carbon steel showed better erosion resistance than thin coatings.

In recent years, nanocomposite coatings (mainly nanocrystalline TiN in a matrix of amorphous Si_3N_4 or nano-composite TiN/Si_3N_4) are actively studied by a number of research groups worldwide using mostly CVD and sometimes PVD (in particular, magnetron sputtering) [6-9] These coating are extremely hard and have shown great wear resistance in laboratories. For tribological applications such as in machine tools, thin coatings of 2-5μm are commonly used. However, in this project, nanocomposite coatings (>20 μm) have been produced and their erosion resistance measured with good success.

Plasma enhanced magnetron sputter (PEMS) deposition is an improved version of conventional magnetron sputtering. It utilizes an electron source and a discharge power supply to generate plasma, independent of the magnetron plasma, in the entire vacuum chamber. The PEMS technology has shown to produce much better TiN coatings for cutting applications [10,

11] and the superior performance is attributed to the very fine grain ($\simeq$ 60 nm) TiN microstructure that is formed due to the heavy ion bombardment [12].

EPRI Research Project Outline

The objective of this Electric Power Research Institute (EPRI) project is to develop erosion resistant nano-technology coatings using the PEMS method to mitigate the erosion problems encountered in gas and steam turbines under SPE and LDE conditions. The technical approach is as follows.

- Apply selected coatings by the plasma enhanced magnetron sputtering (PEMS) method on substrate material used in the turbine blades and vanes
- Conduct screening tests such as, SPE, LDE, harness, adhesion ranking, etc., on small samples to identify the most promising coating(s)
- Compare the properties of these samples with similar results from other commercial techniques
- Develop coating process specification for the selected coatings
- Conduct qualification testing by mechanical property evaluations and thermal exposure tests to simulate field service conditions
- Apply coating(s) to components and conduct field evaluations
- Commercialize the technology

The selected coatings were applied to small disc samples prepared from the various substrate alloys. One of the important tasks is to identify the most effective combination of the processing variables to produce the best coating possible by the PEMS method. Several combinations of the processing variables have been tried initially on Ti-6Al-4V substrate samples to select the optimum combination. Screening tests have been completed on all of the selected coatings and the most promising coatings have been identified. Under Phase II of this project, these coatings will be applied to mechanical and thermal test specimens and tests will be conducted to qualify the coatings prior to field application. Then the selected coatings will be applied to turbine components under Phase II for field service exposure and testing in operating turbines.

Experimental details

Substrate Materials:

Substrate alloys, Ti-6Al-4V alloy, 12Cr (Type 403), 17-4PH, and Custom 450 stainless steels were selected for this study. These materials are used in gas turbine compressors and steam turbine blades and vanes. Some of the samples were directly machined from scrapped turbine blades and some were machined from rod stock. Test samples were machined to 2.5 cm (1 in.) in diameter by 3.2 mm (0.125 in.) thick and then polished using 1μm diamond paste to a surface roughness of ~5 nm Ra. Some of the samples were also ground to 600 grit surface finish and some of the 17-4PH samples were shot peened to evaluate the effects of surface finish on the coating adhesion and properties. They were cleaned with acetone and methanol before entering the PEMS vacuum chamber for coating deposition.

Plasma Enhanced Magnetron Sputtering (PEMS) Process:

Figure 2 shows a schematic of the PEMS system at Southwest Research Institute (SwRI). The PEMS technology utilizes magnetron generated plasma and an additional electron source (a heated filament, for instance) and a discharge power supply to generate plasma. This electron-source generated plasma is independent of the magnetron-generated plasma. There are a number of advantages of this technique. First, during the substrate sputter-cleaning, the magnetrons are not operated, while the electron-source generated plasma alone is sufficient to clean the substrate. In this way, deposition of the target material, which is of concern for conventional magnetron sputtering, will not occur and the cleaning of the sample surface is assured. Second, during the film deposition, the ion bombardment from the electron-source generated plasma is very intense and the current density at the sample surfaces can be 25 times higher than that with the magnetron-generated plasma alone. Consequently, a high ion-to-atom ratio can be achieved in the chamber. Figure 3 shows a schematic diagram of the nano-composite coating microstructure produced using the PEMS processes. Figure 4 shows the improvement in the microstructure of the coating as a function of the ion bombardment intensity. The microstructure consists of hard nano-crystals of TiCN surrounded by amorphous SiCN. The nano-coatings in general have a grain size of less than 100 nm which produces high hardness. X-ray diffraction analysis of the current coating shows a grain size of 10nm a factor of 10 better than what is considered nanostructure coatings.

The nominal deposition procedures in this study included Argon sputter-cleaning of the samples for 60-90 minutes to remove the residual oxide on the surface. Then the samples were coated with a "bond layer" of pure Ti metal for about 5 minutes, which corresponds to 200nm for Ti. After that, deposition of the hard erosion resistant coating is started. In preparing the single-phased nitride of TiN, a solid target of Ti is used in a mixture of $Ar+N_2$ gases. Stellite 6 alloy plate was used to deposit the stellite nano-coating. In preparing the nanocomposite coatings, trimethylsilane ($(CH_3)_3SiH$ or TMS) gas was used as the precursor during sputtering of Ti to form $TiSi_xC_yN_z$. Ti-Si-C-N coating has been prepared using CVD processes [13, 14] but both $TiCl_4$ and $SiCl_4$ were used in all of those studies. Trimethylsilane is much easier and safer to handle and should not cause severe corrosion to the vacuum chamber and pumps.

Tow circular magnetrons of 17.5 cm (6.9 in.) diameter plated on the opposite sides of the coating chamber were used in this process. The major processing variables are as follows:

- The distance between the magnetron and the target
- Power applied to the magnetrons
- Sputter cleaning duration
- Filament current and discharge power supply current
- The bias voltage on the worktable
- Flow rates of argon, nitrogen and TMS gasses
- Duration of coating (controls the thickness)

The sample temperature was measured using a thermocouple embedded in the samples and the steady state temperature was typically about 400°C (752 F). At this temperature and the duration of coating deposition process, the base alloy properties are not expected to change. The various deposition parameter and thus, the deposition rates are carefully controlled to manage the substrate temperature during the costing process.

Characterization of Hard Coatings:

The following laboratory evaluation tests were conducted on the coated samples:

- Nanoindentation to obtain the nano-hardness and elastic modulus
- Rockwell C hardness indentations to qualitatively compare coating adhesion
- Scratch testing for a quantitative assessment of the adhesion strength
- Scanning electron microscopy (SEM) to examine the morphology and microstructure on the cross-section.
- Energy dispersive spectroscopy (EDS) was performed to obtain the Si composition for the nanocomposite coatings
- X-ray diffraction (XRD) to identify the microstructure
- Erosion tests per ASTM Standard G76-04 using a nozzle at an incident angle of 30 degree with respect to the sample surface.

Nano-hardness was measured on the thin coatings with instruments and techniques developed for this purpose. The units are in gaga pascals (GPa) due to the extremely high hardness of these coatings. For reference the hardness of diamond is 100GPa.

Rockwell C hardness indentation is used for a qualitative comparison of the coating adhesion. Scratch testing procedure developed by CSM Instruments

For the SPE testing, the erodent used was 50μm alumina and the back pressure of the nozzle was set at 20 psi as per ASTM Standard G-76-4. A pulsed blast was used to minimize the pressure drop during a long duration of spray. In each test, the pulse was on for 10 seconds; then off for 10 seconds. This constituted one spray cycle. A total number of 10 blast cycles, which constituted one test cycle, were performed before the sample was weighed to measure the weight loss per test cycle.

Results and Discussions

Morphological and Microstructural Analyse

Figures 5 (a, b and c) are the morphological (left) and cross-sectional (right) scanning electron microscope (SEM) images of TiN-coated Ti-6Al-4V alloy substrate for 1.5, 5 and 10 hrs, respectively. The deposition rate of TiN is approximately 4μm/hr under the process variables selected for this deposition cycle. From the micrographs, it is recognized that as the TiN coating thickness increases, the microstructure changes. The thin coating (7μm) looks very dense and featureless, while for the thicker coatings (25 and 43μm), features of V-shaped columnar internal discontinuities appear. No significant difference in the microstructural quality of the coating was observed between the Ti-6Al-4V and stainless steel substrates. The 7μm thick coating looks smooth except for some nodular-like features on the surface (Figure 5a). But when the coating is thicker (25μm and above), the surface becomes rougher indicating the formation of a crystalline structure with preferred orientation (Figure 5b and 5c).

It has been reported that nanocomposite coatings are produced when Si is added to the TiN, and other transition metal nitrides. When the concentration of Si is near 5-10 at.%, the hardness approaches a very high value > 40 GPa [15,16]. In this study, a Ti target was sputtered to obtain Ti, while Si, C and N came from the N_2+TMS environment. To obtain the carbonitrides, initially N_2 flow was started which formed stoichiometric nitrides in these TiN coating trials. Then the TMS flow rate was varied. Based on the EDS data, at a TMS flow rate of 3sccm, about 10% Si was obtained and this flow rate was then selected to prepare the thick coatings.

The surface morphology and cross-sectional microstructure of some samples from TiSiCN coated Ti-6Al-4V substrate are shown Figure 6. The coating thickness was 25 μm in all cases. It is seen in this figure that TiN coating produced by keeping the TMS flow rate at zero, has "cauliflower"-like surface morphology. From an earlier research, it is known that there are "V"-shaped features that grow from inside the coating up to the surface. When a small amount of TMS (3sccm) is added, for sample TiSiCN6, the surface view suggests that the surface features (grain boundaries) are more diffused but the cross-sectional view looks the same as that for the TiN (TiSiCN1). When the TMS flow rate is further increased (to 6 and 9 sccm), the surface morphology becomes a columnar structure is clearly visible in cross section. This condition is not desirable since it weakens the integrity of the coating.

Stellite Coatings. Stellite coating was deposited on to 17-4PH substrate using the PEMS process. Both mono layer and a multi-layer recipes were tried under a given ion bombardment ratio to obtain dense coating. For the multi-layer coating, the flow of nitrogen was turned on and off for specific time periods. Figure 7 shows the microstructure of these coatings. The coatings appear dense with no internal defects and good surface morphology. However, under the solid particle erosion tests (using 50 micron size alumina sand), the coating did not perform well. There is little improvement from the stellite coating on the erosion resistance. The coating yielded similar erosion results as the uncoated solid stellite substrate sample. The test samples are shown in Figure8. The test results are shown in Figure 9 as cumulative mass loss versus the number of test cycles. The erosion rate is linear with the number of test cycles, i.e., the weight of sand used. Further study of the stellite coatings was discontinued due to this lack of improvement in the erosion performance.

Ranking of Microstructure. The microstructure of the coatings produced using several processing parameters were ranked based on extensive SEM evaluations. They were ranked from a subjective scale of 1 though 4, 1 being the best. Some examples are shown in Figure 10. Microstructure ranked 3 (Figure 10a) shows columnar growth in the coating and needle like surface morphology. This condition is undesirable. The other two microstructures are very dense with smoother surfaces.

Nanoindentation Tests

Nanoindentation tests were conducted at the University of Windsor. The nanohardness and elastic modulus of selected samples are shown in Table 1. For comparison, data from three other nano coatings is also included. It is noted that CrN has the lowest hardness while the hardness of

TiSiCN exceeds 40 GPa, well within the "super hard" coating regime [16] It is also noted that the nanocomposite TiSiCN has a good combination of high hardness and lower modulus than single phased TiN. For comparison, the hardness of diamond is 100GPa on the same scale.

Coating Adhesion Tests

Two methods were used to assess the coating adhesion to the substrate. The first one is a qualitative method by comparative ranking using Rockwell C indentation. The second method is a more quantitative technique where a scratch mark is produced on the coated sample by a diamond stylus under increasing load. The details of these methods and the results are summarized below.

Rockwell C Hardness Indentation. Conventional Rockwell C hardness indentations are used for relative comparison for coating adhesion to the substrate. The hardness indenter produces a strong localized strain on the coating and the substrate. After the hardness indentation is made the samples are examined under optical microscope. If the coating has strong adhesion to the substrate, no cracking or spalling will be observed. Otherwise, varying degree of coating cracking and spallation will be seen around the edges of the indentation mark. Examples of results of such tests on the current samples are shown in Figure 11. The samples were coated using different coating parameters in the PEMS chamber. Varying degrees of coating spallation can be seen around the rim of the indentation marks of the three indentations. A qualitative ranking is given from 1 through 6 based on the observed condition of the coating. Rank 1 is the best where no cracking or slight cracking is seen. Moderate cracking and slight delamination is given Rank 2 as in Figure 11b. More severe cracking and coating spallation is given Rank of 3. This ranking in association with the erosion results and microstructural results is used to select the best coating in the series for further evaluation and testing.

Scratch Testing. In this second method, a diamond stylus which is identical to the Rockwell C indenter is drawn across the coated surface of the coated specimen at a constant speed with progressively increasing normal force. This test method is covered by ASTM Specification C1624-05. Recorded test variables are, (a) normal force, (b) frictional force, (c) acoustic emission signal, (d) penetration depth and (e) residual depth. The applied variable is the normal force and the speed of the coated sample with respect to the stylus. The damage along the scratch track is microscopically assessed using an optical microscope or SEM as a function of the applied force. An example of a scratch produced in this test on a TiSiCN coated sample and the associated test variables are shown in Figure 12. Location Lc1 is associated with the start of coating cracking indicating cohesive failure in the coating. Location Lc2 is associated with coating chipping, delamination and spallation of the coating indicating adhesive failure between the substrate and the coating. In this test the load corresponding to location Lc1 is 40N and that at Lc2 is 65N. Results from such results from the various coated samples will be used in conjunction with the Rc indentation results above and erosion tests results to develop an overall coating ranking and quality assessment procedure.

Erosion Testing

Alumina sand erosion data for bare Ti-6Al-4V and 17-4PH samples along with TiSiCN coated samples are shown in Figure 13. Erosion rate results from 30 and 90 degree incident angles and the average values of the two are plotted as. Data from two uncoated substrate alloy samples and 18 other samples coated under various processing conditions are included in this figure. The average values are used to rank the specimens from left to right in this plot, the right most being the best coating. The erosion resistance is affected by the process variables during the coating which produced coatings with different microstructures and thicknesses. The three samples TDOE12, 13 and 14 yielded the best erosion resistance. The relative improvement in the erosion resistances normalized to the uncoated substrates is also shown in Figure 13. These three samples show improvements by factor of one to two orders of magnitude (10 to 100 times). The processing variables corresponding to these three coatings were selected for further development and qualification.

Figure 14 illustrates the behavior of ductile and brittle materials under solid particle erosion conditions. Erosive mass loss as a function of impact angle for aluminum and Al_2O_3 substrates is shown. The variation in response typifies the characteristic definitions of a "Ductile" and "Brittle" response respectively [18]. The ductile materials have better erosion resistance at high angels of incidence and the brittle materials are better at low angles. This tendency is also observed in the current coatings and substrates (Figure 13).

Multilayer Coatings

Some multilayer coatings were also produced in this project with the Ti-TiN and Ti-TiSiCN combinations. Various coating properties were evaluated similar to those discussed above. Figures 15 and 16 show typical microstructure of the two multilayer coatings. Rc indentations indicate that they have good adhesion strengths as shown in Figure 17. The erosion results are summarized in Figure 18 as bar chart. Bare substrates and single (mono)-layer coatings produced to verify the reproducibility of the coating process are also included in this plot for comparison. Even though the multi layer coatings provide significant improvement over the bare substrates, they are not as good as the single layer coatings.

Figure 19 shows the erosion rate comparison of bare substrates, a commercially produced TiN coating, and two nanocomposite coatings. These results are in agreement with the previously published results indicating that the TiSiCN outperforms any other nitrides against erosion [17, 19]. The nanocomposite coatings show significant improvement over the traditional commercial coatings.

Summary and Conclusions

A plasma enhanced magnetron sputtering (PEMS) method has been successful in depositing single phase TiN, Stellite 6 and nanocomposite TiSiCN coatings; and multi-layer coatings on Ti-6Al-4V and stainless steel substrates. Higher coating thicknesses beyond the limit by conventional PVD technologies have been achieved by the PEMS process. Coatings have been deposited using several process variables to identify the optimum process to obtain the best combination of physical and mechanical properties. The selection of trimethylsilane (TMS) gas to produce the nanocomposite coating as the precursor for Si is a novel approach, which allows ones to easily deposit nanocomposite coatings compared to the use of other complex target materials and gasses. Microstructure, hardness, adhesion strength and erosion resistance were used to rank the coatings in their performance. Stellite coating did not show much improvement in the erosion resistance over bare substrates. Specific set pf processing variables have been identified to produce the best TiN and TiSiCN nanocomposite coatings by the PEMS process. Reproducibility of the coatings by the PEMS has been verified for some of the selected coatings. These coatings outperform the single phase TiN nitride coatings and show erosion resistance improvement of over an order of magnitude. Additional deposition of single and multilayer coatings and further screening tests are in progress. Mechanical test specimens will be coated to evaluate the effects of selected coatings on tensile and high-cycle fatigue properties.

ACKNOWLEDGMENT

The work was supported by EPRI and some of the results were generated under a Southwest Research Institute internal research project. The authors wish to thank Mr. Edward Langa for assistance in preparing the coated samples and conducting the erosion tests; and Byron Chapa and Chris Wolfe for the metallographic work.

REFERENCES

1. W. Tabakoff, "Investigation of Coatings at High Temperature for use in Turbomachinery," *Surf. Coat. Technol.*, 39/40 (1989) 97-115.

2. J.Y. DeMasi-Marcin and D.K. Gutpa, "Protective Coatings in the Gas Turbine Engine," *Surf. Coat. Technol.*, 68/69 (1994) 1-9.

3. Turbine Steam Path Damage: Theory and Practice, Volume 2--Damage Mechanisms, Report No. TR-108943-V2, Electric Power Research Institute, 1999

4. V. P. Swaminathan, "Investigation of High-Pressure Compressor Blade Failures in LM6000 Sprint Engines," Unpublished results presented at the Western Turbine Users and Combustion Turbine Operators Task Force Conferences (March 2004 and August 2006).

5. D.S. Rickerby and P.J. Burnett, "The Wear and Erosion Resistance of Hard PVD Coatings," *Surf. Coat. Technol.*, 33 (1987) 191-211.

6. S. Veprek, "New Development in Superhard Coatings: The Superhard Nanocrystlline-Amorphous Composite," *Thin Solid Films*, 317 (1998) 449-454.

7. M. Diserens, J. Patscheider and F. Levy, "Improving the Properties of Titanium Nitride by Incorporation of Silicon," *Surf. Coat. Technol.*, 108-109 (1998) 241-246.

8. J. Musil, "Hard and Superhard Nanocomposite Coatings," *Surf. Coat. Technol.*, 125 (2000) 322-330.

9. L. Rebouta, C.J. Tavares, R. Aimo, Z. Wang, K. Pischow, E. Elves, T.C. Rojas and J.A. Odriozola, "Hard Nanocomposite Ti-Si-N Coatings Prepared by DC Reactive Magnetron Sputtering," *Surf. Coat. Technol.*, 133-134 (2000) 234-239.

10. J.N. Matossian, R. Wei, J. Vajo, G. Hunt, M. Gardos, G. Chambers, L. Soucy, D. Oliver, L. Jay, C. M. Tylor, G. Alderson, R. Komanduri and A. Perry, "Plasma -enhanced, Magnetron-Sputtered Deposition (PMD) of Materials" *Surf. Coat. Technol.*, 108-109 (1998) 496-506.

11. R. Wei, J.J. Vajo, J.N. Matossian, and M.N. Gardos, "Aspects of Plasma-enhanced Magnetron-sputtered Deposition (PMD) of Hard Coatings on Cutting Tools," *Surf. Coat. Technol.*, 158-159 (2002) 465-472.

12. S.V. Fortuna, Y.P. Sharkeev, A.P. Perry, J.N. Matossian, A. Shuleopov, "Microstructural Features of Wear Resistant Titanium Nitride Coatings Deposited by Different Methods," *Thin Solid Films*, 377–378 (2000) 512–517.

13. D.-H. Huo and K.-W. Huang, "A New Class of Ti-Si-C-N Coatings Obtained by Chemical Vapor Deposition, Part I, Part II and Part III: *Thin Solid Films*, (2001 & 2002)

14. D. Ma, S. Ma, H. Dong, K. Xu and T. Bell, "Microstructure and Tribological Behavior of Super-hard Ti-Si-C-N Nanocomposite Coatings Deposited by Plasma Enhanced Chemical Vapor Deposition," *Thin Solid Films*, 494 (2006) 438-444.

15. S. Veprek, P. Nesladek, A. Niederhofer, F. Glatz, M. Jilek, and M. Sima, "Recent Progress in Superhard Nanocrystalline Composites: towards their Industrialation and Understanding of Origin of the Superhardness," *Surf. Coat. Technol.*, 108-109 (1998) 138-147.

16. R. Hauert and J. Patscheider, "From Alloying to Nanocomposites – Improved Performance of Hard Coatings," *Adv. Eng. Mat.*, 2, No. 5 (2000) 247-259

17. R. Wei, E. Langa, C. Rincon and J. Arps, "Solid Particle Erosion Protection of Turbine Blades with Thick Nitrides and Carbonitride Coatings from Magnetron Sputter Deposition," Proc. ASM International Surface Engineering Conference (ISEC), Seattle, WA, May 2006.

18. Finnie, I, 'Some Reflections on the Past and Future of Erosion," Proc. of the 8th Intl. Conf. On Erosion by Liquid and Solid Impact, Eds., I. M. Hutchings and J.A. Little, Cambridge, U.K, Sept 4-8, 1994.

19. V. P Swaminathan, R. Wei and D.W. Gandy, "Erosion Resistant Nano Technology Coatings for Gas Turbine Components, " ASME Turbo Expo 2007, Montréal Canada, May 2007, Paper No. GT2007-27027.

Table 1: Nano-hardness and elastic modulus of selected coatings produced by the PEMS process

	Nano hardness (GPa)	Elastic Modulus (GPa)	H/E
CrN	24.5	278.0	0.09
ZrN	33.8	317.3	0.11
ZrSiCN	29.6	241.2	0.12
TiN	31.7	333.7	0.09
TiSiCN	42.4	299.7	0.14
Diamond	**100**		

(a) LDE damage to 7FA engine R-0 Compressor blade

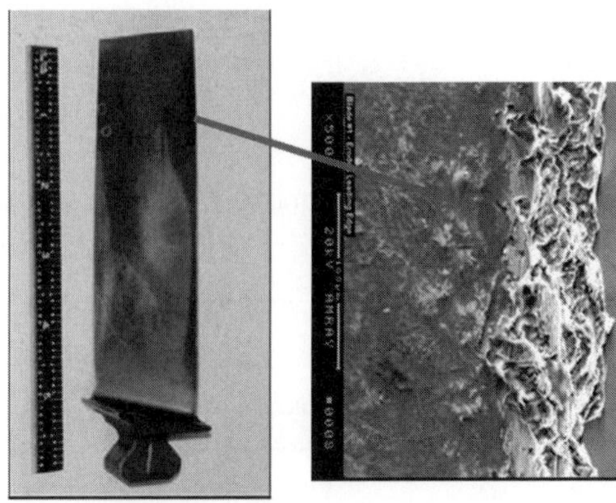

b) LDE damage to an aeroderivative engine blade using water injection

(c) LDE damage to LP steam turbine blades

(d) SPE damage to Steam turbine IP Blade

Figure 1. Examples of solid particle (SPE) and liquid droplet erosion (LDE) damage to gas and steam turbine components

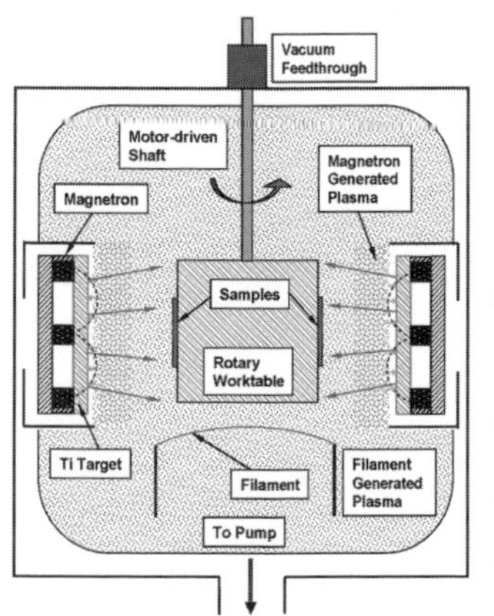

Figure 2. Plasma Enhanced Magnetron Sputtering (PEMS) System with two magnetrons

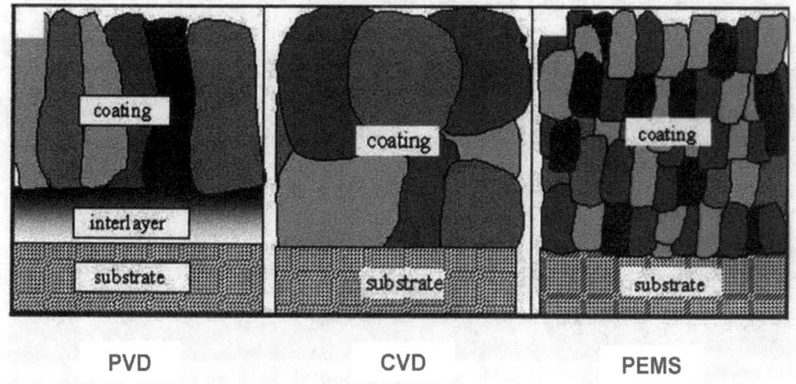

Figure 3. Schematics of the microstructure of TiN coating produced by three different methods. Note the very fine grain size produced by the PEMS process

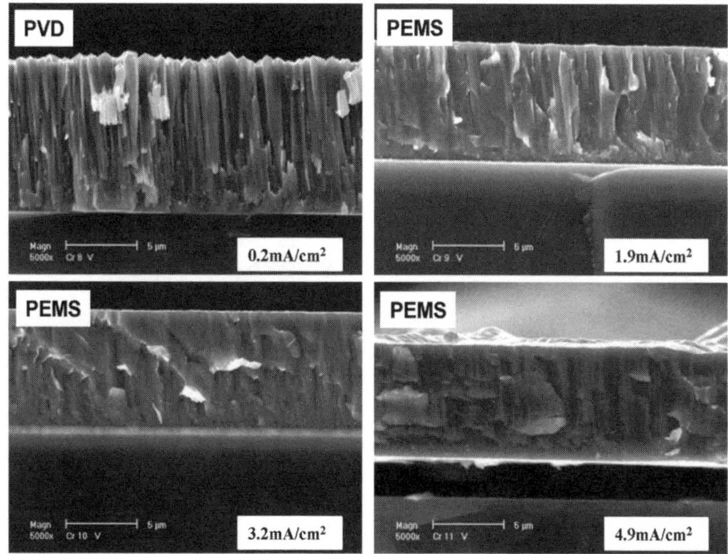

Figure 4. Effect of ion bombardment intensity on the coating microstructure. PEMS method has advantage over other methods in producing very dense and hard coatings

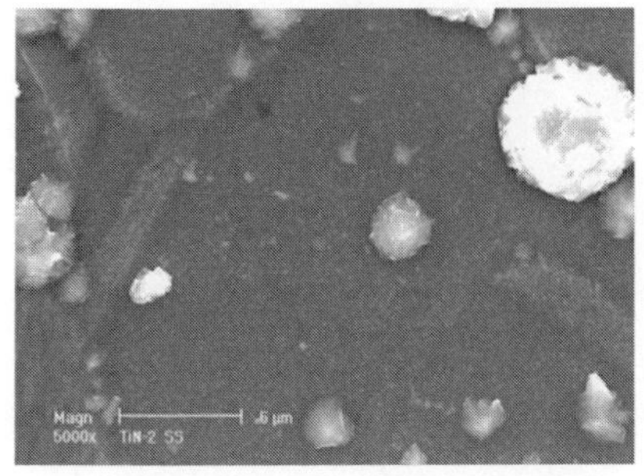

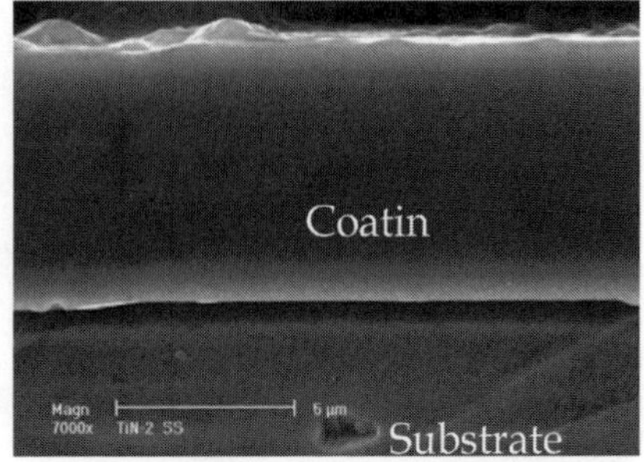

(a) Thickness: 7µm (1.5 hours)

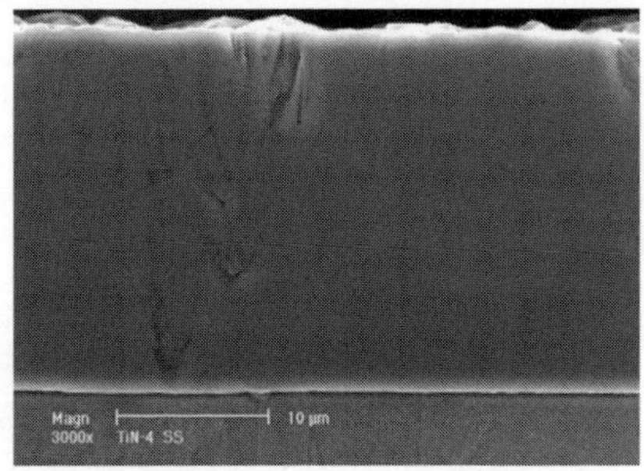

(b) Thickness: 25µm (5 hours)

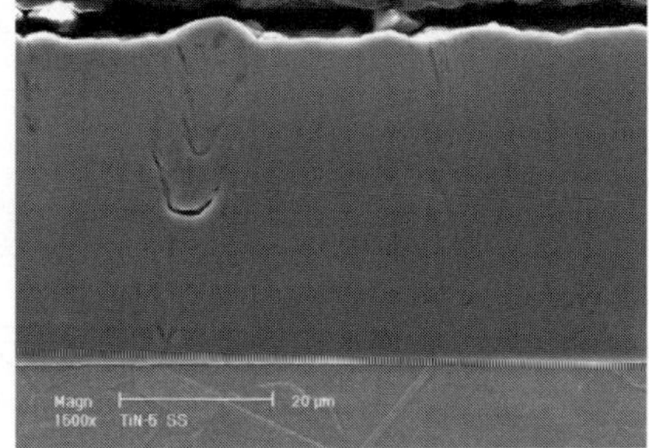

(c) Thickness: 43µm (10 hours)

Figure 5. Morphological (left) and cross sectional (right) SEM images of TiN Coatings deposited for 1.5, 5, and 10 hours on Ti-6Al-4V substrate

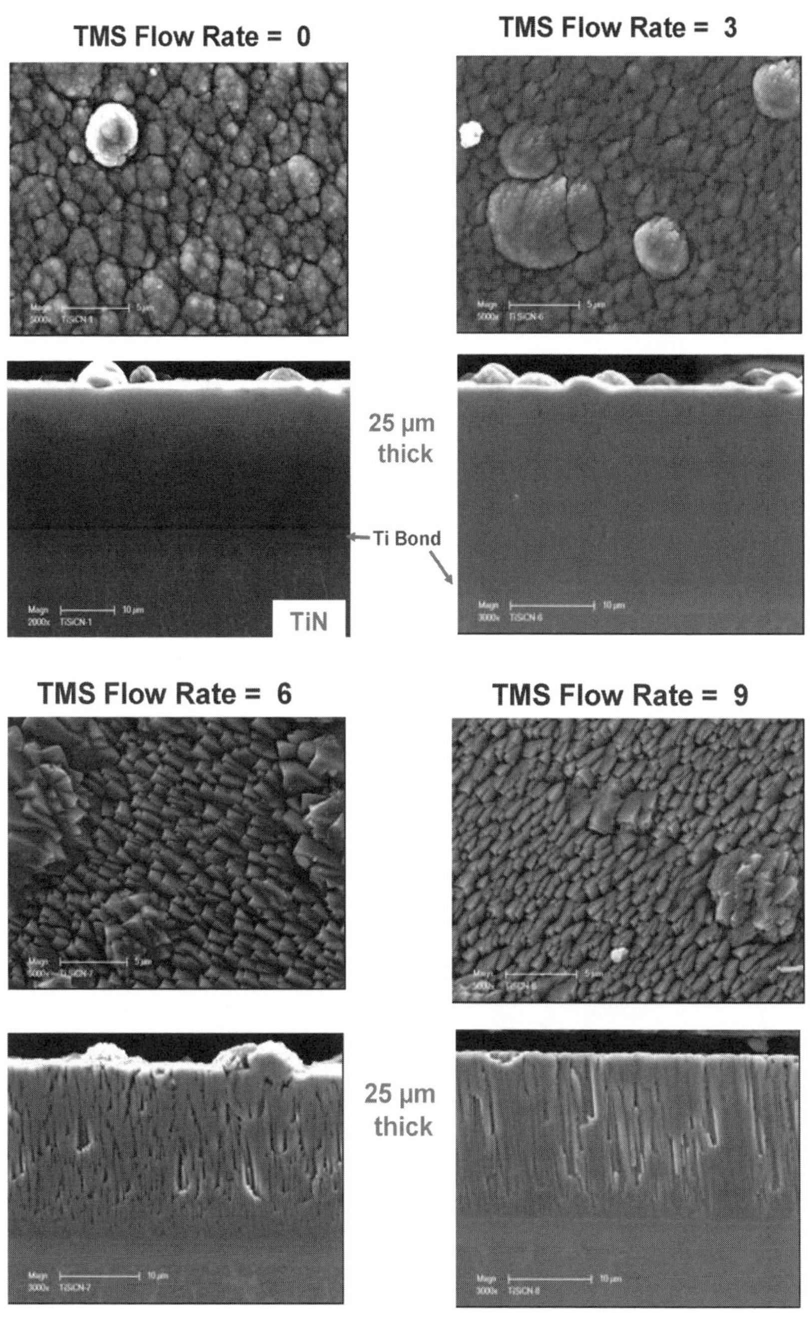

Figure 6. Surface morphology and microstructure of TiSiCN coating on Ti-6Al-4V substrate as a function of trimethylsilane (TMS) gas flow rate (SEM images)

(a) N_2= 0 sccm, 10 A discharge current

(b) N_2= 25 sccm, 10 A discharge current, multi layer (alternate with and without nitrogen)

Figure 7. Topological (left) and cross-sectional (right) images of Stellite coating deposited on 17-4 PH substrate; SEM images. (a) with no nitrogen (straight stellite) and (b) multi-layer coating

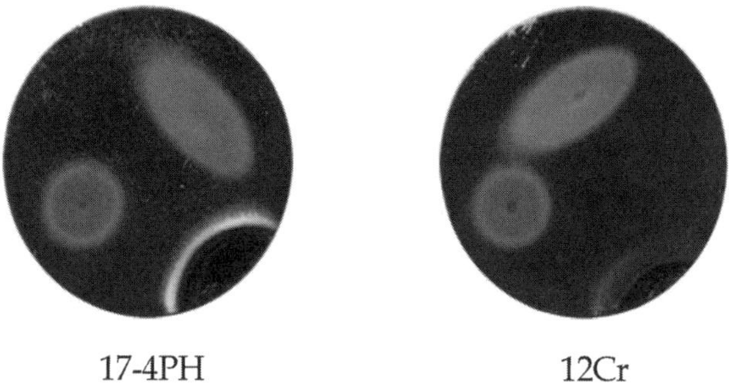

Figure 8. Examples of stellite coated disc specimens subjected to erosion tests at 30 and 90 degree incident angles

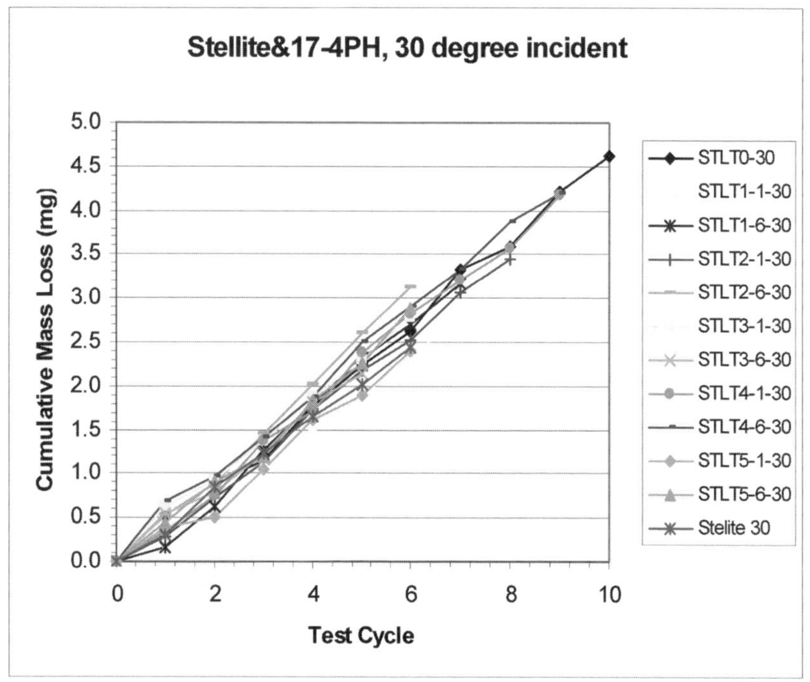

Figure 9. Erosion resistance of solid Stellite vs. Stellite coatings on 17-4PH at 30 degree incident angle showing similar erosion resistance for all the samples. STLT0-30 is the substrate 17-4PH with no coating. Stellite 30 is the sample prepared from the Stellite 6 target plate stock. Results from 90 degree tests are similar

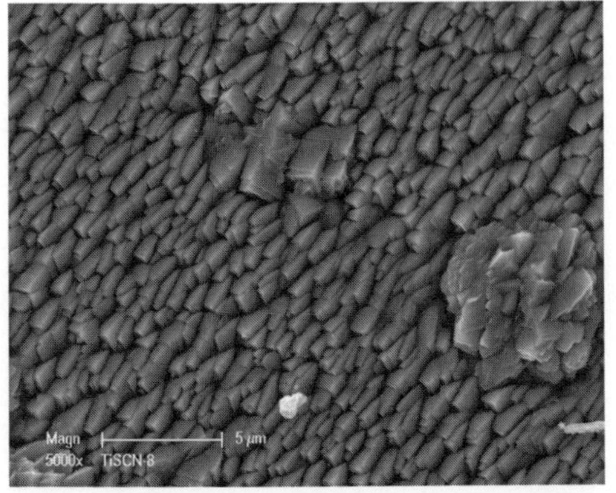

(a) Ti-6-4 Substrate; N_2=25 sccm, TMS=9 sccm (TiSiCN). Microstructure Rank "3"

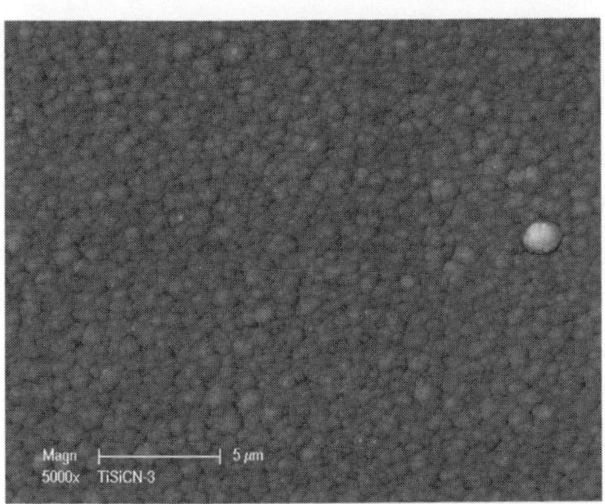

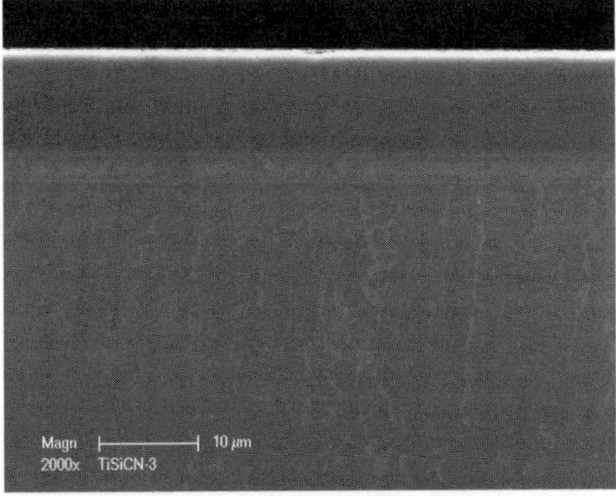

(b) Ti-6-4 Substrate; N_2=50 sccm, TMS=3 sccm (TiSiCN). Microstructure Rank "1"

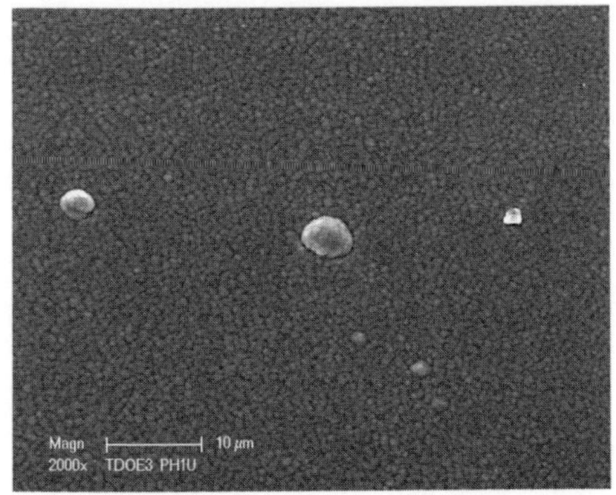

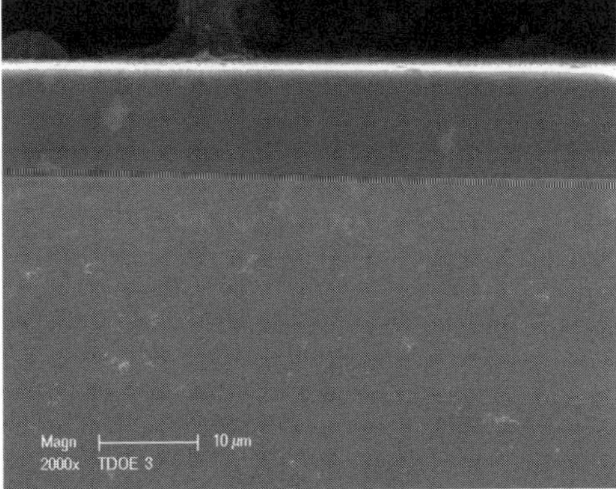

(c) 17-4PH Substrate; N_2=45 sccm, TMS=9 sccm (TiSiCN). Microstructure Rank "1"

Figure 10. Microstructural quality of TiSiCN coating on two substrate at different coating deposition process variables. Rank 1 is the best structure

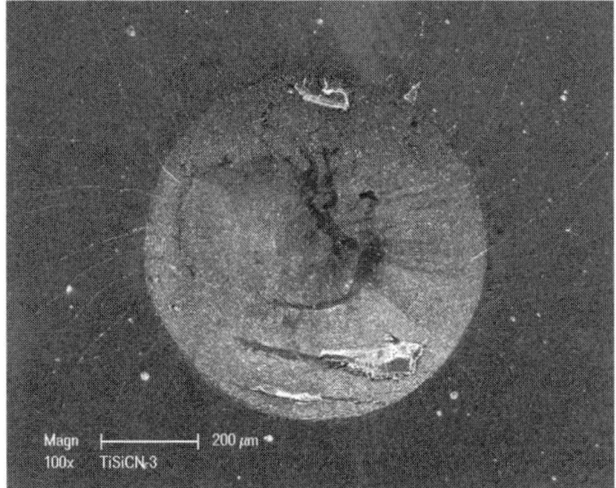

(a) Rank 1 (slight cracking; no delamination)

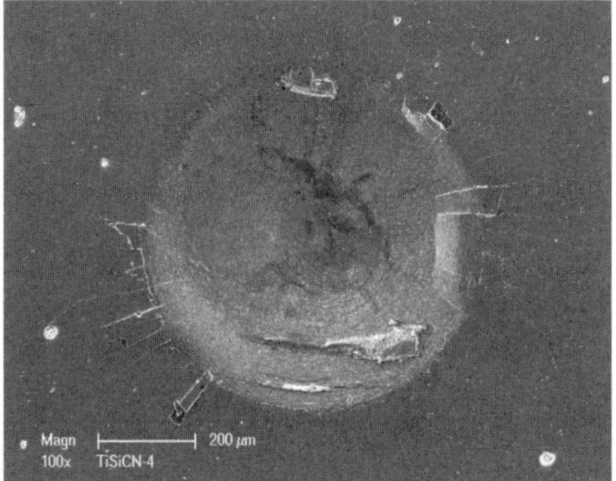

(b) Rank 2 (moderate cracking; some delamination)

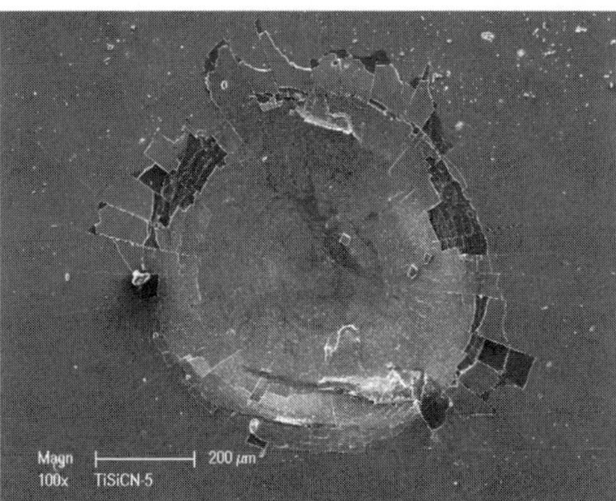

(c) Rank 3 (severe cracking, delamination and spalling)

Figure 11. Example of coating adhesion strength assessment of TiSiCN nano-composite coatings on Ti alloy substrate using R_c Hardness indentations.

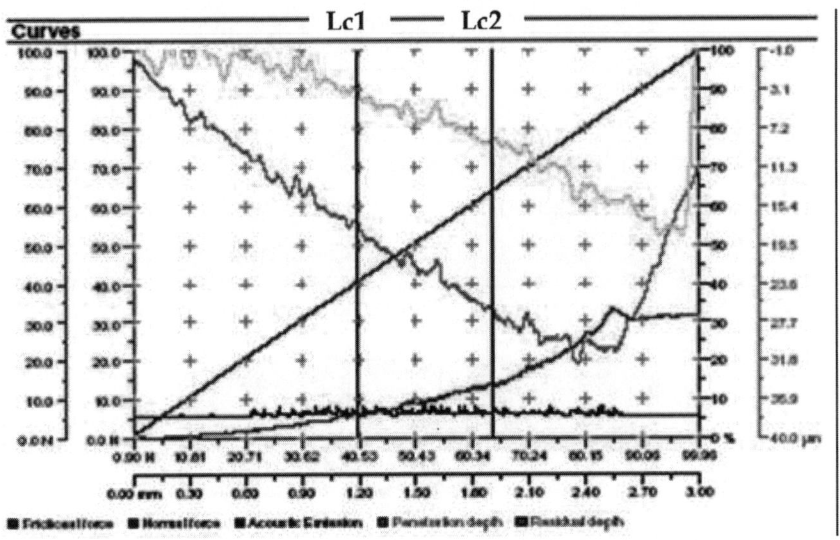

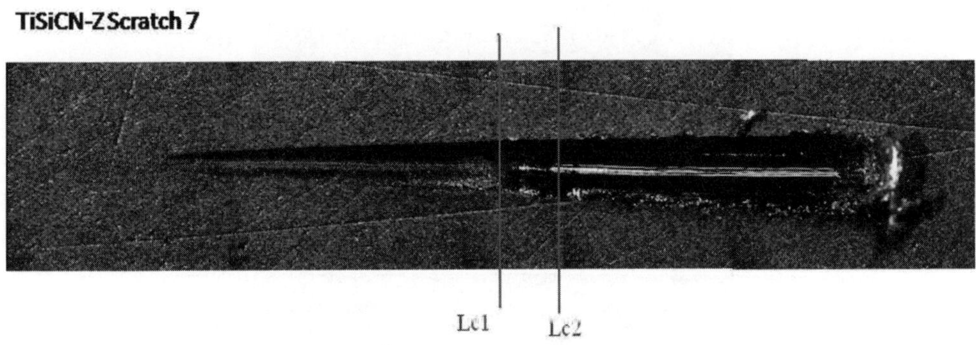

Figure 12. Example of scratch test data for quantitative measurement of the adhesion strength of the coating

Top: Traces of the various parameters recorded during the scratch test

Bottom: Scratch (about 3 mm long) on a sample coated with TiSiCN nano-composite. Lc1 shows the start of coating cracking and at Lc2 coating delamination and spallation occurs

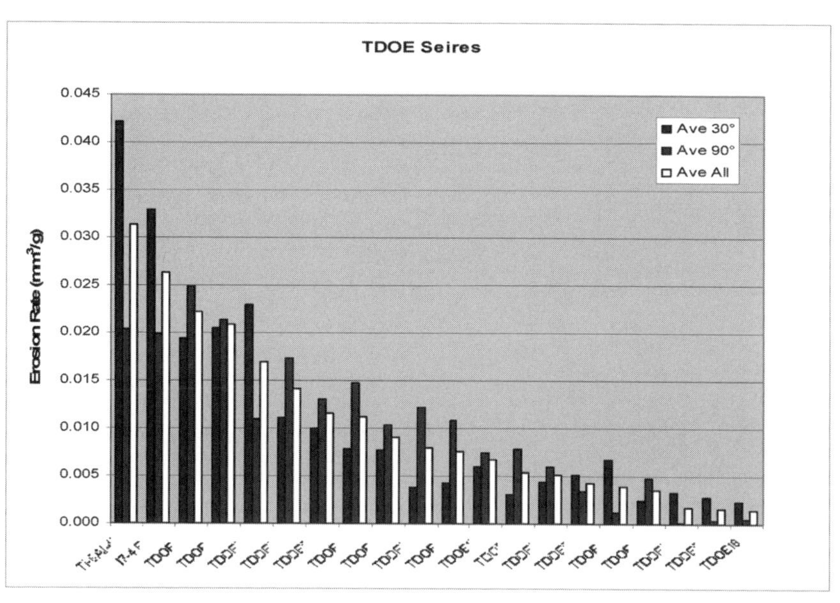

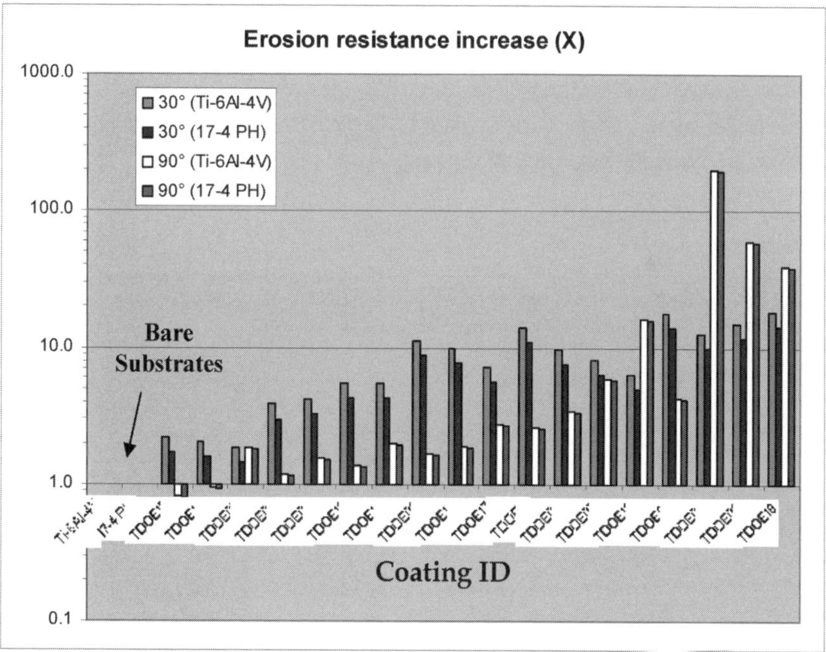

Figure 13. Solid particle erosion (SPE) test results.
Top: Erosion rate comparison of Ti-6Al-4 and 17-4PH samples coated with TiSiCN coating using various processing parameters. Better coatings are towards the right side of the plot with lower erosion rates.
Bottom: Erosion resistance improvement normalized to the uncoated Ti-6Al-4V and 17-4PH substrates showing significant improvement by this coating

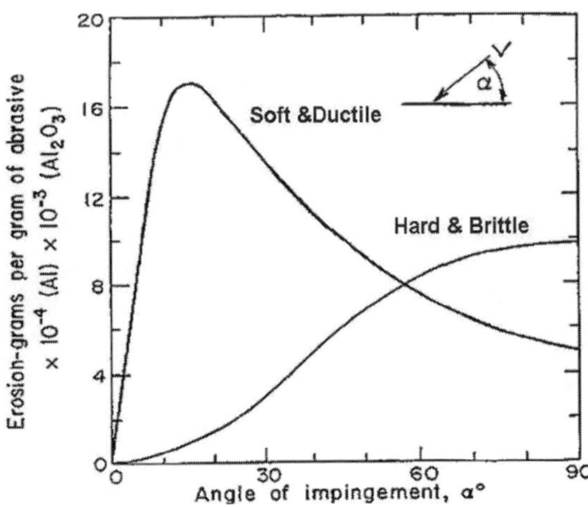

Figure 14. Erosive mass loss as a function if incident angle for ductile (Al) and brittle (Aluminum oxide) materials showing typical 'ductile' and 'brittle' responses to solid particle erosion. Note the variation in the magnitude of erosion in the Y axis label.

Figure 15. SEM images of surface morphology (left) and cross section (right) of a multi-layered TiSiCN coating

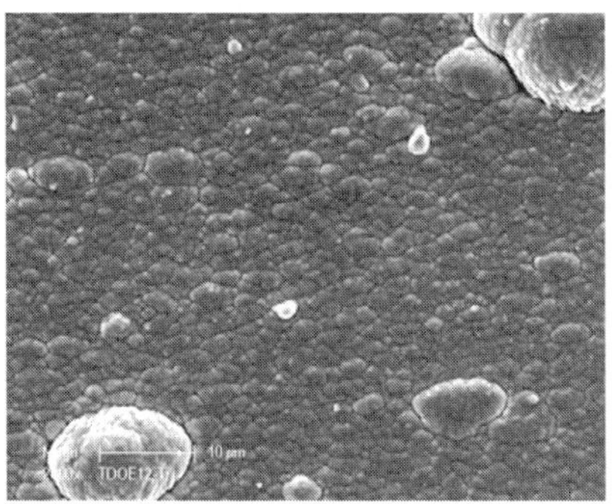

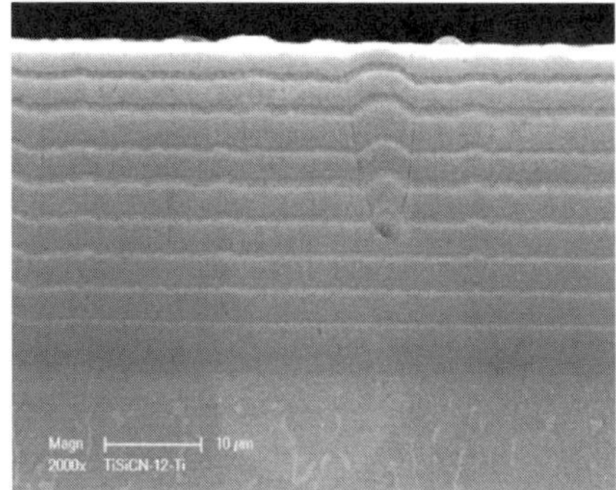

Figure 16. SEM images of surface morphology (left) and cross section (right) of a multi-layered TiN coating

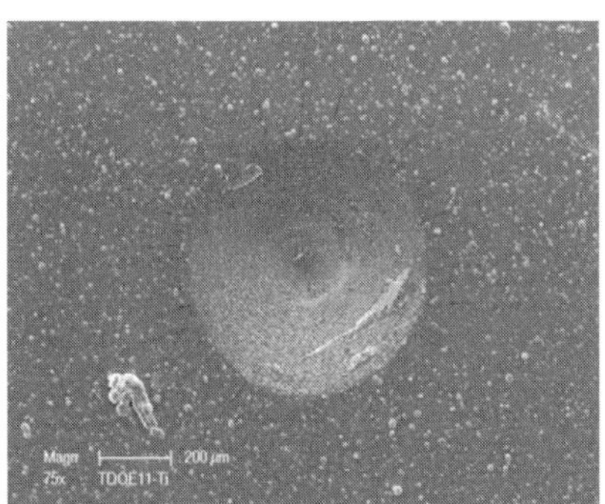

(a) Single Layer TiN Coating

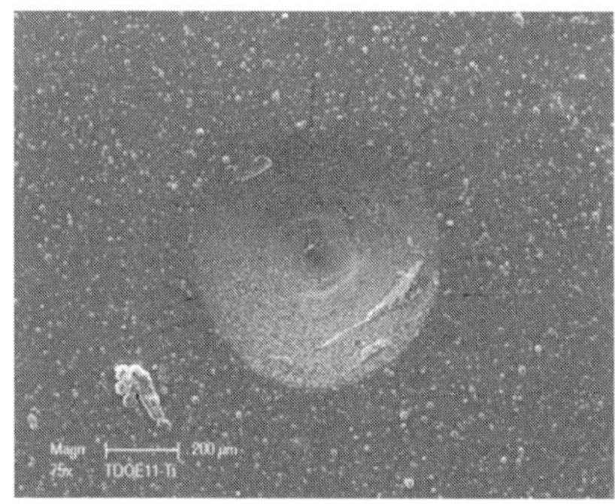

(b) Multi-layer TiN Coating

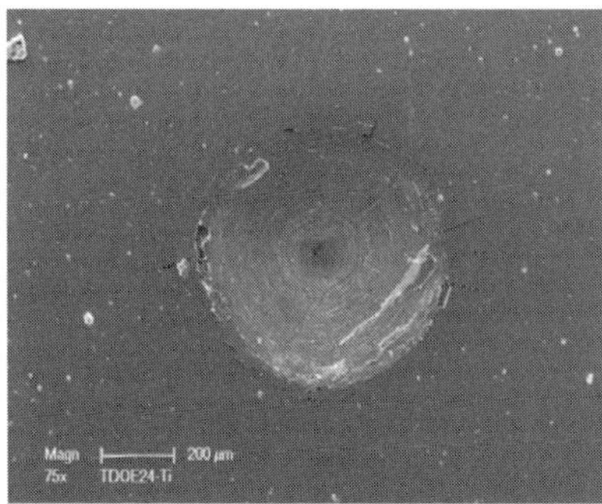

(c) Single layer TiSiCN coating

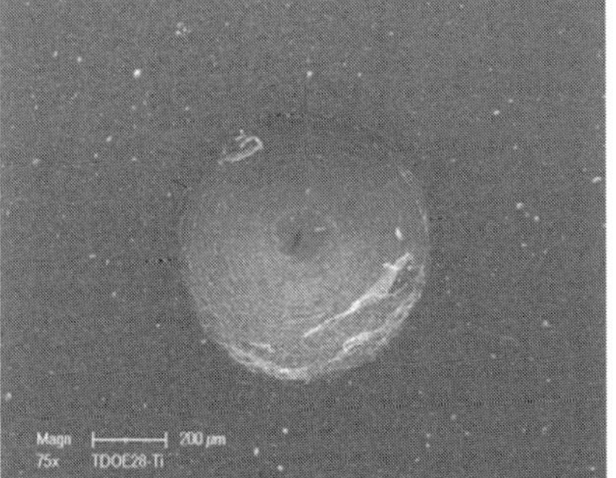

(d) Multi-layer TiSiCN coating

Figure 17. Rockwell C hardness indentations on three different coatings on Custom-450 stainless steel substrate showing good adhesion

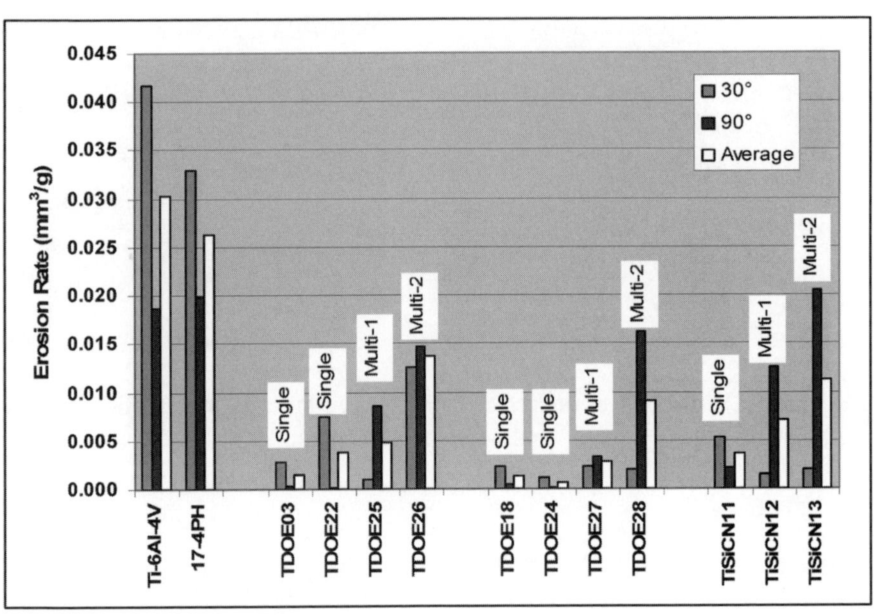

Figure 18. Solid particle erosion test results on single and multi-layer coatings produced under different coating conditions showing significant erosion resistance. Single layer coatings are better than the multilayer coatings under SPE

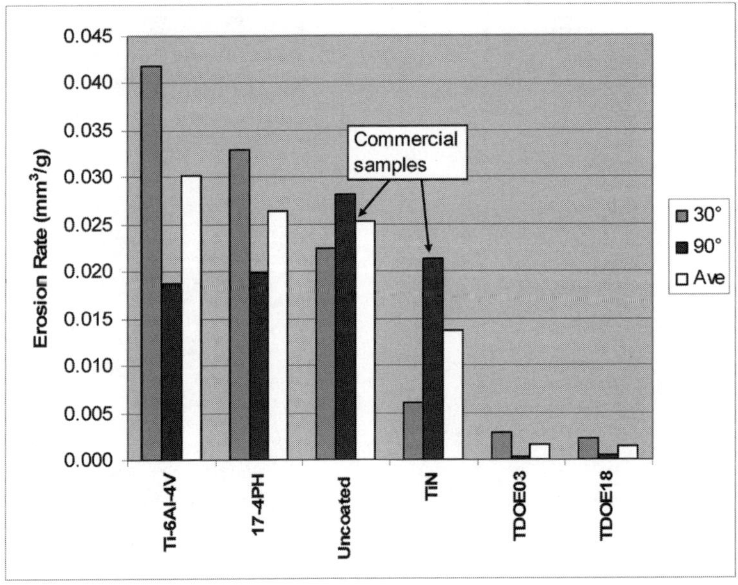

Figure 19. Comparison of erosion rates of commercial TiN coating and the advanced nano-composite coatings show significant improvement by the nanocomposite coatings

Advances in Materials Technology for Fossil Power Plants
Proceedings from the Fifth International Conference
R. Viswanathan, D. Gandy, K. Coleman, editors, p 471-487

DOI: 10.1361/cp2007epri0471

THE STEAMSIDE OXIDATION BEHAVIOR OF CANDIDATE USC MATERIALS AT TEMPERATURES BETWEEN 650°C AND 800°C

J. M. Sarver
The Babcock & Wilcox Company
B&W Research Center
180 South Van Buren Avenue
Barberton, Ohio 44203

J. M. Tanzosh
The Babcock & Wilcox Company
20 South Van Buren Avenue
Barberton, Ohio 44203

Abstract

The U.S. Department of Energy (DOE) and the Ohio Coal Development Office (OCDO) are sponsoring the "Boiler Materials for Ultrasupercritical Coal Power Plants" program. This program is aimed at identifying, evaluating, and qualifying the materials needed for the construction of critical components for coal-fired boilers capable of operating at much higher efficiencies than the current generation of supercritical plants. Operation at ultrasupercritical (USC) conditions (steam temperatures up to 760°C (1400°F)) will necessitate the use of new advanced ferritic materials, austenitic stainless steels and nickel-based alloys. As well as possessing the required mechanical properties and fireside corrosion resistance, these materials must also exhibit acceptable steamside oxidation resistance.

As part of the DOE/OCDO program, steamside oxidation testing is being performed at the Babcock & Wilcox Research Center. More than thirty ferritic, austenitic and nickel-based materials have been exposed for up to 4,000 hours in flowing steam at temperatures between 650°C (1202°F) and 800°C (1472°F). In addition to wrought materials, steamside oxidation tests have been conducted on weld metals, coated materials and materials given special surface treatments. Exposed specimens were evaluated to determine oxidation kinetics and oxide morphology. High chromium ferritic, austenitic and nickel-based alloys displayed very good oxidation behavior over the entire temperature range due to the formation of a dense chromium oxide. With increasing steam temperature, low chromium ferritic materials experienced breakaway oxidation, and low chromium austenitic materials experienced significant oxide exfoliation. Special surface treatments that were applied to these materials appeared to have a beneficial effect on their oxidation behavior.

Introduction

Coal will remain a low-cost fuel for the generation of electricity for many years to come. Increasingly stringent environmental standards necessitate the development of more efficient power plants to keep costs in check. Operating a boiler under ultrasupercritical (USC) conditions increases its efficiency, but the the higher steamside temperatures associated with USC operating conditions requires the utilization of new materials of construction. In the United States, the "Boiler Materials for Ultrasupercritical Coal Power Plants" program (DOE CONTRACT NO. DE-FG26-01NT41175, OHIO COAL DEVELOPMENT OFFICE (OCDO) REF. NO. D-00-20) is being performed to determine the most suitable materials of construction for a USC boiler.

An important consideration in the selection of materials for a USC boiler is steamside oxidation. In a power plant, steamside oxidation has three primary detrimental aspects, all of which are exacerbated at the higher temperatures planned for USC operation: 1) oxidation leads to wall loss which can eventually compromise structural integrity, 2) oxidation can act as an insulating barrier to heat transfer and cause local overheating of tubing, and 3) spalled oxides can plug tube bends and/or erode steam turbines (1). Because the control of steamside oxidation is an important factor in the successful operation of a USC boiler, one task of the DOE/OCDO program is devoted to the study of the steamside oxidation of candidate alloys.

Specifically, the goal of steamside oxidation task in the DOE/OCDO program is to determine the steamside oxidation behavior and temperature limits of currently available ferritic, austenitic, nickel-based and coated materials. Task 3 is also seeking to better understand the fundamental aspects of steamside oxidation to advance the future development of materials for USC service.

Experimental

Since the goal of the DOE/OCDO USC project is to attain a maximum main steam temperature of 732°C, the steamside oxidation tests are being performed on coupons in slowly flowing atmospheric pressure steam at aim temperatures of between 650°C and 800°C. The total exposure time at each temperature is 4,000 hours, with interim specimen removals after 1,000 and 2,000 hours.

The test environment is high purity water with a pH of 8.0-8.5 (adjusted by an addition of 20-70 ppb ammonia) and 100-300 ppb of dissolved oxygen. Inside a furnace, the test solution flashes to steam and passes into a retort that contains the test specimens. The specimens are coupons that are hung from a test frame and oriented parallel to the flow of steam. The temperature spread within the retort was approximately ±15°C at each test temperature. In each exposure, the materials that contained the lowest chromium content were exposed to the lowest temperatures and those with the highest chromium content were exposed to the highest temperatures. The temperature at each specimen location within the retort remained constant during each exposure, so the specimens were not exposed to any temperature cycling during the exposures. The test setup is explained in greater detail in a previous paper (2).

In general, six coupons are tested from each material at each temperature. The coupons are measured and weighed prior to testing. At each shut-down (after 1,000, 2,000 and 4,000 hours), two coupons from each material are removed and weighed to determine the weight change. One of the removed coupons from each material is cross sectioned and metallographically examined with a scanning electron microscope equipped with energy dispersive spectroscopy capabilities (SEM/EDS) to determine oxide morphology and composition. The other coupon from each material is de-scaled (using alkaline permanganate, hot diammonium citrate and, when necessary, hot inhibited hydrochloric acid) and re-weighed to determine the de-scaled weight loss experienced by each material.

The materials that have been tested in this program range from a 2% Cr ferritic alloy up to nickel-based superalloys. The compositions of the test materials and their manufacturers are displayed in Table 1.

Results

Oxidation Kinetics

In order to account for the oxide exfoliation that occurred on some of the samples, the oxidation kinetics in this program were calculated based on the descaled weight loss data instead of non-descaled weight change data. The relative parabolic rate constants (k_p) calculated for the test material at temperatures of 650 and 800°C are displayed in Figure 1.

Oxide Morphology

A description of the steam-formed oxide that formed on the materials as a function of time and temperature is displayed in Table 2. SEM photographs and EDS maps of selected materials are displayed in Figures 2-5.

Oxide Exfoliation

The DOE/OCDO steamside oxidation program was not specifically designed to study oxide exfoliation, so the test specimens were only exposed to temperature cycles when the facility was shut down to remove specimens. Thus, the specimens exposed for 1,000 hours experienced one temperature cycle, specimens exposed for 2,000 hours experienced two temperature cycles, and specimens exposed for 4,000 hours experienced three temperature cycles. After each exposure, the specimens were cooled under slowly flowing argon at a rate of approximately 75°C per hour.

The weight of exfoliated oxide from each material was calculated using the descaled weight loss and the non-descaled weight change of the test coupons. The difference in the weight of the pre-exposed coupon and the descaled post-test coupon represents the total metal lost to oxidation.

Assuming that the oxide formed on ferritic materials was Fe_3O_4, and that the oxide formed on austenitic materials was M_2O_3, where M had a molecular weight of 55.5 g/mole, then the expected weight change can be calculated from the weight loss. The expected weight change can then be compared to the measured weight change to approximate the amount of metal oxide that exfoliated.

The relative exfoliation rates of the materials evaluated in this program are shown in Figure 6. Except for the MARB2 alloy, ferritic materials with <~9% Cr experienced exfoliation at all of the test temperatures. At 650°C, oxide exfoliation was not observed for austenitic materials; however, at 800°C, iron-based austenitics with relatively low Cr levels (<~19 %) experienced significant oxide exfoliation. Austenitic materials containing high Ni and Cr concentrations did not experience exfoliation at temperatures up to 800°C.

Surface Treatments

Steamside oxidation tests were performed on several coupons which had been given surface treatments to improve their oxidation resistance. The surface treatments fell into three catagories: 1) barrier coatings, 2) surface alloying and 3) mechanical surface alteration. The surface treatments employed, base metal, test temperature and steamside oxidation results are displayed in Table 3.

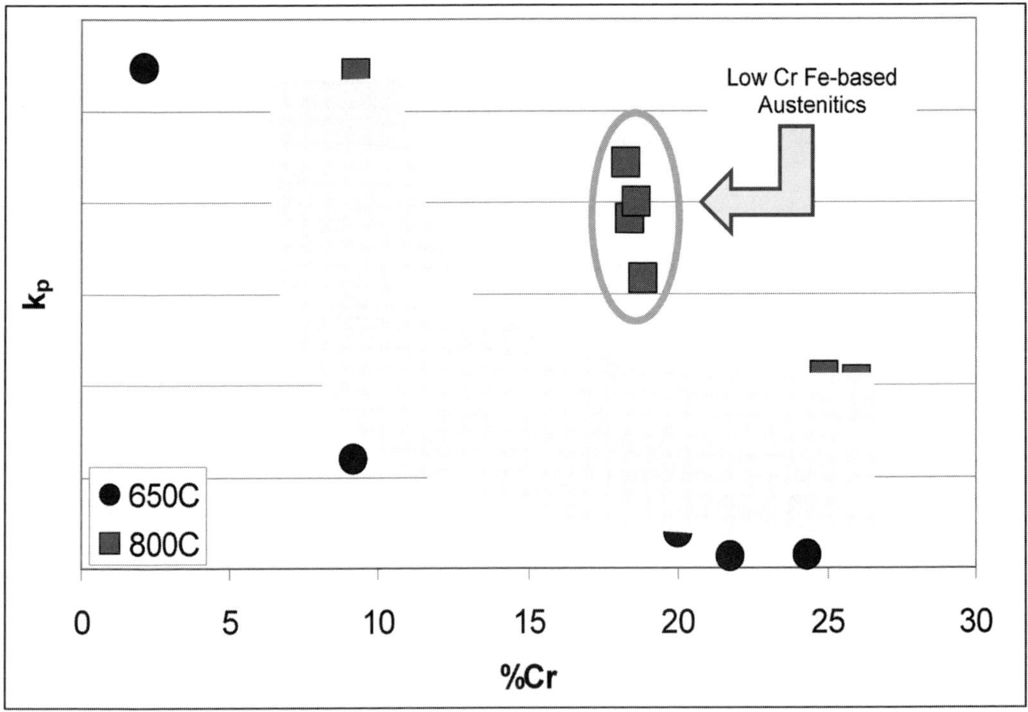

Figure 1. Relative Steam Oxidation Parabolic Rate Constants as a Function of Cr Content

Table 1: Composition of Test Materials

Material	C	Si	Fe	Cr	Ni	Mo	W	Nb	Other
T23	0.070	0.24	Bal	2.09	0.13	0.17	1.7	0.031	0.002 B
P91	0.11	0.37	Bal	8.29	0.14	1.03	0.024	0.068	0.18 Cu
P92	0.11	0.21	Bal	8.93	0.12	0.49	1.65	0.05	0.005 B
MARB2	0.082	0.73	Bal	9.16		<0.01	2.47	0.048	3.3 Co 0.019 B
SAVE 12	0.12	0.28	Bal	9.25			2.92	0.05	2.68Co
VM12	0.12	0.48	Bal	11.37	0.29	0.28	1.44	0.047	1.49 Co
Alloy 214	0.040	0.10	3.52	16.34	Bal	<0.1	<0.1	<0.1	4.43 Al 0.008 Y 0.03 Zr
347HFG	0.091	0.48	Bal	18.61	12.40	0.07		0.87	
304H	0.050	0.45	Bal	18.83	11.0				
SUPER304H	0.080	0.25	Bal	19.10	9.57	0.15		0.50	2.73 Cu 0.11 Co
Alloy 800HT	0.070	0.27	Bal	19.49	32.32				0.56Ti 0.53Al
Alloy 282	0.068	<0.05	0.35	19.63	Bal	8.56	<0.01		10.35 Co 2.21 Ti 1.41 Al
Nimonic 263	0.050	0.07	0.34	20.02	51.17	5.91			19.51 Co 2.16 Ti 0.44 Al
20-25+Nb	0.080	0.36	Bal	20.42	25.83	1.54		0.36	
CCA617	0.059	0.17	0.87	21.73	55.0	8.71	0.26	0.03	11.57Co 1.23Al 0.41Ti
SAVE 25	0.074	0.24	51.25	21.85	19.25	0.12	1.35		4.1Cu
Alloy 230	0.110	0.39	1.25	22.42	Bal	1.31	14.27	0.05	0.22 Co 0.33 Al
HR6W	0.07	0.26	Bal	23.44	44.70		6.0	0.25	0.12Ti
Alloy 740	0.034	0.450	1.020	24.31	49.45	0.520		1.830	0.75 Al 19.63 Co 1.58Ti
RA602CA	0.17	0.1	9.4	25.2	62.5				2.3 Al 0.08 Y 0.08 Zr
HR-120	0.06	0.59	Bal	25.94	36.49	0.38	<0.1	0.66	0.05Al 0.14 Co
310HCbN	0.067	0.34	50.45	25.98	19.97	0.11	<0.001	0.470	

Discussion

As shown in Table 1, many different materials were evaluated in this program, ranging from low chromium ferritic materials to high chromium nickel-based materials. In general, the materials were iron-chromium or iron-nickel-chromium alloys with other alloying elements added to enhance their properties. Many of the materials tested are relatively new alloys, or alloys which have had limited use in fossil boiler systems.

A study of the effect of the various alloying elements on the k_p of the materials studied in this program reveals that the chromium concentration is the single most important alloying element with respect to steamside oxidation. As shown in Figure 1, the k_p of the low chromium materials is high. The k_p decreases sharply at chromium concentrations between 9 and 12% and then becomes relatively stable at chromium concentrations above ~12%.

The low chromium ferritic steels (such as T23) form a multiple layered oxide, as shown in Figure 2. The outermost oxide layers are iron-rich, with chromium-rich inner oxide layers. The demarcation between the iron-rich and chromium-rich oxide layers is the original surface of the material, indicating that metal ions are diffusing outward and oxygen ions are diffusing into the metal.

The decrease of k_p between 9 and 12% chromium corresponds to the inner chromium oxide layer becoming more dense and protective. An example of this transition is shown Figure 3. For SAVE12 at 650°C, some areas of the surface display a thin, dense chromium oxide covered with iron oxide. Other areas (as shown in Figure 3) show an inner oxide that is a mixture of dense chromium oxide and non-protective iron-chromium oxide. Above these less protective inner oxide areas, pronounced iron-rich oxide nodules form. Over time these nodules grow and dominate the surface of the material.

The presence of oxide nodules and incomplete chromium oxide layers within the inner oxide has been reported by other researchers (3). The incomplete chromium oxide layers typically initiate along edges. The surface chromium concentration along edges is expected to be less than at flat surfaces since the chromium in the specimen is diffusing to a greater surface area along edges, as shown in Figure 7. A 3-dimensional image of Figure 3 would be expected to show that the chromium oxide layers are discontinuous and, therefore, are not effective in limiting the diffusion of iron to the surface and oxygen into the base metal. The result is the formation of the observed oxide nodules. Since oxidation is readily occurring at these locations, chromium from areas adjacent to the nodules is incorporated into the discontinuous chromium oxide layers, and the protective chromium oxide layers adjacent to the nodules are destabilized. Thus, the nodules grow laterally and cover the surface of the specimen. This progression is shown in Figure 8.

Materials with chromium concentrations greater than ~12% (at 650°C) and ~19% (at 800°C) form only the dense protective chromium oxides shown in Figure 4, and the k_p of these materials is relatively invariant as the chromium content increrases. This behavior is due to the fact that the chromium oxide remains stable and protective, even at temperatures up to 800°C.

Table 2: Steam-formed Oxide Morphology

Material	%Cr	Morphology
T23	2.09	At 650C very thick multiple layered Fe oxide above a Fe/Cr oxide
P91	8.29	At 650C Fe oxide over Cr-rich oxide nodules at shorter times evolving to very thick Fe oxide over Cr-rich oxide at longer times
P92	8.93	At 650C & 800C nodules and thick Fe oxide over Cr oxide
MARB2	9.16	At 650C a thin Fe oxide over a thin, dense Cr oxide; oxides thicken and nodules form at 800C
SAVE12	9.25	At 650C & 800C multiple layered Fe oxide over mixed Cr/Fe oxide + nodules
VM12	11.37	At 650C dense Cr oxide, oxide thickens with time and temperature; at 800C Fe oxide layer covers Cr oxide
214	16.34	At 800C thin dense Al oxide
347HFG	18.61	At 800C Fe oxide above thick Cr oxide and nodules
304H	18.83	At 650C Cr oxide; at 800C nodules (Fe oxide over Cr/Ni oxide)
S304H	19.1	At 650C thin Cr oxide w/ Cr depleted zone beneath; at 800C Cr/Fe oxide and nodules
20-25+Nb	20.47	At 800C dense Cr oxide above very thin Si layer, minor Al oxide grain boundary penetrations
800HT	19.49	At 650C thin Cr oxide above thin Ti oxide above very thin Al layer
N263	20.02	At 650C Ti oxide above Cr oxide
CCA617	21.73	At 650C thin Cr oxide above very thin Al oxide; at 800C dense Cr oxide mixed with some Ti oxide that thickens with temperature; Al oxide penetrations along grain boundaries
SAVE25	21.85	At 650C thin Cr oxide above Cr depleted zone; at 800C Cr oxide above a very thin Si layer
230	22.42	At 650C Cr oxide above Al layer; at 800C Cr oxide above Al oxide grain boundary penetrations
HR6W	23.44	At 650C thin Cr oxide; at 800C Cr oxide
740	24.31	At 650C thin Cr oxide mixed with Ti oxide; at 800C dense Cr oxide mixed with some Ti oxide above a Ti oxide layer; Al oxide penetrations along grain boundaries
HR120	25.94	At 650C Cr oxide; at 800C Cr oxide above thin Si layer
310HCbN	25.98	At 800C dense Cr oxide that thickens with temperature above a very thin Si layer

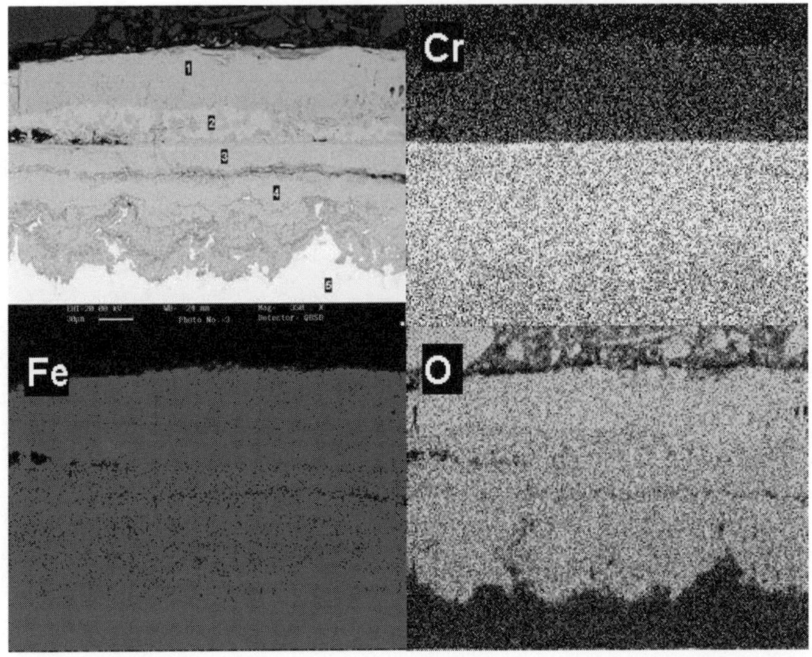

Figure 2. SEM Photograph of Cross Section of Steam-formed Oxide on T23 Tested at 650°C and Associated EDS Maps

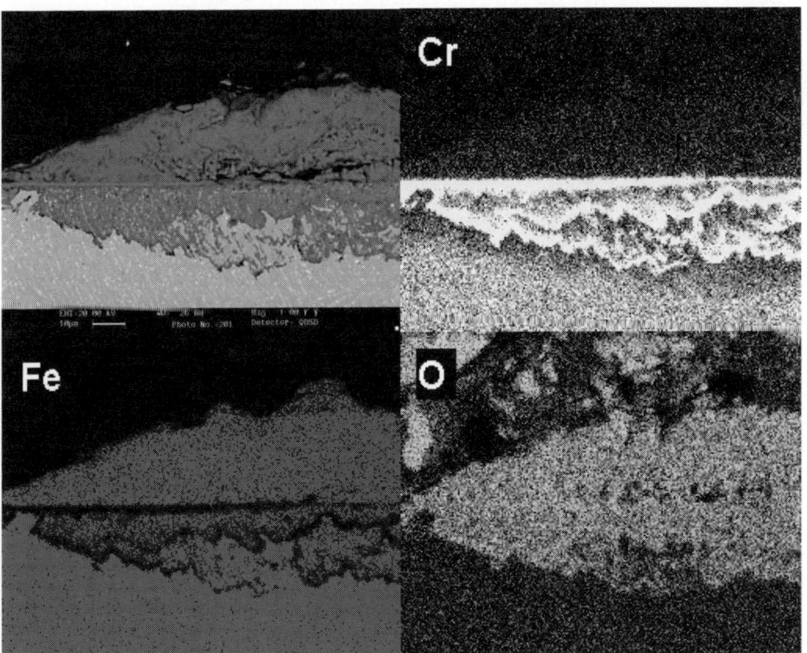

Figure 3. SEM Photograph of Cross Section of Steam-formed Oxide Nodules on SAVE12 Tested at 650°C and Associated EDS Maps

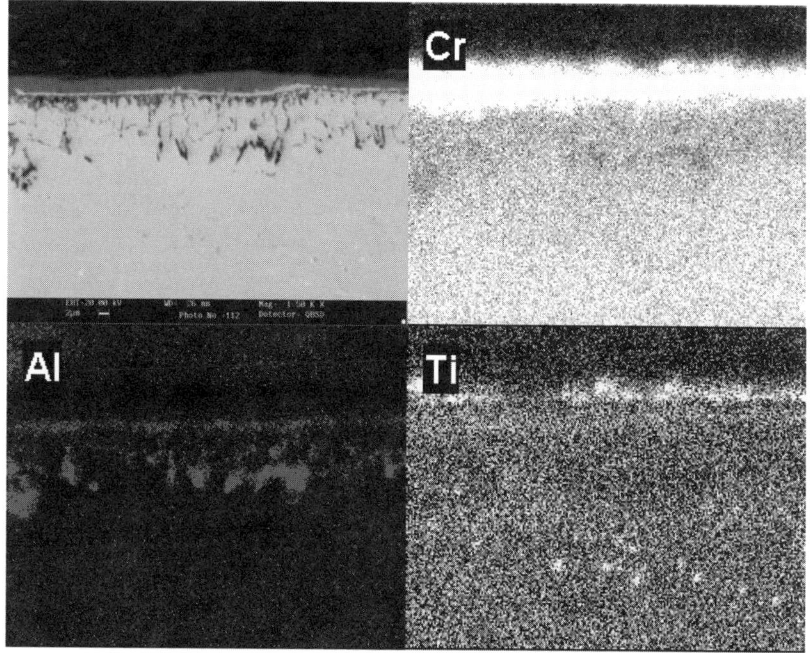

Figure 4. SEM Photograph of Cross Section of Steam-formed Oxide and Grain Boundary Aluminum Oxide Penetrations on CCA617 Tested at 800°C and Associated EDS Maps

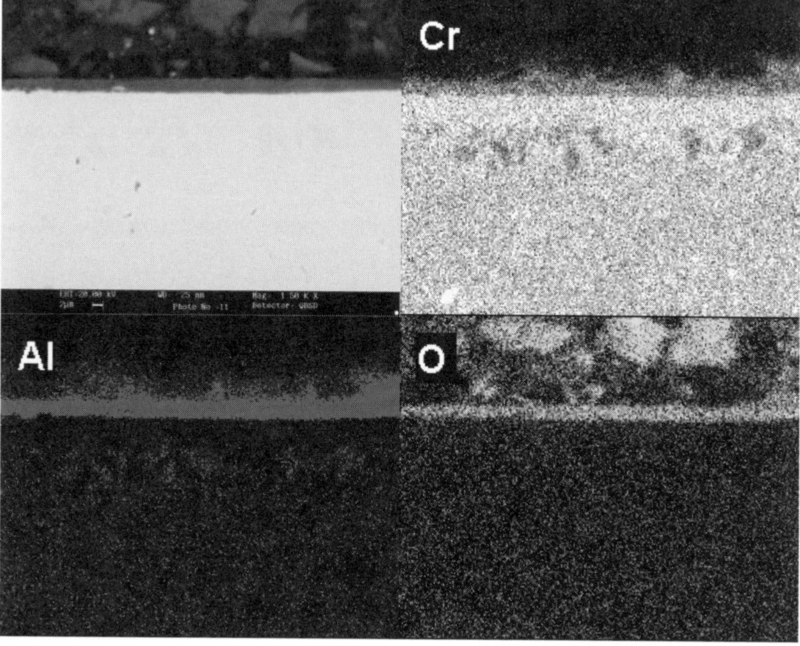

Figure 5. SEM Photograph of Cross Section of Steam-formed Aluminum Oxide on Alloy 214 Tested at 800°C and Associated EDS Maps

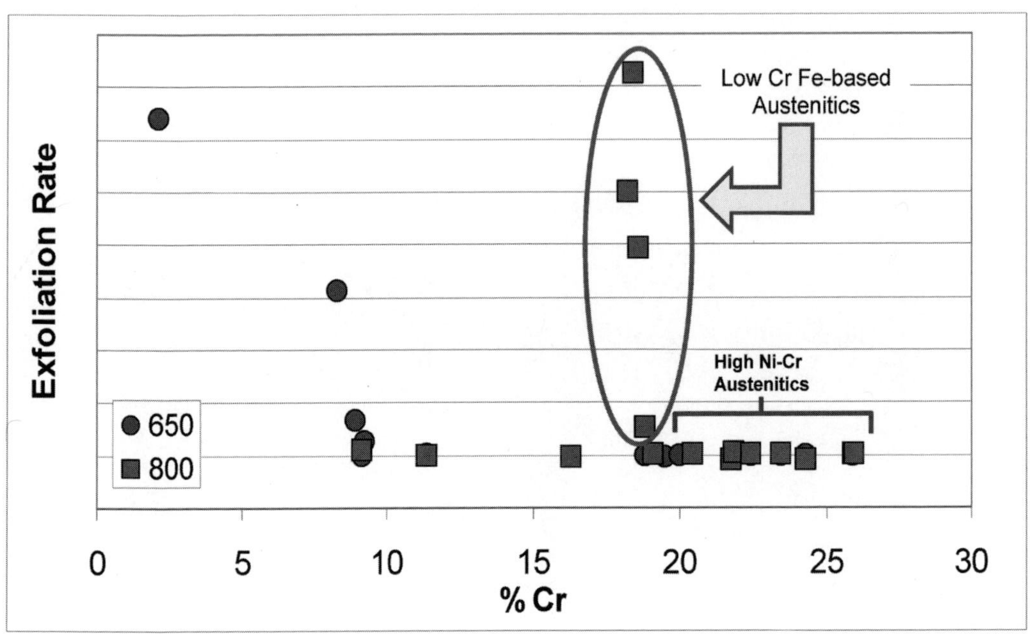

Figure 6. Relative Oxide Exfoliation Rates Experienced During Steam Exposures

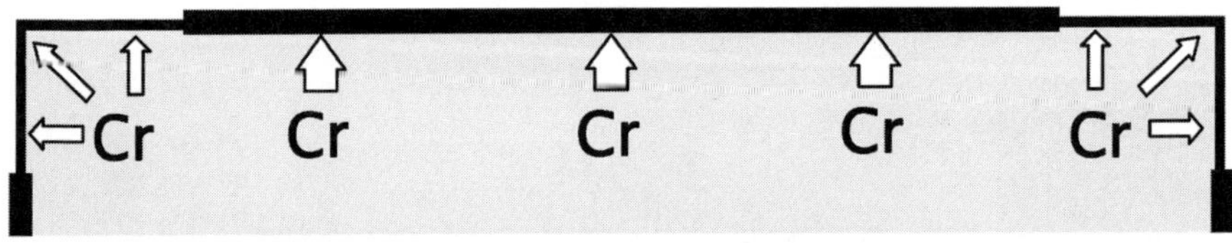

Figure 7. Multiple Diffusion Paths at Edges Leads to Lower Chromium Surface Concentrations and Less Protective Oxides along Edges

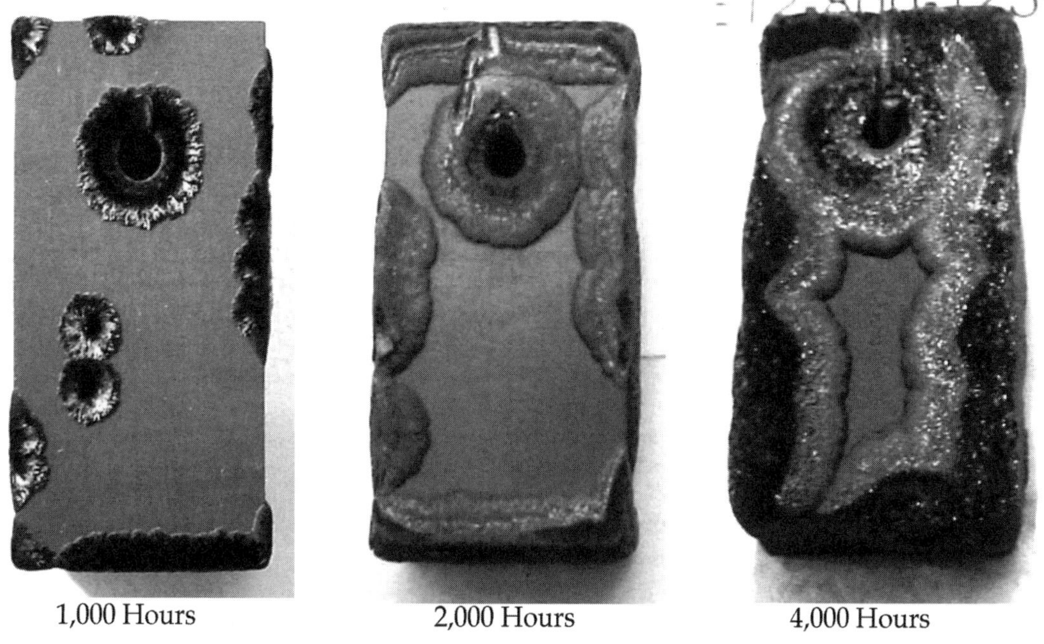

Figure 8. Oxide Nodule Growth on SAVE12 as a Function of Exposure Time to 800°C Steam

While the chromium concentration dominates the steamside oxidation behavior of these materials, other alloying elements also have an impact. In Figure 1, ferritic materials containing between 8.29 and 9.25% chromium display variations in k_p of greater than 4 orders of magnitude. These variations in k_p do not correlate to variations in the chromium concentration. For these ferritic materials, an increase in the silicon content appears to have a beneficial effect on the steamside oxidation behavior of the alloys; an effect reported previously by other researchers (4-6). This beneficial effect of silicon was not obvious for the higher chromium austenitic materials in this program.

From Table 2, the elements other than iron, nickel and chromium that are most often incorporated into the oxides of the austenitic materials are aluminum and titanium. From a thermodynamic standpoint, the reactivity of these elements with oxygen is as follows:

Aluminum > Titanium >Chromium > Iron > Nickel.

Thus, aluminum and titanium will form oxides at lower partial pressures of oxygen than will the other alloying elements.

Table 3: Effectiveness of Surface Treatments in Reducing Steamside Oxidation

Surface Treatment	Base Metal	Result
Barrier		
Ceramic/Glass	P92	Coating exfoliated during first 1,000 hour exposure
Electroless Ni	P92	Coating was non-protective and exfoliated
Surface Alloying		
Chromized	P92 and S304H	Cr oxide forms. Oxidation rate is less than P92; approximately equal to S304H, no exfoliation
Aluminum-Chromized	P92 and S304H	Mixed Cr, Fe, Al, Ni oxide forms above thin Al oxide. Oxidation rate is less than P92, but greater than S304H and other chromized and Si-chromized materials
Silicon-Chromized	P92 and S304H	Cr oxide forms. Oxidation rate is less than P92; approximately equal to S304H, no exfoliation
Mechanical		
Shot Blasting	SUPER304H and 347HFG	Cr oxide forms. k_p of shot blasted materials ~10X less than for non-blasted materials
Deep Rolling	347HFG	Thin Cr oxide forms. k_p of deep rolled materials 10-100X less than for non-deep rolled materials

Of the materials tested in this program, Alloy 214 is the only alloy with sufficient aluminum to form a protective aluminum oxide layer, as shown in Figure 5. Under the test conditions for this program, Alloy 214 exhibited k_p values that were at least 2 orders of magnitude less than any other material tested (Alloy 214 is not included in the data points shown in Figure 1). Since Alloy 214 contains only 16.3% chromium, it is the aluminum content of this alloy, not the chromium content, that is the determining factor in its steamside oxidation behavior.

Aluminum may have had a slightly beneficial effect in the steamside oxidation behavior of austenitic materials that contained between 0.2 and 2.3% aluminum but, in general, the aluminum and titanium additions to these alloys do not improve their steamside oxidation, nor were they intended to do so. Aluminum and titanium aer added to these alloys to improve their mechanical properties, such as high temperature creep. The reactivity of aluminum and titanium are apparent from the oxide that forms on Alloy CCA617 (Figure 4). The outer oxide layer is composed of chromium oxide and titanium oxide, with an inner layer of aluminum oxide. Figure 4 also shows aluminum oxide penetrations along the near-surface grain boundaries. Grain boundary oxide penetrations in aluminum-containing austenitics were not observed after 4,000 hours of exposure at 650°C, but were observed at 800°C after 1,000 hours of exposure.

The presence of aluminum oxide grain boundary penetrations indicates that: 1) oxygen is diffusing into the alloy more rapidly along grain boundaries than in the bulk matrix, 2) the oxygen partial pressure along the grain boundaries is very low since only aluminum (the most reactive element in the alloy) is oxidizing along the grain boundaries, and 3) it appears that aluminum is preferentially diffusing to the near-surface grain boundaries where oxygen penetration is occurring. The grain boundary penetrations may be a concern for these materials since it reduces the cross sectional area of sound material. Over the first 4,000 hours of exposure, the depth of the penetrations increased with time.

One group of data points in Figure 1 does not fit the trend displayed by the remainder of the data. The k_p values for the low chromium iron-based austenitic alloys tested at 800°C are between 1 and 3 orders of magnitude higher than would be expected based on their chromium content. The reason that these materials (304H, 347HFG and SUPER304H) displayed much higher than expected steamside oxidation rates is that they experienced oxide exfoliation during testing, as shown in Figure 6.

Oxide exfoliation is understood to be a combination of oxide growth stresses and thermal stresses resulting from coefficient of thermal expansion (CTE) differences between an oxide and its base metal, or between two different oxides (7). The exfoliation of steamside oxides from low chromium iron-based austenitic alloys in operating boilers at temperatures below 650°C is well documented (8-9), so the observed oxide exfoliation of these alloys at 800°C is not surprising.

The similarity in the shape of the plots in Figures 1 and 6 is also not surprising. For materials that contain sufficient chromium to form a protective oxide, steam-formed oxides grow at a parabolic rate (thickness increases with the square root of time). Freshly exposed materials experience rapid oxidation initially; however, the oxidation rate decreases with time as the oxide thickens and diffusion through the oxide becomes more difficult. When an oxide (typically the outer oxide layer) becomes sufficiently thick, it will exfoliate due to CTE differences, as was explained above. When this occurs, the overall oxide thickness is usually reduced by at least half, the remaining oxide permits diffusion at a rate commesurate with its thickness, and the oxidation rate jumps to a rate that corresponds to the new, thinner oxide thickness. This "re-energizing" of the oxidation process for materials that experience oxide exfoliation leads to the high oxidation rates for these materials observed in Figure 1.

The results presented in Table 2 may provide an explanation for the observed behavior. At 650°C, only a chromium oxide was observed on the 304H and SUPER304H materials after the 4,000 hour test (347HFG was not tested at 650°C). At 800°C, however, an outer iron oxide was observed above a less protective chromium-rich oxide on these alloys. Thus, at temperatures above 650°C, it appears that the chromium oxide that forms on these materials lack the ability to prohibit diffusion through the oxide. As a result a relatively thick duplex oxide forms, the outer layer of which (due to growth stresses and/or CTE mismatch stresses) exfoliates during the test and/or during cooldown. For the specimens tested at 650°C, the same behavior may also occur after longer exposures.

The results in Figures 1 and 6 and Table 2 suggest that the oxidation/exfoliation behavior of austenitic materials is strongly dependent upon the chromium concentration. At chromium concentrations below ~20%, high oxidation/exfoliation rates and oxide nodules are observed for iron-based austenitic materials exposed to 800°C steam. However, materials that possess chromium concentrations of 20% and higher display low oxidation/exfoliation rates and protective chromium oxides.

As fossil power moves toward operating at USC conditions, it is vital that the industry develop a thorough understanding of steamside oxide exfoliation. It is important to learn if this condition is limited to low chromium iron-based austenitic alloys, or if, given sufficient time at temperature, it will also affect higher chromium iron-based or nickel-based austenitic alloys. Knowing the maximum time-temperature conditions to which different alloys can be exposed without experiencing severe oxide exfoliation is also critical in determining locations within the boiler where these alloys can be employed. Finally, any adjustments to the steamside chemistry that could stabilize the chromium oxide and make it more protective should be explored to mitigate this concern.

Certainly one method that can be used to decrease steamside oxidation rates and reduce oxide exfoliation is surface treatments. Surface treatments such as claddings and coatings have already been used for years to improve the fireside corrosion resistance of boiler materials. As shown in Table 3, the barrier coatings that were attempted in this program (coatings designed to stop oxidation or to oxidize in place of the original alloy) were unsuccessful in improving the steamside oxidation behavior of the base alloy. While other barrier-type coatings may be developed for USC materials in the future, their long-term performance and their ability to maintain the heat transfer characteristics of the base alloys to which they are applied must be thoroughly researched.

Treatments involving surface alloying did show promise in this program, particularly chromizing and silicon-chromizing. Application of these techniques permitted a 9% chromium ferritic alloy to develop a protective chromium oxide layer and, thus, greatly reduce its oxidation rate. It is not surprising that these techniques did not significantly reduce the oxidation rate of a low chromium iron-based austenitic material, but the test specimens did not display visible signs of exfoliation, as did the base alloy. The aluminum-chromizing technique appears to be the least effective chromizing technique evaluated. SEM/EDS results indicate that a continuous protective aluminum oxide did not form, nor did a continuous protective chromium oxide. The mixed oxide that did form lacked the protective properties exhibited by the oxides formed using the other chromizing techniques.

For the 4,000 hour exposures in this program, mechanical surface treatments were the most effective treatments in improving the steamside oxidation behavior of candidate USC alloys, as shown in Table 3. Both shot blasting and deep rolling cold work the surface of the materials and produce surface compressive stresses (10-11). The resulting cold worked microstructure permits fast diffusion paths to the surface of the material, allowing a thin, protective chormiumm oxide to form quickly. The low chromium iron-based austenitic materials given a shot blast or deep roll treatment exhibited greatly reduced oxidation rates and no oxide exfoliation. It seems likely, however, that oxide exfoliation will eventually occur on these mechanically treated austenitic materials once their oxides becomes sufficiently thick. Thus, while these techniques show promise, much longer time at temperature studies must be conducted to determine if this improvement has a finite life due to microstructural changes and/or chromium depletion.

For all of the surface treatments discussed above, consideration must be given to the cost of the treatment, the ease of application, the reproducibility of application, the ability to apply a treatment to complex shapes, and the ability for the treatment to provide long-term benefit.

Conclusions

The higher steam temperatures required for USC boiler operation necessitates the use of materials that are new to this industry. The DOE/OCDO program described above has investigated the steamside oxidation behavior of many of these new materials at steam temperatures well in excess of current operating conditions. The results from this program have increased our general understanding of the steamside oxidation process, and have provided important information regarding the expected steamside oxidation behavior of specific alloys. The results suggest that alloys and surface treatments exist that will provide acceptable steamside oxidation behavior at USC steam temperatures.

Operating at USC conditions still presents significant steamside materials challenges that will require additional study. First, accurate temperature limits for different clases of materials and specific alloys must be determined. Second, a thorough understanding of oxide exfoliation must be developed, along with techniques to mitigate exfoliation. Finally, since there is a strong economic driver to utilize some form of surface treatment to improve the steamside oxidation behavior of USC materials, these surface treatments must be thoroughly understood and optimized to provide significant long-term benefit.

Through USC materials programs such as the one sponsored by the DOE/OCDO, and others being performed throughout the world, the materials challenges will be met, and safe, clean USC power generation will become a reality in the near future.

Acknowledgements

The authors would like to thank Sridhar Ramamurthy, Ross Davidson and Brad Kobe at SSW for their expertise in performing the SEM/EDS analyses and their insights in the interpretation of the results. The authors would also like to thank Doug Zeigler, John Jevec and Bob Pelger at B&W for their assistance in performing the oxidation testing.

References

1. J. Sarver, R. Viswanathan and S. Mohamed, Boiler Materials for Ultrasupercritical Coal Power Plants – Task 3, Steamside Oxidation of Materials, NETL/DOE, 2003. USC T-5.

2. J. M. Sarver and J. M. Tanzosh, "Preliminary Results from Steam Oxidation Tests Performed on Candidate Materials for Ultrasupercritical Boilers", EPRI International Conference on Materials and Corrosion Experience for Fossil Power Plants, Isle of Palms, South Carolina, November 18-21, 2003.

3. J. P. Shingledecker and I. G. Wright, "Evaluation of the Materials Technology Required for a 760°C Power Steam Boiler", 8th Liege Conference on Materials for Advanced Power Engineering, Liege, Belgium, September 18-20, 2006.

4. Y. Fukuda, K. Tamura, "Effect of Cr and Si Contents on the Steam Oxidation of High Cr Ferritic Steels," Proceedings from the International Symposium on Plant Aging and Life Prediction of Corrodible Structures, Sapporo, Japan, May 15-18, 1995.

5. K. Tamura, T. Sato, Y. Fukuda, K. Mitsuhata, H. Yamanouchi, "High Temperature Strength and Steam Oxidation Properties of New 9-12% Cr Ferritic Steel Pipes for USC Boilers," Proceedings of the ASM 2nd International Conference on Heat Resistant Materials, ASM, Materials Park, Ohio, 1995, pp 33-39.

6. W.J. Quadakkers, J. Ehlers, V. Shemet, L. Singheiser, "Development of Oxidation Resistant Ferritic Steels for Advanced High Efficiency Steam Power Plants", Materials Aging and Life Management, Kalpakkam, India, October 3-6, 2000.

7. I. G. Wright, M. Schütze, S. R. Paterson, P. F. Tortorelli and R. B. Dooley, "Progress in Prediction and Control of Scale Exfoliation on Superheater and Reheater Alloys", EPRI International Conference on Boiler Tube and HRSG Tube Failures and Inspections, San Diego, California, November 2-5, 2004.

8. O. H. Larsen, R. B. Frandsen and R. Blum, "Exfoliation of Steam Side Oxides from Austenitic Superheaters", VGB PowerTech, July, 2004, pp 89-94.

9. A. Hughes, "Case Study of 347HFG Exfoliation in a Supercritical Boiler", EPRI International Conference on Materials and Corrosion Experience for Fossil Power Plants, Isle of Palms, South Carolina, November 18-21, 2003.

10. T. Sato, Y. Fukuda, K. Mitsuhata and K. Sakai, "The Practical Application and Long-Term Experience of New Heat Resistant Steels to Large Scale USC Boilers", Proceedings from the Fourth International Conference on Advances in Material Technology for Fossil Power Plants, Hilton Head Island, South Carolina, October 25-28, 2004.

11. Ecoroll Product Information, http://www.ecoroll.de/index_e.htm.

LEGAL NOTICE/DISCLAIMER

This report was prepared by The Babcock & Wilcox Company pursuant to a Grant partially funded by the U.S. Department of Energy (DOE) under Instrument Number DE-FG26-01NT41175 and the Ohio Coal Development Office/Ohio Department of Development (OCDO/ODOD) under Grant Agreement Number CDO/D-00-20. NO WARRANTY OR REPRESENTATION, EXPRESS OR IMPLIED, IS MADE WITH RESPECT TO THE ACCURACY, COMPLETENESS, AND/OR USEFULNESS OF INFORMATION CONTAINED IN THIS REPORT. FURTHER, NO WARRANTY OR REPRESENTATION, EXPRESS OR IMPLIED, IS MADE THAT THE USE OF ANY INFORMATION, APPARATUS, METHOD, OR PROCESS DISCLOSED IN THIS REPORT WILL NOT INFRINGE UPON PRIVATELY OWNED RIGHTS. FINALLY, NO LIABILITY IS ASSUMED WITH RESPECT TO THE USE OF, OR FOR DAMAGES RESULTING FROM THE USE OF, ANY INFORMATION, APPARATUS, METHOD OR PROCESS DISCLOSED IN THIS REPORT.

Reference herein to any specific commercial product, process, or service by trade name, trademark, manufacturer, or otherwise, does not necessarily constitute or imply its endorsement, recommendation, or favoring by the Department of Energy and/or the State of Ohio; nor do the views and opinions of authors expressed herein necessarily state or reflect those of said governmental entities.

NOTICE TO JOURNALISTS AND PUBLISHERS: Please feel free to quote and borrow from this report, however, please include a statement noting that the U.S. Department of Energy and the Ohio Coal Development Office provided support for this project.

Advances in Materials Technology for Fossil Power Plants
Proceedings from the Fifth International Conference
R. Viswanathan, D. Gandy, K. Coleman, editors, p 488-506

DOI: 10.1361/cp2007epri0488

EFFECTS OF FUEL COMPOSITION AND TEMPERATURE ON FIRESIDE CORROSION RESISTANCE OF ADVANCED MATERIALS IN ULTRA-SUPERCRITICAL COAL-FIRED POWER PLANTS

Horst Hack
horst_hack@fwc.com; Tel. 973-535-2200; Fax. 973-535-2242

Greg Stanko
greg_stanko@fwc.com; Tel.973-535-2256; Fax. 973-535-2242

Foster Wheeler North America Corp.
12 Peach Tree Hill Road, Livingston, NJ 07039

Abstract

The U.S. Department of Energy (DOE) and the Ohio Coal Development Office (OCDO) are co-sponsoring a multi-year project, managed by Energy Industries of Ohio (EIO), to evaluate candidate materials for coal-fired boilers operating under ultra-supercritical (USC) steam conditions. Power plants incorporating USC technology will deliver higher cycle efficiency, and lower emissions of carbon dioxide (CO_2) and other pollutants than current coal-fired plants. Turbine throttle steam conditions for USC boilers will approach 732°C (1350°F), at 35 Mpa (5000 psi). The materials used in current boilers typically operate at temperatures below 600°C (1112°F) and do not have the high-temperature strength and corrosion resistance required for USC operation. Materials that can meet the high temperature strength and corrosion requirements for the waterwalls and superheater/reheater sections of USC boilers have been tested, and evaluated.

The focus of the current work is the evaluation of the fireside corrosion resistance of candidate materials for use in USC boilers through field testing. These materials include high-strength ferritic steels (SAVE12, P92, HCM12A), austenitic stainless steels (Super304H, 347HFG, HR3C), and high-nickel alloys (Haynes® 230, CCA617, INCONEL® 740, HR6W). Protective coatings (weld overlays, diffusion coatings, laser claddings) that may be required to mitigate corrosion were also evaluated. Corrosion resistance was previously evaluated under synthesized coal-ash and flue gas conditions typical of three North American coals, representing Eastern (mid-sulfur bituminous), Mid-western (high-sulfur bituminous), and Western (low-sulfur sub-bituminous) coal types. Laboratory testing for waterwall materials was performed at temperatures ranging from 455°C (850°F) to 595°C (1100°F), while superheat/reheat materials were tested at temperatures ranging from 650°C (1200°F) to 870°C (1600°F). The laboratory test results were previously presented, which illustrated the effects of temperature and coal type on corrosion of alloys with varying chromium level.

Promising materials from the laboratory tests were assembled on air-cooled, retractable corrosion probes for testing in utility boilers. The probes were designed to maintain metal temperatures using multiple zones, ranging from 650°C (1200°F) to 870°C (1600°F). Three utility boilers, equipped with low NOx burners, were identified that have adequate flue gas temperatures and represent each of the three coal types. This paper presents new fireside corrosion probe results for Mid-western, and Western coal types, after approximately one-year of exposure in the field.

Introduction

Program Background

Power plants in the United States are under increasing pressures to improve efficiency and reduce emissions. The efficiency of conventional pulverized coal power plant cycles is strongly related to operating temperature and pressure. The need to improve efficiency has been a driving force to advance materials suitable for these higher boiler operating temperatures and pressures. The present program, through a unique government and industry consortium, is conducting a multi-year materials evaluation effort to advance the technology in coal-fired power generation. The objective of this program is to allow boiler operation at much higher temperatures and pressures than are presently used in conventional power plants. These higher operating conditions will enable the use of advanced, more efficient ultra-supercritical (USC) steam cycles in coal-fired power generation, which offers the added advantage of reducing carbon dioxide emissions.

The existing fleet of pulverized-coal power plants is typically 35 percent (HHV) efficient, and operates at steam temperatures below 600°C (1112°F). The materials used in these existing plants do not have the high-temperature strength and corrosion properties required for USC operation. By developing improved materials systems that can withstand higher temperatures, pulverized coal power plant efficiencies of up to 47 percent (HHV) are possible. These efficiency gains, alone, would cut the release of carbon dioxide and other emissions by nearly 30 percent. Additionally, USC cycles can be combined with oxycombustion technology to facilitate carbon dioxide capture.

This multi-year program is funded by DOE through the National Energy Technology Laboratory (NETL), co-funded by OCDO, and managed by EIO. A consortium of industry members who are cost-sharing the project (Alstom Power, Riley Power, Babcock and Wilcox, and Foster Wheeler), along with the direct participation of DOE's National Laboratories (NETL and Oak Ridge National Laboratory), are performing the technical aspects of the project. The Electric Power Research Institute (EPRI) provides technical management and direction for the program. The program has been divided into eight tasks, with responsibilities distributed among the consortium members, as described by Viswanathan (1). The focus of this paper is the work being conducted by Foster Wheeler on the "Fireside Corrosion Resistance" task.

Fireside Corrosion

In the first half of the 20th century, the development of pulverized coal boilers evolved gradually, setting the groundwork for the explosion of technology that occurred in the second half. Steam temperatures rose an average of 3.7°C (6.6°F)/yr from 173°C (343°F) in 1903 to about 621°C (1150°F) in 1955. However, the excursion above 538°C (1005°F) was short lived; by 1960 most steam generators were designed for 565°C (1049°F), and by 1970 they were back to 538°C (1005°F). There were several reasons for this retreat, one being a materials problem that, at that time, defied a technical or economic solution. Materials that exhibited strength to withstand the higher temperatures and pressures present in the superheaters of these supercritical advanced boilers were available; but they proved to be especially susceptible to corrosion by certain coals, most notably high-sulfur bituminous coals.

The cause of this type of corrosion, referred to as coal ash corrosion, was discovered in the late 1950's and is now generally accepted to be the presence of liquid alkali iron trisulfates on the surface of the superheater and reheater tubes beneath an overlying ash deposit (2). Approaches to solving this problem have included changing the fuel or providing protective baffling with sheaths of corrosion resistant material around selected tubes. Coal ash corrosion is a widespread problem for superheater and reheater tubes, especially where high-sulfur, high-alkali, and high-chlorine coals are used, and is a critical problem that needs to be resolved before advanced ultra supercritical boilers can be deployed. The installation of low-NOx burners in boilers in the 1980's has, in some cases, resulted in conditions in the superheater and reheater sections that have exacerbated the coal ash corrosion problem. Tube samples obtained from the superheater section in units operating with low-NOx burners contained evidence of carbon carryover and less oxidizing flue gases. This environment has resulted in carburization of the tube surfaces. Carburization ties up the chromium in the alloy, which is the element that provides the most benefit to coal ash corrosion resistance and, as a result, promotes higher wastage rates. In a somewhat related matter, Powder River Basin (PRB) fuels that were previously thought to be non-corrosive have been found to cause very high wastage rates under certain less oxidizing operating conditions (3). There have been a number of literature reviews and recent updates discussing the variables affecting the corrosion mechanism.

As part of a project, "Boiler R&D for Improved Coal Fired Power Plants" sponsored by the Electric Power Research Institute (EPRI), Foster Wheeler Development Corporation (FWDC), and Ishikawajima Harima Heavy Industries Co, Inc. (IHI) performed both laboratory and field-testing to assess coal ash corrosion mechanisms (3, 4). The results of the laboratory tests for twenty two materials, three alkali sulfate levels, three levels of SO2, and four temperatures, indicated that the alkali and sulfur content, along with the metal operating temperature, can have a significant effect on coal ash corrosion. The loss from corrosion increased with higher levels of SO2, alkali sulfates, and temperature. The temperature for the maximum corrosion rate, can be represented by a bell-shaped curve (plotting corrosion loss vs. temperature), the form of which shifts and varies in width as a function of alloy composition, SO2 level in gas, alkali content, and CaO concentration in the coal. Pitting attack was not an active mechanism at low levels of SO2 and alkali, but it became active, especially on stainless steels and chromized coatings at very high levels of SO2 and alkali sulfate concentrations. Also, chromium was found to be the most beneficial alloying element for enhanced corrosion resistance. The corrosion resistance was greatly improved as the chromium levels approached and exceeded 25 percent. Laboratory tests also indicated that standard Type 347 tube material is adequate for mildly-corrosive environments, but Type 310 or Type 310 HCbN (Sumitomo HR3C, ASME code case 2115) is required for more corrosive conditions.

The second phase of the above project included field tests exposing eight alloys on air cooled, retractable corrosion probes for up to 16,000 hours. FWDC and EPRI chose three boilers; two burning Eastern high-sulfur coal and one burning Western low-sulfur sub-bituminous coal, for test probe exposures. The air-cooled, retractable corrosion probes were designed and installed; they exposed eight metal samples for 4000, 12000, and 16000 hours at temperatures generally in the range of 650°C to 700°C (1200°F to 1300°F). In both of the units burning the high-sulfur Eastern coal, the classic alkali-iron trisulfate attack occurred. As expected, chromium was the most beneficial element that improved coal ash corrosion resistance in these tests. Alloys with 25 percent or more chromium generally showed satisfactory corrosion resistance. The boiler burning the low-sulfur Western coal produced atypical reducing conditions on the backside of the corrosion probes. This resulted in high corrosion rates, from sulfidation in the presence of calcium sulfate, on the backside of the tube, rather than at the 2 and 10 o'clock positions on the front of the tube, which are typical for coal-ash corrosion. The results from these field corrosion probe exposures confirmed the performance of alloys in the laboratory corrosion tests. The combination of the laboratory screening and parameter testing and field test results under complex realistic conditions provided a database for alloy selection.

A similar field test, at temperatures typical of an advanced-cycle plant, exposed 10 alloys (including developmental and clad compositions) on air-cooled retractable corrosion probes for up to 16000 hours at temperatures in the range of 620°C to 730°C (1150°F to 1340°F) (5). The results of that Oak Ridge National Laboratory program showed that some of the developmental alloys and claddings had wastage rates that were one fourth of the wastage typical for a 347 superheater tube. In a separate study for EPRI, samples of 304, 347, 800H, NF709 HR3C, CR30A and chromized T22 were exposed for over 45,000 hours at temperatures from 521°C to 685°C (970°F to 1265°F), which covers the first part of the first target for the expected tube metal temperatures for a ultra-supercritical steam plant (2). The active corrosion mechanism was found to be molten salt attack by potassium-iron trisulfate. Additionally, the tube location in the boiler was found to be very important, and in some areas this variable overshadowed the effect of temperature and alloy content.

Under the USC conditions, maximum steam temperatures will approach 760°C (1400°F), which means that metal temperatures will approach 815°C (1500°F) or higher. As a result of these metal temperatures, high-strength, high-nickel alloys and advanced austenitic stainless steels have initially been selected for the superheater and reheater sections of the boiler. An insidious form of corrosion that could be problematic in the high-strength Ni base super-alloys that will be necessary in USC boilers is known as Type II hot corrosion. A temperature of 815°C (1500°F) is high enough that the ashes, even with the fuel and low-NOx burner conditions discussed above, should not result in the classic coal ash corrosion. Rather, the corrosion mechanisms that will be operative are oxidation and sulfidation from the gas and solid deposits. In much of the previous work cited, fireside corrosion data were measured that were generally in the temperature range of 620°C to 727°C (1148°F to 1340°F). More importantly, no field data were obtained on the high-nickel alloys that have been selected for the high-temperature tubular components of the USC plant. Thus, the absence of fireside corrosion data for many alloys above temperatures of 727°C (1340°F) where Type II hot corrosion could be a problem, and particularly for the high-nickel alloys over the full temperature range expected for the USC plant requires additional laboratory and field-testing.

The waterwalls located in the furnace also have fireside exposure, but the operating conditions, deposits and resulting corrosion mechanisms differ from those in the superheater and reheater. In the area of the waterwalls, the introduction of new combustion systems to reduce NOx emissions has led to severe waterwall corrosion for some fuels, especially eastern bituminous coals. The active fireside corrosion mechanism is expected to be sulfidation from a sub-stoichiometric gas with H_2S and deposits containing carbon and iron sulfide. The higher waterwall temperatures of the USC plant will significantly increase corrosion rates vis-à-vis existing plants. Future, more stringent NOx emission regulations will further increase the risk of waterwall corrosion. The utility industry is studying and implementing combustion system modifications to reduce waterwall corrosion, but it is likely that waterwall corrosion will persist in core areas. This will require protective coatings or claddings applied by various means, such as weld overlaying, laser cladding, and various forms of thermal spraying. In response to present waterwall corrosion problems, various promising coating compositions and application techniques are under development, but need thorough evaluation and possibly some more development.

Fireside corrosion evaluation involves a three-task approach. First, laboratory testing is necessary to screen the different alloys in controlled environments where the different variables of alloy content, temperature, fuel/ash and sulfur in the flue gas can be evaluated. Second, the evaluation of the best-performing alloys in the laboratory tests can be made on air-cooled retractable corrosion probes inserted in the superheater and reheater areas of actual operating boilers burning distinctly different fuels. Third is a test of the best-performing alloys under pressure in actual boiler operating conditions, e.g., steam loop, in-line probes. Since the temperature at any given location in an operating boiler is usually not isothermal, because of unit cycling, multiple test locations must be utilized to cover multiple temperatures and different deposit formations. This third task requires some boiler modification because none of the existing boilers are operating at the advanced USC conditions. Some limited higher temperatures can be obtained by the use of flow-restricting orifices in the tubes, but the highest temperatures will require a separate steam circuit. That steam circuit will be at the needed higher temperature but not at a higher pressure than the operating boiler. In the waterwalls the outer diameter metal temperature can be increased by the use of a greater wall thickness.

The objectives of the present fireside corrosion resistance program are as follows:

- To determine the fireside corrosion and Type II hot corrosion resistance of the alloys which have the strength for USC tubing applications, and to select the optimum tubing alloy or coating for superheaters and reheater tubes.

- To determine the fireside corrosion resistance of alloys, overlays and coatings which have the strength for USC waterwall applications and to select the optimum alloy and protection system for the waterwalls.

Field Testing

During the laboratory-testing phase of the fireside corrosion resistance task, experiments were performed on a variety of developmental and commercial alloys and coatings by exposing them to simulated USC superheater and reheater environments (6, 7, 8). While these laboratory tests are a valuable screening tool for down-selecting candidate materials for use in advanced USC power plant applications, there are a variety of variables that cannot be addressed in laboratory tests. Field exposures allow for the evaluation of a number of important environmental parameters that cannot be fully simulated in the laboratory tests. With regard to these parameters:

- The actual composition of the deposits formed on the tubes is more complex than the composition of the simulated ash in laboratory tests.
- The SO3 concentration formed by heterogeneous reaction on cooled surfaces is variable.
- The temperature gradients that occur within the ash deposits are very large.
- The ash and flue gas move past tubes at high velocity; the rate varies with design.
- The composition of the corrosive deposits changes over time.
- The temperatures are not constant.
- The effect of fly ash erosion results in the removal of the protective oxides from the tube and replenishes the metal surface with fresh corrosive, molten sulfate ash.

The alloys and coating materials that indicated the most promising results based upon laboratory testing were selected for field-testing. Some of the materials that were tested in the laboratory were not available in tubular form, and/or in sizes compatible with the probe design. Therefore, the materials selected for testing in the field were a subset of those tested in the laboratory. Table 1 lists the materials that were used in the present field tests. The weld overlays and laser claddings were applied to 230 tubing material. The diffusion coatings were applied to Super 304H tubing material.

Table 1: Materials and Coatings Selected for Field Testing

Wrought	Weld Overlays	Diffusion Coated	Laser Clad
Super 304H	622	FeCr	50/50
347HFG	52	SiCr	
800HT	72		
617	33		
230			
HR6W			
HR3C			
740			

Three utility boilers were selected for use as host sites for installation of the field corrosion probes. Two air-cooled, retractable, corrosion probes have been installed at each utility plant. Each of these boilers was equipped with low-NOx burners and one each was expected to be burning Eastern, Mid-West, and Western coal for the duration of the testing. Promising alloys from the laboratory testing were assembled on air-cooled retractable corrosion probes for testing in the superheater or reheater areas of these three utility boilers. At each site two corrosion probes were intended to be exposed for one and two years, respectively, at each site. The corrosion probes included material samples intended to be exposed at controlled metal temperatures of 650°C, 704°C, 760°C, 815°C, and 871°C, (1200°F, 1300°F, 1400°F, 1500°F, and 1600°F), representing the range of temperatures expected in the superheater and reheater components of the USC plant.

The layout of the material specimens on each probe was based upon knowledge of the fuels burned at the host site, combined with the results of the laboratory testing. For example, the first of the three installations is burning a high-sulfur Mid-western coal. The specimen layout for this site is shown in Table 2.

Table 2: Specimen Layout for Mid-western Utility Site Showing Target Temperatures

Time	Zone 5 - 1200F	Zone 4 - 1300F	Zone 3 - 1400F	Zone 2 - 1500F	Zone 1 - 1600F
2 YR	HR6W	740	50/50 LC	740	50/50 LC
	HR3C	617	230	HR3C	HR6W
	S304H	230	740	72WO/230	52WO/230
	33WO/230	50/50 LC	SiCr/S304H	52WO/230	HR3C
	622WO/230	HR6W	33WO/230	HR6W	72WO/230
1 YR	800HT	HR3C	HR6W	740	72WO/230
	740	S304H	230	800HT	740
	617	800HT	740	50/50 LC	52WO/230
	347HFG	622WO/230	52WO/230	SiCr/S304H	HR6W
	230	347HFG	800HT	230	50/50 LC

The selected materials were machined into uniform tubular specimens, and assembled into probes, each containing 25 specimens. Figure 1 shows one of the corrosion probes, in the extended position, mounted to the retraction mechanism, currently installed at a Mid-western utility site. The probes are instrumented with thermocouples to monitor and control the zone temperatures. Figure 2 shows a close-up view of the same probe, in the retracted position, in which the sample segments are visible. The completed probe extends approximately five feet into the furnace.

The first set of corrosion probes was successfully installed and tested at a Mid-western host utility site in January 2006. Installations at the Western and Eastern utility sites followed thereafter. The corrosion probe system was designed to be independent from the main boiler, to be installed and removed without a boiler outage, and to be fail-safe, retracting the probe in the event of a malfunction. With these safety features, years of testing would not be compromised by a sudden overheating event. Each assembled and instrumented probe is attached to a retraction mechanism that allows automatic removal of the probe in the event of a malfunction. A blower supplies cooling air to maintain the temperatures of each of the five zones within each probe. A control system monitors selected thermocouples for each temperature zone, and modulates the airflow to maintain the desired average metal temperature for each zone. The probes will retract automatically upon failure of the cooling air supply system, loss of data signal, power failure, computer failure, and thermocouple failure. Probe status and temperature data are being continuously monitored and collected remotely. Periodic on-site inspections are performed to remove the probes and visually assess the deposits, corrosion rates, and wall thickness loss of the specimens.

Figure 1: Corrosion Probe Installed at Mid-western Utility

Figure 2: Close-up View Corrosion Probe Installed at Mid-western Utility

The field corrosion probe testing is an ongoing part of this program. At the end of the testing period, post-exposure metallurgical evaluation will be performed on each of the probes. All of the samples from each probe will be disassembled and will be evaluated for thickness loss and sub-surface penetration of the corrosive species. The samples and deposits will be evaluated to determine what corrosion mechanisms are operative at each test condition.

Mid-Western Site Exposure Results

The first probe to complete the field exposure was the one-year corrosion probe at the Mid-Western site. This probe was removed for metallurgical evaluation during April 2007, with approximately 5,200 hours of exposure time to boiler operating temperatures. The initial metallurgical evaluation focused on Zone 2, which had a target temperature of 816°C (1500°F). The median temperature for the exposure time was approximately 732°C (1350°F), with short-term exposures to temperatures as high as 927°C (1700°F). The external surface was mostly covered with relatively thin, hard, dark-colored ash deposits that were non-uniformly distributed across the length of the probe and around the circumference, as seen in Figure 3. The heaviest deposit buildup was generally located between the 3 and 5 o'clock positions (as viewed from the probe tip) where the flue gas was impinging on the probe surface. In several locations along the probe length, however, entire coupon specimens were free of ash deposits, and covered with only thin, black and red oxide layers. The specimens located near the back (coolest) portion of the probe were also covered with streaks of light-colored residue.

Following post exposure evaluation and documentation, the probe was sectioned into five zones, each representing the different temperature regimes that the probe specimens were expected to maintain during testing. Ring sections for the five different materials residing within each zone were cut from the coupons samples and lightly grit blasted for macroscopic and physical evaluation of wall loss, as shown in Figure 4. Visual examination of the samples removed from zones 1 - 4 revealed little to no evidence of significant wall loss for any of the proposed materials. Scattered pits were observed on a few of the sections removed from zones 3 and 4, but the extent of wall loss was relatively shallow. The wall thickness loss on the Zone 2 material samples ranged from 0.4% to 1.2% of the original wall thickness. Photomicrographs of the typical subsurface penetration noted in the samples are presented in Figure 5. Further evaluation and quantification of total metal wastage through wall thickness measurements and metallographic examination of subsurface penetration is required before any definitive statements can be made concerning the condition of the probe samples, and relative ranking of candidate material alloys.

Western Site Exposure Results

The second probe to complete the field exposure was the one-year corrosion probe at the Western site. This probe was removed for metallurgical evaluation during July 2007, with approximately 5,800 hours of exposure time to boiler operating temperatures. The initial metallurgical evaluation focused on Zone 2, which had a target temperature of 816°C (1500°F). This zone exhibited surface attack that was representative for the probe materials exposed at this site. The median temperature for the exposure time was approximately 749°C (1380°F), with short-term exposures to temperatures as high as 893°C (1640°F). The external surface was generally covered with relatively thick, nodular, yellow and orange deposits (and/or corrosion products) that overlaid a dense, black oxide layer, as seen in Figure 6. Locally heavier deposition was noted at the 1 and 5 o'clock positions (as viewed from the probe tip) along the periphery of where the flue gas was impinging on the probe surface. In several locations along the probe length, however, entire coupon specimens were free of ash deposits, and were only covered with a thin, brownish-black oxide layer.

Following post exposure evaluation and documentation, the probe was sectioned into five zones, each representing the different temperature regimes that the probe specimens were expected to maintain during testing. Ring sections for the five different materials residing within each zone were cut from the coupons samples and lightly grit blasted for macroscopic and physical evaluation of wall loss, as shown in Figure 7. Visual examination of the samples removed from zones 1 - 5 revealed some general wall loss and scattered pitting on many of the materials. The average wall thickness loss on the Zone 2 material samples ranged from 0.6% to 2.0% of the original wall thickness; however localized pitting of up to 10% of original wall thickness was noted. The microscopic morphology of subsurface attack in the samples that exhibited notable wastage is illustrated in Figure 8. This preliminary data after one-year of exposure suggests tha the wall loss under Western coal samples was significantly greater than for the Mid-Western coal conditions. This is clearly a departure from the results obtained in the laboratory testing, in which the Western coal samples showed the least wall loss. Further evaluation and quantification of total metal wastage through wall thickness measurements and metallographic examination of subsurface penetration is required before any definitive statements can be made concerning the condition of the probe samples, and relative ranking of candidate material alloys.

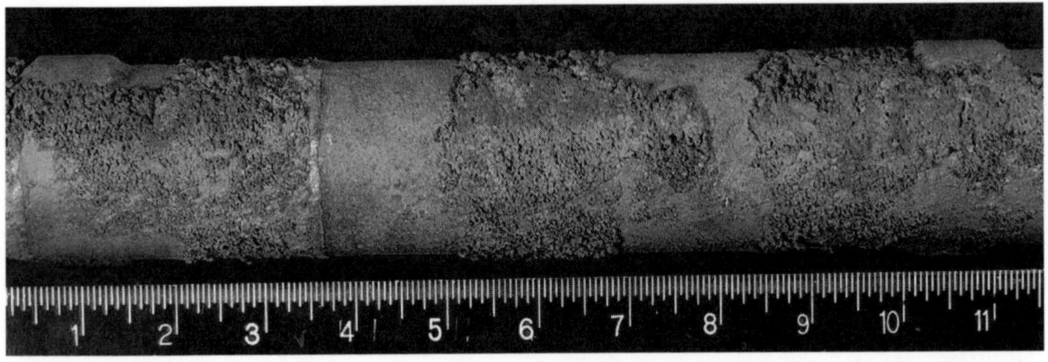

Figure 3: One-year Corrosion Probe Zone 2 - Mid-western Utility Post Exposure (Before Cleaning)

Figure 4: One-year Corrosion Probe Rings from Zone 2 - Mid-western Utility – Showing Post Exposure - Before Cleaning (Top Left) and After Cleaning (Top Right, and Bottom)

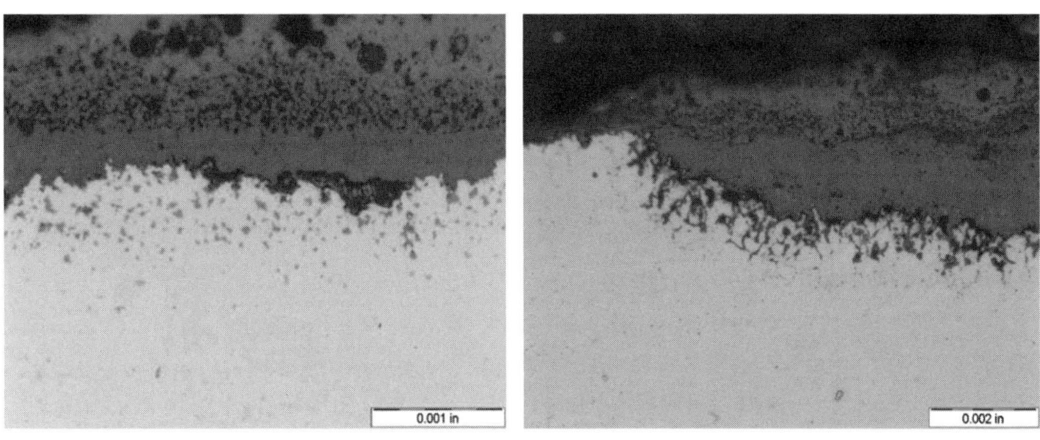

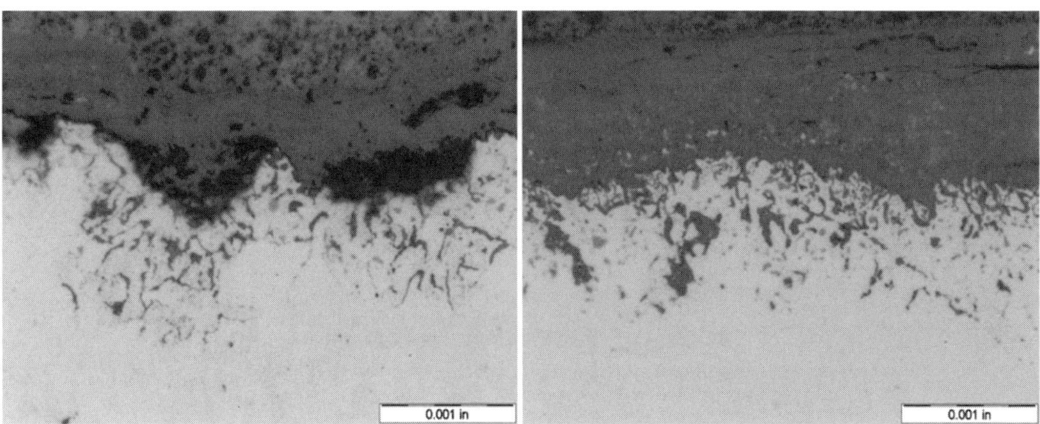

Figure 5: Photomicrographs taken from 1-Year Mid-Western Probe

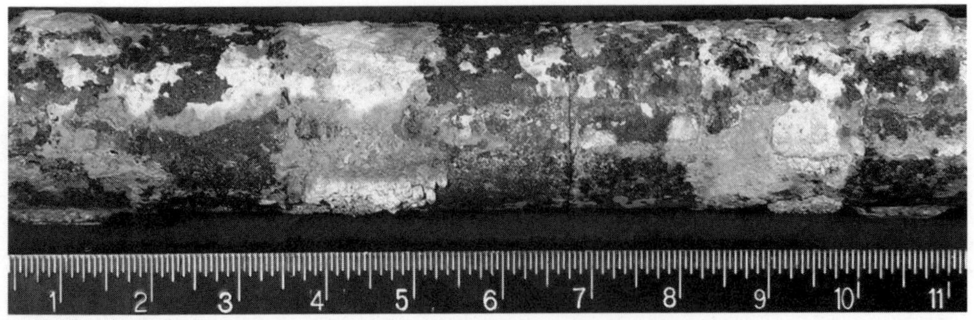

Figure 6: One-year Corrosion Probe Zone 2 - Western Utility Post Exposure (Before Cleaning)

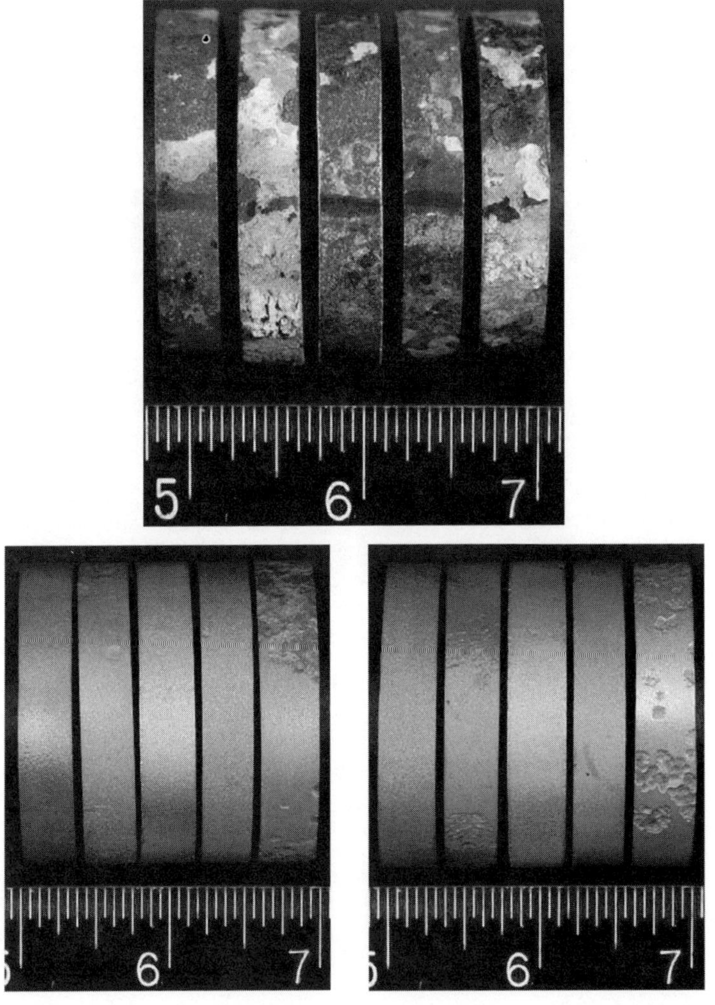

Figure 7: One-year Corrosion Probe Rings from Zone 2 - Western Utility – Showing Post Exposure - Before Cleaning at 12 O'clock Position (Top) and After Cleaning at 12 O'clock (Bottom Left), and 4 O'clock Position (Bottom Right)

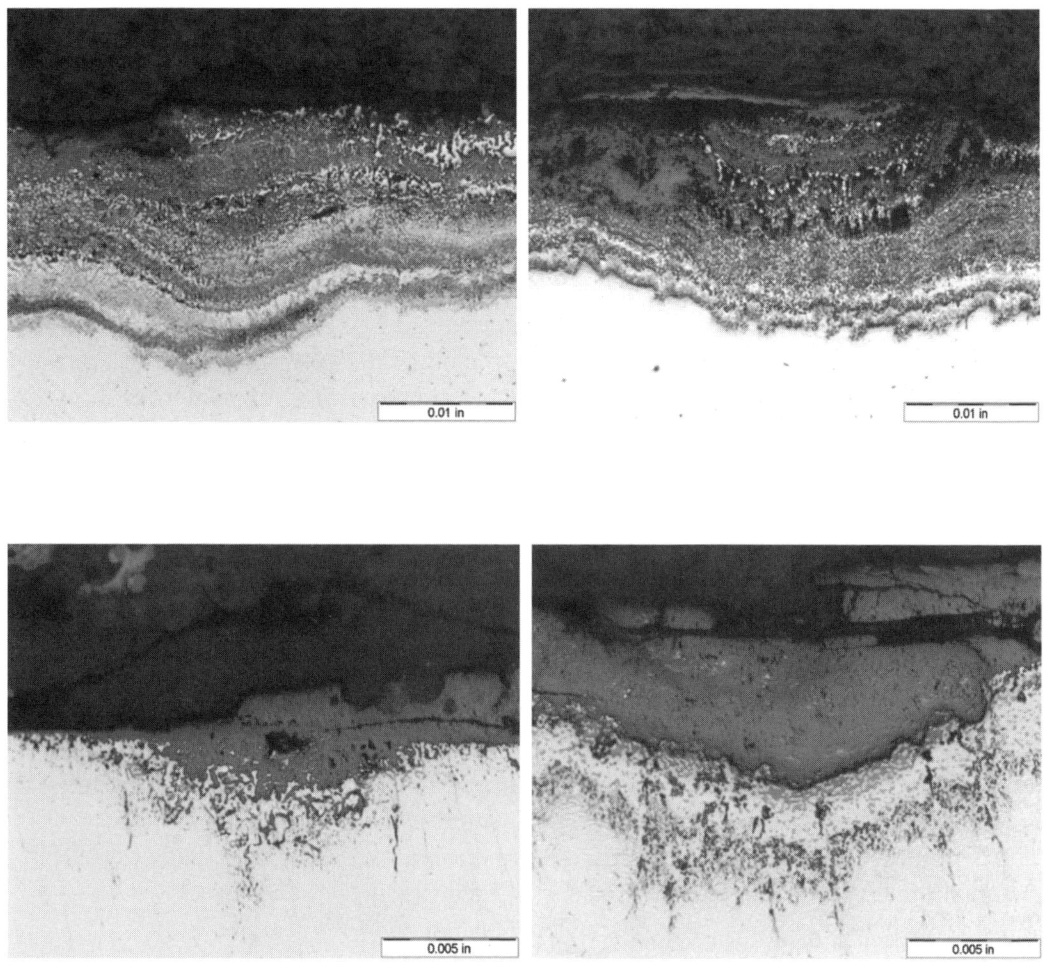

Figure 8: Photomicrographs taken from 1-Year Western Probe

Summary and Conclusions

The status of current work in experimental evaluation of fireside corrosion resistance of candidate materials for use in USC boilers was presented. Corrosion resistance was previously evaluated in laboratory tests under synthesized coal-ash and flue gas conditions typical of three North American coals, representing Eastern (mid-sulfur bituminous), Mid-western (high-sulfur bituminous), and Western (low-sulfur sub-bituminous) coal types. A status report on the ongoing field corrosion probe test phase of this fireside corrosion test program has been presented. Promising materials from the laboratory tests were assembled on probes, and are being exposed in three utility boilers for one and two-year durations. These air-cooled, retractable corrosion probes were designed to maintain metal temperatures using multiple zones, ranging from 650°C (1200°F) to 870°C (1600°F).

Visual examination of the samples removed from the Mid-Western utility site indicated minimal evidence of significant wall loss for any of the tested materials, and the extent of wall loss, especially in Zone 2, was relatively shallow. Visual examination of the samples removed from the Western utility site indicated evidence of wall loss for some of the tested materials, and the extent of wall loss, as presented for Zone 2, was more severe when compared to the Mid-Western samples. Further evaluation and quantification of total metal wastage through wall thickness measurements and metallographic examination of subsurface penetration is required before any definitive statements can be made concerning the condition of the probe samples.

Acknowledgments

The authors wish to thank Dr. Robert Romanosky (NETL), Robert Purgert (EIO), and Dr. R. Viswanathan (EPRI) for their support and guidance during this project.

This work was prepared with the support of the U.S. Department of Energy, under Award No. DE-FG26-01NT41175 and the Ohio Coal Development Office/Ohio Department of Development (OCDO/ODOD) under Grant Agreement Number CDO/D-00-20. However, any opinions, findings, conclusions, or recommendations expressed herein are those of the authors and do not necessarily reflect the views of the DOE and/or the OCDO/ODOD".

LEGAL NOTICE/DISCLAIMER

This report was prepared by Foster Wheeler North America Corp. pursuant to a Grant partially funded by the U.S. Department of Energy (DOE) under Instrument Number DE-FG26-01NT41175 and the Ohio Coal Development Office/Ohio Department of

Development (OCDO/ODOD) under Grant Agreement Number CDO/D-00-20. NO WARRANTY OR REPRESENTATION, EXPRESS OR IMPLIED, IS MADE WITH RESPECT TO THE ACCURACY, COMPLETENESS, AND/OR USEFULNESS OF INFORMATION CONTAINED IN THIS REPORT. FURTHER, NO WARRANTY OR REPRESENTATION, EXPRESS OR IMPLIED, IS MADE THAT THE USE OF ANY INFORMATION, APPARATUS, METHOD, OR PROCESS DISCLOSED IN THIS REPORT WILL NOT INFRINGE UPON PRIVATELY OWNED RIGHTS. FINALLY, NO LIABILITY IS ASSUMED WITH RESPECT TO THE USE OF, OR FOR DAMAGES RESULTING FROM THE USE OF, ANY INFORMATION, APPARATUS, METHOD OR PROCESS DISCLOSED IN THIS REPORT.

Reference herein to any specific commercial product, process, or service by trade name, trademark, manufacturer, or otherwise, does not necessarily constitute or imply its endorsement, recommendation, or favoring by the Department of Energy and/or the State of Ohio; nor do the views and opinions of authors expressed herein necessarily state or reflect those of said governmental entities.

References

1. R. Viswanathan, et al, "U.S. Program on Materials Technology for USC Power Plants", Proceedings from the Fourth International Conference on Advances in Materials Technology for Fossil Power Plants, October 2004.

2. J.L. Blough, G. J. Stanko, M. T. Krawchuk, "Superheater Corrosion in Ultra-Supercritical Power Plants, Long-Term Field Exposure at TVA's Gallatin Station," Palo Alto, CA: Electric Power Research Institute, February 1999. TR-111239.

3. J. L. Blough, M. Krawchuk, G. J. Stanko, W. Wolowodiuk, "Superheater Corrosion: Field Test Results," Palo Alto, CA: Electric Power Research Institute, November 1993. TR-103438.

4. W. Wolowodiuk, S. Kihara, and K. Nakagawa, "Laboratory Coal Ash Corrosion Tests," Palo Alto, CA: Electric Power Research Institute, July 1989. GS-6449.

5. J. L. Blough, W. W. Seitz, A. Girshik, "Fireside Corrosion Testing of Candidate Superheater Tube Alloys, Coatings, and Claddings - Phase II Field testing," Oak Ridge, TN: Oak Ridge National Laboratory, June 1998. ORNL/Sub/93-SM401/02.

6. H. Hack, G. S. Stanko, "Fireside Corrosion Resistance of Advanced Materials for Ultra-Supercritical Coal-Fired Plants," 22nd Annual International Pittsburgh Coal Conference, September 2005.

7. H. Hack, G. S. Stanko, "Update on Fireside Corrosion Resistance of Advanced Materials for Ultra-Supercritical Coal-Fired Power Plants," The 31st International Technical Conference on Coal Utilization & Fuel Systems, May 21-25, 2006, Clearwater, Florida, USA.

8. H. Hack, G. S. Stanko, "Experimental Results for Fireside Corrosion Resistance of Advanced Materials in Ultra-Supercritical Coal-Fired Power Plants", The 32nd International Technical Conference on Coal Utilization & Fuel Systems, June 10 - 15, 2007, Clearwater, Florida, USA

Advances in Materials Technology for Fossil Power Plants
Proceedings from the Fifth International Conference
R. Viswanathan, D. Gandy, K. Coleman, editors, p 507-519

DOI: 10.1361/cp2007epri0507

IMPACT OF STEAM-SIDE OXIDATION ON BOILER HEAT EXCHANGER TUBES DESIGN

P. Billard
Electricite de France, R&D
77818 Moret-surLoing, France

R. Fillon
Electricite de France, Basic Design Dpt
69628 Villeubanne, France

J. Gabrel
V&M Research Center
59620 Aulnoye-Aymeries, France

B. Vandenberghe
V&M Tubes
59880 St Saulve, France

Abstract

In fossil fired boilers, the thermal energy generated by combustion is transferred to the fluid *via* exchanger tubes. On the internal side of these tubes, an oxide layer grows and, due to its low thermal conductivity, the oxide causes a rise of the metal temperature. The phenomenon is self-activated and dependant on the thermodynamic characteristics of the boiler. The increase of the metal temperature leads to a significant reduction of the creep lifetime of the component.

The Maximum Allowable Stresses are given by the codes in accordance with the mechanical properties, mainly creep behavior for the highest temperature components. Even if the oxidation phenomenon is largely described in the literature, this is taken into account in a global way in the construction codes for boilers through the selection of the "design temperature". This approach is no longer suitable for the high performance boilers planned today, with more and more severe steam parameters.

This paper highlights the need to integrate the oxidation behavior in the design of such advanced boiler heat exchanging components.

- The impact of steam oxidation on tube service life is described: Growth of an oxide layer, metal consumption, increase of tube temperature and reduction of creep rupture time. The importance of parameters such as thermal flux is illustrated by examples.
- The behaviors of two 9-12Cr% materials are analyzed: Comparison of creep life of i) grade 92 steel which exhibits good creep properties and ii) VM12 steel which offers good oxidation resistance.

- An expression of the "design temperature" considering performance index in term of oxidation resistance is formulated, in relation with the thermal characteristics of the exchanger.

Finally, the paper concludes on the need for continuing activities for a better knowledge of oxidation laws, thermal properties and the conditions of the exfoliation of oxide layers.

Exchange tubes in a steam generator

Heat exchanger tubes are loaded in different ways:

- mechanically or thermomechanically (pressure, weight, expansion, etc.);
- metallurgically (change in microstructure);
- chemically, with gas-side corrosion and steam-side oxidation.

While the first issue is clearly taken into account by the tube designers in accordance with the construction codes, the same does not apply to those phenomena related to structural changes or the tube environment. This is particularly the case of internal oxidation in the steam phase.
The oxidation of the tubes is a result of interaction between the metal and the steam; it causes a layer of oxide to form at the surface of the metal, resulting in:

- a metal protection more or less efficient;
- metal consumption;
- building of a heat insulating film which causes the tube temperature to increase.

Oxidation of the tubes at the "steam side" not only leads to increased stress (reduced cross-section), but causes a layer with low heat conductivity to form which, by increasing the temperature of the tube, will reduce its creep resistance.

We therefore propose to assess the effect of internal oxidation on the design of the tubes and choice of materials, particularly in the case of 9-12% Cr ferritic steels.

Oxidation mechanisms

For steels containing chromium, the oxide structure formed in a steam environment is made up of layers that will differ according to the type of steel and the temperature [1, 2]. Whatever their formation mechanism and composition, we only consider those oxides formed:

- have a two-layer structure. The inner film at the metal side is dense and protective, the outer layer is porous and greatly reduces the heat conductivity of the tube,
- have a molar volume greater than the volume of the material they replace; the layer formed has a thickness multiplied by the density ratio which for iron and steel is around 2.1.

Growth kinetics

The law according to which the layer grows is parabolic. Experimentation and feedback from experience show that the general formula is as follows:

$$e_{ox} = kt^{1/n} \tag{1}$$

where,			
	e_{ox}:	Thickness of metal lost by oxidation	(mm)
	k:	Oxidation constant	($mm^2.h^{-1}$)
	t:	Exposure time	(h)

The value of n is between 2 and 3, which means that there are other phenomena involved in the oxide growth (and which slow it down). It can differ according to the steels tested and experimental conditions [3].

The oxidation constant k is related to the temperature as follows:

$$k = k_0 \exp(- E/RT) \tag{2}$$

where,			
	E:	Reaction activation energy	($J.mol^{-1}$)
	R:	Constant of perfect gases	($8.314\ J.mol^{-1}.K^{-1}$)
	T:	Reaction temperature	(K)
	k_0:	Constant dependent on the material	

It is however important to note that while there is a lot of data available for the materials used in the steam generators, there is a significant mismatch between the laboratory test results (simulated atmosphere) and the in-situ tests [4].

Exfoliation of oxide layers

Exfoliation is the removal of all or part of the oxide formed on the metal [5]. This results in:

- the release of solid particles of varying size into the steam loop;
- loss of the metal protection which will accelerate the oxidation process and thus further reduce the tube cross-section;
- a reduction in the thickness of the insulating layer of oxide, thus limiting the increase of the tube temperature, which has a beneficial effect on creep life.

Calculation of the "metal temperature"

Expression of the temperature difference

The exchanger tubes are subjected to heat flux through the wall thickness from the flue gas to the steam. The difference in temperature between the outer surface of the tube and the steam can be expressed by the following approached relation [6] (the impact of the outer corroded layer on the metal temperature is of little importance and has not been taken into account):

$$T_{surf} - T_{steam} = Q_{ext} \,.\, (r_3/r_1) \,.\, (r_1/\lambda_{metal} \,.\, \ln(r_3/r_2) + r_1/\lambda_{oxide} \,.\, \ln(r_2/r_1) + 1/\alpha_{steam}) \quad (3)$$

Where (see figure 1 for the representation of the geometrical values of the oxidized tube),

r_1:	Inside radius/oxide;	$r_1 = r_2 - 2{,}1.x$	(m)
r_2:	Inside radius/tube;	$r_2 = r_{20} + x$	(m)
r_3:	Outside radius/tube;	$r_3 = r_{30} - ct$	(m)
Q_{ext}:	Heat flux at the outer surface		$(W.m^{-2})$
T_{steam}:	Steam temperature		(°C)
T_{surf}:	Temperature of the outer metal surface		(°C)
α_{steam}:	Heat transfer coefficient at steam side		$(W.m^{-2}.K^{-1})$
λ_{metal}:	Heat conductivity of the metal		$(W.m^{-1}.K^{-1})$
λ_{oxide}:	Heat conductivity of the oxide		$(W.m^{-1}.K^{-1})$

Heat flux

The heat flux in a steam generator is maximal at the evaporator (a flux of 300 $kW.m^{-2}$ can be encountered). Superheaters and reheaters are exposed to a lower heat flux (approximate values between 5 and 60 $kW.m^{-2}$).

Temperature profile

Relation (3) expresses the difference in temperature between the steam and the outer surface of the tube; figure 2 shows an illustration of the radial temperature profile.

The concept of cumulated lifetime fractions

A simplified consideration of the creep behavior of a material (Robinson's law) results in:

$$\Sigma\, t_i/t_r = 1 \quad (4)$$

Where, $\mathbf{t_i}$ = lapsed time at stress $\boldsymbol{\sigma_i}$ and temperature $\mathbf{T_i}$

$\mathbf{t_r}$ = time at rupture for stress $\boldsymbol{\sigma_i}$ and at temperature $\mathbf{T_i}$

This law consists of formulating the lifetime consumption as cumulated fractions corresponding to the times spent with each stress-temperature pair. Rupture occurs when 100% of lifetime has been consumed.

While this law only partly translates the actual behavior of the material, it has the advantage of allowing incremented stresses to be taken into account (gradual increase in stress and temperature in the calculation of tube life that concerns us...).

Calculation of the impact of oxidation on tube life

The approach: The calculation involves estimating for each unit of time, the stress-temperature pairs applied to the component.

- The term “stress” is calculated from the internal steam pressure and tube dimensions which change according to metal consumption both inside and outside.
- The term “temperature” is deduced i) from the steam temperature and ii) from the ΔT resulting from the gradient inside the tube.

Examples of calculations

The impact of oxidation on the creep life of an exchanger tube was calculated for the following initial conditions:

- Material: T92 (martensitic steel with 9% Cr and 2% W) [7]
- Steam temperature: 580°C (1076°F)
- Steam pressure: 300 bar (4 350 psi)
- Outside diameter of tube: OD = 38 mm
- Heat flux: 50 000 $W.m^{-2}$ (example of a low temperature superheater in a USC power plant)
- Heat transfer coefficient at steam side: 5 000 $W.m^{-2}.K^{-1}$
- Heat conductivity of the metal: 30 $W.m^{-1}.K^{-1}$
- Heat conductivity of the oxide: 1 $W.m^{-1}.K^{-1}$
- Rate of corrosion (outside): 0.005 $\mu m.h^{-1}$
- Test time: 100 000 hrs

Thickness calculation

The wall thickness is calculated using the following relation:

$$e = (P_{design} \times OD) / (2\sigma_{allowable} + P_{design}) \quad (5)$$

With: P_{design}, design pressure (steam pressure + safety ΔP)

When OD = 38 mm, the thickness obtained is 7.8 mm.

Results of the calculation of oxidation effects

Under the conditions indicated above,

- the tube has consumed 29% of its lifetime after 100 000 hrs, while only 5% would be consumed if there was no oxidation or corrosion (figure 3);
- the thickness of the internal oxide layer is 370 μm (corresponding to a loss of metal of 180 μm);
- the circumferential stress in the tube changes over time from 58 to 63 MPa while the temperature of the metal (at mid-thickness of the tube) changes from 605 to 635°C (1120 to 1175°F).

The temperature profile after 100 000 hrs is shown in figure 4.

A test lasting over 200 000 hrs under the same conditions entails rupture after around 170 000 service hours (figure 5) and shows that the degradation accelerates over time.

Impact of the boiler environment

Table I presents the results of the calculations corresponding to various scenarios. The results show the great impact of heat flux and heat transfer coefficient on the temperature of the metal and, consequently, on the reduction in lifetime.

Table I – calculation of the impact of oxidation – impact of parameters

Conditions	Rupture?	Oxide thickness (μm)	Total ΔT K (°F)	ΔT interf. K (°F)	ΔT oxide K (°F)	σ max (MPa)	t_{rupt} without ox. nor cor. (h)
Reference *	Yes (168 000 h)	560	70 (158)	17 (63)	46 (115)	67.7	1 830 000
* but $T_{steam} = 590°C$	Yes (115 000 h)	640	76 (169)	17 (63)	53 (127)	68.1	920 000
* but $Q_{out.} = 30\ 000\ W.m^{-2}$	No (cons. 24%)	470	38 (100)	10.2 (50)	23.3 (74)	67.1	3 690 000
* but $Q_{out.} = 65\ 000\ W.m^{-2}$	Yes (98 400 h)	657	101 (214)	22 (72)	70 (158)	68.2	1 090 000
* but $\alpha_{in.} = 2\ 000\ W.m^{-2}.K^{-1}$	Yes (60 800 h)	780	113 (235)	42 (108)	64 (147)	69.0	317 000
* but $\lambda_{oxide} = 0,7\ W.m^{-1}.K^{-1}$	Yes (112 600 h)	640	99 (210)	17 (63)	68 (154)	68.1	1 835 000
* but $Q_{out.} = 0\ W.m^{-2}$	No (cons. 3%)	370	0	0	0	66.5	1 835 000

* conditions mentioned in the text above

Figure 6 illustrates the "sliding" of the creep curve in case no.2 (steam T = 590°C – 1094°F): the blue curve corresponds to the creep of the steel when the generator was started up (metal temperature 615°C – 1140°F); the red curve corresponds to that of the steel at the metal temperature of 666°C – 1230°F at the time of rupture. The horizontal dot line shows the circumferential stress in the tube; the values calculated in the given example are 58.1 MPa – 8.43 ksi on start up and 68.1 MPa – 9.88 ksi at the time of rupture. It appears:

- that the change in stress in the tube, further to the reduction in cross-section, does not entail a great reduction in creep life;
- that the increase in temperature, combined with the impact of oxidation, causes a great reduction in lifetime;
- that rupture occurs after a time that takes into account the damage undergone at the various temperatures (cumulated lifetime fractions).

Impact of the material

The exercise illustrated in figure 5 corresponds to an environment with a high heat flux in which oxidation has a great effect on the temperature of the tube and therefore on its creep life. In this case, it would appear preferable to use a material with good oxidation resistance, even if its creep resistance is lower than that of Grade 92 used in the example.

The same exercise was carried out using VM12 steel (martensitic steel with 11.5% Cr, 1.5% Co and 1.5%W) and which is particularly oxidation resistant [8, 9]. Figures 7, 8 and 9 illustrate the results of the calculations:

- steel Grade 92 is penalized by its relatively poor oxidation resistance, the safety margin (1.5) used in the design not being sufficient to prevent rupture after 170 000 hours of service;
- VM12 steel has the advantage here of its good behavior in a steam environment. Taking the same tube thickness as for the T92 (low safety margin in this case for the VM12), there is no creep rupture after 200 000 hours of service. If we consider a wall thickness in accordance with the design codes (9 mm instead of 7.8 mm), less service life is of course consumed (under 70% after 200 000 hours);
- We verify that the oxide thickness and tube temperature are lower than in grade VM12 in the long term.

We can therefore see the advantage of using materials with good oxidation resistance in environments with high heat fluxes (in the case of water-walls).

Impact of exfoliation

The calculation results verify in this case that the consumption of metal is increased by oxidation that is periodically "reactivated". This results in increased stress. But the most important parameter is the temperature, which means that exfoliation has a favorable effect on service life (figure 10). Note that in reality, exfoliation only acts on part of the oxide layers and there will always be a residual "insulating layer" to a certain degree.

The exfoliation phenomenon is therefore very important when considering the behavior of tubes exposed to high heat flux. In the example in figure 7, rupture of the Grade 92 tube will be retarded if we consider that the oxide layer is subjected to exfoliation.

Taking oxidation into account in heat exchanger design

The strong impact of oxidation on the lifetime of heat transfer tube has been demonstrated. This effect is taken into account in a global way in the construction codes but this approach is not always suitable for the high performance boilers planned today, with more and more severe steam parameters. Manufacturers do however define "design temperatures" according to the type of component. Table II gives ΔT values given by EN code and examples taken from the design of two recent projects.

Table II – examples of "design" temperatures for the various components of a steam generator

Component	ΔT K (°F) EN 12952-3	Estimated Heat Flux ($kW.m^{-2}$)	ΔT / Project 1 K (°F)	ΔT / Project 2 K (°F)
Evaporator	50 (122)	~ 300	50 (122)	70 (158)
SH & RH / radiation	50 (122)	~ 60	50 (122)	50 (122)
SH & RH / convection	35 (95)	~20	35 (95)	50 (122)
Outlet header	15 (59)	~ 0	15 (59)	15 (59)
Main steam piping	15 (59)	~ 0	5 (41)	5 (41)

It is necessary to find a compromise between mechanical strength (long term creep resistance) and oxidation resistance according to the service conditions of the component, and particularly the heat flux to which it is exposed. We could thus define a "design temperature" that i) takes the application into account and ii) is in accordance with specific oxidation resistance of each type of material (e.g. difference of behavior between 9 and 12%Cr steels). It could be expressed as follows:

$$T_{design} = T_{steam} + \Delta T_{tube} + \Delta T_{oxide} + \text{margin} \quad (6)$$

Where T_{design} : temperature taken into account for the tube dimensions

T_{steam}: temperature of the steam inside the tube

$\Delta T_{tube} = f(Q, \alpha, \lambda_{metal})$,

$\Delta T_{oxide} = f(Q, k, \lambda_{oxide}, t)$

The current construction codes integrate ΔT_{tube} and a margin which covers operating variations and parameters scattering. Besides, this new proposal takes into account ΔT_{oxide} which allows integrating the true oxidation resistance of the material in accordance with the operating condition.

For the "header" application, T_{design} is close to T_{steam} and Grade 92 is favored owing to its good creep resistance. For a "superheater tube" application however, the value of T_{design} for the same steel would be higher than that determined for a grade with 12% Cr (oxidation resistant), thus presenting a "handicap" that would need to be analyzed in accordance with the allowable stresses at the various temperatures.

For the various materials, it is not only necessary to determine the creep characteristics, but also to make further investigations into their oxidation laws, the conductivity of their oxides and their exfoliation abilities.

Conclusion

The tubes used in advanced steam boilers are subjected internally to steam at very high pressure and temperature. The material therefore is gradually deformed by creep, in conjunction with oxidation of the metal in contact with the fluid heated to a high temperature. The oxide film thus formed provides an insulating layer which causes the temperature of the tube to increase under the effect of the heat flux and reduces the creep lifetime of the heat exchanger.

The parameters that have the greatest impact on service life have been discussed and evaluated; they concern the tube environment (mainly steam temperature and heat flux) and the intrinsic properties of the material (oxidation constant, heat conductivity of the oxide, exfoliation law).

The design codes for high temperature heat exchangers only take the creep properties of the materials into account. If the "design temperature" of the component nevertheless takes the gradual heating of the tubes into account, it would appear necessary to explicitly integrate oxidation into the design rules in order to prevent excessive creep and allow less precautions for those materials that offer greater resistance to the effects of the steam.

References

1 J. ZUREK, P.J. ENNIS: "Oxidation behaviour of ferritic and austenitic steels in simulated steam environments" ; 4th International Conference on Advances in Materials Technology for Fossil Power Plants - Hilton Head Island, October 2004.

2 S. OSGERBY, A. FRY: "Assessment of the steam oxidation behaviour of high temperature plant materials" ; 4th International Conference on Advances in Materials Technology for Fossil Power Plants - Hilton Head Island, October 2004.

3 R VISWANATHAN, J. SARVER, JM. TANZOSH, "Boiler materials for Ultrasupercritical Coal Power Plants - Steam side oxidation"; Journal of Material Enginering and Performance - Vol 15(3), June 2006.

4 A. FRY, S. OSGERBY, I.G. WRIGHT: « Oxidation of Alloys in Steam Environments – a Review »; Report of the 'National Physical Laboratory', Septembre 2002.

5 I.G. WRIGHT, M. SCHUTZE et al. : « Progress in Prediction and Control of Scale Exfoliation on Superheater and Reheater Alloys » ; EPRI Conference on Boiler Tube Failures and Inspections, November 2005.

6 N. HENRIKSEN, O. LARSEN, R. BLUM: « Evaluation of Superheater Tube Lifetime »; ESKOM Conference on "Process Water Treatment and Power Plant Chemistry", November 1997.

7 D. RICHARDOT, JC. VAILLANT, A. ARBAB, W. BENDICK, "The T92/P92 Book" - Vallourec & Mannesmann Tubes, 2000.

8 V. LEPINGLE, G. LOUIS, D. PETELOT, B. LEFEBVRE, B. VANDENBERGHE, « Long term exposure of new 12% Cr boiler steels in steam high temperature ». Eurocorr 2005 – Lisbon, 4-8 Sept. 2005.

9 J. GABREL, W. BENDICK, B. VANDENBERGHE, B. LEFEBVRE, Liège conference, "Status of the VM12 for tubular applications in advanced power plants", Liege, 18-20 Sept. 2006.

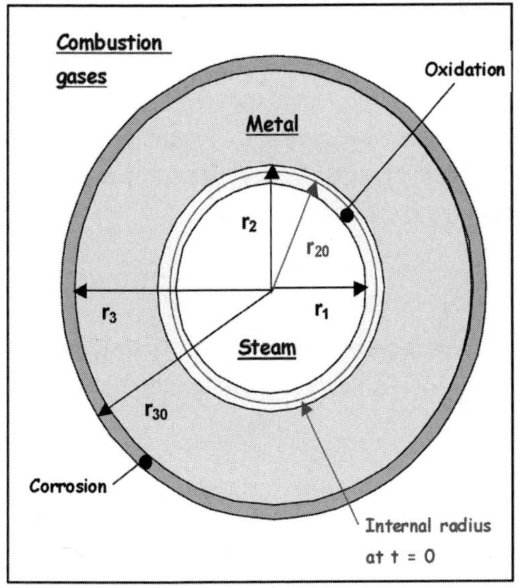

Figure 1. Definition of the geometrical values used in temperature calculations.

Figure 2. Temperatures profile in a tube subjected to heat flux from the outside.

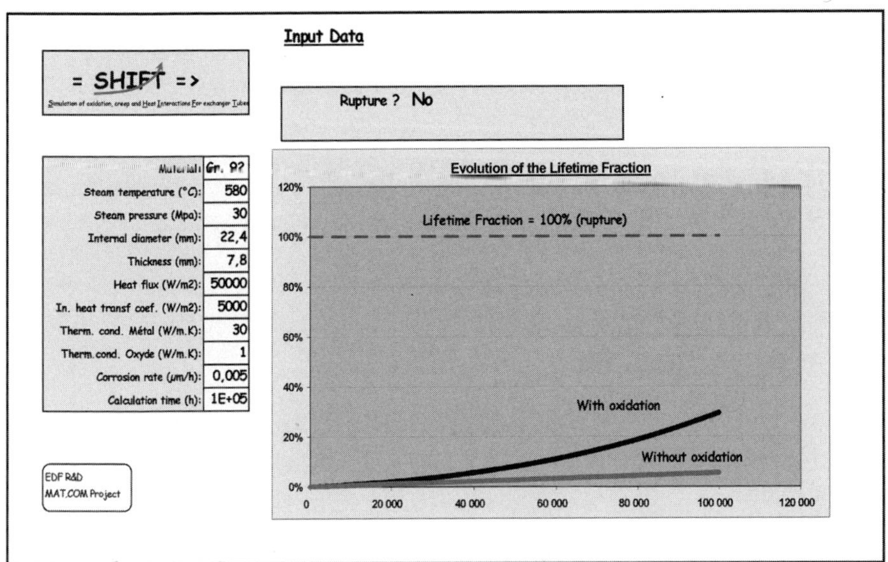

Figure 3. Calculation results of the impact of oxidation on lifetime.

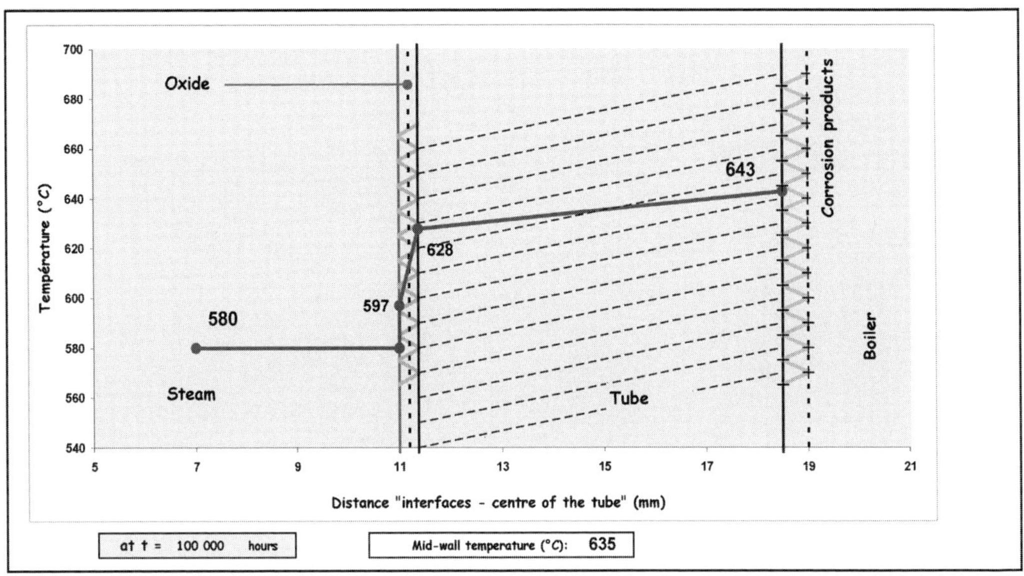

Figure 4. Temperature profile in the tube at t = 100 000 h.

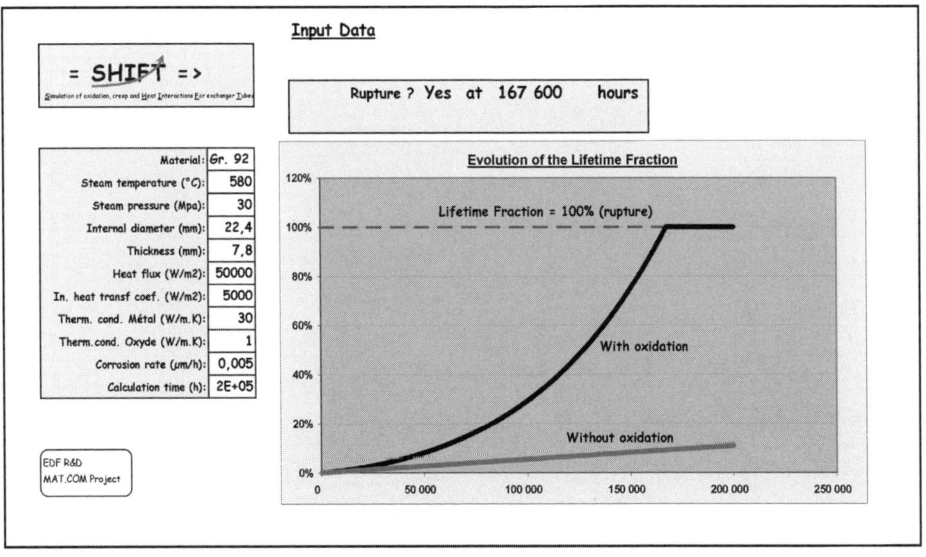

Figure 5. Calculation results of the impact of oxidation on lifetime – analysis over 200 000 hrs

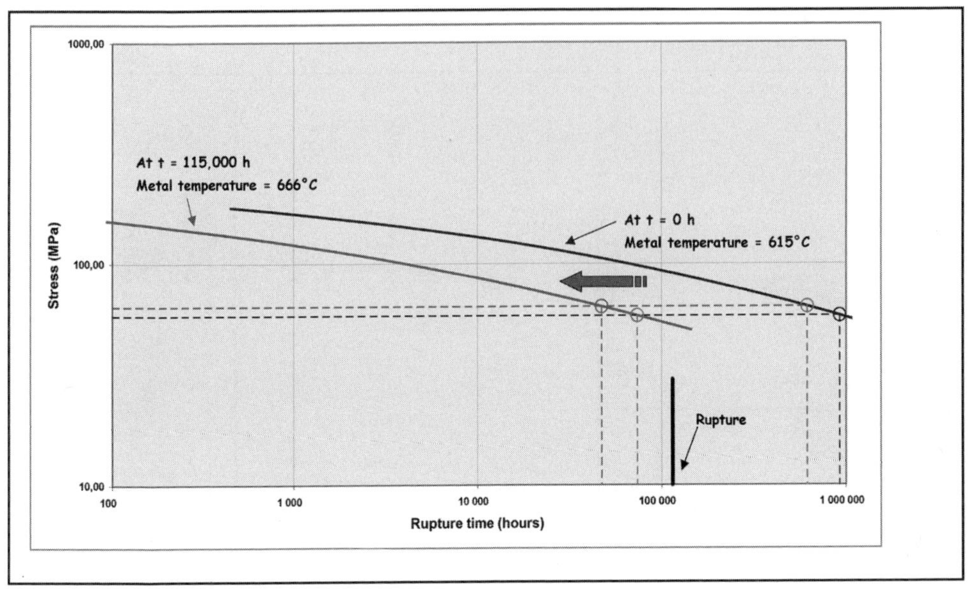

Figure 6. Shift of creep curve representative of the material after oxidation, corresponding to case no.°2 (steam T = 590°C).

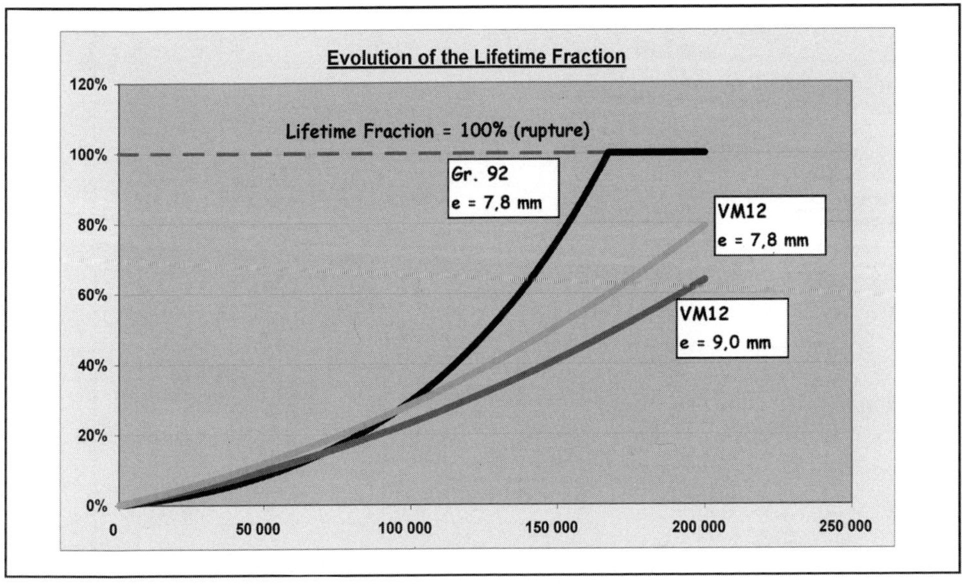

Figure 7. Impact of the material on lifetime consumption – exercise using steels Gr.92 and VM12 under the « reference » conditions of chart I.

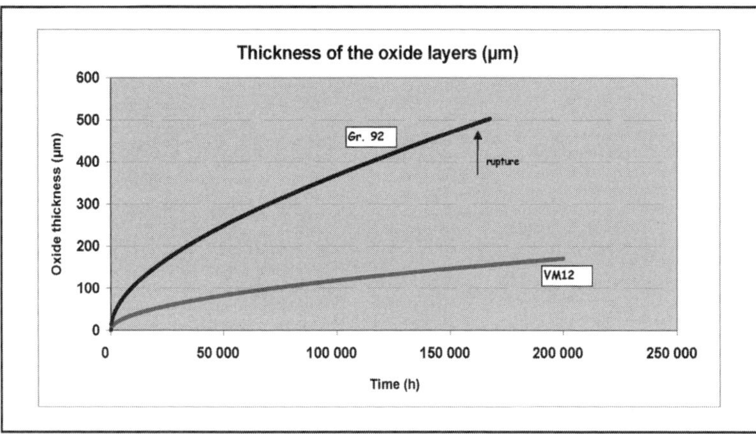

Figure 8. Exercise illustrated by figure 7 – oxide layer thicknesses for both materials.

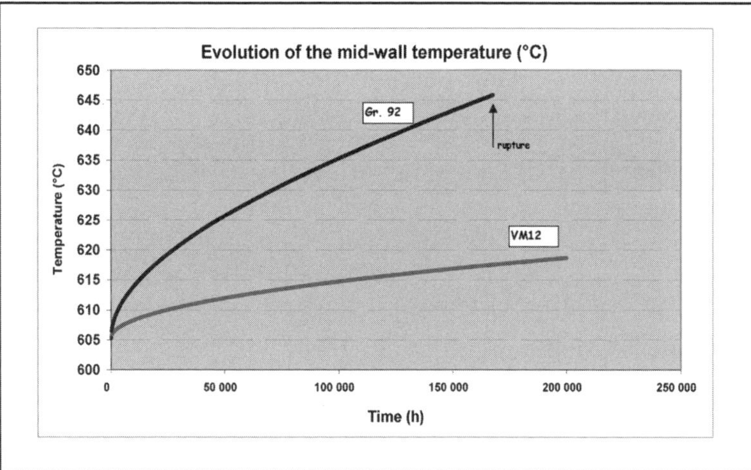

Figure 9. Exercise illustrated by figure 7 – evolution of tube temperature.

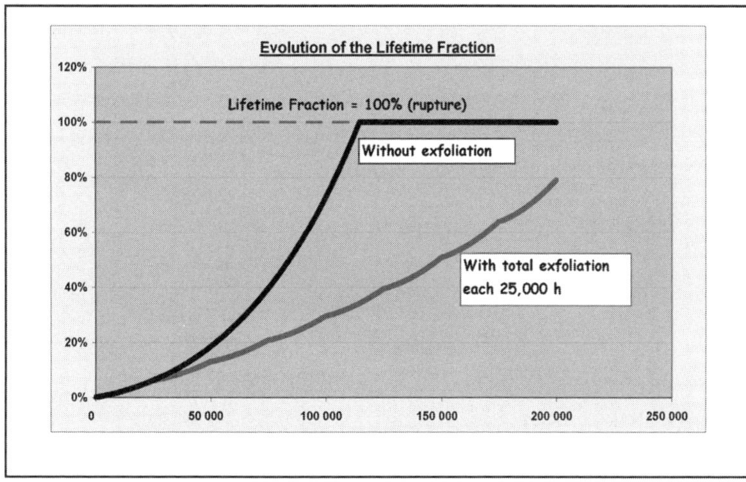

Figure 10. Impact of exfoliation on the lifetime of a tube subjected to heat flux.

Advances in Materials Technology for Fossil Power Plants
Proceedings from the Fifth International Conference
R. Viswanathan, D. Gandy, K. Coleman, editors, p 520-530

DOI: 10.1361/cp2007epri0520

Steamside Oxidation Behavior of Experimental 9%Cr Steels

Ö. N. Doğan
G. R. Holcomb
D. E. Alman
P. D. Jablonski
U.S. Department of Energy
National Energy Technology Laboratory,
Albany, Oregon

Abstract

Reducing emissions and increasing economic competitiveness require more efficient steam power plants that utilize fossil fuels. One of the major challenges in designing these plants is the availability of materials that can stand the supercritical and ultra-supercritical steam conditions at a competitive cost. There are several programs around the world developing new ferritic and austenitic steels for superheater and reheater tubes exposed to the advanced steam conditions. The new steels must possess properties better than current steels in terms of creep strength, steamside oxidation resistance, fireside corrosion resistance, and thermal fatigue resistance.

This paper introduces a series of experimental 9%Cr steels containing Cu, Co, and Ti. Stability of the phases in the new steels is discussed and compared to the phases in the commercially available materials. The steels were tested under both the dry and moist conditions at 650°C for their cyclical oxidation resistance. Results of oxidation tests are presented. Under the moist conditions, the experimental steels exhibited significantly less mass gain compared to the commercial P91 steel. Microstructural characterization of the scale revealed different oxide compositions.

Introduction

Considerable amount of research is being carried out to extent the application temperature range of 9Cr steels to 650°C [1]. Sufficient creep [2], steam side oxidation [3], and fireside corrosion [4] resistance are required at this temperature. Compositional modifications are considered to improve the creep strength by obtaining stable second phase particles and by improving the stability of the tempered martensitic matrix. These chemical compositional modifications may also have significant effects on the surface degradation mechanisms of these steels. In this paper, we will examine the effects of alloying elements on the high temperature oxidation of the 9Cr steels.

Experimental Procedure

The experimental steels were melted in a vacuum induction furnace and poured into cylindrical graphite molds (76 mm diameter). The ingots were fabricated to a 12 mm thick plate through hot forging and rolling. Chemical composition of the experimental steels is listed in Table 1. They primarily differed from each other for their Si and Mn content. Effect of Mn additions on the transformation temperatures was determined using a linear variable differential transducer (LVDT) based dilatometer. The Thermocalc® software in conjuction with the TCFE5 database was utilized to calculate equilibrium phases for each composition. A commercial P91 steel was used as a comparison material in all oxidation tests. All oxidation specimens were cut from these plates in 25 mm x 12 mm x 3 mm dimensions. They were wet-ground using 600 grit SiC abrasive paper.

Three different tests were employed to characterize the oxidation behavior of the experimental steels. The first was a pseudo-cyclical test utilizing a flow of dry air at 650°C. These tests were conducted in a tube furnace with samples hanging on a rack placed in the constant temperature zone of the ceramic tube. The specimens were taken out of the tube in certain time intervals and cooled down to room temperature in ambient air. After weighing, they were placed in the dry air flow at 650°C. These tests ran up to 1160 hours.

The second test was similar to the first test except that it used a flow of moist air. Dry air was bubbled through two columns of distilled water before it was fed into the furnace tube. Through this process, the air was saturated with 3 percent (by volume) water. This test lasted nearly up to 2000 hours at 650°C.

The third test can be described as a true cyclical test. These tests were conducted in air in the presence of steam at atmospheric pressure. The test consisted of 1-hour cycles of heating and cooling (55 minutes in the furnace and 5 minutes out of the furnace) in a vertical tube furnace equipped with a programmable slide to raise and lower the samples. Water was metered into the bottom of the furnace along with compressed air (50% water vapor – 50% air, by volume). The exposure temperature for these tests was 650°C.

The oxidation scales were examined using various analytical techniques including SEM, WDX, and EDS.

Table 1. Chemical composition (weight %) of the steels used in this study.

	Cr	C	Cu	Co	Mo	Ni	Mn	Si	Ti	V	Nb
HR52	9.06	0.08	2.94	3.03	0.69	1.19	0.01	0.07	0.50		
HR58	9.02	0.14	3.03	2.94	0.71	1.13	0.01	0.28	0.54		
HR59	9.01	0.13	3.04	2.95	0.71	1.14	0.24	0.28	0.59		
HR60	9.02	0.11	3.02	2.94	0.71	1.14	0.49	0.28	0.59		
HR61	9.07	0.20	3.06	2.92	0.71	1.14	0.98	0.29	0.53		
P91	8.26	0.08	0.12	0.05	0.92	0.32	0.51	0.34	0.01	0.23	0.08

Figure 1. Optical micrographs of HR58 (a) Bright field image showing the tempered martensite structure and prior austenite grain boundaries. (b) Dark field image illuminating second phase particles. The large particles are the primary TiC forming during solidification. The fine particles are the Cu-rich and TiC precipitates formed in solid state.

Results and Discussion

Microstructure

Matrix of HR alloys was predominantly martensitic as determined using optical microscopy and XRD (Figure 1a). A small amount of delta ferrite was present in all compositions. After the hot rolling, a dispersion of precipitates was observed in the HR alloys as shown in Figure 1b. Two different types of precipitates were identified. They were (i) copper rich precipitates with a higher volume fraction and a larger average size and (ii) titanium carbide (TiC) precipitates with finer size distribution. Figure 2 shows the increase in size of the precipitates with time at 750°C. Generally, the TiC precipitates are smaller (40-60 nm) and they coarsen slower. On the other hand, the Cu-rich precipitates tend to be larger (50-80 nm) and their coarsening kinetics is faster.

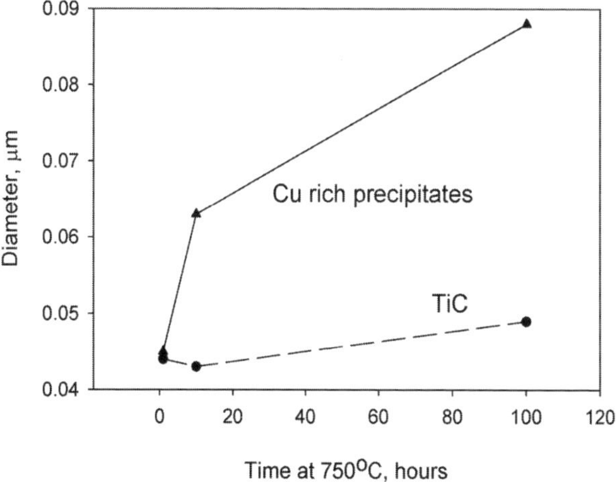

Figure 2. Size and coarsening behavior of Cu-rich and TiC precipitates in HR52.

Effect of Mn additions (HR58 through HR61) on the transformation temperatures was determined experimentally by dilatometry and calculated using Thermocalc®. As shown in Figure 3a, the Mn additions has a significant effect on the A_1 temperature (the $\alpha/(\alpha+\gamma)$ transition). The A_1 temperature decreases with increasing Mn from 0 to 1 mass percent. The effect on the A_3 temperature (the $(\alpha+\gamma)/\gamma$ transition) is less clear. These observations were also supported by the thermodynamic calculations as shown in Figure 3b.

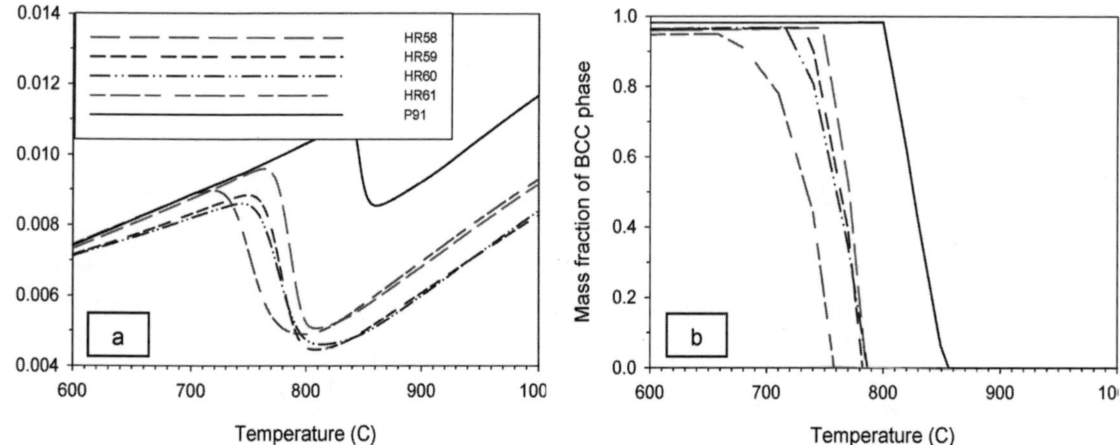

Figure 3. Effect of Mn additions on the A_1 and A_3 temperatures in the 9 Cr steels. (a) Change in length of specimen with temperature during heating as determined by dilatometry, and (b) Predictions calculated using Thermocalc®.

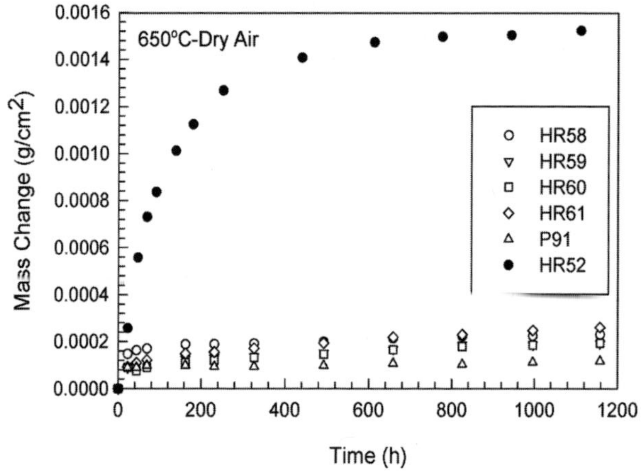

Figure 4. Mass gain of the 9Cr steels during the oxidation test in dry air at 650°C.

Oxidation of 9Cr steels in dry air at 650°C

The mass gain of the 9Cr steels as a function of time is shown in Figure 4 as determined during the pseudo-cyclical tests in a flow of dry air at 650°C. All steels tested gain mass initially at a high rate and then reach to a near steady state. HR52 alloy which is essentially the base composition of the experimental 9Cr steels without Si and Mn gains mass at a high initial rate for a longer time than the rest of the alloys tested. It eventually reaches to a steady state after about 800 hours of exposure. Addition of Si to the HR52 composition significantly reduces the time during which the steel (HR58) gains mass at the initial high rate. As a result, the total mass gain is about 8 times lower. Addition of Mn does not affect the mass gain significantly as shown for HR59, HR60, and HR61 in Figure 4. The commercial steel (P91) gains the smallest amount of mass after exposure to dry air at 650°C for 1160 hours.

The mass change discussed above is due primarily to growth of oxidation scale on the surface of the specimens. Internal oxidation contributes to the mass gain to a smaller extent. No spalling of the scale was detected on any of the specimens during the pseudo-cyclical test.

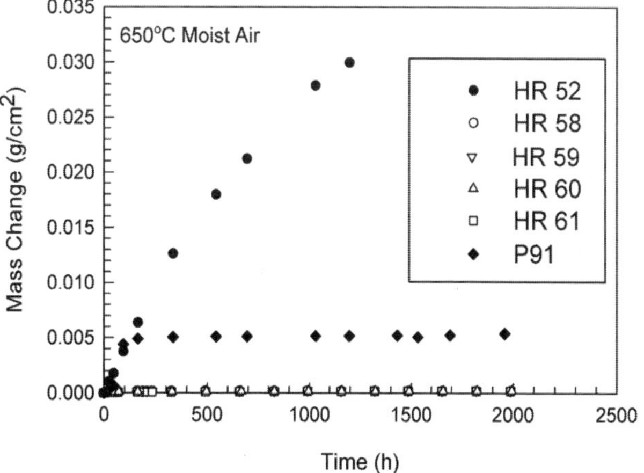

Figure 5. Mass gain of 9Cr steels during the oxidation test in air with 3%H_2O at 650°C.

Oxidation of 9Cr steels in moist (3%H_20) air at 650°C

Mass change of the steels during the pseudo-cyclical oxidation tests in the moist air at 650°C is shown in Figure 5. Introduction of 3%H_2O in the environment changes the oxidation characteristics of the HR52 and P91 steels drastically. HR52 gains mass in moist air about 30 times more than it does in dry air. The increase in the mass gain of the commercial P91 by the introduction of moisture is about 50 times. Interestingly, the

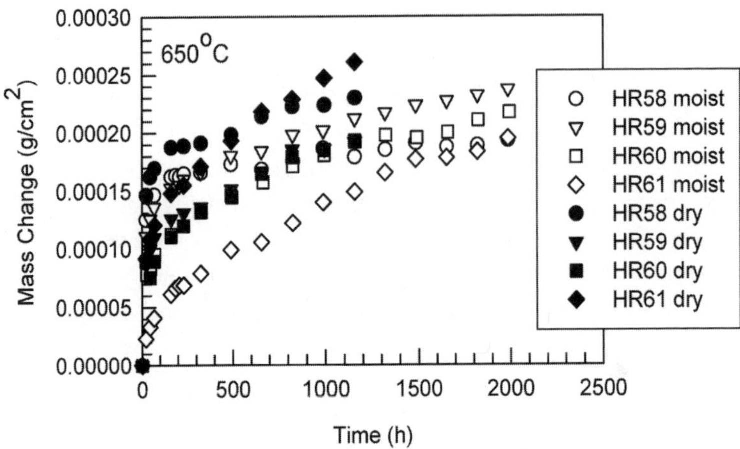

Figure 6. Comparison of oxidation rates of the experimental 9Cr steels in dry and moist air.

HR58, HR59, HR60, and HR61 steels do not follow this trend. In fact, the introduction of moisture does not seem to significantly affect the mass change in these steels (Figure 6). It is clear from these results that addition of Si into the HR52 composition improves the oxidation resistance of 9Cr steels drastically in moist environment. The effect of Si in moist air is much stronger than in dry air. Effect of the Mn addition to the Si containing experimental steels is not significant in moist air as shown in Figure 6. This is similar to the effect in dry air. Although the Si addition appears to be the main factor in reducing mass gain in the experimental 9Cr steels, the effect seems to be a result of synergistic interactions between Si and other elements in the composition since Si alone does not cause the same effect in the commercial P91 steel in moist air.

Cyclical oxidation of 9Cr steels in moist (50%H_20) air at 650°C

The cyclical oxidation tests were performed in air with increased water content (50%H_2O). The mass gain for the test steels is presented as a function of time in Figure 7. The mass gain results obtained from the cyclical oxidation test is similar to the results of the pseudo-cyclical test performed in air with 3%H_2O. The commercial steel (P91) gained about 30 times more mass than the experimental 9Cr steels. The mass gain of the experimental steels as a function of time can be represented by an equation of the form

$$\Delta m = a\, t^b$$

Where m is mass gain (mg/cm^2), t is exposure time (h), a is rate constant (mg^{2}cm^{-4}h^{-1}), and b is reaction order. The rate constants and the reaction orders are listed in Table 2 for the experimental steels.

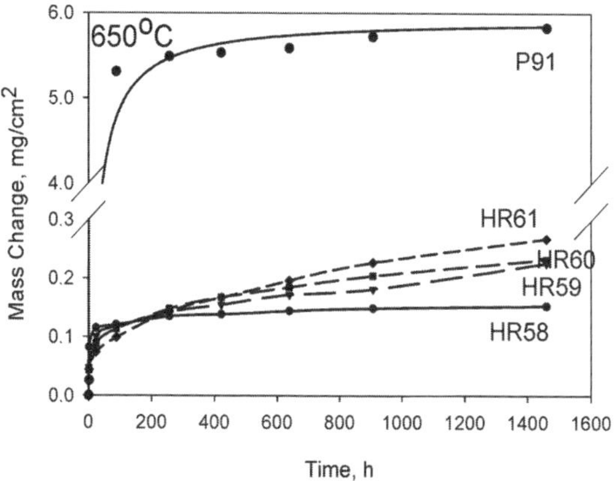

Figure 7. Mass gain of the 9Cr steels during the cyclical oxidation test in air with 50%H_2O at 650ºC.

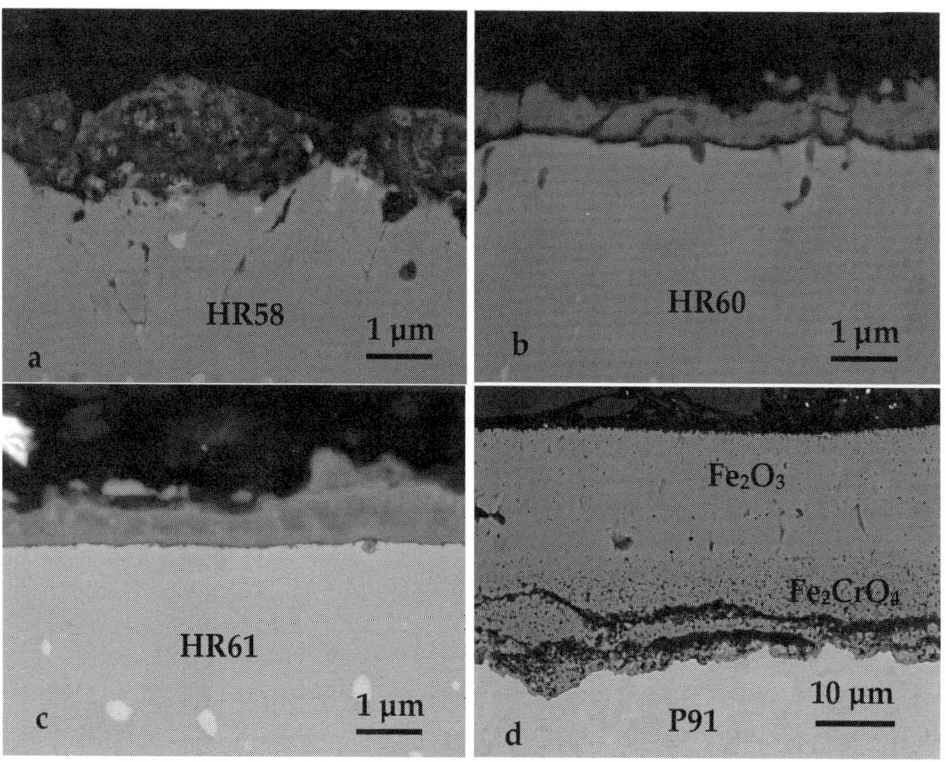

Figure 8. SEM back-scattered electron images of the cross-sections of oxidation specimens of the 9Cr steels (a) HR58, (b) HR60, (c) HR61, and P91. Note the different length scale on P91.

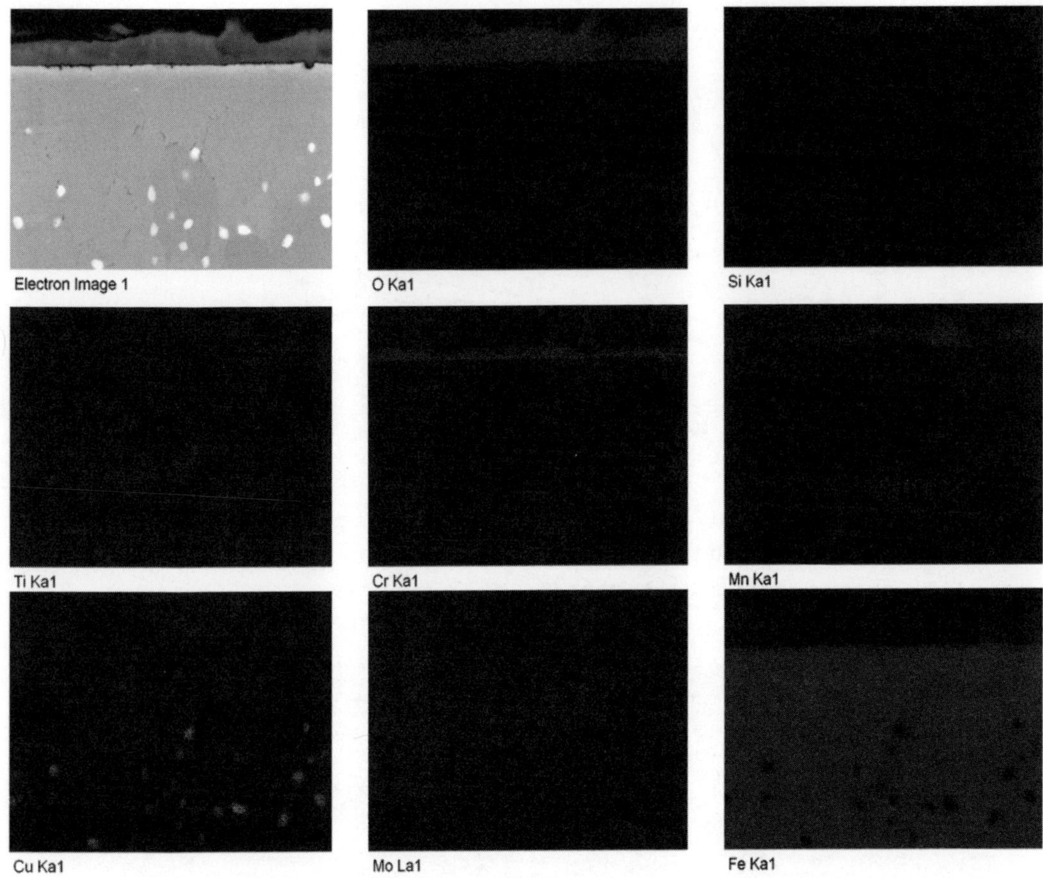

Figure 9. Elemental x-ray maps on the cross-section of HR61 oxidized in air + 50%H_2O at 650°C.

Table 2. Experimentally determined rate constants (a) and reaction orders (b) for the experimental steels.

	a	b	R^2
HR58	0.082	0.09	0.9960
HR59	0.045	0.21	0.9843
HR60	0.038	0.25	0.9989
HR61	0.024	0.33	0.9939

Back-scattered electron images of the cross-sections of the specimens exposed to moist air at 650°C for 2000 hours in one-hour temperature cycles are shown in Figure 8. While the experimental alloys developed a thin scale (about 1 µm thick), the scale on the P91 was about 30µm thick. The thin oxide scale on the experimental 9Cr alloys was very adherent to the substrate steel resulting no detectable spalling.

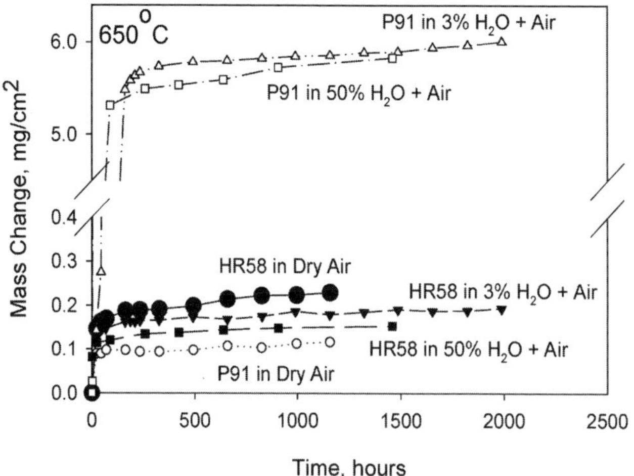

Figure 10. Comparison of oxidation rates of HR58 and P91 in different environments.

X-ray maps in Figure 9 demonstrate distribution of elements on the cross-section of the HR61 specimen after 2000 hour cyclic exposure to moist air at 650°C. O, Cu, and Mn are distributed fairly evenly throughout the scale. Cr and Ti are concentrated in the inner scale near the substrate whereas Si is concentrated in the outer scale. The scale does not appear to contain Fe in the x-ray maps although spot analysis done using WDX indicated some Fe is contained in the scale. Below the scale near the surface of the substrate alloy, a Cu depleted zone is apparent. Evidently, the Cu rich precipitates near the surface dissolve and Cu diffuses to the scale during the exposure. In contrast, P91 scale contains Fe_2O_3 in the outer layer and a more porous Fe_2CrO_4 in the internal scale (Figure 8).

Finally, Figure 10 compares P91 and HR58 under different oxidation conditions. Oxidation rate of P91 is affected significantly by introduction of moisture at 650°C. Both 3% and 50% H_2O in air have similar effects. On the other hand, the oxidation rate of HR58 does not change significantly with the introduction of moisture. In moist air, HR58 is much more resistant to oxidation than P91.

Conclusions

A very protective oxide scale forms on the experimental HR58 through HR61 in both dry and moist air at 650°C. Growth kinetics of this scale is extremely slow. This results in a thin (about 1-2 µm) scale. The scale is also very adherent to the substrate under the cyclical temperature conditions. No spalling was observed during the oxidation tests.

References

1. F. Masuyama, "History of Power Plants and Progress in Heat resistant Steels", ISIJ International, Vol. 41 (2001), No. 6, pp. 612-625.
2. P. J. Ennis, "Creep strengthening mechanisms in 9-12% chromium steels", Proceedings from the Fourth International Conference on Advances in Materials Technology for Fossil Power Plants, October 25-28, 2004, Hilton Head Island, SC, ASM International, 2005, pp. 1146-1159.
3. R. Viswanathan, J. Sarver, and J.M Tanzosh, "Boiler materials for ultra-supercritical coal power plants – Steamside oxidation", J. Mat. Eng. & Per., Vol. 15 (2006), pp. 255-274.
4. D.K. McDonald and E.S. Robitz, "Coal ash corrosion resistant materials testing program", Proceedings from the Fourth International Conference on Advances in Materials Technology for Fossil Power Plants, October 25-28, 2004, Hilton Head Island, SC, ASM International, 2005, pp. 310-322.

Advances in Materials Technology for Fossil Power Plants
Proceedings from the Fifth International Conference
R. Viswanathan, D. Gandy, K. Coleman, editors, p 531-543

DOI: 10.1361/cp2007epri0531

An Investigation of Key Experimental Parameters in Steam Oxidation Testing and the Impact they have on the Interpretation of Experimental Results.

A. T. Fry
National Physical Laboratory, Teddington, Middlesex, TW11 0LW

Abstract

The acceptance of materials for extended duration, safety critical power generation applications usually requires several stages of testing and data generation. The initial sift of candidate materials can be based upon simple, short-term exposures under nominally constant atmospheres and temperatures. These often over simplified exposures can eliminate materials that are grossly unsuitable for the application but do not differentiate between materials that have broadly acceptable properties. After these initial sifts of candidate materials the next step is usually to expose the remaining materials in a pilot plant costing in excess of £100K for test durations of one month. The availability of an intermediate step, involving laboratory testing with some features of plant, would provide a cost-effective route to a more informed decision on material selection and lifetime predictions. Hence, to better differentiate between candidate materials it is desirable to tailor the laboratory test conditions such that they closer replicate in-service conditions. In terms of components that are exposed to steam oxidation degradation mechanisms, this means replicating the steam conditions with an aim of producing oxide scale morphologies similar to that seen in plant.

Key experimental parameters have been identified, including water chemistry, pressure, steam delivery and flow rate, and a series of steam exposure tests on ferritic (P92), austenitic (Esshete 1250) and superalloy (IN740) material conducted to evaluate the effect on the degradation rate and oxide scale morphology.

Results indicate that the oxidation rate of the austenitic, and to a lesser extent the ferritic material, is sensitive to the level of dissolved oxygen in the feed water, producing an increase in the oxidation rate for the austenitic material and a decrease in the oxidation rate for the ferritic material. The propensity for spallation was also affected by the oxygen content. In addition the steam pressure and steam delivery method are shown to affect the oxidation rate and scale morphology for these materials.

The resultant scale morphologies of the oxide scales grown have been compared with service exposed materials, these did not adequately replicate the scale morphology. Cyclic oxidation tests have been conducted and are shown to better replicate service grown scales.

Keywords: Water chemistry, Pressure, Flow rate, Steam oxidation

Introduction

During service in high temperature plant, tubing, piping and headers in boilers are exposed to steam at high pressure flowing at high rates and often with an associated heat flux. Similarly components in steam turbines are exposed to high-pressure steam atmospheres.

Service lifetimes of these components may be limited by creep, fatigue and/or oxidation. Traditionally materials designed for use at high temperature have been developed primarily for their mechanical properties but there is a growing realisation that oxidation may limit lifetime, either directly through metal wastage or indirectly through raising local temperatures due to the lower thermal conductivity of the oxide scale.

The acceptance of materials for high temperature, extended duration applications usually requires several stages of testing and data generation. The initial sift of candidate materials can be based upon simple, short-term exposures under nominally constant atmospheres and temperatures. These often over simplified exposures can eliminate materials that are grossly unsuitable for the application. After these initial sifts of candidate materials the next step is usually to expose the remaining materials in a pilot plant costing in excess of £100K for test durations of one month. Better methods to improve the differentiation between candidate materials are desirable. Hence laboratory test conditions need to be designed such that they closer replicate in-service conditions. In terms of components that are exposed to steam oxidation degradation mechanism, this means replicating the steam conditions with an aim of producing oxide scale morphologies similar to those seen in plant.

Experimental parameters which are controllable within the laboratory have been identified. These are the water chemistry, test pressure, the steam delivery method and the flow rate of the steam. Steam exposure tests have been conducted on low alloy ferritic materials (T22, T23 and P92), austenitic material (Esshete 1250) and super alloy material (IN740) to evaluate the effect of these parameters on the degradation rate and oxide scale morphology of these previously mentioned parameters. Where appropriate the results from these tests have been compared with results from the literature.

Test Methods

Steam exposure tests have been conducted on a range of materials, compositions of which are presented in Table I, using a range of test procedures. In all cases the water used to generate the steam was distilled and deoxygenated by bubbling nitrogen gas through the water reservoir. Tests performed using flowing steam, were conducted in horizontal tube furnaces, with a controlled hot zone of around 30 cm. Samples were nominally 10 cm by 10 cm by 3 mm thick, although in some cases the material was supplied as either tube or finned tube resulting in circular specimen with a nominal diameter of 10 cm. Specimens were placed in alumina boats for testing and in all cases the specimens were placed into hot furnaces. Once the exposure time was completed specimens were removed and left to cool on the laboratory benches.

The high pressure exposures were conducted within a static autoclave. In this case water is added to the pressure vessel along with the specimens prior to heating. Once the exposure is complete the pressure vessel is cooled before the specimens can be removed.

In the case for tests performed using a relatively high level of dissolved oxygen, typically greater than 200 ppb, the steam was pumped into the furnace tubes using a peristaltic pump. Once in the tube the water formed the steam which passed along the tube to vent to atmosphere through a gas bubbler. Tests were also performed using low levels of dissolved oxygen, typically less than 10 ppb as measured using an oxygen meter. In this case the water was distilled and deoxygenated as before, however it was stored and delivered to the furnaces in a more controlled manner, making sure that there could be no oxygen ingress. The deoxygenated water was introduced into the furnace by application of a small positive argon gas pressure.

The effect of using a carrier gas was investigated by comparing the results of tests using 100% flowing steam and 50% Ar + 50% H_2O.

Steam exposures are usually performed under isothermal conditions for the total exposure time. Experience has shown that such isothermal tests rarely, if ever, produce oxide scale morphology consistent with those observed on service-exposed material. These are usually multilayered in form, whilst the laboratory grown oxide usually consists of one or two layers. A small number of exposures have therefore been performed using a thermal cycling exposure method to examine the effect of thermal cycles on the oxidation rate and scale morphology. The cycles consisted of 20 hours hot and 4 hours cold, during which a continuous stream of 50% Ar + 50% H_2O was being passed through the apparatus.

For each test two specimens of the same material were used. After the tests the specimens were weighed and mass change calculations made using the mass of the specimens prior to exposure. One of the specimens was then nickel plated and sectioned prior to mounting for metallographic evaluation which consisted of oxide thickness measurements and assessment of the scale morphology.

Table 1 Composition of the materials used in this study.

Material	Alloy composition, wt% (bal Fe)															
	C	P	Cr	V	Nb	Cu	Si	S	Mo	W	N	B	Mn	Al	Ni	Co	Ti
T22	0.1	0.025	2.25				0.5	0.025	0.54				0.6				
T23	0.07	0.03	2.25	0.25	0.05		0.5	0.01	0.17	1.6	0.03	0.06	0.35	0.03			
P92	0.12		9.37	0.23			0.1		0.45	1.98	0.04		0.61				
9Cr1Mo(1)	0.11	0.02	8.82				0.41	0.004	0.95				0.45			0.019	
9Cr1Mo(2)	0.112	0.018	9.89				0.63	0.009	1.01				0.47			0.024	
9Cr1Mo(3)	0.1	0.006	9.7				0.59	0.007	1				0.5			0.031	
Alloy 122	0.12	0.009	11.1	0.22	0.05	1	0.2	0.007	0.35	1.9	0.06	0.02	0.68	0.005	0.3	0.011	
X20	0.21	0.13	10.9	0.22	0.41		0.33	0.005	0.66		0.05	0.015	0.6	0.007	0.41		
Esshete 1250	0.09	0.024	15.3	0.25	1	0.14	0.56	0.26	1.09		0.03	0.038	6.23		9.9		
IN740	0.034	0.003	24.3		1.83		0.45	0.002	0.52	0.003			0.27	0.75	49.4		1.58

Results

Effect of water chemistry

Tests have been performed on P92, Esshete 1250 and IN740 using steam generated from water containing two levels of dissolved oxygen. These two levels are referred to as 'High' and 'Low' with around 200 ppb and 10 ppb of dissolved oxygen respectively. In all cases the specific mass change and oxide thickness (in micrometers) have been measured, the results of these tests are shown in Figures 1 and 2.

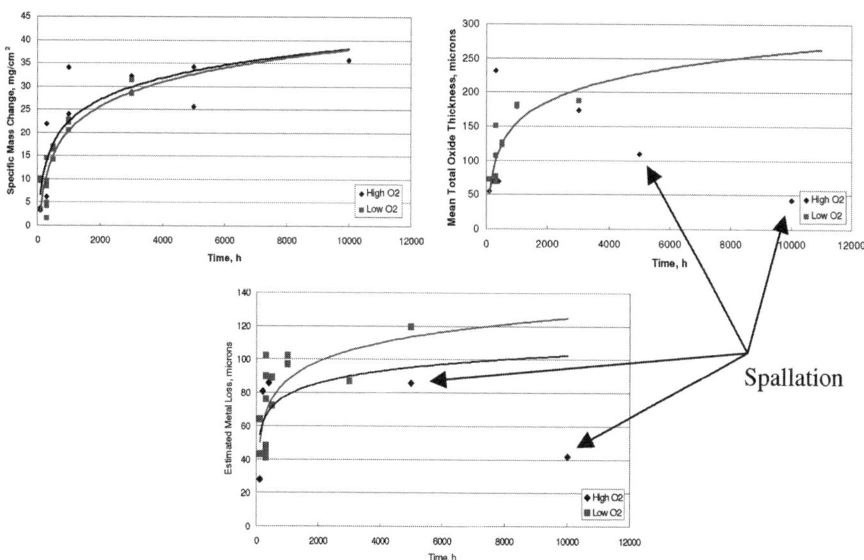

Figure 1 Specific mass change, total oxide thickness and estimated metal loss plots for P92 exposed to flowing steam at 650 °C with high and low levels of dissolved oxygen, log trend lines have been drawn to illustrate the parabolic growth rate

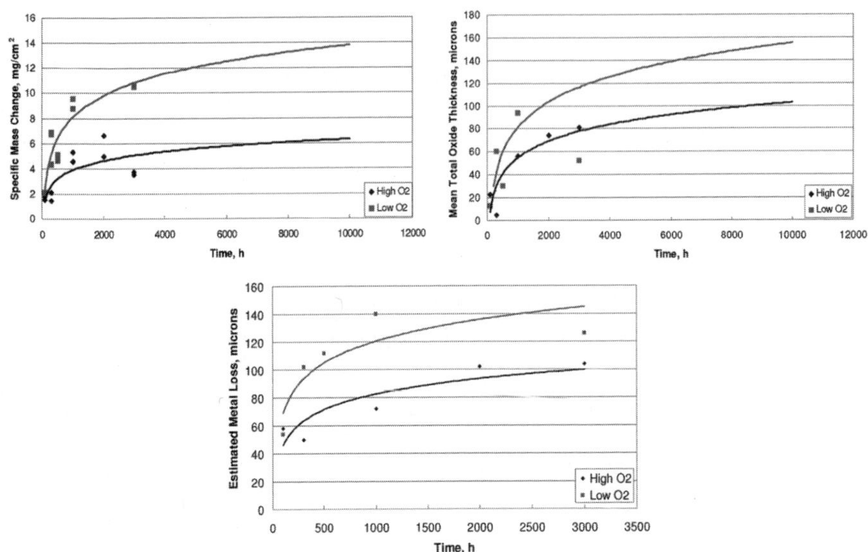

Figure 2 Specific mass change, total oxide thickness and estimated metal loss plots for Esshete 1250 exposed to flowing steam at 700 °C with high and low levels of dissolved oxygen

Log trend lines have been added to Figures 1 and 2 to aid in the comparison of the oxidation rate. The estimated metal loss in Figures 1 and 2 has been calculated using the maximum thickness of the inward growing oxide, and assumes that none of this oxide has spalled away. Recent experimental evidence from the European Komet project shows that this assumption may not reflect reality; it would therefore be better to measure the actual metal loss as just using the inner oxide thickness can result in under estimates as illustrated in Figure 1 after about 5000 hours.

The results show that there is an apparent effect on the oxidation kinetics from the water chemistry for the low alloy P92 material and the austenitic Esshete 1250 material. Results from test on IN720, not shown here, did not show a measurable effect due to the high oxidation resistance of this material. In the case of P92 the effect appears to be quite small and may potentially be within the experimental scatter. However, when plotting the estimated metal loss it does appear that low oxygen containing steam is producing an increase in the oxidation kinetics with a low propensity for scale spallation. The microstructure of the oxide scale for both conditions is shown in Figure 3. This shows that the low oxygen containing steam is producing a more dense oxide scale after 500 hours, with a lower amount of porosity. It is suggested therefore that the lower amount of defects in the scale accounts for the improved spallation properties.

In the case of the austenitic Esshete 1250 material the effect of the oxygen content is more obvious, with an increase in the oxidation rate and the thickness of the final oxide scale. This is also clearly illustrated in Figure 4 which shows the microstructure of the oxide scale formed at 700 °C.

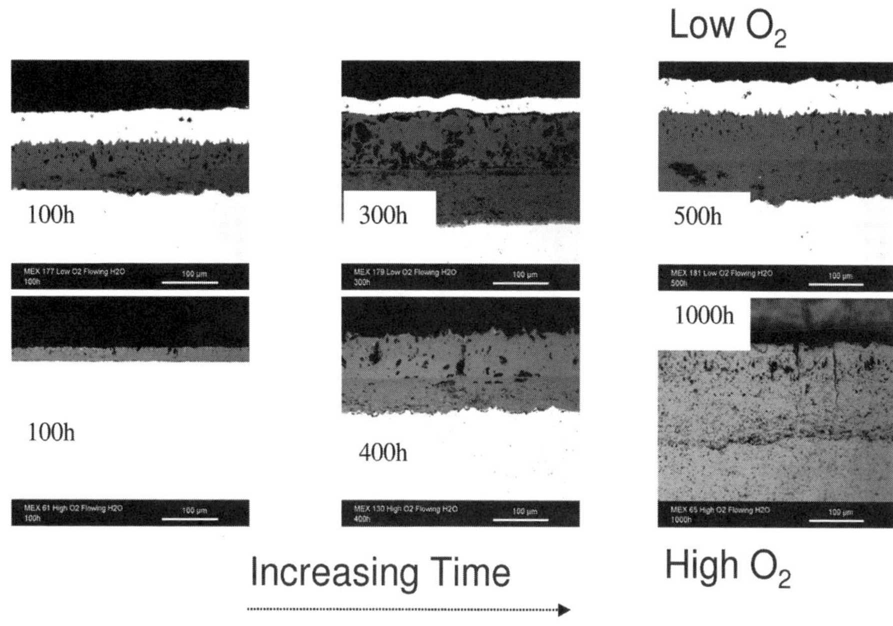

Figure 3 Micrographs showing the morphology of the oxide scale formed on P92 at 650 °C in low and high oxygen steam

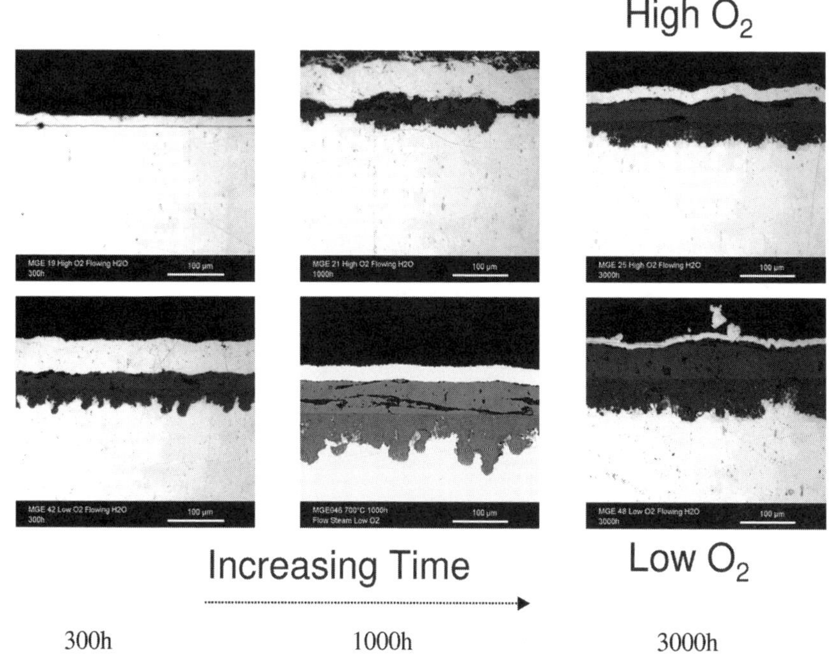

Figure 4 Micrographs showing the morphology of the oxide scale formed on Esshete 1250 at 700 °C in low and high oxygen steam

Effect of test pressure

To examine the effect of pressure on the oxidation rate and the scale morphology, tests have been performed on T22, T23, 9Cr1Mo, P92 and IN740. The results of these tests are shown in Figures 5, 6 and 7. For the low alloy materials T22 and T23 the results show that there is a trend for the oxidation rate and oxide scale thickness to increase with increasing pressure. This is also the case with P92 and IN740, which both show an increase in the oxide thickness with increasing pressure. The 9Cr1Mo materials however do not follow this trend showing a reduction in the oxide thickness with increasing pressure.

When examining the scales in cross section it can be seen that as the pressure increases the density of the oxide scale increases and the presence of defects, in the form of voids and porosity, decreases both in terms of volume and distribution. This will impart higher strengths in the oxide scale and thus reduce the propensity of spallation. This has been discussed and shown qualitatively using Rockwell Hardness tests in earlier work by Osgerby and Fry [1-2].

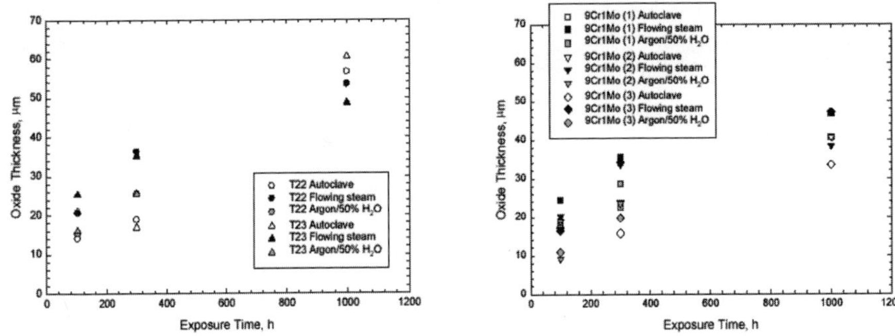

Figure 5 Oxide thickness results for T22, T23 and various 9Cr1Mo alloys exposed to flowing steam, static high pressure (50 bar) steam and a mixture of argon and steam at 550 °C

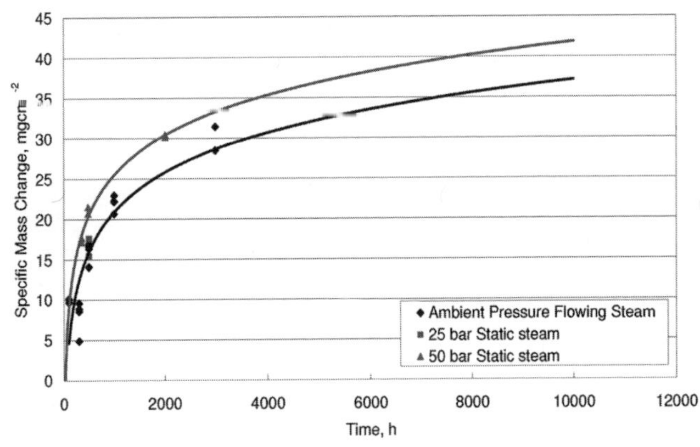

Figure 6 Specific mass change results for P92 exposed to steam at 650 °C under various pressure conditions

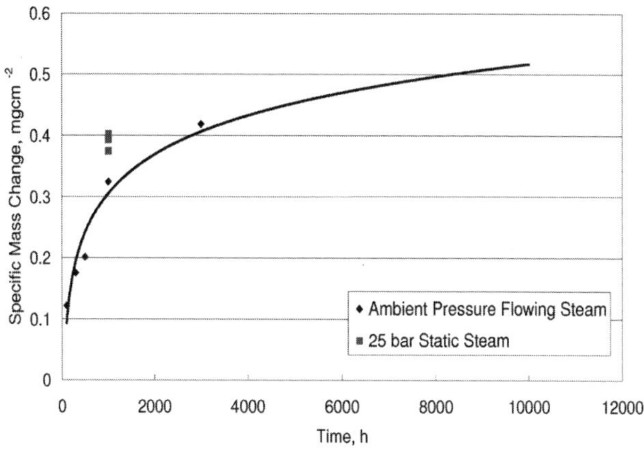

Figure 7 Specific mass change of IN740 at 750 °C in flowing and static steam at 25 bar

Effect of steam delivery method

It is generally observed that the oxidation rate, before spallation (approx. 1000 hours), for materials exposed to steam delivered using a carrier gas, lies approximately midway between the two extremes of 100% flowing steam and static high pressure steam, as shown in Figure 8.

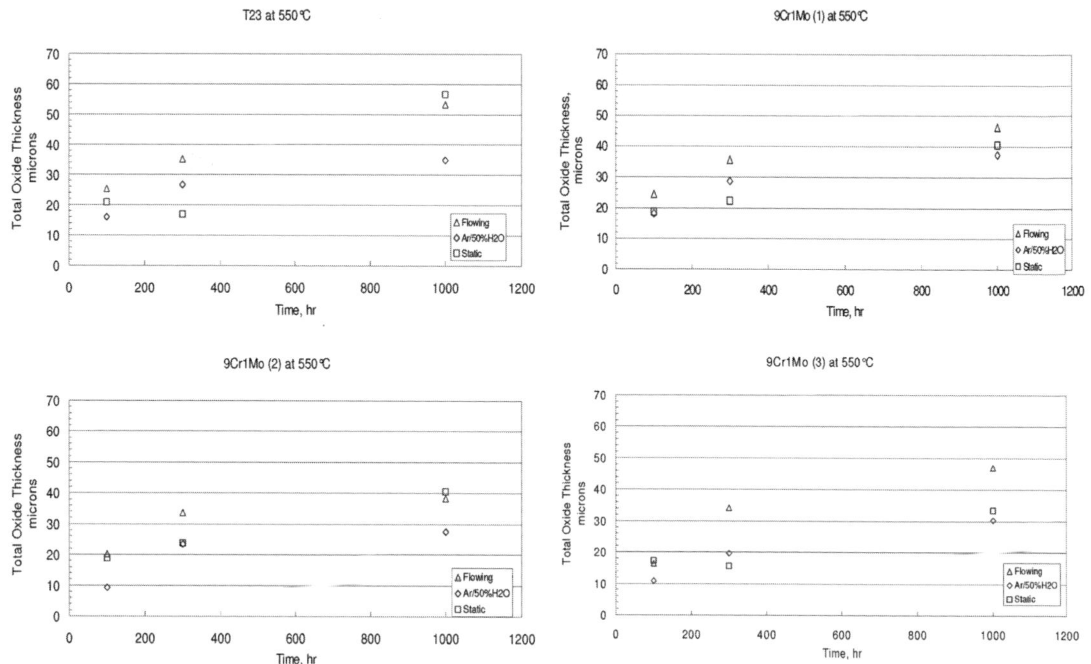

Figure 8 Steam oxidation data for T22 and various 9Cr1Mo alloys

In the tests the three 9Cr1Mo steels showed fairly consistent behaviour with the oxidation rate in 9Cr1Mo(1) greater than that in 9Cr1Mo(2) which in turn was greater than that in 9Cr1Mo(3). This ranking is probably a result of the slightly lower Cr and Si content in 9Cr1Mo(1) in comparison with the other two materials and the lower C content of 9Cr1Mo(3) compared to 9Cr1Mo(2). It is suggested that C affects the oxidation resistance through tying up Cr in the form of carbides: this Cr would otherwise be available for oxidation resistance.

Significant differences in the morphology of the oxide scales grown under different conditions can be observed. The behaviour of the 2¼Cr steels (T22 and T23) is illustrated in Figure 9. Scales grown under Ar/50% H_2O show a compact inner spinel layer with an outer porous magnetite layer. There are traces of haematite at the surface but these are very thin and discontinuous. The scale grown under flowing steam has a compact outer layer of magnetite with an inner spinel layer that contains a plane of massive defects. In contrast the scale grown under high pressure static steam is compact throughout, as described previously, although there is some evidence of roughening at the surface of the outer scale.

Similar diversity in the scales grown under different conditions is observed in the 9Cr1Mo steels, as illustrated in Figure 10. Two additional features in the scales grown on these materials are the presence of internal oxidation below the spinel under all conditions and the development of wavy scale/substrate interface and surface in the scales grown under flowing steam and static high pressure steam.

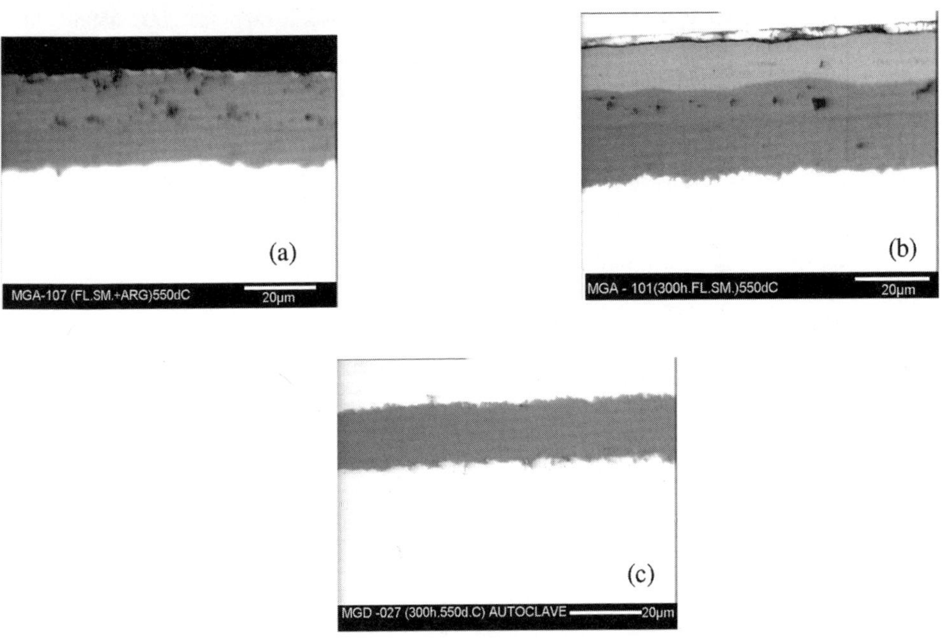

Figure 9 Microstructure of oxide scales grown on T22 ferritic steel at 550 °C for 300 h using different laboratory procedures (a) argon/50% H_2O (b) flowing steam (c) static high pressure steam

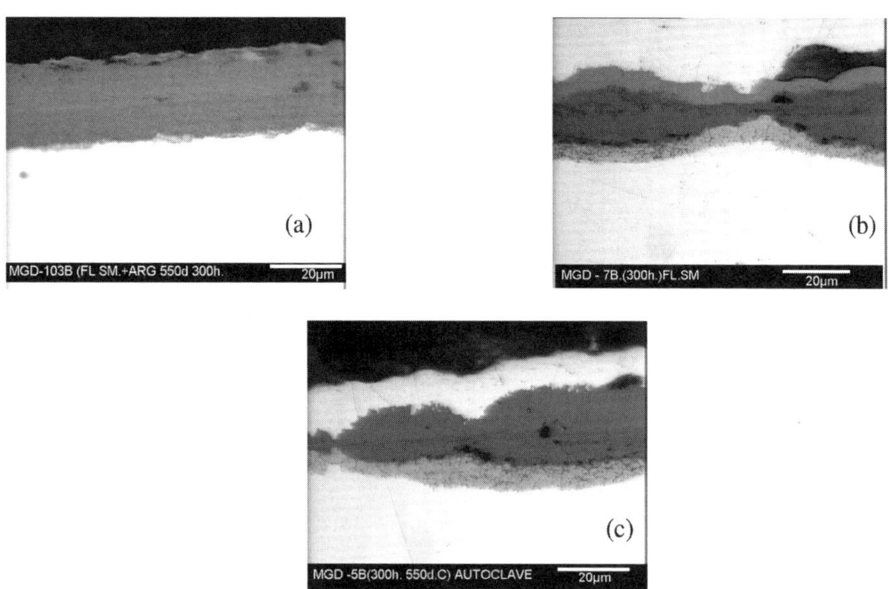

Figure 10 Microstructure of oxide scales grown on 9Cr1Mo(2) steel at 550 °C for 300 h using different laboratory procedures (a) argon/50% H2O (b) flowing steam (c) static high pressure

Figure 11 shows the probability of the 9Cr1Mo(2) alloy developing an oxide of a certain thickness. A horizontal line indicates a uniform scale whilst increasing slope to the data indicates increasing variation in the scale thickness. Two effects are apparent. Firstly the wavy interface develops with time – data from the 100 h exposures (open symbols) follow an almost horizontal path; the slope of the data generally increases with increasing exposure time (black- and grey-filled symbols). Secondly the data from scales grown under argon/50% H_2O are much more homogeneous in thickness than those grown under flowing steam or in the autoclave.

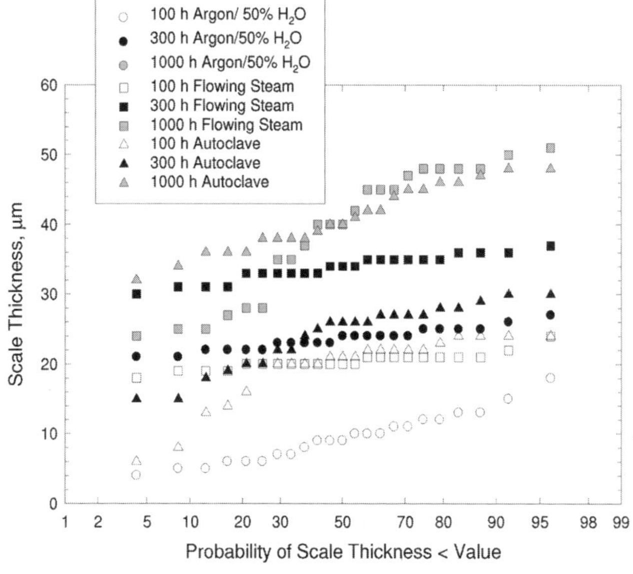

Figure 11 Probability plot of oxide scale thickness for 9Cr1Mo(2) material, exposed to steam at 550 °C using various procedures.

Effect of flow rate

It is believed that the flow rate affects the oxidation rate of materials during testing. To investigate this tests were performed on the P92 material, using low oxygen containing steam, and two flow rates. The high flow rate was twice the low rate, and was at the limit of the current apparatus capability. The results of these limited tests are shown in Figure 12, and indicates qualitatively that by doubling the flow rate the oxidation rate is slightly increased, although with a limited amount of data this is only an indicative observation, but agrees with observations in the literature as discussed later.

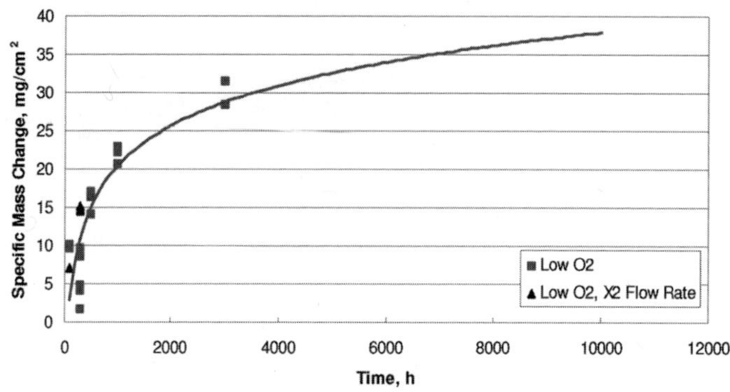

Figure 12 P92 exposed to flowing steam at 650°C, containing low levels of dissolved oxygen and flowing through the furnace at two flow rates

Discussion

Laboratory exposure methods for steam oxidation testing vary from being fairly simple to much more complex. When deciding on which test method to adopt the researcher needs to consider a range of variable test parameters. The decision of which parameters are important and which best replicate the in-service conditions has been shown to effect the results, in terms of both the oxidation rate and the morphology of the resultant oxide scale.

From the literature examined there is no consensus on these points and also little mention of the potential effects of steam chemistry. Results from tests performed within this study indicate that the level of dissolved oxygen in the water can affect the oxidation behaviour of the test specimens. The extent of this effect appears to vary depending on the material composition, although there does appear to a general trend for increased oxidation rate, lower porosity and better spallation properties. (Analysis of these results has shown that some types of data are influenced more by spallation than others.) It would appear that measuring the metal loss is the most robust method, but it is also the most labour intensive and difficult.

Within the literature it is generally reported that an increase in the test pressure results in an increase in the oxidation rate [3]. Otoguro *et al* [4] noted that the resistance to steam oxidation was influenced by four factors; steam pressure, temperature, Cr content and Ni content. Montogomery *et al* [5] suggest that by increasing the pressure the integrity of the scale is decreased, thus increasing the oxidation rate. The oxidation rate will also be increased due to the increased temperature of the underlying metal substrate, caused by voids restricting the heat transfer. Results within this work substantiate this, with a slight increase in the oxidation rate, as shown in Figure 6 and 7. With regards to the defect structure in the scale and the effect on the integrity of the oxide there is evidence that the water chemistry may have some impact on this. In early test on T22, T23 and 9Cr1Mo the data shows spallation of the oxide in the autoclave exposures (Figure 5). These were performed using high oxygen steam. The tests performed using low oxygen steam on P92 and IN740, as shown in Figure 6, did not show this spallation, although this could equally be due to alloy composition.

The gas velocity has been shown to strongly affect the oxidation of austenitic and ferritic chromium steels. In work conducted by Asteman *et al* [6] it was concluded that the vaporization of chromium from the oxide in the form of chromium (VI) oxyhydroxide increased the oxidation rate of X20, 304L and 310 steels in O_2/H_2O environments. They found that at low gas velocities the X20 and 310 showed completely protective behaviour. In contrast the 304L material did not form a completely protective oxide at low gas velocities. For higher gas velocities (>2.5 cms^{-1}), X20 exhibited breakaway corrosion while the oxide on the 304L and 310 was observed to fail locally. Observations within this study agree with the effect on oxidation rate, in that it was seen to increase with increased gas velocity. No evidence as to the precise mechanism is proposed here.

Figure 13 presents micrographs of specimens which were cut from pipes that had been exposed to steam during service. The piece of T22 had been in service for 19000 h at 550 °C and the 9Cr1Mo material had been at 590 °C for 28000 h. The steam pressure during service was unknown. The scales are significantly thicker that those formed in the laboratory, due to the much longer exposure time, and do not appear to have spalled. In both cases the outer magnetite layer shows some porosity whilst the inner spinel has a fully dense structure. Rather surprisingly it is the structure of the scales formed under the argon/50% H_2O mixture that is most similar to that which is formed in service. This procedure is the most dissimilar to the actual service conditions so it would be expected to produce the least representative scales.

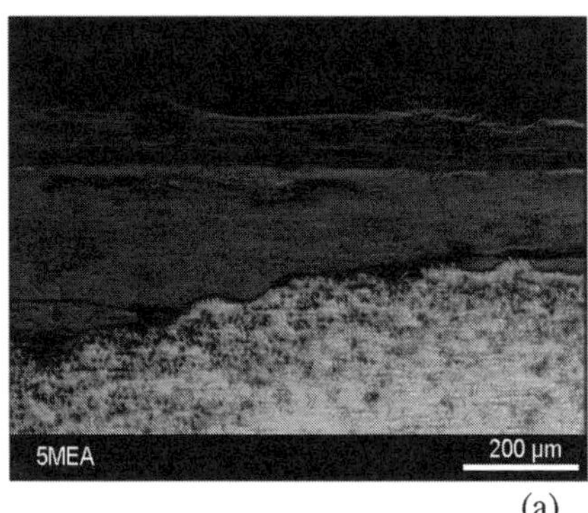

(a)

(b)

Figure 13 Morphology of oxide scales formed in service in a steam environment (a) T22 after 19000 h at 550 °C (b) 9Cr1Mo after 28000 h at 590 °C

Thermal cycling during the exposure is also believed to influence the oxidation rate and scale morphology. Preliminary work looking into this has shown that the multilayered scale morphology found in service exposed materials can be replicated using thermal cycling techniques, as shown in Figure 14.

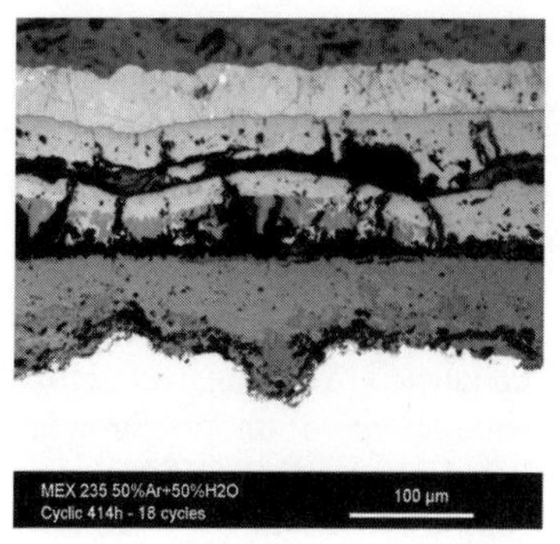

Figure 14 P92 exposed to argon/50% H2O at 650 °C, after 18 thermal cycles or 20 hours hot, 4 hours cold.

Conclusions

It has been shown that a number of test parameters, that can be varied when conducting laboratory based steam oxidation testing, affect the oxidation rate and scale morphology of test specimens. A range of tests have been performed to investigate the effect of water chemistry, steam pressure, flow rate and delivery method.

This has shown that for the materials investigated here: -

- There is an increase in the oxidation rate when the oxygen content in the steam is reduced, in addition there appears to be less defects in the scale structure thereby giving the oxide scale greater strength and better spallation properties.
- Results have shown that measuring the metal loss of the specimens is a better method to use when evaluating the oxidation kinetics, as it is less prone to errors caused by spallation.
- Increasing the pressure and the gas flow velocity of the steam increases the oxidation rate.
- Thermal cycling with an argon/steam mix produces oxide scale morphology that most closely resembles that seen in service exposed materials.

Acknowledgements

This work was carried out as part of the C05 project on “Improved Test Methods and Understanding of Steam Oxidation”, which was part of the Performance Programme, a programme of underpinning research funded by the United Kingdom Department of Trade and Industry. The author extends thanks to Mr David Laing, Mr Dipak Gohil, Mr Jim Banks and Dr Louise Brown, who were responsible for laboratory testing and analysis of the samples.

References

1 S. Osgerby and A. T. Fry, “Steam Oxidation Resistance of Selected Austenitic Steels”, Materials Science Forum, Vols. 461-464 (2004), pp. 1023-1030

2 S. Osgerby and A. Fry, “Simulating Steam Oxidation of High Temperature Plant Under Laboratory Conditions: Practice and Interpretation of Data”, HTCERS Conference 2003.

3 A Fry, S Osgerby and M Wright, "Oxidation of Alloys in Steam Environments – A Review", NPL Report MATC(A)90 July 2002, National Physical Laboratory, UK

4 Y. Otoguro, M. Sakakibara, T. Saito, H. Ito and Y. Inoue,. "Oxidation Behaviour of Austenitic Heat-resisting Steels in a High Temperature and High Pressure Steam Environment". *Transactions ISIJ*, Vol. 28, 1988, pp. 761-768.

5 M. Mongomery and A. Karlsson, "Survey of Oxidation in Steamside Conditions", *VGB Kraftwerkstechnik*, 75, 1995, pp. 235-240.

6 H. Asteman, K. Segerdahl, J. –E. Svensson and L. –G. Johansson. The Influence of Water Vapor on the Corrosion of Chromia-Forming Steel, Materials. *High Temperature Corrosion and Protection of Materials 5, Materials Science Forum* Vls. 369-372 (2001), pp. 277-286.

Advances in Materials Technology for Fossil Power Plants
Proceedings from the Fifth International Conference
R. Viswanathan, D. Gandy, K. Coleman, editors, p 544-550

DOI: 10.1361/cp2007epri0544

Role of Minor Compositional Variation in Oxidation 'Cr-Mo' Steels

R.K. Singh Raman[#,@] and A. Al-Mazrouee[#]

Department of Mechanical Engineering, @Department of Chemical Engineering
Bldg 31 Monash University, Vic 3800, Australia
e-mail: raman.singh@eng.monash.edu.au

Keywords: Chromium-Molybdenum ferritic steels

Abstract: Oxide scale growth kinetics of 1.25Cr-0.5Mo and 2.25Cr-1Mo steels have been investigated and oxide scales have been characterized. In spite of the considerable difference in chromium contents of the two steels, their oxidation kinetics have been found to be similar. The similarity in oxidation behavior has been associated with formation of a protective innermost layer of silicon rich oxide in the case of 1.25Cr-0.5Mo steel and absence of such a layer in the case of 2.25Cr-1Mo steel, suggesting predominating role of Si over Cr.

Introduction

Chromium-Molybdenum ferritic steels (generally referred as Cr-Mo steels) possess a good combination of mechanical properties, formability, weldability and resistance to stress corrosion cracking and other forms of corrosion[1,2]. These attributes make Cr-Mo steels the extensively used family of engineering materials for moderately high temperature applications over several decades, such as construction material for steam generation/handling, petroleum processing/refining, thermal reforming/ polymerisation/cracking systems [1-3]. With increasing demand for suitable materials for supercritical and ultra supercritical steam generators for power plants having much improved efficiencies, Cr-Mo steels have attracted renewed interest for their potential application in low-temperature parts of such steam generators (that will be expected to operate at higher than usual temperatures).

Chromium, aluminum and/or silicon are the common alloying elements, which when present in sufficient amounts in steels can confer oxidation resistance, as a result of the formation of a protective inner layer of Cr_2O_3, Al_2O_3 or SiO_2[4]. A layer of SiO_2 is the most protective of the three oxides[4,5] since this is the most defect-free of the three oxides. Silicon content is well known to influence oxidation resistance also of other iron-chromium alloys[6-8], including Cr-Mo ferritic steels[9,10].

This study investigates the influence of silicon contents of 1.25Cr-0.5Mo and 2.25Cr-1Mo steels in oxidation kinetics and composition of the oxide scales during air-oxidation of the two steels. The paper also presents a brief discussion on the possible inaccuracies in creep data as a result of cross-sectional losses of the specimens during creep testing as well as need for generating oxide scale thickness data for individual steels for the purpose of accurate life predication of components.

Test Materials and Experimental Description

Compositions of the two steels investigated in the present study, as analyzed by optical emission spectroscopy, are given in Table 1. As discussed later, relative silicon contents of the two steels have been found in this study to have considerable influence on their oxidation kinetics and scale characteristics. The accuracy of the silicon analyses was ± 0.02 wt%. Cylindrical specimens (6.5 mm in diameter and 25 mm long) were machined from the rods of the test materials and ground and polished to 1200 grit finish and cleaned ultrasonically before high temperature oxidation tests. Before and after the oxidation tests, specimens were weighed using a digital balance (0.1 mg accuracy) and their diameters and lengths were measured using a digital micrometer (±2μm accuracy). The oxidation tests were carried out using a tubular furnace with a digital temperature controller (accuracy of ±2°C). Several specimens were subjected to air-oxidation in the furnace at 700°C, and removed after different durations, viz., 1, 10, 30, 100, 300, and 600 hours. In order to determine the reduction in diameter due to oxide scaling losses, oxidized specimens were subjected to an ultrasonic cleaning by immersion in a special descaler solution.

Table 1 Chemical compositions (wt%) and the corresponding ASTM specifications of the two Cr-Mo steels

Steels	Mn	C	S	P	Si	Mo	Cr	Fe
1.25Cr-0.5Mo	0.46	0.12	0.006	0.010	0.56	0.54	1.27	BAL.
2.25Cr-1Mo	0.60	0.098	0.008	0.012	0.22	0.98	2.22	BAL.

For morphological and chemical characterization of different scale layers, duly polished cross sections of the oxidized specimens were examined by Scanning Electron Microscopy (SEM), Energy Dispersive X-ray spectroscopy (EDXS) and X-Ray Mapping (XRM).

Results and Discussion

Oxidation kinetics, represented as mass gains per unit area, during air-oxidation of the two steels at 700°C for 600h are compared in Figure 1. Both steels follow parabolic behaviour[4]. It is noted that despite the difference in chromium contents of the two steels, their mass gains during 600h oxidation are similar.

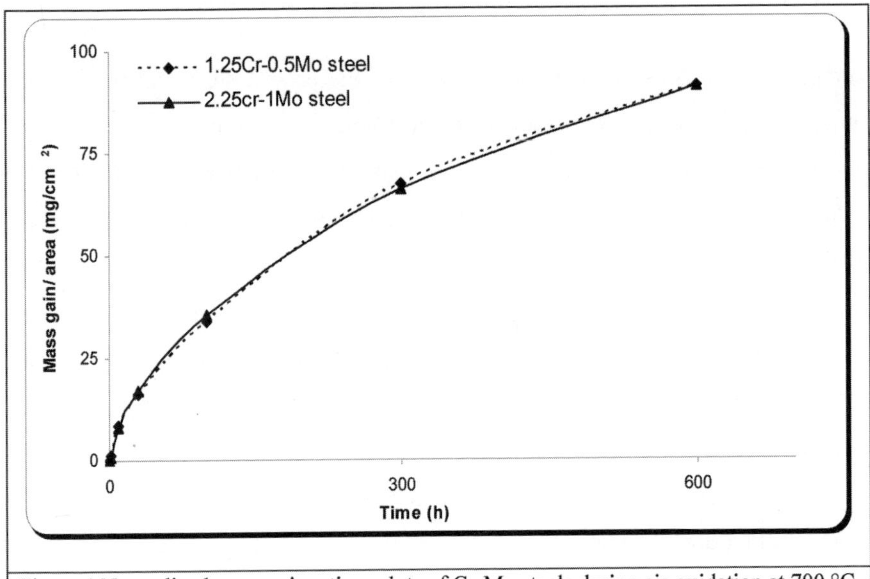

Figure 1 Normalized mass-gain v time plots of Cr-Mo steels during air-oxidation at 700 °C

Similarity of oxidation kinetics of the two steels has been explained on the basis of the chemical characteristics of the inner scales developed during their oxidation. Oxidation resistance of iron-chromium alloys is effected due to a protective inner layer of either chromium oxide (in high Cr alloys) or iron-chromium spinel oxide (in low chromium steels)[4]. However, it is the relative chromium content of the inner layer of Fe-Cr oxides in low Cr steels that has been found to govern the extent of protectiveness and the oxidation kinetics[11].

Multi-layered scales were present in the case of oxidized specimens of both 1.25Cr-0.5Mo and 2.25Cr-1Mo steels, as shown in Figure 2. Chromium contents (determined as Cr/Fe ratio) of the Cr-rich inner oxide layer (which is known to confer oxidation resistance to Fe-Cr alloys) of the scales formed over the two steels were compared. Relative Cr content in the inner layer of the scale developed over 2.25Cr-1Mo steel was considerably greater (maximum: ~15%) than those in similar scale layers over the 1.25Cr-0.5Mo steel (maximum: ~10%). On the basis of the chromium contents of the inner oxide layers of the two steels, 2.25Cr-1Mo steels should show a greater oxidation resistance, unlike the similar oxidation rates of the two steels suggested in Figures 1.

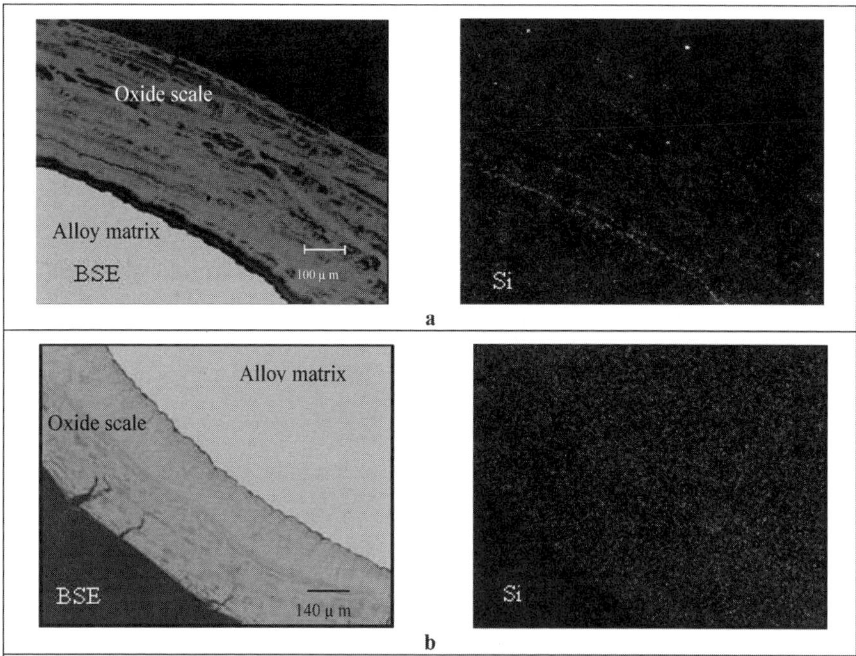

Figure 2 Backscattered electron image (BSE) and X-ray images for silicon in the cross-sections of oxide scales formed during oxidation in air at 700 °C: (a) 1.25Cr-0.5Mo steel and (b) 2.25Cr-1Mo steel.

In order to understand similarity in oxidation rates of the two steels, X-ray mapping (XRM) of the elements that are likely to influence oxidation resistance, viz., Fe, O, Si, S and Cr, has been carried out in the cross-sections of the oxidized specimens of the two steels. Inner scales on both the steels had considerable chromium contents. However, of particular relevance is a thin but distinguishable layer of high Si content exclusively at the alloy-scale interface of the oxidized 1.25Cr-0.5Mo steel. There is no such Si-rich inner layer in the case of scale developed over 2.25Cr-1Mo steel, where Si appears to be relatively evenly distributed across the scale. Formation of the thin Si-rich inner layer, presumably of SiO_2 in the case of 1.25Cr-0.5Mo steel is believed to be related to the higher silicon content of this alloy (Table 1). SiO_2 is known to be more effective as a protective scale[4] than other protective oxides, which may provide the reason for similarity in oxidation rates of 1.25Cr-0.5Mo and 2.25Cr-

1Mo steels, in spite of the lower Cr contents of the former.

Creep testing assumes determination of deformation essentially under the mechanical loading. A reduction in area due to reasons other than deformation under mechanical loading requires to be accounted for during such tests. Therefore, it is necessary to quantitatively account for an enhancement in stress due to reduction in the specimen diameter as a result of high temperature oxidation. In a recent study[12], a Stress Enhancement Factor (SEF) has been quantified for this purpose, which is given by:

$$SEF = \frac{\sigma}{\sigma_0} = \frac{F/A}{F/A_0} = \frac{A_0}{A} = \frac{d_0^2}{(d_0 - y)^2}$$

where, d_0 is the original diameter before oxidation, y is the reduction in diameter as a result of oxidation, A_0 and A are the cross-sectional areas corresponding to the original and reduced diameters, σ_0 and σ, corresponding stresses, and F is the applied load.

Kinetics of reduction in diameter as a result of oxidation losses were similar to those shown for the two steels in Figure 1. On the basis of the relationship shown in the above equation, the reduction of area due to oxidation in 30 hours at 700°C will translate into a stress enhancement by more than 0.03%. The stress enhancement factor for the two steels rises to 16-18% after 600 hours. However, these values violate the recommendation of ASTM E-139 according to which the tolerance should not exceed ±1% for load or stress variation[13].

Similarity in scale thickness of the two steels with different chromium and silicon levels during their oxidation under identical conditions has a great relevance to life assessment of in-service components by oxide scale thickness measurement[14,15]. Undue deterioration in creep life of an elevated temperature component (e.g., steam generator tube bank) is often caused by the component getting subjected to excessively high temperatures due to abnormal events, such as formation of hot spots etc. An excessive temperature will also cause development of a greater-than-average scale thickness, which can be detected by non-destructive testing, and correlated with the excessive temperature experienced that results in the loss of creep life. In order to correlate the thickness of the in-service oxide scale with the temperature, oxide scale growth rate data are generated over a range of temperatures[16]. For the purpose of life assessment, the power industry-based research groups require to generate long-term scale growth data for the steel used[16] over an industrially relevant range of temperatures. Similarity of oxide growth of a chromium-lean steel (1.25Cr-0.5Mo steel) with 2.25Cr-1Mo steel as a result of its slightly higher silicon content (as observed in this study) suggests that it will be necessary to generate the long-term scale growth data for the individual steels used, before such data can be accurately applied for life assessment by scale thickness measurement. Taking the role of minor compositional variations into account in generating industrially relevant scale growth data is particularly important since the ASTM specifications have provision for a wide range for certain elements in the for the two steels, particularly the range for silicon content, which is so critical for oxide scale growth (as established in the present study).

Conclusions

The scale growth kinetics during oxidation of 1.25Cr-0.5Mo and 2.25Cr-1Mo steels have been investigated in air at 700°C for 600 h, using cylindrical specimens, similar to the specimen geometry used in creep testing. Alloy-scale cross-sections have been characterized. The investigation has led to the conclusions as listed below:

(1) 1.25Cr-0.5Mo and 2.25Cr-1Mo steels have similar mass gain kinetics despite the difference in their chromium contents.

(2) The enhanced oxidation resistance for the lower chromium content alloy (1.25Cr-0.5Mo steel) was a result of higher silicon content of this steel than 2.25Cr-1Mo steel. The higher silicon content facilitated formation of silicon rich oxide near steel substrate leading to an improved protection.

References

1. A.K. Khare *(ed.), Ferritic Steels for High Temp. Appl.: ASM Int. Conf., Warren, PA, 1981.*
2. S.F. Pugh and EA. Little (eds.), Proc. *BNES lnt. Conf Ferritic Steels for Fast Reactor Steam Generators, Pub:* BNES, London, 1978.
3. T. Wada and G.T. Eldis, *Ferrinc Steels for High Temperature Applications: ASM Int. Conf. (*A.K Khare (ed.)), *Warren, PA, 1981, p.343-62.*
4. N. Birks and G.M. Meyer, *Introduction to High Temperature Oxidation of Metals*, Pub: Edward Arnold, 1982.
5. A. Rahmel, *Werkstoffe und Korrosion*, 1968, Vol.16, p.837.
6. S.N. Basu and G.J. Yurek, *Oxid Met*, 35 (1991) 441.
7. G.D. Yurek, D. Eisen and A. Garatt Reed, *Metall Trans A*, 13 (1982) 473.
8. M.J. Bannett, J.A. Desport and P.A. Labun, *Oxid Met,* 22 (1984) 291.
9. R.K. Singh Raman and A.K. Tyagi, *Materials Science and Technology*, 10 (1994) 27.
10. R.K. Singh Raman, J.B. Gnanamoorthy and S.K.Roy, *Oxidation of Metals*, 42 (1994) 1.
11. RK. Singh Raman: *Metall Trans A,* 26 (1995) 1847.
12. L. Marino and L.O. Bueno, *J Pressure Vessel Technology, Trans of ASME*, 123 (2001) 88.
13. ASTM E 139-00, Pub: International, West Conshohoken, PA.
14. R. K. Singh Raman, *Metall and Mater Trans A,* 31 (2000) 3101.
15. L.W. Pinder *Corr. Sci.*, 21 (1981) 749.
16. EPRI Report TR-102433-V2, 'Boiler Tube Failure Metallurgical Guide', 1993, Palo Alto, CA.

Advances in Materials Technology for Fossil Power Plants
Proceedings from the Fifth International Conference
R. Viswanathan, D. Gandy, K. Coleman, editors, p 551-563

DOI: 10.1361/cp2007epri0551

Creep Strength of High Cr Ferritic Steels Designed Using Neural Networks and Phase Stability Calculations

F. Masuyama* and H.K.D.H. Bhadeshia**

*Kyushu Institute of Technology
Graduate School of Engineering
1-1, Sensui-cho, Tobata, Kitakyushu 804-8550, Japan

**University of Cambridge
Department of Materials Science and Metallurgy
Pembroke Street, Cambridge CB2 3QZ, U.K.

Abstract

The highest creep rupture strength of recent 9-12% Cr steels which have seen practical application is about 130 MPa at 600°C and 100,000 h. While the 630°C goal may be realized, much more work is needed to achieve steam temperatures up to 650°C. Conventional alloy development techniques can be slow and it is possible that mathematical models can define the most economical path forward, perhaps leading to novel ideas. A combination of mechanical property models based on neural networks, and phase stability calculations relying on thermodynamics, has been used to propose new alloys, and the predictions from this work were published some time ago. In the present work we present results showing how the proposed alloys have performed in practice, considering long term creep data and microstructural observations. Comparisons are also made with existing enhanced ferritic steels such as Grade 92 and other advanced 9-12%Cr steels recently reported.

Introduction

Development of heat-resistant steel for power boilers and turbines has been ongoing for about five decades. This has led to an increase in the thermal efficiency of power plants whenever innovative steels have been commercially implemented. Through this effort, the steam conditions, temperature and pressure of power plants have recently been raised to about 600°C and 35 MPa using ferritic steels for improved efficiency in response to environmental protection and energy conservation requirements(1). Further enhancements in the creep resistance of 9-12%Cr steels used for boiler header/piping and steam turbine rotor applications are vital in order to achieve steam temperatures in excess of 630°C. Ferritic steels developed to date have maximum creep rupture strengths of approximately 130 MPa at 600°C and 65 MPa at 650°C for 100,000 hours. However there is strong demand for the development of ferritic steels having a strength of 100 MPa at 650°C for future advanced plants, including a 700°C class[(2)]. According to conventional alloy development techniques, creep strength can be improved by optimizing chemistry and heat treatment based on experience and test results of creep data obtained after long-term creep tests for several years or more. This can be a very slow way forward. It is possible that we are now at the limits of what can be achieved with ferritic steels. Given this situation, it is possible that mathematical models can provide the most economical path, with a different approach to alloy development[(3)].

A combination of mechanical property models based on neural networks, together with phase stability calculations relying on thermodynamics, has been used to propose new ferritic steels, and the predictions from this work were published some time ago[(4)]. The published work predicted excellent stress rupture strength, double or greater than that of the strongest steel developed to date. However, such predictions remain to be tested experimentally. Accordingly, two of the proposed steels, and one standard steel, were melted to be manufactured as wrought materials, and their creep rupture properties and microstructures were investigated for comparison with the predicted results. Based on this comparison, differences between the predictions and experimental results are discussed.

Analysis and Predicted Properties of the Proposed Steels [(4)]

Neural networks now comprise a general method of non-linear regression analysis in which a mathematical relationship is established between each of the independent input variables, x_j and one or several dependent output variables, y. In linear regression analysis, the sum of all products x_j multiplied by a weight w_j and a constant θ then gives an estimate of the $y = \sum_j w_j x_j + \theta$.

Neural networks are in general non-linear and the non-linearity is achieved by taking a hyperbolic tangent of the right hand side of this equation, and then applying a linear transfer into y. In fact, many hyperbolic tangents and corresponding weights can be added to make the function as complex as is necessary. The relationship is generated by presenting a network with a database consisting of a set of input conditions for which the value is known. The network then learns a relationship between the input conditions and corresponding values for the output in a procedure that is known as training the network. Once the network is trained, output prediction (creep rupture strength in the present study) for any given inputs such as chemical composition and heat treatment condition is very rapid. However, there are no physical models for the output parameters or creep rupture strength. An example of the neural network structure is shown in Figure 1. In the present study the network consisted of 37 input nodes (one for each variable), a number of hidden nodes and an output node representing the creep rupture strength. The hidden nodes are where the mathematical operations described below are carried out. More hidden nodes represent a more complex model. The network was trained using 1033 examples randomly chosen from a total of 2066 available, which are combinations of creep rupture strength and 30

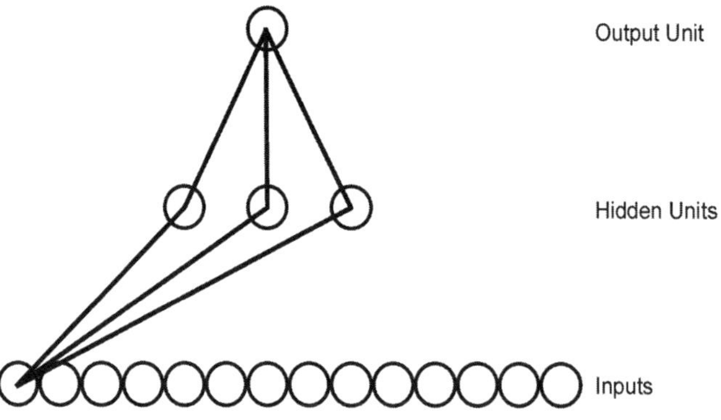

Fig.1 A typical three-layer network in the analysis

original inputs including the time to rupture, chemical composition, and heat treatment in the database compiled from 27 published reports, with the remaining 1033 examples being used as new experiments to test the trained network.

Linear functions of the inputs x_j are operated on by the hyperbolic tangent transfer function:

$$h_i = \tanh\left(\sum_j w_{ij}^{(1)} + \theta_i^{(1)}\right) \tag{1}$$

so that each input contributes to every hidden unit. The bias is designated θ_i and is analogous to the constant that appears in linear regression. The strength of the transfer function is in each case determined by the weight w_{ij}. The transfer to the output y is linear:

$$y = \sum_i w_{ij}^{(2)} + \theta^{(2)} \tag{2}$$

The specification of the network structure, together with the set of weights is a complete description of the formula relating the inputs to the output.

To explore the network, a study was made on the behavior of the model for two classic steels: 2.25Cr-1Mo (Grade 22 steel bainitic) and 10CrMoW (similar in composition to Grade 92 steel martensitic, but containing 0.86%Cu and higher Cr content of 10.61% - all concentrations in wt%). The resulting calculated curves of creep rupture strength as a function of time corresponded well to the actual strength for both steels, and trend curves for the effects of alloying elements and heat treatment conditions on the creep rupture strength were also developed and demonstrated.

Based on the findings by neural network analysis, innovative steels were proposed with creep rupture strengths. Table 1 shows the compositions emerging from the design, as well as the analyzed chemical composition of the manufactured steels. Table 2 shows the designed heat

Table 1 Input parameters on chemical composition and analysis of test steels

(mass%)

Steels		C	Si	Mn	P	S	Cr	Mo	W	Ni	Cu	V	Nb	N	Al	B	Co	Ta	O	Re
A	Input	0.12	0.00	0.48	0.0016	0.001	9.00	0.75	3.00	0.00	0.00	0.21	0.01	0.064	0.000	0.0080	1.25	0.0003	0.01	0.0003
	Analysis	0.11	<0.01	0.49	0.003	0.001	9.04	0.74	2.99	0.01	<0.01	0.20	0.011	0.064	0.004	0.0070	1.25	<0.0001	0.003	0.0003
B	Input	0.13	0.00	0.50	0.0016	0.001	8.70	0.30	3.00	0.00	0.00	0.21	0.01	0.064	0.000	0.0080	0.00	0.0003	0.01	0.0003
	Analysis	0.13	<0.01	0.50	0.002	0.001	8.75	0.30	2.99	<0.01	<0.01	0.20	0.011	0.068	0.003	0.0078	0.01	<0.0001	0.002	0.0006
N	Input	0.12	0.05	0.64	0.016	0.001	10.61	0.44	1.87	0.32	0.86	0.21	0.01	0.064	0.022	0.0022	0.02	0.0003	0.01	0.0003
	Analysis	0.12	0.04	0.64	0.018	0.001	10.65	0.43	1.87	0.32	0.85	0.20	0.011	0.072	0.021	0.0024	0.01	< 0.0001	0.003	0.0006

Table 2 Input parameters on heat treatment conditions applied to test steels

Steels	Normalizing			Tempering			Annealing		
	Temp. (°C)	Duration (h)	Cooling Rate	Temp. (°C)	Duration (h)	Cooling Rate	Temp. (°C)	Duration (h)	Cooling Rate
A	1200	2	AC	800	4	AC	740	4	AC
B	1180	2	AC	800	4	AC	740	4	AC
N	1065	2	AC	770	4	AC	740	4	AC

treatment parameters which were actually applied to the test steels. Steels A and B are proposed steels which are modified from the above-mentioned model or standard steel, 10CrMoW (steel N in Table 1). The first attempt led to the design of steel A, but its long term strength at 650°C barely failed to meet the 100 MPa requirement. Changes were thus made to improve both the mean long term strength and the certainty of prediction, reducing the cobalt, chromium and molybdenum concentrations, with the resulting material designated as steel B. The newly designed steels A and B do not contain any silicon, aluminum, nickel or copper, which are well known to cause deterioration in creep rupture strength. The boron concentration is reduced primarily to reduce the uncertainty of the predictions. There is also an increase in the normalizing temperature, as well as reductions in the manganese and chromium concentrations together with an increase in the level of tungsten. Consequently, the predicted 100,000 h creep rupture strength of each steels were 240-340 MPa, 230-310 MPa and 110-130 MPa at 600°C for steels A, B and N (10CrMoW) respectively, and 80-200 MPa, 130-180 MPa and 40-70 MPa at 650°C for steels A, B and N respectively. The predicted creep rupture strengths of Steels A and B are highly distinguished. The strength of steel N approaches the level of Grade 92, which is the highest among all steels commercially developed to date. If the mean values of the predicted 100,000 h creep rupture strengths of steels A and B are compared to steel N, steels A and B would exhibit approximately 2.5 times the strength of steel N at 600°C and 650°C.

In conjunction with the neural network calculation, phase diagram calculations and kinetic predictions through the calculation of precipitation reactions on the basis of thermodynamic data and kinetic theory were also carried out. As a result of those calculations, it is predicted that the equilibrium fractions are $M_{23}C_6$ and Laves phase in each of the steels, and that the Laves phase only occurs in steel N (10CrMoW) at the very late stage of annealing at 650°C. Steels A and B were not found to exhibit Laves phase precipitation, at least at 10^6 h at 600°C or 650°C.

Test Materials and Experimental Procedures

The above-mentioned steels A and B designed using neural networks and phase stability calculations, proposed as new strong ferritic steels for power plant applications, were melted in a vacuum furnace as well as a standard material consisting of steel N (10CrMoW). Melted ingots weighing 20 kgf were forged to bars with sectional dimensions of 20 mm by 40 mm. Chemical compositions analyzed in comparison with target values, i.e., the input or predicted design parameters and heat treatment conditions of the test steels, are shown in Tables 1 and 2, respectively. Hardness measurement of the test steels (steels A, B and N) indicated 214, 216 and 225 respectively in terms of Vicker's hardness. Tensile strength at room temperature of the test steels showed a trend similar to that of hardness, indicating 659 MPa, 675 MPa and 715 MPa for steels A, B and N respectively, but the elevated temperature tensile strength of steels A and B turned out to be greater than for steel N at the temperature of 700°C. Values were 229 MPa and 233 MPa for steels A and B respectively, against 217 MPa for steel N. Creep tests were carried out at the temperatures of 600°C, 650°C and 700°C for a maximum duration of 23,500 h. Creep strain measurement and microstructural studies were also performed for comparative study of the creep deformation behavior among the test steels, as well as comparison of the precipitation properties with the predicted results.

Creep Properties of the Designed Steels

Figure 2 shows rupture stress versus time to rupture diagrams at the temperatures of 600°C, 650°C and 700°C for steels A, B and N. In the graph, the creep rupture data for Grade 92[5] are also plotted with the average lines obtained by parametric analysis for the purpose of comparison. Steel N is slightly weaker than Grade 92 at 600°C, and the difference in strength between the two becomes greater at 650°C and 700°C with increased temperature. The strengths of steels A and B are nearly the same and lie on the upper bound or on a slightly stronger level than the Grade 92 data band at the temperatures of 600°C and 650°C, but at 700°C steel B shifts to the average line of Grade 92, and steel A shifts to the lower bound of the data band. Figure 3 demonstrates the Larson-Miller plot of the data for the three test steels. A parameter constant of 34.1 was selected as best fitting the average strength of Grade 92, but this value does not provide the best fit to the test steels in the entire parameter range. From the Larson-Miller master curve, the 100,000 h creep rupture strengths of the test steels were predicted as 129 MPa, 133 MPa and 108 MPa at 600°C, and 74 MPa, 76 MPa and 56 MPa at 650°C for steels A, B and N, respectively. Meanwhile, the strength of Grade 92 was predicted as 122 MPa at 600°C and 64 MPa at 650°C. Figure 4 visually illustrates the comparison of creep rupture strength estimated by neural network analysis and measured with respect to the actually melted test specimens, as well as the measured tensile strength in the graph appearing in previously published literature(4). The measured creep rupture strength of steel N is generally observed in the band close to the predicted line with statistical deviation, while the strengths of steels A and B lie on the upper bound or slightly toward high stress at 600°C, but slightly stronger than the predicted line for Steel N at 650°C. The prediction of rupture stress versus time to rupture for steel N corresponds

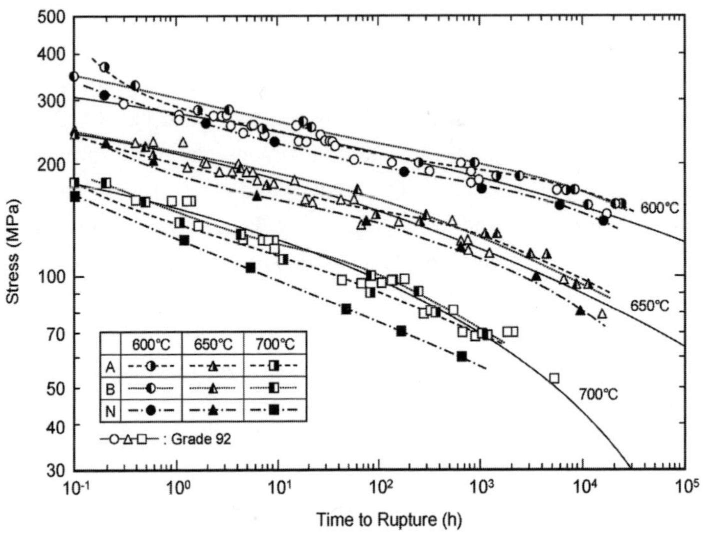

Fig.2 Stress vs. time to rupture diagrams at 600°C, 650°C and 700°C for test steels

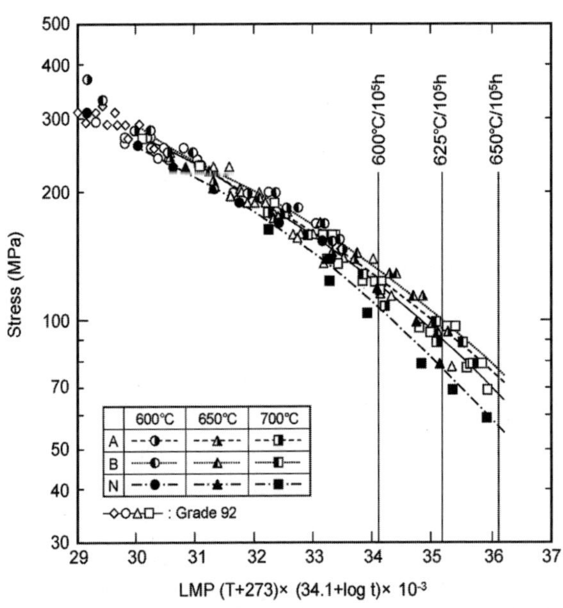

Fig.3 Larson-Miller parameter plots of creep rupture strength of test steels

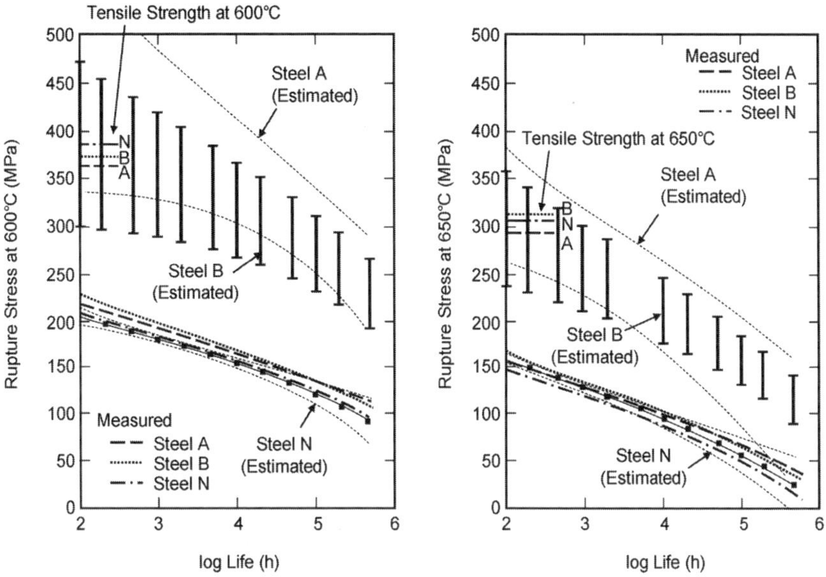

Fig.4 Comparison of creep rupture strength between predictions and measured results

very well to the measured results at 600°C, and shifts slightly in a parallel manner towards the weaker direction at 650°C.

Figure 5 shows the creep deformation behaviors of the test steels at 650°C. The primary creep rates of the test steels start at the same value and decrease along nearly the same line despite the different test stress conditions, and then start tertiary creep at different minimum creep rates

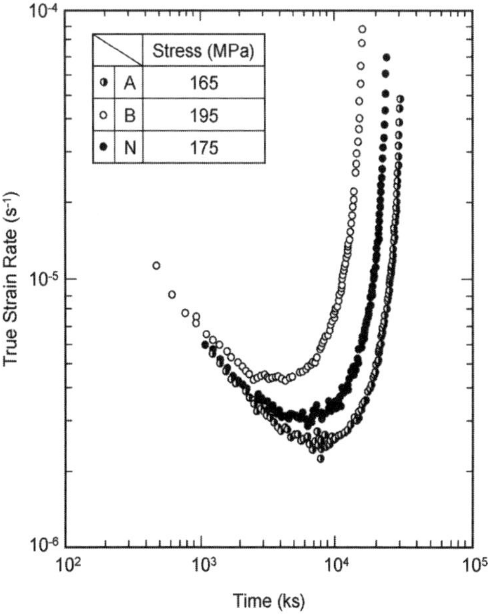

Fig.5 Creep deformation behaviors of test steels at 650°C

depending on the stress and the material. However, the tertiary creep curves of the three steels behave according to a similar trend, suggesting that the same mechanisms govern the creep. Figure 6 shows the average creep rate versus stress relationship for the test steels. It is confirmed that the average creep rate is proportional to the minimum creep rate, with a coefficient of approximately 0.98 x 10^1 across the entire range of stress(6). Norton's law therefore applies to the relationship between average creep rate and stress. At the temperatures of 600°C and 650°C, very high stress exponent values such as 16 and 25 are seen (25 in the case of steel A only, at the lower stress range below the stress which lead 16). This means that the steels are highly strengthened by precipitations promoting a large amount of internal stress. At the temperature of 700°C, the stress exponent is in the range of 8 to 11 with moderate strengthening by precipitation.

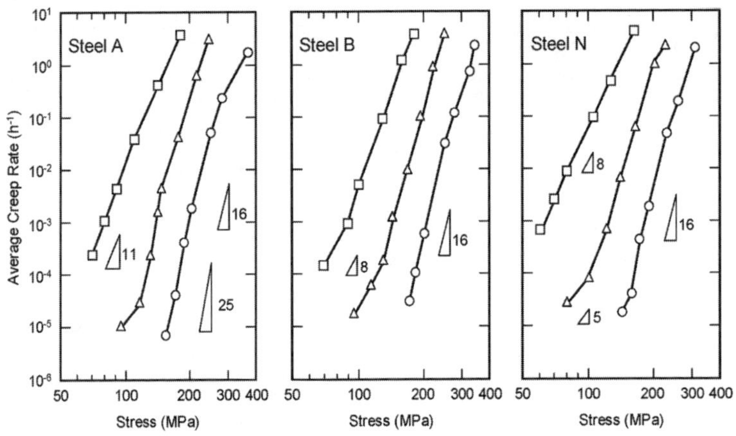

Fig.6 Creep rate vs. stress relationship for test steels

Microstructures of the Designed Steels

Figure 7 shows optical micrographs of the as-manufactured or creep ruptured test steels. Steels A and B exhibit courser prior austenite grain size than steel N, and very tight martensite lath structures are observed in each of the steels before and after the creep test. The crept specimens include recovered lath structures at certain locations in the grains, but it is hard to distinguish the recovered lath in the complicated fine structures. Figure 8 shows highly magnified transmission electron micrographs of the creep ruptured test steels. The martensite lath grain exhibited reduced dislocation density, with recovery to subgrain structures. Precipitations are observed along the grain boundaries and the interior of the recovered subgrain. Figures 9 and 10 show FE-SEM graphs of as-manufactured and creep ruptured test steels respectively. The as-manufactured specimens were observed using secondary electron imaging (SEI) and back scattered electron imaging (BEI), but the creep ruptured specimens were viewed only by BEI. The SEI graph displays all of the precipitation as white points, while the BEI graph displays the precipitates containing heavy elements such as molybdenum and tungsten as bright points, meaning that the bright points correspond to the Laves phase in the present study. The SEM graph of the as-manufactured specimen in Figure 9 shows numerous fine precipitates along the grain boundaries. On the other hand, in the case of the BEI graph, only steel A shows a number of bright points corresponding to the Laves phase, while steels B and N show no bright spots. The BEI graphs of the gauge portion and grip portion of the creep ruptured specimens show numerous Laves phase precipitation along the prior austenite grain boundaries and packet/block boundaries. The fact that there is no notable difference between the gauge and grip portions means that the stress does not influence the Laves phase precipitation. The observation of FE-SEM structures and other element analysis such as EPMA clarified that all the test steels formed precipitates of $M_{23}C_6$ and

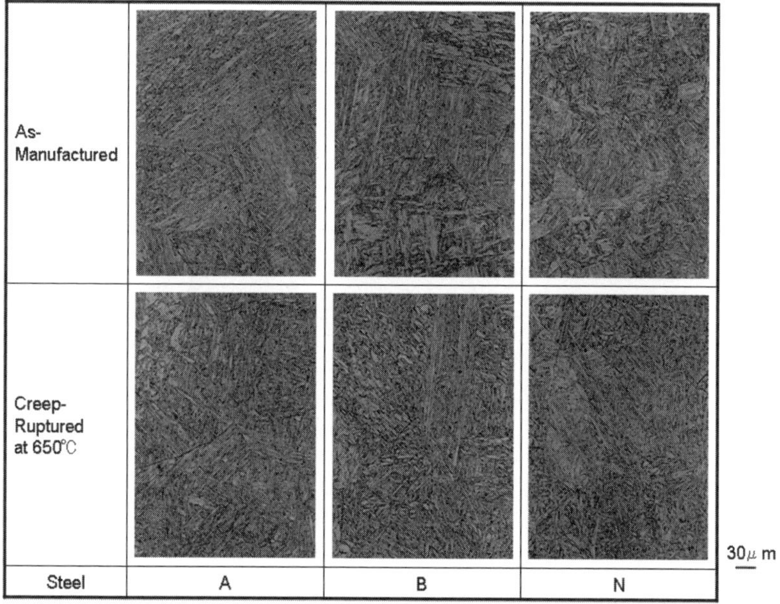

Fig.7 Microstructures of tested steels as manufactured or creep-ruptured

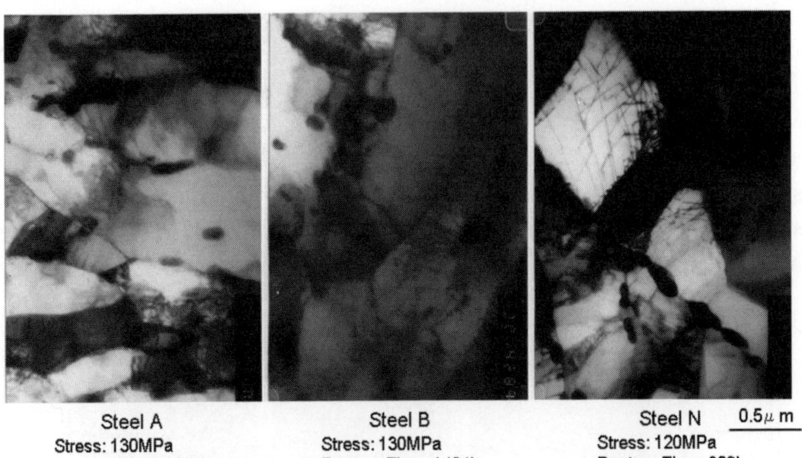

Fig.8 TEM structures of test steels creep-ruptured at 650℃

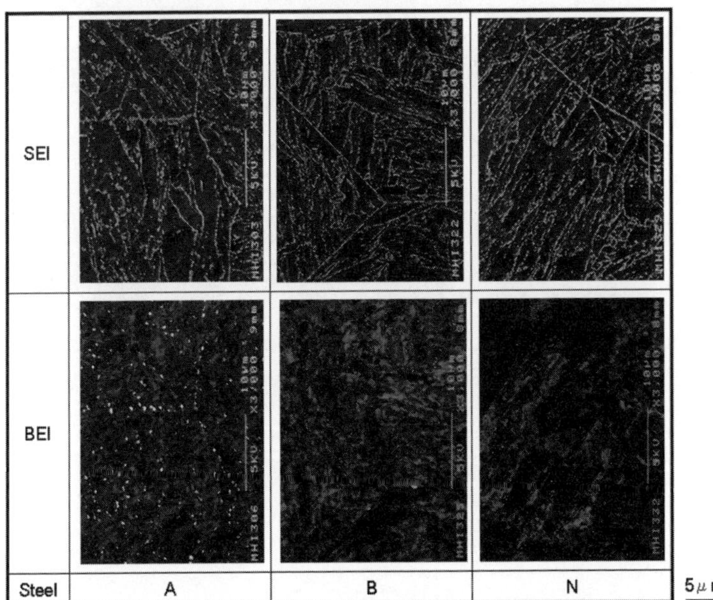

Fig.9 FE-SEM structures of test steels as-manufactured

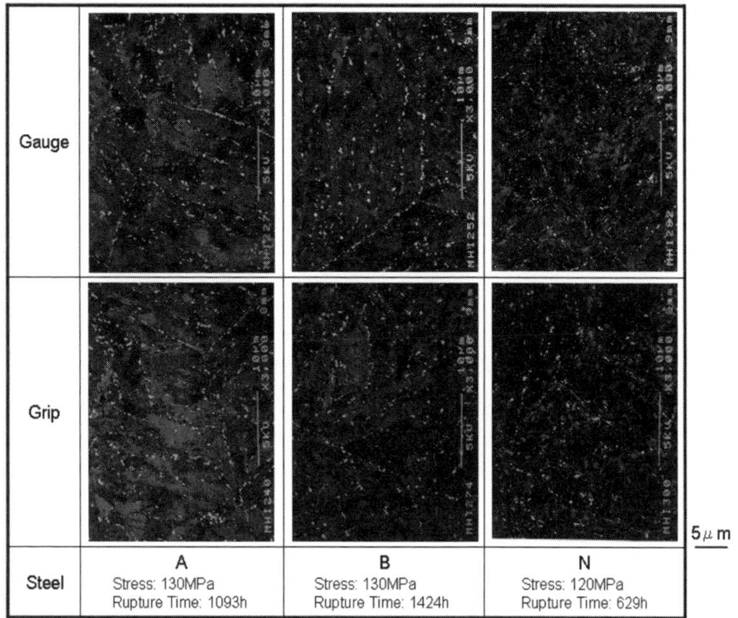

Fig.10 FE-SEM BEI structures of test steels creep-ruptured at 650℃

MX under the as-manufactured conditions (normalized, tempered and annealed). In addition to these two precipitates, Laves phase was heavily generated in steel A, but hardly at all in steels B and N. It is also suggested that Laves phase appears in the all steels due to heating during the creep test, but without any effect on creep stress. The M_2X precipitation predicted by the kinetic calculation was not observed in the experimental specimens in the present work. These results obtained from the actually melted steels differ from the phase prediction based on the thermodynamic and kinetic models mentioned previously.

Discussion

The predictions of the properties of steels designed using neural networks and phase stability calculations were tested experimentally. The major differences between predictions and experimental results were (1) creep rupture strength, and (2) precipitation of Laves phase and M_2X. The measured creep strength of the experimental specimens whose chemical compositions and heat treatment were nearly the same as predicted excepting silicon, nickel, copper and aluminum for steels A and B, and in addition to these cobalt for Steel B, was two fifths of the predicted values, but lay at the upper bound of the strongest existing steel (Grade 92). This was well reproduced by the predicted steel, steel N (10CrMoW), particularly at the temperatures of 600°C and 650°C. In comparison with commercial steel grades, the predicted steels A and B are approximately 5%, 30% and 40% stronger than Grades 92, 122 and 91[(7)] respectively at 600°C and 100,000 h based on data with maximum testing duration of 23.500 h. As indicated above, some alloying elements were input as 0%, despite the fact that actual melting cannot achieve these ideal inputs. The roles of the 0% element are not certain in the designed steel nor is the effect of the lean elements in the melted steels on creep strength reduction. Also, there exists no data on 0% silicon, nickel, copper and aluminum alloyed steels. As shown in Figure 4, elevated tensile strengths at 600°C and 650°C for each of the steels actually melted are at nearly the same level, but the creep rupture strengths of steels A and B are predicted as being extremely high against tensile strength. If the steel N exhibits a reasonable balance of tensile strength and creep strength, steels A and B would have the twice of tensile strength measured or 700 MPa at 600°C and 600 MPa at 650°C, which would be impractical. However, if an unbalanced relationship between tensile strength and creep strength could be accepted, the creep strength mechanism of steels A and B should be expected to differ from that of steel N. Nevertheless, the creep deformation mechanism and strengthening mechanism (precipitation strengthening) were confirmed to be the same in the previous section. It is considered that the lack of data on chemistry effects (particularly 0% elements) causes the prediction of mutation of creep rupture strength if there is no discontinuity in the effect of concentration of elements on the creep strength.

Precipitates observed in the test steels are nearly the same as in existing high strength ferritic steels, such as Grades 91, 92 and 122, excepting Laves phase precipitation. The Laves phase in general is not observed in steels normalized and tempered at temperatures above 730°C, typically 760°C and above, and generally applied as boiler steels. However, steel A showed considerable Laves phase after annealing at 740°C, and no Laves phase was also observed in steels B and N. Annealing followed by air cooling is somewhat effective in Laves phase generation, and steel A contains a larger quantity of molybdenum compared with steels B and N, and large amount of tungsten. Phase stability calculations at the equilibrium state suggested Laves phase formation, and the kinetic model predicted Laves phase after long term heating at 650°C for steel N, but not for steels A and B. Meanwhile, it was predicted that M_2X would form in all steels due to the short time of heating. In fact, the test steels formed $M_{23}C_6$, MX and Laves phase before and after the creep test. It is also not certain whether the 0% elements affect precipitation behavior, or why M_2X (Cr_2N based precipitation) occurs in the kinetic calculation, as opposed to MX in the designed and actually melted steels.

Conclusions

A combination of mechanical property models based on neural networks together with phase stability calculations relying on thermodynamics has proposed new ferritic steels for power applications, and these predictions were tested experimentally. The results obtained from this work are as follows:

(1) The major differences between the predictions and the experimental results were regarding creep rupture strength and the precipitation of Laves phase and M_2X.

(2) The creep strength of the experimental specimens was two fifths of the predictions, but lay at the upper bound of the strongest existing steel (Grade 92), and this was well reproduced by the predicted steel, steel N (10CrMoW), particularly at the temperatures of 600°C and 650°C. In comparison with commercial steel grades, the predicted steels A and B are approximately 5%, 30% and 40% stronger than Grades 92, 122 and 91 respectively at 600°C and 100,000 h.

(3) The experimental specimens formed precipitates consisting of $M_{23}C_6$, MX and Laves phase before and after the creep test despite the prediction of precipitates in the designed steels of $M_{23}C_6$ and Laves phase at the equilibrium state, and of $M_{23}C_6$ and M_2X on the kinetic time scales.

References

(1) F. Masuyama, "History of Power Plants and Progress in Heat Resistant Steels", ISIJ International, Vol. 41, No. 6, (2001), pp. 612 – 625.

(2) F. Masuyama, "Advanced Power Plant Developments and Material Experiences in Japan", J. Lecomte-Beckers, et al. (eds), Materials for Advanced Power Engineering, Part I, Forschungszentrum Juerich GmbH, Germany, (2006), pp. 175 – 187.

(3) H. K. D. H. Bhadeshia, "Neural Networks in Materials Science", ISIJ International, Vol. 39, No. 10, (1999), pp. 966 – 979.

(4) F. Brun, T. Yoshida, J. D. Robson, V. Narayan, H. K. D. H. Bhadeshia and J. C. Mackay, "Theoretical Design of Ferritic Creep Resistant Steels using Neural Network, Kinetic, and Thermodynamic Model", Materials Science and Technology, Vol. 15, (1999), pp. 547 – 554.

(5) H. Mimura, M. Ohgami, H. Naoi and T. Fujita, "Properties of 9Cr-1.8W Steel with High Creep Strength for USC Boiler Piping and Tubing Applications", D. Coutssouradis, et al. (eds.), Materials for Advanced Power Engineering 1994, Part I, Kluwer Academic Publishers, Netherlands, (1994), pp. 361 – 372.

(6) F. Masuyama and N. Komai, "Evaluation of Long-term Creep Rupture Strength of Tungsten-strengthened Advanced 9-12%Cr Steels", Key Engineering Materials, Vol. 171-174, (2000), pp. 179 – 188.

(7) K. Kimura, "Review of Allowable Stress and New Guideline of Long-term Creep Strength Assessment for High Cr Ferritic Creep Resistant Steels", I. A. Shibli, et al. (eds.), Creep & Fracture in High Temperature Components -Design & Life Assessment Issues-, DEStech Publications, Pennsylvania, (2005), pp. 1009 – 1022.

Advances in Materials Technology for Fossil Power Plants
Proceedings from the Fifth International Conference
R. Viswanathan, D. Gandy, K. Coleman, editors, p 564-581

DOI: 10.1361/cp2007epri0564

Ferritic and Austenitic Grades for New Generation of Steam Power Plants

S. Caminada [1)], L. Cipolla [2)], G. Cumino [1)], A. Di Gianfrancesco [2)], Y. Minami [3)], T. Ono [3)]
[1)] Tenaris: Piazza Caduti 6 Luglio 1944 1, Dalmine 24044 (BG) Italy
[2)] Centro Sviluppo Materiali SpA: Material & Product Directorate, Via di Castel Romano 100, 00128 Rome, Italy
[3)] TenarisNKKt R&D, 1-10 Minamiwatarida, Kawasaki, Kanagawa, 210-0855 Japan

Abstract

The steam parameters in the new high efficiency fossil fuel power plants are continuously increasing, requiring new advanced materials with enhanced creep strength able to operate on the most severe temperature and pressure conditions.
Tenaris focused on the development of ferritic-martensitic and austentitic grades for tubes and pipes applications.
The product development in TenarisDalmine for the ferritic-martensitic grades has been focused on:
- low alloyed ASTM Grade 23 as substitute of Grade 22 for components operating at relatively low temperatures, containing 1.5% W and with quite good weldability and creep properties up to 580°C and a competitive cost;
- high alloyed ASTM Grade 92, an improved version of the well known Grade 91 for the superheaters, headers and other parts of the boiler operating at temperatures up to 620°C: its tempered martensitic structure offers very high creep strength and long term stability.

The product development in TenarisNKKt R&D on austenitic grades has been focused on:
- TEMPALOY AA-1 as improved version of 18Cr8NiNbTi with the 3%Cu, showing high creep and corrosion properties,
- TEMPALOY A-3: a 20Cr-15Ni-Nb-N showing good creep behaviour and corrosion properties better than AA-1 due to the higher Cr content.

This paper describes the Tenaris products, the process routes and the main characteristics of these steels, including the effect of shot blasting on steam oxidation properties of the austenitic grades, as well as, the R&D activities in the field of alloy design, creep tests, data assessment, microstructural analysis and damage modelling, conducted with the support of the Centro Sviluppo Materiali.

KEYWORDS: Ferritic Steel; Austenitic Steels; Boiler components; Creep properties, Creep data assessment;

1. Introduction

Severe requirements on strength, corrosion, creep properties and thermal stability during service are requested for high temperature steels for boilers, steam lines and headers in Ultra Super Critical (USC) Power Plants.
Tenaris has accumulated a great experience in the production of C-Mn and Cr-Mo ferritic/martensitic grades in TenarisDalmine and TenarisSilcotub plants, as well as, in the TEMPALOY AA1 and A3 austenitic grades in TenarisNKKt plant.
The development and the industrialization were supported by R&D activities, carried out jointly with Centro Sviluppo Materiali (CSM) and TenarisNKKt R&D.
In particular for the ferritic/martensitic grades, in the last twenty years huge R&D activities were made in the frame of European Coal and Steel Community programs [1] and later in the COST 501, 522 and now 536 programs to develop and optimize high Cr-Mo steels such as Grade 91, 911 and

92 and the new generation of low alloyed Grades 23 [2,3,4,5,6]. At present all the above grades are currently produced, as seamless tubes and pipes, in a wide range of sizes.
The product development in TenarisNKKt for the austenitic grades has been focused on:

- TEMPALOY AA-1 as improved version of 18Cr8NiNbTi with the 3%Cu, showing high creep and corrosion resistance properties,
- TEMPALOY A-3: a 20Cr-15Ni-Nb-N showing good creep behaviour and corrosion properties better than AA-1 due to the higher Cr content.

This paper describes the Tenaris products and the main characteristics of all these steels, including the effect of shot blasting on high temperature corrosion properties of the austenitic grades, as well as, the R&D activities in the field of alloy design, creep tests, data assessment, microstructural analysis and damage modelling, conducted with the effort of the Centro Sviluppo Materiali.

2. Production route

2.1 Ferritic/Martensitic Steels

TenarisDalmine and TenarisSilcotub tubes and pipes are produced from billets, manufactured either directly by continuous casting process or manufactured from hot rolled ingots. All CrMo steels are vacuum degassed to improve the cleanliness and to reduce gas content.
Tubes and pipes are produced in 5 different mills, according to the size of the finished products. The working window for thermomechanical (piercing and rolling) processes is selected on the basis of hot torsion test results [7].
Tubes with OD up to 88.9 mm are manufactured in a continuous mill (FAPI mill); tubes and pipes up to 146mm are produced in Silcotub mandrel mill; pipes with OD up to 406.4 mm are produced in a multistand pipe mill (MPM) with retained mandrel [8], whereas pipes with OD from 406.4 up to 711 mm are manufactured in a rotary expansion mill (Expander mill).
The heavy thickness and large diameter pipes are produced in a pilger mill. Heat Recovery Steam Generator (HRSG) tubes produced in the continuous mill are available in lengths up to 24 meters.
Low and high Cr-Mo Grades currently produced by TenarisDalmine are shown in Table 1. All the compositions comply with the main standards: ASTM, ASME code cases, EN and TÜV.

Grade		C	Mn	P	S	Si	Cr	Mo	W	Nb	V	B	Other
9	min	-	0.30	-	-	0.25	8.00	0.90	-	-	-	-	-
	max	0.15	0.60	0.025	0.025	1.00	10.00	1.10	-	-	-	-	
22	min	0.05	0.30	-	-	-	1.90	0.87	-	-	-	-	-
	max	0.15	0.60	0.025	0.025	0.50	2.60	1.13	-	-	-	-	
23	min	0.04	0.10	-	-	-	1.90	0.05	1.45	0.02	0.20	0.0005	N: 0.03 max
	max	0.10	0.60	0.030	0.010	0.50	2.60	0.30	1.75	0.08	0.30	0.0060	
91	min	0.080	0.30	-	-	0.20	8.00	0.85	-	0.06	0.18	-	N: 0.03-0.07
	max	0.012	0.60	0.020	0.010	0.50	9.50	1.05	-	0.10	0.25	-	
92	min	0.07	0.30	-	-	-	8.50	0.30	1.50	0.04	0.15	0.0010	N: 0.03-0.07
	max	0.13	0.60	0.020	0.010	0.50	9.50	0.60	2.00	0.09	0.25	0.0060	
911	min	0.09	0.30	-	-	0.10	8.50	0.90	0.90	0.06	0.18	0.0003	N: 0.04-0.09
	max	0.13	0.60	0.010	0.010	0.50	10.50	1.10	1.10	0.10	0.25	0.0060	
X20	min	0.17	-	-	-	0.15	10.00	0.80	-	-	0.45	-	Al: <0.040
	max	0.23	1.00	0.030	0.025	0.50	12.50	1.20	-	-	0.55	-	
WB36 (15NiCuMo Nb5-6-4)	min	-	0.80	-	-	0.25	-	0.25	-	0.015	-	-	Ni: 1.0-1.3 Cu: 0.5-0.8
	max	0.17	1.20	0.025	0.020	0.50	0.30	0.50	-	0.045	-	-	

Table 1 – Chemical composition of main low and high Cr-Mo steels in the main international standards

2.2 Austenitic steels

Tenaris austenitic steel grades are shown in Table 2 [9,10,11]. The TEMPALOY AA-1 has been registered as KA-SUS321J2HTB in METI Standards of Japan in 2001, and also as ASME code Case N° 2512 and in ASTM A213-06a as UNS S30434.
TEMPALOY A-3 has been registered as KA-SUS309J4HTB in METI Standards of Japan in 1997.
The production route for austenitics starts from rolled hot forged billets followed by hot extrusion and cold tube forming.

Grade		C	Si	Mn	P	S	Cu	Ni	Cr	Ti	Nb	B	(Ti+Nb/2)C
TEMPALOY AA-1	min	0.07	-	-	-	-	2.5	9.0	17.5	0.1	0.1	0.001	2.0
	max	0.14	1.00	2.00	0.04	0.01	3.5	12.0	19.5	0.25	0.4	0.004	4.0
TEMPALOY A-3	min	0.03	-	-	-	-	-	14.5	21.0	-	0.5	0.001	N: 0.10%
	max	0.10	1.0	2.0	0.04	0.03	-	16.5	23.0	-	0.8	0.005	N: 0.20%

Table 2 – Chemical composition of austenitic steels

3. Microstructure and mechanical properties

3.1 Ferritic/Martensitic steels

In order to obtain a good compromise between creep resistance, toughness and cold formability (i.g. limited hardness values), the CrMo steels are usually supplied in normalized and tempered conditions.
The normalizing heat treatment gives a martensitic (bainitic or ferritic for low alloy grades) structure and provides a good carbides solubilisation into the matrix. The optimum austenitizing temperature during normalizing depends on the chemical composition and on the desired properties: a coarse grain promotes creep resistance, while a finer one increases proof strength and toughness.
In order to set up and tune the heat treatment parameters an extensive data base of CCT diagrams was produced, taking into account different austenitizing temperatures and chemical compositions for the same Grade.
Examples of heat treatment conditions used for Cr-Mo-V steels are listed in Table 3.
The subsequent tempering heat treatment softens the material and promotes the formation of $M_{23}C_6$ and fine MX precipitates, resulting into a better creep resistance. The desired final microstructure is constituted of a tempered martensitic structure with V and Nb carbides and carbonitrides [7].
After tempering $M_{23}C_6$ carbides, which exhibit high thermal stability, nucleate predominantly at prior austenite grain boundaries and along martensite lath boundaries, increasing creep strength by retarding subgrain growth.
All TenarisDalmine products were characterized in terms of:

- microstructure and precipitation by Light (LM), Scanning (SEM) and Transmission Electron Microscopy (TEM/STEM-EDS);
- surface quality: measurement of roughness, decarburization depth and oxide (scale) thickness;
- mechanical properties: tensile tests from room up to service temperature, hardness, impact energy, with FATT evaluation; long term creep behaviour;
- suitability to cold bending and hot induction bending.

Steel Grade	Dimension [mm x mm]	Normalizing		Tempering	
		Temperature (°C)	Time (min)	Temperature (°C)	Time (min)
P23	219x31.75	1070	45	760	90
P91	609 x 17.5	1070	20	780	60

Table 3 - Examples of heat treatment conditions for CrMoV steels

Examples of typical microstructures by LM, SEM and TEM for the different Grades in the final heat treatment conditions are shown in Figures 1 and 2. Figure 3 shows the typical tempered martensitic structure of grade 91.

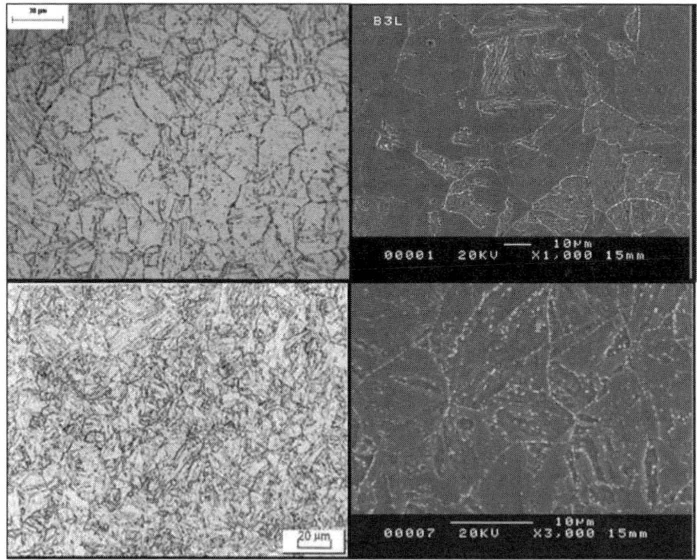

Figure 1 - Tempered bainitic microstructure of a Grade 23 pipe (219 mm OD x 31.75 mm WT) after N+T by LM and SEM

Figure 2 - Tempered martensitic microstructure with $M_{23}C_6$ carbides at the boundary of the martensitic laths for Grade 91 pipe (609 mm OD x 17.5 mm **WT)**

Figure 3 - TEM thin foil of Grade 91 pipe (609 mm OD x 17.5 mm WT) after N+T

The precipitation state in TenarisDalmine Grades 91, 911 and 92 tubes and pipes after normalizing and tempering was investigated by STEM/TEM technique: mean dimensions of MX carbonitrides and $M_{23}C_6$ carbides were measured; the particles in the as-treated Grade 92 tube exhibit comparable dimensions with those in as-treated Grades 911 and 91 (Table 4).

Grade	**92**	**911**	**91**
Heat treatment	1070°C+780°C	1060°C+760°C	1070°C+780°C
AVERAGE DIAMETER (nm)			
MX	45	50	35
$M_{23}C_6$	125	130	139

Table 4 – State of precipitation in advanced 9%Cr steels after quality heat treatment. subgrains; b) fine MX carbonitrides inside the matrix.

3.2 Austenitic steels

TEMPALOY AA-1 and A-3 austenitic steels are supplied after solution heat treatment at T $\geq$ 1160°C with a soaking time of 3 minutes + 1min/mm.

3.2.1 TEMPALOY AA-1

The TEMPALOY AA-1 is characterised by a MC and $M_{23}C_6$ carbides strengthening increased by the Cu-rich phase precipitation. Figures 4, by light microscopy, show the austenitic microstructure without δ-ferrite with a grain size ASTM 4. In figures 5 it is described the microstructure after two year of service in a power plant: $M_{23}C_6$, MC and Cu phase are present.
The tensile properties of TEMPALOY AA-1, as well as, the ductility versus temperature are shown in figures 6.

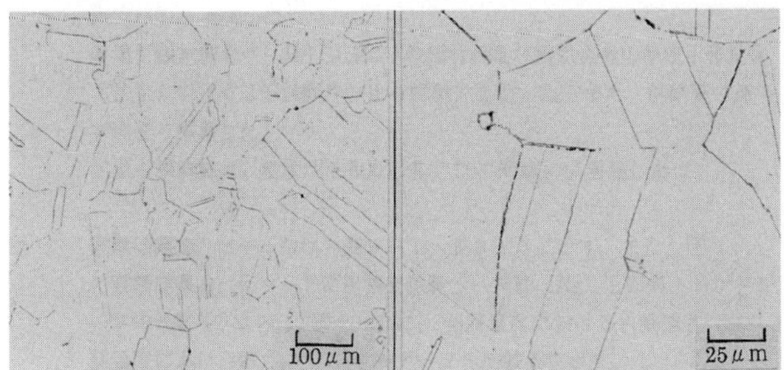

Figure 4: TEMPALOY AA-1 microstructure in as treated condition

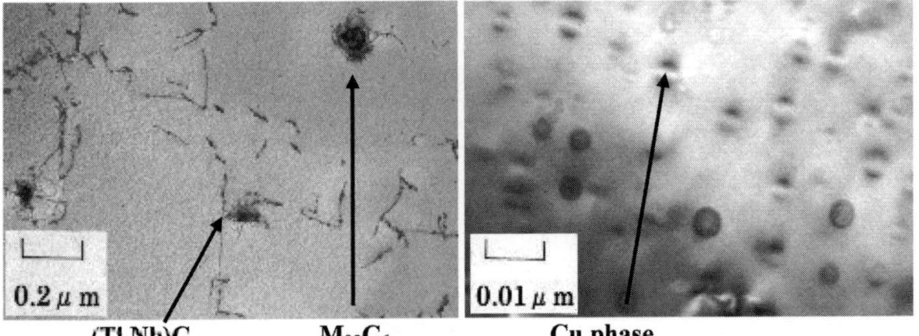

Figure 5: TEM TEMPALOY AA-1 microstructure show $M_{23}C_6$, MC and Cu phase after 2 years operation in actual power plant

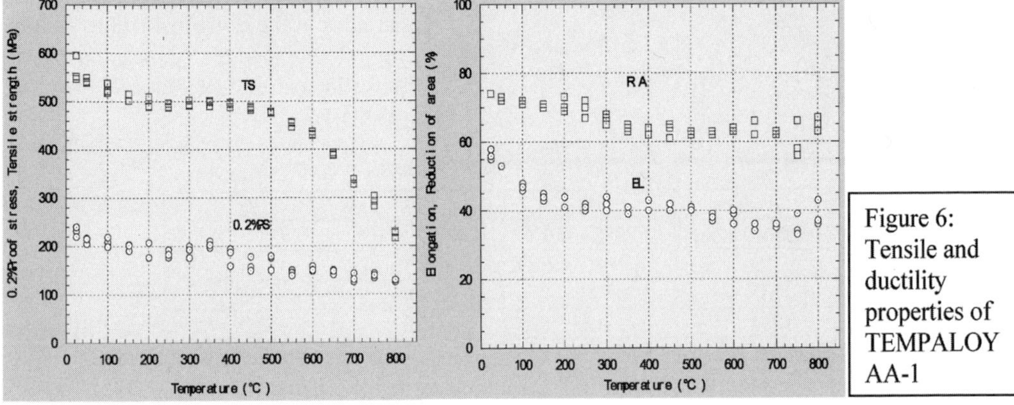

Figure 6: Tensile and ductility properties of TEMPALOY AA-1

TEMPALOY A-3

The TEMPALOY A-3 creep strength is mainly due to the precipitation of MX-Nb(C,N), complex CrNb Nitrides and $M_{23}C_6$ carbides. Figures 7, by light microscopy, show the austenitic microstructure without δ-ferrite. The tensile properties of TEMPALOY A-3, as well as, the ductility versus temperature are shown in figures 8.

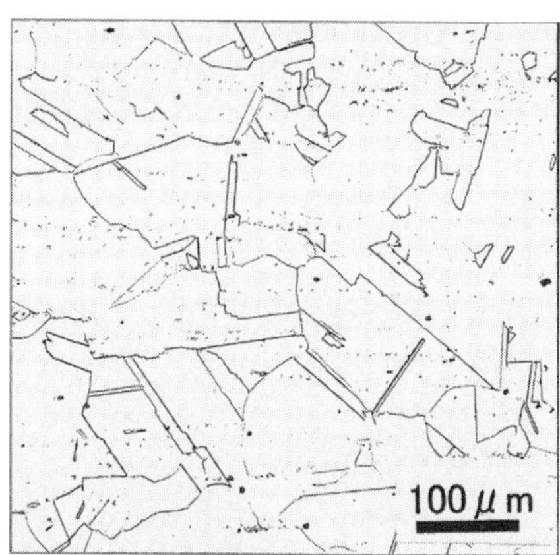

Figure 7: TEMPALOY A-3 microstructure in as treated condition

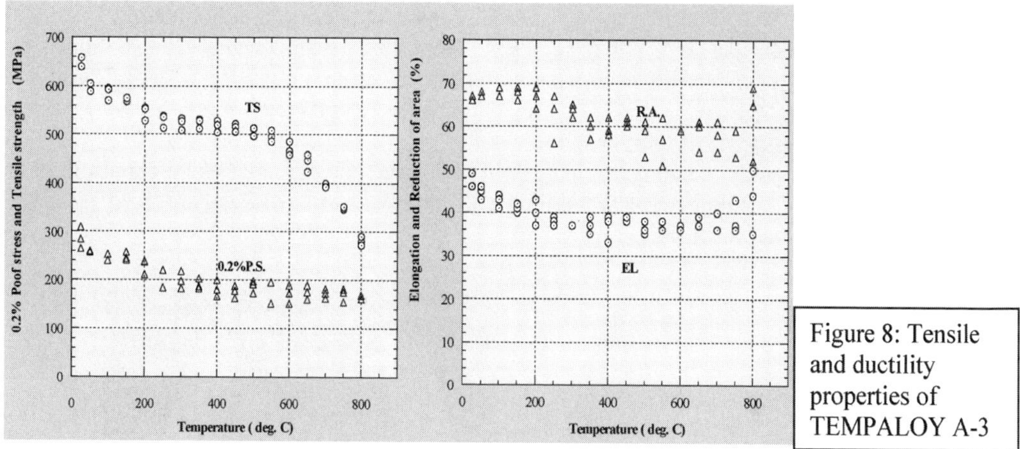

Figure 8: Tensile and ductility properties of TEMPALOY A-3

4. BENDABILITY

Tubes of various Grades, normalized and tempered, were cold bent in the most severe conditions: a 180° bending angle was used, together with critical R/D ratios (up to 1.00). A working window for the cold bending as a function of diameter (D), thickness and bending radius (R) was identified. Microstructural and hardness investigations were performed on the extrados, intrados and neutral axis of the bent tubes. No micro-voids at the extrados were detected by SEM. The influence of Post Bending Heat Treatment (PBHT) on the microstructure and hardness values was investigated: various tempering treatments were performed in agreement with European Standard EN 12592 recommendations.

Figure 9 – Example of bent T91 tube (44.5 mm OD x 3.05 mm WT; R/D=1.40)

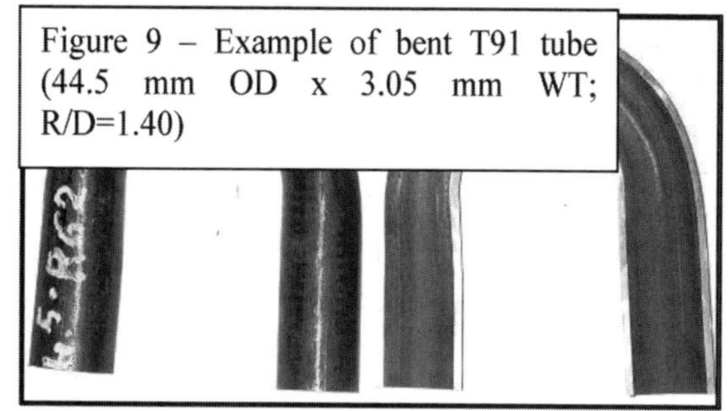

Hot bending tests on P91 pipes of large diameter were also performed using different bending angles and various R/D ratios.

In Figure 10 the hot induction bending operation is shown. After bending the pipes

were normalized and tempered again (PBHT). A working window for hot induction bending was defined. The bent pipe microstructures were investigated by LM and SEM. No micro voids were detected at the extrados position. In Figure 10a, the microstructure of P91 hot-bent pipe (609 mm OD x 17.5 mm WT, R=5D, bending angle 30°), after a new normalizing and tempering treatment is shown. The microstructure is tempered martensite, similar to that of the unbent mother pipe.
Impact and tensile tests were performed at extrados, intrados and neutral axis of the bend. No significant differences between bent portion and original mother pipe were observed.

Figure 10 – Industrial hot bending of P91 (609 mm OD x 17.5 mm WT, R/D=5); (Courtesy of Simas-Bassi Luigi Group)

At present CSM and TenarisDalmine are running a creep program on specimens taken from extrados, intrados and neutral portion of hot-bent P91 and P23 pipes. From first results, within 20.000 hours, no differences of creep properties between bent and unbent pipes were obtained.

Also for the austenitic grades bending test have been performed: the appearance and the result of colour check after cold bending are shown in Figure 11: no cracks or flaws were observed. Excellent formability was verified.

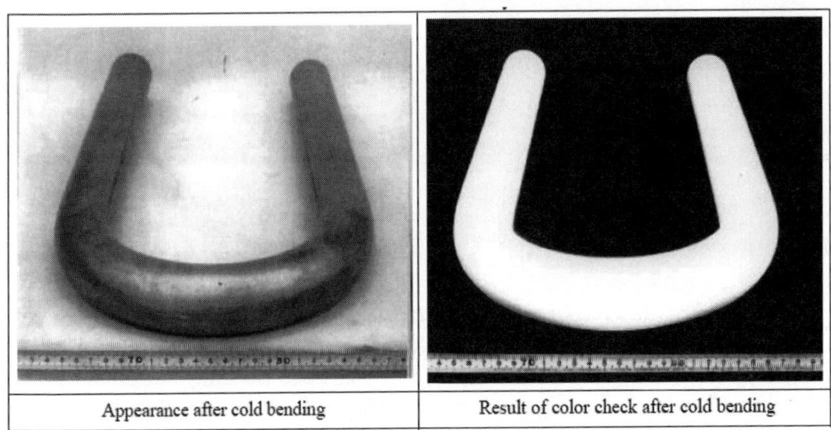

Appearance after cold bending	Result of color check after cold bending

Figure 11: Bending test on TEMPALOY A-3

6. WELDABILITY

TenarisDalmine in cooperation with Air Liquide Welding and CSM has investigated the weldability of Grade 91, 92 and 23 [12] tubes and pipes (butt welds and T-Joints, similar and dissimilar weldments), in order to identify suitable welding procedure specifications (WPS).
Welded joints underwent a microstructural and mechanical characterization, including long term creep tests that are ongoing.
Weldability tests on austenitic alloys TEMPALOY AA-1 and A-3 were performed using GTAW and GTAW+SMAW processes with specific similar consumables developed by Kobelco. Mechanical, microstructural and creep properties together with the soundness of the weld and the properties of welded joints were investigated for both austenitic alloys. Figure 12 shows an example of the hardness profile of the TIG welded joint for TEMPALOY AA-1.

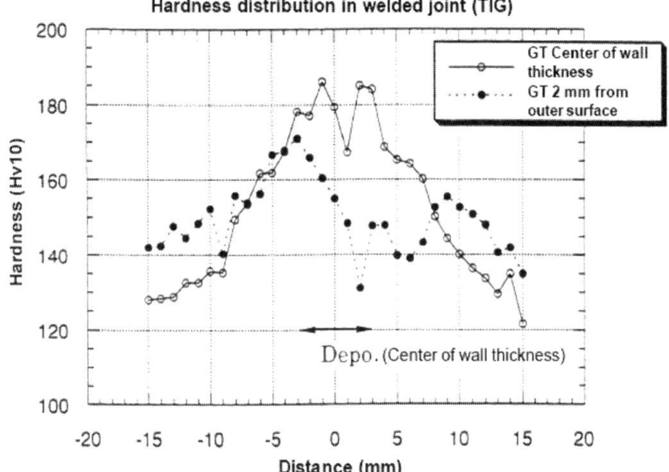

Figure 12: Results of hardness distribution measurement in TEMPALOY AA-1

7. CREEP BEHAVIOUR

7.1 Creep behaviours of ferritic/martensitic steels

A wide creep test program is characterisation performed for on the TenarisDalmine high temperature materials. As an example, Tenaris Grade 91 database contains more than 750 test results from tubes and pipes from more than 35 different heats, with outside diameter from 38 to 660 mm and wall thickness from 4.6 to 84 mm. The applied stresses range from 35 to 400 MPa and more than 7 million creep hours were accumulated. Several tests are still running and some of them have reached more then 99.000 hours at 600 and 650°C with foreseen time to rupture over more 100.000 hours. Up to now the longest creep test has time to rupture of 110,301h at 550°C with 150MPa stress [13]. All data are completed by information about material pedigree: chemical composition, heat treatment details and mechanical properties.

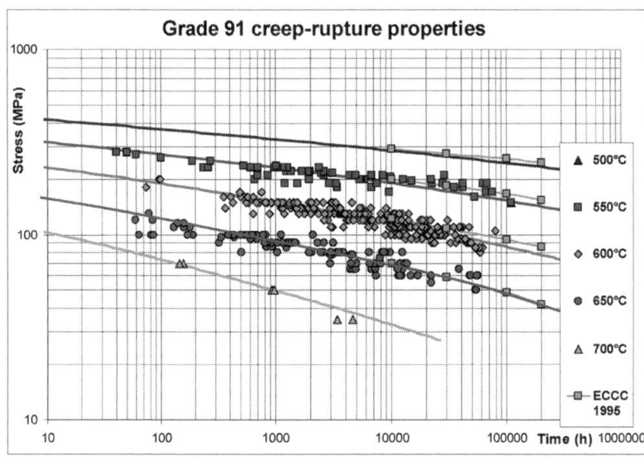

Figure 13 – Tenaris Grade 91 creep-rupture curves (main testing temperatures)

Following European Creep Collaborative Committee (ECCC) Recommendations, the assessments of the creep-rupture data were performed and the Post Assessment Tests (PATs) were successfully passed [14,15,16,17]. The calculated creep-rupture curves of Grades 91 are plotted in figure 13 for the main testing temperatures. The stresses for rupture at 10^5 hours for Grade 91 and also for Grade 911 are compared in Table 5 [15-17] with ECCC official material datasheets.

Due the availability of large and accurate database, it was possible to perform reliable creep-rupture extrapolations at very long times: 600°C/78MPa/200.000h for Grade 91. It is important to notice that a reduction of creep-rupture strength in comparison to ECCC official extrapolation, dated 1995, was obtained for Grade 91. The difference is notable especially at 600°C. Tenaris Grade 91 database contains longer creep data than those used for extrapolations in 1995 by ECCC working group.

Stress values (MPa) to rupture into 10^5 hours				
Temperature (°C)	Grade 91		Grade 911	
	TenarisDalmine	ECCC assessment [18]	TenarisDalmine	ECCC [19]
550	153	150	176	173
575	117	116	135	134
600	86	86	101	98
625	62	62	71	71
650	46	44	50	/

Table 5 – Predicted stresses to rupture for TenarisDalmine Grades 91 and 911 compared with 2005 ECCC new assessments [15-16]

Similar databases are available for the all the older grades; the database for the new grades 23 and 92 base materials, as well as for the similar welded joint, are under construction and at present the longest tests are in the range of 30.000 hours in the Tenaris/CSM laboratories.

An assessment of TenarisDalmine Grade 91 creep-strain data was performed and the stresses to reach 1% elongation were extrapolated at different temperatures: at the stress value of 74MPa the steel should reach 1% strain into 100.000 hours and should break after 300.000 creep hours.

7.2 Creep behaviours of austenitic steels

The creep behaviours of the TEMPALOY AA-1 and TEMPALOY A-3 have been evaluated in the range temperature of 600-800°C. Specimens of TEMPALOY AA-1 are still running above 100kh. The figures 14 and 15 show the master curves for both steels.
Figure 16 shows the creep behaviour of the TIG and SMAW welded joint for both austenitic grades compared with the base materials. The welded joint shown creep resistance very similar to the base material in all the test temperature range investigated (600-800°C)

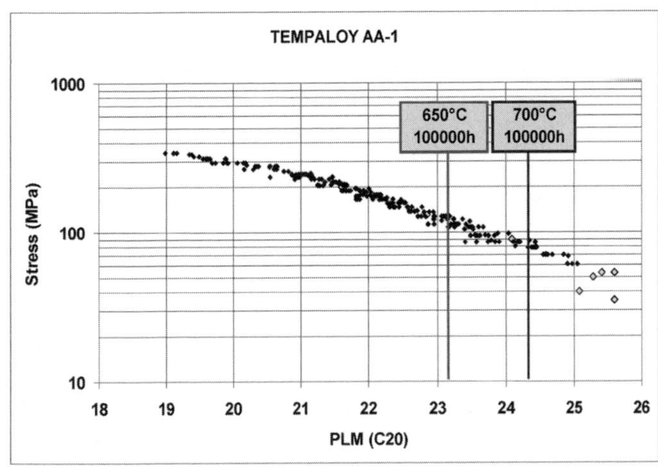

Figure 14: Master curve of TEMPALOY AA-1 (open point = test running)

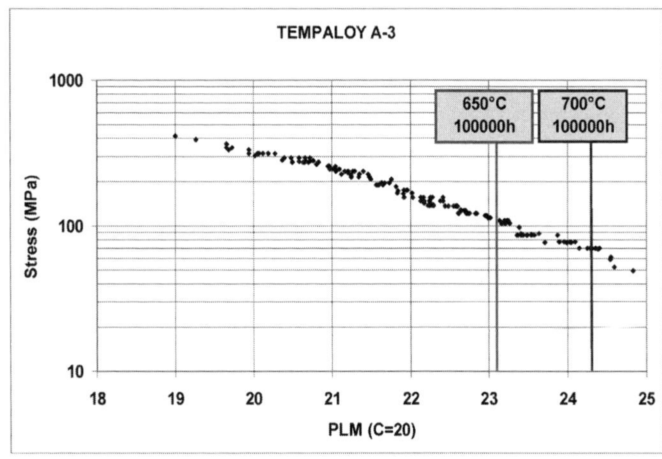

Figure 15: Master curve of TEMPALOY A-3

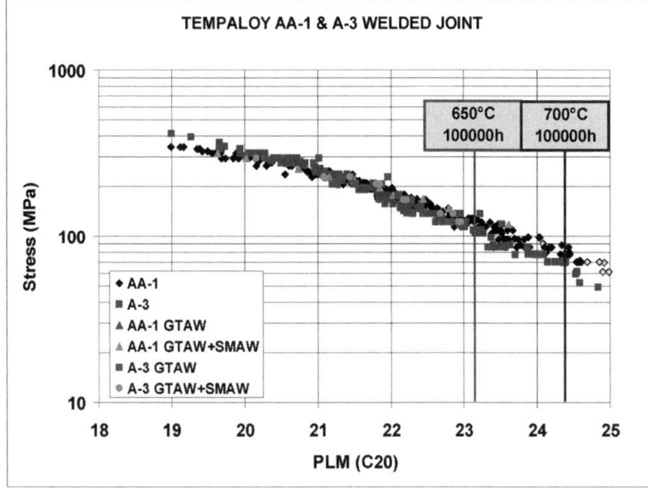

Figure 16: creep behaviour of the TIG and SMAW welded joint for both austenitic grades compared with the base material (open point are test still running).

The allowable stresses for the TEMPALOY AA-1 are 30% superior to the ones of grade TP347H as shown in figure 17. Also TEMPALOY A-3 allowable stresses are better than Incoloy 800H and TP310S (figure 18).

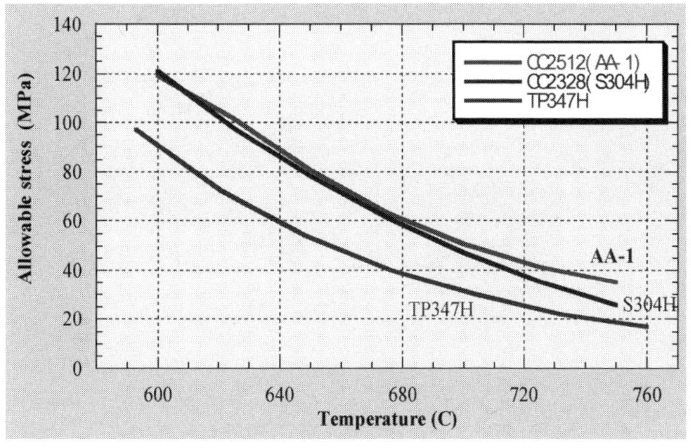

Figure 17: Allowable stress for TEMPALOY AA-1 and competitor materials

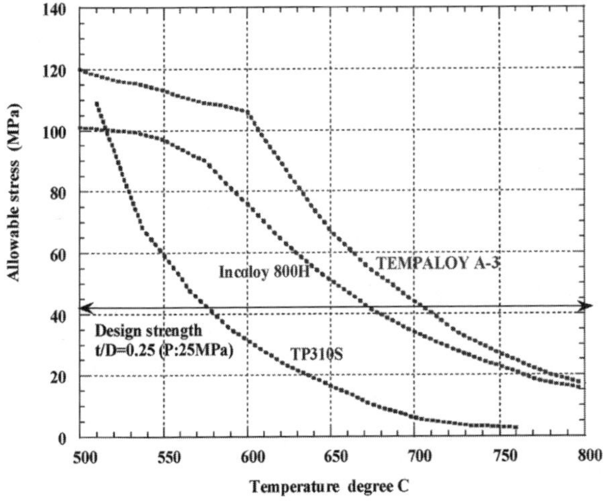

Figure 18: Allowable stress for TEMPALOY A-3 and benchmark materials

7.3 Hot Corrosion behaviour of austenitic grades

High-temperature corrosion tests were performed in conformity with "JSPS Method of High-temperature Corrosion Test by V_2O_5/Na_2SO_4 Synthetic Ash Coating" prescribed by the 123rd Committee (University-Industry Research Cooperation Committee on Heat Resisting Metals and Alloys) of the "Japan Society for the Promotion of Science (JSPS)"[20].
The test specimens were cut out from both the base metal and welded joint.
The base metal of the TP321HTB tube was selected as a comparative material. The tests were performed in two types of environments, each simulating coal ash combustion and heavy oil ash combustion. The testing temperatures were 600°C, 700C°, and 800°C for both types of environments. The duration of the test was 100h at each temperature. The test results are graphically summarized in figure 19.
No grain boundary corrosion was observed on any of the test specimens subjected to the high-temperature corrosion test. The amounts of high-temperature corrosion of the base metal and weld metal are nearly equivalent at each temperature to those of the comparative material (TP321HTB tube). Presumably, this is because both materials have nearly same levels of Cr content.

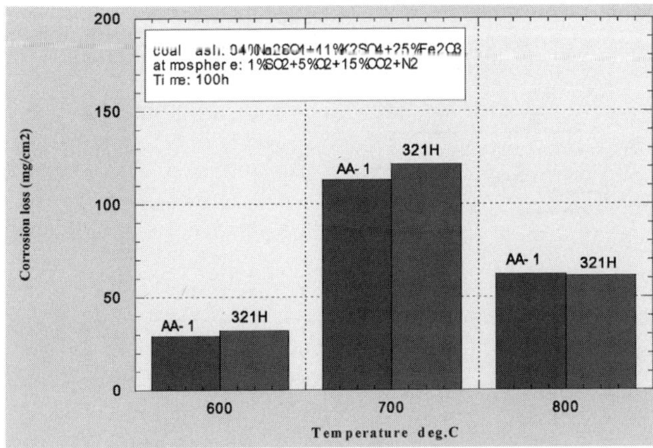

Figure 19: results of corrosion on TEMPALOY AA-1 test in simulating coal ash combustion.

The steam oxidization property was also evaluated using the inner surface of the tube as the test surface. The testing temperatures were 600°, 650°, 700°, and 750°C for 1000 hours.

For TEMPALOY AA-1 the test was also performed before and after shot-blasting: also in the case of benchmark material was a TP321HTB tube. The results of the steam oxidization test are shown in figure 20.

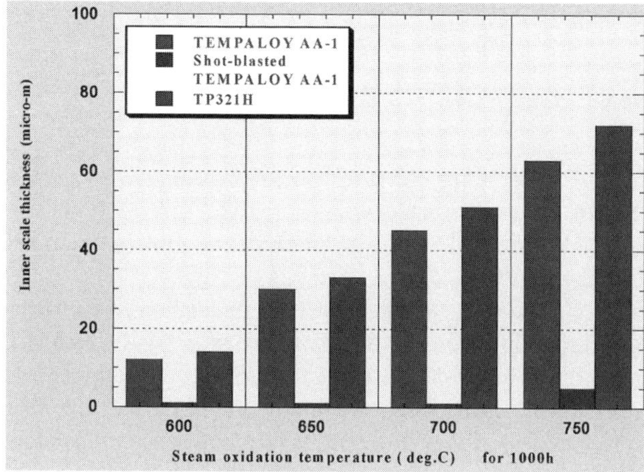

Figure 20: results of steam oxidation tests TEMPALOY AA-1

The base metal exhibited steam corrosion resistance nearly equivalent to that of the TP321HTB tube. Presumably, this is because both materials have same levels of Cr content. Shot blasting is proved to be extremely effective for suppressing the generation of steam-oxidized scale.

Similar tests have been performed also on TEMPALOY A-3 in the same condition of TEMPALOY AA-1 and the figure 21 shows the obtained results.

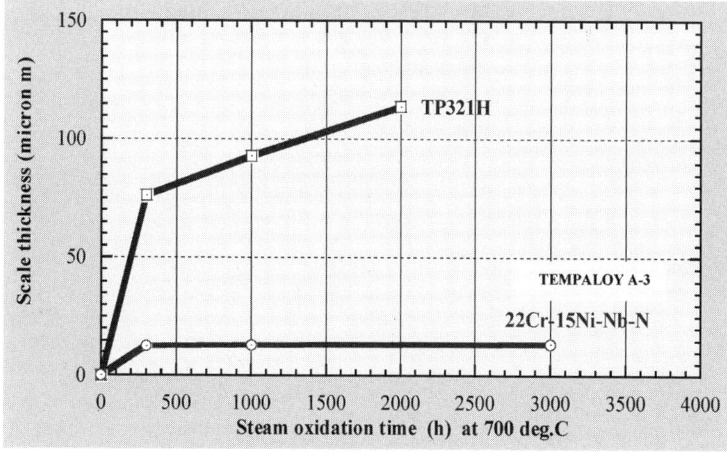

Figure 21: results of steam oxidation tests for TEMPALOY A-3

7.4 Exposure test in Power Plant

Both TEMPALOY AA-1 and A-3 have been exposed to actual service condition in several power plants for long time:

- TEMPALOY AA-1:
    - Ogishima Thermal Power Plant Unit 1, Japan: Steam temperature: S/H-541°C, R/H-541°C, in 1997
    - VESTKRAFT Unit No.3, Denmark, in 1997
- TEMPALOY A-3:
    - Eddystone No.1 Boiler, Exelon, USA, in 1991
    - Ogishima Thermal Power Plant Unit No.3, NKK, Japan, in 1992

- VESTKRAFT Unit No.3, Denmark, in 1997

The TEMPALOY A-3 has been successfully installed in:
- Lamma Unit No.4, Hong Kong Electric Company, Hong Kong, in 1992
- Tomatoh Atsuma Unit No.3 Unit, Hokkaido Electric Power Company, Japan, in1994

8 MICROSTRUCTURAL EVOLUTION OF MARTENSITIC STEELS

The microstructure of Grades 91, 911 and 92 after long creep tests was investigated to give a metallurgical explanation of the reduction of creep resistance with increasing service time and temperature. The crept samples were observed by LM and SEM too; afterwards the microstructures were investigated by STEM/TEM technique [21,22,23,24,25].
Microstructural evolution of 9%Cr steels during creep exposure consists of nucleation and growth of second phases, followed by coarsening of these ones, after growth reaction ends: $M_{23}C_6$ and MX precipitates are present since after tempering, Laves and Z-phase, nucleate during creep service.
The coarsening of precipitates in Grade 91 during aging is shown in Figure 22, where the size frequency of precipitates (referring to all precipitates) of four different crept samples is compared as function of Larson-Miller Parameter (LMP): in the as-treated Grade 91 tube, the precipitation mainly consisted of small particles with 100 nm average diameter; after 53,318 hours at 650°C the whole population of precipitates suffered a shift towards larger particles, some of them even with micrometric diameter.

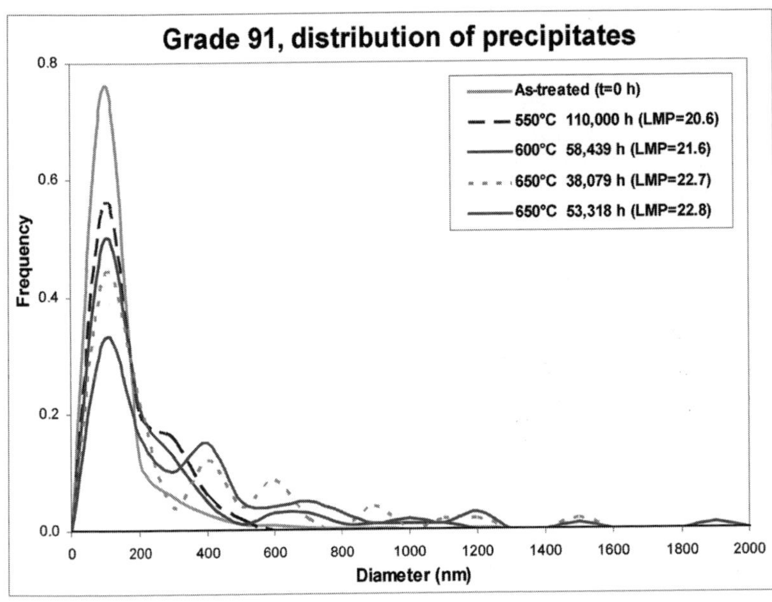

Figure 22 – Size distribution of precipitates in Grade 91 with increasing time

8.1 Evolution of carbides

The growth and coarsening of $M_{23}C_6$ and MX carbonitrides in Grades 91, 911 and 92 was also investigated by STEM/TEM technique: those particles appeared stable enough against coarsening; in particular MX precipitates are only slightly affected by ageing, and their equivalent diameter remains below 100 nm even over 100,000 hours at high temperature; equivalent diameter of $M_{23}C_6$ carbides remains below the critical diameter of 250 nm at 600°C too (Figure 23).
The most evident changes in the microstructure of 9%Cr steels occur with the nucleation of Laves-phase as well as the nucleation of Z-phase at longer times.

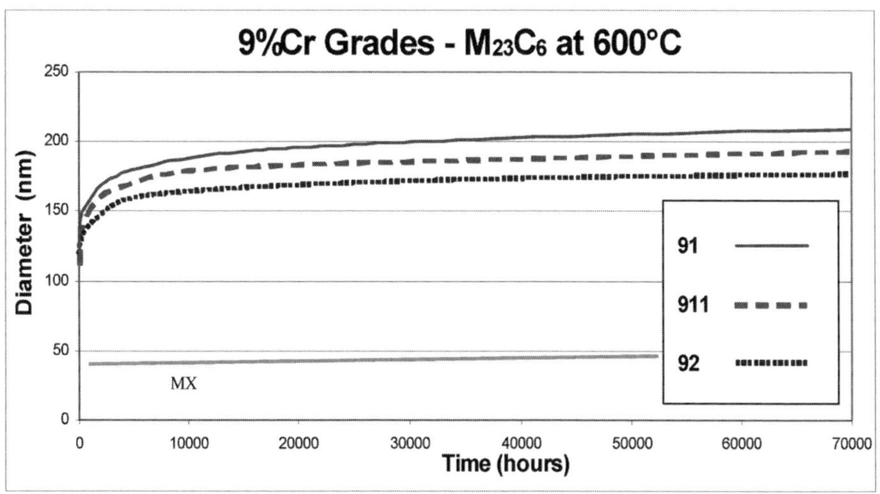

Figure 23 - Evolution of MX and $M_{23}C_6$ carbides in Grades 91, 911 and 92.

3.1 Evolution of Laves phase particles

Increasing of creep strength of 9%Cr steel has been achieved by a change of the Mo content and/or the addition of W to the steels with respect to ASTM Grade 9 base chemical composition [16].
These changes in Mo and W contents lead to the formation of intermetallic Laves phase during thermal service, $(Fe,Cr)_2(Mo,W)$, which nucleates mostly on subgrain boundaries and on prior austenite grain boundaries (PAGB). The presence of Laves phase on crept materials was investigated both by SEM and STEM/TEM technique, adopting automatic image analysis (AIA). SEM back scattered electron (BSE) images were used to distinguish $M_{23}C_6$ particles (dark contrast) from Laves particles (bright contrast).
This technique enables quantitative characterization of the precipitation and coarsening of Laves phase during creep, selecting the particles by shape, contrast and EDS punctual analysis. Size distribution of Laves precipitation was determined investigating 3 mm^2 of each crept sample.
The Laves phase nucleates and grows relatively fast during approximately the first 1,000-10,000 hours in the temperature range 600°C-650°C: below 600°C this phase remains relatively small even at very long time, but it can reach very large dimensions at 650°C especially in Grades 91 and 911 (equivalent diameter up to 2μm).
Figure 24 shows the growth of Laves radius with increasing time at 600°C in Grades 91, 911 and 92. Presuming that Laves phase particle coarsening obeys the Ostwald ripening law:

$$d^3 - d_0^3 = k_d \cdot (t - t_0) \qquad (1)$$

with d_0 and d the mean precipitate diameter at times t_0 and t, respectively, the rate constant K_d can be obtained by fitting Eq.1 to the measured values in different aged samples.
Based on these results, the coarsening rate and the mean dimension of Laves phase are evidently effected by the Mo and W content in the steel: Grades 91 and 911 with higher Mo content than Grade 92, show larger Laves phases. This means that W promotes higher stability against coarsening. The volume fraction of Laves phase in aged samples was found to be similar among Grades 91, 911 and 92 at same time and temperature ageing (Figure 25).

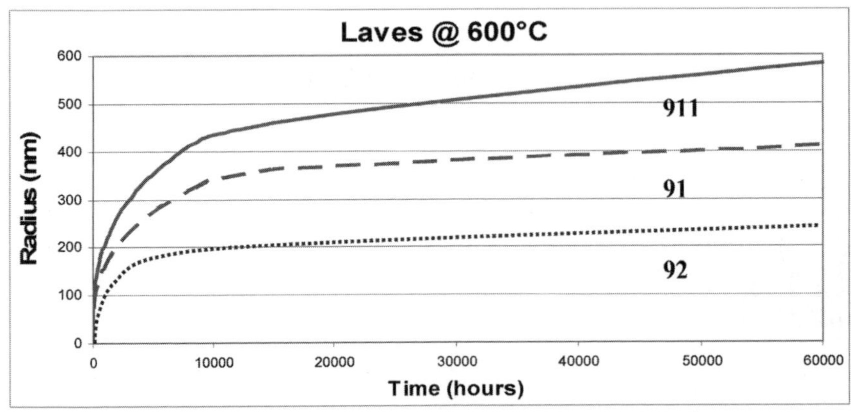

Figure 24 – Laves phase size evolution in Grades 91, 911 and 92.

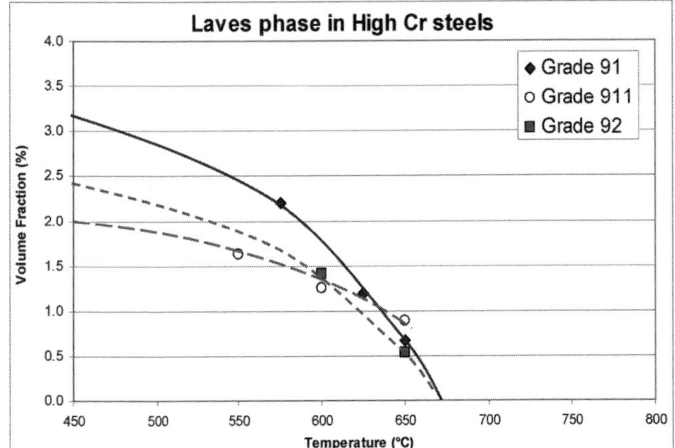

Figure 25 – Volume fraction of Laves phase in 9%Cr steels as a function of time

By the creep strengthening point of view the precipitation of Laves phase has two aspects. On one hand, high amounts of Mo and W contents are incorporated in this phase, causing a depletion of these elements from the solid solution and thus a reduction of their contribution to the overall creep resistance. On the other hand, the increased volume fraction of secondary phases leads to a higher precipitation strengthening during the first precipitation phase: at the beginning, the precipitation of fine Laves phase increases the creep resistance. However, if the coarsening rate is not taken under control, the mean diameter of these particles quickly reaches micrometric dimensions with a detrimental effect on creep behaviour. Laves phase precipitates in Grade 92 after 57,715 hours at 650°C are shown in Figure 26, together with $M_{23}C_6$ carbides, by TEM thin foil.

3.2 Evolution of Z-phase particles

Like Laves particles, modified Z-phase is an intermetallic phase which nucleates in the range of 600-650°C during creep service. This phase is common to all 9-12%Cr steels with high Nb and V content; MX carbonitrides during thermal exposure slowly transform in favour of a more stable composition, Cr(V,Nb)N, which is called modified Z-phase. The particles of this phase form thin plates within the matrix.

It is experimentally proved [26,27,28,29] that high amount of Z-phase nucleates and grows in large number and size mostly in 12%CrMoV steels where it causes the so-called "sigmoidal" shape of isothermal creep curve, with a dramatic drop of creep resistance. MX carbonitrides are one of the most effective strengthening precipitates in the 12%Cr advanced ferritic steels. However, Z-phase is formed at the expense of MX particles during high temperature exposure. It grows rapidly and cannot contribute to strengthening. The premature creep breakdown in 12%Cr steels has been explained by the disappearance of MX carbonitrides and the fast coarsening of Z-phase.

Figure 26 – Laves phases in Grade 92 after 57,715 hours at 650°C

Z-phase was recognized also in 9%Cr steels, but in far smaller amount: no dramatic drop in volume fraction of MX was observed in association to the nucleation of this phase, therefore it is believed that the modified Z-phase does not affect negatively creep properties of Grades 91, 911 and 92. In Figure 27 two Z-phase particles in Grade 92 crept specimen after 57,715h at 650°C are shown: their average diameter is about 1 μm and the mean phase composition in mass percent was measured as 44%Cr-2%Mo-23%V-19%Nb-2%W-6%Fe. Evolution of Z-phase in Grade 92 is shown in Table 6, together with size evolution of MX, $M_{23}C_6$ as well as Laves particles.

		Equivalent diameter (nn)			
Temperature (°C)	Time (h)	MX	$M_{23}C_6$	Laves	Zeta
	0	40	130	0	0
600	33,051	60	135	420	50 (a)
650	57,715	70	190	655	310

Table 6 – Precipitates evolution in Grade 92. ((a) to be verified by diffraction pattern)

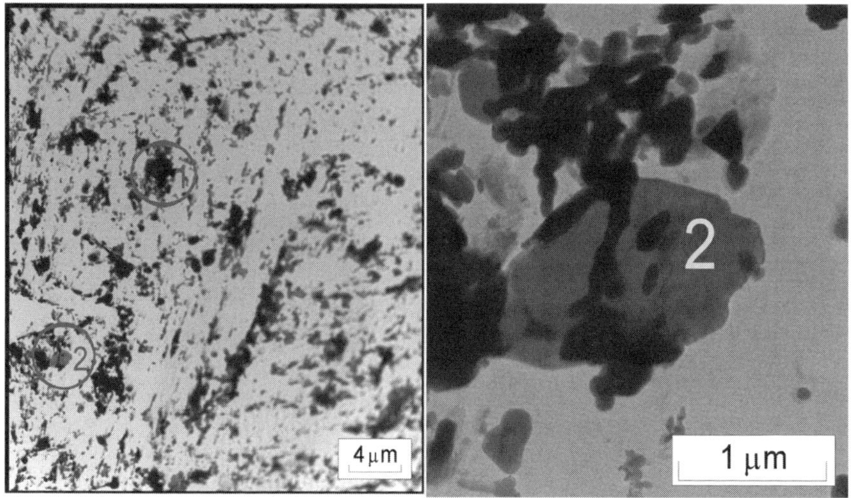

Figure 27 – Z phase particles in Grade 92 crept specimen (650°C/57,715h).

9. CONCLUSIONS

The production route, the main microstructural features and properties of low and high Cr-Mo ferritic/martensitic steels and of austenitic grades TEMPALOY AA-1 and A-3 have been described. The current R&D activities of Tenaris and CSM on advanced creep steels have been presented.
For Grade 91 the long term creep behaviour has been deeply investigated. TenarisDalmine creep-rupture dataset was assessed, following ECCC recommendations: a stress value of 86MPa at 600°C/100.000h has been obtained, in agreement with recent ECCC assessments. The values for grade 91 and the other 9%Cr steels is guaranteed by the analysis of microstructural evolution carried out, showing that the only few Z-phase particles appear after very long term aging at high temperature, without relevant effect on creep strength.

The austenitic stainless steel TEMPALOY AA-1 has been developed. The allowable tensile stress of this steel is more than 30% higher compared with that of ASME SA-213 Grade TP347H in the temperature range 600-700°C. This superior creep rupture strength is caused by the precipitation of MC and $M_{23}C_6$ carbides, and Cu-rich phase. The hot corrosion resistance and long term aging properties of this steel are at the same level as 18Cr-8Ni stainless steels. The steam oxidation resistance can be significantly improved with the internal shot-blasting method.
The practical service as superheater tubes in a boiler has revealed that the mechanical properties, creep and environmental resistance of this steel are excellent as boiler material. This steel has been already approved by METI standard in Japan as SUS321J2HTB; it is also approved by ASME as code Case N° 2512 and by in ASTM A213-06a as UNS S30434.

Also TEMPALOY A-3 has been developed for more aggressive environment; it has been registered as KA-SUS309J4HTB in METI Standards of Japan in 1997.

Tenaris is now able to supply a wide range of tubular product for new advanced Ultra Super Critical Boiler, from C steel to high CrMo steel with W up to the austenitic grades for superheater applications.

REFERENCES

1. Di Gianfrancesco A., Cumino G., Roffin A., ECSC Final Report EUR 15589 IT, 1996
2. Sikka V.K., Ward C.T., Thomas K.C.: Ferritic Steel for High-Temperature Applications, ASM Int. Conf. on Production, Fabrication, Properties and Applications of Ferritic Steels for High-Temperature Applications, 6-8 October 1981, USA, Ed. Ashok K. Khake, ASM 1983
3. Naoi H., Mimura H., Oghami M., Morimoto H., Tanaka T., Yazaki Y., Fujita T., Proc. New Steels for Advanced Plant up to 620°C°C, Ed. E. Metcalfe, London, May 1995, EPRI, USA, 1995, p. 8
4. Sawarahi Y., Ogawa K., Masuyama F., Yokoyama T., Proc. New Steels for Advanced Plant up to 620°C, Ed. E. Metcalfe, London, May 1995, EPRI, USA, 1995, p. 45
5. R.W. Vanstone, Alloy design and microstructural control for improved 9-12%Cr power plant steels, Annex A, COST 522 Steam Power Plant, Final Report, 1998-2003
6. Foldyna V., Microstructural Stability of Ferritic 9-12% Chromium Steels, COST Contract No. 94-0076-CZ (DG 12CSMS)
7. A. Poli, S. Spigarelli, Analysis of effect of chemical composition on hot forming operations of P91 steel: Proceding of Super High Strenght Steels Conf, 2-4 November 2005, Rome Italy
8. Anelli E., Cumino G., Gonzalez G.C., Proceedings from Materials Solutions '97, Indianapolis 15-18 Sept. 1997, p. 67

9. Minami Y, Thoyama A., Hayakawa H.: Properties And Experience With A New Austenitic Stainless Steel (TEMPALOY AA-1) For Boiler Tube Application; 8th Liege Conference, 18-20 September 2006
10. Minami Y., Thoyama A.: High Temperature Characteristics of High Cr Austenitic Stainless Steel Boiler Tubes for Ultra Super Critical Pressure Thermal Power Plants; First Int. Conf. on Heat Resistant Materials, 23-26 Sept. 1993, p. 533, Fontana, Wisconsin USA
11. Minami Y., Thoyama A.: Development of the High Temperature Materials for Ultra Super Critical Boilers (NKK TEMPALOY Series), Int. Conf. Advanced Heat Resistant Steels for Power Generation, April 1998, San Sebastian, Spain
12. Poli, S. Caminada, C. Rosellini, A. Bertoni, G. Liberati: Fabrication and weldability of grade 23 tubing and piping: Proceeding of Super High Strenght Steels Conf, 2-4 November 2005, Rome Italy
13. S. Caminada, G. Cumino, L. Cipolla, A. Di Gianfrancesco: Long term creep behaviour and microstructural evolution of ASTM Grade 91 steel; EPRI Fourth International Conference on Advanced in Materials Technology for Fossil Power Plants; October 25-28, 2004, Hilton Head Island, South Carolina USA
14. Creep data validation and assessment procedures, ECCC Recommendations Volumes (2005). Published by European Technology Development.
15. ECCC Data Sheets 2005. Published by European Technology Development
16. S.R. HOLDSWORTH, C.K. BULLOUGH and J. ORR, BS PD6605 creep rupture data assessment procedure. ECCC Recommendations 2005, Volume 5, Appendix D3
17. J. ORR, ECCC ISO CRDA procedure document. ECCC Recommendations 2005, Volume 5, Appendix D1a
18. L. CIPOLLA, J. GABREL: New creep rupture assessment of grade 91, 1° Super High Strength Steel Conference, Rome (2005) Italy
19. W. BENDICK, J. GABREL: Proc. ECCC Creep Conference, London (2005), UK, pp. 406- 418
20. Minami Y. Minoru Y.: TEMPALOY A-3 Tubes with High Corrosion Resistance for High Temperature Use; NKK Technical review N° 61 (1991)
21. A. Di Gianfrancesco, L. Cipolla, F. Cirilli, G. Cumino, S. Caminada: Microstructural stability and creep data assessment of Tenaris Grades 91 and 911: Proceeding of Super High Strenght Steels Conf, 2-4 November 2005, Rome Italy
22. H. BHADESHIA, Design of Ferritic Creep-resistant Steels, ISIJ Int. 41, (2001), p 626
23. Korcakova L., Microstructure Evolution in High Strength Steel for Power Plant Application: Microscopy and Modelling, Tech. Uni. of Denmark, August 2002
24. Hald J., NIMS-MPA workshop, 17th March 2004, Tsukuba, JP
25. J. HALD, Proc. ECCC Creep Conf., London (2005), UK, pp. 20-30
26. K. Sawada, H. Kushima, K. Kimura, M. Tabuchi; ISIJ International, vol.47 (2007) No. 5 pp 733-739,
27. Hald J., Korcakova L., "Precipitate Stability in Creep Resistant Ferritic Steels – Experimental Investigations and Modelling, Submitted by ISIJ International (2002)
28. Kimura K. Sawada K. Kushima H. Long-term Creep Strength Prediction of 9Cr Steels. Proc of NIMS-MPA workshop 2004;1-14. Japan: Tsukuba
29. Abe F. NIMS-MPA workshop, 17th March 2004, Tsukuba, JP

Advances in Materials Technology for Fossil Power Plants
Proceedings from the Fifth International Conference
R. Viswanathan, D. Gandy, K. Coleman, editors, p 582-589

DOI: 10.1361/cp2007epri0582

The Use of Advanced Materials on Large Steam Turbines in Supercritical Steam Cycles in the Czech Republic

L. Prchlik
V. Polivka
M. Kapic
K. Duchek
Skoda Power
Tylova 57
316 00 Plzen, CZECH REPUBLIC

Abstract

The paper summarizes several years of research on the application of modern materials in the design of large steam turbines operating at high temperatures. The use of 9-12% chromium steels on main steam turbine components, the application of abradable coatings in seals and the seize/corrosion protection of selected components by modern surfacing techniques are presented. Results of materials long-term testing supported by the field application at elevated steam temperatures were used to verify the new material solutions and manufacturing techniques.
The second section of the paper presents the design of a new 660 MW supercritical power plant to be built in the Czech Republic between 2008 and 2010. The unit parameters and steam cycle characteristics are presented together with the visualization of the new block. The steam turbine design is discussed with respect to the application of advanced materials.

High temperature materials in the Czech power generation - history

The first Czech supercritical steam turbine was designed and built in the State Research Institute of Machine Design in Bechovice in the early sixties of the last century. The parameters of the turbine were fully comparable to similar projects in Europe and overseas [1-2]. The steam inlet conditions were 38 MPa and 650°C. The HP part of this test turbine is depicted on Figure 1. The research turbine was build from austenitic materials and its power output was 2 MW. Technical difficulties associated with high pressure and temperature eventually led together with a relatively low coal price to the termination of this ambitious research program. Rather than rapid change in the steam inlet temperature, a slow ongoing increase in steam admission pressures and temperatures could be seen in the Czech Republic.

In the following decades the creep properties and strength of ferritic low and medium alloyed steels had been gradually improved through collaboration between research institutes and leading steel suppliers. Several modifications of 10-12% Cr creep/corrosion resistant steels were developed, yet eventually the Poldi steel R-M-AK2MV (similar to X22CrMoV12-1) has been most widely used. Low alloyed rotor steels also received a significant attention. A unique steel Skoda T56 (CSN15335) based on the composition 0.25C1.25Cr0.5Mo0.5W0.6VTi has been successfully developed and applied on 100 MW turbines in early sixties. Its creep properties were fully comparable to X22CrMoV12-1, while its oxidation behavior was naturally inferior. The steel development exceeded the needs of the power generation and highly creep resistant rotor steels had been only rarely applied. Nowadays, the inlet steam temperature does not exceed 540°C on most large Czech power plants. Currently, no supercritical steam turbine is operated in the Czech Republic.

In the last decade of the twentieth century the application of modern 9-12% chromium steels began. An extensive project on welding qualification has been started in Skoda Power and Vitkovice Research. Skoda Research took active part in COST 501 and 522 EU funded projects. P91 casting trials were conducted in Vitkovice Steel and ZDAS. Creep testing was an inherent part of all development efforts. Currently 9%Cr advanced steel application is supported by over a decade of experience and optimization and forms a basic prerequisite for USC steam turbine construction.

Application oriented materials research

Skoda Power is a medium size company with less than one thousand employees having independent development of solutions for power generation industry. The development of manufacturing techniques is often completed on collaborative basis together with equipment suppliers. The basic properties of new 9-10% ferritic steels have been established in EU funded COST 501, 522 projects. A need therefore existed to adopt these materials in the manufacturing process and verify their properties on real or near-to-real components.

Application of modern 9-10% steels. Traditionally, the Czech steel 15128 has been used for power plant high temperature piping until mid nineties. The low alloyed steel 15128 possessed very good creep properties considering its relatively simple composition (1/2%Cr-1/2%Mo-1/4%V) An excellent operational experience has been gained over several decades of its application. The V-content of this steel was not exceeding the limit of 0.35% unlike in equivalent steel 14MoV6-3 used in Germany, where several failures of pipes have been reported. In spite of its low alloyed character, the steel requires very careful handling during preheat, welding and PWHT as opposed to 21/4%Cr-1%Mo type steel (i.e.P22) that can be handled more easily.

As the result, the transition from 15128 to P91 in the Czech Republic was in some sense similar to the transition from 12%Cr steel X22 taking place in Western Europe. In both cases welders were accustomed to stringent process control. This led to very good mechanical properties obtained already in first welding trials with P91 in Skoda. In fact, the first application of advanced 9-10%Cr steel in Skoda Power was a replacement of a 15128 piping section with P91 in mid nineties. Together with the welding qualification for homogenous welds, heterogeneous joints with 14MoV6-3, X22CrMoV121 and P22 were successfully completed. Since then, P91 steel has been used for new installation and repair, where advantage was taken of its favorable thickness compared to other steels. During the application of P91 other specialized manufacturing techniques have been adopted. These included special resistive preheat and heat treatment, bending, TIG orbital welding, cast component repair and localized cold repairs of turbine components machined to final dimensions.

Creep testing. Extensive verification of weld creep properties followed basic tests required by EN288-3. This helped to further support the positive experience with P91 manufacturing and application. The results of creep testing for dissimilar P91-P22 welds have been reported earlier [3]. Figure 2 shows the overview of creep testing for homogenous welds of forged and cast components. It can be noticed that the creep properties were nearly identical to the P91 baseline for shorter exposure times and lower temperatures. A drop below 20% of the values for the basic material could be observed for longer exposure times and higher temperatures.

New power plants for central European power grid

The feasibility study of the largest Czech power generation utility showed that a new power source output should not exceed the 660 MW limit in order to satisfy requirements on power grid stability. In addition, the state environmental policy and other European related regulations must be respected. The fundamental requirements for the power plant were defined as follows.

- The net block efficiency at least 42%
- Emission limits according to BAT (SO_2<150 mg/Nm^3, NO_X<200 mg/Nm^3 , solid particles<20 mg/Nm^3, CO<20 mg/Nm^3) for north bohemian brown coal
- Minimized use of cooling water, minimized waste water production
- Maximized scheduled service intervals, minimized service time
- Minimized unscheduled shutoffs (all critical plant section backed up)
- Minimize the property required for power plat construction (use existing power plant site)
- Recuperate the heat of exhaust gasses for the feed water preheat
- Use steam extraction for district heating

These requirements resulted in a new power plant design and pushed for the design of a new ultrasupercritical steam turbine.

New ultrasupercritical steam turbine

The development of a new 660 MW unit that started in 2003 and represents a qualitatively new step in the Skoda Power turbine portfolio taking full advantage of previous experience with subcritical 500 MW and 1000 MW units. To guarantee the high reliability of the new 660 MW turboset, its design is based on field-proven 500 MW turbines that have been operating successfully in the Czech Republic and abroad over two decades. The basic configuration is similar; the turboset consist of one HP, one IP and two LP sections. Most turbine components are the modification of proven design features. The high admission steam parameters are enabled by the use of modern creep resistant materials from leading European suppliers. Figure 3 illustrates the change in thermal cycle compared to the original 500 MW machine. The turbine cross-section and overall arrangement of valves and connection ducts are depicted on Figure 4. Table 1 gives an overview of the turbine operating parameters.

Valve design and the anti-seize protection. The design of HP/IP turbine valves used was derived from the verified concept of control and stop valves and is presented on Figure 5. Two HP inlet valves are designed of cast G-X12CrMoVNbN91 steel with selected inner parts manufactured from P92 forgings. Four integrated control-stop valves are used at the inner into the IP section. Optimized NiCr-Cr_2C_3 HVOF coating in combination with stellite cladding is applied to critical valve surfaces. Fatigue and corrosion testing at elevated temperature verified the concept.

Turbine IP section blading. The material used for IP section blading is the corrosion resistant X22CrMoV12-1 steel, which Czech equivalent has been used for HP blading for over two decades. Its creep properties are well known from both in-house and ECCC creep standards. The use of this material for the inlet steam temperature of 605°C is possible due to the double flow design of the IP section and the low reaction blading used. The temperature drop on the first row of for the impulse type guide vanes is significantly higher than that for a comparable reaction type blading. The low-reaction rotating blade row is exposed to temperature nearly by 20°C lower compared to an equivalent reaction type blading. This results in higher permissible stress for the same material. In addition, the length of the first-row impulse-type buckets is at least 40% smaller than that for the reaction type blading. This length reduction significantly reduces the resulting centrifugal-force in spite of a slightly larger root diameter for the impulse-type blades. This approach can be used only limited power output of the turbine since the increased flow rate leads to blade length increase and correspondingly elevated creep loading of rotating blades. In addition, a vortex passive cooling will be used in the central part of the IP rotor.

Improved sealing. NiCrAlY/Bentonite coating has been optimized and applied to rotor seals. Figure 6 shows the application of abradable coatings on labyrinth seal section. Fail-safe approach with minimized coating thickness allows operation with acceptable leakage even with local coating delamination. Honeycombs are used in labyrinths as stationary rims with rotating fins from austenitic steels on blade shroud. Both sealing technologies have been extensively used on turbines up to 500 MW output. Special anti-swirl arrangement is used upstream of rotor shrouds to improve the rotor stability.

Bearing journals overlay welds. To improve the bearing journal tribological properties Cr-Mn hardfacing overlay is applied to their surface. An intermediate layer and double heat treatment procedure has been applied. SAW process has been used. In the first layer low carbon 2.25% Chromium steel with PWHT at 720°C has been applied. The second Cr-Mn layer of a larger thickness has been heat treated at approximately 650°C to achieve optimum surface hardness.

Conclusions

A robust design of a supercritical steam turbine has been prepared based on the experience with steam turbines of a similar power output. The turbine uses proven design features while several crucial improvements in both the thermal cycle and the steam turbine components have been applied. The steam turbine design relies on modern creep resistant 9% steels for rotors and turbine casings. The advantage is taken of the low reaction blading that allows to use a verified blading material for the first IP stage. The increase in steam admission parameters together with other features such us full 3D blading and improved seals significantly improves net plant efficiency and reduces CO_2 emissions.

References

1. American Society of Mechanical Engineers. *Eddystone Station, 325 MW Generating Unit, 1960.* Brochure of an ASME Mechanical Engineering Heritage Site, 2003.

2. American Society of Mechanical Engineers. *Philo 6, Steam-Electric Generating Unit, 1957.* Brochure of an ASME Mechanical Engineering Heritage Site, Columbus, Ohio, 2003.

3. L. Prchlik, E. Folkova, P. Hranek, "Long-term Mechanical Properties of Dissimilar Welds between P91 Grade an Low Alloyed Steels,", presented at the Fourth International Conference on Advances in Materials Technology for Fossil Power Plants, Hilton Head Island, SC (October 2004).

Figures and Tables

Table 1. The characteristics of the new ultrasupercritical steam turbine.

Skoda USC 660 steam turbine/power plant parameters

Electric output	660 MW
HP admission steam pressure	26.4 MPa
Live steam temperature	600°C
Reheat steam temperature	610 °C
Cooling water temperature	18.5°C (max. 28.5°C)
Speed	3000 1/min

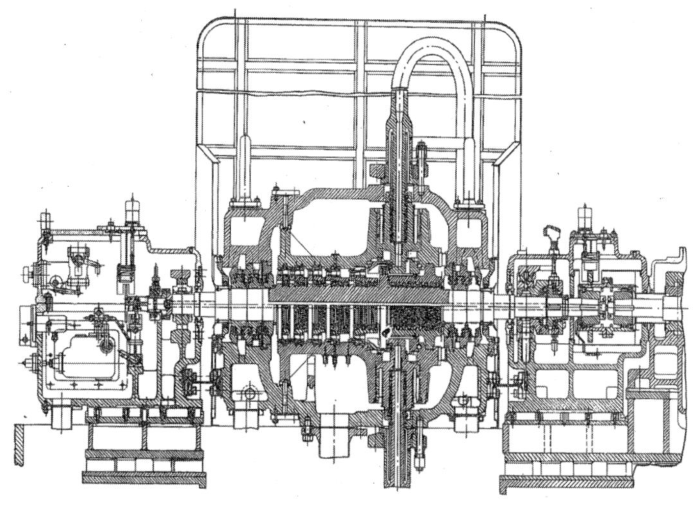

Figure 1. USC research turbine designed in the Czech Republic in early sixties.

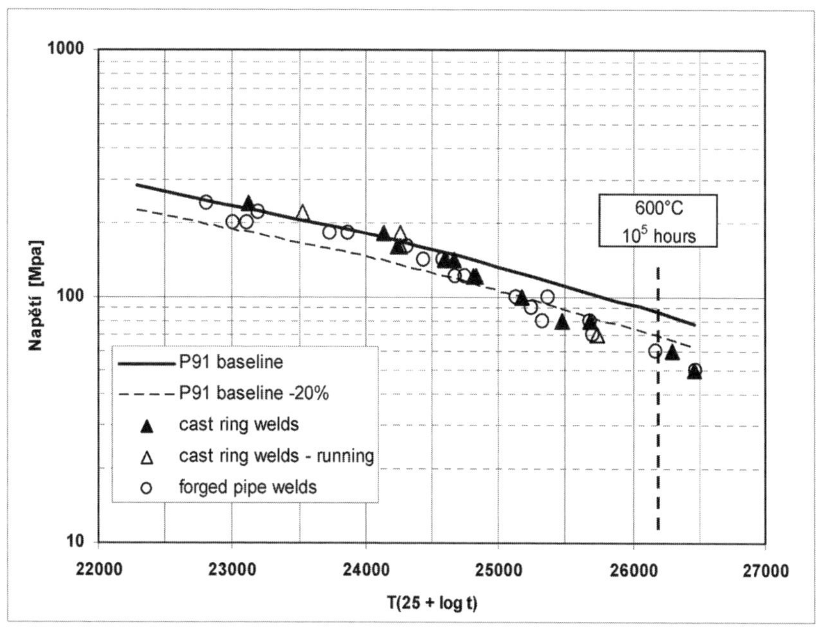

Figure 2. The results of creep testing for forged and cast components. The details of the manufacturing procedure can be found in ref. [1].

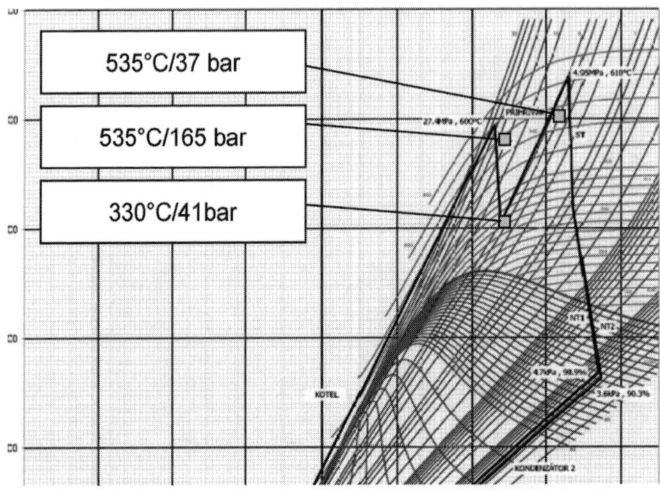

Figure 3. The solid line represents the thermal cycle of 600 MW block. Live and reheat steam parameters and feed water temperature of the 500 MW turbine are shown for comparison.

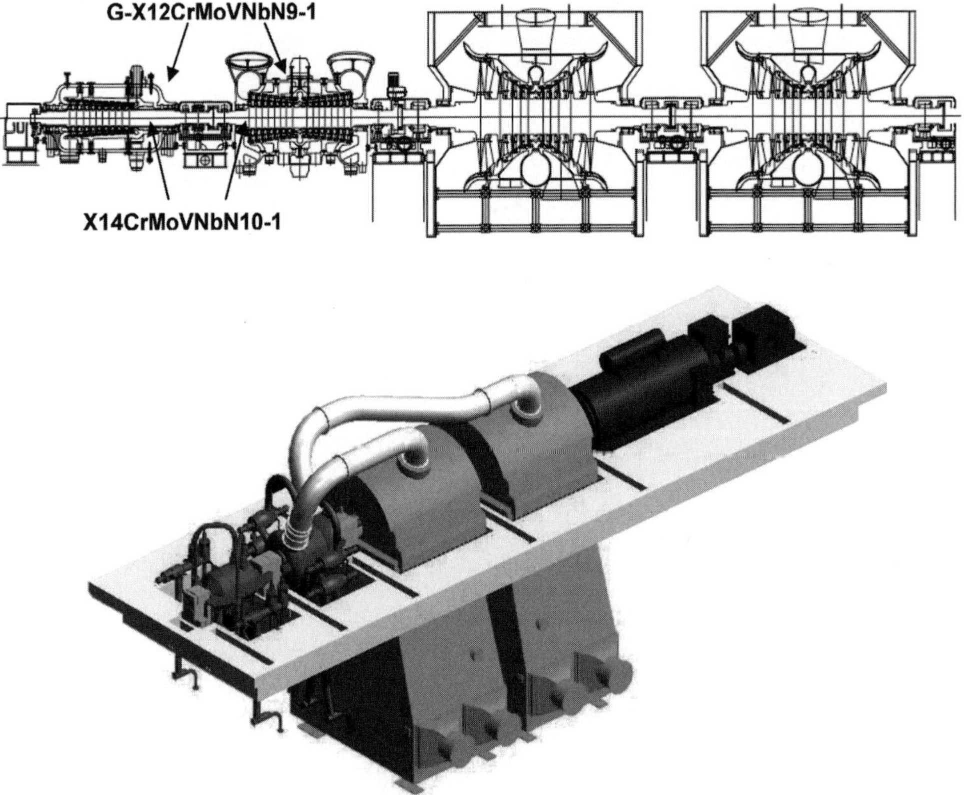

Figure 4. The cross-section and the 3D view of the turbine island for the new 660 MW USC steam turbine.

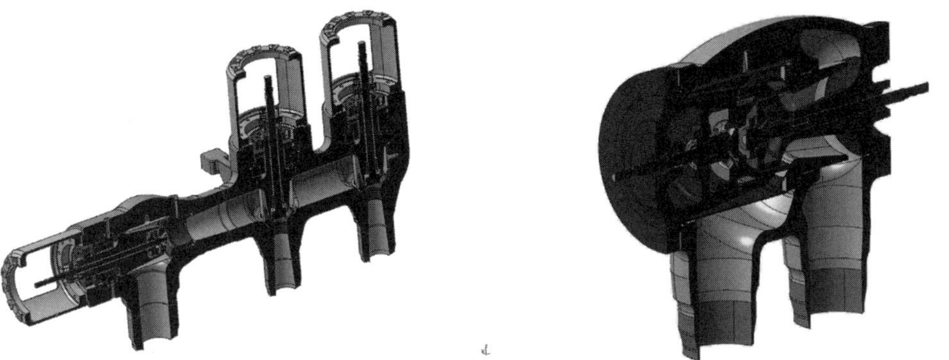

Figure 5. The design of HP and IP valves used on the new USC steam turbine.

Figure 6. Examples of components used in the USC turbine: optimized IP blading section and abradable coatings applied onto rotor seal.

Figure 7. The visualization of the new block on one of the construction sites considered.

Advances in Materials Technology for Fossil Power Plants
Proceedings from the Fifth International Conference
R. Viswanathan, D. Gandy, K. Coleman, editors, p 590-600

DOI: 10.1361/cp2007epri0590

Service Experience with A Retrofit Modified 9Cr (Grade 91) Steel Header

S J Brett
RWE Power International

Plant Life & Integrity, RWE npower
Windmill Hill Business Park
Whitehill Way, Swindon SN5 6PB, UK
e-mail: steve.brett@rwenpower.com

Abstract

In 2004 extensive Type IV cracking was found on branch and attachment welds on a modified 9Cr (grade 91) header after 58kHrs service. The header was a retrofit component installed on a 500MW unit in 1992. Early inspection of the header was undertaken because it had been established that a number of low nitrogen to aluminum (N:Al) ratio components had been incorporated in its construction. This had been identified as a factor common to earlier premature plant failures in this grade of steel elsewhere in the UK, investigations subsequently showing both the presence of coarse aluminum nitride (AlN) precipitates, a depleted VN-type MX precipitate population, and low parent and Type IV creep strength in grade 91 material with low N:Al ratio. The cracking found in the present case was overwhelmingly on the header barrel sides of the welds, in material which, while meeting the chemical compositional requirements of the ASTM specification for this grade of steel, was confirmed to have low N:Al ratio.

This paper summarises the inspection history of the header, describing the distribution of cracks found in 2004 and at a later outage in 2006. The results are discussed against the background of the shortfall of Type IV creep life in grade 91 and its relationship to chemical composition. The wider implications for other grade 91 components in service are considered.

Introduction

Following its introduction in the late 1980s, modified 9Cr (grade 91) steel found widespread use for retrofit header applications in the UK power industry. Although it has only been in service for a relatively short period of time, currently approximately 80,000 hours, a number of problems have arisen. Many of these are related to incorrect fabrication procedures and failure to produce a correctly tempered martensitic structure. However service experience has also shown that grade 91 material which has been produced in a fully martensitic condition, and apparently correctly tempered, can fail relatively early in service by Type IV cracking in the fine grained region towards the parent side edge of the heat affected zone. Investigations of failed components have shown both the presence of coarse AlN precipitates, a depleted VN-type MX precipitate population, and low parent and Type IV creep strength in grade 91 material with low N:Al ratio (1)(2).

In 2004 extensive Type IV cracking was found on branch and attachment welds on a modified 9Cr (grade 91) header after 58kHrs service (3). The header was a retrofit component installed on a 500MW unit in 1992. Early inspection of the header was undertaken because it had been established that a number of low N:Al ratio components had been incorporated in its construction.

The header is shown schematically in Fig.1. It was designed to BS1113:89 with a design pressure of 17.58 MPa and a design temperature of 580°C to supply steam at 568°C. It was constructed from six ASTM A335 P91 cylindrical barrel sections (450mm OD x 50mm t), Barrels 1 and 6 being 2.3m long, Barrels 2 and 5 being 4.9m long and Barrels 3 and 4 being 3.5m long. Each barrel was separated from its neighbor by an ASTM A182 F91 forged T-piece, except the middle two barrels which were separated by a central circumferential butt weld. The ends of the header were closed by forged domed ends and the header was fitted with four ASTM A182 F91 safety valve branches (190mm OD x 57mm t), one ASTM A182 F91 main steam atmospheric pass out branch (210mm OD x 54mm t), and two much smaller pressure tapping branches. A total of 408 ASTM A213 T91 stubs (54mm OD x 8mm t) were distributed along the header body, grouped mainly in 68 elements of 6 stubs (A-F) each. Most of the stubs were attached to the barrel sections with a smaller number on the forged T-pieces. On the barrels the six stubs in each element were arranged at 50° intervals around the circumference between 55° and 305° from top dead centre position. A number of attachment welds were also present in the form of centralizing restraint brackets, main hanger supports and anti-rotation lugs.

The cracking was investigated in some detail and confirmed to be Type IV (Fig.2) with an analysis of the cracking distribution leading to the conclusion that it was a result mainly of poor Type IV strength. There was no evidence of high loading on the header from, for example, inadequate support or systems loads (3).

An analysis of thermocouple data from the four steam lines coming out of the header for the ten years prior to 2005, and therefore covering most of the period the header had been in service, showed an average steam temperature of 570°C and a maximum leg steam temperature of 572°C. The header had therefore delivered steam at a temperature only slightly higher than the expected 568°C. Although there was evidence from stub oxide thickness measurements of more cracking on those parts of the header body operating at the highest temperature, the cracking could not be explained by this factor alone (3). Fig.3 shows the temperature profile on the header with the relative positions of the T-pieces.

The cracking found in the present case was overwhelmingly associated with the three barrel sections having, on the basis of their material test certificates, the lowest N:Al ratio. The two worst cracked, Barrels 2 and 5, come from the same cast and have N:Al ratio <1.5. The third cracked barrel, Barrel 4, has a N:Al ratio of 2.8, significantly higher than found in earlier problem casts. The adjacent Barrel 3, with only very limited cracking, and remaining Barrels 1 and 6, had much higher N:Al ratios.

All large branch welds cracked were found to have cracking on both flank positions on the header barrel side. The largest surface length of cracking found was 185mm and the largest through-wall extent was 8mm. For cracked stubs the crack position was the same, on the header body side flank weld toe. Cracking on attachment welds was generally confined to header side weld toes with an orientation parallel to the long axis of the header. The longest surface length of cracking here was 310mm. Significantly, the cracking orientation in virtually all cases corresponded to a driving force controlled primarily by the hoop stress.

Action taken in 2004

Because the cracks were found at a relatively early stage of development, it was found possible to grind them out and return to service without the need for weld repair. For stub cracks, where it was calculated that early recracking would be most likely, finite element analysis was used to define ground profiles which left the surface Type IV zones subjected to operating stress levels no higher than those on the Type IV zones of the original profiles.

In the expectation that Type IV cracking would eventually reappear in the ground areas and extend to other header areas, the decision was taken to replace the header. This was planned for a short outage in 2006 with the recognition that replacement might have to wait until the next major outage in 2008. To allow for the eventuality that the header would have to remain in service until 2008, but that cracking might be present in 2006 that could not be contained by further grinding, a weld repair option was developed. Prior to the 2006 outage a weld repair procedure to be used without post weld heat treatment, capable of maintaining the header in operation between the 2006 and 2008 outages, was approved.

Cracking found in 2006

During the short outage in 2006 the header was re-inspected and some further stub cracking found. Almost all the cracks were confined to Barrels 2 and 5, the sections which had been most badly cracked in 2004. Cracks had appeared on stubs previously uncracked and also on stubs previously cracked and ground. Significantly, virtually all recracking occurred adjacent to the ground areas, with only a single example of recracking within a ground area. No new cracks or recracking were found on Barrel 4 and only a single new cracked stub on Barrel 3. No new cracks or recracking were found on any of the larger branches and attachment welds inspected. Because of the limited extent of cracking on the innermost four barrels, the outermost two, operating a little colder, were not inspected in 2006.

The total distribution of cracked stubs, including those cracked in 2006, is shown in Fig.4. All the cracks and recracking found in 2006 were ground out within the previously agreed grinding limits and the header was returned to service without the need for weld repair. It is now planned to replace the header in 2008.

Aluminum Level and Aluminum Nitride

As with earlier failures in the UK the cracking on the header occurred in material with a relatively high aluminum level and a low N:Al ratio (1). It was therefore considered worth investigating how these factors relate to the presence of observable AlN, on which the weakening effect depends. Investigations were carried out on samples collected from a number of components, including other UK retrofit headers.

In measuring aluminum level it is important to distinguish between total aluminum and soluble aluminum. In the present context soluble aluminum was considered the more relevant measurement since this will be more directly related to the amount of aluminum available to form AlN. The most convenient way of measuring aluminum content, where sufficient material is available, is by optical emission spectroscopy (OES) in which the sample surface is vaporized and the content analyzed spectrographically. Where the amount of material available is not sufficient for this measurement has to be by soluble methods, in which the sample is first dissolved in acid.

Two soluble techniques were used on samples investigated by the present author: a "standard" soluble method, using hydrochloric acid, and a microwave digestion method (MDM) using a mixture of strong acids heated in a microwave oven. The results for a range of grade 91 samples are shown in Fig.5 where soluble and MDM values are normalized by the corresponding OES value. It can be seen that soluble and OES values are very close, differing only by about 5%. It appears that OES gives near soluble aluminum levels as measured by the standard technique. Rather surprisingly the MDM values are significantly and consistently higher. It appears that this method, although clearly an alternative soluble technique, is probably measuring total aluminum level.

On the basis of this it was decided that, for the present investigations, the OES and soluble methods were preferred and could be used interchangeably. In Fig.6 the results of subsequent investigations for the presence of AlN particles in the samples are shown semi-quantitatively against nitrogen and aluminum content using only OES or standard soluble aluminum values. The results appear to indicate that AlN will always be present in grade 91 steel unless the aluminum can be kept to a very low level (<0.01wt%), substantially below the ASME limit of 0.04wt% current at the time the UK retrofit headers were manufactured or indeed the more recently revised ASME limit of 0.02wt%.

The effect of AlN can be inferred from a modified graph shown in Fig.7. This shows the nitrogen and aluminum levels present in RWE npower plant items, mainly in this case taken from available materials test certificates, and highlights those compositions associated with early cracking in terms of operating hours. It can be seen that the earliest cracking is largely confined to the lowest N:Al ratio materials but that, with increasing operating time, material with higher N:Al ratio has also been affected. Because the cracking observed to date has occurred at such an early stage, there is unfortunately ample scope for this process to continue with even material with quite high N:Al ratio potentially cracking within the design life (typically 150kHrs).

The Role of Aluminum Nitride in Type IV Cracking

Although the role of AlN in weakening parent material creep strength by suppressing the formation of the finer VN precipitation can now be regarded as well established, its specific role in the Type IV zone is less clear. Three general possible weakening effects can be suggested for grade 91:

- The grain size in the Type IV zone is small which, for any material failing in creep by a grain boundary cavitation mechanism, will reduce creep strength.
- Following on from this, the small grain size inhibits the formation of the martensitic lath structure typical of the parent and any strengthening contribution from this will be suppressed.
- The limited welding thermal cycle in the Type IV zone will not be sufficient for significant resolution and re-precipitation of precipitates which might provide a strengthening mechanism. The temperature cycle will simply coarsen existing precipitates with some corresponding loss of strength.

The last point is also relevant to AlN which, once formed, will not re-dissolve in the Type IV zone to release nitrogen for the formation of new VN precipitates. The role of AlN can therefore be seen as indirectly influencing Type IV strength by lowering the strength of the parent from which it forms.

Some support for this hypothesis is provided by Type IV rupture data emerging recently from tests being carried out by the UK High Temperature Power Plant Forum (UKHTPPF) (4). The early results indicate that for a given grade 91 material, the Type IV strength within the Type IV scatter band reflects that of the parent strength within the parent scatter band.

Wider Implications

The earliest use of grade 91 in the UK, for retrofit headers on large coal-fired plant, occurred at a time when electricity in England and Wales was supplied by the Central Electricity Generating Board. As a result the cracked header is one of a family of approximately 100 similar grade 91 components constructed with broadly the same design philosophy. Whereas in the case of the earlier grade 91 failures in the UK some contribution from high stress or high temperature could not be ruled out, this is less true of this header. Extensive investigations led to the conclusion that the early appearance of Type IV cracking could not be explained by high operating loads or excessive operating temperature. The cracking appears to have been driven primarily by the hoop stress in the casts with the worst N:Al ratio. In terms of operating conditions therefore the header has to be considered representative of the general population of UK retrofit headers, although probably containing material at the bottom of the creep strength range.

Unfortunately for plant operators another provisional conclusion from the UKHTPPF program is that, while lower aluminum increases rupture life for both parent material and the Type IV zone, the improvement is limited for the latter case (4). The difference in rupture life between Type IV failures in high and low aluminum material is relatively small. The implication is that other UK headers of this type with better composition could still be vulnerable to cracking within the design life.

For pipework systems operating at pressures and temperatures comparable to the UK retrofit headers Type IV cracking may also appear within the design life. The most vulnerable welds will be those containing Type IV zones perpendicular to the hoop stress, primarily large branches or seam welds. Girth welds experiencing system stresses which increase axial loads towards the level of the hoop stress may also be affected.

For headers or pipework systems operating at lower temperatures, the incidence of Type IV cracking will depend on the extent to which design philosophy has taken advantage of the lower temperature to increase operating stress. The major increase in parent strength at lower temperature will not necessarily be reflected in the Type IV zones and design stress levels may not fully take account of the weakness of these regions.

Conclusions

Early Type IV cracking of a grade 91 header, along with increasing evidence that the range of Type IV strength in this material is relatively narrow, indicates a general problem for retrofit grade 91 headers in the UK and possibly for grade 91 more generally.

Grade 91 casts with the highest aluminum level and lowest nitrogen:aluminum ratio are likely to exhibit the earliest Type IV cracking. However lowering aluminum level and increasing nitrogen:aluminum ratio, while beneficial in retarding the onset of Type IV cracking, may not eliminate it completely.

Unless the relative weakness of the Type IV zone has been fully accounted for at the design stage, even material with more favorable composition may suffer Type IV cracking within the design life.

Acknowledgements

This paper is published with the permission of RWE npower. The author would also like to acknowledge useful technical discussions with colleagues within the UK High Temperature Power Plant Forum and in particular Dr David Allen, E.ON UK.

References

1. S. J. Brett, "In-Service Failures of Modified 9Cr (Grade 91) Components," IMechE Seminar: Forensic Investigation of Power Plant Failures, One Birdcage Walk, London, March 2, 2005.

2. S. J. Brett, J. S. Bates & R. C. Thomson, "Aluminum Nitride Precipitation in Low Strength Grade 91 Power Plant Steels," EPRI - 4th International Conference on Advances in Materials Technology for Fossil Power Plants, South Carolina, October 25 – 28, 2004.

3. S. J. Brett, D. L. Oates and C. Johnston, "In-Service Type IV Cracking on a Modified 9Cr (Grade 91) Header," ECCC Conference: Creep & Fracture in High Temperature Components – Design & Life Assessment Issues, IMechE, London, September 12 – 14, 2005.

4. S. J. Brett, D. J. Allen and L. W. Buchanan, "The Type IV Creep Strength of Grade 91 Materials," 3rd International Conference on the Integrity of High Temperature Welds, IOM3, London, April 24-26, 2007.

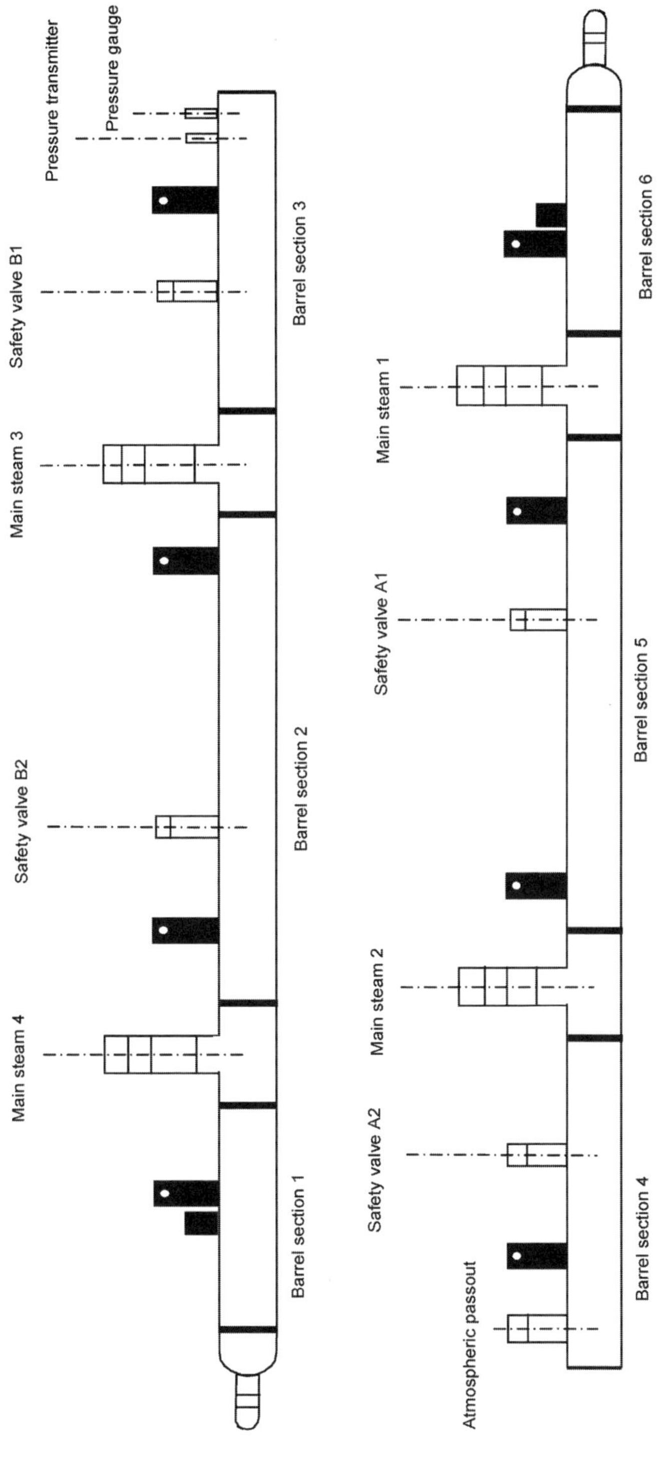

Fig.1. Schematic arrangement of the cracked header.

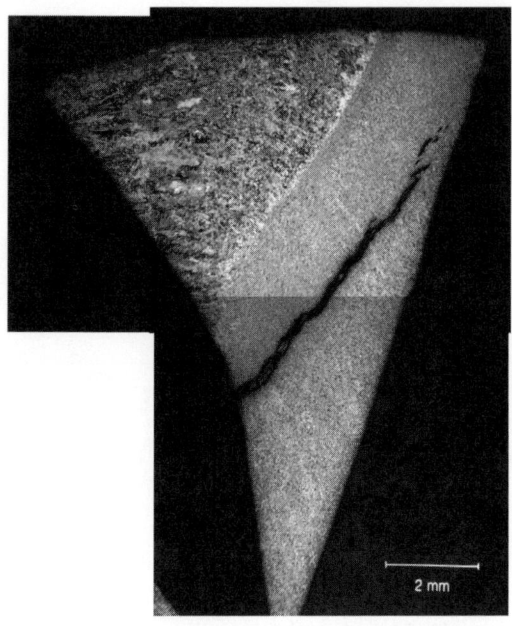

Fig.2. Example of a Type IV crack sampled from one stub.

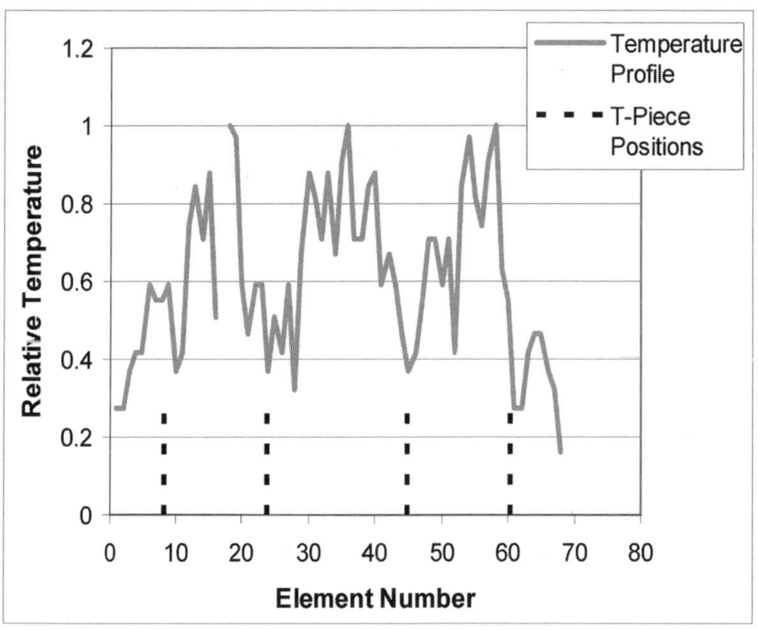

Fig.3. Relative operating temperature profile along the header estimated from stub oxide thickness data.

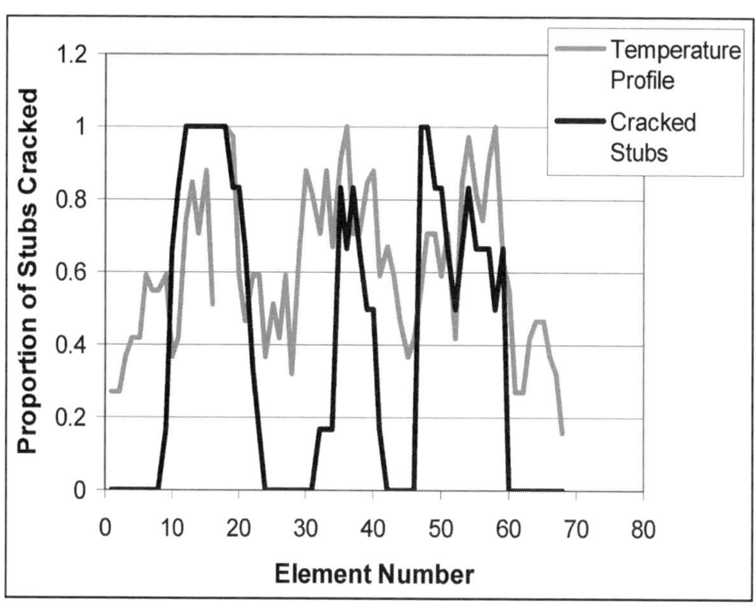

Fig.4. Distribution of cracked stubs along the header superimposed on the temperature profile shown in Fig.3.

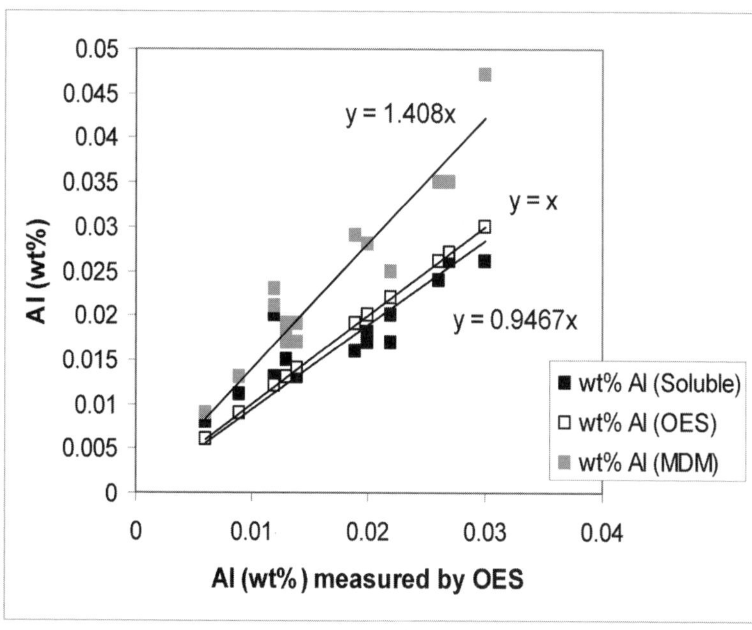

Fig.5. Relationship between aluminum levels measured by three different methods.

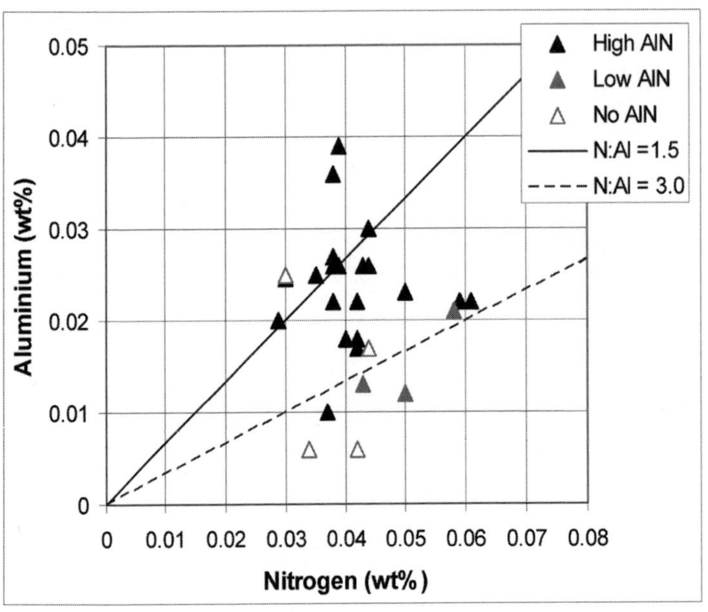

Fig.6. Qualitative levels of aluminum nitride found in samples with varying levels of nitrogen and aluminum (as measured by standard soluble or OES methods only).

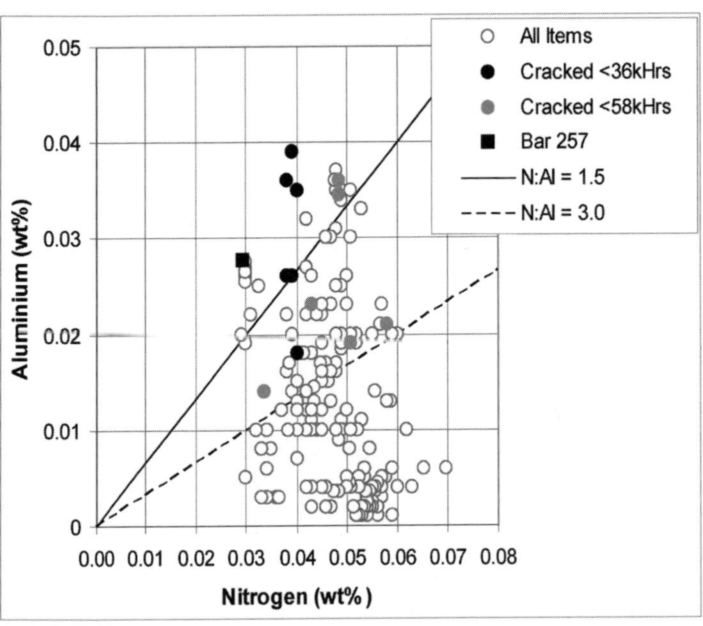

Fig.7. Variation of observed cracking incidence with operating hours for varying levels of nitrogen and aluminum (mainly from material test certificates).

Advances in Materials Technology for Fossil Power Plants
Proceedings from the Fifth International Conference
R. Viswanathan, D. Gandy, K. Coleman, editors, p 601-615

DOI: 10.1361/cp2007epri0601

STRESS DEPENDENCE OF DEGRADATION AND CREEP RUPTURE LIFE OF CREEP STRENGTH ENHANCED FERRITIC STEELS

K. Kimura
K. Sawada
H. Kushima
Y. Toda
National Institute for Materials Science
1-2-1 Sengen, Tsukuba
Ibaraki 305-0047, Japan

Abstract

Overestimation of long-term creep strength of creep strength enhanced ferritic steels is caused by change in stress dependence of creep rupture life with decrease in stress. Creep rupture strength of those steels has been re-evaluated by a region splitting analysis method and allowable tensile stress of the creep strength enhanced ferritic steels regulated in METI (Ministry of Economy, Trade and Industry) Thermal Power Standard Code in Japan has been reduced in December 2005 and July 2007. A region splitting analysis method evaluates creep rupture strength in the high stress and low stress regimes individually, which is separated by 50% of 0.2% offset yield stress. Change is stress dependence of the minimum creep rate is observed at 50% of 0.2% offset yield stress, which roughly corresponds to 0% offset yield stress, on ASME Grade 122 type steels. Stress dependence of the minimum creep rate in the high stress regime is equivalent to that of flow stress observed in tensile test, and a magnitude of stress exponent, n, in the high stress regime decreases with increase in temperature from 20 at 550°C to 10 at 700°C. On the other hand, n value in the low stress regime is 4 to 6 for tempered martensite single phase steel, however, remarkably small value of 2 to 4 is observed in the low stress regime of the dual phase steel containing delta ferrite. Large stress dependence of creep rupture life and minimum creep rate in the high stress regime is caused by contribution of considerable plastic deformation due to applied stress higher than a proportional limit. Creep deformation in the low stress regime is considered to be governed by diffusion controlled phenomena and dislocation climb as a rate controlling mechanism.

Introduction

Creep strength enhanced ferritic (CSEF) steels have been widely used for high temperature structural components in modern thermal power plant, and energy efficiency of the power plant has been improved by increasing steam temperature and pressure. It has contributed to reduce both fuel consumption and emission of carbon dioxide which is regarded to be a green house gas. However, a risk of overestimation of long-term creep strength has been pointed out (1-3) and premature failure due to Type IV cracking has been also found on branch and attachment welds of CSEF steels (4). Allowable tensile stress of the CSEF steels regulated in Thermal Power Standard Code in Japan has been reviewed (5, 6) with a region splitting analysis method (7-9) and allowable tensile stress of the CSEF steels has been reduced (10, 11). Since remarkable reduction of creep strength has been observed on the weldment of the CSEF steels, creep strength reduction factor for weldment has been also investigated by means of a region splitting analysis method (12, 13), as well as allowable tensile stress of the parent materials.

Accuracy of creep rupture strength evaluation is remarkably improved by a region splitting analysis method (7-9), which evaluates creep rupture data in the high stress and low stress regimes individually. Two regimes are divided by 50% of 0.2% offset yield stress, and availability of the method is attributable to change in stress dependence of creep rupture life. In this paper, influence of stress on creep deformation property is investigated on ASME Grade 122 type steels, in order to understand an influence of stress on creep strength property of CSEF steels and to obtain a theoretical basis of a region splitting analysis method.

Experimental Procedure

The steels used in the present study are ASME Grade 122 type steels and chemical composition of the steels is shown in Table 1. Heat treatment condition and microstructure of the steels are shown in Table 2. Three steels of those are ASME Grade 122 type pipe (P122), plate (Gr122) and tube (T122) materials, and chemical composition of those three steels is essentially the same. The other tube steel (12CR) contains slightly higher chromium of 12.10mass%, and it is higher than the upper limit of ASME Grade 122, that is 11.5mass%. Microstructure of the steels is tempered martensite, except for 12CR which is dual phase containing about 5 vol% of delta ferrite due to higher chromium concentration.

Tensile test was conducted over a range of temperatures from 550 to 700°C under a constant nominal strain rate of 5×10^{-5} s^{-1} up to 2 to 3% of total strain, and 1.25×10^{-3} s^{-1} beyond that. Strain rate of the tensile test was controlled by differential transformer whose resolution is 1μm, with an extensometer attached to the gauge portion of the specimen. Flow stress was evaluated under a constant nominal strain rate of 5×10^{-5} s^{-1}, as well as 0% and 0.2% offset yield stresses, in addition to tensile strength, that is a flow stress under a constant nominal strain rate of 1.25×10^{-3} s^{-1}. The former flow stress is hereinafter denoted as FS, in contrast to TS for the latter one. Creep test was conducted over a range of temperatures from 550 to 700°C.

Table 1: Chemical composition (mass%) of the steels studied

Steels	C	Si	Mn	Ni	Cr	Mo	Cu	W	V	Nb	Al	B	N
P122	0.12	0.30	0.60	0.32	10.65	0.34	0.85	1.89	0.19	0.05	0.007	0.0029	0.054
Gr122	0.12	0.24	0.63	0.36	10.73	0.38	0.97	1.97	0.22	0.06	0.006	0.0039	0.072
T122	0.13	0.31	0.60	0.36	10.65	0.33	0.86	1.87	0.19	0.05	0.007	0.0024	0.057
12CR	0.11	0.27	0.59	0.33	12.10	0.34	0.82	1.82	0.19	0.06	0.016	0.0030	0.066

Table 2: Heat treatment condition and microstructure of the steels studied

Steels	Product form	Normalizing	Tempering	Microstructure
P122	pipe	1050°C x 60min / AC	770°C x 6h / AC	Tempered martensite
Gr122	plate	1050°C x 100min / AC	770°C x 6h / AC	Tempered martensite
T122	tube	1050°C x 10min / AC	770°C x 6h / AC	Tempered martensite
12CR	tube	1050°C x 10min / AC	790°C x 1h / AC	Tempered martensite + Delta ferrite (5vol%)

Results and Discussion

Creep Strength

Stress vs. time to rupture curves at 550, 600, 650 and 700°C of the steels are shown in Figure 1. Slope of the curves at 550°C is almost constant from short-term to long-term, although creep rupture strength of Gr122 is slightly lower than those of the others. It becomes steeper, however, in the low stress regime at 600, 650 and 700°C. In the high stress regime where slope of the curve is gentle and stress dependence of creep rupture life is large, creep rupture life of the four steels are almost the same. On the other hand, dual phase steel of 12CR containing delta ferrite indicates further steep slope than the other delta ferrite free steels in the low stress regime, and creep rupture life of 12CR in the low stress regime is remarkably shorter than those of the other steels.

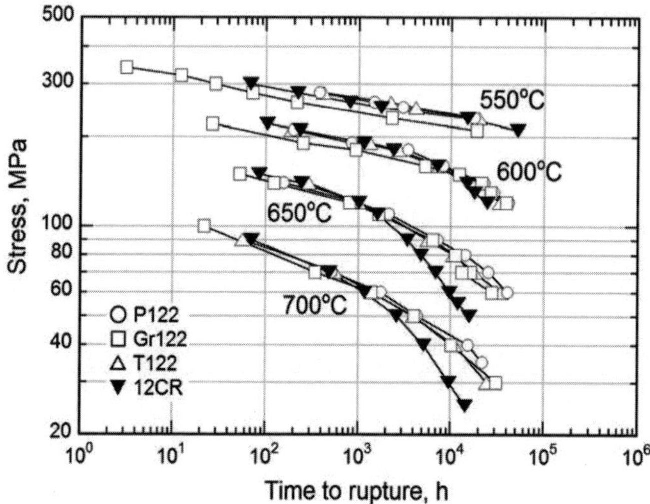

Figure 1. Stress vs. time to rupture curves at 550, 600, 650 and 700°C of the steels.

Stress vs. minimum creep rate curves at 550, 600, 650 and 700°C of the steels are shown in Figure 2. The curves at 550°C are close to linear in both logarithmic scales for all the steels in the investigated stress range. However, steeper slope of the curves in the high stress regime decreases with decrease in stress at 600, 650 and 700°C, and change in slope of the stress vs. minimum creep rate curve is observed notably on dual phase steel of 12CR. Minimum creep rate of 12CR in the low stress regime is remarkably larger than those of the other single phase steels, although no significant difference in minimum creep rate is observed on these four steels in the high stress regime. Stress dependence of the minimum creep rate is corresponding to that of creep rupture life shown in Fig. 1.

Minimum creep rate of the steels is plotted against time to rupture and shown in Figure 3. The parameters of A and B obtained by regression analysis for individual steel with the following equation (1) are shown in the figure.

$$\dot{\varepsilon}_m = A\, t_R^{B} \tag{1}$$

where $\dot{\varepsilon}_m$ is a minimum creep rate and t_R is a time to rupture. Differences in parameters of the three steels of P122, Gr122 and T122 are small, and those of dual phase steel of 12CR are obviously different from those of the other steels. Scatter band of the data in the short-term is narrow, and it tends to expand with increase in time to rupture, since 12CR indicates shorter time to rupture than those of the other steels for the same minimum creep rate in the long-term.

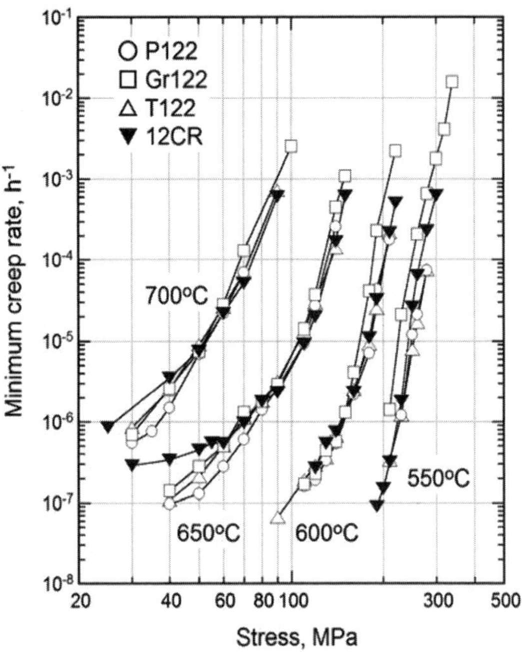

Figure 2. Stress vs. minimum creep rate curves at 550, 600, 650 and 700°C of the steels.

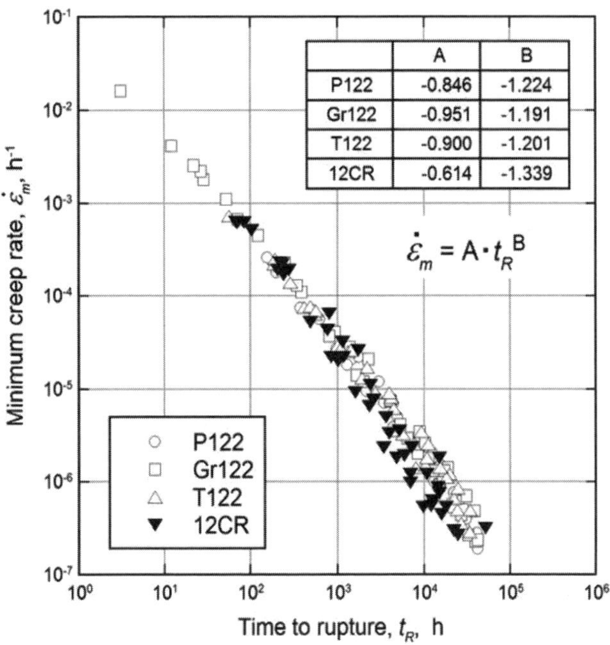

	A	B
P122	-0.846	-1.224
Gr122	-0.951	-1.191
T122	-0.900	-1.201
12CR	-0.614	-1.339

Figure 3. Monkman-Grant relationship of the steels.

Creep rate vs. time curves over a range of stresses from 60 to 150MPa at 650°C of Gr122 are shown in Figure 4. Creep deformation consists of transient and accelerating creep stages and no obvious steady state creep stage is observed. With decrease in stress, time to rupture increases in accordance with decrease in minimum creep rate as expected from a linear relationship of the Monkman-Grant plot shown in Fig.3. For the single phase steels of P122, Gr122 and T122, essentially the same creep deformation property as shown in Fig.4 is observed.

Creep rate vs. time curves over a range of stresses from 60 to 120MPa at 650°C of T122 and 12CR are shown in Figure 5. Creep deformation consists of transient and accelerating creep stages for both steels, and creep deformation behavior of both steels at high stresses of 110 and 120MPa are resemble each other. In the low stress condition, however, increase in creep rate in an accelerating creep stage of 12CR is faster than that of T122 and creep rupture life of 12CR is shorter than that of T122. A magnitude of such difference in accelerating creep behavior and creep rupture life increases with decrease in stress, although difference in minimum creep rate is relatively small. Shorter creep rupture life in the low stress regime and different Monkman-Grant relationship of 12CR from the other steels is caused by a faster increase in creep rate in an accelerating creep stage. From the above results, it has been found that stress dependence of creep rupture life and minimum creep rate is large in the high stress regime, and it decreases in the low stress regime. Influence of delta ferrite phase is observed only in the low stress regime, as a result of faster increase in creep rate in an accelerating creep stage.

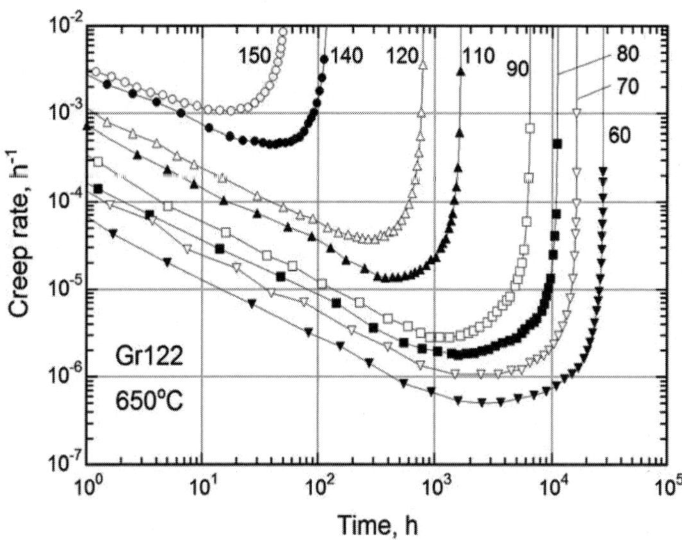

Figure 4. Creep rate vs. time curves of Gr122 at 650°C. Numerical values in the figure indicate stress in MPa.

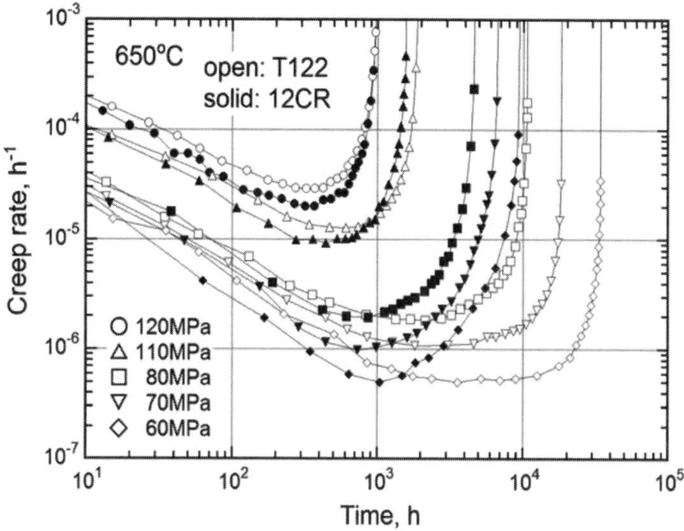

Figure 5. Creep rate vs. time curves of T122 and 12CR at 650°C.

Tensile Strength Property

Stress vs. strain curve of T122 at 600°C under a constant nominal strain rate of 5 x 10^{-5} s^{-1} is shown in Figure 6. A 0.2% offset yield stress of the steel was evaluated to be 321MPa, and a maximum stress of 362MPa was obtained as a flow stress under a constant nominal strain rate of 5 x 10^{-5} s^{-1} (FS). Magnitude of plastic strain was estimated by subtracting elastic strain evaluated by linear relationship between stress and strain from total strain, and it was used to estimate a 0% offset yield stress. Measurement accuracy of a 0% offset yield stress was about ±10MPa.

Temperature dependence of tensile strength (TS), flow stress under a constant nominal strain rate of 5 x 10^{-5} s^{-1} (FS), 0.2% offset yield stress (0.2% YS) and 0% offset yield stress (0% YS) of the steels are shown in Figure 6 for individual steels. An error bar of 0% offset yield stress indicates measurement accuracy of ±10MPa, and 50% of 0.2% offset yield stress is described by dashed line. These stress values decreases with increase in temperature monotonously for all the steels. Good correspondence between 0% offset yield stress and 50% of 0.2% offset yield stress is observed over a range of temperatures from 550 to 700°C for all the steels. It has been shown that 50% of 0.2% offset yield stress is regarded to be a proportional limit stress of the steels at the temperatures from 550 to 700°C.

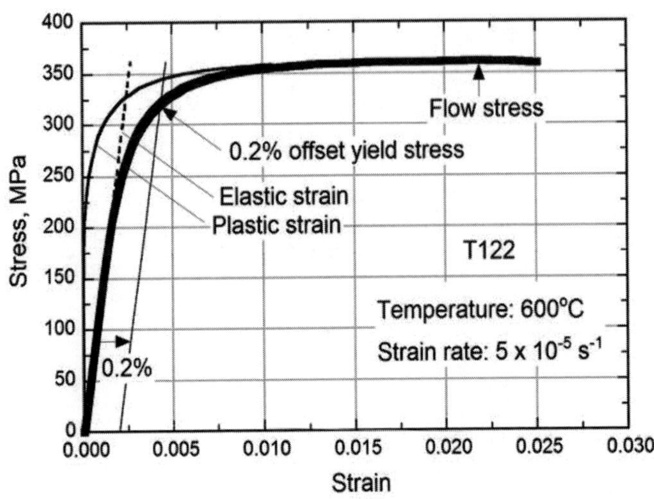

Figure 6. Stress vs. strain curve of T122 at 600°C under constant strain rate of 5×10^{-5} s^{-1}.

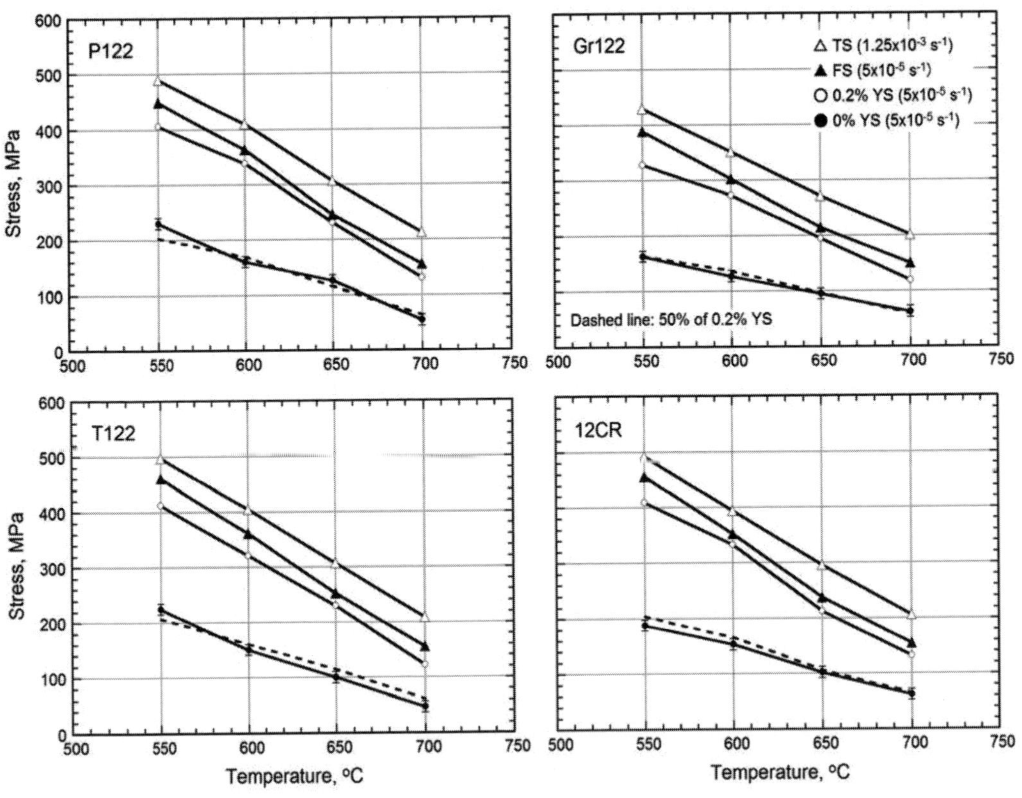

Figure 7. Temperature dependence of tensile strength (TS), flow stress under a strain rate of 5×10^{-5} s^{-1} (FS), 0.2% offset yield stress (0.2% YS) and 0% offset yield stress (0% YS) of the steels. Dashed line indicates 50% of 0.2% offset yield stress

Stress Dependence of Minimum Creep Rate

Stress vs. minimum creep rate curves of the steels at 600°C are shown in Figure 8. Since tensile strength is a flow stress under a constant nominal strain rate of 1.25×10^{-3} s^{-1} (4.5 h^{-1}), tensile strength and flow stress under a constant nominal strain rate of 5×10^{-5} s^{-1} (1.8×10^{-1} h^{-1}) (FS) are also plotted in the same figure. Tensile strength and FS are observed on an extension of the stress vs. minimum creep rate curves in the high stress regime. Slightly higher minimum creep rate of Gr122 than the other steels corresponds to lower tensile strength and FS.

Stress vs. minimum creep rate curves of Gr122 over a range of temperatures from 550 to 700°C are shown in Figure 9, together with tensile strength and FS. It has been found that stress dependence of minimum creep rate in the high stress regime is equivalent to that of flow stress evaluated by tensile test. Magnitude of stress exponent, n, in the high stress regime is larger than that in the low stress regime, and it decreases with increase in temperature from 19 at 550°C to 10 at 700°C. On the other hand, n value in the low stress regime is in a range of 3.8 to 5.3.

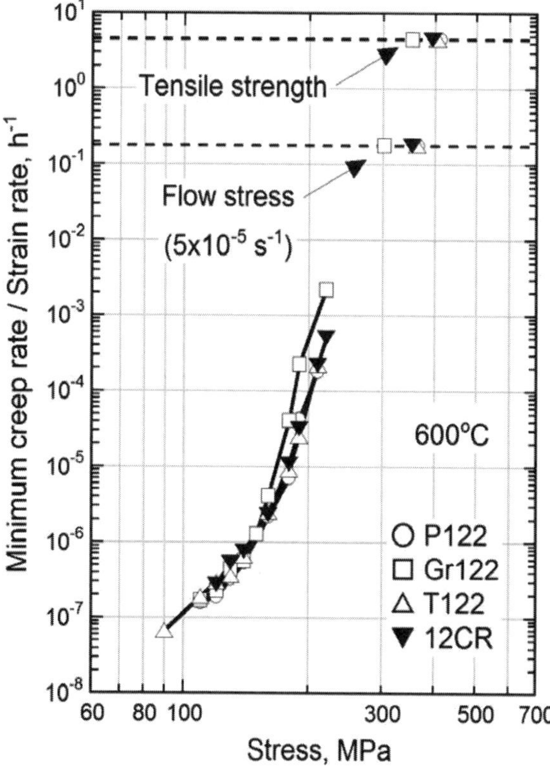

Figure 8. Stress vs. minimum creep rate curves of the steels at 600°C. Tensile strength and flow stress under a strain rate of 5×10^{-5} s^{-1} is plotted at the strain rate tested.

Similar plots of minimum creep rate, tensile strength and FS for P122, T122 and 12CR to Fig.9 are shown in Figures 10, 11 and 12, respectively. Stress dependence of the minimum creep rate is clearly divided into two groups of high stress and low stress regimes, and that of minimum creep rate in the high stress regime, tensile strength and FS is expressed by a common linear relationship in both logarithmic scales. Magnitude of stress exponent, n value in the high stress regime is almost the same for all the steels, regardless of delta ferrite phase, and it decreases with increase in temperature from about 20 at 550°C to about 10 at 700°C.

On the other hand, magnitude of stress exponent, n value in the low stress regime is in a range of 3.8 to 5.6 for single phase steel of P122, Gr122 and T122, however, that of 12CR is in a range of 2.0 to 3.6 except for 6.7 at 600°C and it is significantly smaller than that of single phase steels. Stress exponent, n value of 3.8 to 5.6 corresponds to that of creep deformation controlled by dislocation climb. Smaller stress exponent, n value of 2.0 to 3.6 for dual phase steel corresponds to that of creep deformation controlled by dislocation glide. However, it is hard to explain the smaller stress exponent value of 12CR by different rate controlling mechanism, since about 95vol% of 12CR is same tempered martensite as that of the other single phase steels.

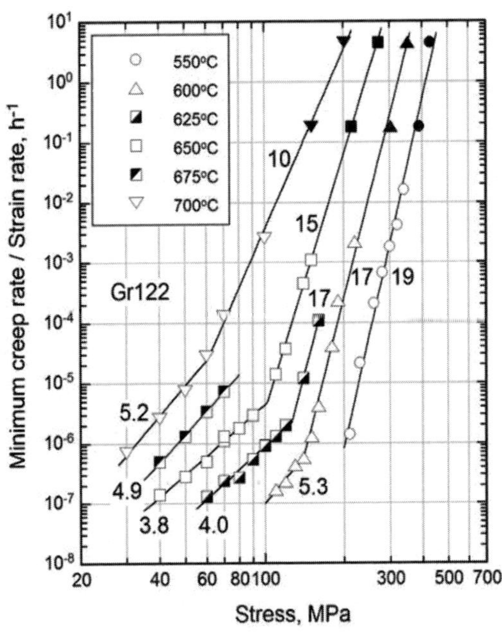

Figure 9. Stress vs. minimum creep rate curves, tensile strength and flow stress under a strain rate of 5×10^{-5} s^{-1} of Gr122. Numerical values in the figure indicate stress exponent, n value.

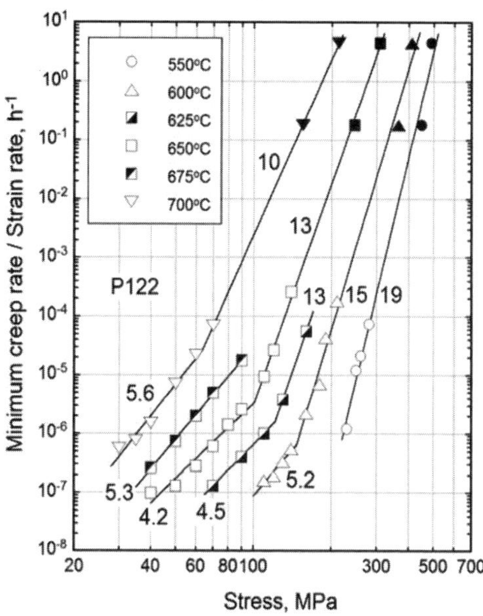

Figure 10. Stress vs. minimum creep rate curves, tensile strength and flow stress under a strain rate of 5×10^{-5} s^{-1} of P122. Numerical values in the figure indicate stress exponent, n value.

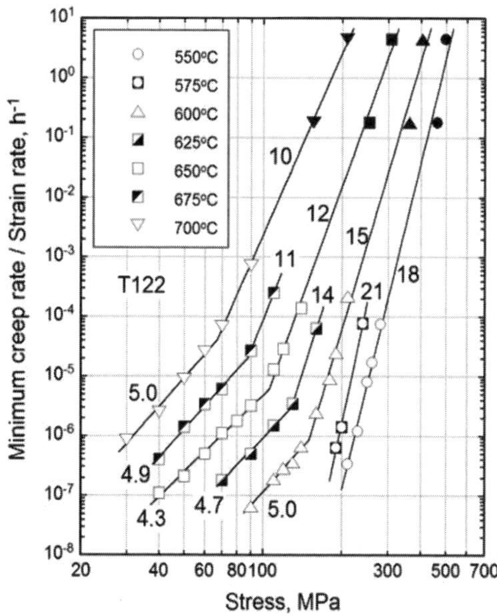

Figure 11. Stress vs. minimum creep rate curves, tensile strength and flow stress under a strain rate of 5×10^{-5} s^{-1} of T122. Numerical values in the figure indicate stress exponent, n value.

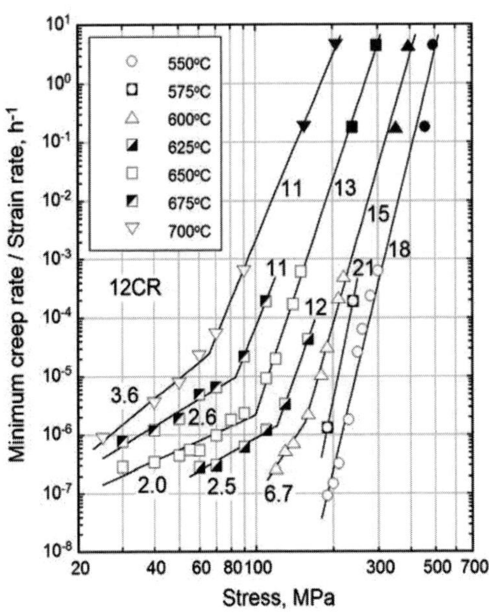

Figure 12. Stress vs. minimum creep rate curves, tensile strength and flow stress under a strain rate of 5 x 10^{-5} s^{-1} of 12CR. Numerical values in the figure indicate stress exponent, n value.

According to change in stress exponent, a boundary stress between high stress and low stress regimes is evaluated from the Figs. 9 to 12. Relation between 0% offset yield stress and a boundary stress between high stress and low stress regimes is shown in Figure 13. Good correspondence between both stresses is observed, and it indicates that the low stress regime where a stress exponent value is small is equivalent to an elastic range below proportional limit, and the high stress regime corresponds to a plastic range beyond proportional limit. Consequently, large stress dependence of creep rupture life and minimum creep rate in the high stress regime is caused by contribution of considerable plastic deformation due to applied stress higher than a proportional limit. On the other hand, creep deformation in the low stress regime is considered to be governed by diffusion controlled phenomena and dislocation climb as a rate controlling mechanism. A smaller stress exponent value of 12CR in the low stress regime is speculated to be caused by an enhanced diffusion due to composition partitioning as a result of a presence of delta ferrite (9).

Change in stress dependence of creep deformation is clearly observed at a proportional limit stress, which corresponds to 50% of 0.2% offset yield stress. Long-term creep strength, therefore, should be discussed on the creep strength at the stresses below proportional limit, since it is different from that in the high stress regime and high temperature components are operated under stress condition below proportional limit.

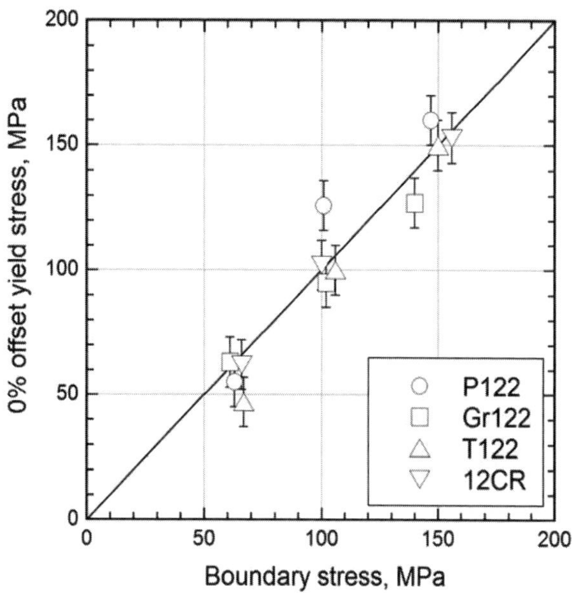

Figure 13. Relation between 0% offset yield stress and a boundary stress between high stress and low stress regimes whose stress exponent values are different at 600, 650 and 700°C.

Conclusion

Influence of stress on creep deformation property is investigated on ASME Grade 122 type steels and theoretical basis of a region splitting analysis method is discussed. The obtained results are as follows.

1. Difference in stress dependence of creep rupture life and minimum creep rate is observed in the high stress and low stress regimes.

2. A half of 0.2% offset yield stress is regarded to be a proportional limit stress of the steels.

3. Stress dependence of minimum creep rate in the high stress regime, tensile strength and flow stress under a strain rate of 5×10^{-5} s^{-1} is expressed by a common linear relationship in both logarithmic scales. Magnitude of stress exponent, n value in the high stress regime is almost the same for all the steels, regardless of delta ferrite phase, and it decreases with increase in temperature from about 20 at 550°C to about 10 at 700°C.

4. Large stress dependence of creep rupture life and minimum creep rate in the high stress regime is caused by contribution of considerable plastic deformation due to

higher stress than a proportional limit. On the other hand, creep deformation in the low stress regime is considered to be governed by diffusion controlled phenomena and dislocation climb as a rate controlling mechanism.

5. Change in stress dependence of creep deformation is clearly observed at a proportional limit stress, which corresponds to 50% of 0.2% offset yield stress. It has been concluded that long-term creep strength should be discussed on the creep strength at the stresses below proportional limit.

Acknowledgments

A part of this study was financially supported by the Budget for Nuclear Research of the Ministry of Education, Culture, Sports, Science and Technology, based on the screening and counseling by the Atomic Energy Commission.

References

1. V. Foldyna, Z. Kubon, A. Jakobova and V. Vodarek, "Development of Advanced High Chromium Ferritic Steels", *Microstructural Development and Stability in High Chromium Ferritic Power Plant Steels,* The Institute of Materials, 1997, pp. 73-92.

2. A. Strang and V. Vodarek, "Microstructural Stability of Creep Resistant Martensitic 12%Cr Steels", *Microstructural Stability of Creep Resistant Alloys for High Temperature Plant Applications,* The Institute of Materials, 1998, pp. 117-133.

3. K. Kimura, H. Kushima and F. Abe, "Heterogeneous Changes in Microstructure and Degradation Behaviour of 9Cr-1Mo-V-Nb Steel During Long Term Creep", *Key engineering Materials,* 171-174, 2000, pp.483-490.

4. S.J. Brett, D.L. Oates and C. Johnston, "In-Service Type IV Cracking in a Modified 9Cr (Grade 91) Header", *Creep and Fracture in High Temperature Components - Design and Life Assessment Issues,* DEStech Publications, Inc., 2005, pp. 563-572.

5. K. Kimura, "Assessment of Long-term Creep Strength and Review of Allowable Stress of High Cr Ferritic Creep Resistant Steels", PVP2005-71039, presented at the 2005 ASME Pressure Vessels and Piping Division Conference, Denver, CO (July 2005).

6. K. Kimura, "Creep Strength Assessment and Review of Allowable Tensile Stress of Creep Strength Enhanced Ferritic Steels in Japan", PVP2006-ICPVT11-93294, presented at the 2006 ASME Pressure Vessels and Piping Division Conference, Vancouver, Canada (July 2006).

7. K. Kimura, H. Kushima and F. Abe, "Degradation and Assessment of Long-term Creep Strength of High Cr Ferritic Creep Resistant Steels", presented at the International Conference on Advances in Life Assessment and Optimization of Fossil Power Plant, Orland, FL (March 2002).

8. K. Kimura, K. Sawada, K. Kubo and H. Kushima, "Influence of Stress on Degradation and Life prediction of High Strength Ferritic Steels", PVP2004-2566, presented at the ASME Pressure Vessels and Piping Division Conference, San Diego, CA (July 2004).

9. K. Kimura, H. Kushima, K. Sawada and Y. Toda, "Region Splitting Analysis on Creep Strength Enhanced Ferritic Steels", PVP2007-26406, presented at the Eighth International Conference on Creep and Fatigue at Elevated Temperatures, San Antonio, TX (July 2007).

10. Ministry of Economy, Trade and Industry. Nuclear and Industrial Safety Agency, *Thermal Power Standard Code,* Tokyo, Japan, 2005.

11. Ministry of Economy, Trade and Industry. Nuclear and Industrial Safety Agency, *Thermal Power Standard Code,* Tokyo, Japan, 2007.

12. M. Tabuchi and Y. Takahashi, "Evaluation of Creep Strength Reduction Factors for Welded Joints of Modified 9Cr-1Mo Steel (P91)", PVP2006-ICPVT11-93350, presented at the ASME Pressure Vessels and piping Division Conference, Vancouver, Canada (July 2006).

13. Y. Takahashi and M. Tabuchi, "Evaluation of Creep Strength Reduction Factors for Welded Joints of HCM12A (P122)", PVP2006-ICPVT11-93488, presented at the ASME Pressure Vessels and Piping Division Conference, Vancouver, Canada (July 2006).

Advances in Materials Technology for Fossil Power Plants
Proceedings from the Fifth International Conference
R. Viswanathan, D. Gandy, K. Coleman, editors, p 616-626

DOI: 10.1361/cp2007epri0616

The Estimation of Residual Life of Low-Alloy Cast Steel Cr-Mo-V Type after Long-Term Creep Service

Zieliński A.*, Dobrzański J.*
Renowicz D.**, Hernas A.**
*Institute for Ferrous Metallurgy
44-100 Gliwice ul. K. Miarki 12, Poland
** Silesian University of Technology
Department of Material Engineering and Metallurgy

Abstract

A three-way pipe of fresh steam pipeline made of low-alloy cast Cr-Mo-V steel after long-term creep service longer than 100,000 hours was examined. In the article, results of microstructure investigation and mechanical properties (at room and elevated temperatures) are presented in comparison to initial state. Impact transition temperatures of materials in initial state and after long term service were determined. Presented results are applied to the diagnostic of materials used for pressure installations in power stations. Shortened creep tests at constant stress and variable test temperature and constant test temperature at variable stress were performed. On the base of results of many years of authors investigations of Cr-Mo-V low-alloy as well as high alloyed 12Cr-Mo-V steels usability above mentioned methods for estimation life and residual life in the practice were determined. Methods of selection of parameters for long term creep tests were developed on the basis of shortened creep tests. Residual life of low-alloy Cr-Mo-V cast steel after long term service on the basis of results shortened creep tests were determined. Moreover, low-cycle isothermal fatigue and thermal fatigue were done. On the basis of test results the degree of material properties degradation were estimated.

1. Introduction.

A review of the specialist literature and gained experiences show that until now there has been no satisfactory description of low-alloy cast steel Cr-Mo-V components being operated in fatigue and creep conditions. In particular, the mechanisms of crack initiation and the influence of material structure degradation processes onto cast steel components service life have hardly been made familiar.

Nowadays in Poland the majority of the power plants have already reached or even exceeded considerably the calculated service life, i.e. 100,000 or 200,000 hours of operation. In view of the above, the necessity to develop diagnostic examinations of boiler critical components, in order to allow their further safe and trouble-free operation, is well justified [1,2].

The actual service life of power installations is often considerably different from the calculated service life. This is mainly due to often higher resistance to creep than the average value according to standards, which is applied in design calculations. The material structure condition, which changes owing to the overall long-term temperature and load influence thus affecting the service life of a structural component to a large extent, plays a vital part in those processes. Creep resistance of the components operated at such temperature conditions, i.e. above the limit temperature, forms the main material feature when estimating further capability of material for further operation [2]. Both a design engineer and an end user are mostly interested in the time period of safe and trouble-free operation, also called "available service life". In reality, the available service life of structural components is most often higher, even many times, than their calculated service life [2].

When making prognoses of the residual life, the most reliable results can be obtained on the basis of the material creep tests, both in the initial state and after a long-term operation. Such tests still constitute the most important source of information when evaluating service life and residual life of materials operated above the limit temperature T_g [3].

Since the duration of long-term creep tests ranges from minimum several thousand to several dozen hours, the time for coming up with the creep test results is minimum several years. In order to shorten the period of performing creep tests and evaluation of service life or residual life, in the engineering practice so-called "accelerated" creep tests are carried out, lasting from several dozen to several thousand hours. This allows to obtain the test results within maximum a dozen months or so, enabling to evaluate the residual life of a component [4÷8].

Below some test results of structural changes and strength properties at room temperature and elevated temperature are presented, in comparison with the initial condition of the tested cast steel. The material transition temperature into brittle state for material in initial state and following a long-term operation has been determined [9,10]. The principles of shortened creep test execution at constant stress and variable testing temperature have also been described. Based on our own experiences in testing low-alloy steels type Cr-Mo-V, the usefulness of the aforementioned method was determined with regard to evaluating the service life and residual life in practice. The residual life of Cr-Mo-V low-alloy cast steel has been determined after a long-term operation and based on the accelerated creep tests. Additionally, complex thermal fatigue tests have been executed. Based on the obtained results, the degree of properties degradation for a tested T-connection material has been determined [11,12].

2. Material, scope and methodology of testing.

The tested object was constructional material of boiler pressure part components in the form of T-connections made of Cr-Mo-V low-alloy cast steel in the initial state and following 100,000 hours of operation in creep and fatigue conditions.

This paper includes the results of the following tests:

- Mechanical properties tests at room and elevated temperatures:
  - Static tensile test with the objective to determine tensile strength R_m, yield point R_e, R_e^t, elongation A_5 and area reduction of the specimen Z,
  - Impact test using CHARPY V specimens,
- Microstructure examinations.
- Shortened creep tests.
- Thermal fatigue examinations.

3. Tests results

3.1. Testing of mechanical properties.

The testing of strength properties in the tensile test at room temperature was carried out on Zwick company's machine, having the maximum load of 200 kN, using the load range of 100 and 50kN, whereas the tests of strength properties at elevated temperature T_b=392, 572, 752, 842, 932 and 1022°F were carried out on Amsler company's machine, having the maximum load of 200 kN but applying the loads of 20 and 50kN. The test results of strength properties of the examined materials of the components which were made of Cr-Mo-V low-alloy cast steel at room temperature and at elevated temperature are presented in Table No. 1.

Table No. 1. Strength properties at room and elevated temperatures of material in initial state and after 100,000 hours of operation in creep conditions.

Test temperature [°F]	R_m^t [MPa]		R_e^t [MPa]		A_5 [%]		Z [%]	
	i.s.	a.s	i.s.	a.s.	i.s.	a.s.	i.s.	a.s.
68	568	556	403	323	21.5	15.0	69.2	47.9
392	557	454	363	250	-	16.2	25.1	38.0
572	555	475	311	246	13.3	11.2	30.2	25.1
752	517	481	290	232	17.3	16.7	43.7	41.8
842	502	416	283	225	21.6	12.0	53.7	16.3
932	431	378	252	212	26.6	26.5	70.0	69.8
1022	374	320	247	196	23.3	20.5	81.6	50.7

▭ incompliant with requirements acc. to standard PN-89/H-83157,
i.s. – initial state,
a.s. – following 100,000 hours of operation.

The impact strength tests were carried out at temperature range from 68 to 284°F. Impact strength changes in relation to the test temperature are shown in the form of a diagram in Fig. 1. The nil ductility transition temperature of Cr-Mo-V cast steel in initial state has the value of approx. 113°F. The long-term operation lasting 100,000 in creep conditions resulted in elevation of the nil ductility transition temperature up to approx. 140°F.

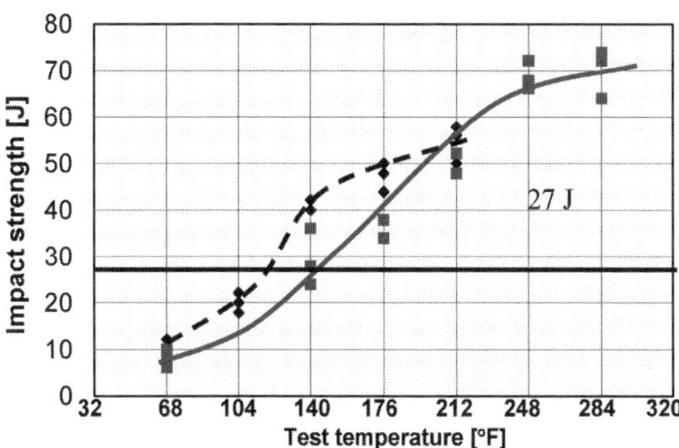

Fig. 1. Impact strength changes in relation to test temperature and brittle state transition temperature of Cr-Mo-V cast steel in initial state and following approx. 100,000 hours of operation.

3.2. Microstructure testing.

The metallographic tests were carried out with Philips XL30 scanning electron microscope, at magnifications 500, 1000, 2000 and 3000X. The material of Cr-Mo-V cast steel in initial state has ferrite structure with degraded pearlite areas. The degraded pearlite are the areas in the structure where phase transition is on the boundaries of pearlite and bainite transition. Inside the ferrite grains, uniformly located and very fine releases can be observed. Singular very fine releases on the grain boundaries can also be seen (Fig. 2). The difference in the structure appearance between the initial state and that following approx. 100,000 hours of operation can only be seen after at magnifications ≥ 2000X. The structure of Cr-Mo-V cast steel following 100,000 hours of operation in creep conditions can be characterised by disintegration of pearlite / bainite areas, which is manifested by coagulation and increase of releases in those areas and on grain boundaries, where carbides locally form release chains. Moreover, inside the ferrite grains impoverishment in fine dispersion releases can be observed. However, larger, non-uniformly located coagulated releases are also formed (Fig. 3).

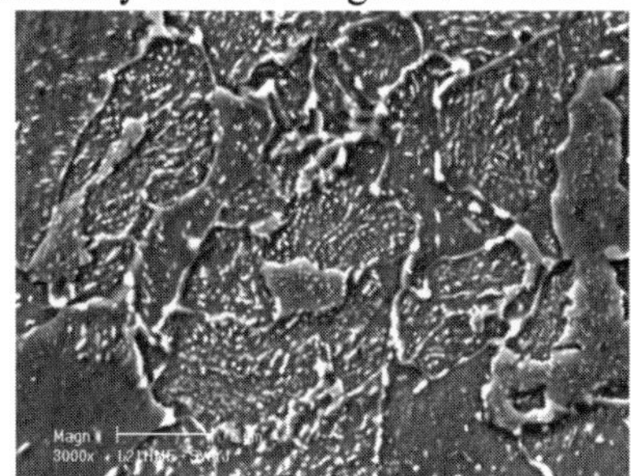

Fig. 2. Cr-Mo-V low-alloy cast steel structure in initial state; hardness 181 HV10.

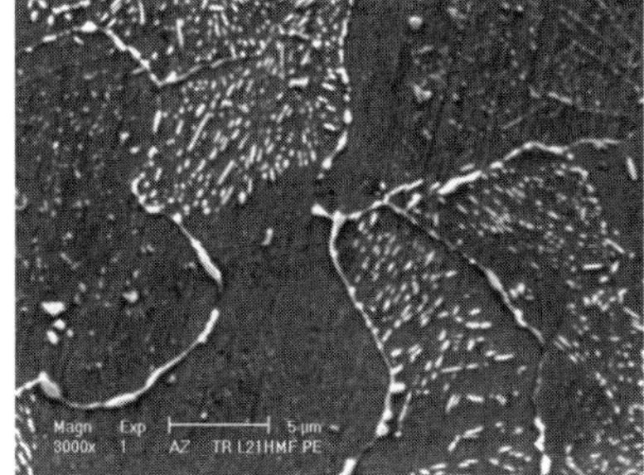

Fig. 3. Cr-Mo-V low-alloy cast steel structure after 100,000 hours of operation in creep conditions; hardness 171 HV10.

3.3. Shortened creep tests.

The accelerated creep tests for the tested material of live steam installation T-connections in the initial state as well as after a long-term operation have been carried out at testing stresses of σ_b = 50, 70 and 100 MPa and testing temperature of T_b = 1112, 1148, 1184, 1202, 1220, 1256 and 1292°F.

The obtained results enabled to determine, using the graphical extrapolation method, the residual life for three levels of test stresses σ_b, i.e. 50, 70 and 100 MPa at temperatures of 932, 977, 1022, 1067 and 1112°F. Based on the determined residual life of the examined components, creep resistance characteristics have been drawn up in the system $\log\sigma = f(\log t_r)$ for temperatures 932, 977, 1022, 1067 and 1112°F. The next stage in the execution of the accelerated creep tests was to determine the residual creep resistance $R_{ze/10000}$ and $R_{ze/30000}$, based on the creep curves characteristics. The achieved values of $R_{ze/t}$ are presented in the form of parametrical time curve of creep resistance in the form of log σ = f(L-M) in Fig. 4. The parametrical curve L-M may be used to determine the long-term test parameters, including anticipated time until rupture. The final outcome of the accelerated creep tests is to determine the available residual service life for the working parameters of further operation (p_r, T_r), which in turn determines the safe operation time of the tested component (Table No. 2). A graphical presentation of the elaborated way to determine the anticipated creep resistance values in function log σ = f(L-M) and based on the obtained results of the shortened tests for some selected test stress levels σ_b and test temperature T_b is shown in Fig. 5.

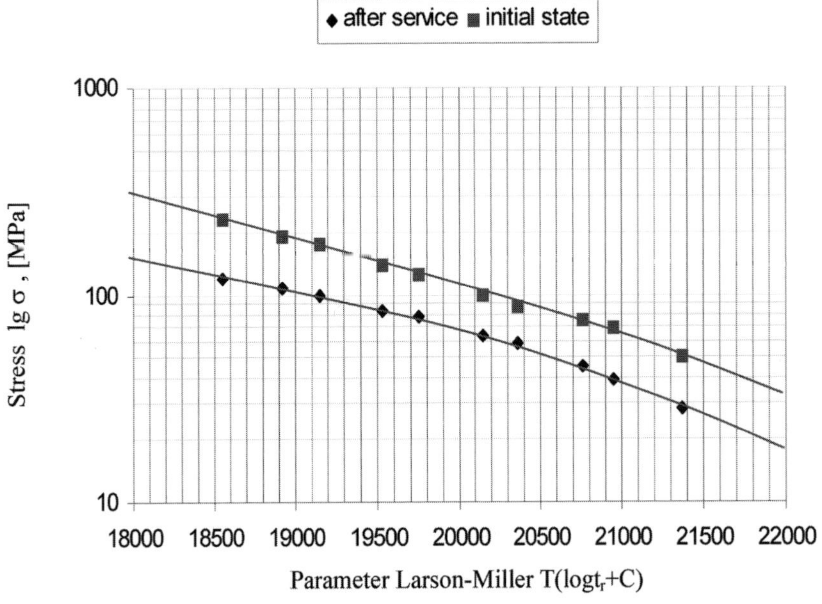

Fig. 4. Residual creep resistance curves in function log σ = f(L-M) of T-connection material in initial state and following 100,000 hours of operation.

Table No. 2. Anticipated residual life of the tested T-connection material made of Cr-Mo-V low-alloy cast steel on the basis of accelerated creep tests.

Component	Assumed working stress σ_r [MPa]	Assumed temperature of further operation T_r [°F]	Anticipated service life [hrs]	
			residual	available residual
T-connection after 100,000 hours of operation	50	1004	103,000	56,500
	70		30,000	16,500
	100		7,500	4,125
T-connection in initial condition	50		500,000	275,000
	70		300,000	165,000
	100		50,000	27,500

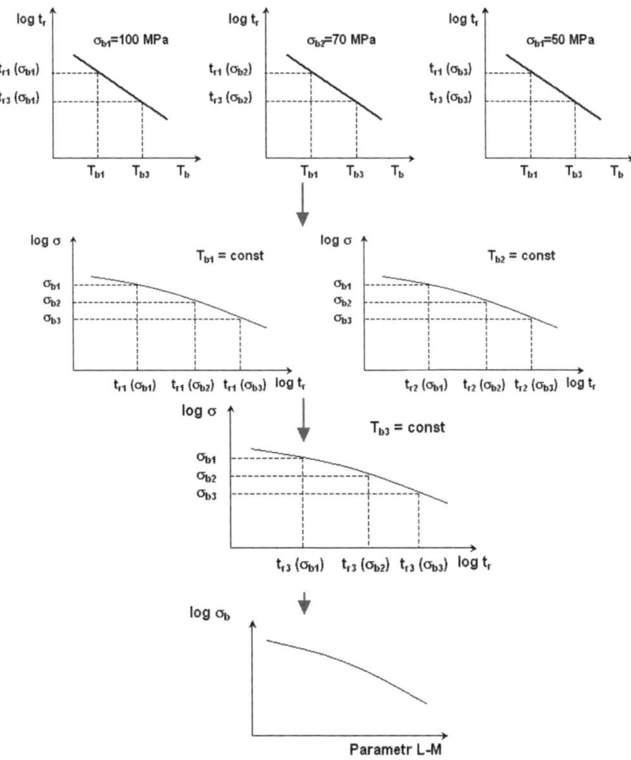

Fig. 5. Graphical presentation of the elaborated way of selecting long-term creep test parameters based on the obtained results of the shortened tests for some selected test stress levels σ_b and test temperature T_b [13].

3.4 Thermal fatigue testing.

The objective of the performed thermal fatigue tests was to determine the influence of thermal deformations, which result in time-variable thermal stresses onto material service life, as well as to determine the limit working conditions for a plant when variable thermal-mechanical stresses are activated in T-connection material, made of L21HMF cast steel, following 100,000 hours of operation. Therefore, an analysis of the tested T-connection working conditions was carried out and the operation time until failure, i.e. when the T-connection is to be replaced, was also determined. This analysis allowed to determine the thermal cycle parameters, which should be simulated for specimens at the test workplace.

During the fatigue tests, the values of axial force, temperature and total deformation ε_c were recorded. The tests were run using stiff control of MTS machine, owing to which in the course of the fatigue tests the load stiffness coefficient was $K=\varepsilon_m/\varepsilon_t=1$ and the total deformation was $\varepsilon_c = 0$. That meant that the thermal deformation of specimen ε_t. in cyclic temperature variation conditions was transformed into mechanical deformation ε_m.

The fatigue tests executed on 4 test specimens and at the aforementioned assumed parameters enabled to determine the fatigue characteristics of the tested T-connection material in function of number of thermal-mechanical load change cycles. These tests allowed to determine dissipation of deformation energy W_k, accumulated in the specimen during the thermal-mechanical fatigue process, which was determined from hysteresis loop surfaces, recorded when the specimen was under load. Some exemplary curves in the fatigue test for specimen No. 1 are shown in Figures 6÷8.

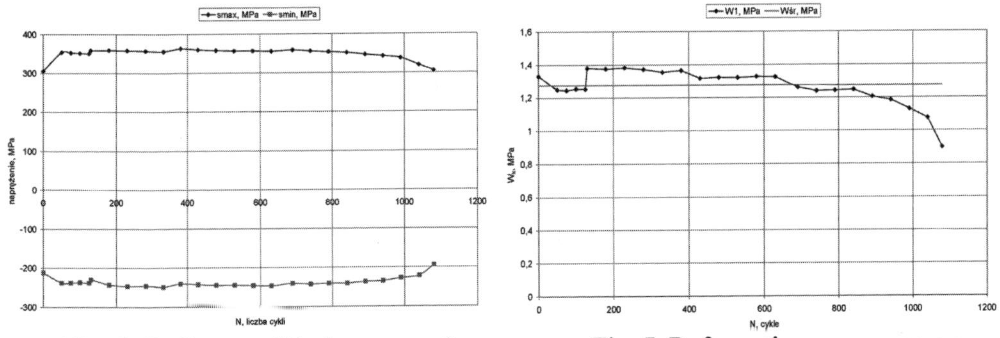

Fig. 6. Cyclic consolidation curves for specimen No. 1.

Fig. 7. Deformation proper energy dissipation curves for specimen No. 1.

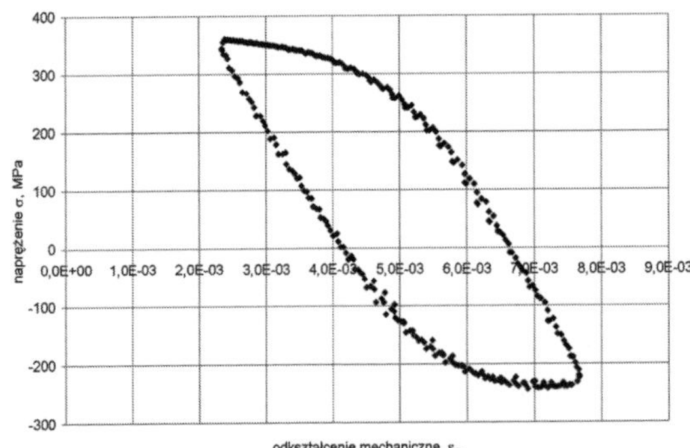

Fig. 8. Exemplary hysteresis loop for specimen No. 1 in cycle No. N=530.

In order to determine the value of deformation energy W_i required to initiate cracks on the specimen measuring length, thermal fatigue tests were carried out for 1080, 700, 635 and 550 numbers of cycles N_f. When a given number of load change cycles was executed, the test was stopped and the specimen was cut alongside its axis and thoroughly examined in a light microscope. The purpose of this was to measure the length of the largest crack on the specimen measuring length. The results of the calculated dissipation energy and the lengths of the maximum cracks measured are presented in Table No. 3.

Table No. 3. Calculation results of dissipation energy and maximum crack lengths in tested specimens.

Specimen No.	**N_f [cycles]**	**W_k [MPa]**	**l_{max} [μm]**
1	1,080	1,376	362
2	700	794	196
3	635	769	44
4	550	674	107

The obtained maximum crack lengths in the tested specimens in relation to the calculated accumulated energy W_k for a given number of cycles allowed to determine the value of deformation energy W_i. The way of determining the crack initiation energy is shown in Fig. 9.

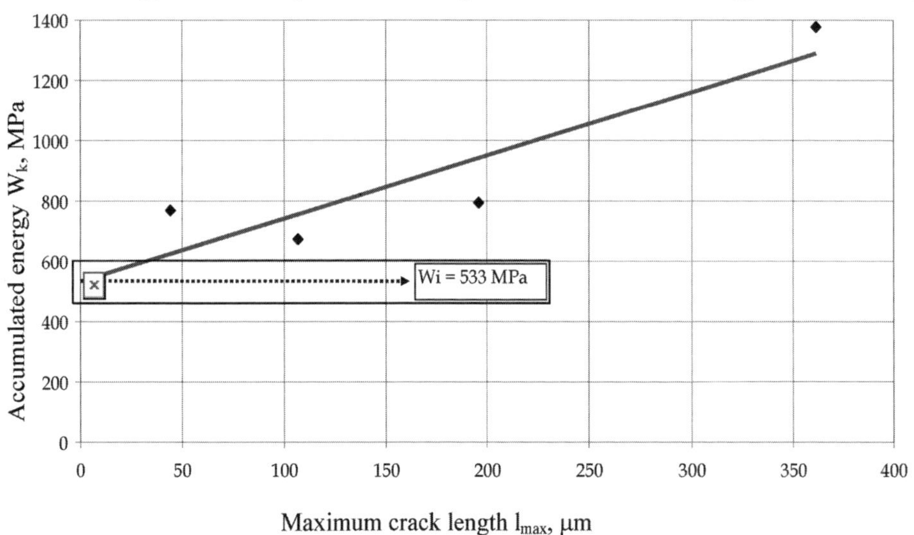

Fig. 9. Way of determining crack initiation energy in L21HMF low-alloy cast steel specimens after 100,000 hours of operation.

Based on the obtained results of thermal fatigue tests on T-connection material after 100,000 hours of operation in creep and fatigue conditions, the number of boiler start-ups, after which cracks may occur, was determined. In this case, the critical number of the T-connection load change cycles was:

$$N_f = \frac{W_i}{W_{K,av}} = \frac{533}{1,265} = 421 \quad \text{cycles}$$

This calculated number of cycles determines the T-connection service life. When the value of boiler start-ups N_f is exceeded, there may be initiation of cracks in the T-connection internal surfaces.

On the basis of the theory of linear crack cumulation, the available operation time for the T-connection material was determined in relation to:

$$t_r = (t/N_{odst})\cdot N_r = (100\,000/1612)\cdot 421 = 26\,117 \text{ hours}$$

where: t_r – T-connection available operation time,
t – T-connection operation time so far,
$N_{odst.}$ – number of pipeline shutdowns,
N_r – available number of pipeline shutdowns,
1,612- number of boiler start-ups and shutdowns.

5. Service life evaluation.

The operational conditions of such equipment as steam pipelines, power boiler components, etc., generate complex stress conditions resulting from variable loads influence and non-stationary heat flow through their walls. In such cases, the criteria of any plant material condition have to be widened by thermal-mechanical fatigue criteria. Concentrating on the tested T-connection following 100,000 hours of operation, the way of evaluating material condition after long-term operation was developed based on the performed tests. Additionally, such tests enable to determine appropriate weight coefficients which are helpful in determining the intensity of thermal-mechanical creep and fatigue influence onto material wear degree. The residual life can be calculated in accordance with the following formula:

$$t_r = t_p \cdot n_p + t_{zc} \cdot n_{zc}$$

where:
t_P – time or number of cycles until rupture (number of shutdowns) in creep conditions,
t_{ZC} - time or number of cycles until rupture (number of shutdowns) in thermal-mechanical fatigue conditions,
n_P, n_{ZC} – weight coefficients, which take into consideration the intensity of influence of wear process in creep and fatigue conditions respectively.

6. Conclusions:

1. The metallurgical tests and examinations performed have confirmed general degradation of the T-connection material after a long-term operation. The T-connections were made of L21HMF cast steel material. The following facts have contributed to such state of affairs:
   - the material does not meet R_e^t requirements at elevated temperature,
   - the low-cycle tests of the material after operation have shown over-excessive influence of temperature onto changes in mechanical properties,
   - the nil ductility transition temperature has been shifted to temperature +140°F,
   - diversified degree of microstructure degradation has been detected.
2. The executed accelerated creep tests of T-connection material in initial state and following 100,000 hours of operation have enabled to determine the available residual service life for working parameters of further operation. At working stress of 50 MPa and temperature of 1004°C, the available residual service life is 56,500 hours for a T-connection following 100,000 hours of operation. With the same parameters, the service life of a T-connection in initial state is 275,000 hours.
3. The T-connection service life, as determined in the thermal fatigue tests, enabled to calculate the operation time till crack initiation in solid material, which is 26,117 hours.
4. The test results have proven without a doubt that any evaluation of service life and residual life of cast steel T-connections (being operated in undetermined working conditions) is not sufficient or satisfactory when based solely on creep criteria. Therefore, all standard criteria as well as diagnostic procedures have to be broadened by values which are characteristic for fatigue processes [14].

References:

1. Hernas A.: *Żarowytrzymałość stali i stopów. Monografia 1*, Wydawnictwo Politechniki Śląskiej, Gliwice 2000 (in Polish).
2. Hernas A., Dobrzański J.: *Trwałość i niszczenie elementów kotłów i turbin parowych. Monografia 2*, Wydawnictwo Politechniki Śląskiej, Gliwice 2003 (in Polish).
3. Dobrzański J., Zieliński A. „*Ocena trwałości eksploatacyjnej stali energetycznych pracujących powyżej temperatury granicznej w oparciu o skrócone próby pełzania*" IX Seminarium Naukowo – Techniczne. *Badania materiałów na potrzeby Elektrowni i Przemysłu Energetycznego*. Zakopane 2002, pages 97÷106 (in Polish).
4. Dobrzański J.: *Zmodyfikowanie wybranych nieniszczących metod oceny stanu i trwałości resztkowej elementów z niskostopowych stali Cr-Mo-(V) po długotrwałej eksploatacji*. Sprawozdanie IMŻ Nr 103/91/BM, niepublikowane (in Polish).
5. Dobrzański J., Miliński P.: *Sprawdzenie przydatności skróconych prób pełzania do oceny stanu stali typu Cr-Mo-(V)*. Sprawozdanie IMŻ Nr 53/91/BM, niepublikowane. (in Polish).
6. Tokarz A., Zieliński A., Maciosowski A. „*Automatyzacja badań pełzania, zbierania, archiwizowania i przetwarzania wyników*". Materiały IX Seminarium naukowo – technicznego pt.: „Badania materiałowe na potrzeby elektrowni i przemysłu energetycznego", Zakopane, 2002, p. 77 (in Polish).

7. Bołd T., Dobrzański J., Miliński P. *„Metody oceny trwałości resztkowej elementów urządzeń kotłowych do stosowania w krajowej energetyce"*, Zeszyty Naukowe Politechniki Śląskiej, Energetyka , z. 120, 1994, pp. 69-80 (in Polish).
8. Dobrzański J. *„Diagnostyka materiałowa w ocenie stanu i prognozie czasu eksploatacji poza obliczeniowy rurociągów parowych pracujących w warunkach pełzania"*, Energetyka nr 12, 2002 (in Polish).
9. Zieliński A., Dobrzański J., Krztoń H.: *Structural changes in low alloy cast steel Cr-Mo-V after long time creep service.* Journal of Achievements in Materials and Manufacturing Engineering. Vol. 25, Issue 1, p. 33-36. August 2007.
10. Zieliński A.: *Wpływ długotrwałej eksploatacji w warunkach pełzania na własności mechaniczne staliwa Cr-Mo-V.* VIII Międzynarodowa Konferencja Naukowa – Nowe technologie i osiągnięcia w metalurgii i inżynierii materiałowej, Częstochowa 2007 Wydawnictwo Politechniki Częstochowskiej, tom II pp. 717-721 (in Polish).
11. Renowicz. D. Cieśla M.: *Crack initiation in Steel Parts Working under Thermo-Mechanical Fatigue Conditions, International Journal of Computational Materials Science and Surface Engineering* , published by Inderscience Publishers – w druku.
12. Hernas A., Renowicz D., Cieśla M., Mutwil K., Zieliński A.: *Materiałoznawcze i wytrzymałościowe kryteria trwałości elementów rurowych pracujących w warunkach pełzania i zmęczenia,* Raport końcowy z realizacji projektu badawczego nr 3T08A 04127, 2007 – niepublikowane (in Polish).
13. Zieliński A., Dobrzański J.: *Trwałość resztkowa niskostopowego staliwa typu Cr-Mo-V po długotrwałej eksploatacji w warunkach pełzania.* Sprawozdanie IMŻ nr S0-574/BE/2006 – niepublikowane (in Polish).
14. Renowicz. D, Hernas A., Cieśla M., Mutwil K.: *Degradation of the cast steel parts working in power plant pipelines.* Procedings od the 15th Scientific Inernational Conference. "Achivements in Mechanical and Materials Engineering" AMME 2006, Gliwice-Wisła Poland Vol. 18, Issue 1-2, 2006.

Advances in Materials Technology for Fossil Power Plants
Proceedings from the Fifth International Conference
R. Viswanathan, D. Gandy, K. Coleman, editors, p 627-644

DOI: 10.1361/cp2007epri0627

APPLICATION OF A COMPREHENSIVE R&D CONCEPT TO IMPROVE LONG-TERM CREEP BEHAVIOUR OF MARTENSITIC 9-12% Cr STEELS

H. Cerjak
I. Holzer
P. Mayr
C. Pein
B. Sonderegger
E. Kozeschnik
Institute for Materials Science, Welding and Forming
Graz University of Technology,
Kopernikusgasse 24, A-8010 Graz, Austria.
Ph. 43-316-873-7181 Fax 43-316-873-7187

Abstract

The research activities on modern martensitic 9-12% Cr steels for the application in environmentally friendly power plants at the Institute for Materials Science, Welding and Forming (IWS) are represented by numerous interacting projects. Focusing on mechanical properties of base and weld metal, microstructural characterisation of creep and damage kinetics, weldability, microstructure analysis in the course of creep, modelling of precipitation and coarsening kinetics, simulation of complex heat treatments and the deformation behaviour under creep loading, a comprehensive picture of the material behaviour can be drawn. The individual projects are briefly described and the conceptual approach towards a quantitative description of the creep behaviour of 9-12% Cr steels is outlined.

Keywords: 9-12% chromium steels, creep strength, microstructure, microscopy, modelling, simulation

Introduction

For many components in thermal power plants, creep resistant martensitic-ferritic 9-12% Cr steels are used because of their good creep resistance, oxidation behavior and low thermal expansion coefficient. The main challenge in the development of these steels are the increase of the creep resistance at temperatures greater than 600°C, for both, base material and weldments. Final goal is to reach a creep resistance for loads of 100MPa and exposing times of more than 10^5h. In the past, often short time creep tests were used in order to extrapolate the lifetime of the components. Although many steels show optimal microstructure after the heat treatment, but show a drop in the creep resistance after about 10^4h. Mainly, two reasons are responsible for this deterioration: First, creep mechanisms change when applying different creep loads. Usually, these mechanism changes are not considered when extrapolating lifetimes of components. Second, the microstructure of the material undergoes an evolution: precipitates coarsen and dissolve, other precipitates form, martensite laths broaden etc. This change in microstructure is not incorporated into macroscopic extrapolation methods. Both aspects are considered in the comprehensive approach, which is presented in the following.

Metallographic investigations

Several microscopical techniques combined in order to overlap their information content. Following microstructural features were examined: Size of martensite laths and subgrains; the size distribution, number density, type and location of precipitates and the size and location of pores.

Many works on these topics have been published so far: Hättestrand and Andren [1-4], Hofer et al. [5-9], Hofer and Cerjak [10-13], Strang and Vodarek [14, 15], Kimura et al. [16], Papst [17-19], Danielsen and Hald [20] and Sonderegger [21, 22] measured size and chemical composition of precipitates via TEM/EFTEM or atom probe. Korcakova, Hald and Somers [23] and Dimmler [24-26] investigated Laves-phase precipitates by SEM. Investigations of size and distribution of creep pores were also carried out by Dimmler. Martensite lath width and size of subgrains were measured by Sklenicka [27], Cerri [28], Sawada [29, 30], Sonderegger [21, 31] and others. However, the isolated investigation of different features is not sufficient, only the combination of techniques and the implementation of the results into numerical simulation tools will enhance the knowledge and understanding sufficiently.

Sub-μm precipitates: TEM/EFTEM

The following work mainly concentrates on the creep resistant 9-12% Cr COST steel variant CB8. The chemical composition is given in Tab. 1.

Table 1: Chemical composition of the steel CB8 (wt%).

Al	B (ppm)	C	Co	Cr	Mn	Mo	N	Nb	Ni	Si	V
0.028	0.0112	0.17	2.92	10.72	0.20	1.40	0.0319	0.06	0.16	0.27	0.21

Transmission-electronmicroscopy (TEM) and energy-filtered TEM (EFTEM) are ideal methods to measure size, shape and number density of small (<1µm) precipitates. In order to determine the chemical composition, EELS (Electron Energy Loss Spectroscopy) and EDX is applied. Additional methods like CBED (Convergent Beam Electron Diffraction) or SAD (Selected Area Diffraction) are necessary to get informations on the crystal structure of the precipitates.

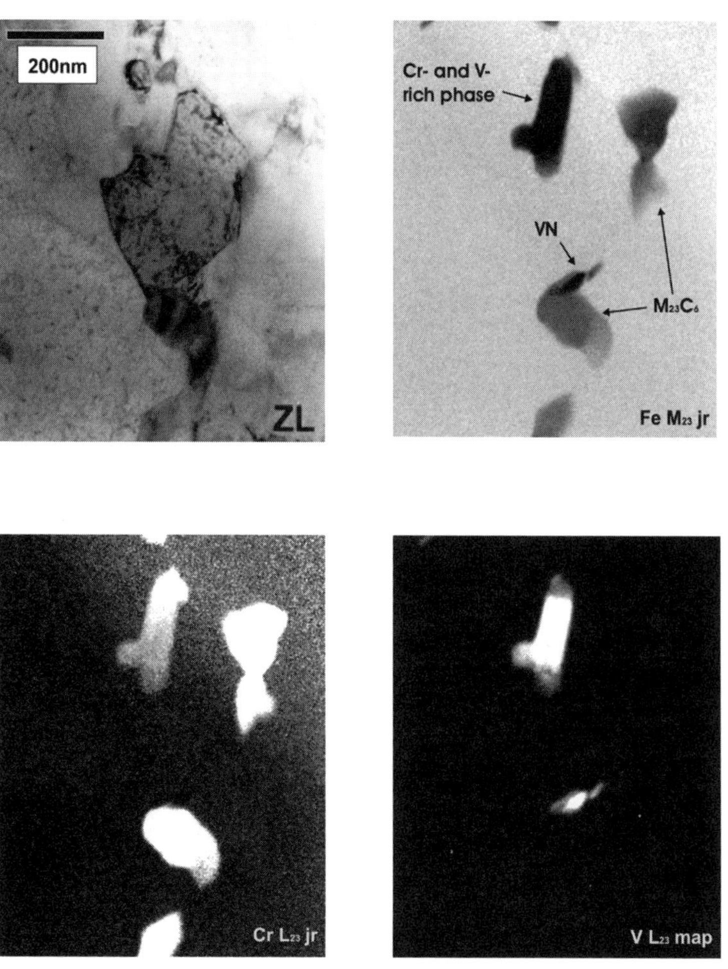

Figure 1: Cr-carbides ($M_{23}C_6$), V-nitrides (VN) and a Cr and V-rich phase, measured by EFTEM jump ratio images and elemental maps

Figure 1 shows EFTEM micrographs of different precipitate types in a thermally aged sample of the creep resistant 9-12% Cr steel CB8. The combination of EFTEM micrographs (indicating the local composition) with zero loss images (indicating matrix boundaries, dislocations etc.) gives valuable information on the local microstructure near the precipitates. Additionally, the size and number density can be measured as a function of creep loading and tempering time [3, 22, 32]. Figure 2 shows the results on thermally aged samples of the material CB8:

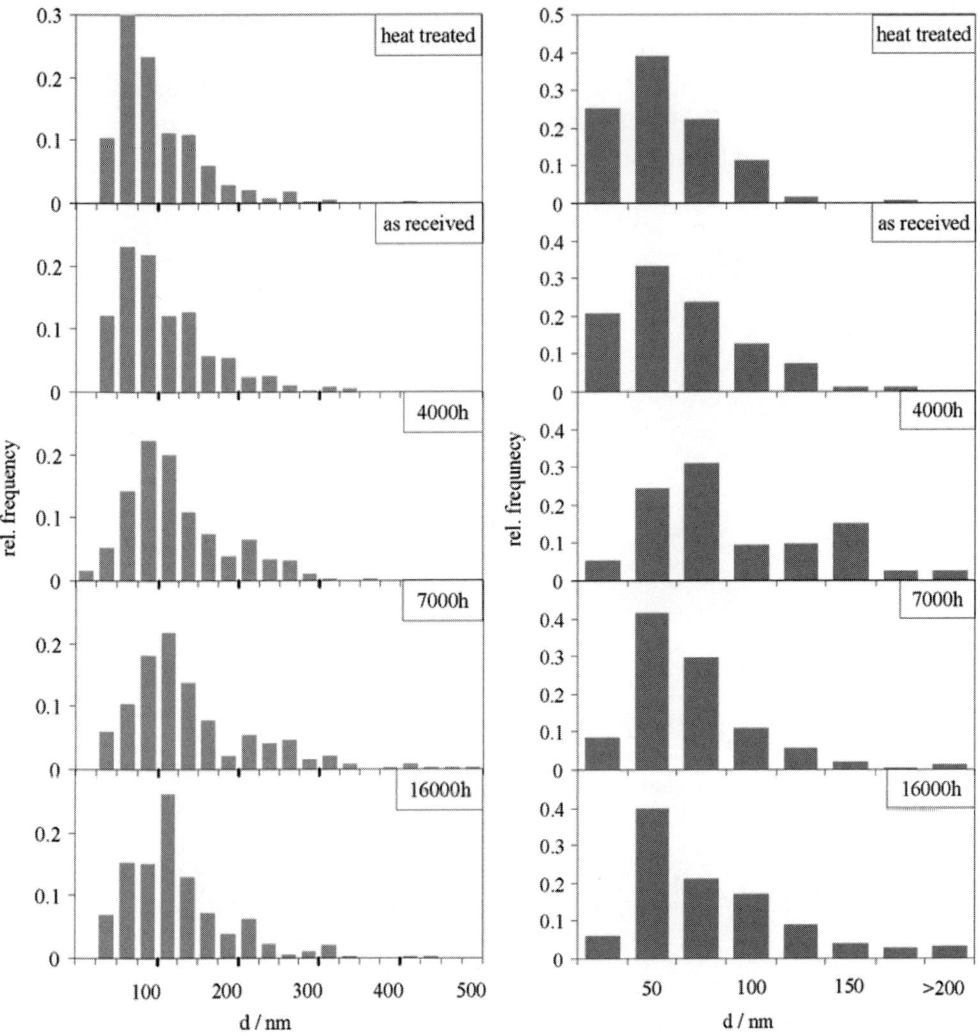

Figure 2: Size distributions of $M_{23}C_6$ precipitates (left diagram) and VN (right) after one heat treatment cycle (heat treated), as received condition, and several durations (4000h, 7000h, 16000h) of overageing at 650°C

Precipitates greater than 0.5-1μm

For precipitates greater than 0.5-1μm, Scanning Electron Microscopy (SEM) is much more convenient. Particles of this size usually provide lower number densities, which are unsuitable for investigations in TEM. According to Reuter [33], precipitates can be visualized due to their Z-contrast relative to the matrix, thus the chemistry must be sufficiently different. Table 2 shows calculated contrasts of several precipitate types relative to the matrix of a 9-12% Cr steel. Laves phase has the best combination of great diameters and good Z-contrast due to its high Mo- or W- content.

Table 2: Chemical composition, calculated Z-contrast and mean diameters of several precipitate types in 9-12% Cr steels [34]

Precipitate type	Z-contrast	d [nm]
$M_{23}C_6$	-2%	50-200
VN	-5%	10-100
Nb(C,N)	16%	10-100
Laves phase	28%	200-2000
Z phase	-4%	50- 1000

Creep pores

View damage investigations can be found in literature [35], but in order to understand the creep behavior in the tertiary regime, also the evolution of creep pores have to be considered. Figure 3 shows the spatial distribution of creep pores in the vicinity of the fracture surface of a broken creep sample. Following steps are necessary for the metallographic investigations:

- Special preparation techniques for even very small creep pores [25]
- Automatic image processing and evaluation
- Reconstruction of position oft he pores relative to other microstructural components, such as precipitates or grain boundaries

These systematic evaluations lead to an improved understanding for the formation of creep pores up to macroscopic damage of the material in the tertiary creep regime.

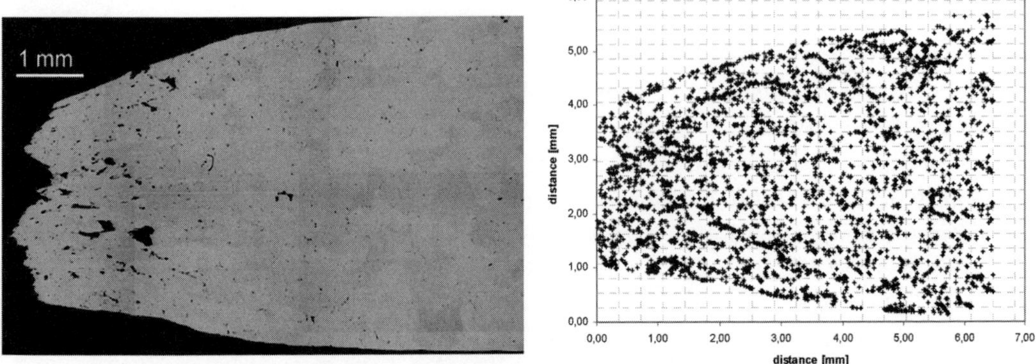

Figure 3: SEM micrograph of a broken creep sample (CB8, left image) plus reconstruction of the position of creep pores (right image)

Martensite laths and grains: EBSD

By EBSD (Electron Backscatter Diffraction) in SEM, grains, subgrains and martensite laths can be differed via measuring crystal orientations. The spatial resolution reaches down to approximately 200nm and an orientational resolution of 1-1,5°. Not only the size and shape of grains can be measured - also informations regarding the matrix boundary are available, like misorientation or deviation from an ideal twin boundary. This, EBSD also provides indirect information on boundary energy or mobility. Figure 4 shows an EBSD map of steel CB8; single martensite laths are shown by different brightness. Figure 5 shows the evaluation of a number of thermally aged or creep loaded samples regarding the subgrain size. It is clearly visible, that thermally ageing (at 650°C) alone does not produce a significant amount of subgrain coarsening, whereas additional creep loading (80-110MPa) does.

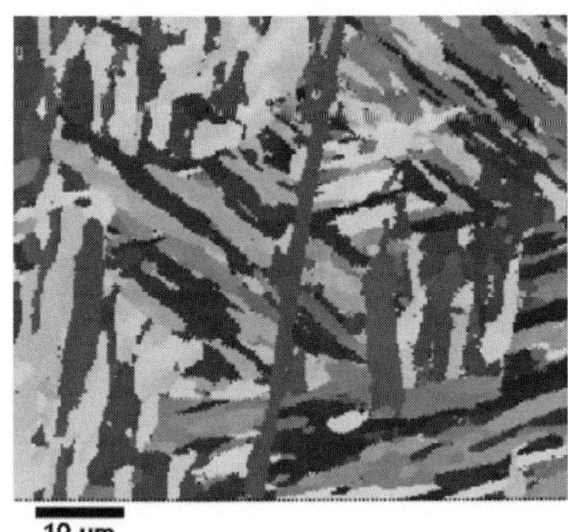

Figure 4: EBSD micrograph of CB8, thermally aged

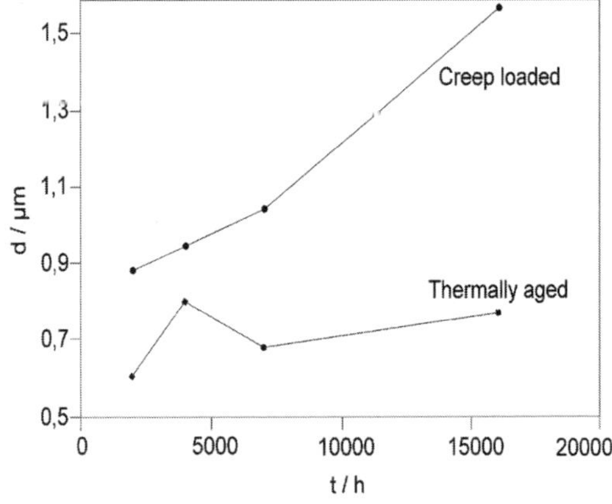

Figure 5: Mean Subgrain diameters of thermally aged or creep loaded samples

Weldability assessment

For a successful service application and acceptance in practice, the weldability and the long-term behaviour of welds of the newly developed materials is one of the key issues. Therefore, the weldability of several 9-12%Cr steels is investigated at the IWS. Microstructural evolution during welding is studied by heat affected zone (HAZ) simulation using a Gleeble thermo-mechanical simulator and subsequent metallographic investigations applying most advanced electron microscopic methods. The creep strength of base materials, weld metals and cross-welds is studied by long-term creep testing at service temperature.

Creep strength of cross-welds is decreasing with increasing testing duration compared to the mean base material creep strength. This decrease in creep strength is independent of the selected weld metal. Figure 6 shows results of creep tests at 600°C of cross-welds fabricated with three different filler metals differing in weld metal creep strength. After 10.000 hours of testing cross-weld creep strength deviates from the mean line of base material creep strength. After 30.000 hours of testing, creep rupture strength of cross-welds is already approximately 30% below the base material creep strength independently of the weld filler. The mismatch in creep rupture strength between base material and cross-welds is expected to increase further with increasing testing duration.

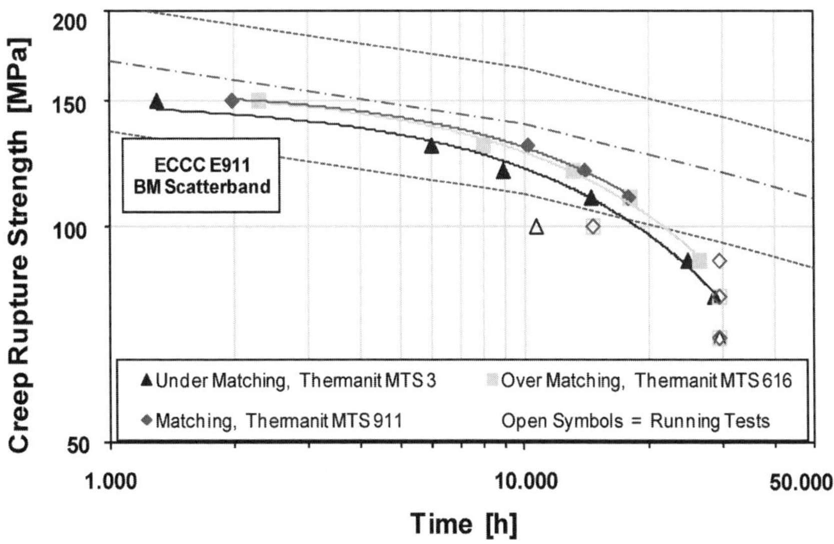

Figure 6: Creep rupture data at 600°C for welds of G-X11CrMoWVNb 9-1-1 (E911). Comparison of the cross-weld creep strength to the mean creep strength of E911 base material.

Type IV cracking in the fine-grained HAZ has been identified as major reason for the decrease of cross-weld creep strength. In Figure 7, the fracture location investigation of a creep tested cross-weld of G-X11CrMoWVNb 9-1-1 (E911) parent cast material welded with a strength matching filler material is shown. Failure takes place in the fully refined region of the HAZ (see Figure 8). Creep damage by formation of voids is limited to a very narrow region parallel to the weld fusion line adjacent to the unaffected base material.

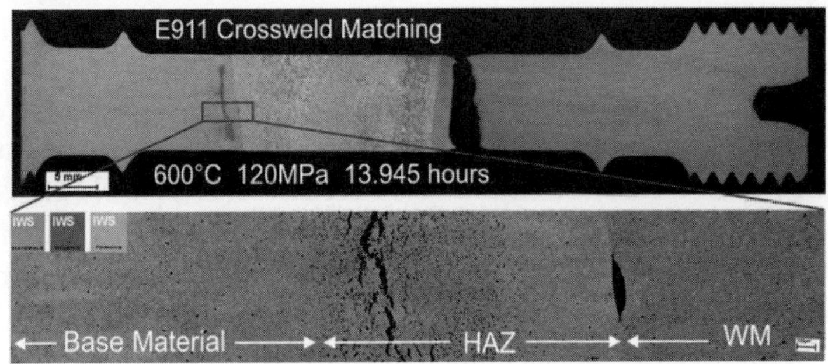

Figure 7: Fracture location investigation of a creep tested cross-weld sample. E911 pipe material welded with Thermanit MTS 911 (matching creep strength level). Excessive pore formation in a 1 mm wide zone adjacent to the unaffected base material.

Creep at low stress levels as a diffusional problem is enhanced in a microstructure containing a high volume fraction of prior austenite grain boundaries. All diffusion driven processes, e.g. recovery or coarsening of precipitates are proceeding much faster in this region of the HAZ compared to the weld metal, unaffected base material or other HAZ regions.

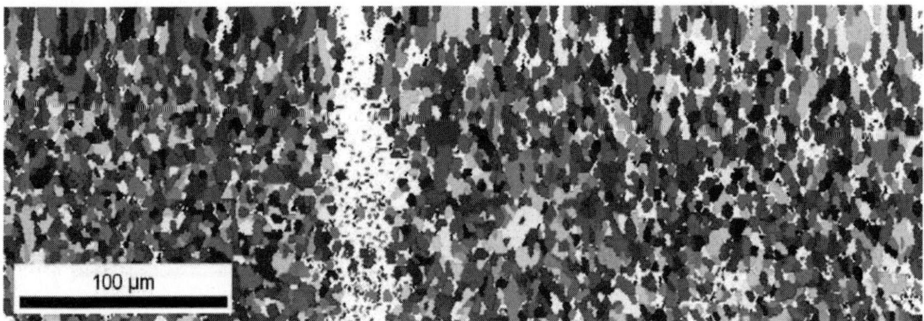

Figure 8: EBSD grain mapping at the location of fracture in specimen shown in Figure 7. Crack and creep voids are represented by white areas within the image.

Figure 9 shows a comparison of the initial microstructure in the grain-refined region of a cross-weld after post-weld heat treatment and the same region after 14.000 hours of creep exposure at 600°C. The number density of precipitates decreases during creep and the average diameter of the precipitates increases. Both results in a decrease of creep strength in this part of the HAZ.

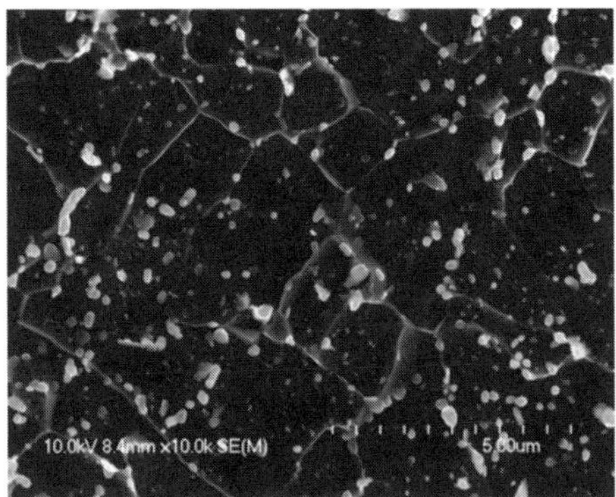

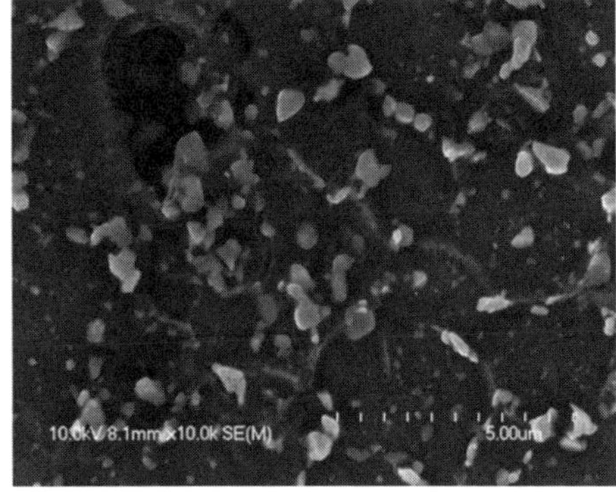

Figure 9: Microstructural evolution during creep in the fine-grained HAZ of E911 cross-welds tested at 600°C. Precipitates of the initial microstructure (left) show extensive coarsening during 14.000 hours of creep exposure.

Type IV cracking has been identified as the major end-of-life failure mechanism in creep exposed welded structures of 9-12% Cr steels. So far, the reduced cross-weld creep strength can only be considered by the introduction of a weld strength factor during the design of components. Further research at the IWS heads into the direction of the suppression of a fully refined microstructure in the HAZ by modifications of base material chemistry with controlled addition of boron and nitrogen. This allows increasing the creep strength of cross-welds to the level of the base material and eliminating Type IV cracking

Thermodynamic and kinetic simulations

In the last years thermodynamic and kinetic simulations have become increasingly important when optimizing the chemical composition and production process [36]. Since the long-term stability of the microstructure of 9-12% Cr-steels is closely related to the stability of the precipitate microstructure, it is very important to describe and predict the evolution of each precipitate population during the entire lifetime of a component. Based on a recently developed theoretical approach for the simulation of the precipitation kinetics of multi-component multi-phase materials [37, 38], the evolution of the precipitate microstructure during heat treatment and service of the COST steel CB8 has been simulated on the computer.

Thermodynamic equilibrium analysis

The thermodynamic equilibrium analysis is an important step in a comprehensive material characterization. Figure 10 shows the calculated phase diagram for the steel CB8 as a function of carbon content. For the calculation a slightly modified version of the thermodynamic database TCFE3 was used, e.g. to account for the stabilizing effect of silicon on the Laves phase [24]. Moreover, although not considered in the phase diagram below, a revised thermodynamic description for the modified Z-phase [42] has been added to this database, which is a further development of the initial assessment of Danielsson and Hald [20]. These values have been used in the kinetic simulations presented in the following section.

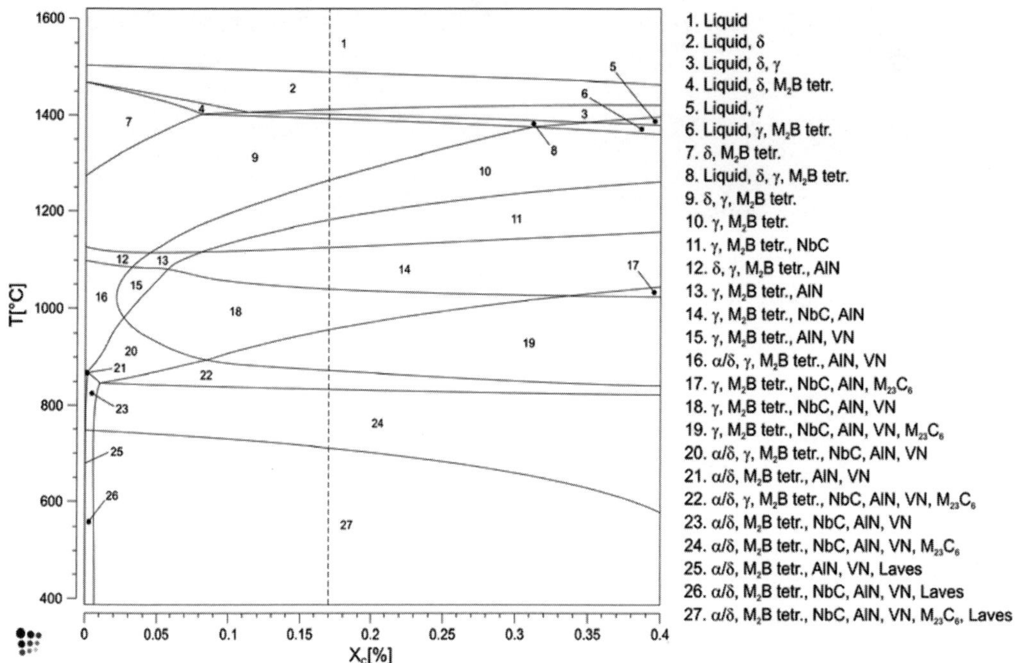

Figure 10: Calculated phase diagram of the COST alloy CB8.
A modified version of the TCFE3 database was used.

Kinetic simulation

Beside thermodynamic equilibrium calculations also kinetic simulations have to be carried out to reach a better understanding of the effect of precipitate evolution. Therefore the industrial heat treatment for the COST alloy CB8 was simulated with the kinetic simulation software MatCalc [41]. The underlying theory and model implementation is described in refs. [37- 40]. For the simulation a modified version of the database TCFE3 and the diffusion database Mobility_v21 from ThermoCalc AB, Stockholm, Sweden were used. The adaptions were made based on the experimental observations in [21]. The simulation result for the heat treartment is shown in Figure 11.

The top figure shows the temperature history during heat treatment. The simulation starts closely below the solidus temperature of this material, i.e. 1400 °C. Cooling proceeds down to 350°C. At this temperature, the austenite matrix decomposes into martensite. It is assumed that no precipitation reactions occur below this temperature due to the sluggish diffusion. In the next simulation step, the temperature is increased again. The simulation is performed in a ferritic matrix up to 847°C, the A_1 temperature of this steel. Then, the matrix is changed to austenite again and austenitization takes place at 1080 °C. The three quality heat treatment cycles take place again in a ferritic matrix. The two lower graphs show the results of the simulation of the time temperature sequence explained above. The phase fraction of the $M_{23}C_6$ and Laves phase is divided by a factor of ten to get a more concise diagram. These results are compared to experimental data measured by one of the authors [21]. Good agreement is observed between simulation and experiment. A more detailed description of the heat treatment simulation is given in [43].

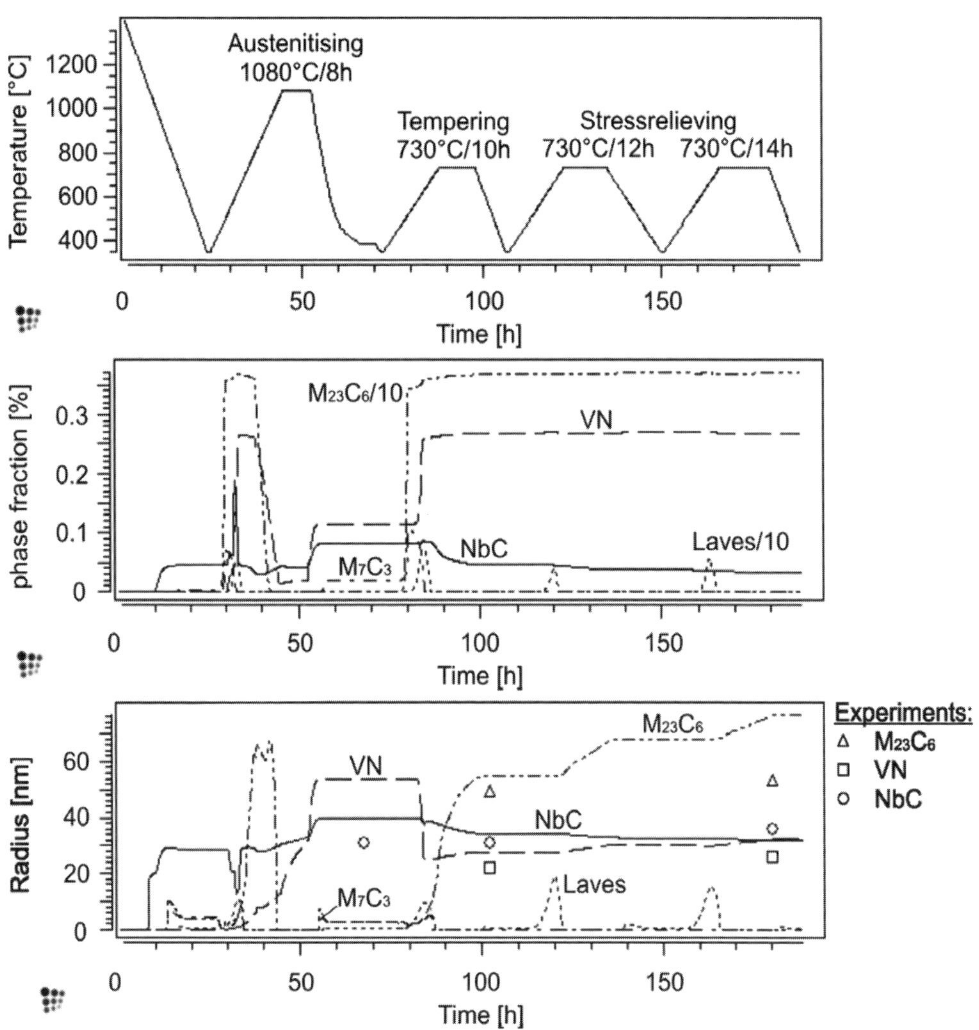

Figure 11: Results of the heat treatment simulation of COST alloy CB8

Microstructure modelling

In oreder to predict the behavior of creep resistant steels, the full complexity of the creep process can only be described with models that can cover all physical mechanism that interact in the material. A comprehensive microstructural model of creep must include, for instance, diffusional creep, dislocation dynamics as well as all interactions between the microstructural components, such as dislocations, grain boundaries and precipitates. It was also proposed that local effects must be taken into account, because any weak spot in the microstructure can cause ultimate failure of the macroscopic component [45]. Therefore, to give a short overview about the ongoing modelling activities, two different approaches are outlined in the following which are developed or applied at the IWS: the back-stress concept and a newly developed spatially resolved model for the evolution of the local microstructure.

Back-stress concept

If an external force is acting on a microstructure, it is frequently assumed that the external load σ_{ex} is counteracted by heterogeneous internal microstructural constituents, such as precipitates and interfaces. Consequently, not the entire external load can be assumed to represent the driving force for the creep process; only this part of the external stress σ_{ex}, which exceeds the amount of inner stress σ_i from the counteracting microstructure, effectively contributes to the creep process. Since the inner stress reduces the effect of the external stress, this approach is commonly denoted as back-stress concept. The effective creep stress σ_{eff} can be expressed as

$$\sigma_{eff} = \sigma_{ex} - \sigma_i \quad (1)$$

In a recent treatment by Dimmler [24], the inner stress σ_i has been expressed as a superposition of individual contributions from dislocations and precipitates. When also taking into account the contribution from subgrain boundaries, the inner stress is

$$\sigma_i = M\tau_i = M(\tau_{disl} + \tau_{prec} + \tau_{sgb}), \quad (2)$$

where M is the Taylor factor (usually between 2 and 3, see ref. [24]) and τ is the shear stress. The subscripts in the bracket term denote contributions from dislocations, precipitates and subgrain boundaries, respectively. The concept of the effective and inner stress to describe the creep behavior is shown in Figure 12. The input parameters are the applied stress σ, temperature T and time t [24].

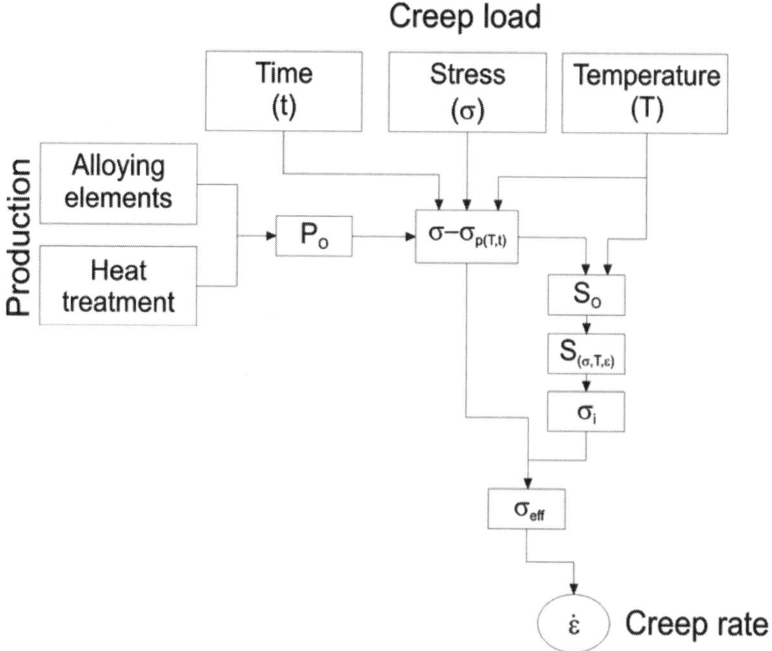

Figure 12: Back-Stress concept, scematic [47]

The combination of the back-stress concept with the simulation software MatCalc allows therefore determining at every point the inner stress, which influences the plastic deformation, as a function of the microstructural evolution. First results and more detailed information about the theoretical approach are given in an accompanying paper [46].

Microstructure simulation

If an external stress is applied on a poly-crystalline, multi-phase microstructure, the resulting microscopic stress field within the specimen will usually show a complex pattern of areas with alternating compressive and tensile stresses. This pattern strongly depends on the local microstructure and its elements as precipitates, pores, grainboundaries etc. For this reason phenomenological and statistical approaches cannot capture the essential microstructural changes that occur in the material on a local scale. In our model these local phenomena shall be considered, including self-diffusion of vacancies, motion of dislocations, kinetics of precipitation and other effects, as well as interactions among them.

In our opinion, phenomenological and statistical approaches cannot capture the essential microstructural changes that occur in the material on a local scale. Important physical processes in creep, such as nucleation and growth of micro-voids or dissolution of precipitates, operate on a local basis, and, therefore, an appropriate model must take these local events into account. Only by these means modelling and simulation can support our conception of creep.

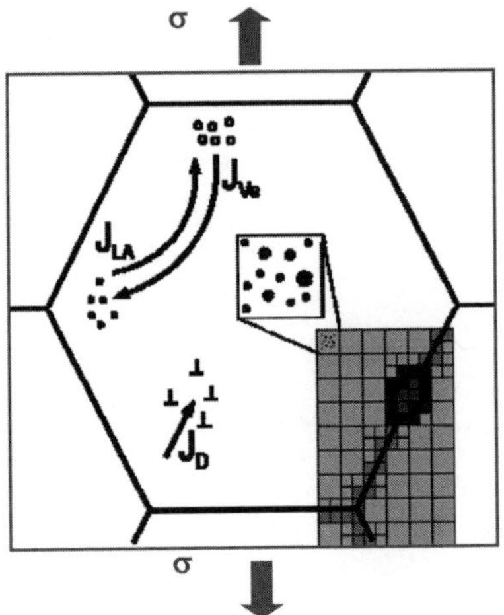

Figure 13: Microstructural model

Figure 13 presents a sketch of our local microstructure approach illustrating the basic mechanisms and interactions. The flux of vacancies which is caused by local microscopic stresses produces mass transport in the opposite direction. Motion of dislocation is activated by the local stress gradient. Moving boundaries interact with precitpitates within the matrix and vice versa. Depending on these mechanisms creep can be divided into diffusional creep, creep caused by dislocations and other effects. The model development now is focused on the correct specification of diffusional creep, which is the predominant process at low stresses [48].

There are two main mechanisms which are causing diffusional creep, namely diffusion via the bulk and diffusion via grain boundaries.

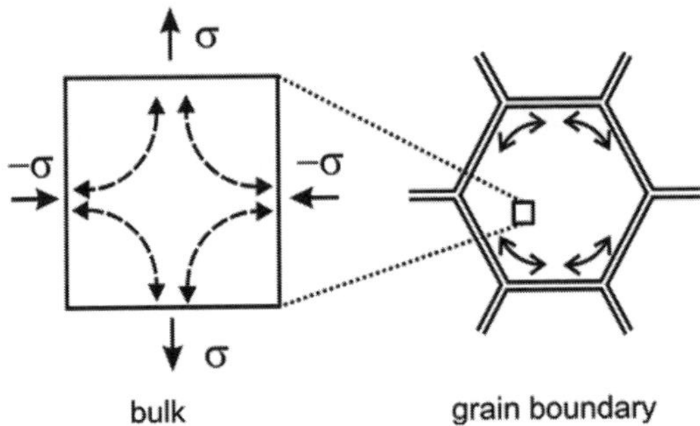

Figure 14: Finite element model for the simulation of diffusional creep

Two boundary conditions are relevant for the description of the flux in the matrix: The characterisation of the flux itself (Ficks Law) and the calculation of the concentration of vacancies in equilibrium, dependent of the local stress distribution. The output of the calculations is the spatially distributed flux of vacancies and the resulting time dependent deformation of the material.

The behaviour of grain boundaries differs from that of the grain itself. The main reason is the different reaction on shear stress - two grains glide along the grain boundary when shear stress is applied. This behaviour is modelled with a stong anisotropic tensor of the Elastic Modulus.

In our simulation, the microstructure is discretized with rectangular cells, each of them corresponding to a representative volume of the simulation domain. Each element can be one single phase, e.g. matrix, carbide or a grain boundary and, accordingly, material properties such as Young's modulus, shear modulus etc. are assigned to it.

As a first example for our microstructure model, we consider a hard lens-shaped precipitate within a homogeneous soft matrix. An external stress in the y axis is applied, which causes a local stress distributiuon and deformation. Figure 15 shows first results of our simulation.

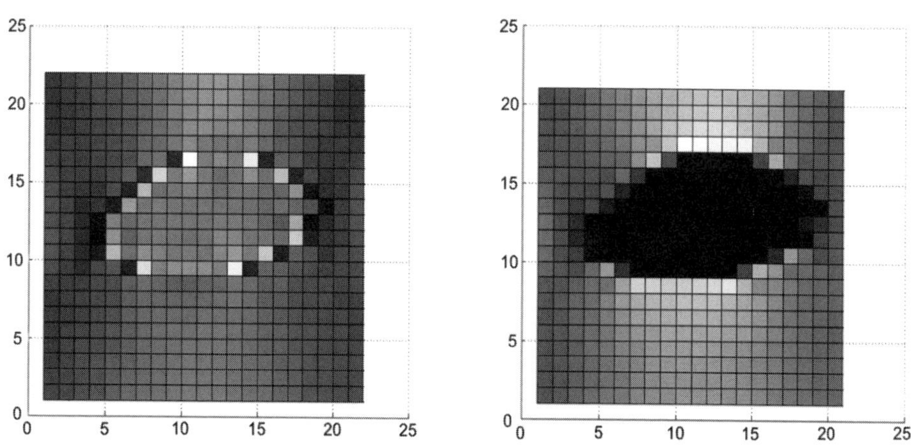

Figure 15: Elastic stress field near a hard precipitate.
Von Mises stress (left) and elastic deformation (right)

Conclusion and Outlook

Material development of creep resistant 9-12% Cr steels in thermal power plants has become increasingly complex. Many steel variants show superior creep resistance within the first year of service, but a pronounced drop at longer exposure times. In order to overcome this degradation, it is necessary to investigate the material by different methods with overlapping information content, understand the evolution processes, and simulate the microstructural evolution. Models, which have been used in the past, like estimating the lifetime by short term creep experiments, show the danger of being too optimistic regarding the estimated lifetime of the components. The comprehensive approach presented in this paper overcomes these difficulties and shows the potential of reducing development costs and time.

References

1 M. Hättestrand, H. O. Andrén, Materials Science and Engineering. A270 (1999), 33-37.

2 M. Hättestrand, M. Schwind, H. O. Andrén, Materials Science and Engineering. A250 (1998), 27- 36.

3 M. Hättestrand, H. O. Andrén, Micron 32 (2001), 789- 797.

4 M. Hättestrand, H. O. Andrén, Acta mater. 49 (2001), 2123- 2128.

5 F. Hofer, P. Warbichler: Ultramicroscopy, 63 (1), 1996, 21-25.

6 F. Hofer et al., J. Microscopy, 204 (2), (2001), 166.

7 P. Warbichler, F. Hofer, P. Hofer, E. Letofsky, H. Cerjak, Micron (1998), 63-72

8 F. Hofer, P. Warbichler, B. Buchmayer, S. Kleber, Journal of Microscopy 184 (1996), 163-174

9 F. Hofer et al., Metallurgical and materials transactions A 31A (2000), 975

10 P. Hofer. PhD thesis, Technische Universität Graz, 1999.

11 P. Hofer, H. Cerjak, B. Schaffernak, Steel research 69 (1998) 8, 343-348

12 P. Hofer, H. Cerjak, P. Warbichler, Materials Science and Technology 16 (2000), 1221-1225

13 H. Cerjak, P. Hofer, B. Schaffernak, P. Warbichler, Praktische Metallographie 27 (1997)

14 A. Strang, V. Vodarek, Proc. Microstructural development and Stability in high Chromium ferritic power plant steels, Cambridge, 1995, 31-51.

15 V. Vodarek, A. Strang, Materials for advanced power engineering, Liege, 2002, 1223-1231.

16 K. Kimura et al., Proc. Materials for advanced power engineering, Liege, 2002, 1171-1180.

17 I. Papst et al., Metallographietagung Dortmund, Germany, 6.-10. September 1999

18 I. Letofsky-Papst et al., Zeitschrift für Metallkunde 95 (2004) 1, 18-21

19 I. Letofsky-Papst et al., Praktische Metallographie 41 (2004) 7, 334-343

20 H. Danielsen, J. Hald, ISSN 0282-3772, Värmeforsk Service AB, Stockholm, 2004

21 B. Sonderegger, PhD thesis, Technische Universität Graz, 2005

22 B. Sonderegger, Ultramicroscopy 106 (2006) 10, 941-950

23 L. Korcakova, J. Hald, M. Somers, Materials Characterisation 47 (2001), 111- 117.

24 G. Dimmler, PhD thesis, Technische Universität Graz, 2003

25 G. Dimmler et al., Praktische Metallographie, 39 (2002) 12, 619-633

26 G. Dimmler et al., Materials Characterization 51 (2003), 341-352

27 V. Sklenicka, K. Kucharova, A. Dlouhy, J. Krejci, Proc. Materials for Advanced Power Engineering 1994. Liége (1994), 435- 444.

28 E. Cerri, E. Evangelista, S. Spigarelli, P. Bianchi, Materials Science and Engineering. A245 (1998), 285- 292.

29 K. Sawada. et al., Materials Science and Engineering. A267 (1999), 19- 25.

30 K. Sawada, K. Kubo, F. Abe, Materials Science and Engineering. A319- 321 (2001), 784- 787.

31 B. Sonderegger, S. Mitsche, H. Cerjak, Materials Characterization, in print

32 E. E. Underwood: Quantitative Stereology, London, 1970, 174

33 W. Reuter. In: Shinoda G, Kohra K, Ichinokawa T, editors. Proceedings of the 6th International Congress on X-ray Optics and Microanalysis. Tokyo: University of Tokyo Press (1972), 121.

34 P. Weinert et al., proc. MPA Seminar (2002)

35 W. Poeppel, U. Hildebrandt, Praktische Metallographic; 26 (1989); 141-153.

36 J. O. Andersson, T. Helander, L. H. Höglund, P. F. Shi, B. Sundman, CALPHAD, vol. 26, Nr. 2 (2002), 273-312

37 J. Svoboda, F. D. Fischer, P. Fratzl, E. Kozeschnik, Mater. Sci. Eng. A, vol. 385, Nrs. 1-2, (2004), 166-174

38 E. Kozeschnik, J. Svoboda, F. D. Fischer, CALPHAD, vol. 28, Nr. 4, (2005), 379-382

39 J. Svoboda, I. Turek, F. D. Fischer, Phil. Mag., vol. 85 (2005), 3699-3707

40 L. Onsager, Physical Review, I. vol. 37, pp. 405-426, (1931), II. vol. 38 (1931), 2265-2279

41 Software package, MatCalc, http://matcalc.tugraz.at.

42 H. Danielsen, private communication

43 E. Kozeschnik, B. Sonderegger, I. Holzer, J. Rajek, H. Cerjak; Materials Science Forum Vols. 539-543 (2007), pp. 2431-2436

44 J. Rajek, PhD thesis, Graz University of Technology, 2005.

45 P. Weinert: Microstructural Modelling of Creep in Ferritic/Martensitic 9-12% Cr-Steels, PhD thesis, Graz University of Technology, 2001.

46 I. Holzer, E. Kozeschnik, H. Cerjak, this issue

47 B. Reppich: Zeitschrift für Metallkunde, 73 (11), (1982), 697-805

48 F.R.N. Nabarro, Met. Mater. Trans., Vol 33A, (2002), 213-218.

Advances in Materials Technology for Fossil Power Plants
Proceedings from the Fifth International Conference
R. Viswanathan, D. Gandy, K. Coleman, editors, p 645-657

DOI: 10.1361/cp2007epri0645

Novel Hafnium-Containing Steels for Power Generation

R.J.Grice
R.G.Faulkner
Y.Yin
Institute of Polymer Technology and Materials Engineering
Loughborough University
Leicestershire
UK, LE11 3HH

Abstract

There is clear evidence that creep damage in power plant steels is associated with grain boundary precipitates. These particles provide favourable nucleation sites for grain boundary cavities and micro-cracks. The formation of $M_{23}C_6$ carbides as grain boundary precipitates can also lead to grain boundary chromium depleted zones which are susceptible to corrosive attack. Such precipitates are the causing loss of creep life in the later stages of creep because of their very high coarsening rate.

Through Monte Carlo based grain boundary precipitation kinetics models, combined with continuum creep damage modelling (CDM), it is predicted that improvements in creep behavior of power plant steels can be achieved by increasing the proportion of MX type particles. Studies of a Hafnium-containing steel have produced improvements in both creep and corrosion properties of 9%Cr steels. Hafnium has been ion-implanted into thin foils of a 9 wt% Cr ferritic steel to study its effect on precipitation. Two new types of precipitates are formed, hafnium carbide, (an MX type precipitate) and a Cr-V rich nitride, with the formula M_2N. The hafnium carbide particles were identified using convergent beam diffraction techniques, and micro-analysis. The nano-sized particles are present in much higher volume fractions when compared to VN volume fractions in conventional power plant ferritic steels. Furthermore it is confirmed that the Hf causes the removal of $M_{23}C_6$ grain boundary precipitates. This has led to an increased concentration of Cr within the matrix, reduced chromium depleted zones at grain boundaries, and increased resistance to inter-granular corrosion cracking.

1. Introduction

Corrosion resistance and oxidation is at current, hindering the temperature at which some of the most common ferritic steels can achieve in service. Thus for both austenitic and ferritic steels, a improvements in the corrosion protection and creep properties of the material, will enable plant design and operation to be more efficient and reliable (1). In order to obtain good creep properties in a chromium containing steel, the primary requirement is to obtain a suitable dispersion of small MX type particles within the grains of the matrix and on the grain boundaries. These retard dislocation motion and grain growth under creep conditions. However, coarsening(Ostwald ripening), reduces this strengthening effect in the later stages of creep. By the removal of grain boundary precipitates, commonly $M_{23}C_6$, there is a reduction in the number of chromium depleted zones, and failure by intergranular corrosion cracking (IGCC) can be reduced.(2,3)

This report investigates the effect of Hafnium as a potential addition to stainless steels to improve creep properties, and corrosion resistance. This is achieved through the use of precipitation and creep modelling, and experimental analysis of Hf containing steels.

2. Experimental

2.1. Materials

The metals used in this study consist of two stainless steels, one ferritic and one austenitic. The composition of the ferritic stainless steel, E911 is shown in Table 1, It is normalised at 1060°C for 1 hour and then air cooled.

Table 1: Chemical Composition of E911 (wt%, Fe Balance)

C	Si	Mn	P	S	Cr	Mo	Ni	V	Al	Nb	W	N
0.115	0.19	0.35	0.007	0.003	9.10	1.00	0.22	0.23	0.006	0.069	0.98	0.069

The austenitic stainless steel used is grade 316L with a composition shown in Table 2.

Table 2: Chemical Composition of E911 (wt%, Fe Balance)

C	Si	Mn	P	S	Cr	Mo	Ni	Cu	Nb	B	N
0.115	0.19	0.35	0.007	0.003	9.10	1.00	0.22	0.23	0.006	0.069	0.98

2.2. MTData Modelling

A thermodynamcial modelling programme (4) was used to predict the effects hafnium would have upon the phases present with the system. In order to determine the amount of hafnium needed to significantly diminish the $M_{23}C_6$ phase, various computations were conducted over a range of hafnium additions. The programme requires inputs of the composition of the system, in this case E911 and 316L as indicated in Tables 1 and 2 respectively. A balance of Fe was used in both cases, and the amount of hafnium varied between 0 wt% and 6 wt %. Computations used the MTsol database, a database compiled by NPL, the creators of the programme, over temperatures ranging from 800 – 1200K in steps of 10K. To easily identify the effect of hafnium, some phases are removed and miscibility gaps accounted for. The phases present for the resulting plots were the Liquid, BCC, FCC, HCP_A3, $M_{23}C_6$, Sigma, and Laves phase.

2.3. Precipitation and Creep Modelling

2.3.1. Simulation of Precipitation Kinetics. The details of the simulation of precipitation kinetics used in this study have been reported elsewhere.(5,6) Here, the essentials are summarised.

The simulation starts with the establishment of the simulation cell, a representation of the material where the phase transformations will be followed. Random nucleation sites are then generated within the cell and particle growth dependent upon time, and an initial size, which if too small is discarded back into the system. The heat treatment and service (ageing) time is divided into smaller time intevals, over which the calculation steps are repeated.

$$\Delta N(t) = Z\beta^* \left(\frac{N}{x_\theta} \right) \exp\left(-\frac{\Delta G^*}{kT} \right) \Delta t \tag{1}$$

where $\Delta N(t)$ is the number of nuclei generated in the simulation cell during the time interval $t \sim t + \Delta t$. As Z, β^* and ΔG^* are different between grain boundaries and the matrix, the number of nuclei generated during each time interval in grain boundaries, $\Delta N_{GB}(t)$ and in the matrix, $\Delta N_M(t)$ is different. N is the number of a particular type of atomic site, x_θ is the molar fraction of solute atoms in the nucleus phase, k the Boltzmann constant, T the absolute temperature, ΔG^* the energy required to form the critical nucleus, and t the time. The two coefficients Z and β^* are as follows

$$Z = \frac{V_{\theta a} (\Delta G_V)^2}{8\pi \sqrt{kTK_j \gamma_{\alpha\theta}^3}} \tag{2}$$

$$\beta^* = \frac{16\pi \gamma_{\alpha\theta}^2 D x_\alpha L_j}{a^4 (\Delta G_V)^2} \tag{3}$$

where $V_{\theta a}$ is the volume occupied by one atom in the nucleus, ΔG_V the free energy change per unit volume of nucleus, $\gamma_{\alpha\theta}$ the interfacial free energy, D the diffusivity, x_α the solute concentration in the matrix, a the lattice parameter, L_j and K_j are shape factors.

The inter-granular and the intra-granular nuclei are randomly located within the grain boundaries and the matrix (excluding grain boundaries) respectively. At present, we assume a nucleation is successful when the generated nucleus does not overlap with others. No particular size distribution of the nuclei is introduced here. The nucleation rate decreases with time because the matrix solute concentration decreases due to the formation and growth of precipitates.

The growth and coarsening mechanism are dependant upon solute concentration. An average concentration is calculated from the initial content and the amount that has been precipitated. The concentration at the surface of each particle is then calculated from the Gibbs-Thomson Equation (4).

$$C_r = C_\infty \exp \frac{2\gamma_{\alpha\theta} V_\theta}{RTr} \tag{4}$$

This can then be used to estimate a concentration gradient for each particle, with both grain boundary and volume diffusion mechanism being taken into account for grain boundary particles. The volume change of each particle is then calculated using the diffusion law (5).

$$\Delta V = D_V S g \frac{\rho_\theta}{C_\theta \rho_\theta - C_r \rho_\alpha} \Delta t \tag{5}$$

These steps are repeated over the required time period to produce results including average size, distributions, and inter-particle spacing, as a function of time.

2.3.2. Continuum Creep Damage Mechanics Modelling. The continuum creep damage mechanics model in this paper has been discussed in detail elsewhere (6). The model described in the previous section allows the position of each precipitate particle at any time to be recorded during the precipitation kinetics simulation. The evolution of the average inter-particle spacing for both inter- and intra-granular particles can also be calculated. This can then be fed into the CDM modelling code, negating the application of the Livshitz-Wagner model. In addition, this new approach allows the study of the effects that different precipitates and alloy compositions may have upon the creep behaviour of the material.

2.4. Ion Implantation

These implantations were carried out at Hokkaido University, Japan, using a ULVAC 400kV Ion Accelerator. Hafnium targets supplied by the Institute of Pure Chemicals, Japan, of 99.99% purity were used for the implantation. The ion current was kept at approximately 1 μA (10^{-6} Amperes), for 30 and 60 minutes. These two implantation treatments are roughly equivalent to 1.5 and 3 wt.% Hf implantation respectively. The foils produced were then tempered to 700°C for 2 hours in the case of the austenitic steel, and 760°C for 1 hour with the ferritic steel using the in-situ furnace in the high voltage TEM (JEM-ARM1300) at Hokkaido University, Japan. Once cooled, the samples were investigated with the FEI Tecnai F20 Field Emission Gun TEM at Birmingham University, to determine microstructural changes, the presence of phases and elements at grain boundaries using EDX methods. The diffraction patterns obtained using convergent beam diffraction relating to the structures of hafnium containing particles were used to determine their type. Once the d-spacing's had been calculated, reference was made to powder diffraction records to determine the structure and identity of the material.

3. Results and Discussion

3.1. MTData Modelling

The theory surrounding the alloying of hafnium is fundamentally described through the use of MTData and the thermodynamic models this programme produces when describing a system containing the elements of the steels in question with the addition of hafnium. Figures 1 and 2 show the resulting plot of E911 (composition as Table 1) with and without hafnium, noting the different phases that become apparent with the addition of hafnium.

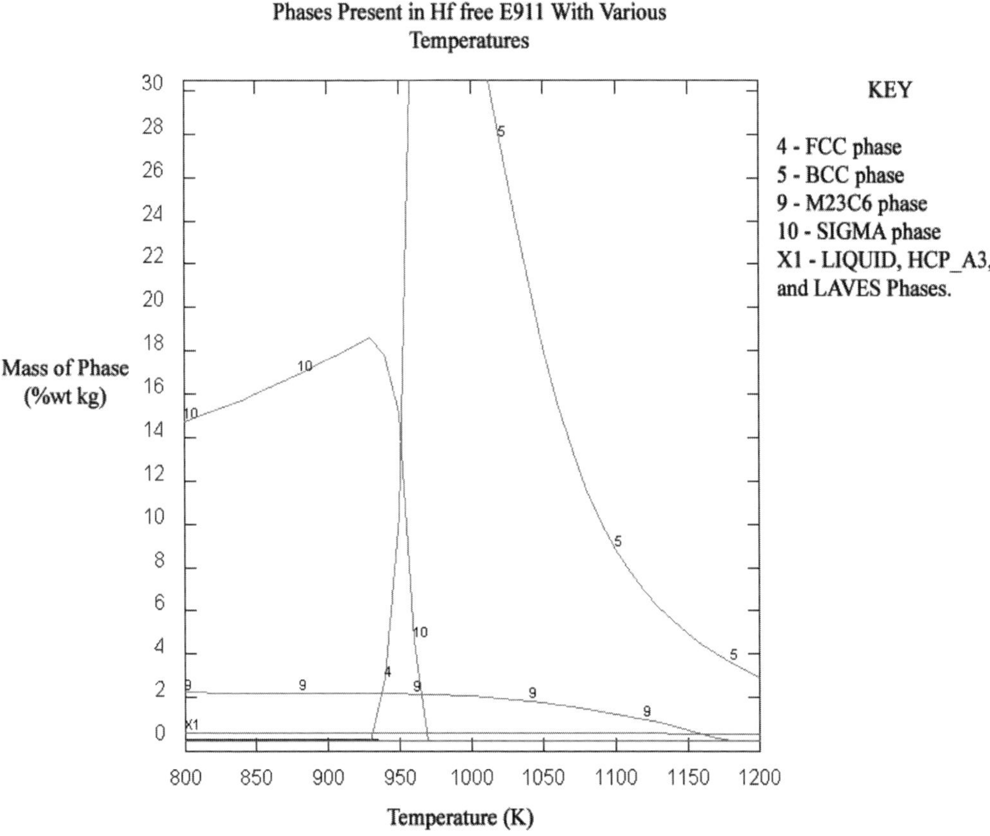

Figure 1. MTData plot of E911 without Hf, line 9 represents the $M_{23}C_6$ phase.

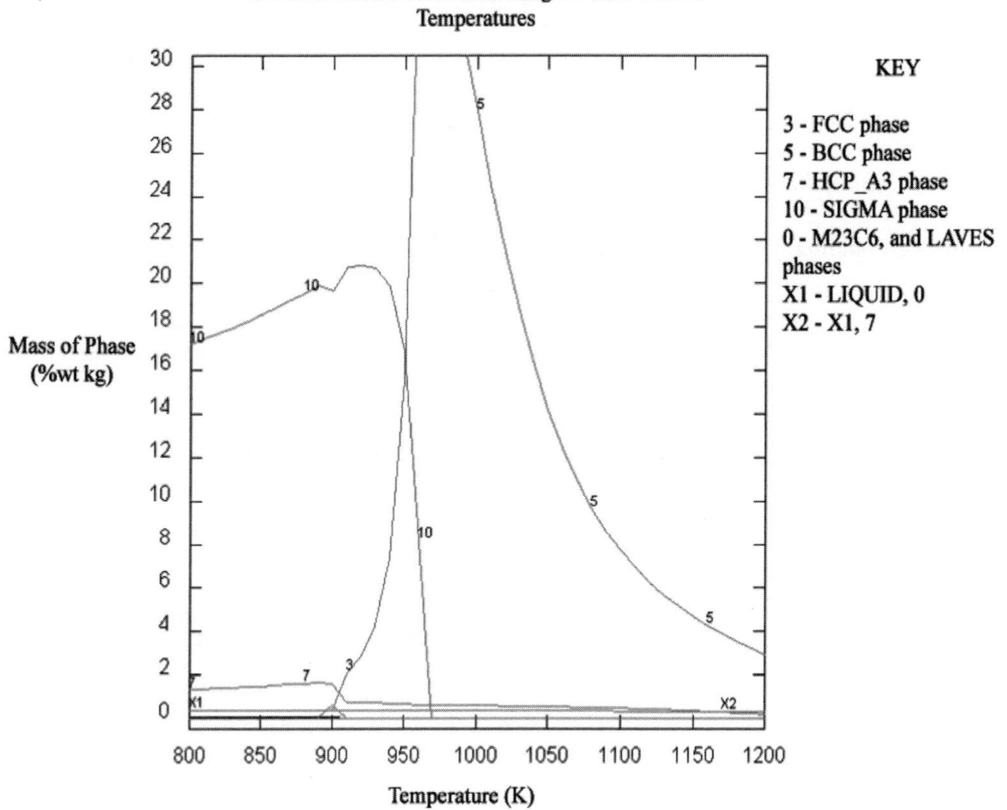

Figure 2. MTData plot of E911 with Hf, the $M_{23}C_6$ phase has been replace with line 7 and 3 (HCP_A3 and FCC respectively).

With the ability to vary the amount of hafnium added to the system it is possible to determine the lowest amount of hafnium required to initiate the removal of the $M_{23}C_6$ phase in both cases. Through these calculations it was proven that as little as a 0.9 %wt of hafnium would significantly diminish the $M_{23}C_6$ phase.

Two new phases have become apparent with the addition of Hf (figure 2), an FCC and a HCP_A3 phase. Further calculations using MTData show that the FCC is mainly HfC with a very small amount of VN at lower temperatures. The HCP_A3 phase has the composition M2N, where M mainly includes Cr, V, Nb, and Mo. The depletion of the M23C6 phase is due to the lower enthalpy of HfC, which results in the $M_{23}C_6$ being replaced by the more stable HfC phase.

3.2. Precipitation and Creep Modelling

The effects of two new phases on the creep behavior are shown in figure 3, together with that of the VN in the raw E911 material.

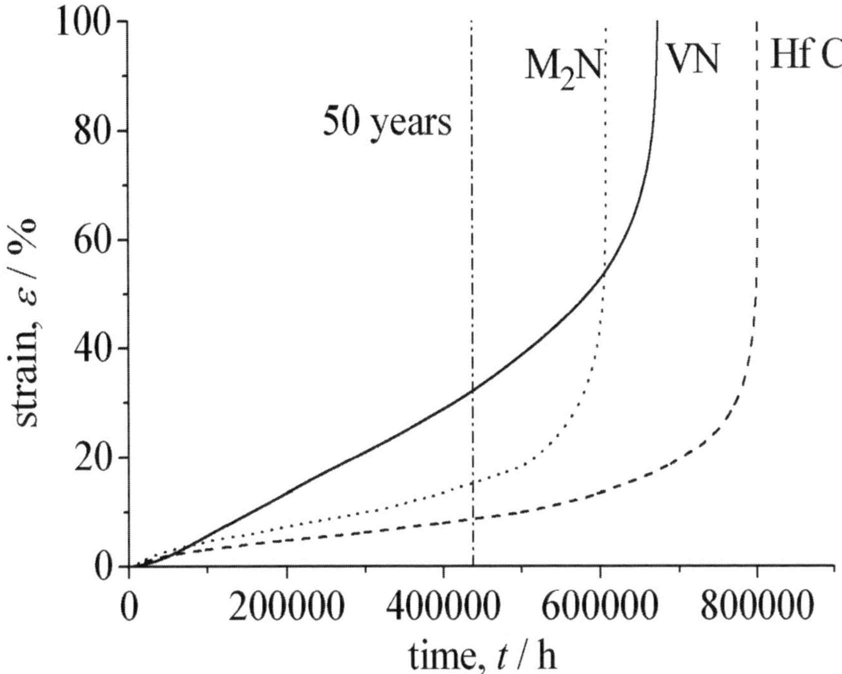

Figure 3. Creep life calculations of materials containing various MX type particles.

Both new phases promote slower creep rates than VN, therefore give much better creep rupture properties. It is clear that by the removal of the M23C6 phase, the creep properties of the material increase with regard to the later stages.

3.3 Ion Implantation

The ion implantation of hafnium into both ferritic and austenitic steels produced the expected results suggested by the thermodynamic modelling of the MTData programme. Micrographs of the microstructures prior to implantation show clear sensitisation and through the use of diffraction, have been clarified as the common $M_{23}C_6$ particles.

3.3.1. E911 Samples Without Hafnium Implantation. The ferritic samples were all heat treated at 700°C for 2 hours, regardless of hafnium implantation, this was necessary to produce the effects of sensitisation within the microstructure.

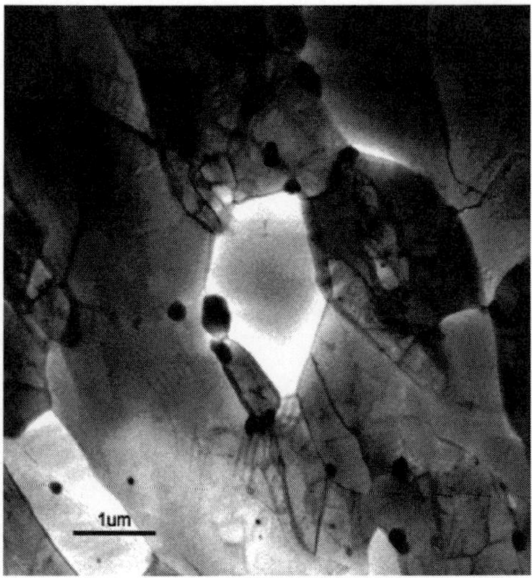

Figure 4. TEM micrograph of heat treated Hf free E911.

The grain boundary particles seen within this microstructure (figure 4) were also analysed using EDX (figure 5) to further clarify its identification as an $M_{23}C_6$ type carbide. The austenitic steel samples were also heat treated in to 760°C for 1 hour.

3.3.2. Hafnium Implanted Samples. Both the ferritic and austenitic steels produced the predicted results of the MTData programme, following the implantation of hafnium and consequent heat treatments.

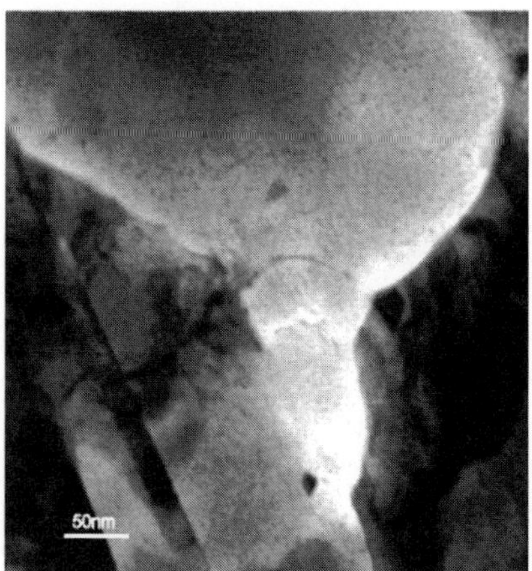

Figure 6. TEM mircograph of E911 with Hf implantation.

Figure 6 clearly shows the reduced, if not complete removal of the amount of carbides at grain boundaries. Further analysis of the grain boundaries confirmed the thermodynamic predictions of the removal of the $M_{23}C_6$ phase.

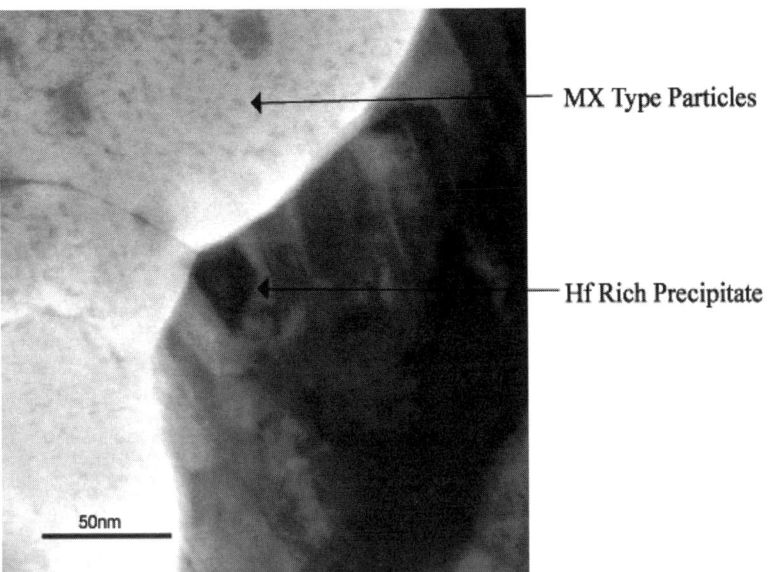

Figure 7. High magnification of Hf implanted E911 showing the presence of small MX type particles within the matrix, and a larger Hf rich precipitate at the grain boundary.

The larger grain boundary precipitates (Figure 7) were analysed using EDX methods (figure 8) and determined to be hafnium containing. These particles were smaller in size and volume fraction (65-70nm, and 1.5%) than commonly found M23C.

The smaller particles within the matrix were clearly in the order of 5-10 nm, and hafnium rich, with a volume fraction of 1.9%. These particles are similar to VN found in E911, but have a higher volume fraction than that of VN (0.3%). Convergent beam analysis of the matrix particles produced diffractions patterns that allow the determination of their crystal structures.

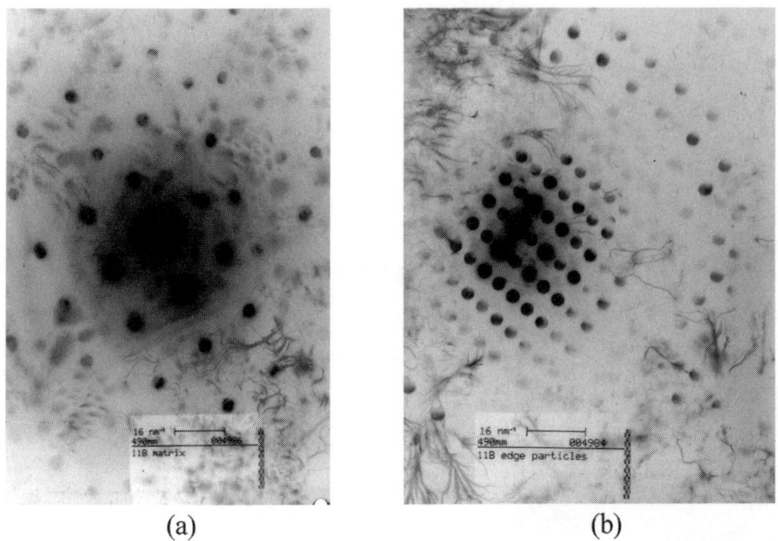

Figure 9. Convergent beam diffraction micrographs of a) the matrix, and b) the smaller matrix particles

The patterns shown in Fig. 9(a) and 9(b) were measured to find planer spacings and inter-planer angles as show in Fig. 10(a) and 10(b) respectively. The structures of the matrix, Figure 9(a), and the smaller hafnium containing particles in Fig. 9(b), were determined from these patterns. The results concurred that the particles were hafnium carbide with an FCC structure.

For the austenitic steel, the observations were similar in respect to the removal of phases as described by MTData calculations. Figure 11 clearly shows the difference in observed microstructures of 316 stainless steel with and without hafnium implantation.

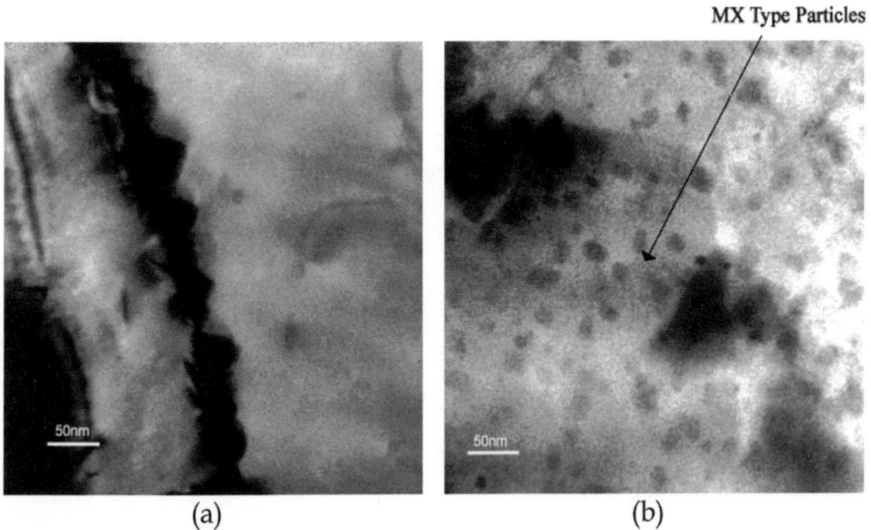

Figure 11. TEM micrograph of 316L microstructure; a) Hf free, and b) with Hf implantation.

Within the altered microstructure of the hafnium implanted steel, a similar effect can be seen with respect to the ferritic steels earlier, however there is a slight difference in the matrix particles in that clusters can be seen.

The particles within this matrix are a mixture of fine particles of <5nm in size, and areas of clustered particles or perhaps large particles in the range of 10-20nm. Primarily the use of EDX and HAADF (High-Angle Annular Dark-Field) methods within the STEM produced micrographs confirming the presence of the heavy hafnium element within these particles, as is observed by the brighter colour in Figure 12. Further analysis using convergent beam techniques also produced concurrent evidence of the presence of hafnium carbide within the matrix of the 316 stainless steel.

4. Conclusions.

Literature shows that grain boundary precipitates, most commonly M23C6, in most Cr containing steels used in power generation is detrimental to the creep properties of the material in the later stages of creep. Not only is this particular carbide damaging to the creep strength, but also to the corrosion resistance of the material is reduced because of chromium depleted zones. Hafnium additions to ferritic and austenitic steels used in power plant design have changed the microstructure markedly. The formation of two new phases bring improved creep rupture properties by increasing the volume fraction of MX type particles with in the matrix. The HfC particles found within the matrix are the fundamental constituents of this improvement, with the additional effect of having depleted significantly the grain boundary $M_{23}C_6$ phase. The near complete removal of the $M_{23}C_6$ phase can improve the resistance to inter-granular corrosion cracking (type IV cracking), by the stabilisation of chromium levels across grain boundary regions. Cast alloys of Hf containing E911 have been produced, studies have been undertaken to determine similar microstructural effects, corrosion resistance, and creep life and rupture strengths. The combination of improved creep and corrosion resistance by the relatively small additions of hafnium is very useful in the development and production of new power plant alloys.

Acknowledgments

The Worshipful Company of Tin Plate Workers Alias Wire Workers for their sponsorship of the research.

References

1. *Fabrication of Steels for Advanced Power Plant.* M. Barrie, Mitsui Babcock Energy Ltd. 2003. R226/ DTI/Pub URN 02/1508.

2. V.K. Sikka, C.T. Ward, and K.C. Thomas, "Production, Fabrication, Properties and Applications of Ferritic Steels for High Temperature Applications," presented at the ASM International Conference, Warren, PA (October 1981).

3. V. Cihal. *Intergranular Corrosion of Steels and Alloys.* Elsevier Publication, 1994, p80-83.

4. MTDATA

5. Y.Yin, R.G. Faulkner, *Materials Science and Technology,* 2003, vol 19, 91.

6. Y.Yin, R.G. Faulkner, *Power Technology,* Vol. 21. Forschungszentrum Julich GmbH; 2002 pp.1247-1256

Figures and Tables

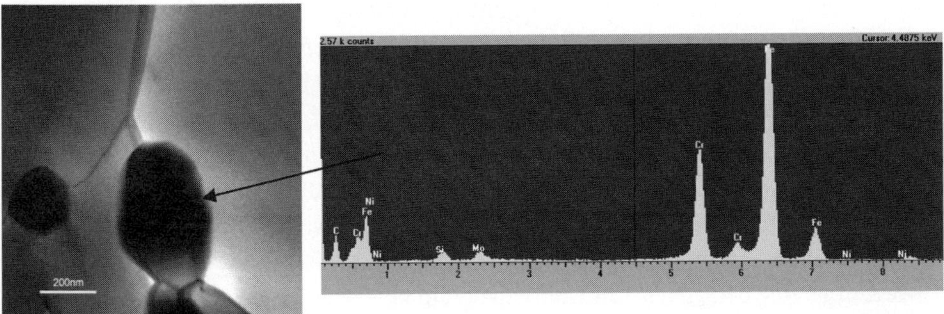

Figure 5. Grain boundary precipitate of $M_{23}C_6$ type, with its respective EDX graph showing Cr content.

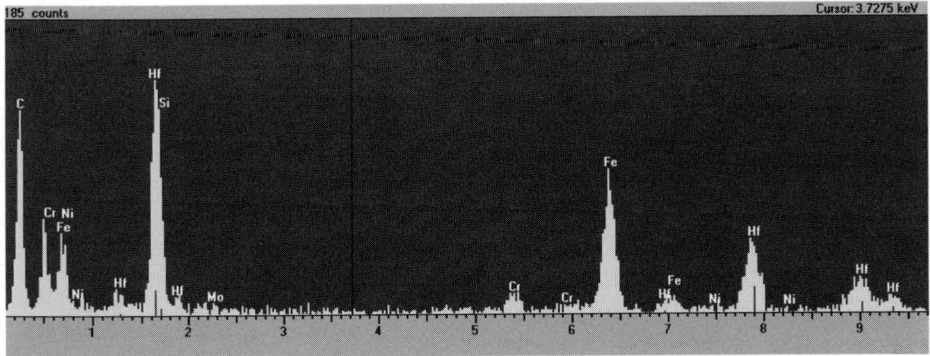

Figure 8. EDX analysis of grain boundary particle shown in Figure 7.

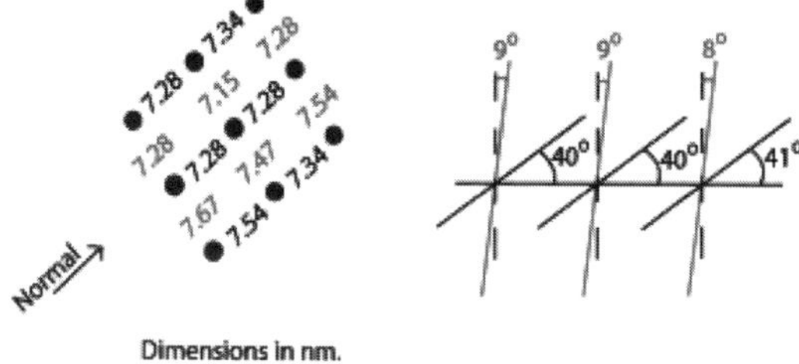

Figure 10(a). Simplified diagram of the measurements taken from Figure 9(a).

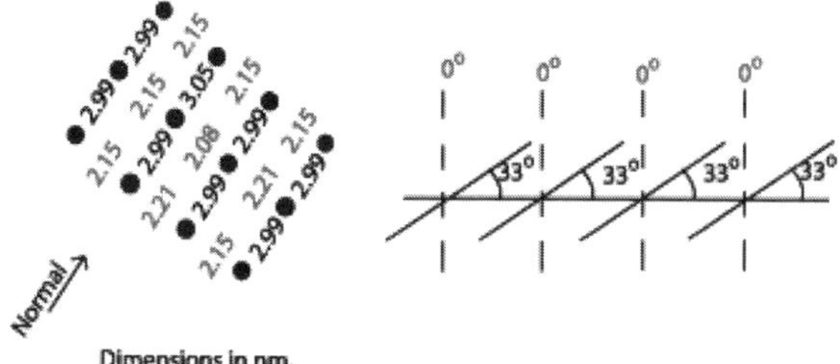

Figure 10(b). Simplified diagram of the measurements taken from Figure 9(b)

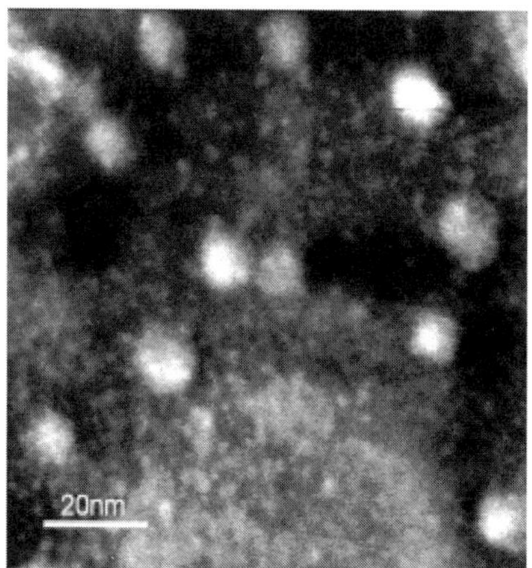

Figure 12. HAADF imaging of 316 Hf implanted matrix, the brighter areas showing heavier atoms, i.e. those containing hafnium.

Advances in Materials Technology for Fossil Power Plants
Proceedings from the Fifth International Conference
R. Viswanathan, D. Gandy, K. Coleman, editors, p 658-674

DOI: 10.1361/cp2007epri0658

The Role of Creep-Fatigue in Advanced Materials

J. Parker (Structural Integrity Associates, Inc)
2904 South Sheridan Way, Suite 303
Oakville, Ontario, L6J 7L7

D. Gandy (EPRI)
1300 W Wt Harris Blvd
Charlotte, NC 28262-8550

Abstract

Improved knowledge of creep and fatigue interactions is necessary today more than ever as power-generating plants are exposed to cyclic operation. Understanding factors that impact damage initiation and propagation, as well as technologies to predict accumulation of damage in systems and components are required.

A major EPRI initiative to rationize apparently conflicting evidence in the field of creep fatigue started in 2006. This project has successfully engaged key industry experts from around the world to do the following:

- Assess and document current creep-fatigue test methods
- Evaluate analytical methodologies with respect to crack initiation and growth
- Discuss life prediction methodologies for different applications
- Assess deficiencies that exist in the area of creep-fatigue damage
- Identify future research and development requirements

The present paper summarizes selected information from this project with particular emphasis on the performance of creep strengthened Ferritic steels such as Grade 91 and 92 steels.

Introduction

Creep Fatigue knowledge is necessary for meaningful reliability assessment of existing power generating plant because cyclic operation of an aging power plant fleet requires:

- Better understanding of factors affecting damage initiation and propagation, as well as
- Technologies and methods to predict damage accumulation for individual components and overall safe/economic life.

Utilities typically adopt operating practices which involve more severe cycles, either in terms of number or magnitude or both, so that there is, and will continue to be, a requirement to assess the creep fatigue performance of traditional boiler and turbine components.

The recent installation of large numbers of combustion turbines and the associated heat recovery steam generators (HRSG's) has significantly increased the number of materials and components which are operating under cyclic conditions at high temperatures. Indeed, the materials and component types susceptible to creep-fatigue is increasing even more since much of the new high efficiency coal plant will be operating with steam/metal temperatures much higher than those of traditional plant. Indeed, recent expert review has reemphasized that with most practical scenarios regarding future generation mix indicates that fossil plant will continue to be important to the international generation mixture, Table 1.

Table 1 - Exajoules of Coal Use (EJ) and Global CO_2 Emissions (Gt/yr) in 2000 and 2050 with and without Carbon Capture and Storage* (1)

	BUSINESS AS USUAL		LIMITED NUCLEAR 2050		EXPANDED NUCLEAR 2050	
	2000	**2050**	**WITH CCS**	**WITHOUT CCS**	**WITH CCS**	**WITHOUT CCS**
Coal Use: Global	100	448	161	116	121	78
U.S.	24	58	40	28	25	13
China	27	88	39	24	31	17
Global CO_2 Emissions	24	62	28	32	26	29
CO_2 from Coal	9	32	5	9	3	6

* Universal, simultaneous participation, High CO_2 prices and EPPA-Ref gas prices.

Thus, there is the expectation that:

- Creep -Fatigue damage will increase in occurrence, and affect an increased range of components.

- Different challenges will be faced since the new components will involve a greater range of alloys, methods of manufacture, and types of operation than conventional fossil generating stations.

It is therefore apparent that operation under cyclic conditions at temperatures where creep can occur is, and will continue to be, a matter of significant concern. The complexity associated with understanding damage development is further increased because there will be many different types of operating cycles (indeed for a single component location the types of cycle could vary with time) and in some circumstances environmental factors can influence crack initiation and growth. Some examples of the different types of cycle which can occur and thus need to be considered include:

- Temperature increase/decrease in phase or out of phase with pressure , Load changes
- Where in the cycle the hold period occurs
- Stress cycle considering both primary and thermal stresses
- Residual Stress Effects

Damage Morphology Issues

In many cases laboratory evaluations concentrate on generating data rather than performing detailed characterization of the morphology of damage. This then raises issues for identification of early stages of damage in service components. Additional complexities occur particularly for thick section components such as headers and rotors since there will frequently be the need to take into account gradients in temperature and/or stress.

The generally accepted view of how fatigue and creep damage appears is shown schematically in Figure 1 (2). This figure indicates that Fatigue is typically fully transgranular and creep is typically fully intergranular. It should be noted that while with traditional alloys exhibiting equiaxed microstructure creep voids generally develop on grain boundaries. These voids then increase in number and size leading to microcrack formation. It should be emphasized that with advanced steels creep damage can also take the form of precipitate coarsening and microstructural changes. These changes may in some circumstances be described as strain softening and will clearly significantly change materials strength.

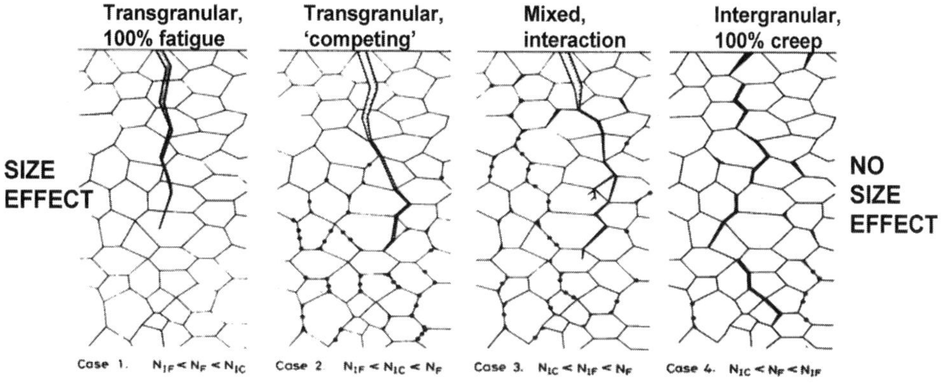

Figure 1.

Schematic diagram illustrating different types of damage which can develop in high temperature components, (2)

When operating conditions are such that both forms of damage can occur there are at least 2 different options for damage development. One is where the 2 forms of damage both occur but act independently of each other, an example of a sample removed from a low alloy steel weld showing inside surface connected fatigue cracking and midwall creep damage is shown in Figure 2. The other condition is where damage is mixed and there is an interaction (3). A typical photograph showing damage in a laboratory sample is shown in Figure 3 (4). An alternative photomicrograph from an ex-service component, shown in Figure 4 (5), illustrates that damage in service components maybe significantly more difficult to identify as creep fatigue. In this case, the cyclic stresses were relaxed by creep which occurred in the Type IV region of the weld HAZ. Thus, the cyclic loading resulted in higher stress conditions than would have been experienced from pressure alone and it was the local damage developed from the stress relaxation which resulted in cracking. It is worth noting that while this example is a special case, in service damage frequently occurs at welds in part because welds typically exhibit heterogeneous microstructure and properties. It is well understood that relaxation of high stress may cause cracking in brittle regions, e.g. as is the case with stress relief cracking. However, systems or cyclic loads can be concentrated in the weaker regions of a weld leading to accelerated creep damage and crack formation, e.g. as with Type IV cracking.

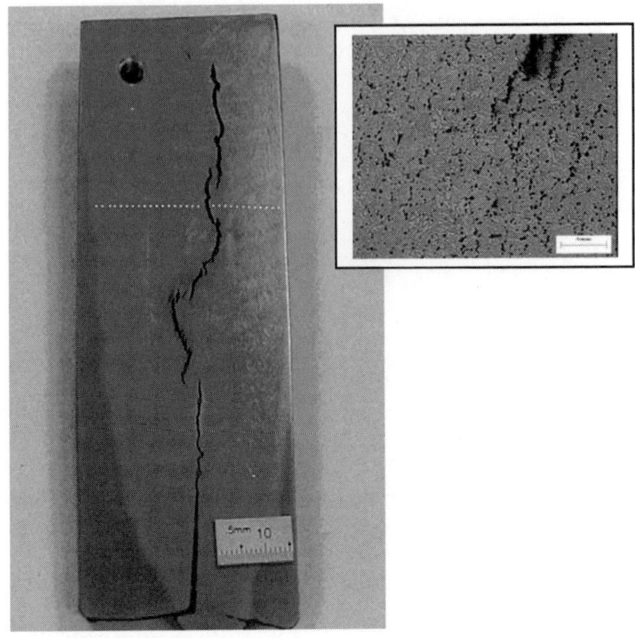

Figure 2.

Example of cracking from an exservice weld.The defect A initiated at the inside surface and appears to be related to fatigue. The midwall damage in transgranular and is typical of creep, see detailed micrograph.

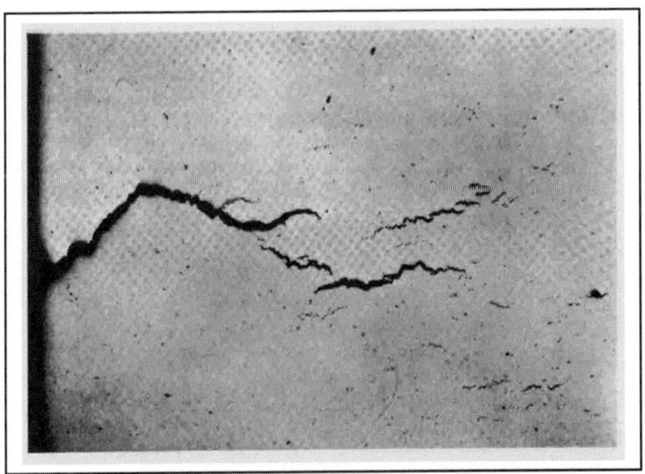

Figure 3.

Example of creep fatigue damage observed in a laboratory test specimen (4).

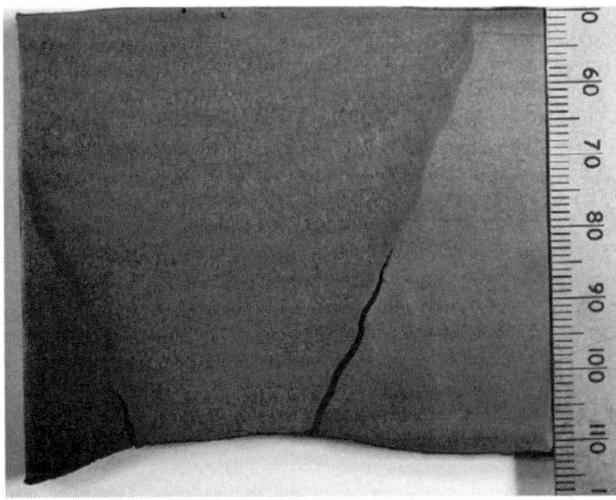

Figure 4.

Example of damage developed in a low alloy steel weldment after elevated temperature operation with both steady and cyclic stresses present (5).

Laboratory Testing Methodologies

Creep-fatigue tests are typically carried out to generate material property input data for use in models for assessment of component behaviour. In some cases these tests are performed under idealised cycles since, in general, relatively simple cycles facilitate subsequent analysis. These tests lead to the development of models such as cyclic/hold deformation and endurance data models. Alternatively, testing maybe performed under conditions believed to directly represent service. Thus, by application of a repeated typical, representative service - related cyclic/hold, it may be possible to assess life empirically. Furthermore, these tests provide results which can be used to verify the effectiveness of creep-fatigue assessment procedures in close-to-service conditions.

It is important to show that the methods used to correlate laboratory data with plant behavior are relevant. In that regard, component targeted feature-specimen tests were shown to be essential. These allow both benchmarking of design methods and evaluation of remaining life predictions made using available creep-fatigue assessment procedures. Since it is not always straightforward to establish the specifics of microdamage development which occurs prior to the initiation of a defect, through - test monitoring of damage would be an advantage.

Unfortunately, this type of evaluation is rarely performed. As noted earlier, all too frequently no post test laboratory assessment of specimens is carried out. This is particularly unfortunate since this type of examination permits direct assessment of damage and comment on the applicability of the data to in-service assessment. It was agreed that post test inspection to characterize the mechanism of damage/cracking and to quantify the extent of any dimensional instability should be an integral part of any creep-fatigue testing campaign. To facilitate documentation and reporting of damage the terminology used should be rationalized (e.g. through agreed definitions of damage).

While there are documents providing guidance on Low Cycle fatigue Testing, there are no National or International Standards or Codes of Practice covering creep-fatigue testing. This means that cross comparison of results is difficult if not impossible. The results of a major European standardisation project seeking to provide a formalised approach for thermal fatigue testing have shown that:

- The measurement uncertainty associated with the determination of the number cycles corresponding to the 10% load drop on any individual curve is about ±3%.
- The average in house repeatability of test results was about ±13%.
- Even for partners involved in developing the standard some test results were unacceptable and discarded as outliers.
- The scatter at the 95% confidence level derived from data sets excluding outliers were about ±67% for tests performed by partners and about ±45% for tests performed by others.
- A realistic target tolerance of ±5°C should be considered as being acceptable for the control of temperature throughout a TMF test at the heating rate of 5K/s.

In general, for tests involving cycling, lower scatter was noted for samples with greater cross sectional area and with smooth surface. Even so it was noteworthy that no universally agreed specimen geometry exists even for relatively simple tests on homogeneous material such as parent. Where tests are carried out on heterogeneous samples such as in welds, it appears that samples are generally dimensionally similar to those used for parent to accommodate test rig requirements etc.

Some general points regarding creep-fatigue testing include:

- Ensure the accuracy, stability and calibration of equipment
- To reduce the scatter of the data, the tests should preferably be carried out under axial strain control, isothermal conditions, and by cylindrical smooth specimens.

- Tension hold time tests are useful for ferritic steels as well as austenitic steels to simulate the thermal fatigue.

- Ductility normalized strain range partitioning method is useful because the lives depend on the ductility and the strength of the materials.

Analytical Methodologies to estimate Crack Initiation

Methodologies for performing analytical evaluation of crack initiation included:

- Time fraction

- Strain fraction/Ductility Exhaustion

  - Simplified methods for total strain
  - Estimate creep strain with time for applicable stress/temperature combinations
  - Strain to Failure, elongation/Reduction of Area or other method for different conditions

The behavior of different steels has been reported in major studies as well as in individual papers. It should be emphasized that full analysis requires knowledge of a range of materials information. In many cases, full data has not been developed for an alloy under consideration the analysis is based on certain assumptions. In addition, in most cases laboratory assessments have concentrated on parent properties because less data exists for weld metals, heat affected zone (HAZ) or overall weldment performance.

Review of recent developments associated with application of the British Energy R5 code with regard to estimation of crack initiation demonstrates that:

- A 'stress modified' ductility exhaustion approach has been derived from consideration of cavity nucleation and growth. This approach gives less pessimistic predictions of failure for cycles where the creep dwell is positioned at a stress below the peak in the cycle and when the strain range is low.

- Models developed for the effects of multiaxial states of stress.

- The majority of creep-fatigue problems are associated with welds. The current advice for creep-fatigue initiation for welds is complicated and can be excessively conservative. Thus, changes have been proposed to R5 Vol.

2/3 App. A4. These changes involve replacement of Fatigue Strength Reduction Factors (FSRF) by:

- o Weld strain enhancement factor (WSEF), which takes account of material mismatch and geometry and
- o Weldment endurance reduction (WER), which takes account of the reduced endurance of a weldment.

Although in combination these give a similar result to a FSRF so the fatigue damage is unchanged when only the WSEF is used as in the determination of the start of dwell stress, the calculated creep damage fraction is reduced.

In addition, there are several proposed modifications to account for the effects of compressive dwells. These can be summarized as:

- Fatigue strain range is enhanced for the presence of a creep dwell.
- The fatigue damage should be calculated by removing the nucleation phase of fatigue endurance. This makes allowance for the physical processes that reduce fatigue endurance in the presence of a compressive dwell.
- The creep damage per cycle is zero.

Future considerations for developments of the R5 methodology include:

- Engineering or True Strain at Failure?
- Further validation of 'stress modified' approach for Type 316H parent (testing is advanced) and for Type 316H HAZ (creep-fatigue testing to start, creep tests in progress). Evaluation is also planned for 1CrMoV rotor steel (using existing data).
- Improved prediction of stress relaxation
- Improved prediction of thermo-mechanical cycle and start of dwell stress
    - o Modified elastic modulus methods, which provide an approximate computational means of estimating more accurate values of start-of-dwell stress, elastic follow-up factor and strain range than are obtained using a basic shakedown analysis.
    - o Inelastic analysis including the development of constitutive models for austenitic parent and weld metals.
        - Further work on the influence of high residual stresses in weldments.

- Application of the weldment procedure to creep-fatigue features tests.
- Derivation of the WSEFs for ferritic materials.

Materials Information

An evaluation of available creep fatigue information has highlighted concerns regarding a lack of metallurgical data, in particular, concerning:

- Long term creep behaviour
- Mechanical behaviour
- Behaviour of welds
- Fabricability / manufacturing

It should also be noted that in many cases the results of a test program are limited to creep rupture lives and endurance data either for different ramp rates or for different hold times. It is apparent that material creep ductility will exhibit a critical effect on the type of damage developed and how that damage will influence creep fatigue interaction. The data presented in Figure 5 provide an

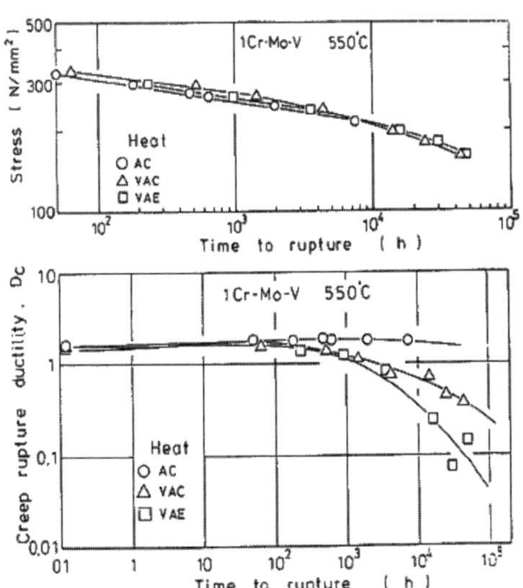

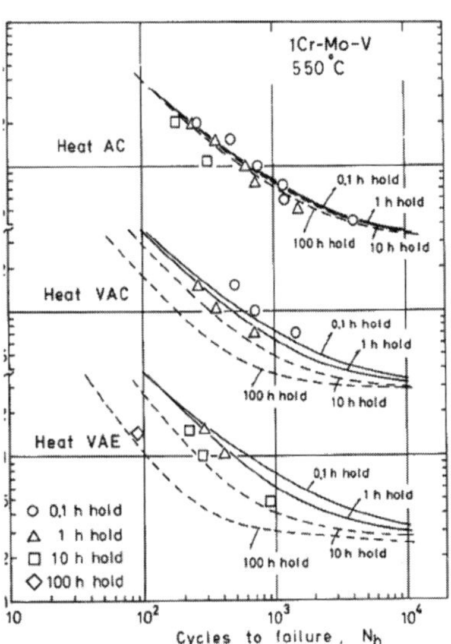

Figure 5.

Creep rupture strength and ductility for three 1CrMoV steels (a) and cyclic endurance curves for these steels showing the different effect of hold time (b). (6)

illustration of this effect. Three turbine low alloy steels were tested and all exhibited similar creep rupture strength. However, the creep ductility of the 3 aqlloys was different and this difference was a factor in the level of life reduction noted in cyclic tests with hold time. Thus, the alloy with the highest ductility exhibited very little life reduction when hold times were included in the cycle. Conversely, creep brittle material revealed a significant reduction in cycles to failure for cycles with hold time (6).

Creep fatigue in Creep Strengthened Ferritic Steels

The behavior of creep strengthened ferritic steels such as Grade 91 and Grade 92 has been evaluated by a number of investigators. Typically this type of evaluation involves performing low cycle fatigue tests at different ramp rates or performing cyclic tests which include hold time. In both cases data are normally compared to the results of tests performed with rapid cyclic ramps. Typical results for Grade 91 steel are illustrated in Figure 6. Here both the tests with slow ramps and hold times show reduced lives. Interestingly tests where the hold was in tension or compression or both show similar lives.

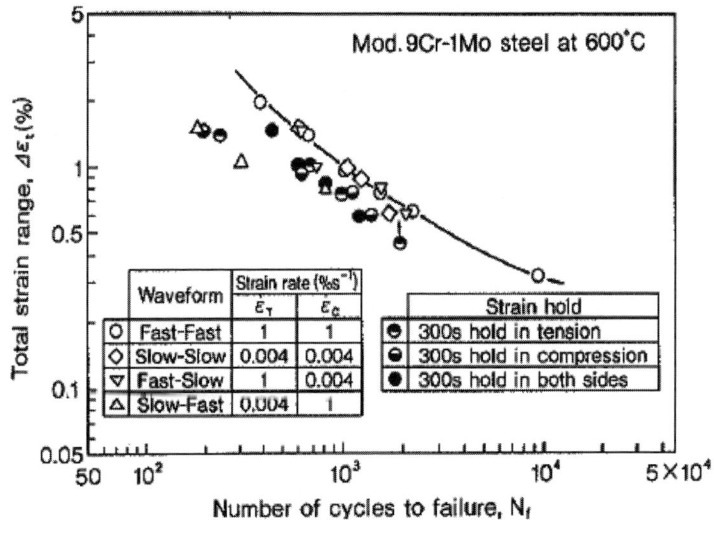

Figure 6.

The effect of different ramp rates and hold times on the number of cycles to failure of Grade 91 steel tested at 600°C, (7).

This observation differs in detail to results from an alternative study considering creep – fatigue – oxidation interactions in grade 91 steel. Since this steel typically enters service with a tempered martensitic structure it is susceptible to microstructural changes with exposure time at high temperature. Detailed microstructural studies have shown that cyclic work softening was accelerated by hold times at peak temperature. Thus, marked differences in microstructure were found for creep fatigue exposure compared to similar samples cycled without hold, for example Figure 7. In this study the most extreme life reductions for creep fatigue exposure were noted in samples which had experienced compressive holds. This behavior was consistent with the greater incidence of surface cracking observed under these conditions, Figure 8. The cracking observed was consistent with scale disruption and internal oxidation which promoted defect initiation in the steel substrate.

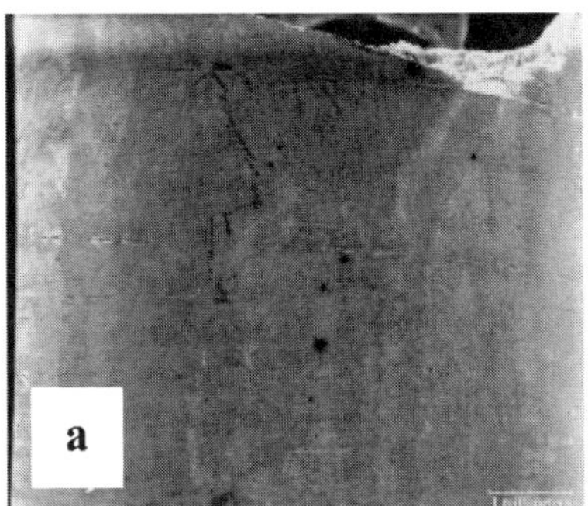

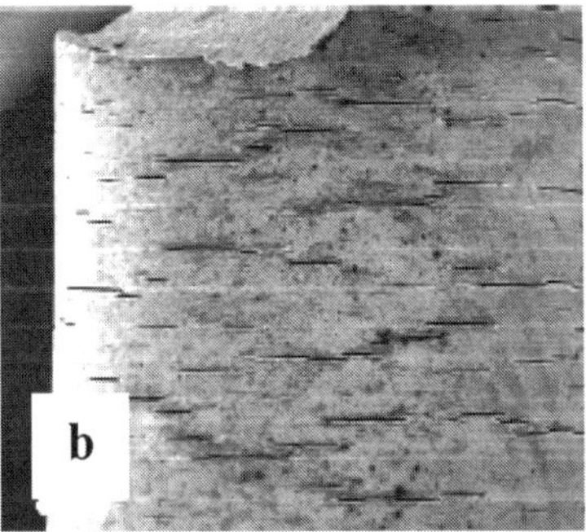

Figure 7.

Comparison of the specimen surface of Grade 91 samples tested under creep fatigue conditions. The sample after holding on the tensile stress side, shown in (a), exhibited limited surface cracking while the sample after compressive holds (b) revealed many surface cracks and microcracks, (8)

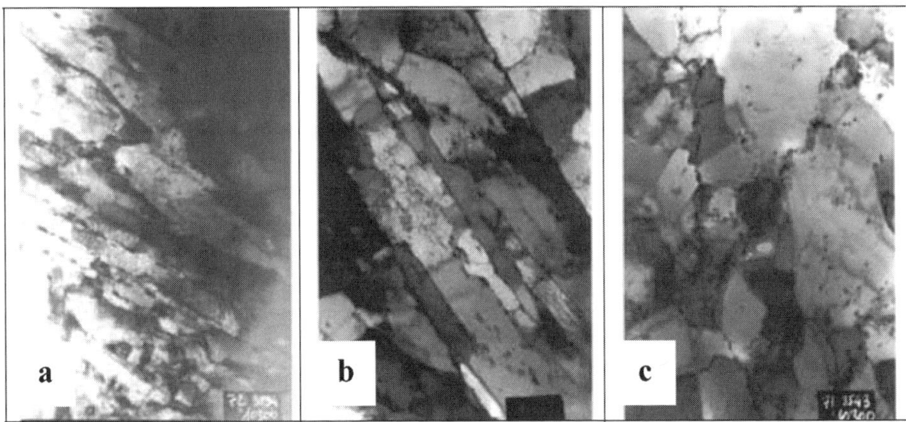

Figure 8.

Microstructures observed in Grade 91 steel showing the pretest structure (a), the microstructure after cycling at elevated temperature without hold (b), and the microstructure after cycling at elevated temperature with hold times (c) (9).

The additional complications involved with assessment of creep fatigue behavior of strengthened Ferritic steels arise in part because:

- the initial microstructure is complicated and can be varied by the fabrication process used, these variations may or may not be identified by optical microscopy
- the creep strength and ductility depend on both the detailed chemical composition and processing details. These differences in strength cannot simple be identified using optical metallography.
- surface effects in general and scale formation in particular can also be involved in crack initiation. The details of how the scales form, whether they crack and if so whether these cracks initiate defects in the underlying substrate requires a level of understanding not presently established.

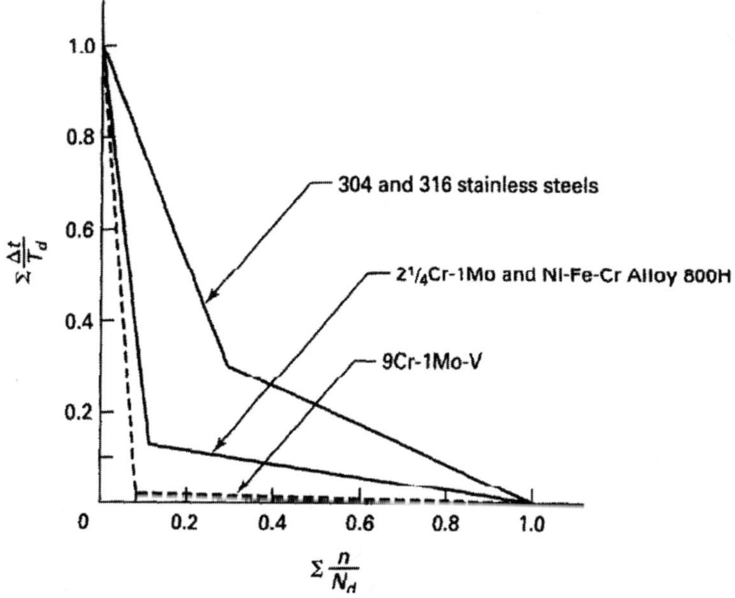

Figure 9.

Creep fatigue diagram suggested by the ASME section NH (10).

In view of these and other complications in understanding and analysis of creep fatigue behavior it is not surprising that when data are analyzed using the traditional approaches the results can appear very pessimistic. For example, assessment of data in the ASME section NH applied a very conservative interpretation to the expected creep behavior for Grade 91 steel, Figure 9. As shown this conservative approach implied that the creep fatigue endurance for Grade 91 steel was markedly worse than that for conventional boiler steels. This conservatism is likely a function of the fact that the analysis approaches did not take into account the complications associated with variable strength, microstructural changes, strain softening and the influence of surface scale.

Work performed by CRIEPI in Japan has summarized analysis of many creep-fatigue tests carried out on the austenitic alloy 316FR and the 9 and 12 % Cr steels Grade 91, Grade 122, and TMK1. These test results included data from several long term tests. The main findings of this work were summarized as:

- Time fraction type stress based creep damage is insufficient for predicting life reduction due to holding. This is especially the case at the small strain ranges of practical interest. This can not be covered by any interaction diagram.
- Calculation of ductility exhaustion type strain based creep damage tends to overestimate creep fraction used per cycle when all inelastic strain during hold period is counted.
- By decomposition of inelastic strain or extracting ductility exhausting effect using a rate dependent overall ductility, life reduction can be estimated with reasonable accuracy (within factor of 2).
- Low cycle fatigue endurance can be described using the Coffin Manson method but creep fatigue damage analysis should recognize that when there is interactive damage there is an acceleration of fatigue damage because of ductility reduction as a consequence of the accumulated creep damage

Review of data generated in Japan demonstrated that the new steels P23, P92 and P122 have superior creep strength, and comparable low-cycle fatigue and creep-fatigue strength compared with conventional steel P22 and P91. However, plotting creep fatigue data on the conventions life fraction interaction diagram indicates very short life expectancies. It appears that the excessive life reduction in creep-fatigue testing of laboratory specimens can be mainly attributed to the strain concentration in the specimen due to the deformation, such as bulging and necking induced during creep-fatigue test. The detailed measurements performed in this evaluation are of a type rarely undertaken. The fact that post test laboratory evaluation provides key data to assist interpretation of results was noted and offers further weight to the earlier recommendation regarding the need for laboratory test programs to include post test examination.

It was also noted that the creep-fatigue endurance life of P23, P92 and P122 can be evaluated by using an inelastic strain range including creep strain and Coffin-Manson relationship obtained by low-cycle fatigue test without a creep effect. However, it should be noted that Yamaguchi showed that good relationships required that the creep strain was normalized by the appropriate value of creep ductility.

Especially for boiler materials uncertainty exists about extrapolation from short term tests to the long-term creep range so that there is reduced confidence that the expected performance will have the normal factors of safety. Since in - service experience is not available regarding damage mechanisms (e.g. corrosion, creep damage) for these materials and operating conditions it will be necessary to adopt a series of measures to evaluate performance. These include:

- Additional instrumentation to permit measurement of key factors (e.g. temperature, pressure, creep deformation, displacement) at critical component locations.
- Storage and retrieval of information in component data bases to allow more accurate life time consumption calculation, and
- Generation of additional laboratory materials data, in parallel to service operation so that the accuracy of extrapolation methods can be reviewed. In addition, by conducting tests on sections used to manufacture actual components it will be possible to develop heat specific data which allows the determination of the location within the scatter band.

Concluding Remarks

In July 2006, EPRI organized an Expert Workshop on *Creep-Fatigue Damage Interaction* to bring together key industry experts from around the world. Discussions took place over 3 days with sessions covering:

- Introduction
- Laboratory Testing Methodologies
- Materials Information
- Analytical Methodologies – crack initiation
- Analytical Methodologies – crack propagation
- Application of Creep - Fatigue data to Component Design and Assessment

- Concluding Discussion and Summary

The Expert Workshop succeeded in presenting up-to-the-minute information concerning the current state-of-knowledge of creep-fatigue damage interaction. The discussion of key factors resulted in the development of an agreed listing of key issues for future consideration. The second workshop held in 2007 summarized progress on the short term issues and agreed a schedule for the future. Current work is focused in part on detailed assessment of data for selected alloys used in fossil boilers and turbines.

Based on assessment of information from Grade 91 steel it is apparent that the advantages offered by creep strengthened ferritic steels can only be realized with careful control of fabrication methods. Since ASME codes provide general information further engineering judgement is frequently of benefit. An Integrated Life Management Strategy for CSF steels should include review and quality assurance. However, even with good acceptance and validation checks effort will be needed during service to evaluate component performance.

Particular emphasis should be placed on assessment of locations where there are complexities, these include:

- Geometry
- Microstructural variations
- Surface Scale
- Loading, including cycling

A comprehensive approach to evaluation of creep fatigue damage development requires integration of knowledge from laboratory programmes with detailed appreciation of in-service conditions.

References

1. MIT Report *"The Future in Coal"*, Massachusetts Institute of Technology, 2007

2. R. Hales. A quantitative metallographic assessment of structural degradation of type 316 stainless steel during creep-fatigue, *Fatigue Eng. Mater. Struct.*, 1980, 3, 339-359

3. R.P. Skelton. *Technology Innovation: Creep-Fatigue Damage Accumulation and Interaction Diagram Based on Metallographic Interpretation of Mechanisms,* EPRI, Palo Alto, CA 2007. 1014837

4. S.J. Brett. *"In-service cracking mechanisms affecting 2CrMo welds in ½ CrMoV steam pipework systems"*, International Conference on Integrity of High Temperature Welds, 1998, pp. 3

5. R.P. Skelton. Growth of short cracks during high strain fatigue and thermal cycling, in *"Low Cycle Fatigue and Life Prediction"* (eds C. Amzallag et al.,) ASTM STP 770, Philadelphia, 1982, pp. 337-381

6. K. Yamaguchi, K. Ijima, K. Kobayashi and S. Nishijima. *Creep –fatigue of 1 Cr-Mo-V Steels under Simulated Cyclic Thermal Stresses.* ISIJ International, Vol 31 (1991), No. 9, pp. 1001-1006

7. K. Sonoya, I. Nonaka and M.Kitagawa, *"Prediction of creep – fatigue lives of Cr - Mo steels with Diercks equation"*, ISIJ Internatrional, Vol 31, No 12. (1991), pp. 1424 - 1430

8. A. Pineau, *"Creep-Fatigue-Oxidation Interactions with Tensile and Compressive Hold Times on a 9Cr Steel at 550°C"*. Proceedings of the Expert Workshop on Creep-Fatigue Damage Interaction, 2006

9. M.T.Cabrillat, L. Allais, M. Mottot, B. Riou and C. Escaravage, *"Creep fatigue behavior and damage assessment for mod 9Cr-1Mo steel"*, ASME ICPVT11, Vancouver, July 2006.

10. ASME : Boiler and Pressure Vessel Code, Part III, Division 1, Subsection NH-Class 1 Components in Elevated Temperature Services, ASME, New York, 2004.

Advances in Materials Technology for Fossil Power Plants
Proceedings from the Fifth International Conference
R. Viswanathan, D. Gandy, K. Coleman, editors, p 675-688

DOI: 10.1361/cp2007epri0675

MICROSTRUCTURAL CHARACTERIZATION OF MODERN MARTENSITIC STEELS

K. Maile
F. Kauffmann
A. Klenk
E. Roos
MPA University of Stuttgart
Pfaffenwaldring 32
70569 Stuttgart, Germany

S. Straub
ALSTOM Power Generation AG
68309 Mannheim, Germany

K.H. Mayer
ALSTOM Power Service GmbH
90208 Nürnberg, Germany

Abstract

This paper deals with results from microstructural investigations (TEM), characterization of precipitation and long term development of precipitates, mainly done within the framework of COST 536.

Introduction

In the development of modern steam power plants, it is the aim to increase the inlet temperature of the steam from about 540 °C - 550 °C up to 600 °C - 650 °C. This increase would lead to a significantly improved rate of efficiency and reduced CO_2 emission. To enable such an increase in temperature, new ferritic-martensitic 9-12 wt. % Cr steels are developed [1]. Initially the addition of Nb, N and in some cases W to the traditionally used X20(2)CrMoV-12-1 helped to improvement the high temperature stability of these modern steels. The modified steels yield a significantly enhanced creep resistance, which has its origin in thermally stabilised $M_{23}C_6$ and vanadium and niobium carbonitrides. This development step has already found its transition to the industrial application and is applied in nearly 30 power stations, operating at maximum steam inlet temperatures of 610 °C. Typical examples for these steels are the X10CrMoVNbN-9-1 (P91), X12CrMoVNbN-10-1 and X12CrMoWVNbN-10-1-1 (E911) [1].

In a second step towards the optimisation of these steels, Boron has been added for a further stabilization of the $M_{23}C_6$ precipitates. These boron steels are suited for applications up to 625 °C [1]. The most prominent alloy of this development step is X10CrWMoVNbNB-9-2 (P92). Recent research results suggest that increasing the boron content and adjusting the nitrogen content accordingly can further improve the creep properties. This should lead to an increased amount of fine vanadium carbonitrides and prevent the formation of boron nitride at the same time [2, 3].

The potential for the improvement by alloying boron is demonstrated by the development of the 10.5 wt% Cr TAF steel by Fujita and co-workers about 30 years ago [4, 5]. The drawback of the TAF steel is the high B content of about 300 to 400 ppm. In accordance with the current experiences it reduces for example the hot workability of large turbine components dramatically. Nevertheless, the quantitative investigations of the microstructure of this steel with modern analysis techniques are of great importance for the development of new 9-12 wt. % Cr steels with moderate boron contents. The available data for the microstructure of the TAF steel, dated from 1982, give only a qualitative view of the observed phases [6].

Microstructural Characterization

Optical microscopy and scanning electron microscopy

The optical microscopy (OM) analysis was done using a Leitz Aristomet. The specimens were prepared by standard metallographic methods such as cutting, grinding and polishing. To develop the microstructure the specimen were etched by 3% HNO3. Scanning electron microscopy (SEM) investigations were carried out using a JEOL JSM 6400 operated at 10 kV for the analytical measurements and at 25 kV for acquiring images. For the analytical measurements the SEM is equipped with an energy-dispersive X-ray spectrometer (EDX) system (TN5500 from NORAN) which allows the detection of elements with an atomic number ≥ 5 (Boron).

Transmission electron microscopy

Conventional transmission electron microscopy (TEM) investigations were performed using a JEOL JEM 2000 FX operated at 200 kV to determine the subgrain size and dislocation density. For analytical measurements the TEM is equipped with an EDX detector (Kevex Sigma 1 from NORAN, capable for detection of elements with an atomic number ≥ 11) and with a GATAN energy filter. The latter was used for acquiring elemental maps. A detailed description of energy filtered TEM (EFTEM) measurements can be found for example in [7, 8]. For the quantitative evaluation of the precipitate state (type, size, shape and distribution) a digital image processing system (SEM-IPS 500, Kontron) was used to analyze the EFTEM images and/or TEM bright field micrographs. For each specimen an area of about 40 μm^2 was quantitatively analyzed.

For the conventional TEM measurements metal foils were prepared by mechanical cutting, grinding and chemical etching using a Tenupol-3 from Struers. Most of the analytical TEM investigations were performed on formvar extraction replica. Detailed information about the two applied preparation techniques is given in [9]. Each precipitation type was characterised by electron diffraction to determine the crystallographic structure and by EDX measurements to determine the chemical composition. EDX was done on approximately 80% of all particles visible in the TEM micrographs. The electron diffraction experiments were performed on selected particles using either a parallel beam for selected area diffraction or by convergent beam electron diffraction (CBED). A detailed description of the applied methods can be found in [10].

Test material

For our own investigations, we used a TAF test melt from Saarschmiede in Völklingen, produced in the framework of the COST 501 research project. Long-term creep tests have been performed on this material. Some of the tested specimens were characterized in detail within the framework of the COST 536 project and the results of the experiments at 650 °C are compared with the results of the COST steels B0 and FB8, which are different in the B, Cr, Co and C content as well as in the heat treatment condition. The chemical composition of the investigated melts of the TAF, the B0 and the FB8 alloys are given in Table 1.

The investigated creep specimen of the TAF steel has been tested at 650 °C with a load of 100 MPa which lead to failure after 26 931 hours. Additionally, the specimen head, which has seen the thermal load of 650 °C for the test duration of 26931 h, but no mechanical load, has been investigated. As no initial material was available for the microstructural characterization of the TAF specimen, the original heat treatment has been applied again to the second specimen head. The hardness values obtained after heat treatment are in accordance with the tensile strength during the initial tests, which suggests that the initial state was successfully reproduced.

The investigated B0 material has been tested at 650 °C for 18 788 h with a load of 80 Mpa and the FB8 steel specimen has been tested at 650 °C for 19436 h with a load of 70 MPa. The heat treatment data and the 0.2 limit at RT after heat treatment of the samples TAF, B0 and FB8 are presented along with the creep data for these three alloys in Figure 1. The heat treatment of the TAF steel is typical for bars that are used for the manufacturing of blades and bolts. The heat treatment of the Steels B0 and FB8 represents the heat treatment procedure of large forgings.

Results

Microstructure

The initial state of the TAF steel as observed by OM is shown along with the creep tested condition in Figure 2. This steel has a ferritic-martensitic microstructure. Prior austenite grain boundaries and martensite laths can be observed. The martensitic lath structure and the grain size did not change visibly. For the creep tested specimen a slightly enlarged grain size was observed along with several creep pores. Similar investigations showed no visible influence of the thermal exposure on the microstructure. For these investigations, electron microscopy methods are needed.

The former austenite boundaries and also to a lesser extent the subgrain boundaries are decorated with precipitates, which can be observed by SEM (Figure 3). The particles were analyzed by EDX in the SEM and most of them show Cr enrichment compared to the neighboring matrix. Additionally, several primary niobium containing particles were observed. Large Laves phase particles containing mainly iron and molybdenum were found in the creep tested and also in the thermally loaded condition. The thermal load did influence the precipitates in the material, but for a quantification of the effect, and a comparison to the influence of creep experiments on the microstructure and the precipitation state, transmission electron microscopy investigations are a necessity.

Figure 4 shows a montage of four low magnification TEM micrographs of an extraction replica of the TAF material in the initial state. By comparison with figure 3, one can see that the density and spatial distribution of the precipitates has been well preserved during the preparation of the replica. The analysis of the particle sizes and size distributions has been carried out at extraction replica. For the analysis of the subgrain sizes and the dislocation densities metal foils have been investigated in the TEM (Figure 5).

The subgrain size and shape of the TAF steel was not significantly influenced by the exposure at the test temperature. The subgrains remained elongated with an average size of about 300 nm. In contrast, during the creep test not only the shape of the subgrains changed, but also the size increased by a factor of almost 2 (Figure 5b). The B0 showed very comparable evolution of the subgrain size, but the subgrains have grown faster in the case of the FB8 material, where the final subgrain size reached 750 nm. The dislocation density after creep was very similar for all three investigated steels, but the TAF material had a significantly higher initial dislocation density. During the thermal exposure at test temperature the dislocation density of the TAF material decreased, but slower as during creep. The results of the analysis of the subgrain size and the dislocation density have been summarized in figure 6a. The evolution of these parameters with increasing strain are given in figure 6b and 6c. These diagrams clearly show the importance of plastic strain for the increase of the subgrain size.

Precipitation analysis

The analysis of the particle sizes and size distributions has been carried out at extraction replica. In Figure 7 a TEM bright-field image and the corresponding elemental map showing the chromium (red) and the vanadium (green) distribution of the particles of the FB8 steel after creep are depicted. The Cr-rich particles have been identified by EDX and electron diffraction experiments as $M_{23}C_6$ phase and the V-rich particles as MX-phase. A common occurrence of Cr and V appears yellow and can indicate particles of the M_2X type or Z-Phase. In the case of Figure 7b, the particles have bees identified as Z-Phase particles by EDX measurements and by selected area diffraction. Some particles are not visible in the two elemental maps. These particles have all been analysed by EDX and can be attributed to either Nb(C, N), or to the Laves phase (particles containing mainly iron and molybdenum; present only after thermal exposure or creep testing). The particle sizes and their distributions have been analysed from 10 of these EFTEM images, corresponding to an analysed area of 40 μm^2, for each condition.

The results of the quantitative precipitation analysis of the three different steels are presented in Figure 8. The fractions of the different precipitate types observed are given in relation to the total precipitation number. In all samples, the most prevalent precipitates are from the $M_{23}C_6$ type. The initial size of these particles lies in the range of 75 nm for the TAF steel up to a relatively high number of 98 nm in the case of the FB8 steel. Of crucial importance for the creep resistance is the coarsening rate of these particles. In case of the TAF steel, they grew to 138 nm after 26 931 hours. This slow growth prevents the creep properties from degrading. A further positive effect on the creep strength is the dynamic precipitation of the fine VN particles during creep. This dynamic precipitation is also found in the specimen head, with only the thermal load applied. In this thermally loaded condition, an increase of the $M_{23}C_6$ particles diameter to a size of 107 nm after 26 931 hours at 650 °C was observed. The B0 steel exhibited a very comparable precipitation behaviour in relation to the TAF steel, but the absolute values are all slightly in favour of the TAF steel. The $M_{23}C_6$ particles in the B0 steel also had a low initial diameter of about 80 nm. The growth rate was higher and resulted in a particle diameter of 143 nm after 18 788 hours, more than 9 000 hours less than in the case of the TAF steel. In accordance with the results of the TAF steel, the B0 steel exhibits also a dynamic precipitation of VN particles, but also at a higher growth rate. The analysis of the FB8 steel yielded a different precipitation behaviour. The initial size of the $M_{23}C_6$ was relatively high with 98 nm, and they coarsened quickly. But of greater importance is the emergence of Z-phase in this steel. This relatively large phase consumed all fine MX particles during creep, which is very detrimental for the creep stability. In all steels, Laves-phase has been found after creep.

Figure 9a gives an overview of the growth rate of $M_{23}C_6$ at 650 °C in different steels developed within the COST projects along with data for the P92 steel [12]. According to this graph, both, the absolute size and the growth rate are quite low for the B0 steel, but even lower for the TAF test materials. This is usually attributed to the relatively high boron content of this material. The boron is built-in in the $M_{23}C_6$ particles and reduces their rate of Oswald ripening (e.g. [2, 13]).

Additionally to the slowly coarsening $M_{23}C_6$ precipitates, MX-type vanadium nitrides precipitate during the heat treatment and even more during creep of the TAF steels. The initial size and the growth of these particles are summarized in Figure 9b for the three test steels and other COST steels. The MX particles in the TAF steel exhibit the smallest initial size of less then 20 nm in the initial state and still only 41 nm after creep testing. Additionally, the amount of MX precipitates increased significantly during exposure and even more during the creep testing of the TAF material. These fine MX-type particles increase further the creep resistance of the material. The present niobium carbonitrides have no significant role in the strengthening of the material. The effect of the precipitation of the Laves phase on the creep properties is probably small due the low number and relatively large size of the precipitates in comparison to the $M_{23}C_6$ and VN precipitates. The same effects found for the TAF steel are also true for the B0 steel, but to a lesser extent. The main reason may be the lower boron content of this steel, which leads to a lower creep strength, but it also enhances the hot workability of the steel. The decrease of the mean diameter in the cases of FB8 and FB6 is due to the nucleation of the Z-phase in these steels. This Z-phase coarsens quickly and consumes the fine MX particles in this process. In the case of the CB6 steel, the slight decrease in MX size can also be attributed to the beginning nucleation of the Z-phase (fraction 2 %).

Conclusions

It is on one side the small initial size and slow coarsening of the $M_{23}C_6$ precipitates and on the other side the dynamic precipitation of small V(C, N) particles along with the absence of Z-phase that strengthen the more creep resistant TAF steel during creep. The origin of the slow coarsening of the $M_{23}C_6$ precipitates can be mainly attributed to the high boron content of this material. The reason for the dynamic precipitation of the V(C, N) particles could be the relatively short tempering time of 2 hours at 700 °C and the high austenitizing temperature of the TAF material (1150 °C) which promote probably the dynamic precipitation during the creep process.

The B0 steel exhibits a comparable evolution of the microstructure, but always with slightly lessened properties than the TAF material with regard to creep resistance. Higher initial precipitation diameters and a faster growth rate (although both at very good values) lead to lower creep strength as evidenced by Figure 1.

The 11.1 wt.% Cr FB8 steel, alloyed in addition with 2.94 % Co, reveals more coarsening of the $M_{23}C_6$ precipitates and forms Z-phase during the creep process in expense of the beneficial fine MX particles. Both effects are responsible for the low long time creep strength. These worse creep properties are also supported by the faster increase of the subgrain sizes, which indicate that they have been less stabilized than in the case of the TAF and the B0 steels.

The dislocation densities did not yield quantitative information on the creep resistance, but the very high initial value for the TAF material is in good correlation to the high yield strength and short annealing times of this steel. At the time of failure, all three alloys had about the same dislocation density.

Acknowledgement

The authors wish to thank their partners of the COST projects for the creep test specimens and Dr. Wang from the University of Darmstadt for the additional heat treatment for reproducing the initial state of the TAF material. Thanks are also extended to the German COST partners and to the German Government for the financial support of the investigations (project number 0327705B).

Literature

1. T.-U. Kern, K.H. Mayer, C. Berger, G. Zies, M. Schwienheer: „Stand der Entwicklungsarbeiten in COST 522 für Hochtemperatur-Dampfturbinen," 27. Vortragsveranstaltung FVHT, VDEH, 26. 11.2004, Düsseldorf.

2. F. Abe: „Alloy Design of Creep and Oxidation Resistant 9Cr Steels for Thick Section Boiler Components Operating at 650°C", 4th EPRI International Conference on Advanced in Materials Technology for Fossil Power Plants", Hilton Oceanfront Resort, Hilton Head Island, SC, USA, October 25-28, 2004, 273-283

3. K. Sakuraya, H. Okada, F. Abe: „Coarse Size BN Type Inclusions formed in Boron Bearing High Cr Ferritic Heat Resistant Steels", 4th EPRI International International Conference on Advanced in Materials Technology for Fossil Power Plants", Hilton Oceanfront Resort, Hilton Head Island, SC, USA, October 25-28, 2004, 1325-1338

4. T. Fujita, N. Takahashi: The Effect of Boron on the Long Period Creep Rupture Strength of the modified 12%Cr heat-resisting Steel", Transaction of ISIJ 18, 1978, 702-711

5. T. Fujita: „Twenty-first Century Electricity Generation Plants and Materials", International Workshop of Advanced Heat-resisting Steels, Yokohama, Japan, 8. November 1999

6. I-M. Park, T. Fujita: „Long Term Creep Rupture Properties and Microstructure of 12% Cr Heat resisting Steels", Transaction of ISIJ 22, 1982, 830-837

7. F. Hofer, P. Warbichler, W. Grogger: „Imaging of nanometer-sized precipitates in solids by electron spectroscopic imaging," Ultramicroscopy 59, 1995, 15-31

8. F. Hofer, W. Grogger, G. Kothleitner, P. Warbichler: „Quantitative analysis of EFTEM elemental distribution images," Ultramicroscopy 67 (1997) 83-103

9. J.S. Brammer, M.A.P. Dewey: „Specimen Preparation for Electron Metallography," Blackwell Scientific Publications, Oxford, 1966

10. C. Scheu, F. Kauffmann, G. Zies, K. Maile, S. Straub, K.H. Mayer: „Requirements for microstructural investigations of steels used in modern power plants," Metallkunde 06, 2005, 653-659

11. G. Zies, K. Maile, A. Klenk, S. Straub and K.H. Mayer: "Determination of Microstructural Parameters Influencing Creep Behaviour of 9-12%Cr Steels", 27. MPA-Seminar, Stuttgart, 4. and 5. October 2001

12. K.H. Mayer: „Vergleichende Bewertung des Einflusses der Mikrostruktur auf die Kriechfestigkeit der borlegierten COST-Stähle mit erhöhtem Cr- und Co-Gehalt," Interner COST 536-Bericht, 14. 9. 2004

13. L. Lundin: „High Resolution Microanalysis of Creep Resistant 9-12%Cr Steels," Doctoral Thesis for the Degree of Doctor of Philosophy, Department of Physics, Chalmers University of Technology and Gothenbourg University, Sweden, 1995

Table 1. The chemical composition of the TAF test melt F35 by Saarschmiede and of the B0 and the FB8 COST test melts.

Test melt	C	Si	Mn	Cr	Co	Mo	Al	Ni	V	Nb	N	B
TAF (F35)	.21	.33	.87	10.5	-	1.54	.014	<.02	.24	.18	.017	.03
B0	.14	.20	.53	9.2	-	1.43	.008	.11	.23	.06	<.02	.018
FB8	.17	.09	.09	11.1	2.94	1.46	.005	.20	.21	.07	.023	.010

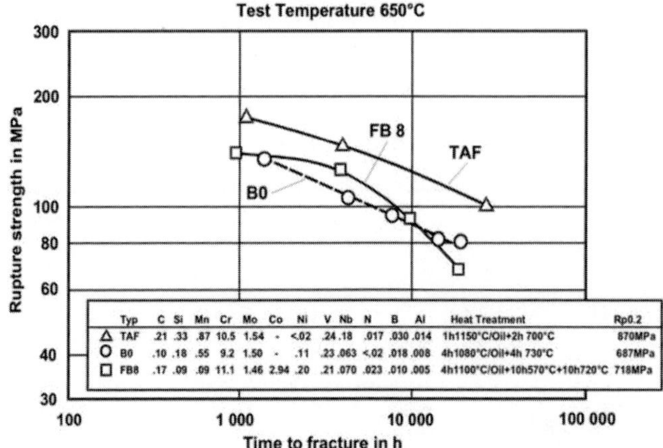

	Typ	C	Si	Mn	Cr	Mo	Co	Ni	V	Nb	N	B	Al	Heat Treatment	Rp0.2
△	TAF	.21	.33	.87	10.5	1.54	-	<.02	.24	.18	.017	.030	.014	1h1150°C/Oil+2h 700°C	870MPa
○	B0	.10	.18	.55	9.2	1.50	-	.11	.23	.063	<.02	.018	.008	4h1080°C/Oil+4h 730°C	687MPa
□	FB8	.17	.09	.09	11.1	1.46	2.94	.20	.21	.070	.023	.010	.005	4h1100°C/Oil+10h570°C+10h720°C	718MPa

Figure 1. Creep resistance of the TAF, the B0 and the FB8 steels tested at 650 °C.

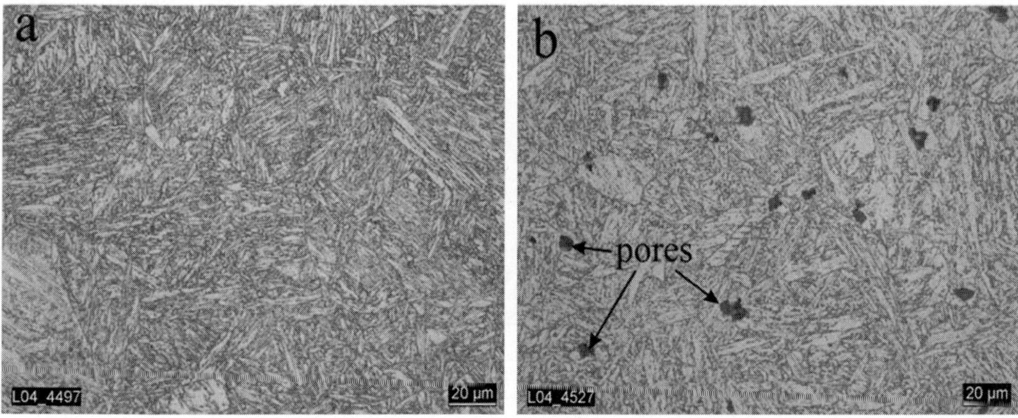

Figure 2. Optical microscope images of the TAF steel in the initial state and after creep.

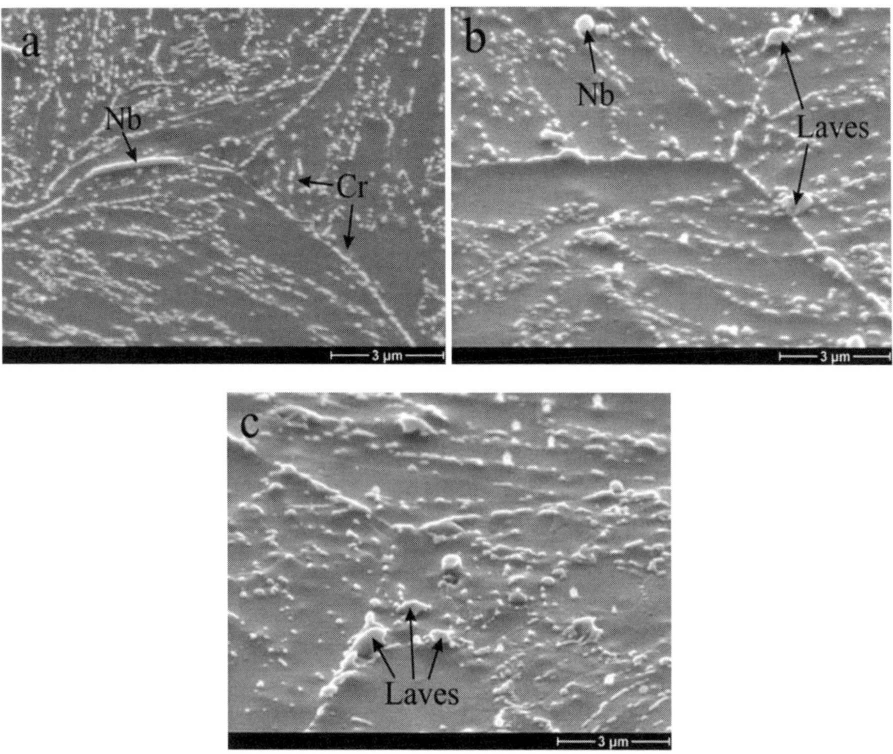

Figure 3. Scanning electron microscopy images of the TAF steel in the initial state (a), after thermal exposure (b) and after creep (c)

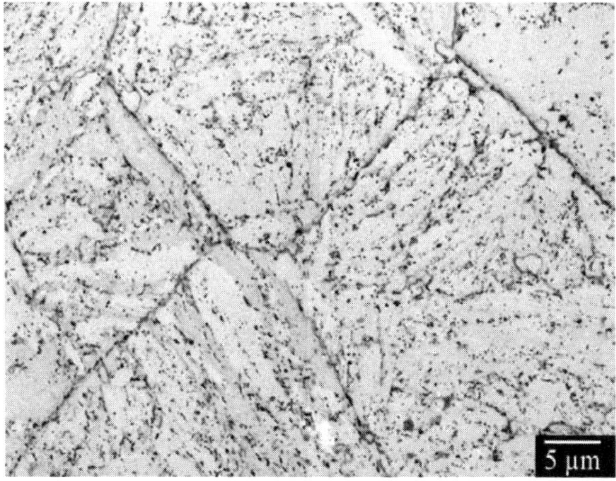

Figure 4. Transmission electron microscopy image of a extraction replica sample of the TAF steel after creep.

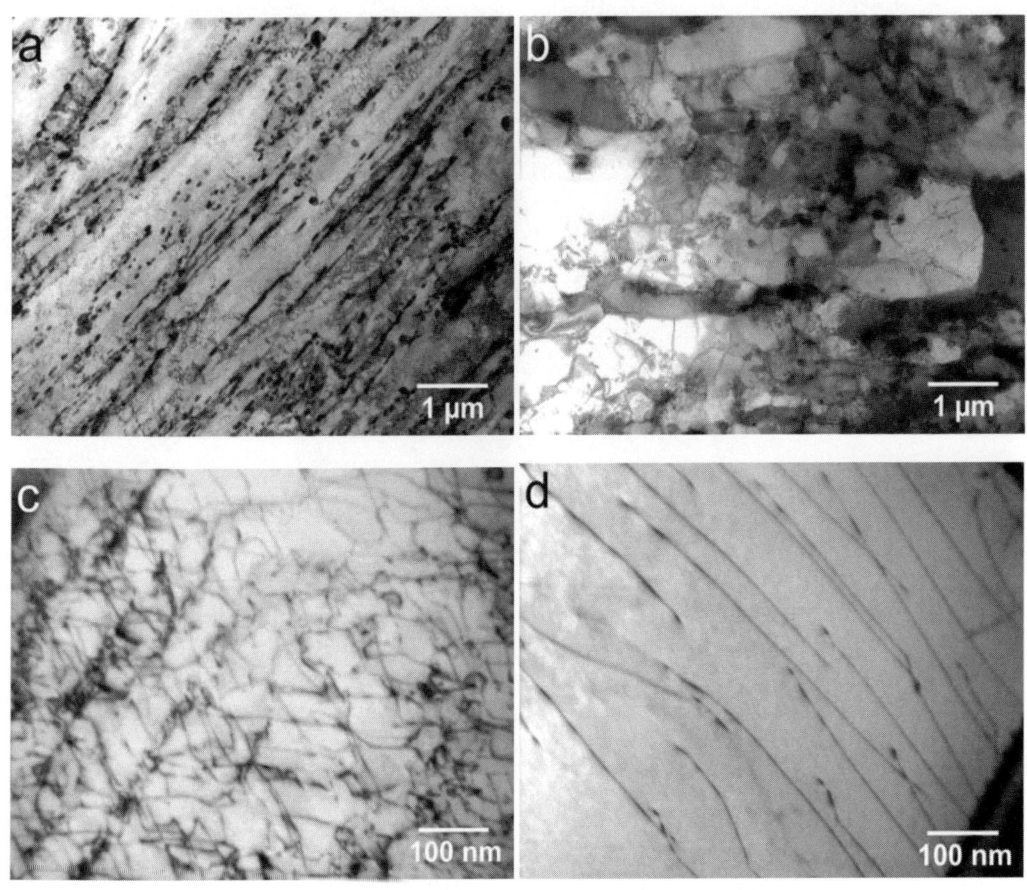

Figure 5. Transmission electron microscopy images of the investigated metal foils of the TAF steel prior (a,c) and post (b,d) creep

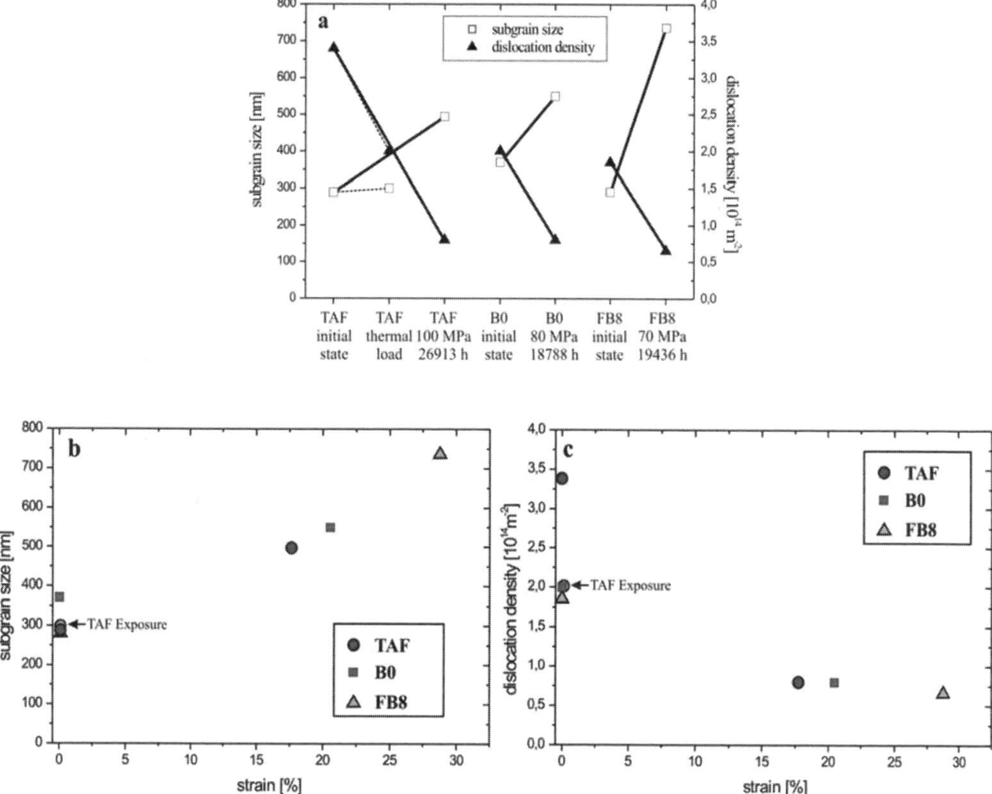

Figure 6. Subgrain sizes and dislocation densities of the investigated steels in relation to creep conditions and strain.

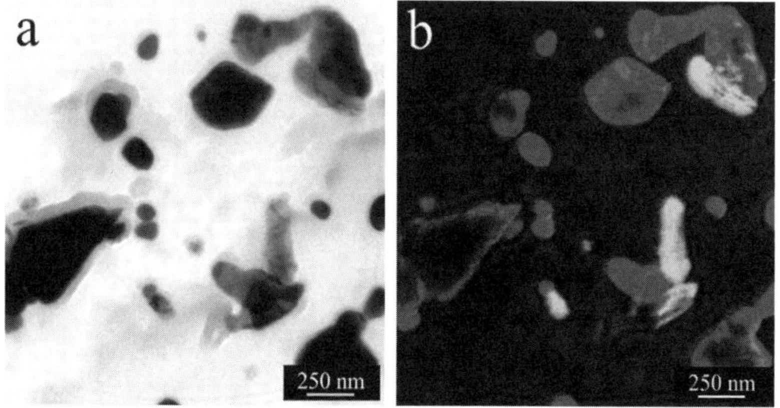

Figure 7. TEM bright field image and elemental distribution map (red: Cr; green:V) of the FB8 steel after creep

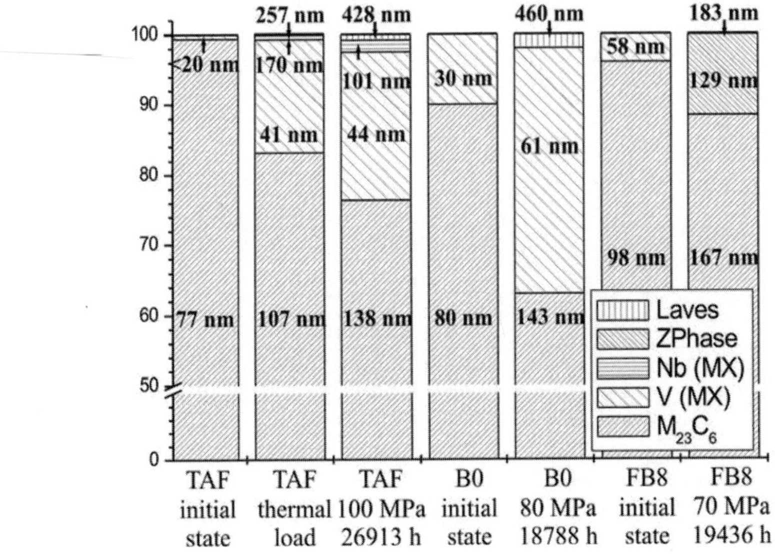

Figure 8. Precipitations smaller than 1 μm present in the investigated steels.

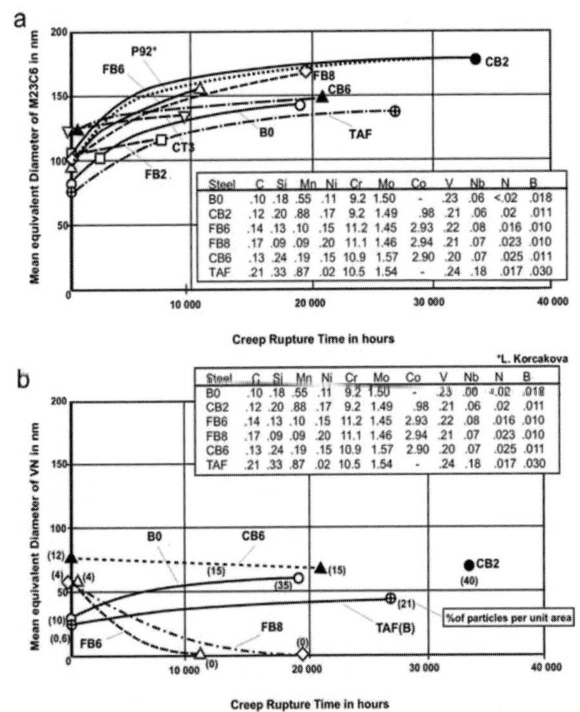

Figure 9. Evolution of the particle size with creep time for $M_{23}C_6$ (a) and MX (b) particles at 650 °C

Advances in Materials Technology for Fossil Power Plants
Proceedings from the Fifth International Conference
R. Viswanathan, D. Gandy, K. Coleman, editors, p 689-701

DOI: 10.1361/cp2007epri0689

Towards A Standard for Creep Fatigue Testing

S R Holdsworth
EMPA, Swiss Federal Laboratories
for Materials Testing & Research
Überlandstrasse 129
CH-8600 Dübendorf
Switzerland

D. Gandy
EPRI
1300 West WT Harris Blvd
Charlotte, NC 28262
USA

ABSTRACT

Procedures for the assessment of components subject to cyclic loading at high temperature require material property input data which characterise creep-fatigue deformation response and resistance to cracking. While there are a number of standards and codes of practice defining test procedures for the acquisition of low cycle fatigue (LCF) and creep properties, there is no formal guidance for the determination of creep-fatigue data. The paper reviews the results of a worldwide survey which has been conducted by EPRI to underpin the content of a new draft testing procedure in preparation for submission to ASTM and ultimately to ISO standards committees.

The survey comprised a review of the guidance contained in all relevant national and international standards, and the responses to a questionnaire circulated to high temperature testing specialists in Europe, North America and Japan. Standards relating to the calibration of load, extension and temperature measuring devices, were also reviewed. The questionnaire responses provided feedback relating to testpiece geometry and test machine, the control and measurement of load, extension and temperature, and data acquisition practices. The paper provides background to the guidance under consideration for inclusion in the new standard.

INTRODUCTION

Creep-fatigue testing is typically performed at elevated temperatures and involves the sequential or simultaneous application of the loading conditions necessary to generate cyclic deformation/damage enhanced by creep deformation/damage and (or) vice versa, with the purpose of such tests being to determine material properties for:

a) assessment input data for the deformation and damage condition analysis of engineering structures operating at elevated temperatures
b) material characterisation (ranking), and/or
c) the verification of constitutive deformation and damage model effectiveness.

The requirements of *a)* to *c)* may be satisfied by a test procedure prescribing for example a standard testpiece geometry and cycle waveform, and a standard crack initiation criterion. However if the results for *c)* are required to be related to a specific practical application, in principle a more generic testing standard is needed. These potentially contrasting requirements are considered in the following paper.

While there are a number of testing standards and codes of practice which cover the determination of low cycle fatigue (LCF) deformation and cyclic endurance properties, some of which provide guidance for testing at high temperatures (1-9), there is no single standard which specifically prescribes a procedure for creep-fatigue testing. The rectification of this omission was taken as an action item from an EPRI workshop on Creep-Fatigue Damage Interaction which took place in Amsterdam in July-2006 (10).

The following paper reviews the results of a worldwide survey of creep-fatigue testing practices conducted in advance of, and to underpin a new draft testing procedure in preparation for submission to ASTM and ultimately to ISO standards committees.

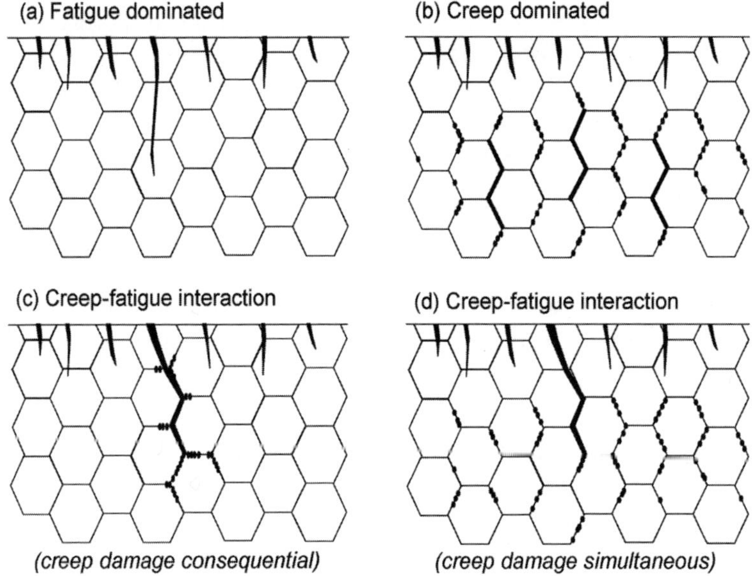

Figure 1. Creep-fatigue failure mechanisms: (a) Fatigue dominated, (b) Creep dominated, (c) Creep-fatigue interaction (due to consequential creep damage accumulation), (d) Creep-fatigue interaction (due to simultaneous creep damage accumulation)

CREEP-FATIGUE TEST

The development of creep-fatigue damage is influenced by temperature, strain amplitude, strain rate and hold time, and the creep strength and ductility of the material. For example, in the absence of a significant hold time (and/or at relatively high strain rates) in tests on steels, crack initiation and growth is fatigue-dominated, even at temperatures high in the creep regime for the material (e.g. Figure 1a). With increasing hold time (and/or decreasing strain rate) and decreasing strain amplitude at high temperatures, the creep damage condition within the testpiece becomes increasingly influential, to the limit beyond which crack development becomes fully creep-dominated (e.g. Figure 1b). At intermediate hold times and strain amplitudes, fatigue cracking interacts with creep damage developing consequentially or simultaneously resulting in accelerated crack growth (e.g. Figure 1c,d), the extent of any interaction increasing with decreasing creep ductility. Creep-fatigue tests are performed to characterise the deformation behaviour and crack initiation endurances of materials subject to loading conditions responsible for an interaction between creep and fatigue deformation/damage.

SURVEY

Prior to preparing a first draft standard document, a questionnaire was circulated to high temperature fatigue testing specialists in Europe, North America and Japan to survey current creep-fatigue testing practices concerning:
- the types of test employed
- testpiece geometries and dimensions
- testing machines and loading
- strain measurement
- temperature measurement, and
- data acquisition

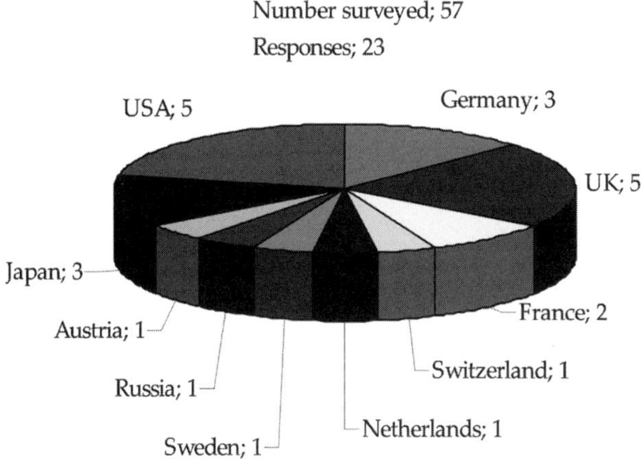

Figure 2. Distribution of responses to survey

The questionnaire was circulated to 57 specialists in 13 countries. 23 responses were received from the countries shown in Figure 2.

Type of Test

Creep-fatigue damage may be generated in tests involving sequential blocks of creep and fatigue loading, e.g. (11). However from the results of the survey, it was more common to apply a waveform shape responsible for the generation of both static and transient loading within the same cycle, e.g. Figure 3. Creep-fatigue tests were performed in both load and strain control, although more commonly in strain control (Table 1).

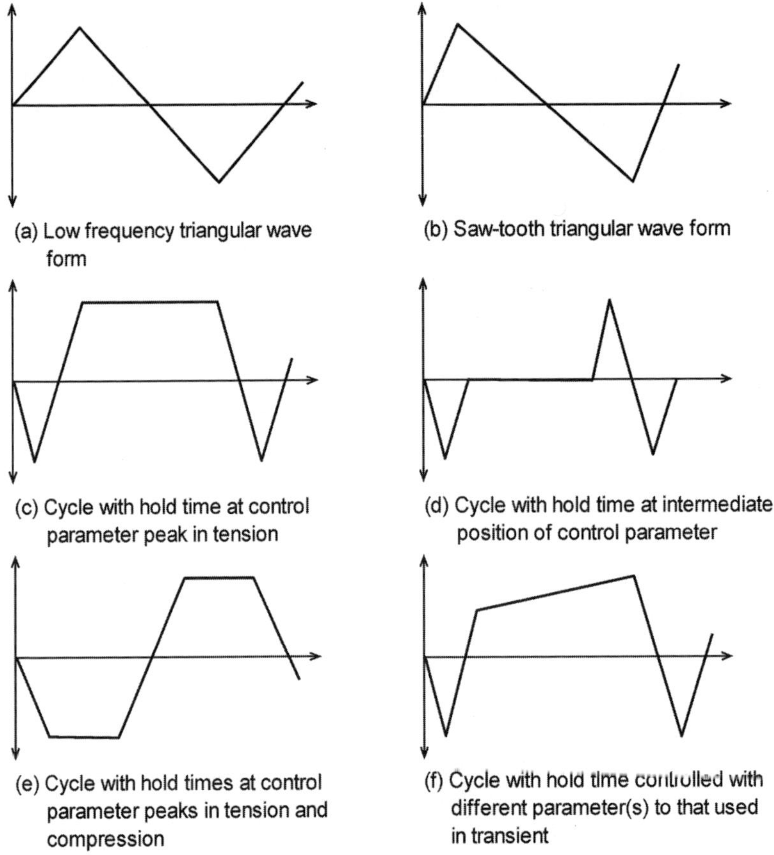

Figure 3. Examples of creep-fatigue cycle waveforms

The most commonly adopted creep-fatigue waveform was a cycle involving one or more hold times (e.g. Figure 3c-f), where hold periods could be anything between 1min and 24h (with an extreme case of 90 days).

With the recent improvements in test machine technology and digital control systems, thermo-mechanical fatigue (TMF) testing is increasingly being adopted for characterising material response to creep-fatigue loading. Procedures for TMF testing are well covered by existing standards (12,13).

Table 1: Survey Indicated Summary of Waveform Usage for Creep-Fatigue Testing

Waveform	Load Control (User %)	Strain Control (User %)
Low frequency triangular (isothermal)	14	59
Saw tooth triangular (isothermal)	27	68
Cyclic hold (isothermal)	32	86
Thermo-mechanical fatigue (TMF) (without and with hold time)	14	68

Testpieces (Specimens)

The most widely used testpiece type was the uniform parallel gauge section specimen (without ridges for extensometer fixation), Figure 4. All questionnaire respondees employed a version of this testpiece configuration, although other specimen types were used for special circumstances (see below). However, while there was a general consensus concerning testpiece type, predictably a range of gauge section dimensions and end connections were employed.

Gauge section diameters (*d*) were typically in the range 5-10mm, with respondees often adopting more than one size to cover a range of applications. Gauge lengths (l_o) were in the range 8-25mm, but typically between 12 and 15mm. Adopted l_o/d ratios varied between 2 and 4, with the upper limit being determined by users wishing to avoid buckling during compression loading.

Examples of grip-end connection types are shown in Figure 4 and include:
- screw threaded connections (77%)
- button-end connections (36%), and
- smooth cylindrical connections (for use with hydraulic jaws) (50%)

For testpieces subject to through-zero loading, threaded and button-ended grip arrangements invariably incorporated features to ensure a smooth transition from tension to compression and vice-versa.

In addition to uniform parallel gauge section testpieces, hour-glass, tubular and notched geometries were also employed. Hour-glass testpieces were generally only used for high strain amplitude tests when the risk of buckling a parallel gauge section testpiece was high. The use of a diametral extensometer was required to control or monitor strain with this testpiece geometry.

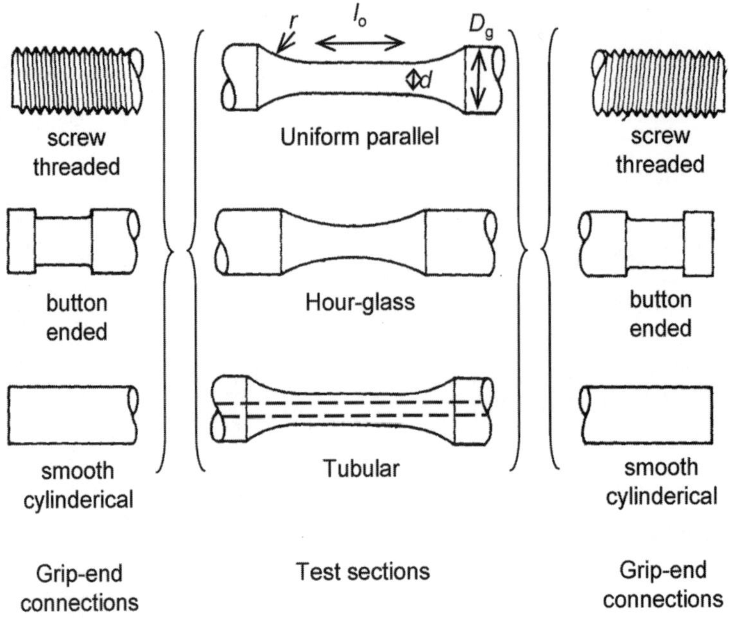

Figure 4. Testpiece configurations

A variant to the uniform parallel gauge section testpiece was the tubular testpiece. As a generality, such testpieces were only used for creep-fatigue testing to minimize thermal gradients when heating was by induction, e.g. in TMF testing (12,13).

Circumferential notched round bar testpieces may be used to determine the effect of triaxiality on creep-fatigue crack initiation endurance or to measure the rate of short crack growth from an application related stress concentrating feature.

Testing Machine

Machine Type. Servo-controlled tension-compression testing machines involving hydraulic or electro-mechanical based drive systems were commonly used, with respondees to the questionnaire indicating a wider use of hydraulic machines (82%) compared with electro-mechanical machines (64%). Some laboratories used both types of servo-controlled machine depending on the cycle type under investigation. Electro-mechanical machines perform well at low frequencies, but increasingly hydraulic machines are being used for low frequency as well as higher frequency testing.

Load Cell Classification. Existing fatigue (and creep) testing standards required that load cells conform to the Class 1 requirements of standards such as ISO 7500-1 (14). Consequently most load cells were calibrated to meet these requirements, although a small number of respondees (10%) chose to calibrate to Class 0.5.

Alignment. Testpiece bending due to misalignment is generally caused by:
- an angular offset of the testpiece grips
- a lateral offset of the loading bars (or testpiece grips) in an ideally-rigid system, and/or
- an offset in the load-train assembly with respect to a non-rigid system.

Most respondees conducted alignment checks according to one of the published practices (e.g. (15,16)) or to their own internal procedure. The check was typically performed every 12 months, or after an extraordinary loading incident or before the start of a new test series. A small number of laboratories (10%) checked alignment more frequently (i.e. every 6 months).

Strain Measurement

Extensometer Type. All respondees used side entry contact extensometers to control or monitor strain. Axial extensometers mounted to ridges on the testpiece gauge length were no longer used except by one respondee, and only then in certain circumstances for deformation testing and never for crack initiation endurance testing. Diametral extensometers were used by 3 respondees but only for strain measurement and control with hour glass testpieces or notched testpiece geometries.

Classification. Axial extensometers were most commonly calibrated to at least meet the Class 0.5 requirements of ISO 9513 (17) or equivalent. Two respondees calibrated to Class 1 while one calibrated to Class 0.2. Diametral extensometers could only be calibrated to Class 2. The recommendation in the new standard will be for a minimum axial extensometer requirement of Class 0.5.

Temperature Measurement and Control

Heating equipment. A range of heating systems was employed by the respondees, including:
- resistance furnace (59%),
- radiant furnace (14%),
- resistance heating (5%),
- induction heating (64%), and
- inert gas or liquid heating (5%)

Most commonly, induction heating and resistance furnace heating were used, often by the same respondees when their laboratories conducted both TMF and isothermal tests.

Temperature sensing devices. Four types of thermocouple were commonly employed for temperature measurement and control, i.e.
- Type K - Nickel-chromium/nickel-aluminium (59%)
- Type N - Nickel-chromium-silicon/nickel-silicon (nicrosil/nisil) (18%)

- Type R – Platinum-13%rhodium/platinum (50%)
- Type S – Platinum-10%rhodium/platinum (23%)

While Type K thermocouples were widely used, their repeated use is not recommended, in particular for tests above 400°C.

Thermocouples were either tied with heat resistant cord to the gauge section or spot welded to the blend radii or testpiece shoulders. Thermocouples were never spot welded to the gauge section of testpieces to be used for crack initiation endurance testing to avoid premature cracking. In one case, temperature measurement was by means of embedded thermocouples located in axial centre holes.

One respondee referred to the use of pyrometry. However, optical pyrometers are not usually recommended for testpiece temperature measurement, without the use of supportive observations from thermocouples attached to the shoulders, in particular when the test material is prone to oxidation.

Less than one third of the respondees referred to sensing device calibration practice. Of these, two employed an external organisation to confirm traceability to the international unit (SI) of temperature (18a), while two used internal laboratory procedures. The remaining three relied on the quality assurance of the thermocouple manufacturers (implying no recalibration strategy).

Temperature Tolerances. A wide spectrum of tolerances was adopted for the control of temperature *i)* at a single gauge section location during the course of test, and *ii)* along the gauge length at any given time during test. Adopted tolerances, where cited, ranged from ≤±1°C to ≤±10°C, with values in the range ≤±2°C to $\leq\pm0.01.T_{max}$ being the most frequent. Significantly, no qualification was given of whether the values cited were instrument indicated temperatures only or temperatures including all uncertainties.

On the basis of this feedback, it was concluded that temperature measurement and control was a topic requiring specific attention in the new standard.

Test Results

Crack Initiation Criteria. A range of crack initiation (failure) criteria were adopted, varying between 2% and 25% reductions in steady state maximum stress. Notably, the most commonly adopted criteria was a 10% reduction in maximum stress (45%) compared with the anticipated outcome of a 2% reduction (22%).

Data Acquisition. With one exception, all respondees now gathered their creep-fatigue test data by digital recording. It is normal practice to record and archive peak and user-defined key-event values of stress and strain for all cycles. However, full stress-strain hysteresis loop records are usually only retained for selected cycles based on a linear or logarithmic sampling strategy. The number of data points saved to characterize respective stress-strain hysteresis loops was surveyed in the questionnaire.

There was consensus concerning the number of data points archived to record the cyclic part of a given stress-strain hysteresis loop. Irrespective of how many co-ordinates were recorded during test, almost all respondees permanently stored ~200 data points per cycle.

The practices adopted for hold time data storage were more varied, ranging from 200-300 data points to the total from a collection rate of 1-2 per sec. (apparently without reduction), irrespective of hold time duration. Original data collection strategies (prior to data reduction) varied from simply gathering $t(\sigma,\varepsilon)$ co-ordinates at a uniform rate to 'logarithmic + level-crossing'.

DISCUSSION

Of the three purposes identified in the Introduction for which creep-fatigue testing is performed, a fully prescriptive procedure recommending for example a standard testpiece geometry, a common waveform and a single crack initiation criterion, could be proposed for the determination of material properties for:

a) assessment input data for the deformation and damage condition analysis of engineering structures operating at elevated temperatures, and

b) material comparison (ranking).

Such a standard procedure could also cover requirements for the determination of material properties for:

c) the verification of constitutive and damage model effectiveness,

although in practice the information required for this purpose is usually service application specific. This is the main reason for the large variety of creep-fatigue waveforms adopted worldwide (only a small sample of which is given in Figure 3). For many users, creep-fatigue tests are a form of benchmark test. Consequently for *c)*, it may therefore only be possible to specify generic test conditions. For example, such details as those relating to cycle shape, strain rates, and the duration and position within the cycle of hold time(s) are likely to be dictated by the service conditions for which verification is required.

In preparing a creep-fatigue testing procedure, it will be necessary to clarify whether the scope is to provide specific guidance for the determination of standard material property data or to provide more generic guidance for a wide range of standard and benchmark creep-fatigue testing practices.

The results of the survey overwhelmingly indicated that the subject requiring the most attention in a new standard concerned specifically with creep-fatigue testing is that of temperature measurement and control.

Temperature is debatably the most important control parameter in creep- fatigue testing. A significant difference between existing standard procedure documents covering creep and stress relaxation testing, e.g. (19,20), and LCF testing at elevated temperatures, e.g. (1-6), concerns the temperature control requirements in terms of *i)* variation at a single gauge section location during the course of test and *ii)* variation along the gauge length at any given time during test. Whereas the latest creep and stress relaxation testing standards require temperature tolerances to include all uncertainties, those for LCF testing (apart from (7)) do not.

There are a number of uncertainties associated with the temperature indicated on a digital display or gathered by the data acquisition system. These include:

- the calibration tolerance on the thermocouple (even when traceable to the international unit (SI) of temperature (18b)),
- thermocouple drift during use
- temperature difference between testpiece and thermocouple
- measurement device tolerance (including uncertainties due to electrical connections and their temperature sensitivity and cold junction compensation)

These can exceed ±2°C (21). Permissible indicated temperature variations of ±3°C can therefore mean actual temperature variations of ~±5°C.

For creep-fatigue testing, the temperature tolerances need to be more consistent with those specified for creep and stress relaxation testing, and tighter than those currently specified for LCF testing in international standards. Current creep and relaxation standards require temperature tolerances to be less than ±3°C (for $T \leq 600°C$), including all uncertainties, e.g. (19,20). In practice, this means a tolerable indicated temperature variation of ≤~±1°C.

A similar anomaly exists for measurement sensor calibration. Whereas creep and stress relaxation standards are prescriptive in terms of thermocouple calibration frequency, procedure and traceability to the international unit (SI) of temperature (18), existing LCF testing standards are not. It is therefore also strongly recommended that the new creep-fatigue testing standard considers thermocouple calibration practice to at least the same level of detail as that prescribed for creep and stress relaxation testing. There are undoubtedly practical difficulties associated with calibration of the spot-welded and ribbon type thermocouples used for creep-fatigue testing but these must be addressed.

With the almost universal use of digital data acquisition, it is essential that calibration of not only the measurement sensor but also the entire temperature measurement and control system is traceable to the international unit (SI) as defined by (18).

CONCLUDING REMARKS

In response to an identified user requirement for a creep-fatigue testing procedure, a worldwide survey of current practices has been conducted by EPRI as a pre-cursor to the preparation of a draft standard for consideration by ASTM and ultimately ISO standards committees. The results of this survey are reviewed.

In preparing a creep-fatigue testing procedure, it will be necessary to clarify whether the scope is to provide specific guidance for the determination of standard material property data or to provide more generic guidance for a wide range of standard and benchmark creep-fatigue testing practices.

A main consideration for a new standard procedure covering creep-fatigue testing concerns temperature measurement and control. In particular, temperature control tolerances must be more consistent with those currently required by the latest creep and stress relaxation standards. Importantly, calibration of the entire temperature measurement and control system must be traceable to the international unit (SI) as defined by (18).

ACKNOWLEDGEMENTS

The authors acknowledge the valuable contribution of those specialists who responded to the EPRI Creep-Fatigue Testing Questionnaire, the findings of which form the basis of this paper.

REFERENCES

1 G.B. Thomas, G.B. et al, 1989, "A code of practice for constant-amplitude low cycle fatigue testing at elevated temperature", Fatigue, Fract. of Engng. Mater. Struct., 1989, 12, 2, 135-153.

2 A 03-403: 1990, "Produits métalliques: Pratique des essais de fatigue oligocyclique", Normalisation Français.

3 BS 7270: 2000, "Method for constant amplitude strain controlled fatigue testing" British Standards Institution.

4 ASTM E606, 2004, "Standard practice for strain-controlled fatigue testing", ASTM Standards, Vol. 03.01.

5 PrEN 3874, 1998, "Test methods for metallic materials - constant amplitude force-controlled low cycle fatigue testing", European Standards, Aerospace series.

6 PrEN 3988, 1998, "Test methods for metallic materials - constant amplitude strain-controlled low cycle fatigue testing", European Standards, Aerospace series.

7 ISO 12106, 2003, "Metallic materials - Fatigue testing - Axial strain-controlled method", International Organisation for Standardisation.

8 M. Sakane & K. Yamaguchi, "High temperature low cycle fatigue standard testing - JSMS recommendation", Proc. 4th Japan-China Bilateral Symp. on High Temperature Strength of Materials, NIMS, Tsukuba, 11/13- June-2001, 179-184.

9 R. Hales et al., "A code of practice for the determination and interpretation of cyclic stress-strain data", Materials at High Temperatures, 2002, 19, 4, 165-186.

10 D.W. Gandy, "Creep-fatigue damage", Proc. Expert Workshop, Amsterdam, 11/13-July-2006, EPRI Report 1014482, 2006, December.

11 M.M. Leven, "The interaction of creep and fatigue for a rotor steel", Experimental Mechanics, 1973, September, 353-372.

12 ASTM E2368, 2004, "Standard practice for strain controlled thermomechanical fatigue testing", ASTM Standards, Vol. 03.01.

13 P. Hähner et al., 2006, "Validated code-of-practice for strain-controlled thermo-mechanical fatigue testing", JRC EUR 22281 EN.

14 ISO 7500-1, 2004, "Metallic materials - Verification of static uniaxial testing machines - Part 1. Tension/compression testing machines - Verification and calibration of the force measuring system", International Standards Organisation.

15 Kandil, F.A., 1998, "Measurement of bending in uniaxial low cycle fatigue testing", NPL Measurement Good Practice Guide No. 1 NPLMMS001 (IBSN 0946754 16 0).

16 ASTM E1012, 1999, "Standard practice for verification of specimen alignment under tensile loading", ASTM Standards Vol.03.01.

17 EN ISO 9513, 1999, "Metallic materials - Calibration of extensometers used in axial testing", International Standards Organisation.

18 EN 60584-1, 1996, *a)* "Thermocouples - Reference tables (IEC 584-1)", *b)* "Thermocouples - Tolerances (IEC 584-2)", European Committee for Standardisation (CEN).

19 EN 10291, 2000, "Metallic materials - Uniaxial creep testing in tension - Methods of test", European Committee for Standardisation (CEN).

20 EN 10319, 2003, "Metallic materials - Tesnsile stress relaxation testing - Part 1: Procedure for testing machines", European Committee for Standardisation (CEN).

21 ECCC Recommendations Vol.3, 2005, "Acceptability criteria for creep, creep rupture, stress rupture and stress relation data", J. Granacher & S.R. Holdsworth (eds), publ. ECCC/ETD, *www.ommi.co.uk/etd/eccc/open.htm*.

Advances in Materials Technology for Fossil Power Plants
Proceedings from the Fifth International Conference
R. Viswanathan, D. Gandy, K. Coleman, editors, p 702-717

DOI: 10.1361/cp2007epri0702

ECCC Rupture Data for Austenitic Stainless Steels – Experiences Gained with Demanding Data Analyses

M W Spindler† and H Andersson*
†British Energy, Barnett Way, Barnwood, Gloucester, GL4 3RS, UK.
*KIMAB, Drottning Kristinas väg 48, SE-114 28 Stockholm, Sweden.

Abstract

The European Creep Collaborative Committee Working Group on Austenitic stainless steels (WG3B) has performed; (i) reviews of existing rupture strength values, (ii) data collations for existing and new grades of austenitic steels and (iii) new assessments of rupture strength for existing and new grades of austenitic steels. In particular, new creep rupture data assessments have been performed for TP316L, TP316, Alloy 800H, Alloy 800HT, 253MA, Esshete 1250, HR3C, NF709 and NF709R. A variety of different data analyses methods were applied by different assessors. Interestingly, most of these assessments encountered similar problems when attempting to simultaneously fit; high-stress, low-temperature data and low-stress, high-temperature data. These problems are similar to those encountered by others when assessing modified 9Cr and 12Cr alloys. This paper summarises WG3B's experience with such demanding datasets and offers a number of possible solutions to these difficulties.

INTRODUCTION

The European Creep Collaborative Committee, ECCC, Working Group on Austenitic Steels (WG3B) brings together European industries and research institutes, which have an interest in the creep properties of austenitic steels and weldments. During the first phase of ECCC (1992 to 1996) WG3B reviewed the creep rupture strength values contained in different national standards for the common grades of austenitic steels, the 300 series and Alloy 800 grades. The results of this review have been reported in [1], also WG3B performed collations of the available European, American and Japanese data on Types 316, 316L, Alloys 800H, 800HT and 253MA. During the second phase of ECCC (1997-2001) WG3B focused upon austenitic steel weldments and weld-metals and new austenitic steels for service at 650 and 700°C. The new higher strength austenitic steels that were considered included NF709, NF709R, Super 304H, HR3C and TP347HFG (fine grained). In addition, the currently available high strength austenitic steels Esshete 1250, AC66, Type 316L(N) and Type 316LNB were also considered. However, as with the first stage of ECCC only creep rupture strength was considered. The third phase of ECCC (2001-2005) "Advanced Creep" widened the range of properties covered and work was performed on creep ductility, creep deformation, cold work and multiaxial effects, in addition to continuing work on the rupture strength of new austenitic steels and their weldments. The work of WG3B on creep rupture strength has been published in the ECCC data sheets and the work on the advanced creep properties have been published in the scientific literature [2,3]. In this paper WG3B has taken the opportunity to describe the experiences gained whilst performing some of these creep rupture data assessments. In particular, most of these assessments encountered similar problems when attempting to simultaneously fit; high-stress, low-temperature data and low-stress, high-temperature data. These problems are similar to those encountered by others when assessing modified 9Cr and 12Cr alloys. This paper summarises WG3B's experience with such demanding datasets and offers a number of possible solutions to these difficulties.

Power stations currently use ferritic or martensitic creep resistant steels for steam generation at temperatures of up to 600°C, with only limited use of austenitic steels as boiler tubing. In order to raise the efficiency in future plant it is necessary to raise steam and metal temperatures. The steels used in the past (P22, 1CrMoV, 12CrMoV and P91) and the new modified 9Cr steels (P23, P92 and E911) neither have sufficient rupture strength nor are able to resist oxidation at temperatures above 650°C. The relative 100,000 hour creep strengths at 650°C of the austenitic steels are shown in Figure 1. In contrast Ni alloys such as Alloys 617, 740 and 263, which promise rupture strength and oxidation resistance at temperatures above 700°C, are increasingly expensive. Therefore, austenitic steels such as AC66, Esshete 1250, Type 347HFG, Super 304H, HR3C and NF709 will be increasingly used in both boiler tubes but also perhaps pipes and headers. Furthermore, newly developed austenitic stainless steels such as Save 25, Sanicro 25 and COST 522's BGA4 will be used once creep rupture data assessments have been preformed.

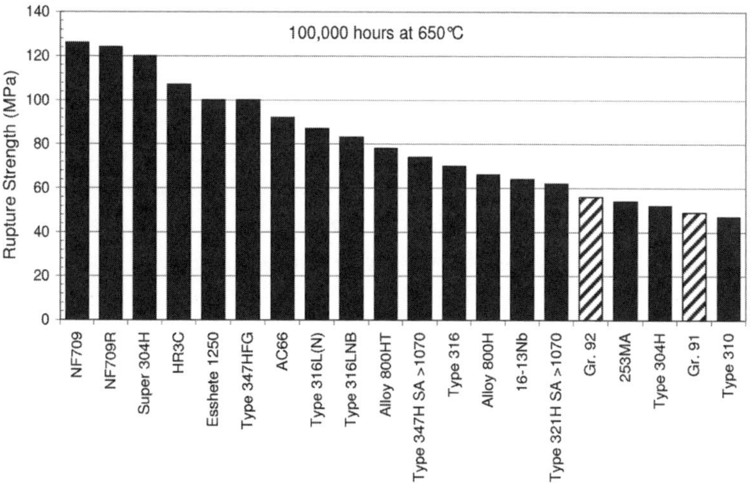

Figure 1 The 100,000 hour Creep Rupture Strength of Austenitic Stainless Steels.

Creep rupture data assessments of austenitic stainless steels

Type 316

One of the first creep rupture data assessments (CRDA) preformed by WG3B was for Type 316. This analysis was a very large (2255 failed data) multiheat (123 heats) data set with long term data at 50°C intervals from 500 to 800°C. The analysis was conducted using the PD6605 procedure [4]. This procedure uses the maximum likelihood method that utilises the information from unfailed tests, which are often associated with long duration tests. The models fitted include the Soviet models and Minimum Commitment Equation (A=0), which are parametric equations, and a number of polynomials including a simplified Mendelson-Roberts-Manson, Manson-Haferd, Larson-Miller and Orr-Sherby-Dorn. The most suitable creep rupture model was chosen on the basis of its deviance and the results of the ECCC Post Assessment Tests (PAT) 1.1, 1.2, 1.3 and a modified form of PAT 3.2 [5]. PATs 1.1 and 1.2 test the physical realism of the model, PAT 1.3 tests whether the strength falls away too quickly at low stress and PAT 3.2 tests the repeatability and stability of the model in extrapolation. For the Type 316 analysis the Soviet Model 1 (SM1) was chosen which is given by

$$\log_e(t_u^*) = \beta_0 + \beta_1 \log_{10}(T) + \beta_2 \log_{10}(\sigma_0) + \frac{\beta_3}{T} + \beta_4 \frac{\sigma_0}{T} \qquad (1)$$

where t_u^* is the predicted rupture time in hours, σ_0 is the stress in MPa, T is the temperature in Kelvin and β_0 to β_4 are the constants determined by the fitting process. In common with many PD6605 assessments a parametric equation was chosen rather than a polynomial equation. This is because polynomials tend to be unstable in extrapolation and often suffer from turn-back or points of inflection. However, the SM1 model and all of the other models suffered from a similar fault which was that the data at the lowest temperature, 500°C, were poorly fitted. In the case of the Type 316 data this occurred in the long term data with durations greater than 10,000 hours for which the SM1 model is non-conservative (see Figure 2). For design purposes this non-conservatism has no effect since at these low temperatures the design strength is determined by the tensile properties. However, rupture strength at 500°C is important to the life assessment of components operating at this temperature and since eq. (1) is non-conservative at 500°C the life assessments would be non-conservative as well. Within British Energy this has been overcome by performing a separate analysis of the creep rupture data at 500, 525 and 550°C alone. Nevertheless, it would clearly be preferable to have an analysis route that can simultaneously fit all of the data.

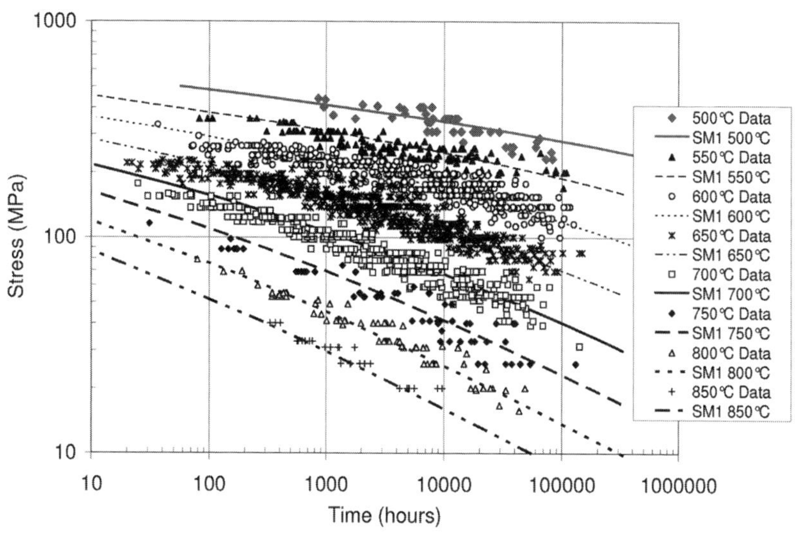

Figure 2 Creep Rupture Strength of Type 316

HR3C

The creep rupture data for HR3C were provided by Sumitomo Metal Industries Ltd, the data were at temperatures of 600 to 750°C. Three creep rupture data assessments were carried out. The first of the CRDA were carried out by Robertson [6] using the PD6605 procedure with the standard set of models as used for Type 316. As for the Type 316 assessment one of the parametric equations was chosen in preference to a polynomial, albeit in this case it was the Soviet Model 2 (SM2), which met the requirements of the PD6605 procedure. However, in the ECCC and PD6605 procedures, the model realism check (PAT 1.1) is somewhat subjective and although Robertson noted that the SM2 model did not fit the short term data at 600°C (see Figure 3), the SM2 model was chosen by Robertson [6] because it was the most stable in extrapolation. Nevertheless, all of the official ECCC and PD6605 PATs were passed. In particular, the goodness of fit tests PAT 2.1, which examines the predicted versus observed plots for all the data, and PAT 2.2, which examines this at the main test temperature (700°C), the minimum (650°C) and maximum (750°C) temperatures at which more than 10% of the data exist, were passed. However, it was noted that if PAT 2.2 were applied at 600°C, which is currently an important temperature for the use of these steels in supercritical coal stations then owing to the poor fit at this temperature then this test would have been failed.

The second assessment was by Schwienheer [7] and used the DESA procedure [8], in which the time temperature parameter can be chosen from the Larson-Miller, Sherby-Dorn, Manson-Haferd and Manson-Brown parameters and the polynomial can be based on a stress function of either $\log(\sigma_0)$ or $\sigma_0{}^m$, where m can be between 0.1 and 1. The model chosen by Schwienheer used the Manson-Brown time temperature parameter with a third order polynomial in stress to the power of 0.25 which is given by

$$\frac{\log_{10}(t_u^*) - \log ta}{[(T-G)/1000]^R} = \beta_0 + \beta_1 \sigma_0^{\,0.25} + \beta_2 \left(\sigma_0^{\,0.25}\right)^2 + \beta_3 \left(\sigma_0^{\,0.25}\right)^3 \tag{2}$$

where log*ta*, G and R are the constants in the time temperature parameter. The assessment made by Schwienheer [7] passed all of the ECCC PATs and gave a good fit the data at all temperatures, including 600°C (see Figure 3).

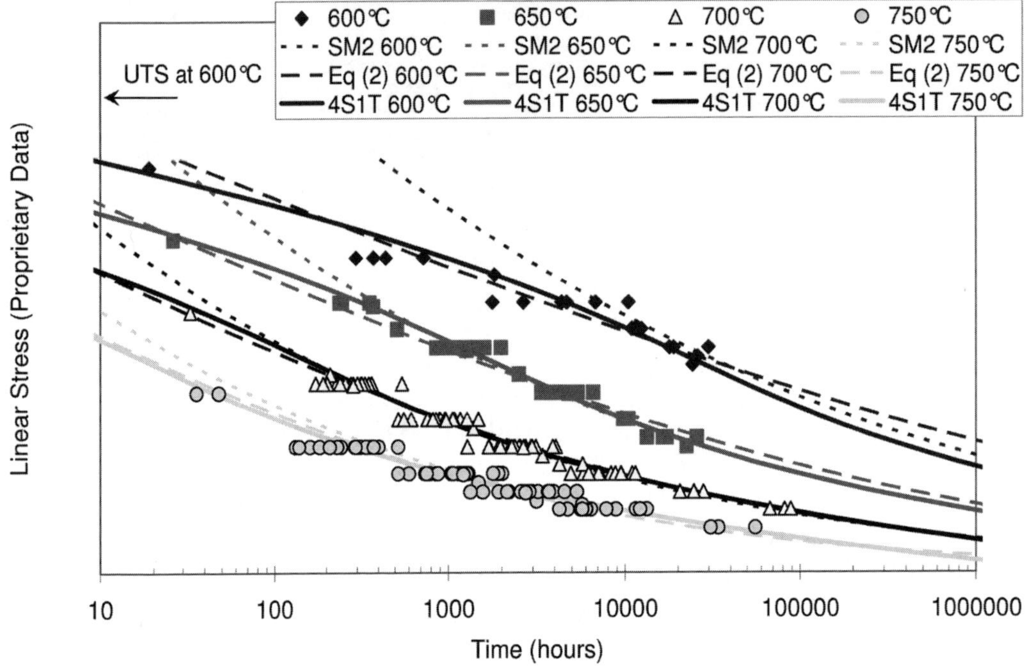

Figure 3 Creep Rupture Strength of HR3C

The third assessment [9] also used the PD6605 procedure but included eight additional user defined creep rupture models. The best of these eight models was chosen because of its goodness of fit and reliability on extrapolation. In particular, these user defined models examined whether polynomials in stress rather than log(σ_0) would give improved fits to the data. It was found that a polynomial in stress with a simple 1/T temperature term, which is given by

$$\log_e(t_u^*) = \beta_0 + \beta_1\sigma_0 + \beta_2\sigma_0^{\,2} + \beta_3\sigma_0^{\,3} + \beta_4\sigma_0^{\,4} + \beta_5/T \tag{3}$$

gave the most reliable fit in terms of both goodness of fit and stability in extrapolation. In particular, the simple change from the use of log(σ_0) to stress gave a good fit at 600°C (see Figure 3). The model given by equation (3) is refereed to as 4S1T, since it is a fourth order polynomial in stress and has one temperature term.

It was then necessary for WG3B to choose between the three assessments for the data sheet on HR3C. Since both equations (2) and (3) passed all of the PAT tests and gave similarly good fits at 600°C, and the other test temperatures, it was decided that the decision should be between these two equations only. The final choice being made to select equation (3) it was more conservative at 600 and 650°C (see Table 1), which are important temperatures for the commercial applications of HR3C.

Notwithstanding the differences between each of these assessments it is interesting to note that the predicted 100,000 hour rupture strengths from the two assessments at the main test temperature (700°C), the minimum (650°C) and maximum (750°C) temperatures are within less than 10% (see Table 1).

Table 1: Comparison Between the Creep Rupture Strength Values for HR3C Parent.

Temp.	100,000 hour Rupture Strength (MPa)		
(°C)	Schwienheer [7]	4S1T	Sumitomo Data Sheets
600	198	178.7	184
650	116	106.8	114
700	63.7	65.2	66
750	34	37.1	38

Esshete 1250

The creep rupture strength of Esshete 1250 has been analysed previously and is reported in PD6525 [10]. This analysis used the ISO 6303 procedure [11]. In addition, Baker [12] has performed a more recent analysis using the PD6605 procedure [4], which produced an Orr-Sherby-Dorn 4th order polynomial (OSD4). However, both of these assessments experienced difficulties with fitting the data over the whole temperature range. In the ISO 6303 assessment only the data at 650, 700, 750 and 800°C were used. However, the temperatures of interest to British Energy are in the range 550 to 600°C, thus the assessment by Baker included all of the data from 550 to 900°C. Nevertheless, the assessment by Baker also experienced difficulties in fitting the whole data set. In particular, at 550°C the long term strength appeared overly optimistic and at 650°C the model failed to describe the sigmoidal rupture behaviour (see Figure 4). Hence, a third assessment was conducted [3], although this assessment differed from the others in that it used a multi-region approach. This was done to overcome the two problems with the two previous assessments; which were that (i) the data at 550 and 600°C appear to converge at long times (see Figure 4) and (ii) that the data at 650 and 700°C show sigmiodal behaviour, whereas this sigmiodal behaviour is not shown at 550, 600, 750 or 800°C.

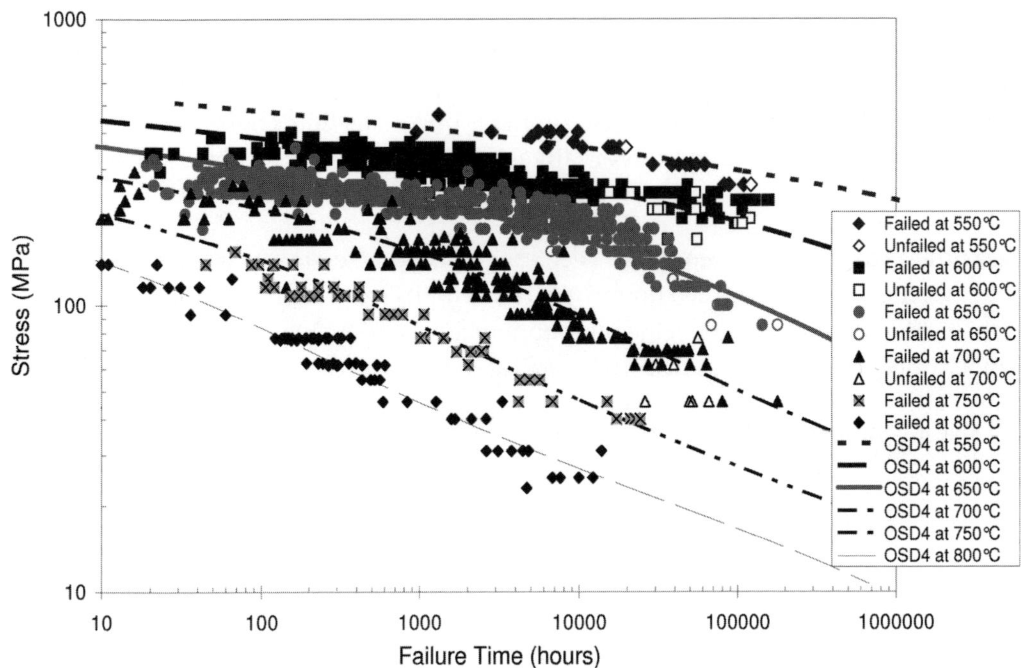

Figure 4 Creep Rupture Strength of Esshete 1250 compared with the OSD4 Model

This multi-region approach modelled the rupture strength using only three failure mechanisms, which should be sufficient to describe the sigmoidal rupture behaviour of Esshete 1250. Each of the three parts of the model varies with temperature according to an Arrhenius expression and varied with stress according to a power law, in which the power is a function of temperature. Thus, the rupture strength for each mechanism is given by

$$t_u^* = A_i \exp(Q_i/T)\sigma^{(B_i.T+C_i)} \tag{4}$$

where A_i, B_i, C_i, and Q_i are the material constants in each of the three failure mechanisms. The three mechanisms were combined using logical statements such that the sigmoidal rupture behaviour can be reproduced. This gave

$$t_u^* = MIN\left\{ \begin{matrix} MAX\left[A_P \exp(Q_P/T)\sigma_0^{(B_P.T+C_P)},\ A_{Cons.} \exp(Q_C/T)\sigma_0^{(B_C.T+C_{C.})}\right], \\ A_D \exp(Q_D/T)\sigma_0^{(B_D.T+C_{D.})} \end{matrix} \right\} \tag{5}$$

Fitting to the above model was carried out using non-linear regression with $\log_{10}(t_u^*)$ as the dependent variable. One of the main difficulties with non-linear regression is in identifying realistic starting values for the constants. In [3] these were determined by performing three initial analysis to each of the three mechanisms independently. Thus, judgements were made regarding which of the data failed by each mechanism and these three sets of data were fitted independently to give the starting values that were used in the final analysis. The advantages of using the final analysis, eq. (5), is that it avoids any subjectivity and gives a better fit to the data than the independent fits. It can be seen from Figure 5 that eq. (5) gives a good fit, within the range of the data, and that it resolves the two problems with the previous analyses. Namely, that the data at 550 and 600°C appear to converge at long times and that the data at 650 and 700°C show sigmiodal behaviour, whereas this behaviour is not shown at 550, 600, 750 or 800°C. In addition, it can be seen in Figure 5 that eq. (5) is stable on extrapolation, which contrasts with many polynomial expressions, which turn back or have inflections on extrapolation.

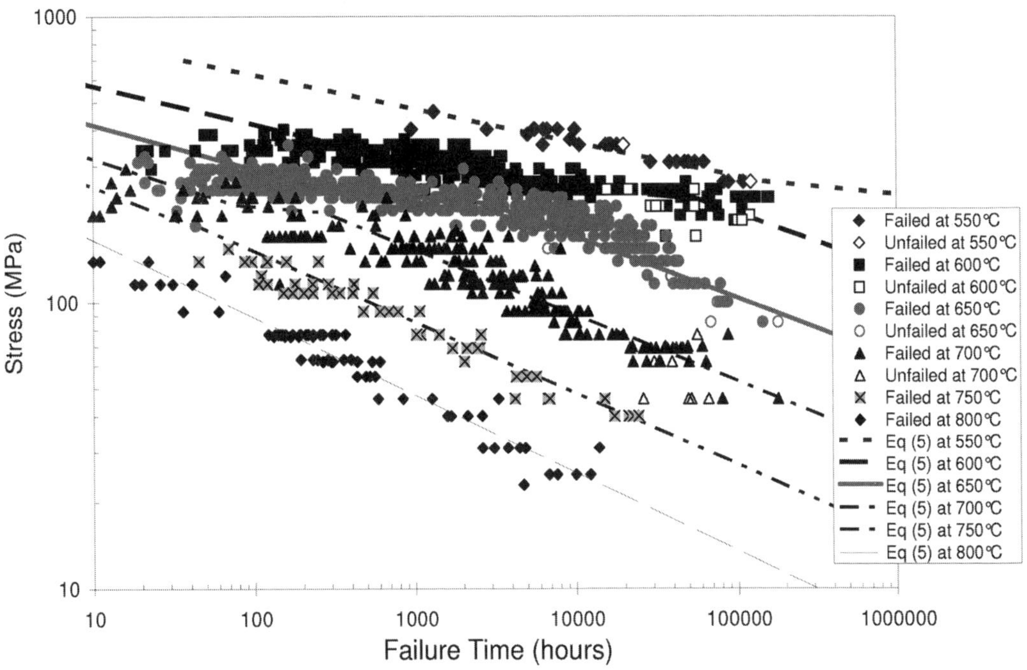

Figure 5 Creep Rupture Strength of Esshete 1250 Compared to the equation (5)

The multi-region model for Esshete 1250 passed all of the ECCC PATs [3]. In particular, the multi-region model showed very good repeatability and stability on extrapolation (PATs 3.1 and 3.2). Indeed, the differences between the 300,000 rupture strengths for the new model and the fit based on a 50% cull of the longest time data (PAT 3.1) are all less than 1%.

Nevertheless, it is interesting to note that the OSD4 model also passed the ECCC PATs and that the 300,000 hour rupture strengths for the multi-region model and the OSD4 model are relatively close to one another (4.8%, 0.4%, 0.8% and 6.4% difference at 550, 600, 650 and 700°C respectively).

NF709R

The WG3B has performed three creep rupture data assessments on the rupture data for NF709R austenitic stainless steel. The majority of the data (on five heats) being provided by Nippon Steel Corporation, with additional data on one heat being provided by ECCC's own test programme. The heat tested by ECCC has previously been tested by Nippon Steel hence the data are for a total of five heats. The first CRDA was carried out by Andersson and Sandström [13] using the 'free temperature model'. The second CRDA was carried out by Nespoli [14] using the ISO 6303 proceedure. A third CRDA has been performed and is reported here for the first time. In addition, to these three CRDAs a new multi-region modelling approach, which has been proposed by Wilshire [15], has been investigated. NF709R is a subset of grade NF709 which are both propriety steels from Nippon Steel Corporation. The composition of NF709R falls within the overall specification for NF709, but has a tighter specification with lower carbon content to; increase the resistance to intergranular corrosion, and increased Cr content to; (i) increase the corrosion resistance in sulphate rich environments such as those in waste incineration boilers and (ii) to increase steam oxidation resistance for high-temperature boilers. The purpose of the CRDAs performed by ECCC is to produce rupture strength values for NF709R that are distinct from those for NF709.

The assessment by Andersson and Sandström [13] used a novel model called the 'free temperature model'. In the free temperature model $\log_{10}(\sigma_0)$ is treated as the dependent variable, which is different from other fitting methods that usually treat $\log_{10}(\sigma_0)$ as an independent variable. $\log_{10}(\sigma_0)$ is then fitted to a complex time temperature parameter which has two polynomials in temperature, one of which is multiplied by $\log_{10}(t_u^*)$. These two polynomials can be chosen to have different orders, typically third second or first order polynomials are used. For NF709R the chosen 'free temperature model' was

$$\begin{aligned}
&\log_{10}(\sigma_0) = a_1\,TTP^2 + a_2\,TTP + a_3 \\
&\text{where}\quad TTP = w(T) + v(T).\log_{10}(t_u^*) \\
&\text{where}\quad w(T) = c_1(T/1000)^3 + c_2(T/1000)^2 + c_3(T/1000) + c_4 \\
&\text{and}\quad v(T) = b_1(T/1000) + b_2
\end{aligned} \tag{6}$$

Thus, the 'free temperature model' has a very flexible time temperature parameter but a stable description of stress. Nevertheless, in common with many austenitic steels the fit at the lowest temperature, in this case 600°C, is not particularly good (see Figure 6).

The assessment by Nespoli [14] followed a more traditional approach and used the ISO6303 method. However, whereas ISO6303 usually considers polynomials up to fourth order Nespoli investigated the use of higher order polynomials. The model chosen was a Larson-Miller model with a fifth order polynomial in $\log_{10}(\sigma_0)$.

Thus, this model has a simple temperature dependence and a very flexible stress dependence. This model fitted the 600°C data a little better than the 'free temperature model' (see Figure 6).

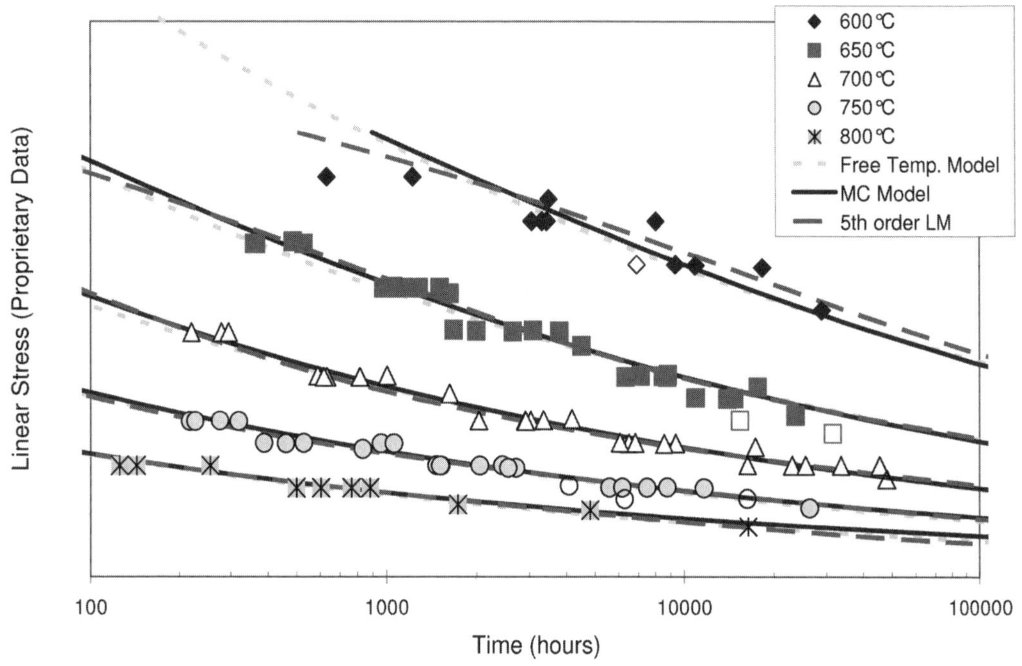

Figure 6 Creep Rupture Strength of NF709R

The third assessment followed a similar approach to that used for HR3C [9]. Once again the PD6605 procedure was used initially and eighteen models were fitted to the data. The PD6605 procedure identified the Minimum Commitment, MC, as the preferred model for NF709R. However, at 600°C the curvature of the data was opposite to that of the MC model and was opposite to that of the data at 650°C and above (see Figure 6), thus, it was decided to investigate alternative rupture strength models, with the aim of improving the fit at 600°C. A second problem with the rupture models in PD6605 (when used to fit NF709R) was that all of the models had one or more coefficients that showed large standard errors. Indeed, all of the models had one or more coefficients that would fail the Student's t-test at a 5% level. Within the PD6605 [4] procedure there is the facility for the user to enter additional models to the eighteen that are used routinely. The selection procedures that were used to choose between these models involved the standard PD6605 tests but in addition the significance of the coefficients was tested using the Student's t-test at a 5% level. This enabled a backward elimination, stepwise regression procedure to identify the preferred model. Two starting models were used; the first included a polynomial in $\log(\sigma_0)$ and the second included a polynomial in σ_0. Both models included polynomials in $1/T$. These equations can be written as;

$$\log_e(t_u^*) = \beta_0 + \sum_{k=1}^{4} \beta_k (\log(\sigma_0))^k + \sum_{m=1}^{4} \chi_m (1/T)^k \qquad (7a)$$

and

$$\log_e(t_u^*) = \beta_0 + \sum_{k=1}^{4} \beta_k \sigma_0^{\ k} + \sum_{m=1}^{4} \chi_m (1/T)^k \qquad (7b)$$

It was found from the Post Assessment Tests and the Student's t-test that a model of the form

$$\log_e(t_u^*) = \beta_0 + \beta_1\sigma_0 + \beta_2\sigma_0^{\ 2} + \beta_3\sigma_0^{\ 3} + \chi_1/T + \chi_2/T^2 \qquad (8)$$

gave the most reliable rupture strength predictions. The model given by equation (8) is referred to as 3S2T, since it is a third order polynomial in stress and a second order polynomial in 1/T. This model was chosen because the polynomials in terms of σ_0 gave significantly lower deviancies than the polynomials in terms of $\log(\sigma_0)$. In both cases (eqs (7a) and (7b)) the $1/T^3$ and $1/T^4$ terms were aliased by the software, i.e. these parameters were found to be linearly dependent on the preceding parameters and were automatically set to zero. In the case of the 4th order polynomial in stress, with a second order polynomial in 1/T, the coefficients β_3 and β_4 failed the Student's t-test. Thus, the $\sigma_0^{\ 4}$ term was removed and the data set was refitted, which resulted in the 3S2T model, eq. (8). The deviance of the 3S2T model was 1499.8, which is better than that of the MC model at 1504.1, and the model gave a good prediction of the short term data at 600°C (see Figure 7) and passed all of the ECCC PATs.

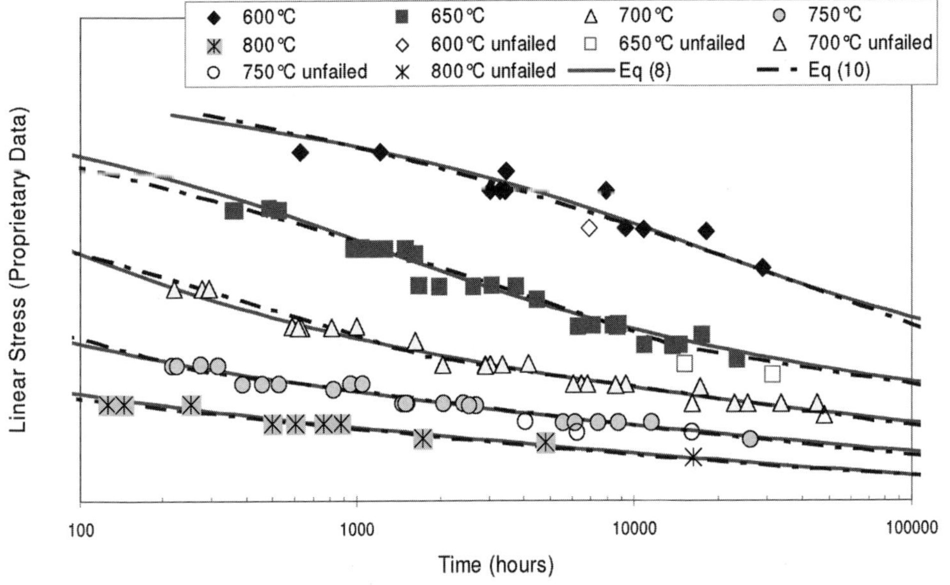

Figure 7 Creep Rupture Strength of NF709R

The final assessment followed a novel multi-region approach which is based on the work of Wilshire and Scharning [15], who have proposed that rupture strength can be described by a simple equation given by

$$(\sigma_0 / \sigma_{UTS}) = \exp\{-k_1[t_u^*.\exp(-Q_c^* / RT)]^u\} \tag{9}$$

where σ_{UTS} is the ultimate tensile strength at the test temperature, R is the gas constant and k_1 and u are material constants and Q_c^* is the activation energy for lattice diffusion in the alloy steel matrices. Furthermore, Wilshire and Scharning [15] have noted that in some materials, which exhibit a change in failure mechanism, it is necessary to use this equation in conjunction with a multi-region approach in which the data were split manually into two different regions. However, it is also possible to statistically fit the data in one step using piecewise non-linear regression. Thus, two equation (9)s can be rearranged and written as

$$\log_e(t_u^*) = \max\begin{bmatrix} \frac{1}{u_1}.\log_e(1/k_1) + \frac{1}{u_1}.\log_e(-\log_e(\sigma_0/\sigma_{UTS})) + \frac{Q_c^*}{RT}, \\ \frac{1}{u_2}.\log_e(1/k_2) + \frac{1}{u_2}.\log_e(-\log_e(\sigma_0/\sigma_{UTS})) + \frac{Q_c^*}{RT} \end{bmatrix} \tag{10}$$

Wilshire and Scharning assert that Q_c^* should be assumed to be the measured activation energy for lattice diffusion in the alloy steel matrices, which for Fe in austenite is often taken to be 280kJ/mol. Nevertheless, it is equally possible to allow the regression routine to determine best fit values for Q_c^* in each of the two regions. Thus, the regression routine returns values for u_1, u_2, k_1, k_2, Q_{c1} and Q_{c2}. Interestingly, for NF709R it was found that Q_{c1}, which covers the high stress region was 295.5kJ/mol, and Q_{c2}, for the low stress region was 276.4kJ/mol. Both of these values are close to the actual value for lattice diffusion of Fe in austenite of 280kJ/mol. It was found that this multi-region approach gave an accurate fit to the data (Figure 7). Indeed, it was more accurate, as measured by the standard error, than the other 3 preceding assessments. It also passed the ECCC post assessment tests for model realism (PATs 1.1, 1.2 and 1.3) and the goodness of fit tests (PATs 2.1 and 2.2). However, it failed one of the tests for stability on extrapolation namely PAT 3.2, which culls the lowest stress data, 10% at each temperature, and compares the predictions for the culled and full data sets. The other test PAT 3.1, which culls half of the longest term data, those with durations greater than $t_{u[max]}/10$, was passed. The failure of PAT 3.2 is slightly surprising in that eqs (9) and (10) are intrinsically stable on extrapolation. Nevertheless, this is only a single application of the Wilshire and Scharning approach to a relatively small and short term data set. Hence, it can be concluded that the Wilshire and Scharning approach shows promise and is worthy of investigation on other data sets. In particular, for data sets where the high stress data approach the ultimate tensile strength.

Notwithstanding the differences between each of these four assessments it is interesting to note that the predicted 100,000 hour rupture strengths are within 10% or less (see Table 2).

Table 2: 100,000 hour Rupture Strength of NF709R

Temperature (°C)	100,000 hour Rupture Strength (MPa)				
	Nippon Steel NF709	Free Temperature Model	Larson Miller 5th Order Model	3S4T Model	Multi-Region Model
600	187	194	201	189	181
650	124	124	125	123	119
700	84	78.7	82.5	80.9	77.4

Discussion and Conclusions

ECCC Working Group 3B on Austenitic stainless steels has conducted a number of creep rupture data assessments on different steels. Because of the diverse membership of ECCC these assessments have been conducted using a variety of different methods. These methods have included 'traditional' creep rupture models such as polynomials in log stress with time temperature parameters such as Mendelson-Roberts-Manson, Manson-Haferd, Larson-Miller and Orr-Sherby-Dorn and parametric equations such as the Soviet models and Minimum Commitment Equation (A=0). In addition, a number of new methods have been tried out including multi-region approaches.

A comparison between the many assessments shows that it is possible to develop models for creep rupture that provide better fits to the data and more reliable long term predictions than traditional creep rupture models. Nevertheless, it is also clear that these models also often predict similar values of 100,000 and 300,000 hours rupture strength, which are used in pressure vessel design. Thus, if these values were the only purpose for performing a creep rupture data assessment then there might be little reason for improving the models. However, creep rupture data are also used for predictions of creep-fatigue endurance and in life assessments of defective components subjected to both steady and cyclic loading.

The time fraction approach to calculate creep damage during a creep-fatigue cycle and component life assessments require creep rupture models to give good estimates of duration for a much wider range of conditions than pressure vessel design codes. In particular, at the high stresses that can arise in cyclic hardening materials such as austenitic stainless steels. However, it has been shown that the 'traditional' creep rupture models do not perform well in austenitic steels at the lowest temperatures and highest stresses. This is particularly, true when the stresses are well above the proof stress and indeed as the stresses approach the ultimate tensile strength (as in Type 316,

HR3C and Esshete 1250). Nevertheless, many of the new creep rupture models give improved fits at the lowest temperatures and highest stresses. For example the use of polynomials in σ_0, such as those used in the 4S1T and 3S2T models and the use of σ_0^m, in the DESA procedure, give consistently better fits than polynomials in $\log(\sigma_0)$, particularly at high stress (see HR3C and NF709R). Multi-region modelling would also be a benefit when rationalising high and low stress data (see Esshete 1250 and NF709R).

In creep rupture, high stresses go hand in hand with low temperatures and it is equally clear that the 'traditional' time temperature parameters are not flexible enough to cope with changes in activation energy. Methods that increase the flexibility in the temperature term; include the free temperature model, polynomials in 1/T (see NF709R) and ultimately multi-region approaches (see Esshete 1250 and NF709R).

The most demanding problems associated with changes in behaviour with stress and temperature are exemplified by the 'sigmoidal' creep rupture behaviour that is shown not only by Esshete 1250 but by many ferritic and martensitic steels. Whilst, high order polynomials in σ_0 and $\log(\sigma_0)$ can display sigmoidal behaviour the fits and stability on extrapolation are often found to be poor. The ultimate solution to these problems are multi-region approaches such as those used on Esshete 1250 and NF709R. Multi-region approaches have also been used by Kimura [16] and Maruyama *et al* [17]. The main differences between these approaches are; the rupture equations used, the methods used to determine which data lie in which region and the methods used to fit the models to the data. In the case of the multi-regions approaches used here these have used different but equally stable rupture equations. Nevertheless, the splitting of the data into regions and fitting of the model has always been done using in a single step by the statistical analysis software using logical statements and non-linear regression. This avoids any subjectivity and will give a better fit to the data than independent fits.

It should be noted that owing to the diverse number of creep rupture modelling approaches tried by ECCC WG3B, no one approach can be recommended but it is suggested that once a number of different approaches have been tried the ECCC PATs offer a powerful tool to chose between the best of these for the particular data set being fitted.

Acknowledgement

This paper is published by permission of British Energy Generation Ltd and the European Creep Collaborative Committee Management Board.

References

1 L Linde, R Sandstrom, R Gommans, M W Spindler & A Fairman, Evaluation of Creep Rupture Data for the New European Standard for Stainless Steels-Edition 2, ECCC Doc. No. 0509/WG3.3/7, Sept. 1998.

2 M W Spindler, The Multiaxial Creep Ductility of Austenitic Stainless Steels, Fatigue Fract. Engng. Mater. and Struct. Vol. 27, Issue 4, pp 273-281, 2004.

3 M W Spindler and S L Spindler, Creep Deformation, Rupture and Ductility of Esshete 1250, Creep & Fracture in High Temperature Components Conference Proceedings pp. 452-464, DEStech Publications, Inc., Lancaster, USA, 2005.

4 PD6605-1998, Guidelines on Methodology for Assessment of Stress Rupture Data, British Standards Institute, 1998.

5 ECCC, Creep Data Validation and Assessment Procedures, Pub. European Tech. Develop. Ltd on behalf of ECCC Management Committee, 2005.

6 D G Robertson, Assessment of Creep Rupture Data for HR3C Steels and Weldments, ECCC Report AC/WG3B/8, 2001.

7 M Schwienheer, Assessment of Creep Rupture Data for Parent and Weldment Dataset for Austenitic Steel HR3C, ECCC Report AC/WG3B/53, 2004.

8 Granacher J. Monsees M. Hillenbrand P & Berger C. Software for the assessment and application of creep and rupture data. Nuclear Engineering and Design.1999;190(3):273-285.

9 M W Spindler, An Assessment of the Creep Rupture Strength of HR3C Parent and Weldments, ECCC Report AC/WG3B/54, 2005.

10 PD6525, Elevated Temperature Properties for Steels for Pressure Purposes Part 1: Stress Rupture Properties, British Standards Institute, 1990.

11 ISO 6303. Pressure Vessel Steels Not Included in ISO 2604, Parts 1 to 6 - Derivation of Long-Time Stress Rupture Properties First Edition, 1981.

12 Baker A. J. Creep Rupture of Esshete 1250, BEGL Report E/REP/ATEC/0049/GEN/02, 2002.

13 H C M Andersson and R Sandström, Assessment of creep rupture data for the stainless steel NF709R, ECCC Doc. No. AC/WG3B/56, SIMR Report IM-2004-534, 2004.

14 N Nespoli, Assessment of Stress Rupture Data for Austenitic Steel NF709R, ECCC Doc. No.AC/WG3B/48, ISB Report 03067/IS/R02, 2003.

15 B Wilshire and P J Scharning, Design Data Prediction for Grade 92 Steel, Proc. CREEP8: 8th Int. Conf. on Creep and Fatigue at Elevated Temperatures, July 22-26, 2007, San Antonio, USA.

16 K Kimura, Creep Strength Assessment and Review of Allowable Tensile Stress of Creep Strength Enhanced Ferritic Steels In Japan, Proc. PVP2006, ASME Pressure Vessels and Piping Division Conference, July 23-27, 2006, Vancoucer, Canada.

17 K Maruyama, H G Armaki and K Yoshimi, Multiregion Analysis of Creep Rupture Data of 316 Stainless Steel, Int. J. Press. Vess. & Piping, Vol. 84, 2007, pp. 171-176.

Advances in Materials Technology for Fossil Power Plants
Proceedings from the Fifth International Conference
R. Viswanathan, D. Gandy, K. Coleman, editors, p 718-732

DOI: 10.1361/cp2007epri0718

Long-Term Crack Behavior under Creep and Creep-Fatigue Conditions of Heat Resistant Steels

A. Scholz, C. Berger, F. Mueller
Institute of Materials Technology, Darmstadt University of Technology
Grafenstraße 2, 64283 Darmstadt, Germany

A. Klenk
Materialpruefungsanstalt Universitaet Stuttgart
Pfaffenwaldring 32, 70569 Stuttgart, Germany

ABSTRACT

High temperature components with notches, defects and flaws may be subject to crack initiation and crack propagation under long-term service conditions. To study these problems and to support an advanced remnant life evaluation, fracture mechanics procedures are required. Since a more flexible service mode of power plants causes more start up and shut down events as well as variable loading conditions, creep-fatigue crack behavior becomes more and more decisive for life assessment and integrity of such components.

For steam power plant forged and cast components, the crack initiation time and crack growth rate of heat resistant steels were determined in long-term regime up to 600 °C. Component-like double edge notched tension specimens have been examined. The results are compared to those obtained using the standard compact tension specimen. Crack initiation time and crack growth rate have been correlated using the fracture mechanics parameter C*. The applicability of the stress intensity factor K_I to describe the creep crack behavior is also being assessed. A modified Two-Criteria-Diagram was applied and adapted in order to recalculate crack initiation times under creep-fatigue conditions. Recommendations are given to support the use of different fracture mechanics parameters in order to describe the long-term crack behavior under creep and/or creep-fatigue conditions.

KEYWORDS

high temperature crack behavior, crack initiation, crack growth, creep and/or creep-fatigue conditions, stress intensity factor K_I , parameter C*

INTRODUCTION

Standards, which refer to the behavior of crack-like defects/flaws and unavoidable notches, are required for the design and surveillance of high temperature power plant components in long-term service. In case of unavoidable notches, creep or creep-fatigue cracks may be initiated or propagated under static or cyclic high-temperature loading. In both cases a quantitative description is based on a sufficiently large experimental basis, which consists of corresponding long term tests.

In the course of over two decades a solid database for conventional 1- and 12CrMoV-steels [1, 2] and modern 10Cr-steels [3, 4] was established which comprises a large number of long-term tests up to 35 000 h. Testing methods were optimized continuously. For evaluation of the tests, standard methods have been applied and new evaluation methods have undergone continuous development. In order to obtain reliable results, validity criteria have been defined. This experience was applied in the evaluations described in the following.

EXPERIMENTAL PROCEDURES

The test materials (4 different modern 10Cr-steels, with and without Tungsten, forged and cast conditions) were taken from industrial components. Their chemical composition is listed in Table 1. Component dimensions are given in Table 2. In this study side-grooved C(T)- and DEN(T)-specimens were used, Figure 1. A spark eroded crack starter was used.

Long-term tests up to 20 000 h under cyclic tension load with load ratio $R = F_{min}/F_{max} = 0.1$, 0.6 and 1.0 were performed with hold times at maximum load of 0.3, 1.0 and 3.0 h. Most of the tests were carried out at 600 °C.

The small specimens of type Cs25 (side grooved compact tension specimens with 25 mm thickness) were tested using the interrupted test technique [5]. For each stress level a series of up to 10 specimens was tested under the same loading condition in a multi-specimen machine. The specimens were unloaded after predetermined time proportions from 10 % to 80 % of the estimated "rupture time" have been exceeded. During each interruption the load line displacement of all specimens was measured and one specimen was fractured at low temperature. The crack length of the specimen was determined fractographically.

The component-like specimens Ds60 (side grooved double edge notched tension specimens with 60 mm thickness) were tested in servo-hydraulic testing machines [5]. During these tests the load line displacement was measured online by means of capacitive high temperature strain gauges. The crack propagation was monitored online by using the alternating current potential drop (ACPD) technique. At the end of each test the potential drop signal was calibrated with the final crack length measured on the fractured specimens.

ANALYSIS OF CRACK DATA

Crack initiation and crack growth under creep and/or creep-fatigue conditions are described by the stress intensity factor K_I and the parameter C*. Creep Crack Initiation (CCI)-data, Creep Crack Growth (CCG)-rate and Load Line Displacement (LLD)-rate were determined based on the measured crack length a and load line displacement v_{LLD}.

Determination of the parameter C*

The parameter C* is valid for stationary creep in the crack tip environment. The parameter C* can be determined by [6]:

$$C^* = \dot{v}_c \cdot \frac{F}{B_n \cdot (W-a)} \frac{n}{n+1} \cdot \eta \tag{1}$$

with η = [2+0.522(1-a/W)] for C(T)-specimens and η = 1 for DEN(T)-specimens, the applied load F, the specimen width W, the specimen thickness between side grooves B_n and the creep part of measured load line displacement (LLD) rate $\dot{v}_c$ [6].

In case of absence of load line displacement rate, the parameter C* can be estimated using the reference stress method [7]

$$C^*_{ref} = \mu \cdot \sigma_{ref} \cdot \dot{\varepsilon}_{ref} \cdot (K_I/\sigma_{ref})^2 \tag{2}$$

where μ = 1 represents plane stress conditions and μ = 0.75 for plane strain conditions. The reference strain rate is given by:

$$\dot{\varepsilon}_{ref} = A_u/t_u \tag{3}$$

where A_u is the uniaxial creep ductility.

Determination of the Stress Intensity Factor K_I

The stress intensity factor K_I is valid for linear elastic behavior only [8], but it can be used as an approximation if the plastic zone near the crack tip is limited [9]. For side-grooved C(T)-specimens the stress intensity factor is calculated as [10]:

$$K_I = \frac{F}{\sqrt{B \cdot B_n \cdot W}} \cdot (2 + a/W) \cdot f(a/W) \tag{4}$$

$$f(a/W) = \frac{0.886 + 4.64 \cdot (a/W) - 13.32 \cdot (a/W)^2 + 14.72 \cdot (a/W)^3 - 5.6 \cdot (a/W)^4}{(1 - a/W)^{3/2}} \tag{5}$$

For side-grooved DEN(T)-specimens the stress intensity factor is calculated as [10]:

$$K_I = \frac{F}{\sqrt{B \cdot B_n \cdot 2 \cdot W}} \cdot f(a/W) \tag{6}$$

$$f(a/W) = 1.4 \cdot (a/W)^{1/2} + 0.2556 \cdot (a/W)^{3/2} - 1.5 \cdot (a/W)^{5/2} + 2.42 \cdot (a/W)^{7/2} . \tag{7}$$

Validity Criteria

Validity criteria are specified for the use of a fracture mechanics parameter as correlating parameter for crack behavior. The criteria "transition time" and "ratio $\dot{v}_c/\dot{v}$ " are originally defined for crack growth under creep conditions [6] but these validity criteria can also be used for creep crack initiation data and creep-fatigue crack initiation/growth data. The transition time t_1 is given by [6]:

$$t_1 = \frac{K_I}{E' \cdot (n+1) \cdot C^*} . \tag{8}$$

For test times $t >> t_1$ the parameter C* is valid.

Further, the data can be classified as being creep-ductile if $\dot{v}_c/\dot{v} \geq 0.5$ and the crack behavior may be characterized by the parameter C*. For creep-brittle situations (i.e. $\dot{v}_c/\dot{v} \leq 0.25$) the stress intensity factor may be used to describe the crack behavior.

The German Creep Crack Group "W14" has considered further validity criteria [5]. These validity criteria are aimed to determine in general validity of a creep crack test, without consideration which parameter (C* or K_I) is applied. At first, validity is restricted to a crack tip opening displacement, which is small compared to the specimen geometry. This criterion is defined as:

$$a/50 < \frac{v}{1 + 3 \cdot \left(a/(W-a)\right)} \tag{9}$$

for C(T)-specimens and

$$a/50 < v/2.7 \tag{10}$$

for DEN(T)-specimens.

The second criterion is aimed to ensure reasonable use of the creep fracture mechanics parameters by limiting the net stress in the ligament to a value smaller than creep rupture strength:

$$\sigma_{n\,pl} < R_{u/t/T} \; @ \; t_i \tag{11}$$

with

$$\sigma_{n\,pl} = \frac{F}{B \cdot (W-a)} \cdot \left(1 + 2 \cdot \frac{W+a}{W-a}\right) \tag{12}$$

for C(T)-specimens and

$$\sigma_{npl} = \frac{F}{2 \cdot B \cdot (W - a)} \tag{13}$$

for DEN(T)-specimens.

RESULTS AND DISCUSSION

For the description of Creep Crack Initiation (CCI) and Creep Crack Growth (CCG), predictive models and methods have been applied, which, among other goals, assist the industry in safe operation of power plant components. For example, the steady state CCG rate da/dt may be approximated as [11]

$$da/dt^{NSW\text{-}APP} = \frac{3 \cdot (C^*)^{0.85}}{A_u{}^*} \tag{14}$$

where da/dt and C* are in mm/h and MPam/h, respectively. $A_u{}^*$ is taken as the uniaxial strain after fracture, A_u, for plane stress conditions and $A_u/30$ for plane strain conditions [11]. For CCI (Δa_i = 0.5 mm) this model can be taken as [11]:

$$t_i^{NSW\text{-}APP} = \frac{\Delta a_i \cdot A_u{}^*}{3 \cdot C^{*0.85}} \,. \tag{15}$$

An established method for prediction of CCI is the so-called Two-Criteria-Diagram (2CD) [9]. A brief description of this method is given below.

Based on data for Creep and Fatigue Crack Behavior methods were evaluated that allow estimating Creep-Fatigue Crack Initiation (CFCI) and Creep-Fatigue Crack Growth (CFCG).

Creep Crack Initiation (CCI)

In the Two-Criteria-Diagram (2CD) the nominal stress $\sigma_{n\,pl}$ considers the stress situation in the ligament, i.e. in the far field of the creep crack and the fictitious elastic parameter K_{I0} at time zero characterizes the crack tip situation (Figure 2). These loading parameters are normalized in the 2CD by the respective time and temperature dependent values, which indicate material resistance against crack initiation. The normalized parameters are the stress ratio $R_\sigma = \sigma_{n\,pl}/R_{u/t/T}$ for the far field and the stress intensity ratio $R_K = K_{I0}/K_{Ii}$ for the crack tip. The value $R_{u/t/T}$ is the creep rupture strength of the material and the parameter K_{Ii} characterizes the CCI of the material. This parameter has to be determined from specimens with high ratio $K_I/\sigma_{n\,pl}$, preferably side grooved C(T)25-specimens. The 2CD distinguishes three fields of damage mode separated by lines of constant ratio R_σ/R_K. Above R_σ/R_K = 2.0 ligament damage is expected, below R_σ/R_K = 0.5 crack tip damage is expected and between these lines a mixed damage mode is observed. Crack initiation is expected above a boundary line only. The 2CD has been developed as a way of practice to transfer CCI data from specimens with different sizes to larger components with similar far field and crack tip situation and the applicability of the method was proven by the results of more than 100 small and large scale specimens with artificial and natural defects. Past experience has shown that the boundary line is dependent on creep ductility and specimen size [12]. The "adapted" boundary line (dashed line) in Figure 2 considers the size influence but not the ductility effect. The estimate of the CCI for large specimens with the 2CD is conservative.

The parameter C* is shown in Figure 3 against the CCI time t_i for the investigated 10Cr-steels at 550 and 600 °C. The plane strain and plane stress initiation time predictions, given by Eqn. (15) are included. The plane strain NSW-APP Model provides a very conservative estimate of the time to CCI. Applying the formula for plane stress a better agreement between CCI data and the approximate NSW Model were observed.

Creep Crack Growth (CCG)

The CCG-rate da/dt is plotted against C* in Figure 4, for the investigated 10Cr-steels at 550 and 600 °C. Data points for $\Delta a < 0.5$ mm and $t < t_1$ have been removed. For these data an almost linear correlation between da/dt and C* on a log-log scale can be observed. It may also be observed that the data from the DEN(T)- and C(T)-specimens lie within a relatively tight data band. Furthermore, in Figure 4 the data are compared with the approximate Nikbin-Smith-Webster-Model (NSW-APP) [11]. All data fall close to the plane stress line. The NSW-APP plane strain line provides a very conservative prediction for the data.

If the linear elastic stress intensity factor K_I is used for description of CCG-rate da/dt, surprisingly a mostly linear correlation between da/dt and K_I on a log-log scale can be observed (Figure 5). Data points for $\Delta a < 0.5$ mm have been removed. Further, only data with ratio $K_I/\sigma_{n\,pl} > 3.5$ (i.e. crack tip damage is dominated) are included in Figure 5. K_I and $\sigma_{n\,pl}$ are in $N/mm^{3/2}$ and N/mm^2, respectively. As expected, the CCG-rate in dependence of K_I is temperature dependent. Nevertheless, the stress intensity factor K_I can be used for CCG descriptions in consideration of the mentioned restrictions.

For several component-like Ds60-specimens made of 10Cr-cast steel (Figure 6a) and 10Cr-forged steel (Figure 6b), the creep crack initiation and creep crack growth was recalculated with the NSW-Model. For the calculation of the needed parameter C* the reference stress method was used. If the creep crack growth $\Delta a_c < 5$ mm (i.e. crack incubation and stable crack growth under "c" - creep conditions) the calculation yields a conservative prediction.

Creep-Fatigue Crack Initiation (CFCI) and Growth (CFCG)

For the prediction of crack initiation under creep-fatigue conditions a modified Two-Criteria-Diagram (M2CD) has been introduced and validated on 1CrMoV-steels [13]. The original method was developed for crack initiation under creep only and is described above.

The modification considers the creep-fatigue crack initiation behavior due to hold time t_H and load ratio R. With increasing hold times at constant load ratio R the CFCI-data approach the CCI-data (Figure 7a). The test with higher frequency present shorter crack initiation times, the decrease is due to the influence of fatigue. Further, with increasing R-values at constant hold times the CFCI-data approaches the CCI-data (Figure 7b). The influence of hold time changes with the R-ratio. For practical applications, a new time dependent parameter $K_{Ii\ cf}(t_{i\ cf})$ has to be used ("c" - creep, "cf" - creep-fatigue). This parameter is reduced against the parameter $K_{Ii\ c}(t_{i\ c})$ according to a reduction of the crack initiation time from $t_{i\ c}$ to $t_{i\ cf} = 0.6 \cdot t_{i\ c}$. This reduction is dependent of hold-time and load ratio R (Figure 7c).

All other details concerning the 2CD for CCI remain unchanged. The modified 2CD for 10Cr-cast steel is shown in Figure 8. The validity of the boundary line is confirmed by the results of creep-fatigue crack initiation tests. The prediction of CFCI-time is conservative.

For intermediate loading conditions an accumulative crack growth is assumed, which can be determined from increments of creep crack growth and fatigue crack growth. The fatigue crack propagation per cycle can be described in terms of $\Delta K_{I\ eff}$ (Figure 9) and the creep crack growth rate in terms of C* (Figure 4). The creep-fatigue crack propagation per cycle is then given by an accumulation in form of

$$(da/dN)_{cf} = (da/dN)_f + 1/f\ (da/dt)_c \ . \qquad (16)$$

Further a creep crack initiation time $t_{i\ c}$ to achieve a technical crack initiation length of $\Delta a_{i\ c}$ = 0.5 mm was taken (Figure 3). Beginning from the initial crack length a_0 the creep crack growth increments were only accumulated when the time increments exceeded the creep crack initiation time $t_{i\ c}$ or the crack increments exceeded the creep crack initiation length Δa_i = 0.5 mm.

The results of such a calculation are represented in Figure 10. The calculated creep-fatigue crack length $\Delta a_{cf}'$, which is composed of the fatigue portion and the creep portion, agrees relatively well with the measured values Δa_{cf}, whereas the results are consistently conservative.

CONCLUDING REMARKS

The crack initiation and crack propagation behavior of modern 10Cr-steels in forged and cast conditions have been investigated under creep and creep-fatigue conditions at 550 and 600 °C. The evaluation of the obtained experimental results was used subsequently for the assessment of material behavior, taking into account defects and crack-like stress concentrations. The application of this database to component behavior was validated by a number of component-like large-scale specimens. For practical use a software system including a data base with relevant creep crack initiation and growth data and calculation tools representing methods for the calculation of creep and/or creep-fatigue crack initiation and growth for components was developed [14].

The creep crack initiation results were successfully evaluated with the Two-Criteria-Diagram (2CD), which is based on the linear elastic parameter K_I. This analysis always results in conservative predictions. The approximate Nikbin-Smith-Webster-Model (NSW) in terms of the creep fracture parameter C* shows also a conservative approximation of creep crack initiation. For predicting creep-fatigue crack initiation times the 2CD was modified, adapted and verified.

In order to characterize the creep crack growth rate data, the stress intensity factor K_I as well as the creep fracture mechanics parameter C* have been successfully applied for different steels. Using the approximate NSW-Model similar to crack initiation conservative approximations for creep crack growth rates could be evaluated.

An accumulation of increments of creep crack growth and fatigue crack growth delivers satisfactorily prediction of creep-fatigue crack growth.

Summarizing, the development of advanced design methods and the establishment of new materials demands the generation of a long-term database. As a consequence of this work, a reduction of technical risk of highly loaded components is going to be achieved.

ACKNOWLEDGEMENT

Thanks are due to the "Forschungsvereinigung der Arbeitsgemeinschaft der Eisen und Metall verarbeitenden Industrie e.V.", the "Arbeitsgemeinschaft industrieller Forschungsvereinigung", the "VDEh-Gesellschaft zur Förderung der Eisenforschung mbH" and the "Forschungsvereinigung Verbrennungskraftmaschinen" for their financial and technical support. A special thanks is due to Dr. J. Ewald (Convenor of the German working group "Creep Crack Behavior") for useful discussions and contributions to this work.

REFERENCES

1) Kloos K.H., Kußmaul K., et al.: *Rissverhalten warmfester Kraftwerksbaustähle im Kriech- und Kriechermüdungsbereich*, Final report of IfWD and MPAS, AiF 7251, 1992.
2) Berger C., Roos E., et al.: *Kriechrissverhalten ausgewählter Kraftwerksstähle in erweitertem, praxisnahem Parameterbereich*. Final report of IfWD and MPAS, AVIF A78, 1999.
3) Berger C., Roos E., et al.: *Hochtemperaturrissverhalten der 600 °C-Stähle für Wellen und Gehäuse von Dampfturbinen*. Final report of IfWD and MPAS, AVIF Nr. A127, 2002.
4) Roos E., Berger C., et al.: *Kriech- und Kriechermüdungsrissverhalten moderner Kraftwerkstähle im Langzeitbereich*, Schlussbericht zum AVIF-Vorhaben Nr. A178 der Materialprüfungsanstalt, Universität Stuttgart und des Instituts für Werkstoffkunde der TU Darmstadt, 2006.
5) ECCC WG1.2. *Recommendations for Creep Crack Initiation Assessments*. Issue 1, 2003.
6) ASTM E 1457-00. *Standard Test Method for Measurement of Creep Crack Growth Rates in Metals*, 2000.
7) Webster G.A., Ainsworth R.A.: *High Temperature Component Life Assessment*, Chapman & Hall, 1994.
8) Irwin G.R.: Analysis of stresses and strains near the end of a crack traversing a plate. Trans. ASME, J. Appl. Mech., 24, 1957: 361-364.
9) Ewald J., Keienburg K.-H.: *A Two-Criteria-Diagram for creep crack initiation*. Int Conf on creep, Tokyo, 1986: 173-78.
10) Fett T., Munz D.: *Stress Intensity Factors and Weight Functions*. Computational Mechanics Publications, advances in Fracture Series, Southampton UK and Boston USA, 1997.
11) Nikbin K.M., Smith D.J., Webster G.A.: *Influence of Creep Ductility and State of Stress on Creep Crack Growth*, in Advances in Life Prediction Methods at Elevated Temperatures ASME, New York, pp.249/58, 1983.
12) Ewald J.: *Zwei-Kriterien-Diagramm für Kriechrisseinleitung: Berücksichtigung des Kriech-Verformungsvermögens (für ferritische Werkstoffe)*, Vortrag auf der 26. Vortragsveranstaltung der Arbeitsgemeinschaft für warmfeste Stähle und der Arbeitsgemeinschaft für Hochtemperaturwerkstoffe, Düsseldorf, 2003.
13) Ewald J., Sheng S., Schellenberg G.: *Engineering guide to assessment of creep crack initiation on components by Two-Criteria-Diagram*, 2nd Inter-national HIDA Conference, 4 - 6 Oct. 2000, Stuttgart, Germany, S5-2-1/5-2-20, 2000.
14) Ewald J., Mao T., Müller F., Scholz A., Machalowska-Tracz M., Klenk A.: *Programmgestützte fortschrittliche Kriech- und Kriechermüdungsrissbeschreibung für typische langzeitbeanspruchte Kraftwerksbauteile*. 28. Vortragsveranstaltung der Arbeitsgemeinschaft für warmfeste Stähle und der Arbeitsgemeinschaft für Hochtemperaturwerkstoffe "Langzeitverhalten warmfester Stähle und Hochtemperaturwerkstoffe", Düsseldorf, S. 70-85, 25. November 2005.

FIGURES

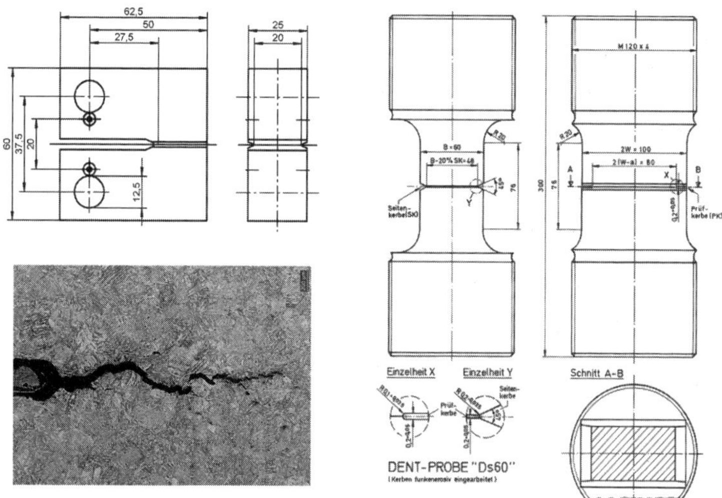

Figure 1. Standard Compact Tension Cs25-specimen and component-like Ds60-specimen

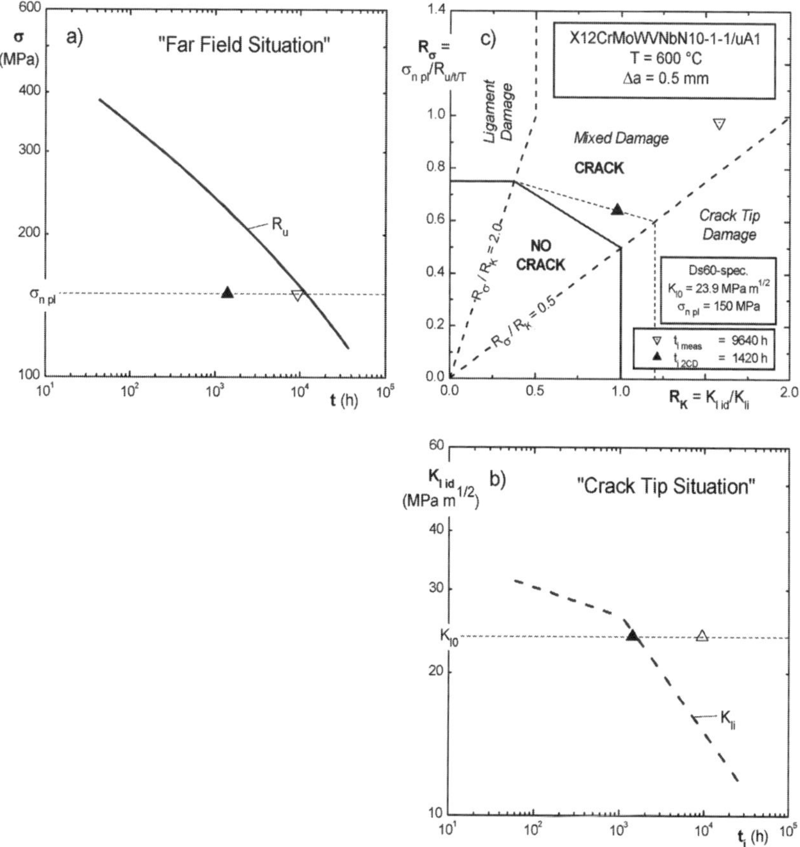

Figure 2. Determination of Creep Crack Initiation Time of a component-like Ds60-specimen using Two-Criteria-Diagram

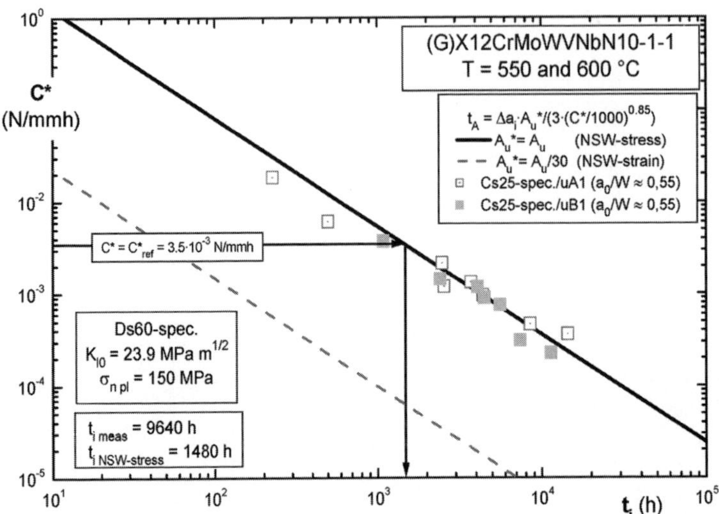

Figure 3. Comparison of measured Creep Crack Initiation Times on Cs25-specimen and NSW-Model, Determination of Creep Crack Initiation Time on a component-like Ds60-specimen by using the NSW-Model, 10Cr-steels, 550 °C and 600 °C

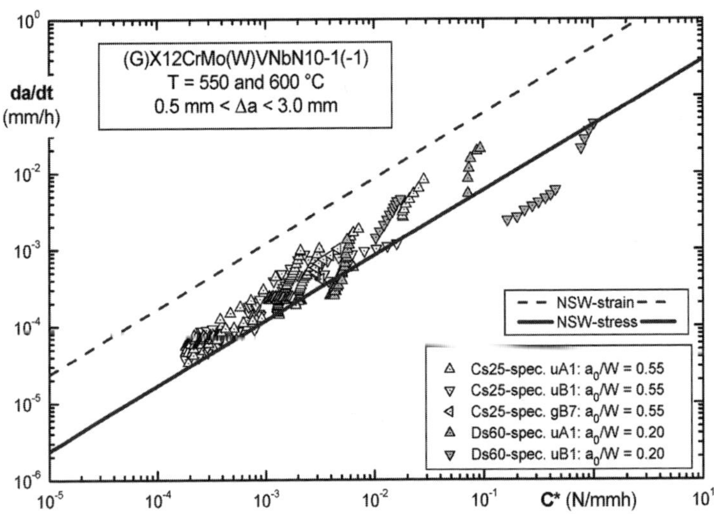

Figure 4. Comparison of measured Creep Crack Growth Rate on Cs25- and Ds60-specimens and NSW-Model, 10Cr-steels, 550 °C and 600 °C

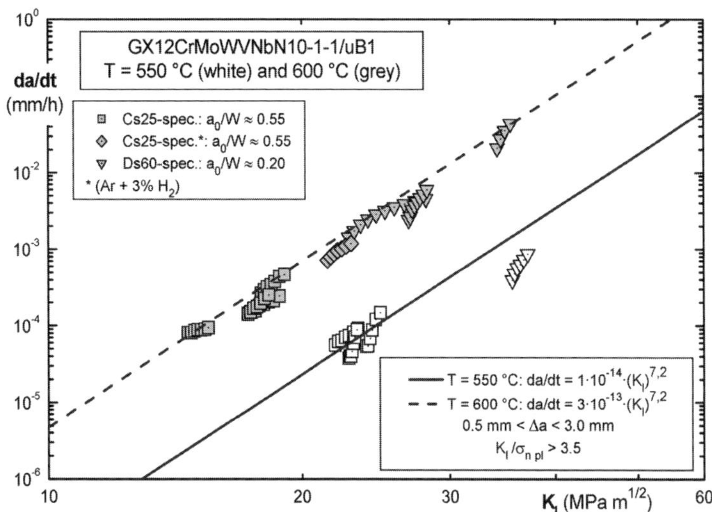

Figure 5. Creep Crack Growth Rate vs. K_I, 10Cr-cast steel, 550 °C and 600 °C

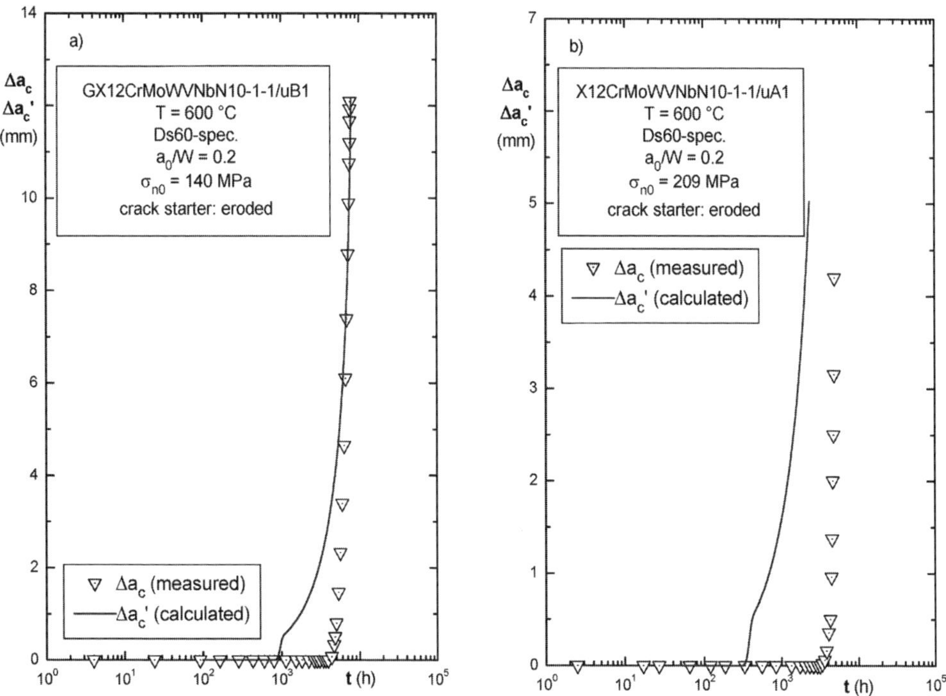

Figure 6. Recalculation of Creep Crack Initiation and Growth on component-like Ds60-specimens made of 10Cr-cast steel (a) and 10Cr-forged steel (b)

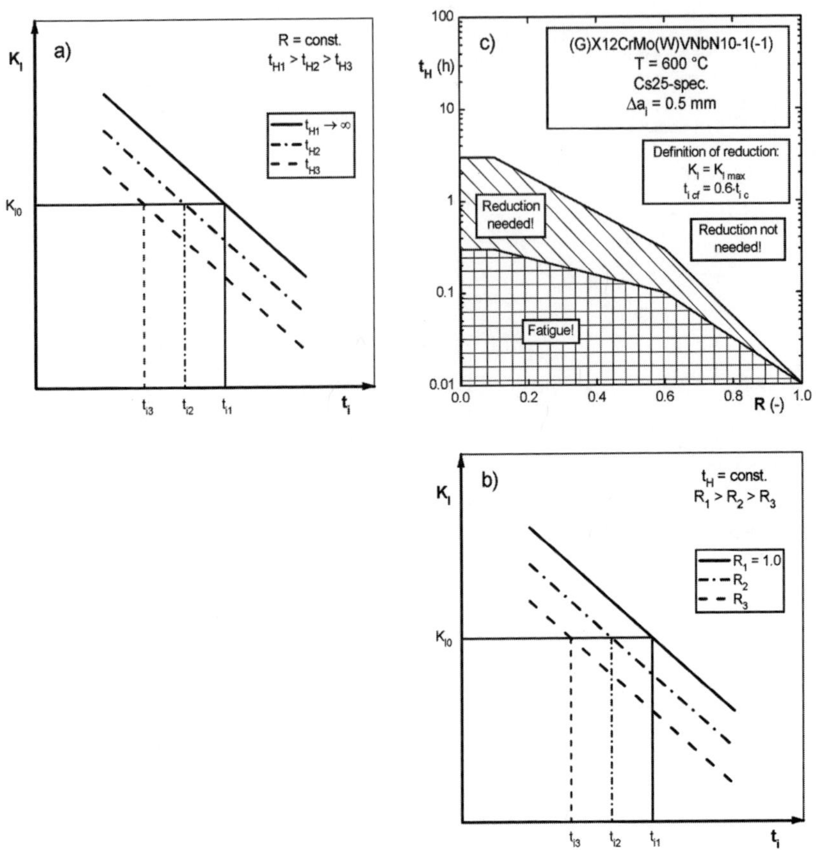

Figure 7. Modification of Two-Criteria-Diagram for Creep-Fatigue Loading, schematic

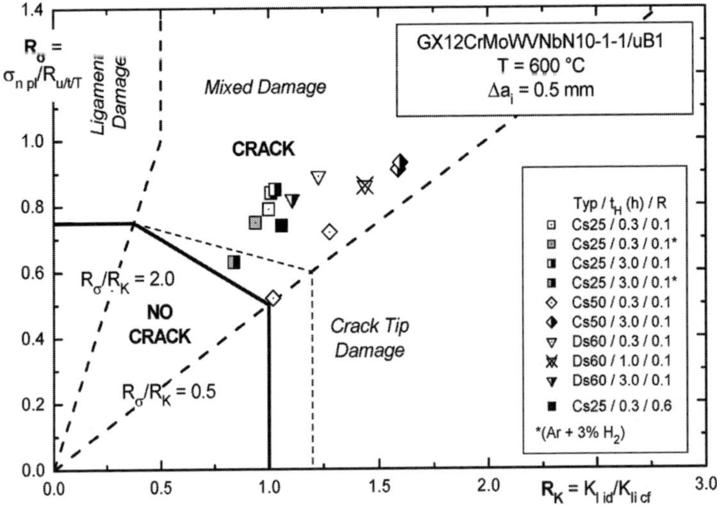

Figure 8. Modified Two-Criteria-Diagram for Creep-Fatigue Loading, 10Cr-cast steel, 600 °C

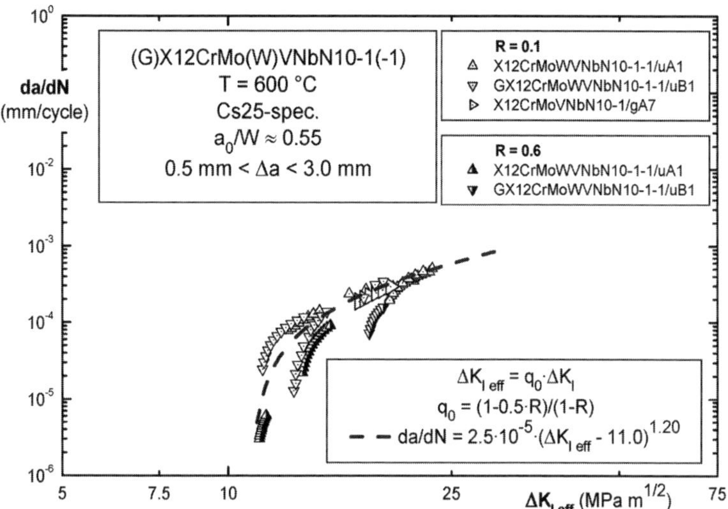

Figure 9. Fatigue Crack Growth Rate vs. $\Delta K_{I\ eff}$, 10Cr-steels, 600 °C

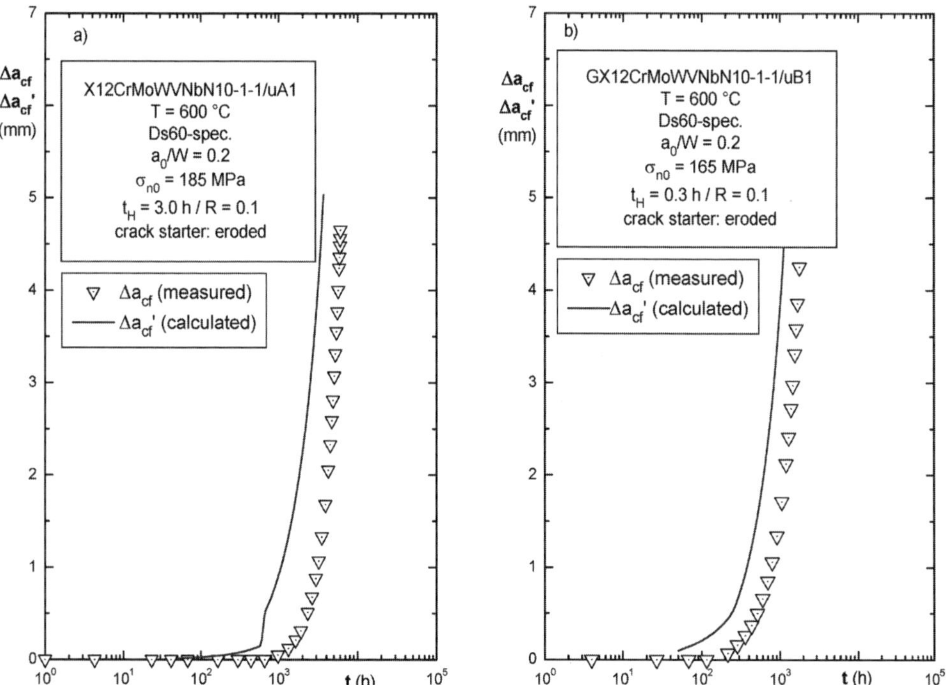

Figure 10. Recalculation of Creep-Fatigue Crack Initiation and Growth on component-like Ds60-specimens made of 10Cr-cast steel (a) and 10Cr-forged steel (b)

TABLES

	Weigth - %											
Material	C	Si	Mn	P	S	Cr	Mo	Nb	Ni	N	V	W
X12CrMoWVNbN10-1-1 uA1	0,12	0,10	0,42	0,007	0,001	10,70	1,04	0,050	0,76	0,056	0,16	1,04
GX12CrMoWVNbN10-1-1 uB1	0,13	0,29	0,82	0,014	0,005	9,51	1,02	0,059	0,52	0,041	0,19	1,02
X12CrMoVNbN10-1 gA7	0,12	0,05	0,53	0,010	0,003	10,16	1,46	0,052	0,55	0,046	0,18	-
GX12CrMoVNbN9-1 gB7	0,13	0,24	0,55	0,004	0,004	9,00	0,93	0,070	0,36	0,046	0,21	-

Table 1. Chemical composition of the test materials

Material	Product form, Dimensions
X12CrMoWVNbN10-1-1 uA1	rotor, ∅ 400 mm, L = 6 750 mm
GX12CrMoWVNbN10-1-1 uB1	block, 500 × 340 × 300 mm^3 block, 500 × 340 × 150 mm^3
X12CrMoVNbN10-1 gA7	1/2 disk (rotor), D_a = 1 200 mm, D_i = 840 mm, B = 60 mm
GX12CrMoVNbN9-1 gB7	ring, D_a = 620 mm, D_i = 420 mm B = 100 mm

Table 2. Component dimensions of the test material

Advances in Materials Technology for Fossil Power Plants
Proceedings from the Fifth International Conference
R. Viswanathan, D. Gandy, K. Coleman, editors, p 733-747

DOI: 10.1361/cp2007epri0733

A DESIGN PERSPECTIVE OF ELEVATED TEMPERATURE MATERIAL BEHAVIOUR

Robert Jetter
Consultant
1106 Wildcat Canyon Rd.
Pebble Beach, CA 93953

Abstract

The purpose of this paper is to discuss elevated temperature materials behavior from the perspective of the component designer/stress analyst and the developer of elevated temperature design rules and criteria. Some issues associated with the design and structural integrity evaluation process are discussed including application of criteria supporting elevated temperature design of nuclear components to non-nuclear power and petrochemical components, particularly those subjected to significant cyclic service. The fundamental problem is how to relate specimen data gathered over limited time, loading conditions and geometry to complex structures subject to intermixed long and short term variable loading. Finally, a path forward is suggested for the development of elevated temperature design criteria for power and petrochemical plant components operated in cyclic service in the creep regime.

Introduction

In the last several years, there has been increasing interest in elevated temperature design criteria; in large part due to the interest in the design and deployment of the gas cooled Very High Temperature Reactor (VHTR). There has likewise been interest in extending the permitted operating temperatures for some of the materials of construction in Section VIII Div 2 and 3 and in the development of cyclic design procedures in the creep regime for Section I construction. Many of the issues associated with these interests have been discussed in the Subgroup Elevated Temperature Design (SG-ETD) (SC-D) and there is currently a DOE/ASME Materials Project to address some of the VHTR issues.

Within the ASME B&PV Code, the only rules for quantitative assessment of cyclic loading of vessels at elevated temperatures in the creep regime are in Subsection NH "Class 1 Components in Elevated Temperature Service" (1). Restrained thermal expansion stresses in piping systems are addressed at elevated temperatures in B31.1, "Power Piping" (2) and B31.3, "Process Piping" (3) but neither address the effects of stresses due to local thermal gradients. Sections I (4) and VIII, Div 1 (5) apply in the creep regime but neither have quantitative rules for cyclic loading conditions. Sections VIII, Div 2 (6) and Div 3 (7) do have quantitative rules for cyclic loading but they are not applicable where creep effects are significant. However, the general consensus of the non-nuclear B&PV Code Committees, when briefed on the rules of Subsection NH, has been that those rules are too complex and conservative to be used for non-nuclear components even though there are, apparently, numerous cases where individual subparts of NH have been used for specific design issues. The following discussion will highlight a number of these and related issues along with some thoughts on potential approaches to elevated temperature design criteria for power and petrochemical applications.

Design Criteria for Load Controlled Stress

One of the difficult areas in current elevated temperature design standards, shared with criteria for loading below the creep regime, is the difficulty and potential conservatism involved in separating primary, load controlled stresses from secondary and peak, displacement controlled stresses. A related issue is the problem of load determination in structures with redundant load paths, i.e. statically indeterminate structures.

The conceptual basis for the current design by analysis methodology was established several decades ago based on a methodology wherein elastically calculated stresses are sequentially categorized based on the relevant failure mode. Primary stresses, those which normally determine wall thickness, are first determined by separating the structure into simpler segments (free bodies) in equilibrium with external loads. Next, secondary and peak stresses (which in combination with primary stresses normally determine cyclic life) are determined from stresses at discontinuities and induced thermal stresses. Different allowable stresses are assigned to the different stress categories based on the failure mode of concern. This sequential methodology works well with "hand analysis" but difficulties can arise with today's wide spread use of Finite Element Analyses (FEA). These difficulties are a result of the need to deconstruct the results of FEA in order to determine which portion of the total stress from the FEA falls within the various stress categories for comparison with their appropriate allowable values.

Plastic analysis, and, in particular, limit load analysis is a method to avoid the process of deconstructing FEA results and to credit the reserve load capacity of redundant load paths. Plastic analysis and limit loads are permitted in Subsection NB (8) for temperatures below the creep regime and are the preferred approach for the newly drafted rules for Section VIII Div 2, also applicable below the creep regime.

There is a procedure applicable in the creep regime, analogous to limit load procedures below the creep range, usually referred to as the reference stress approach. In its basic formulation, the reference stress is given by:

$$\sigma_R = (P/P_L)\sigma_y \tag{1}$$

where σ_R and σ_y are the reference stress and yield strength respectively, P is the load on the structure and P_L is the limit load expressed as a function of the geometry of the structure and its yield strength. While the reference stress approach has not been incorporated into the ASME Code, it is used in the British code for elevated temperature nuclear systems, R5, and in the recent European Standard, EN 132445. However, the British code recommends an adjustment factor for design of 1.2 times the basic reference stress given by Eq. (1) and the EN standard cautions against the use of the reference stress method by those not familiar with its use.

One potential cause for concern can be illustrated with the following example. Consider an indeterminate structure consisting of two parallel bars of equal area, A, lengths L_1 and L_2, constrained in the lateral direction and loaded by P. The stationary creep stress solution for the stress in the bars can be expressed as:

$$\sigma_1 = [L_2^{1/n}/(L_1^{1/n} + L_2^{1/n})]P/A \tag{2}$$

$$\sigma_2 = [L_1^{1/n}/(L_1^{1/n} + L_2^{1/n})]P/A \tag{3}$$

where σ_1 and σ_2 are the stress in bars 1 and 2, respectively, and n is the stress exponent in the familiar power law representation of the steady creep rate. If $n = 1$, Eq. (2) and (3) become the elastic stress distribution. As $n \rightarrow \infty$, Eq. (2) and (3) become the reference stress solution given by Eq. (1).

Table (1) shows solutions for the stress in bars 1 and 2 as a ratio to the NH methodology, which presumes a single designated load path for primary loads. The first row is the single load path solution. The second row is the solution based on the elastic deflection characteristics of both bars. The third row is the stationary creep solution for $n = 3$ and the last row provides the reference stress solution. The columns headed $L_1 = L_2/8$ represent an unbalanced system with one bar much longer than the other and the columns headed $L_1 = L_2/2$ represent a more balanced system. The advantage of accounting for load carrying redundancy is apparent; a factor of two increase in loading capacity if the evaluation is based on the reference stress.

However, the difference between the balanced and unbalanced configurations highlights a potential problem. As is the case for limit loads, the reference stress is an instability criterion, not a damage based criterion. Inherent to the concept is that there is enough ductility for the structure to deform and redistribute the internal loads prior to local failure. This is a reasonable premise below the creep regime where code approved materials can be presumed to have substantial ductility prior to failure. This may not be the case in the creep regime where ductility prior to rupture can be considerably lower, particularly for the low allowable stress levels and strain rates associated with sustained loading conditions.

Table (1) Stress Distribution in a Two Bar Model

	$L_1 = L_2/8$		$L_1 = L_2/2$	
	Bar 1	**Bar 2**	**Bar 1**	**Bar 2**
Single load path	1.0	-	1.0	-
Elastic	0.89	0.11	0.67	0.33
Stationary creep, n = 3	0.67	0.33	0.56	0.44
Reference stress	0.50	0.50	0.50	0.50

In the above unbalanced example, the strain in the shorter, more highly loaded bar is a factor of 8 greater than the longer, more lightly loaded redundant bar. Particularly in the creep regime, there is, thus, the potential for failure of the more highly loaded bar and subsequent premature load transfer to the redundant load path before the advantage of the reference stress load redistribution is realized. If the more highly loaded bar fails prematurely, the load in the more flexible redundant bar will increase by a factor of two greater than predicted from the reference stress criterion. In the more balanced system, the strains are not so different and the stationary creep stress distribution is reasonably close in the two bars, roughly within the 1.2 design factor recommended by R5.

The above discussion highlights a potential problem with the application of reference stress methods to unbalanced structures. However, Penny and Marriott (9) have reported a theoretical and experimental study of a pressurized sphere/cylinder intersection that showed a good correlation between the predicted rupture time using the reference stress and experimental results. Two aluminum vessels were tested and the reference stress predictions of rupture time using small deflection theory were conservative by factors of approximately 0.35 to 0.8. By comparison, the failure life prediction using the initial elastically calculated stress was conservative by factors of 0.01 to 0.04; again demonstrating the advantage of the reference stress approach where stress/strain redistribution is taking place. It may be that the sphere/cylinder intersection is an example of a well balanced system. What is lacking is a quantitative method for differentiating unbalanced systems.

Allowable Stresses

Time Dependent Stress Criteria

ASME Standards Technology LLC (ST- LLC) recently issued Requests for Proposals (RFPs) for a number of issues including development of time dependent allowable stress values for use in Section VIII, Div 1 construction. There are two aspects of this endeavor. The first is to accommodate loading conditions occurring at higher temperature and shorter time durations than those covered by Design Conditions based on normal operation. Typically, this might be some sort of "bake-out" cycle in a petrochemical plant. The other aspect would be to have a more quantitative method for evaluating the design margin for operating conditions that extend well beyond 100, 000hr; perhaps, even several 100,000hr.

There are precedents for accommodating both these issues. Subsection NH has design procedures to accommodate different time duration loadings at different temperatures. For primary, load controlled stresses, the procedure is a use fraction summation. For peak stresses associated with cyclic loading, the procedure is a linear life fraction summation where fatigue damage and creep damage are summed separately and their interaction evaluated by an allowable combined damage factor.

The criteria for establishing allowable stress values in NH are somewhat different than those used for non nuclear applications, e.g. for each specified time, t, the allowable time dependent stress is the lesser of 67% of the minimum stress to cause rupture, 80% of the minimum stress to cause initiation of third stage creep and 100% of the average stress for a total strain of 1%. The criteria for Section VIII, Div 1 and Section I allowable stress values are based on creep properties referenced to 100,000hr and are given by the lesser of 67% of the average or 80% of the minimum stress to cause rupture and 100% of the minimum stress to cause a minimum (secondary) creep rate corresponding to 1% in 100,000hr.

There is also a precedent for time dependent allowable stresses in the in the European EN standard, EN 132445. In the EN standard, the allowable stresses in the creep range are based on the mean stress to cause creep rupture in time, t, and the mean stress to cause a creep strain of 1% in time, t. There is a factor of safety applied to the mean creep rupture stress equal to 2/3 if there is no in-service monitoring and 0.8 if there is. There is no safety factor on the strain. However, if there is in-service monitoring, the strain limit does not apply, but strain monitoring is required.

Using the average creep rupture strength as representative of the mean value and assuming that the ratio of the minimum to average value is represented by their respective allowable stress criteria for Section I/Section VIII, Div 1 applications, the relative conservatism of these approaches to allowable stress values can be determined. For purposes of illustration, it will be assumed that the allowable stress is governed by the average creep rupture strength for a reference time of 100,000hr. Based on these assumptions, the Subsection NH allowable stresses are a factor of 1.2 *more* conservative than those of Sections I/VIII Div 1; whereas, the EN allowable stresses, with in-service monitoring, are a factor of 1.2 *less* conservative than Sections I/VIII Div 1. Without in-service monitoring, the EN allowable stresses are equivalent to Sections I/VIII Div 1.

However, there are additional considerations. The time dependent allowable stresses under NH are used in the evaluation of operating conditions, not Design Conditions as is the case for Sections I/VIII, Div 1 and, presumably, the EN standard. Since Design Conditions are typically less severe than operating conditions, Subsection NH is in the position of applying more conservatively defined allowable stresses to less conservatively defined design parameters. Thus, the net effect is that the inherent conservatism of Subsection NH more nearly approximates Sections I and VIII, Div 1 than would be deduced from the allowable stress values alone.

In the case of the EN allowable stress values, the higher values are used in conjunction with in service monitoring which should ensure the component is not exposed to unexpected overloads and, based on the required strain measurements, the structural response falls within predicted limits. From that perspective, the rationale for higher allowable stress values in conjunction with in-service monitoring is consistent with the rationale for higher allowable stresses in Section VIII, Div 2 as compared to Div 1 based on more rigorous and thorough analysis procedures and explicit consideration of cyclic loading. In both cases, higher allowable stresses are rationalized, in part, by use of more rigorous requirements elsewhere in the overall process of design, construction and operation.

Extension of Allowable Stress Values

One of the tasks in the afore mentioned DOE/ASME Materials Project is to address allowable stress values for Alloy 800H and 9Cr-1Mo-V (Grade 91) steel; both base metal and weldments. Also of interest is the extension of the allowable stress values for 800H to 900 C for time durations to 500,000hr. In reviewing the data to extend the allowable stresses for 800H it was determined that there was enough information to extrapolate the stress rupture parameters. It was also found that for longer times and higher temperatures the allowable stress values were governed by the 1% total strain criteria instead of the creep rupture criteria which governed at lower temperatures. Additional data acquisition is recommended to sort out the influence of diffusion creep at long times, low stresses and very high temperatures.

Design Criteria for Displacement Controlled Stress

Unlike the primary stress evaluation criteria in NH for load controlled stresses, which are based on elastic analysis results, the criteria for evaluation of displacement controlled stresses, important for ratcheting and fatigue due to cyclic loading, are based on either elastic analysis or inelastic analysis. Neither the elastic nor inelastic analysis route is without its unique challenges.

Inelastic Analysis Criteria

Conceptually, the inelastic analysis methodology is the most straightforward. Using appropriate constitutive material models, the stress and strain history at the location of interest are determined for a representative segment of the total component design life. First, the accumulated strain summed over the whole life is compared to the strain limits; 1% membrane, 2% bending and 5% total. Next, the creep-fatigue damage due to cyclic loading is calculated using linear life fraction summations where fatigue damage and creep damage are summed separately and their interaction evaluated by an allowable combined damage factor. Although the actual evaluation uses readily available properties, fatigue life and creep rupture data; there are issues in the development of the material behavior models to predict the stress and strain history and the determination of the creep-fatigue interaction factor.

Subsection NH doesn't include constitutive models for the permitted materials of construction, only general guidance on what key features should be considered. There was an extensive Department of Energy (DOE) program that was undertaken to define and validate constitutive models for application to the Liquid Metal cooled Fast Breeder Reactor (LMFBR). The modeling guidance from that program was a contractual requirement for the LMFBR. It is not clear how or even if a similar effort could be accomplished for materials of primary interest to the petrochemical and power industries.

Creep-Fatigue Interaction. A second, related issue is the determination of the allowable creep-fatigue damage factor. A number of tests with different combinations of hold time, cycle magnitude and temperature are required. Interpretation of the results is further clouded by the fact that the creep damage is a calculated quantity that can not be measured directly. Thus, it is subject to the material modeling complexity described above. Also, as has been increasingly apparent in the recent DOE sponsored evaluation of the creep-fatigue properties of 9Cr-1M0-V (Grade 91 steel), it is necessary to take into account the change in material properties due to cycling and aging. The computed creep damage for Grade 91 is particularly sensitive to cyclic softening effects. Revisions to the current creep-fatigue criteria for Grade 91 are being explored in the aforementioned DOE/ASME Materials Project.

Elastic Analysis Criteria

In the late 1960s and 1970s, when rules in Subsection NH were developed, FEA capabilities were much more limited and much more time consuming and expense than now. In particular, even if the appropriate material models were available, the routine application of inelastic analyses to design would have been prohibitively expensive and time consuming. As an alternate, methods were developed that relied only on the results of elastic analyses. Conceptually, the idea was that analyst would first attempt to satisfy the presumably simpler rules based on the results of elastic analyses and only resort to inelastic analyses if it was not practical to either modify the design or operating conditions such that the elastic analysis rules could be satisfied.

Conceptually, the rules for elastic analysis are based on simplified mechanistic models, such as the familiar elastic shakedown model, which can be used to approximate and/or limit inelastic response. A detailed description of the models employed in Subsection NH is beyond the scope of this discussion but more information and further references can be found in Chapter 12 of the background document "Companion Guide to the ASME Boiler & Pressure Vessel Code" (10). The problem with rules based on elastic analyses is that if they are relatively simple, the approximations involved tend to be overly conservative and if they are more accurate, it comes with the price of increasing complexity. An example of the former is Tests A-1 and A-2 in subsection NH whose objective is to ensure that the maximum primary stress intensity plus the range of secondary stress intensity does not exceed the yield strength. An example of the latter is the elastic analysis rules for creep-fatigue damage in Subsection NH. The starting point for the NH creep-fatigue evaluation procedure, which involves a number of very specific steps and deconstruction of FEA results, is the output of the Section VIII, Div 2/Subsection NB fatigue evaluation. Although other international codes differ in the specifics of their approach, the complexity of procedures based on the results of elastic analyses is a common problem.

The complexity of the NH elastic analysis creep fatigue procedure is largely due to the need to account for time dependent strain and stress redistribution due to creep. The basic problem is that creep-fatigue data are generated from compact specimens tested under uniaxial, isothermal, strain controlled conditions whereas, in actual design environments, none of the above are generally the case; thus, the afore mentioned need to use simplified, mechanistic models to bound the inelastic response due to creep.

Alternate Approach to Creep-Fatigue. To avoid the difficulties associated with separating creep-fatigue damage into to separately calculated parts with their attendant approximations and conservatisms, an alternate approach has been proposed, Jetter(11), based on the use of hold time creep-fatigue test data which includes stress and strain redistribution effects representative of real structures. In this approach, called the

Simplified Model Test (SMT), the test is designed to have follow-up characteristics conservatively bounding the stress and strain redistribution that occurs in structural components of interest. In this approach, creep and fatigue damage are not calculated separately and the results of elastic analyses are used directly in a design evaluation procedure similar to the current methods in Section VIII, Div 2 and Subsection NB.

The SMT approach is shown conceptually in Figure 1. Hold time creep-fatigue data are generated using a test specimen which includes representative stress and strain redistribution characteristics. From that data, cyclic design curves are generated analogous to the current design fatigue curves which do not include the effects of creep. An important point is that the data and resultant design curve are based on the elastically calculated strain in the test specimen so that the elastically calculated component strain can be used directly with the cyclic design curve. In effect, the component is being replicated with a simplified model, tested to the same elastically calculated strain range. As noted in the referenced paper, although some limited supporting analyses and test data were presented, additional analyses and testing are required to verify the concept

Exemption from Fatigue Analysis. An important feature of Section VIII, Div 2 and Subsection NB is the criteria which permit exemption from detailed fatigue analyses. These criteria set a limit on the number and type of pressure cycles based on a fixed limit, 1000 for integral parts, or as a function of the allowable cycles from the design fatigue curve as a function of the allowable stress for the material of construction. Because of the effects of hold time on strain controlled cyclic life and stress and strain redistribution due to creep, these criteria do not provide the same assurance in the temperature regime where creep is significant as they do at temperature below the creep regime. For that reason, Subsection NH does not have criteria comparable to that of Section VIII, Div 2 and Subsection NB for exempting detailed fatigue analysis. In discussions at SG-ETD meetings it has been pointed out by Koves (12) and others that for many power and petrochemical applications there are relatively few significant cycles and that exemption from fatigue criteria would provide a practical alternative for many applications.

Negligible Creep Criteria. Although Subsection NH does not provide exemption from fatigue requirements in the creep regime, it does permit use of rules in Subsection NB for evaluation of cyclic service if the NH criteria for negligible creep are satisfied. The current negligible creep criteria in NH are based on two considerations. The first consideration is that the creep damage at the flow stress should be less than 10% over the design life of the component and the other consideration is that the accumulated strain due to creep should be less than 0.2%. It should be noted that these criteria, which are based on local stresses, result in lower temperatures for negligible creep than criteria which are based on primary stress levels only. That is because primary stresses are limited to lower levels than localized secondary and peak stresses and, since creep is stress dependent, creep effects become significant at lower temperatures for the higher localized stresses. The negligible creep criteria in Subsection NH are currently under review as part of the DOE/ASME Materials Project. An important observation by Riou (13) is that the NH criteria have factors to account for cyclic strain hardening that are inappropriate for strain softening materials such as Grade 91 steel. Smoothed curves for time-temperature limits based on the above NH criteria are shown in Appendix E of Code Case N-253 for a limited number of materials.

Welds and Weld Strength Reduction Factors

There are several ways that Subsection NH deals with welds; weld strength reduction factors are provided to account for the degradation of creep rupture strength of weld metal as compared to base metal at higher temperature and longer times, the allowable strain at welds is half that permitted for base metal and the and the allowable cycles from the design fatigue curve for weld metal is a factor of two less tan that for the corresponding base metal. Also, the weld strength reduction factors are provided for specific weld rods and processes, thus restricting Subsection NH to those specific combinations. EN 132445 also provides for the use of weld strength reduction factors as does B31.3, Process Piping.

Suggestions for Power and Petrochemical Component Design Criteria in the Creep Regime

Based on the preceding discussion, the following is presented as a suggested path forward to the development of elevated temperature design criteria for power and petrochemical plant components operated in cyclic service in the creep regime. Both near term and longer term options are presented. The goal of these recommendations, both near and longer term, is to have rules that are no more complex to implement than the current design-by-analysis methods of Section VIII, Div 2 and Subsection NB.

Allowable Stresses and Required Wall Thickness

For the near term, the required wall thickness would be based on the thickness formulas in Sections I and VIII, Div 1 as applicable. If in-service monitoring is invoked, particularly if it is in conjunction with explicit consideration of cyclic damage, then it would be appropriate to reduce the required wall thickness to a percentage of that required by the Section I and VIII thickness formulas. Based on the EN 132445 precedent for allowable stress values, a reduction to 83% of the previously required thickness would be appropriate. Time dependent allowable stresses should be developed as should weld strength reduction factors. In both cases, the approaches in Subsection NH, EN 132445 and B31.3 should be considered.

For the longer term, incorporation of the reference stress methodology should be considered provided that issues related to adequate creep ductility and unbalanced systems are resolved.

Cyclic Loading Evaluation

There are two main aspects of cyclic load evaluation; (a) strain limits and ratcheting which are analogous to the primary plus secondary stress limits in Section VIII, Div 2 and Subsection NB, and (b) creep-fatigue damage rules.

Strain Limits and Ratcheting. There are two approaches in Subsection NH for satisfying strain limits using elastic analysis results. The more conservative approaches in T-1320 require the maximum primary stress plus the secondary stress range to be below the average yield strength. If satisfied, a strain calculation is not required. The less conservative but more complex approaches in T-1330 evaluate strain limits and ratcheting based on the Bree (14) model of a pressurized cylinder with cyclic thermal gradients and extensions based on the work of O'Donnel and Porowski (15) and Sartory (16) to bound accumulated strain. For the near term, one approach would be to simplify and streamline these procedures; however, the result would still be more complex and require more FEA deconstruction than the analogous primary plus secondary stress limits below the creep regime.

An alternate approach, based on creep modified shakedown is suggested. This useful concept is shown in Figure 2. The criteria for shakedown in the creep regime, i.e. no yielding in subsequent load applications, is that the applied strain range doesn't exceed the yield stress at the cold (or short duration) end of the cycle plus the stress remaining after full relaxation of the yield stress over the life of the component. If both ends of the cycle are in the creep regime, then the shakedown criteria are based on the sum of the relaxation strength at either end of the cycle. Interestingly, satisfaction of this criterion is already a requirement in Subsection NH for evaluation of creep-fatigue using elastic analyses; see T-1431*(2)*. Thus, using the notation of Section VIII, Div 2, the limit becomes:

$$(P_L + P_b + Q)_R \leq S_{PS} \tag{4}$$

where $S_{PS} = 1.5S_m + S_{rH}$ if one end of the cycle is below the creep regime and $S_{rH} + S_{rL}$ if both ends of the cycle are within the creep regime. S_{rH} and S_{rL} = relaxation strengths associated with the hot and cold ends of the cycle, respectively. The yield stremgth is approximated by $1.5S_m$.

This approach has the advantage of relative simplicity corresponding to established criteria at lower temperatures and it would be generally applicable without some of the restrictions of the current criteria in Subsection NH. It is also consistent with the use of elastic analyses for evaluation of creep-fatigue.

For the longer term, methods using FEA of the specific geometry and cyclic loading to determine shakedown characteristics and strain accumulation might be considered. They would have the advantage of being less conservative than the creep modified shakedown approach and probably more accurate and without some of the restriction of the current criteria; however, it is not clear that these methods would achieve the objective of complexity no greater than the existing procedures applicable below the creep regime.

Creep-Fatigue. Development of criteria for evaluating creep-fatigue that are not either overly conservative or unduly complex is a difficult task. For the near term, perhaps the best that can be done is to start with the current creep-fatigue evaluation criteria in Subsection NH that are based on elastic analyses. As currently shown in NH, the rules have several optional paths of varying complexity and conservatism, factors which must be determined to account for such effects as multiaxiality, Poisson ratio variability, average as opposed to minimum properties, and, in the case of creep damage determination, a multistep process to sort out variable hold times and the effect of sustained loads. In addition, there are a number of cases of cross referencing to other paragraphs to define specific parameters. It is suggested that the existing rules could be streamlined and simplified to eliminate some of the options, develop constants to approximate some of the variable factors, assume uniform hold times and eliminate cross referencing. While this would greatly simplify the rules, they would still be significantly more complex than existing procedures where creep is not a consideration.

There are several other approaches that might be used to address the complexity and/or the conservatism of the current rules for evaluation of creep-fatigue damage in Subsection NH. This complexity and conservatism can be attributed to two main factors; (a) the need to consider the effects of inelasticity due to creep in the determination of the strain range and (b) the need to separately account for creep damage including the effects creep on stress and strain redistribution. Two approaches that deal directly with these issues are (a) the Simplified Model Test (SMT) approach as previously discussed and (b) the use of isochronous stress curves to estimate the strain range to be used in conjunction with creep-fatigue design curves based on strain controlled hold time data representing saturation of the hold time effect.

With either the SMT approach or the use of isochrnous stress-strain curves, the allowable number of cycles would be based on a design fsatigue curve adjusted for the effects of creep damage. In the case of the SMT approach, the effect of stress and strain redistribution due to creep are built into the design curve through the use of a test specimen which models those effects. In the isochronous stress-strain approach, the stress and strain redistribution during the hold time is estimated by the calculated strain at the end of the cycle. Experience with isochronous stress-strain calculations has shown quite good agreement with the results of detailed inelastic FEA for monotonic loading cycles. This makes the approach most suitable for gradual startup and shutdown cycles where transient stress and strain values do not exceed sustained levels.

Exemption from Fatigue Analysis. Development of criteria for exemption from fatigue analysis is a high near term priority. The suggested approach is to use bounding assumptions similar to those used in the criteria for temperatures below the creep regime with the following modifications. The strain amplitude should be increased using the procedures for elastic creep-fatigue evaluation and the fatigue curve should be reduced by bounding assumptions on the effects of hold time and elastic follow-up. This will quite likely result in fewer permitted cycles and, perhaps, tighter restrictions on thermal stress magnitudes; however, these criteria should still provide considerable relief from the requirements for detailed analysis.

References

1. 2004 ASME Boiler and Pressure Vessel Code. Section III, Division 1, Subsection NH, *Class 1 Components in Elevated Temperature Service*; The American Society of Mechanical Engineers

2. ASME B31.1, *Power Piping*; The American Society of Mechanical Engineers

3. ASME B31.3, *Process Piping*; The American Society of Mechanical Engineers

4. 2004 ASME Boiler and Pressure Vessel Code, Section I, *Rules for Construction of Power Boilers;* The American Society of Mechanical Engineers

5. 2004 ASME Boiler and Pressure Vessel Code, Section VIII, Div 1, *Rules for Construction of Pressure Vessels;* The American Society of Mechanical Engineers

6. 2004 ASME Boiler and Pressure Vessel Code, Section VIII, Div 2, *Alternative Rules;* The American Society of Mechanical Engineers

7. 2004 ASME Boiler and Pressure Vessel Code, Section VIII, Div 3, *Alternative Rules for Construction of High Pressure Vessels;* The American Society of Mechanical Engineers

8. 2004 ASME Boiler and Pressure Vessel Code. Section III, Division 1, Subsection Nb, *Class 1 Components;* The American Society of Mechanical Engineers

9. R. K. Penny and D. Marriott, *Design for Creep;* McGraw-Hill Book Co., Ltd, London (1971)

10. *Companion Guide to the ASME Boiler & Pressure Vessel Code, Second Edition*; K. R. Rao, Editor, The American Society of Mechanical Engineers (2006)

11. R. I. Jetter, "An Alternate Approach to Evaluation of Creep-Fatigue Damage for High Temperature Structural Design Criteria" PVP-Vol. 5, *Fatigue, Fracture, and High Temperature Design Methods in Pressure Vessels and Piping*; Book No. H01148 - 1998, The American Society of Mechanical Engineers

12. W. Koves, personal communication

13. B. Riou, "Improvement of ASME NH for Grade 91 (negligible creep)" Areva Technical Data Report, 1/24/07

14. J. Bree, "Incremental Growth Due to Creep and Plastic Yielding of Tubes Subjected to Internal Pressure and Cyclic Thermal Stresses" *Journal of Strain Analysis*, Vol. 3, No. 2, 1968

15. W. J. O'Donnell and J. S. Porowski, "Upper Bounds for Accumulated Strains Due to Creep Ratcheting" *Trans ASME, Journal of Pressure Vessel Technology*, Vol. 96, 1974

16. W. K. Sartory, "Effect of Peak Thermal Stress on Simplified Ratcheting Analysis Procedures" PVP-Vol. 163, *Structural Design for Elevated Temperature Environments – Creep, Ratchet Fatigue and Fracture*, Book No. H00478 - 1989

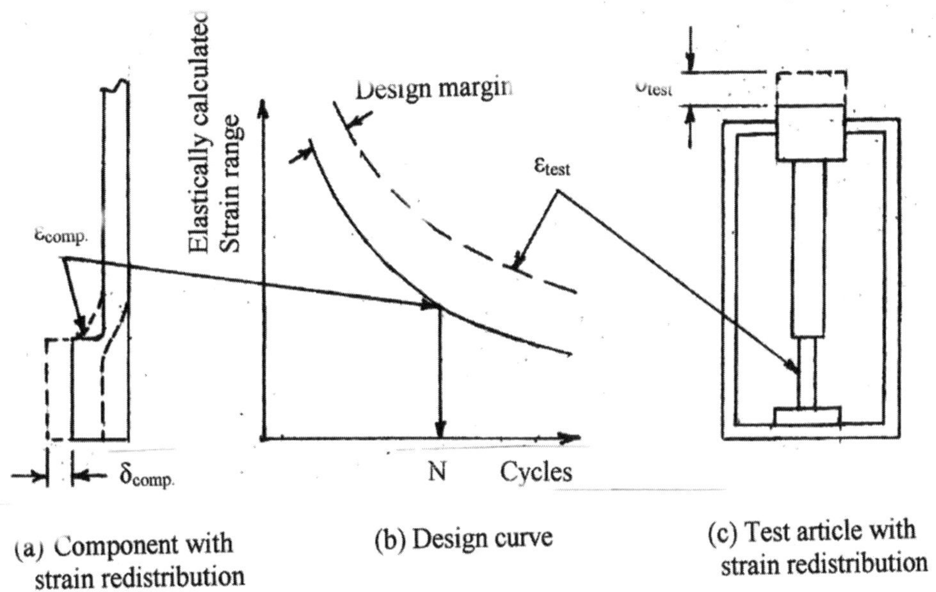

Figure 1. SMT Methodology

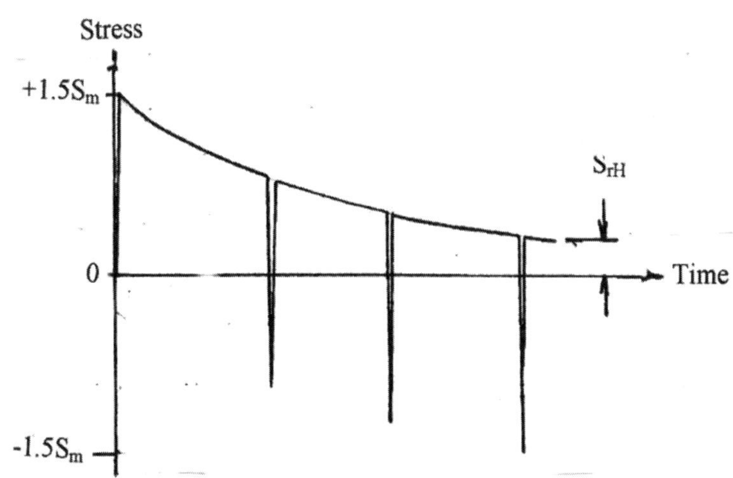

Figure 2. Creep Modified Shakedown

Advances in Materials Technology for Fossil Power Plants
Proceedings from the Fifth International Conference
R. Viswanathan, D. Gandy, K. Coleman, editors, p 748-761

DOI: 10.1361/cp2007epri0748

Improved Methods of Creep-Fatigue Life Assessment of Components

A. Scholz, C. Berger
Institute of Materials Technology (IfW), Darmstadt University of Technology
Grafenstrasse 2, 64283 Darmstadt, Germany

Abstract

Improved life assessment methods contribute to an effective long term operation of high temperature components, reduces technical risk and increases high economical advantages.

Creep-fatigue at multi-stage loading, covering cold start, warm start and hot start cycles in typical loading sequences e.g. for medium loaded power plants, was investigated here. At hold times creep and stress relaxation, respectively, lead to an acceleration of crack initiation. Creep fatigue life time can be calculated by a modified damage accumulation rule, which considers the fatigue fraction rule for fatigue damage and the life fraction rule for creep damage. Mean stress effects, internal stress and interaction effects of creep and fatigue are considered.

Along with the generation of advanced creep data, fatigue data and creep fatigue data as well scatter band analyses are necessary in order to generate design curves and lower bound properties inclusive. Besides, in order to improve lifing methods the enhancement of modelling activities for deformation and life time are important. For verification purposes, complex experiments at variable creep conditions as well as at creep fatigue interaction under multi-stage loading are of interest. Generally, the development of methods to transfer uniaxial material properties to multiaxial loading situations is a current challenge.

Further, a constitutive material model is introduced which is implemented as a user subroutine for Finite Element applications due to start-up and shut-down phases of components. Identification of material parameters have been achieved by Neural Networks.

Keywords

advanced 9-12% Cr–steels, 1%Cr-steel, creep fatigue, multiaxiality, fatigue scatter band assessment, stress relaxation, multi-stage loading, life time assessment, simulation

Introduction

Life time of components of power plants depends in most cases on variable loading conditions. This concerns fatigue and creep-fatigue as well as crack initiation and crack propagation. The normal variable service conditions which contain phases of start-up, full load, partial load and shut-down cause variable stress-strain distributions and temperature transients and lead to a large variety of combined static (primary load) and variable loading (secondary load) situations.

A wide range of loading parameters have been introduced in order to simulate complex loading of components. Conventional testing of temperature induced loading deals with fatigue experiments representing the cyclic loading at the heated surface of components, while creep experiments represent the quasi-static loading phases of components.

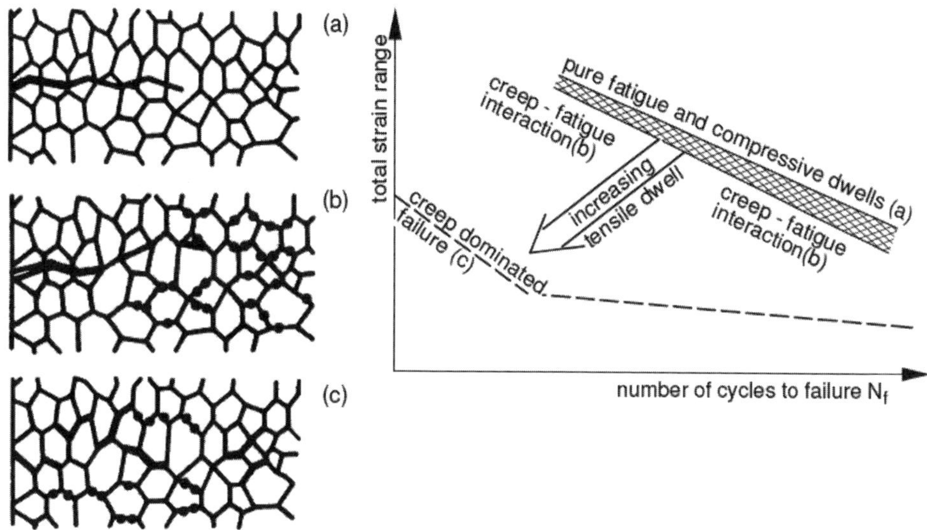

Figure 1. Failure modes at fatigue (transgranular crack path a), creep fatigue superposition (trans- intergranular b) and creep (intergranular c), schematically [2]

A key problem deals with the identification and interpretation of different physical damage mechanisms. Therefore, phenomenological solutions were traditionally introduced and put into practical use. Fatigue cracks occur at the surface of a component (Figure 1), but creep damage is initiated by creep cavities and micro cracks preferably at grain boundaries. Their interaction is intended for conventional heat resistant steels, but their consideration in life assessment methods is not realized to satisfaction. A clear influence on endurances due to the superposition of fatigue and creep has been observed [1]. Increasing tensile hold time decrease the fatigue endurance limit. Hereby, failure mode changes from fatigue dominant to creep dominant condition. At the high strain regime damage is characterized as fatigue dominated. With decreasing strain creep damage predominates. It is known from the literature [2] that hold times at tension and compression stage can have a material individual influence on endurance. Additionally, oxidation effects can contribute significantly to the reduction of endurance limits [2]. Summarizing, either fatigue damage or creep damage prevail, or they may interact under variation of strain range, tension and/or compression hold time, frequency, temperature and ductility of material.

The initial stage of fatigue failure is characterized by dislocation processes which lead to surface defects. This is followed by growth processes by bulk deformation, and is completed by tearing up a small remaining ligament. Creep damage due to nucleation is much more difficult to define. It is assumed that microstructure damage is nucleated in the creep life and can be interpreted as the onset of tertiary creep.

Creep-Fatigue

Generally, conventional mechanical experiments are needed to investigate the material properties in order to provide a consistent basis for quality control purposes and for design data. In addition, complex experiments may represent service conditions and contribute to verification purposes of life prediction methods.

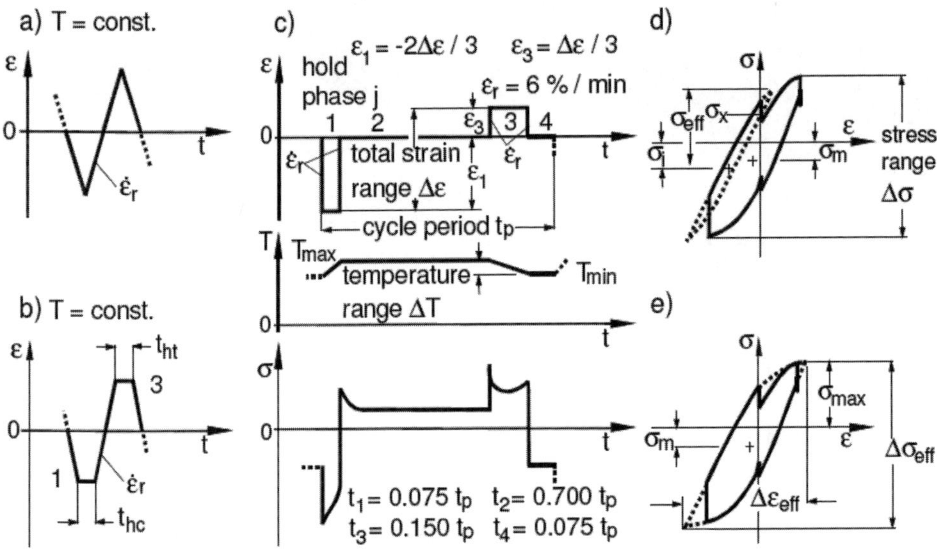

Figure 2. Different strain cycles simulating the conditions at the heated surface of heavy components: standard cycle without (a) and with (b) hold times as well as service-type cycle (c), the stress-strain path (d and e) show mean stress σ_m, maximum stress σ_{max}, internal stress σ_i , effective stress σ_{eff} and corresponding effective values [1]

The creep fatigue behaviour at the heated surface of heavy components, such as turbine rotors, normally is investigated by conventional low cycle fatigue experiments with standard (LCF-) cycles (Figure 2a) and creep-fatigue cycles with dwell periods at maximum and minimum strain (Figure 2b). In contrast, a single-stage service-type strain cycle (Figure 2c to e) was developed [1, 3, 4], which is characterized by a compressive strain hold phase 1 simulating start-up condition, a zero strain hold phase 2, with approximate temperature equilibration at constant loading, a tensile strain hold phase 3, simulating shut-down conditions and an additional zero strain hold phase 4, which characterizes a zero loading condition.

Non-isothermal experiments (type "an") approximate the service conditions more closely than isothermal experiments. They were carried out up to failure times of 8 000 h and gave only insignificantly smaller numbers of cycles to failure than comparable isothermal tests (type "iso"). Considering the design life of power plants of up to 200 000 h or more, long term strain cycling is of interest. This could be realized by an isothermal package-type testing procedure ("pa") (Figure 3). It is composed of packages of strain cycles with short hold times which are periodically inserted into much longer creep packages. Maximum test durations of 70 000 h have been achieved at a turbine rotor steel of type 1CrMoNiV [4].

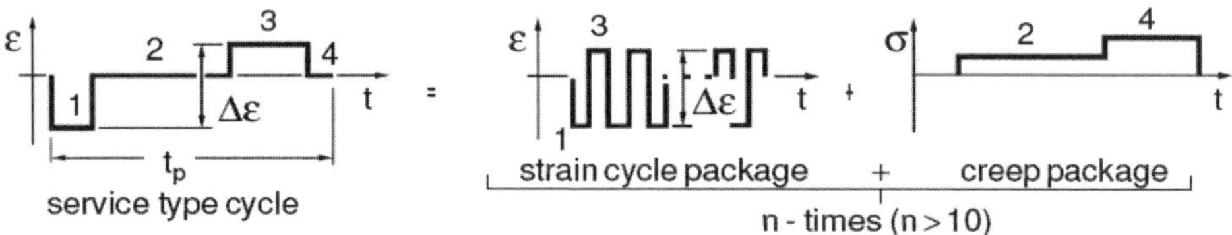

Figure 3. Simulation of long term isothermal service-type testing by isothermal package type tests [3, 4]

The stress-strain path of the service-type experiments were analysed in order to develop a creep-fatigue life assessment procedure. Deformation analyses led to the determination of an effective stress concept, $\sigma_{eff} = \sigma - \sigma_i$ (Figure 2d). The internal stress σ_i for any time of the measured hysteresis loop, e.g. stress σ_{eff}, is defined as the centre of the hypothetical elastic-plastic flank curve loop which is inserted in the flank curve loop enveloping the whole measured loop. The flank curves are derived from a cyclic or quasi-static yield curve by multiplying the latter by a factor of two. The cyclic yield curve can be experimentally determined by a strain cycle without hold times which is inserted into the service-type strain cycling. Due to the different relaxation phases during the hold times, the mean stress σ_m varies from zero (figure 2e) and has to be considered for the life time assessment. For this purpose a mean stress factor v_σ which includes the Smith-Watson-Topper-Parameter [1, 3, 4, 6] can be used.

Life assessment

For life assessment under creep fatigue loading the life fraction rule [1, 5, 7, 8] is widely used. Failure is determined by the summation of fatigue damage D_f as a cycle fraction and creep damage D_c as a time fraction up to a critical creep fatigue value D, whereby D depends on material. In this paper D is used in the sense of life fraction.

The damage mechanisms during creep-fatigue conditions are mainly influenced by microstructural coarsening [7] and cavity growth. Typical mechanisms of cavity growth have been identified in 1CrMoNiV steels during stress relaxation [8]. Grain boundaries may be cavitated in tensile hold time, but cavitation is usually not found in a balanced cycle containing hold time of equal duration. At service-type strain cycling cavitation dominates (Figure 2c and 3).

As indicated in [5], high deformation rate can be associated with transgranular damage, while low deformation rate leads to intergranular damage. Cavities were found at the transition of transgranular damage to mixed-mode damage on a 1CrMoNiV steel at 525°C. At long-term creep-fatigue loading microstructural analyses confirm that the superposition of fatigue and creep which is underlying the damage accumulation hypothesis (Figure 4).

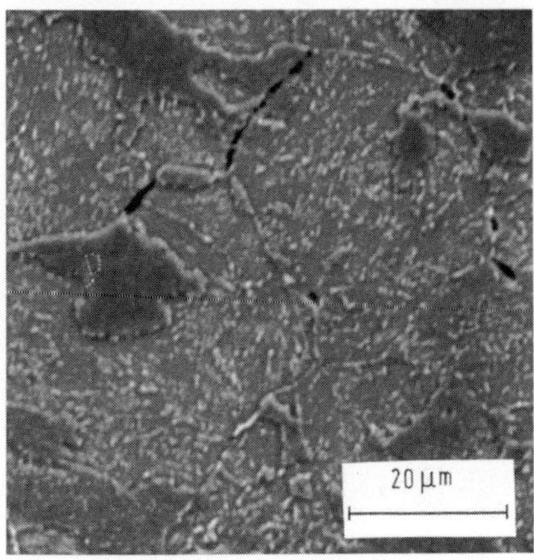

Figure 4. Microstructure with pores and micro cracks beyond of the main crack tip, isothermal service-type strain cycling according to Figure 2b, crack initiation time 31.300 h (T = 525°C, Δε = 0.54%, t_p = 32 h), 1CrMoNiV – steel

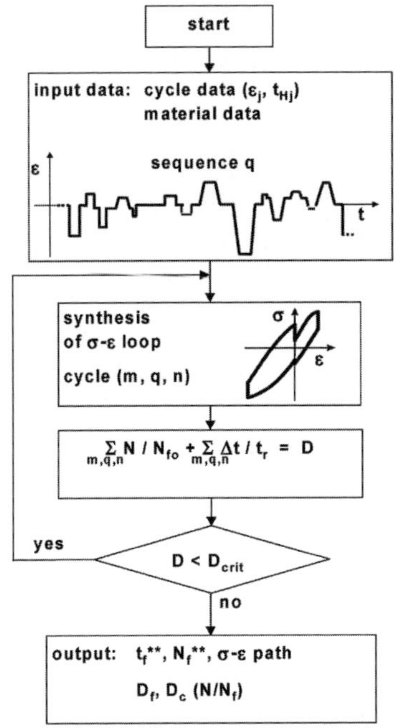

Figure 5. Structure of user program SARA for creep-fatigue life assessment [4, 5]

A prevalent rule for creep-fatigue life analysis in the long-term region is the generalized damage accumulation rule

$$\sum_k \sum_j \left(\Delta t_j / t_{uj}\right) + \sum_k \left(N_k / N_{fok}\right) = D \tag{1}$$

which combines the Miner rule for fatigue damage and the life fraction rule for creep damage [1, 3, 5, 8]. The damage summation over all cycles k including damage at hold times j =1 to 4 leads to a creep-fatigue damage D. For the calculation of creep damage the ratio of time increments t_j and rupture time t_{uj} is considered.

The structure of this relationship is relatively simple and therefore suitable for life assessment (Figure 5) and in practical use for life monitoring systems [9].

This rule was modified empirically with some features in order to cover physical aspects (Figure 6). The reference value of the number of cycles to failure N_{fo}

$$N_{fo} = N_f(\Delta\varepsilon, 2t_{H\,sta}) \cdot \nu_\sigma \tag{2}$$

is taken from standard strain-cycling tests, with tension and compression hold times $t_{h\ sta}$ (Figure 2b). Failure is determined here as a crack depth of 0.5 to 1 mm of a specimen diameter of 10 mm. In the case of a visco-elastic stress-strain path, at low strain amplitudes where plastic deformation disappears, creep damage dominates due to intergranular damage, and fatigue damage was calculated by a fatigue life curve (t_p = 0 h, Figure 2a). In the full elastic-plastic regime at high values of total strain range, creep-fatigue damage dominates due to transgranular damage. Here fatigue damage is calculated on the basis of a failure life curve based on symmetrical creep fatigue experiments with symmetric hold times (Figure 2b). The hold time $t_{H1} = t_{H\ sta}$ at compression strain is fully considered as creep-fatigue damage while creep damage is derived from hold phase 2 and 4 and the remaining time of hold phase 3 ($t_{H3} - t_{Hsta}$). The reference value of rupture time t_u (eq. (1)) is taken from a creep rupture curve for time dependent stresses $\sigma_{eff}(t)$ and is also affected by the hold times $t_{H\ sta}$ (Figure 2b and 6). Hereby, cyclic softening effects are taken into account. In addition, preloading effects can have an influence on the reference value N_{fo} and t_{uo} [1, 5].

Furthermore, N_{fo} depends on a mean stress factor ν_σ:

$$\nu_\sigma = N_f(\Delta\sigma_{max}, \Delta\varepsilon_{eff}) / N_f(\Delta\sigma_{eff}/2, \Delta\varepsilon_{eff}) \qquad (3)$$

This factor ν_σ is derived from the Smith-Watson-Topper parameter (P_{SWT}) [2, 5] and considers a mean stress $\sigma_m \neq 0$ of the cycle. Herein the values σ_{max}, $\Delta\sigma_{eff}$ and $\Delta\varepsilon_{eff}$ are taken from the flank curve loop (Figure 2e).

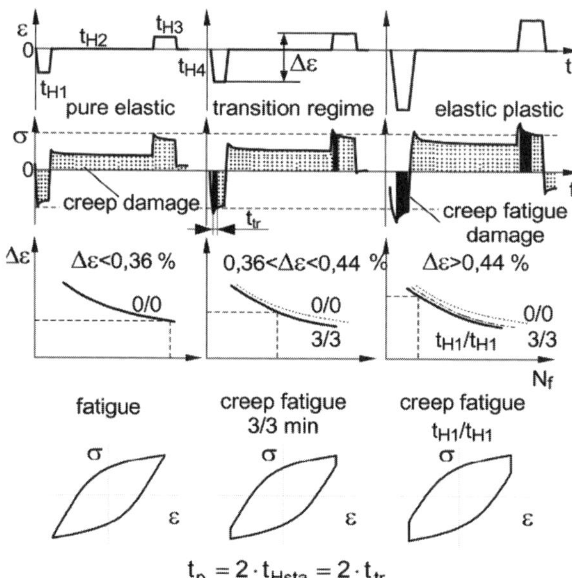

Figure 6. Association of elastic-plastic deformation and failure mechanisms at stress relaxation for the calculation of creep damage and fatigue damage [5]

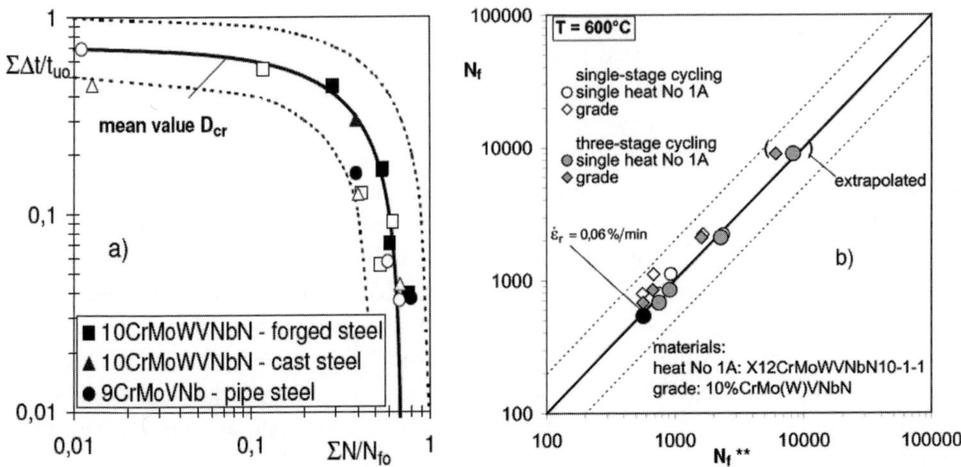

Figure 7. Results of creep-fatigue life analysis of service-type strain cycling experiments 10CrMoWVNbN forged and cast steels, 9CrMoVNb pipe [3] (a) , measured numbers of cycles to failure N_f of isothermal service-type strain cycling tests versus predicted numbers of cycles to failure $N_f{}^{**}$(b), single heat and material grade of 10%CrMo(W)VNbN

The main result of a creep-fatigue life analysis as described above is a creep-fatigue damage mean value D_{cr} . At the conventional European ferritic 1CrMoNiV rotor steel, mean values of D_{cr} = 0.54 at 500 °C and D = 0.52 at 525 °C were identified [3, 5]. Higher mean values were obtained with the martensitic 12CrMoV steel, whereby as there were D_{cr} = 0.75 at 550 °C and D_{cr} = 0.93 at 600 °C. Finally, for the modern 600 °C steel of type 10CrMoWVNbN and its cast version mean values D_{cr} = 0.71 and D_{cr} = 0.65 were found at 600 °C [4]. As a result, all features introduced within this creep-fatigue interaction concept yielded to the smallest scatter band in the creep fatigue life diagram (Figure 7a).

In order to transfer knowledge of deformation behaviour and damage assessment obtained in several research programs, a software tool SARA (Figure 5) was developed for life time studies and industrial application as a life-cycle counter in power plants [9]. The numerical simulation by SARA envelopes the input of cycle and material data, the synthesis of stress-strain hysteresis loops according to cycle counting methods, the individual assessment of fatigue damage, creep-fatigue damage and finally an output of the predicted number of cycles-to-failure N_f**, failure time t_f** as well as cycle specific results. Life assessment on the basis of single heat data and material grade data by the usage of creep-fatigue life mean values D_{cr} given above lead to an acceptable result (Figure 7b).

Three-stage service-type strain cycling (Figure 8) demonstrates typical service loading conditions as cold start, warm start and hot start. Such loading sequences are of specific design interest. The frequency is typical for a medium loaded power plant. Three-stage cycling tests have reached longest times to failure up to $1.5x10^4$ h for the 1CrMoNiV steel and 10^4 h for the 10CrMoWVNbN steel.

The creep fatigue damage analysis of the three-stage service-type experiments leads to values D within the scatter band of the single-stage service-type tests (Figure 7a). This important result confirms the ability of the damage accumulation rule (eq. (1)) for multi-stage service-type loading. Further, life assessment method by SARA can also applied to multi-stage loading.

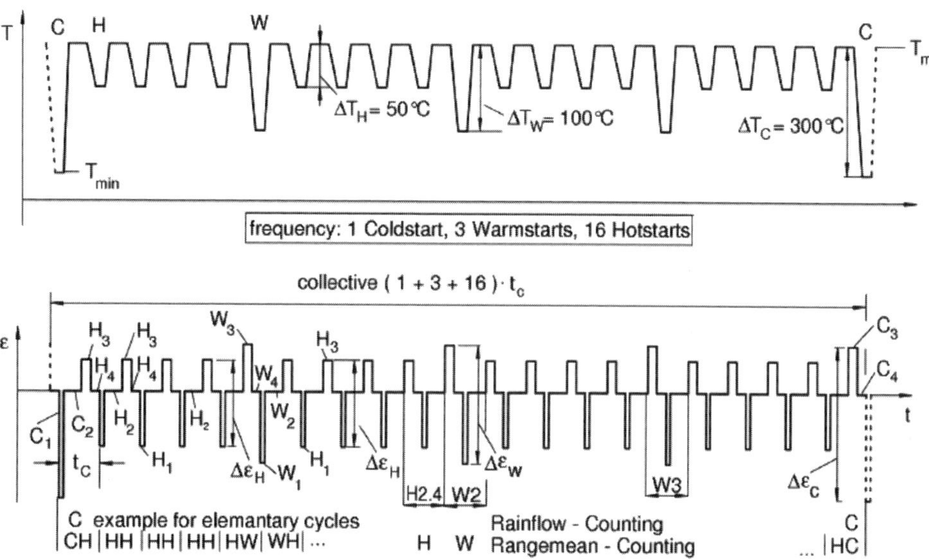

Figure 8. Three-stage service-type strain cycling and principle of cycle counting [4, 5]

Scatter band analysis

A new procedure has been developed in order to describe temperature dependent fatigue life behaviour. This procedure is based on a cycle-temperature-parameter

$$P_{NT} = \tau \cdot (C + \lg N_f) \ , \tag{4}$$

whereby $\tau = (T + 273) \cdot 10^{-3}$ is a temperature function and C is a material constant (Figure 9) [10]. Experimental basis is the irreversible work derived form the hysteresis loop at mid life. The necessary data can be derived from standard LCF experiments (Figure 2a).

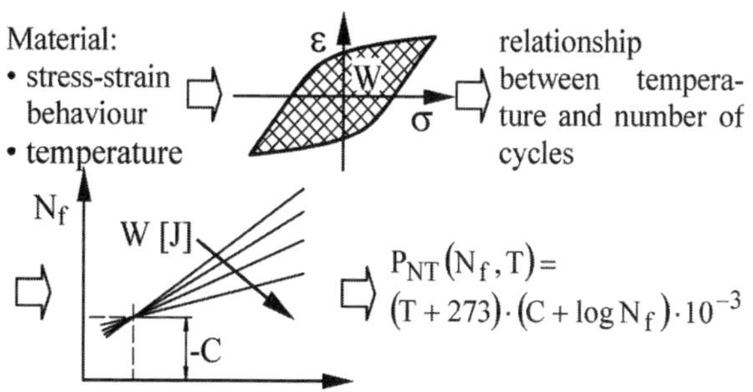

Figure 9. Principal of determination of constant C [10].

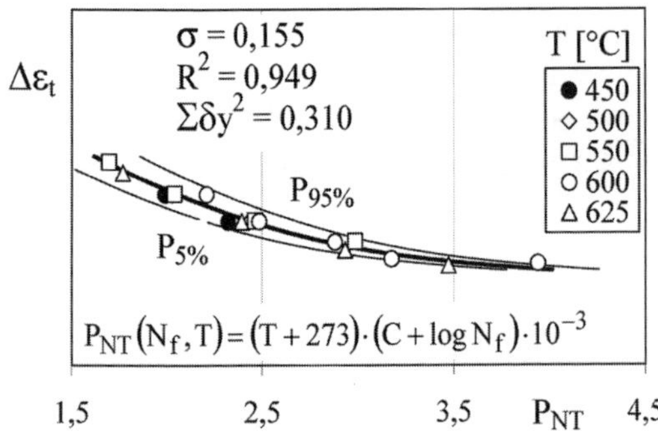

Figure 10. Mean curve with parameter P_{NT}, 10CrMoWVNbN heat 1 [10]

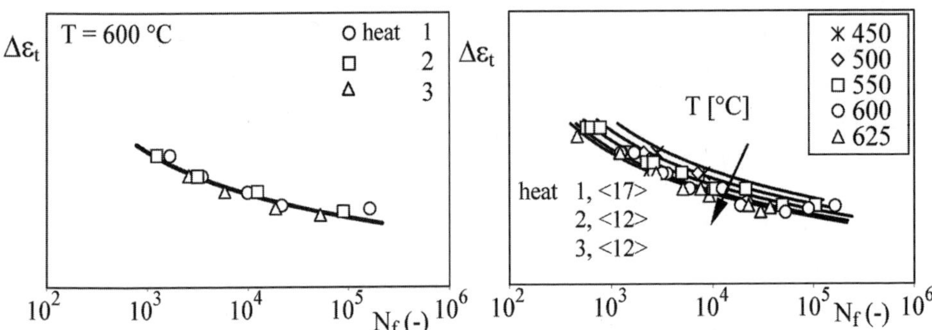

Figure 11. Result of data assessment by the temperature dependent method at T = 600 °C and for temperature range from 450 to 625 °C , 10CrMoWVNbN steel [10], < > number of experiments [10]

The new procedure is suitable for the temperature range 500-600 °C, but not for temperatures out of this range (Figures 10 and 11). It can be assumed that the investigated material does not show a dominated creep deformation for temperatures ≤ 450°C. On the other hand the relatively high application temperature of 625°C leads to microstructural instabilities [11]. Summarizing it can be stated that an analytic procedure was developed which considers the influence of temperature in a suitable way. Further, this procedure allows to determine mean curves and scatter band parameters in a reproducible manner. Future work will focus on the influence of hold times and of strain rates.

Besides fatigue, for the design of components both the creep strain behaviour as well as the creep rupture behaviour are also of special interest. Here a number of different graphical and numerical scatter band assessment methods were currently evaluated within **ECCC** (European Creep Collaborative Committee) [12, 13]. Numerical scatter band assessments will often be made on the basis of a time-temperature parameter P(T, t) (e.g. Larson Miller Parameter P_{LM}). It is supposed to permit a combination of the isothermal time rupture and time-to-strain curves for different temperatures each to one master curve. For a numerical solution a model function for the master curve must be found. A decisive problem concerning assessments of this type is the evaluation of the quality of a model function which concerns the interpretation of the basis test results. A software tool ***DESA*** has been developed which provides a versatile and robust methodology for multi-heat creep rupture data and supports the user to come to an optimum model function with moderate effort. This DESA method is also described in the ECCC Recommendations [13]. In addition Post Assessment Tests (PAT's) have been developed.

Other frequently used scatter band assessment procedures are the ISO 6303 method and the BS PD6605 procedure as an alternative to ISO 6303 method [12]. The Graphical Averaging and Cross-Plotting Method, preferably in usage in Germany, fits graphically isothermal curves of creep rupture data or time to specific strain values. Generally, versatile and robust methodologies are necessary for the assessment of multi-heat data.

Multiaxial behaviour

Life assessment methods either of conventional type or of advanced type require suitable multiaxial experiments for verification purposes. Additionally to tension/torsion or internal pressure experiments developed in the past, experiments with cruciform test pieces (Figure 12) are of high interest due to its plane testing zone.

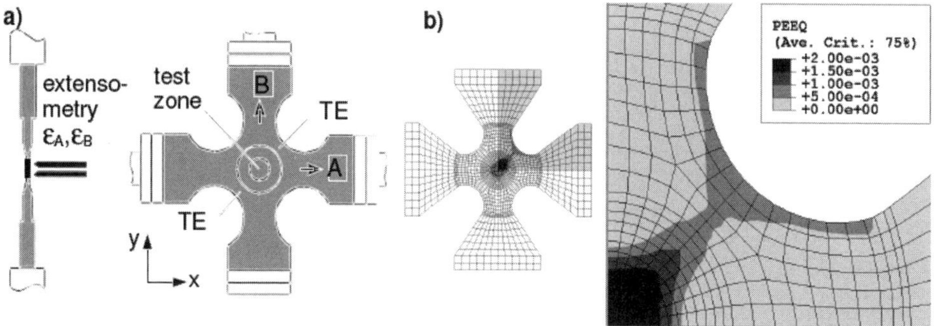

Figure 12. Scheme of the cruciform specimen (a) and elastic-plastic Finite Element calculation showing equivalent plastic strain $\varepsilon_{p,eq}$ distribution and maximum deformation in the test zone (b)

Investigations by Ohnami [14] have demonstrated the large influence of loading ratio on number of cycles to crack initiation. Pure fatigue tests on a 1%Cr-steel show a factor of 10 in life between biaxial strain ratio $\phi_\varepsilon = -1$ and $\phi_\varepsilon = +1$ (Figure 13a, solid lines).

Long-term service-type creep-fatigue experiments on a 1%CrMoNiV rotor steel with four hold times (cycle period t_p) are performed on the cruciform testing system. Experiments run under strain controlled mode with long hold times. The strain ratio is given to $\phi_\varepsilon = 0.5$ and $\phi_\varepsilon = 1$. A total of five biaxial service-type creep-fatigue experiments were performed on the 1%CrMoNiV rotor steel with relevant test durations up to about 2 000 hours. As a first result at strain ratio $\phi_\varepsilon = 1$, a clear influence of superimposed creep at hold times of a factor of two can be observed. Secondly, a strain ratio of $\phi_\varepsilon = 0.5$ leads to an increase of number of cycles to crack initiation N_i of factor 1.5 compared to $\phi_\varepsilon = 1$. This result confirms the pure fatigue biaxial experiment results [14].

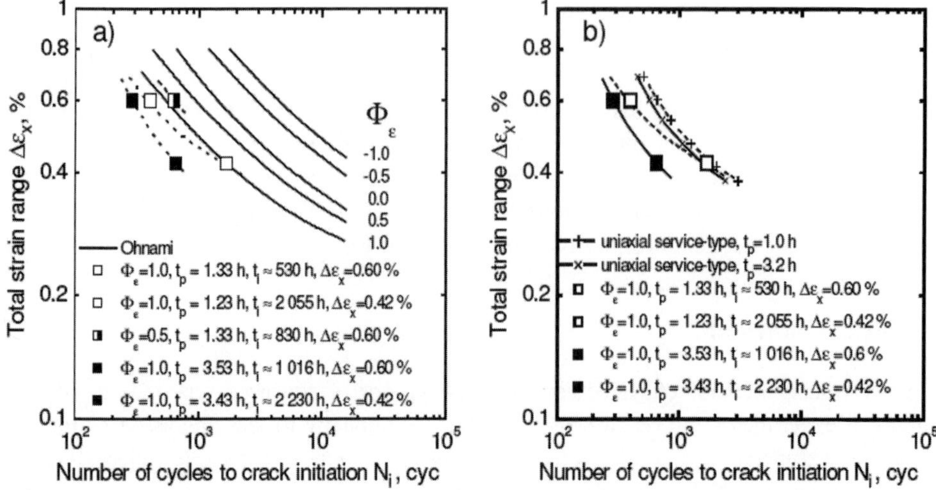

Figure 13. Influence of biaxial strain ratio ϕ_ε on the number of cycles to crack initiation N_i at fatigue testing, 1CrMoV, T = 550 °C [16] and first results of long term service-type creep-fatigue experiments (Figure 2c) (ϕ_ε = 1.0 and ϕ_ε = 0.5), dε / dt = 0.06 % / min (a), comparison with uniaxial experiments (b) , t_p periodic time, t_i time to crack initiation corresponding to a crack depth of about 0.2 mm, 1CrMoNiV, T = 525 °C [16]

Further, biaxial strain controlled experiments lead to a reduction of number of cycles to crack initiation N_i up to factor 3 compared to uniaxial service-type creep fatigue experiments (Figure 13b). On the one hand, an increase of total strain range leads to a larger difference of N_i values. But on the other hand, increasing hold time has a more significant influence.

A constitutive material model describes elastic-viscoplastic behaviour for small deformations and was introduced by Tsakmakis [15]. A key feature is the combination of effective stress with a generalised energy equivalence principle. An undamaged fictitious material is described by means of effective variables which are the basis of the constitutive material model. A damage variable is defined with an approach proposed by Lemaitre additionally to another set of variables for the damaged real material. The known behaviour of the undamaged fictitious material is then mapped to the unknown behaviour of the real material with damage. This step is done by substitution of the effective variables using relations which implicate the damage variable.

Within the current research work on creep-fatigue, the material parameters of the constitutive model were determined by a two-step approach with a combination of the Neural Networks method and the optimisation method by Nelder-Mead. The Neural Networks method is already established in similar non-linear problems and can deliver a "global" solution. The method by Nelder-Mead is a direct search method without the need of numerical or analytical gradients and leads only to a "local" solution.

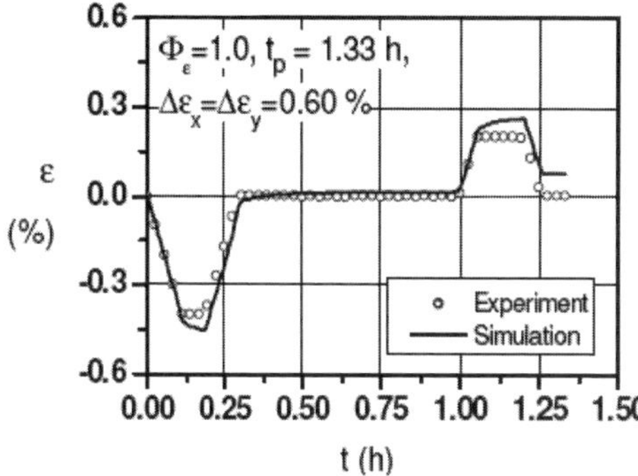

Figure 14. Finite Element simulation of biaxial service-type creep fatigue experiments (see Figure 13a), comparison of strain vs. time curves from experiment to those of FE model calculated with the constitutive material model and the material parameters determined, ϕ_ε = 1.0, 1CrMoNiV steel, T = 525 °C, dε / dt = 0.06 % / min [16]

Results of experiments with cruciform specimens can be employed in combination with Finite Element (FE) simulations in order to verify various material models. These simulations were performed with ABAQUS whereby the constitutive material model is implemented in user defined subroutine UMAT. The example in Figure 14 demonstrates the applicability of the material model introduced.

Concluding remarks

A phenomenological life assessment procedure and an advanced material model have been investigated and applied to heat resistant steels. Knowledge on cyclic deformation and creep fatigue damage assessment has been obtained here.

- Long-term experiments both, uniaxial as well as multiaxial are necessary in order to verify advanced life prediction methods for components.
- Stress strain path and creep fatigue life can be predicted by user program *SARA* on the basis of rules for deformation, relaxation and cyclic stress strain behaviour including internal stress and mean stress effects.
- A creep-fatigue interaction concept which covers physical interpretations of deformation and failure mechanisms has been applied to a 1CrMoNiV steel and 10CrMo(W)VNbN steels as well.
- Scatter band assessment procedures provide the generation of mean curves and lower bound curves for creep and fatigue properties.
- The simulation of creep-fatigue behaviour at multi-axial and multi-stage loading under start-up and shut-down conditions with constitutive models is a current challenge.
- Advanced lifing methods and knowledge of materials as well as methodologies enables the reduction of efforts for design.
- Consequently, an increase of design quality and efficiency can be achieved.

Acknowledgement

Thanks are due to the Forschungsvereinigung der Arbeitsgemeinschaften der Eisen und Metall verarbeitenden Industrie e.V. (AVIF), Projects No. A 165 and A 166, the FKM Forschungskuratorium Maschinenbau e.V., Projects No. 052500 and No. 052510, the Arbeitsgemeinschaft industrieller Forschungsvereinigungen (AiF) and the VDEh-Gesellschaft zur Förderung der Eisenforschung mbH, Project No. 11200 N and the Deutsche Forschungsgemeinschaft (DFG), Projects No. BE 1890, 13-1 and 16-1 for financial support and the working groups of Power Plant Industry for supply of material.

References

[1] J. Granacher, A. Scholz, and C. Berger, "Creep fatigue behaviour of heat resistant turbine rotor steels under service-type strain cycling", Proc. of the Fourth Int. Charles Parsons Turbine Conf., Newcastle upon Tyne, Eds. A. Strang, W.M. Banks, R. D. Conroy and M. J. Goulette, The Institute of Materials, London (1997) pp. 592-602.

[2] D. A. Miller, and R. H. Priest, "*Materials Response to Thermal-Mechanical Strain Cycling*", High Temperature Fatigue: Properties and Prediction, ed. R. P. Skelton, Elsevier Appl. Science Publ. Ltd, 1987, pp. 113-176.

[3] J. Granacher, and A. Scholz, "Creep fatigue behaviour of heat resistant steels under service-type long term conditions", Proc. of the Third Int. Conf. "Low Cycle Fatigue and Elasto-Plastic Behaviour of Materials", Berlin, Ed.: K.-T. Rie, Elsevier Appl. Sc., London, 1992, pp. 235-41.

[4] A. Scholz, H. Haase, and C. Berger, "Simulation of Multi-Stage Creep-Fatigue Behaviour", Fatigue 2002, Proc. of the Eighth Int. Fatigue Congress, A. F. Blom (Ed.), Volume 5/5, Stockholm, June 2002, pp. 3133-40.

[5] A. Scholz, and C. Berger, "Deformation and life assessment of high temperature materials under creep fatigue loading", First Symposium on Structural Durability in Darmstadt, June 9-10, 2005, Proc. Editor: C. M. Sonsino, Fraunhofer IRB Verlag, 2005, pp. 311 – 328, ISBN 3-8167-6788-5.

[6] K. N. Smith, P. Watson, and T. H. Topper, "A Stress-Strain Function for the Fatigue of Metals", J. of Materials, Vol. 4 (1970) pp. 767–778.

[7] J. S. Dubey, H. Chilukuru, J. K. Chakravartty, M. Schwienheer, A. Scholz, and W. Blum, "Effects of cyclic deformation on subgrain evolution and creep in 9 – 12 % Cr-Steels", Materials Science und Engineering A, 406 (2005), pp. 152-159, ISSN 0921-5093.

[8] K. Yagi, O. Kanemaru, K. Kubo, and C. Tanaka, "Life prediction of 316 stainless steel under creep-fatigue loading", Fat. & Fract. of Engng. Mat. & Struct., Vol. 9, No. 6 (1987) pp. 395-408.

[9] B. J. Cane, "Life Management of Ageing Steam Turbine Assets", Proc., Ed. by the Institute of Materials, London, 1997, pp. 554–574.

[10] C. Berger, A. Scholz, M. Schwienheer, and R. Znajda, "A new scatter band assessment procedure for multi heat fatigue life data", Proc. Creep & Fracture in High Temperature Components – Design & Life Assessment Issues, ECCC Creep Conf., September 12 – 14, 2005, London, UK. Edited by I. A. Shibli, S. R. Holdsworth, G. Merckling. Lancaster, PA: DEStech Publications, 2005, pp. 628/637, ISBN 1-932078-49-5.

[11] M. Schwienheer, H. Haase, A. Scholz, and C. Berger, "Long Term Creep and Creep Fatigue Properties of the martensitic Steels of Type (G)X12CrMoWVNbN10-1-1", Proc. of the 7th COST Conf. "Materials for Advanced Power Engineering 2002", Part III, Lecomte-Beckers, Carton, Schubert, Ennis (Ed), 2002, Liège, Belgium, pp. 1409-1418.

[12] C. Berger, and A. Scholz, "Perspectives on improved life assessment methods for new plants", Advances in Materials Technology for Fossil Power Plants, Proc. Fourth Int. Conf., October 25-28, 2004, Hilton Head Island, South Carolina, EPRI Report Number 1011381, Eds: R. Viswanathan, D. Gandy, K. Coleman. Materials Park, OH: ASM Int., 2005, pp. 653 – 671, ISBN 0-87170-818-3.

[13] ECCC Recommendations - Volume 5 Part I [Issue 4], Guideline for the Exchange and Collection of Creep Rupture, Creep Strain-Time and Stress Relaxation Data for Assessment Purpose, Appendix B1, ECCC-Document, Ed.: S.R. Holdsworth, May 2001.

[14] M. Itoh, M. Sakane, and M. Ohnami, "High Temperature Multiaxial Low Cycle Fatigue of Cruciform Specimen", Trans. ASME, JEMT, Vol.116, No.1, 1994, Jan., pp. 90-98.

[15] C. Tsakmakis, and D. Reckwerth, "The Principle of Generalized Energy Equivalence in Continuum Damage Mechanics", 2003, Deformation and Failure in Metallic Materials, Springer Verlag, ISBN 3-540-00848-9

[16] A. Samir, A. Simon, A. Scholz, and C. Berger, "Service-type creep-fatigue experiments with cruciform specimens and modelling of deformation", Int. J. of Fatigue, Volume: 28, Issue: 5-6, May - June, 2006, pp. 643-651.

Advances in Materials Technology for Fossil Power Plants
Proceedings from the Fifth International Conference
R. Viswanathan, D. Gandy, K. Coleman, editors, p 762-782

DOI: 10.1361/cp2007epri0762

FAILURE BEHAVIOR OF HIGH CHROMIUM STEEL WELDED JOINTS UNDER CREEP AND CREEP-FATIGUE CONDITIONS

Y. Takahashi
Central Research Institute of Electric Power Industry
2-11-1 Iwado Kita, Komae-shi, Tokyo 201-8511 Japan

M. Tabuchi
National Institute of Materials Science
1-2-1 Sengen, Tsukuba-shi, Ibaraki 305-0047, Japan

Abstract

Strength of welded joints of high chromium steels are of highly important concern for operators of ultra supercritical thermal power plants. A number of creep-fatigue tests with tensile strain hold have been carried out for the welded joints of two types of high chromium steels widely used, i.e. Grade 91 and 122 steels. It was found that failure occurred in fine grain heat-affected zone in all the creep fatigue tests, even at a relatively low temperature and fairly short time where failure occurred in plain base metal region in simple creep testing. Four procedures were used to predict failure lives and their results were compared with the test results. A newly proposed energy-based approach gave the best estimation of failure life, without respect of the material and temperature.

Introduction

Ferritic steels containing high chromium contents such as Grades 91, 92 and 122 are widely used in recent ultra supercritical fossil power plants as well as combined cycle plants due to their high temperature strength and superior oxidation resistance. Many tests have been done for determining allowable stress values of these materials (1) and developing the procedures for evaluating life consumption under the superposition of creep and fatigue damages (2-4).

It is now well-recognized from service experiences and laboratory testing that welded joints of these steels are more prone to creep failure than the base metals because of deterioration of creep strength by a temperature history experienced during the welding process. Introduction of weld strength reduction factor has been attempted recently in order to cope with this finding (5-6).

Creep-fatigue interaction constitutes another important failure mechanism to be addressed in the assessment of high-temperature components subjected to cycles consisted of transient and steady-state periods. However, only very limited data have been obtained concerning creep-fatigue behavior of the welded joints and no systematic effort has been made for developing design and life evaluation procedures for them.

The authors recently initiated a study for developing a procedure to properly make a creep-fatigue assessment of these welded joints as a part of a government-sponsored program aiming at promotion of the employment of these materials in the future nuclear power plants (7). It was found in this study that type-IV cracking occurs in creep-fatigue loading easier than pure creep loading and this leads to large life reduction due to the introduction of hold time even at relatively low temperature and short total elapsed time where type-IV cracking hardly occurs under simple creep conditions. This paper presents updated results of this work, including test data of both Grade 91 and 122 steels as well as the application of several creep-fatigue life prediction procedures.

Tested materials and welded joints

Both Grade 91 and 122 steels have been tested in the study. They were produced as a plate form of either 25 mm or 30 mm thickness. Chemical compositions are shown in Table 1. The plates were connected by Gas Tungsten Arc Welding (GTAW) using the weld consumables which have similar chemical compositions as the base metals. Post-weld heat treatment was given to the welded plates by keeping at 745C, for 60 minutes for Grade 91 and 75 minutes for Grade 122.

Cross-sectional views of the welded joints are shown in Figure 1. Weld metal width varied between 10 and 15 mm depending on the position along the thickness while the width of the heat-affected zone (HAZ) was constantly about 3 mm. The heat-affected zone can be subdivided by microstructural difference into a coarse grain heat-affected zone (CGHAZ) and a fine grain heat-affected zone (FGHAZ). The weld metal and CGHAZ are harder than the base metal whereas the narrow region of approximately 1 mm width designated as FGHAZ is softer than those as shown in Figure 2. The poor creep strength of FGHAZ is considered as a main cause of the type-IV failure. In order to grasp mechanical behavior of FGHAZ, simulated FGHAZ material was also produced by heating the base metal up to 900°C for grade 91 and 950°C for grade 122 and cooled at a speed of 10°C /s, followed by the post-weld heat treatment at the above conditions.

Summary of creep tests

Creep tests for cross-weld specimens taken from the both welded joints as well as those fabricated by similar procedures were conducted at three temperatures covering the range of practical interest, i.e. 550, 600 and 650°C. Most of the tests were performed using round bar specimens with 6 mm diameter and 30 mm gauge length but the plate specimens with 17.5 mm width and 5 mm thickness were also tested. Difference between the rupture times of both specimens was found to be small. Creep tests were also carried out for simulated FGHAZ specimens made by the above heat treatment.

Creep rupture data obtained for both kinds of materials are displayed in Figure 3. At the lowest test temperature, i.e. 550°C, most of the specimens ruptured at base metal remote from the welded joints but rupture occurred in the heat-affected zone at 600°C and 650°C except very short-term tests. Weakness of the welded joints in comparison with the base metals apparently becomes clearer as the temperature increases and/or the applied stress decreases in both steels. Predictions by equations representing average trend of many existing creep rupture data taken in Japan are also drawn in the figure. It can be seen that rupture times of the present welded joints tend to be even shorter than the predictions in low stress region.

It was also found that the FGHAZ material showed weaker creep strength than the welded joints whereas the weld metals have higher creep strength than the base metals. Although the absolute strength levels of the both kinds of steel differ to some extent, the above observations on the relative strength of the welded joints and their constituents against their base metals commonly hold for both materials and this suggests that mechanism of creep strength reduction is shared by them.

As indicators of ductility of the materials, rupture elongation and reduction of area are plotted as a function of rupture time in Figures 4 and 5, respectively. Both of them of the welded joint specimens tend to decrease with the rupture time, down to quite small values even in the case of grade 91 where the base metals and FGHAZ maintained a quite high value regardless of temperature and rupture time.

Summary of creep-fatigue tests

Round bar specimens having a test section of 8 mm diameter shown in Figure 6 were machined from the welded joints in a way that the center of the specimens coincides with the boundary of weld metal and the heat-affected zone (fusion line) as shown in Figure 7. As the gauge length for controlling the "nominal" strain was 12 mm, half of the gauge length was occupied by the weld metal and another half was constituted by the heat-affected zone and the plain base metal.

Pure fatigue tests and creep-fatigue tests were conducted at the same three temperatures as in the creep tests. The nominal strain for the gauge length was changed at the rate of 0.1%/s during ascending and descending portions of the strain wave. In the pure fatigue tests, strain was cycled in a triangular wave whereas it was controlled in a trapezoidal waveform with a hold period between 10 minutes to 10 hours at tensile peak of the strain. Due to cycling softening behavior of these steels, stress range at each cycle decreased with cycles as in the case of the base metal. Failure was judged to occur when the maximum stress during the cycle became less than 75 % of the extrapolated trend of the previous record before the initiation of rapid drop.

Relation between the hold time and the number of cycles to failure is shown in Figure 8. It can be seen that reduction of life due to hold time is more pronounced at a smaller strain range and a higher temperature. Grade 91 and 122 exhibited similar failure lives in contrast to the difference in static creep strength, presumably due to the difference between stress versus strain control. Main cracks were found in the plain base metal in the pure-fatigue tests but all specimens of creep-fatigue tests were failed by cracks in FGHAZ zone without any exception even at the lowest temperature, where type-IV failure occurs only at long term tests under static creep condition.

Figure 9 shows a comparison with the test results for the base metals (2). Although the failure lives of the welded joints were shorter, to some extent, than the base metal even under pure-fatigue conditions, difference between them grew as the hold time increased, up to a factor of 10. Although some consideration would be possible such as the effect of cyclic softening, detailed evaluations of the stress distribution and its effect on damage accumulation would be required to make the reason for this clearer.

Procedure for creep-fatigue life prediction

As an initial step to develop creep-fatigue life prediction method suitable for life evaluation for these welded joints, existing creep-fatigue life prediction methods were applied and the results were compared with the test results. Specifically, the time fraction rule and ductility exhaustion approach were chosen as they are used in design and/or assessment (8-9) widely. In addition, a modified ductility exhaustion method recently proposed by one of the authors (2) was also applied. An approach focused on the strain energy, instead of strain, was newly developed and also applied.

Time Fraction Rule. Time fraction rule regards the stress as a principal indicator for creep damage and calculates the creep damage per cycle by the following equation using the stress value, σ during the hold time:

$$D_c = \int_0^{t_H} \frac{dt}{t_R(\sigma, T_{abs})}, \tag{1}$$

where t_H is the hold time in each cycle and t_R is rupture time at the stress, σ, and the absolute temperature, T_{abs} in pure creep condition

The following best-fit creep rupture equation with constants listed in Table 2 was used to calculate the rupture time as a function of stress and temperature.

$$\log_{10} t_R = \left[a_0 + a_1 \log_{10} \sigma - a_2 (\log_{10} \sigma)^2\right] / T_{abs} - C \tag{2}$$

Ductility Exhaustion Approach. According to the ductility exhaustion approach, creep damage per cycle in creep-fatigue test is calculated from the creep strain rate, $\dot{\varepsilon}_c$, as:

$$D_c = \int_0^{t_H} \frac{\dot{\varepsilon}_c}{\varepsilon_f(\dot{\varepsilon}_c)} dt, \tag{3}$$

where ε_f is the strain limit usually called "creep ductility". Quantities such as rupture elongation, reduction of area, and true rupture strain, have been used as its value, depending on researchers and/or materials. Moreover, its dependency on the creep strain rate was taken into account in some cases, but not in others. In this study, the values of rupture elongation obtained in the creep tests were plotted as a function of average creep strain rate calculated by ε_f / t_R. Afterwards, the following equation was used to fit the data as shown in Figure 10.

$$\begin{aligned} \varepsilon_f &= \varepsilon_{f0} \quad when\ \dot{\varepsilon}_{in} > \overline{\dot{\varepsilon}_{in}} \\ &= 10^{b_0 + b_1 \log \dot{\varepsilon}_{in} + b_2 (\log \dot{\varepsilon}_{in})^2} \quad when\ \dot{\varepsilon}_{in} \le \overline{\dot{\varepsilon}_{in}} \end{aligned} \tag{4}$$

where ε_{f0} is the upper bound ductility whereas b_0, b_1 and b_2 are fitting parameters. These values for each welded joint are summarized in Table 2.

Modified Ductility Exhaustion Approach. In an attempt to improve the life predictability of ductility exhaustion method, one of the authors developed the following equation for the estimation of creep damage, by redefining creep damage as an amount of loss of ductility caused by inelastic deformation,:

$$D_c = \int_0^{t_H} \left(\frac{1}{\varepsilon_f(\dot{\varepsilon}_c)} - \frac{1}{\varepsilon_{f0}} \right) \dot{\varepsilon}_c dt \tag{5}$$

where δ_0 is the ductility under sufficiently fast loading thus permitting the use of ductility values obtained from conventional short-term tensile tests.

It should be noted that the new equation always gives smaller creep damage than the classical approach and that the creep damage is estimated to be zero even in the creep tests when no decrease of ductility is seen as the rupture time increases. On the contrary, when decrease in the ductility in the creep tests is significant (i.e. ε_f is much smaller than ε_{f_0}), difference between the conventional and new approaches becomes small. In this way, the modified approach separates creep damage leading to ductility reduction from creep deformation.

Energy-based Approach. It was found that rupture elongation showed peculiar behavior at 550C when plotted against the average inelastic strain rate; taking a maximum value at an intermediate strain rate, rather than monotonically decreasing with the decrease in strain rate. Strain energy per unit section area given to rupture would be a better parameter to represent the ductility of the material. Following the similar process as in the modified ductility exhaustion approach led to the equation:

$$D_c = \int_0^{t_H} \left(\frac{1}{W_f(\dot{W}_{in})} - \frac{1}{W_{f0}} \right) \dot{W}_{in} dt \tag{6}$$

where $\dot{W}_{in} = \sigma \dot{\varepsilon}_{in}$ is the work rate per unit section area whereas W_f is the total work accumulated to rupture ($= \sigma \varepsilon_f$ in the case of creep tests), with its saturated value, W_{f0}, at a sufficiently fast loading.

Figure 11 shows the relationship between the average work rate calculated as W_f / t_R and W_f. Smoother relations were obtained than the case of rupture elongation and the data were fitted by a power-law relation with the upper bound value obtained in the fast tensile tests as

$$W_f = \min(\alpha \dot{W}_{in}^{\ n}, W_{f0}), \tag{7}$$

where α and n are constants given in Table 2.

Whichever the method is used, stress and/or strain rate during strain hold is required to calculate creep damage. In order to obtain these values, stress relaxation during holding in the mid-life cycle was approximated by the following equation:

$$\sigma = c_1 + c_2 \log t + c_3 \left(\log t\right)^2, \tag{8}$$

where t denotes the time from the start of strain holding and c_1, c_2, c_3 are parameters to fit the recorded stress data.

Fatigue damage per cycle, D_f, was simply calculated as a reciprocal of the failure life, N_{f0}, obtained in the pure fatigue test at the same strain range and temperature as in the creep-fatigue test to evaluate.

$$D_f = \frac{1}{N_{f0}} \tag{9}$$

Finally, the following equation based on the assumption of linear damage summation was commonly applied to predict the failure lives.

$$N_f = \frac{1}{D_f + D_c} \tag{10}$$

Results of Life Prediction

The predicted numbers of cycles to failure are compared with the test results in Figures 12 through 15. Failure lives predicted by the time fraction rule and the modified ductility exhaustion approach showed a relatively good agreement with the test results for most conditions but the failure life of grade 122 at the lowest temperature tended to be excessively over-predicted. The classical ductility exhaustion approach under-predicted the failure life for all conditions, more pronouncedly with the decrease in strain range. Finally, the energy-based approach seems to have brought about the best agreement with the test results as a whole, although a few points lie outside a factor of 2 band on conservative side.

Conclusion

A number of creep-fatigue tests with tensile strain hold have been carried out for the welded joints of Grade 91 and 122 steels. It was found that type-IV failure occurred more easily under creep fatigue loading than pure creep condition at the same temperature and test period. Four procedures have been applied to predict failure lives in creep-fatigue tests and their results were compared with the experimental life. It was found that the newly proposed energy-based approach gave better estimation of failure life than the time fraction or ductility exhaustion approaches, regardless of material and temperature.

Acknowledgment

Present study is the result of "Development of damage prevention technology for welded structures in the next-generation high temperature nuclear plant" entrusted to Central Research Institute of Electric Power Industry by the Ministry of Education, Culture, Sports, Science and Technology of Japan (MEXT).

References

1. e.g. K. Kimura, "Assessment of long-term creep strength and review of allowable stress of high Cr ferritic creep resistant steels," Proceedings of PVP2005, ASME PVP2005-71039.

2. Y. Takahashi, "Study on creep-fatigue evaluation procedures for high chromium steels, Part I : Test results and life prediction based on measured stress relaxation," Submitted to Int. J. Pressure Vessel and Piping.

3. Y. Takahashi, "Study on creep-fatigue evaluation procedures for high chromium steels, Part II : Life prediction based on calculated deformation," Submitted to Int. J. Pressure Vessel and Piping.

4. Y. Takahashi, "Accuracy and margin in creep-fatigue evaluation for modified 9Cr-1Mo steel," Proceedings of PVP2007-CREEP8, PVP2007-26695.

5. M. Tabuchi, Y. Takahashi, "Evaluation of creep strength reduction factors for welded joints of modified 9Cr-1Mo steel (P91)," Proceedings of PVP2006-ICPVT-11, PVP2006-ICPVT11-93350

6. Y. Takahashi and M. Tabuchi, Evaluation of creep strength reduction factors for welded joints of HCM12A (P122), Proceedings of PVP2006-ICPVT-11, PVP2006-ICPVT11-93488.

7. Y. Takahashi, "Study on type-IV damage prevention in high-temperature welded structures of next-generation reactor plants, part I fatigue and creep-fatigue behavior of welded joints of modified 9Cr-1Mo steel," PVP2006, Proceedings of PVP2006-ICPVT-11, PVP2006-ICPVT11-93081

8. American Society of Mechanical Engineers, Boiler and Pressure Vessel Code, Section III, subsection-NH 2005.

9. British Energy, Assessment procedure for the high temperature response of structures R5 Issue 3; 2003.

Table 1 Chemical compositions of tested materials

	C	Si	Mn	P	S	Cu	Ni	Cr	Mo	V	Nb
Grade 91 Base metal	0.1	0.25	0.43	0.006	0.002	0.01	0.06	8.87	0.93	0.19	0.07
Grade 91 Weld metal	0.08	0.14	0.90	0.005	0.004	0.10	0.58	9.01	0.94	0.18	0.04
Grade 122 Base metal	0.11	0.26	0.64	0.016	0.002	1.03	0.39	10.87	0.31	0.20	0.054
Grade 122 Weld metal	0.08	0.22	0.36	0.006	0.004	0.03	0.41	9.66	0.94	0.20	< 0.01

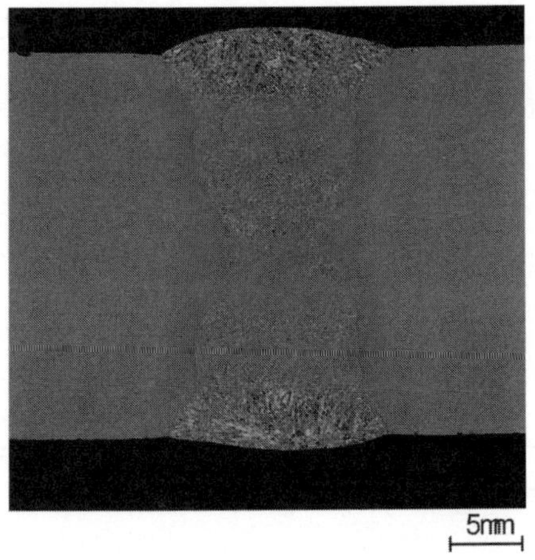

(a) Grade 91

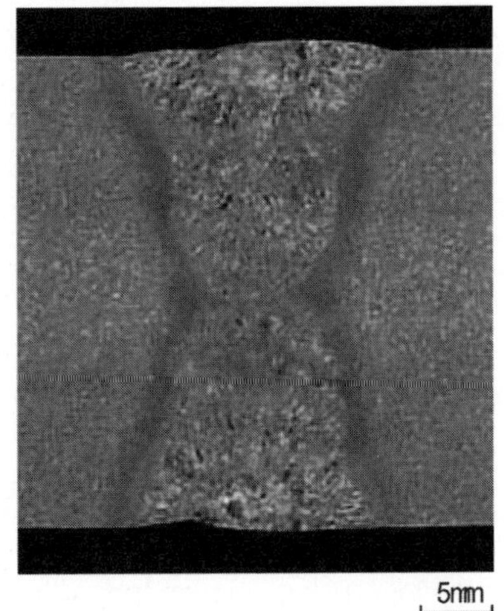

(b) Grade 122

Figure 1 Cross-sectional view of welded joint

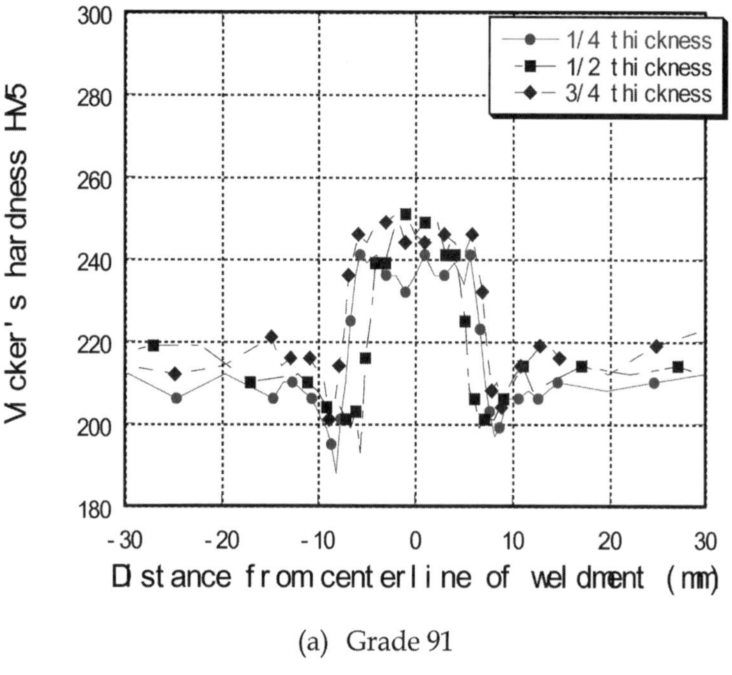

(a) Grade 91

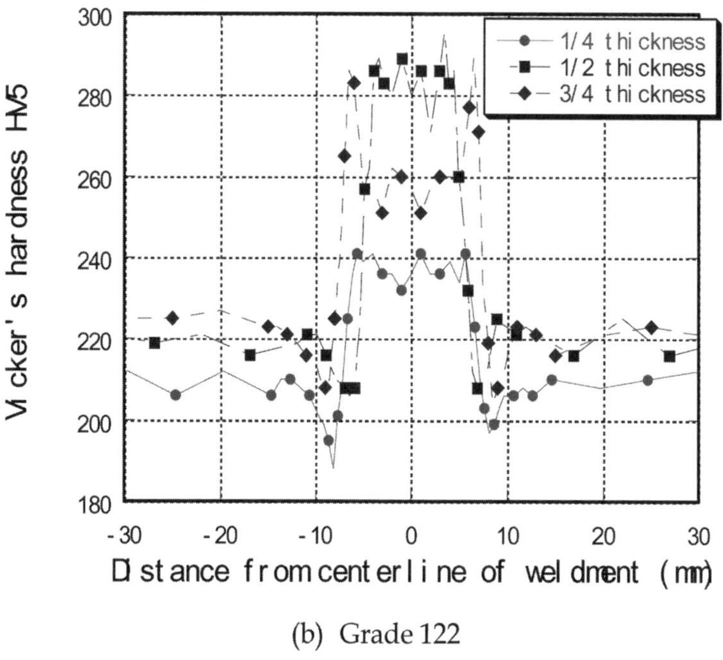

(b) Grade 122

Figure 2 Distribution of hardness across the welded joint

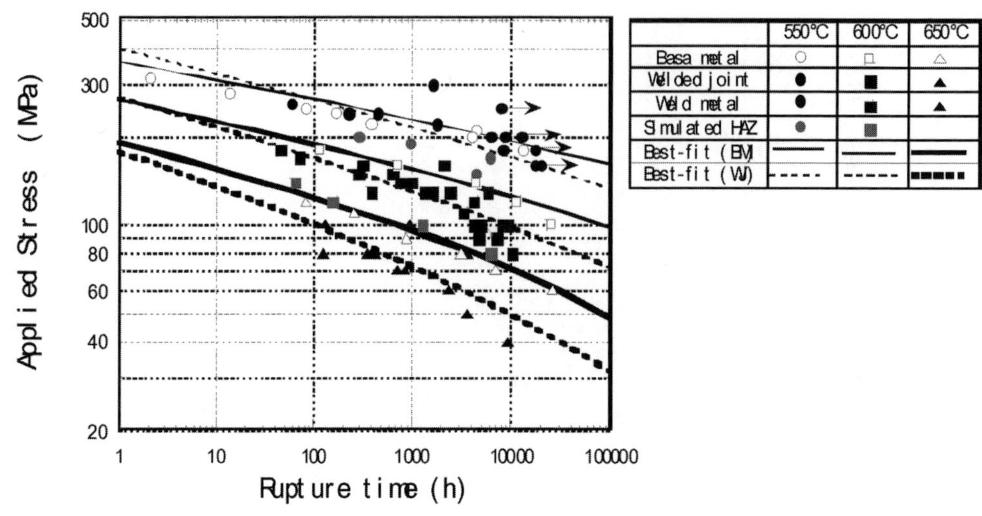

(a) Grade 91

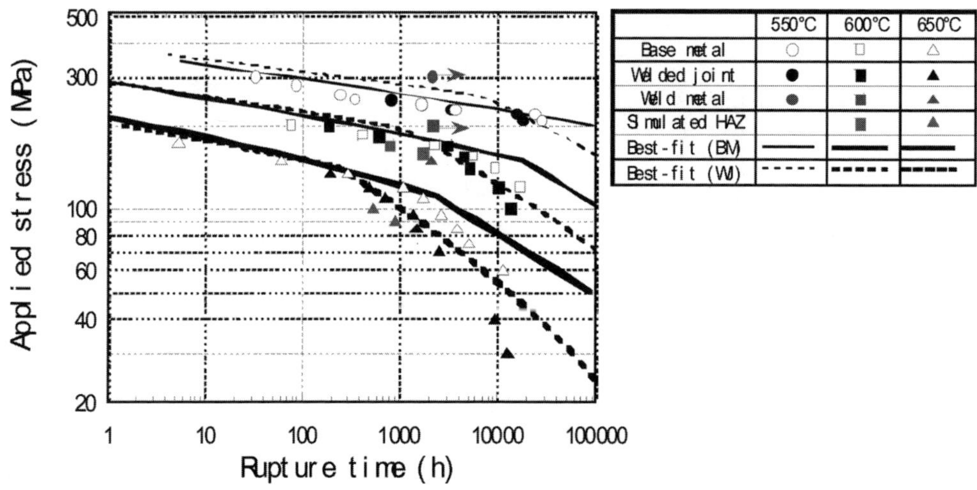

(b) Grade 122

Figure 3 Creep rupture data of welded joints and their constituents

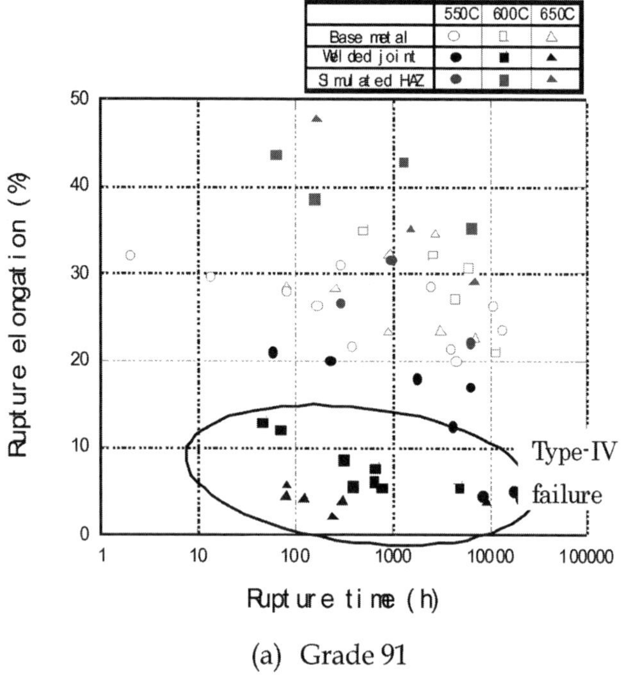

(a) Grade 91

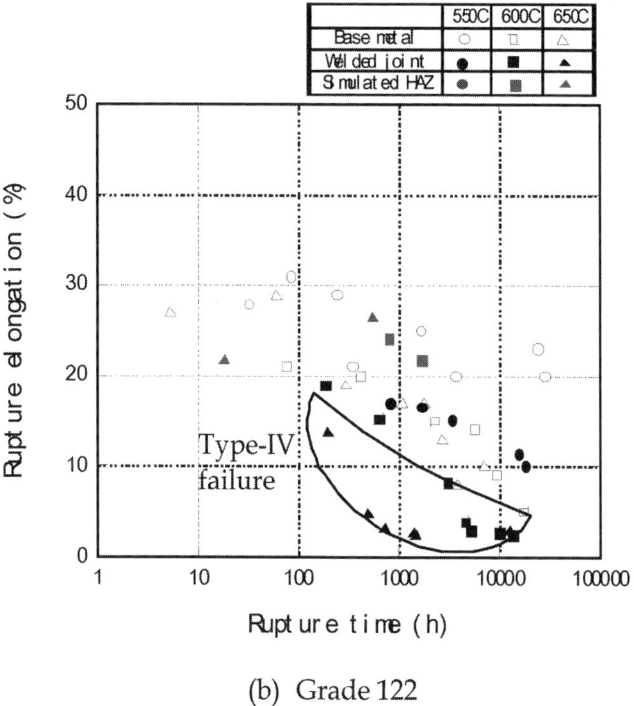

(b) Grade 122

Figure 4 Variation of rupture elongation with rupture time

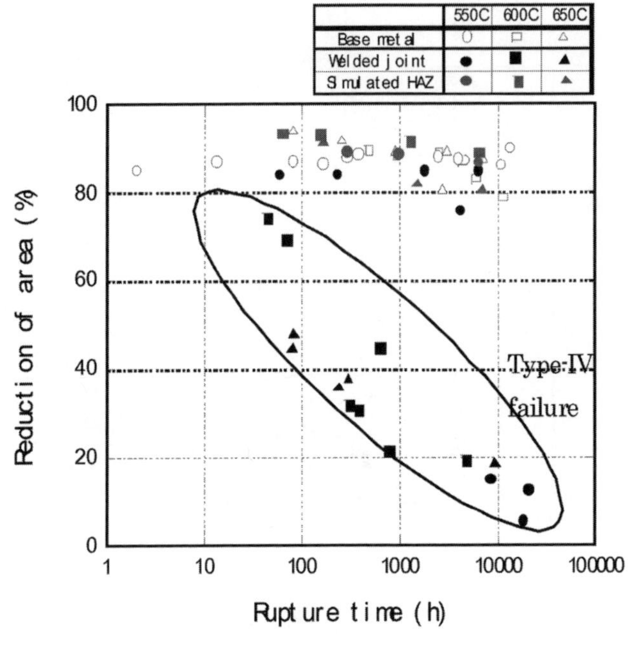

(a) Grade 91

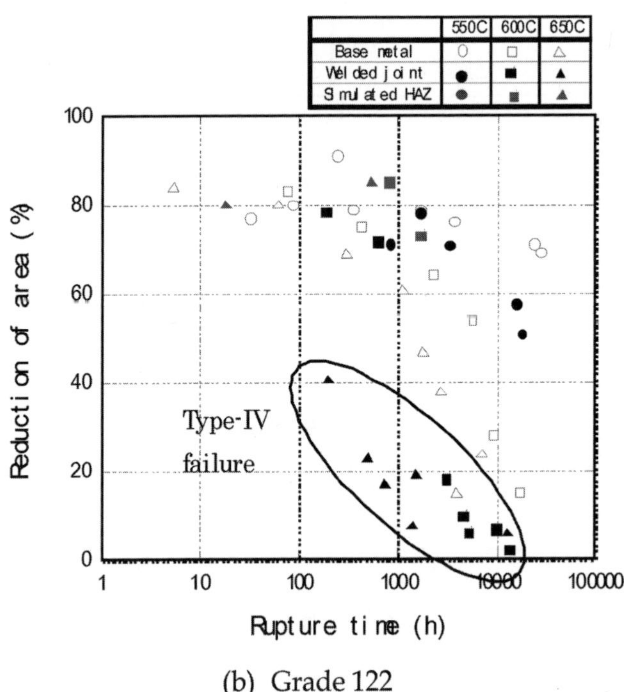

(b) Grade 122

Figure 5 Variation of reduction of area with rupture time

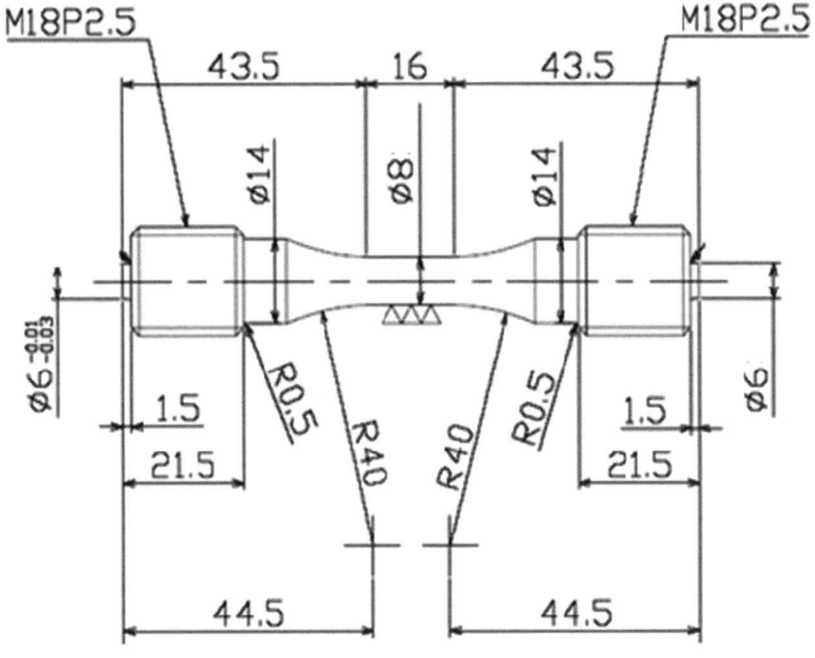

Figure 6 Geometry of creep-fatigue test specimens

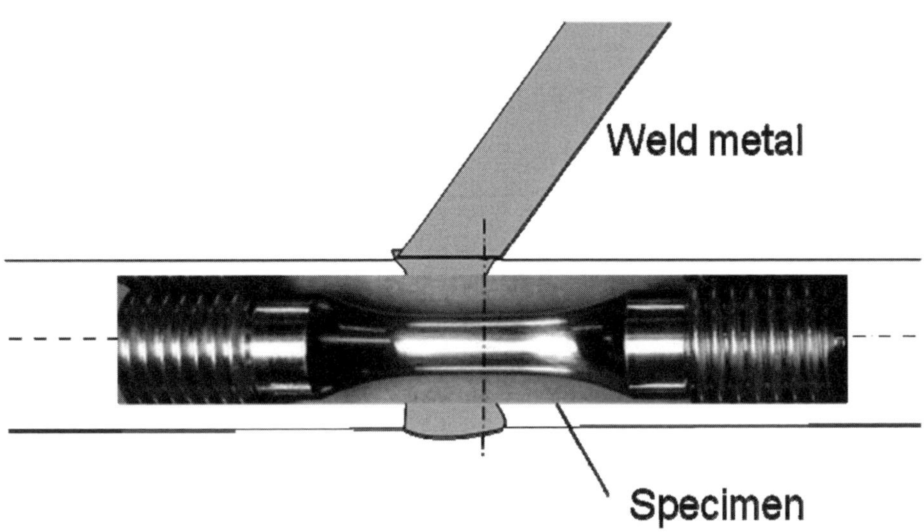

Figure 7 Specimen sampling method

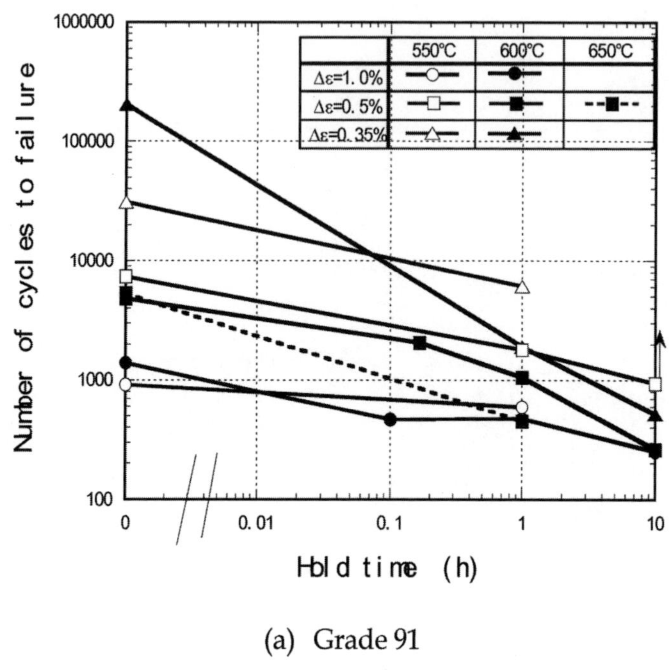

(a) Grade 91

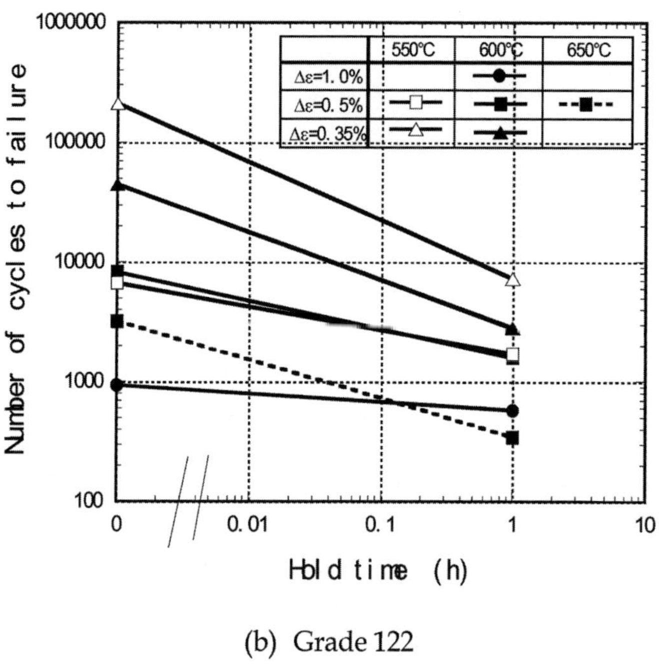

(b) Grade 122

Figure 8 Relation between hold time and number of cycles to failure

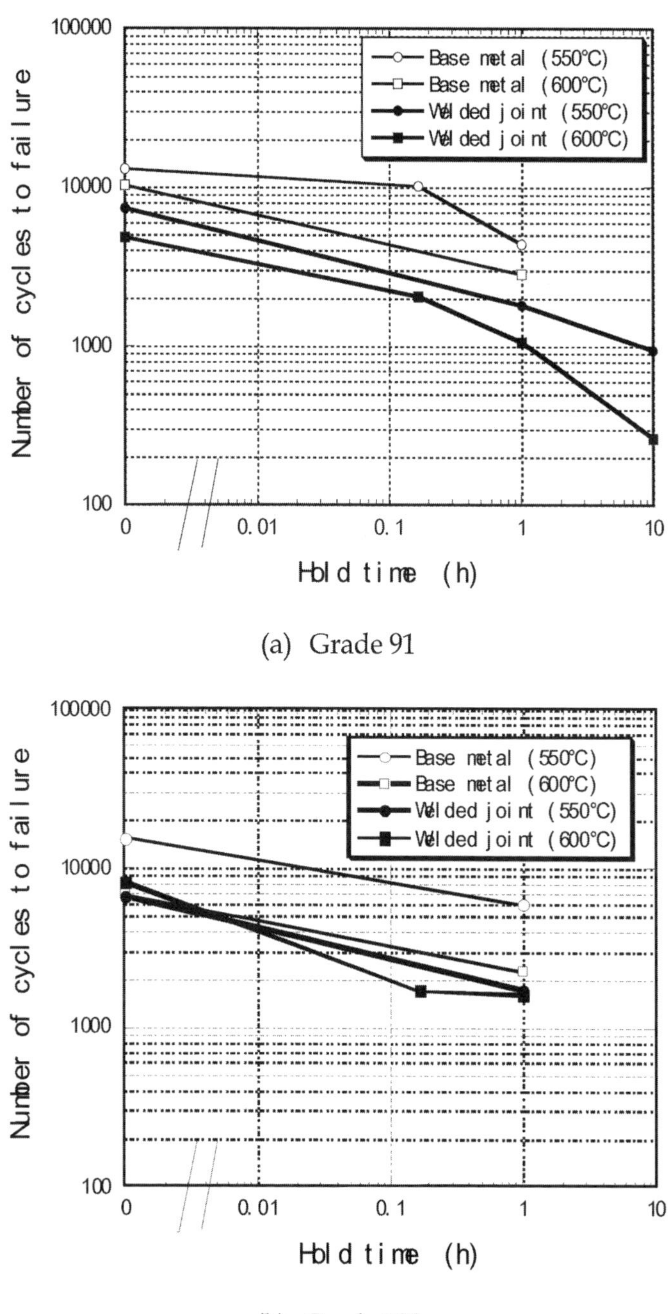

(a) Grade 91

(b) Grade 122

Figure 9 Comparison of failure life of base metal and welded joint specimens

Table 2 Constants for creep-fatigue life prediction

Procedure	Constant	Grade 91			Grade 122		
		550C	600C	650C	550C	600C	650C
Time fraction rule	a_0	34154			24670		
	a_1	3494			1225.3		
	a_2	-2574			-1237.8		
	C	31.4			21		
Ductility exhaustion approach	b_0	1.384	1.453	1.098	1.031	1.455	2.239
	b_1	0.0249	0.550	0.548	-0.252	0.099	1.174
	b_2	-0.0354	0.099	0.144	-0.079	-0.054	0.194
	ε_{f0}	19	20	24	17	20	22
Energy-based approach	α	79.9	29.8	6.81	75.9	91.1	35.6
	n	0.277	0.285	0.162	0.180	0.414	0.392
	W_{f0}	54	51	50	65	60	51

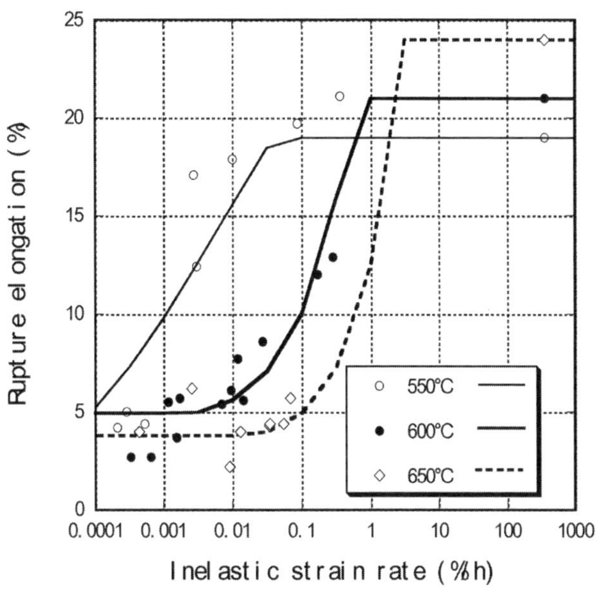

(a) Grade 91

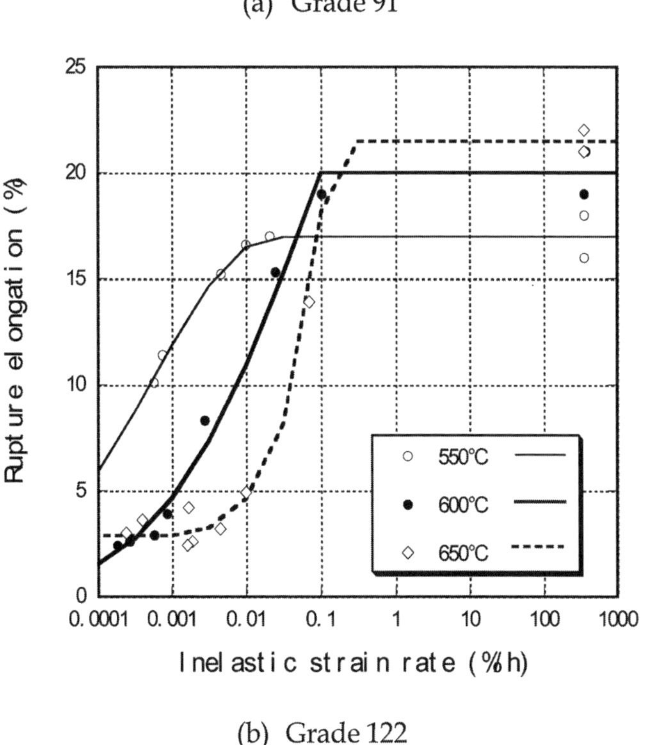

(b) Grade 122

Figure 10 Relation between average inelastic strain rate and rupture elongation

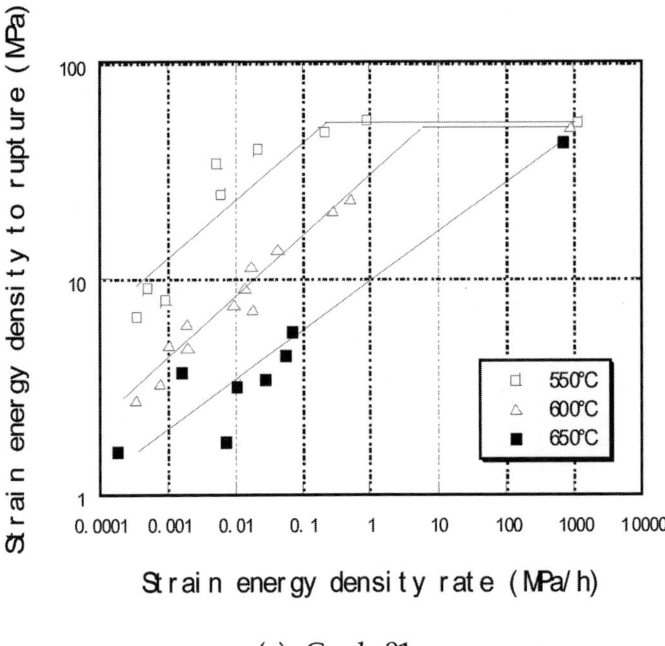

(a) Grade 91

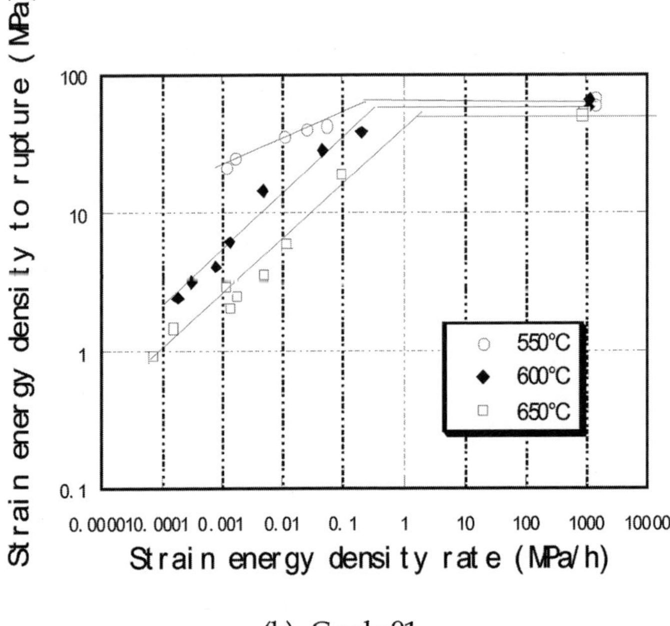

(b) Grade 91

Figure 11 Relation between average work rate and accumulated work to rupture

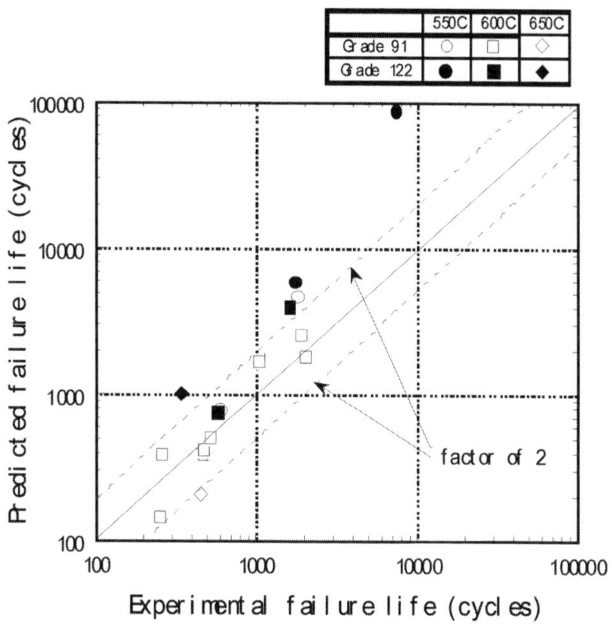

Figure 12 Comparison of predicted life by time fraction rule with test results

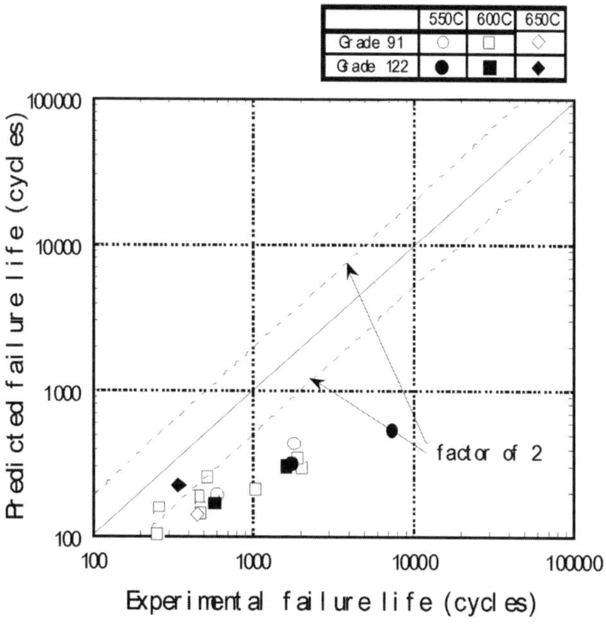

Figure 13 Comparison of predicted life by ductility exhaustion approach with test results

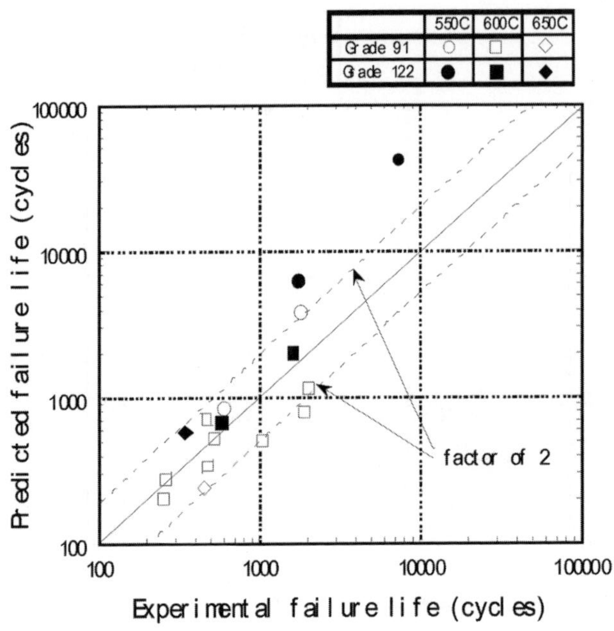

Figure 14 Comparison of predicted life by modified ductility exhaustion approach with test results

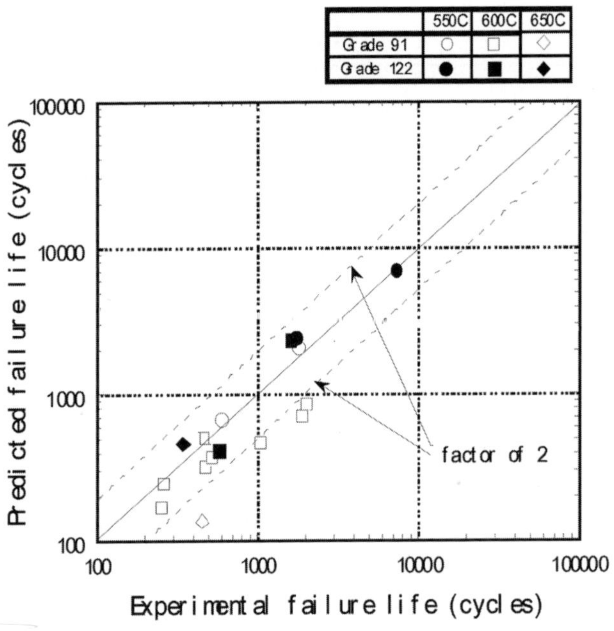

Figure 15 Comparison of predicted life by energy-based approach with test results

Advances in Materials Technology for Fossil Power Plants
Proceedings from the Fifth International Conference
R. Viswanathan, D. Gandy, K. Coleman, editors, p 783-789

DOI: 10.1361/cp2007epri0783

Prediction of In-Service Stress States of Single Crystal Superalloys Based on Mathematical Analyses of γ/γ' Microstructural Morphologies

Masakazu. OKAZAKI*
Motoki. SAKAGUCHI*
Nagaoka University of Technology,
Tomioka, Nagaoka 940-2188, Japan.

Abstract:

It has been well recognized that the morphology of γ/γ' microstructures of single crystal superalloys may be changed during their in-service period, associated with a reasonable manner following material science rules. This implied, it may be possible for us to use this phenomenon for failure analysis, through analysis of the morphological change in γ/γ' microstructure. In order to explore this possibility, the change of γ/γ' microstructure of a single crystal Ni-base superalloy, CMSX-4, was studied in this work, when the alloy was subjected to several modes of external lodgings. The experimental variables in the loading mode in this work were: level of stress in tension and compression, loading temperature, loading rate, monotonic/cyclic loadings and multi-axial level in stress states. The experimental results clearly demonstrated that the γ/γ' microstructures were changed, associated with a very sensitive manner depending on the above loading modes. A new image analysis method has developed to quantitatively analyze the morphological changes. These experimental results are summarized into a two-dimensional map so that it can be easily applied for failure analysis and other engineering applications.

1. Introduction

Single crystal Ni-based superalloys have regularly arrayed composite microstructure consisting of cuboidal γ' precipitates surrounded by narrow channels of γ matrix. During the gas turbine's operation period, on the other hand, the microstructure may change accompanying with severe directional coarsening, so-called rafting, of the initially cuboidal γ' precipitates to the plate-like or needle-like structure by the creep stress due to the centrifugal force [1,2]. Nowadays, several investigations have indicated that the directional coarsening could occur by many types of loading conditions: monotonic loading associated with plastic strain [3,4] and the high temperature fatigue or thermo-mechanical fatigue loadings [5]. A series of relating works strongly indicate the morphologies of the γγ' microstructure should be formed and reformed under any metallurgical kinetics and interactions between internal and external stresses, temperature and loading histories. Accordingly, once the understanding on the kinetics as well as the influencing factors are achieved, it may be possible to estimate the stress state and the remaining life of superalloy components in service, through a quantitative analysis of their γ/γ' microstructures. Primary objective in this work is to make clear experimentally the effects of the monotonic and cyclic (fatigue) loadings on the geometrical γ/γ' morphologies. The second is to propose and explore a new quantitative method to analyze the effects of thermo-mechanical histories from the γ/γ' morphology.

2. **Experimental Procedure**

The material used in this work is a second generation single crystal superalloy, CMSX-4, to which the recommended 8-stages solution and 2-stages aging treatments were given. After the heat treatment, the CMSX-4 revealed very regular γ/γ' microstructure, involving cubic γ' precipitate. It is thought that there is negative misfit between γ/γ' interfaces [1].

From the master material, the solid cylindrical specimens were machined for monotonic and cyclic straining tests, where all specimen's axes lie within 5° from <001> crystallographic orientation. The straining tests were carried out according to the test program summarized schematically in Fig.1, where all tests were conducted under uni-axial loading condition in air. The Specimen 1~6 were subjected to plastic strain(s) under the room temperature and received the subsequent heat treatments (1080C° for 20hrs. in vacuum), where experimental variables were the plastic strain magnitudes, directions and their histories. The Specimen 7 and 8 were subjected to stress controlled cyclic loadings at 1000C° for 150h. For all specimens tested, sections parallel to the loading axis were carefully polished and observed in the SEM.

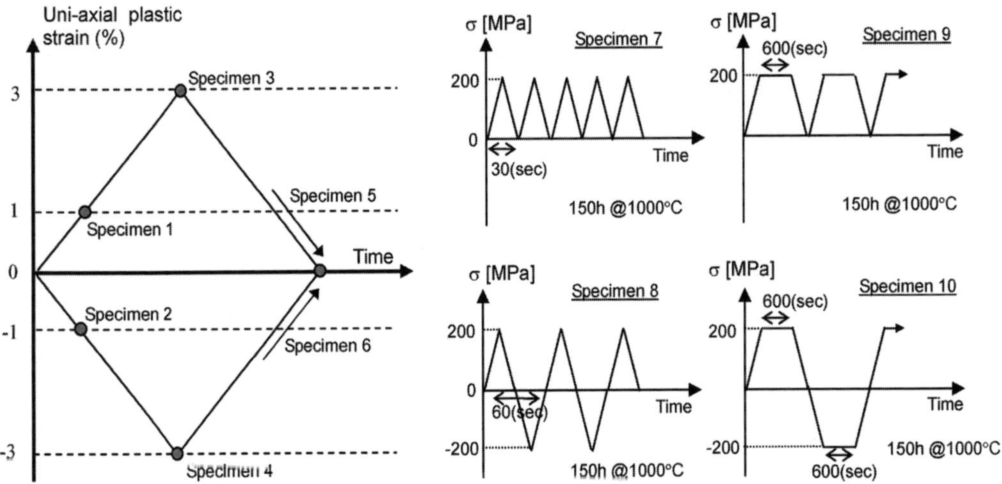

Fig.1 Schematic illustration for the monotonic and cyclic straining process. The Specimens 1-6 were subjected to plastic strain at R.T., while the Specimens 7-10 were to fatigue or creep-fatigue loading cycles at 1000°C. The post heat treatments at 1080°C were performed to all specimens after pre-straining.

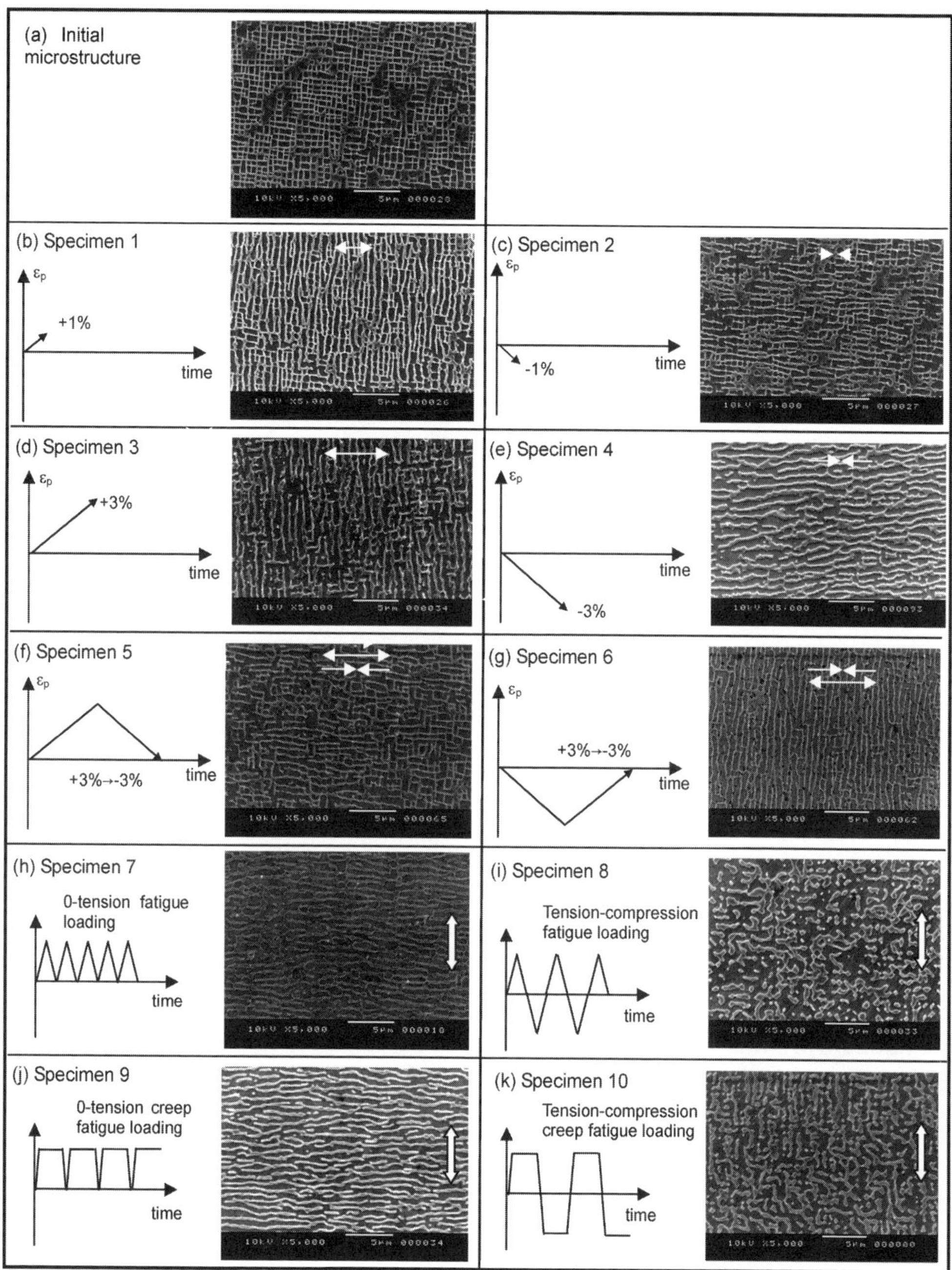

Fig.2 SEM images of initial microstructure and the Specimen 1-10. Arrows in figures represent the uni-axial loading directions

3. Experimental Results

Fig.2 summarizes the SEM micrographs of the Specimen 1~10. The following characteristics should be pointed out from Fig. 2.

(1) On the effect of sign of monotonic plastic straining:

The γ/γ' microstructures in the Specimen 1 and 3 reveal the directional coarsening normal to the loading axis (see Fig. 2(b) and (d)) similar to the plate-type rafted structure by creep. On the other hand, in the Specimen 2 and 4, the directional coarsening occurred parallel to the compressive strain axis (see Fig. 2(c) and (e)). These results clearly indicate the γ/γ' microstructure may be reconstructed under an influence of the sign of straining. These relationships between the plastic strain and the coarsening directions are same to that in the rafting phenomenon during creep for negative misfit alloys. M.Fahrman et al. [3] and M. Veron et al. [4] have also obtained the similar results from the pre-straining tests for single crystals.

(2) On the effect of plastic strain magnitude:

Comparing between the Specimen 1 and 2 (see Fig. 2(a) and (b)), and between the Specimen 3 and 4 (see Fig. 2(c) and (d)), it is found that the extents of the coarsening are significantly affected by magnitudes of plastic strain. It suggests that there would be a critical strain to reconstruct the γ/γ' morphology.

(3) On the effect of plastic strain history.

It is clear that the microstructure in the Specimen 5 (Fig. 2(f)) directionally coarsens parallel to the strain axis, while that in the Specimen 6 (Fig. 2(g)) is normal to the axis (i.e. compressive in Specimen 5, while tensile in the Specimen 6). Note that the final level of plastic strain reached was same between these two specimens, but along with different loading histories. Accordingly, the γ/γ' microstructure is reformed depending on loading sequence, as well.

(4) Regarding the effects of stress-controlled cyclic loading on the γ/γ' microstructure

The specimen 7 (Fig.2(h)), subjected to the 0-tension cyclic stress (R_σ=0), shows the directional coarsening normal to the uni-axial stress axis similar to the rafted structure during tensile creep and plastic-strain-influenced coarsening shown in Fig.2(d). On the other hand, the specimen 8, to which the tension-compression cyclic stress (R_σ=-1) were applied, reveals the unique microstructure coarsened to both normal and parallel to the stress axis (see Fig. 2(i)). Thus, the comparison between the Specimens 7-10 clearly shows that the γ/γ' morphologies are reconstructed under an influence of stress ratio and stress waveform. It is also worthy to note that the cyclic lading could also produce the rafted structure, similar to that by creep.

4. Quantitative evaluations for the γ/γ' morphologies

The investigation in the Sec. 3 is not beyond phenomenological discussions, thus it becomes inevitable for the engineering application to analyze the γ/γ' morphologies in a quantitative manner. In this work, two quantitative analytical methods are tried. One is a so-called ***"intercept method"***, and the other is an original method named by ***"Cluster-Method"*** in this work.

4.1 Intercept method

4.1.1 Evaluating procedure *The intercept method* is a traditional method that is usually used to evaluate grain size. In this method, some straight lines were drawn parallel and perpendicular to the loading axis and then the number of the intersections between each line and the γ' phases was counted. In this work, at first arbitrary 15×15 (μm) regions were selected from Fig. 2(a)~(e), (k) and 30 lines were drawn parallel and perpendicular to the loading axis, where the interval of each line was every 0.5 μm. Then, the number of intersections between the lines and the γ' phase were counted.

4.1.2 Evaluated results In this paper, the number of the intersections between the normal lines and the γ' phases are denoted "*H*" and that between the parallel lines and γ' phases are denoted "V", respectively. The evaluated results shown in Fig. 2(a)~(e), (i) are summarized in Table 1. Here, values of H and V are normalized by the values for the initial microstructure, H_0 and V_0, respectively. Following characteristics are found from the value of H/H_0, V/V_0 and their relative ratio ($R=log\{(V/V_0)/(H/H_0)\}$.

(1) The values of R in the directional coarsened microstructure normal to strain axis (Fig.2(b),(d)) reveal the negative values, while those in the parallel coarsened ones (Fig.2(c),(e) are positive.

(2) The more significant coarsened microstructures (Fig. 2(d), (e)) have larger absolute values of R compared to the less coarsened ones (Fig. 2(b), (d)).

Thus, the sign and the absolute value of R can quantitatively evaluate the coarsening directions and the extent of coarsening, respectively. However, it is difficult to distinguish the Fig. 2(i) from the initial microstructure, in which the directional coarsening occurred in both normal and parallel directions.

Table 1 Summary of the values of V/V_0, H/H_0 and their relative ratios (R) for actual microstructures in Fig.2.

Fig. number	V/V_0	H/H_0	$R = \log(\frac{V/V_0}{H/H_0})$
Fig.2(a)	1	1	0
Fig.2(b)	0.558	1.23	-0.345
Fig.2(c)	1.03	0.615	0.225
Fig.2(d)	0.371	0.929	-0.399
Fig.2(e)	0.906	0.315	0.459
Fig.2(i)	0.644	0.651	-4.69×10^{-3}

4.2 Cluster-Method

4.2.1 Evaluation Procedure

In this method, the actual microstructures were represented by ***the cluster*** consisting of minute square elements as schematically shown in Fig. 3. Here, the microstructural morphologies are numerically represented by following two parameters;

s; the number of the elements which consist the cluster (black region in Fig. 3(b); s=23),

t; the number of the elements which surround the cluster (dashed region in Fig. 3(b), t=29).

The values of **s** and **t** represent the size of microstructures (e.g., the extent of randomly coarsening) and the extent of directional coarsening, respectively.

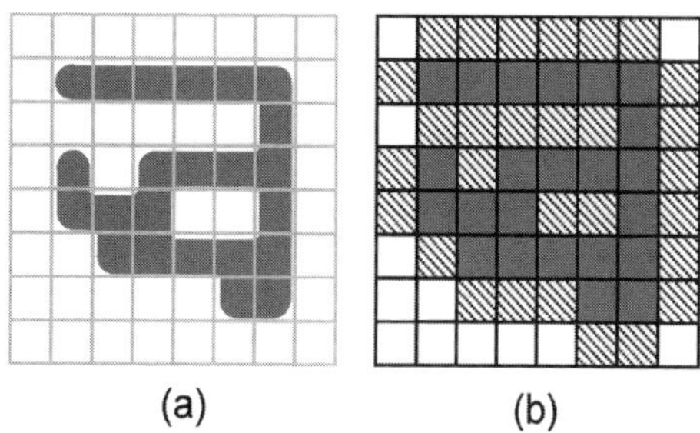

Fig.3 Schematic illustration of the "Cluster-Method. The values of **s** and **t** are counted by 23 and 29 for this geometry, respectively.

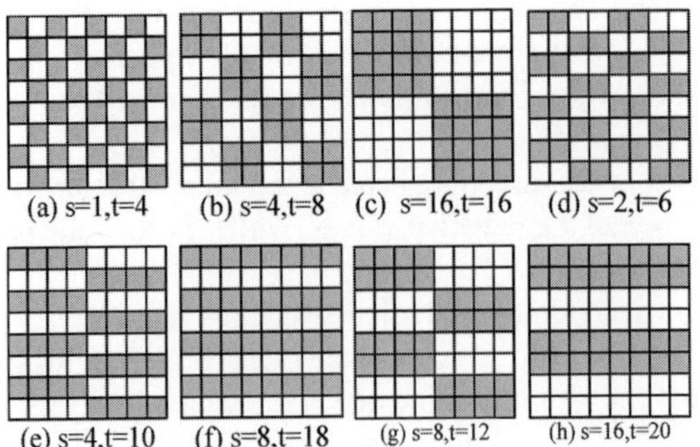

Fig.4 Typical morphologies in the Cluster-Method.

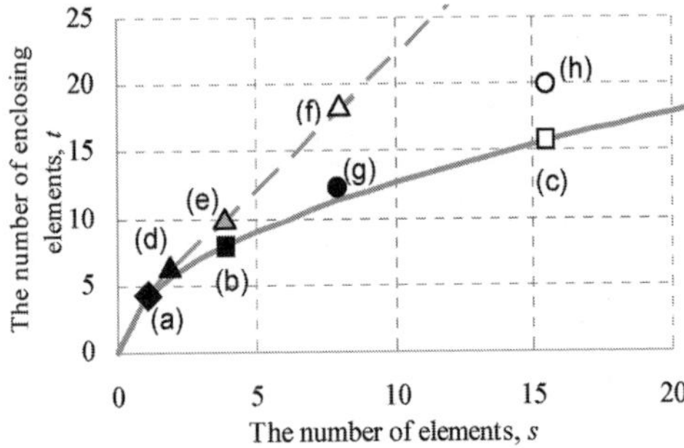

Fig.5 Relationship between *s* and *t* for some typical morphologies in Fig. 4. (a)~(h) corresponds to the micrograph in Fig. 4.

Fig. 4 shows some models with typical morphologies and their values of ***s*** and ***t***. are plotted in Fig. 5. It is found from Fig. 5 that the square shape clusters such as Fig, 4(a),(b),(c) lie on the solid line ($t = 4\sqrt{s}$), while the dashed line (;$t = 2s + 2$) corresponds to the case the elements array in a line. The morphologies such as Fig. 4(g) or (h) distribute in the region between the solid and dashed line in Fig. 5. Note that the well coarsened (large) morphologies such as Fig. 4(c) or (h) lie in the region far from the origin, while the directional coarsened types leave from the solid line to dashed line.

4.2.2 Evaluated results Fig. 6 shows the relationship between *s* and *t* for the actual microstructure in Fig. 2(a)~(e) and (i), where the γ/γ' microstructures were represented by 0.1x0.1 (μm) elements. In Fig. 6, the values of *s* and *t* are put into the special coordinate system in which the solid and dashed lines in Fig. 5 are transformed to the horizontal and vertical axes for easily understanding. From this transformation, the size of the coarsened γ' phase can be represented by the distance from the origin (***r***), while the angle from the *s'* axis (***ϕ***) corresponds to the level of the directional coarsening. It is found from fig.6 that the well directional coarsened structures such as Fig. 2(d) or (e) have larger values compared with fig.(b) or (c). In addition, the differences between the initial microstructure and the Fig. 2(i) is obviously evaluated, which could not distinguished by the Intercept-Method in section 4.1.2. These results suggest that the Cluster-Method proposed in this work can lead the quantitative evaluation for the various morphological changes in the γ/γ' microstructure.

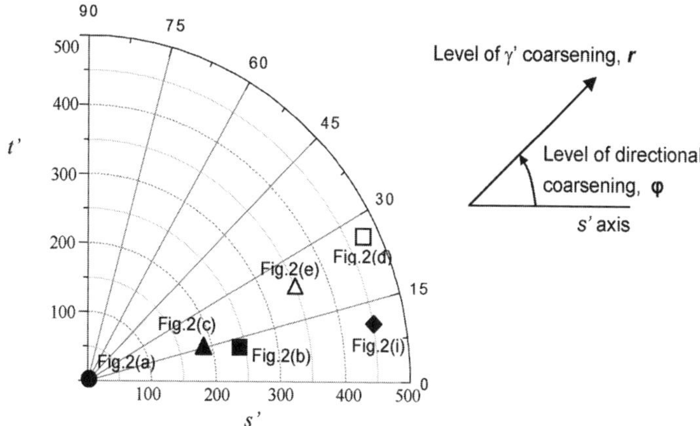

Fig. 6 The analysis of the γ/γ' morphologies by the Cluster-Method. On this 2-dimensional diagram, the degree of the γ' coarsening and that of the γ/γ' directional coarsening are represented by the distance from the origin, ***r***, and the argument angle, ***φ***, respectively.

5. Summary

The present experiments clearly have demonstrated that an application of the monotonic straining at R.T. and the subsequent heat treatment at high temperature resulted in a uni-directional, or bi- and tri-axial coarsenings of γ/γ' microstructure, depending on the magnitude of strain and the histories. A critical strain was found to be needed to build up the microstructural changes. Not only monotonic but also cyclic loadings could induce the similar changes in the γ/γ' microstructure. These changes have been summarized in the table. In order to quantitative expression of the γ/γ' morphologies, a new method, named by "***Cluster-Method***", was proposed, where the degree of the coarsening and the γ/γ' directional rafting can be displayed on an original 2-dimensional diagram. This method successfully evaluate various types of the γ/γ' microstructures. Furthermore, this method may provide us new potentials for both the stress field detection and the remaining life analysis of superalloy components in service.

References

[1] T.M. Pollock & A.S. Argon, *Acta Metall Mater.* Vol.42, No.6, (1994), pp.1859-1874
[2] S.Socrate & D.M.Parks, *Acta Metall Mater.,* Vo.41, No.7, (1993), pp.2185-2209
[3] M.Fahmann, E.fahmann, O.Paris, P.Tratzl and T.M.Pollock, *Superalloys* 1996, (1996), pp.191-210
[4] M.Veron and P.Batie, *Acta Mater.,* Vol.45, No.8, (1997), pp.3277-3282
[5] V.Brien and B.Decamps. *Mater. Sci. Eng. A*, Vol.316, (2001), pp.18-31
[6] M.Okazaki, T.Hiura and T.Suzuki, *Superalloys 2000*, (2000), pp.505-514.

Advances in Materials Technology for Fossil Power Plants
Proceedings from the Fifth International Conference
R. Viswanathan, D. Gandy, K. Coleman, editors, p 790-808

DOI: 10.1361/cp2007epri0790

Advances in Welded Creep Resistant 9-12% Cr Steels

D. J. Abson
J. S. Rothwell
B. J. Cane
TWI
Granta Park
Cambridge, CB21 6AL, UK

Abstract

Research on high chromium ferritic materials for high temperature power plant components generally concentrates on the properties of the parent steel. Weldments, however, are often the weak link, leading to premature failures and associated forced outages and high maintenance spend. Clearly, consideration of the creep performance of weld metals and associated heat-affected zones (HAZs) in these materials is important. Despite this, relevant weldment creep rupture data are not commonly available, and weldment creep rupture "strength reduction factors" are not always known. This paper provides comment on the available information on parent materials, and highlights the need for the assessment of the creep performance of weldments. Strategies for increasing HAZ creep rupture strength are reviewed, and some available weldment data are considered. Less conventional welding processes (GTA/TIG variants and EB welding) appear to provide improved creep performance of weldments. They therefore merit further study, and should be considered for welding the new steel grades, particularly in supercritical and ultra-supercritical applications.

Introduction

World electricity demand is increasing relentlessly, with many parts experiencing very rapid growth. At the same time, since power plants are a major source of greenhouse gases, there is growing pressure to reduce emissions per unit of electricity generated. Whilst nuclear power and alternative or renewable energy generating systems are likely to be increasingly introduced, IEA projections of new plant-builds indicate a reliance on fossil fuels over the next 20 to 30 years. Electricity generated from coal currently accounts for around 40% of the world's generating capacity and, because of its abundance in many of the world's developing countries, it is likely to remain the dominant fuel in growing economies such as China and India for the foreseeable future.

With this situation, there is a clear incentive to increase the efficiency of coal-fired plants. Sub-critical plants typically operate at efficiencies of 35%, whereas supercritical (SC) plants have efficiencies up to 46%, and ultra-supercritical (USC) plants achieve efficiencies around 50%. Such efficiency improvements provide significant reduction in greenhouse gas emissions. More than 240 high efficiency SC plants are in operation worldwide. World Coal Institute figures for 2005 indicate that China, for example, has 22 SC units, supplying around 14 GW of electricity. 24 USC plants also exist, with units in Denmark, Germany, the Netherlands, Japan and the U.S.A., and many are at the planning stage elsewhere. The concept of carbon dioxide capture is now adding further importance to the need for higher efficiencies, since the integration of carbon capture plant gives an efficiency penalty of around 10%. These SC and USC plants (with provision for integration of carbon-capture) are thus clear candidates for the next generation of coal-fired plants.

SC and USC plants operate at higher temperatures and pressures than conventional sub-critical plants, as shown in Table 1 (1). High temperature materials technology has consequently become the major challenge, particularly for USC plant. The reality of USC plant has been made possible by major developments in Europe, U.S.A. and Japan in 9 to 12% Cr ferritic steels for operation up to 650°C. The evolution of steels for power plants is shown in Table 2 (2), and also in Figure 1 (3).

The opportunities offered by the advanced steels in terms of USC plants, however, may not fully materialise if plant reliability suffers. Poor confidence in the long term performance of welds, particularly in components such as steam piping, headers and valves, is a major safety and economic concern. The phenomenon of type IV cracking is recognised as a primary failure mechanism that limits the life of components operating at elevated temperatures.

Type IV cracking has plagued the power industry for around 30 years. Unfortunately, with increasing precipitate-strengthening in the advanced ferritic steels, the effect on lifetime may be exacerbated. In other words, the weldment-parent creep strength ratio, WPCSR (commonly called the weld creep rupture "strength reduction factor") can decrease, i.e. worsen, with increasing parent metal creep strength. Type IV cracking can thus erode the benefits achieved in parent metal creep strength. Unless the problem can be reduced or overcome, for example by placing welds in lower stress or lower temperature locations, there will need to be more conservative design philosophies and remnant life assessments based on parent metal strength, which realistically account for long-term WPCSR values.

This paper considers possible solutions to the problem of type IV cracking, including improvements in the welding process, and the need for long-term weldment data as a prerequisite for any new high temperature design approach using advanced ferritic steels.

Parent Steel Developments

Whilst substantial improvements have been made over the last four decades in the development of 9-12%Cr creep-resistant steels (4), the steam temperatures in recent and projected thermal power stations are now approaching the upper temperature limit for ferritic steels. An understanding of the microstructural factors that control long term creep rupture strength is vital for further improvements in creep rupture strength. Two basic alloy design philosophies are possible. One is to adopt nominally 9%Cr as the base, recognising that the alloys will have inadequate steam oxidation resistance, and will require cladding in order to operate at temperatures of the order of 650°C. Alternatively, the poorer creep rupture strength of 10.5%Cr steels (5) might be improved through judicious choice of alloying. The improved oxidation resistance of higher chromium steels would eliminate the need for cladding. In all cases, it is essential to avoid the formation of delta ferrite, since its presence is detrimental to creep rupture strength (6).

In designing further alloys based on a nominally 9%Cr composition in the European COST 536 programme, Hald (7) stated the basic design ideas as:

- Avoid Z-phase formation (with low Cr contents)
- Achieve Laves phase strengthening
- Add other nitride formers (Ti, Ta, Zr and Hf)
- Optimize the boron and nitrogen contents

Creep damage in power plant steels is associated with grain boundary precipitates. In most Cr-containing ferritic steels, grain boundary precipitates, such as $M_{23}C_6$, provide favourable nucleation sites for creep damage, in the form of grain boundary cavities and micro-cracks. An increase in the proportion of intragranular MX type particles, such as VN, is beneficial. Hald (7) reviewed the development of 9-12%Cr steels. He noted that the poor long term creep rupture strength of some of the recently-developed alloys demonstrated that the ideas behind the alloy design have been based on a lack of understanding of the long-term microstructure stability of this class of alloys. Microstructural explanations for the improved creep strength of the new 9-12%Cr steels should be found in mechanisms which retard the migration of dislocations and sub-grain boundaries, and thus delay the accumulation of creep strain with time.

Among recent developments, including some on experimental alloy systems, Yin and Faulkner (8) showed that ion implantation of hafnium into thin foils of a 9 wt-%Cr ferritic steel can markedly improve the creep properties of the material. The addition of hafnium markedly increased the volume fraction of MX precipitates, and completely eliminated the formation of the common grain boundary $M_{23}C_6$ particles.

Semba and Abe (9) showed the beneficial effect on creep rupture strength of a progressive increase in boron up to 0.0139% in their 0.08%C-9%Cr-3%W-3%Co-0.2%V-0.05%Nb-0.008%N alloys. Steels with boron or nitrogen contents at higher levels lead to the detrimental formation of coarse BN precipitates. The solubility limit of BN at 1150°C is expressed by the equation:

$$\log(\%B) = -2.45\log(\%N) - 6.81 \qquad (10)$$

Figure 2 shows how this equation defines a boundary beyond which coarse precipitates form. Although nitrogen is an important addition for the formation of fine MX precipitates, which enhance creep strength, high solution temperatures are required during steel processing, leading to extra costs for the steel maker. Taneike et al (11), employed a solution treatment temperature of 1,300°C to take titanium carbo-nitrides into solution, but observed a substantial reduction in the minimum creep rate at 650°C in 8.4%Cr steels containing 0.047%Ti, 0.13%C and 0.006%N. Optimizing the solution treatment temperature is an important aspect of producing modern B-containing steels.

Type IV Failure and its Consequences for Creep Rupture Life

Creep rupture studies on modified 9-12%Cr steels have been largely focused on the rupture strength of the parent steels. Much less attention has been given to the behaviour of weldments. The failure of high chromium steels in power plant components is often associated with failure in the "type IV" region. In short term, high stress laboratory tests on welded joints, failure generally occurs in the parent steel or weld metal. However, at lower stresses and after longer times, the failure location switches to the outer region of the visible HAZ. This has been classified as a type IV failure, and the location identified as the type IV region. The classification of the different failure locations has been discussed by Brear and Fleming (12) and Francis et al. (4). Type IV cracking is the most troublesome in-service failure classification facing the industry today.

In 9%Cr steels, Cerjak and Letofsky (13) found that the transition to type IV cracking occurred after test durations of ~10,000h. In their article on 91 grade steel weldments, Brear and Fleming (12) examined equations to describe parent steel and type IV creep rupture failures in grade 91 steels. The intersection of these two equations predicts the transition from parent metal failure (at high stresses and short times) to HAZ failure (at lower stresses and longer times). They found good agreement between experimental data and the equations describing HAZ behaviour due to Nath and Masuyama (14).

Abe (15) carried out creep rupture tests on simulated HCM12A specimens that had been heated to temperatures in the range 800 to 1,000°C, to simulate different parts of the HAZ. He reported that, while hardness had its minimum value after heating to temperatures near the Ac_1, the creep rupture time in tests carried out at 650°C had its minimum value in specimens heated to temperatures near the Ac_3. From their results, it appears that at high stresses (140MPa), failure preferentially occurred in specimens heated slightly below the Ac_3 i.e. intercritically heated, whilst at lower stresses (<120MPa), the minima shifted to specimens heated to just beyond the Ac_3 i.e. the fully transformed fine grained specimens. This apparent shift in failure position with stress may explain why both regions of the outer HAZ have been associated with type IV failure.

It has been reported that type IV cracking of the weldments can reduce creep life, compared with parent design life, by as much as a factor of five (16-18). When considering creep rupture strength, Nath and Masuyama (14) reported that, at 600°C, 10,000h extrapolated data for grade 91 weldments indicated a 37% disparity between the parent steel and the HAZ, i.e. a WPCSR of 0.63. However, it should be recognised that the extent of the disparity increases progressively with increasing log (exposure time), as shown by Sawaragi (19) for 2.25%Cr-1%Mo weldments. In creep rupture tests at 538°C on tubular modified 9%Cr-1%Mo specimens, a progressive reduction in WPCSR from 0.94 to 0.93 to 0.90 for rupture times of 10^1, 10^3 and 10^5h, respectively, was reported by Blass et al. (20). In a study of grades E911, 91, 92, and 122 weldments, in the UK-funded Fourcrack programme, it was concluded that the weldment creep rupture strength "falls towards a floor value of about 60% of the parent strength in the longer term" (21).

Parent steel and cross-weld creep rupture data for 92 grade steel, tested at 650°C, have been presented by Abe (22), Figure 3. Similar data have also been presented by Abe and Tabuchi (23) for grade 122 steel. The important features of this figure are that, for long exposure times, the parent steel creep rupture strength falls below a linear extrapolation of the short term data, and that the disparity between the parent steel and the cross-weld creep rupture strength increases progressively with increasing life, i.e. the WPCSR decreases.

The first issue, i.e. the parent steel creep rupture strength, has been addressed by Japanese researchers, who incorporated an increased slope for a second straight line fitted to long term creep data, obtained at stress levels less than 50% of the 0.2% offset yield strength (24,6). They have called this method 'region splitting'. It is clearly essential, when extrapolating short-term parent steel test data to projected long service lives, to ensure that appropriate allowance is made for the increasing fall-off in parent steel creep rupture strength with time.

Concerning the second issue, i.e. the weldment behaviour, WPCSR values estimated from the data in Figure 3, are plotted in Figure 4, along with similar data obtained for grades 91 and 122 from a variety of sources. The linear extrapolation suggests that the WPCSR appropriate for a 30 year design life at 650°C for this steel is of the order of 0.27! Furthermore a linear extrapolation may be optimistic. This is clearly at variance with, and much more alarming than, the conclusion of the Fourcrack programme cited above, even though 92 and 122 steel grades were included in both studies. The information in Figure 4 reveals that, in addition to measures needed for extrapolation of parent material data, allowance must also be made for the progressive decrease in WPCSR, when extrapolating short term data to longer times.

In all cases, it should also be recognised that, particularly for small diameter test specimens, at least part of the apparent fall-off in creep rupture strength at long exposure times will be attributable to the loss of cross-section through oxidation, unless a suitable protective atmosphere is provided (27).

Masuyama (28) and Kimura et al. (29) noted that changing the constant in the Larsen-Miller parameter has a strong influence on the value of the parameter, and thus on the estimated creep rupture life at a different temperature. Kimura et al. (29) have shown how more accurate fitting of data can be achieved in the type IV region by using a different constant. Therefore, also included in Figure 4, on a second abscissa is the equivalent time at 600°C, estimated using the Larsen-Miller parameter with an assumed constant of 36 (Figure 3a) and 20 (Figure 3b). This demonstrates the strong effect of changing the Larsen-Miller constant on the lifetimes of weldments at temperatures different from that of the test data available. Clearly, the test temperature and the creep rupture life must be reported whenever WPCSR values are quoted. If predictions are made for temperatures other than the test temperature, there must be confidence in the applicability of the parametric constant chosen.

In agreement with this, data obtained by Kuboň and Sobotka (30) for grade 91 steel show that the ratio was 0.77 after 10,000h at 600°C, but decreased with increasing time and temperature. Even with this modest diminution of extrapolated creep rupture strength, for a 91 grade parent steel creep rupture life of 10,000h, there is a dramatic reduction in the HAZ (type IV) rupture life to only 1,700h.

The sizing of component thicknesses in the ASME boiler and pressure vessel construction codes (31-33), is normally based on closed form equations which include an allowable stress for the material of construction and an efficiency factor (E). The allowable stresses are supplied by ASME (2004), and are based on the parent material properties at the design temperature. Most commonly for creep service the allowable stress is taken to be 67% of the average creep rupture stress of the parent material at 100,000 h. The allowable stress does not consider the loss of strength due to the presence of weldments. The efficiency factor varies between the different ASME construction codes. However, Section I (2004) states that E is 1 for seamless and welded cylinders or the ligament efficiency. Hence, for components designed to ASME section I with no ligaments, there will be no allowance for weld creep rupture strength reduction.

It appears from the information presented above that premature weldment failure is likely wherever weldments in fabrications from 91 grade and similar steels operating at temperatures around 600°C are subjected to stresses that are of similar magnitude to the parent steel design stresses. It is likely that utilities that have used these steels in high pressure steam service at such temperatures with inadequate allowance for the presence of weldments are "sitting on a time-bomb", and that monitoring and perhaps in-service stress measurements are required urgently to avoid catastrophic premature failures.

Improving the WPCSR and Avoiding Type IV Failure

Through Parent Steel Developments

In view of the importance of failure in the type IV region during long-term service, it is inevitable that increasing attention will be given to the creep rupture properties of the HAZ. One approach to improving the creep rupture strength of the type IV region is to modify the parent steel composition.

For 91 grade steel tested t 650°C, Iseda et al. (34) reported an improvement in both the parent steel creep rupture strength and the WPCSR, as nitrogen was increased, for times up to 40,000h. However, lower nitrogen levels would clearly be required if the beneficial effects of increased boron levels are to be exploited.

The single most effective step towards the elimination of type IV failure and its negative effect on the WPCSR has been the addition of boron. As well as improving the parent steel creep properties, boron has been shown to improve the WPCSR substantially (35, 22, 34). Tabuchi et al. (35) added from 90 to 180ppm B to low nitrogen (<0.0017%), 9%Cr steels. They tested cross-weld specimens, and reported that 90-130ppm was the optimum level to improve both parent steel and HAZ creep rupture strength. A similar conclusion was reached by Kondo et al. (36), who studied welded joints and simulated HAZs. The beneficial effect of boron was attributed to it combining within $M_{23}C_6$, and suppressing its coarsening, thereby reducing the minimum creep rate (37-38). It also prevented the formation of a fine-grain HAZ, which adds to the evidence pertaining to this region being the most susceptible to type IV cracking.

The substitution of tungsten for Mo has also been reported to be very effective in enhancing the creep rupture strength of the HAZ (39). However, this route of enquiry does not appear to have been addressed by other investigators.

Through Welding Process Development

A novel approach to improving the creep rupture strength of the type IV region, as noted by Bell (40), is to carry out a partial tempering of the parent steel prior to welding, and completing the tempering treatment as the post-weld heat treatment, as described by Sikka (41). This approach almost doubled the creep life, compared with conventionally heat treated 9%Cr steel weldments in the short term (<10,000 hours) (42); it would be interesting to see the effect over a longer test period. Other process variables suspected to have a strong influence on the type IV creep life of weldments have been highlighted by Francis et al. (43). Among the variables shown to influence type IV cracking, as determined by Baysian neural network analysis, were pre-heat temperature and PWHT time.

A further strategy, which could presumably be combined with the half tempering approach, is to join modified 9% Cr steels by welding processes that create a very narrow HAZ (and therefore a shorter thermal cycle, and less time for over ageing of precipitates). Abe (15) reported that electron beam welding, which produces a very narrow HAZ, had a beneficial effect on the creep rupture performance of the type IV region, with more recent data confirming the behaviour (23); see Figure 5. The approximate HAZ widths of electron beam and GTA/TIG welds were 0.5mm and 2.5mm, respectively, and the cross-weld creep rupture life of EB welds was approximately twice that of GTA/TIG welds (15, 23). However, consideration should be given to any adverse effects on toughness of EB welds in such steels (44). A number of alternative processes, including EB, narrow gap GTA/TIG, and friction welding are currently being investigated in one of the creep research programmes at TWI within the European COST programme.

Albert et al. (45) have shown that simulated fine grain HAZ and real weldments behave significantly differently in terms their creep strain. This observation was attributed to the triaxiality introduced by the different creep properties in the various regions of the HAZ. They pointed out that changing the weld preparation angle can alter the stress state of the joint, and influence creep results significantly. They concluded that 'By reducing HAZ width or the groove angle of the joint, this stress state can be altered to achieve significant improvement in rupture life of the weld joints'.

Weld metal development

Welding consumables have been developed with matching or near-matching properties, for welding 92 grade 9%Cr steels by manual metal arc and submerged arc welding (46-48) and flux-cored wire welding (47-48). However, where creep rupture properties are presented, they are generally marginally inferior to those of the 92 grade parent steel. In view of the lower creep rupture strength of the HAZ at long durations, the creep rupture strength of the weld metal is probably of little consequence.

For their series of experimental weld metal compositions, Vanderschaeghe et al. (47) showed the detrimental effect of increasing oxygen and also boron content on impact toughness. However, few long-term creep rupture test data are available, and so it will be of interest to see how the various consumables perform in long-term tests.

Limited creep rupture data, from tests at 600°C have been presented for experimental rutile flux-cored wires, for which the optimum Ti addition was ~0.06% (49). For welds containing 9%Cr-1%Co~0.06%Ti~0.003%B, the maximum test duration was ~9,000h, with a creep rupture strength ~40% greater than that of 92 grade parent steel. Despite the short test duration, it is a dramatic improvement in creep rupture strength that warrants further qualification. An unusual feature of the failure was the extensive ductility (~80% reduction of area) displayed. The reason for the substantial enhancement of creep rupture strength from the Ti addition has not yet been elucidated, but is assumed to result from a change in the character of some of the precipitating phases.

Whilst it would be undesirable to employ a weld metal that over-matched the parent steel creep rupture strength by such a large margin, it is nevertheless vital to press ahead with the development of weld metals, in order to be able to weld the higher creep strength parent steels that are currently emerging. The development of improved weld metals will become particularly important as the problem of type IV cracking diminishes.

Concluding remarks

In view of the decrease in WPCSR values with increasing creep rupture life, the design of existing plant that operate at temperatures close to 600°C should be reviewed, and appropriate action taken if the assumed WPCSR values were non-conservative. Where the deployment of new steel grades is concerned, long term (>10,000h) test results should be taken into account not only for the parent steel, but also for the weldments to be used. Extrapolation of short term data obtained outside the type IV regime is greatly misleading, and possibly dangerous. The change in WPCSR over time needs to be predicted accurately for the type IV region, so that appropriate life-time projections can be made. It is clear that the test temperature and the creep rupture life must be reported whenever WPCSR values are quoted, and that care must be exercised in manipulating data determined at different temperatures.

Research should be aimed at increasing the WPCSR and increasing further our understanding of the mechanisms by which type IV degradation occurs. To this effect, revised parent steel compositions will need to be combined with alternative welding processes to achieve optimal performance. Results for EB welds appear to give promising improvements in the HAZ creep rupture strength ratio, and should be explored further, together with other alternative welding processes. R & D programmes to address these issues are being initiated at TWI and work is ongoing within the European COST program, with emphasis on commercial fabrication, as well as performance improvement.

As continued improvements are made to the creep strength of parent steels and of the HAZ region, it is clear that further development of welding consumables is required, to ensure that the weld metal is not the creep weak region in long-term service.

Conclusions

1. WPCSR values (commonly called weld creep rupture "strength reduction factors") are not constant, but decrease with increasing creep rupture life.

2. There is evidence that type IV behaviour can be improved by incorporating boron.

3. Type IV behaviour can be improved by welding with processes producing a short thermal cycle, thereby producing a narrow HAZ, together with attention to weld preparation, preheat and PWHT.

Acknowledgements

The Authors would like to Acknowledge Paul Woollin and Ian Partridge of TWI for their contributions to the paper.

References

1. World Coal Institute web site, June 2007, www.msm.cam.ac.uk/phase-trans/2005/link/188.pdf.

2. Viswanathan R, Purget R, Rao U: "Materials technology for advanced coal power plants", www.msm.com.ac.uk/phase-trans/2005/link/188.pdf.

3. Masuyama F: "Advance power plant developments and materials experience in Japan", Proc 8th Liege Conference September 2006, Liege, Belgium, Research Centre Julich, 2006, Eds. J Lecompte-Beckers et al., Vol. 53 Part I 175-187.

4. Francis J A, Mazur W and Bhadeshia H K D H: "Type IV cracking in ferritic power plant steels", Mat. Sci. & Tech. 2006, 22(12), 1387-1395.

5. Hald J: Microstructure and long-term creep properties of 9-12%Cr steels", Proc. ECCC Creep Conf. on Creep and fracture in high temperature components - design & life assessment issues, 12-14 September 2005, London, U.K., Eds. I A Shibli et al., European Creep Collaborative Committee, 2005, 20-30.

6. Igarashi M, Yoshizawa M, Iseda A, "Long-tem creep strength degradation in T122/P122 steels for USC power plants", Proc 8th Liege Conf. September 2006, Liege, Belgium, Research Centre Jölich, 2006, Eds. J Lecompte-Beckers et al., Vol. 53, Pt II, 1095-1104.

7. Hald J: "Metallography and alloy design in the COST 536 action", Proc 8th Liege Conf. September 2006, Liege, Belgium, Research Centre Jölich, 2006, Eds. J Lecompte-Beckers et al., Vol. 53 Part II, 917-930.

8. Yin Y F and Faulkner R G: "Creep damage and grain boundary precipitation in power plant metals", Mater. Sci. Technol. 2005, 21(11), 1239-1246.

9. Semba H and Abe F: "Alloy design and creep strength of advanced 9%Cr USC boiler steels containing high boron", Proc 8th Liege Conf. September 2006, Liege, Belgium, Research Centre Jölich, 2006, Eds. J Lecompte-Beckers et al., Vol. 53, Pt II, 1041-1052.

10. Sakuraya K, Okada H, and Abe F: "BN type inclusions formed in high Cr ferritic heat resistant steel", Energy Materials, 2006, 1(3), 158-166.

11. Taneike M, Fujitsuna N and Abe F: "Improvements of creep strength by fine distribution of TiC in 9Cr ferritic heat resistant steel", Materials Science and Technology 2004, 20(11), 1455-1461.

12. Brear, J M and Fleming A: "Prediction of P91 life under plant operating conditions", ETD Int. Conf. on High Temperature Plant Integrity and Life Extension, 14-16 Apr 2004, Robinson College, Cambridge,.UK.

13. Cerjak H: and Letofsky E: "The behaviour of weldings in Large 9% Cr Alloy Castings", Proc. Fifth In. Charles Parsons Turbine Conf. on Advanced materials for 21st century turbines and power plant, 3-7 July 2000, Churchill College, Cambridge, UK, Eds. A Strang et al. The University Press, Cambridge 2000, 386-398.

14. Nath B and Masuyama F: "Materials comparisons between NF616 HCM12A and TB12M - 1: Dissimilar metal welds", Proc. EPRI/National Power Conference on New steels for advanced plant up to 620°C, 11 May 1995, London, Society of Chemical Industry, London, UK, Ed. E Metcalfe, 1995, 114-134.

15. Abe F: "R & D of advanced ferritic steels for 650°C USC boilers", in Ultra-steel 2000, Proc. Int. Workshop on the innovative structural materials for infrastructure in 21st century, 12-13 January 2000, Tsukuba, Japan, National Research Institute for Metals, 2000, 119-129.

16. Ellis F V and Viswanathan R: "Review of type IV cracking in piping welds" Proc. Int. Conf. on Integrity of high-temperature welds, Institution of Mechanical Engineers, 1998, 125-134.

17. Tabuchi M, Watanabe T, Kubo K, Matsui M, Kinugawa J and Abe F: "Creep crack growth behaviour in the HAZ of weldments of W containing high Cr steel", Int. J. Pressure vessels and piping, 2001, 78, 779-784.

18. Albert K, Matsui M, Watanabe T, Hongo H, Kubo K and Tabuchi M: "Microstructural investigations on type IV cracking in high Cr steel", ISIJ International 2002 42(12) 1497-1504.

19. Sawaragi Y: "Creep rupture properties of welds (1): Low alloy steels, high-Cr ferritic steels, Cu-Ni base steels", Welding Int. 1995, 9(7), 515-519.

20. Blass J, Battiste R L and O'Connor D G: "Reduction Factors for Creep Strength and Fatigue Life of Modified 9Cr -1 Mo Steel Weldments: conformation by Axial or Torsional tests of tubular specimens with longitudinal or circumferential welds", PVP Vol. 213, Pressure Vessel integrity ASME 1991, 253-259

21. Allen D J, Harvey B and Brett S J: " 'Fourcrack' - An investigation of the creep performance of advanced high alloy steel welds", Proc. ECCC Creep Conf. on Creep and fracture in high temperature components - design & life assessment issues, 12-14 September 2005, London, Eds. I A Shibli et al., European Creep Collaborative Committee, London, 2005, 772- 782, and "Fourcrack- Advanced coal-fired power plant steels -Avoidance of premature weld failure by type IV cracking", Project summary 304, 2005, available from the U.K. Department of Trade and Industry web site <http://www.dti.gov.uk/files/file20068.pdf>.

22. Abe F: "High performance creep resistant steels for 21st century power plants", Proc Eurocorr Exhibition, 2-4Nov. 2005, Italy (Keynote paper in the High temperature materials - boilers session)

23. Abe F and Tabuchi M: "Microstructure and creep strength of welds in advanced ferritic power plant steels", Science and Technology of Welding, 2004, 9(1), 22-30.

24. Kimura K, Sawada K, Kushima H and Toda Y: Degradation behaviour and long-term creep strength of 12Cr ferritic creep resistant steels", Proc 8th Liege Conf. September 2006, Liege, Belgium, Research Centre Jülich, 2006, Eds. J Lecompte-Beckers et al., Vol. 53, Pt II, 1105-1116.

25. American Society of Mechanical Engineers, 2004: ASME Boiler and Pressure vessel code Section III: "Rules for construction of nuclear facility components".

26. American Society of Mechanical Engineers, Code Case N-253-14, Construction of Class 3 Components for Elevated Temperature Service, Section III, Division 1, 2006, Appendix C, Table C-1.3 (e).

27. Cane B J: "Creep life assessment techniques – Decade of progress". ASME, Pressure Vessels and Piping Division, Special Publication, New York, 1993.

28. Masuyama F: "Creep strength reduction and risk in high strength ferritic steel welds", Proc. IOM Conf. on Integrity of high temperature welds, 24-26 April 2007, London, IOM Communications, 2007, 497-506.

29. Kimura K, Sawada K, Kushima H and Toda Y: "Influence of Composition Partitioning on Creep Strength of High Chromium Ferritic Creep Resistant Steels", Proc. IOM Conference on Integrity of high temperature welds 24-26 April 2007, London, IOM Communications, 2007, 497-506

30. Kuboň Z and Sobotka J: "Assessment of long-term creep properties, creep failure and strength reduction factor of boiler tube and pipe weldments", Proc. Baltica VI , VTT Symposium 234 on Life management and maintenance for power plants, 8-10 June 2004, Helsinki, Eds. J Veivo and P Auerkari, VTT, 2004, 477-490.

31. American Society of Mechanical Engineers, 2004: ASME Boiler and pressure vessel code Section I: "Rules for construction of power boilers".

32. American Society of Mechanical Engineers, 2004: ASME Boiler and pressure vessel code Section VIII Division 1:' Rules for construction of pressure vessels'.

33. American Society of Mechanical Engineers, 2004: ASME Boiler and pressure vessel code Section VIII Division 2:' Alternative rules for construction of pressure vessels'.

34. Iseda A, Sawaragi Y and Natori A: "Properties of welded joint of 9%Cr-1%Mo-V-Nb steel", Proc. Symp. on Joining of Materials for 2000 AD, 12-14 Dec.1991, Tiruchirapalli, India, Publ.: Tiruchirapalli, India; Indian Institute of Welding; 1991, 465-470.

35. Tabuchi M, Masayuki K, Kubo K and Albert S K: "Improvement of type IV creep cracking resistance of 9Cr heat resisting steels by boron addition", OMNI 2004, 3(3), 1-11.

36. Kondo M, Tabuchi M, Tsukamoto S, Yin F and Abe F: "Suppressing type IV failure via modification of heat affected zone microstructures using high boron content in 9Cr heat resistant steel welded joints", Science and Technology of Welding and Joining, 2006, 11(2), 216-223.

37. Abe F, Okada H, Wanikawa S, Tabuchi M, Itagaki T, Kimura K, Yamaguchi K and Igarashi M: "Guiding principles for development of advanced ferritic steels for 650°C USC boilers", Proc. 7th Liege Conf. On Materials for advanced power engineering, 2002, Eds. J Lecompte-Beckers et al., European Commission, 2002, 1397-1406.

38. Abe F: "Metallurgy for long-term stabilization of ferritic steels for thick section boiler components in USC power plants", Proc 8th Liege Conf. September 2006, Liege, Belgium, Research Centre Jülich, 2006, Eds. J Lecompte-Beckers et al., Vol. 53, Part II, 965-980.

39. Otoguro Y, Matsubara M, Itoh I and Nakazawa T: "Creep rupture strength of heat affected zone for 9Cr ferritic heat resistant steels", Nuclear Engineering and Design, 2000, 196, 51-61.

40. Bell K: 'An Analysis of published creep rupture data for modified 9% Cr steel weldments' TWI Research Report for Industrial Members, 598/1997, March 1997.

41. Sikka V K: "Method for welding chromium molybdenum steels" US Patent No. 4, 612, 070, September 1986.

42. Coussement C, de Witte M, Dhooge A, Dobbelaere R and van der Donckt E: "High-temperature properties of improved 9% Cr steel weldments", Revue de la Soudure 1990, 1, 58-63.

43. Francis J A, Mazur W and Bhadeshia HKDH: "Trends in welding research". Pine Mountain, CA, USA, May 2005, ASM International. From review article "Type IV Cracking in Ferritic Steels".

44. Panton-Kent R: "Weld metal toughness of MMA and electron beam welded modified 9Cr-1Mo steel", TWI Research Report for Industrial Members, 429/1990, November 1990.

45. Albert S K, Tabuchi M, Hongo H, Watanabe T, Kubo K and Matsui M: "Effect of welding process and grave angle on type IV cracking behaviour of weld joints of a ferritic steel". Science and Technology of Welding and Joining 2005, vol 10, No.2, 149-156.

46. Heuser H and Jochum C: "Properties of matching filler metals for P91, E911 and P92", Proc. 3rd EPRI Conf. on Advances in materials technology for fossil power plants, 5-6 April 2001,

Swansea, U.K. Ed. R. Viswanathan, W. T. Bakker, J. D. Parker, Institute of Materials, 2001, 249-265.

47. Vanderschaeghe A, Gabrel J and Bonnet C: "Mise au point des consommables et procedures de soudage pour l'acier grade 92", (in French), ("Development of welding consumables and procedures for steel grade 92"), Proc. ESOPE Conf. 23-25 October 2001, Paris.

48. Metrode Products Limited, "P92 welding consumables for the power generation industry", Metrode Products Limited web site 2007.

49. Abson D J, Rothwell J S and Woollin P: "The Influence of Ti, Al and Nb on the Toughness and Creep Rupture Strength of Grade 92 Steel Weld Metal" Proc IOM Conf. on Integrity of high temperature welds, 24-26 April 2007, London, IOM Communications, 2007, 129-138. (See also Abson D J: "The influence of Ti and Al on the toughness and creep rupture strength of grade 92 steel weld metal", TWI Research Report for Industrial Members, 833/2005, September 2005.)

Tables and Figures

Table 1: Temperatures and pressures for different types of coal-fired plant (1).

	Temperature (°C)	Pressure (bar)
Sub-critical	538	167
Supercritical	540 - 566	250
Ultra-supercritical	580 - 620	270 - 285

Table 2: Evolution of creep rupture strength for ferritic steels (2).

Years	Alloy Modification	600^0C /10^5h Creep Rupture Strength, MPa	Example Alloys	Maximum Metal Use Temp., ^{0}C
1960-70	Addition of Mo or Nb, V to simple 12Cr and 9Cr steels	60	EM12, HCM9M, HT9, Tempaloy F9, HT 91	565
1970-85	Optimization of C, Nb, V	100	HCM12, T91, HCM2S	593
1985-95	Partial substitution of W for Mo	140[1]	P92, P122, P911 (NF 616, HCM12A)	620
Emerging	Increase W and addition of Co	180[1]	NF12, SAVE 12	650[2]

[1] In a later paper by Masuyama (3), 140MPa was revised to 130MPa and 180MPa was revised to 150MPa.
[2] Masuyama (3) added MARN and MARB2 steels (10^5 h creep rupture strength 150MPa).

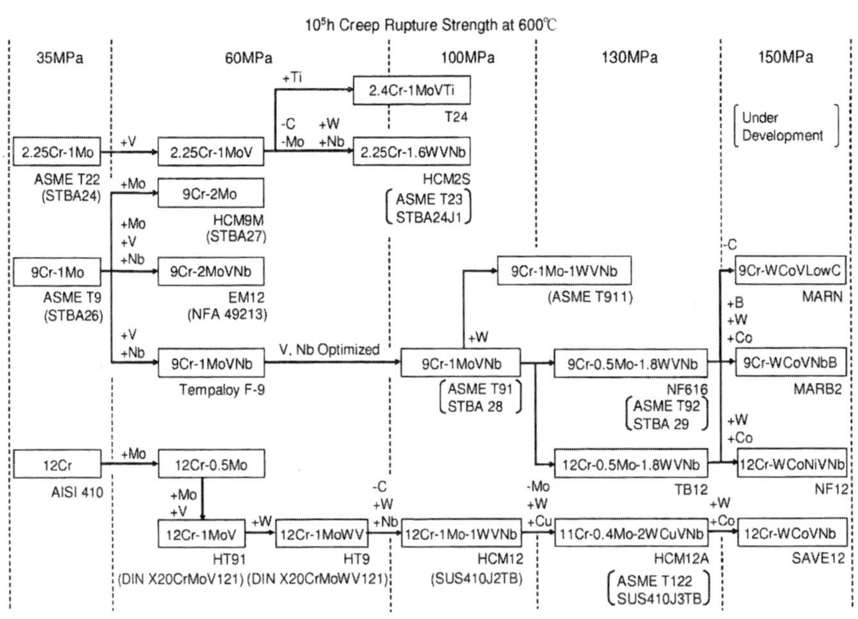

Figure 1. Chart showing the progressive development of 9-12%Cr steels (3)

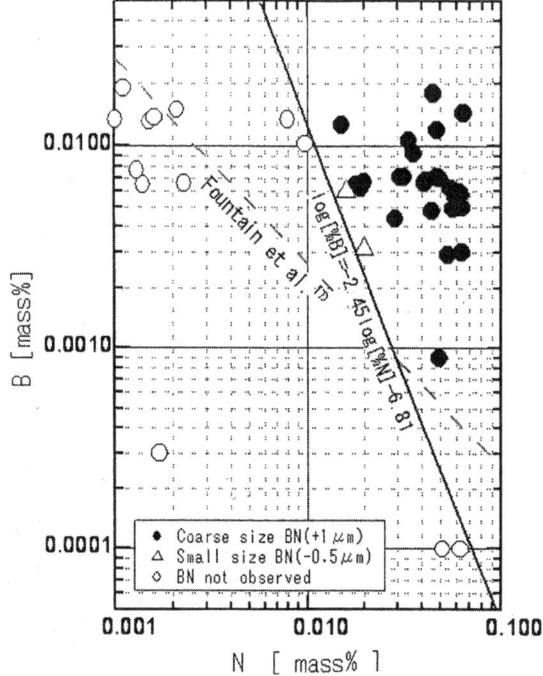

Figure 2. Solubility limit for BN at 1150°C for high Cr steel, with the corresponding information for the Fe-B-N system from Fountain et al. superimposed (9).

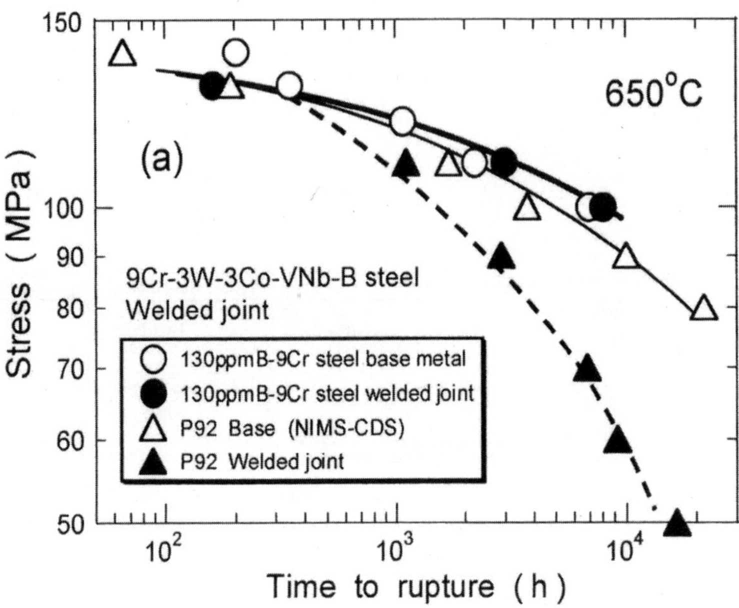

Figure 3. Creep rupture data of Abe (22) for parent steel and cross-weld specimens of 92 grade and a boron-containing steel tested at 650°C.

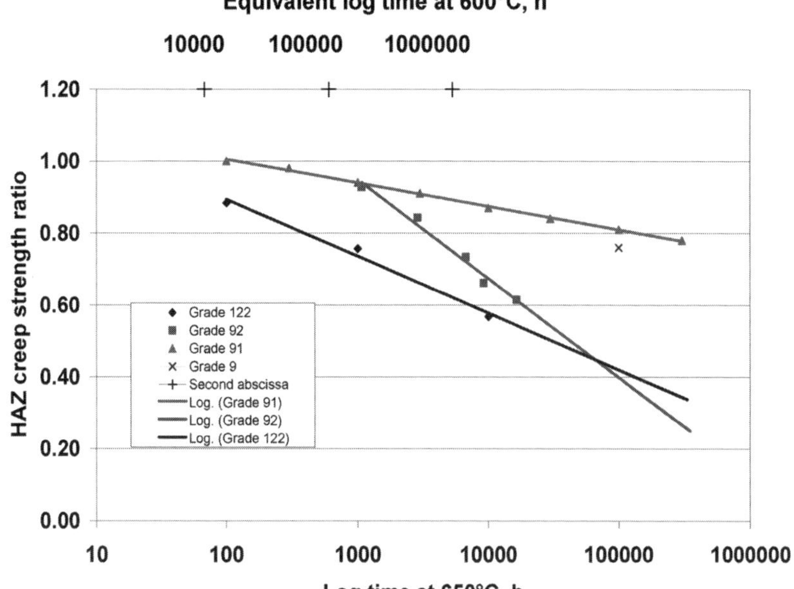

(a)

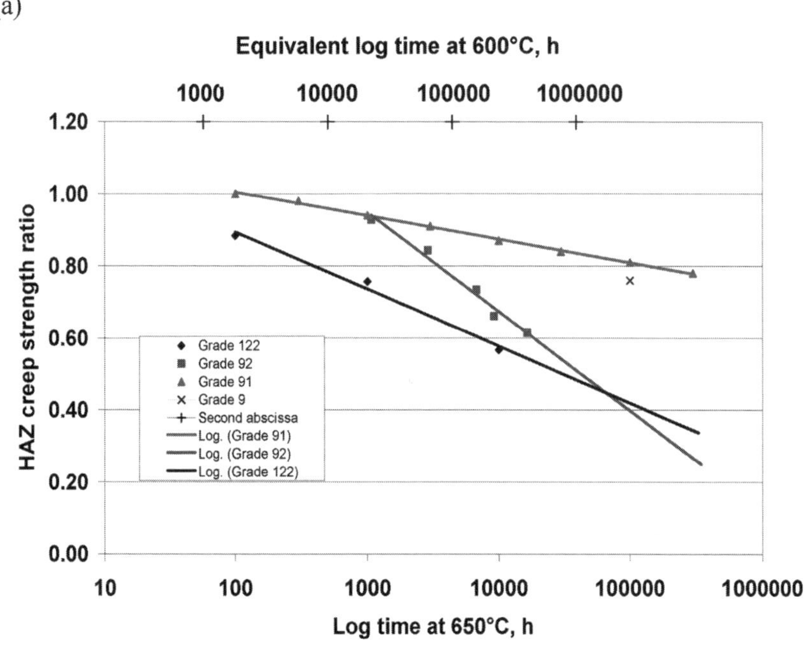

(b)

Figure 4. WPCSR values derived from creep rupture data of cross-weld or simulated fine-grain HAZ specimens tested at 650°C for grade 122 (23), grade 92 (22) and grade 91 (25). A linear extrapolation is tentatively assumed, up to a 30 year life. A single data point for grade 9, from ASME code case N-253-14 (26), is represented here as a 100,000h value. Equivalent times at 600°C are shown, using a Larson-Miller constant of (a) 36 and (b) 20.

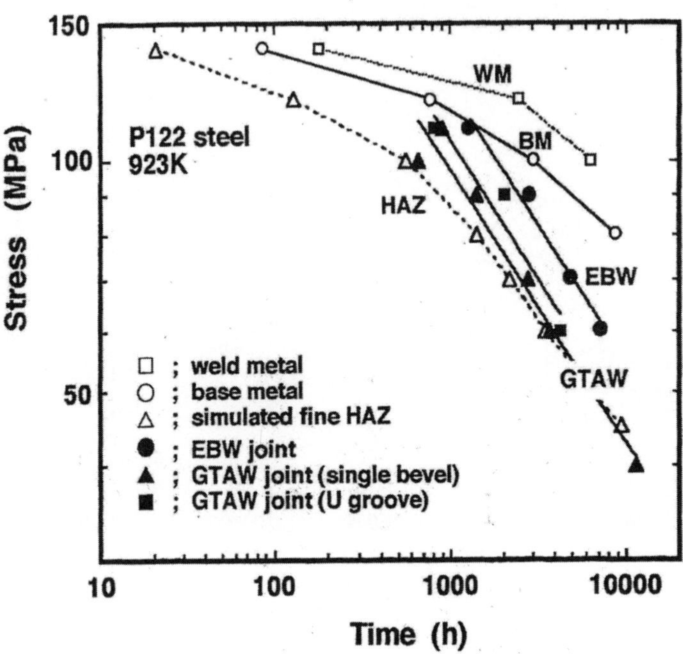

Figure 5. Creep rupture data for grade 122 weldments produced with different welding processes (23).

Advances in Materials Technology for Fossil Power Plants
Proceedings from the Fifth International Conference
R. Viswanathan, D. Gandy, K. Coleman, editors, p 809-817

DOI: 10.1361/cp2007epri0809

16-8-2 and Other Weld Metal Compositions that Utilize Controlled Residual Elements to Enhance and Maintain Elevated Temperature Creep Strength

William F. Newell, Jr., PE, IWE
Euroweld, Ltd.
255 Rolling Hill Road
Mooresville, NC 28117

Abstract

Achieving high temperature creep strength while maintaining rupture ductility in weld metal for austenitic stainless steel weldments has always been challenging. In the late 1940's and early 1950's, independent work in both Europe and the USA resulting in what is known today as the 16-8-2 (nominally16% chromium -8% nickel -2% molybdenum) stainless steel weld metal. Philo 6 and shortly thereafter at Eddystone used the alloy to construct the first supercritical boilers and piping in the USA. Concurrent with domestic boiler and piping fabrication, the US Navy was also using this material for similar supercritical applications. Over the decades, enhanced performance has evolved with variations of the basic composition and by adding specific residual elements. Controlled additions of P, B, V, Nb and Ti have been found to greatly enhance elevated temperature as well as cryogenic behavior. The history of these developments, example compositions and areas of use as well as mechanical property results are presented.

Introduction

When austenitic stainless steel alloys are chosen for elevated temperature service, specific problems may occur. These include formation of intermetallic phases (sigma and chi), decreased ductility, and reduced corrosion or fatigue resistance. Other factors such as heat affected zone (HAZ) cracking, crater cracking, solidification cracking or adverse ferrite content are normally associated with welding or the welding consumable used.

There is a proven and nearly forgotten weld filler metal and formulation approach which offers solutions to both high and low temperature pressure service condition problems associated with austenitic stainless steel. This weld filler metal is identified by its major alloying elements 16-8-2 [nominally, 16% chromium (Cr)-8% nickel (Ni)-2% molybdenum (Mo)]. This alloy was originally developed by Babcock and Wilcox in the mid 1950's under contract to the U.S. Navy, Bureau of Ships, to evaluate 18-8Mo welding alloys for fabricating steam lines and pressure vessels in Navy components. (1) In a parallel and perhaps earlier effort, Murex (United Kingdom) offered an alloy (ARAMAX GT) for similar purposes in the late 1940's. (2)

Babcock & Wilcox developed 16-8-2 to mitigate cracking associated with the type 347H alloy for use in service up to 565°C (1050°F). The British ARAMAX GT product was subsequently replaced by Metrode Products with a 17-8-2 formulation. Further, a lean 16-8-2 product with restricted Mo was reintroduced in the last decade. (3)

The first known commercial use of 16-8-2 was by Babcock & Wilcox for Philo 6, the first 31 MPa (4,500 psi) ultra supercritical boiler in the United States. By 1955, Babcock and Wilcox had manufactured and used nearly 14 metric tons (30,000 lbs) of the 16-8-2 composition to fabricate austenitic boiler and piping components for high temperature service. On a parallel path, Combustion Engineering (now Alstom Power) in 1957 repaired Sultzer high pressure valves at the ultra supercritical Eddystone Power Plant located in Eddystone, PA, which is still in operation today. In the mid 1960's, Combustion Engineering's version of 16-8-2 was again used at Eddystone, for repairs and reinstallation of high pressure piping. (4-5)

Work continued in the following decades to further adjust the composition to increase performance. Many variants evolved, especially with respect to controlling or adding specific residual elements.

16-8-2 Weld Filler Metal

The 16-8-2 welding alloy is a dilute hybrid between E308H and E316H. However, rather than just matching any particular base material, it can be used to weld almost any 3XX or 3XXH series of austenitic stainless steel. (2) Until its development, welding of stainless steels used in high temperature/high pressure service conditions were often plagued with a loss of high temperature ductility, fatigue resistance, and corrosion resistance. Due to the ferrite content associated with existing austenitic weld filler metals, problems were often encountered from the formation of intermetallic phases and the resulting embrittlement. 16-8-2 typically exhibits satisfactory properties up to a service temperature approaching 800°C (1472°F). (2)

It has been demonstrated that limiting the Mo content to about 1.2 wt. % enhances rupture ductility and reduces thermal fatigue issues. The 16-8-2 "lean version" is formulated and used in many current applications. (3) Most oil companies specify the "lean version" of the electrode to weld 304H, 316H, 321H, and 347H alloys in piping systems for high temperature/high pressure service conditions. The "lean version" is also used in the fabrication of catalytic cracking units in the petroleum refining process, which are predominantly fabricated entirely out of 304H stainless steel. Current formulations of the 16-8-2 filler metal family place rigorous controls on the residual elements including: boron (B), niobium (Nb), phosphorus (P), titanium (Ti) and vanadium (V). Control of these residual elements has been found to enhance elevated temperature creep strength while maintaining rupture ductility. The "lean version" of the 16-8-2 coupled with residual element control offers a combination of alloying elements to avoid high temperature/high pressure service condition problems.

16-8-2 Mechanical Properties

16-8-2 does not exhibit extraordinary mechanical properties until elevated temperatures are reached, then it surpasses almost all other austenitic stainless steel filler metals, at least where creep ductility is concerned. Figures 1,2 and 3 illustrate mechanical properties of selected weld metal under high temperature conditions. Figures 2 and 3 show the superior ductility offered by the 16-8-2 composition.
Low total Cr + Mo with controlled carbon and ferrite content (2 – 5 FN) ensure high resistance to thermal embrittlement caused by formation of intermetallic phases (sigma and chi)" [4]. By limiting the amount of Mo, resistance to additional creep ductility and thermal fatigue is provided. The as-deposited weld metal has an excellent hot ductility properties which permit the weld joint to move, thus preventing in service joint cracking. Figure 4. compares creep performance of selected shielded metal arc and flux cored arc welding consumables with various high temperature austenitic stainless steel alloys. The weld metal performs equal to and normally better than the base materials. In addition, 16-8-2 offers good low service temperature toughness, down to -196°C (-320°F). (6)

Resistance to solidification cracking is another area where the 16-8-2 alloy performs well. An example of this was demonstrated by Fluor Canada, Ltd in 2001 when building a new oil refinery in Alberta, Canada. The client had requested an economical stainless steel which would withstand 100,000 hrs of service life in a naphthenic acid environment. This led to a new stainless steel being developed which would be called TP 316 Cb for use on heater tubes inside the firebox.(7) The first attempt at welding this material was performed with type 316 filler metal, but resulted in cracking problems. After careful consideration, 16-8-2 was selected as the filler metal of choice for the root and hot pass with the balance welded with filler material similar in composition to the base material (TP 316 Cb); the result was crack free welds. Success was attributed to the 16-8-2's ability to yield before the base material and heat affected zone during solidification, yet exhibit adequate strength in service.

A similar approach is being utilized to mitigate HAZ embrittlement issues in Type 347H piping. Type 347H weld metal and defective HAZ base material is removed and the entire weld grooves are rewelded with 16-8-2 consumables. (8)

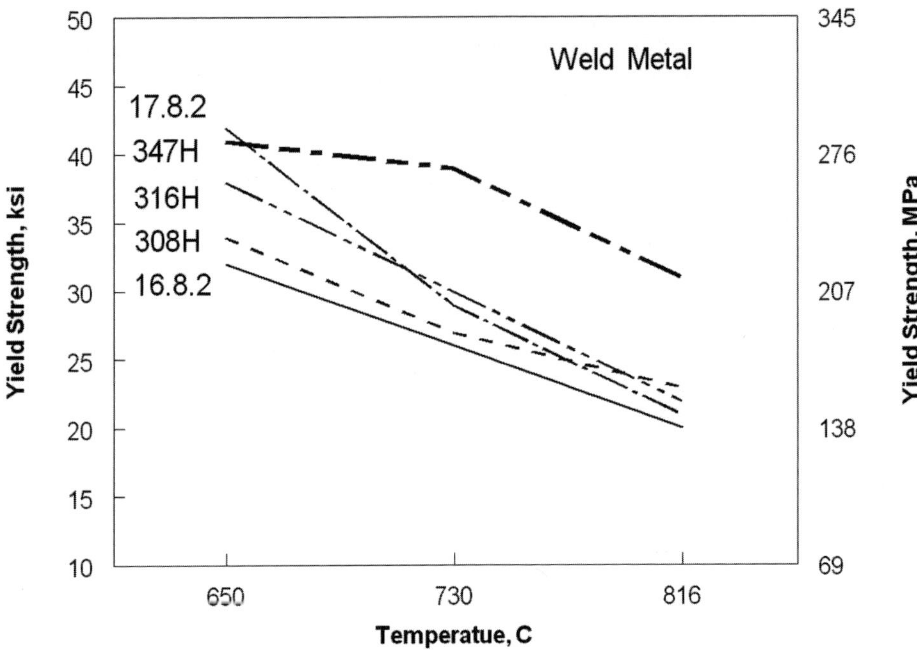

Figure 1. Weld metal yield strength versus temperature for selected alloys. (2-6)

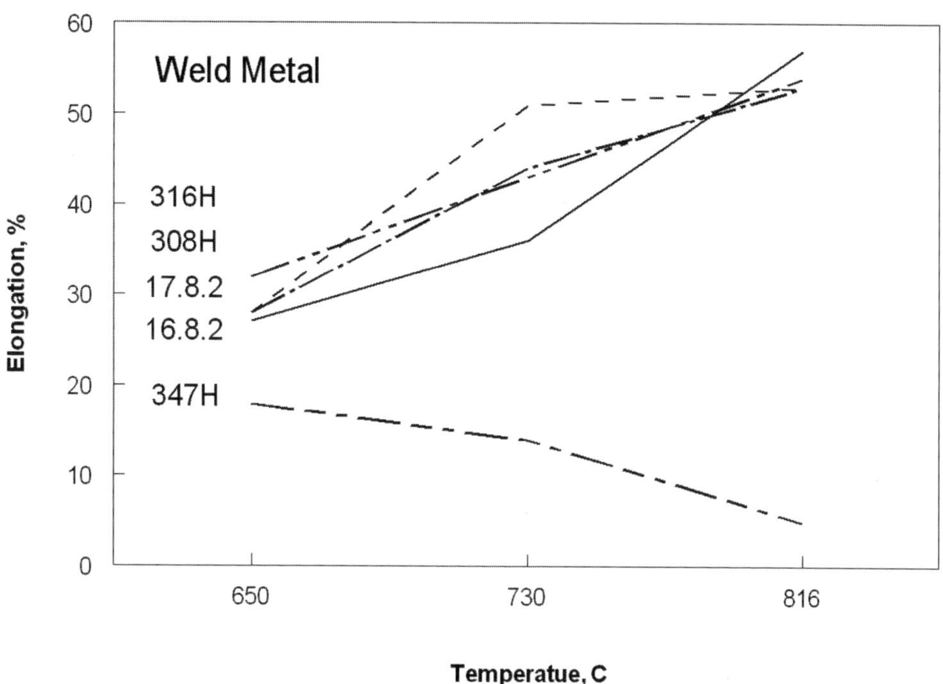

Figure 2. Weld metal elongation versus temperature for selected alloys. (2-6)

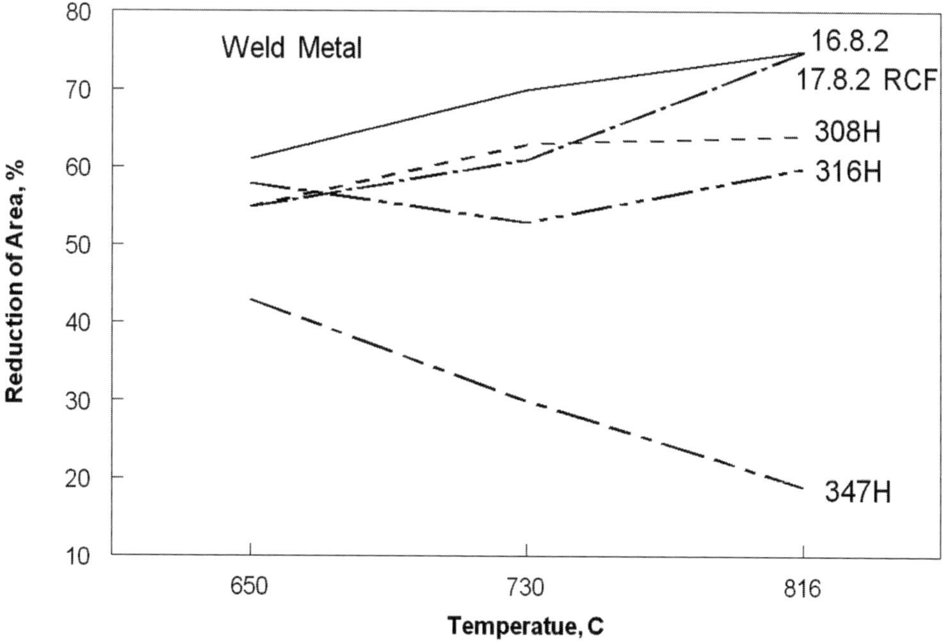

Figure 3. Weld metal reduction of area versus temperature for selected alloys. (2-6)

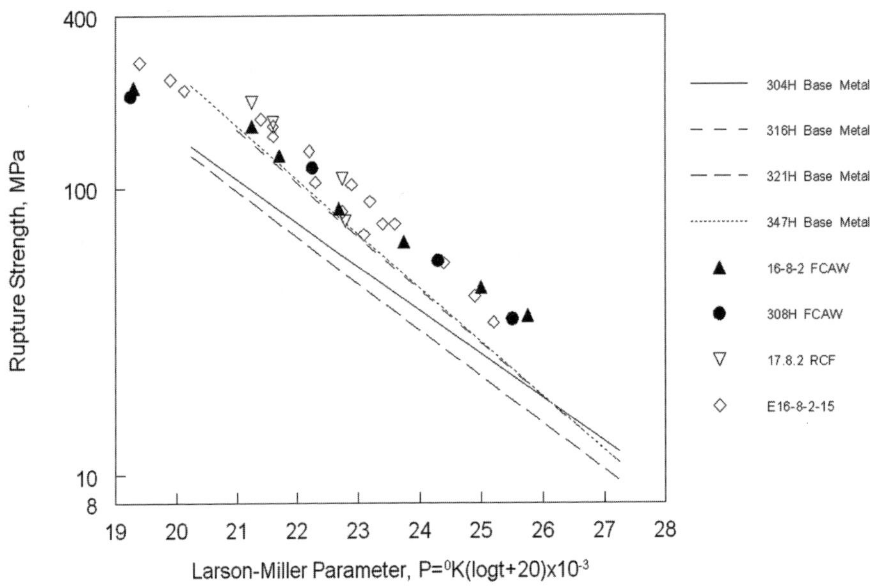

Figure 4. Rupture Stress versus Larson-Miller Parameter comparing selected welding consumables and base metals. (2-3)

16-8-2 Chemical and Microstructure Properties

The specified combination of chromium (Cr), nickel (Ni), and molybdenum (Mo) is what makes the 16-8-2 alloy unique. The lean chemical composition and low ferrite typically found in 16-8-2 type weld metals provide excellent microstructual stability and ductility retention for service at elevated temperature. The nominal composition is 15.5 Cr, 8.5 Ni, and 1.5 Mo, percent by weight. Typically, weld deposits have a ferrite number ranging from 2 to 5 FN. The weld metal is very tolerant to welding process, welder technique and cooling rate issues that typically affect the deposited FN of other austenitic weld metals. (2-7)

Deposited ferrite content for this filler material helps to increase resistance or even eliminate microfissuring and solidification cracking during fabrication. There is a widely held view that weld metal must have above 3 FN for optimum resistance to hot cracking in austenitic stainless steel welds. However, past experience has shown that 16-8-2 weldments, even with only1-2 FN, are always sound. (3) Exceptional resistance to microfissuring has been demonstrated in E16-8-2 welds exhibiting as low as 0.7 - 1.2 FN in comparison with many 300 series weld metals. Experimentation and field testing of selected solid wire has been found to have no more than 0.5 - 1.5 FN.(3) Historically 16-8-2 is one of the most reliable alloys manufactured when trying to predict the final percentage of ferrite.

16-8-2 Variants and Controlled Residual Elements

Over the decades, enhanced performance was obtained by varying the basic composition and by adding specific residual elements. Controlled additions of P, B, V, Nb and Ti have been found to greatly enhance elevated temperature as well as cryogenic behavior for selected 300-series and 16-8-2 variants. Increases in rupture ductility, corrosion resistance and thermal fatigue resistance plus decreased thermal embrittlement tendencies are typically achieved. Controlled Residual Element (CRE) variations are shown in Table 1.

Table 1. Electrode variations and typical applications. (2-6)

Electrode[1]	Application
E308H-15 CRE	Fast Flux Test Facility; Hanford, WA
E310H-15 CRE	Handcuff Tube Panel Hangers
E316H-15 CRE	Fast Flux Test Facility; Hanford, WA
16-8-2-15 CRE	Eddystone 1 and 2; Petro-chem; Cryogenic Applications
15-9-4-15 CRE	Forerunner to 17-9-1; High Temp Piping Repairs
17-9-1-15 CRE	Floor Supports for 304H Economizer Panels
17-8-2-16 RCF[2]	High Temperature Power Piping

1. SMAW electrodes shown. Most compositions were available in SAW and GTAW also. 16-8-2 is available and used with SMAW, SAW, GTAW and FCAW processes.
2. The "RCF" designation indicates both controlled ferrite and residual elements.

Most 16-8-2 consumables manufactured today utilize controlled residual element formulation approaches but the extent is typically proprietary. Typical increase in creep rupture strength from using a CRE formulation approach is shown in Figure 5 and has been observed with those alloys noted in Table 1. (4-5)

Perhaps one of the most dramatic examples of applying CRE technology is shown in Figure 6 for a 17-9-1-15 consumable. This CRE alloy was tested at 1202F (650C), 28 ksi (193 MPa) and performed 18,926 hours before rupture. Reduction of area of 71.3% and 19.1% (50mm) elongation substantiate that this alloy exhibited significant creep ductility and surpasses most high temperature austenitic stainless steel base materials. (4-5)

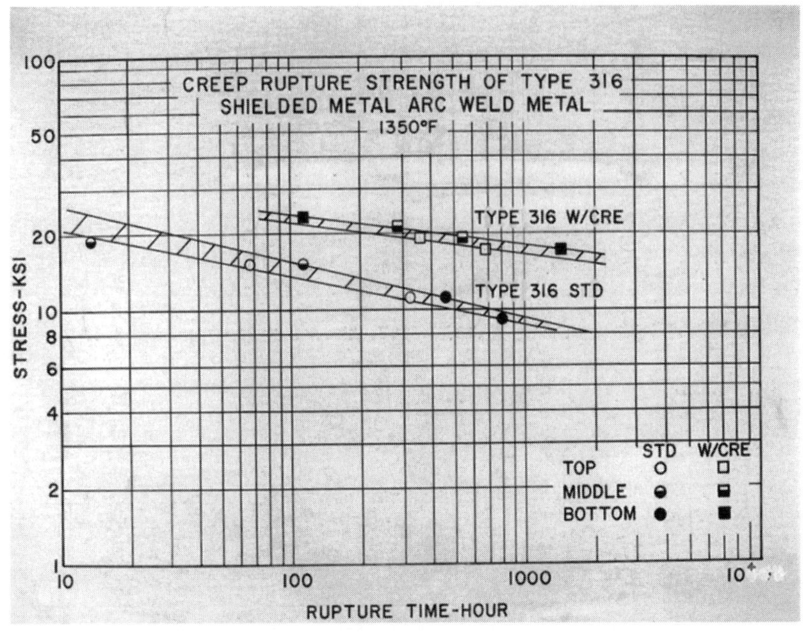

Figure 5. Increase in Creep Rupture Strength with a CRE formulation. (4-5)

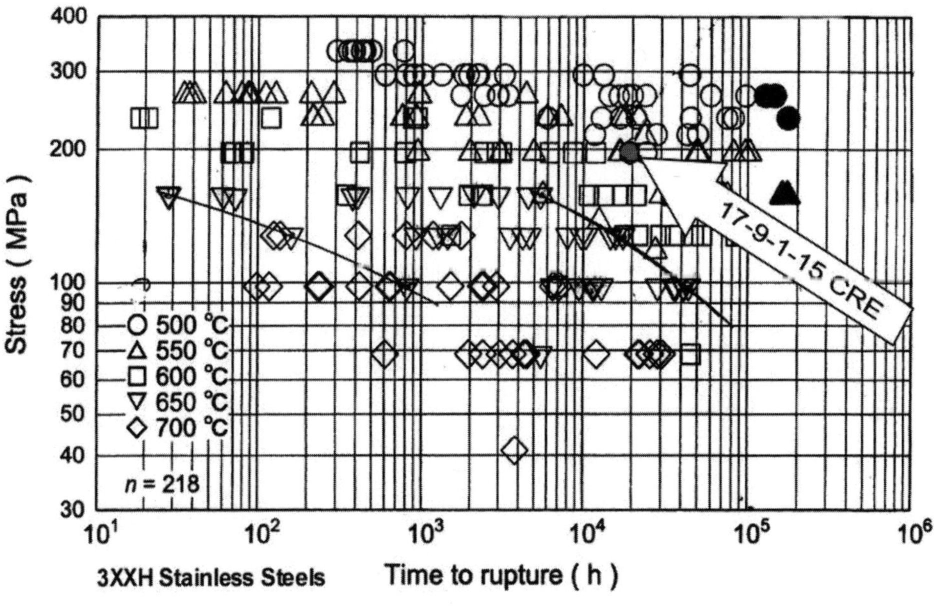

Figure 6. 17-9-1-15 CRE tested at 1202 F (650C), 28 ksi (193 MPa); 18,926 hours until rupture. (4-5)

Conclusion

16-8-2 weld filler metal has the attributes necessary to solve nearly any austenitic 3XXH stainless steel high temperature high pressure welding problem. The combination of the filler metals' high creep rupture strength, high creep ductility, and corrosion resistance make it very appealing. It not only has applications at high temperature, but also at low temperatures down to -196°C (-320°F). This suggests an operating temperature range for this filler metal from -196°C (-320°F) to 800°C (1472°F). 16-8-2 has proven that it can be used even with new base metals such as TP 316 Cb [5] or for making existing alloys such as 347H more serviceable. Variants such as the 17-8-2 RCF and 17-9-1 CRE or other 3XX CRE formulations can offer significant mechanical properties over traditional high temperature austenitic stainless steel welding consumables. The 16-8-2 alloy was developed over 50 years ago and continues today to offer a creative way to weld stainless steels used in high pressure and high temperature applications.

References

1. Carpenter, O R and Wylie, R D: "16-8-2 Cr-Ni-Mo for welding electrode", Met. Prog. 1956, 70 (5), 65-73.

2. Metrode Welding Consumables: "16.8.2 For High Temperature 3XXH Stainless Steels", Metrode products limited, Data Sheet, 2002, 1.

3. A W Marshall, J C M Farrar: "Lean Austenitic type 16-8-2 Stainless Steel Weld Metal", Metrode Products Limited, 1-3.

4. Private Notes. C. T. Ward (formerly with Combustion Engineering, Chattanooga, TN)

5. Private Notes. J. F. Turner, Electrode Engineering Inc., Harrison, TN.

6. Metrode Welding Consumables: "308H & 16-8-2 Stainless Steel Weld Metals For Oil Refinery Catalytic Cracking Units", Metrode Products Limited, 5-7.

7. D. Dove, B. Messer, T. Phillips: "An Austenitic Stainless Steel, Resistant To High Temperature Creep and Naphthenic Acids Attack In Refinery Environments", Corrosion, 2001.

8. J. Mitchell, Syncrude Canada, Ltd.: "Welding Challenges and Opportunities in the Heavy Oil Industry", Euroweld 4th Annual Welding Seminar, Edmonton, Alberta, Canada, 25-26 April 2006.

9. W. Newell, Jr. and J. Oley: "16-8-2 History, Production and Use", 16 August 2007.

Advances in Materials Technology for Fossil Power Plants
Proceedings from the Fifth International Conference
R. Viswanathan, D. Gandy, K. Coleman, editors, p 818-829

DOI: 10.1361/cp2007epri0818

WELDABILITY INVESTIGATION OF INCONEL® ALLOY 740 FOR ULTRASUPERCRITICAL BOILER APPLICATIONS

John M. Sanders
The Babcock & Wilcox Company
1562 Beeson Street
Alliance, OH 44601

Jose E. Ramirez
Edison Welding Institute
1250 Arthur E. Adams Drive
Columbus, OH 43221

Brian A. Baker
Special Metals Corporation
3200 Riverside Drive
Huntington, WV 25705

Abstract

INCONEL[1] alloy 740 has been identified as a prime candidate for the severe operating conditions of USC boilers. It exhibits the highest stress rupture strength and corrosion resistance at projected operating temperatures approaching 760 degrees Centigrade. Alloy 740 is a precipitation-hardenable nickel-chromium-cobalt alloy with a niobium addition. In general, the issues of welding precipitation-hardenable alloys include heat-affected zone (HAZ) liquation cracking, ductility-dip cracking (DDC), and PWHT cracking. Alloy 740 is a new alloy with no fabrication history that was derived from NIMONIC®[1] alloy 263. The base chemistry of alloy 740 is different enough to require study of its weldability. This paper describes the weldability investigation now underway for this alloy. Tomorrow's USC boiler will require the use of thin section superalloys previously used in aerospace applications. Lessons learned from that industry are being applied and expanded to boiler applications up to three inches in thickness. The workscope of this project includes basic material characterization studies to determine cracking sensitivity relative to other well known aerospace alloys such as Waspalloy and INCONEL alloy 718. This information is being used in welding trials to determine acceptable welding processes and techniques which will result in sound thick section, crack free welds. Several weld processes were evaluated relative to welding the INCONEL alloy 740. Gas tungsten arc welding and pulsed Gas Metal Arc welding have emerged as the two most favorable processes.

---

[1] INCONEL® and NIMONIC® are registered trademarks of the Special Metals family of companies.

Introduction

Coal fired power generation facilities can gain efficiency and reduction of undesirable byproducts by operating at higher temperatures and pressures. Traditional materials such as ferritic and austenitic stainless steels aren't suited for the projected operating conditions of tomorrow's supercritical boilers. The U.S. Department of Energy and the Ohio Coal Development Office have sponsored a consortium including boiler manufacturers to evaluate the material requirements of advanced ultrasupercritical boilers. This paper describes the approach taken to evaluate the weldability of a newly developed alloy known as INCONEL alloy 740. It was derived from an alloy used previously in the aerospace industry known as NIMONIC alloy 263. The base chemistry of INCONEL alloy 740 resulted from modifications to enhance coal ash corrosion resistance. INCONEL alloy 740 has been identified as the only alloy presently commercially available which possesses the high temperature rupture strength required for plant operating conditions of 5000 PSI steam pressure at 1400 degrees Fahrenheit. INCONEL alloy 740 is a high strength heat resisting alloy which is strengthened by precipitation of a metallurgical phase during an aging heat treatment similar to nickel base "superalloys" utilized in the aerospace industry.

The approach taken on this project was to review the methods developed by the aerospace industry for quantifying/classifying weldability attributes of high strength heat resisting alloys. The "aerospace" approach was used to define various properties of the alloy and to rank it among the alloys with known fabrication databases. The results of this material characterization have been dovetailed into an ongoing welding process evaluation study to determine suitable welding techniques for section thicknesses approaching three inches.

This paper is divided into the main sections of historic background from the aerospace industry, material characterization studies performed by The Edison Welding Institute, and welding process evaluations now being performed by The Babcock & Wilcox Company.

Historical Background

Figure 1 is a plot of the allowable stress of several alloys versus operating temperature for boiler applications. It can be seen that INCONEL alloy 740 stands alone at the highest projected operating temperatures. Future ultrasupercritical boilers are projected to operate at 1400 degrees Fahrenheit.

The chart is also a useful reference relative to today's most utilized alloys. There might be some other nickel base alloys with some potential for the highest temperatures but as of this writing INCONEL alloy 740 appears to be the most viable. The weldability issues would be similar for all nickel base alloys.

Boiler Materials for USC Plant

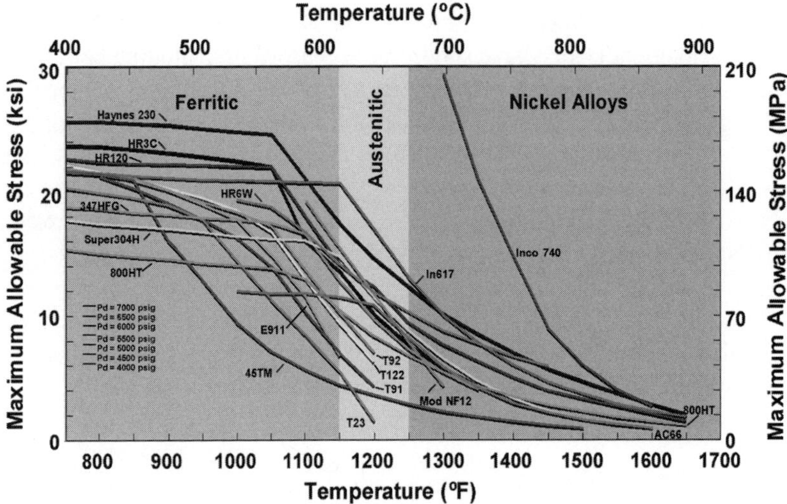

Figure 1. Allowable Stress for various boiler materials (Source: Brian Vitalis , Riley Power)

The aerospace industry has been welding nickel base alloys for decades. The precipitation hardenable alloys are susceptible to three types of cracking known as liquation, strain age, or ductility dip cracking. These three mechanisms will now be described in general terms.

Liquation Cracking

This occurs typically in the heat affected zone and is related to the low melting point constituents remaining liquid while the weld puddle is freezing and contracting. Figure 2 shows a photomicrograph of weld and heat affected zone liquation cracking found in a nickel base superalloy. These liquation cracks have been shown to cause bend test failures in ASME Section IX weld procedure qualification testing.

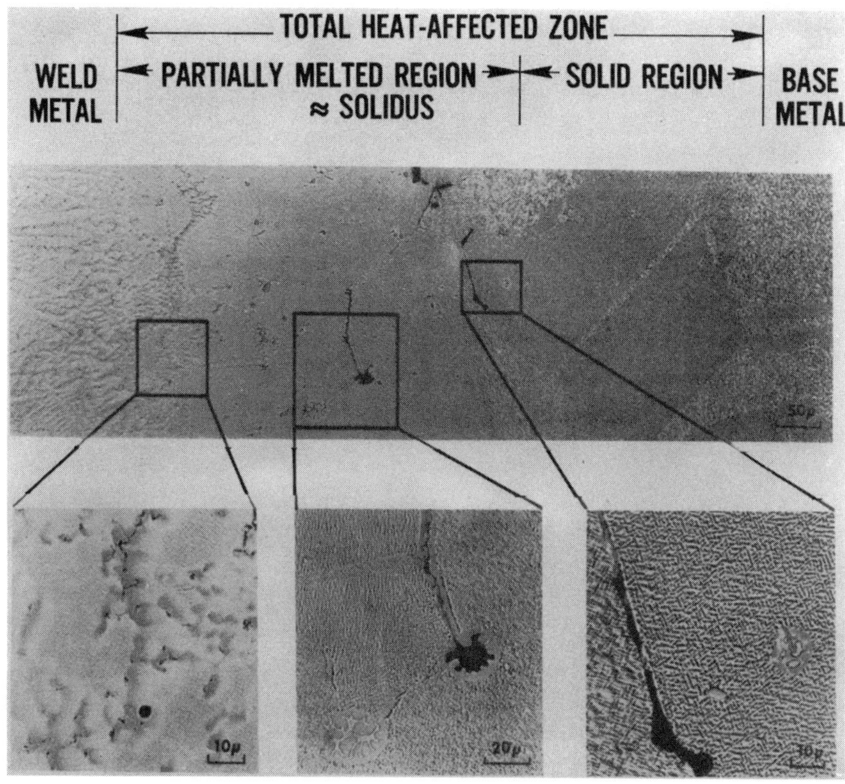

Figure 2. Photomicrograph of Typical Liquation Cracking (2)

Strain Age Cracking

The "aging" heat treatment which causes the precipitation strengthening in some nickel base alloys produces several metallurgical phases. Production of these phases, notably gamma prime, creates shrinkage forces in the matrix. The summation of the welding residual stresses and this precipitation related stress can be enough to crack the weldment as shown in Figure 3. Note that the cracks aren't in the actual weld or its heat affected zone.

Figure 4 depicts the influence of the actual aging heat treatment on the strained weldment which can cause cracking.

Figure 5 is an empirical representation of the relative strain age cracking sensitivity of various aerospace superalloys which are precipitation strengthened. INCONEL alloy 740 would lie in the region of INCONEL alloy 718 due to the aluminum and titanium content.

Ductility Dip Cracking

Some nickel base superalloys exhibit a loss of ductility at elevated temperatures known as a ductility dip. This lack of ductility can lead to cracking. Figure 6 is a chart which represents elongation for a superalloy at various temperatures. There is a noticeable drop in elongation (ductility dip) in the temperature region of 600C to 900C.

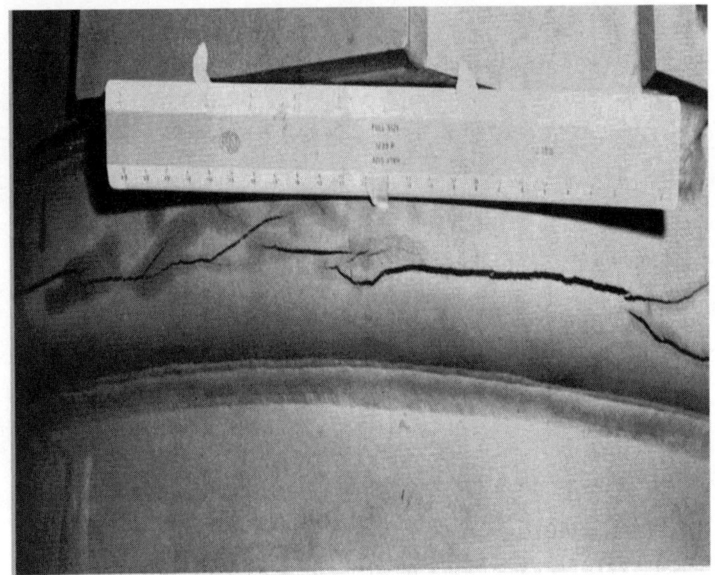

Figure 3. Strain Age Cracking (2)

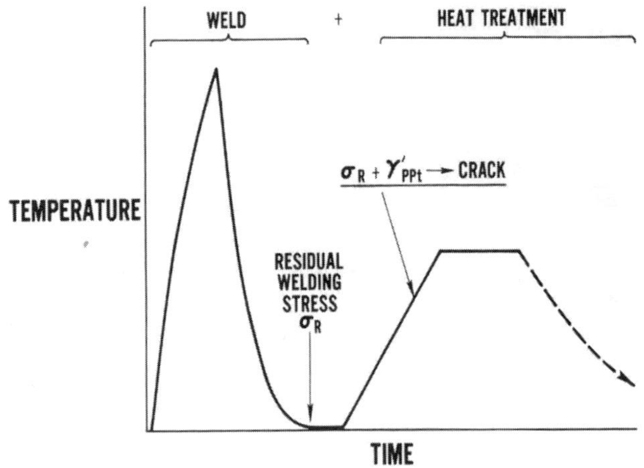

Figure 4. Two contributors to Strain Age Cracking (2)

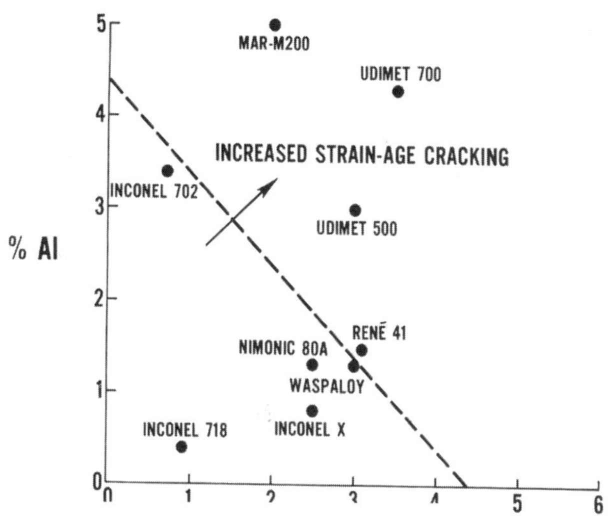

Figure 5. Effect of Aluminum and Titanium on Strain Age Cracking Sensitivity (2)

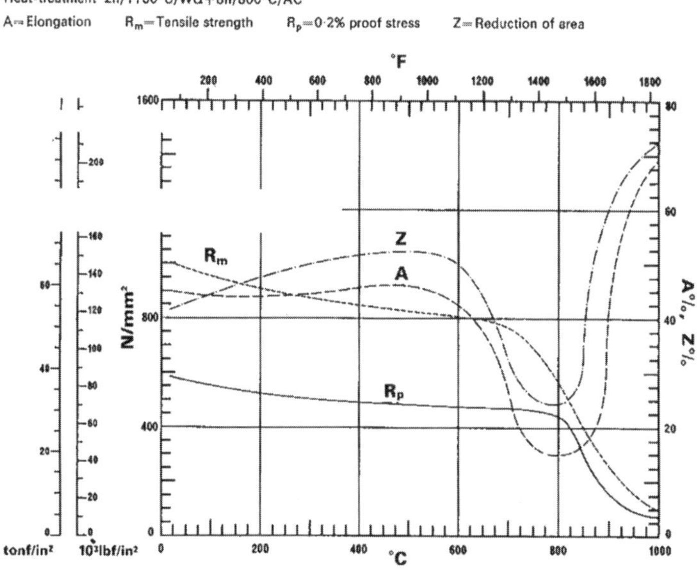

Figure 6. Temperature versus Ductility for NIMONIC alloy 263 (4)

Weldability of INCONEL Alloy 740

Table 1 details the chemical composition of INCONEL alloy 740, heat BLT2819 used in this study.

Table 1: Chemical Composition of INCONEL Alloy 740 (Heat BLT2819) Used in This Testing Program (3)

Element	**C**	**Ni**	**Cr**	**Co**	**Nb**	**Ti**	**Al**
Wt. Pct.	0.029	Balance	24.4	20.0	2.0	1.8	1.0
Element	**Si**	**Mn**	**P**	**S**	**Ag**	**B**	**Bi**
Wt. Pct.	0.53	0.26	<0.005	<0.001	<0.1	0.0045	<0.1
Element	**Cu**	**Fe**	**Mo**	**Pb**	**V**	**Zn**	**Zr**
Wt. Pct.	0.02	0.45	0.50	0.6 ppm	<0.01	1 ppm	0.225

The weldability of INCONEL alloy 740 was investigated by the Edison Welding Institute by quantifying its tendency toward the three kinds of cracking defined above. This was accomplished through a series of tests including spot varestraint, along with Gleeble determinations of elevated temperature properties such as nil ductility range, ductility dip temperature range, and strain age cracking simulation. EWI compared the relative weldability of INCONEL alloy740 to alloys with know fabrication histories in the aerospace industry, namely Waspalloy and INCONEL alloy 718. The results of their investigations show that INCONEL alloy 740 exhibits weldability closer to INCONEL alloy 718 (more weldable) than it does to Waspalloy (less weldable). It does show sensitivity to liquation cracking which has been verified by welding trials. Figure 7 is a schematic representation of welds in alloys with differing sensitivity to liquation cracking. The region shown as the zero ductility plateau represents the temperature zone where the material has no ductility because of liquated grain boundaries. When the material is cooled below this range ductility returns. If strain from weld shrinkage is present while the material is cooling through this zero ductility temperature range liquation cracking occurs as shown in Figure 7b. INCONEL alloy 740 falls into this category. Special Metals is in the process of determining the optimum chemistry to minimize this zero ductility range. Elements which are thought to influence this are carbon, silicon, boron, niobium, titanium, and aluminum. In this program liquation cracking has caused bend test failures during some weld procedure qualification tests to ASME Section IX requirements.

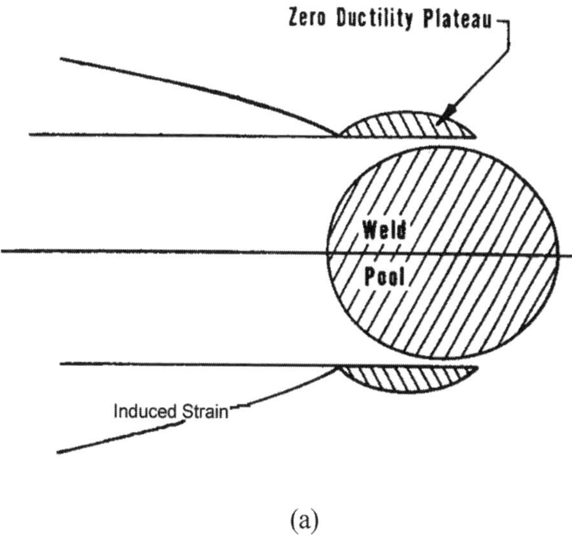

(a)

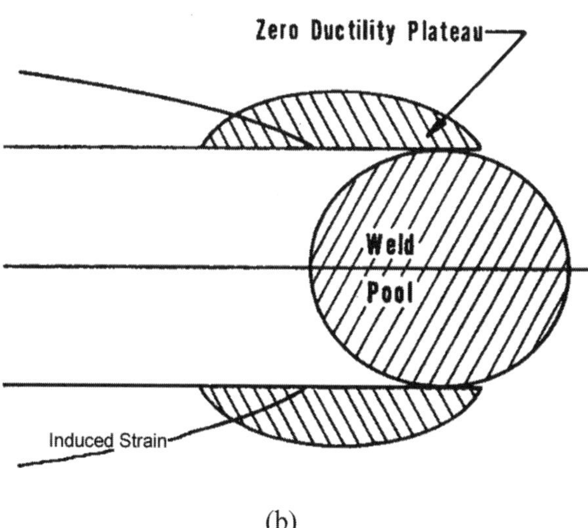

(b)

Figure 7. Schematic Representation of the Susceptibility to Liquation Cracking for Alloy with a Small Zero-Ductility Plateau (lower cracking susceptibility) (a), and Alloy With a Large Zero-Ductility Plateau (higher cracking susceptibility) (b). (Reference 3)

Weld Process Evaluation

Figure 8 is a representation of the interplay of the separate influences which will result in liquation cracking. The material weldability has been described previously in the section on EWI testing. The main influences here are material chemistry and processing effects such as hot working and thermal treatment. One of the unknowns regarding use of aerospace superalloys in boiler projects is the pushing of the envelope to greater thicknesses than previously applied. In Figure 8 this is the influence represented by the Joint Restraint circle. As weld joints become thicker the residual stress increases. Also shown in the figure is the third influence which is Weld Process. Superalloys are known to be heat input sensitive relative to weldability. Early studies with INCONEL alloy 740 have shown that some traditional boiler fabrication welding processes will not be an option for this alloy. Processes such as Submerged Arc Welding or Shielded Metal Arc welding are either not chemically compatible or have excessively high heat input.

FACTORS AFFECTING
MICROFISSURE FORMATION

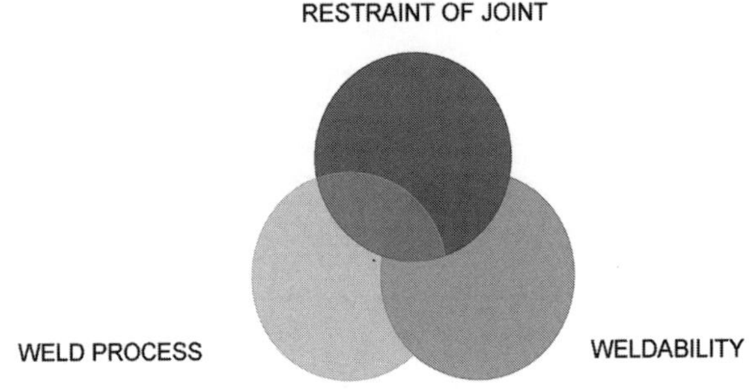

Figure 8. Schematic representation of factors affecting welding induced microfissuring (Reference 3)

Microfissures, such as those shown in Figure 9, have been shown to cause some bend test failures. Typically aged specimens that have liquation microfissures will pass a 4T radius bend but not a 2T bend. Processes such as GTAW or Pulsed GMAW have shown the most promise to date in producing welds with a minimal number of microfissures.

Figure 10 shows a cross section of a GTAW weld in1.75" thick INCONEL alloy 740, which exhibited only a few microfissures. Parameters of prospective weld processes which affect the heat input must be evaluated. These include travel speed, amperage, voltage, shielding gas, filler wire diameter and feed speed, interpass temperature, and interbead cleaning. To date welding procedure qualifications have been accomplished using GTAW in sections 5/8" in thickness. Most procedure qualification attempts were successful in the as welded condition but when the weld was aged to full strength bend test failures occurred. Some welds were successfully tested in the aged condition if the section thickness was 5/8ths inch or less and the weld was performed manually with the GTAW process.

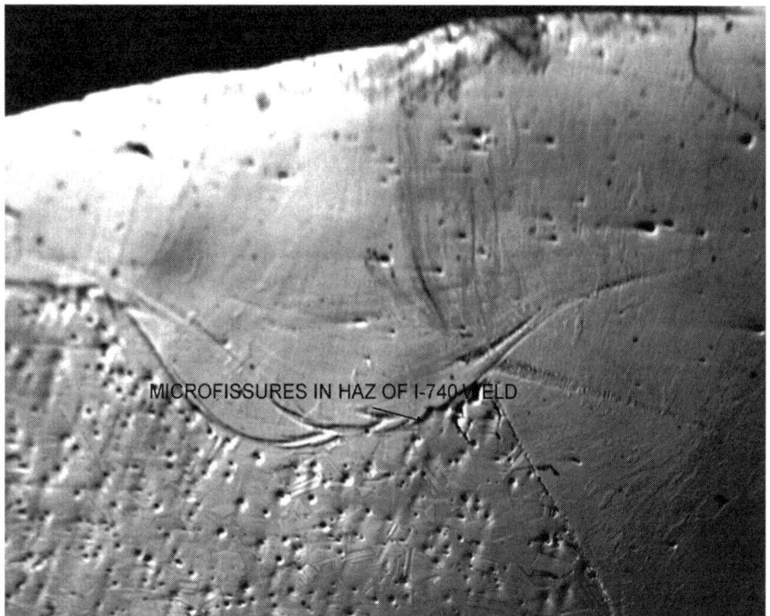

Figure 9. Microfissures near cap bead of P-GMAW in INCONEL alloy 740 weld which opened up in bend testing.

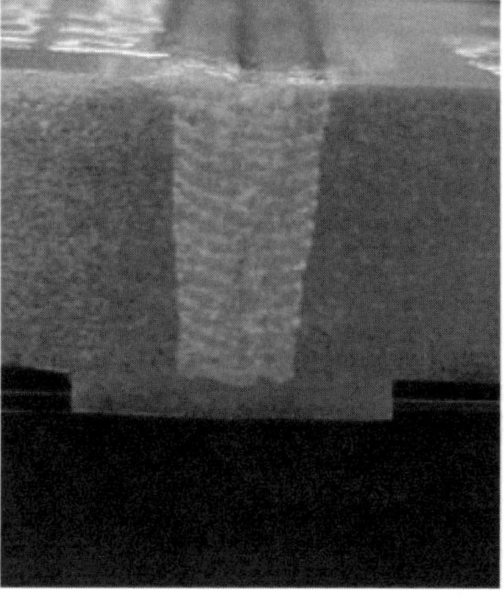

Figure 10. GTAW weld in 1.75" thick INCONEL alloy 740

Conclusions and Recommendations

1. INCONEL alloy 740 is susceptible to liquation cracking. These microfissures are formed in the heat affected zone of a weld very close to the fusion line. Sometimes they encroach into the actual weld bead. This was predicted by EWI gleeble testing and confirmed by welding trials. As heat input is increased the number of fissures also increases. Bend test specimens have failed due to the presence of liquation cracks. Elimination of these microfissures is the top priority of an ongoing investigation into the weldability of INCONEL alloy 740.

2. Gleeble studies have shown that the alloy should not be sensitive to strain age or ductility dip cracking. It should be noted that for boiler applications weldments will be thicker than those produced to date in superalloys. As the thickness increases, so does the propensity for cracking due to the additional restraint. The precipitation strengthened alloys might exhibit uneven mechanical properties due to the thickness. Some heat treatments such as solution annealing and quenching are time dependent and thicker plates might not respond to the heat treatments uniformly.

3. Welding procedure qualification testing of thick welds to the requirements of ASME Section IX has been hampered by microfissuring due to liquation cracking. Most welds produced to date have exhibited some microfissuring in the heat affected zone. Welds in the as welded condition will typically pass the 2T radius bend test. Once the weld is aged to obtain full strength, bend test specimens will pass a 4T radius bend but tend to fail the 2T bend test although some welds have passed. Work is ongoing in this area to obtain consistently acceptable 2T bend test results in the fully aged condition. Welds in thinner materials such as tubing may not be affected by microfissuring. Early results from thin section manual welding of INCONEL alloy 740 were very positive for both as welded and post-weld aged tests.

4. Recommendations include more weld process development to minimize microfissuring in thick section welding of INCONEL alloy 740. This will include welding base metal in the aged condition.

5. Evaluation of heat to heat variability should be investigated. This has been problematic in the aerospace industry for these nickel base superalloys.

Acknowledgements

This paper was prepared with the support of the U.S. Department of Energy, under Award No. DE-FG26-01NT41175 AND the Ohio Coal Development Office of the Ohio Air Quality Development Authority under Grant Agreement Number CDO/D-0020. However, any opinions, findings, conclusions, or recommendations expressed herein are those of the author(s) and do not necessarily reflect the views of the DOE and/or the Ohio Coal Development Office of the Ohio Air Quality Development Authority.

References

1) M. Prager, and C.S. Shira. "Welding of Precipitation-Hardening Nickel-Base Alloys." *Welding Research Council Bulletin* No. 128, New York, 1968.

2) William A. Owczarski. "Process and Metallurgical Factors in Joining Superalloys and Other High Service Temperature Materials." *Superalloys Source Book.* American Society for Metals, Metals Park, Ohio, 1984.

3) Jose E. Ramirez. "Evaluation of Weldability of INCONEL Alloy 740 for the Ultra-Suprecritical Boiler Project." Edison Welding Institute report for project 47067GTH, sponsored by Energy Industries of Ohio.

4) NIMONIC alloy 263 product literature, Special Metals Corp, Huntington, West Virginia.

5) R. Viswanathan, R. Purgert, and U. Rao. "Materials For Ultra-Supercritical-Coal Fired Power Plant Boilers." *Materials For Advanced Power Engrg.* 2002, Vol 21, Part II, p 1109-1130, Ed: J. Lecomte-Bekers, et al, ISBN 3-89336-312-2, Proc. of the Seventh Liege Conference.

Legal Notice/Disclaimer

This report was prepared by The Babcock & Wilcox Company and EdisonWelding Institute pursuant to a Grant partially funded by the U.S. Department of Energy (DOE) under instrument Number DE-FG26-01NT41175 and the Ohio Coal Development Office of the Ohio Air Quality Development Authority under Grant Agreement Number CDO/D-0020. NO WARRANTY OR REPRESENTATION, EXPRESS OR IMPLIED, IS MADE WITH RESPECT TO THE ACCURACY, COMPLETENESS, AND/OR USEFULNESS OF INFORMATION CONTAINED IN THIS REPORT. FURTHER, NO WARRANTY OR REPRESENTATION, EXPRESS OR IMPLIED, IS MADE THAT THE USE OF ANY INFORMATION, APPARATUS, METHOD, OR PROCESS DISCLOSED IN THIS REPORT WILL NOT INFRINGE UPON PRIVATELY OWNED RIGHTS. FINALLY, NO LIABILITY IS ASSUMED WITH RESPECT TO THE USE OF, OR FOR DAMAGED RESULTING FROM THE USE OF, ANY INFORMATION, APPARATUS, METHOD OR PROCESS DISCLOSED IN THIS REPORT.

Advances in Materials Technology for Fossil Power Plants
Proceedings from the Fifth International Conference
R. Viswanathan, D. Gandy, K. Coleman, editors, p 830-862

DOI: 10.1361/cp2007epri0830

Behavior of New Pipe Steels and Their Welds in Modern High Efficiency Power Stations with High Steam Parameter

W. Bendick,
Salzgitter Mannesmann Forschung GmbH,
Ehinger Str. 200
47259 Duisburg, Germany

Bernd Hahn
Vallourec & Mannesmann Deutschland GmbH
Rather Kreuzweg 106
40472 Duesseldorf, Germany

Herbert Heuser
Böhler Thyssen Schweisstechnik Deutschland GmbH
Unionstrasse 1
59067 Hamm, Germany

Russel Fuchs
Böhler Thyssen Welding USA Inc.
10401, Greenbough Drive
Stafford, Texas 77477

PART I: BASE METALS

1. Introduction

Today's, the increased energy demand results in a boom in the construction of new high-efficiency power stations with high steam parameters. As is shown by the national and international projects, fossil fuels will still form the basis for the generation of power in the next years, even despite of the great efforts and the progress which is made with regard to the utilization of alternative energy sources. The increased economic demands and the protection of the climate give reasons for a more cost-efficient and ecologically justifiable provision of energy. All this can only be achieved by lowering the specific fuel and heat consumption for the generation of one kilowatt hour, it is essential to further increase the efficiency of the new power stations in comparison to the modern plants which were started in Germany from 1992 to 2002. New construction and procedural solutions are only part of the possibilities. The main influencing factors on the efficiency increase are the further increase of the steam parameters "pressure" and "temperature". Steam temperatures ranging from 605 °C (live steam) to 625 °C (hot reheat steam) as well as pressures of 300 bar (live steam) and 80 bar (hot reheat steam) as design parameters have become an important factor for the issue of a building and operating license in Germany. The new evaluations of the creep strength for the steel T/P92 (X10CrWMoVNb9-2) which were performed on the European level by the European Creep Collaborative Committee (ECCC) in 2005 have currently limited the further increase of the steam temperature to more than 625 °C [1].

In addition to the requirement of high creep strength, the use of the currently available martensitic 9-12 % Cr steels in the previously mentioned range is limited by their high temperature corrosion resistance. The properties, the restrictions concerning their use and the processing conditions are the focus of this article. The new steel VM12-SHC will also be included since it helps to close the gap to the austenitic steels when being used as a boiler tube.

But on the other hand, the strength resources of the previously used membrane tube wall steels like T1 (15Mo3) and T12 (13CrMo4-4) are exhausted by the further increase of the steam parameters to the temperature range of 600/625 °C for the new generation of power stations. Thus, the development of new membrane tube wall steels was essential and resulted in the development of the steels T/P23 and 7CrMoVTiB10-10 (T/P24). In the following, the focus will also be on these low-alloy steels to show the variety of applications not only for the new generation of power stations but also for the improved efficiency and modernization of existing power stations which is of enormous technical and economic importance.

2. Characteristics and recommended use of the martensitic 9-12 % Cr steels

2.1 General material characteristics of the 9 % Cr steels

The martensitic Cr steels can only be safely processed when having a basic knowledge of their metallurgical properties.

The development of the steels T/P91 (X10CrMoVNb9-1) and T/P92 (X10CrWMoVNb9-2) is based on the martensitic 9 % Cr steel T/P9 (see also **Table 1**). The German steel X20CrMoV11-1 which also belongs to the group of 9 – 12 % Cr-steels was included into the considerations for comparative purposes as well.

The increase of the creep strength of the steel X20CrMoV11-1 in comparison to the steel T/P9 is generally based on the complex impact of the Cr content in connection with Mo and V as well as the resulting martensitic structure which is stabilized by the precipitation of $M_{23}C_6$ carbides. A further increase of strength concerning the steel T/P91 was achieved by precipitation hardening through finely distributed V/Nb carbonitrides of the type MX. An additional increase of strength concerning the steel T/P92 could be achieved by adding tungsten to the alloy. The influence of tungsten on the increase of the creep strength characteristics is very complex and has not yet been understood in detail. For example, the so-called Laves phase $Fe_2(Mo,W)$ will be precipitated after a period of operation. While precipitation hardening of Laves phase will increase the strength, it decreases the original solution hardening effect of tungsten. Eventually the coarsening of Laves phase may decrease creep strength after long term operation.

All previously mentioned, Cr steels are characterized by the same transformation behavior, as can be seen from the comparison of the CCT diagrams for the steels X20CrMoV11-1, T/P91 and T/P92 shown in the **Figures 1 to 3**. One characteristic is the ferrite region which has been shifted into the direction of the long cooling times. A complete martensite transformation will be accomplished even at relatively long cooling times from normalization temperature. Due to the lower carbon contents in comparison to the steel X20CrMoV12-1, the 9 % Cr steels are characterized by the formation of martensite which contains less carbon; this results in a finish martensite temperature which has increased by approx. 100 °C. At the same time, the martensite hardness will be reduced by approx. 150 HV units. These facts have a positive influence on the processability (hot forming and welding).

Basically, secured long-term strength properties for the mentioned 9 % Cr steels will only be achieved by the correct quenching and tempering treatment, which means that cooling from normalization temperature has to be accomplished with a sufficiently high cooling rate depending on the components dimension. It is essential that the steel will be cooled down safely below the finish martensite temperature to ensure a complete martensite transformation and to obtain only tempered martensite through the following tempering procedure. Air cooling is sufficient for seamless tubes with wall thicknesses up to approx. 130 mm; this has been shown by comprehensive tests performed by V&M. Forgings of larger section sizes should undergo oil quenching or a cooling in similar emulsions from normalization temperature to ensure the martensite transformation though thickness.

The influence of different tempering temperatures on the mechanical properties for the steel T/P91 is shown in the **Figure 4**. The optimum material properties will be achieved with tempering temperatures ranging from 750 to 780 °C. The steel T/P92 shows the same behavior [2]. As detailed in **Table 2**, the PWHT specifications of these steels are the same and thus facilitate handling in terms of manufacturing and construction.

The high tempering resistance of the martensitic steels in comparison to the group of the low-alloy Cr-Mo-V steels must be emphasized here as well.

The new 9 % Cr steels do not only show improved creep strength characteristics in comparison to the existing steel X20CrMoV11-1, but also a clearly higher level of toughness. While the impact values at room temperature of the steel X20CrMoV11-1 in its initial condition reach average values ranging from 60 to 70 Joule, these values are lower bound values depending on process route and deformation grade as well as for very large wall thicknesses; the average values are normally clearly above 100 Joule. This positive behavior ensures a safe processing especially during the design of the PWHT processes of dissimilar welds with the low-alloy ferritic steels. The toughness of the heat affected zone on part of the 9 % Cr steel does not represent a failure criterion when using a low-alloy welding filler and a PWHT for such connections which has been adapted to the temperature of the low-alloy steel [3].

2.2 Creep and corrosion behavior

The basis for the construction of power stations with higher steam parameters are secured long-term strength values. To achieve these values, it is required to perform assessments on the available creep tests at certain time intervals, especially with regard to the new 9 % Cr steels. This task has been accomplished for more than 50 years by the working committee for heat resistant steels and high-temperature materials AGW/AGHT (Arbeitsgemeinschaft für warmfeste Stähle und Hochtemperaturwerkstoffe). The long-term strength values which are specified in the German codes are mainly based on the evaluations of this working committee. On the European level, the European Creep Collaborative Committee (ECCC) was founded in 1991 and pursues similar goals.

An evaluation of the steel T/P92 on the European level was accomplished for the first time in 1999. With regard to the steel T/P92, only the design values according to ASME were available which had been determined by Nippon Steel on the basis of a Larson-Miller evaluation. The first evaluation of the ECCC showed that the creep strength values for the steel T/P92 specified in the ASME-Code Case 2179 [4] were too high. A 100,000 h creep strength value of 132 MPa at a temperature of 600 °C was modified with an ECCC value of 123 MPa [5].

The activities performed with regard to the planning of new, supercritical power stations in Europe and especially in Germany were the reason for the start of a second evaluation of the long-term behavior of this steel which was performed by the ECCC group in the middle of 2004. With regard to this second evaluation, Nippon Steel and V&M provided comprehensive datasets concerning the relevant temperature range between 550 and 650 °C with long term tests. **Figure 5** shows the result of the new ECCC evaluation of 2005 [6] concerning the steel T/P92 with the results of the ECCC evaluation of 1999 and the underlying creep strength values of the ASME-Code Case 2179.

The following conclusions have to be drawn on the basis of these new evaluations:

- The steel T/P92 shows a significant decrease of the creep strength after long periods of operation which is caused by microstructural changes, especially by the formation of new phases (Laves phase and Z phase); this decrease was not clearly shown in the creep strength test results until a test period of more than 50,000 hours. This means that a sufficient extrapolation security of the creep strength of these steels is only provided after having achieved these test periods.

- The ongoing microstructural processes have only been insufficiently investigated and understood so far, especially with regard to their kinetics and the influence of alloy composition, and must therefore be further investigated. The respective research activities have been initiated by V&M.

- T/P92 can still be characterized by a common scatter band (± 20%) with regard to their creep behavior and will cover the higher part of the scatter band.

- The steel T/P92 should be preferred regarding the construction of new power stations with steam parameters up to 625 °C. Thus, the design of thick-walled components (steam headers, joints) currently reaches a technological limit, since these components have wall thicknesses that would significantly limit a flexible start-up or shut down of the power station due to the arising thermal stresses if the steam parameters would be higher.

In addition to the requirement of a sufficient creep behavior, the high temperature corrosion behavior of the steels to be used is also becoming a service life-limiting factor with regard to high steam parameters. Due to their average Cr content of 9 %, the steel T/P92 is inferior to the steel X20CrMoV11-1 with regard to their high temperature corrosion behavior; thus, they do not present an alternative for the substitution of this steel in the heated area of the boiler.

2.3 Boiler tubes made of the steel VM12-SHC – Information concerning the first application

To close the gap between the creep behavior and the high temperature corrosion behavior of the steel X20CrMoV12-1 and the 9 % Cr steels, V&M Tubes initiated the development of a 12 % Cr steel with the company designation VM12 for use at temperatures up to 650 °C [7-9]. This steel will be described in detail in another article of this conference [10] so that only the characteristics which are specific for its use and its further processing will be mentioned at this point.

The following development objectives were specified:

- Sufficiently high scale resistance for safe use at temperatures up to 650 °C
- Better or comparable short-term and long-term strength characteristics than those of the steels T/P911 and T/P92
- Good hot formability and welding properties. The steel T/P91 is considered as the optimum steel concerning these issues.
- The new steel should have a martensitic structure.
- The chrome content should be higher than 11 % to secure good scale resistance.
- Similar contents of vanadium, niobium and nitrogen compared to the 9 % Cr steels should to secure sufficient creep strength by precipitation hardening in the form of MX particles.
- By optimizing the content of tungsten, cobalt and boron, their strength-increasing influence should be used as well.

The current state of the developments has resulted in an alloy composition which is characterized by very good high temperature corrosion properties and corresponds to the service requirements of the new Germany power station projects in the temperature range up to 610 °C. Although the development of the steel VM12 has not yet been completed, an earlier application of the comprehensively tested alloy modification with the name VM12-SHC was required since no other 12 % Cr steel meets the demanded service requirements, namely a sufficiently safe and high creep strength level in connection with high demands concerning the high temperature corrosion resistance.

The chemical composition of VM12-SHC in comparison to the known 9-12 % Cr steels is shown in **Table 3**. The main differentiating factors with regard to the steel T/P92 is the slightly increased carbon content, an average chrome content higher than 11 % and an average cobalt content >1.0 %.

As can be seen in the Schaeffler diagram presented in **Figure 6**, the phase stability of the steel T/P92 will be influenced if the chrome content is increased from 9 % to 11 %. To achieve a completely martensitic structure, the nickel equivalent has to be increased accordingly. To achieve this, cobalt was added to the alloy, and the carbon content was slightly increased in comparison to the 9 % Cr steels. The steel VM12-SHC is located close to the point in which martensite, austenite and ferrite meet each other (as can be seen in the Figure). According to this diagram, insignificant shares of δ ferrite are possible; however, these shares are lower than 2 % (according to the practical experiences which have been gathered so far). The diagram shows the typical compositions of the considered steels.

The transformation behavior of the steel VM12-SHC is not different from the transformation behavior of the known 9 % Cr steels (**Figure 7**).

The steel will be delivered in normalized and tempered condition. The maximum tempering temperature is limited by the point A_{c1b}, which may be between 810 and 830 °C depending on the chemical analysis. Normalization is accomplished with temperatures ranging from 1040 to 1080 °C. The optimum properties will be achieved with tempering temperatures ranging from 750 to 800 °C. **Figure 8** shows the microstructure of a VM12-SHC boiler tube (dimensions: 38 x 7.1 mm) after the quenching and tempering treatment. The structure consists of tempered martensite. In this case, no δ ferrite could be detected.

Table 4 shows the specified mechanical properties at room temperature. The strength values at elevated temperatures can be seen in **Table 5**.

According to the German codes, creep tests of several melts with running periods of at least 30,000 hours are required for the qualification of a new steel for service. These testing periods will be reached this year. To allow for the safety wishes of the power station operators, however, the design values used in the current applications are not based on the actually determined creep strength which can be found in the upper scatter band region of the steel T/P911 for temperatures between 550 to 610 °C, but on the creep strength of the steel T/P91 [10]. **Figure 9** illustrates this issue.

The excellent scale resistance is shown as an example in **Figure 10** by means of a comparison with the steel T/P92. This figure shows the weight increase per unit of area caused by oxidation.

Currently, superheater heating surfaces are being produced from VM12-SHC boiler tubes for five utility steam generators with capacities between 800 and 1100 MW. Qualifying examinations for these applications were performed by the TÜV Rheinland [*German Inspection Authority*].

2.4 Recommended use and current state of European standardization

As a consequence of the ECCC creep strength evaluation, the effective application range of the steel T/P92 (X10CrWMoVNb9-2) has been changed. The steel T/P92 has to be used preferably for temperatures ranging between 600 and 625 °C.

The inclusion of the steel X10CrWMoVNb9-2 (T/P92) into the European standard EN10 216-2 [11] has already been initiated 3 years ago by V&M. A draft dated April 2005 is available and includes the results of the ECCC creep strength evaluations so that the future use of these steels according to this European standard is secured as well.

In addition, X10CrWMoVNb9-2 has been approved TÜV Rheinland in order to ensure the use of this steel in Europe according to the valid technical regulations until the new EN standard 10216-2 comes into force. The material requirements for X10CrWMoVNb9-2 (T/P92) are available as material data sheets prepared by the TÜV Rheinland [12]

The development of the steel VM12, with the objective to cover the temperatures ranging from 625 to 650 °C with a steel which does not only show a high oxidation resistance but also high creep strength values, has not yet been finished. In the boiler area, there is a gap between the steel X20CrMoV11-1 and the austenitic steels which cannot be closed by using the 9 % Cr steels due to insufficient oxidation resistance. The temperature range up to 610 °C is now being covered with the VM12 modification VM12-SHC. The legal requirements for this application have been checked and approved by the qualifying examination which has been performed by a notified body (TÜV Rheinland).

3. New low-alloy steels for boiler tubes/membrane tube walls and thick-walled piping components

3.1 Metallurgical basis and material properties

The previously used membrane tube wall steels T1 (15 Mo3) and T12 (13 CrMo4-4) can no longer be used for the vaporizer area of the new boilers with increased steam parameters since the new design requirements are reaching far into the creep range. Using the known higher alloy steels T22 (10CrMo9-10) or T91 (X10CrMoVNb9-1) is not possible since heat treatment following weld fabrication is not intended from a technical perspective and is very difficult to perform. Thus, the development of new membrane tube wall steels was clearly essential and resulted in the development of the steel T/P23 in Japan and 7CrMoVTiB10-10 (T/P24) at Vallourec & Mannesmann in Germany.

The basis for the development of the steels T/P23 and T/P24 was formed by the known steel T/P22, as can be seen in **Table 6**.

The creep strength of these steels was strongly increased by adding the carbide-forming elements vanadium, niobium and titanium. In the case of the steel T/P23, the chemical element tungsten was also added and the molybdenum content was reduced. A special characteristic of both steels is their reduced carbon content; the effect of this reduced carbon content can be seen in the CCT diagram (**Figure 11**) for the steel T/P24. Even very fast cooling in the martensite region results in maximum hardness values of only 350 to 360 HV10. This behavior is a decisive factor in the that no post-weld heat treatment has to be done for wall thicknesses s < 10 mm in the thin-walled boiler tubes. Both steels are used in normalized and tempered condition only. The normalization procedure for the steel T/P23 is normally performed at 1060 °C ± 10 °C, and the normalization procedure for the steel T/P24 is done at 1000 °C ± 10 °C. To achieve optimum material properties at wall sizes > 10 mm, it is required to cool the steel more quickly (water quenching) from the austenisation temperature. The tempering procedure for the steel T/P23 will be performed at temperatures in the range 760 °C ± 15 °C. In the case of the steel T/P24, the tempering procedure at 750 °C ± 15 °C has been determined as the optimum procedure.

With regard to their general material properties, the steels T/P23 and T/P24 largely correspond to the steel T/P22 which has been the basis for their development. This applies for the physical properties and the scale resistance as well as for the notched bar impact strength which is on a similarly high level. However, a shift of the A_V-T curve can be determined depending on the cooling rate from normalization temperature; this is illustrated in **Figure 12** for the different quenching and tempering conditions and wall sizes (T: Tube, P: Pipe). To ensure high impact strength values at room temperature, it is essential to perform a quench and tempering treatment on thick-walled components for both steels.

Compared to the steel T/P22, the steels T/P23 and T/P24 show significantly increased strength properties (see overview **Table 7**).

Figure 13 shows an exemplary comparison of the minimum values of the 0.2 % proof strength for thin-walled tubes. As it is shown by the aging tests for the steel T/P24 in **Figure**

14, neither the strength values nor the ductility values show any significant changes in the long term. A similar behavior can also be determined with regard to the steel T/P23.

The most defining characteristic of high temperature steel is its creep strength. As can be seen in **Figure 15**, there are also clear differences compared to the steel T/P22. In the lower temperature range, the long-term values of the steel T/P24 are only slightly below the values of the steel T/P91. The values for the steel T/P23 are a little lower, but they are significantly above the level of the steel T/P22. In the temperature range of 575 °C, they meet the creep strength curve of the steel T/P24 and come close to the curve of the steel T/P91 at a temperature of 600 °C. Due to the limited scale resistance, which can be attributed to the low chrome content, it is not recommended to use both steels for long periods at temperatures of more than 575 °C.

The steels T/P23 and T/P24 are showing good processing properties which are similar to the steel T/P22. The same regulations as for the steel T22 have to be applied to the cold bending procedure of thin-walled boiler tubes. In the case of inductive bending or other hot forming procedures of thick-walled pipes, a new quenching and tempering procedure with water is required in any case.

3.2 Recommended use and current state of standardization

With regard to new power stations with steam temperatures higher than 585 °C, the use of the steels T23 or T24 (7CrMoVTiB10-10) in the boiler area is essential from a technical perspective to satisfy the increased requirements concerning the operation parameters. The steels are mainly used for membrane tube walls. In the area of thick-walled components, the use of the steels P23 and P24 (7CrMoVTiB10-10) does not only provide a good alternative to the known low-alloy steel grades for boiler headers but also for entire piping systems both for plants and for refurbishments in the temperature range between 500 and 550 °C. As can be seen in **Figure 15**, both new steels show a clear advantage concerning their use at temperatures ranging from 500 to 550 °C compared to the steel 10CrMo9-10 (P22) which has been traditionally used for these applications. Their creep strength values can only be exceeded by the new martensitic tube steels like the steels P91, P911 and P92. The advantages of these new steels in the high pressure piping area will be shown on the basis of the results of the wall size calculations of representative components of a 350 MW block, which was originally designed for the steel P22 according to ASME B31.1 [13]. **Tables 8 and 9** demonstrate this issue by giving the example of the wall size calculation for the straight pipe of the steam line and a fitting integrated into this system.

Similarly, the same effects which are achieved when reducing the wall size are also achieved when dimensioning the hot reheat steam pipe. Basically, a wall size reduction by 35 % up to 50 % compared to the low-alloy steels which are normally used in these applications can be assumed in the temperature range between 500 and 545 °C, an obvious application area for both steels with regard to pipes. Furthermore, there is an advantage over the use of the high-quality 9 % Cr steel X10CrMoVNb9-1 (P91) concerning the welding fabrication which corresponds to the welding fabrication of the steel 10CrMo9-10.

The steel T/P23 is currently standardized in ASTM A 213 (Code Case 2199) [15] and ASTM A 335. A respective Code Case is currently being prepared with regard to the steel T24. Under the designation 7CrMoVTiB10-10 (material number 1.7378) an approval by TÜV Rheinland is available in form of a material data sheet [16].

Both steels have been introduced by V&M into the revision of the EN standard 10216-2 similar to the steel X10CrWMoVNb9-2 (T/P92) and they have been included into the present draft from April 2005 [11] with the designations 7CrWVMoNb9-6 (T/P23) and 7CrMoVTiB10-10 (T/P24).

PART II. FILLER METALS

4. Development of new, high-temperature welding fillers for bainitic and martensitic materials

During the implementation of new steels, there will also be the requirement for the respective and appropriate welding fillers. Thus, it is important that, at the same time of the development of new base materials, the respective welding filler materials, with sufficient creep characteristics, are developed as well. It will normally be assumed that the analyses of the welding filler materials are close to the analyses of the base materials. In the following, the welding fillers for the bainitic steels T/P23 and T/P24 as well as for the martensitic steels P92 and VM12 are presented for the common practical processes GTAW, SMAW and SAW. In the course of the (steel) development, it could be determined that both bainitic steels cannot only potentially be used as boiler tubes (as originally intended), but that they also present an appropriate alternative for the martensitic 9 % Cr steels in the thick-walled high pressure pipe area at temperatures of 500 to 550 °C.

5. Matching filler metals for the bainitic steel T/P23

First, a matching filler metal for the GTAW-process has been developed for girth welding of thin-walled tubes for the production of membrane tube walls. In this case, no PWHT is required after welding since the low carbon content in the base material and in the weld metal does not cause an increase in hardness with values above 350 HV10 in the weld or in the HAZ. Since this material can also be processed as thick-walled component, stick electrodes and SA welding fillers are required as well in addition to GTAW. Depending on the wire diameter and the welding parameters, the toughness of the all GTA weld metal is very high, regardless of with or without PWHT. Sometimes, values of more than 200 J could be determined. The hardness without any PWHT is approx. 270 HV10 and 250 HV10 after PWHT. However, PWHT at a temperature of 740 °C has to be performed after the welding process for the slag producing welding procedures (**Figure 16**), since without PWHT the toughness of the weld is only on the low level of 20 J.

The PWHT has to be adapted carefully to the construction circumstances. Thin components without huge wall thickness transitions have to be cooled below a temperature of 250 °C after welding to prevent the possible formation of martensite. Coming directly from the welding procedure, thick-walled components, especially for wall thickness transitions, have to undergo an intermediate stress relief at 500 - 550 °C before the components are heated to the tempering temperature. This procedure is supposed to prevent "stress relief cracking" (**Figure 16**).

Table 10 summarizes the chemical composition of the welding filler metal and the mechanical properties of the all weld metal for the individual processes.

Table 11 outlines the mechanical properties of a thick-walled girth weld at a tube (Ø 219.1 x 30 mm). The root was welded in a single pass by GTAW. The first and the second filler pass were welded with the stick electrode (Ø 2.5 and 3.2 mm), and the other passes were welded by SAW. The PWHT which is obligatory for this wall thickness was performed at a temperature of 740 °C for 1 hour. Even this short PWHT resulted in a toughness level between 125 J and 150 J. The fracture of the transverse tensile specimen occurred in the base material. The hardness of the weld metal was max. 250 HV10. The bend test shows no defects at a bending angle of 180°.

6. Matching filler metals for the bainitic steel T/P24 (7CrMoVTiB10-10; 1.7378)

With regard to the steel T/P24, welding fillers have been developed for GTAW, SMAW and SAW. The analysis limits for the all weld metal are summarized in **Table 12**. Due to the high oxygen affinity of the element titanium (Ti), a more or less visible Ti burn out occurs during the welding process compared to the initial values of the wire. This applies both for GTA welding, where the arc should be optimally insulated by inert gas, and especially for welding with an electrode or submerged arc welding. This fact led to the result that the welding fillers were alloyed with niobium instead of titanium. Both niobium and titanium are carbide-forming elements and contribute to the creep strength like of the martensitic steels P91 and P92.

Table 12 lists also the mechanical values for the all weld metal. Like in the case of GTA-welding of thin-walled tubes made of T23, T24 tubes also do not require PWHT after welding. The hardness is 322 HV10 for the non-tempered GTAW weld metal. A PWHT which is necessary on weldments of SMA and SA at wall thicknesses >10 mm at a temperature of 740 °C for 2 hours reduces the hardness by 100 units to 230 HV10.

A thick-walled P24 joint (tube diameter 159 x 20 mm) was welded in the scope of a qualification. The root was welded in a single pass by GTAW followed by 2 passes which were welded with the stick electrode. The rest of the seam was filled by submerged arc welding. **Table 13** summarizes the results of this girth weld. The transverse tensile test resulted in fracture of the base material. The toughness of the weld metal with values around 250 J is on a comparably high level with the all weld metal. The bend test did not result in any defects at a bending angle of 180°, and the hardness was below 250 HV10.

Figure 17 shows schematically the temperature cycle during welding and the following PWHT at wall thicknesses > 10 mm. The PWHT can be carried out from the interpass temperature ≤ 250 °C.

Furthermore, **Table 13** shows the results from a boiler tube weld. When having a wall thickness of 6.3 mm, notched bar impact tests can only be performed up to a size of max. 5 mm. Many fabricators are surprised when boiler tubes are GTA-welded for the first time and when the determined notched bar impact values are compared with the values from thick-walled connections. **Figure 18** shows a schematic overview specifying which welding technology must be used to ensure sufficiently high notched bar impact values in the as welded condition (boiler tube connections do not require any PWHT). It is important to weld thin beads and that especially the top layer is welded very thin. This bead should "temper" the lower bead and should not be located in the area where the notched bar impact test specimen will be removed.

Transverse samples were taken from the welds described in **Table 13** to determine the long-term properties of these connections at a temperature of 550 °C (see also chapter 10).

7. Matching filler metals for the martensitic material P92 (Nf 616)

Increasing the steam parameters to increase the efficiency means that higher requirements must be met with regard to the materials and concerning the properties according to the intended use. Thus, it is obvious that the manufacturers of welding filler materials can already prove sufficient creep characteristics for their materials when using new steels. Furthermore, a sufficiently high toughness level at room temperature is required to save costs concerning the heating of water during the pressure test.

In general, the toughness of the martensitic welding fillers is clearly lower than the toughness of the ferritic or bainitic welding fillers. The minimum requirements for weld metals are geared to the requirements of the base materials and should at least be 41 J at room temperature. The reasons for low notched bar impact values can be attributed to the alloy contents of C, Nb, V, N and W in the martensitic structure. These elements are necessary to ensure the required creep strengths, but at the same time, a higher content of these elements reduces the toughness of the weld metal. **Table 14** includes the analysis limits as well as the mechanical properties of P92.

The negative toughness influence of these elements is partly compensated by an increased nickel content of the weld metal compared to the base material. While the nickel (Ni) content of the base material P92 is limited to a maximum of 0.4 %, the upper limit of nickel in the weld metal was set to ≤ 1 %. The niobium (Nb) content of the weld metal was reduced to a minimum of 0.04 %. At the same time, the manganese (Mn) content in the weld metal had to be adjusted to comply with the requirement of Mn + Ni ≤ 1.5 %. Both manganese and nickel have an influence on the lower critical point Ac_{1b}. The analyses of the weld metals were set in a way so that the δ ferrite content was less than 1 %.

Welding the martensitic steel P92 is accomplished in the martensitic temperature range, i.e. between 200 and 300 °C. Due to the martensitic structure, the temperature during the welding process and the post-weld PWHT has to be controlled very carefully.

For wall thicknesses up to approx. 80 mm, a pure martensitic transformation will occur in the case of cooling in air. In the case of P92, this transformation starts at a temperature around 400 °C. The individual cooling speeds should be determined in a way so that a complete martensitic transformation will be achieved.

After a PWHT procedure at 760 °C for 2 hours, the hardness of the martensite in the weld metal is approx. 250 HV10. After the welding process (before the PWHT), the hardness of the weld metal of approx. 400 HV10. The danger of intercrystalline stress corrosion cracking after the warm bending procedure or after welding P92 is therefore clearly reduced. At the same time, also the danger of cold cracking is reduced so that less stressed components can be cooled down to room temperature after welding.

As described before, the operators of power plants and the monitoring organizations attach great importance to the best possible toughness properties of the used weld metals. However, the possible metallurgical scope for the martensitic metals is not wide enough to set the notched bar impact value for weld metals welded with electrodes or by submerged arc welding to a significantly higher level than 41 J. To a limited extent, the toughness of these weld metals can be influenced by the choice of the welding parameters and by the conditions of the PWHT. Here, it is important that the welded connection cools down below the martensite finish temperature (Mf) prior to the PWHT to allow a complete tempering of the martensite through the PWHT. The Mf-temperature is approx. 150 °C for the weld metals matching P92 so that a cooling to at least 100 °C is required. Protection against hydrogen-induced cold cracking can be achieved with a soaking procedure directly from the welding heat (2–3 hours at a temperature of 250 – 300 °C).

Figure 19 shows a schematic overview of the heat control during the welding procedure and the following PWHT. By welding "thin" passes, the toughness can be systematically increased. The thinner the individual passes are welded, the higher is such a tempering effect. This has to be taken into account especially during a SMA weld.

Figure 20 shows the dependency of the notched bar impact value on the conditions of the PWHT for the weld metal matching the base material P92.

Different pipe joints were welded in the scope of the weld metal development and qualification measures. **Figure 21** shows an example of the welding sequence of a P92 pipe weld. The root was welded in a single pass by GTAW. The lower part of the wall was welded with SMAW, and the upper part was welded by SAW. After welding, the parts were slowly cooled to room temperature. The following PWHT was performed at a temperature of 760 °C. For this, the connections were cut into halves first so that different PWHTs could be performed for 2 hours as well as for 4 hours.

The analyses of the weld metals correspond to the analyses of the base materials (with the exception of nickel and niobium). The strength values of the matching welding fillers meet the requirements both at room temperature and at higher temperatures (600 °C). **Tables 15 and 16** summarize the results of the procedure qualification tests.

A PWHT with an extended hold time of at least 4 hours at a temperature of 760 °C is recommended for the all weld metal for submerged arc welding. **Table 16** shows the respectively different toughness values of the weld metal (all weld metal) after a too short (2 hours) or a longer (4 hours) PWHT. In terms of the girth welds, the differences are not that significant. This can be attributed to the different structure of the individual passes.

The technical data sheet for the base material P92 specifies a temperature range from 730 to 780 °C for the PWHT after the welding procedure. However, it could be determined that a temperature of 760 °C is required for the welding filler materials to restrict the required hold time to an economically reasonable duration of two to four hours (see also **Figure 19**). However, low temperatures can be selected, but they reduce the toughness in the weld metal or they require substantially longer hold times, respectively, to reach the required level again. Temperatures which are significantly higher than 760 °C can cause the Ac_{1b} temperature to be exceeded. However, it is not critical to exceed the temperature only to a minor extent and only for a short time.

8. Matching filler metals for a new martensitic material with 12 % chrome – VM12-SHC

According to the latest knowledge, the previously described base material P92 is only used for temperatures up to a maximum of 620 °C. The reason for this is the insufficient scaling resistance at higher temperatures. Here, materials with higher chromium content have to be used. The European research program „COST 536" currently optimizes a 12 % chrome steel (VM12) developed by Vallourec & Mannesmann for applications including temperatures of up to 650 °C. Böhler Thyssen Schweisstechnik is also involved in this project and develops welding fillers matching VM12.

The developments for thick-walled components have not yet been finished. However, the basic tests are already on a very advanced level. Due to the existing alloying concept, the weld metal is characterized by high mechanical strength properties. At the same time, however, there is a toughness level in the welding fillers showing values around 40 J. Thus, these values are below the values of the previously mentioned welding fillers for the 9-10 % Cr steels. This alloy has also shown that a nickel/niobium modification of the weld metal which could otherwise successfully be used does not result in any improvements. Nevertheless, the toughness level is sufficient enough since the required minimum values are currently at > 27 J. However, it is necessary to exercise reasonable care during the welding process (only low heat input, right choice of the electrode diameter, compliance with the admissible pass size etc.). The PWHT should be performed at a temperature of 770 °C since sufficiently high toughness values cannot be guaranteed at a temperature of 760 °C (**Figure 22**). The A_{c1b} temperature of the weld metal and the base material is at 800 °C. **Table 17** shows the results for the all weld metal for GTAW and SMA welding.

Compared to P92, this alloy has a higher content of chromium. Since this causes a formation of ferrite, this must be balanced with an austenite-forming element. For this, the chemical element cobalt (Co) will be used, since cobalt does not have any influence on the A_{c1b} point (compared to nickel).

Compared to the continuous development of the base material for thick-walled components, this material will be used in new power plant projects for wall sizes up to a maximum of 10 mm. This tube material was qualified for wall sizes up to 10 mm under the denomination VM12-SHC and is welded using the GTA- and SMA-welding processes. **Table 18** includes the results from qualification welding procedures. The time for the PWHT in the GTA-welding procedure was only 30 minutes which can be regarded as the lowest limit. The PWHT for welding procedures with SMAW is specified with a holding time of at least 2 hours.

9. Creep rupture tests

To use high-temperature steels effectively, it must be ensured at an early stage that the respective weld metals and the welded connections show almost the same creep strength characteristics than the base materials. Respective long-term tests concerning the all weld metal and the welded connections have thus been performed for the welding filler materials presented in this article.

10. Creep rupture tests for the bainitic welding fillers

Creep rupture tests at 525 °C, 550 °C and 575 °C were performed for the welding fillers matching T/P23 and T/P24.

The longest current tests (for fillers matching T/P23) are performed over a period of 14.000 hours, and the stress level is selected in a way so that 10.000 hours (already exceeded) or 30.000 hours are intended. With regard to the welding fillers matching T/P23 (SMAW and SAW), the rupture points of the all weld metal and the rupture points of the welded connections are slightly below the mean value curve of the base material. The ruptures in the welded connections mainly occurred in the HAZ or in the base material.

In the case of the welding fillers matching T/P24, the creep properties of the welding fillers cannot be evaluated so easily. It is noticeable that the matching titanium-alloyed electrodes do not fulfill the requirements of the creep strength. The rupture points for girth welds are already below the lower scatter band curve of the base material with values around 1.000 hours, and the ruptures are in the weld metal and not, as expected, in the base material. In contrast to this, GTA-welded connections have already reached running periods of more than 10.000 hours. Also the SMA weld metal does not show consistent and satisfying results. The reasons for this can obviously be found in the burn out behavior of certain elements, especially with regard to the important alloy elements like titanium and boron. As described before, it is difficult to warrant a sufficiently high content of titanium and boron in the welded connections produced under welding conditions in line with actual requirements, especially for tube connections with different welding parameters. Due to this reason, welding fillers matching T/P24 have been developed with a changed alloying concept, i.e. they were alloyed with niobium. Previously achieved results with regard to creep rupture tests confirm the correctness of this analysis modification.

11. Creep rupture tests for the martensitic welding fillers

Creep rupture tests at 600 °C, 625 °C and 650 °C were performed for the welding fillers matching P92. The creep characteristics were determined both for the all weld metals for SMAW and SAW and for P92 tube welding seams.

The longest tests for P92 weld joints are for more than 60.000 hours.

During the tests of the connections for P92, the rupture points occurred at a temperature of 600 °C within the scatter band of the base material. In contrast to this, the rupture points at higher temperature were partly outside the scatter band, and there could be detected both ruptures in the weld metal as well as in the HAZ. During long term tests, the creep ruptures occurred in the HAZ. At a temperature of 650 °C, ruptures could exclusively be detected in the HAZ.

12. Summary

The intention of this article was to draw the attention to material developments for modern power plants. These materials are of great economic importance for the optimization of existing power plants as well as for new power plants with increased steam parameters. The attempt to use martensitic steels in power plants even for temperatures higher than 620 °C, resulted in an initiative of a European research program „COST 536“ which includes more than 15 European countries. From today's perspective, the developments head for steels alloyed with boron and cobalt. The chrome content of the materials will be increased to approx. 11 – 11.5 % to be able to warrant a sufficient corrosion resistance also at high temperatures. Other alloying modifications, especially with regard to the increased commodity prices, are currently being discussed. The development of welding fillers matching these steels is accomplished at the same time like the development of the base materials.

It remains a challenging task to warrant an optimal balance between long-term characteristics up to a temperature of 650 °C, sufficient strengths in the weld metal with the shortest possible PWHT and sufficient scaling resistances with further developments. It is also possible to cut back the requirements with regard to the toughness of the all weld metal (e.g. from more than 41 J to 27 J at room temperature) to no longer be forced to accept clearly longer tempering periods of the welded connections which have been common up to now.

13. Literary references

[1] ECCC-Creep Conference „Creep & Fracture n high temperature components – Design & Life Assessment"; 12.-14. September 2005, London

[2] „T92/P92-Book"; V&M Tubes, edition 2000

[3] Verbändevereinbarung FDBR-VGB-VdTÜV „Mischverbindungen an Warmfesten Stählen"; Januar 1995

[4] Code Case of ASME Boiler and Pressure Vessel Code, Code Case 2179-3, October 1999

[5] Hald, J.; „Assessments by the European Creep Collaborative Committee of steel grade 92"; Proc. International Colloquium "High Temperature Materials and Design – Objectives and Achievements of Joint Research Activities"; German Iron and Steel Institute, Düsseldorf 1999, Tagungsband S. 142-146

[6] ECCC data sheet „Steel ASTM Grade P92 (X10CrWMoVNb9-2)" Working Group WG3A, April 2005

[7] Bendick, W., Gabrel, J., Vaillant, J-C.; Vandenberghe, B.: „Development of a new 12 % Cr-Steel for tubes and pipes in power plants with steam temperatures up to 650 °C", 28th MPA-Seminar, Stuttgart 2002

[8] Bendick, W., Hahn, B., Vandenberghe, B.: „Entwicklung eines neuen 12%Cr-Stahles für Rohre in Kraftwerksanlagen mit Dampftemperaturen bis 650 °C"; 26. Vortragsveranstaltung der AG Warmfeste Stähle und AG Hochtemperaturwerkstoffe, Düsseldorf 2003

[9] Bendick, W., Hahn, B., Vandenberghe, B.: „Stand der Entwicklung von Werkstoffen für den Hochtemperaturbereich bis 650 °C" VGB-Werkstofftagung 2004

[10] J. Gabrel, B., B. Vandenberghe, C. Zakine,
"VM12 - a new 12 %Cr steel for application at high temperature in advanced power plants - status of development"; 5th Int. Conf. on Advances in Materials Technology for Fossil Power Plants"; October 3-5, 2007

[11] prEN 10 216-2:2002/prA2:2004 „Seamless steel tubes for pressure purpose – Technical delivery conditions – Part 2: Non alloy and alloy steel tubes with specified elevated temperature properties“; Draft April 2005

[12] TÜV-Rheinland Werkstoffblatt V&M 06-01: Warmfester Stahl X10CrWMoVNb9-2 – Werkstoff-Nr. 1.4901; Ausgabe 12/06

[13] Power Piping: ASME Code for Presssure Piping B31; ASME B31.1-2001

[14] EN 13 480-3 „Metallische industrielle Rohrleitungen – Teil 3: Konstruktion und Berechnung“; August 2002

[15] Code Case of ASME Boiler and Pressure Vessel Code,
Code Case 2199, June 1995

[16] TÜV Rheinland Werkstoffblatt V&M 06-003: Warmfester Stahl 7CrMoVTiB10-1 - Werkstoff-Nr.17378; Ausgabe 12/2003

Table 1: Comparison of the chemical composition of the steels T/P9, T/P91, T/P911, T/P92 and X20CrMoV12-1

	T/P9	T/P91 X10CrMoVNb9-1	T/P92 X10CrWMoVNb9-2	X20CrMoV11-1
C	max. 0,15	0,08 – 0,12	0,07 – 0,13	0,17 – 0,23
Si	0,25 - 1,00	0,20 – 0.50	max. 0.50	0,15 – 0,50
Mn	0,30 - 0,60	0,30 – 0,60	0,30 – 0,60	max. 1,00
P		max. 0,020	max. 0,020	max. 0,025
S		max. 0,010	max. 0,010	max. 0,020
Al		max. 0,020	max. 0,020	max. 0,040
Cr	8,00 - 10,00	8,00 – 9,50	8,50 – 9,50	10,0 – 12,50
Ni	-	max. 0,040	max. 0,040	0,30 – 0,80
Mo	0,09 - 0,1,10	0,85 – 1,05	0,30 – 0,60	0,80 – 1,20
W	-	-	1,50 – 2,00	-
V	-	0,18 – 0.25	0,15 – 0,25	0,25 – 0,35
Nb	-	0,06 –0,10	0,04 –0,09	-
B	-	-	0.0010 – 0.0060	-
N	-	0.030 – 0.070	0.030 – 0.070	-

Table 2: PWHT parameters of the 9 % Cr steels

Steel grade	Normalizing temperature °C	Tempering temperature °C
X20CrMoV11-1	1020 to 1070	730 to 780
T/P91 (X10CrMoVNb9-1)	1040 to 1080	750 to 780
T/P92 (X10CrWMoVNb9-2)	1040 to 1080	750 to 780

Table 3: Chemical composition of the steel VM12-SHC in comparison to the other 9-12 % Cr steels

	X20CrMoV11-1	T/P91 X10CrMoVNb9-1	T/P92 X10CrWMoVNb9-2	VM12-SHC
C	0,17 – 0,23	0,08 – 0,12	0,07 – 0,13	0,10 – 0,14
Si	0,15 – 0,50	0,20 – 0.50	max. 0.50	0,40 – 0,60
Mn	max. 1,00	0,30 – 0,60	0,30 – 0,60	0,15 – 0,45
P	max. 0,025	max. 0,020	max. 0,020	max. 0,020
S	max. 0,020	max. 0,010	max. 0,010	max. 0,010
Al	max. 0,040	max. 0,020	max. 0,020	max. 0,020
Cr	10,0 – 12,50	8,00 – 9,50	8,50 – 9,50	11,0 – 12,0
Ni	0,30 – 0,80	max. 0,040	max. 0,040	0,10 – 0,40
Mo	0,80 – 1,20	0,85 – 1,05	0,30 – 0,60	0,20 – 0,40
W	-	-	1,50 – 2,00	1,30 – 1,70
V	0,25 – 0,35	0,18 – 0.25	0,15 – 0,25	0,20 – 0,30
Nb	-	0,06 –0,10	0,04 –0,09	0,03 – 0,08
B	-	-	0.0010 – 0.0060	0,0030 – 0,0060
N	-	0.030 – 0.070	0.030 – 0.070	0,030 – 0,070
Co	-	-	-	1,40 – 1,80

Table 4: Mechanical properties of the steel VM12-SHC at room temperature

Delivery condition	YS MPa	TS MPa	Elongation A_5 %		Impact energy J
			long	trans	trans
annealed	≥ 450	620 – 850	≥ 19	≥ 19	≥ 27

Table 5: Mechanical properties at higher temperatures

Temperature °C	100	200	250	300	350	400	450	500	550
YS (min) MPa	412	390	383	376	367	356	342	319	287
TS (min) MPa	565	526	510	497	490	479	448	406	338

Table 6: Chemical composition of modern low alloy high-temperature steels and their predecessors

	T/P22	T/P23	T/P24
C	max. 0,15	0,04-0,10	0,05-0,10
Si	max. 0,50	max. 0,50	0,15-0,45
Mn	0,30-0,60	0,10-0,60	0,30-0,70
P	max. 0,025	max. 0,030	max. 0,020
S	max. 0,025	max. 0,010	max. 0,010
Ni	---	---	---
Cr	1,90-2,60	1,90-2,60	2,20-2,60
Mo	0,87-1,13	0,05-0,30	0,90-1,10
W	---	1,45-1,75	---
Ti	---	---	0,05-0,10
V	---	0,20-0,30	0,20-0,30
Nb	---	0,02-0,08	---
Al	---	max. 0,030	max. 0,020
N	---	max. 0,030	max. 0,010
B	---	0,0005-0,0060	0,0015-0,0070

Table 7: Mechanical properties of modern, low-alloy, high temperature steels in comparison to the steels 10CrMo9-10 and X10CrMoVNb9-1

Steel grade	YS MPa	TS MPa	Elongation A_5 %
10CrMo9-10	min. 280	480 - 630	min. 20
T/P23	min. 400	min. 510	min. 20
7CrMoVTiB10-10	min. 450	585 - 840	min. 17
X10CrMoVNb9-1	min. 450	620 - 850	min. 17

Table 8: Wall size calculation for the straight pipe according to EN standard 13 480-3 [14] and ASME B31.1 [13]; design pressure p = 191 bar, design temperature T = 545°C, pipe diameter D_i = 450 mm

Steel grade	Wall thickness e acc. to EN 13480-3 [mm]	Wall thickness t acc. to ASME B31.1 [mm]
P22	105,2	96,2
P23	58,6	55,7
P24	52,3	50,0
P91	41,3	39,8

Table 9: Wall size calculation for a T fitting according to EN standard 13 480-3 [14] and ASME B31.1 [13]; design pressure p = 191 bar, design temperature T = 545°C, basic pipe D_i = 450 mm/connection D_i = 300 mm

Steel grade	Wall thickness e acc. to EN 13480-3 [mm]	Wall thickness t acc. to ASME B31.1 [mm]
P22	155,0 / 103,0	135,0 / 96,0
P23	91,0 / 61,0	91,0 / 61,0
P24	83,0 / 55,0	84,0 / 55,0
P91	67,0 / 45,0	68,0 / 45,0

Table 10. Chemical composition and mechanical properties all weld metal for matching filler metal to T/P23

		C	Si	Mn	P	S	Cr	Mo	Ni	V	W	Nb	B	N	Al
Base metal T/P23	min.	0,04	-	0,10	-	-	1,90	0,05	-	0,20	1,45	0,02	0,0005	-	-
	max.	0,10	0,50	0,60	0,030	0,010	2,60	0,30	-	0,30	1,75	0,08	0,0060	0,030	0,030
Filler metal (GTA, SMA, SA)	min.	0,04	0,15	0,30	-	-	2,00	-	-	0,18	1,3	0,02	-	-	-
	max.	0,10	0,60	0,80	0,020	0,015	2,80	0,40	0,60	0,3	1,8	0,10	0,007	0,050	0,020

Weld process	Temp. °C	PWHT °C/h	YS MPa	TS MPa	Elongation %	CVN J	Hardness HV10
Base metal requirements		730 – 760	≥ 450	585 – 840	≥ 17	≥ 27	≤ 250
GTAW Ø 2,4 mm	+20	---	639	818	21,4	228 230 268	270
	+20	740/2	520	620	20,2	261 286 299	250
	+550	740/2	426	449	17,4	---	---
SMAW Ø 4,0 mm	+20	740/2	509	625	19	128 136 140	227
	+20	740/15	421	553	25	156 156 160	192
	+550	740/15	302	350	26,4	---	---
SAW wire-Ø 4,0 mm flux: UV 430 TTR-W	+20	740/2	615	702	18,1	187 204 208	237

Table 11. Results of a P23 tube round seam welding with matching welding fillers

Pipe dimension: Ø 219 mm x 30 mm Filler metal: Root GTAW 2 Filler Pass SMAW 10 Filler Pass SAW; wire-Ø 3,2 mm Preheating temperature: 250 °C; Interpass temperature: max. 300 °C. SAW: Amperage = 450 A (=/+); Voltage = 28 V; Welding speed = 52 cm/min.; E = 14,5 kJ/cm							
PWHT °C/h	**Test temp. °C**	**TS (MPa)**	**Location of fracture**	**Bending angle**	**A_v (ISO-V) J**		**Hardness HV10**
					WM	HAZ	
740/1	+20	580	BM	180°	124 150 153	92 156 223	< 250
740/1	+600	333	BM	---	---	---	---

Table 12. Chemical composition and mechanical properties all weld metal for matching filler metal to T/P24

		C	Si	Mn	P	S	Cr	Mo	Ni	V	Ti	Nb	B	N	Al
Base metal	min.	0,05	0,15	0,30	-	-	2,20	0,90	-	0,20	0,05	-	0,0015	-	-
T/P24	max.	0,10	0,45	0,70	0,020	0,010	2,60	1,10	-	0,30	0,10	-	0,0070	0,010	0,020
Filler metal	min.	0,04	0,15	0,30	-	-	2,10	0,90	-	0,20	0,03		0,001	-	-
(GTA, SMA, SA)	max.	0,10	0,50	0,80	0,020	0,015	2,70	1,20	0,40	0,30	0,10		0,007	0,050	0,020

Weld process	**Temp. °C**	**PWHT °C/h**	**YS MPa**	**TS MPa**	**Elongation %**	**CVN J**	**Hardness HV10**
Base metal requirements		730 – 760	≥ 450	585 – 840	≥ 17	≥ 27	≤ 250
GTAW	+20	---	664	803	19,2	> 250	≤ 322
Ø 2,4 mm	+20	740/2	595	699	20,3	> 250	≤ 230
SMAW Ø 4,0 mm	+20	740/2	577	689	18,1	150	≤ 233
SAW wire-Ø 4,0 mm flux: UV 430 TTR-W	+20	740/2	495	600	23,8	260	≤ 206

Table 13. Mechanical properties of girth welds at T/P24 (transverse specimen)

P24 (Pipe dimension: Ø 159 x 20 mm)						
Weld process	**PWHT °C/h**	**TS MPa**	**LoF**	**CVN (ISO-V) J**	**Hardness HV10**	**Bending angle**
Base metal	730 – 760	585 – 840	---	> 27	175 – 260	---
GTA, SMA, SA*	740/2	597	BM	200 – 290	< 250	180° o.k.
*) Root: GTAW; 2 Fill passes with stick electrodes, 7 Fill passes SAW						
T24 (Tube dimension: Ø 38 x 6,3 mm)						
Weld process	**PWHT °C/h**	**TS MPa**	**LoF**	**CVN (ISO-V) J**	**Hardness HV10**	**Bending angle**
Base metal	730 – 760	585 – 840	---	> 27	175 – 260	---
GTAW	---	639	BM	60 – 76	≤ 336	180° o.k.

Table 14. Requirements to chemical composition and mechanical properties of matching filler metals for T/P92

a) Chemical composition (weight-%)								
	C	**Cr**	**Mo**	**V**	**W**	**Nb**	**N**	**Mn+Ni**
min.	0,08	8,5	0,3	0,18	1,5	0,04	0,04	-
max.	0,13	9,5	0,6	0,25	2,5	0,08	0,08	1,5
A_{c1b}-temp.: > 760 °C								

b) Mechanical properties				
Test temperature °C	**YS MPa**	**TS MPa**	**Elongation A_5 %**	**CVN (ISO-V) J**
+20	$\geq$ 450	620 – 850	$\geq$ 17	$\geq$ 41
+600	231	267	---	---
Creep rupture strength at 10^5h and 600 °C: 108 MPa				

Table 15. Mechanical properties of SMA-weld metal Thermanit MTS 616 and a girth weld at P92-pipe

El-Ø 4,0 mm, 120 – 140 Amps., T_p 250 °C, T_i 270 °C

Chemical composition (weight-%); all weld metal

C	Si	Mn	P	S	Cr	Ni	Mo	V	W	Nb	N
0,11	0,27	0,65	0,018	0,008	8,95	0,7	0,53	0,19	1,72	0,044	0,045

Mechanical properties, a) all weld metal

PWHT °C/h	Test temperature °C	TS MPa	Elongation A5 %	CVN (ISO-V) J
760/2	+20	800	17,6	54
760/2	+600	585	12	---

b) girth weld; pipe P92, pipe-Ø 336 mm, wall thickness: 40 mm

PWHT °C/h	Test temperature °C	TS MPa	Location of fracture	CVN (ISO-V) J	Hardness HV10
760/2	+20	665	BM	60	236 – 262
760/2	+600	349	BM	---	---

Table 16. Mechanical properties of SA-weld metal Thermanit MTS 616 / Marathon 543 and a girth weld at P92-pipe

Wire-Ø 3,2 mm, 380 – 420 Amps., 28 – 30 Volts, v: 5 m/min., T_p 250 °C, T_i 300 °C									
Chemical composition (weight-%); all weld metal									
C	**Si**	**Mn**	**Cr**	**Ni**	**Mo**	**V**	**W**	**Nb**	**N**
0,09	0,36	0,60	8,45	0,73	0,41	0,17	1,59	0,043	0,059

Mechanical properties, a) all weld metal

PWHT °C/h	**Test temperature °C**	**YS MPa**	**TS MPa**	**Elongation A5 %**	**CVN (ISO-V) J**
760/2	+20	678	789	19,8	35
760/4	+20	621	742	20,8	53

b) girth weld; pipe P92, pipe-Ø 300 mm, wall thickness: 40 mm

PWHT °C/h	**Test temp. °C**	**TS MPa**	**Elong. A_5 %**	**Location of fracture**	**CVN (ISO-V) J**	**Hardness HV10**
760/2	+20	674	16,5	BM	92	238 – 250
760/4	+20	678	15	BM	89	234 – 249
760/4	+600	355	18	BM	---	---

Table 17. Chemical composition and mechanical properties of matching filler metals for VM12-SHC (all weld metal)

		C	Si	Mn	P	S	Cr	Mo	Ni	W	V	Nb	B	Co	N	Al
Base metal VM12-SHC	min.	0,10	0,40	0,15	-	-	11,0	0,20	0,10	1,30	0,20	0,03	0,0030	1,40	0,030	-
	max.	0,14	0,50	0,45	0,020	0,010	12,0	0,40	0,40	1,70	0,30	0,08	0,0060	1,80	0,070	0,020
Filler metal (GTA, SMA)	min.	0,08	-	0,20	-	-	10,5	0,20	0,20	1,40	0,15	0,03	0,0010	1,3	0,02	-
	max.	0,17	0,60	0,80	0,020	0,015	12,3	0,40	0,80	2,00	0,30	0,09	0,0060	2,0	0,07	0,020

Weld process	**Temp. °C**	**PWHT °C/h**	**YS MPa**	**TS MPa**	**Elongation %**	**CVN (ISO-V) J**	**Hardness HV10**
Base metal requirements		760 – 800	≥ 450	620 – 850	≥ 17	27 / 40	≤ 260
GTAW: Ø 2,4 mm	+20	770/2	684	822	18,5	44	< 297
SMAW: Ø 4,0 mm	+20	770/2	689	832	17,2	44	< 281

Table 18. Girth welding of VM12-SHC; Mechanical properties on transverse specimen

Weld process	**PWHT °C/h**	**TS MPa**	**LoF**	**CVN (ISO-V) J**	**Hardness HV10**
Base metal	740 – 780	620 – 850	---	27 / 40	---
GTAW	770/0,5	745	BM	37 – 73	≤ 351
SMAW	770/2	728	BM	31 – 51	≤ 322

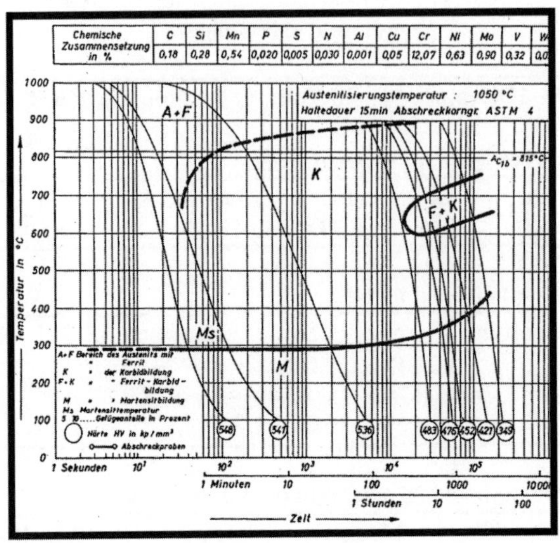

Figure 1: CCT diagram of the steel X20CrMoV11-1

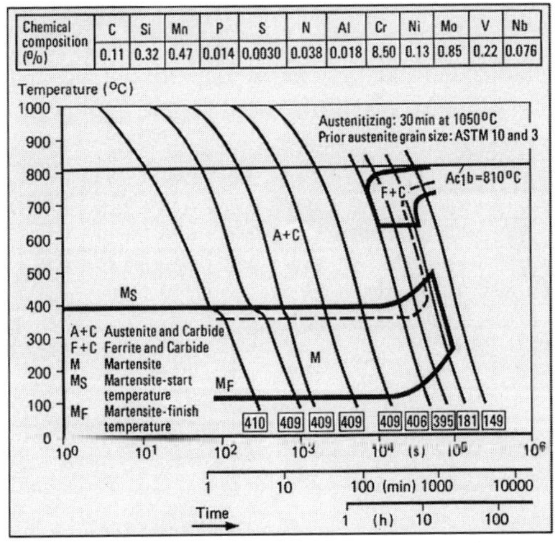

Figure 2: CCT diagram of the steel T/P91 (X10CrMoVNb9-1)

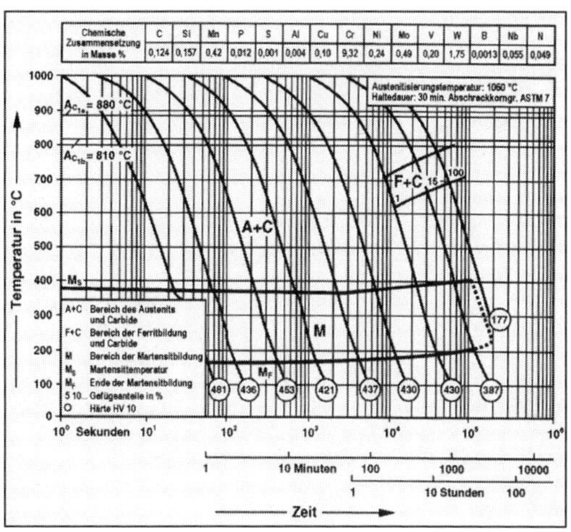

Chemische Zusammensetzung in Masse %	C	Si	Mn	P	S	Al	Cu	Cr	Ni	Mo	V	W	B	Nb	N
	0,124	0,157	0,42	0,012	0,001	0,004	0,10	9,32	0,24	0,49	0,20	1,75	0,0013	0,055	0,049

Figure 3: CCT diagram of the steel T/P92 (X10CrWMoVNb9-2)

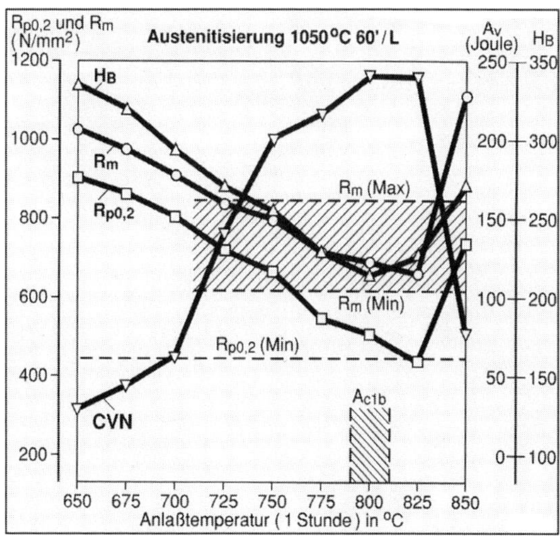

Figure 4: Tempering behavior of the steel T/P91

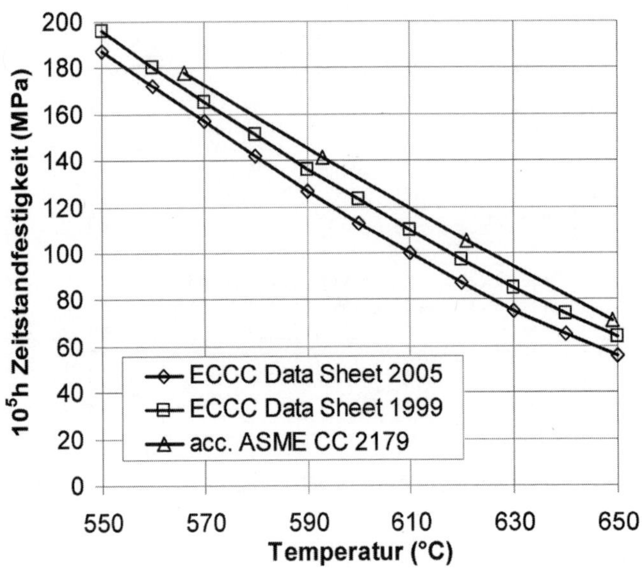

Figure 5: Comparison of the results of the ECCC creep strength evaluations of 1999 and 2005 with the values specified in the ASME-Code Case 2179 for the steel T/P92.

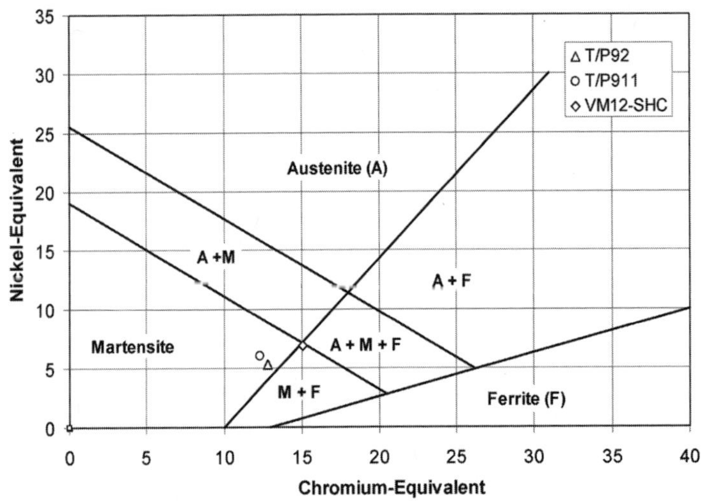

Figure 6: Schaeffler diagram modified according to H. Schneider

$Cr_{Äq} = Cr + 2Si + 1.5\ Mo + 5\ V + 5.5\ Al + 1.75\ Nb + 1.5\ Ti + 0.75\ W$

$Ni_{Äq} = Ni + Co + 0.5\ Mn + 30\ C + 0.3\ Cu + 25\ N$

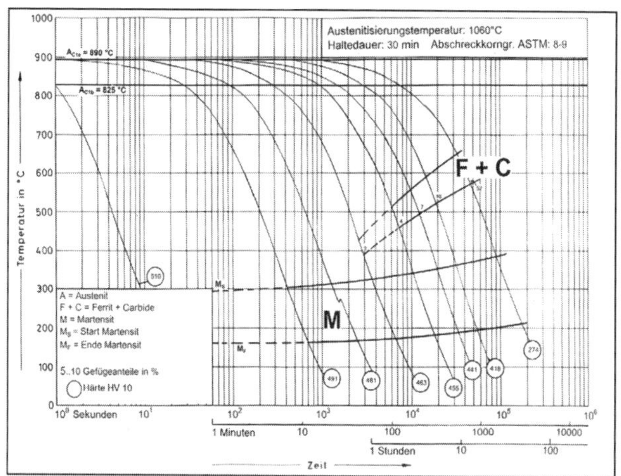

Figure 7: Time-temperature transformation diagram of the steel VM12-SHC

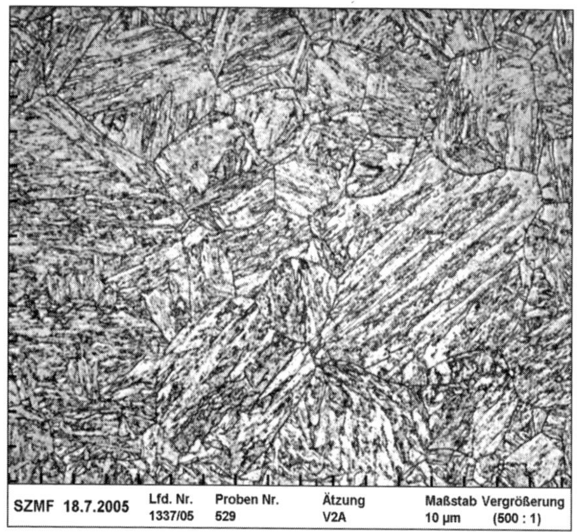

Figure 8: Microstructure of the steel VM12 after the quenching and tempering treatment (normalization 1060°C/20 min., air cooling; tempering treatment 790°C/1 h, air cooling)

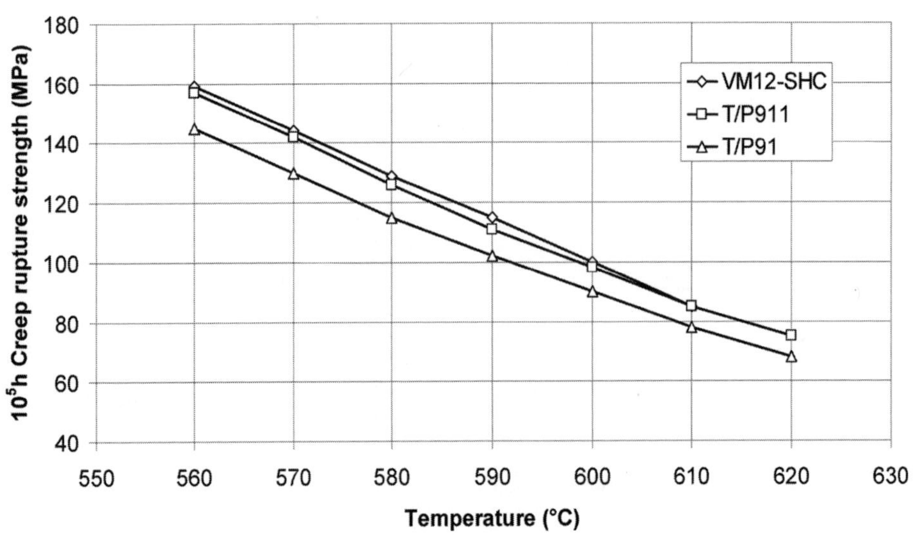

Figure 9: Creep strength of the steel VM12-SHC on the basis of a Larson-Miller evaluation

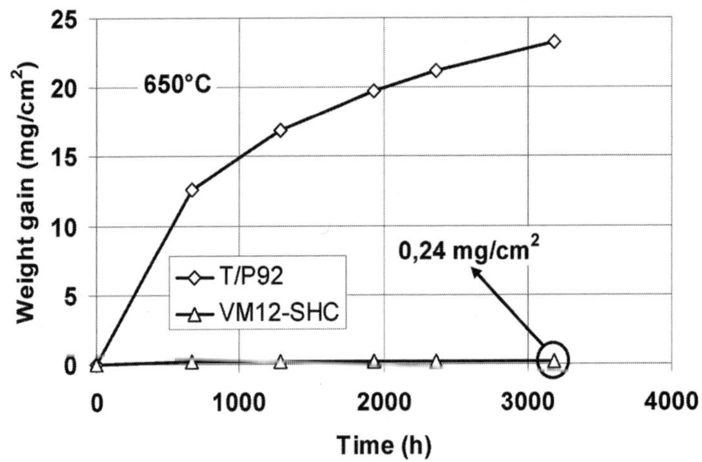

Figure 10: Oxidation behavior of the steel VM12-SHC at a temperature of 650°C in a water steam atmosphere in comparison to the steel T/P92 (source: ALSTOM)

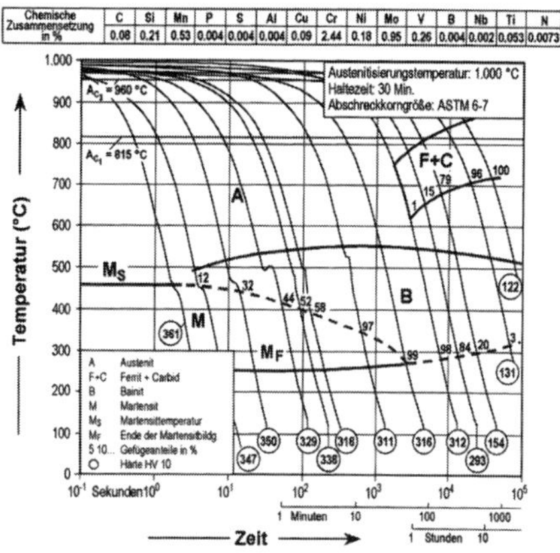

Figure 11: CCT diagram T/P24

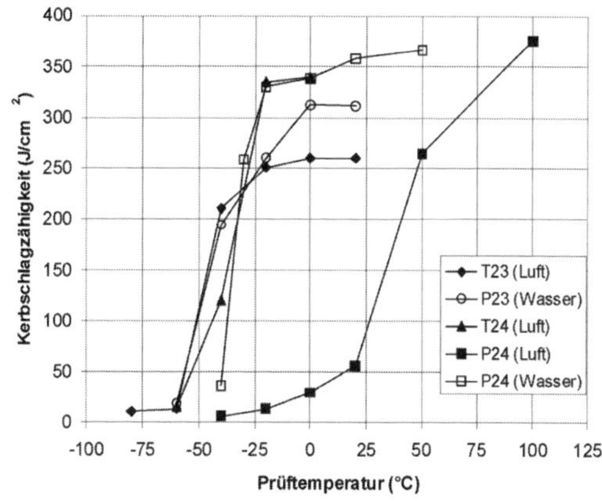

Figure 12: A_V-T curves of T23 (L), P23 (L), P23 (W), T24 (L) and P24 (W)

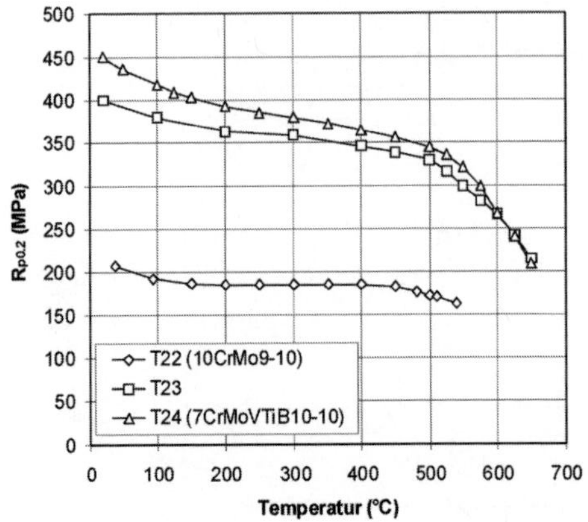

Figure 13: Minimum values of the 0.2 % proof strength depending on the temperature for T22, T23 and T24

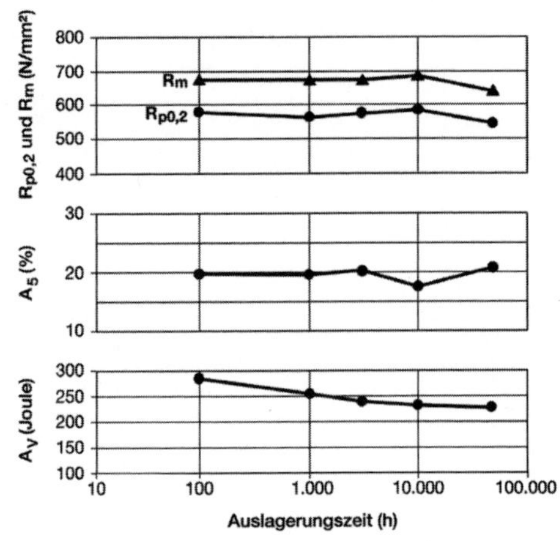

Figure 14: Mechanical parameters of the steel T/P24 after a long-term aging high temperature creep tests at a temperature of 550°C

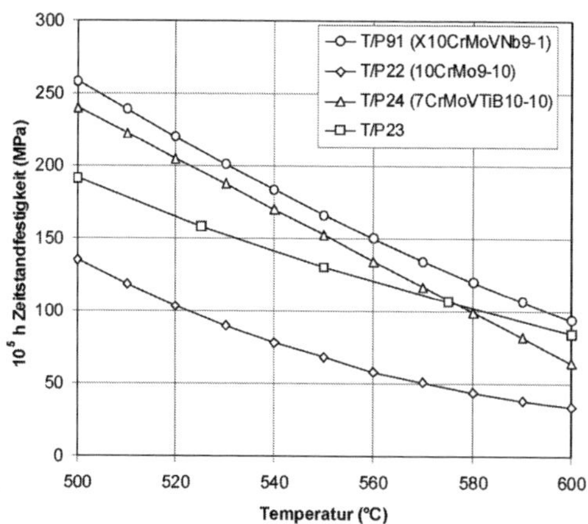

Figure 15: 10^5 h Creep strength values depending on the temperature for the steels T/P22, T/P23, T/P24 and T/P91

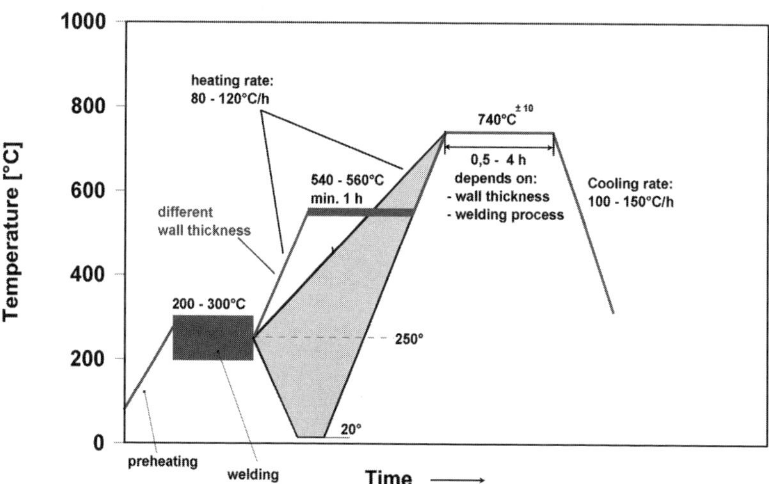

Figure 16. Schematic diagram of the temperature cycle during the welding and PWHT of T/P23

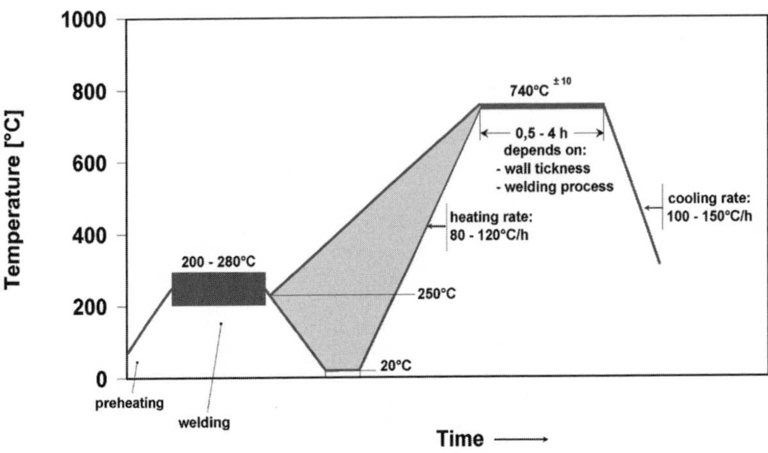

Figure 17. Schematic diagram of the temperature cycle during the welding and PWHT of T/P24

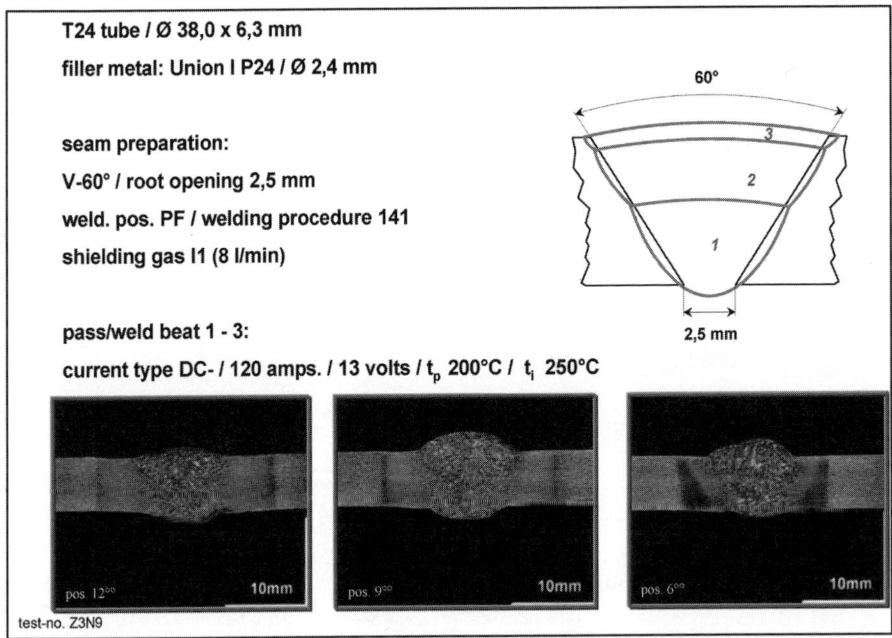

Figure 18: Optimization of welding technique with respect to high CVN-values

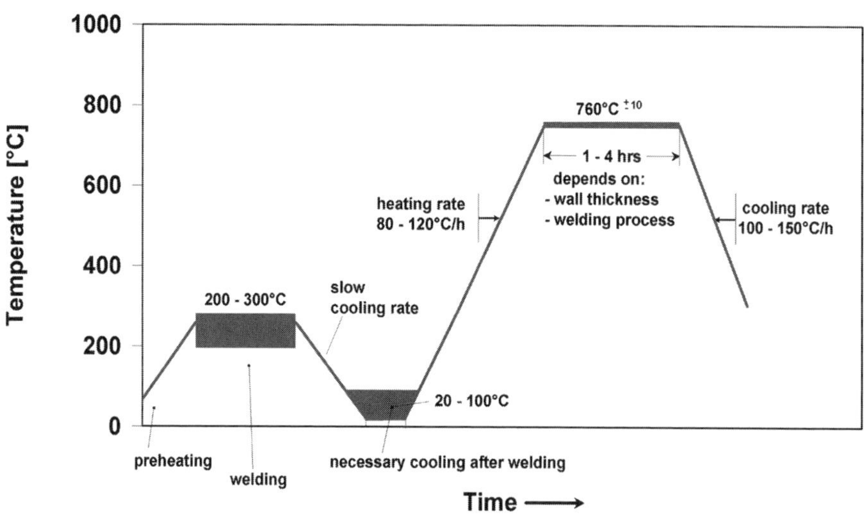

Figure 19. Schematic diagram of the temperature cycle when welding P92

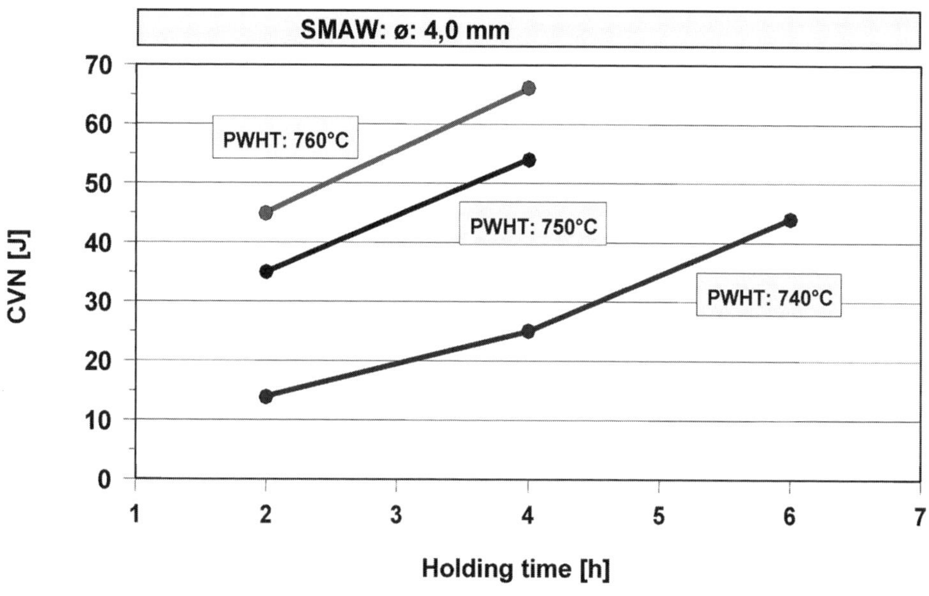

Figure 20. Influence of the temperature on the toughness of the weld metal; all weld metal matching P92

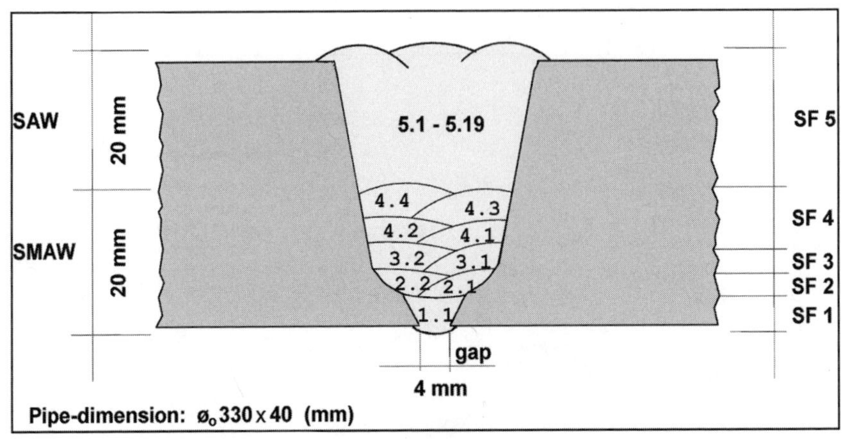

Figure 21. Base metal (pipe): P92

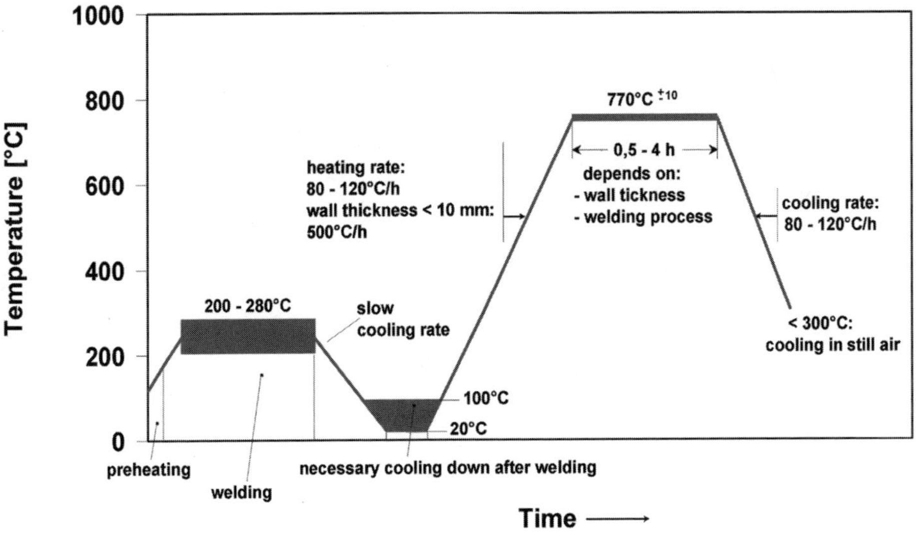

Figure 22. Schematically temperature cycle during welding of 12 % Cr-steel VM12-SHC

Advances in Materials Technology for Fossil Power Plants
Proceedings from the Fifth International Conference
R. Viswanathan, D. Gandy, K. Coleman, editors, p 863-873

DOI: 10.1361/cp2007epri0863

Welding of Dissimilar Joints of New Power Plant Steels

H. Heuser, C. Jochum
Böhler Thyssen Schweisstechnik Deutschland GmbH
Unionstrasse 1
59067 Hamm, Germany

W. Bendick
Salzgitter Mannesmann Forschung
Ehinger Strasse 200
47259 Duisburg, Germany

B. Hahn
Vallourec & Mannesmann Deutschland GmbH
Rather Kreuzweg 106
40472 Düsseldorf, Germany

R. Fuchs
Böhler Thyssen Welding USA Inc.
10401, Greenbough Drive
Stafford, Texas 77477

In conventionally fired power plants, appropriate materials are required which correspond to the different temperature and oxidizing conditions in the boiler and in the superheater sections. Pipe steels with 2 ¼ Cr (P22; T23; T24) must be welded to 9 - 12 % Cr steels (P91; E911; P92; VM12). In this area, the choice of the appropriate welding filler material is vital for the quality of the weldment. This report highlights the possibilities for achieving optimal properties in differing dissimilar metal welds under conditions of reduced carbon diffusion.

1. Introduction

The increase of the steam temperature and pressure in fossil fuel fired power plants and the related improvements to levels of efficiency only can be realized by the use of new materials. The choice of the materials depends on the steam parameters in the particular sections. Alloying concepts with respect to the materials in the steam turbines, including the turbine housing, the steam generator (boiler) and the superheater must be reconsidered. New materials are available for all sections which allow both higher temperature and increased pressure. In many cases, conventional materials such as

16Mo3, 10CrMo9-10 or 13CrMo4-5 are no longer satisfying the demanding stress levels. For this reason, new materials are now being considered in the construction of power plants. Table 1 shows the bainitic and martensitic steels that are being used in the new generation of power plants.

During the fabrication and construction of the different components, dissimilar metal welds will occur. Due to the differing Cr-contents of the steels, carburization or decarburization will be present. Much has been reported on this subject [1 - 3]. However, there is very little research data available about dissimilar metal welds between the new steels.

The aim of this report is to take a specific look at a few of the more common dissimilar metal welds in these new steels, providing insight into the properties of the weldments.

2. Presentation of problem

The problems concerning dissimilar metal welds between materials with distinctly differing Cr contents primarily exist in the form of carbon diffusion. The carbon diffuses to the material with the higher chrome content during post weld heat treatment. As a result, a carbon depleted zone develops in the material with the lower chrome content, and a zone with enriched carbon, the so-called carbide seam, develops in the material with greater amount of chrome. The severity of these zones depends on the time at temperature. Unless the welding is carried out with nickel-based welding filler metals, this is generally unavoidable.

Figure 1 shows a schematic representation of the carbon diffusion in a joint between 10CrMo9-10 (P22) to X20CrMoV11-1, using different welding fillers metals.

The structural changes influence the material behavior (toughness and strength) of the weldment in the area of the de-carburized and carburized zones.

On dissimilar metal welds between X20CrMoV11-1 and 10CrMo9-10, welded with either of the two matching materials, the toughness requirements in the C-depletion zone and the carbide seam are often not easily achieved (wide scattering of the individual values) [1,2]. Additional analyses show that fracture initiation and propagation are limited in the notched bar impact test to the softer de-carburized zone [3]. However, there is also no cause for concern in operation as sufficiently high toughness levels occur at these temperatures. Creep damage in these dissimilar metal welds has not yet been recognized in service, as demonstrated by weldments with far in excess of 100,000 h in operation [3]. Even using creep tests with grooves in the de-carburized zone, it was not possible to observe any premature break [2].

In the case of weldments between martensitic steels such as X20CrMoV11-1 and X10CrMoVNb9-1 (P91) for example, , it is to be concluded that carbon diffusion does not occur at all, or only to a very negligible degree, irrespective of the filler material chosen, due to the small differences between both materials in terms of Cr-content.

In contrast, with dissimilar metal welds between 10CrMo9-10 and X10CrMoVNb9-1, the carburization and/or partial de-carburization are considerably more prevalent, irrespective of the welding filler used. The weak point of the connection is either in the partially de-carburized HAZ of the 10CrMo9-10 (welded with matching filler metal to

P91) or in the de-carburized weld metal (welded with matching filler metal to 10CrMo9-10, see Figure 2). On the basis of extensive tests it has been proven that during a notched bar bending test, with these dissimilar metal welds, the initiation of fissures generally occurs in the C-depleted and hence softer zones [3].

3. The task

In the previous examples, reducing the rate of C-diffusion in the material 10CrMo9-10 (P22) was not possible due to a lack of special carbide forming elements such as Nb, V or Ti. With the newer bainitic materials, T/P23 and T/P24, such elements are important alloy elements which significantly improve the degree of creep strength. It is to be expected that these carbide forming elements also have a beneficial effect with regard to C-depleted zones on dissimilar metal welds with the martensitic materials that have a higher Cr content, T/P91, E911, T/P92 and VM12. This expectation should be substantiated through corresponding experiments.

In order to evaluate this effect, different dissimilar metal welds were welded and tested.

4. Test Procedure

The following material combinations were tested:

T23 / T91 tubes
T24 / T91 tubes
P23 / P92 pipes

It should be noted that for the T24 / T91 combination, material was not available with approximately the same diameters.

Table 1 shows the analyses and mechanical properties of the base materials. Matching welding filler metals to both pipe materials were used. Table 2 contains the analyses and mechanical properties of the welding filler metals used.

The thin-wall boiler pipe welds were made using GTAW. With the heavy-wall weldment P23 / P 92, the root was welded using GTAW and additional fill passes were welded using SMAW. The joint preparations and the welding parameters can be seen in Figures 3 and 4.

For the tube welds, the preheat and interpass temperature was 150°C. The heavy-wall P23 / P92 welds were preheated to 200°C, with an interpass temperature of 270 °C maximum. All welds were post weld heat treated.

5. Test results

5.1 Mechanical properties

The mechanical properties of the welded joints were tested in the as welded condition as well as after post weld heat treatment in order to determine whether the location of the failure varies, depending on the thermal treatment of the weldment.

The strength of the weldments were determined using flat bar cross weld tension specimens, tested at room temperature and at 500°C. All failures occurred in the lower alloy base material, well above the minimum specified tensile strength of the T23, T24 and P23.

The highest levels of toughness were achieved using the matching weld filler metals to the lower alloyed base material.

Tables 3 - 5 contain the mechanical properties determined. Figures 5 and 6 clearly indicate that the matching weld metal to P92 and to P23 demonstrates a higher level of strength than the base material P23.It can be concluded from this that no significant de-carburization is present that influences the strength of the weldments.

This was substantiated by metallographic testing.

5.2 Metallographic tests, hardness and alloy distribution

Metallographic examination of the fusion zones of the weldments was done to determine the extent of de-carburization and carburization.

Figures 7 – 10 show the results of the metallographic examinations. The hardness of the weld metal and HAZ does not fall below the hardness of the base material in any case. The carbide forming elements in the matching weld metals to T/P23, T24 and P92 prevent a strong C-diffusion, as is present in weld metal on 10CrMo9-10 (Figure 2). Electron beam microprobe tests of the alloy distribution confirm these this. In this way, it is possible that the creep strengths of such dissimilar metal welds are comparable with weldments made using matching base materials to the lower alloy component. Tests that are now in progress should confirm this.

6. Summary

The characteristics of the dissimilar metal welds of the new power plant steels T23 / T91; T24 / T91 and P23 / P92 were tested.

In comparison with the previous dissimilar metal welds P22 / P91 the special carbide forming elements V, Nb and Ti either prevent or reduce the degree of C-diffusion, irrespective of whether the matching weld metals are chosen for the low alloyed steels or for the higher alloyed materials. This should have a positive effect on the creep strength of dissimilar metal welds. Corresponding tests have been started to verify this.

With the weld metals matching T/P23 and T/P24, welding fillers are available which should also lead to a reduction of the C-diffusion with dissimilar metal welds in steels in which one member does not contain any special carbide forming elements, which are advantageous for the P22 / P91 dissimilar metal welds, for example.

Additional tests with the new, low alloyed welding filler metals presented here should support their application advantage when compared to the customary welding solutions with dissimilar metal welds.

The product trade names of the weld filler materials used in this investigation can be found in Table 6.

7. Literary References

[1] Düren, C.; Jahn, E.; Langhardt, W.; Schleimer, W.: Kerbschlagbiegeversuche an Schweißverbindungen zwischen den Stählen X20CrMoV12-1 und 10CrMo9-10; Stahl und Eisen 102 (1982); *(Notched bar impact tests between the steels X20CrMoV12-1 und 10CrMo9-10; Stahl und Eisen 102 (1982); pages 479 - 483).*

[2] Blind, D.; Kaes, H.; Weber, H.: Eigenschaften von betriebsbeanspruchten Schweißverbindungen zwischen den warmfesten Stählen X20CrMoV12-1 und 10CrMo9-10; VGB-Werkstofftagung „Werkstoffe und Schweißtechnik im Kraftwerk"; Essen (1985). *(The qualities of operationally used weld connections between the heat resistant steels X20CrMoV12-1 und 10CrMo9-10; VGB materials conference: "Materials and welding technology in power plants": Essen, Germany, 1985.)*

[3] Bendick, W.; Niederhoff, K.; Haarmann, K.; Wellnitz, G.; Zschau, M.: 9 %-Chromstahl X10CrMoVNb9-1 - Ein Rohrleitungswerkstoff für Hochtemperaturkraftwerke. *(The 9% chrome steel X10CrMoVNb9-1 - a pipe material for high-temperature power plants.)*

Table 1. New types of bainitic and martensitic materials compared to tested creep resistant materials 10CrMo9-10; X20 and P91

Designation	Chemical composition in weight-% C	Si	Mn	Cr	Ni	Mo	V	W	Nb	Others	Working temp. °C [1)]
Bainitic Steels											
T/P22 (10CrMo9-10)	0,08-0,14	≤ 0,50	0,40-0,80	2,0-2,5	-	0,90-1,10	-	-	-	-	≤ 550
T/P23: ASTM A213, code case 2199-1 ASTM A335 (7CrWVNb9-6)	0,04-0,10	≤ 0,50	0,10-0,60	1,9-2,6	-	0,05-0,30	0,20-0,30	1,45-1,75	0,02-0,08	N ≤ 0,03 B 0,0005-0,0060	≤ 550
T/P24: ASTM A213 (7CrMoVTiB10-10)	0,05-0,10	0,15-0,45	0,30-0,70	2,20-2,60	-	0,90-1,10	0,20-0,30	-	-	N ≤ 0,010 B 0,0015-0,0070 Ti 0,05-0,10	≤ 550
Martensitic Steels (9 - 12 % Cr-Stähle)											
X20CrMoV11-1 1.4922	0,17-0,23	< 0,50	< 1,0	10,0-12,5	0,30-0,80	0,80-1,20	0,25-0,35	-	-	-	≤ 560
T/P91: ASTM A213; A335 (X10CrMoVNb9-1)	0,08-0,12	0,20-0,50	0,30-0,60	8,0-9,5	< 0,40	0,85-1,05	0,18-0,25	-	0,06-0,10	N 0,03-0,07	≤ 585
T/P911: ASTM A213, code case 2327 ASTM A335 (X11CrMoWVNb9-1-1)	0,09-0,13	0,10-0,50	0,30-0,60	8,50-9,50	0,10-0,40	0,90-1,10	0,18-0,25	0,90-1,10	0,06-0,10	N 0,05-0,09	≤ 600
T/P92: ASTM A213, code case 2179-3 ASTM A335 (X10CrWMoVNb9-2)	0,07-0,13	< 0,5	0,30-0,60	8,5-9,5	< 0,40	0,30-0,60	0,15-0,25	1,5-2,0	0,04-0,09	N 0,03-0,07 B 0,001-0,006	≤ 620
VM12-SHC	0,10-0,14	0,40-0,60	0,15-0,45	11,0-12,0	0,10-0,40	0,20-0,40	0,20-0,30	1,30-1,70	0,03-0,08	Co 1,40-1,80 N 0,030-0,070 B 0,0030-0,006	≤ 620

[1)] constructive depicted limit of working temperature in power stations

Mechanical properties at RT

	YS MPa (min)	TS MPa	Elongation % (min)	CVN (ISO-V) J (min)
T/P22	310	480-630	18	40
T/P23	400	≥ 510	20	-
T/P24	450	585-840	17	41
X20 (1.4922)	500	700-850	16	39
T/P91	450	620-850	17	41
T/P911	450	620-850	17	41
T/P92	440	620-850	17	27
VM12-SHC	450	620-850	17	27

Table 2. Composition and mechanical properties of the filler metals used

Chemical composition wire resp. all-weld-metal (weight-%)

Filler metal	C	Si	Mn	Cr	Mo	Ni	Nb	N	V	W	Cu	B	Ti
GTAW; similar to P23 Ø 2,4 mm	0,061	0,45	0,53	2,02	0,03	0,13	0,04	0,01	0,22	1,78	0,10	0,002	0,005
GTAW; similar to P24 Ø 2,4 mm	0,073	0,26	0,45	2,32	0,92	0,09	0,01	0,006	0,25	<,002	0,17	0,002	0,086
SMAW; similar to P23 Ø 3,2 mm	0,057	0,23	0,62	2,20	0,03	0,05	0,04	0,022	0,20	1,59	0,06	0,002	<,001
SMAW; similar to P24 Ø 3,2 mm	0,113	0,35	0,74	8,97	0,56	0,61	0,06	0,038	0,22	1,57	0,03	0,005	0,007

Mechanical properties all-weld-metal; test temp. +20 °C

Filler metal	PWHT [°C/h]	YS [MPa]	TS [MPa]	Elongation [%]	CVN, ISO-V [J]
GTAW; similar to P23 Ø 2,4 mm	740/2	621	708	21,0	256 / 207 / 242
GTAW; similar to P24 Ø 2,4 mm	740/2	595	699	20,5	264 / 286 / 292
SMAW; similar to P23 Ø 3,2 mm	750/2	523	633	20,8	100 / 137 / 144
SMAW; similar to P24 Ø 3,2 mm	750/2	691	810	19,0	54 / 60 / 65

Table 3. Dissimilar T23 / T91; GTA welded

Filler metal: matching T23, Ø 2,4 mm; Tube material T23 (44,5 x 7,6 mm) to T91 (44,5 x 7,14 mm)

PWHT [°C/min]	Test temp. + [°C]	TS [MPa]	Location of fracture	CVN centre WM [J/cm²] at +20 °C	Side bend test
740/30	20	563	BM T23	73/95/158	180° passed
	500	436	BM T23		

Table 4. Dissimilar joint T24 / T91; GTA welded

Filler metal: matching P24 (Ti/B-alloyed), Ø 2,4 mm; Tube material T24 (38,3 x 6,3 mm) to T91 (44,5 x 7,14 mm)

PWHT [°C/min]	Test temp. + [°C]	TS [MPa]	Location of fracture	CVN centre WM [J/cm²] at +20 °C	Side bend test
740/30	20	574	BM T24	135/152/148	180° passed
	500	464	BM T24		

Table 5. Dissimilar joint P23 / P92; SMA-welded; Root pass GTAW

Filler metal: matching P23, Ø 3,2 / 4,0 mm; Pipe dimension: both 219,10 x 20 mm

PWHT [°C/h]	Test temp. + [°C]	TS [MPa]	Location of fracture	CVN [J] at + 20°C	Side bend test
740/2	20	613	BM P23	138/136/132 132/135	180° passed
		598			
	500/550	432/386			

Filler metal: matching P92, Ø 3,2 / 4,0 mm; Pipe dimension: both 219,10 x 20 mm

PWHT [°C/h]	Test temp. + [°C]	TS [MPa]	Location of fracture	CVN [J] at + 20°C	Side bend test
740/4	20	589	BM P23	40/46/44	180° passed
		590			
	500/550	419/385			

Table 6. Matching filler metals

	matching			
	T/P23	T/P24	T/P91	P92
GTAW	Union I P23	Union I P24	Thermanit MTS 3	Thermanit MTS 616
SMAW	Thermanit P23	Thermanit P24	Thermanit MTS 3	Thermanit MTS 616

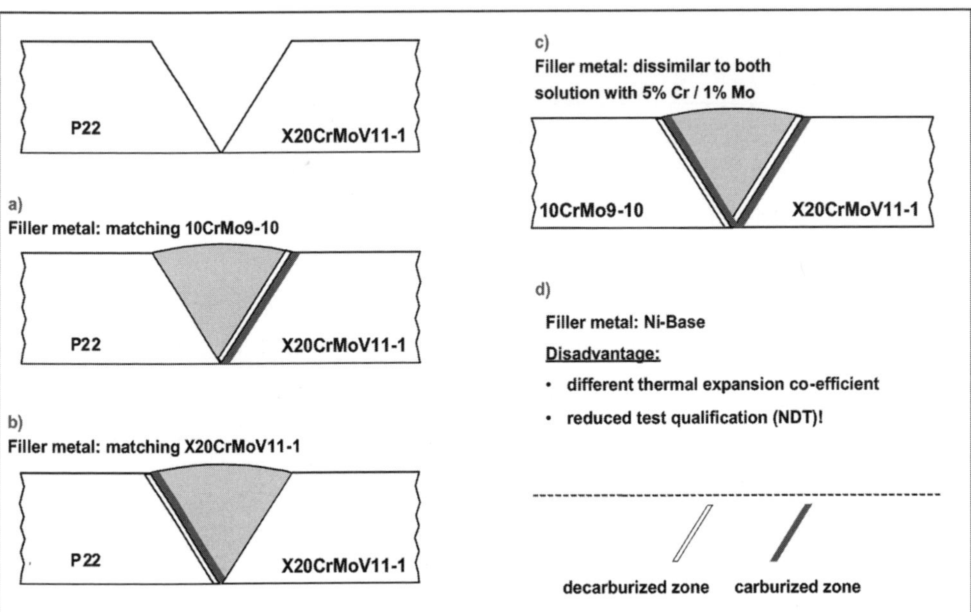

Figure 1. Schematic representation of C-diffusion of the dissimilar joint 10CrMo9-10 - X20CrMoV11-1

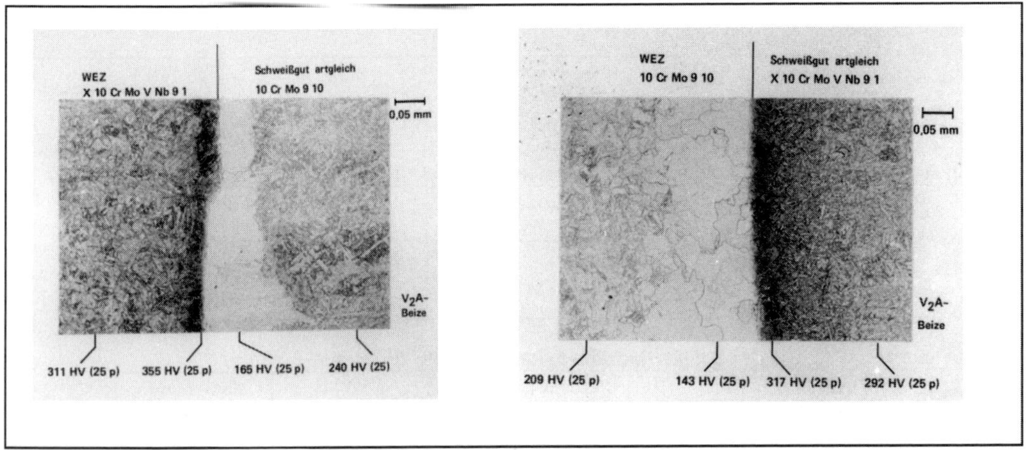

Figure 2. Decarburized zone in dissimilar joints P91 / P22; similar welded to P22 resp. to P91 [3]

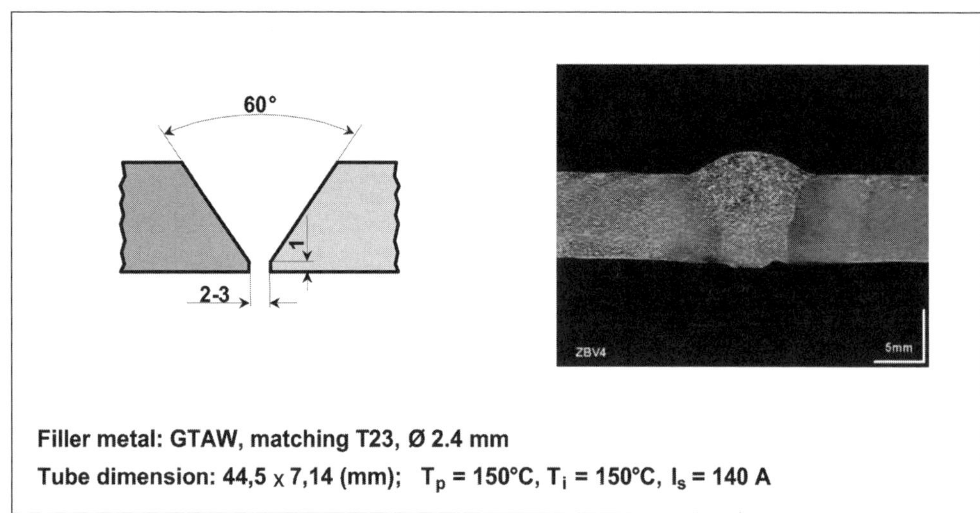

Figure 3. Joint preparation and welded joint T23 / T91

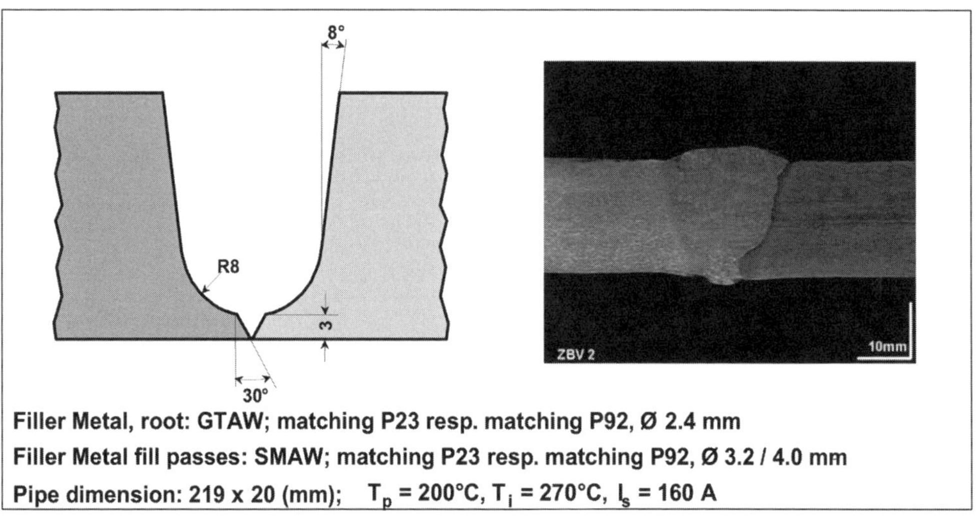

Figure 4. Joint preparation and welded joint P23 / P92

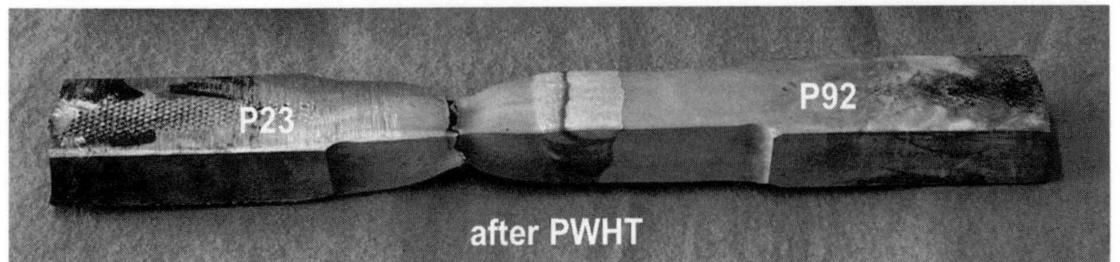

Figure 5. Tensile test specimen of the joint P23 / P92, filler metal matching P23

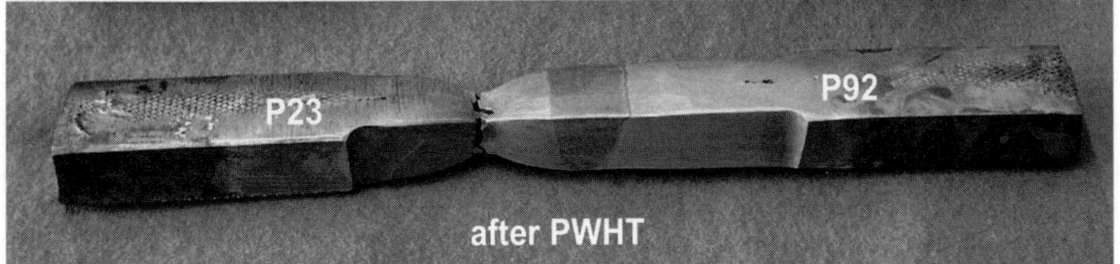

Figure 6. Tensile test specimen of the joint P23 / P92, filler metal matching P92

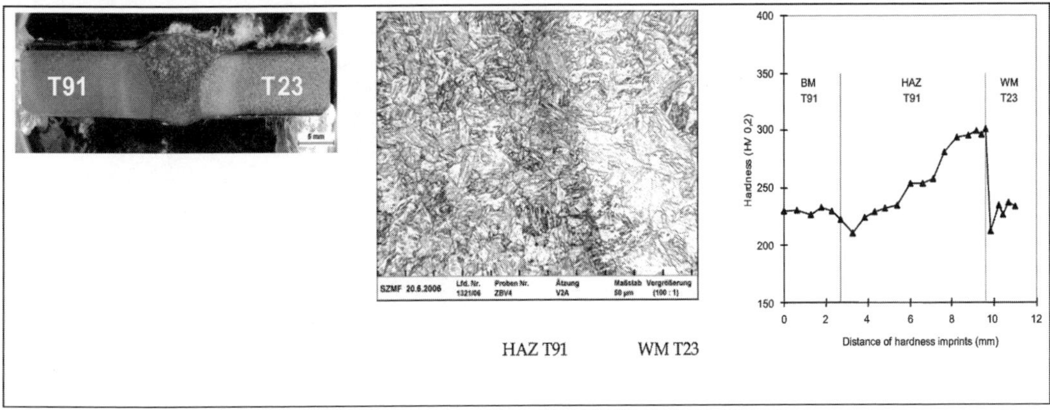

Figure 7. Dissimilar joint T91 / T23; filler metal matching T23 (PWHT 740 °C / 30 min.)

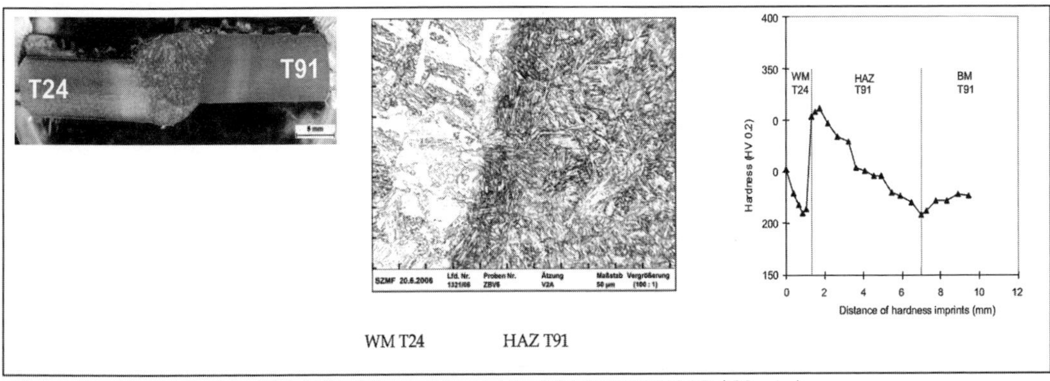

Figure 8. Dissimilar joint T24 / T91; filler metal matching T24 (PWHT 740 °C / 30 min.)

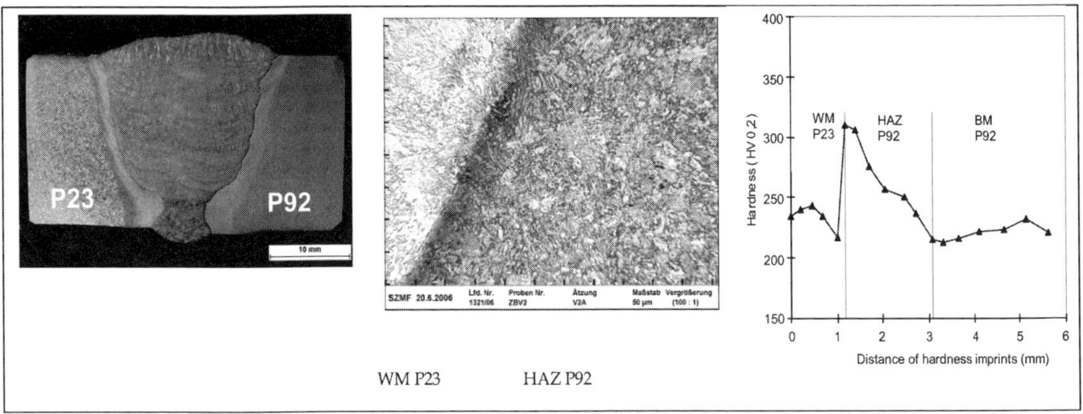

Figure 9. Dissimilar joint P23 / P92; filler metal matching P23 (PWHT 740 °C / 2 h)

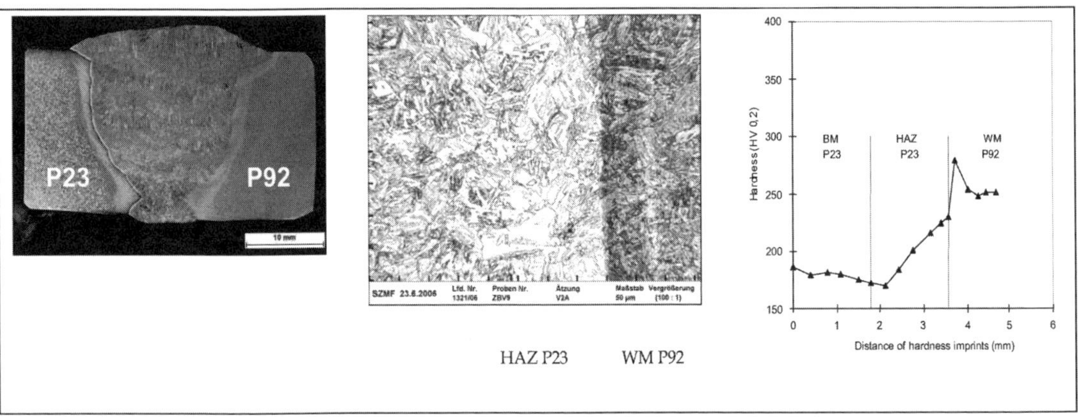

Figure 10. Dissimilar joint P23 / P92; filler metal matching P92 (PWHT 740 °C / 4 h)

Advances in Materials Technology for Fossil Power Plants
Proceedings from the Fifth International Conference
R. Viswanathan, D. Gandy, K. Coleman, editors, p 874-883

DOI: 10.1361/cp2007epri0874

Improvement of Creep Rupture Strength of 9Cr1MoNbV Welded Joints by Post Weld Normalizing and Tempering

T. Sato, K. Tamura
Kure Research Laboratory, Babcock-Hitachi K.K.
5-3 Takaramachi, Kure, Hiroshima, 737-0029 Japan
K. Mitsuhata, R. Ikura
Kure works, Babcock-Hitachi K.K.
6-9 Takaramachi, Kure, Hiroshima, 737-8508 Japan

Abstract

Recent years high strength 9Cr1MoNbV steel developed in USA has been major material in boiler high temperature components with the increase of steam parameters of coal fired thermal power plants. As the microstructure of this steel is tempered martensite, it is known that the softening occurs in HAZ of the weldment. In the creep rupture test of these welded joints the rupture strength is lower than that of the parent metal, and sometimes this reduction of strength is caused by TypeIV cracking. To develop an effective method to improve the rupture strength of welded joint, advanced welding procedure and normalizing-tempering heat treatment after weld was proposed. 9Cr1MoNbV plates with thickness of 40-50mm were welded by 10mm width automatic narrow gap MAG welding procedure using specially modified welding material. After normalizing at 1,050°C and tempering at 780°C, material properties of the welded joints were examined. Microstructure of HAZ was improved as before weld, and rupture strength of the welded joints was equal to that of the parent metal. The long term rupture strength of the welded joints was confirmed in the test exceeded 30,000hours. This welding procedure has been applied to seam weld of hot reheat piping and headers in USC boilers successfully.

Introduction

Ferritic steels containing 9 to 12% chromium have been studied for extending the operating temperature range up to 600°C and higher. For this purpose, Mod.9Cr-1Mo, the high strength 9% chromium steel containing niobium and vanadium, was developed in the U.S. and standardized as T/P91 in ASTM in the early 1980s(1). Mod.9Cr-1Mo steel has now become a major material for the high pressure and temperature components of large capacity boilers in Japan (2). Mod.9Cr-1Mo steel also has been applied to seam-welded pipes with large diameter and relatively thin wall, because of difficulty in manufacturing as hot extruded seamless pipe.

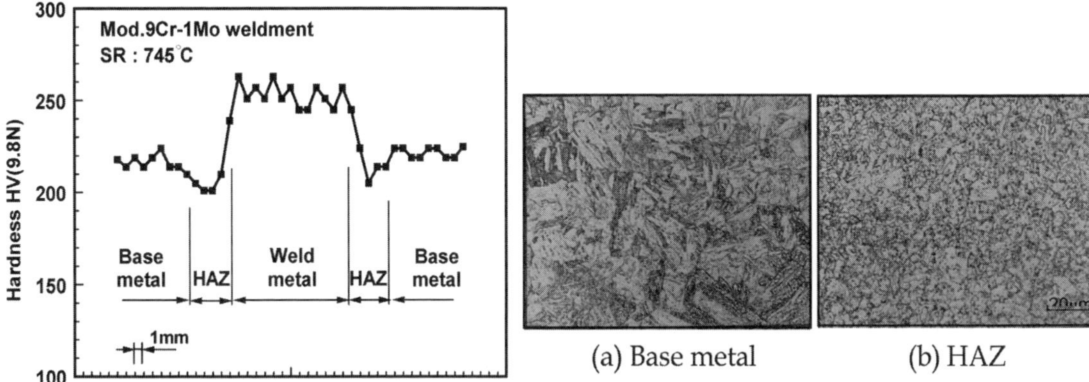

(a) Base metal (b) HAZ

Figure 1. Hardness profile and microstructures of a conventional Mod.9Cr-1Mo weldment heat treated for SR

It is well known, however, that the particular region within a HAZ of a weld joint undergoes softening by weld heat cycles, as this steel is a tempered martensitic material. Figure 1 shows the hardness profile, microstructures of a typical Mod.9Cr-1Mo weldment which was stress relieved (SR) at 745°C. The softening occurs at the fine-grained region heated around the AC_3 temperature during welding. Its main mechanism has been considered to be the coarsening of vanadium nitrides (3).

In the creep rupture test of Mod.9Cr-1Mo weldments given an SR treatment, the obtained rupture strength is lower than that of the base metal, and the rupture occurs in the HAZ in Figure 1. The strength of welded joint was evaluated in the U.S. in past (4) and also discussed in Japan recently (5). It is not a critical problem for circumferential weld of pipes as the stress applied to HAZ perpendicularly is small compared to the hoop stress. But in the case of seam-welded pipes, the strength reduction is an important issue. From the metallurgical standpoint, the HAZ softening can be recovered by an appropriate normalizing and tempering heat treatment after welding. But welding material matched with normalizing and tempering heat treatment has not been developed in past.

In the present study, the effect of normalizing-tempering heat treatment after welding on the properties of weldments was investigated to develop an effective method to improve the rupture strength of martensitic steel weldments based on our new concept.

New Concept of Seam Welding for Mod.9Cr-1Mo Steel

The new concept of a seam welding for Mod.9Cr-1Mo steel is shown in Figure 2. It consists of three important factors. 1) Welding procedure is narrow gap MAG to get high welding quality and high welding efficiency. 2) Welding filler material is newly optimized one to maintain creep rupture strength of weld metal after normalized and tempered. 3) As post weld heat treatment, normalizing and tempering is applied to improve HAZ softening. By the combination of these three factors weld joints with excellent creep rupture strength should achieved. Figure 3 shows the narrow gap welding procedure. Its mechanics is very simple and one pass welding for each layer is possible even for thick walled plate or pipe. It was originally developed by Babcock-Hitachi K.K.(6)(7) and its practicability is proven by experience of application to low alloy steels and Mod.9Cr-1Mo steel for long years.

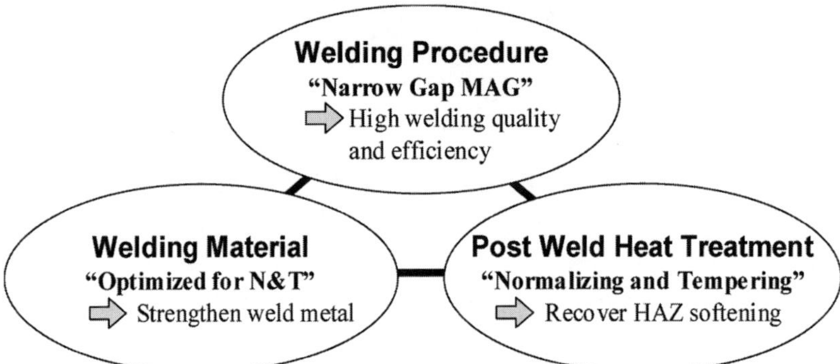

Figure 2. New concept of seam welding for Mod.9Cr-1Mo steel

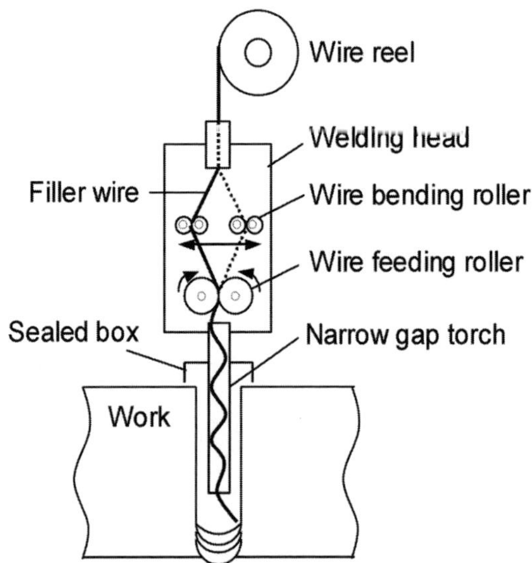

Figure 3. Narrow gap MAG welding procedure

Experimental Details

Materials and Welding

Materials used in this study are Mod.9Cr-1Mo steel plates with thickness of 40 to 52 mm. The chemical compositions of the plates and the filler wire are shown in Table 1. Table 2 lists the welding conditions. These plates were welded by a proprietary narrow gap metal active gas (MAG) welding procedure (7). The filler wire used here was newly produced to satisfy the requirement of creep rupture strength by increase of niobium, nitrogen and reduction of manganese compared with conventional filler wires. This wire was selected by screening test from several candidates. These weldments were then normalized at 1,050°C and tempered at 780°C for 1 to 2.5 hours (hereafter called as NT weldment). In actual fabrication of boiler components constructing of various welded parts, such NT weldments could be subjected to additional heat treatments for stress relieving. Considering this, after the post-weld normalizing and tempering heat treatment, an SR heat treatment was conducted at 745°C for 2 to 5.5 hours for each weldment.

Table 1: Chemical compositions of materials tested in this study

(wt%)

Material	C	Si	Mn	P	S	Ni	Cr	Mo	V	Nb	N
Base metal A (t=40mm)	0.10	0.34	0.45	0.004	0.001	0.06	8.44	0.99	0.19	0.08	0.051
Base metal B (t=52mm)	0.11	0.35	0.45	0.007	0.001	0.10	8.54	0.99	0.20	0.08	0.054
Base metal C (t=44mm)	0.09	0.35	0.44	0.006	0.001	0.14	8.44	0.98	0.20	0.08	0.047
Filler wire (φ=1.2mm)	0.08	0.39	1.28	0.004	0.004	0.45	8.77	0.88	0.17	0.08	0.050

Table 2: Welding condition and post weld heat treatment

	Welding process	Shield gas	Preheating	Number of pass	Arc current	Arc voltage	Welding speed
Face run	Narrow gap MAG	Ar + 20%CO_2	>200°C	10-15	240-260A	26-27V	21-24cm/min
Back run				1-5	240-250A	27-28V	21-22cm/min

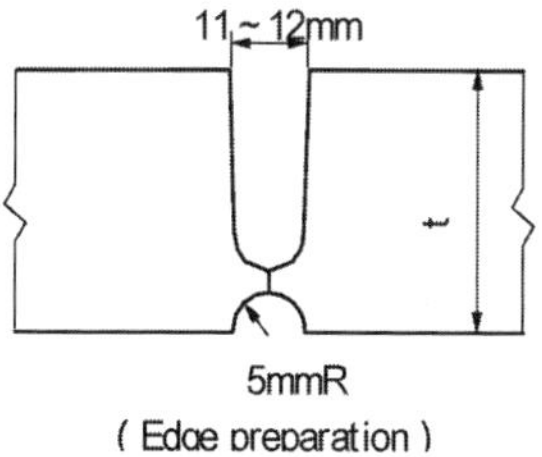

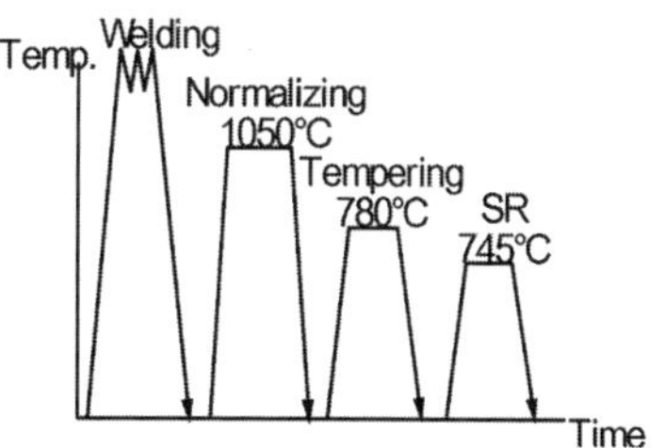

Creep Rupture, Tensile and Impact Tests

Creep rupture specimens were machined to a conventional geometry (6 mm in diameter and 30 mm long gauge) from each weldment transverse to the welding direction, and the fusion line was located at the centre of the gauge length. Creep rupture tests were conducted at 600, 650 and 700°C. Tensile test specimens were of the same type as that of creep rupture specimens. The tensile tests were performed at room temperature, 550, 600 and 650°C. Standard 10x10x55 mm Charpy V notch impact specimens were machined from the weldments. The samples were taken transverse to the welding direction and given a 2 mm deep V notch in the centre of the weld metal. The Charpy impact test was carried out at 0°C.

Results and Discussion

Metallurgical Observation

A typical macrostructure and microstructures of the representative locations of a normalized and tempered (NT) weldment are shown in Figure 4. The microstructure of the area corresponding to the previous HAZ of the as-welded steel is a tempered martensite similar to that of the base metal, and the fine-grained structure such as mentioned in Figure 1 has almost diminished. Figure 5 illustrates the hardness profiles at the mid-cross section of weldments A and B. The hardness of the weld metal of each weldment is close to that of the base metal. It means that the strength and metallurgical discontinuity in the NT weldments are rather mild in comparison with the conventional weldments given only an SR treatment after welding.

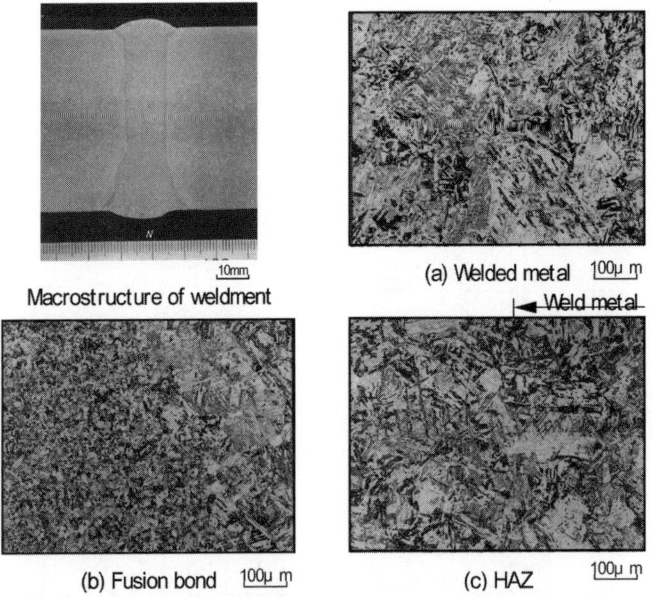

Figure 4. Macro- and microstructures of Mod.9Cr-1Mo NT weldment

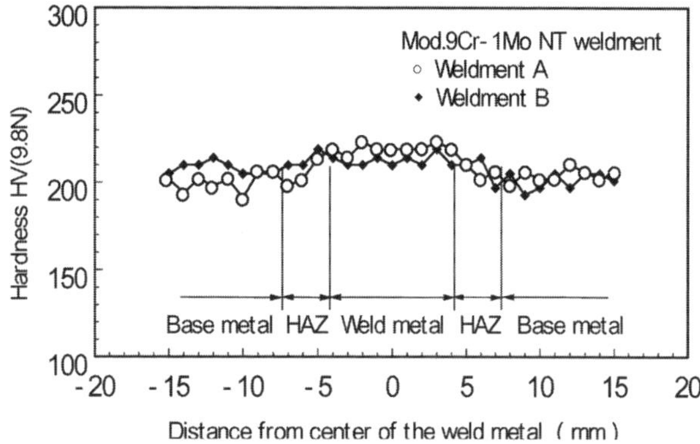

Figure 5. Hardness profiles of Mod.9Cr-1Mo NT weldments

Creep Rupture Strength of Weldments

Figure 6 presents cross sectional views of typical specimens. Some specimens ruptured in the base metal, and others ruptured in the weld metal. No particular tendency in failure location was observed. This result indicates that the strength of the weld metal after the present NT treatment is well balanced with the base metal.

Figure 7 shows the results of creep rupture testing. The solid and dotted lines in this figure are the average and minimum (ave.-1.65SEE) strengths of Mod.9Cr-1Mo steel reported by Brinkman (4), which were recalculated for the present test temperatures using his proposed equation. The rupture strengths of the tested weldments were equal to that of the base metal.

These rupture data are re-plotted in the form of a parameter, logtr-31080/T, in Figure 8. The solid and dotted lines in this figure are the average strengths of the Mod.9Cr-1Mo base metal and weldment calculated by the Brinkman's equation (4). It is estimated that the NT weldment has high rupture strength equal to that of the base metal even at the condition of 650°C/100,000 hours. Considering the applicable temperature range of this steel for boiler components, it seems that the NT weldment has sufficient creep rupture strength.

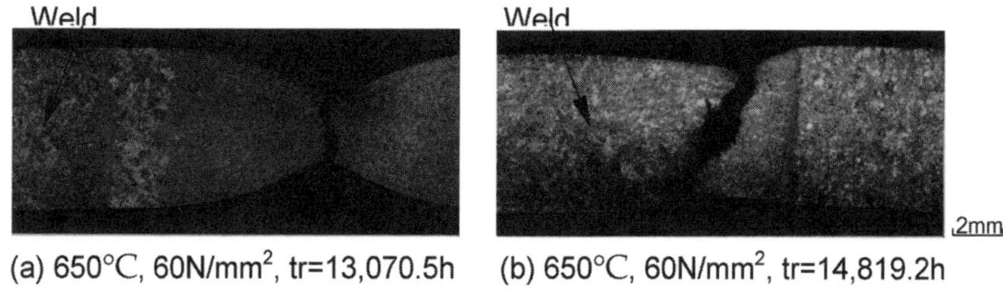

(a) 650°C, 60N/mm^2, tr=13,070.5h (b) 650°C, 60N/mm^2, tr=14,819.2h

Figure 6. Cross sectional views of creep ruptured specimens

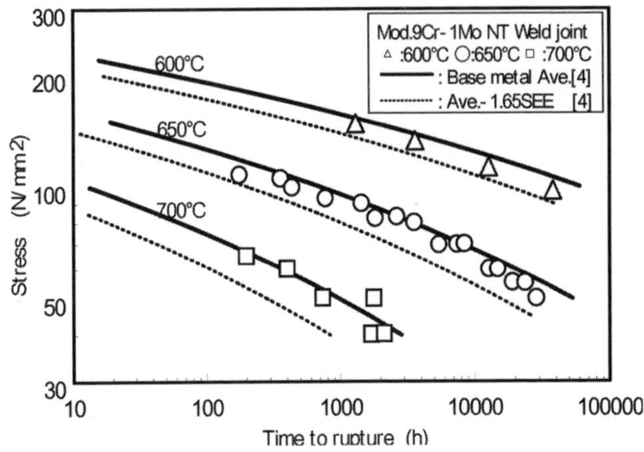

Figure 7. Creep rupture strength of Mod.9Cr-1Mo NT weld

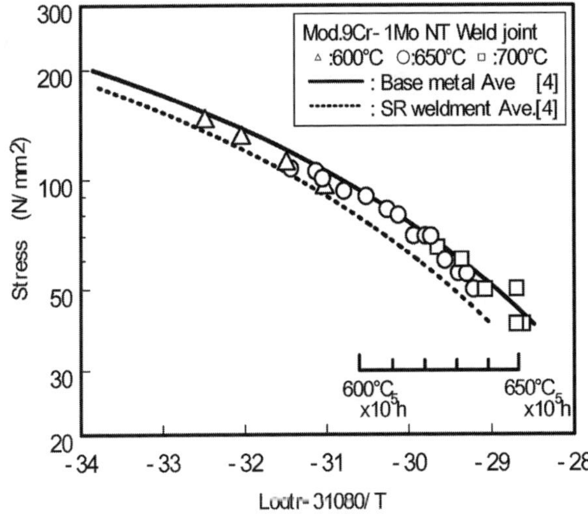

Figure 8. Parametric plot of creep rupture data of Mod.9Cr-1Mo NT weld

Tensile and Impact Properties

Figure 9 shows the tensile properties of the NT weldments. The tensile and 0.2% proof strengths satisfied the values specified in the ASME Code. All the specimens fractured in the base metal and had the strength identical to that of the base metal.

Charpy impact test results of the weld metals from different NT weldments are presented in Figure 10. The average absorbed energy of each NT weld metal was 80 joules or above, which was higher than that of the conventional weld material only stress relieved after welding

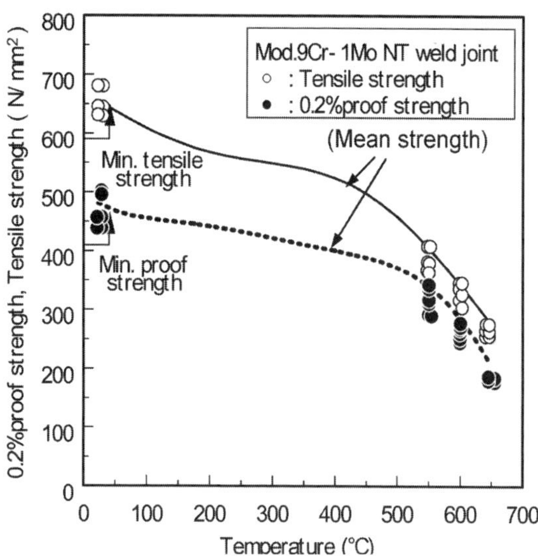

Figure 9. Tensile properties of Mod.9Cr-1Mo NT weld joints

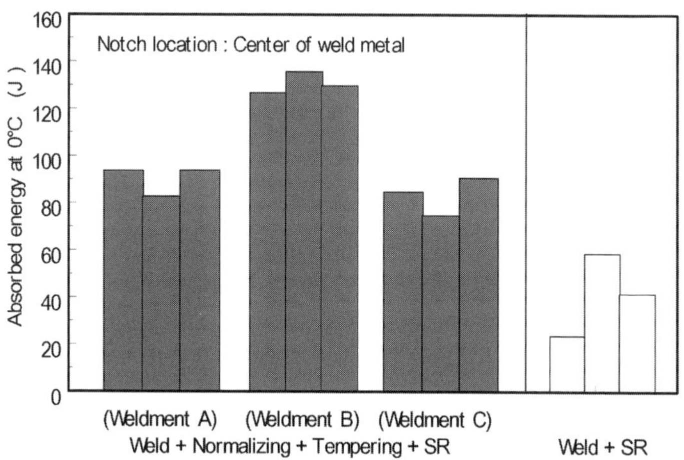

Figure 10. Absorbed energy of Mod.9Cr-1Mo weld metals

Application to Boiler Hot Reheat Piping and Header

The NT welding procedure has been applied to the seam weld of boiler hot reheat piping and headers in several USC plants successfully. Figure 11 shows the narrow gap MAG seam welding process of actual boiler hot reheat pipe. The operation time of first plant reached approximately 50,000 hours and no defects were found by nondestructive inspection.

(a) Set up of bent plates

(b) Seam welding

Figure 11. Narrow gap MAG seam welding process of boiler hot reheat pipe

Conclusion

New seam welding procedure with the modified filler material, narrow gap MAG welding process and normalizing-tempering heat treatment after welding was investigated as an effective fabrication procedure for Mod.9Cr-1Mo steel.

(1) The improvement of creep rupture strength of the weldment was clearly confirmed by long-term creep rupture test. The rupture strengths of the weldments are equal to that of the base metal.

(2) The tensile strength of weldment is identical to that of the base metal.

(3) The absorbed energy of impact test of the welded metal is higher than that of the conventional weld metal.

(4) This procedure has already applied to hot reheat piping and headers of 1,000MW class power boilers successfully.

References

1. A DOE Newsletter "A Modified 9Cr-1Mo Steel Shows Excellent Potential for Steam Generator Applications" DOE/FE-0053/41, Dec. (1982)

2. A. Iseda et al. "Application and Properties of Modified 9Cr-1Mo Steel Tubes and Pipes for Fossil-fired Power Plants" The Sumitomo Search, No.36, p.17, May (1988)

3. Y. Tsuchida et al. "Study on Creep Rupture Strength in Heat Affected Zone of 9Cr-1Mo-V-Nb-N Steel by Welding Heat Cycle Simulation" Welding Research Abroad, 43, 8/9 p.27 (1997)

4. C.R.Brinkman et al. "Modified 9Cr-1Mo Steel for Advanced Steam Generator Applications" ASME Paper 90-JPGC/NE-8 (1990)

5. M. Tabuchi, Y. Takahashi "Evaluation of Creep Strength Steel(P91)" ASME, PVP 2006-ICPVT11-93350 (2006)

6. M.Kawahara, Isao-Asano, "BHK Type Narrow Gap Welding Process", Narrow Gap Welding, The Japan Welding Society (1986)

7. S. Sawada et al. "Application of Narrow-Gap Process", Welding Journal, 58-9 p.17 (1979)

Advances in Materials Technology for Fossil Power Plants
Proceedings from the Fifth International Conference
R. Viswanathan, D. Gandy, K. Coleman, editors, p 884-896

DOI: 10.1361/cp2007epri0884

SELECTION OF EROSION RESISTANT MATERIALS IN THE SEVERE ENVIRONMENT OF COAL FIRED POWER PLANTS

Chris Harley
Senior Applications Engineer
Conforma Clad
501 Park East Boulevard
New Albany, IN 47150

ABSTRACT

Competitive pressures throughout the power generation market are forcing individual power plants to extend time between scheduled outages, and absolutely avoid costly forced outages. Coal fired power plant owners expect their engineering and maintenance teams to identify, predict and solve potential outage causing equipment failures and use the newest advanced technologies to resolve and evade these situations.

In coal fired power plants, erosion not only leads to eventual failure, but during the life cycle of a component, affects the performance and efficiency due to the loss of engineered geometry. "Wear" is used very generally to describe a component wearing out; however, there are numerous "modes of wear." Abrasion, erosion, and corrosion are a few of the instigators of critical component wear, loss of geometry, and eventual failure in coal fired plants.

Identification of the wear derivation is critical to selecting the proper material to avoid costly down-times and extend outage to outage goals. This paper will focus on the proper selection of erosion resistant materials in the severe environment of a coal fired power plant by qualifying lab results with actual field experiences.

INTRODUCTION

Power generation utilities and holding companies goals are to extend times between major planned boiler outages. Systems types and configurations, the age of the plant, their specific operating demands and both preventative and general maintenance philosophies can dictate the accomplishment of these goals. Extending time between major outages two, four, and even five years is resulting in increased forced outages due to erosion related equipment failures – pushing the equipment beyond their intended useful life cycle. Additional contributing factors to increased erosion are:

- Frequent/increased soot blowing – lances, cannons
- Fuel Changes – higher ash content and lower BTU coals
- Fuel Firing Equipment – Low No_x burners
- Abrupt changes in material flow paths
- Ash plugging reducing engineered gaps and openings

The Electric Power Research Institute has generated an in-depth report titled Tube Repair and Protection from Damage Caused by Sootblower Erosion 10080837 March 2004, which will be referenced in the **Lab Testing** section of this paper.

Babcock Power completed a comprehensive engineering study to identify erosion resistant materials suitable for the harsh erosive environment of burner tips titled Abrasive Wear Testing Development Project # 310003, which will also be summarized in the following pages.

EROSION

Examining two definitions, we find that erosion is caused by the impact, cutting action, or abrasive wear of small solid particles freely immersed in the direction of fluid flow that frequently undercut portions of the material they strike [1]. Erosion is also defined as the progressive loss of original material from a solid surface due to mechanical interaction between that surface and the impinging fluid or solid particles [2].

If high erosion-resistant particles such as tungsten carbide exist in low erosion resistant or soft matrix, the impacting particles can undercut and remove portions of the material (Figure 1). However, if the high erosion resistant particles are densely packed in a matrix material that causes the impacting particles to impinge on a greater percent of the hard particle, the erosion resistance increases dramatically (Figure 2).

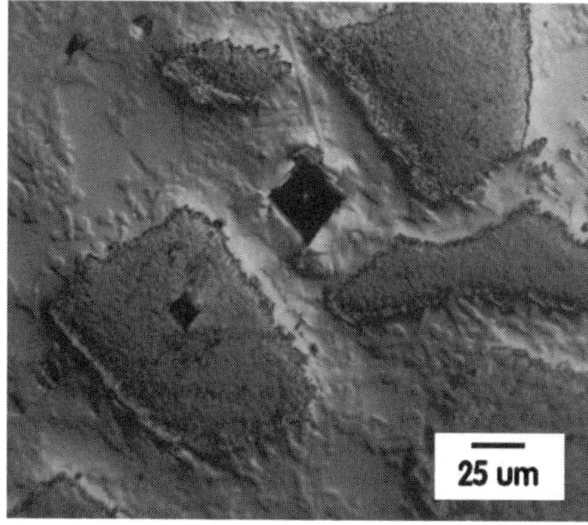

Figure 1

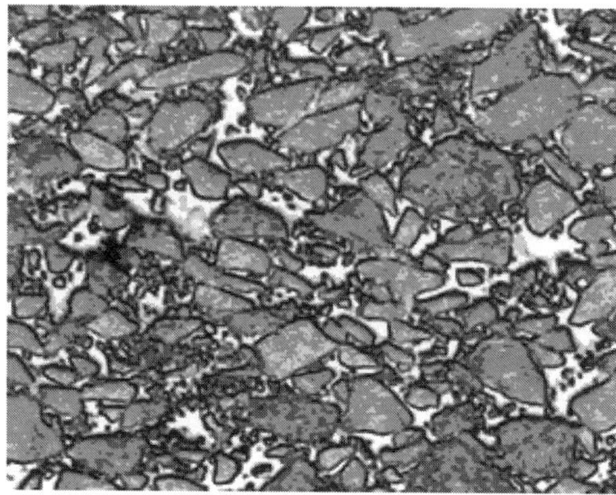

Figure 2

When evaluating the relative erosion resistance of materials, a number of factors must be considered. The obvious factors are temperature, velocity of the impacting particles, their size and shape, and the impacting or impinging angle. These factors can be controlled in standardized testing, but combining their range of variability to comprehensively evaluate performance is limited.

LAB TESTING

Standardized testing procedures, such as ASTM G76 Figure 3, reduce a number of the variables with the intent of providing a common baseline for comparison. This test method utilizes a repeated impact erosion approach involving a small nozzle delivering a stream of gas containing abrasive particles which impact the surface of the test specimen. A standard set of test conditions is described. However, deviations from some of the standard conditions are permitted if described thoroughly. These test methods can be used to rank the erosion resistance of materials under the specified conditions.

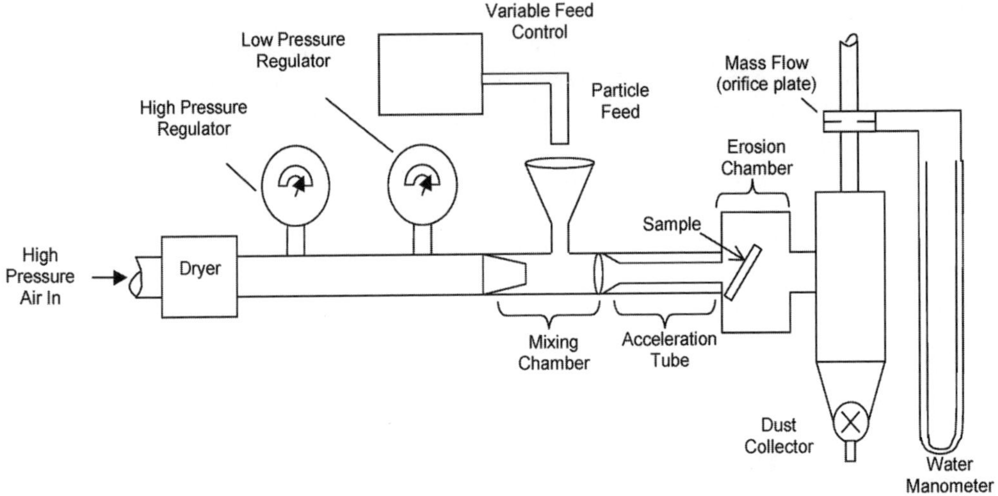

Figure 3

Additional testing methods both ASTM standardized and mathematical erosion models are described in detail in Tube Repair and Protection from Damage Caused by Sootblower Erosion 10080837, March 2004.

TEST 1

Hot Erosion Testing:

Electric Power and Research Institute. Tube Repair and Protection from Damage Caused by Sootblower Erosion 10080837 March 2004.

Erodent:

High temperature erosion tests were carried out using the bed ash from an operating boiler as the erodent material. The particle morphology was a mixture of both round and angular with a mean particle size of 556 microns and mean particle density of 164.4 LB_m/ft^3. The erodent material particles were comprised of high concentrations of silicon and calcium with minor concentrations of aluminum, magnesium, sulfur, iron, phosphorus, titanium and chlorine.

Reporting:

Test results are typically reported as both a weight loss and a thickness loss for each of the tested specimens. However, since the weight measurements included the material erosion wastage (-), oxide scale (+), ash deposit (+), and different densities, the weight loss scheme was not a desirable approach for predicting the erosion rate. Therefore, the thickness loss was determined to be a more valid method for determining the erosion rates of the tested alloys. Material testing was performed at two temperatures; 900 and 1100 degrees and at two different impingement angles.

Materials Tested:

Twelve alloys were selected for high temperature erosion testing:

- SA387 Grade 11 alloy steel
- Nickel alloy 52 – GMAW
- Nickel alloy 622 – GMAW
- Nickel alloy 602CA – GMAW
- WC200 braze alloy – infiltration brazed
- Duocor coating – TWAS
- 309L stainless steel – GMAW
- Nickel alloy 72 – GMAW
- Nickel alloy 625 – GMAW
- 312 stainless steel – GMAW
- Cr3C-NiCr coating – HVOF
- LMC-M WC blend coating – HVOF

The list was generated through a combination of those applied in the industry and those found from erosion and/or erosion data of carbon steels, stainless steel alloys, nickel-based alloys, tungsten carbide claddings and thermal spray coatings, as presented in Section 3.3.4 Tube Repair and Protection from Damage Caused by Sootblower Erosion 10080837, March 2004. Details regarding the selected material descriptions, specifications, chemical composition, thermal properties, application processes, and cost estimates are also available in the aforementioned Electric Power Institute report.

The base material for all test samples was SA387 grade 11 alloy.

High Temperature Test Results

Chart 1-1 shows graphically the results omitting the Duocor coating.

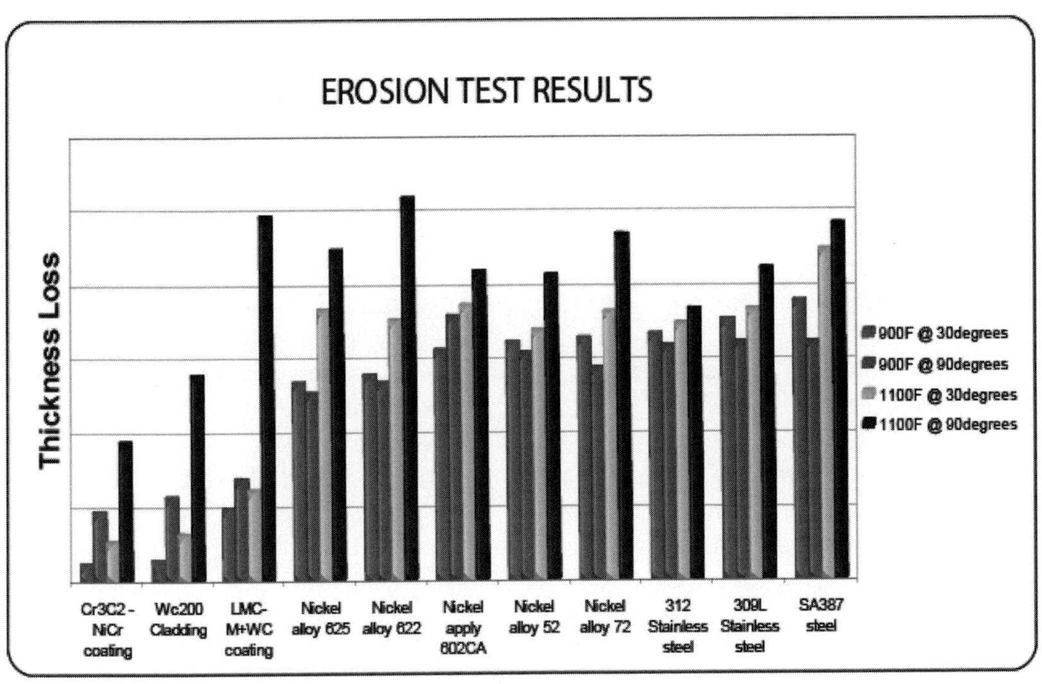

Chart 1-1

Hot Erosion Test Summary

The results indicate that among the twelve alloys tested, the materials with the highest density of erosion-resistant particles i.e., Tungsten carbide and Chrome carbide showed the highest erosion resistance. The Cr_3C_2-NiCr HVOF - applied coating showed the highest erosion resistance followed closely by the infiltration brazed WC 200 material both with erosion resistance particle percentages of close to 70%.

Additional detailed information regarding the lab test summarized above can be found in The Electric Power Research Institute's technical report - Tube Repair and Protection from Damage Caused by Sootblower Erosion 10080837 March 2004.

TEST 2

Erosion Testing:

Babcock Power. Abrasive Wear Testing Development Project # 310003

Erodent:

Erosion tests were carried out using Black Beauty coal slag abrasive as the erodent material. The particle morphology was angular with a particle hardness of 6-7on the moh's scale. The erodent material particles were comprised of high concentrations of Silicon dioxide, Aluminum oxide, and Ferric oxide with minor concentrations of Titanium oxide, Calcium oxide, Magnesium Oxide, Potassium oxide and Sodium oxide with a bulk density of 87 LB/ft^3.

Reporting:

Test results are typically reported as both a weight/volume loss adjusted by materials original density. All tests were performed at room temperature at a 90 degree impingement angles. The velocity of the erodent was 240 ft/second for a duration of 30 minutes. Figure 4.

Materials Tested:

Eight materials were selected for testing. The materials tested were of those currently in service as well as materials and coatings know for their ability to withstand high erosion and also good thermal response to high temperature gradients. Details regarding the selected material descriptions, specifications, chemical composition, thermal properties, application processes are available for review.

Silicon Carbide SiC
Stellite 6
Stellite 31
Chrome carbide
Conforma Clad WC219 infiltration brazed
Stellite 12
Stoody 101
50Cr/50Ni (A560)

The base material for all test samples was ASTM A560.

Figure 4. Test Fixture with Sample

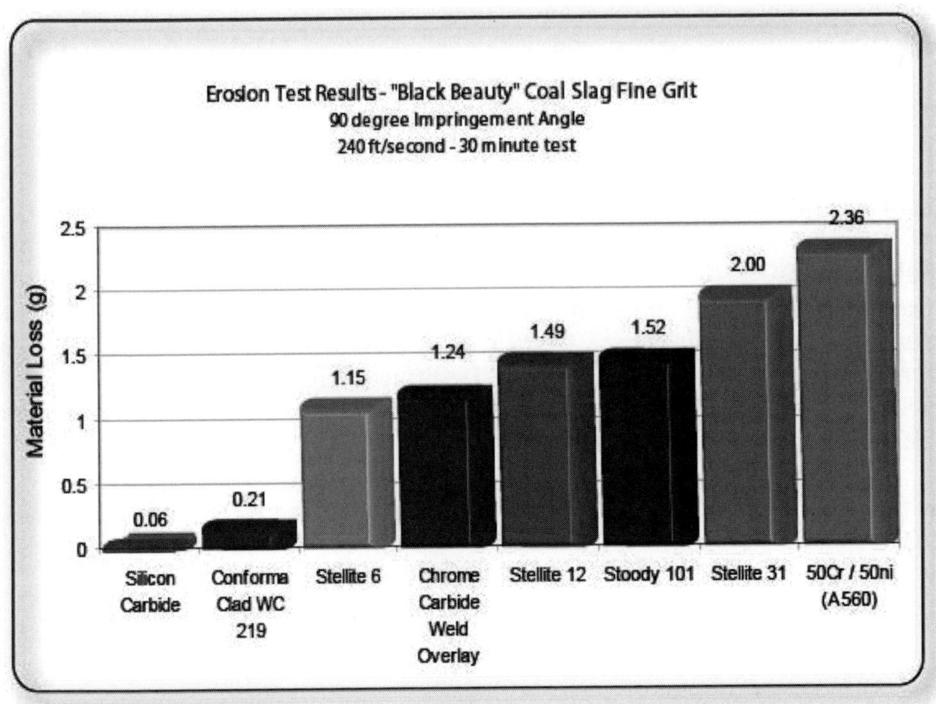

Chart 2.2

Erosion Test Summary

The results indicate that the Silicon carbide material had the lowest volume loss in these tests based on the 90 degree impingement angle followed closely by the infiltration brazed Conforma Clad Wc219 (chart2.3.

Continuing Tests – Field Application

Erosion resistance is complex, combining the many variables to actually duplicate, recreate, field environments is next to impossible in laboratory tests. Additional environmental factors such as thermal shock resistance, erosion resistant material bond strength, as well as many others come into play. The following field tests will compare the laboratory qualified erosion resistant materials to other industry accepted methods of erosion protection.

FIELD TEST 1

Tennessee Valley Authority, Shawnee Fossil Plant

7900 Metropolis Lake Road
Paducah, KY 42086

Plant overview

The Shawnee Plant generates its 1750 MW with 10 boilers supplying over 580,000 area homes with power. The plant's unit 10 is the nation's first commercial scale atmospheric bubbling fluidized bed unit designed in the early 1980's as a test unit for advanced coal firing technologies. This unit started up in October of 1988 and began a demonstration period until May 1991 when the unit was turned over to the TVA generating group for normal commercial operations.

Unit 10 atmospheric bubbling fluidized bed

There are 3 evaporator sections in the boiler fed in parallel from the boiler feed pumps. Each section is a vertical 4 pass arrangement with the bottom tube being pass 1. The in-bed tubes are submerged in a mixture of coal, limestone and recycled ash in the bottom of the furnace where combustion takes place reaching maintained temperatures of 1450°F to 1600°F for optimum sulfur capture. Combustion air enters the bed through air nozzles located in the furnace floor causing the materials to become fluidized and completely cover the in-bed tubes to a depth of 40"-42". Large bubbles of air form and carry the burning bed material up through the tube matrix at a velocity of 7.5 ft/second. The bed material, with less than 3% combustible material, is made up of calcium sulfate, calcium oxide and coal ash.

Evaporator tube history - early erosion rates

The original evaporator tubes were 2.25" OD x .220" SA178C rifled tubes. The bottom row of tubes (pass 1) and all evaporator bends were protected with Extendalloy flame sprayed and fused 45% tungsten in a nickel matrix.

From December 1988 – October 1991 the maximum erosion rates ranged from .001 - .002"/1000 hours near the recycle feed nozzles. During the Mid 90's changes in fuel and operating conditions increased tube erosion resulting in numerous failures (Figure 5). Due to the cost and availability, a mixture of pet coke and coal was burned in the unit. The pet coke reduced the amount of ash the unit produced resulting in the need to add additional limestone to the bed to control temperature.

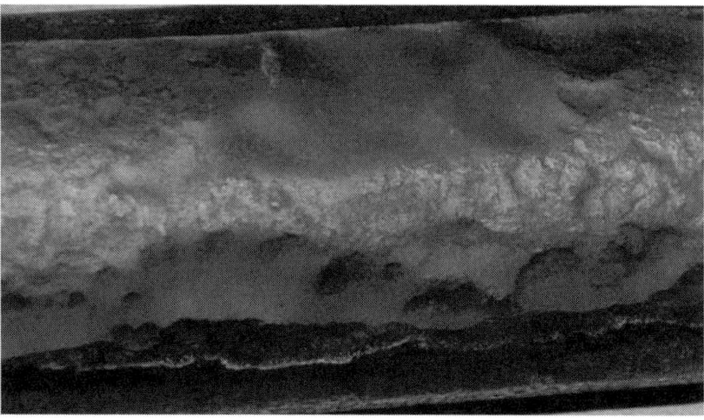

Figure 5

By 1996 tubes leaks had become a serious problem resulting in replacements of tubes in evaporator 2 and half of evaporator 1. The replacement tubes were rifled 2.25" x .220" SA210A1 coated 360 degrees all passes with Extendalloy. The remaining tubes in evaporator section 1 and 3 were flame sprayed on-site with the same Extendalloy material. However, continuing erosion caused tube failures resulted in all evaporator tube replacements during December 1999 outage.

Evaporator tests

Replacement in-bed tubes installed during the 1999-2000 outage were coated 100% - 360 degrees with another spray and fuse material of similar composition as Extendalloy. However, due to the past poor performance of this type of coating, boiler engineers from TVA installed test materials in high erosion areas for evaluation. Each test area consisted of an approximately 6" section coated with the test material separated by the current spray and fuse material. The materials tested were a Stoody 140 weld overlay, NiCr-3 and NiCrMo-3 HVOF, 312 and 309 stainless steel weld overlay, and a Chrome carbide weld overlay. In January 2002, both stainless steel weld overlay test samples had worn through and resulted in a tube failure. After inspections and noticeable erosion on the remaining test sections, all segments were removed.

Evaporator tube continued testing

With spray and fuse tube failures continuing to plague the unit, TVA boiler engineers continued the search for a suitable erosion resistant material for tube protection. Due to successful erosion prevention applications in other areas of TVA's fleet, Conforma Clad was contacted. Conforma Clad supplied in-bed testing; infiltration brazed 70% tungsten carbide coating applied to the high erosion areas of tube bends (Figure 6). The Conforma Clad coated test sections were installed during a November 2003 outage. Arrangements were made between TVA engineers and Conforma Clad to perform inspections during unit availability to track erosion. Due to the non-magnetic characteristics of the infiltration brazed NiCr matrix coating, eddy-current measurements could be taken for accurate measurements (Figure 7)

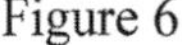

Figure 6

Figure 7

Evaporator tube inspection results

Evaporator 2 Tube inspections Conforma Clad		
Date	**Thickness**	**Material Loss**
Nov-03	.036"	as supplied
Nov-04	.036"	NA
Apr-05	.036"	NA
Sep-05	.0348"	.0012"*
Apr-06	.0342"	.0018"
Feb-07	.0334"	.0026"

* Material loss measured 1.5" x .750 area directly in-line with nozzle.

Table 1

Summary

Although there are differences in materials from the lab test to actual field tests, high erosion resistant particles densely packed in a matrix material characteristic of infiltration brazed Conforma Clad coating, withstand the impinging particles of these environments substantially better than other materials. Additional factors, as mentioned in these discussions, were thought by Tennessee Valley Authority engineering to play a role in the success or failure of erosion resistant coatings. Material bond strengths to the base substrate were thought to play a role in the failure of the spray method coatings. Bond strengths of only approximately 40MPa for the spray methods were unable to withstand thermal cycling along with the simple handling and installations. However, the bond strength of the infiltration brazing is 483MPa and easily withstands the environmental requirements. Due to the low erosion rate displayed by the Conforma Clad cladding (table 1), Tennessee Valley Authority replaced all 3 sections of the evaporator with Conforma Clad infiltration brazed cladding in the scheduled 2007 outage.

FIELD TEST 2

We Energies, Valley Station

333 West Everett Street, Milwaukee, WI 53203

Plant overview

The Valley Station generates 280 megawatts with four boilers. This plant occupies more than 22 acres adjacent to Milwaukee between the south Menomonee canal and the Menomonee River. The pulverized coal fired at this station during a portion of this test was blended with approximately 9% petroleum coke. The plant discontinued this addition in April of 2003.

Unit 2, boiler 3

New low swirl coal spreaders were installed into the existing CCV® low NO_x burners at We Energies Valley Station, Unit 2, Boiler 3 in February of 2003 as part of the normal maintenance schedule. Several of the materials tested in the laboratory were supplied for a direct comparison. They were installed in burners fed by the same mill. Three (3) low swirl coal spreaders were supplied for installation; one of Riloy 74 clad with Conforma Clad infiltration brazed tungsten carbide, and two of cast silicon carbide.

Burner Erosion History

Due to relatively high nozzle velocity at full load of approximately 87 ft/sec, combined with the high silica and alumina content in the coal, this burner application is considered to be a moderately high erosive environment. This is evident from the wear that can be seen on the standard burner components. (See Figures 8 and 9).

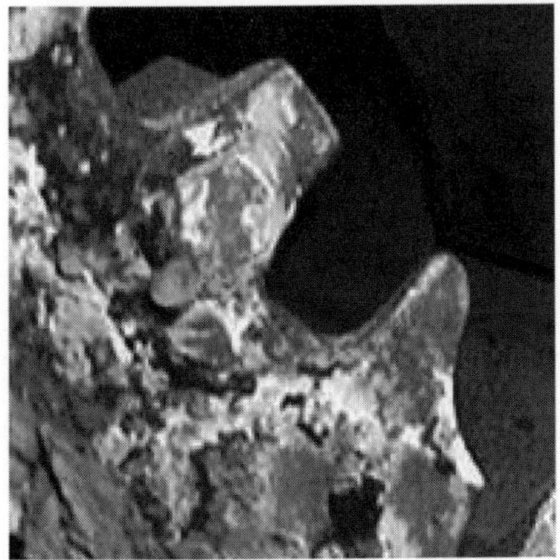

Figure 8. Unit #2, Boiler #3 unprotected burner component showing typical wear after 22 months of service

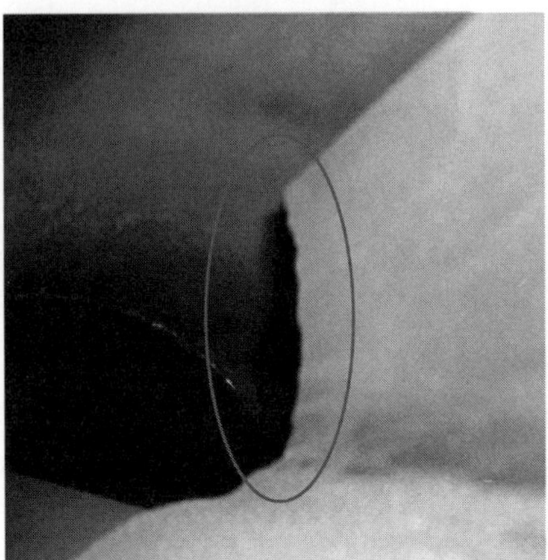

Figure 9. Stellite 31 Weld Overlay on the leading edge shows approximately 1 ½" off vane leading edge after 22 months of service

Burner Tests:

As part of development and evaluation of the selected erosion resistant materials, burner designers chose to install coal spreaders to confirm performance in operation. One prototype coal spreader test piece was protected with 0.040" thickness of Conforma Clad WC219 applied directly to the leading edge of the spreader base material using a proprietary infiltration brazing process. Two other coal spreaders test pieces were of the tested cast silicon carbide. The prototype coal spreaders were installed in Unit 2, Boiler 3 D1 burner location on February of 2003 along with the balance of coal spreaders being supplied with Stellite weld overlay on the leading edges. After approximately 9 months of continuous service, the prototype test pieces were inspected on October 20, 2003.

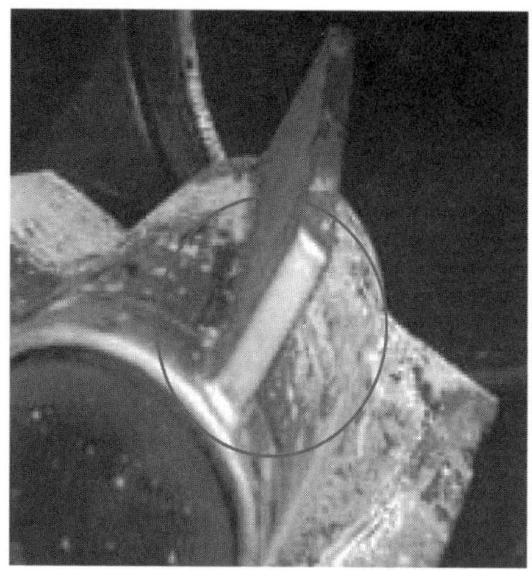

Figure 10. Conforma Clad infiltration brazed tungsten carbide spreader measured .007" material erosion loss

Figure 11. Silicon Carbide spreader experienced .080" leading edge erosion and had significant cracking

	BASE	MIDDLE	TIP	
LEADING EDGE				
VANE 1	.042"	.036"	.039"	
VANE 2	.040"	.033"	.043"	
VANE 3	.038"	.036"	.039"	
VANE 4	.039"	.037"	.040"	
BODY				
LOCATION	1	2	3	4
	.038"	.039"	.037"	.036"

Table 2

Summary

The standard Stellite 31 weld overlay protected coal spreader shown in Figure 9, has approximately 1-1/2" of the coal spreader vane eroded. The prototype test spreader of cast silicon carbide had to be removed due to thermal cracking (Figure 11). The prototype test spreader, protected with tungsten carbide cladding was visually inspected and showed no visible signs of erosion (Figure 10). Due to the non-magnetic nature of the cladding protection, it was possible to measure actual remaining cladding thickness using an Elcometer eddy current thickness gauge. Measurements showed that the maximum extent of erosion was 0.007", or less than 20% of the total protective layer thickness (Table 2). From the measured results the predicted life of the coal spreader protected by the tungsten carbide coating was estimated at approximately 5 years. Discussions with the plant, as recent as August 2007, confirm that the Conforma Clad protected coal spreaders are still in service.

CONCLUSION

While erosion caused failures is only one of the many reasons for forced outages. Returns on the initial investment of preventative maintenance programs involving high erosion prone components can have a payback in as little as one forced outage avoidance. Utilizing today's modern erosion resistant technologies for component protection is getting plants one step closer to achieving the new outage-to-outage goals.

"We can't solve problems by using the same kind of thinking we used when we created them!"
—A. Einstein

References

Tube Repair and Protection for Damage Caused by Sootblower Erosion, K Colman, D.Overcash Fossil Repair Applications Center (FRAC), EPRI, Charlotte, NC

Babcock Power Development Project #310003 – May 7, 2002 Bonnie Coutemanche, Engineer Fuel Equipment Design

B. Wang, A Comparison of Erosion Resistance of Twelve Different Materials. Technical and Research Memorandum, Nov 13, 2003. LBW001-1

[1] Metals Handbook. H. Boyer and T. Gall, eds. American Society for Metals, Metals Park, OH, 1992
[2] Annual Book of ASTM Standards, Vol. 03.02 Wear and Erosion; Metal Corrosion. American Society for Testing and Materials, Philadelphia, PA 2002

Advances in Materials Technology for Fossil Power Plants
Proceedings from the Fifth International Conference
R. Viswanathan, D. Gandy, K. Coleman, editors, p 897-913

DOI: 10.1361/cp2007epri0897

Alloys for Advanced Steam Turbines—Oxidation Behavior

G. R. Holcomb
National Energy Technology Laboratory
U. S. Department of Energy
1450 Queen Avenue SW
Albany, Oregon 97321

Abstract

Advanced or ultra supercritical (USC) steam power plants offer the promise of higher efficiencies and lower emissions. Current goals of the U.S. Department of Energy (DOE) include power generation from coal at 60% efficiency, which would require steam temperatures of up to 760°C. Current research on the oxidation of candidate materials for advanced steam turbines is presented with a focus on a methodology for estimating chromium evaporation rates from protective chromia scales. The high velocities and pressures of advanced steam turbines lead to evaporation predictions as high as 5×10^{-8} kg $m^{-2}s^{-1}$ of $CrO_2(OH)_2(g)$ at 760°C and 34.5 MPa. This is equivalent to 0.077 mm per year of solid Cr loss.

Introduction

Current goals of the U.S. Department of Energy's Advanced Power Systems Initiatives include power generation from coal at 60% efficiency, which would require steam conditions of up to 760°C and 35MPa, so called ultra-supercritical (USC) steam conditions. This is in comparison to conventional sub-critical steam plants which operate at about 37% efficiency (steam at 540°C-14.5 MPa) and advanced plants that are currently just being introduced into the market that operate at 40 to 45% efficiency (steam at 600°C-28MPa). The importance of increased efficiency is because it is estimated that for each 1% raise in plant efficiency will eliminate approximately 1,000,000 tons of CO_2 over the lifetime of an 800MW coal fired plant (1). The overarching limitation to achieving the DOE goal is a lack of cost effective metallic materials that can perform at these temperatures and pressures (2). Improving alloy resistance to high temperature corrosion is one key in developing new, efficient and clean coal-fired ultra-supercritical (USC) steam plants (2).

For the USC application, both turbine and boiler materials will operate at higher temperatures and pressures than in conventional plants. However, the development of creep strength in alloys is often obtained at the expense of corrosion and oxidation resistance. Therefore, the strategies to confer corrosion resistance may be needed if ever increasing cycle temperatures are to be achieved in advanced plants. To identify or develop alloys and strategies that can meet these performance requirements, it is critical to understand the degradation mechanisms that will occur during operation.

A critical aspect of materials usage in USC steam turbines is oxidation behavior. Oxidation can result in several adverse conditions: general section loss from material thinning, deep and localized section loss from internal oxidation along grain boundaries, dimensional changes that are critical in airfoils, and downstream erosion from oxide spallation. Evaporation of protective chromia scales may also be an issue at the higher temperatures and pressures of USC steam turbines. The evaporation of chromia scales is the focus of the research presented here.

Chromia Evaporation

The oxidation of alloys protected by the formation of Cr_2O_3 (chromia formers) can undergo scale loss due to reactive evaporation of chromium containing gas species. Water vapor increases the evaporation loss by allowing the formation of $CrO_2(OH)_2(g)$, which for the same conditions has a higher vapor pressure than $CrO_3(g)$. $CrO_3(g)$ is the primary Cr gas specie in dry air or oxygen. A generalized reaction equation for Cr evaporation from Cr_2O_3 is

$$\tfrac{1}{2}Cr_2O_3(s) + nH_2O(g) + mO_2(g) = CrO_{1.5+n+2m}H_{2n}(g) \qquad (1)$$

For $CrO_2(OH)_2(g)$, n=1 and m= ¾ so Eq. 1 becomes:

$$\tfrac{1}{2}Cr_2O_3(s) + H_2O(g) + \tfrac{3}{4}O_2(g) = CrO_2(OH)_2(g) \qquad (2)$$

Evaporation can change the overall oxidation kinetics from parabolic behavior to linear kinetics or even to breakaway oxidation. Linear kinetics can arise after scale growth from oxidation, which decreases with increasing scale thickness, matches the scale loss from reactive evaporation. The change in scale thickness, x, with time, t, can be described in terms of the parabolic rate constant, k_p, and the linear reactive evaporation rate, k_e, as:

$$\frac{dx}{dt} = \frac{k_p}{x} - k_e \qquad (3)$$

At long times or high reactive evaporation rates, a limiting scale thickness, x_L, arises that is given by:

$$x_L = \frac{k_p}{k_e} \qquad (4)$$

In this case metal loss rates are linear, but still involve diffusion through a protective scale. Rapid metal loss can occur when reactive evaporation of Cr depletes the scale (and sometimes the substrate metal) of Cr (3-4). Decreased Cr in the scale or metal can lead to the formation of less protective oxides, such as Fe-Cr oxides in Fe-Cr base alloys. Unprotective scales can lead to rapid metal loss, or "break-away" oxidation.

A methodology for calculating evaporation rates in a high pressure steam turbine is presented. Experimental results were used to validate the methodology.

Methodology

Evaporation

One way to determine evaporation rates is to assume that volatility is limited by the transport of the volatile specie through a boundary layer in the gas phase. For flat plate geometry with laminar flow, the evaporation rate can be calculated by Eq. 5 (5-6):

$$k_e\left(\tfrac{kg}{m^2 s}\right) = 0.664\, Re^{0.5} Sc^{0.343} \frac{D_{AB}\rho}{L} \tag{5}$$

Where Re and Sc are the dimensionless Reynolds and Schmidt numbers, D_{AB} is the gaseous diffusion coefficient between the Cr gas specie and the solvent gas (m^2s^{-1}), ρ is the density ($kg\ m^{-3}$) of the evaporative specie in the gas, and L is the length (m) in the flow direction of the flat plate. Equation 5 is valid for Sc numbers between 0.6 and 50 (5). Assuming ideal gas behavior and a reaction described by Eq. 1, this can be expanded to:

$$k_e\left(\tfrac{kg}{m^2 s}\right) = 0.664\, Re^{0.5} Sc^{0.343} \frac{D_{AB} M_i P_T}{LRT} P^n_{H_2O} P^m_{O_2} \exp\left(-\frac{\Delta G}{RT}\right) \tag{6}$$

Where P_T is the total pressure (Pa), P_i is the partial pressure of gas specie i, M_i is the molecular mass ($kg\ mol^{-1}$) of gas specie i (in this case i is the Cr-containing gas specie), and ΔG is the Gibbs energy of Eq. 1 ($J\ mol^{-1}$). The dimensionless Reynolds and Schmidt numbers are defined as:

$$Re = \frac{\rho_s v L}{\eta} \tag{7}$$

$$Sc = \frac{\eta}{\rho_s D_{AB}} \tag{8}$$

Where ρ_s is the density of the solvent gas ($kg\ m^{-3}$), η is the absolute viscosity ($kg\ m^{-1}s^{-1}$) and v is the gas velocity ($m\ s^{-1}$).

For turbulent flow ($Re > 5\times10^5$), the equation equivalent to Eq. 6 is (5):

$$k_e\left(\tfrac{kg}{m^2 s}\right) = 0.0592\, Re^{4/5} Sc^{1/3} \frac{D_{AB} M_i P_T}{LRT} P^n_{H_2O} P^m_{O_2} \exp\left(-\frac{\Delta G}{RT}\right) \qquad (9)$$

Each of the parameters in Eqs. (6-9) that require additional commentary are described.

Diffusion Coefficient, D_{AB}

Estimation of the diffusion coefficient, D_{AB}, between the Cr gas species and the solvent gas is the most tenuous of the parameters. Tucker and Nelken (7) compared several different methods for estimation of D_{AB} and of two recommended choices, the one developed by Fuller *et al.* (8) was used here because it contains fewer parameters that themselves need to be estimated. After conversion to SI units, the estimation equation is:

$$D_{AB} = \frac{(3.203\times10^{-4})T^{1.75}}{P_T (v_A^{1/3} + v_B^{1/3})^2} \sqrt{\frac{1}{M_A} + \frac{1}{M_B}} \qquad (10)$$

Here v_i is the diffusion volume of species i ($m^3\ mol^{-1}$). Diffusion volumes, as given by Fuller *et al.* (8), were from a fit of Eq. 10 with an extensive list of diffusion data measurements of various A-B pairs. Diffusion volumes of Cr gas species are not available and so were estimated based on a molecule with a radius of 1.6×10^{-10} m, then converted to molar volume. The value of 1.6×10^{-10} m comes from a density functional theory estimation of the length of a Cr-O bond in CrO (9). The values of v_i for the mixtures other than air were based on a weighted average of the component v_i values.

An additional consideration for supercritical steam turbine environments is that D_{AB} can diverge from the inverse pressure relationship of Eq. 10 at high pressures (10). The method in Bird *et al.* (10) was used to approximate this divergence using reduced temperature and pressure, T_r and P_r. Reduced temperature and pressure are equal to $T/T_{critical}$ and $P/P_{critical}$. For water $T_{critical}$ is 647.25 K and $P_{critical}$ is 218.25 MPa. As an example, consider the conditions of 760°C and 34.5 MPa. In this case T_r is 1.60 and P_r is 1.56, which from Bird (10) reduces the value of D_{AB} obtained from Eq. 10 by a factor of 0.88. This is an approximation because this method was developed for self-diffusivity using Enskog kinetic theory and fragmentary data (11).

Absolute Viscosity, η

The absolute viscosity of non-polar gases, for example O_2, N_2, Ar, and air, can be calculated from the following equation (7-8):

$$\eta = 8.44\times10^{-25} \frac{\sqrt{MT}}{\sigma^2 \Omega_\eta} \qquad (11)$$

Where σ is a characteristic diameter of the molecule (m) and Ω_η is the dimensionless collision integral. Literature values of σ and Ω_η were used (5-6).

The absolute viscosity of water, in the temperature range of interest, can be found using Eq. 12, which was obtained from linear portions of absolute viscosity curves as functions of temperature and pressure (12). Equation 12 is for the temperature range 811K-1089K and pressures up to 3.45×10^7 Pa where the linear fit has a correlation coefficient (R^2) of 0.994. Absolute viscosity for temperatures below 811K, nearer to the critical point of water, are decidedly non-linear and Eq. 12 should not be used.

$$\eta = 3.701\times10^{-8} + 3.080\times10^{-7}T + 1.144\times10^{-13}P_T \tag{12}$$

For gas mixtures, the absolute viscosity of each component gas was combined using Eqs. 13-14 (5-6):

$$\eta_{mix} = \sum_{i=1}^{n} \frac{x_i \eta_i}{\sum_{j=1}^{n} x_j \Phi_{ij}} \tag{13}$$

$$\Phi_{ij} = \frac{1}{\sqrt{8}}\left(1+\frac{M_i}{M_j}\right)^{-\frac{1}{2}}\left[1+\left(\frac{\eta_i}{\eta_j}\right)^{\frac{1}{2}}\left(\frac{M_j}{M_i}\right)^{\frac{1}{4}}\right]^2 \tag{14}$$

Solvent Gas Density, ρ_s

The density of the solvent gas is found by assuming ideal gas behavior, which allows Eq. 15:

$$\rho_s = \frac{P_T M_{Ave}}{RT} \tag{15}$$

where M_{Ave} is the average molecular weight of the solvent gas mixture.

Gibbs Energy, ΔG

The two primary Cr gas species for reactive evaporation are $CrO_3(g)$ in either dry conditions or moist conditions at higher temperatures, and $CrO_2(OH)_2(g)$ in moist conditions at most of the temperatures of interest here. Equation 1 describes the evaporation reaction. It is necessary to know the Gibbs energy of formation for each of the products and reactants in Eq. 1 to obtain the ΔG of the reaction used in Eqs. 6 and 9. The ΔG of Eq. 1 is given by:

$$\Delta G_1 = -RT\ln\frac{P_{CrO_{1.5+n+2m}H_{2n}}}{a^{\frac{1}{2}}_{Cr_2O_3}P^{m}_{O_2}P^{n}_{H_2O}} \tag{16}$$

The ΔG_f for $CrO_2(OH)_2(g)$ is not well established. Opila (13) has reviewed the literature and has found that using data based on Glusko (14) results in much lower calculated partial pressures of $CrO_2(OH)_2(g)$ than using data based on Ebbinghaus (15). Glusko (14) is the source of $CrO_2(OH)_2(g)$ data for the ITVAN (16) and HSC (17) thermodynamics programs. The experimental data of Gindorf *et al.* (18) lie between that predicted by Glusko (14) and Ebbinghaus (15) (in terms of log $P_{CrO2(OH)2}$). Ebbinghaus (15) used estimates of molecular parameters to formulate thermodynamic information. Gindorf *et al.* (18) used transpiration experiments to measure the partial pressure of $CrO_2(OH)_2(g)$. It is unclear (13) how the Glusko (14) data was generated. Table 1 shows ΔG_f values for compounds and species of interest.

Table 1 – Gibbs Energy of Formation for Species of Interest. Calculated from Roine (17) Unless Otherwise Indicated.

T (K)	ΔG_f $Cr_2O^3(s)$ (J mol^{-1})	ΔG_f $H_2O(g)$ (J mol^{-1})	ΔG_f CrO_3 (J mol^{-1})	ΔG_f $CrO_2(OH)_2$ (J mol^{-1}) (Glusko 14)	ΔG_f $CrO_2(OH)_2$ (J mol^{-1}) (Gindorf 18)	ΔG_f $CrO_2(OH)_2$ (J mol^{-1}) (Ebbinghaus 15)
500	-998,700	-219,100	-292,200	-616,600	-632,100	-644,700
573	-979,300	-215,400	-287,400	-599,000	-615,100	-629,500
600	-972,300	-214,000	-285,700	-592,500	-608,900	-624,000
673	-954,300	-210,300	-280,900	-574,900	-592,500	-608,900
700	-947,400	-208,900	-279,200	-568,400	-586,300	-603,400
773	-928,700	-205,000	-274,400	-550,900	-569,400	-588,300
800	-921,800	-203,600	-272,600	-544,400	-563,300	-582,800
873	-903,300	-199,700	-267,800	-526,900	-546,400	-567,700
900	-896,500	-198,200	-266,100	-520,500	-540,200	-562,200
973	-878,100	-194,200	-261,200	-503,000	-523,400	-547,200
1000	-871,300	-192,700	-259,500	-496,500	-517,200	-541,700

The partial pressures of $CrO_3(g)$ and $CrO_2(OH)_2(g)$ over pure Cr_2O_3 (activity of 1) were found for conditions of atmospheric pressure, $P_{O2} = 0.20$, and $P_{H2O} = 0.03$ (air plus 3%

H_2O) and are shown in Fig. 1a. Figures 1b and 1c also show partial pressures of CrO_3(g) and $CrO_2(OH)_2$(g) over pure Cr_2O_3, but in these cases for H_2O with 180 ppb dissolved O_2 (DO). Figure 1b is at atmospheric pressure and Fig. 1c is at 30.0 MPa.

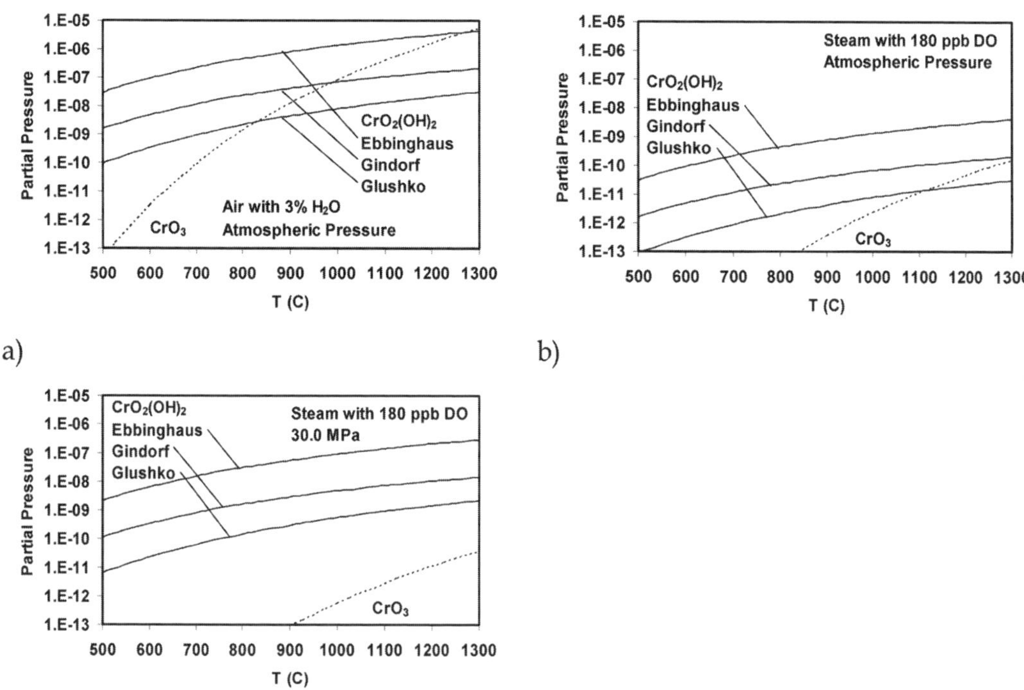

c)

Figure 1. Partial pressures of CrO_3(g) and $CrO_2(OH)_2$(g) over pure Cr_2O_3 (activity of 1) for a) 3% H_2O in air at atmospheric pressure, b) steam with 180 ppb DO at atmospheric pressure, and c) steam with 180 ppb DO at 30.0 MPa.

Gibbs energies for partial pressures above atmospheric were calculated by adjusting ΔG_f of each compound or specie using:

$$\Delta G_{f,P} = \Delta G_f^\circ + \left(n_{gas\ prod} - n_{gas\ react}\right) RT \ln \frac{P_T}{P_T^\circ} \tag{17}$$

Equation 17 assumes that all changes in ΔG_f due to pressure are from the gas phase and follow ideal gas behavior. In Eq. 17, ($n_{gas\ prod} - n_{gas\ react}$) is the change in the number of moles of gas to form the compound or specie. P_T° and ΔG_f° are P_T and ΔG_f at atmospheric pressure.

Since the formation of $CrO_2(OH)_2$(g) reduces the total moles of gas, higher pressures increase its partial pressure (Figs. 1b-1c). Conversely the formation of CrO_3(g) increases the total moles of gas, so higher pressures decrease its partial pressure (Figs. 1b-1c).

Experimental Comparison

Experiments best suited to verify the methodology have situations where a steady-state scale thickness (Eq. 4) is quickly established and mass loss due to reactive evaporation of Cr_2O_3 can be found from mass change with time measurements. Several such tests are described below. Otherwise it would be necessary to separate the effects from scale growth from oxidation and scale thinning from evaporation by the integration of Eq. 3, from which it can be difficult to obtain reliable k_p and k_e values.

Cyclic oxidation experiments on Haynes 230 (UNS NO6230) (19) and Inconel 625 (UNS NO6625) were conducted in air in the presence of steam at atmospheric pressure. The compositions of these alloys are given in Table 2. This was designed to examine the adhesion and spallation behavior of protective oxides. The tests consisted of 1-hour cycles of heating and cooling (55 minutes in the furnace and 5 minutes out of the furnace) in a tube furnace equipped with a programmable slide to raise and lower the samples, Fig. 2. Periodically (between cycles) the samples were removed for mass measurements and then returned for more exposure. The suspension of the samples as shown in Fig. 2 allowed the passage of the gas steam to flow unimpeded across the samples. Water was metered into the bottom of the furnace along with compressed air. Two total gas flow rates were used with rates of 1.9×10^{-3} m s^{-1} (38% water vapor and air, by volume) and 7.6×10^{-3} m s^{-1} (37% water vapor and air, by volume). The exposure temperature for these tests was 760°C. There was no evidence of scale spallation during these tests. In similar tests on certain other alloys, for example with TP347HFG, there was evidence of scale spallation from visible scale debris from handling during mass measurements.

Table 2 – Alloy Compositions as Found by X-ray Florescence (XRF) for the Nickel Alloys or the Nominal Composition for 304L Stainless Steel.

Alloy	Type	Fe	Cr	Ni	Mo	Nb	Mn	Si	Cu	Al	Other
Haynes 230	XRF	1.3	22.6	58.8	1.3		0.5	0.3	0.04	0.4	14.3 W
Inconel 625	XRF	4.4	21.4	61.0	8.4	3.4	0.1	0.4	0.3	0.2	0.3 Ti 0.01 V 0.07 Co
304L	Nom	Bal	19.0	10.0							

Note that the reactive evaporation rates calculated from the preceding methodology are on a $CrO_2(OH)_2(g)$ basis and the experimental mass losses were from chromia scale evaporation and so are on a $Cr_2O_3(s)$ basis. To compare the two rates on the same Cr_2O_3 basis, the following conversion was used:

$$k_e[Cr_2O_3\,basis] = \frac{M_{Cr2O3}}{2M_{CrO2(OH)2}} k_e[CrO_2(OH)_2\,basis] = 0.644 k_e[CrO_2(OH)_2\,basis] \tag{18}$$

Figure 2. Cyclic oxidation apparatus for testing in atmospheric pressure steam/air mixtures.

Results from these tests are shown in Fig. 3 in comparison with predicted slopes from reactive evaporation of $Cr_2O_3(s)$ to $CrO_2(OH)_2(g)$ using the Gindorf *et al.* (18) data for $CrO_2(OH)_2(g)$. The agreement is close, suggesting that the reactive evaporation methodology is validated for this case.

Figure 4 is a backscattered electron micrograph of Haynes 230 (UNS NO6230) after exposure at 760°C in moist air for 2000 cycles. It shows a very thin oxide scale, approximately 1 µm thick. Aluminum was internally oxidized.

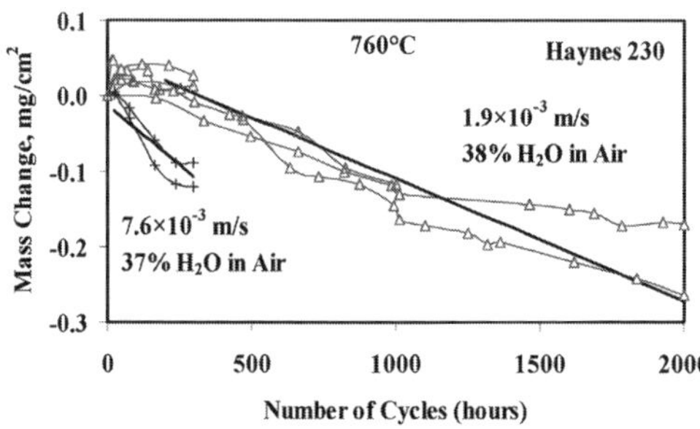

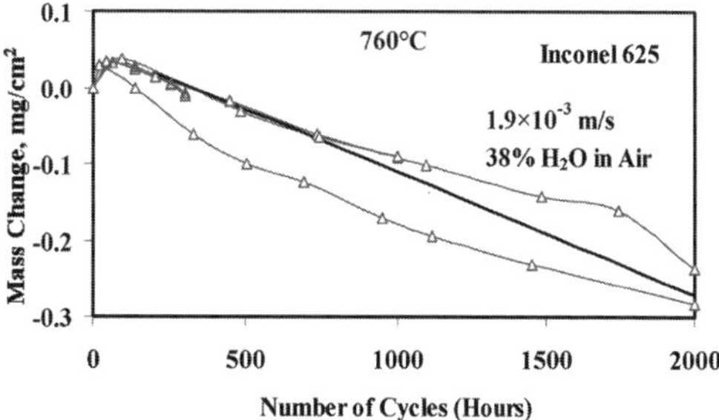

Figure 3. Cyclic oxidation of Haynes 230 (UNS NO6230) (19) and Inconel 625 (UNS NO6625) at 760°C in moist air. Triangle data points for 1.9×10^{-3} m s^{-1} and plus data points for 7.6×10^{-4} m s^{-1}. Straight solid lines are the predicted slopes (on a Cr_2O_3 basis) from reactive evaporation of $Cr_2O_3(s)$ to $CrO_2(OH)_2(g)$ using the Gindorf *et al.* (18) data for $CrO_2(OH)_2(g)$.

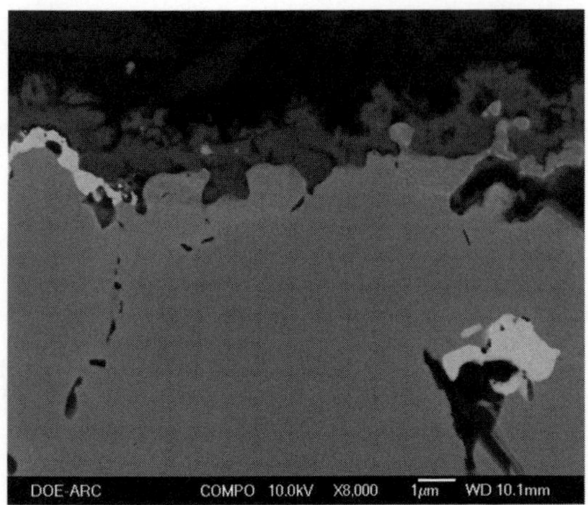

Figure 4. Micrograph using backscattered electrons of Haynes 230 (UNS NO6230) after exposure at 760°C in moist air for 2000 cycles (19). The scale is predominately Cr_2O_3. The bright second phase is W-rich.

Table 3 shows a comparison of the experimental slopes (after 200 hours for the 1.9×10^{-3} m s^{-1} data and after 24 hours with the 7.7×10^{-3} m s^{-1} data) and the predicted reactive evaporation rates using the three sets of ΔG_f data for $CrO_2(OH)_2(g)$. The evaporation model results using Gindorf ΔG values are in good to excellent agreement with experimental values.

Table 3 also compares experimental results from Asteman *et al.* (5) for 304L (UNS S30403, composition in Table 2) at 600°C in 10% H_2O in O_2. In this case evaporation was evident from 72 to 168 hours of exposure. Once again there is good agreement between experiment and evaporation rates calculated using the Gindorf *et al.* (18) data for $CrO_2(OH)_2(g)$.

For the applications and experimental design discussions that follow, the Gindorf *et al.* (18) data for $CrO_2(OH)_2(g)$ is used.

Asteman *et al.* (4) also showed breakaway oxidation for 304L at 600°C in 40% H_2O in O_2. For a gas mixture of H_2O and O_2, the equilibrium constant in Eq. 2 can be used to show that the partial pressure of $CrO_2(OH)_2(g)$ is at a maximum with $P_{H2O} = 4/7P_{O2}$ (~57% H_2O). So increasing the water content in Asteman *et al.* (4) from 10% to 40% increased the evaporation rate, which led to breakaway oxidation due to a depletion of Cr in the scale.

Supercritical Steam Turbine Environments

Development of advanced steam turbines is underway in much of the world to improve the efficiency of power generation from coal. While much of the alloy development involves improving high temperature creep strength, steam oxidation resistance is also of importance. Current U.S. DOE research programs are aimed at 60% efficiency from coal generation, which would require increasing the operating conditions to as high as 760°C and 37.9 MPa for the high pressure (HP) turbine. Current technology limits operation to about 620°C. Above 650°C, it is expected that nickel-base alloys will be required based on creep strength limitations of ferritic and austenitic stainless steels.

Table 3 - Comparison of Experimental and Predicted Evaporation Rates (All on a Cr_2O_3 basis).

Alloy and Conditions	Experimental Slope (kg $m^{-2}s^{-1}$)	Evaporation based on Glusko (14) $CrO_2(OH)_2(g)$ data (kg $m^{-2}s^{-1}$)	Evaporation based on Gindorf (18) $CrO_2(OH)_2(g)$ data (kg $m^{-2}s^{-1}$)	Evaporation based on Ebbinghaus (15) $CrO_2(OH)_2(g)$ data (kg $m^{-2}s^{-1}$)
Haynes 230 UNS NO6230 760°C 38% H_2O in air 1.9×10^{-3} m s^{-1}	-3.46×10^{-10} (19)	-6.50×10^{-11}	-7.00×10^{-10}	-1.38×10^{-08}
Haynes 230 UNS NO6230 760°C 37% H_2O in air 7.6×10^{-3} m s^{-1}	-1.11×10^{-9} (19)	-1.27×10^{-10}	-1.37×10^{-9}	-2.69×10^{-08}
Inconel 625 NS NO6625 760°C 37% H_2O in air 1.9×10^{-3} m s^{-1}	-4.13×10^{-10}	-6.50×10^{-11}	-7.00×10^{-10}	-1.38×10^{-08}
304L UNS S30403 600°C 10% H_2O in O_2 2.5×10^{-2} m s^{-1}	-5.68×10^{-10} (4)	-3.79×10^{-11}	-5.23×10^{-10}	-1.03×10^{-08}

Since candidate alloys for this application are all chromia formers, reactive evaporation could be an important degradation mechanism. Representative environments for current and advanced steam turbines were chosen as: temperatures of 540, 600, 680, 720, 740, and 760°C, pressures of 16.5, 20.0, 31.0 and 34.5 MPa, steam velocity of 300 m s^{-1} (calculated from 60 Hz, 3600 revolutions per minute, and 0.8 m rotor + blade radius), and characteristic length of 0.05 m. This is turbulent flow, so Eq. 9 was used. The values used for the partial pressure of oxygen were based off of oxygenated feedwater that is typical of once-through supercritical power plants, *i.e.*, dissolved oxygen (DO) of 150-180 ppb and a pH of 8.0-8.5 controlled with ammonia additions (12). By the time the feedwater enters the boiler, most of the DO has been removed to less than 1 ppb (20). However, at high temperatures, water undergoes dissociation to O_2 and H_2 to levels above 1 ppb. To estimate the DO at temperature and pressure, the program FactSage (21) was used to first determine the amount of NH_3 required for a pH of 8.25 at 25°C: 34.5 ppb. This agreed well with the reported (12) 20-65 ppb NH_3 used for pH control to 8.0 to 8.5. Next FactSage was used to find the value of P_{O2} for each temperature and pressure combination from water with 34.5 ppb NH_3. A minimum of 1 ppb of DO was used for cases where the dissociation pressure of O_2 was less than 1 ppb. Output from FactSage included the fugacities of H_2O and O_2, so these were used in place of PH_2O and PO_2 in Eq. 9. The use of fugacities made only a minor difference because the fugacity adjustments tended to cancel each other out in Eq. 9. Results are shown in Fig. 5 and Table 4.

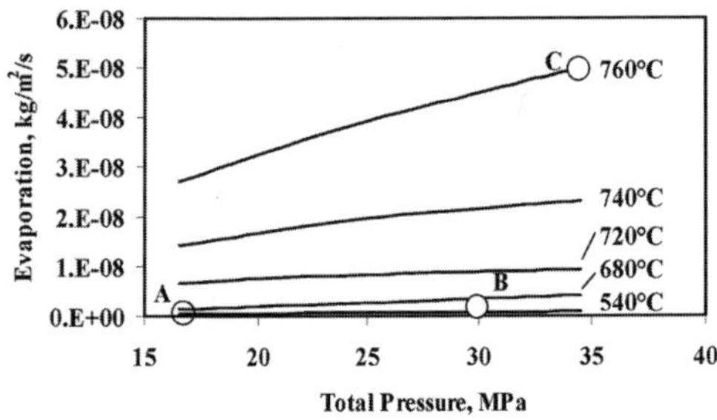

Figure 5. Predicted evaporation rates in supercritical steam turbine conditions with DO set by the greater of the dissociation of water or 1 ppb, 300 m s^{-1} flow rate, and a characteristic length of 0.05 m. Region "A" is typical for current power plants, "B" is for current advanced power plants, and "C" is the U.S. DOE target conditions.

These predicted rates are large compared to the experimental tests at atmospheric pressure and low gas velocities (Table 3). The highest value in Table 4 (for 760°C and 34.5 MPa) of 4.98×10^{-8} kg $m^{-2}s^{-1}$ is equivalent to 0.077 mm per year of solid Cr loss (assumes a metal density of 9 g cm^{-3} and a conversion to a Cr basis in a manner similar to Eq. 18). This is a large value for metal loss for a component expected to operate many years, and it may be larger if the scale losses enough Cr to become non-protective. Current state-of-the-art steam turbines operate at approximately 600°C and 31 MPa, with a predicted evaporation rate of 1.72×10^{-9} kg $m^{-2}s^{-1}$. Typical subcritical steam power plants operate at 538°C and 16.5 MPa, with a predicted evaporation rate of about 3.8×10^{-10} kg $m^{-2}s^{-1}$. These later two cases should have lower evaporation rates in practice because the ferritic-martensitic steels used usually form Fe-Cr spinel outer scales instead of chromia scales. This lowers the activity of chromia in the scale, which lowers the partial pressure of $CrO_2(OH)_2(g)$ (Eq. 16) and thus lower the evaporation rate.

Table 4 - Predicted Partial Pressures of $CrO_2(OH)_2$ and Evaporation Rates in Supercritical Steam Turbine Conditions with DO Set by the Greater of the Dissociation of Water or 1 ppb. There is also 34.5 ppb of NH_3.

T, °C	P_T, MPa	DO, ppb	O_2 fugacity coefficient	H_2O fugacity coefficient	$P_{CrO2(OH)2}$	k_e, kg m^{-2} s^{-1}
540	16.5	1	1.136	0.892	2.27E-12	3.85E-10
540	20.0	1	1.167	0.871	2.61E-12	5.03E-10
540	31.0	1	1.270	0.807	3.58E-12	8.28E-10
540	34.5	1	1.304	0.788	3.86E-12	9.58E-10
600	16.5	1	1.118	0.917	4.30E-12	7.35E-10
600	20.0	1	1.145	0.901	4.96E-12	9.54E-10
600	31.0	1	1.233	0.850	6.88E-12	1.72E-09
600	34.5	1	1.263	0.835	7.45E-12	2.00E-09
680	16.5	1	1.101	0.942	8.88E-12	1.49E-09
680	20.0	1	1.124	0.930	1.03E-11	1.94E-09
680	31.0	1	1.198	0.893	1.44E-11	3.68E-09
680	34.5	1	1.223	0.882	1.56E-11	4.30E-09
720	16.5	4.94	1.094	0.951	4.04E-11	6.71E-09
720	20.0	3.95	1.115	0.941	3.96E-11	7.58E-09
720	31.0	2.12	1.184	0.910	3.50E-11	9.14E-09
720	34.5	1.78	1.207	0.901	3.32E-11	9.35E-09
740	16.5	11.26	1.091	0.955	8.71E-11	1.43E-08
740	20.0	9.39	1.112	0.946	8.81E-11	1.68E-08
740	31.0	5.65	1.178	0.918	8.48E-11	2.21E-08
740	34.5	4.87	1.200	0.909	8.25E-11	2.31E-08
760	16.5	21.70	1.089	0.959	1.64E-10	2.72E-08
760	20.0	18.67	1.108	0.951	1.70E-10	3.25E-08
760	31.0	12.35	1.173	0.925	1.76E-10	4.62E-08
760	34.5	10.94	1.194	0.917	1.75E-10	4.98E-08

The presence of Cr evaporation taking place in the superheater tubes prior to the turbine may to some degree saturate the steam with $CrO_2(OH)_2(g)$, thereby reducing the driving force for evaporation. The lower steam velocity in the superheater tubes (10-25 m s^{-1} is typical (12)) will result in lower evaporation rates than in the steam turbine (~300 m s^{-1}), but there is considerable length of superheater tubing at the high temperature and pressure of the high pressure turbine that could allow some build up of $CrO_2(OH)_2(g)$. Evaporation in the superheater may move the problem upstream and reduce it in the turbine.

Laboratory Experimentation

Laboratory corrosion tests generally seek to mimic the process environment as closely as possible. In cases where this is difficult, then one seeks to establish conditions where the corrosion mechanisms are the same. For steam turbines, laboratory tests with the same combination of temperature, pressure, gas velocities, and steam chemistry are extremely difficult and expensive. Therefore tests sacrifice one or more of the conditions—usually pressure or gas velocity.

For examining the effects of Cr-evaporation as a corrosion mechanism, laboratory tests may be best served with much higher oxygen partial pressures so as to increase the evaporation rate. A comparison of the evaporation rates from Tables 3 and 4 show that even with air and water vapor mixtures, experimental tests (Table 3) fail to achieve the predicted evaporation rates at high pressures and gas flows (Table 4) by several orders of magnitude. As discussed earlier, for O_2+H_2O mixtures, a P_{H2O} equal to $4/7P_{O2}$ (~57% H_2O) should maximize the evaporation rate. The same holds true (albeit at a lower evaporation rate) for air+H_2O mixtures with the maximum also at 57% H_2O. Laboratory tests in steam at atmospheric pressure will have extremely small evaporation rates due to the low partial pressure of oxygen. This is all illustrated in Fig. 6 for predictions made at 760°C. In Fig. 6 the advanced steam turbine points are from the 760°C data in Table 4. The representative laboratory curves are a function of the partial pressure of O_2 in either air+H_2O or O_2+H_2O atmospheres. The laboratory curves were all calculated at atmospheric pressure, a relatively large laboratory gas velocity of $v = 0.02$ m s^{-1}, and $L = 0.02$ m. The right-hand-side of the laboratory curves drop sharply as P_{H2O} approaches zero. The right-hand-side of the laboratory curves are limits. Reactive evaporation in drier O_2 or drier air would switch at that point from $CrO_2(OH)_2(g)$ being the dominate gas specie to $CrO_3(g)$, and would not drop further with less H_2O.

Efforts to improve laboratory tests for higher evaporation rates would include testing in either O_2+H_2O or Air+H_2O at 57% H_2O, increasing the gas velocity (k_e is proportional to $v^{1/2}$), increasing the sample size (k_e is proportional to $L^{1/2}$), or increasing the total pressure (moving the reaction of Eq. 2 to the right).

Conclusions

A methodology was developed to calculate Cr evaporation rates from Cr_2O_3 with a flat planar geometry. As part of this calculation, the interdiffusion coefficient, absolute viscosity, and the Gibbs energy of reaction were determined. The major variables include temperature, total pressure, gas velocity, and gas composition. Experimental verification was done at atmospheric pressure in moist air and moist oxygen. It was concluded that the Gindorf *et al.* (18) data for $\Delta G_{f, CrO_2(OH)_2}$ gave a close match with observed evaporation rates, and so was used for further calculations.

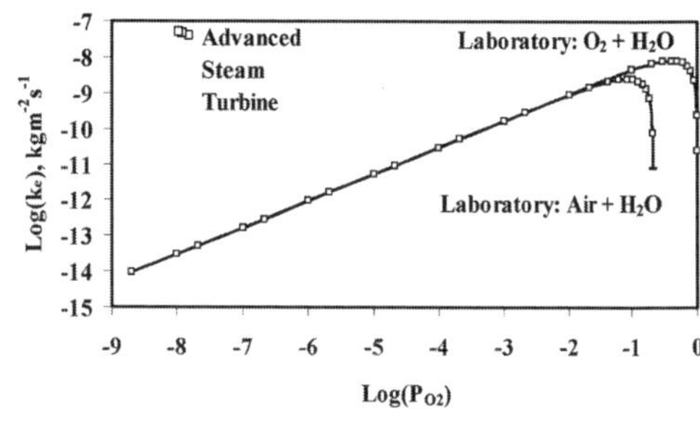

Figure 6. Predicted evaporation rates at 760°C for advanced steam turbines (DO set by dissociation, P_T of 25.0, 31.0, and 34.5 MPa, v = 300 m/s, L = 0.05 m) compared with atmospheric pressure laboratory tests (v = 0.02 m/s, L = 0.02 m) as a function of the partial pressure of O_2 for tests in either O_2+H_2O or Air+H_2O.

The methodology was also applied to advanced steam turbines conditions. The high velocities and pressures of the advanced steam turbine led to evaporation predictions as high as 4.98×10^{-8} kg $m^{-2}s^{-1}$ at 760°C and 34.5 MPa. This is equivalent to 0.077 mm per year of solid Cr loss. Should this Cr loss be too large to maintain sufficient Cr for a protective oxide scale, then much higher oxidation rates could result. Chromium evaporation is expected to be an important oxidation mechanism with the types of nickel-base alloys proposed for use above 650°C in advanced steam boilers and turbines. Chromium evaporation is of less importance for the ferritic and austenitic alloys used in current steam boilers and turbines due to their relatively large oxidation rates with respect to evaporation rates.

It was shown that laboratory experiments, with much lower steam velocities and usually much lower total pressure than found in advanced steam turbines, would best reproduce chromium evaporation behavior with atmospheres that approach either O_2+H_2O or Air+H_2O at 57% H_2O instead of with oxygenated steam.

References

1. Robert Swanekamp, "Return of the Supercritical Boiler," *Power*, 146 (4), 2002, pp. 32-40.

2. R. Viswanathan, J. F. Henry, J. Tanzosh, G. Stanko, J. Shingledecker, B. Vitalis, R. Purgert, "U.S. Program on Materials Technology for Ultra-Supercritical Coal Power Plants," *Journal of Materials Engineering and Performance*, 14, 2005, pp. 281-292.

3. H. Asteman, J.-E. Svensson, L.-G. Johansson, M. Norell, "Indication of Chromium Oxide Hydroxide Evaporation During Oxidation of 304L at 873 K in the Presence of 10% Water Vapor," *Oxidation of Metals*, 52, 1999, pp. 95-111.

4. H. Asteman, J.-E. Svensson, M. Norell, L.-G. Johansson, "Influence of Water Vapor and Flow Rate on the High-Temperature Oxidation of 304L; Effect of Chromium Oxide Hydroxide Evaporation," *Oxidation of Metals*, 54, 2000, pp. 11-26.

5. David R. Gaskell, *An Introduction to Transport Phenomena in Materials Engineering*, New York, New York: Macmillan Publishing, 1992, pp. 78-89, 569-578.

6. G. H. Geiger, D. R. Poirier, *Transport Phenomena in Metallurgy*, Reading, Massachusetts: Addison-Wesley Publishing, 1973, pp. 7-13, 463-467, 529-537.

7. William A. Tucker, Leslie H. Nelken, "Diffusion Coefficients in Air and Water," in *Handbook of Chemical Property Estimation Methods*, Eds. Warren J. Lyman, William F. Reehl, David H. Rosenblatt Washington DC: American Chemical Society, 1990, pp. 17.1-17.25.

8. Edward N. Fuller, Paul D. Schettler, J. Calvin Giddings, "A New Method for Prediction of Binary Gas-Phase Diffusion Coefficients," *Industrial & Engineering Chemistry*, 58, 1966, pp. 19-27.

9. S. Veliah, K.-H. Xiang, R. Pandey, J. M. Recio, J. M. Newsam, "Density Functional Study of Chromium Oxide Clusters: Structures, Bonding, Vibrations and Stability," *Journal of Physical Chemistry B*, 102, 1988, pp. 1126-1135.

10. R. Byron Bird, Warren E. Stewart, Edwin D. Lightfoot, *Transport Phenomena*, New York, New York: John Wiley & Sons, 1960, pp. 504-506.

11. J. C. Slattery, R. B. Bird, "Calculation of the Diffusion Coefficient of Dilute Gases and of the Self-Diffusion Coefficient of Dense Gases, "American Institute of Chemical Engineers Journal, 4, 1958, pp. 137-142.

12. *Steam*, 40th ed., Eds. S. C. Stultz, J. B. Kitto, Barberton, Ohio: Babcock & Wilcox, 1992, pp. 3.8-3.9, 42.11.

13. Elizabeth J. Opila, "Volatility of Common Protective Oxides in High-Temperature Water Vapor: Current Understanding and Unanswered Questions," *Materials Science Forum*, 461-464, 2004, pp. 765-774.

14. Glusko Thermocenter of the Russian Academy of Sciences—Izhorskaya 13/19, 127412, Moscow, Russia: IVTAN Association, 1994.

15. B. B. Ebbinghaus, "Thermodynamics of Gas Phase Chromium Species: The Chromium Chlorides, Oxychlorides, Fluorides, Oxyfluorides, Hydroxides, Oxyhydroxides, Mixed Oxyfluorochlorohydroxides, and Volatility Calculations in Waste Incineration Processes, *Combustion and Flame*, 93, 1993, pp. 119-137.

16. V. S. Yungman, V. A. Medvedev, I. V. Veits, G. A. Bergman, IVTANTHERMO—A Thermodynamic Database and Software System for the Computer, Boca Raton, Florida: CRC Press and Begell House, 1993.

17. A. Roine, HSC Chemistry 5.11, Pori, Finland: Outokumpu Research Oy, 2002.

18. C. Gindorf, K. Hilpert. L. Singheiser, "Determination of Chromium Vaporization Rates of Different Interconnect Alloys by Transpiration Experiments, *Solid Oxide Fuel Cells (SOFC VII)*, Eds. H. Yokokawa, S. C. Singhal, Proceedings Vol. 2001-16, Pennington, New Jersey: Electrochemical Society, 2001, pp. 793-802.

19. G. R. Holcomb, M. Ziomek-Moroz, D. E. Alman, "Oxidation of Alloys for Advanced Steam Turbines," *Proceedings of the 23rd Annual Pittsburgh Coal Conference*, Pittsburg, Pennsylvania: University of Pittsburg, 2006.

20. Steven C. Kung, The Babcock & Wilcox Company, Private Communication, June 2007.

21. C. W. Bale, A. D. Pelton, W. T. Thompson, G. Eriksson, K. Hack, P. Chartrand, S. Decterov, J. Melançon, S. Petersen, FactSage 5.5, Thermfact and GTT-Technologies, 1976-2007.

Advances in Materials Technology for Fossil Power Plants
Proceedings from the Fifth International Conference
R. Viswanathan, D. Gandy, K. Coleman, editors, p 914-926

DOI: 10.1361/cp2007epri0914

THE HEAT AFFECTED ZONE OF BORON ALLOYED CREEP RESISTANT 9% CHROMIUM STEELS AND THEIR SUSCEPTIBILITY TO TYPE IV CRACKING

P. Mayr
H. Cerjak
Institute for Materials Science, Welding and Forming
Graz University of Technology
Kopernikusgasse 24
8010 Graz, Austria
peter.mayr@tugraz.at

Abstract

In thermal power plants, weldments of all currently used martensitic 9% chromium steels suffer from Type IV cracking in the fine grained region of the heat affected zone (HAZ). Japanese researchers presented a new martensitic steel for ultra super critical (USC) steam conditions which showed resistance against Type IV cracking. Within this work, a modified type of this boron-nitrogen balanced advanced 9Cr-3W-3Co steel is compared in terms of weldability to the most promising 9%Cr steel (CB2) developed within the European research activity COST. The HAZ has been investigated by applying the so-called 'Heat Affected Zone Simulation' technique using a Gleeble 1500 thermo-mechanical simulator. Basic optical microscopy was supported by most modern electron microscopic investigations methods like energy filtered TEM (EFTEM), electron energy loss spectroscopy (EELS), electron backscatter diffraction (EBSD) and energy dispersive X-ray analysis (EDX). Phase transformations in the HAZ were directly observed using in situ X-ray diffraction using synchrotron radiation at Advanced Photon Source (APS) of Argonne National Laboratory, IL, USA. Although both steels showed similar transformation behaviour the resulting microstructure due to the applied weld thermal cycle was significantly different. At peak temperatures higher than 1200°C delta ferrite was formed and remained stable down to room temperature by the fast cooling cycle in both steels. While CB2 showed conventional formation of coarse-grained (CG), fine-grained (FG) and intercritical HAZ, boron-nitrogen balanced 9Cr steel showed no formation of a fine-grained HAZ. As Type IV cracking is a phenomenon of the FGHAZ, this type of alloy shows high potential of eliminating Type IV cracking as major life limiting factor in heat resistant steel weldments.

Introduction

Martensitic 9-12% chromium steels have been identified as favored steel grades for thick-walled components in thermal power generation [1- 4]. Arc welding is the major joining and repair technology for such components [5]. The weld thermal cycle alters the microstructure of the base material in a narrow zone close to the fusion line. In this heat affected zone (HAZ) processes like phase transformations, dissolution of precipitates or recrystallisation lead to a microstructure different to the unaffected base material [6, 7]. Components loaded at high temperatures for several years are operated in the regime of creep. Long-term service experience and creep tests of welded structures have shown that the HAZ of 9-12% chromium steels is the weakest part and results in premature failures [8-10]. The dominant failure mechanism is Type IV cracking in the refined region of the HAZ. Good overviews on Type IV cracking are given in Refs. [11-13]. Physical HAZ simulation using a Gleeble thermo-mechanical simulator allows reproducing HAZ microstructure and studying its formation.

Experimental

Experimental

The CB2 material used in this study was part of a 4.3 ton pilot test valve (Heat Nr. 37514) produced by PHB Stahlguss International, Germany. CB2 is a boron containing 9 wt.% chromium steel with 1.5 wt.% molybdenum and 1 wt.% cobalt. The NPM1 material was produced as a test melt of 20 kg in a vacuum induction furnace and forged to 50 x 50 mm cross section. NPM1 is based on 9 wt.% chromium with the addition of 3 wt.% tungsten and 3 wt.% cobalt. Boron and nitrogen level was adjusted to suppress the formation of boron nitrides. The exact chemical composition in wt. % of the steels is given in Table 1.

Table 1: Chemical composition (in wt. %) of the investigated boron containing 9 % chromium steels.

Analysis (Heat)	C	Si	Mn	Cr	Mo	W	Ni	Co	V	Nb	B	N
CB2 (37514)	0.11	0.28	0.86	9.12	1.46	-	0.22	1.02	0.19	0.060	0.0115	0.0211
NPM1	0.074	0.29	0.44	9.26	-	2.84	0.06	2.95	0.21	0.056	0.0120	0.0130

The quality heat treatment to adjust the final base material microstructure differs for both steels. Heat treatment of CB2 steel is characteristic for large cast components and consists of normalising at 1100°C for 9 hours followed by tempering at 730°C for 10 hours. NPM1, foreseen as thick-walled pipe material, was normalised at 1150°C for one hour after forging followed by tempering at 770°C for four hours. Post-weld heat treatment of CB2 was performed at 730°C for 24 hours, while PWHT of NPM1 material was performed at 740°C for four hours. Exact heat treatment parameters are listed in Table 2.

Table 2: Parameters of quality heat treatment and post-weld heat treatment for CB2 and NPM1 steel.

	CB2		**NPM1**	
	Temperature	**Duration**	**Temperature**	**Duration**
Austenitising	1100°C	9 h/AC	1150°C	1 h/AC
Tempering	730°C	10 h/AC	770°C	4 h/AC
PWHT	730°C	24 h/AC	740°C	4 h/AC

AC ... Air Cooling

Physical simulation of the HAZ microstructure was performed using a Gleeble thermo-mechanical simulator. Weld thermal cycles acting as input data for the physical HAZ simulation were calculated using the software 'Tempcycle' developed at the authors organisation. Welding parameters are listed in Table 3. For welding simulation of CB2, typical parameters of shielded metal arc welding process (SMAW) were applied. The heat input (Q)

$$Q\left[\frac{kJ}{cm}\right] = \frac{U[V] \times I[A] \times 60}{v\left[\frac{cm}{min}\right] \times 1000} \times k$$

was evaluated with 11.3 kJ cm^{-1}. The characteristic cooling time between 800°C and 500°C was 18 seconds. Three different peak temperatures representing different regions of the HAZ were simulated. For the NPM1 welds, the gas tungsten arc welding process (GTAW) was simulated. The heat input of 10.8 kJ cm^{-1} brought characteristic cooling times of 19 seconds. Twelve different peak temperatures ranging from 800°C to 1300°C were simulated.

Table 3: Input data for calculation of weld thermal cycles for physical HAZ simulation.

Process	Preheat Temp. [°C]	Current I [A]	Voltage U [V]	Welding Speed [cm/min]	k	HC	Heat Input [kJ/cm]	$t_{8/5}$ [s]
SMAW (CB2)	200	100	26	11	0.8	3D	11.3	18
GTAW (NPM1)	200	160	15	8	0.6	3D	10.8	19

k ... process efficiency HC ...heat conduction

$t_{8/5}$ ... characteristic cooling time 3D ... 3 dimensional

The calculated weld temperature cycles for CB2 and NPM1 steel are shown in Figure 1.

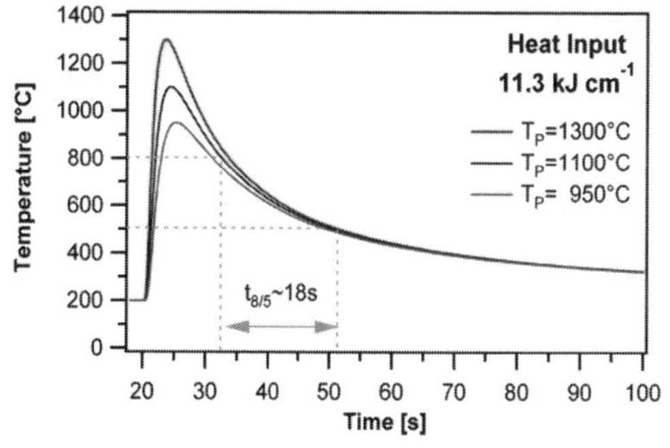

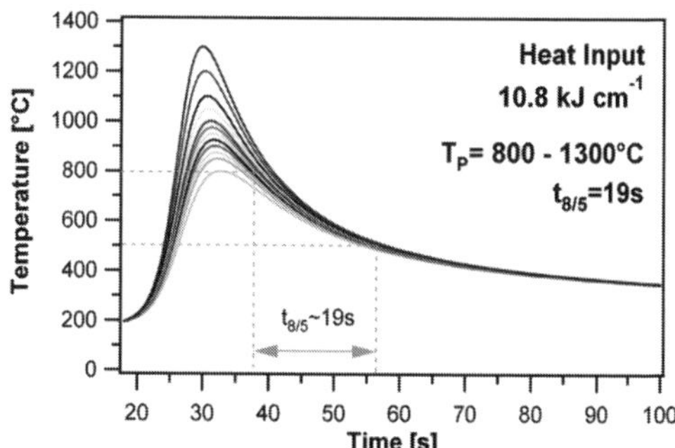

Figure 1: Calculated weld time-temperature cycles for the Gleeble HAZ simulation of CB2 (left) and NPM1 (right). The heat input in the range of 11 kJ cm^{-1} resulted in a characteristic cooling time of about 18 seconds. Peak temperatures were selected to simulate different regions of the HAZ, i.e. CGHAZ (~1300°C), FGHAZ (~1100°C) and ICHAZ (~950°C).

For light microscopic investigations, specimens were polished to a 1 µm surface finish and etched either with modified picric acid (100 ml distilled water, 10 g picric acid, 5 ml hydrochloric acid) to reveal prior austenite grain boundaries or modified LBII etchant (100 ml distilled water, 0.75 g ammonium hydrogen fluoride, 0.9 g potassium disulfide) to reveal the martensitic lath structure. Optical microscopy was supported by transmission electron microscopy (TEM) and analytical TEM investigations.

Electron backscatter diffraction (EBSD) was applied to reveal prior austenite grain and martensite lath morphology.

Equilibrium phase transformation temperatures were calculated using the software package Matcalc [14]. Transformation behaviour of both steels during Gleeble weld simulation was recorded with an attached dilatometer. In-situ X-ray diffraction using Synchrotron radiation was performed to determine the phase fractions of martensite, austenite and delta ferrite at all temperatures of the weld thermal cycles. The experimental setup of the XRD experiments is given in Ref. [15].

Results and Discussion

Base Material Characterisation

Both steels showed a tempered martensitic microstructure in "as-received" condition, i.e. in normalised and tempered condition (Figure 2 and 3). No delta ferrite was observed in both steels.

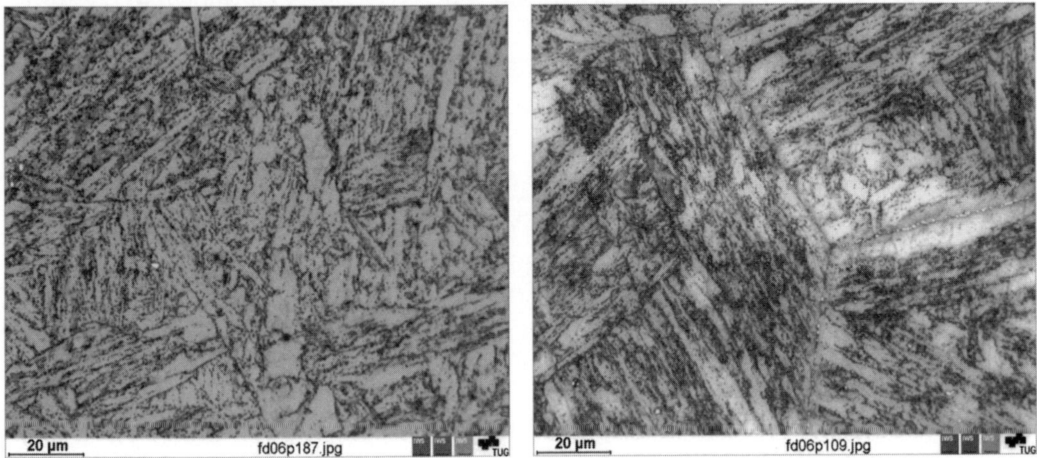

Figure 2: Optical micrographs of base material microstructure of CB2 (left) and NPM1 (right) revealing tempered martensitic microstructure with precipitates along prior austenite grain boundaries and martensite lath boundaries.

Figure 3: EBSD image of NPM1 base material showing tempered martensitic lath microstructure with an average prior austenite grain size of 250 μm.

Transformation behaviour

Equilibrium phase transformation temperatures were calculated with the software Matcalc and based on the thermodynamic database TCFE3 [16] (Table 4). The beginning of ferrite (tempered martensite) to austenite transformation (A_{e1}), as well as full austenitisation (A_{e3}) under equilibrium conditions, was taking place at slightly higher temperatures for the NPM1 steel. Nucleation of delta ferrite (A_{e4}) in steel NPM1 was calculated to start at about 80°C lower than in CB2 and a fully delta ferritic matrix is only obtained in NPM1 at 1411°C (A_{e5}). Measured transformation temperatures (A_{c1}, A_{c3}) using dilatometry are more than 100°C higher than the calculated equilibrium transformation temperatures. This is attributed to the high heating rates, which represent far from equilibrium conditions and results in a superheating of ferrite and austenite before transformation. Formation of delta ferrite was not observed by dilatometry, since this method is not sensitive enough to record this phase transformation. Martensite start temperature is approximately 30°C higher in the NPM1 material, which is mainly caused by the increased carbon content of CB2 steel. Higher carbon content is known to shift the onset of martensitic transformation to lower temperatures.

Table 4: Measured and calculated phase transformation temperatures for CB2 and NPM1 steel.

	CB2	**NPM1**
A_{C1} (A_{e1})	915°C - 930°C (792°C)	930°C - 970°C (828°C)
A_{C3} (A_{e3})	975°C - 1030°C (854°C)	1010°C - 1030°C (881°C)
(A_{e4})	(1257°C)	(1183°C)
(A_{e5})	-	(1411°C)
M_S	350°C - 400°C	400°C - 430°C

The calculated equilibrium phase diagrams, as a function of the chromium content, show that both steels solidify primarily ferritic (Figure 4). At the nominal chemical composition, both steels are predicted to be free of delta ferrite. At chromium contents higher than 12.2 wt.% for CB2 and 11.1 wt.% for NPM1, delta ferrite is thermodynamically stable down to room temperature and the final microstructure is expected to contain martensite and delta ferrite.

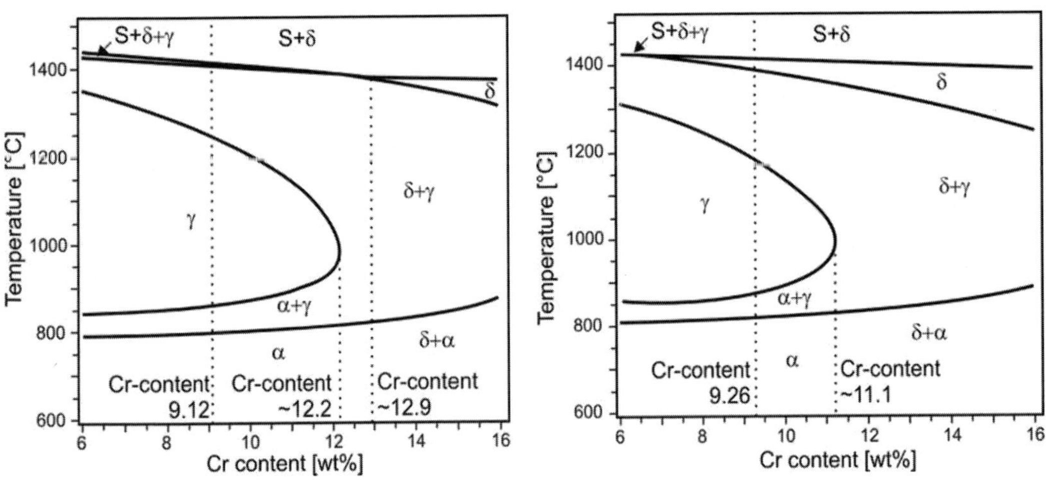

Figure 4: Calculated phase diagram for CB2 (left) and NPM1 (right) for the chemical composition given in Table 1.

Recorded phase transformations by in-situ X-ray diffraction [17] revealed a more complex microstructure of NPM1 steel after a single cycle weld simulation (see Figure 5 and 6). The initial tempered martensitic microstructure transformed very fast into a 100% austenitic microstructure (at position 45 seconds). Increasing temperature produced an increased ferrite fraction (start of delta ferrite formation), which reached its maximum of 45% close to the peak temperature of 1300°C. On cooling, the majority of delta-ferrite transformed back into austenite, except for approximately 4 vol.% of retained delta ferrite. After about 170 seconds at a temperature of 419°C, martensite transformation started and the virgin martensite fraction steadily increased while the austenite fraction decreased. After the completed weld thermal cycle, the microstructure consisted of martensite with approximately 4 vol.% of retained delta ferrite and 4 vol.% of retained austenite. For CB2, a microstructure consisting of martensite with approximately 4 vol.% of retained delta ferrite and 4 vol.% retained austenite after a single cycle weld simulation was identified by XRD.

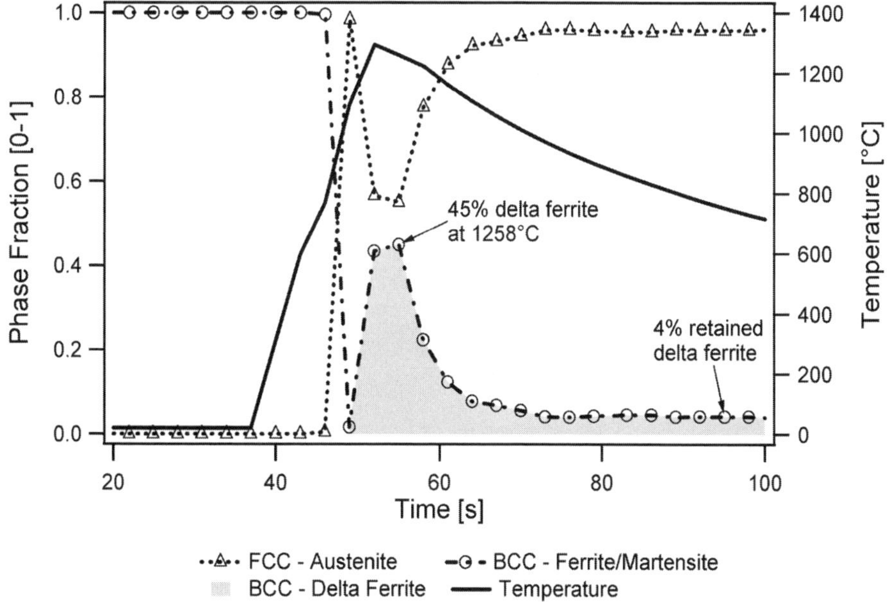

Figure 5: In-situ X-ray diffraction using synchrotron radiation revealing phase transformations in NPM1 steel during a single cycle weld simulation (T_P=1300°C, $t_{8/5}$=40s).

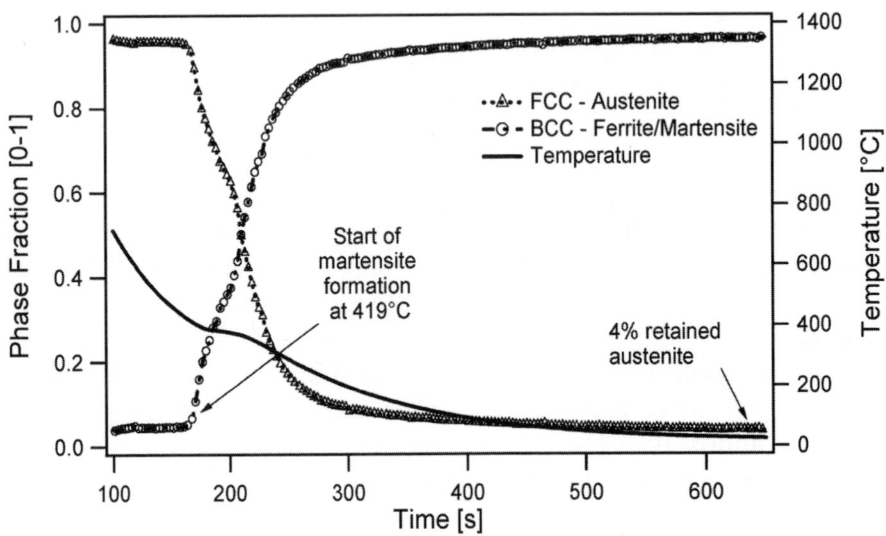

Figure 6: In-situ X-ray diffraction using synchrotron radiation revealing phase transformations in NPM1 steel during a single cycle weld simulation (T_P=1300°C, $t_{8/5}$=40s).

HAZ Microstructure

The HAZ simulated microstructure was investigated by optical microscopy (Figure 7). The cast steel variant CB2 showed the conventional formation of a heat affected zone. Coarse-grained microstructure was observed after simulation with highest peak temperatures (1300°C).

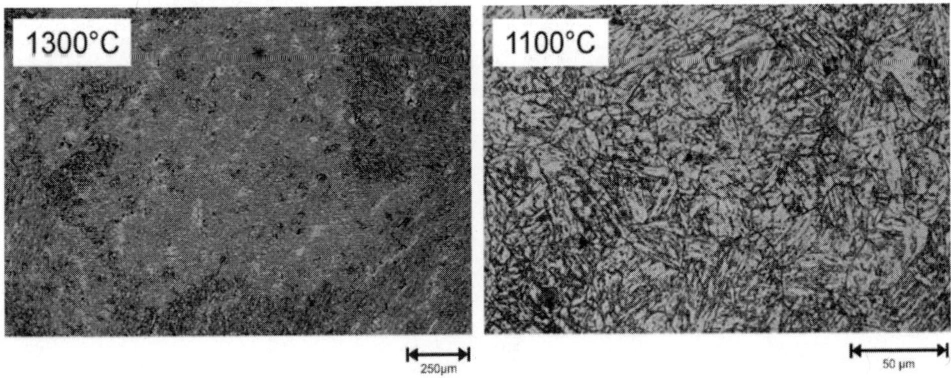

Figure 7: Weld simulated microstructure of steel CB2 showing grain refinement applying a peak temperature of 1100°C and grain coarsening after weld simulation with a peak temperature of 1300°C.

After simulation with peak temperatures close to the A_{C1} transformation temperature (T_P=1100°C) a fully refined microstructure existed.
The forged steel grade NPM1 showed an exceptional formation of the HAZ microstructure. Up to the highest simulated peak temperatures no complete refinement of the microstructure was observed. Grain structure visible in the base material persisted after weld simulation although the steel had undergone several phase transformations. Figure 8 shows a prior austenite grain in NPM1 base material before weld simulation (left) and exactly the same location after weld simulation with a peak temperature of 1100°C and subsequent post-weld heat treatment (right). After weld simulation, the prior austenite grain had still the same shape as observed before simulation. Even martensite lath packages in the grain interior had the same appearance before and after weld simulation although the steel had undergone full austenitisation and re-transformation to martensite.

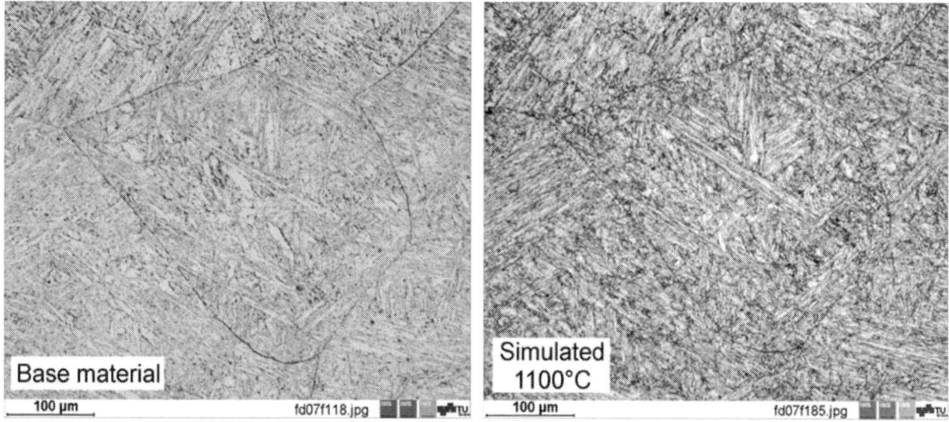

Figure 8: Comparison of the NPM1 microstructure at the same location before and after HAZ simulation applying a peak temperature of 1100°C revealing no grain refinement during the weld thermal cycle.

At higher magnifications, additionally formation of fine grains close to the prior austenite grain boundaries was observed (Figure 9). Increasing the peak temperature of the weld simulation led to an increased formation of small grains close to the initial grain boundaries but still larger areas with the initial martensite lath packages persisted. Throughout all simulated peak temperatures no complete refinement of the NPM1 HAZ microstructure, as in CB2, was observed.

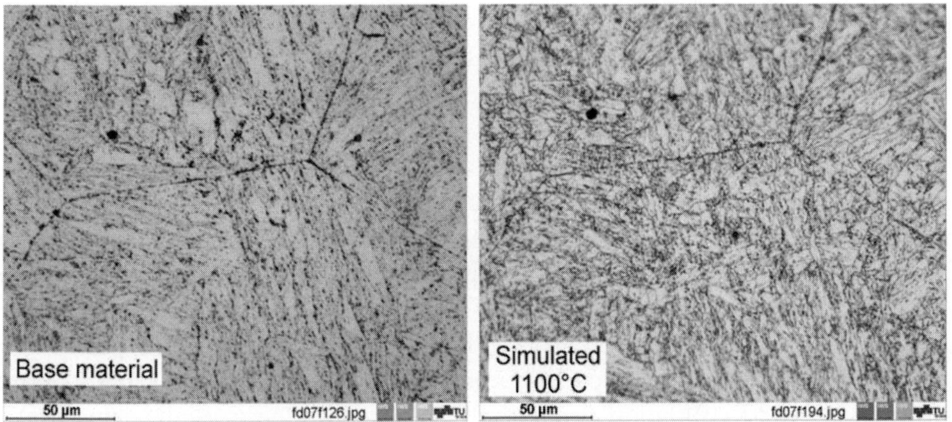

Figure 9: Comparison of the NPM1 microstructure at a grain boundary region at the same location before and after HAZ simulation revealing preservation of prior austenite grain boundary structure.

Summary and Conclusions

The microstructure of the heat affected zone of two boron alloyed 9% chromium steels was investigated by physical HAZ simulation. Both steels showed a tempered martensitic microstructure in as-received condition. The transformation temperatures observed by dilatometry and in-situ X-ray diffraction where similar for both steels. At highest peak temperatures the formation of delta ferrite was observed. The fast weld thermal cycle resulted in an incomplete re-transformation of the delta ferrite to austenite and finally of the austenite to martensite on cooling. After completed weld simulation, the microstructure consisted of martensite with certain amounts of retained delta ferrite and austenite.

While after simulation with a peak temperature of 1100°C, CB2 showed a complete refinement of the microstructure, NPM1 steel was characterized by the same microstructure as observed in the as-received condition. NPM1 showed the formation of some small grains close to the initial grain boundaries but never showed a complete refinement of the microstructure. As Type IV cracking is, per definition, limited to the fine-grained region of the HAZ, weldments of NPM1 steel show high potential to overcome a strength loss during creep caused by Type IV cracking. The mechanism of the suppression of a complete refinement in NPM1 HAZ microstructure is not fully understood yet and still under investigation.

Acknowledgments

This work was part of the Austrian research cooperation "ARGE ACCEPT - COST 536" and was supported by the Austrian Research Promotion Agency (FFG) which is gratefully acknowledged. The X-ray diffraction work carried out at the APS Synchrotron was supported by Dr. John Elmer and Dr. Todd Palmer who are gratefully acknowledged. Special thanks for their continuous support to all the members of the Heat Resistant Design Group at the National Institute for Materials Science (NIMS), Tsukuba, Japan under the supervision of Dr. Fujio Abe.

References

[1] Armor A F and Viswanathan R, "*Supercritical fossil steam plants: Operational issues and design needs for advanced plants*", in 4th int conf Advances in Materials Technology for Fossil Power Plants, Hilton Head Island, ASM International, 2005.

[2] Viswanathan R, Henry J F, Tanzosh J, Stanko G, Shingledecker J and Vitalis V, "*U.S. Program on materials technology for USC power plants*", in 4th int conf Advances in Materials Technology for Fossil Power Plants, Hilton Head Island, ASM International, 2005.

[3] Kern T U, Wieghardt K and Kirchner H, „*Material and design solutions for advanced steam power plants*", in 4th int conf Advances in Materials Technology for Fossil Power Plants, Hilton Head Island, ASM International, 2005.

[4] Masuyama F, "*Alloy development and material issues with increasing steam temperature*", in 4th int conf Advances in Materials Technology for Fossil Power Plants, Hilton Head Island, ASM International, 2005.

[5] Cerjak H, "*Welding of steam turbine components*", Study report of the COST 505 Welding Group, Directorate General Science, Research and Development, Brussels, 1992.

[6] Easterling K (1983), *Introduction to the Physical Metallurgy of Welding*, London, Butterworths.

[7] Granjon H (1991), *Fundamentals of Welding Metallurgy*, Cambridge, Abington Publishing.

[8] Tabuchi M and Takahashi Y, "*Evaluation of creep strength reduction factors for welded joints of modified 9Cr-1Mo steel (P91)*", 2006 ASME Pressure Vessels and Piping Division Conference, Vancouver, ASME, 2006.

[9] Takahashi Y and Tabuchi M, "*Evaluation of creep strength reduction factors for welded joints of HCM12A (P122)*", in 2006 ASME Pressure Vessels and Piping Division Conference, Vancouver, ASME, 2006.

[10] Schubert J, Klenk A and Maile K, "*Determination of weld strength factors for the creep rupture strength of welded joints*", in int conf Creep and Fracture in High Temperature Components - Design & Life Assessment Issues, London, DEStech Publications, 2005.

[11] Middleton C J, Metcalfe E, „*A review of laboratory Type IV cracking data in high chromium ferritic steels*", in int conf Steam Plants for the 1990's, London, IMechE, 1990.

[12] Ellis F V and Viswanathan R, "*Review of Type IV cracking in piping welds*", in 1st int conf Integrity of High Temperature Welds, London, IOM, 1998.

[13] Francis J A, Mazur W and Bhadeshia H K D H, "Type IV cracking in ferritic power plant steels", *Mater Sci Techno,* 2006 **22(12)** 1387-95.

[14] Kozeschnik E and Buchmayr B, "*MatCalc – A simulation tool for multicomponent thermodynamics, diffusion and phase transformations*", in "Mathematical Modelling of Weld Phenomena 5", Eds. H. Cerjak and H.K.D.H. Bhadeshia, IOM Communications, London, book 738, 2001, 349-361.

[15] Mayr P, Palmer T A, Elmer J W, Cerjak H, "In situ Observation of Phase Transformation and their Effects in 9-12% Cr Steels During Welding", *Advanced materials research,* 2007 15-17, 1014 - 1019.

[16] TCFE3 thermodynamic database, ThermoCalc AB, Stockholm, Sweden.

[17] Elmer J W, Palmer T A, Babu S S and Specht E D, „In situ observations of lattice expansion and transformation rates of α and β phases in Ti-6Al-4V", *Materials Science and Engineering A,* 2005 A391, 104 – 113.

Advances in Materials Technology for Fossil Power Plants
Proceedings from the Fifth International Conference
R. Viswanathan, D. Gandy, K. Coleman, editors, p 927-939

DOI: 10.1361/cp2007epri0927

Simplified Methods for High Temperature Weld Design and Assessment for Steady and Cyclic Loading

P.Carter
Alstom Power Inc.,
2000 Day Hill Road,
Windsor, CT 06095

Abstract

Simplified or reference stress techniques are described and demonstrated for high temperature weld design and life assessment. The objective is the determination of weld life under steady and cyclic loading in boiler headers and piping systems. The analysis deals with the effect of cyclic loading, constraint and multiaxiality in a heterogeneous joint. A common thread that runs through most high temperature weld reports and failure analyses is the existence of a relatively creep-weak zone somewhere in the joint. This paper starts with the assumption that the size and creep strength of this zone are known, in addition to parent metal properties. Life prediction requires an efficient analysis technique (such as the reference stress method), which separates the structural and material problems, and does not require complex constitutive models. The approach is illustrated with a simple example of an IN617 main steam girth weld, which could be present in an advanced plant concept with 700°C steam temperature.

Introduction

Welded high-energy boiler components (steam lines and headers) that operate in the creep range have a finite life and significant consequences of failure. Managing this risk involves complex technical questions. At the design stage, relatively simple calculations are used to select materials and component geometry. However the literature on high temperature weld failures is characterized by a level of complexity, which is clearly not dealt with in current design procedures. The ASME Section I Boiler and Pressure Vessel design code [1] merely requires that in room temperature tests, the weld exhibits at least as much strength and ductility as the parent material. For high temperature welded components in service, the complexity problem has driven the industry to a high dependence on inspection. Successful plant life management generally uses combinations of analysis, inspection and testing. The three disciplines are complementary and each provides information which the others cannot. This paper describes developments in predictive techniques for high temperature weld life, which may lead to improved design procedures and life assessment.

The literature on high temperature weld failure analysis and prediction is broadly divided into two groups, based, respectively on weld processes and metallurgy, and on mechanical properties of weld and parent materials.

The second group has seen significant developments in detailed modeling of continuum damage mechanics (CDM) of heterogeneous welded joints. Examples and reviews may be found in [2-5]. Here creep and damage models are used which are complex enough to capture the essential features of creep-rupture data. Parent, weld metal and weak HAZ materials are modeled. Agreement with vessel and mechanical tests has shown that reasonably complex structural behavior is understandable with this approach. The models predict both the initiation and growth of cracks and volumetric damage in multi-material models of welded joints and have focused on steady loading.

This approach taken in this paper follows the second group, and aims to reduce the complexity of the mechanical models and analysis required. Predicting weld properties from the weld process is still a daunting problem, but design and risk management of a weld is in principle simpler. Whether welds are weaker than parent metal, or if there is a weak Type IV or Type IIIa zone in the HAZ-parent region, weld life should be predictable if the properties of the weakest, (and, as it happens, the strongest) material in the joint are known. This is a question of obtaining reliable rupture data and using it in an appropriate model of the joint.

Although the literature is overwhelmingly focused on ferritic welds, the connection with nickel-based welded components is that there is likely to be a strength mismatch between filler and parent materials. The position and size of the weak zone is of secondary importance as far as the assessment methodology is concerned. The analysis methods required for ferritic type IV welds turn out to be the same as for heterogeneous nickel-based welds.

In addition to basic parameters such as stress and temperature, weld creep failures depend on:

- i) Heterogeneous creep material properties.
- ii) Constraint and multiaxial stress state.
- iii) Weld geometry.
- iv) Cyclic effects.

This paper describes effective analysis techniques which address these effects.

Reference Stress Assessment of Weld Life

Reference stress and approximate analysis techniques are widely used in design and assessment codes [6-8]. The critical feature of a reference stress technique is that the stress analysis and material properties are separated. This is essential for practical design and life assessment problems. The reference stress calculation is a hand calculation or a finite element limit load calculation. For steady load, we seek the lowest value of yield stress such that the structure does not collapse. In practice this is achieved by a single analysis which determines the collapse load for an arbitrary yield stress. The reference stress is then obtained for the real load from a simple calculation of ratios. The R5 [8] reference stress approach for steady loading in heterogeneous structures is to define material strength for the reference stress calculation in proportion to creep strength. So the analysis of a welded joint will generally have more than one yield stress, and the overall strength of the joint may involve stress redistribution leading to lower stresses in weaker zones. The reference stress for each material is in proportion to its yield stress, reflecting this stress re-distribution. In R5 this approach also deals with cyclic loading by replacing limit load analysis with shakedown analysis.

For welds it has been known for some time that life may be conservatively defined by the maximum principal stress (MPS), [9]. From the limit analysis it is straightforward to obtain the integration point MPS at the limit load, and to use it for life assessment. Combinations of von Mises (effective) stress and MPS have been used in detailed continuum damage (CDM) calculations, [3-5]. Use of a combination of von Mises stress and MPS with a reference stress could be termed a "modified reference stress" approach.

Key to the use of approximate methods for welds, is understanding the competing mechanisms driving the failure, namely

- Stress redistribution which tends to protect weaker materials.
- Multiaxiality which tends to increase the ratio between maximum principle stress and Mises stress as redistribution occurs.

The final result will reflect the difference between these significant effects, so the approximate technique cannot afford to ignore or make significant errors with either of the processes.

In a recent study undertaken by the Boiler Materials for Ultrasupercritical Coal Power Plants Program [10], comparisons between detailed CDM calculations and modified reference stress calculations were undertaken for typical main steam girth weld geometries. To provide continuity with previous work, Type IV zones adjacent to typical Grade 22 welded joints were considered. In addition weak welds as expected in nickel-based welded pipe were considered. Figure 1 shows the two cases with weak material in a strip representing the HAZ, and as the complete weld.

The loading cases are high and low pressure with associated axial loads, and a case of axial load alone. In practice we expect to see either pressure dominated cases, or combinations of pressure and system loads. The analyses show that for girth welds, pressure and axial loading can produce different levels of constraint and multiaxiality. So, realistic cases must deal with specific combinations of pressure and system loads.

Typical continuum damage mechanics (CDM) relationships for creep strain rate and damage rate in a uniaxial form are:

$$\dot{\varepsilon}_c = A.\exp(-Q_1 / RT)\sinh^n\{\sigma/(1-D)/\sigma_D\}$$

$$\dot{D} = B.\exp(-Q_2/ RT)\sinh^n\{\sigma/(1-D)/\sigma_R\}(\sigma_{mps}/\sigma)^m$$

where A, B, Q_1, Q_2, σ_D, σ_R, n, m are material constants, σ is effective stress, σ_{mps} is maximum principle stress. The problem of obtaining adequate data to model welded joints in this way is evident.

In the study, use was made of data in the literature to represent a typical power plant steel.

This data was used to calculate reference yield stresses for parent, weld and weak zones as outlined above.

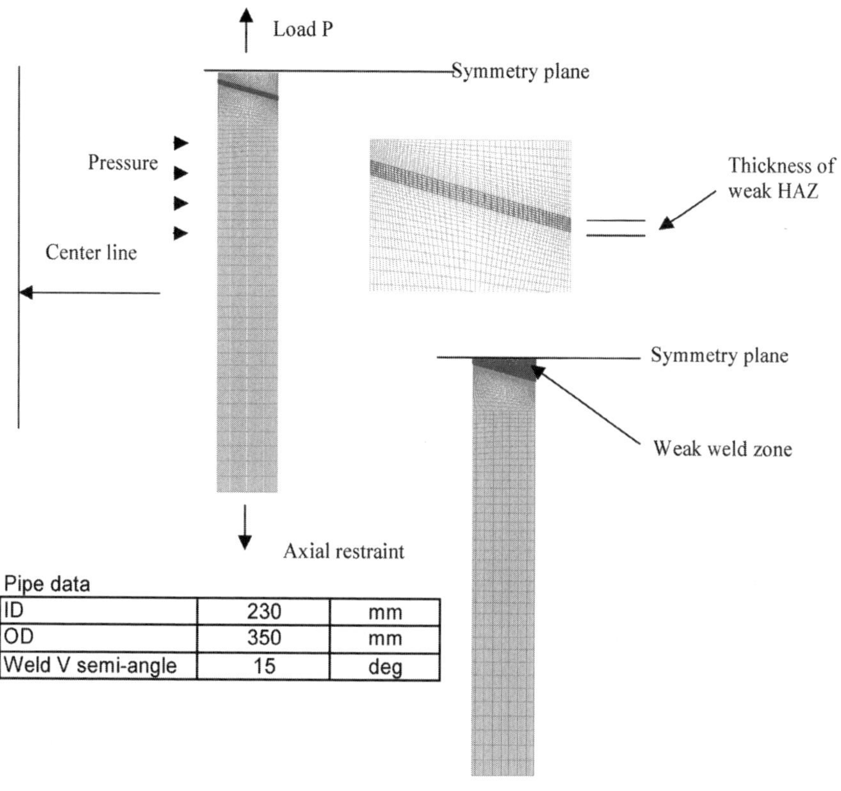

ID	230	mm
OD	350	mm
Weld V semi-angle	15	deg

Figure 1. Finite element models of weld and base metal

Combinations of the following cases were analyzed:

Pressure loads with stress of 50 and 25 MPa.
Axial loads with stress of 50 MPa.
Weak (HAZ) width: 1mm, 4mm, full weld, homogeneous.
Weak (HAZ) properties: weak-ductile and weak-brittle.

Figures 2 and 3 show typical results of CDM analyses.

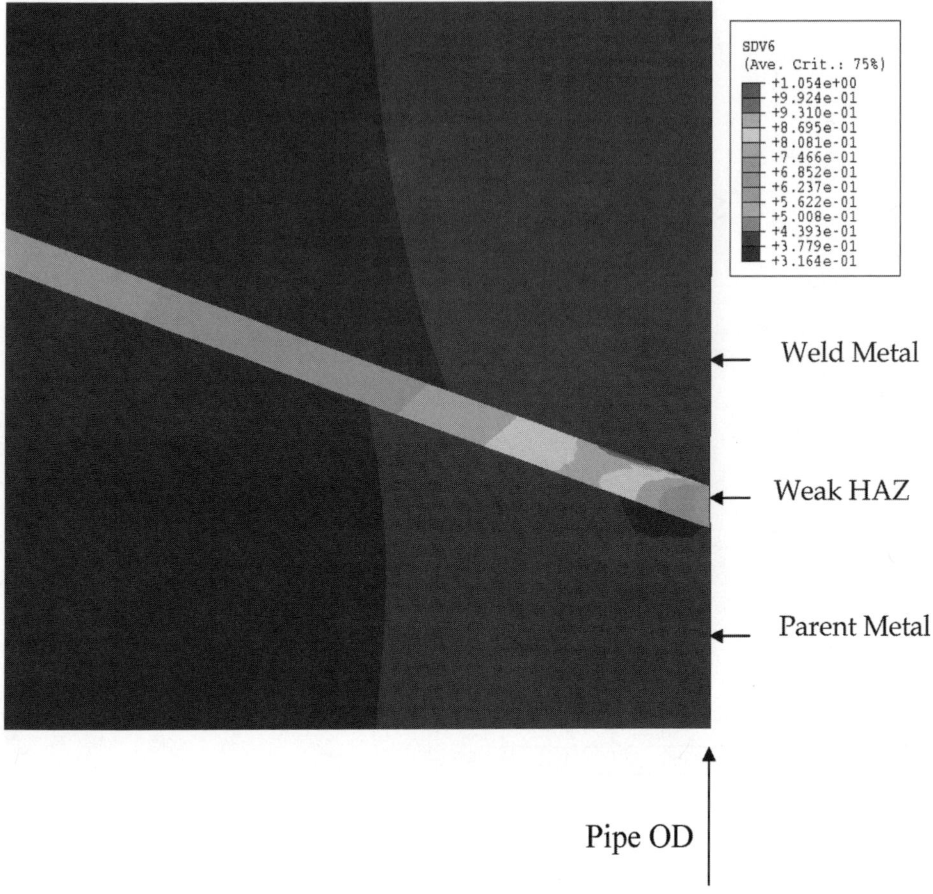

Figure 2. Damage Contours Showing Sub-surface Crack Initiation Near Pipe OD for 1mm HAZ Wide Weak Ductile Material, 50 MPa Pressure at 7500 hours

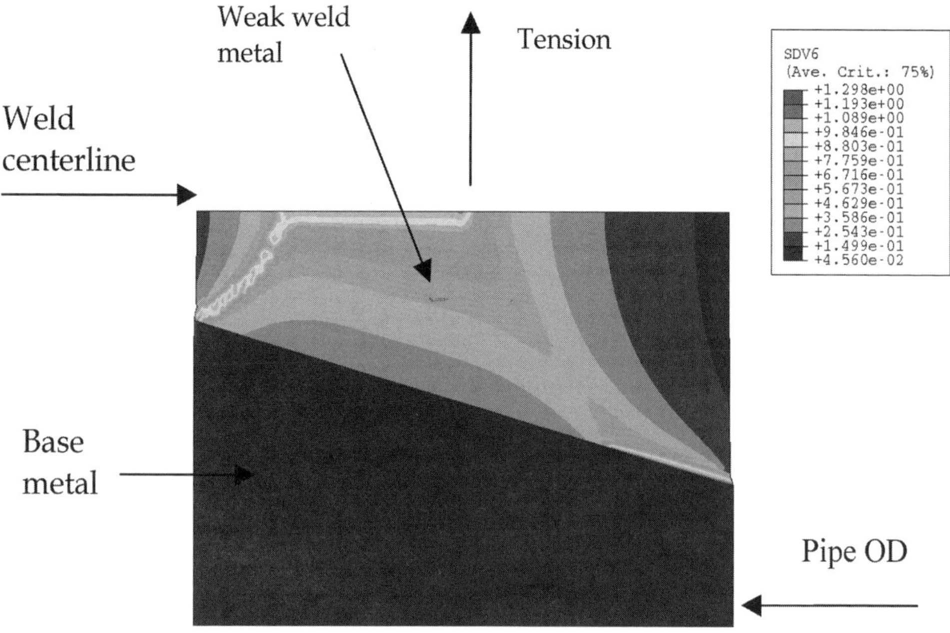

Figure 3. Contours of damage showing multiple cracks for weak weld under 50 MPa axial tension at 73,000 hours

Figure 4 is a summary of the calculations. It shows the accuracy of the modified reference calculations as a reference stress ratio. The denominator is the reference stress which has the same time to crack initiation or failure as the CDM calculation. In addition to the modified reference calculations, the results of other reference stress or design approaches are shown. The conservative results (BS6539, ASMEIII NH) are obtained by defining the strength of the component by the weakest material. The non-conservative results (R5) take material strength variations into account, but do not consider multiaxiality.

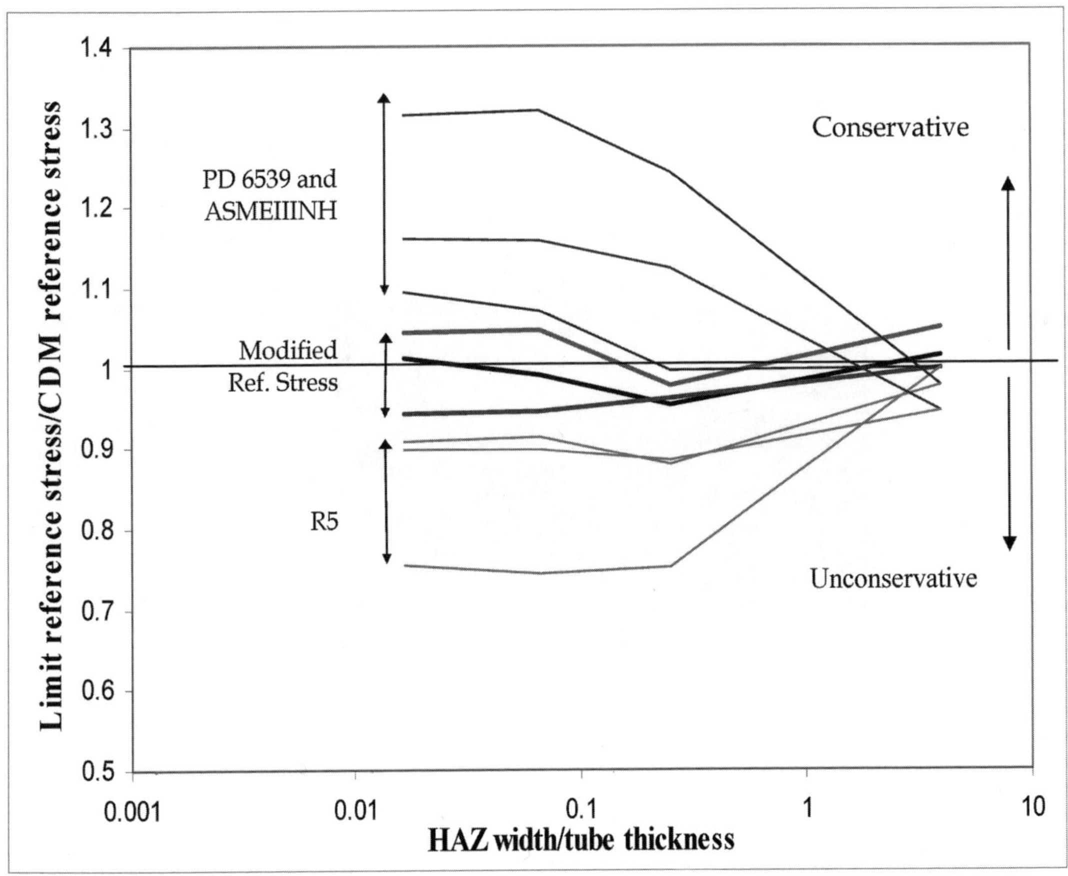

Figure 4. Reference stress/CDM reference stress ratio

Effect of Cyclic Loading: Pressure and Piping Loads on Girth Welds

The R5 [8] procedure describes the use of the shakedown reference stress for cyclic high temperature damage assessment. Reference stress calculations use a fictitious yield (or reference) stress. As noted above, for steady load, we seek the lowest value of yield stress such that the structure does not collapse. For cyclic loading, the shakedown reference stress is the lowest value of yield stress such that the structure achieves shakedown, that is, eventually behaves elastically. Calculation of this parameter is possible using commercial finite element software in different ways, depending on the capabilities. Unlike the steady load reference stress calculation, trial and error is required for cyclic loading. Since temperature varies over the cycle, the method has to take temperature-dependent yield stress into account.

The physical process we are seeking to model is as follows. During the hot, steady load part of the cycle, the welded structure experiences creep under steady load and temperature. During this time, stresses relax from the initial elastic distribution with residual stresses associated with welding, erection, etc, to the creep steady state distribution. Depending on how long this takes, a number of start-up shut-down cycles may occur before the structure can be said to be in its steady state condition. If no yielding occurs during the cycle, then the favourable steady state residual stresses are not changed by the cycle and component life is defined by steady load conditions. If yielding (resetting) does occur, then component life is reduced compared to the steady load case. Then life prediction has to take the load cycle into account.

The shakedown reference stress approach embodies the rapid cycle assumption [11,12]. That is that creep damage calculated from a fictitious elastic cycle with an arbitrary constant residual stress would be higher or more conservative than the real damage. Therefore the rapid cycle approach does not consider relaxation during the cycle, and there may be some conservatism as a result. Since finite element codes can take temperature-dependent yield stress into account with no more difficulty than constant values, we may take full advantage of the fact that on shut-down, the yield stress is generally higher than at operating temperature. The stress rapid cycle changes from the hot, steady state to the cold residual stress state. We aim to see if this is possible without yielding. We define a fictitious temperature-dependent "yield" stress as follows. For a given target life, define the fictitious yield stress is the lesser of the yield stress and the rupture stress. Figure 5 shows yield and rupture curves for IN617 material for 1 x 10^6 hours.

This data is used in a piping model with a local weld/parent detail model such as in Figure 6. This is a simple geometry which can demonstrate the problem of cyclic pressure and thermal loading. It has two straight sections and a long radius elbow, bother ends s are fully constrained. The load cycle consists of 0 - 30 - 0 MPa pressure, and 300 - 700 - 300 ^{0}C uniform temperature.

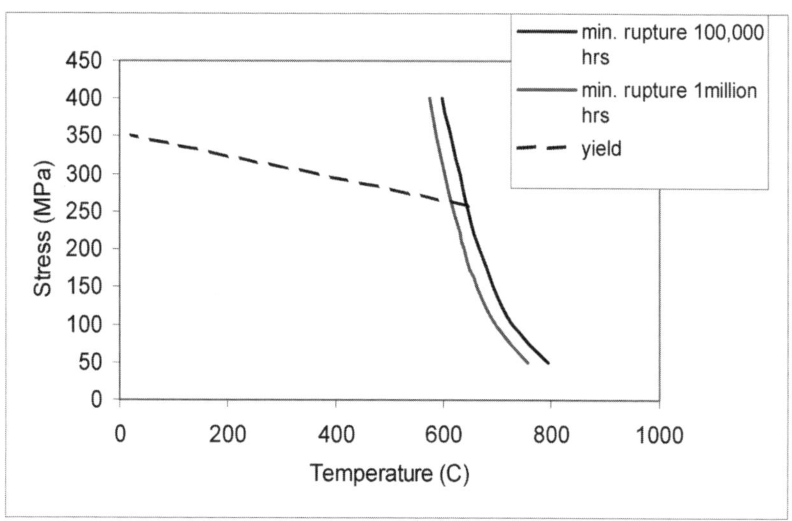

Figure 5. Definition of temperature-dependent "yield" stresses: IN617 data for 100,000 and 1 million hours minimum rupture life.

Weld properties may be represented by a strength reduction factor, or by specific creep rupture data if it is available. Initially we analyse the component using the Figure 5 10^6 hour "yield" properties for weld and parent material, for a combined load - temperature cycle. If the component shakes down, that is, it eventually does not yield over a cycle, then based on the effective stress, the cyclic life of the component would be at least 10^6 hours. Figure 7 shows the distribution of plastic strain on the weld centreline for the 10^6 hour data, which does not change after the first three cycles. We conclude that the weld shakes down, and will not fail in less than 10^6 hours. (The minimum life under steady load for this design is 9 x 10^6 hours.)

The weld strength should be determined by testing. Using, as an example, a weld strength factor of 0.9, we find that the weld fails between 10^5 and 10^6 hours. Figure 8 shows the weld plastic strain history for the 10^6 hours case, which does not shakedown.

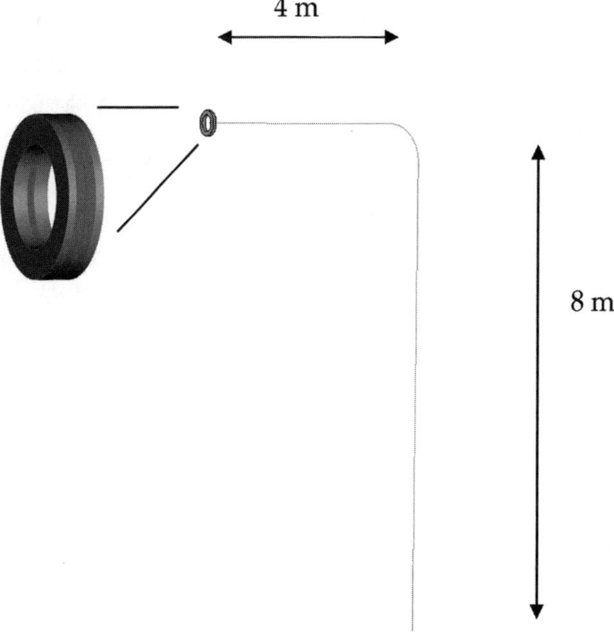

Figure 6. 440 mm OD x 80 mm thick piping model with solid weld/parent metal detail. Ends are constrained.

PEEQ
Angle = 0.0000, Radius = 140.0000, Section Point = 1
(Avg: 75%)
+2.444e-02
+2.240e-02
+2.037e-02
+1.833e-02
+1.629e-02
+1.426e-02
+1.222e-02
+1.018e-02
+8.146e-03
+6.110e-03
+4.073e-03
+2.037e-03
+0.000e+00

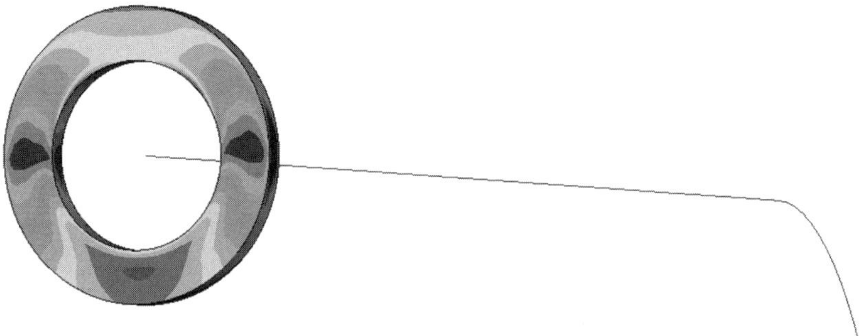

Figure 7. Distribution of plastic strain on weld centreline

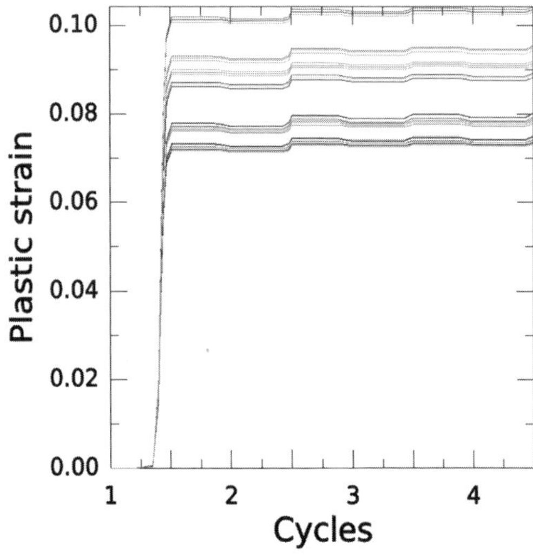

Figure 8. History of weld plastic strain indicating shakedown not achieved.

Conclusions

Weld life assessment has been demonstrated using a piping model with a girth weld having inferior creep rupture properties. Bounds to the effect of cyclic loading on piping life were calculated.

The method requires yield and rupture properties for the parent material and the weakest material is the weld metal/HAZ system. The use of unmodified crossweld test data is expected to be conservative for this purpose.

References

1. ASME Boiler and Pressure Vessel Code, Section I, 2001 Edition, The American Society of Mechanical Engineers, New York, NY.

2 D.J. Gooch and S.T. Kimmins, "Type IV Cracking in ½Cr½Mo¼V / 2¼Cr1Mo Weldments", Proc. 3rd Int. Conf on Creep and Fracture of Engineering Materials and Structures, Eds. Wilshire and Evans, Swansea, 1987.

3. F.R. Hall and D.R. Hayhurst, "Continuum damage mechanics modeling of high temperature deformation and failure in a pipe weldment", Proc. R. Soc. Lond. A, 433, 383-403, 1991.

4. D.R. Hayhurst, I.W. Goodall, R.J. Hayhurst and D.W. Dean, "Lifetime predictions for high-temperature low-alloy ferritic steel weldments", J. Strain Analysis, 40, 7, 2005.

5. T.H. Hyde, W. Sun and J.A. Williams "Creep analysis of pressurized circumferential pipe weldments - a review", J. Strain Analysis, 38, 1, 2003.

6. API 579 Recommended Practice: Fitness for Service, American Petroleum Institute, New York, 2001.

7. PD 6539. Guide to Methods for the assessment of the influence of crack growth on the significance of defects in components operating at high temperatures. British Standards Institution, 1994.

8. R5 Assessment Procedure for the High Temperature Response of Structures, Issue 3, British Energy Generation Ltd. 2003.

9. A.T. Price and J.A. Williams, "The influence of welding on the creep properties of steels", Recent Advances in Creep and Fracture of Engineering Materials and Structures, Eds Wilshire and Owen, Pineridge Press, Swansea, 1982.

10. Boiler Materials for Ultrasupercritical Coal Power Plants - Task 8, Design Methods and Data, Subtask F, Weld Analysis and Assessment, NETL/DOE, 2006.

11. P. Carter, D.L. Marriott and M.J. Swindeman, "Cyclic analysis for high temperature design", ASME/JSME PVP Conference, San Diego, 2004.

12. H.F. Chen, A.R.S. Ponter and R.A. Ainsworth, "The linear matching method applied to the high temperature life integrity of structures. Part 1. Assessments involving constant residual stress fields", Int. J. Pressure Vessels and Piping, 83, 123-135, 2006.

LEGAL NOTICE/DISCLAIMER

Advances in Materials Technology for Fossil Power Plants
Proceedings from the Fifth International Conference
R. Viswanathan, D. Gandy, K. Coleman, editors, p 940-967

DOI: 10.1361/cp2007epri0940

Alternative Filler Materials for DMWs Involving P91 Materials

K. Coleman
D. Gandy
EPRI
1300 Harris Blvd
Charlotte, NC 28262

Abstract

In the late 1980's the domestic utility industry suffered from dissimilar weld (DMW) failures between low alloy ferritic tubing and austenitic tubing in superheaters and reheaters. EPRI performed extensive research into the problem and found that nickel based filler metals provided significant service life improvements over 309 SS filler metals. Improved joint geometries and additional weld metal reinforcement were determined to provide added service life. A new nickel-based filler metal was also developed that provided similar thermal expansion properties to the low alloy base metal. The new filler metal exhibited low chromium content that would result in a smaller carbon denued zone than currently available fillers. This new filler metal was never commercialized because of a tendency to microfissure, which resulted in less than desired service life. This paper discusses a further investigation of the filler metal microfissuring and looks at long term testing to determine suitability of the filler for high temperature applications.

Introduction

Weld transitions between ferritic and austenitic tubing in superheater and re-heater sections have a history of failing before design lives have been reached (1). There are several reasons for these premature failures including:

1. Thermal stresses caused by differenced in coefficient of thermal expansion

2. Carbon denuded zones caused by carbon migration from low alloy base metals to higher alloyed filler metals

A new filler metal (HFS6) was developed in earlier EPRI work that would solve these problems (2). It contained low chromium content that would result in a smaller carbon denued zone than available nickel-based and austenitic fillers (3). The high nickel content of the alloy resulted in thermal expansion similar to low alloy ferritic tube materials. However, the new filler metal was never commercialized because of a tendency to microfissure, which resulted in less than desired service life. The chemical composition from the filler material specification and from actual electrode and deposited weld pads is listed in Table 1.

Table 1: Chemical composition of HFS6 electrode from reference 1.

Heat	C	Si	Mn	P	S	Cr	Mo	Fe	Al	Nb	Ti	Ni
Specification	0.04	0.75	0.9-1.5	0.015	0.015	7.5-9.5	1.5-2.5	Bal	0.1	0.1	0.1	43-50
DI2956	0.09	0.13	0.82	0.0015	-	8.05	1.79	Bal	-	-	-	42.0
9561-014	0.04	0.47	1.25	0.001	0.005	8.80	1.98	Bal	-	-	-	44.32
6568-P1950 [a]	0.03	0.54	1.26	0.005	0.003	8.26	1.87	Bal	0.01	-	0.03	43.36
6568-P1950 [b]	0.04	0.60	1.45	0.014	0.003	9.25	2.28	Bal	0.04	-	0.011	47.65

a Analysis from a weld pad.
b This was a check analysis on the same heat as (a) but from a weld plate.

With the development and use of higher strength alloys in new power installations, specifically Grade 91, EPRI's Materials and Repair Program saw a need for further research into this filler metal. If the microfissuring could be eliminated, the filler metal would offer substantial benefits in weld joints between Grade 91 pipes and tubes, as well as joints between Grade 91 and low alloy ferritic or austenitic pipes and tubes.

The objective of the present project targets the design and manufacture of an experimental creep-resisting nickel-based SMAW electrode to be used to produce microfissure-free weld metal deposits. The criteria established for the weld metal includes:

- Exhibit chromium content less than 10 percent
- Demonstrate thermal expansion properties similar to ferritic steels
- Develop good toughness in the as welded condition (>20 ft-lbs, 27 J)
- Provide acceptable stress rupture/creep properties

In the earlier investigation, the project focus was primarily to find a weld metal suitable for dissimilar welding applications involving 300H series stainless steels (304H, 316H, 321H, 347H) and P11, P22 or P9 CrMo materials. The scope of the potential applications has now increased to not only include P91, but also the possibility of using the new filler metal for producing stub pieces that could be used in the field to make as-welded joints in P91. This type of application would require the weld metal to demonstrate sufficient strength and toughness in the as welded condition and retain sufficient strength to match the P91 base material following a subcritical or full normalize and temper heat treatment (N+T). As a result, preliminary investigation of tensile properties following N+T was carried out as part of the current work

The current project is separated into three tasks to be performed over a three-year period. The tasks are as follows:

1. Development of microfissure-free electrode
2. Preparation of sample coupons
3. Metallurgical, physical and long term rupture testing

This paper describes the results of Task 1, along with some of the early development in Task 2.

Task 1: DEVELOPMENT OF MICROFISSURE-FREE ELECRODE

Alloy Development

Owing to the promising nature of the results on the original electrodes, the original composition was used as the starting point for work in this project. It was considered important to reproduce compositions of good technical purity close to the original so that its susceptibility to microfissuring could be confirmed, and against which improvements could be calibrated. The experimental plan involved producing numerous heats of electrodes in which the variations were primarily carbon, manganese and niobium. The effect of carbon was investigated up to ~0.12%, the effect of manganese up to ~2.3% and the effect of niobium up to ~2.0%. Over fifty-five experimental heats were produced and evaluated, however specific chemistries will not be presented in this paper because evaluation is on going (4).

All of the experimental heats were used to deposit all-weld metal pads, which were then analyzed for microfissuring tendencies. The analysis pads were deposited onto carbon steel blocks using stringer beads essentially in accordance with AWS A5.11. Sufficient layers were deposited to produce an undiluted all-weld metal composition for testing.

Microfissuring

An effective test method for assessing and ranking the microfissuring susceptibility of each experimental heat of electrodes was deemed necessary to evaluate microfissuring tendency. Microfissuring occurs in the heat affected zones (HAZs) of weld beads reheated by subsequent beads. Typical microfissures can be seen in Figure 1. The technique used in the present work rearranges the beads to form a stack of single beads in a vertical build-up similar to that used as a reference method for ferrite measurements in AWS A5.4 (Figure 2)(5).

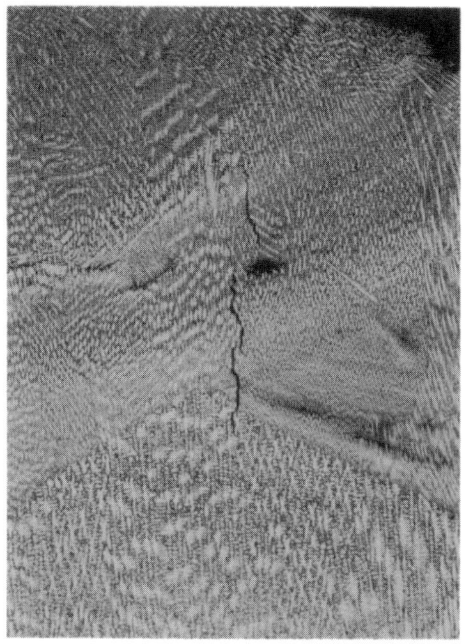

Figure 1:Example Micrograph of Microfissuring

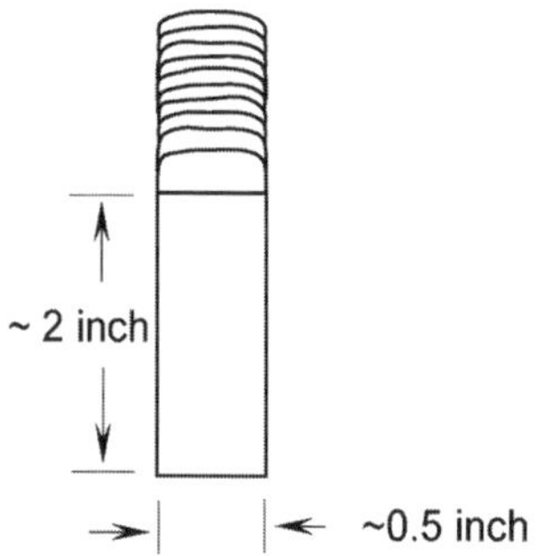

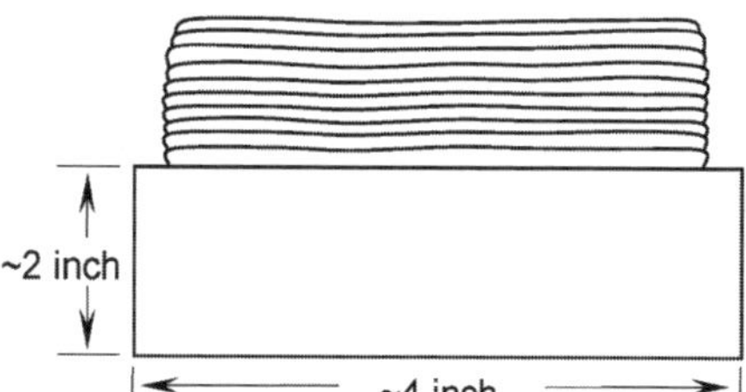

Figure 2: Microfissuring assessment build-ups.

The HAZ were therefore in underlying beads, and any microfissures were free to propagate through layers if inclined to do so. The sides of the blocks were then ground flat and any microfissures that may have been present were detected using dye penetrant. The specimens were not preheated and were kept to a nominal interpass temperature of 300°F (150°C) maximum. All welding was performed using DC+ polarity at ~110A.

Over fifty-five different chemical compositions of filler metals were manufactured and evaluated for microfissuring tendencies. The filler metals were grouped into four major groups, A through D, and given unique identifying numbers that are used throughout this paper (for example D724). The filler metals were produced through controlled additions of sixteen different elements including carbon, silicon, manganese, phosphorus, sulfur, chromium, moly, iron, vanadium, tungsten, copper, aluminium, cobalt, niobium, tin, and nickel.

A wide range of microfissuring susceptibility was observed. For the purposes of evaluation, a microfissuring index (MI) was allocated which allowed the experimental heats to be divided into five groups. The coupons were ranked from 1 to 5, where a number one ranking indicates a very low incidence of microfissuring and a ranking of 5 demonstrates extensive microfissuring. In some cases fissures propagated almost through the full depth of the build-up (Figure 3). The allocation of a heat into a particular MI level was judged by visual examination, based on the general extent and amount of microfissuring seen in the dye penetrant test.

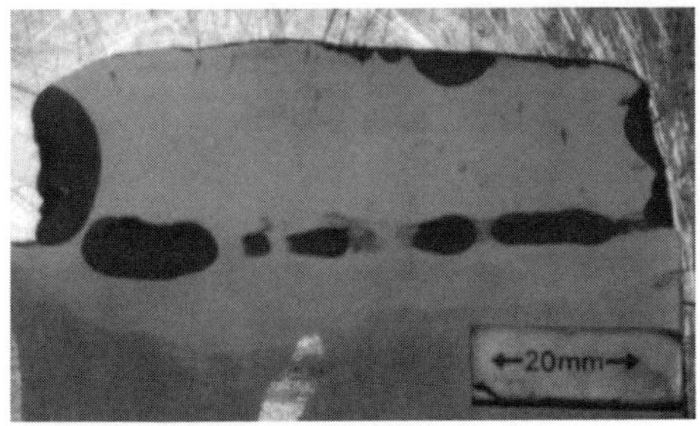

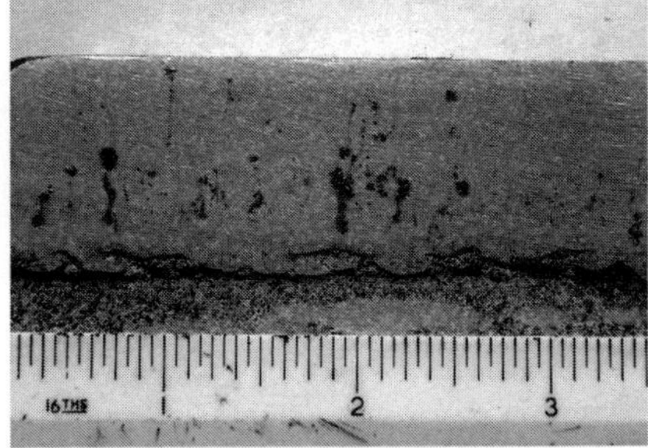

Figure 3: Extent of microfissuring found on the longitudinal face of test build-ups when ground and dye penetrant tested.

Ignoring at this stage any interaction between the various alloying additions investigated, graphs illustrating the microfissuring index (MI) versus alloying additions were plotted for each element. An example for manganese can be seen in Figure 4. Although this graph does not show a strong effect of manganese by itself, some of the graphs of other elements in combination with manganese indicated an optimum level.

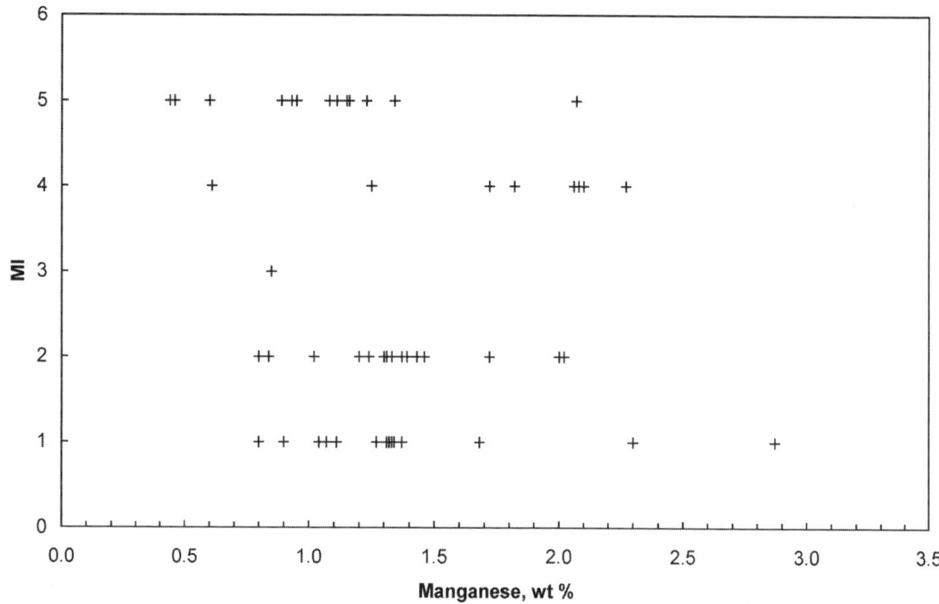

Figure 4: Microfissuring index plotted against alloy content.

After comparing the effect of each chemical on the microfissuring tendencies of the filler metal, a microfissuring factor (MF) was developed. An empirical MF formula was conceived that accounts for compositional effects of each element and provides a quick calculation to the microfissuring tendencies of each filler metal.

Filler Metal Composition

Final production chemistry was determined from the above microfissuring test and is shown in Table 2. A trade name for the new filler metal was also determined- **EPRI P87**.

Table 2: Chemical composition of EPRI P87 electrode.

	C	Si	Mn	P	S	Cr	Mo	Fe	Nb	Ni
Specification	0.1	0.3	1.5	0.008	0.008	9	2	38	1	Bal

Evaluation of As Welded Condition

It was assumed that the filler metal would be used in one of three different heat treatment conditions. These include: 1) as welded, 2) heat treated to the base metal subcritical heat treatment (PWHT) temperature, and 3) heat-treated to the base metal normalized and tempered temperatures (N&T).[1] Most of the early testing was

[1] Note, these heat treatments are based on the normal heat treatment temperatures for the base metal.

performed without any heat treatment, except for a few select tests where in normalization procedure for Grade 91 base metals was used.

The following provides the results of testing performed on as-welded coupons.

Tensile Testing. The room temperature tensile properties of heat D724 are given in Table 3. Heat D724, was selected because it was represented a filler metal heat with a chemical analysis within the optimum range that yielded a low microfissuring index.

Table 3: Room Temperature Strength for heat D724.

Temperature °F (°C)	0.2% proof ksi (MPa)	UTS ksi (MPa)	4d elongation %	Reduction of area %
68 (20)	52 (359)	81.5 (562)	34	49

The strength was comparable to what would be expected from a high nickel alloy austenitic weld metal and it exceeds the tensile requirements for P11/P22 base material given in ASME SA-335. It does not meet the tensile requirements for P91 at room temperature (Table 4). Additional discussion of this issue is provided in the next section.

Table 4: Comparison of room temperature tensile strength new alloy to various alloys (Based on ASME minimums)(6)

Alloy	UTS Ksi (MPa)	Yield Strength Ksi (MPa)
Heat D724	81.5 (562)	52 (359)
IN 182	80.0 (550)	NA
SA 335-P91	85.0 (586)	60.0 (414)
SA 376 TP304H	75.0 (517)	30.0 (207)
SA 335-P22	60.0 (414)	30.0 (207)

To evaluate the strength of the filler material at operating temperatures, two types of high temperature tests were carried out, hot tensile tests (ASTM E21 (7)) and short-term stress rupture tests (aiming for failure in 100 hours). All of the high temperature testing was carried out at 1100°F (593°C). This is the same temperature at which previous stress-rupture tests were carried out (1), and it is within the potential service temperature range for P91.

Stress rupture tests were carried out according to ASTM E139 (8) on longitudinal all-weld metal samples. All of the stress rupture tests were carried out at 593°C (1100°F) and the load for the first set of tests, 43ksi (297MPa) was selected to be about 90-95% of the hot tensile proof stress, aiming for failure to occur in a relatively short time. Based on the hot tensile properties and the result of the first set of stress rupture tests a second series of tests will be carried out for longer duration in Task 3.

Hot Tensile Testing. Hot tensile tests were carried out at 593°C (1100°F). The experimental heats tested showed quite a variation in hot strength ranging from 35.5ksi (245MPa) to 50.3ksi (347MPa) 0.2% proof stress; and 52.7ksi (363MPa) to 76.9ksi (530MPa) ultimate tensile strength.

The tensile strength and proof stress achieved with the heats containing niobium and carbon all exceed the requirements of P91 base material at 1100°F (593°C). Even though the weld metal does not meet the room temperature strength requirement of P91, it will exceed the P91 base material strength at typical operating temperatures. Table 5 and Figures 5 and 6 provide comparisons of the room temperature and elevated temperature (1100°F/593°C) tensile and yield strength of heat D724 (note the room temperature properties from Table 3 have been repeated here). The primary area of interest is above 900F as noted by the circles. These results show that over this temperature range there is only about a 20-25% drop in strength compared to the 30-50% drop seen for ferritic steels. This comparatively low drop in strength means that although the weld metal does not match the room temperature strength of P91 it exceeds its strength at 1100°F (593°C).

Table 5: Comparison of room and elevated temperature strength for heat D724 and P91 base metal at 1100F.

Temperature °F (°C)	Yield Strength Ksi (MPa)	UTS Ksi (MPa)	4d elongation %	Reduction of area %
68 (20) D724	52 (359)	81.5 (562)	34	49
1100 (593) D724	42 (290)	61 (420)	13.5	26
1100 (593) P91	32.7(226)	41.3(285)	--	--

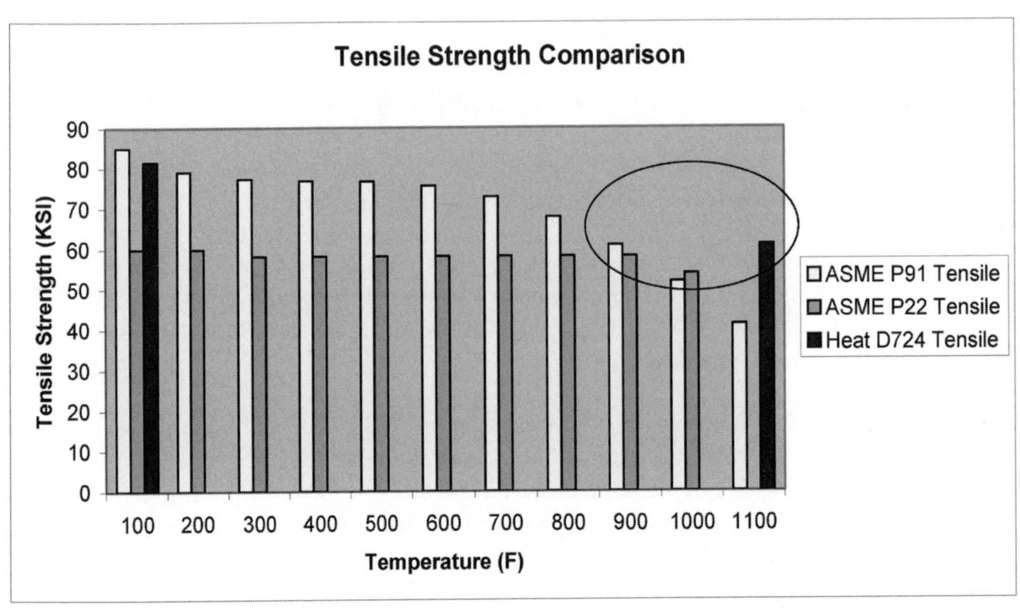

Figure 5: Tensile strength versus temperature for heat D724 compared to P91 and P22 ASME data. Note: the 1100F tensile strength for heat D724 exceeds that of both P91 and P22.

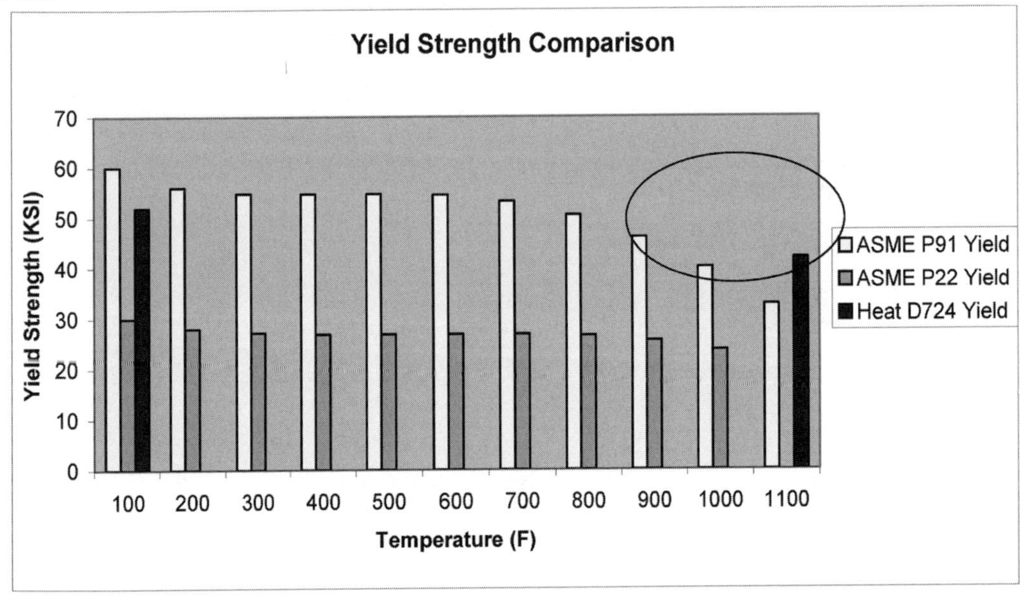

Figure 6: Yield Strength versus temperature for heat D724compared to P91 and P22 ASME data. Note: the 1100F yield strength for heat D724 exceeds that of both P91 and P22.

Toughness Testing. One of the goals for the program was to develop a filler metal that would demonstrate toughness of at least 20 ft-lbs (27J) in the as welded condition. To evaluate this, Charpy tests were carried out on heat D724. At 68°F (20°C) an average of 60 ft-lb (82J) in the as-welded condition was considered to be more than adequate for any of the intended applications. See Table 6 for the individual Charpy values and lateral expansion.

Table 6: Room temperature impact properties of heat D724

Test temperature °F (°C)	Charpy energy ft-lb (J)	Lateral expansion mils (mm)
68 (20)	67, 58, 57 (91, 78, 77)	56, 49, 48 (1.41, 1.23, 1.22)

Short Term Stress Rupture Tests. Four heats (D927, D923, C269 and C273) were selected for initial stress rupture tests, including one heat containing boron (C273). The load for the first series of tests, 43ksi (297MPa), was about 90-95% of the proof stress of the weld metal at 1100°F (593°C) and was selected with the intention of achieving a rupture life of around 100 hours. Table 7 shows the results of the stress rupture tests. The original development work on the HSF6 filler metal was tested at a much lower stress (35ksi) and failed in a substantially shorter time (15 hours). **These results show a significant improvement over the original HSF6 filler metal.**

Table 7: Stress rupture test results

Heat	Test temperature °F (°C)	Load ksi (MPa)	4d rupture elongation %	Rupture reduction of area %	Time to rupture hours
D923	1100 (593)	43 (297)	8.8	33.2	155
	1400 (760)	12 (83)	6.0	25.8	354
D927	1100 (593)	43 (297)	9.0	27.2	247
	1400 (760)	12 (83)	1.8	6.2	692
C269	1100 (593)	43 (297)	11.7	39.0	204
	1400 (760)	12 (83)	2.3	10.9	401
C273	1100 (593)	43 (297)	8.5	28.9	216
	1400 (760)	12 (83)	6.1	38.3	694

Figure 7 shows a Larson Miller Plot for this alloy compared to other nickel based alloys.

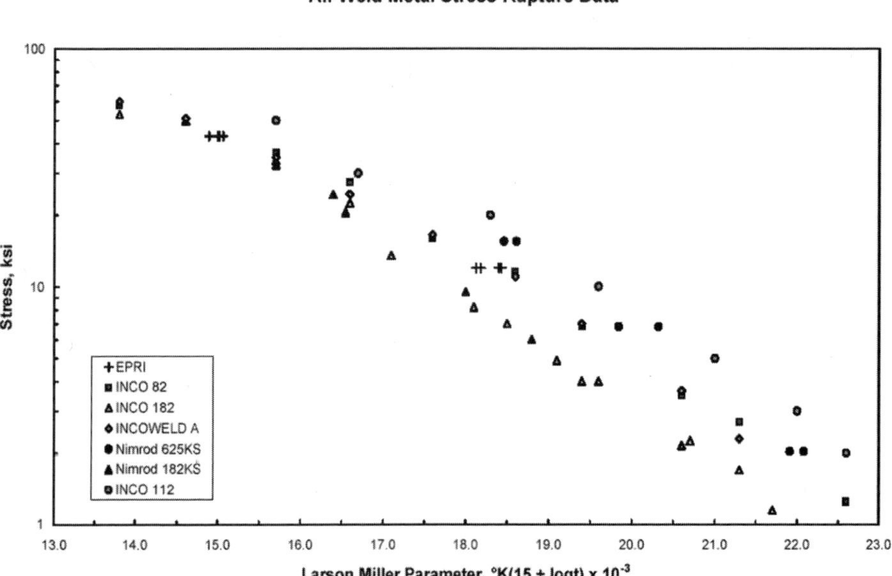

Figure 7: Larson Miller Plot of Short Term Rupture Tests

Evaluation of Normalize and Temper Weldment

A proposed application for the dissimilar weld metal was to allow for buttering of weld preparations in P91 components in a shop location. These components could then be normalised and tempered (N+T) prior to leaving the shop to eliminate the Type IV damage location and offer greater service lives than joints made with conventional B9 filler metal. An added benefit is that the components could be joined in the field without field PWHT.

As a preliminary check to establish whether there may be any loss of weld metal strength in this application, a hot tensile test was carried out on a representative heat of electrodes following a N+T heat treatment of 1940°F (1060°C) for one hour which was air cooled followed by 1400°F (760°C) temper for two hours and air cooled. This N+T heat treatment was selected as being typical of that applied to P91 base material.

Hot tensile. Heat D925 which had previously been subject to a hot tensile test and produced 49.5ksi (341MPa) 0.2%proof stress was subject to another all-weld metal hot tensile test but following a normalize and temper (N+T) heat treatment. Following a N+T heat treatment the 0.2% proof stress at 1100°F (593°C) was reduced to 33ksi (226MPa); details of results are given in Table 8. This equates to a reduction in proof stress of about a third following N+T and a reduction in UTS of about one sixth. The N+T heat treatment was selected because it was typical of that applied to P91 base material (normalize: 1940°F/1060°C one hour air cool + temper: 1400°F/760°C two hours air cool).

Table 8: Comparison of as-welded and N+T hot strength (1100°F/593°C) of heat D925

Condition	0.2% proof ksi (MPa)	UTS ksi (MPa)	4d elongation %	Reduction of area %
As-welded	49.5 (341)	76.5 (528)	21	24
N+T	33 (226)	63 (435)	24.5	33

N+T: 1940°F (1060°C) / 1 hour AC + 1400°F (760°C) / 2 hours AC

These properties were just in excess of the minimum 0.2% proof stress requirement of 211MPa (30.6ksi) at 593°C (1100°F) for P91 base material according to BS EN 10222-2:2000. These properties may require further evaluation but they were considered satisfactory for initial strength validation as a buttering layer subject to N+T.

Thermal Expansion

Thermal expansion of the filler metal was also tested. The difference between ER309 SS filler metal and low alloy steels was one of the contributing factors to premature failures in DMW's. It was desirable for the filler metal to demonstrate thermal expansion close to Grade 22 and Grade 91. Thermal expansion of EPRI P87 is shown in Table 9.

Table 9: Thermal Expansion of EPRI P87

Mean Thermal Expansion Tests	
Material	CTE in/in/F 70F-1000F
CS-Gr 22	8.1
Gr 91	7.0
In 626	7.9
P87	7.5

Task 2: PREPARATION OF COUPONS FOR LONG TERM STRESS RUPTURE TESTS

Following the short-term stress rupture test, the most promising alloy composition (EPRI P87) was used to prepare 220 lb (100KG) of SMAW filler metal. To evaluate how well the new filler metal will perform in service a series of stress rupture test was proposed. The tests were originally proposed to be cross weld tests but this joint geometry would introduce two fusion zones and HAZ which would make comparison of results difficult. Instead tests were composite rupture samples containing one half base metal and one half weld metal with the fusion line centered in the gage length (Figure 8).

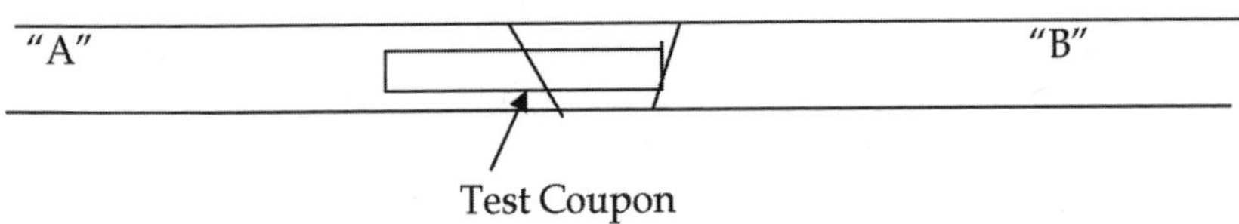

Figure 8: Coupon Location to Weld Joint

The first set of coupons (Coupons 1, 2, and 3) was prepared to evaluate the new filler metal in the standard PWHT condition. All coupons will utilize Gr 91 on the "B" side of the weldment. The "A" side was varied between three materials; Gr 22, Gr 91, and 304 SS with coupon #1 using Gr 91, coupon #2 using Gr 22, and coupon #3 using 304 SS. Coupons were prepared using 1 ½" thick plate. The coupons on side "A" are detailed below.

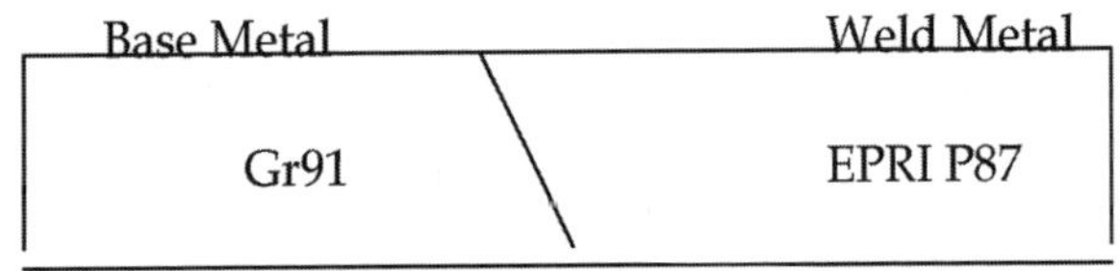

Figure 9: Coupon #1 using Gr 91 base metal

The first coupon (Figure 9) investigates a standard Gr 91 to Gr 91 weldment to determine if the newly developed filler metal shows any improvement over standard 9081-B9 filler usually utilized in these joints. Of particular concern is if the filler metal will demonstrate at least the same life in high stress short-term tests as B9 and to see if long term/low stress tests result in a type IV failure. The coupon was welded with a 400F (200C) preheat, cooled to room temperature and then subjected to a PWHT at 1400F (760C). A picture of the completed block and a cross section are shown in Figure 10.

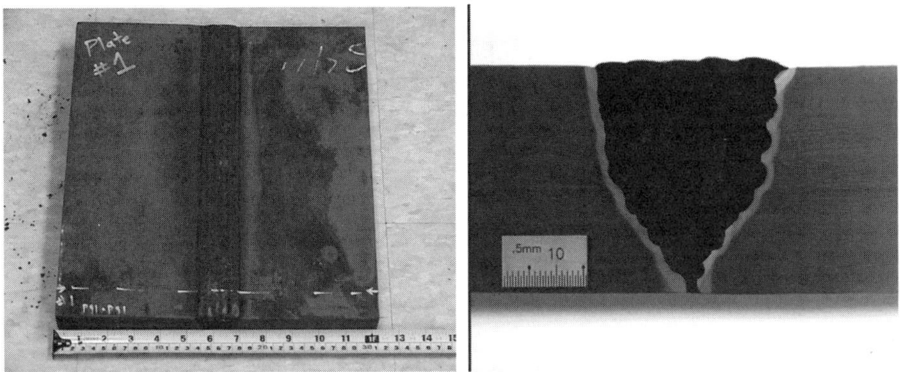

Figure 10: Completed Coupon #1 and cross section of weld.

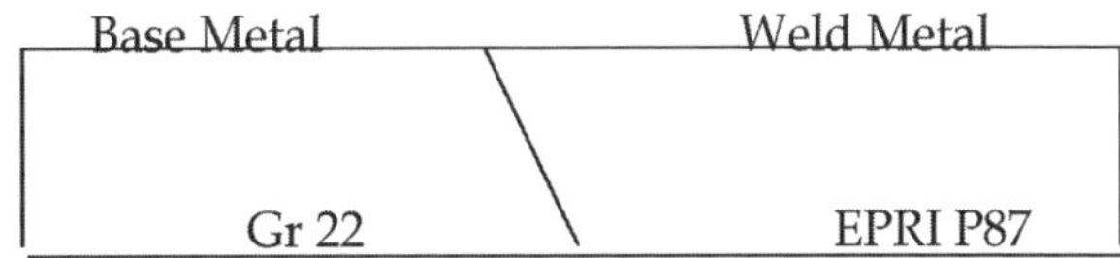

Figure 11: Coupon #2 using Gr 22 base metal

The second coupon (Figure 11) was the same as coupon number 1 except was prepared using SA 387 Gr22 base metal and was subjected to 1350F (732C) PWHT. The purpose of this coupon is to determine if any problems would develop with the filler when welding to lower alloy base metals. Carbon migration tests will also be performed on this coupon as the project continues.

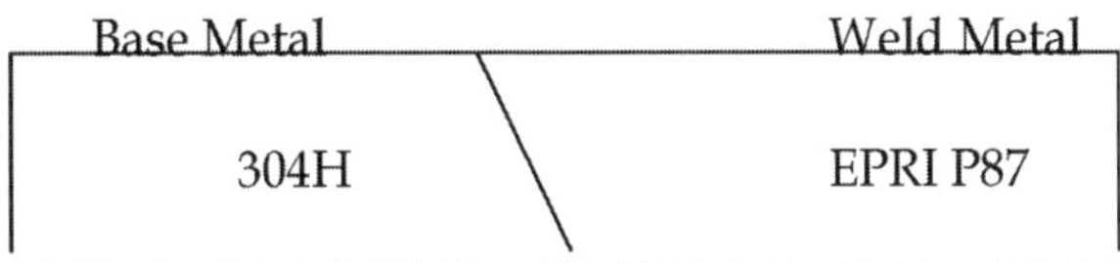

Figure 12: Coupon #3 using 304H base metal

Stainless steel (Type 304H) was used for coupon #3 (Figure 12). No PWHT was utilized. This coupon demonstrates the suitability of the filler metal being used with austenitic base metals in the as welded condition.

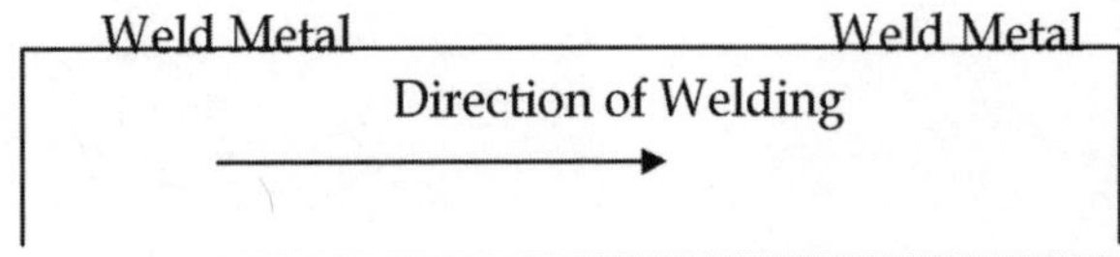

Figure 13: All weld metal Coupon #4

The previous derivative of this filler metal indicated a potential for hot cracking transverse to the weld. This test (Figure 13) was performed using a weld metal coupon with the sample removed parallel to the direction of welding. The samples received a 1400F (760C) PWHT because this is the condition that would be used on P91 joints, which are the major concern of this development.

Effect of Buttering

One of the goals of this filler metal was to eliminate field PWHT by allowing for buttering of the component followed by a standard sub-critical heat treatment in the factory. The components would then be shipped to the construction site for erection. To evaluate if a buttering followed by heat treatment would provide good service two coupons were prepared and tested.

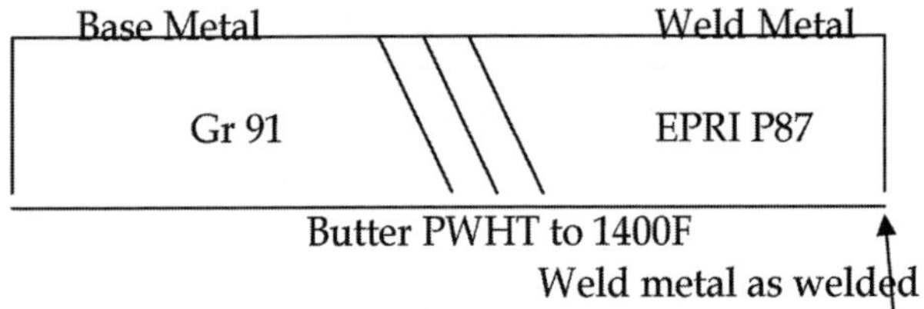

Figure 14: Coupon #5 using Gr 91 with weld butter

Coupon #5 (figure 14) has a two layer butter welded to SA387-Gr91 with a 400F (200C) preheat, allowed to cool and then PWHT to 1400F (760C). The coupon was machined and then the weld was completed without PWHT. Testing will be performed with most of the weld in the "as welded" condition (except for the butter area).

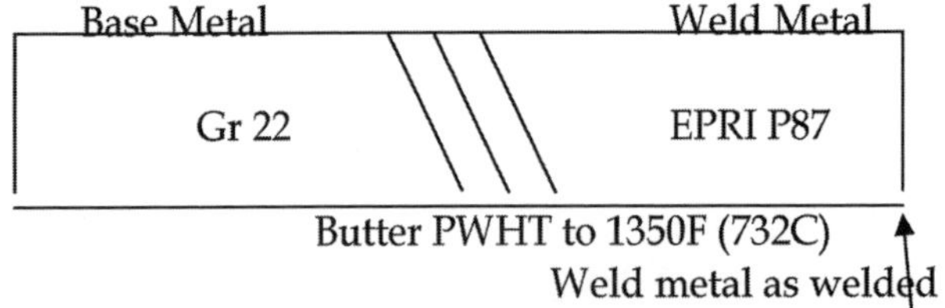

Figure 15: Coupon #6 using Gr 22 with weld butter

Coupon #6 (Figure 15) was made like coupon #5 above except the base material used was SA387-Gr22 and a 1350F (732C) PWHT was utilized on the butter. Figure 16 shows the buttered side after welding and heat treatment prior to completing the joint.

Figure 16: Coupon #6 after buttering and heat treatment prior to completion of welding.

Elimination of Type IV Failure Location

One of the problems associated with the manufacture of components is that the weldments fail in the Type IV location (fine grain heat affected zone) sooner than ruptures would occur in the weld metal or base metal (9, 10, 11). The heat from the welding process heats the metal in this zone above the lower critical temperature of the metal. P91 has been shown to loose its long-term rupture properties when heated to this level and hence the premature failures.

These failures are generally called weld failures because they are associated with welds but they are not associated with the filler metal. Developing a new filler metal will not by itself solve this problem. One goal of the current research program was to develop a welded joint what would have rupture strength equal to or better than the base metal. This would result in longer service lives of welded components.

The only way to restore properties to the HAZ is through a normalization heat treatment. Unfortunately this requires being able to join the components together after the heat treatment without creating a new HAZ. The proposed development involves buttering the P91 base metal with the newly developed alloy (EPRI P87) sufficiently thick to insulate the base metal from the heat of subsequent welding. Then a normalization and temper heat treatment would be performed to eliminate the HAZ.

Then the components could be joined without any additional heat treatment. This requires a filler metal that does not require PWHT to develop adequate properties for service. The filler metal developed during this project offers this type of improvement in performance.

To evaluate the filler metal in this type of joint, coupon #7 was produced (Figure 17).

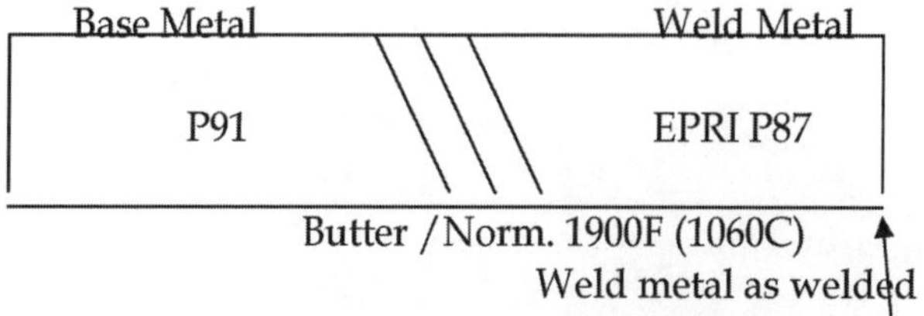

Figure 17: Coupon #7 using a butter that was normalized.

To simulate buttering in the shop followed by normalization and field erection w/o PWHT, halves of the coupon were preheated to 400F (200C), buttered with two layers of filler metal, and allowed to cool. They were then normalized at 1900F (1060C) and allowed to cool to room temperature. The blocks were then tempered at 1400F (760C) prior to machining a new weld bevel. The coupons were then joined using room temperature preheat and allowed to cool before machining and rupture testing. No subsequent heat treatment was performed.

Stress Rupture Testing

The seven coupons above will be subjected to stress rupture testing with planned failures occurring at 250, 500, 750, 1000, 2500, and 5000 hours resulting in 48 tests.

Figure 18 shows the results from Coupon #1 using standard PWHT and Grade 91 plate. The solid line is the Grade 91 base metal minimum line. The dashed line is from other EPRI work using the same heat of base metal as used in this project but welded with E9018-B9 filler metal and PWHT for two hours at 1400F (760C). All failures with the new filler metal, EPRI P87, were in the base metal in the type IV location. No significant benefit was seen form the new filler metal but it was at least as good as a B9 filler metal.

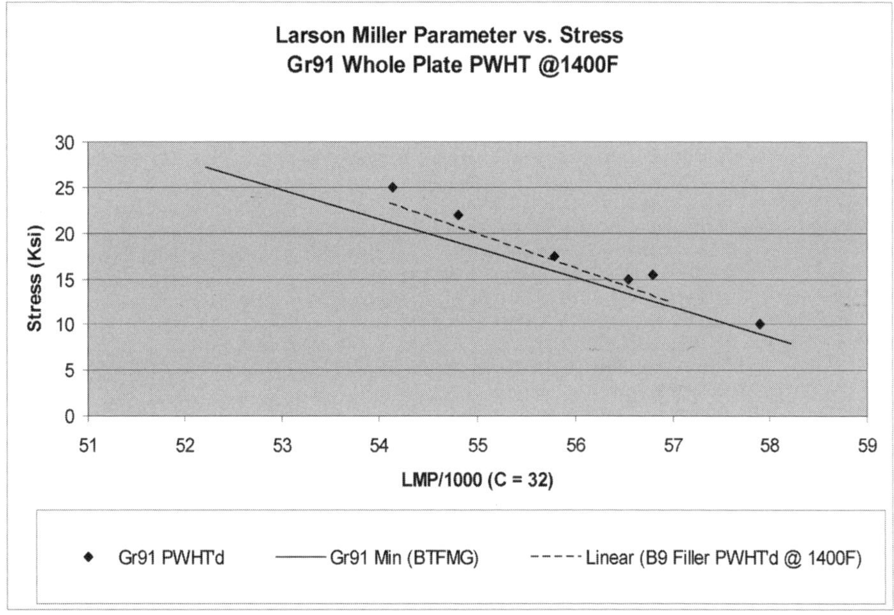

Figure 18: Rupture data for Coupon #1

Figure 19 shows the results from Coupon #2 using standard PWHT and Grade 22 plate. The solid line is the Grade 22 base metal minimum line used in the ASME Code. All failures with the new filler metal, EPRI P87, were in the base metal. No significant benefit was seen form the new filler metal but it was at least as good as a B3 filler metal.

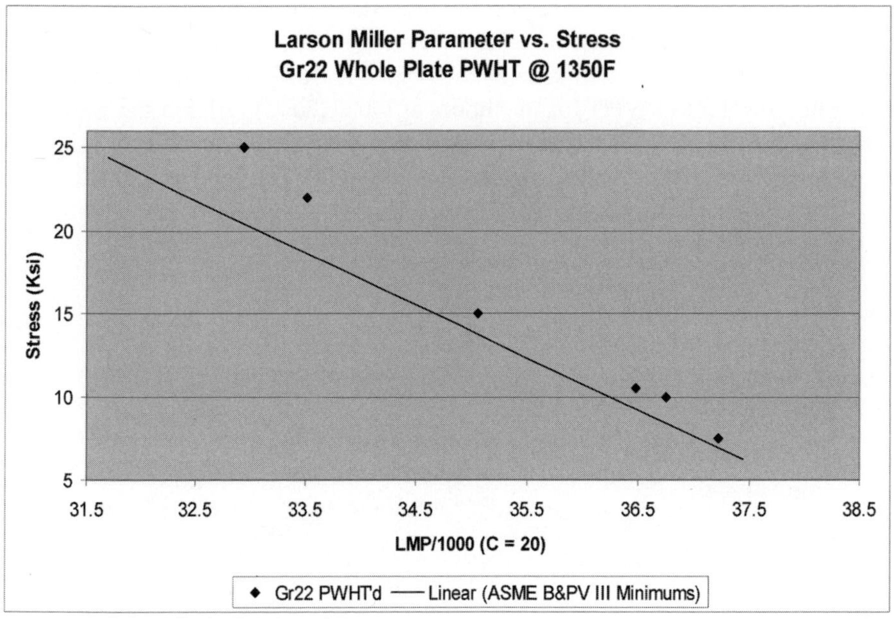

Figure 19: Rupture data for Coupon #2

Figure 20 shows the results from Coupon #3 using no PWHT and 304H SS plate. The solid line is the 304H SS base metal minimum line from EPRI's Boiler Tube Failure manuals. All failures with the new filler metal, EPRI P87, were in the base metal. No significant benefit was seen form the new filler metal but it was at least as good as an ER309 filler metal.

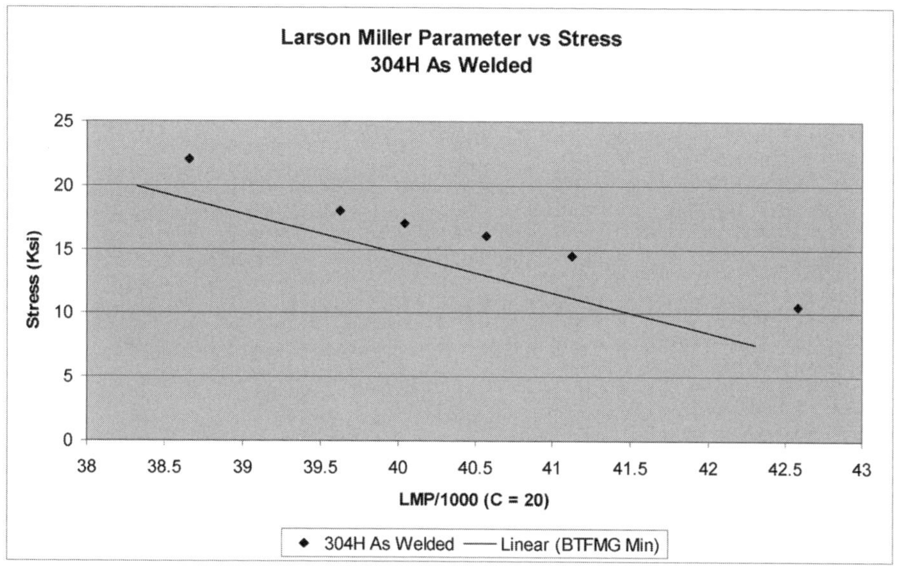

Figure 20: Rupture data for Coupon #3

Figure 21 shows the results from Coupon #4 which was the all weld metal coupon. The purpose of this test was to determine the actual creep rupture strength of EPRI P87 filler metal. The solid line on the left is for Grade 22 base metal, the dotted line is for 304H base metal and the dashed line is for Grade 91 base metal. The filler metal clearly demonstrates significant creep strength improvements over common filler metals used in high temperature joints with the materials tested in this program.

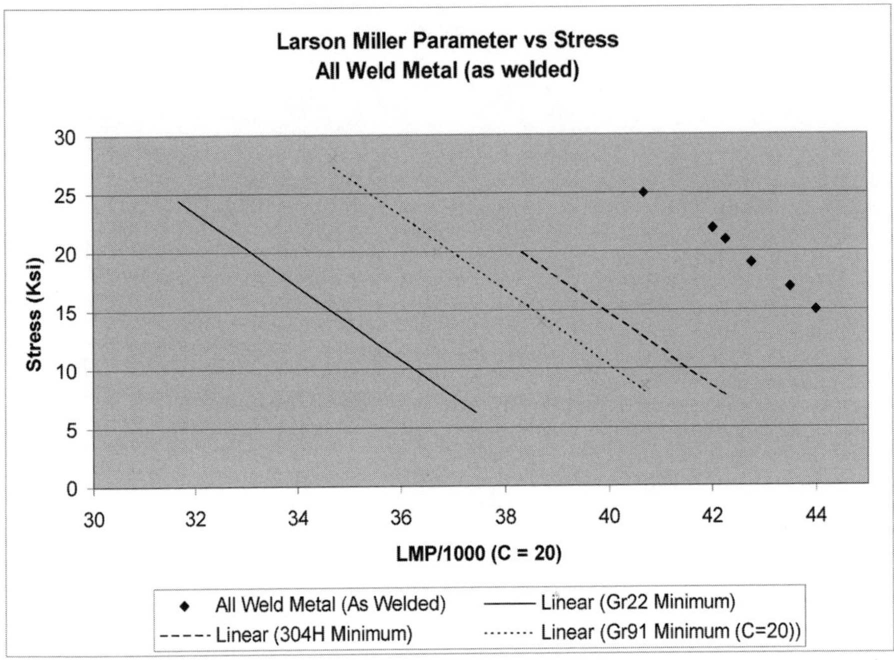

Figure 21: Rupture data for Coupon #4

Figure 22 shows the results from Coupon #5 which was tested to demonstrate if Grade 91 base material could be buttered and PWHT'ed in the shop followed by erecting in the field with no subsequent PWHT. The solid line on the left is for Grade 91 base metal, the squares are from coupon #1 and the diamonds are the results for this coupon. Again, all failures were in the Type IV location. No benefit was demonstrated in rupture life for this process over a standard joint using B9 filler and PWHT although, no determent was found for performing PWHT in the shop. This demonstrates that significant time savings may be realized by batch heating the components in an oven in the shop and then erecting with out any additional PWHT in the field. The reader is cautioned though that residual stresses should be considered when using this option.

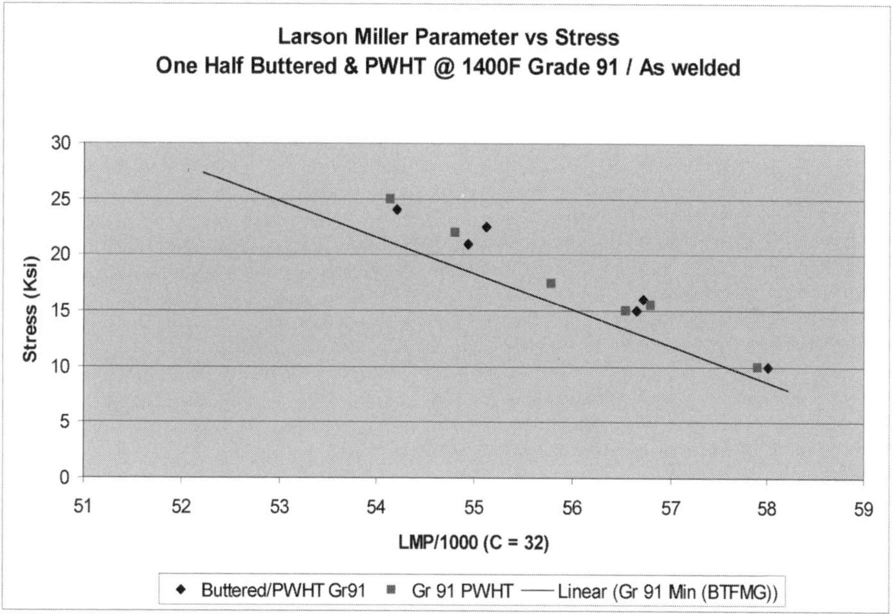

Figure 22: Rupture data for Coupon #5

Figure 23 shows the results from Coupon #6 which was tested to demonstrate if Grade 22 base material could be buttered and PWHT'ed in the shop followed by erecting in the field with no subsequent PWHT. The solid line on the left is for Grade 22 base metal, the squares are from coupon #2 and the diamonds are the results for this coupon. No benefit was demonstrated in rupture life for this process over a standard joint using B3 filler and PWHT although, no determent was found for performing PWHT in the shop. This demonstrates that significant time savings may be realized by batch heating the components in an oven in the shop and then erecting with out any additional PWHT in the field. The reader is cautioned though that residual stresses should be considered when using this option.

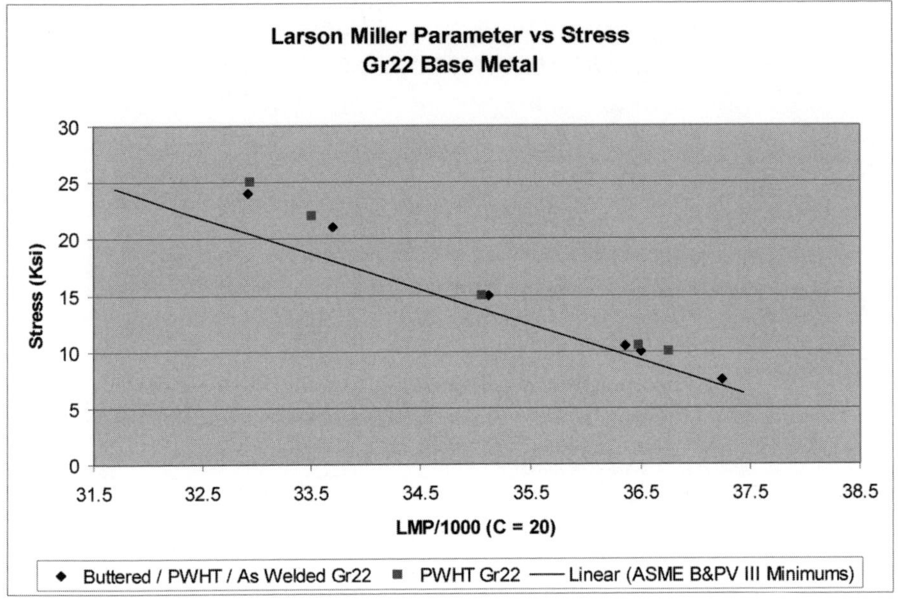

Figure 23: Rupture data for Coupon #6

Figure 24 shows the results from Coupon #7 which was tested to demonstrate if additional rupture life could be developed by eliminating the Type IV location through a N&T heat treatment of the Grade 91 base metal after buttering the coupons. The solid line on the left is for Grade 91 base metal, the dashed line if from Coupon #1 data and the diamonds are the results for this coupon. Significant benefit was shown in rupture life from elimination of the HAZ. Failures occurred in the base metal significantly away from the Type IV location or close to the fusion line.

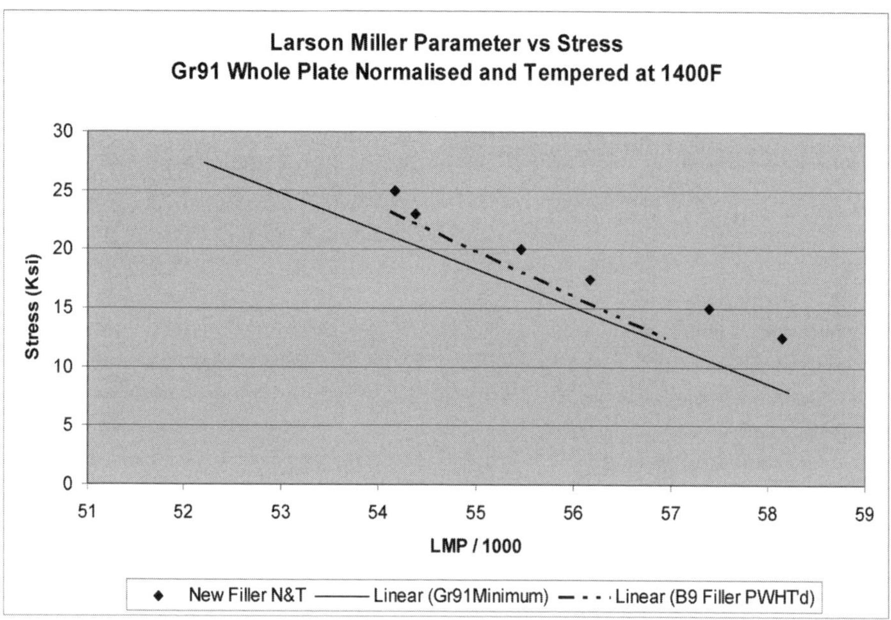

Figure 24: Rupture data for Coupon #7

Carbon Migration

Failure mechanisms determined to limit life in DMW joints was the migration of carbon and formation of carbides at the fusion line. The driving force for this carbon migration is the different levels of chromium across the weld joint with traditional nickel based or austenitic alloys. Carbon migration was one of the primary reasons this filler metal was developed. As shown in Table 2, EPRI P87 only contains 9% chromium. Carbon migration was first tested by running single bead on plate coupons. One plate was used and three different beads were placed on the plate. The first was 309SS filler metal, the second used In 625 filler, and the third used EPRI P87 filler metal. The chromium content for ER 309 is 22% to 25%. The chromium content of In 625 is 20% to 23%. As noted in Table 2, EPRI P87 only contains 9% chromium.

After welding, the plate was normalized at 1900F for two hours and then tempered at 1400F (760C) for two hours. Figures 25, 26, and 27 show results of carbon migration testing for these three filler metals. Figure 28 shows carbon migration from one of the rupture samples from Coupon #7. This coupon had a 2 hr normalization heat treatment at 1900F (1060C), followed by two hrs at 1400F (760C), and finally exposed to 4400 hrs at 1150F (620C) during rupture testing. No carbon migration was noted in this coupon.

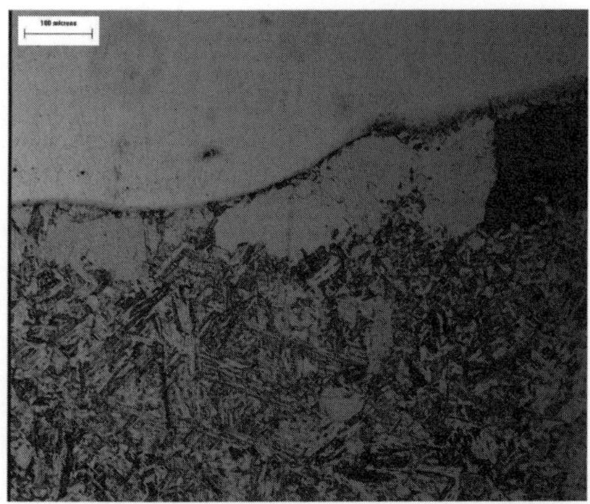

Figure 25: Carbon migration using ER 309 filler metal

Figure 26: Carbon migration using In 625 filler metal

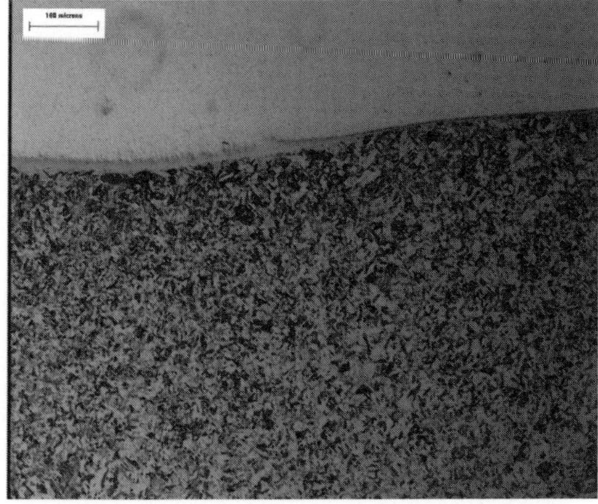

Figure 27: Carbon migration using EPRI P87 filler metal

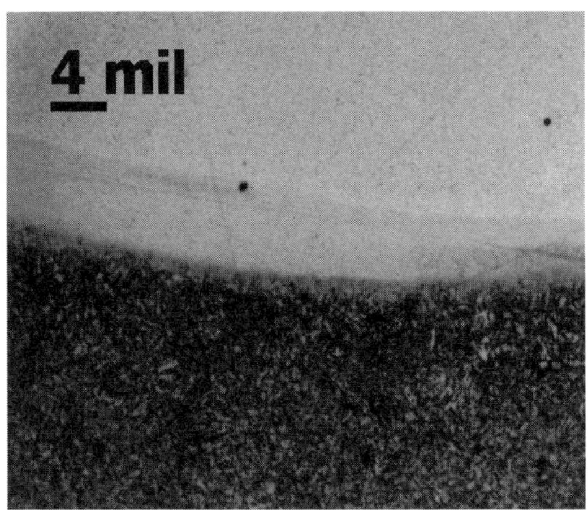

Figure 28: Carbon migration using EPRI P87 filler metal after rupture testing

AVAILABILITY

EPRI P87 is currently available for the SMAW process in common sizes and will soon be available for GTAW and GMAW process. Metrode Products LTD. has a current license to manufacturer and sell the filler metal. Figure 29 shows packaging of the SMAW filler metal in 3/32", 1/8", and 5/32" (2.5mm, 3.2mm, and 4.0mm).

Figure 29: EPRI P87 SMAW filler metal

CONCLUSIONS

- A novel microfissuring test was used which was proved to be a simple and effective method for discriminating microfissuring susceptibility. This test was used successfully to rank the behavior of different variants to a baseline composition (HFS6) in a large matrix of experimental SMAW electrodes.
- A baseline composition similar to the HFS6 composition established in the earlier EPRI work was produced. This composition developed microfissuring in much the same manner as the HFS6 alloy.
- Modifications to the baseline alloy composition were completed yielding an alloy that was virtually microfissure free in the deposited condition.
- Ambient temperature tensile properties will not match P91 base material but will match lower alloyed CrMo base material e.g. P11/P22. However, properties at operating temperature will surpass the P91 base metal strength.
- The elevated temperature yield and tensile strength of the new alloy comfortably exceeds P11, P22 and P91 base material requirements at 1100°F (593°C). Hot ductility of the modified composition was determined to be satisfactory.
- Carbon migration and carbide formation was virtually eliminated through the use of low chromium content.
- Rupture testing demonstrated no significant benefit or degradation compared to standard joints in Grade 91, Grade 22, or 304H base metals and common filler metals.
- All weld metal rupture tests demonstrate that the new filler metal, EPRI P97, is significantly stronger than any of the base metals tested.
- For some applications, buttering of weld joints followed by shop PWHT and subsequent erecting in the field without additional PWHT may be acceptable.
- Significant joint rupture life improvements may be possible with this filler through a butter/N&T/Erection process.

ACKNOWLEDGEMENTS

The authors of this paper would like to thank G B Holloway, A W Marshall, and J A Sanderson of Metrode Products Ltd and W F Newell of Euroweld Ltd. for conducting the evaluation of the new filler metal this report is based upon.

We would also like to thank the R. Viswanathan for his vision and the Metal Properties Council, Inc and GA Technologies for their support in the earlier EPRI work to develop the HSF6 filler metal that was used as a starting point for the current research.

REFERENCES AND BIBLIOGRAPHY

1. Dissimilar weld failure analysis and development program, Volume 1: Executive Summary; EPRI Report CS-4252, volume 1, Research project 1874-1, November 1985.
2. Dissimilar weld failure analysis and development program, Volume 9: Optimised filler metal development; EPRI Report CS-4252, volume 9, Research project 1874-1; draft final report, March 1987.
3. C. Lundin et al, Carbon Migration in Cr-Mo Weldments Effect on Metallurgical Structure and Mechanical Properties, WRC Bulletin 407, December 1995
4. G. Holloway, et al, Development of Advanced Methods of Joining Low Alloy Steel, February 2003.
5. ANSI/AWS A5.4-92, Specification for Stainless Steel Electrodes for Shielded Metal Arc Welding
6. ASME Section II, Part D, 2001
7. ASTM E21, Standard Test Methods for Elevated Temperature Tension tests of Metallic Materials, 1992
8. ASTM E139, Standard Practice for Conducting Creep, Creep-Rupture, and Stress-Rupture Tests of Metallic Materials, 1990
9. R. Viswanathan, Review of Type IV Cracking in Piping Welds, EPRI Report TR-108971, October 1997
10. C. Middleton et al, An Assessment of the Risk of Type IV Cracking in Welds to Header, Pipework, and Turbine Components Constructed from the Advanced Ferritic 9% and 12% Chromium Steels, presented at the 3rd Conference on Advances in Material Technology for Fossil Power Plants, University of Wales Swansea, April 2001.
11. D. Gandy et al, Performance Review of P/T91 Steels, EPRI Report 1004516, November 2002

Advances in Materials Technology for Fossil Power Plants
Proceedings from the Fifth International Conference
R. Viswanathan, D. Gandy, K. Coleman, editors, p 968-981

DOI: 10.1361/cp2007epri0968

Overview of Oxy-Combustion Technology for Utility Coal-Fired Boilers

Brian Vitalis
Riley Power Inc.
(a Babcock Power Inc. company)
5 Neponset St, PO Box 15040, Worcester, MA, USA 01615-0040

Abstract

With nearly half of the world's electricity generation fueled by coal and an increasing focus on limiting carbon dioxide emissions, several technologies are being evaluated and developed to capture and prevent such emissions while continuing to use this primary fossil energy resource. One method aimed at facilitating the capture and processing of the resulting carbon dioxide product is oxy-combustion. With appropriate adjustments to the process, the approach is applicable to both new and existing power plants.

In oxy-combustion, rather than introducing ambient air to the system for burning the fuel, oxygen is separated from the nitrogen and used alone. Without the nitrogen from the air to dilute the flue gas, the flue gas volume leaving the system is significantly reduced and consists primarily of carbon dioxide and water vapor. Once the water vapor is reduced by condensation, the purification and compression processes otherwise required for carbon dioxide transport and sequestration are significantly reduced.

As an introduction to and overview of this technology, the paper summarizes the basic concepts and system variations, for both new boiler and retrofit applications, and also serves as an organized review of subsystem issues identified in recent literature and publications. Topics such as the air separation units, flue gas recirculation, burners and combustion, furnace performance, emissions, air infiltration issues, and materials issues are introduced.

Introduction

With the goal of mitigating global warming, much research is focused on ways to capture the carbon dioxide (CO_2) produced from fossil fuels use. Roughly 85% of the world's energy needs are supplied by fossil fuels. While coal use represents only a quarter of worldwide energy release for all purposes, it is favored for use in large power plants and fuels roughly half of the electricity generated in the USA and worldwide. Being large stationary sources of emissions, power plants are excellent candidates for application of highly effective emissions control systems, and significant progress has been achieved in reducing the emissions of particulate, nitrogen oxides, sulfur dioxide, and mercury from coal-fired plants. Conceptually, there are three main approaches to capturing CO_2 from combustion of fossil fuels: 1. Pre-combustion, or decarbonization in IGCC applications with water-gas shift reaction, 2. Post-combustion, such as amine scrubbing, and 3. Oxy-combustion, to produce CO_2-rich flue gas. Oxy-combustion is attractive because it essentially combines advantages of the other two: the cost of CO_2 capture and plant efficiency are comparable to IGCC, and it can be applied as a retrofit to many existing coal-fired plants. However, in all cases, capture of CO_2 will pose significant energy and economic penalties, and the pursuit of emissions capture should be balanced with sensitivity to the associated increase in direct heat release to surroundings, as well as the accelerated consumption of resources.

Oxy-combustion itself is not a new concept, in fact it was used for high temperature applications such as welding and metal cutting in the 1940's and moved into the aluminum, cement, and glass industries in the 1960's [1, 2]. In the 1980's, Abraham proposed its use specifically to produce CO_2 for Enhanced Oil Recovery (EOR) [1] and in the 1990's it gained further attention for NO_x reduction [3]. Until now, it has not been widely adopted because the energy requirement for oxygen separation makes implementation for these uses too expensive. However, now with the potential valuation of limiting CO_2 emissions, the economics are being changed again and oxy-combustion is seen as a way to produce a CO_2-rich flue gas that is easier to process for sequestration than conventional air-fired flue gas [4].

Oxy-Combustion – Basic Concepts

In conventional combustion, air (21% oxygen, balance mainly nitrogen) is used as the convenient oxygen source for burning of fuel. The nitrogen from the air is mostly inert in this process (a trace of it is oxidized to NO_x) and ends up mixed with the combustion products. As shown in Figure 1, typical flue gas from burning bituminous coal with air contains about 74% nitrogen, 14% carbon dioxide, and 8% water vapor, with other species. At the tremendous quantities of flue gas involved in power generation, if carbon dioxide is to be captured and sequestered, practical limitations dictate that it be separated from the other primary components considered benign. The basic premise of oxy-combustion for power plant application is to separate the nitrogen from the oxygen in the air rather than having to separate it from the carbon dioxide in the flue gas. Without the air-borne nitrogen diluting the flue gas, a concentrated CO_2 flue gas results which is also much less in volume. As shown in Figure 1, by using nearly pure oxygen as the oxidant instead of air, the resulting flue gas is primarily CO_2: 61% CO_2, 30% water vapor. Due to the relative absence of nitrogen, the net volume of flue gas (after any recycling) is reduced by 80% compared to the dilute, air-firing condition. After condensation of water vapor, the flue gas is further reduced in volume and about 88% CO_2. This stream is then much easier to process for CO_2 purification and compression than dilute, air-fired flue gas that would require significant scrubbing processes.

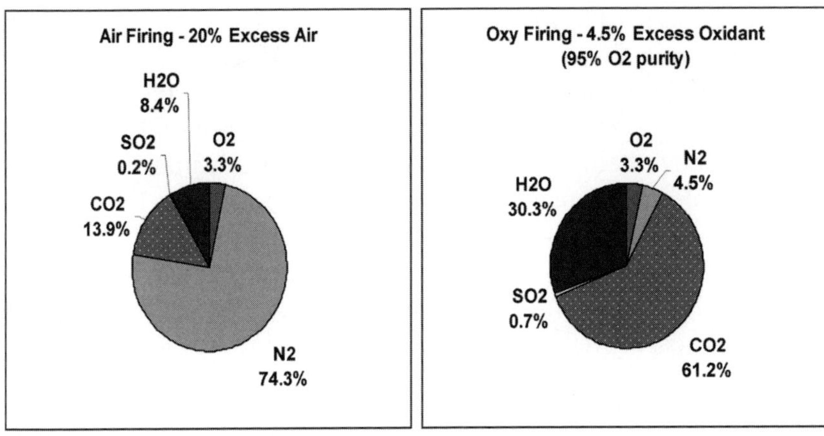

Figure 1: Comparison of Flue Gas Compositions

However, simply combining fuel with oxygen results in very hot flames and, at the scale of coal-fired power generation, some means of controlling furnace temperature is desired to permit use of economical materials and conventional designs. Retrofit of existing boilers is also a consideration, and is discussed later in this paper. In the near term, the use of various diluents (in place of nitrogen from air) is being considered. By far the most popular choice right now is recycled flue gas, which introduces no additional need for flue gas species separation and provides a means for coal drying and transport, though it does require some processing to avoid problems. Other means to control the flame temperature include water or steam injection [5], or more sophisticated burner zone oxygen staging can permit a reduced level of flue gas recirculation (FGR) [6].

A basic schematic of the oxy-combustion system with FGR is shown in Figure 2. In conventional air-firing, the combustion air is preheated in a regenerative heater, cooling the flue gas for heat recovery. In oxy-combustion systems, the air entering the air separation unit is not available for cooling of the flue gas since the air separation unit needs to begin with cool air and has a significant heat rejection load itself. Instead, significant plant integration of feedwater and process heating is needed, taking heat from both the flue gas stream and the air separation unit process. Oxygen is mixed with a diluent - generally FGR as shown here - and burns the fuel. Most proposed schemes have an Electrostatic Precipitator (ESP) handling the combined flue gas stream, so that the recycled gas is relatively clean. Downstream of the recycle take-off, the flue gas volume is much less than that from air-firing, and system equipment requirements vary primarily according to the purity requirements of the ultimate CO_2 sequestration use and destination. In some cases, and for low sulfur coal, conventional FGD and SCR systems may not be required at all and the minor pollutants are sequestered along with the CO_2.

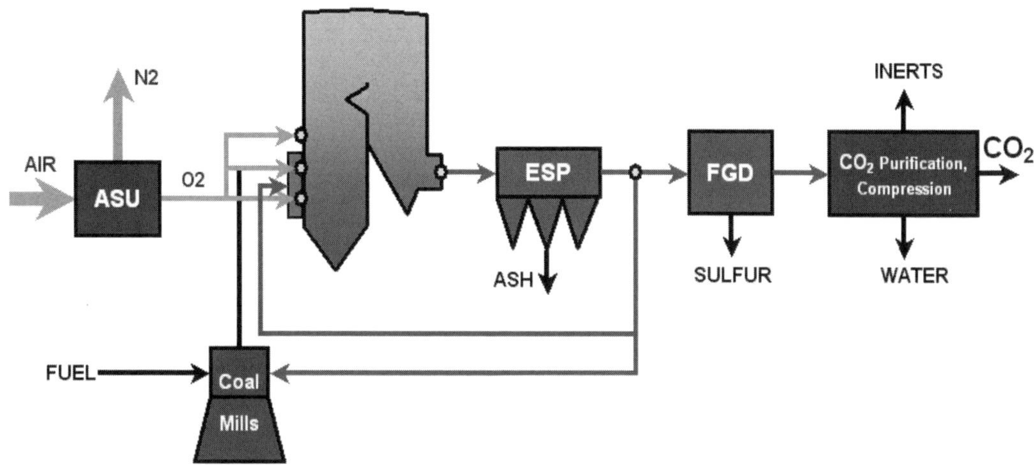

Figure 2: Basic Schematic of Oxy-Combustion Concept

Oxygen Source

At the quantities required for utility power production, the only commercially available oxygen production technology is cryogenic separation. Using this technology, the system is often referred to as an Air Separation Unit (ASU). A typical ASU consists of an air compressor, precooling system, purification unit, heat exchangers, and distillation column. Though it's a well-proven technology, the power involved (210-220 kWhr/ton O_2) is one of the primary barriers to implementation of oxy-combustion [7, 2, 4]. Development continues for several novel, lower-power alternatives such as ceramic membranes (Oxygen Transport Membranes, Ion Transport Membranes, Mixed Conducting Membranes), Ceramic Autothermal Recovery systems, and Chemical Looping, with DOE funding for some of these activities, targeting commercialization by 2010.

Ideally, nitrogen should be entirely eliminated, but ASU energy requirements rise sharply when oxygen purity needs to be above 98% [3, 1], since argon-oxygen separation then becomes a factor. The operating economics must not only be optimized for the oxygen equipment alone, but require consideration of the ultimate CO_2 purity requirements and optimization with removal of the inerts by the CO_2 product recovery system. In general, when recognizing that at least some nominal amount of air infiltration to the boiler system is inevitable, most studies have determined that 95% purity oxygen provides the best economic balance when using an ASU [3, 8, 2, 1].

Flue Gas Recirculation (FGR)

Recycle flue gas is used to moderate the combustion temperatures and to maintain overall heat transfer characteristics, enabling oxy-combustion to be easily retrofitted to existing boilers, and permitting conventional design for new units. For new unit designs, the convective heating surfaces and flue gas handling equipment could be dramatically reduced if FGR could be omitted or reduced, however there are certain practical considerations for the flame and furnace that limit reductions in FGR, as further described in the section **Combustion and Furnace Conditions**, below.

When the recycled flue gas is mixed with the oxidant before entering the burners, the recycle rate has a corresponding diluting effect on the oxygen concentration at the burner. This is an interesting new variable available to the burner designer, as it opens up possibilities for different stoichiometries and oxygen concentrations in various parts of the burner and/or for various parts of the furnace, enabling further control of the flame and furnace conditions. By using such techniques, aided by CFD modeling and pilot-scale testing, there is the possibility of reducing overall FGR rates and the size of the boiler, while still controlling combustion conditions within practical limits.

Wet vs. Dry Recycle

While most of this and other introductory discussions presume that the recycled flue gas is the same condition and composition as that flowing from the boiler, the issue of water vapor removal should be mentioned. *Wet* vs. *Dry* FGR is illustrated in Figure 3 and described below.

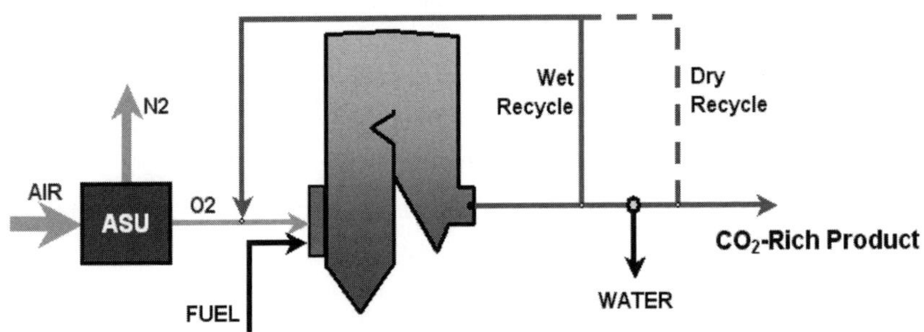

Figure 3: Wet vs. Dry Flue Gas Recirculation

Wet FGR: The recycle gas contains the full water vapor content of the main flue gas. Sometimes referred to as "the practical" approach to FGR, but the high moisture content may cause problems in the FGR duct, fan, and pulverizers.

Dry FGR: Some of the water vapor is removed from the recycle gas stream.

More development is needed in the area of pulverizer performance under a range of FGR conditions. The effect of wet vs. dry recycle does not end at the pulverizer or FGR circuit, but impacts furnace and overall boiler performance by altering the density, heat capacity, radiation, and convection properties of the flue gas mixture.

Regenerative Heating

An interesting combination of solutions is described in a 2005 report sponsored by the International Energy Administration (IEA Report 2005/9) [1]. As shown in Figure 4, the system utilizes a regenerative gas-gas heat exchanger to cool the combined flue gas from 644°F to 518°F in order to optimize cost vs. efficiency of the ESP. The heat is transferred back to the recycled flue gas. The FGR is treated as two separate streams. The primary recycle, serving the pulverizers, gets cooled and dried, while the secondary recycle is "wet", and both are reheated by the gas-gas heat exchanger. Oxygen is introduced to the secondary recycle stream to bring the mixture oxygen content up to 23% oxygen, and the balance of oxygen is fed to the windbox / burner separately.

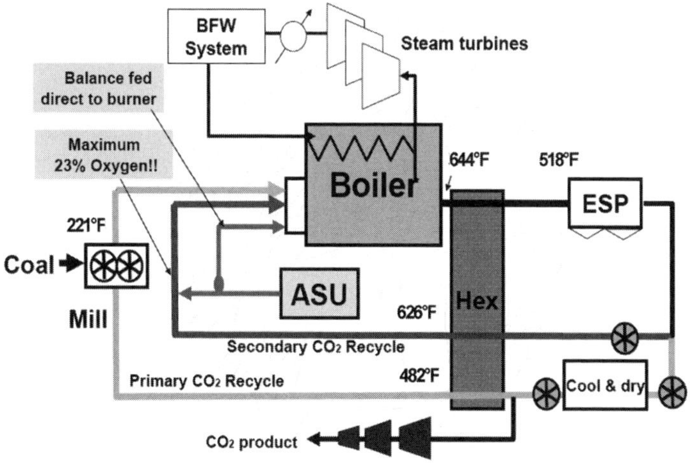

Figure 4: Oxy-Combustion System Schematic with G-G Heat Exchanger

(Figure adapted from (IEA Report 2005/9) via ref. [1]

Combustion and Furnace Conditions

Most combustion science and conventional furnace and heat transfer designs have been developed on the basis of air-firing. Research and test programs have started and are continuing to investigate fundamentals of oxy-combustion environments. At the macro level, oxy-combustion flame stability, overall heat transfer, and thermodynamic performance have been found to be comparable to well-known air-fired conditions [3, 1], confirming good potential for retrofit of existing units, but there are some differences that should be taken into account.

Excess Oxidant

Generally, the basic mechanics of ensuring complete combustion continue to require a modest amount of excess oxygen in the flue gas. The concept of "excess air" may be extended to "excess oxidant", but in any case the proven target of about 3-3.5% oxygen left over in the flue gas is a reasonable amount. Selective use of oxygen concentrations and stoichiometry staging at various burner levels may permit some reduction in the overall excess oxidant level. Experiments are continuing to determine how close an oxy flame can come to stoichiometric [4, 6]. This has significant appeal considering that a primary penalty of oxy-combustion is the energy required to separate the oxygen from air in the first place.

Flame Temperature

Some initial insight to the differences in performance between the firing conditions can be gained by review of relevant gas properties. Figure 5 is a comparison of some of the relevant properties of the gas species; recall that N_2 is the dominant presence in air-firing, and CO_2 and H_2O dominate oxy-firing. In order to use conventional materials and designs - especially a requirement for retrofit applications, it is desired to achieve a conventional adiabatic flame temperature. If the air-fired and oxy-fired gas properties were the same, one might expect that a similar flame temperature would be produced if the oxygen concentration into the burner were diluted (via FGR) to about 21% (such as in air). But since the specific heat of CO_2 is greater than that of N_2, the adiabatic flame temperature is suppressed and less FGR dilution effect is needed. Several studies have shown that to match the flame temperature from air-firing, the oxygen at the burner should only be diluted to about 30%, requiring about 60% FGR [9, 1, 10].

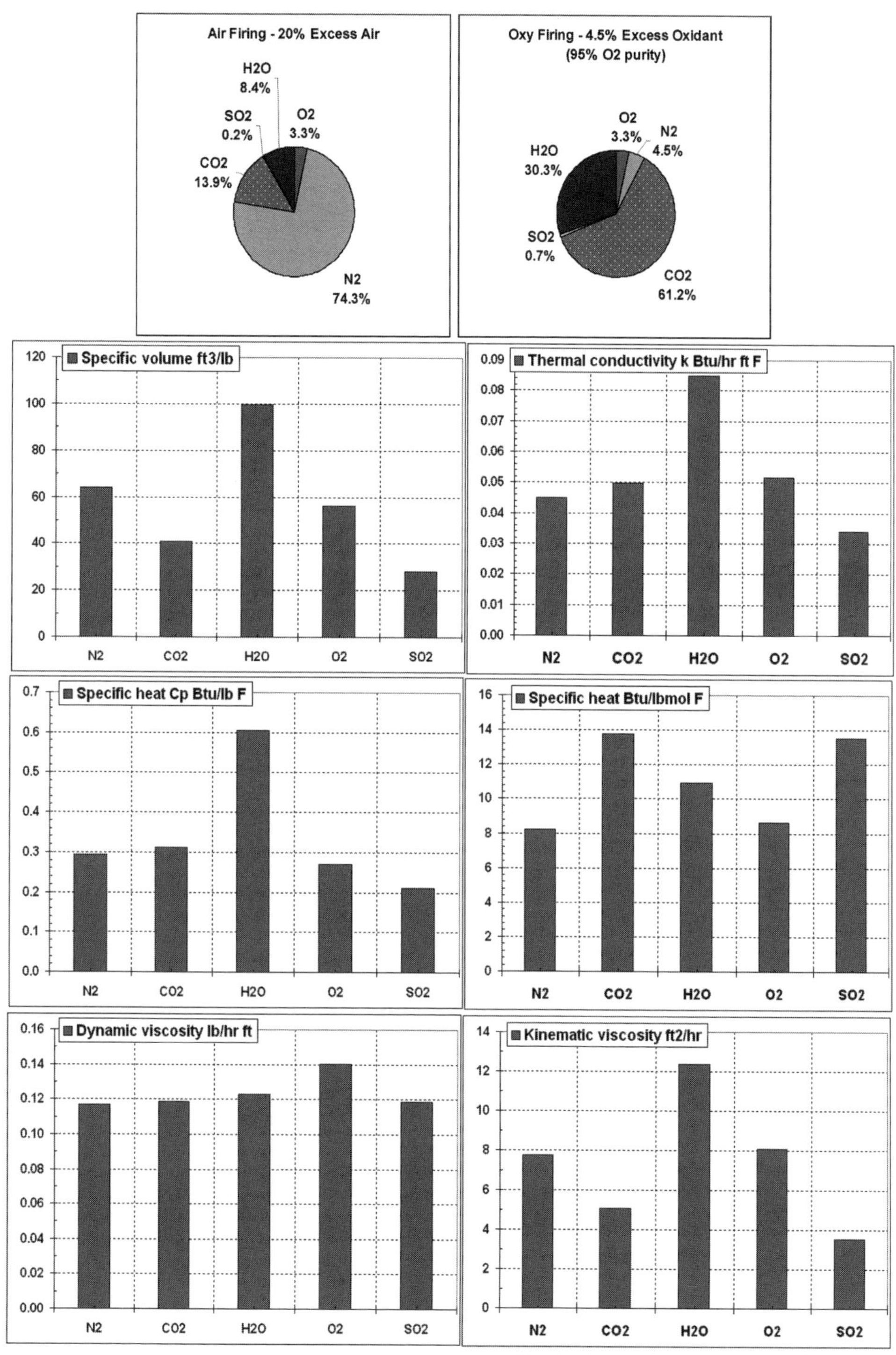

Figure 5: Properties of Gases at 2000F, 1 atm

Radiation and Overall Furnace Performance

Another key furnace design parameter is the furnace exit gas temperature. The temperature at the furnace exit must be controlled below the particular coal's ash softening point in order to prevent significant ash fouling in the convection section. As discussed above for the flame temperature consideration, radiative heat transfer differences can be identified by review of gas properties and have been confirmed by lab and pilot-scale testing. Strongly radiating gases, CO_2 and H_2O, dominate the oxy-fired environment and this suggests that some further tempering of the flame temperature with FGR is appropriate. Several studies conclude that an oxygen concentration of 25-27% at the burner would result in matching the radiation heat transfer rate of air firing [9, 10, 7]. The enhanced radiative properties of the oxy-fired gases must be accounted for in evaluating existing furnace wall materials in retrofits, and for properly predicting furnace wall temperatures in advanced UltraSuperCritical units, especially sliding pressure designs which require high furnace outlet steam enthalpies.

While the conditions described above match the radiation heat transfer rate in the burner zone, the furnace exit gas temperature will be less for oxy-combustion. With the denser flue gas and somewhat less mass flow from oxy-firing conditions, the residence time is increased and furnace exit gas temperature has been found to be 70-90 F° less than for air [9]. Further evaluations of furnace performance and variations of wet vs. dry FGR and selective staging of burner levels will be facilitated by Computational Fluid Dynamic (CFD) modeling. Fundamental issues of gas modeling methods, coal devolatilization, and char burnout must be further investigated to support detailed modeling. Regarding unburned carbon, University of Newcastle trials suggest 20-60% reduction in unburned carbon loss, which might be explained since the combustion process begins at higher oxygen concentrations at the burner, and furnace residence time is longer due to reduced flue gas volume [9].

Emissions

NO_x

Various test programs have shown oxy-fired NO_x levels to be significantly lower than those from air-firing. With very little molecular nitrogen available at the flame, thermal NO_x is reduced. Further, some of the fuel NO_x in the recycled flue gas is reduced back to molecular nitrogen when it passes through the flame again. Tests with both staged and unstaged burning of coal indicate oxy-fired NO_x about 50-70% less than air-fired, suggesting that post-combustion controls would not be required even if the flue gas were to be released [3, 9, 11, 7]. But at such low levels, any NO_x may ultimately be co-sequestered with the CO_2.

SO_x

In general, the fuel sulfur will be fully oxidized as in air-firing so there may not be any significant change due to oxy-firing mode, though the potential for SO_2 to SO_3 conversion still needs to be investigated. However, oxy-firing poses some opportunities for avoiding the cost of dedicated FGD equipment. First, use of low sulfur coal may bring the uncontrolled SO_2 level to within acceptable limits for co-sequestration with CO_2. In fact, small amounts of SO_x and H_2S help the EOR process by improving oil miscibility [11]. Second, Buhre et al. [9] suggest the oxy-combustion environment may permit in-furnace desulfurization due to inhibition of $CaSO_4$ decomposition at higher temperatures and recirculation of flue gas [9].

Mercury

Some literature reports lower mercury emissions are possible with oxy-combustion [1]. Thermodynamic analysis by CANMET [11] suggests that both oxidized and elemental mercury could be captured in the condensed CO_2 product and therefore not admitted to the atmosphere or handled separately.

Air Infiltration

Air infiltration may be a significant challenge depending on the CO_2 capture and sequestration requirements. Ochs et al. [4], report that if the tramp nitrogen can be tolerated in the sequestration scheme, then there should be no problem in the compression and delivery of the mixture, and that initial experiments suggest that geologic sites should be tolerant of minor constituents including SO_x. However, if high purity CO_2 is required and air infiltration is significant, then an additional distillation process is required to purify the CO_2, requiring significant energy and limiting the amount of CO_2 that can be economically captured [4].

A second aspect of air infiltration is its potential impact on NO_x production. Without infiltration, several studies and test programs indicate that NO_x from oxy-combustion can be very low and would generally be tolerated for most sequestration sites. However, further research is required to determine the impact of varying amounts of tramp nitrogen on the oxy-fuel NO_x production, and NO_x treatment systems could be required at some point.

Retrofit vs. New Unit

For new unit builds, the integration aspects described previously - especially for heat recovery from the ASU and the CO_2 recovery train - are important aspects to help offset these systems' considerable energy penalties. Further, the capital cost of new plants can take advantage of the potential reduced size of the boiler and omission of conventional, discrete emissions control equipment for both criteria and trace pollutants. The term "capture-ready" is sometimes used to refer to a plant that has essentially been fully designed for CO_2 capture (via oxy-combustion technology or other means), but some of the equipment is not initially purchased or installed and the plant operates in more conventional mode prior to CO_2 regulation or policy being firmly established.

In the case of retrofit to an existing coal-fired boiler, opportunities for full heat recovery and system integration may be limited, and the oxy-combustion system may need to be somewhat simplified. It is anticipated that many boilers will be able to be converted without making significant changes to the heat transfer surfaces or other pressure parts. Working with the existing furnace enclosure and arrangement, burner retrofits may be advisable. The significant additions to the plant would be the oxygen supply system, flue gas recirculation system (if not already present or sufficient), and of course the CO_2 product recovery system. In general, the existing air system would be left in place (and tied-into) to permit startup on air and transition to oxy-combustion at a stable load.

Materials Issues

Simply by nature of the absence of diluent nitrogen from the air, the concentration of all other combustion products are elevated in oxy-combustion. The comparison of air-fired and oxy-fired flue gas analyses of Figure 1 helps to illustrate this point. Based on the same 2% sulfur coal, and regardless of flue gas recirculation, the concentration of sulfur dioxide increases from 0.2% to 0.7% (vol. wet). More concentrated SO_2, SO_3, H_2O, and other trace species in the flue gas could result in greater corrosion rates.

The higher concentration of SO_2 and the higher concentration of oxygen passing through the burner may suggest the likelihood of greater SO_2 to SO_3 conversion, where SO_3 is a greater concern for corrosive attack. SO_3 reacts with metal surfaces to form low melting point components, such as alkali-iron trisulfates, that cause molten salt attack [12]. However, the overall oxygen concentration from oxy-combustion will be the same or possibly less that that in air-fired flue gas, so the variations and potential for excess oxidant reductions should be considered in terms of corrosion potential as well as overall system optimization. Further research is required to confirm the detailed behaviors of SO_3 and various trace species in the denitrified environment.

As described in the furnace radiation discussion above, with high CO_2 and H_2O there is the possibility of increased radiative heat flux in the furnace resulting in slightly higher tube metal temperatures. This may aggravate the situation of high furnace metal temperatures generally present in sliding pressure supercritical and UltraSuperCritical boilers. Despite previous reasoning and separate test results indicating lower unburned carbon with oxy-firing, others suggest that the CO_2-rich environment could inhibit burnout and lead to CO and elevated carbon in the ash. The combination could result in more reducing conditions underneath the ash deposits, resulting in accelerated corrosion of boiler tubes [12, 13].

In several publications, flue gas recirculation (FGR) is blamed for accumulation of corrosive species. This requires some clarification. If the FGR is recycled "wet" - without any preferential separation or reduction of species other than removal of flyash - then it is of the same composition as the flue gas resulting from the combination of fuel and oxidant in the furnace. Mixing gases of the same composition will result in a mixture of the very same composition. However, in "dry" recirculation, much of the moisture is removed from the recycle stream, and a concentrating effect similar to that described for nitrogen above takes place. As an upper bound on this concentrating effect, Figure 6 shows the effect of removing all of the moisture from the recycle stream. Iterative calculations can then predict the feedback effects of this continual mixing and separation scheme. The interactions and path history of certain trace species require special review depending on specific system designs.

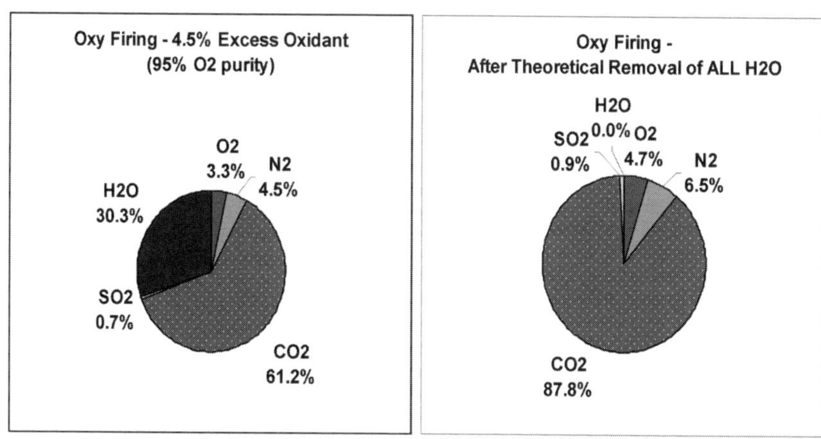

Figure 6: Concentrating Effect of Moisture Removal

Until more is known about the performance of materials in the oxy-combustion environment, several studies propose including sulfur-removing scrubbers (FGD) in the recycle path. This poses a significant capital cost as well as heat rate penalty, and the issue adds to the need to better understand the materials issues under the denitrified conditions [4].

Riley Power is involved in an industry consortium partly sponsored by the US DOE and Ohio Coal Development Office, investigating advanced materials issues for UltraSuperCritical (USC) coal-fired power plants. The group is beginning a second phase of the work, extending considerations into the application of oxy-combustion mode to USC power plants. The specific combinations of oxy-combustion gas and ash conditions and elevated metal temperatures will be investigated.

Closing Remarks

Efforts to mitigate global warming should first be applied to two areas: improved economy - reduced use of energy throughout society - and higher efficiency. Gains in efficiency should span from the use of supercritical and ultrasupercritical steam cycles in central power stations, to improvements in vehicles and household appliances. Improvement in these areas not only prevents unnecessary production of CO_2, but also reduces the often-overlooked direct heat release into the environment and slows the depletion of key resources. Building on that presumed foundation, several technologies are being developed to capture and permanently store CO_2 emissions that result from responsible use of fossil fuel resources. Oxy-combustion is proposed for coal-fired boilers in order to yield a concentrated CO_2 flue gas that is relatively easy to capture, transport, and sequester. It is a promising technology in that it can be readily applied to the existing base of fossil-fired power plants, is relatively straightforward, and can offer significant cost advantages for new units including the potential avoidance of discrete emissions control equipment.

References

1. S Santos, M Haines, "Oxy-Fuel Combustion Application for Coal Fired Power Plant", IEAGHG International Oxy-Fuel Combustion Network Workshop, Cottbus, Germany, November 2005.

2. M Simmonds, I Miracca, K Gerdes, "Oxyfuel Technologies for CO2 Capture: A Techno-Economic Overview", The 7th International conference on Greenhouse Gas Control Technologies, Vancouver, Canada, September 2004.

3. H Farzan, S Vecci, D McDonald, K McCauley, P Pranda, R Varagani, F Gauthier, "State of the Art of Oxy-Coal Combustion Technology for CO_2 Control from Coal-Fired Boilers", 32nd International Technical Conference on Coal Utilization & Fuel Systems, Clearwater, FL, USA, June 2007.

4. T Ochs, D Oryshchyn, J Ciferno, C Summers, "Ranking of Enabling Technologies for Oxy-Fuel Based Carbon Capture", 32nd International Technical Conference on Coal Utilization & Fuel Systems, Clearwater, FL, USA, June 2007.

5. K Zanganeh, C Salvador, A Shafeen, "Pilot-Scale Evaluation of Coal Combustion in Hydroxy-Fuel Mode", 32nd International Technical Conference on Coal Utilization & Fuel Systems, Clearwater, FL, USA, June 2007.

6. V Becher, A Goanta, S Gleis, H Spliethoff, "Controlled Staging with Non-Stoichiometric Burners for Oxy-Fuel Processes", 32nd International Technical Conference on Coal Utilization & Fuel Systems, Clearwater, FL, USA, June 2007.

7. V Sethi, K Omar, P Martin, T Barton, K Krishnamurthy, "Oxy-Combustion Versus Air-Blown Combustion of Coals", 32nd International Technical Conference on Coal Utilization & Fuel Systems, Clearwater, FL, USA, June 2007.

8. M Shah, "Oxy-Fuel Combustion for CO2 Capture from New and Existing PC Boilers", presented at Electric Power Conference, May 2007.

9. BJP Buhre, LK Elliott, CD Sheng, RP Gupta, TF Wall, "Oxy-Fuel Combustion Technology for Coal-Fired Power Generation", Progress in Energy and Combustion Science, 31, p. 283-307, 2005.

10. SP Khare, AZ Farida, TF Wall, Y Liu, B Moghtaderi, RP Gupta, "Factors Influencing the Ignition of Flames from Air Fired Swirl PF Burners Retrofitted to Oxy-fuel", 32nd International Technical Conference on Coal Utilization & Fuel Systems, Clearwater, FL, USA, June 2007.

11. L Zheng, Y Tan, R Pomalis, B Clements, "Integrated Emissions Control and Its Economics for Advanced Power Generation Systems", 31st International Technical Conference on Coal Utilization & Fuel Systems, Clearwater, FL, USA, May 2006.

12. B Covino, S Matthes, S Bullard, "Corrosion in Oxyfuel / Recycled Flue Gas-Fired vs. Air-Fired Environments", 32nd International Technical Conference on Coal Utilization & Fuel Systems, Clearwater, FL, USA, June 2007.

13. J Henry, J Nava, "The Changing Face of Corrosion in Coal-Fired Boilers", 2005 Conference on Unburned Carbon on Utility Fly Ash, NETL, April 2005.

14. Energy Information Administration (EIA), "International Energy Annual 2004 (May-July 2006)", website www.eia.doe.gov/iea; Report # DOE/EIA-0484(2007)

15. M Raindl, S senthoorselvan, H Spliethoff, G Haberberger, "H2O/CO2 Condensation - Heat and Mass Transfer Coefficient Analysis for Oxy-Fuel Cycles", 32nd International Technical Conference on Coal Utilization & Fuel Systems, Clearwater, FL, USA, June 2007.

16. D Stopek, "Options and Economics for CO2 Control Technologies", Electric Power Conference, May 2007.

Advances in Materials Technology for Fossil Power Plants
Proceedings from the Fifth International Conference
R. Viswanathan, D. Gandy, K. Coleman, editors, p 982-992

DOI: 10.1361/cp2007epri0982

Fireside Corrosion Study Using B&W Clean Environment Development Facility for Oxy-Coal Combustion Systems

S. C. Kung
J. M. Tanzosh
D. K. McDonald

The Babcock & Wilcox Company
20 South Van Buren Avenue
Barberton, OH 44203

Abstract

The development of oxy-fuel combustion technology for coal-based power generation may produce combustion products different from those typically found in traditional boilers. In particular, the enrichment of CO_2 and perhaps SO_3 could alter the chemical equilibrium to favor the formation of certain carbonates and sulfates in the deposit. Higher concentrations of these gases would also increase the potential for condensation of carbonic and sulfuric acids in lower-temperature areas of the boiler. To address these concerns, B&W has instituted a comprehensive research program to better understand the effect of oxy-coal combustion on fireside corrosion. The scope of this program includes gas and deposit analyses of actual combustion products sampled from B&W's Clean Environment Development Facility (CEDF) during the oxy-coal combustion of three commercial coals. The sampling locations consist of regions representing the lower furnace, superheater bank, and pulverizer outlet. Following the gas and deposit analyses, a series of laboratory corrosion tests will be performed to expose candidate alloys and coatings to conditions simulating the oxy-coal combustion environments. The technical approaches and results of the fireside corrosion program obtained to date are discussed.

Introduction

Carbon dioxide (CO_2) emissions are believed to be a major contributor to climate change and global warming. One of the more significant sources of CO_2 emissions comes from combustion of fossil fuels in power plants. Currently, the US has about 310 GW of coal-fired power generating capacity, producing about two trillion kilowatt-hours of electricity.[1] This capacity represents about 50% of the electricity generated annually and about one third of total CO_2 emissions in the US. In recent years, public awareness and legislation have led to a policy of reduction of the greenhouse gases, and regulations are being considered both locally and nationally.

While the specifics of regulations and economic incentives are being debated, equipment suppliers and power generators are moving ahead and developing several efficient technologies to capture and dispose of the CO_2 from power generation with coal. Attention has also been given to CO_2 utilization and storage. Among these technologies, IGCC (integrated gasification combined cycle), oxy-coal combustion, and post-combustion are considered the front runners.[2] Recent cost studies by DOE and others have compared the three technologies with somewhat varying results but all close in effectiveness within the accuracy band of the estimates.[3,4] It is widely felt that only large scale demonstrations will determine which technology is most economical.

Currently, B&W, with Air Liquide, is focusing on the development of oxy-coal combustion technology due to its significant potential for low cost, high efficiency, and improved reliability.[5] In addition, this technology can be deployed to both new and retrofit units and provide the same fuel flexibility as is available to conventional coal-fired plants. Research studies have shown that oxy-coal combustion is the only technology capable of near zero emissions.[6]

In pursuit of a large scale demonstration, B&W has decided to convert its $30MW_{th}$ Clean Environment Development Facility (CEDF) located in Alliance Ohio to be capable of operating in full oxy-coal combustion mode. The CEDF was originally built in 1993 with funding from B&W, US Department of Energy, and the Ohio Coal Development Office. The unit is fired by a single full size burner delivering 100 MBtu/h heat input. Over the years, it has been used to support several pivotal programs, including Combustion 2000 and Advanced Emission Control Development Program. Several generations of burners and wet and dry scrubbers have also been developed using this facility. Once fully converted in the fall of 2007, the CEDF will become the largest oxy-coal combustion facility in the world. Immediately following the conversion, a B&W-Air Liquide test program will commence, which includes the materials study discussed in this paper for oxy-coal combustion systems.

Materials Consideration

The unique features of oxy-coal combustion are the use of oxygen to eliminate nitrogen and incorporation of flue gas recirculation to the system for CO_2 enrichment. After sulfur is scrubbed through a wet flue gas desulfurizer (WFGD), a large portion of the flue gas is redirected back to the pulverizers and burners. The recirculation allows the accumulation of over 80% CO_2 by volume in the flue gas, making it suitable for subsequent purification, sequestration, and storage.

Concerns have been raised that the enrichment of CO_2 in the combustion gas may promote the formation of alkali carbonates as part of the ash constituents in boilers. In air-fired boilers, alkali carbonates do not typically form in a meaningful quantity because the presence of a relatively low CO_2 concentration in the gas phase. In addition, alkali metals tend to preferentially react with sulfur in coal and form much more stable sulfate compounds in the furnace. However, in oxygen-fired units, the excess CO_2 may shift the chemistry of the combustion products sufficiently, causing the alkali sulfates and carbonates to coexist in the ash. Where such a mixture condenses on boiler tubes, aggressive fireside corrosion can take place, as evidenced by the high metal wastage in process recovery boilers for the pulp and paper industry.

Furthermore, as the flue gas is scrubbed through the WFGD, some of the sulfuric acid may be carried over by the gas stream returning to the boiler as fine mist. Part of the acid mist could condense in the ductwork beyond the WFGD. It is not known if the amount of acid condensate will be in sufficient quantity to cause aqueous corrosion in the ductwork. The remaining mist will eventually be heated and dissociate to form moisture and SO_3 prior to entering the pulverizer. Because of the moisture content in coal, additional water vapor is introduced to the gas stream in the pulverizer. It is unclear if the dew point of sulfuric acid, resulting from the returning SO_3 and an increased total moisture content, would be significantly raised. If the dew point of sulfuric acid is greater than the gas temperature, dew point corrosion could be anticipated in the pulverizer and coal piping. Finally, the returning SO_3 from the WFGD could eventually enter the furnace and lead to the formation of more alkali sulfates, thus creating a higher propensity for fireside corrosion on boiler tubes.

The high CO_2 concentration in the flue gas, along with a relatively high moisture content, can also favor a condition where condensation of carbonic acid is possible. This is of particular concern for the pulverizer and coal pipe where additional moisture is introduced from the coal. In general, the dew point of carbonic acid is much lower than that of sulfuric acid. Therefore, dew point corrosion due to sulfuric acid condensation is usually more problematic for lower-temperature boiler components in which the minimum metal temperatures are not properly maintained. However, it has been well documented in the oil and gas industry that, when carbonic acid is allowed to form, accelerated corrosion attack can take place.[7,8]

With the above concerns in mind, B&W is taking a proactive approach to quantify the potential of fireside corrosion in coal-based oxycombustion systems. A two-year comprehensive research program was instituted in 2007, which will address all of the key fireside corrosion issues unique to oxy-coal combustion. The research program is divided into four tasks, which include investigation of (1) dew point corrosion in pulverizers due to carbonic acid condensation, (2) thermodynamic prediction of fireside corrosion, (3) gas and deposit sampling in CEDF, and (4) long-term laboratory corrosion testing. This paper summarizes the planned development effort and some of the test data available to date. The majority of the planned activities is still ongoing, and results will be reported later as they become available. However, a brief description of the workscope for each task is presented here.

Task 1 – Dew Point Corrosion in Pulverizers Due to Carbonic Acid

The presence of CO_2 and a relatively high moisture content in pulverizers can significantly increase the dew point of carbonic acid. Once the dew point exceeds the operating temperature, condensation of carbonic acid can occur, leading to rapid corrosion attack on alloys. In B&W's current system design, Shand lignite is one of the model coals for oxycombustion. Lignite typically consists of a high moisture content (>30 wt.%), thus representing an extreme challenge for the management of carbonic acid condensation and dew point corrosion. According to the mass balance for lignite, the gas phase in the pulverizers will be approximately 50 vol.% CO_2 and 30 vol.% H_2O. Note that the CO_2 concentration is reduced from 80% to 50% due to the additional moisture content from coal. The maximum outlet operating temperature of the pulverizers is set at approximately 170°F.

Accurate prediction of the dew point of carbonic acid based on the CO_2 and moisture concentrations has not been precisely established. However, it is of general consensus that the dew point of carbonic acid follows closely and is just slightly above the dew point of water. Therefore, the temperature at which carbonic acid starts to condense can be estimated based solely on the moisture content in the gas phase. Using this approach, the dew point of carbonic acid in the pulverizers is assessed to be 10-40°F below the maximum operating temperature, depending on the moisture content. This assessment suggests that condensation of carbonic acid may occur when large temperature cycles are encountered in operation.

Extensive research has been performed in the past several decades dealing with aqueous corrosion resulting from carbonic acid condensation. Findings of this research have led to the development of a predictive model that is widely used by the oil and gas industry for component design.[9,10] The model is capable of estimating the pH value of the condensate as well as the corrosion rate of carbon steel based on the CO_2 and moisture concentrations in the gas and metal temperature. Using this model, pH of the condensate in the pulverizers of B&W's oxy-coal combustion system has been predicted to be quite low (approximately 4.2) and the corrosion rate of carbon steel quite high (in mm).

However, direct application of this model to the pulverizers of oxy-coal combustion system can be misleading, as the model only considers the corrosion of carbon steel in contact with carbonic acid. In reality, coal contains a large amount of ash that will also be present in the pulverizers. Depending upon its constituents, the ash may provide a significant buffering effect on the acid condensate, thus preventing a low pH to be reached on the metal surface. If the pH of the condensate is buffered to near neutral, its corrosivity can be much reduced. No research data are available in the open literature pertinent to the role of coal ash in CO_2 corrosion. Therefore, Task 1 is intended to address this unexplored area and demonstrate the potential effect of coal ash on carbonic acid corrosion.

Task 2 - Thermodynamic Prediction of Fireside Corrosion

The focus of this task is to determine the likelihood of different corrosion mechanisms operating on the furnace walls and superheaters/reheaters in oxy-fired utility boilers. In air-fired boilers without staging, the combustion gas is largely oxidizing with 2-3% excess oxygen throughout the lower furnace. Under this condition, the predominant corrosion mechanism operating on the furnace walls is oxidation. The only exception is that, in certain areas of the lower furnace walls where the combustion gas is reducing and sulfidizing due to improper air/coal mixing, sulfidation can become the dominant corrosion mechanism.[11] For staged combustion, the majority of the lower furnace walls are subjected to sulfidation.

In the upper furnace, however, coal ash corrosion is responsible for the highest metal loss on the outlet section of superheaters and reheaters. Coal ash corrosion, also known as hot corrosion, refers to the presence of a low-melting phase of sulfates formed adjacent to the boiler tubes.[12] When molten, the sulfates can attack the metal rapidly via the processes of dissolution and fluxing. Chlorides can also deposit on the furnace walls and superheaters/reheaters and cause severe wastage if the coal chlorine concentration is high. All of the mechanisms encountered in air-fired boilers have been extensively studied and are reasonably understood.

For oxy-coal combustion, the primary concern is whether alkali carbonates are also stable in the furnace due to an increased CO_2 concentration in the flue gas. If formed, alkali carbonates can condense on the boiler tubes along with other ash constituents and substantially shift the corrosion mechanisms away from those mentioned above for air-fired units. Based on the operating experience of process recovery boilers, the addition of carbonates to the ash constituents can result in accelerated fireside corrosion on all boiler components.

Therefore, Task 2 is intended to evaluate the potential of carbonate formation in oxy-fired boilers by means of thermodynamic calculations. A commercially available software package, HSC Chemistry 5.1, was used for the analysis.[13] Consistent with Task 1, Shand lignite was selected as the model coal. The approach was to calculate the equilibrium compositions of gaseous and condensed phases under both air and oxy-fired boiler conditions at different temperatures typical for the furnace walls and superheater/reheaters. For corrosion evaluation, the emphasis was placed especially on the resulting deposit chemistry. Both sets of calculations were performed at 3% excess oxygen so that a direct comparison of the air and oxy-fired boilers can be made. It is assumed that the corrosion mechanisms would remain unchanged if similar chemistry of the combustion products is predicted.

Task 3 - Gas and Deposit Sampling in CEDF

A total of three model coals will be evaluated for oxycombustion during the CEDF testing scheduled to start in September 2007. These coals consist of Eastern bituminous (Mahoning 7), lignite (Shand), and PRB (Black Thunder). For each of the test coal, gas and deposit samples will be collected through ports located in the lower furnace walls and secondary superheater bank of the CEDF. A number of gaseous species of interest will be extracted at various locations. These include SO_2, SO_3, NO, NO_2, CO, CO_2, O_2, HCl, H_2S, H_2, and H_2O. Online instrumentation will be available for in-situ gas analysis, which consists of gas chromatography (GC) with flame photometric detection and thermal conductivity detection and Fourier transform infrared spectroscopy (FTIR). The EPA methods for sampling and analysis will be followed.

For deposit sampling, HVT (high velocity thermocouple) probes will be used at selected locations of furnace walls and the secondary superheater bank. The probes will be air-cooled to facilitate the collection of ash deposit and minimize reactions between the deposit and probe body. After sufficient deposit is formed, the probes will be retrieved and samples carefully removed. The samples will then be chemically analyzed, with special emphasis given to the inner layer that is in direct contact with the probe surfaces and responsible for corrosion.

The intent of Task 3 is to determine the realistic conditions existing at the furnace walls and superheater bank of oxy-fired boilers through gas and deposit sampling. Once the corrosive environments are defined for different coals, the information will be compared to the thermodynamic prediction from Task 2 and transferred to the laboratory for better simulation of the boiler conditions.

Task 4 - Long-Term Laboratory Corrosion Testing

If different corrosion mechanisms are suspected to prevail in oxy-coal fired boilers, as determined mainly from the outcome of Task 3, a series of laboratory tests will be performed simulating the lower and upper furnace conditions. For each test, coupons of candidate alloys and coatings will be exposed isothermally to the simulated condition for up to 1000 hours. The alloys and coatings will consist of commercially available boiler-tube materials as well as experimental weld overlays under development. At the end of the test, corrosion resistance of the coupons will be metallurgically examined and compared. Details of the laboratory test conditions and materials selection will be finalized at a later date.

Results and Discussion

For Task 1, a sample of Shand lignite coal was chemically analyzed for its proximate and ultimate analyses. Subsequently, the sample was dried and ground to a fineness of 70% through 200 mesh, typical for pulverized coal. High purity water was then added to the pulverized coal at two concentration levels for pH measurements in air at room temperature. The first level returned the total moisture content of the coal to its original quantity, i.e., 33.4%, and the second level resulted in additional moisture to the mixture, i.e., ~53%. The second level with a higher moisture content would simulate some degree of water condensation taking place in the pulverizers. The pH was determined before and after mixing of the slurry to a paste-like consistency. Both a pH meter and litmus paper were used for the measurements.

Results of the measurements indicate that pH of the slurry falls within a narrow range of 7 to 8.2. Mixing did not seem to have a profound effect on the pH readings. The results clearly suggest that the ash in Shand lignite behaves as a weak base at room temperature in air. Such a behavior is highly desirable for the concern of carbonic acid condensation.

Following the pH determination in air, a laboratory apparatus was designed to better simulate the operating conditions of pulverizers in an oxy-coal combustion system. Figure 1 illustrates the laboratory setup capable of sparging a coal slurry (50% pulverized lignite + 50% water) with a mixed gas of 55% CO_2 + 30% O_2 + 15% N_2. The slurry temperature was maintained within 165-170°F on a hot plate.

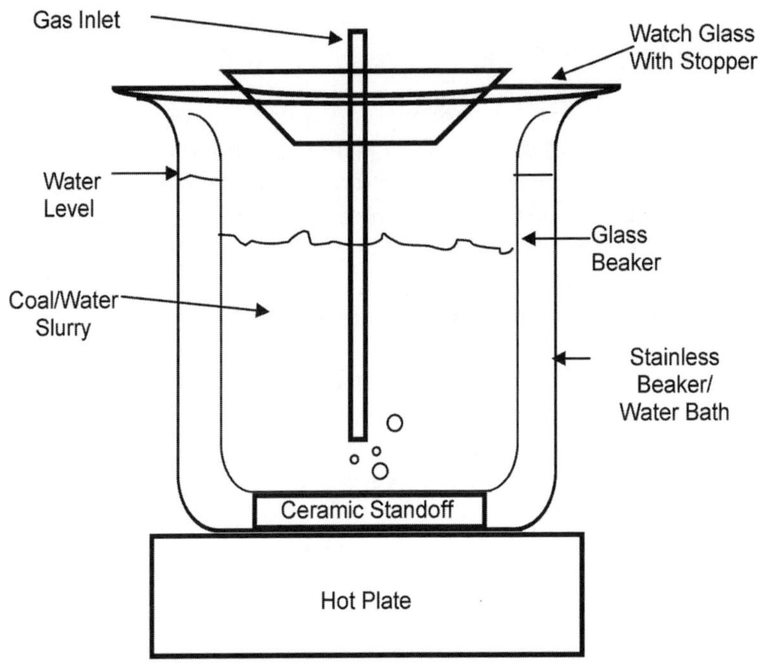

Figure 1 – Illustration of the Experimental Setup for Coal Slurry pH Measurement

Before the start of the test, the slurry was allowed to stand in air at room temperature for a period of approximately 100 hours. After this period, the pH of the mixture was determined to be 7.0, which is consistent with the earlier room temperature results. Subsequently, the slurry was heated to 165-170°F, followed by sparging with the CO_2-containing mixed gas. A decrease in pH to a value of 6.46 was observed a few hours after the sparging. However, continuing monitoring of the system for a longer time did not reveal any further change in pH. Such an observation indicates that the coal slurry was able to equilibrate with the CO_2 quickly to form some acid, most likely bicarbonic acid. However, the buffering effect of the coal ash was sufficiently adequate to prevent the pH from reaching the theoretical value of ~4.2 even after several days. Therefore, the presence of coal ash is expected to significantly reduce the propensity for dew point corrosion in the pulverizers (and coal pipe) of an oxycombustion system by maintaining a near neutral pH in the condensate.

For Task 2, thermodynamic calculations were performed for both air and oxy-fired boilers burning Shand lignite coal in a wide range of temperatures. In addition, the scenarios of staged and unstaged combustion were also explored. To improve accuracy of the calculations, the input files required for the thermochemical program were structured to contain all of the major coal elements as well as the ash constituents. Furthermore, mixtures of various phases were allowed to form, which consist of gas, oxide, sulfide, chloride, sulfate, and carbonate. The chloride phase was included for completeness even though the chlorine concentration in the coal is very low (i.e., 23 ppm). All phases were assumed to be ideal solutions due to the lack of activity coefficient data.

Results of the calculations indicate that no measurable difference in the combustion products is expected between air and oxy-fired boilers. In other words, the increase of CO_2 concentration in the flue gas from oxy-coal combustion would not alter the corrosion mechanisms known to air-fired units. Table 1 gives the deposit chemistry predicted for 3% excess oxygen at 1112°F, a condition pertinent to superheaters/reheaters. For ease of comparison, any species predicted to be less than 1 mole% in the condensed phase is omitted. It can be seen that only oxide and sulfate phases are expected to form on the superheater/reheater tube surfaces, and no other condensed phases are thermodynamically stable. The results demonstrate that the deposit constituents are essentially identical in both types of boilers. Consequently, corrosion in the upper furnace of oxy-coal fired boilers would not change and should still be dominated by the classic coal ash mechanism.

Table 1
Comparison of Predicted Deposit Chemistry Condensed on Superheaters/Reheaters at 1112°F (in mole%)

Phase	Species	Air-Fired	Oxy-Fired
Oxide	$Ca_3Al_2Si_3O_{12}$	27.2	27.2
	Fe_2O_3	15.5	15.5
	SiO_2	55.6	55.7
	TiO_2	1.2	1.3
Sulfate	$CaSO_4$	16.7	16.7
	$MgSO_4$	44.6	44.6
	$K_2SO_4 \cdot 2MgSO_4$	1.8	1.8
	Na_2SO_4	36.6	36.6

Table 2 compares the deposit chemistry expected to exist on the furnace walls at 842°F under oxidizing (unstaged) condition. In addition to the oxide and sulfate phases, some chloride is also predicted to condense on the walls. Because of the low concentration of chlorine in Shand lignite, the actual quantity of chloride as Na_2AlCl_6 is very small. However, due to the fact that Na_2AlCl_6 is the only species present in the chloride phase, a 100 mole% is indicated. Despite the addition of chloride to the deposit mix, Table 2 shows that the condensed combustion products in air and oxy-fired boilers are again nearly identical. Therefore, no change in the corrosion mechanism is anticipated for the furnace walls. Similar findings were also obtained for furnace walls exposed to reducing conditions in boilers burning coal substoichiometrically.

Based on the results of Task 2, it is postulated that carbonates would not form in oxy-fired boilers and thus, existing knowledge on the classic fireside corrosion mechanisms operating in air-fired boilers is applicable to the oxy-coal combustion system as well.

Table 2
Comparison of Predicted Deposit Chemistry Condensed on Furnace Walls at 842°F (in mole%), Unstaged

Phase	Species	Air-Fired	Oxy-Fired
Chloride	Na_2AlCl_6	100	100
Oxide	Al_2O_3	9.1	9.2
	$Ca_3Al_2Si_3O_{12}$	6.9	7.0
	Fe_2O_3	11.0	11.0
	$NaAlSi_3O_8$	2.0	2.1
	SiO_2	69.6	69.4
Sulfate	$CaSO_4$	55.6	55.3
	$MgSO_4$	27.1	27.0
	$K_2SO_4 \cdot 2MgSO_4$	1.2	1.2
	Na_2SO_4	16.0	16.5

At the writing of this paper, facilities required for Tasks 3 and 4 are still under construction. Therefore, no immediate results are available for reporting. Further update of the B&W materials program is planned for future meetings.

Conclusions

A comprehensive research project has been instituted at B&W to address several material concerns relative to the oxy-coal combustion technology. As part of a large-scale, first-of-a-kind oxy-coal combustion demonstration program, the materials project is intended to (1) evaluate the potential of dew point corrosion in pulverizers due to carbonic acid condensation, (2) predict the fireside corrosion mechanisms via thermodynamic calculations, (3) determine actual corrosive environments through gas and deposit sampling in the CEDF, and (4) perform long-term laboratory tests to compare the corrosion resistance of candidate alloys and coatings under oxy-coal combustion environments.

The project has not been completed at the writing of this paper. However, based on the findings obtained to date, it appears that dew point corrosion in the pulverizers (and coal pipe) of an oxy-coal combustion system firing Shand lignite would not be severe due to the presence of neutralizing coal ash. It was demonstrated that the coal ash is capable of buffering the pH of the condensate formed under a high CO_2 environment at 170°F to a value approximately 6.4, instead of a theoretical value of 4.2. Such a pH increase represents a reduction of more than 100 times in condensate acidity.

Results of the thermodynamic calculations revealed that similar deposit would form on the boiler tubes in air and oxy-fired boilers with nearly identical compositions. Carbonates were found to be unstable under the conditions of CO_2-rich flue gas investigated and thus would not cause additional corrosion. Consequently, no change in the fireside corrosion mechanisms is anticipated for the furnace walls and superheaters/reheaters, and the existing knowledge for air-fired boilers is directly applicable to the oxy-coal combustion system.

References

1. Annual Energy Outlook 2007, http://www.eia.doe.gov/oiaf/aeo/index.html, (Access Date: August 16, 2007).

2. T. F. Wall, "Combustion Processes for Carbon Capture," Proceedings of the Combustion Institute, Vol. 31, No. 1, pp. 31-47, January 2007.

3. J. P. Ciferno, "Advanced Pulverized Coal Oxyfuel Combustion," Proc. 5th Annual Conference on Carbon Capture & Sequestration and 31st International Coal Utilization and Fuel Systems Conference, p. 63, May 2006.

4. A.H. Seltzer, and J. Fan, "An Optimized Supercritical Oxygen-Fired Pulverized Coal Power Plant for CO2 Capture," 31st International Coal Utilization and Fuel Systems Conference, p. 51, May 2006.

5. B.J.P. Buhre, L.K. Elliott, C.D. Sheng, R.P. Gupta and T.F. Wall, "Oxy-fuel combustion technology for coal-fired power generation," Progress in Energy and Combustion Science, Vol. 31, No. 4, pp. 283-307, 2005.

6. EPRI Report, "Review of CO_2-Capture Development Activities for Coal-Fired Power Generation Plants," EPRI Project Manager: J. Wheeldon, Report No. 1012239, March 2007.

7. F. Todt, Korrosion und Korrosionsschutz, 2nd ed., Walter de Gruyter & Co., Berlin, p. 861, 1961.

8. F. M. Song, D. W. Kirk, J. W. Graydon. And D. E. Cormack, "CO_2 Corrosion of Bare Steel under an Aqueous Boundary Layer with Oxygen," J. Electrochem. Soc., Vol. 149, No. 11, pp. B479-B486, 2002.

9. C. DeWaard and D. E. Milliams, "Carbonic Acid Corrosion of Steel," Corrosion, Vol. 31, No. 5, pp. 177-181, May 1975.

10. CO_2 Corrosion Rate Calculation Model, Norsok Standard M506, Rev. 2, 2005.

11. S. C. Kung, "Fireside Corrosion in Coal- and Oil-Fired Boilers," ASM Handbook, Vol. 13C, pp. 477-481, 2006.

12. R. A. Rapp and Y. S. Zhang, "Hot Corrosion of Materials: Fundamental Studies," J. Met., p. 47, Dec. 1994.

13. HSC Chemistry for Windows, Outokumpu Research Oy, Finland, ver. 5.1, 2002.

Advances in Materials Technology for Fossil Power Plants
Proceedings from the Fifth International Conference
R. Viswanathan, D. Gandy, K. Coleman, editors, p 993-1000

DOI: 10.1361/cp2007epri0993

DESIGN CONSIDERATIONS FOR ADVANCED MATERIALS IN OXYGEN-FIRED SUPERCRITICAL AND ULTRA-SUPERCRITICAL PULVERIZED COAL BOILERS

Horst Hack
horst_hack@fwc.com; Tel. 973-535-2200; Fax. 973-535-2242

Andrew Seltzer
andrew_seltzer@fwc.com; Tel. 973-535-2542; Fax. 973-535-2242

Greg Stanko
greg_stanko@fwc.com; Tel.973-535-2256; Fax. 973-535-2242

Foster Wheeler North America Corp.
12 Peach Tree Hill Road, Livingston, NJ 07039

ABSTRACT

As the demand for worldwide electricity generation grows, pulverized coal steam generator technology is expected to be a key element in meeting the needs of the utility power generation market. The reduction of greenhouse gas emissions, especially CO_2 emissions, is vital to the continued success of coal-fired power generation in a marketplace that is expected to demand near-zero emissions in the near future. Oxycombustion is a technology option that uses pure oxygen, and recycled flue gas, to fire the coal. As a result, this system eliminates the introduction of nitrogen, which enters the combustion process in the air, and produces a highly-concentrated stream of CO_2 that can readily be captured and sequestered at a lower cost than competing post-combustion capture technologies.

Oxycombustion can be applied to a variety of coal-fired technologies, including supercritical and ultra-supercritical pulverized coal boilers. The incorporation of oxycombustion technology in these systems raises some new technical challenges, especially in the area of advanced boiler materials. Local microclimates generated near and at the metal interface will influence and ultimately govern corrosion. In addition, the fireside corrosion rates of the boiler tube materials may be increased under high concentration oxygen firing, due to hotter burning coal particles and higher concentrations of SO_2, H_2S, HCl and ash alkali, etc. There is also potential to experience new fouling characteristics in the superheater and heat recovery sections of the steam generator. The continuous recirculation of the flue gases in the boiler, may lead to increasing concentrations of deleterious elements such as sulfur, chlorine, and moisture.

This paper identifies the materials considerations of oxycombustion supercritical and ultra-supercritical pulverized coal plants that must be addressed for an oxycombustion power plant design.

INTRODUCTION

Since the onset of the industrial age, CO_2 concentrations in the Earth's atmosphere have increased by about 1-2 ppm per year. This represents a 35% increase in the atmospheric CO_2 concentration in less than 200 years (see Figure 1). This increase may have a profound effect in causing global climate change. Increased atmospheric CO_2 concentrations may not only cause solar energy to be trapped in the atmosphere but may also increase the acidity of the ocean due to increased CO_2 dissolution [1].

CO_2 capture technologies are based upon three general concepts: post-combustion capture, pre-combustion capture, and oxygen based combustion. Post-combustion refers to capturing CO_2 from a flue gas after a fuel has been combusted in air. Pre-combustion refers to a process where a hydrocarbon fuel is gasified and water-gas shifted to form a mixture of hydrogen and CO_2 and the CO_2 is captured from the synthesis gas before it is combusted. Oxycombustion (or O_2-fired combustion) is an approach where a hydrocarbon fuel is combusted in a mixture of oxygen and carbon dioxide, rather than air, to produce an exhaust mixture of CO_2 and water vapor.

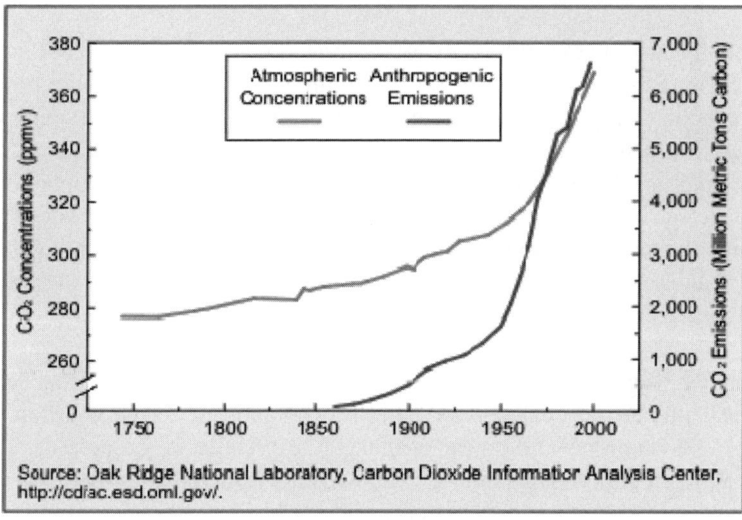

Figure 1 - Atmospheric Concentration of CO2

Although post-combustion capture can appear to be the simplest method of CO_2 removal, the associated chemical absorption process is both expensive and energy intensive. As an alternative to post-combustion scrubbing, the combustion process can be accomplished with oxygen rather than air. With air nitrogen eliminated, a CO_2-water vapor rich flue gas is generated. After partial removal of the water vapor, a portion of the flue gas is recirculated back to the boiler to control the combustion temperature and the balance of the CO_2 is processed for pipeline transport. This oxygen-fired combustion process eliminates the need for the CO_2 removal/separation process and, despite the expense and power consumption of air separation, reduces the cost of CO_2 capture. Furthermore, oxycombustion is a simple, low risk, high efficiency technology. Adapting oxycombustion to retrofit applications is especially important since coal firing is currently the dominant means of power generation (40% worldwide, 51% USA).

OXYCOMBUSTION PROCESS

In an oxycombustion boiler the combustion air is separated into O_2 and N_2 and the boiler uses the O_2, mixed with recycled flue gas, to combust the coal. The products of combustion are thus only CO_2, water vapor, and some small impurities. The water vapor is readily condensed, yielding a nearly-pure CO_2 stream ready for sequestration. The CO_2 effluent is compressed at high pressure (greater than 2000 psia) and is piped from the plant to be sequestered in geologic formations (depleted oil and gas reservoirs, unmineable coal seams, saline formations, and shale formations) as shown in Figure 2.

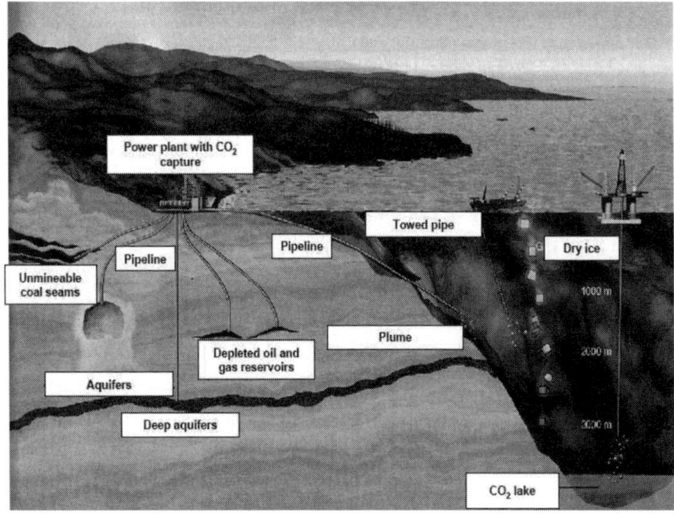

Figure 2 - Sequestration Options (Ref.: US National Energy Technology Laboratory)

The coal is combusted in the furnace or radiant section of the boiler where the oxidizer consists of a mixture of O_2 and recycled flue gas, which contains primarily CO_2 gas. The furnace is specially adapted to oxycombustion by optimizing the designs of burners, ports, internal radiant surfaces, tube materials, and water/steam circuitry. Air is separated into O_2 and N_2 using techniques such as cryogenic air separation or the more efficient membrane-based techniques. Recycling of the flue gas (55% – 80% of the total flue gas flow) is utilized to control flame temperature in the furnace to maintain acceptable waterwall temperatures and reduce fuel NOx release. NOx formation is minimized by advanced NOx control technologies such as combustion staging and low-NOx burners. Oxygen-firing of a steam boiler is a low-risk, high-efficiency technique for CO_2 capture in a power plant. The steam boiler type can either be pulverized coal (PC) or circulating fluidized bed (CFB) depending on which is more suitable to the type of fuel being combusted. Oxycombustion can be readily adapted to a new power plant or to retrofit applications.

In a retrofit application it is important to minimize the power plant derating and CO_2 removal penalty resulting from the oxycombustion power plant conversion. In a greenfield application it is important to minimize the overall plant heat rate and CO_2 removal penalty of the oxycombustion power plant. By optimizing the power plant cycle, the reduction in cycle efficiency and net power due to CO_2 removal can be theoretically limited to approximately 7-10% points and 10-12%, respectively.

The efficiency and cost-effectiveness of carbon sequestration of greenfield oxycombustion boilers can rival competing gasification plants by specifically tailoring boiler design by appropriate surface location, combustion system design, material selection, furnace layout, and water/steam circuitry. Boiler size can be substantially reduced due to higher radiative properties of O_2-combustion versus air-combustion, under certain conditions. Furthermore, a wider range of fuels can be burned due to the high oxygen content of the combustion gas and potential for high coal preheat. Any boiler size reduction may be limited by fuel ash fusion temperatures and required mechanical clearances.

The gaseous environment in an oxycombustion boiler has substantially higher partial pressures of CO_2 and H_2O than in an air-fired boiler. This significantly alters the equilibrium concentrations of CO, sulfur compounds, and chlorides which impacts material carburization, sulfidation, and chloride attack. In a PC the dominant mode of heat transfer is radiation from CO_2, H_2O, and coal particles. Consequently, the radiant gas emissivity of an oxygen-fired furnace is substantially greater than that of an air-fired furnace. This combined with potential higher oxygen concentration can create considerable increases in water wall heat fluxes. This can potentially reduce the size of the furnace for new boilers that burn fuels with a high ash fusion temperature. However, for oxygen-fired retrofit applications it is desirable to maintain similar peak and average heat fluxes by reducing flame temperatures to compensate for the higher gas emissivity. Figure 3 presents the furnace O_2, flame temperature, and equilibrium CO concentration as functions of flue gas recycle flow.

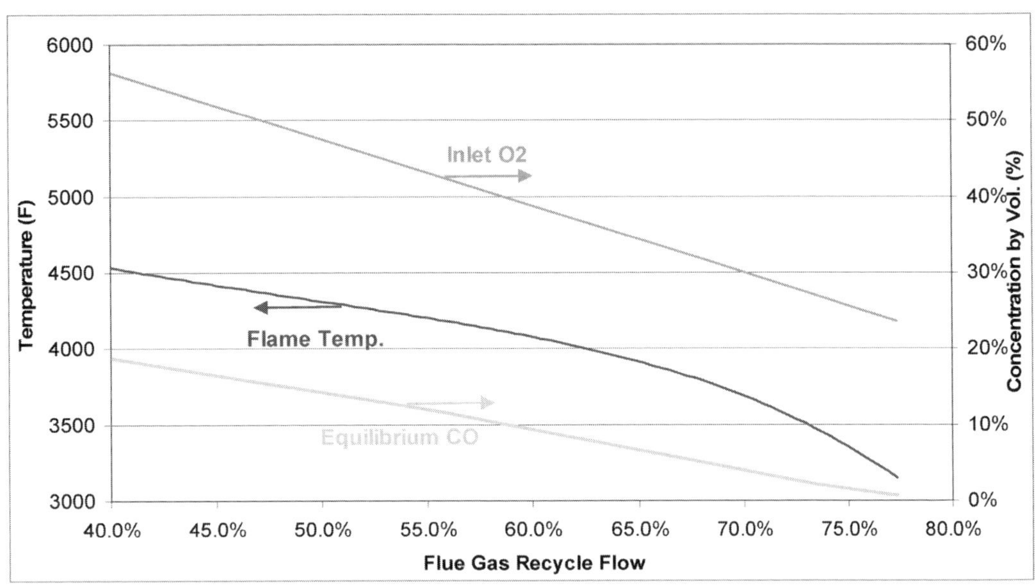

Figure 3 - Temperature and Species Versus Flue Gas Recycle Flow

CONCEPTUAL DESIGN STUDIES

Conceptual design studies have been performed for both a greenfield and retrofit of a 475 MWe (gross) supercritical PC boiler plant burning Illinois #6 coal.

In the oxygen-fired retrofit design, the quantity of flue gas recirculation flow was selected to maintain the same boiler geometry, materials, and burners. Such a design is termed a "low oxygen" (i.e. 23 – 27% O2) rather than the greenfield "high oxygen" (35-45%) design. If the sulfur compounds are not removed (i.e. by FGD) before the flue gas is recycled, the concentration of sulfur compounds increases by a factor of four (due to a 75% flue gas recycle flow) and will greatly increase water wall tube corrosion. In the air fired furnace a H_2S concentration of 1600 ppm was predicted resulting in a gas phase corrosion of 12 mil/yr. In the oxygen-fired furnace without an FGD, this increased to a H_2S concentration of 6400 ppm resulting in a gas phase corrosion of 30 mil/yr. Consequently either an FGD must be provided or weld overlay (e.g. 622 alloy) needs to be applied. Maximum CO concentration in both the air-fired and O_2-fired furnaces was similar (5-10%) as both used over-fire air/gas for NOx reduction. O2-fired Heat Recovery Area (HRA) tube temperatures and materials are similar to the air-fired design.

In the oxygen-fired greenfield design, the flue gas recirculation flow rate was selected to minimize the boiler cost. The lower is the recycled flow rate, the smaller is the required boiler. However, as the recycled flow rate is reduced, the water wall temperatures increase, and more expensive materials are required. Consequently there appears to be an economic optimally O_2

concentration of approximately 35-40% (55-60% flue gas recycle flow). At such an O_2 level, heat flux is increased by a factor of 1.5 to 2.5 increasing water wall temperature by 150-250°F and requiring significant material upgrade (e.g. from T2 to T92). Due to the higher flame temperature and CO_2 concentration of the oxygen-fired furnace, the maximum local CO level is substantially higher than in an air-fired furnace (e.g. 25-30% in an O_2-PC vs. 5-10% in an air-PC). If the sulfur compounds are removed (i.e. by FGD) before the flue gas is recycled, the concentration of sulfur compounds is expected to be similar to that of an air-fired furnace (5000 – 10000 ppm). However, due to the high CO level, weld overlay may be required to minimize corrosion. HRA tube temperatures and materials would be similar to the air-fired design.

ECONOMICS

The levelized cost of electricity (COE) is made up of contributions from the capital cost, operating and maintenance costs, consumables, and fuel costs. The levelized COE was calculated to be 5.0 ¢/kWh for the air-fired reference plant and 6.6 ¢/kWh for the oxycombustion plant. The CO_2 mitigation cost (MC) of the oxycombustion plant was calculated at 20 $/tonne (18 $/ton).

A comparison of the COE and MC of the oxycombustion plant versus other competing CO_2 removal technologies is presented in Figure . The increase in the cost of electricity of oxycombustion is approximately 30% greater than an air-fired plant with no CO_2 capture [2].

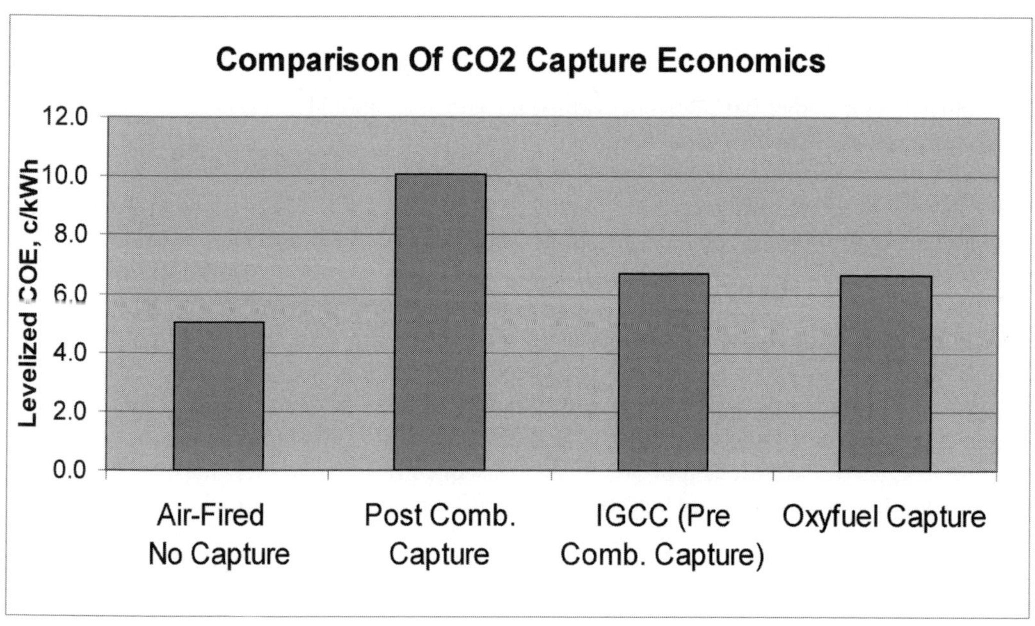

Figure 4 – Economic Comparison of CO_2 Capture Economics

CURRENT WORK

Work is being conducted by Foster Wheeler in the DOE sponsored Oxy-USC Boiler Materials Program to identify the corrosive conditions that exist in the furnace and superheater/reheater sections of both new and retrofitted oxygen-fired pulverized coal boilers and to select representative conditions for material selection and testing. The corrosive microclimates existing in the oxygen-fired furnace are predicted by CFD simulations. Two coals will be simulated: 1) high-sulfur bituminous and 2) low sulfur bituminous. Two oxygen concentrations will be simulated: 1) 23% (typical for retrofit) and 2) 35% (typical for a new plant). Each simulation will be reviewed and a representative furnace microclimate and representative furnace exit gas conditions will be selected that will serve as a design basis for selecting furnace and superheat/reheater tube materials and to generate coal ashes. Coal ashes will be generated in a drop tube furnace. The ash will be collected and analyzed to determine its composition with particular attention being paid to contaminants that will be corrosive to boiler tube materials. Based on the analyses, materials will be selected and test specimens will be fabricated. Candidate waterwall and superheater/reheater materials will then be exposed to synthetic gas and ash conditions for 1000 hour laboratory tests after which corrosion penetration depth and material wastage rates will be determined.

Two demonstration projects (CIUDEN and Jamestown) are planned which will address several needs including the following:

- Scale-up validation for burner design
- Slagging and fouling characteristics of ash
- Impact on radiant and convective heat transfer
- Startup and shutdown considerations
- Effect of plant trips on system
- Long-term material corrosion issues

The CIUDEN demonstration project will be a 20 MWt carbon capture demonstration facility, located in Spain. Several fuels will be combusted in a range of low O_2 to high O_2. Operation of the facility is scheduled for mid-2009.

The Jamestown demonstration project facility is a proposal for a 50 MWe oxygen-fired CFB power plant to be operated by the Jamestown Board of Public Utilities, in Jamestown, New York.

CONCLUSION

The oxycombustion power plant completely removes all CO_2 generated in the combustion process and generates zero ambient air pollutants. It avoids costly CO_2 separation processes, which are limited by equilibrium to only 90% CO_2 removal efficiency. Of the CO_2 sequestration-ready technologies, the oxycombustion requires the least modification of existing proven designs, and requires no special chemicals for CO_2 separation. Furthermore, the cost-effectiveness of the oxycombustion cycle will benefit from several currently promising advanced O_2 separation techniques such as membrane separation.

Conceptual design studies have identified material related issues created by the potentially higher heat flux, increase in corrosiveness of the furnace environment, and changes in the ash slagging and fouling characteristics. It is expected that these issues will be less severe for the low O_2 retrofitable design than for the high O_2 new plant design.

REFERENCES

1. US DOE, NETL, Carbon Sequestration Technology Roadmap and Program Plan – 2005

2. Seltzer, Andrew, Fan, Zhen, and Hack, Horst, "Oxyfuel Coal Combustion Power Plant System Optimization", 7th Annual COAL-GEN Conference Milwaukee, WI, August 2007.

Author Index

K

L

M

N

O

P

R

S

Subject Index

A

B

C

D

E

F

G

H

I

L

O

P

R

S

T

U

V

W